Lecture Notes in Computer Science   2136

Edited by G. Goos, J. Hartmanis and J. van Leeuwen

Springer
Berlin
Heidelberg
New York
Barcelona
Hong Kong
London
Milan
Paris
Singapore
Tokyo

Jiří Sgall   Aleš Pultr   Petr Kolman (Eds.)

Mathematical Foundations of Computer Science 2001

26th International Symposium, MFCS 2001
Mariánské Lázně, Czech Republic, August 27-31, 2001
Proceedings

Springer

Series Editors

Gerhard Goos, Karlsruhe University, Germany
Juris Hartmanis, Cornell University, NY, USA
Jan van Leeuwen, Utrecht University, The Netherlands

Volume Editors

Jiří Sgall
Mathematical Institute, AS CR
Žitná 25, 115 67 Praha 1, Czech Republic
E-mail: sgall@math.cas.cz

Aleš Pultr
Petr Kolman
Charles University, Faculty of Mathematics and Physics
Institute for Theoretical Computer Science (ITI)
Malostranské náměstí 25, 118 00 Praha 1, Czech Republic
E-mail:{pultr/kolman}@kam.ms.mff.cuni.cz

Cataloging-in-Publication Data applied for

Die Deutsche Bibliothek - CIP-Einheitsaufnahme

Mathematical foundations of computer science 2001 : 26th international
symposium ; proceedings / MFCS 2001, Mariánské Láznÿée, Czech Republic,
August 27 - 31, 2001. Jiÿérí Sgall ... (ed.). - Berlin ; Heidelberg ; New York ;
Barcelona ; Hong Kong ; London ; Milan ; Paris ; Tokyo : Springer, 2001
   (Lecture notes in computer science ; Vol. 2136)
   ISBN 3-540-42496-2

CR Subject Classification (1998): F, G.2, D.3, C.2, I.3

ISSN 0302-9743
ISBN 3-540-42496-2 Springer-Verlag Berlin Heidelberg New York

Springer-Verlag Berlin Heidelberg New York
a member of BertelsmannSpringer Science+Business Media GmbH

http://www.springer.de

© Springer-Verlag Berlin Heidelberg 2001
Printed in Germany

Typesetting: Camera-ready by author, data conversion by DA-TeX Gerd Blumenstein
Printed on acid-free paper   SPIN 10840127   06/3142   5 4 3 2 1 0

Foreword

This volume contains papers selected for presentation at the 26th International Symposium on Mathematical Foundations of Computer Science – MFCS 2001, held in Mariánské Lázně, Czech Republic, August 27 – 31, 2001.

MFCS 2001 was organized by the Mathematical Institute (Academy of Sciences of the Czech Republic), the Institute for Theoretical Computer Science (Charles University, Faculty of Mathematics and Physics), the Institute of Computer Science (Academy of Sciences of the Czech Republic), and Action M Agency. It was supported by the European Research Consortium for Informatics and Mathematics, the Czech Research Consortium for Informatics and Mathematics, and the European Association for Theoretical Computer Science. We gratefully acknowledge the support of all these institutions.

The series of MFCS symposia, organized on a rotating basis in Poland, Slovakia, and the Czech Republic, has a well-established tradition. The aim is to encourage high-quality research in all branches of theoretical computer science and bring together specialists who do not usually meet at specialized conferences. Previous meetings took place in Jablonna, 1972; Štrbské Pleso, 1973; Jadwisin, 1974; Mariánské Lázně, 1975; Gdańsk, 1976; Tatranská Lomnica, 1977; Zakopane, 1978; Olomouc, 1979; Rydzina, 1980; Štrbské Pleso, 1981; Prague, 1984; Bratislava, 1986; Karlovy Vary, 1988; Porąbka-Kozubnik, 1989; Banská Bystrica, 1990; Kazimierz Dolny, 1991; Prague, 1992; Gdańsk, 1993; Košice, 1994; Prague, 1995; Kraków, 1996; Bratislava, 1997; Brno, 1998; Szklarska Poręba, 1999; and Bratislava, 2000.

It is our pleasure to announce that at the opening of MFCS 2001, Dana Scott (Carnegie-Mellon Univ., Pittsburg, PA, U.S.A.) was awarded the Bolzano Honorary Medal of the Academy of Sciences of the Czech Republic for his contribution to the development of theoretical computer science and mathematics in general and his cooperation with Czech scientists in particular.

The MFCS 2001 proceedings consist of 10 invited papers and 51 contributed papers. We are grateful to all the invited speakers for accepting our invitation and sharing their insights on their research areas. We thank the authors of all submissions for their contribution to the scientific program of the meeting.

The contributed papers were selected by the Program Committee out of a total of 118 submissions. All submissions were evaluated by three or four members of the committee, with the assistance of referees, for a total of more than 450 reports. After electronic discussions, the final agreement was reached at the selection meeting in Prague on May 11–12, 2001 (the program committee members denoted by * in the list below took part in the meeting). We thank all the program committee members and referees for their work which contributed to the quality of the meeting. We have tried to make the list of referees as complete and accurate as possible and apologize for any omissions and errors.

Special thanks go to Jochen Bern who provided a reliable software system used for electronic submissions and reviewing (tested before on STACS 1999 through 2001), and volunteered a fair amount of his night-time to maintain and further improve the system according to our unrealistic specifications and expectations.

Finally, we would like to thank Milena Zeithamlová, Lucie Váchová, and Andrea Kutnarová from Action M Agency for their excellent work on local arrangements.

We wish MFCS 2002, to be held in Warsaw, success and many excellent contributions.

June 2001

Jiří Sgall
Aleš Pultr
Petr Kolman

Program Committee

Manfred Broy (TU Munich)
Harry Buhrman (CWI, Amsterdam)
Anne Condon (Univ. of British Columbia, Vancouver)
Peter van Emde Boas (Amsterdam Univ.)
Martin Grohe (Univ. of Illinois, Chicago)
Petr Hájek * (Academy of Sciences, Prague)
Juraj Hromkovic * (RWTH Aachen)
Russell Impagliazzo (UC San Diego)
Achim Jung (Univ. of Birmingham)
Juhani Karhumäki * (Univ. of Turku)
Matthias Krause * (Univ. Mannheim)
J. A. Makowsky * (Technion, Haifa)
Tadeusz Morzy (Poznan Univ. of Technology)
Aleš Pultr * (Charles Univ., Prague, *co-chair*)
Giuseppe Rosolini (Univ. of Genova)
Branislav Rovan * (Comenius Univ., Bratislava)
Don Sannella (Univ. of Edinburgh)
Jiří Sgall * (Academy of Sciences, Prague, *chair*)
Gábor Tardos (Academy of Sciences, Budapest)
Igor Walukiewicz (Warsaw Univ.)
Ingo Wegener * (Univ. Dortmund)
Peter Widmayer (ETH, Zurich)
Gerhard Woeginger * (Technical Univ. Graz)
Moti Yung (Columbia Univ., New York).

Referees

E. Allender
A. Ambainis
K. Ambos-Spies
G. Ankoudinov
P. Auer
R. Babilon
J. Baldwin
K. Balinska
R. Beigel
A. Benczur
T. Bernholt
M. Bläser
J. Bloemer
H. L. Bodlaender
B. Bollig
D. Bongartz
A. Brodsky
H. Broersma
R. Bruni
P. Buergisser
H.-J. Böckenhauer
P. Callaghan
C. S. Calude
O. Carton
B. Chlebus
K. Chmiel
J. Chrzaszcz
M. Cieliebak
J. Csirik
L. Csirmaz
F. Cucker
C. Damm
S. V. Daneshmand
G. Delzanno
J. Díaz
S. Dobrev
W. Doerfler
M. Drmota
S. Droste
L. Epstein
M. Escardo
Z. Esik
J. Esparza
S. Fekete

J. Fiala
E. Fischer
S. Fishman
T. Fleiner
J. Flum
J. Forster
L. Fortnow
K. Friedl
W. Gasarch
S. Gerke
O. Giel
A. Goerdt
M. Goldmann
V. Grolmusz
P. Hajnal
V. Halava
T. Harju
J. Hastad
P. Hell
U. Hertrampf
V. Heun
J. Hillston
M. Hirvensalo
J.-H. Hoepman
T. Hofmeister
J. Honkala
H. Hoogeveen
R. Jacob
A. Jakoby
T. Jansen
M. Jerrum
S. Jukna
M. Kaminski
J. Kari
M. Karpinski
H. Klauck
B. Klinz
T. Knapik
P. Koiran
P. Kolman
J. Krajíček
I. Kramosil
A. Królikowski
Z. Królikowski

V. Kůrková
M. Kutylowski
J. van Leeuwen
H. Lefmann
M. Lenisa
S. Leonardi
A. Lepistö
Z. Liptak
L. P. Lisovik
S. Lucks
H. Lötzbeyer
S. P. Mansilla
J. Manuch
M. Mareš
J. Marino
J. Matoušek
R. Mayr
J. McKinna
W. Merkle
S. Merz
M. Michal
B. Moeller
T. Morzy
A. Muscholl
R. Neruda
T. Nipkow
D. Niwinski
T. Noll
J. C. Oostveen
L. Paolini
M. Parente
P. Penna
P. Jeavons
A. Petit
T. Petkovic
J.-E. Pin
T. Pitassi
T. Polzin
C. Pomm
K. Pruhs
P. Pudlák
R. Raz
O. Regev
R. Reischuk

Special thanks go to Jochen Bern who provided a reliable software system used for electronic submissions and reviewing (tested before on STACS 1999 through 2001), and volunteered a fair amount of his night-time to maintain and further improve the system according to our unrealistic specifications and expectations.

Finally, we would like to thank Milena Zeithamlová, Lucie Váchová, and Andrea Kutnarová from Action M Agency for their excellent work on local arrangements.

We wish MFCS 2002, to be held in Warsaw, success and many excellent contributions.

June 2001

Jiří Sgall
Aleš Pultr
Petr Kolman

Program Committee

Manfred Broy (TU Munich)
Harry Buhrman (CWI, Amsterdam)
Anne Condon (Univ. of British Columbia, Vancouver)
Peter van Emde Boas (Amsterdam Univ.)
Martin Grohe (Univ. of Illinois, Chicago)
Petr Hájek * (Academy of Sciences, Prague)
Juraj Hromkovic * (RWTH Aachen)
Russell Impagliazzo (UC San Diego)
Achim Jung (Univ. of Birmingham)
Juhani Karhumäki * (Univ. of Turku)
Matthias Krause * (Univ. Mannheim)
J. A. Makowsky * (Technion, Haifa)
Tadeusz Morzy (Poznan Univ. of Technology)
Aleš Pultr * (Charles Univ., Prague, *co-chair*)
Giuseppe Rosolini (Univ. of Genova)
Branislav Rovan * (Comenius Univ., Bratislava)
Don Sannella (Univ. of Edinburgh)
Jiří Sgall * (Academy of Sciences, Prague, *chair*)
Gábor Tardos (Academy of Sciences, Budapest)
Igor Walukiewicz (Warsaw Univ.)
Ingo Wegener * (Univ. Dortmund)
Peter Widmayer (ETH, Zurich)
Gerhard Woeginger * (Technical Univ. Graz)
Moti Yung (Columbia Univ., New York).

Referees

E. Allender	J. Fiala	V. Kůrková
A. Ambainis	E. Fischer	M. Kutylowski
K. Ambos-Spies	S. Fishman	J. van Leeuwen
G. Ankoudinov	T. Fleiner	H. Lefmann
P. Auer	J. Flum	M. Lenisa
R. Babilon	J. Forster	S. Leonardi
J. Baldwin	L. Fortnow	A. Lepistö
K. Balinska	K. Friedl	Z. Liptak
R. Beigel	W. Gasarch	L. P. Lisovik
A. Benczur	S. Gerke	S. Lucks
T. Bernholt	O. Giel	H. Lötzbeyer
M. Bläser	A. Goerdt	S. P. Mansilla
J. Bloemer	M. Goldmann	J. Manuch
H. L. Bodlaender	V. Grolmusz	M. Mareš
B. Bollig	P. Hajnal	J. Marino
D. Bongartz	V. Halava	J. Matoušek
A. Brodsky	T. Harju	R. Mayr
H. Broersma	J. Hastad	J. McKinna
R. Bruni	P. Hell	W. Merkle
P. Buergisser	U. Hertrampf	S. Merz
H.-J. Böckenhauer	V. Heun	M. Michal
P. Callaghan	J. Hillston	B. Moeller
C. S. Calude	M. Hirvensalo	T. Morzy
O. Carton	J.-H. Hoepman	A. Muscholl
B. Chlebus	T. Hofmeister	R. Neruda
K. Chmiel	J. Honkala	T. Nipkow
J. Chrzaszcz	H. Hoogeveen	D. Niwinski
M. Cieliebak	R. Jacob	T. Noll
J. Csirik	A. Jakoby	J. C. Oostveen
L. Csirmaz	T. Jansen	L. Paolini
F. Cucker	M. Jerrum	M. Parente
C. Damm	S. Jukna	P. Penna
S. V. Daneshmand	M. Kaminski	P. Jeavons
G. Delzanno	J. Kari	A. Petit
J. Díaz	M. Karpinski	T. Petkovic
S. Dobrev	H. Klauck	J.-E. Pin
W. Doerfler	B. Klinz	T. Pitassi
M. Drmota	T. Knapik	T. Polzin
S. Droste	P. Koiran	C. Pomm
L. Epstein	P. Kolman	K. Pruhs
M. Escardo	J. Krajíček	P. Pudlák
Z. Esik	I. Kramosil	R. Raz
J. Esparza	A. Królikowski	O. Regev
S. Fekete	Z. Królikowski	R. Reischuk

Table of Contents

Table of Contents

Table of Contents

A New Category for Semantics

Dana S. Scott

Carnegie Mellon University,
Pittsburgh, PA, USA

Abstract. Domain theory for denotational semantics is over thirty years old. There are many variations on the idea and many interesting constructs that have been proposed by many people for realizing a wide variety of types as domains. Generally, the effort has been to create categories of domains that are cartesian closed (that is, have products and function spaces interpreting typed lambda-calculus) and permit solutions to domain equations (that is, interpret recursive domain definitions and perhaps untyped lambda-calculus).

What has been missing is a simple connection between domains and the usual set-theoretical structures of mathematics as well as a comprehensive logic to reason about domains and the functions to be defined upon them. In December of 1996, the author realized that the very old idea of partial equivalence relations on types could be applied to produce a large and rich category containing many specific categories of domains and allowing a suitable general logic. The category is called Equ, the category of equilogical spaces.

The simplest definition is the category of T_0 spaces and total equivalence relations with continuous maps that are equivariant (meaning, preserving the equivalence relations). An equivalent definition uses algebraic (or continuous) lattices and partial equivalence relations, together with continuous equivariant maps. This category is not only cartesian closed, but it is locally cartesian closed (that is, it has dependent sums and products). Moreover, it contains as a full subcategory all T_0 spaces (and therefore the category of sets and the category of domains).

The logic for this category is intuitionistic and can be explained by a form of the realizability interpretation, as will be outlined in the lecture. The project now is to use this idea as a unifying platform for semantics and reasoning. In the last four years the author has been cooperating with faculty and students at Carnegie Mellon on this program, namely Steven Awodey, Andrej Bauer, Lars Birkedal, and Jesse Hughes. A selection of papers (in reverse chronological order) follows. These (and futute papers) are available via the WWW at http://www.cs.cmu.edu/Groups/LTC.

J. Sgall, A. Pultr, and P. Kolman (Eds.): MFCS 2001, LNCS 2136, pp. 1–2, 2001.

References

1. S. Awodey, A. Bauer. Sheaf Toposes for Realizability. April 2001. Preprint CMU-PHIL-117.
2. A. Bauer, L. Birkedal, D. S. Scott. Equilogical Spaces. September 1998. Revised February 2001. To appear in Theoretical Computer Science.
3. A. Bauer The Realizability Approach to Computable Analysis and Topology. Ph.D. Thesis. September 2000.
4. S. Awodey, J. Hughes. The Coalgebraic Dual of Birkhoff's Variety Theorem. Preprint. October 2000.
5. L. Birkedal, J. van Oosten. Relative and Modified Relative Realizability. Preprint 1146, Department of Mathematics, Universiteit Utrecht. March 2000.
6. A. Bauer, L. Birkedal. Continuous Functionals of Dependent Types and Equilogical Spaces. Proceedings of Computer Science Logic Conference 2000.
7. S. Awodey, L. Birkedal, D. S. Scott. Local Realizability Toposes and a Modal Logic for Computability. January 2000. To appear in Math. Struct. in Comp. Sci.
8. L. Birkedal. A General Notion of Realizability. December 1999. Proceedings of LICS 2000.
9. L. Birkedal. Developing Theories of Types and Computability via Realizability. PhD-thesis. Electronic Notes in Theoretical Computer Science, 34, 2000. Available at http://www.elsevier.nl/locate/entcs/volume34.html. December 1999.
10. S. Awodey and L. Birkedal. Elementary Axioms for Local Maps of Toposes. Technical Report No. CMU-PHIL103. November 1999. To appear in Journal of Pure and Applied Algebra.
11. S. Awodey. Topological Representation of the Lambda Calculus. September 1998. Mathematical Structures in Computer Science(2000), vol. 10, pp. 81–96.
12. L. Birkedal, A. Carboni, G. Rosolini, and D. S. Scott. Type Theory via Exact Categories. LICS 1998. July 1998.
13. D. S. Scott. A New Category?: Domains, Spaces and Equivalence Relations. Unpublished Manuscript. December 1996.

On Implications between P-NP-Hypotheses: Decision versus Computation in Algebraic Complexity

Peter Bürgisser

Dept. of Mathematics and Computer Science, University of Paderborn,
D-33095 Paderborn, Germany
buergisser@upb.de

Abstract. Several models of NP-completeness in an algebraic framework of computation have been proposed in the past, each of them hinging on a fundamental hypothesis of type P≠NP. We first survey some known implications between such hypotheses and then describe attempts to establish further connections. This leads us to the problem of relating the complexity of computational and decisional tasks and naturally raises the question about the connection of the complexity of a polynomial with those of its factors. After reviewing what is known with this respect, we discuss a new result involving a concept of approximative complexity.

1 Introduction

Algebraic complexity theory is the study of the intrinsic difficulty of computational problems that can be posed in an algebraic or numerical framework. Instead of basing this study on the model of the Turing machine, it uses "algebraic models" of computation. Besides the fact that these models form a natural theoretical framework for the algorithms commonly known to solve the problems under consideration, the main motivation for this choice of model is the idea that various methods from pure mathematics like algebraic geometry or topology can be employed to establish lower bounds in this more structured world. This project has been successful for problems of polynomially bounded complexity, as illustrated by the large body of work presented in the textbook [14].

The theory of NP-completeness is one of the main cornerstones of computational complexity, although still resting on the unproven hypothesis that P ≠ NP. In seminal works by Valiant [45,47] and Blum, Shub and Smale [9] models of NP-completeness in an algebraic framework of computation have been proposed, motivated by the hope to prove the elusive separation of P and NP in these frameworks. So far, this hope has not been fulfilled except in rather trivial cases. However, there have been successful attempts in relating these different models to each other and in establishing implications between the various P-NP-hypotheses presently studied. The known "transfer theorems" either relate such hypotheses of the same type refering to different fields, or provide implications

J. Sgall, A. Pultr, and P. Kolman (Eds.): MFCS 2001, LNCS 2136, pp. 3–17, 2001.

from the separation of P and NP in the classical bit model to a corresponding separation in an algebraic model of computation. As an exception to this rule, Fournier and Koiran [20] (see also [31]) recently proved an implication in the reverse direction for a restricted algebraic model. This has challenged the hope that a P-NP-separation in the algebraic model might be easier to prove than in the bit model. We review some of the known transfer results in Section 3.

In Section 4 we will discuss an attempt [12] to relate the P-NP hypothesis in Valiant's framework to the one in the Blum, Shub, Smale (BSS) framework, as well as an attempt [42] to establish a connection to the complexity of certain univariate polynomials. This leads us to the problem of relating the complexity of computational and decisional tasks and naturally raises the question about the connection of the complexity of a polynomial with those of its factors. This relationship is not well understood and turns out to be the bottleneck in both attempts.

In Section 5 we review what is known about the complexity of factors. We mention a result by Kaltofen [27], which seems to be widely unknown in the community (and has been independently discovered by the author). It states that the complexity of an irreducible factor g of a polynomial f can be polynomially bounded in the complexity of f, the degree of g, and the multiplicity of g. We believe that the dependence on the multiplicity is not necessary, but we have been unable to prove this. Instead, we present a new result [10], which states that the dependence on the multiplicity can be avoided when replacing complexity by the related notion of "approximative complexity", which is introduced in Section 6.

As a major application, we obtain the following relative hardness result about decision complexity over the reals: Checking the values of polynomials forming complete families in Valiant's sense cannot be done with a polynomial number of arithmetic operations and tests, unless the permanent has a p-bounded approximative complexity, which seems unlikely. This hardness result extends to randomized algorithms with two-sided error, formalized by randomized algebraic computation trees.

2 Algebraic Models of NP-Completeness

2.1 Blum-Shub-Smale Model

Blum, Shub, and Smale [9] have extended the classical theory of NP-completeness to a theory of computation over arbitrary rings, the three most interesting cases being the field of real or complex numbers, and the finite field $\mathbb{F}_2$. In the latter case, the classical theory of computation is recovered. As usual in algebraic complexity theory, a basic computational step is an arithmetic operation, an equality test, or a $\leq$-test if the field is ordered, and we make the idealizing assumption that this can be done with infinite precision. Moreover a uniformity condition is assumed to be satisfied. For a recent account see [8]. Poizat [40] describes an elegant approach to a P-NP-framework over general structures.

Many ideas of discrete structural complexity can be extended to such a framework; in particular the complexity classes P, NP and the notion of NP-completeness. In this framework, a natural NP-complete problem turns out to be the feasibility problem to decide for a given system of polynomials whether they have a common root. In all three settings ($\mathbb{F}_2, \mathbb{R}, \mathbb{C}$) it is a fundamental open problem whether P $\neq$ NP is true. It has been shown [7] that this question has the same answer over all algebraically closed fields of characteristic zero. Over the reals, no corresponding transfer theorem is known.

2.2   Valiant's Algebraic Model

In [45,47] Valiant proposed an analogue of the theory of #P-completeness in a framework of algebraic complexity, in connection with his famous hardness result for the permanent [46]. This theory features algebraic complexity classes VP and VNP as well as VNP-completeness results for many families of generating functions of graph properties, the most prominent being the family of permanents. For a comprehensive presentation of this theory, we refer to [21,14] and to the recent account [12].

While the complexity classes in the BSS-modell capture decision problems, the Valiant classes deal with the computational problem to evaluate multivariate polynomials. Only straight-line computations are considered and uniformity is not taken into account. The basic object studied is a p-family over a fixed field k, which is a sequence (f_n) of multivariate polynomials such that the number of variables as well as the degree of f_n are polynomially bounded (p-bounded) functions of n. The complexity class VP over k consists of the p-families (f_n) which are p-computable, which means that the straight-line complexity of f_n is p-bounded in n. Note that although X^{2^n} can be computed with only n multiplications, the corresponding sequence is not considered to be p-computable, as the degrees grow exponentially. A p-family (f_n) is called p-definable iff there exists $(g_n) \in$ VP such that for all n

$$f_n(X_1, \ldots, X_{v(n)}) = \sum_{e \in \{0,1\}^{u(n)-v(n)}} g_n(X_1, \ldots, X_{v(n)}, e_{v(n)+1}, \ldots, e_{u(n)}).$$

The set of p-definable families form the complexity class VNP. The class VP is obviously contained in VNP, and Valiant's hypothesis claims that this inclusion is strict. It has been shown in [12, § 4.1] that over algebraically closed fields, this hypothesis depends at most on the characteristic of the field.

3   Known Implications between P-NP-Hypotheses

In each of the algebraic models discussed before we have raised the fundamental question whether P $\neq$ NP. Recently, a considerable amount of research has been directed towards establishing "transfer theorems" that provide implications from the separation of P and NP in the classical bit model to a corresponding separation in an algebraic model of computation. These results rely on various techniques to eliminate the real or complex constants that may be used by a computation in the algebraic model on a Boolean input. One of the first results of this kind was established by Koiran [29] for the additive BSS-model over the reals, which allows additions and subtractions as the only arithmetic operations. Koiran showed that

$$\text{P} \neq \text{NP nonuniformly} \implies \text{P} \neq \text{NP over } \mathbb{R} \text{ as an ordered group.} \tag{1}$$

Here, the elimination of constants is based on polyhedral geometry, in particular on the fact that a nonempty polyhedron defined by a system of inequalities with small coefficients has a small rational point. Quite astonishingly, the converse of the above implication is also true, as recently established by Fournier and Koiran [20]. The proof of the converse of (1) relies on Meyer auf der Heide's [37,39] construction of small depth linear decision trees for locating points in arrangements of hyperplanes.

For the unrestricted BSS-model over the reals (with order) no implication in either direction is known. Over the complex numbers, we know the following about the unrestricted BSS-model

$$\text{P} \neq \text{NP nonuniformly} \implies \text{P} \neq \text{NP over } \mathbb{C}, \tag{2}$$

as noticed independently by several researchers. The essential point here is that using modular arithmetic, it is possible to test in random polynomial time whether an integer given by a straight-line program equals zero. Actually, the left-hand side can be replaced by the weaker statement $\text{NP} \not\subseteq \text{BPP}$. (A proof can be found in Cucker et al. [18], although somewhat hidden.) We do not know how to efficiently test for positivity of an integer given by a straight-line program, and this seems to be the main obstacle for establishing an implication analogous to (2) over the reals.

In [13] we have shown the following implication for Valiant's model over $\mathbb{F}_2$:

$$\text{NC}^2 \neq \oplus \text{P nonuniformly} \implies \text{VP} \neq \text{VNP over } \mathbb{F}_2. \tag{3}$$

More specifically, by interpreting families of polynomials over $\mathbb{F}_2$ as Boolean functions, we can assign to an algebraic class $\mathcal{C}$ its Boolean part $\text{BP}(\mathcal{C})$. It turns out that over $\mathbb{F}_2$ we have

$$\text{NC}^1/\text{poly} \subseteq \text{BP}(\text{VP}) \subseteq \text{NC}^2/\text{poly}, \quad \text{BP}(\text{VNP}) = \oplus\text{P}/\text{poly},$$

which immediately implies (3). Note that the class VNP over $\mathbb{F}_2$ is by definition closely related to $\oplus$P. On the other hand, the Boolean function corresponding to a p-computable family lies in NC^2/poly, due to the fact [6] that an algebraic straight-line program of size $n^{O(1)}$ computing a polynomial of degree $n^{O(1)}$ can be efficiently parallelized, i.e., simulated by a straight-line program of size $n^{O(1)}$ and depth $O(\log^2 n)$. A challenging open problem is to find out to what extent implication (3) might be reversed: does $\text{BP}(\text{VP}) = \text{BP}(\text{VNP})$ imply that $\text{VP} = \text{VNP}$ over $\mathbb{F}_2$? As a first step in this direction, it seems rewarding to locate

BP(VP) exactly in the hierarchy of known complexity classes between NC^1/poly and NC^2/poly.

For Valiant's model in characteristic zero, we have proved in [13] the following implication similar to (3):

$$FNC^3 \neq \#P \text{ nonuniformly} \implies VP \neq VNP \text{ in characteristic zero,}$$

conditional on the generalized Riemann hypothesis. Besides the ideas mentioned for the field $\mathbb{F}_2$, the proof is based on some elimination of constants, which is achieved by a general result about the frequency of primes p with the property that a system of integer polynomial equations solvable over $\mathbb{C}$ has a solution modulo p.

In Section 4.2 we will address the question whether $VP \neq VNP$ implies $P \neq NP$ over $\mathbb{C}$. The difficulty here is not to eliminate constants, but to link the complexity of decisional problems to the complexity of computational problems.

4 Attempts to Establish Further Connections

4.1 Linking Decisional to Computational Complexity

Are there functions, whose value can be checked in polynomial time but which cannot be computed in polynomial time? In fact, this is a basic assumption in cryptography, since it turns out to be equivalent to the existence of one-way functions [24,41]. We ask now this question for a real or complex polynomial function g. More specifically, we ask whether deciding $g(x) = 0$ can be done considerably faster than just by computing g at input x?

In order to make this formal, we introduce the computational complexity $L(g)$ and the decisional complexity $C(g)$. The first one refers to the straight-line model of computation and counts all arithmetic operations (divisions are not allowed for simplicity). The decisional complexity refers to algebraic computation trees and counts besides arithmetic operations also branchings according to equality tests (and $\leq$-tests over the reals). Clearly, $C(g) \leq L(g) + 1$. In both cases, we allow that any real or complex numbers may be used as constants.

Assume that g is the product of m real linear polynomials: $g = h_1 \cdots h_m$. In the case $n = 1$ we have obviously $C(g) = O(\log m)$ using binary search. The result by Meyer auf der Heide [37,39] and Meiser [36] extends this to higher dimensions and states that it is possible to locate a given point $x \in \mathbb{R}^n$ in the hyperplane arrangement given by $h_1, \ldots, h_m$ by an algebraic computation tree with depth $(n \log m)^{O(1)}$. (In fact, linear decision trees are sufficient for this, see [19].) We therefore have $C(g) = (n \log m)^{O(1)}$. On the other hand, one can show that the complexity for evaluating g equals $\Theta(mn)$ if the h_i are in general position [14, Chap. 5]. This example shows that computational and decisional complexity may differ dramatically. In Section 7.1 we will provide strong evidence that this phenomenon does neither occur over the complex numbers nor over the reals if g is irreducible and has a degree polynomially bounded in its complexity.

The following well-known lemma [16,8] provides a link from decisional to computational complexity. It naturally leads to the question of relating the complexity of a polynomial to those of its multiples, which will be investigated systematically in Section 5.

Lemma 1. *There exists a nonzero multiple f of g such that $L(f) \leq C(g)$. Over $\mathbb{C}$ this is true without additional assumption, over $\mathbb{R}$ we have to require that g is irreducible.*

4.2 Does Valiant's Hypothesis Imply the BSS Hypothesis over $\mathbb{C}$?

In [12, §8.4] we have conjectured that Valiant's hypothesis implies the BSS hypothesis over $\mathbb{C}$. Loosely speaking, this means that if the permanent is intractable, then solving systems of polynomial equations over $\mathbb{C}$ is intractable as well. The following reasonings from [12], similar as in Heintz and Morgenstern [25], have lead us to this conjecture.

The real, weighted cycle cover problem is the following decision problem: given a real $n \times n$ matrix $[w_{i,j}]$ of weights and a real number s, one has to decide whether there exists some permutation $\pi \in S_n$ such that $\sum_{i=1}^{n} w_{i,\pi(i)} = s$. (Note that a permutation can be visualized as a cycle cover.) By the above mentioned result [37,39], this decision problem can be solved by algebraic computation trees over the reals with depth $n^{O(1)}$. We reformulate this problem now as follows: let $x_{i,j} = 2^{w_{i,j}}$ and $y = 2^{-s}$. Then the above question amounts to test whether the (reducible) polynomial

$$g_n := \prod_{\pi \in S_n} (1 - Y X_{1,\pi(1)} \cdots X_{n,\pi(n)})$$

vanishes for a given (positive) real matrix $x = [x_{i,j}]$ and $y > 0$.

If we expand g_n according to powers of Y, we see that the permanent of the matrix $[X_{i,j}]$ is the coefficient of Y: in fact, $g_n = 1 - Y \operatorname{PER}_n(X) + O(Y^2)$. A variant of a well-known result on the computation of homogeneous parts implies $L(\operatorname{PER}_n) \leq 4L(g_n)$ (compare [14, § 7.1]). Therefore, g_n is hard to compute if Valiant's hypothesis is true.

It is easy to see that the problem to check whether $g_n(x, y) = 0$ gives rise to a problem in the BSS complexity class NP over the reals. Indeed, for a given matrix x and $y \in \mathbb{R}$ it suffices to guess the permutation π and to check that $y x_{1,\pi(1)} \cdots x_{n,\pi(n)} = 1$.

We present now an attempt to deduce P $\neq$ NP over $\mathbb{C}$ from Valiant's hypothesis, based on the decision problem $g_n(x, y) = 0$ over the complex numbers. Assume that P $=$ NP over $\mathbb{C}$. Then the above decision problem would lie in P and therefore could be solved by algebraic computation trees of depth polynomially bounded in n, hence $C(g_n) = n^{O(1)}$. By Lemma 1 there is a nonzero multiple f_n of g_n for each n such that $L(f_n) \leq C(g_n)$, hence $L(f_n) = n^{O(1)}$. If we could derive from this that the factor g_n has a complexity polynomially bounded in n, then we could conclude $L(\operatorname{PER}_n) = n^{O(1)}$, which contradicts Valiant's hypothesis, as the the permanent family is VNP-complete.

4.3   Univariate Polynomials and P $\neq$ NP over $\mathbb{C}$

For given $n \in \mathbb{N}$ consider the problem ("Twenty Questions") to decide for a given complex number x whether $x \in \{1, 2, \ldots, n\}$. The complexity of this problem in the model of computation trees over $\mathbb{C}$ thus equals $C(p_n)$, where

$$p_n(X) := (X - 1)(X - 2) \cdots (X - n)$$

are the Pochhammer-Wilkinson polynomials. Shub and Smale [42] made the following reasoning similar to the one in Section 4.2. By considering the parameter n also as an input, the above decision problem is easily seen to lie in the class NP over $\mathbb{C}$, when we consider the pair (n, x) as an instance of size $\ell = \lceil \log n \rceil$. The point is that one can check whether x is an integer in the range $0, 1, \ldots, 2^\ell - 1$ by guessing complex numbers $w_0, \ldots, w_{\ell-1}$ and checking that $x = \sum_{i=0}^{\ell-1} w_i 2^i$ and $w_i(w_i - 1) = 0$ for all i.

Assume that P $=$ NP over $\mathbb{C}$. Then the above decision problem would lie in P and we would have $C(p_n) = n^{O(1)}$. By Lemma 1 there would exist a nonzero multiple f_n of p_n with complexity $L(f_n) \leq C(p_n) = n^{O(1)}$. This argument can be refined: by eliminating the finite set of complex constants used by a BSS-machine, it is possible to achieve that the multiple f_n of p_n is an integer polynomial computed by a straight-line program of length $n^{O(1)}$ using 1 as the only constant. We call the minimal length of a straight-line program satisfying this additional condition the τ-complexity $\tau(f)$. Clearly, $L(f) \leq \tau(f)$.

Shub and Smale [42] set up the so-called τ-conjecture, which claims the following connection between the number $z(f)$ of distinct integer roots of a univariate integer polynomial f and its τ-complexity:

$$z(f) \leq (1 + \tau(f))^c ,$$

where $c > 0$ is a universal constant. The τ-conjecture thus implies P $\neq$ NP over the field of complex numbers.

In order to illustrate that the τ-conjecture is of a number theoretic quality, let us ask a more general question. Let k be a field, f be a polynomial in n variables over k, and $d \in \mathbb{N}$. We write $N_d(f)$ for the number of irreducible factors of f over k having degree at most d, not counting multiplicity. For a fixed field k, we raise the following question:

$$\exists c > 0 \; \forall n, d \; \forall f \in k[X_1, \ldots, X_n] : N_d(f) \leq (L(f) + d)^c. \tag{4}$$

It is clear that this statement over $\mathbb{Q}$ implies the τ-conjecture: the difference being that we take $n = d = 1$, count all rational roots of f, and measure complexity with L instead of τ. Refering to question (4), we observe the following:

(i) If (4) is true over some field k, then it must be true over the rationals.
(ii) Question (4) is false over finite fields, real or algebraically closed fields, and p-adic fields.
(iii) Over number fields one may equivalently take $n = 1$ in (4).

The proof is simple: first note that property (4) is inherited by subfields of k. A counterexample over $k = \mathbb{F}_q$ is provided by the factorization of $f = X^{q^d} - X$ into the product of all monic irreducible polynomials over $\mathbb{F}_q$ whose degree is a divisor of d. This example can be lifted to the p-adics. Over $\mathbb{R}$ and $\mathbb{C}$ one may use $f = X^n - 1$ or the Chebychev polynomials. The proof of (iii) is an immediate consequence of the Hilbert irreducibility theorem [32]. (We note that a corresponding conclusion with τ instead of L is not clear.)

It is interesting to note that the Pochhammer-Wilkinson polynomial p_n can be evaluated with $O(\sqrt{n}\log^2 n)$ arithmetic operations. Indeed, assume $n = m^2$ and write for $x \in \mathbb{R}$

$$p_{m^2}(x) = \prod_{\substack{0 \le q < m \\ 0 \le r < m}} (x - qm - r) = \prod_{q=0}^{m-1} h_m(q),$$

where $h_m(Y) := \prod_{r=0}^{m-1}(x - mY - r) = \sum_{i=0}^{m} a_i(x)Y^i$. Using FFT-based fast arithmetic (cf. [14, Chap. 2]) one can compute on input x all coefficients $a_i(x)$ and then use multiple evaluation to obtain $h_m(q)$ for all $q < m$ using only $O(m \log^2 m)$ arithmetic operations. A more detailed reasoning yields $\tau(p_{m^2}) = O(m \log^2 m \log \log m)$. We remark that a similar idea was first formulated for the efficient computation of factorials [44].

We do not know whether the sequence of Pochhammer-Wilkinson polynomials (p_n) is hard to compute in the sense that $L(p_n) \ge n^\epsilon$ for some $\epsilon > 0$. However, we can make the following interesting observation:

$$p_n(X^2) = q_n \cdot \bar{q}_n := \prod_{j=1}^{n}(X - \sqrt{j}) \cdot \prod_{j=1}^{n}(X + \sqrt{j}).$$

Using techniques of algebraic complexity theory, Heintz and Morgenstern [25] were able to prove that each of the sequences (q_n) and $(\bar{q}_n)$ is hard to compute (see also Baur [4]).

If we were able to extend the hardness proof for q_n to all of its nonzero multiples, then we had proved that all nonzero multiples of p_n are hard and thus $P \ne NP$ over $\mathbb{C}$. The problem to relate the complexity of a polynomial with its nonzero multiples appears here closely related to the P-NP problem over the complex numbers.

We remark that Aldaz at al. [1] showed that $\prod_{i=1}^{n}(X - 2^{2^i})$ is hard to compute and Baur and Halupczok [5] proved a corresponding lower bound for all the nonzero multiples of these polynomials.

5   Complexity of Factors

We proceed now with a systematic investigation of the relationship in complexity of a polynomial f with those of its factors. The first question to ask is whether the complexity of a factor g can always be polynomially bounded in the complexity of f. Our developments in Section 4.3 already indicate that the answer to this question is negative, since a positive answer would provide a proof of $P \ne NP$ over $\mathbb{C}$. The answer is indeed negative, as first discovered by Lipton and Stockmeyer [35]. The simplest known example illustrating this is as follows: consider $f_n = X^{2^n} - 1 = \prod_{j<2^n}(X - \zeta^j)$, where $\zeta = \exp(2\pi i/2^n)$. By repeated squaring we get $L(f_n) \leq n+1$. On the other hand, one can prove that for almost all $M \subseteq \{0, 1, \ldots, 2^n - 1\}$ the random factor $\prod_{j \in M}(X - \zeta^j)$ has a complexity which is exponential in n, cf. [14, Exercise 9.8]. A similar reasoning can be made over the rationals based on the factorization into the cyclotomic polynomials. This idea yields reducible factors of high complexity.

Problem 1. Construct a sequence of *irreducible* polynomials g_n with complexity exponential in n, which have a nonzero multiple f_n with complexity polynomially bounded in n.

In the above example, the degree of the factor g is exponential in the complexity of f. We restrict now our attention to factors having a degree polynomially bounded in the complexity of f. A well-known result by Kaltofen [28] describes a randomized polynomial time algorithm for factoring a multivariate polynomial f given by a straight-line program. Hereby, the upper bound is polynomial in the straight-line complexity of f and the degree of f. In a widely unknown paper, Kaltofen [27] also proved that the complexity of any irreducible factor g is polynomially bounded in the complexity of f, the degree of g and the multiplicity of g. This result, stated explicitly below, was independently found by the author, compare [12, Thm. 8.14]. Hereby, the notation $M(d)$ stands for an upper bound on the complexity for the multiplication of two univariate polynomials of degree d over k, e.g. $M(d) = O(d \log d)$ if the field k "supports" fast Fourier transforms.

Theorem 1. *Assume $f = g^e h$ with polynomials $g, h \in k[X_1, \ldots, X_n]$ which are coprime. Let $d \geq 1$ be the degree of g and suppose that k is a field of characteristic zero. Then we have*

$$L(g) = O\big(M(d^3 e)(L(f) + d \log e)\big).$$

In [12, Conj. 8.3] we have conjectured that the dependence on the multiplicity e can be omitted, that is, we think that the complexity of the factor g is polynomially bounded in the complexity of f and the degree d of g.

In Section 7.1 we will present a new result [10] stating that the dependence on the multiplicity in Theorem 1 can indeed be omitted when replacing complexity by the related notion of "approximative complexity", to be defined next.

The fact that a computation with a polynomial number of steps may produce intermediate results of exponential degree is well-known to cause considerable complications. Actually, the so-called weak BSS model of computation [30] was defined in order to cope with this phenomenon, simply by forbidding such an exponential growth of degree. The fact that the P-NP separation in the weak model is trivial to obtain clearly shows that this model is a large oversimplification. We remark that the Valiant model also excludes an exponential growth of degrees during a computation. However, one can show [43] that this is no restriction for the study of polynomials of p-bounded degree, like permanents. Note that resultants of systems of polynomial equations have a huge degree and are not captured by Valiant's framework.

6 Approximative Complexity

The concept of approximative complexity has been systematically studied in the framework of bilinear complexity (border rank) and there it has turned out to be one of the main keys to the currently best known fast matrix multiplication algorithms [17]. For computations of polynomials or rational functions, approximative complexity has been investigated in less detail. Griesser [22] generalized most of the known lower bounds for multiplicative complexity to approximative complexity. Lickteig [33] as well as Grigoriev and Karpinski [23] employ the notion of approximative complexity for proving lower bounds. We refer to [14, Chap. 15] and the references there for further information.

6.1 Algebraic Definition and Topological Characterization

Let $f = f_q(X_1, \ldots, X_n)Y^q + f_{q+1}(X_1, \ldots, X_n)Y^{q+1} + \ldots$ be the expansion of a polynomial f with respect to the variable Y. We do not know whether the complexity of the leading coefficient f_q can be polynomially bounded in the the the complexity of f. However, we can make the following observation. For the moment assume that k is the field of real or complex numbers. We have $\lim_{y \to 0} y^{-q} f(X, y) = f_q(X)$ and $L(f(X, y)) \leq L(f)$ for all $y \in k$. Thus we can approximate f_q with arbitrary precision by polynomials having complexity at most $L(f)$. We could say that f_q has "approximate complexity" at most $L(f)$.

We will formalize this in an algebraic way; a topological interpretation will be given later. In what follows, $K := k(\epsilon)$ is a rational function field in the indeterminate ϵ over the field k and R denotes the local subring of K consisting of the rational functions defined at $\epsilon = 0$. We write $F_{\epsilon=0}$ for the image of $F \in R[X]$ under the morphism $R[X] \to k[X]$ induced by $\epsilon \mapsto 0$.

Definition 1. *Let $f \in k[X_1, \ldots, X_n]$. The approximative complexity $\underline{L}(f)$ of the polynomial f is the smallest natural number r such that there exists F in $R[X_1, \ldots, X_n]$ satisfying $F_{\epsilon=0} = f$ and $L(F) \leq r$. Here the complexity L is to be interpreted with respect to the larger field of constants K.*

Even though L refers to division-free straight-line programs, divisions will occur implicitly since our model allows the free use of any elements of K as constants. In fact, the point is that even though F is defined with respect to the morphism $\epsilon \mapsto 0$, the intermediate results of the computation may not be so. Note that $\underline{L}(f) \leq L(f)$.

We remark that the assumption that any elements of K are free constants is made for achieving conceptual simplicity. We could as well require to build up the needed elements of K from ϵ, ϵ^{-1} and elements of k. One can show that this would not significantly change our main conclusions.

The topological characterization of approximative complexity, to be presented next, shows that this is a very natural notion from a mathematical point of view. Assume k to be an algebraically closed field. There is a natural way to put a Zariski topology on the polynomial ring $A_n := k[X_1, \ldots, X_n]$ as a limit of the Zariski topologies on the finite dimensional subspaces $\{f \in A_n \mid \deg f \leq d\}$ for $d \in \mathbb{N}$. If k is the field of complex numbers, we may define the Euclidean topology on A_n in a similar way. If $f \in A_n$ satisfies $\underline{L}(f) \leq r$, then it easy to see that f lies in the closure (Zariski or Euclidean) of the set $\{f \in A_n \mid L(f) \leq r\}$. Alder [2] has shown that the converse is true and obtained the following topological characterization of the approximative complexity.

Theorem 2. *The set $\{f \in A_n \mid \underline{L}(f) \leq r\}$ is the closure of the set $\{f \in A_n \mid L(f) \leq r\}$ for the Zariski topology. If $k = \mathbb{C}$, this is also true for the Euclidean topology.*

6.2   Computation of p-adic Coefficients

Let $f, p \in k[X_1, \ldots, X_m][Y]$ and assume p to be monic of degree $d \geq 1$ in Y. Let $f = \sum_i f_i p^i$ be the p-adic representation of f, that is, $f_i \in k[X, Y]$ and $\deg_Y f_i < d$. Using the idea in [14, § 7.1] it is not hard to see that the complexity of the coefficient polynomial f_i of Y^i can be polynomially bounded in d, i, and $L(f)$. The following observation shows that the dependence on the degree i cannot be avoided in general.

Proposition 1. *The complexity of coefficient polynomials in a p-adic representation of a polynomial f is not polynomially bounded in $L(f)$ and $d = \deg p$, unless Valiant's hypothesis is false.*

Consider the Y-adic representation of the following polynomial f_n of complexity $L(f_n) = O(n^2)$

$$f_n := \prod_{i=1}^n \left(\sum_{j=1}^n X_{ij} Y^{2^{j-1}} \right) = \sum_i f_{n,i}(X) Y^i$$

and observe that the coefficient $f_{n,2^n-1}(X)$ equals the the permanent of the matrix $[X_{ij}]$. This already provides the proof of Proposition 1.

Assume now that the p-adic representation $f = f_\ell p^\ell + f_{\ell+1} p^{\ell+1} + \ldots$ starts at order ℓ, $f_\ell \neq 0$. We can consider f_ℓ as the leading coefficient of f with respect to the basis p. By contrast with Proposition 1, we can say the following about the approximative complexity of the leading coefficient in relation to the complexity of f (cf. [10]).

Proposition 2. *The approximative complexity of the leading coefficient f_ℓ is polynomially bounded in d and $\underline{L}(f)$: we have*

$$\underline{L}(f_\ell) = O(M(d)\underline{L}(f)).$$

7   Decision versus Computation

7.1   Approximative Complexity of Factors

Here is the result from [10], which eliminates the dependence on the multiplicity in Theorem 1 by switching to approximative complexity. The number $2 \leq \omega \leq 2.38$ denotes the exponent of matrix multiplication (cf. [14, Chap. 15]).

Theorem 3. *Assume that g is an irreducible factor of degree d of a polynomial $f \in \mathbb{R}[X_1, \ldots, X_n]$. We assume that the zeroset of g is a hypersurface in $\mathbb{R}^n$. Then we have for any $\epsilon > 0$ that*

$$\underline{L}(g) = O\big(M(d^4)\underline{L}(f) + d^{2\omega+\epsilon}M(d)\big).$$

A corresponding result holds over $\mathbb{C}$. Moreover, we remark that in the special situation, where g is the generator of the graph of a polynomial function, we obtain considerably better bounds, valid over any infinite field of characteristic zero.

The idea of the proof of Theorem 3 is as follows: After a suitable coordinate transformation, one can interpret the zeroset of the factor g locally around the origin as the graph of some analytic function φ. In order to cope with a possibly large multiplicity of g, we apply a small perturbation to the polynomial f without affecting its complexity too much. This results in a small perturbation of φ. We compute now the homogeneous parts of the perturbed φ by a Newton iteration up to a certain order. Using efficient polynomial arithmetic, this gives us an upper bound on the approximative complexity of the homogeneous parts of φ up to a predefined order. In the special case, where the factor g is the generator of the graph of a function, we are already done. Otherwise, we view the factor g as the minimal polynomial of φ in the variable $Y := X_n$ over the field $k(X_1, \ldots, X_{n-1})$. We show that the Taylor approximations up to order $2d^2$ uniquely determine the factor g and compute the bihomogeneous components of g with respect to the degrees in the X-variables and Y by fast linear algebra.

7.2   Applications to Decision Complexity

By combining Theorem 3 with Lemma 1 we immediately get the following result stating that the approximative complexity of a polynomial g can be bounded polynomially in the decision complexity and the degree of g.

Corollary 1. *Let g be the generator of an irreducible hypersurface in $\mathbb{R}^n$ or in $\mathbb{C}^n$, $d = \deg g$. Then we have for any $\epsilon > 0$ that*

$$\underline{L}(g) = O\big(M(d^4)C(g) + d^{2\omega+\epsilon}M(d)\big).$$

It is quite natural to incorporate the concept of approximative complexity into Valiant's algebraic framework of NP-completeness.

Definition 2. *An* approximatively p-computable family *is a p-family (f_n) such that $\underline{L}(f_n)$ is a p-bounded function of n. The complexity class $\underline{\mathrm{VP}}$ comprises all such families over a fixed field k.*

It is obvious that $\mathrm{VP} \subseteq \underline{\mathrm{VP}}$. If the polynomial f is a projection of a polynomial g, then we clearly have $\underline{L}(f) \leq \underline{L}(g)$. Therefore, the complexity class $\underline{\mathrm{VP}}$ is closed under p-projections. We remark that $\underline{\mathrm{VP}}$ is also closed under the polynomial oracle reductions introduced in [11].

At the moment, we know very little about the relations between the complexity classes VP, $\underline{\mathrm{VP}}$, and VNP.

Problem 2.   1. Is the class VP strictly contained $\underline{VP}$?
2. Is the class $\underline{VP}$ contained in VNP?

Since the class VP is closed under p-projections, the following strengthening of Valiant's hypothesis is equivalent to saying that VNP-complete families are not approximately p-computable.

Conjecture 1. The class VNP is not contained in the class $\underline{VP}$.

This conjecture should be compared with the known work on polynomial time deterministic or randomized approximation algorithms for the permanent of non-negative matrices [34,3,26]. Based on the Markov chain approach, Jerrum, Sinclair and Vigoda [26] have recently established a fully-polynomial randomized approximation scheme for computing the permanent of an arbitrary real matrix with non-negative entries. We note that this result does not contradict Conjecture 1 since the above mentioned algorithm works only for matrices with *non-negative* entries while approximative straight-line programs work on all real inputs.

Under the hypothesis VNP $\not\subseteq \underline{VP}$, we can can conclude that checking the values of polynomials forming VNP-complete families is hard, even when we allow randomized algorithms with two-sided error, formalized by randomized algebraic computation trees. This follows for deterministic computations easily from Corollary 1. The extension to randomized trees is straight-forward using [38,15,18].

Corollary 2. *Assume* VNP $\not\subseteq \underline{VP}$ *over* $\mathbb{R}$. *Then for any* VNP-*complete family* (g_n), *checking the value* $y = g_n(x)$ *over the reals cannot be done by deterministic or randomized algebraic computation trees with a polynomial number of arithmetic operations and tests in* n.

By applying Corollary 1 to the permanent polynomial, we see that Conjecture 1 implies the following separation of complexity classes in the BSS-model of computation (cf. [8]).

Corollary 3. *If* VNP $\not\subseteq \underline{VP}$ *is true, then we have* P $\neq$ PAR *in the BSS-model over the reals.*

References

1. M. Aldaz, J. Heintz, G. Matera, J. L. Montaña, and L. M. Pardo. Time-space tradeoffs in algebraic complexity theory. *J. Compl.*, 16:2–49, 2000. 10
2. A. Alder. *Grenzrang und Grenzkomplexität aus algebraischer und topologischer Sicht.* PhD thesis, Zürich University, 1984. 13
3. A. I. Barvinok. Polynomial time algorithms to approximate permanents and mixed discriminants within a simply exponential factor. *Random Structures and Algorithms*, 14:29–61, 1999. 15
4. W. Baur. Simplified lower bounds for polynomials with algebraic coefficients. *J. Compl.*, 13:38–41, 1997. 10

5. W. Baur and K. Halupczok. On lower bounds for the complexity of polynomials and their multiples. *Comp. Compl.*, 8:309–315, 1999. 10

6. S. Berkowitz, C. Rackoff, S. Skyum, and L. Valiant. Fast parallel computation of polynomials using few processors. *SIAM J. Comp.*, 12:641–644, 1983. 6

7. L. Blum, F. Cucker, M. Shub, and S. Smale. Algebraic Settings for the Problem "$P \neq NP$?". In *The mathematics of numerical analysis*, number 32 in Lectures in Applied Mathematics, pages 125–144. Amer. Math. Soc., 1996. 5

8. L. Blum, F. Cucker, M. Shub, and S. Smale. *Complexity and Real Computation*. Springer, 1998. 4, 8, 15

9. L. Blum, M. Shub, and S. Smale. On a theory of computation and complexity over the real numbers. *Bull. Amer. Math. Soc.*, 21:1–46, 1989. 3, 4

10. P. Bürgisser. The complexity of factors of multivariate polynomials. Preprint, University of Paderborn, 2001, submitted. 4, 11, 13

11. P. Bürgisser. On the structure of Valiant's complexity classes. *Discr. Math. Theoret. Comp. Sci.*, 3:73–94, 1999. 14

12. P. Bürgisser. *Completeness and Reduction in Algebraic Complexity Theory*, volume 7 of *Algorithms and Computation in Mathematics*. Springer Verlag, 2000. 4, 5, 8, 11

13. P. Bürgisser. Cook's versus Valiant's hypothesis. *Theoret. Comp. Sci.*, 235:71–88, 2000. 6, 7

14. P. Bürgisser, M. Clausen, and M. A. Shokrollahi. *Algebraic Complexity Theory*, volume 315 of *Grundlehren der mathematischen Wissenschaften*. Springer Verlag, 1997. 3, 5, 7, 8, 10, 11, 12, 13

15. P. Bürgisser, M. Karpinski, and T. Lickteig. On randomized semialgebraic decision complexity. *J. Compl.*, 9:231–251, 1993. 15

16. P. Bürgisser, T. Lickteig, and M. Shub. Test complexity of generic polynomials. *J. Compl.*, 8:203–215, 1992. 8

17. D. Coppersmith and S. Winograd. Matrix multiplication via arithmetic progressions. *J. Symb. Comp.*, 9:251–280, 1990. 12

18. F. Cucker, M. Karpinski, P. Koiran, T. Lickteig, and K. Werther. On real Turing machines that toss coins. In *Proc. 27th ACM STOC, Las Vegas*, pages 335–342, 1995. 6, 15

19. H. Fournier and P. Koiran. Are lower bounds easier over the reals? In *Proc. 30th ACM STOC*, pages 507–513, 1998. 7

20. H. Fournier and P. Koiran. Lower bounds are not easier over the reals: Inside PH. In *Proc. ICALP 2000*, LNCS 1853, pages 832–843, 2000. 4, 6

21. J. von zur Gathen. Feasible arithmetic computations: Valiant's hypothesis. *J. Symb. Comp.*, 4:137–172, 1987. 5

22. B. Griesser. Lower bounds for the approximative complexity. *Theoret. Comp. Sci.*, 46:329–338, 1986. 12

23. D.Yu. Grigoriev and M. Karpinski. Randomized quadratic lower bound for knapsack. In *Proc. 29th ACM STOC*, pages 76–85, 1997. 12

24. J. Grollmann and A. L. Selman. Complexity measures for public-key cryptosystems. *SIAM J. Comp.*, 17(2):309–335, 1988. 7

25. J. Heintz and J. Morgenstern. On the intrinsic complexity of elimination theory. *Journal of Complexity*, 9:471–498, 1993. 8, 10

26. M. R. Jerrum, A. Sinclair, and E. Vigoda. A polynomial-time approximation algorithm for the permanent of a matrix with non-negative entries. *Electronic Colloquium on Computational Complexity*, 2000. Report No. 79. 15

27. E. Kaltofen. Single-factor Hensel lifting and its application to the straight-line complexity of certain polynomials. In *Proc. 19th ACM STOC*, pages 443–452, 1986. 4, 11
28. E. Kaltofen. Factorization of polynomials given by straight-line programs. In S. Micali, editor, *Randomness and Computation*, pages 375–412. JAI Press, Greenwich CT, 1989. 11
29. P. Koiran. Computing over the reals with addition and order. *Theoret. Comp. Sci.*, 133:35–47, 1994. 6
30. P. Koiran. A weak version of the Blum, Shub & Smale model. *J. Comp. Syst. Sci.*, 54:177–189, 1997. 11
31. P. Koiran. Circuits versus trees in algebraic complexity. In *Proc. STACS 2000*, number 1770 in LNCS, pages 35–52. Springer Verlag, 2000. 4
32. S. Lang. *Fundamentals of Diophantine Geometry*. Springer Verlag, 1983. 10
33. T. Lickteig. On semialgebraic decision complexity. Technical Report TR-90-052, Int. Comp. Sc. Inst., Berkeley, 1990. Habilitationsschrift, Universität Tübingen. 12
34. N. Linial, A. Samorodnitsky, and A. Wigderson. A deterministic polynomial algorithm for matrix scaling and approximate permanents. In *Proc. 30th ACM STOC*, pages 644–652, 1998. 15
35. R. J. Lipton and L. J. Stockmeyer. Evaluation of polynomials with superpreconditioning. *J. Comp. Syst. Sci.*, 16:124–139, 1978. 11
36. S. Meiser. Point location in arrangements of hyperplanes. *Information and Computation*, 106:286–303, 1993. 7
37. F. Meyer auf der Heide. A polynomial linear search algorithm for the n-dimensional knapsack problem. *J. ACM*, 31:668–676, 1984. 6, 7, 8
38. F. Meyer auf der Heide. Simulating probabilistic by deterministic algebraic computation trees. *Theoret. Comp. Sci.*, 41:325–330, 1985. 15
39. F. Meyer auf der Heide. Fast algorithms for n-dimensional restrictions of hard problems. *J. ACM*, 35:740–747, 1988. 6, 7, 8
40. B. Poizat. *Les Petits Cailloux*. Number 3 in Nur Al-Mantiq War-Ma'rifah. Aléas, Lyon, 1995. 4
41. A. L. Selman. A survey of one-way functions in complexity theory. *Math. Systems Theory*, 25:203–221, 1992. 7
42. M. Shub and S. Smale. On the intractability of Hilbert's Nullstellensatz and an algebraic version of "NP ≠ P?". *Duke Math. J.*, 81:47–54, 1995. 4, 9
43. V. Strassen. Vermeidung von Divisionen. *Crelles J. Reine Angew. Math.*, 264:184–202, 1973. 11
44. V. Strassen. Einige Resultate über Berechnungskomplexität. *Jahr. Deutsch. Math. Ver.*, 78:1–8, 1976. 10
45. L. G. Valiant. Completeness classes in algebra. In *Proc. 11th ACM STOC*, pages 249–261, 1979. 3, 5
46. L. G. Valiant. The complexity of computing the permanent. *Theoret. Comp. Sci.*, 8:189–201, 1979. 5
47. L. G. Valiant. Reducibility by algebraic projections. In *Logic and Algorithmic: an International Symposium held in honor of Ernst Specker*, volume 30, pages 365–380. Monogr. No. 30 de l'Enseign. Math., 1982. 3, 5

Playing Games with Algorithms: Algorithmic Combinatorial Game Theory

Erik D. Demaine

Department of Computer Science, University of Waterloo,
Waterloo, Ontario N2L 3G1, Canada
eddemaine@uwaterloo.ca

Abstract. Combinatorial games lead to several interesting, clean problems in algorithms and complexity theory, many of which remain open. The purpose of this paper is to provide an overview of the area to encourage further research. In particular, we begin with general background in combinatorial game theory, which analyzes ideal play in perfect-information games. Then we survey results about the complexity of determining ideal play in these games, and the related problems of solving puzzles, in terms of both polynomial-time algorithms and computational intractability results. Our review of background and survey of algorithmic results are by no means complete, but should serve as a useful primer.

1 Introduction

Many classic games are known to be computationally intractable: one-player puzzles are often NP-complete (as in Minesweeper), and two-player games are often PSPACE-complete (as in Othello) or EXPTIME-complete (as in Checkers, Chess, and Go). Surprisingly, many seemingly simple puzzles and games are also hard. Other results are positive, proving that some games can be played optimally in polynomial time. In some cases, particularly with one-player puzzles, the computationally tractable games are still interesting for humans to play.

After reviewing some of the basic concepts in combinatorial game theory in Section 2, Sections 3–5 survey several of these algorithmic and intractability results. We do not intend to give a complete survey, but rather to give an introduction to the area. Given the space restrictions, the sample of results mentioned here reflect a personal bias: results about "well-known" games (in North America), some of the results I find interesting, and results in which I have been involved. For a more complete overview, please see the full version of this paper [12].

Combinatorial game theory is to be distinguished from other forms of game theory arising in the context of economics. Economic game theory has applications in computer science as well, most notably in the context of auctions [11] and analyzing behavior on the Internet [33].

J. Sgall, A. Pultr, and P. Kolman (Eds.): MFCS 2001, LNCS 2136, pp. 18–33, 2001.
© Springer-Verlag Berlin Heidelberg 2001

2   Combinatorial Game Theory

A *combinatorial game* typically involves two players, often called *Left* and *Right*, alternating play in well-defined *moves*. However, in the interesting case of a *combinatorial puzzle*, there is only one player, and for *cellular automata* such as Conway's Game of Life, there are no players. In all cases, no randomness or hidden information is permitted: all players know all information about gameplay (*perfect information*). The problem is thus purely strategic: how to best play the game against an ideal opponent.

It is useful to distinguish several types of two-player perfect-information games [3, pp. 16–17]. A common assumption is that the game terminates after a finite number of moves (the game is *finite* or *short*), and the result is a unique winner. Of course, there are exceptions: some games (such as Life and Chess) can be *drawn* out forever, and some games (such as tic-tac-toe and Chess) define *ties* in certain cases. However, in the combinatorial-game setting, it is useful to define the *winner* as the last player who is able to move; this is called *normal play*. If, on the other hand, the winner is the first player who cannot move, this is called *misère play*. (We will normally assume normal play.) A game is *loopy* if it is possible to return to previously seen positions (as in Chess, for example). Finally, a game is called *impartial* if the two players (Left and Right) are treated identically, that is, each player has the same moves available from the same game position; otherwise the game is called *partizan*.

A particular two-player perfect-information game without ties or draws can have one of four *outcomes* as the result of ideal play: player Left wins, player Right wins, the first player to move wins (whether it is Left or Right), or the second player to move wins. One goal in analyzing two-player games is to determine the outcome as one of these four categories, and to find a strategy for the winning player to win. Another goal is to compute a deeper structure to games described in the remainder of this section, called the *value* of the game.

A beautiful mathematical theory has been developed for analyzing two-player combinatorial games. The most comprehensive reference is the book *Winning Ways* by Berlekamp, Conway, and Guy [3], but a more mathematical presentation is the book *On Numbers and Games* by Conway [8]. See also [21] for a bibliography. The basic idea behind the theory is simple: a two-player game can be described by a rooted tree, each node having zero or more *left* branches correspond to options for player Left to move and zero or more *right* branches corresponding to options for player Right to move; leaves corresponding to finished games, the winner being determined by either normal or misère play. The interesting parts of combinatorial game theory are the several methods for manipulating and analyzing such games/trees. We give a brief summary of some of these methods in this section.

2.1   Conway's Surreal Numbers

A richly structured special class of two-player games are John H. Conway's *surreal numbers* [8], a vast generalization of the real and ordinal number systems.

Basically, a surreal number $\{L \mid R\}$ is the "simplest" number larger than all Left options (in L) and smaller than all Right options (in R); for this to constitute a number, all Left and Right options must be numbers, defining a total order, and each Left option must be less than each Right option. See [8] for more formal definitions.

For example, the simplest number without any larger-than or smaller-than constraints, denoted $\{\mid\}$, is 0; the simplest number larger than 0 and without smaller-than constraints, denoted $\{0 \mid\}$, is 1. This method can be used to generate all natural numbers and indeed all ordinals. On the other hand, the simplest number less than 0, denoted $\{\mid 0\}$, is -1; similarly, all negative integers can be generated. Another example is the simplest number larger than 0 and smaller than 1, denoted $\{0 \mid 1\}$, which is $\frac{1}{2}$; similarly, all dyadic rationals can be generated. After a countably infinite number of such construction steps, all real numbers can be generated; after many more steps, the surreals are all numbers that can be generated in this way.

What is interesting about the surreals from the perspective of combinatorial game theory is that they are a subclass of all two-player perfect-information games, and some of the surreal structure, such as addition and subtraction, carries over to general games. Furthermore, while games are not totally ordered, they can still be compared to some surreal numbers and, amazingly, how a game compares to the surreal number 0 determines exactly the outcome of the game. This connection is detailed in the next few paragraphs.

First we define some algebraic structure of games that carries over from surreal numbers. Two-player combinatorial games, or trees, can simply be represented as $\{L \mid R\}$ where, in contrast to surreal numbers, no constraints are placed on L and R. The *negation* of a game is the result of reversing the roles of the players Left and Right throughout the game. The (*disjunctive*) *sum* of two (sub)games is the game in which, at each player's turn, the player has a binary choice of which subgame to play, and makes a move in precisely that subgame. A partial order is defined on games recursively: a game x is *less than or equal to* a game y if every Left option of x is less than y and every Right option of y is more than x.

Note that while $\{-1 \mid 1\} = 0 = \{\mid\}$ in terms of numbers, $\{-1 \mid 1\}$ and $\{\mid\}$ denote different games (lasting 1 move and 0 moves, respectively), and in this sense are *equal* in *value* but not *identical* symbolically or game-theoretically. Nonetheless, the games $\{-1 \mid 1\}$ and $\{\mid\}$ have the same outcome: the second player to move wins.

Amazingly, this holds in general: two equal numbers represent games with equal outcome (under ideal play). In particular, all games equal to 0 have the outcome that the second player to move wins. Furthermore, all games equal to a positive number have the outcome that the Left player wins; more generally, all positive games (games larger than 0) have this outcome. Symmetrically, all negative games have the outcome that the Right player wins (this follows automatically by the negation operation).

There is one outcome not captured by the characterization into zero, positive, and negative games: the first player to move wins. An example of such a game is $\{1 \,|\, 0\}$; this fails to be a surreal number because $1 \geq 0$. By the claim above, $\{1 \,|\, 0\} \,\|\, 0$. Indeed, $\{1 \,|\, 0\} \,\|\, x$ for all surreal numbers x, $0 \leq x \leq 1$. In contrast, $x < \{1 \,|\, 0\}$ for all $x < 0$ and $\{1 \,|\, 0\} < x$ for all $1 < x$. In general it holds that a game is fuzzy with some surreal numbers in an interval $[-n, n]$ but comparable with all surreals outside that interval.

For brevity we omit many other useful notions in combinatorial game theory, such as additional definitions of summation, super-infinitesimal games $*$ and $\uparrow$, mass, temperature, thermographs, the simplest form of a game, remoteness, and suspense; see [3,8].

2.2   Sprague-Grundy Theory

A celebrated result in combinatorial game theory is the characterization of impartial two-player perfect-information games, discovered independently in the 1930's by Sprague [39] and Grundy [25]. Recall that a game is *impartial* if it does not distinguish between the players Left and Right. The Sprague-Grundy theory [39,25,8,3] states that every finite impartial game is equivalent to an instance of the game of Nim, characterized by a single natural number n. This theory has since been generalized to all impartial games by generalizing Nim to all ordinals n; see [8,38].

Nim [5] is a game played with several *heaps*, each with a certain number of tokens. A Nim game with a single heap of size n is denoted by $*n$ and is called a *nimber*. During each move a player can pick any pile and reduce it to any smaller nonnegative integer size. The game ends when all piles have size 0. Thus, a single pile $*n$ can be reduced to any of the smaller piles $*0$, $*1$, $\ldots$, $*(n-1)$. Multiple piles in a game of Nim are independent, and hence any game of Nim is a sum of single-pile games $*n$ for various values of n. In fact, a game of Nim with k piles of sizes n_1, n_2, $\ldots$, n_k is equivalent to a one-pile Nim game $*n$, where n is the binary XOR of n_1, n_2, $\ldots$, n_k. As a consequence, Nim can be played optimally in polynomial time (polynomial in the encoding size of the pile sizes).

Even more surprising is that *every* impartial two-player perfect-information game has the same value as a single-pile Nim game, $*n$ for some n. The number n is called variously the *G-value*, *Grundy-value*, or *Sprague-Grundy function* of the game. It is easy to define: suppose that game x has k options $y_1, \ldots, y_k$ for the first move (independent of which player goes first). By induction, we can compute $y_1 = *n_1$, $\ldots$, $y_k = *n_k$. The theorem is that x equals $*n$ where n is the smallest natural number not in the set $\{n_1, \ldots, n_k\}$. This number n is called the *minimum excluded value* or *mex* of the set. This description has also assumed that the game is finite, but this is easy to generalize [8,38].

The Sprague-Grundy function can increase by at most 1 at each level of the game tree, and hence the resulting nimber is linear in the maximum number of moves that can be made in the game; the encoding size of the nimber is only logarithmic in this count. Unfortunately, computing the Sprague-Grundy function for a general game by the obvious method uses time linear in the number of possible states, which can be exponential in the nimber itself.

Nonetheless, the Sprague-Grundy theory is extremely helpful for analyzing impartial two-player games, and for many games there is an efficient algorithm to determine the nimber. Examples include Nim itself, Kayles, and various generalizations [27]; and Cutcake and Maundy Cake [3, pp. 26–29]. In all of these examples, the Sprague-Grundy function has a succinct characterization (if somewhat difficult to prove); it can also be easily computed using dynamic programming.

2.3   Strategy Stealing

Another useful technique in combinatorial game theory for proving that a particular player must win is *strategy stealing*. The basic idea is to assume that one player has a winning strategy, and prove that in fact the other player has a winning strategy based on that strategy. This contradiction proves that the second player must in fact have a winning strategy. An example of such an argument is given in Section 3.1. Unfortunately, such a proof by contradiction gives no indication of what the winning strategy actually is, only that it exists. In many situations, such as the one in Section 3.1, the winner is known but no polynomial-time winning strategy is known.

2.4   Puzzles

There is little theory for analyzing combinatorial puzzles (one-player games) along the lines of two-player theory summarized in this section. We present one such viewpoint here. In most puzzles, solutions subdivide into a sequence of moves. Thus, a puzzle can be viewed a tree, similar to a two-player game except that edges are not distinguished between Left and Right. The goal is to reach a position from which there are no valid moves (normal play). Loopy puzzles are common; to be more explicit, repeated subtrees can be converted into self-references to form a directed graph.

A consequence of the above view is that a puzzle is basically an impartial two-player game except that we are not interested in the outcome from two players alternating in moves. Rather, questions of interest in the context of puzzles are (a) whether a given puzzle is solvable, and (b) finding the solution with the fewest moves. An important open direction of research is to develop a general theory for resolving such questions, similar to the two-player theory. For example, using the analogy between impartial two-player games described above, the notion of sums of puzzles makes sense, although it is not clear that it plays a similarly key role as with games.

3   Algorithms for Two-Player Games

Many nonloopy two-player games are PSPACE-complete. This is fairly natural because games are closely related to boolean expressions with alternating quantifiers (for which deciding satisfiability is PSPACE-complete): there exists a move for Left such that, for all moves for Right, there exists another move for Left, etc. A PSPACE-completeness result has two consequences. First, being in PSPACE means that the game can be played optimally, and typically all positions can be enumerated, using possibly exponential time but only polynomial space. Thus such games lend themselves to a somewhat reasonable exhaustive search for small enough sizes. Second, the games cannot be solved in polynomial time unless P = PSPACE, which is even "less likely" than P equaling NP.

On the other hand, loopy two-players games are often EXPTIME-complete. Such a result is one of the few types of true lower bounds in complexity theory, implying that all algorithms require exponential time in the worst case.

In this section we briefly survey some of these complexity results and related positive results, ordered roughly chronologically by the first result on a particular game. See also [18] for a related survey. Because of space constraints we omit discussion of games on graphs, as well as the board games Gobang, Shogi, and Othello. For details on these and other games, please refer to the full version of this paper [12].

3.1 Hex

Hex [3, pp. 679–680] is a game designed by Piet Hein and played on a diamond-shaped hexagonal board; see Fig. 1. Players take turns filling in empty hexagons with their color. The goal of a player is to connect the opposite sides of their color with hexagons of their color. (In the figure, one player is solid and the other player is dotted.) A game of Hex can never tie, because if all hexagons are colored arbitrarily, there is precisely one connecting path of an appropriate color between opposite sides of the board.

Fig. 1. A 5 × 5 Hex board

Nash [3, p. 680] proved that the first player to move can win by using a strategy stealing argument (see Section 2.3). In contrast, Reisch [35] proved that determining the outcome of a general position in Hex is PSPACE-complete.

3.2 Checkers (Draughts)

The standard 8 × 8 game of Checkers (Draughts), like many classic games, is finite and hence can be played optimally in constant time (in theory). The complexity of playing in a general $n \times n$ board from a natural starting position, such as the one in Fig. 2, is open. However, deciding the outcome of an arbitrary configuration is PSPACE-hard [22]. If a polynomial bound is placed on the number of moves that are allowed in between jumps (which is a reasonable generalization of the drawing rule in standard Checkers [22]), then the problem is in PSPACE and hence is PSPACE-complete. Without such a restriction, however, Checkers is EXPTIME-complete [37].

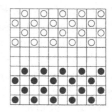

Fig. 2. A natural starting configuration for 10 × 10 Checkers, from [22]

On the other hand, certain simple questions about Checkers can be answered in polynomial time [22,13]. Can one player remove all the other player's pieces in one move (by several jumps)? Can one player king a piece in one move? Because of the notion of parity on $n \times n$ boards, these questions reduce to checking the existence of an Eulerian path or general path, respectively, in a particular directed graph; see [22,13]. However, for boards defined by general graphs, at least the first question becomes NP-complete [22].

3.3   Go

Presented at the same conference as the Checkers result in the previous section (FOCS'78), Lichtenstein and Sipser [32] proved that the classic oriental game of Go is also PSPACE-hard for an arbitrary configuration on an $n \times n$ board. This proof does not involve any situations called ko's, where a rule must be invoked to avoid infinite play. In contrast, Robson [36] proved that Go is EXPTIME-complete when ko's are involved, and indeed used judiciously. The type of ko used in this reduction is shown in Fig. 3. When one of the players makes a move shown in the figure, the ko rule prevents (in particular) the other move shown in the figure to be made immediately afterwards.

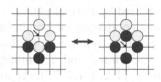

Fig. 3. A simple ko

Recently, Wolfe [41] has shown that even Go endgames are PSPACE-hard. More precisely, a *Go endgame* is when the game has reduced to a sum of Go subgames, each equal to a polynomial-size game tree. This proof is based on several connections between Go and combinatorial game theory detailed in a book by Berlekamp and Wolfe [2].

3.4   Chess

Fraenkel and Lichtenstein [23] proved that a generalization of the classic game Chess to $n \times n$ boards is EXPTIME-complete. Specifically, their generalization has a unique king of each color, and for each color the numbers of pawns, bishops, rooks, and queens increase as some fractional power of n. (Knights are not needed.) The initial configuration is unspecified; what is EXPTIME-hard is to determine the winner (who can checkmate) from an arbitrary specified configuration.

3.5   Hackenbush

Hackenbush is one of the standard examples of a combinatorial game in *Winning Ways*; see e.g. [3, pp. 4–9]. A position is given by a graph with each edge colored either red (Left), blue (Right), or green (neutral), and with certain vertices marked as *rooted*. Players take turns removing an edge of an appropriate color (either neutral or their own color), which also causes all edges not connected to a rooted vertex to be removed. The winner is determined by normal play. Chapter 7 of *Winning Ways* [3, pp. 183–220] proves that determining the *value* of a red-blue Hackenbush position is NP-hard.

3.6 Domineering (Crosscram)

Domineering or *crosscram* [3, pp. 117–124] is a partizan game involving placement of horizontal and vertical dominoes in a grid; a typical starting position is an $m \times n$ rectangle. Left can play only vertical dominoes and Right can play only horizontal dominoes, and dominoes must remain disjoint. The winner is determined by normal play.

The complexity of Domineering, computing either the outcome or the value of a position, remains open. Lachmann, Moore, and Rapaport [31] have shown that the winner and a winning strategy can be computed in polynomial time for $m \in \{1, 2, 3, 4, 5, 7, 9, 11\}$ and all n. These algorithms do not compute the value of the game, nor the optimal strategy, only a winning strategy. We omit discussion of the related game *Cram* [3, pp. 468–472].

3.7 Dots-and-Boxes and Strings-and-Coins

Dots-and-Boxes [1], [3, pp. 507–550] is a well-known children's game in which players take turns drawing horizontal and vertical edges connecting pairs of dots in an $m \times n$ subset of the lattice. Whenever a player makes a move that encloses a unit square with drawn edges, the player is awarded a point and must then draw another edge in the same move. The winner is the player with the most points when the entire grid has been drawn. See Fig. 4 for an example of a position.

Fig. 4.
A   Dots-
and-Boxes
endgame

A generalization arising from the dual of Dots-and-Boxes is *Strings-and-Coins*. This game involves a sort of graph whose vertices are *coins* and whose edges are *strings*. The coins may be tied to each other and to the "ground" by strings; the latter connection can be modeled as a loop in the graph. Players alternate cutting strings (removing edges), and if a coin is thereby freed, that player collects the coin and cuts another string in the same move. The player to collect the most coins wins.

Winning Ways [3, pp. 543–544] describes a proof that Strings-and-Coins endgames are NP-hard. Eppstein [18] observes that this reduction should also apply to endgame instances of Dots-and-Boxes.

3.8 Amazons

Amazons is a game invented by Walter Zamkauskas in 1988, containing elements of Chess and Go. Gameplay takes place on a 10×10 board with four *amazons* of each color arranged as in Fig. 5 (left). In each turn, Left [Right] moves a black [white] amazon to any unoccupied square accessible by a Chess queen's move, and fires an arrow to any unoccupied square reachable by a Chess queen's move from the amazon's new position. The arrow (drawn as a circle) now occupies its square; amazons and shots can no longer pass over or land on this square. The winner is determined by normal play.

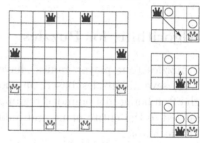

Gameplay in Amazons typically split into a sum of simpler games because arrows partition the board into multiple components. In particular, the *endgame* begins when each component of the game contains amazons of only a single color. Then the goal of each player is simply to maximize the number of moves in each component. Buro [7] proved that maximizing the number of moves in a single component is NP-complete (for $n \times n$ boards). In a general endgame, deciding the outcome may not be in NP because it is difficult to prove that the opponent has no better strategy. However, Buro [7] proved that this

Fig. 5. The initial position in Amazons (left) and black trapping a white amazon (right)

problem is *NP-equivalent* [24], i.e., the problem can be solved by a polynomial number of calls to an algorithm for any NP-complete problem, and vice versa.

It remains open whether deciding the outcome of a general Amazons position is PSPACE-hard. The problem is in PSPACE because the number of moves in a game is at most the number of squares in the board.

3.9 Phutball

Conway's game of *Philosopher's Football* or *Phutball* [3, pp. 688–691] involves white and black stones on a rectangular grid such

Fig. 6. A single move in Phutball consisting of four jumps

as a Go board. Initially, the unique black stone (the *ball*) is placed in the middle of the board, and there are no white stones. Players take turns either placing a white stone in any unoccupied position, or moving the ball by a sequence of *jumps* over consecutive sequences of white stones each arranged horizontally, vertically, or diagonally. See Fig. 6. A jump causes immediate removal of the white stones jumped over, so those stones cannot be used for a future jump in the same move. Left and Right have opposite sides of the grid marked as their *goal lines*. Left's goal is to end a move with the ball on or beyond Right's goal line, and symmetrically for Right.

Phutball is inherently loopy and it is not clear that either player has a winning strategy: the game may always be drawn out indefinitely. One counterintuitive aspect of the game is that white stones placed by one player may be "corrupted" for better use by the other player. Recently, however, Demaine, Demaine, and

Eppstein [13] found an aspect of Phutball that could be analyzed. Specifically, they proved that determining whether the current player can win in a single move ("mate in 1" in Chess) is NP-complete. This result leaves open the complexity of determining the outcome of a given game position.

4   Algorithms for Puzzles

Many puzzles (one-player games) have short solutions and are NP-complete. However, several puzzles based on motion-planning problems are harder, although often being in a bounded region, only PSPACE-complete. However, when generalized to the entire plane and unboundedly many pieces, puzzles often become undecidable.

This section briefly surveys some of these results, following the structure of the previous section. Again, because of space constraints, we omit discussion of several puzzles: Instant Insanity, Cryptarithms, Peg Solitaire, and Shanghai. For details on these and other puzzles, please refer to the full version of this paper [12].

4.1   Sliding Blocks

The Fifteen Puzzle [3, p. 756] is a classic puzzle consisting of 15 numbered square blocks in a 4×4 grid; one square in the grid is a hole which permits blocks to slide. The goal is to order the blocks as increasing in English reading order. See [29] for the history of this puzzle.

A natural generalization of the Fifteen Puzzle is the $n^2 - 1$ *puzzle* on an $n \times n$ grid. It is easy to determine whether a configuration of the $n^2 - 1$ puzzle can reach another: the two permutations of the block numbers (in reading order) simply need to match in *parity*, that is, whether the number of inversions (out-of-order pairs) is even or odd. However, to find a solution using the fewest slides is NP-complete [34]. It is also NP-hard to approximate within an additive constant, but there is a polynomial-time constant-factor approximation [34].

A harder sliding-block puzzle is *Rush Hour*, distributed by Binary Arts, Inc. Several 1×2, 1×3, 2×1, and 3×1 rectangles are arranged in an $m \times n$ grid. Horizontally oriented blocks can slide left and right, and vertically oriented blocks can slide up and down, provided the blocks remain disjoint. The goal is to remove a particular block from the puzzle via an opening in the bounding rectangle. Recently, Flake and Baum [20] proved that this formulation of Rush Hour is PSPACE-complete.

A classic reference on a wide class of sliding-block puzzles is by Hordern [29]. One general form of these puzzles is that rectangular blocks are placed in a rectangular box, and each block can be moved horizontally and vertically, provided the blocks remain disjoint. The goal is to re-arrange one configuration into another. To my knowledge, the complexity of deciding whether such puzzles are solvable remains open. A simple observation is that, as with Rush Hour, they are all in PSPACE.

4.2 Minesweeper

Minesweeper is a well-known imperfect-information computer puzzle popularized by its inclusion in Microsoft Windows. Gameplay takes place on an $n \times n$ board, and the player does not know which squares contain mines. A move consists of uncovering a square; if that square contains a mine, the player loses, and otherwise the player is revealed the number of mines in the 8 adjacent squares. The player also knows the total number of mines.

There are several problems of interest in Minesweeper. For example, given a configuration of partially uncovered squares (each marked with the number of adjacent mines), is there a position that can be safely uncovered? More generally, what is the probability that a given square contains a mine, assuming a uniform distribution of remaining mines? A different generalization of the first question is whether a given configuration is *consistent*, i.e., can be realized by a collection of mines. A consistency checker would allow testing whether a square can be guaranteed to be free of mines, thus answering the first question. A final problem is to decide whether a given configuration has a unique realization.

Kaye [30] proved that testing consistency is NP-complete. This result leaves open the complexity of the other questions mentioned above.

4.3 Pushing Blocks

Similar in spirit to the sliding-block puzzles in Section 4.1 are *pushing-block puzzles*. In sliding-block puzzles, an exterior agent can move arbitrary blocks around, whereas pushing-block puzzles embed a *robot* that can only move adjacent blocks but can also move itself within unoccupied space. The study of this type of puzzle was initiated by Wilfong [40], who proved that deciding whether the robot can reach a desired target is NP-hard when the robot can push and pull L-shaped blocks.

Since Wilfong's work, research has concentrated on the simpler model in which the robot can only push blocks and the blocks are unit squares. Types of puzzles are further distinguished by how many blocks can be pushed at once, whether blocks can additionally be defined to be *unpushable* or *fixed* (tied to the board), how far blocks move when pushed, and the goal (usually for the robot to reach a particular location). Dhagat and O'Rourke [17] initiated the exploration of square-block puzzles by proving that PUSH-*, in which arbitrarily many blocks can be pushed at once, is NP-hard with fixed blocks. Bremner, O'Rourke, and Shermer [6] strengthened this result to PSPACE-completeness. Recently, Hoffmann [28] proved that PUSH-* is NP-hard even without fixed blocks, but it remains open whether it is in NP or PSPACE-complete.

Several other results allow only a single block to be pushed at once. In this context, fixed blocks are less crucial because a 2×2 cluster of blocks can never be disturbed. A well-known computer puzzle in this context is *Sokoban*, where the goal is to place each block onto any one of the designated target squares. This puzzle was proved PSPACE-complete by Culberson [10]. A simpler puzzle, called PUSH-1, arises when the goal is simply for the robot to reach a particular position. Demaine, Demaine, and O'Rourke [14] have proved that this puzzle is NP-hard, but it remains open whether it is in NP or PSPACE-complete.

A variation on the PUSH series of puzzles, called PUSHPUSH, is when a block always slides as far as possible when pushed. The NP-hardness of these versions follow from [14,28]. Another variation, called PUSH-X, disallows the robot from revisiting a square (the robot's path cannot cross). This direction was suggested in [14] because it immediately places the puzzles in NP. Recently, Demaine and Hoffmann [16] proved that PUSH-1X and PUSHPUSH-1X are NP-complete. Hoffmann's reduction for PUSH-* also establishes NP-completeness of PUSH-*X without fixed blocks.

4.4   Clickomania (Same Game)

Clickomania or *Same Game* [4] is a computer puzzle consisting of a rectangular grid of square blocks each colored one of k colors. Horizontally and vertically adjacent blocks of the same color are considered part of the same *group*. A move selects a group

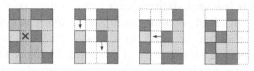

Fig. 7. The falling rules for removing a group in Clickomania. Can you remove all remaining blocks?

containing at least two blocks and removes those blocks, followed by two "falling" rules; see Fig. 7 (top). First, any blocks remaining above created holes fall down in each column. Second, any empty columns are removed by sliding the succeeding columns left.

The main goal in Clickomania is to remove all the blocks. Biedl et al. [4] proved that deciding whether this is possible is NP-complete. This complexity result holds even for puzzles with two columns and five colors, and for puzzles with five columns and three colors. On the other hand, for puzzles with one column (or, equivalently, one row) and arbitrarily many colors, they show that the maximum number of blocks can be removed in polynomial time. In particular, the puzzles whose blocks can all be removed are given by the context-free grammar $S \to \Lambda \mid SS \mid cSc \mid cScSc$ where c ranges over all colors.

Various cases of Clickomania remain open, for example, puzzles with two colors, and puzzles with $O(1)$ rows. Richard Nowakowski suggested a two-player version of Clickomania, described in [4], in which players take turns removing groups and normal play determines the winner; the complexity of this game remains open.

4.5   Moving Coins

Several coin-sliding and coin-moving puzzles fall into the following general framework: re-arrange one configuration of unit disks in the plane into another configuration by a sequence of moves, each repositioning a coin in an empty position that touches at least two other coins. Examples of such puzzles are shown in Fig. 8. This framework can be further generalized to nongeometric puzzles involving movement of tokens on graphs with adjacency restrictions.

Coin-moving puzzles are analyzed by Demaine, Demaine, and Verrill [15]. In particular, they study puzzles as in Fig. 8 in which the coins' centers remain on either the triangular lattice or the square lattice. Surprisingly, their results for deciding solvability of puzzles are positive.

For the triangular lattice, nearly all puzzles are solvable, and there is a polynomial-time algorithm characterizing them. For the square lattice, there are more stringent constraints. For example, the bounding box cannot increase by moves; more generally, the set of positions reachable by moves given an

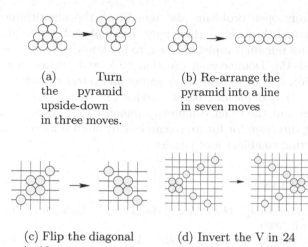

(a)   Turn the pyramid upside-down in three moves.

(b) Re-arrange the pyramid into a line in seven moves

(c) Flip the diagonal in 18 moves.

(d) Invert the V in 24 moves.

Fig. 8. Coin-moving puzzles in which each move places a coin adjacent to two other coins; in the bottom two puzzles, the coins must also remain on the square lattice. The top two puzzles are classic, whereas the bottom two puzzles were designed in [15]

infinite supply of extra coins (the *span*) cannot increase. Demaine, Demaine, and Verrill show that, subject to this constraint, there is a polynomial-time algorithm to solve all puzzles with at least two extra coins past what is required to achieve the span. (In particular, all such puzzles are solvable.)

5   Cellular Automata and Life

Conway's *Game of Life* is a zero-player cellular automaton played on the square tiling of the plane. Initially, certain cells (squares) are marked *alive* or *dead*. Each move globally evolves the cells: a live cell remains alive if it between 2 and 3 of its 8 neighbors were alive, and a dead cell becomes alive if it had precisely 3 live neighbors. Chapter 25 of *Winning Ways* [3, pp. 817–850] proves that no algorithm can decide whether an initial configuration of Life will ever completely die out. In particular, the same question about Life restricted within a polynomially bounded region is PSPACE-complete. Several other cellular automata, with different survival and birth rules, have been studied; see e.g. [42].

6   Open Problems

Many open problems remain in combinatorial game theory. Guy [26] has compiled a list of such problems (some of which have since been solved). An example of a difficult unsolved problem is Conway's angel-devil game [9].

Many open problems also remain on the algorithmic side, and have been mentioned throughout this paper. Examples of games and puzzles whose complexities remain completely open, to my knowledge, are Toads and Frogs [19], [3, pp. 14–15], Domineering (Section 3.6), and rectangular sliding-block puzzles (Section 4.1). For many other games and puzzles, such as Dots and Boxes (Section 3.7) and pushing-block puzzles (Section 4.3), some hardness results are known, but the exact complexity remains unresolved. More generally, an interesting direction for future research is to build a more comprehensive theory for analyzing combinatorial puzzles.

References

1. E. Berlekamp. *The Dots and Boxes Game: Sophisticated Child's Play*. A. K. Peter's Ltd., 2000. 25

2. E. Berlekamp and D. Wolfe. *Mathematical Go: Chilling Gets the Last Point*. A. K. Peters, Ltd., 1994. 24

3. E. R. Berlekamp, J. H. Conway, and R. K. Guy. *Winning Ways*. Academic Press, London, 1982. 19, 21, 22, 23, 24, 25, 26, 27, 30, 31

4. T. C. Biedl, E. D. Demaine, M. L. Demaine, R. Fleischer, L. Jacobsen, and J. I. Munro. The complexity of Clickomania. In R. J. Nowakowski, ed., *More Games of No Chance*, 2001. To appear. 29

5. C. L. Bouton. Nim, a game with a complete mathematical theory. *Ann. of Math. (2)*, 3:35–39, 1901–02. 21

6. D. Bremner, J. O'Rourke, and T. Shermer. Motion planning amidst movable square blocks is PSPACE complete. Draft, June 1994. 28

7. M. Buro. Simple Amazons endgames and their connection to Hamilton circuits in cubic subgrid graphs. In *Proc. 2nd Internat. Conf. Computers and Games*, 2000. 26

8. J. H. Conway. *On Numbers and Games*. Academic Press, London, 1976. 19, 20, 21

9. J. H. Conway. The angel problem. In R. J. Nowakowski, ed., *Games of No Chance*, pp. 1–12. Cambridge University Press, 1996. 30

10. J. Culberson. Sokoban is PSPACE-complete. In *Proc. Internat. Conf. Fun with Algorithms*, pp. 65–76, Elba, Italy, June 1998. 28

11. S. de Vries and R. Vohra. Combinatorial auctions: A survey. Manuscript, Jan. 2001. http://www-m9.ma.tum.de/~devries/comb_auction_supplement/comauction.pdf. 18

12. E. D. Demaine. Playing games with algorithms: Algorithmic combinatorial game theory. Preprint cs.CC/0106019, Computing Research Repository. http://www.arXiv.org/abs/cs.CC/0106019. 18, 23, 27

13. E. D. Demaine, M. L. Demaine, and D. Eppstein. Phutball endgames are NP-hard. In R. J. Nowakowski, ed., *More Games of No Chance*, 2001. To appear. http://www.arXiv.org/abs/cs.CC/0008025. 24, 27

14. E. D. Demaine, M. L. Demaine, and J. O'Rourke. PushPush and Push-1 are NP-hard in 2D. In *Proc. 12th Canadian Conf. Comput. Geom.*, pp. 211–219, 2000. http://www.cs.unb.ca/conf/cccg/eProceedings/26.ps.gz. 29

15. E. D. Demaine, M. L. Demaine, and H. Verrill. Coin-moving puzzles. In *MSRI Combinatorial Game Theory Research Workshop*, Berkeley, California, July 2000. 30

16. E. D. Demaine and M. Hoffmann. Pushing blocks is NP-complete for noncrossing solution paths. In *Proc. 13th Canadian Conf. Comput. Geom.*, 2001. To appear. 29

17. A. Dhagat and J. O'Rourke. Motion planning amidst movable square blocks. In *Proc. 4th Canadian Conf. Comput. Geome.*, pp. 188–191, 1992. 28

18. D. Eppstein. Computational complexity of games and puzzles. http://www.ics. uci.edu/~eppstein/cgt/hard.html. 23, 25

19. J. Erickson. New Toads and Frogs results. In R. J. Nowakowski, ed., *Games of No Chance*, pp. 299–310. Cambridge University Press, 1996. 31

20. G. W. Flake and E. B. Baum. Rush Hour is PSPACE-complete, or "Why you should generously tip parking lot attendants". Manuscript, 2001. http://www. neci.nj.nec.com/homepages/flake/rushhour.ps. 27

21. A. S. Fraenkel. Combinatorial games: Selected bibliography with a succinct gourmet introduction. *Electronic Journal of Combinatorics*, 1994. Dynamic Survey DS2, http://www.combinatorics.org/Surveys/. 19

22. A. S. Fraenkel, M. R. Garey, D. S. Johnson, T. Schaefer, and Y. Yesha. The complexity of checkers on an $N \times N$ board - preliminary report. In *Proc. 19th IEEE Sympos. Found. Comp. Sci.*, pp. 55–64, 1978. 23, 24

23. A. S. Fraenkel and D. Lichtenstein. Computing a perfect strategy for $n \times n$ chess requires time exponential in n. *J. Combin. Theory Ser. A*, 31:199–214, 1981. 24

24. M. R. Garey and D. S. Johnson. *Computers and Intractability: A Guide to the Theory of NP-Completeness.* W. H. Freeman & Co., 1979. 26

25. P. M. Grundy. Mathematics and games. *Eureka*, 2:6–8, Oct. 1939. 21

26. R. K. Guy. Unsolved problems in combinatorial games. In R. J. Nowakowski, ed., *Games of No Chance*, pp. 475–491. Cambridge University Press, 1996. 30

27. R. K. Guy and C. A. B. Smith. The G-values of various games. *Proc. Cambridge Philos. Soc.*, 52:514–526, 1956. 22

28. M. Hoffmann. Push-* is NP-hard. In *Proc. 12th Canadian Conf. Comput. Geom.*, pp. 205–210, 2000. http://www.cs.unb.ca/conf/cccg/eProceedings/13.ps.gz. 28, 29

29. E. Hordern. *Sliding Piece Puzzles.* Oxford University Press, 1986. 27

30. R. Kaye. Minesweeper is NP-complete. *Math. Intelligencer*, 22(2):9–15, 2000. 28

31. M. Lachmann, C. Moore, and I. Rapaport. Who wins domineering on rectangular boards? In *MSRI Combinatorial Game Theory Research Workshop*, Berkeley, California, July 2000. 25

32. D. Lichtenstein and M. Sipser. GO is polynomial-space hard. *J. Assoc. Comput. Mach.*, 27(2):393–401, Apr. 1980. 24

33. C. H. Papadimitriou. Algorithms, games, and the Internet. In *Proc. 33rd ACM Sympos. Theory Comput.*, Crete, Greece, July 2001. 18

34. D. Ratner and M. Warmuth. The $(n^2 - 1)$-puzzle and related relocation problems. *J. Symbolic Comput.*, 10:111–137, 1990. 27

35. S. Reisch. Hex ist PSPACE-vollständig. *Acta Inform.*, 15:167–191, 1981. 23

36. J. M. Robson. The complexity of Go. In *Proceedings of the IFIP 9th World Computer Congress on Information Processing*, pp. 413–417, 1983. 24

37. J. M. Robson. N by N Checkers is EXPTIME complete. *SIAM J. Comput.*, 13(2):252–267, May 1984. 23

38. C. A. B. Smith. Graphs and composite games. *J. Combin. Theory*, 1:51–81, 1966. 21

39. R. Sprague. Über mathematische Kampfspiele. *Tôhoku Mathematical Journal*, 41:438–444, 1935–36. 21

40. G. Wilfong. Motion planning in the presence of movable obstacles. *Ann. Math. Artificial Intelligence*, 3(1):131–150, 1991. 28

41. D. Wolfe. Go endgames are PSPACE-hard. In R. J. Nowakowski, ed., *More Games of No Chance*, 2001. To appear. 24

42. S. Wolfram. *Cellular Automata and Complexity: Collected Papers*. Perseus Press, 1994. 30

Some Recent Results on Data Mining and Search

Amos Fiat

Department of Computer Science, Tel Aviv University

Abstract. In this talk we review and survey some recent work and work in progress on data mining and web search. We discuss Latent Semantic Analysis and give conditions under which it is robust. We also consider the problem of collaborative filtering and show how spectral techniques can give a rigorous and robust justification for doing so. We consider the problems of web search and show how both Google and Klienberg's algorithm are robust under a model of web generation, and how this model can be reasonably extended. We then give an algorithm that provably gives the correct result in this extended model. The results surveyed are joint work with Azar, Karlin, McSherry and Saia [2], and Achlioptas, Karlin and McSherry [1].

1 A General Data Mining Model and Applications

We begin by presenting a general model that we believe captures many of the essential features of important data mining tasks. We then present a set of conditions under which data mining problems in this framework can be solved using spectral techniques, and use these results to theoretically justify the prior empirical success of these techniques for tasks such as object classification and web site ranking. We also use our theoretical framework as a foundation for developing new algorithms for collaborative filtering. Our data mining models allow both erroneous and missing data, and show how and when spectral techniques can overcome both.

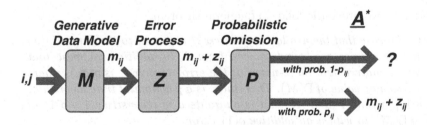

Fig. 1. The data generation model

The data mining model we introduce assumes that the data of interest can be represented as an object/attribute matrix. The model is depicted in Figure 1

J. Sgall, A. Pultr, and P. Kolman (Eds.): MFCS 2001, LNCS 2136, pp. 33–36, 2001.

which shows how three fundamental phenomena combine to govern the process
by which a data set is created:

1. **A probabilistic model of data** M: We assume that there exists an underlying set of probability distributions that govern each object's attribute values (in the degenerate case, these values could be deterministically chosen).
   These probability distributions are captured by the probabilistic model M in
   the figure, where the random variable describing the ith attribute of the j-th
   object is denoted $M_{i,j}$. The actual value of this attribute is then obtained
   by sampling from the distribution $M_{i,j}$; we denote the resulting value m_{ij}.
   We assume that the M_{ij}'s are independent.

2. **An error process** Z: We assume that the data is noisy and error-ridden.
   The error process Z describes the manner by which the error is generated. We
   assume that the data value m_{ij} is corrupted by the addition of the error z_{ij}.

3. **An omission process** P: Some of the data may not be available to the data
   miner. In our model, we assume that there is a probability distribution P
   governing the process by which data is omitted or made available. In particular, the value $m_{ij} + z_{ij}$ is available to the data miner with probability p_{ij},
   and is omitted from the data set (which we represent by the presence of a
   "?") with probability $1 - p_{ij}$. We denote by A^* the resulting data set (which
   is then input to the data mining algorithm).

The goal of the data mining algorithm: given A^* as input (and no knowledge of M, P or Z), obtain meaningful information about M. In particular,
we are interested in obtaining information about the matrix $\mathrm{E}(M)$, whose (i, j)
entry is the expectation of the random variable M_{ij}.

Clearly, without any assumptions about M, Z and P, it is hopeless to achieve
the data mining goal just laid out. We present a general set of conditions under
which it is possible to efficiently retrieve meaningful information about $\mathrm{E}(M)$.
In essence, our results show that if the underlying data model is sufficiently
"structured", then the randomness of a probabilistic process, the addition of
error and the fact that a significant fraction of the data may be missing will
not prevent the data miner from recovering meaningful information about the
"true" data.

More formally, we prove the following general theorem:

Theorem 1. *Suppose that the availability matrix P is known to the data mining
algorithm, and its entries are bounded away from 0. In addition, suppose that
$\mathrm{E}(M)$ is a rank k matrix, and the 2-norm of the error matrix Z is $o(\sigma_k)$, where
σ_k is the k-th singular value of $\mathrm{E}(M)$. Then there is a polynomial time algorithm,
that takes as input only P and A^*, that is guaranteed to reconstruct $1 - o(1)$ of
the entries of $\mathrm{E}(M)$ to within an additive $o(1)$ error.*

Many of the data mining problems can be expressed as special cases of the
above theorem. We can model Latent Semantic Analysis, (LSA) pioneered by
Deerwester et al. [4], via this model. We can also extend the work of Papadimitriou, Raghavan, Tamaki and Vempala [7] on latent semantic indexing to deal
with arbitrary matrices rather than perturbed block matrices.

A fundamental problem in data mining, usually referred to as collaborative filtering (or recommendation systems) is to use partial information that has been collected about a group of users to make recommendations to individual users. To our knowledge, there has been very little prior theoretical work on collaborative filtering algorithms other than the work of Kumar, Raghavan, Rajagopalan and Tomkins who took an important first step of defining an analytic framework for evaluating collaborative filtering [5].

We model the collaborative filtering problem within the framework of our general data mining model as follows: We assume that the utility of product j for individual i is given by a random variable M_{ij} and which data is missing is determined by a probabilistic omission process P.

Kleinberg's seminal work on web *hubs* and *authorities* has had a true impact on the real world [6].

It is an immediate consequence of our results that Kleinberg's definition of importance is robust in the sense that the important sites will remain important (almost) irrespective of the actual random choices made when the "real world" is constructed.

2   Web Search

We present a generative model for web search which captures in a unified manner three critical components of the problem: how the link structure of the web is generated, how the content of a web document is generated, and how a human searcher generates a query. The key to this unification lies in capturing the correlations between each of these components in terms of proximity in latent semantic space. Given such a combined model, the correct answer to a search query is well defined, and thus it becomes possible to evaluate web search algorithms rigorously. We present a new web search algorithm, based on spectral techniques, and prove that it is guaranteed to produce an approximately correct answer in our model. The algorithm assumes no knowledge of the model, and is well-defined regardless of the accuracy of the model.

We like to think of the task at hand for web search as being the following:

1. Take the human generated query and determine the topic to which the query refers. There is an infinite set of topics on which humans may generate queries.
2. Synthesize a perfect hub for this topic. The perfect hub is an imaginary page, no page remotely resembling this imaginary hub need exist. On this imaginary hub the authorities on the topic of the query will be listed in order of decreasing authoritativeness.

We present a new algorithm that is a multi-dimensional generalization of both Kleinberg's algorithm and of Google [3], which provably gives the correct result in the model.

References

1. D. Achlioptas, A. Fiat, A. R. Karlin, and F. McSherry. Web Search vis Hub Synthesis. Manuscript, 2001. 33
2. Y. Azar, A. Fiat, A. R. Karlin, F. McSherry, and J. Saia. Spectral analysis of data. In *STOC 2001*, 2001. 33
3. Sergey Brin and Lawrence Page. The anatomy of a large-scale hypertextual Web search engine. *Computer Networks and ISDN Systems*, 30(1–7):107–117, 1998. 35
4. S. Deerwester, S. T. Dumais, T. K. Landauer, G. W. Furnas, and R. A. Harshman. Indexing by latent semantic analysis. *Journal of the Society for Information Science*, 41(6):391–407, 1990. 34
5. S. Kumar, P. Raghavan, S. Rajagopalan, and A. Tomkins. Recommendation Systems: A Probabilistic Analysis. In *Foundations of Computer Science*, pp. 664-673, 1998. 35
6. Jon M. Kleinberg. Authoritative sources in a hyperlinked environment. *Journal of the ACM*, 46(5):604–632, 1999. 35
7. C. Papadimitriou, P. Raghavan, H. Tamaki, and S. Vempala. Latent Semantic Indexing: A Probabilistic Analysis. In *Proceedings of ACM Symposium on Principles of Database Systems*, 1997. 34

Hypertree Decompositions: A Survey

Georg Gottlob[1], Nicola Leone[2], and Francesco Scarcello[3]

[1] Information Systems Institute, TU-Wien
Vienna, Austria
gottlob@acm.org
[2] Dept. of Mathematics, Univ. of Calabria
Rende (CS), Italy
leone@unical.it
[3] D.E.I.S., Univ. of Calabria
Rende (CS), Italy
scarcello@acm.org

Abstract. This paper surveys recent results related to the concept of hypertree decomposition and the associated notion of hypertree width. A hypertree decomposition of a hypergraph (similar to a tree decomposition of a graph) is a suitable clustering of its hyperedges yielding a tree or a forest. Important NP hard problems become tractable if restricted to instances whose associated hypergraphs are of bounded hypertree width. We also review a number of complexity results on problems whose structure is described by acyclic or nearly acyclic hypergraphs.

1 Introduction

One way of coping with an NP hard problem is to identify significantly large classes of instances that are both recognizable and solvable in polynomial time. Such instances are often defined via some structural property of a graph $G(I)$ that is associated in a canonical way with the instance I. For example, many problems that are NP complete in general become tractable for instances I whose associated graph has bounded treewidth (cf. Sect. 4). Treewidth is a measure of the degree of cyclicity of a graph. Note that instances of bounded treewidth are also easy to recognize given that deciding whether the treewidth of a graph is at most k is decidable in linear time for each constant k.

The structure of a large number of problems is, however, more faithfully described by a *hypergraph* than by a graph. Again, several NP complete problems become tractable if restricted to instances with acyclic hypergraphs. In order to obtain larger tractable instance-classes of hypergraph-based problems, we thus investigated measures of hypergraph cyclicity that play a similar role for hypergraphs as the concept of treewidth does for graphs. In particular, an appropriate notion of hypergraph width (and an associated method of hypergraph decomposition) should fulfil both of the following conditions:

1. Relevant hypergraph-based problems should be solvable in polynomial time for instances of bounded width.

J. Sgall, A. Pultr, and P. Kolman (Eds.): MFCS 2001, LNCS 2136, pp. 37–57, 2001.

2. For each constant k, one should be able to check in polynomial time whether a hypergraph is of width k, and, in the positive case, it should be possible to produce an associated decomposition of width k of the given hypergraph.

Existing measures for hypergraph cyclicity we were aware of do either not fulfil one of these two conditions (e.g. recognizing hypergraphs of bounded *query width* is NP complete, cf. Section 4.3), or are not general enough (such methods are mentioned in Section 10). In particular, various notions of hypergraph width can be obtained by first transforming a hypergraph into a graph (there are several ways of doing so, see Section 4.2) and then considering the treewidth of that graph. However, it is not hard to see that such measures of cyclicity are not very significant due to a loss of structural information caused by the transformation of the hypergraph to a graph (cf. Sect. 4.2). In summary, it appeared that a satisfactory way of determining the degree of cyclicity of a hypergraph was missing, on the basis of which large tractable instances of relevant NP-hard problems could be defined.

Consequently, after a careful analysis of the shortcomings of various hypergraph decomposition methods, we introduced the new method of *hypertree decomposition* and the associated notion of *hypertree width*. To our best knowledge, the method of hypertree decomposition is currently the most general known hypergraph decomposing method leading to large tractable classes of important problems such as *constraint satisfaction problems* or *conjunctive queries*. The notion of hypertree decomposition and the associated notion of hypertree width are the main topics of the present survey paper. However, we will also report on a number of other closely related issues, such as the precise (parallel) complexity of acyclic database queries, and the notion of *query decomposition*. Our results surveyed here are mainly from the following sources, where formal proofs, details, and a number of further results can be found:

- Reference [17], where the precise complexity of acyclic Boolean conjunctive queries (ABCQs) is determined, and where highly parallel database algorithms for solving such queries are presented. (In the present paper, we will not discuss parallel database algorithms and refer the interested reader to [17] and [22]).
- Reference [19], where we first study *query width*, a measure for the amount of cyclicity of a query introduced by Chekuri and Rajamaran [6], and where we define and study the new (more general) concept of *hypertree width*.
- Reference [21], where we establish criteria for comparing different CSP decomposition methods and where we compare various methods including the method of hypertree decomposition. The comparison criteria and the results of the comparison are reported in Section 10 of the present paper.
- Reference [24], where hypertree width is compared to Courcelle's notion of *clique width* [7,8].
- Reference [23], where we give a game theoretic and a logical characterization of hypertree width. These results are reported in Sections 8 and 9 of the present paper, respectively.

This paper is organized as follows. In Section 2 we define a number of important hypergraph-based problems. In Section 3 we discuss the complexity of acyclic instances of these problems. In Section 4, discuss graph treewidth and a generalization termed *query width*. In Section 5 we define the concepts of hypertree decomposition and hypertree width. In Section 6, we show how hypergraphs of bounded hypertree width can be recognized in polynomial time. In Section 7, we show how CSPs can be solved in polynomial time for instances of bounded hypertree-width. In Section 8, we describe the *Robber and Marshals* game which characterizes hypergraphs of bounded hypertree width. In Section 9, we briefly describe our logical characterization of queries of bounded hypertree-width. In Section 10, we give a brief account on how the notion of hypertree decomposition compares to other related notions. Finally, in Section 11, we state some relevant open problems.

2   Hypergraph-Based Problems

A *relational vocabulary* (short: *vocabulary*) consists of a finite nonempty set of relation symbols $P, Q, R \ldots$, with associated arities. A *finite relational structure* (short: *finite structure*) C over a vocabulary τ consists of a finite *universe* U_C and for each k-ary relation symbol R in τ a relation $R^C \subseteq U_C^k$. For a tuple $(c_1, \ldots, c_k) \in R^C$ we often write $R^C(c_1, \ldots, c_k)$. We denote the vocabulary of a structure C by $vo(C)$.

Let A and B be two finite structures such that $vo(A) \subseteq vo(B)$. Then a mapping $h : A \longrightarrow B$ is a *homomorphism* from A to B if for each relation symbol $R \in vo(A)$ it holds that whenever $(c_1, \ldots, c_k) \in R^A$ for some elements $c_1, \ldots, c_k \in U_A$, then it also holds that $(h(c_1), \ldots, h(c_k)) \in R^B$. The following is a fundamental computational problem in Algebra:

Definition 1 (The Homomorphism Problem HOM). *Given two finite structures A and B, decide whether there exists a homomorphism from A to B. We denote such an instance of HOM by HOM(A, B)*

It is well-known (cf. [12]) that HOM is an NP-complete problem. For example, checking whether a graph (V, E) is three colorable amounts to solve the $HOM(A, B)$ problem for structures A and B over a vocabulary with a unique binary relation symbol R, where $R^A = E$ and $R^B = \{(red, blue), (blue, red), (red, green), (green, red), (blue, green), (green, blue)\}$.

In [12,30] it was observed that HOM is equivalent to (and actually, in essence, the same as) the important *constraint satisfaction problem* (CSP) of Artificial Intelligence [9], which, in turn, is equivalent to the database problem BCQ of evaluating Boolean conjunctive queries:

Definition 2 (The Constraint Satisfaction Problem CSP.). *Given a finite set Var of variables, a finite domain U of values, a set of constraints $C = \{C_1, C_2, \ldots, C_q\}$, where each constraint C_i is a pair (S_i, r_i), and where S_i is* *a list of variables of length m_i, called the constraint scope, and r_i is an m_i-ary relation over U, called a constraint relation, decide whether there is a substitution $\vartheta : \text{Var} \longrightarrow U$, such that, for each $1 \leq i \leq q$, $S_i \vartheta \in r_i$.*

Definition 3 (The Boolean Conjunctive Query Problem BCQ.). *A relational database is formalized as a finite relational structure D. A Boolean conjunctive query (BCQ) on D is a sentence of first-order logic of the form: $\exists X_1, \ldots, X_r \; R_1(t_1^1, t_2^1 \ldots, t_{\alpha(1)}^1) \; \wedge \; \cdots \; \wedge R_k(t_1^k, t_2^k \ldots, t_{\alpha(k)}^k)$, where, for $1 \leq i \leq k$, R_i is a relation symbol from $vo(D)$ of associated arity $\alpha(i)$, and for $1 \leq i \leq k$ and $1 \leq j \leq \alpha(i)$, each t_j^i is a term, i.e., either a variable from the list $X_1, \ldots, X_r$, or a constant element from U_D. The decision problem BCQ is the problem of deciding for a pair $\langle D, Q \rangle$, where D is a database and Q is a Boolean conjunctive query, whether Q evaluates to true over D, denoted by $D \models Q$.*

Given that all variables occuring in a BCQ are existentialy quantified, we usually omit the quantifier prefix and write a BCQ as a conjunction of *query atoms*. For example, let emp denote a relation containing (employee#,project#) pairs, and let rel be a relation containing a pair (n_1, n_2) if n_1 and n_2 are numbers of distinct employees who are relatives, then the BCQ $\text{emp}(X, Z) \wedge \text{emp}(Y, Z) \wedge \text{rel}(X, Y)$ expresses that there are two employees who are relatives and work on the same project.

Note that by simple logspace operations (selections and projections on the corresponding relations) one can always eliminate constants occurring in BCQ atoms. We thus assume w.l.o.g. that query atoms contain only variables as arguments. By this assumption, CSP and BCQ are exactly the same problem, where each constraint scope corresponds to a query atom, each constraint relation corresponds to a database relation, and vice-versa. Now each CSP (or BCQ) instance I can in turn be identified with $\text{HOM}(A, B)$, where A is the structure whose universe U_A consists of the set Var of all variables of I and whose relations contain constraint scopes (or query atoms) as tuples, and where B is the structure whose universe U_B is the finite domain U of I and whose relations are just the constraint relations (or the database relations). In this sense, we can speak about instances $\text{CSP}(A, B)$ and $\text{BCQ}(A, B)$, where A is a structure representing the constraint scopes or the query, and B denotes the set of constraint relations, or the database, respectively. On the other hand, each instance $I = \text{HOM}(A, B)$ of HOM can be identified in the obvious way with a CSP instance (or a BCQ instance) by interpreting the elements of U_A as variables and those of U_B as domain elements of the constraint (or database) relations.

Thus all three problems HOM, CSP, and BCQ are the same and are NP-complete (for BCQ this was first shown in [5]). Therefore, it is important to find large classes of instances that can be evaluated in polynomial time. Such classes can be defined by imposing structural restrictions on the problem instances. In particular, for $\text{HOM}(A, B)$, $\text{CSP}(A, B)$, or, equivalently, $\text{BCQ}(A, B)$, one could impose restrictions on the structure A, or on the structure B, or on both (cf. [35,33,30]). In this paper we are interested in restrictions on the structure A. In database terms, we can recast this by saying that we are interested in the structure of the query, rather than on the properties of the database content.

If $\mathcal{A}$ denotes a set of structures, then $\mathrm{HOM}(\mathcal{A})$, $\mathrm{CSP}(\mathcal{A})$, and $\mathrm{BCQ}(\mathcal{A})$ denote the restrictions of HOM, CSP, and BCQ to instances $\mathrm{HOM}(A, B)$, $\mathrm{CSP}(A, B)$, and $\mathrm{BCQ}(A, B)$ respectively, where $A \in \mathcal{A}$.

Each finite structure C over universe U_C defines a hypergraph $\mathcal{H}(C) = (V, H)$ as follows: The set V of vertices V of $\mathcal{H}(C)$ coincides with U_C; the set of hyperedges H of $\mathcal{H}(C)$ consists of all sets $\{c_1, \ldots, c_k\}$ such that there exists a relation R in $voc(C)$ and $(c_1, \ldots, c_k) \in R^C$.

To each problem instance $I = \mathrm{HOM}(A, B)$ or $I = \mathrm{CSP}(A, B)$, or $I = \mathrm{BCQ}(A, B)$, we define the associated hypergraph $\mathcal{H}_I$ by $\mathcal{H}_I = \mathcal{H}(A)$. In particular, this means, that for an instance I of CSP, $\mathcal{H}_I$ denotes the hypergraph whose vertices are the variables of I and whose hyperedges are all sets $\{X_1, \ldots, X_k\}$ such that there exists a constraint scope $S = (X_1, \ldots, X_k)$ belonging to I. For an instance $I = (D, Q)$ of BCQ, $\mathcal{H}_I$ denotes the hypergraph whose vertices are all the variables occurring in Q and whose hyperedges are the sets $var(\alpha)$ of variables occuring in α, for each query atom α.

Example 4. Figure 1(a) shows $\mathcal{H}_{I_1}$ for an instance I_1 of BCQ having query Q_1 :
$a(S, X, T, R) \wedge b(S, Y, U, P) \wedge f(R, P, V) \wedge g(X, Y) \wedge c(T, U, Z) \wedge d(W, X, Z) \wedge e(Y, Z)$

It is furthermore easy to see that HOM, BCQ, and CSP are all equivalent (via logspace transformations) to the following fundamental problems in database theory and artificial intelligence [17]: The *Query Output Tuple Problem:* Given a conjunctive query Q, a database **db**, and a tuple t, determine whether t belongs to the answer $Q(\mathbf{db})$ of Q over **db**. The *Conjunctive Query Containment:* Decide whether a conjunctive query Q_1 is contained in a conjunctive query Q_2. Query Q_1 is contained in query Q_2 if, for each database instance **db**, the answer $Q_1(\mathbf{db})$ is a subset of $Q_2(\mathbf{db})$. The *Clause Subsumption Problem:* Check whether a (general) clause C subsumes a clause D, i.e., whether there exists a substitution ϑ such that $C\vartheta \subseteq D$. A (general) clause is a disjunction of (positive or negative) literals,

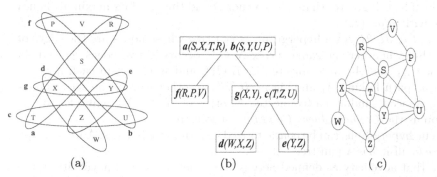

(a)   (b)   (c)

Fig. 1. (a) Hypergraph $\mathcal{H}_{I_1}$; (b) a width 2 hypertree decomposition of $\mathcal{H}_{I_1}$; and (c) the primal graph of $\mathcal{H}_{I_1}$

possibly containing function symbols. Note that subsumption is an extremely important technique used in clause-based theorem proving [1].

Just for the sake of presentation, we will focus in the rest of this paper on the constraint statisfaction problem (CSP).

While, as we will see, many interesting structural properties of a CSP instance $I = \text{CSP}(A, B)$ can be identified by looking at the associated hypergraph $\mathcal{H}_I = \mathcal{H}(A)$, which in AI is called the *constraint hypergraph*, some structural properties of I may be also detected using its *primal graph*, i.e., the primal graph of the hypergraph associated to A, which coincides with the Gaifman graph of A [15]. Let $\mathcal{H}_I = (V, H)$ be the constraint hypergraph of a CSP instance I. The *primal graph* of I is a graph $G = (V, E)$, having the same set of variables (vertices) as $\mathcal{H}_I$ and an edge connecting any pair of variables $X, Y \in V$ such that $\{X, Y\} \subseteq h$ for some $h \in H$. Note that, if the vocabulary of A contains only binary predicates, then all constraints of I are binary and its associated hypergraph is identical to its primal graph. The primal graph of the hypergraph of query Q_1 (and of the equivalent CSP instance) is depicted in Fig. 1(c).

Since in this paper we always deal with hypergraphs corresponding to CSP or BCQ instances, the vertices of any hypergraph $\mathcal{H} = (V, H)$ can be viewed as the variables of some constraint satisfaction problem or of some conjunctive query. Thus, we will often use the term *variable* as a synonym for vertex, when referring to elements of V. For the hypergraph $\mathcal{H} = (V, H)$, $var(\mathcal{H})$ and $edges(\mathcal{H})$ denote the sets V and H, respectively. When illustrating a decomposition, we will usually represent hyperedges of the hypergraph $\mathcal{H}_I$ of a BCQ or CSP instance I by their corresponding query atoms or constraint scopes.

3   Acyclic Instances

The most basic and most fundamental structural property considered in the context of CSPs and conjunctive queries is *acyclicity*. It was recognized in AI and database theory that *acyclic* CSPs or conjunctive queries are polynomially solvable. A CSP instance I is *acyclic* if its associated hypergraph $\mathcal{H}_I$ is acyclic.

A hypergraph $\mathcal{H}$ is acyclic if and only if its primal graph G is chordal (i.e., any cycle of length greater than 3 has a chord) and the set of its maximal cliques coincide with $edges(\mathcal{H})$ [2].

A *join tree* $JT(\mathcal{H})$ for a hypergraph $\mathcal{H}$ is a tree whose nodes are the edges of $\mathcal{H}$ such that whenever the same vertex $X \in V$ occurs in two edges A_1 and A_2 of $\mathcal{H}$, then A_1 and A_2 are connected in $JT(\mathcal{H})$, and X occurs in each node on the unique path linking A_1 and A_2 in $JT(\mathcal{H})$. In other words, the set of nodes in which X occurs induces a (connected) subtree of $JT(\mathcal{H})$. We will refer to this condition as the *Connectedness Condition* of join trees.

Acyclic hypergraphs can be characterized in terms of join trees: A hypergraph $\mathcal{H}$ is *acyclic* iff it has a join tree [3,2,32].

Note that acyclicity as defined here is the usual concept of acyclicity in the context of database theory and AI. It is referred to as α-acyclicity in [11]. This is the least restrictive concept of hypergraph acyclicity among all those defined in the literature.

Acyclic CSPs and conjunctive queries have highly desirable computational properties:

1. Acyclic instances can be efficiently solved. Yannakakis provided a (sequential) polynomial time algorithm solving BCQ on acyclic queries[1] [41].
2. Acyclicity is efficiently recognizable, and a join tree of an acyclic hypergraph is efficiently computable. A linear-time algorithm for computing a join tree is shown in [37]; an L^{SL} method has been provided in [17] (L^{SL} denotes logspace relativized by an oracle in symmetric logspace; this class could also be termed "functional SL").
3. The result of a (non-Boolean) acyclic conjunctive query Q can be *computed* in time polynomial in the combined size of the input instance and of the output relation [41].
4. *Arc-consistency* for acyclic CSP instances can be enforced in polynomial time [9,10].

Intuitively, the efficient behavior of acyclic instances is due to the fact that they can be evaluated by processing any of their join trees bottom-up by performing upward semijoins, thus keeping small the size of the intermediate relations (which could become exponential if regular join were performed).

We have recently determined the precise computational complexity of BCQ, and hence of HOM, CSP, and all their equivalent problems. It turned out all these problems are highly parallelizable on acyclic structures, as they are complete for the low complexity class LOGCFL [17]. This is the class of all decision problems that are logspace-reducible to a context-free language. Note that $NL \subseteq LOGCFL \subseteq AC^1 \subseteq NC^2 \subseteq P$ where NL denotes nondeterministic logspace and AC^1 and NC^2 are logspace-uniform classes based on the corresponding types of Boolean circuits (for precise definitions of all these complexity classes, cf. [29]). Let $\mathcal{AH}$ be the set of all finite acyclic relational structures.

Theorem 5 ([17]). CSP($\mathcal{AH}$) *is* LOGCFL-*complete.*

Moreover, the functional version of these problems belongs to the functional version of LOGCFL, i.e., a solution for a CSP instance can be computed in L^{LOGCFL}, i.e., functional logspace with an oracle in LOGCFL. Efficient parallel algorithms – even for non-Boolean queries – have been proposed in [22]. They run on parallel database machines that exploit the *inter-operation parallelism* [40], i.e., machines that execute different relational operations in parallel.

The important speed-up obtainable on acyclic instances stimulated several research efforts towards the identification of wider classes of queries and constraints having the same desirable properties as acyclic CQs and CSPs.

---

[1] Note that, since both the database **db** and the query Q are part of an input-instance of BCQ, what we are considering is the *combined complexity* of the query [38].

4   Treewidth, Query Width, and Hypertree Width

4.1   Tree Decompositions and Treewidth of Graphs

The treewidth of a graph is a well-known measure of its tree-likeness introduced by Robertson and Seymour in their work on graph minors [34]. This notion plays a central role in algorithmic graph theory as well as in many subdisciplines of Computer Science.

Definition 6. A *tree decomposition* of a graph $G = (V, E)$ is a pair $\langle T, \chi \rangle$, where $T = (N, F)$ is a tree, and χ is a labeling function associating to each vertex $p \in N$ a set of vertices $\chi(p) \subseteq V$, such that the following conditions are satisfied: (1) for each vertex b of G, there exists $p \in N$ such that $b \in \chi(p)$; (2) for each edge $\{b, d\} \in E$, there exists $p \in N$ such that $\{b, d\} \subseteq \chi(p)$; (3) for each vertex b of G, the set $\{p \in N \mid b \in \chi(p)\}$ induces a (connected) subtree of T.

The *width* of the tree decomposition $\langle T, \chi \rangle$ is $\max_{p \in N} |\chi(p) - 1|$. The *treewidth* of G is the minimum width over all its tree decompositions. The treewidth of a CSP instance is the treewidth of its associated primal graph.

The notion of treewidth is a generalization of graph acyclicity. In particular, a graph is acyclic if and only if its treewidth is one [34].

Checking whether a graph has treewidth at most k for a fixed constant k, and in the positive case computing a k-width tree decomposition, is feasible in linear time [4]. Moreover, this task is also parallelizable. Indeed, Wanke [39] has shown that, for a fixed constant k, checking whether a graph has treewidth k is in LOGCFL. By proving some general complexity-theoretic results and by using Wanke's result, the following was shown in [18]:

Theorem 7 ([18]). *For each constant k, there exists an L^{LOGCFL} transducer T_k that behaves as follows on input G. If G is a graph of treewidth $\leq k$, then T_k outputs a tree decomposition of width $\leq k$ of G. Otherwise, T_k halts with empty output.*

Thus, a tree decomposition of width at most k can be also computed in (the functional version of) LOGCFL, and thus by logspace uniform AC^2 and NC^2 circuits.

An important feature of treewidth is that many NP-complete problems are decidable in polynomial-time on structures having bounded treewidth, i.e., having treewidth at most k for some fixed constant $k > 0$. In particular, Courcelle proved that every property expressible in monadic second order logic is decidable in linear time over bounded treewidth graphs [7].

4.2   Treewidth of Hypergraphs

As mentioned in the previous section, many NP-complete problems become tractable on bounded treewidth graphs. In order to exploit this nice feature for CSP, BCQ, and their equivalent problems, many researchers in the AI and the database communities considered the primal graph (of the hypergraph) associated to the relational structure. Let TW[k] be the set of all finite relational structures whose associated primal graph has treewidth at most k. It has been shown that CSP(TW[k]) is solvable in polynomial time [14] and has the same properties of CSP($\mathcal{AH}$), including its precise computational complexity.

Theorem 8 ([17]). CSP(TW[k]) *is LOGCFL-complete.*

Note that considering the primal graph associated to a hypergraph is not the one possible choice. Given a CSP instance I, the *dual graph* [9,10,32] of the hypergraph $\mathcal{H}_I$ is a graph $G_I^d = (V, E)$ defined as follows: the set of vertices V coincides with the set of (hyper)edges of $\mathcal{H}_I$, and the set E contains an edge $\{h, h'\}$ for each pair of vertices $h, h' \in V$ such that $h \cap h' \neq \emptyset$. That is, there is an edge between any pair of vertices corresponding to hyperedges of $\mathcal{H}_I$ sharing some variable.

The dual graph often looks very intricate even for simple CSPs. For instance, in general, acyclic CSPs do not have acyclic dual graphs. However, it is well known that the dual graph G_I^d can be suitably simplified in order to obtain a "better" graph G' which can still be used to solve the given CSP instance I. In particular, if I is an acyclic CSP, G_I^d can be reduced to an acyclic graph that represents a join tree of $\mathcal{H}_I$. In this case, the reduction is feasible in polynomial (actually, linear) time. (See, e.g., [32].) However, in general, it is not known whether there exists an efficient algorithm for obtaining the best simplified graph G' with respect to the treewidth notion, i.e., the simplification of G_I^d having the smallest treewidth over all its possible simplifications (see [30] for a formal statement of this open problem and [21] for a comparison of this notion with some hypergraph-based notions).

Another possibility is considering the so called *incidence graph* [6]. Given a CSP instance I, the incidence graph $G_I^i(\mathcal{H}_I) = (V', E)$ associated to the hypergraph $\mathcal{H}_I = (V, H)$ has a vertex for each variable and for each hyperedge of $\mathcal{H}_I$. There is an edge $\{x, h\} \in E$ between a variable $x \in V$ and and hyperedge $h \in H$ whenever x occurs in h.

The class of all CSP instances whose dual graphs (resp. incidence graphs) have bounded treewidth are solvable in polynomial time and, actually, they are LOGCFL-complete. However, note that none of these classes of CSP instances generalize the class CSP($\mathcal{AH}$). Indeed, there are families of acyclic hypergraphs whose associated primal graphs, dual graphs (without considering simplification), and incidence graphs have unbounded treewidth.

Note that by results of [25] bounded treewidth is most likely the best and most general structural restriction for obtaining tractable CSP and BCQ instances, when the structure of a CSP or BCQ is described via a *graph* (e.g. the primal graph), rather than by a hypergraph. Further interesting material on BCQ and treewidth can be found in [13].

4.3   Query Decompositions and Query Width

A more general notion that generalizes hypergraph acyclicity is *query width* [6]. The notion of bounded query-width is based on the concept of *query decomposition* [6]. We next adapt this notion to the more general setting of hypergraphs, while it was originally defined in terms of queries. Roughly, a query decomposition of a hypergraph $\mathcal{H}$ consists of a tree each vertex of which is labelled by a set of hyperedges and/or variables. Each variable and hyperedge induces a connected subtree (*connectedness condition*). Each hyperedge occurs in at least one label. The width of a query decomposition is the maximum of the cardinalities of its vertices. The *query-width* $qw(\mathcal{H})$ of $\mathcal{H}$ is the minimum width over all its query decompositions.

Example 9. Consider the CSP instance I_2 having the following constraint scopes:

$$a(S, X, X', C, F), b(S, Y, Y', C', F'), c(C, C', Z), d(X, Z), e(Y, Z),$$
$$f(F, F', Z'), g(X', Z'), h(Y', Z'), j(J, X, Y, X', Y')$$

The query-width of $\mathcal{H}_{I_2}$ is 3. Figure 2 shows a query decomposition of $\mathcal{H}_{I_2}$ of width 3. W.l.o.g. we represent hyperedges by the corresponding constraint scopes or query atoms in such decompositions.

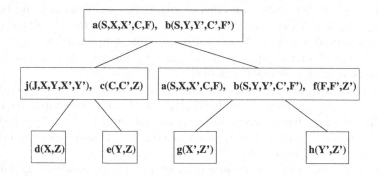

Fig. 2. A 3-width query decomposition of $\mathcal{H}_{I_2}$

Each hypergraph whose primal graph has treewidth at most k has query width at most k, too. The converse does not hold, in general. Moerover, this notion is a true generalization of the basic concept of acyclicity: A hypergraph is acyclic iff it has query width 1.

Let k be a fixed constant. Chekuri and Rajaraman [6] proved that, given a BCQ instance I and a query decomposition of $\mathcal{H}_I$ having width at most k, I is solvable in polynomial time. In [17] it was shown that this problem is LOGCFL-complete (and thus highly parallelizable).

However, when the notion of query-width was defined and studied in [6], no polynomial algorithm for checking whether a hypergraph has query-width at most k was known, and Chekuri and Rajaraman [6] stated this as an open problem. This problem is solved in [19], where it is shown that is unlikely to find an efficient algorithm for recognizing instances of bounded query-width.

Theorem 10 ([19]). *Determining whether the query-width of a hypergraph is at most 4 is NP-complete.*

Fortunately, it turned out that the high complexity of determining bounded query-width is not, as one would usually expect, the price for the generality of the concept. Rather, it is due to some peculiarity in its definition. In the next section, we present a new notion that does not suffer from such problems. Indeed, this notion generalizes query width (and hence acyclicity) and is tractable.

5   Hypertree Decompositions and Hypertree Width

A new class of tractable CSP instances, which generalizes the class $CSP(\mathcal{AH})$ of CSP instances having an acyclic hypergraph, has recently been identified [19]. This is the class of CSPs whose hypergraph has a bounded-width hypertree decomposition [19].

A *hypertree* for a hypergraph $\mathcal{H}$ is a triple $\langle T, \chi, \lambda \rangle$, where $T = (N, E)$ is a rooted tree, and χ and λ are labeling functions which associate to each vertex $p \in N$ two sets $\chi(p) \subseteq var(\mathcal{H})$ and $\lambda(p) \subseteq edges(\mathcal{H})$. If $T' = (N', E')$ is a subtree of T, we define $\chi(T') = \bigcup_{v \in N'} \chi(v)$. We denote the set of vertices N of T by $vertices(T)$, and the root of T by $root(T)$. Moreover, for any $p \in N$, T_p denotes the subtree of T rooted at p.

Definition 11. A *hypertree decomposition* of a hypergraph $\mathcal{H}$ is a hypertree $HD = \langle T, \chi, \lambda \rangle$ for $\mathcal{H}$ which satisfies all the following conditions:

1. for each edge $h \in edges(\mathcal{H})$, there exists $p \in vertices(T)$ such that $var(h) \subseteq \chi(p)$ (we say that p *covers* h);
2. for each variable $Y \in var(\mathcal{H})$, the set $\{p \in vertices(T) \mid Y \in \chi(p)\}$ induces a (connected) subtree of T;
3. for each $p \in vertices(T)$, $\chi(p) \subseteq var(\lambda(p))$;
4. for each $p \in vertices(T)$, $var(\lambda(p)) \cap \chi(T_p) \subseteq \chi(p)$.

Note that the inclusion in Condition 4 is actually an equality, because Condition 3 implies the reverse inclusion.

An edge $h \in edges(\mathcal{H})$ is *strongly covered* in HD if there exists $p \in vertices(T)$ such that $var(h) \subseteq \chi(p)$ and $h \in \lambda(p)$. We then say that p strongly covers h.

A hypertree decomposition HD of hypergraph $\mathcal{H}$ is a *complete decomposition* of $\mathcal{H}$ if every edge of $\mathcal{H}$ is strongly covered in HD.

The *width* of a hypertree decomposition $\langle T, \chi, \lambda \rangle$ is $max_{p \in vertices(T)} |\lambda(p)|$. The *hypertree width* $hw(\mathcal{H})$ of $\mathcal{H}$ is the minimum width over all its hypertree decompositions. A c-width hypertree decomposition of $\mathcal{H}$ is *optimal* if $c = hw(\mathcal{H})$.

The acyclic hypergraphs are precisely those hypergraphs having hypertree width one. Indeed, any join tree of an acyclic hypergraph $\mathcal{H}$ trivially corresponds to a hypertree decomposition of $\mathcal{H}$ of width one. Furthermore, if a hypergraph $\mathcal{H}'$ has a hypertree decomposition of width one, then, from this decomposition, we can easily compute a join tree of $\mathcal{H}'$, which is therefore acyclic [19].

It is worthwhile noting that from any hypertree decomposition HD of $\mathcal{H}$, we can easily compute a complete hypertree decomposition of $\mathcal{H}$ having the same width in $O(\|\mathcal{H}\| \cdot \|HD\|)$ time.

Intuitively, if $\mathcal{H}$ is a cyclic hypergraph, the χ labeling selects the set of variables to be fixed in order to split the cycles and achieve acyclicity; $\lambda(p)$ "covers" the variables of $\chi(p)$ by a set of edges.

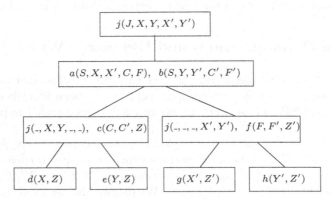

Fig. 3. A Hypertree decomposition of $\mathcal{H}_2$

Example 12. Figure 3 shows an hypertree decomposition HD_2 of the cyclic hypergraph $\mathcal{H}_2$ associated to the CSP instance in Example 9. Each node p in the tree is labeled by a set of hyperedges representing $\lambda(p)$; $\chi(p)$ is the set of all variables, distinct from '_', appearing in these hyperedges. Thus, the anonymous variable '_' replaces the variables in $var(\lambda(p)) - \chi(p)$.

Using this graphical representation, we can easily observe an important feature of hypertree decompositions. Once an hyperedge has been covered by some vertex of the decomposition tree, any subset of its variables can be used freely in order to decompose the remaining cycles in the hypergraph. For instance, the variables in the hyperedge corresponding to constraint j in $\mathcal{H}_2$ are jointly included only in the root of the decomposition. If we were forced to take all the variables in every vertex where j occurs, it would not be possible to find a decomposition of width 2. Indeed, in this case, any choice of two hyperedges per vertex yields a hypertree which violates the connectedness condition for variables (i.e., Condition 2 of Definition 11).

Figure 1(b) shows a complete hypertree decomposition of width 2 of the hypergraph $\mathcal{H}_{I_1}$ in part (a) of the figure. Note that this decomposition also happens to be a query decomposition of width 2.

Let k be a fixed positive integer. We say that a CSP instance I has k-bounded hypertree width if $hw(\mathcal{H}_I) \leq k$, where $\mathcal{H}_I$ is the hypergraph associated to I.

6   Computing Hypertree Decompositions

Let $\mathcal{H}$ be a hypergraph, and let $V \subseteq var(\mathcal{H})$ be a set of variables and $X, Y \in var(\mathcal{H})$. Then X is $[V]$-adjacent to Y if there exists an edge $h \in edges(\mathcal{H})$ such that $\{X, Y\} \subseteq h - V$. A $[V]$-path π from X to Y is a sequence $X = X_0, \ldots, X_\ell = Y$ of variables such that X_i is $[V]$-adjacent to X_{i+1}, for each $i \in [0 \ldots \ell\text{-}1]$. A set $W \subseteq var(\mathcal{H})$ of variables is $[V]$-connected if, for all $X, Y \in W$, there is a $[V]$-path from X to Y. A $[V]$-component is a maximal $[V]$-connected non-empty set of variables $W \subseteq var(\mathcal{H}) - V$. For any $[V]$-component C, let $edges(C) = \{h \in edges(\mathcal{H}) \mid h \cap C \neq \emptyset\}$.

Let $HD = \langle T, \chi, \lambda \rangle$ be a hypertree for $\mathcal{H}$. For any vertex v of T, we will often use v as a synonym of $\chi(v)$. In particular, $[v]$-component denotes $[\chi(v)]$-component; the term $[v]$-path is a synonym of $[\chi(v)]$-path; and so on. We introduce a normal form for hypertree decompositions.

ALTERNATING ALGORITHM k-decomp
Input: A non-empty Hypergraph $\mathcal{H}$.
Result: "Accept", if $\mathcal{H}$ has k-bounded hypertree-width; "Reject", otherwise.

Procedure k-decomposable(C_R: SetOfVariables, R: SetOfHyperedges)
begin
1)   **Guess** a set $S \subseteq edges(\mathcal{H})$ of k elements at most;
2)   **Check** that all the following conditions hold:
     2.a) $\forall P \in edges(C_R), (var(P) \cap var(R)) \subseteq var(S)$ and
     2.b) $var(S) \cap C_R \neq \emptyset$
3)   **If** the check above fails **Then Halt** and **Reject**; **Else**
     Let $\mathcal{C} := \{C \subseteq var(\mathcal{H}) \mid C \text{ is a } [var(S)]\text{-component and } C \subseteq C_R\}$;
4)   **If, for each** $C \in \mathcal{C}$, k-decomposable(C, S)
         Then Accept
         Else Reject
end;

begin(* MAIN *)
     Accept if k-decomposable($var(\mathcal{H}), \emptyset$)
end.

Fig. 4. A non-deterministic algorithm deciding k-bounded hypertree-width

Definition 13 ([19]). A hypertree decomposition $HD = \langle T, \chi, \lambda \rangle$ of a hypergraph $\mathcal{H}$ is in *normal form* (*NF*) if, for each vertex $r \in vertices(T)$, and for each child s of r, all the following conditions hold:
1. there is (exactly) one $[r]$-*component* C_r such that $\chi(T_s) = C_r \cup (\chi(s) \cap \chi(r))$;
2. $\chi(s) \cap C_r \neq \emptyset$, where C_r is the $[r]$-component satisfying Condition 1;
3. $var(\lambda(s)) \cap \chi(r) \subseteq \chi(s)$.

Intuitively, each subtree rooted at a child node s of some node r of a normal form decomposition tree serves to decompose precisely one $[r]$-*component*.

Theorem 14 ([19]). *For each k-width hypertree decomposition of a hypergraph $\mathcal{H}$ there exists a k-width hypertree decomposition of $\mathcal{H}$ in normal form.*

This normal form theorem immediately entails that, for each optimal hypertree decomposition of a hypergraph $\mathcal{H}$, there exists an optimal hypertree decomposition of $\mathcal{H}$ in normal form.

Importantly, NF hypertree decompositions can be efficiently computed. Figure 4 shows the algorithm k-decomp, deciding whether a given hypergraph $\mathcal{H}$ has a k-bounded hypertree-width decomposition. k-decomp can be implemented on a logspace ATM having polynomially bounded tree-size, and therefore entails LOGCFL membership of deciding k-bounded hypertree-width.

Theorem 15 ([19]). *Deciding whether a hypergraph $\mathcal{H}$ has k-bounded hypertree-width is in LOGCFL.*

From an accepting computation of the algorithm of Figure 4 we can efficiently extract a NF hypertree decomposition. Since an accepting computation tree of a bounded-treesize logspace ATM can be *computed* in (the functional version of) LOGCFL [18], we obtain the following.

Theorem 16 ([19]). *Computing a k-bounded hypertree decomposition (if any) of a hypergraph $\mathcal{H}$ is in $\mathrm{L}^{\mathrm{LOGCFL}}$, i.e., in functional LOGCFL.*

As for sequential algorithms, a polynomial time algorithm opt-k-decomp which, for a fixed k, decides whether a hypergraph has k-bounded hypertree width and, in this case, computes an optimal hypertree decomposition in normal form is described in [20]. As for many other decomposition methods, the running time of this algorithm to find the hypergraph decomposition is exponential in the parameter k. More precisely, opt-k-decomp runs in $O(m^{2k}v^2)$ time, where m and v are the number of edges and the number of vertices of $\mathcal{H}$, respectively.

7   Solving CSP Instances of Bounded Hypertree Width

Figure 5 outlines an efficient method to solve CSP instances of bounded Hypertree Width. The key point is that any CSP instance I having k-bounded hypertree width can be efficiently transformed into an equivalent acyclic CSP instance (Step 4.), which is then evaluated by the well-known techniques defined for acyclic CSPs (see Section 3). Let HW[k] be the set of all finite relational structures whose associated hypergraph has hypertree width at most k.

ALGORITHM
Input: A k-bounded hypertree width CSP instance I.
Result: A solution to I, if I is satisfiable; "No", otherwise.

begin
1) Build the hypergraph $\mathcal{H}_I$ of I.
2) Compute a k-width hypertree decomposition HD of $\mathcal{H}_I$ in normal form.
3) Compute from HD a complete hypertree decomposition $HD' = (T, \chi, \lambda)$ of $\mathcal{H}_I$.
4) Compute from HD' and I an acyclic instance I^* equivalent to I.
5) Evaluate I^* employing any efficient technique for solving acyclic CSPs.
6) If I^* is satisfiable, then return a solution to I^*;
   Else Return "No"
end.

Fig. 5. An algorithm solving CSP instances of k-bounded hypertree-width

Theorem 17 ([21]). *Given a CSP instance $I \in \mathrm{CSP}(\mathrm{HW}[k])$ and a k-width hypertree decomposition of $\mathcal{H}_I$ in normal form, I is solvable in $O(\|I\|^{k+1} \log \|I\|)$ time.*

We have also determined the precise computational complexity of solving CSP instances having bounded hypertree-width.

Theorem 18 ([19]). $\mathrm{CSP}(\mathrm{HW}[k])$ *is LOGCFL-complete.*

8 Game Theoretic Characterization of Hypertree Width

In [36], graphs G of treewidth k are characterized by the so called *Robber-and-Cops game* where $k+1$ cops have a winning strategy for capturing a robber on G. Cops can control vertices of a graph and can jump at each move to arbitrary vertices. The robber can move (at infinite speed) along paths of G but cannot go over vertices controlled by a cop. It is, moreover, shown that a winning strategy for the cops exists, iff the cops can capture the robber in a *monotonic* way, i.e., never returning to a vertex that a cop has previously vacated, which implies that the moving area of the robber is monotonically shrinking. For more detailed descriptions of the game, see [36] or [23].

In order to provide a similarily natural characterization for hypertree-width, we defined in [23] a new game, the *Robber and Marshals game (R&Ms game)*. A marshal is more powerful than a cop. While a cop can control a single vertex (=variable) only, a marshal controls an entire hyperedge. In the *R&Ms* game, the robber moves on vertices just as in the robber and cops game, but now marshals instead of cops are chasing her. During a move of the marshals from the set of hyperedges E to to the set of hyperedges E', the robber cannot pass through the vertices in $B = (\cup E) \cap (\cup E')$, where, for a set of hyperedges F, $\cup F$ denotes the union of all hyperedges in F. Intuitively, the vertices in B are those not released by the marshals during the move. As in the monotonic robber and cops game, it is required that the marshals capture the robber by monotonically shrinking the moving space of the robber. The game is won by the marshals if they corner the robber somewhere in the hypergraph.

Example 19. Let us play the robber-and-marshals game on the hypergraph $\mathcal{H}_{I_1}$ of query Q_1 of Example 4 (see Fig 1). We can easily recognize that two marshals can always capture the robber and win the game by using the following strategy: Independently of the initial position of the robber, the two marshals initially move on edges $\{a,b\}$, and thus control the vertices (=variables) T, X, S, R, P, Y, U, as shown in Figure 6.A. After this move of marshals, the robber may be in V, in Z or in W. If the robber is on V, then the marshals move on edge f, and capture the robber, as shown in Figure 6.B (note that the robber cannot escape from V during this move, as both P and R – the only possible ways to leave V – are kept under the marshals' control during the move). Otherwise, if the robber is on W or on Z, then the marshals move on $\{g, c\}$ (see Figure 6.C). Since they keep the control of X, Y, T, and U during the move, then the robber can escape only to vertex W. Therefore, a further move on edge d allows the marshals to eventually capture the robber, as shown in Figure 6.D.

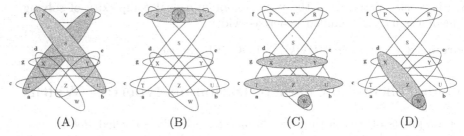

(A)   (B)   (C)   (D)

Fig. 6. (A) The first move of the marshals playing the game on $\mathcal{H}_{I_1}$; (B) move of the marshals if the robber stands on V (capture position); (C) move of the marshals if the robber stands on W or on Z; (D) the marshals capture the robber in W

In [23] we prove that there is a one-to-one correspondence between winning strategies for k marshals and hypertree decompositions of width at most k in a certain normal form.

Theorem 20 ([23]). *A hypergraph $\mathcal{H}$ has k-bounded hypertree width if and only k marshals have a winning strategy for the R&Ms game played on $\mathcal{H}$.*

9   Logical Characterization of Hypertree Width

Denote by L the fragment of first-order logic (FO) whose connectives are restricted to existential quantification and conjunction (i.e., $\neg$, $\vee$, and $\forall$ are disallowed). Kolaitis and Vardi [30] proved that the class of all queries having treewidth $< k$ coincides in expressive power with the k-variable fragment L^k of L, i.e., the class of all L formulas that use k variables only. Se also [13].

In [23], we characterize HW[k] in terms of a guarded logic. We show that HW[k] = GF$_k$(L), where GF$_k$(L) denotes the k-guarded fragment of L. The 1-guarded fragment coincides with the classical notion of guardedness, where existentially quantified subformulas φ of a formula are always conjoined with a *guard*, i.e., an atom containing all free variables of φ. In the k-guarded fragment, up to k atoms may jointly act as a guard (for precise definitions, see [23]). For the particular case $k = 1$, this gives us a new characterization of the acyclic queries stating that the acyclic queries are precisely those expressible in the guarded fragment of L. In order to prove these results, we played the robber and marshals game on the appropriate query hypergraphs.

10   Comparison of Hypertree Width with Other Methods

We report about results comparing the Hypertree decomposition method with other methods for solving efficiently CSPs and conjunctive queries, which are based only on the structure of the hypergraph associated with the problem (we consider *tractability due to restricted structure*, as discussed in Section 2). We call these methods *decomposition methods (DM)*, because each one provides a decomposition which transforms any hypergraph to an acyclic hypergraph. For each decomposition method D, this transformation depends on a parameter called D-*width*. Let k be a fixed constant. The tractability class $C(D, k)$ is the (possibly infinite) set of hypergraphs having D-width $\leq k$. D ensures that every CQ or CSP instance whose associated hypergraph belongs to this class is polynomial-time solvable.

The main decomposition methods considered in database theory and in artificial intelligence are: *Treewidth* [34] (see also [30,17]), *Cycle Cutset* [9], *Tree Clustering* [10], *Induced Width (w*)* cf. [9], *Hinge Decomposition* [27,26], *Hinge Decomposition + Tree Clustering* [26], *Cycle Hypercutset* [21], *Hypertree Decomposition*. All methods are briefly explained in [21]. Here, we do not consider the notion of query width, because deciding whether a hypergraph has bounded query width is NP-complete. However, recall that this notion is generalized by hypertree width, in that whenever a hypergraph has query width at most k, it has hypertree width at most k, too. The converse does not hold, in general [19]. For comparing decomposition methods we introduce the relations $\preceq$, $\triangleright$, and $\prec\!\!\!\prec$ defined as follows:

$D_1 \preceq D_2$, in words, D_2 *generalizes* D_1, if $\exists \delta \geq 0$ such that, $\forall k > 0$, $C(D_1, k) \subseteq C(D_2, k + \delta)$. Thus $D_1 \preceq D_2$ if every class of CSP instances which is tractable according to D_1 is also tractable according to D_2.

$D_1 \triangleright D_2$ (D_1 *beats* D_2) if there exists an integer k such that $\forall m\ C(D_1, k) \not\subseteq C(D_2, m)$. To prove that $D_1 \triangleright D_2$, it is sufficient to exhibit a class of hypergraphs contained in some $C(D_1, k)$ but in no $C(D_2, j)$ for $j \geq 0$. Intuitively, $D_1 \triangleright D_2$ means that at least on some class of CSP instances, D_1 outperforms D_2.

$D_1 \prec\!\!\!\prec D_2$ if $D_1 \preceq D_2$ and $D_2 \rhd D_1$. In this case we say that D_2 *strongly generalizes* D_1.

Mathematically, $\preceq$ is a *preorder*, i.e., it is reflexive, transitive but not antisymmetric. We say that D_1 *is $\preceq$-equivalent to* D_2, denoted $D_1 \equiv D_2$, if both $D_1 \preceq D_2$ and $D_2 \preceq D_1$ hold.

The decomposition methods D_1 and D_2 are *strongly incomparable* if both $D_1 \rhd D_2$ and $D_2 \rhd D_1$. Note that if D_1 and D_2 are strongly incomparable, then they are also incomparable w.r.t. the relations $\preceq$ and $\prec\!\!\!\prec$.

Figure 7 shows a representation of the hierarchy of DMs determined by the $\prec\!\!\!\prec$ relation. Each element of the hierarchy represents a DM, apart ¿from that containing the three $\preceq$-equivalent methods *Tree Clustering, Treewidth,* and w^*.

Theorem 21 ([21]). *For each pair D_1 and D_2 of decompositions methods represented in Figure 7, the following holds. There is a directed path from D_1 to D_2 iff $D_1 \prec\!\!\!\prec D_2$, i.e., iff D_2 strongly generalizes D_1. Moreover, D_1 and D_2 are not linked by any directed path iff they are strongly incomparable. Hence, Fig. 7 completely describes the relationships among the different methods.*

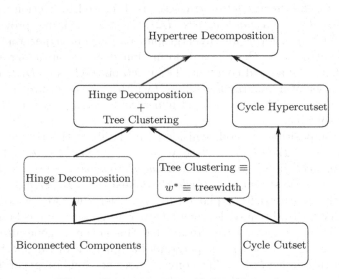

Fig. 7. Constraint tractability hierarchy

Recently, a comparison between hypertree width and Courcelle's concept of *clique-width* [7,8] was made [24]. Given that clique-width is defined for graphs, it had to be suitably adapted to hypergraphs. Defining the clique-width of a hypergraph $\mathcal{H}$ as the cliquewidth of its primal graph makes no sense in the context of CSP-tractability, because then CSPs of bounded clique-width would be intractable. Therefore, in [24], the clique-width $\mathcal{H}$ is defined as the clique-width of its incidence graph $G_I^i(\mathcal{H})$. With this definition it could be shown in [24] that

(a) CSP's whose hypergraphs have bounded clique-width are tractable, and (b) bounded hypertree width strongly generalizes bounded clique-width.

11   Open Problems

Several questions are left for future research. In particular, it would be interesting to know whether the method of hypertree decompositions can be further generalized. For instance, let us define the concept of *generalized hypertree decomposition* by just dropping condition 4 from the definition of hypertree decomposition (Def. 11). Correspondingly, we can introduce the concept of *generalized hypertree width* $ghw(\mathcal{H})$ of a hypergraph $\mathcal{H}$. We know that all classes of Boolean queries having bounded ghw can be answered in polynomial time. But we currently do not know whether these classes of queries are polynomially recognizable. This recognition problem is related to the mysterious *hypergraph sandwich problem* [31], which has remained unsolved for a long time. If the latter is polynomially solvable, then also queries of bounded ghw are polynomially recognizable. Another question concerns the time complexity of recognizing queries of bounded hypertree width. Is this problem fixed-parameter tractable such as the recognition of graphs of bounded treewidth?

Acknowledgements. Research supported by *FWF (Austrian Science Funds)* under the project Z29-INF (Wittgenstein Award), and by *MURST* under project COFIN-2000 "From Data to Information (D2I)"

References

1. L. Bachmair, Ta Chen, C. R. Ramakrishnan, and I. V. Ramakrishnan. Subsumption Algorithms Based on Search Trees. *Proc. CAAP'96*, Springer LNCS Vol.1059. 42

2. C. Beeri, R. Fagin, D. Maier, and M. Yannakakis. On the Desiderability of Acyclic Database Schemes. *Journal of ACM*, 30(3):479–513, 1983. 42

3. P. A. Bernstein, and N. Goodman. The power of natural semijoins. *SIAM J. Computing*, 10(4):751–771, 1981. 42

4. H. L. Bodlaender. A linear-time algorithm for finding tree-decompositions of small treewidth. *SIAM J. Computing*, 25(6):1305-1317, 1996. 44

5. A. K. Chandra and P. M. Merlin. Optimal Implementation of Conjunctive Queries in relational Databases. *Proc. STOC'77*, pp.77–90, 1977. 40

6. Ch. Chekuri and A. Rajaraman. Conjunctive Query Containment Revisited. *Theoretical Computer Science*, 239(2):211–229, 2000. 38, 45, 46, 47

7. B. Courcelle. Graph Rewriting: an algebraic and logic approach. Chapter 5 in *Handbook of Theor. Comp. Sci., vol. B*, J. Van Leeuwen ed., 1990. 38, 44, 54

8. B. Courcelle: Monadic second-order logic of graphs VII: Graphs as relational structures, in Theoretical Computer Science, Vol 101, pp. 3-33 (1992). 38, 54

9. R. Dechter. Constraint Networks. In *Encyclopedia of Artificial Intelligence*, second edition, Wiley and Sons, pp. 276-f285, 1992. 39, 43, 45, 53

10. R. Dechter and J. Pearl. Tree clustering for constraint networks. *Artificial Intelligence*, pp. 353–366, 1989. 43, 45, 53

11. R. Fagin. Degrees of acyclicity for hypergraphs and relational database schemes. *J. of the ACM*, 30(3):514–550, 1983. 42

12. T. Feder and M. Y.Vardi. The Computational Structure of Monotone Monadic SNP and Constraint Satisfaction: A Study through Datalog and Group Theory. *SIAM J. Comput.*, 28(1):57–104, 1998. 39

13. J. Flum, M. Frick, and M. Grohe. Query Evaluation via Tree-Decomposition. In *Proc. of ICDT'01*, Springer LNCS, Vol. 1973, pp.22–38, 2001. 45, 53

14. E. C. Freuder. Complexity of K-Tree Structured Constraint Satisfaction Problems. *Proc. of AAAI'90*, 1990. 45

15. H. Gaifman. On local and nonlocal properties. In *Logic Colloquium '81*, pp. 105–135, J. Stern ed., North Holland, 1982. 42

16. M. R. Garey and D. S. Johnson. *Computers and Intractability. A Guide to the Theory of NP-completeness*. Freeman and Comp., NY, USA, 1979.

17. G. Gottlob, N. Leone, and F. Scarcello. The Complexity of Acyclic Conjunctive Queries. *Journal of the ACM*, 48(3), 2001. Preliminary version in FOCS'98. 38, 41, 43, 45, 46, 53

18. G. Gottlob, N. Leone, and F. Scarcello. Computing LOGCFL Certificates. *Theoretical Computer Sciences*, to appear. Preliminary version in ICALP'99. 44, 50

19. G. Gottlob, N. Leone, and F. Scarcello. Hypertree Decompositions and Tractable Queries. JCSS. to appear. Preliminary version in PODS'99. 38, 47, 48, 50, 51, 53

20. G. Gottlob, N. Leone, and F. Scarcello. "On Tractable Queries and Constraints," in *Proc. DEXA'99*, Florence, 1999, LNCS 1677, pp. 1-15, Springer. 50

21. G. Gottlob, N. Leone, and F. Scarcello. A Comparison of Structural CSP Decomposition Methods. *Artificial Intelligence*, 124(2):243–282, 2000. Preliminary version in *IJCAI'99*. 38, 45, 51, 53, 54

22. G. Gottlob, N. Leone, and F. Scarcello. "Advanced Parallel Algorithms for Processing Acyclic Conjunctive Queries, Rules, and Constraints," *Proc. SEKE00*, pp. 167–176, KSI Ed., Chicago, USA, July 6-8, 2000. 38, 43

23. G. Gottlob, N. Leone, and F. Scarcello. "Robbers, Marshals, and Guards: Game-Theoretic and Logical Characterizations of Hypertree Width," in *Proc. PODS'01*. 38, 51, 52, 53

24. G. Gottlob and R. Pichler. Hypergraphs in Model Checking: Acyclicity and Hypertree-Width Versus Clique-Width. *Proc. ICALP 2001*, to appear. 38, 54

25. M. Grohe, T. Schwentick, and L. Segoufin. When is the Evaluation of Conjunctive Queries Tractable? Proc. ACM STOC 2001. 45

26. M. Gyssens, P. G. Jeavons, and D. A. Cohen. Decomposing constraint satisfaction problems using database techniques. *Artificial Intelligence*, 66:57–89, 1994. 53

27. M. Gyssens, and J. Paredaens. A Decomposition Methodology for Cyclic Databases. In *Advances in Database Theory*, vol.2, 1984. 53

28. P. Jeavons, D. Cohen, and M. Gyssens. Closure Properties of Constraints. *Journal of the ACM*, 44(4):527–548.

29. D. S. Johnson. A Catalog of Complexity Classes. In J. van Leeuwen, editor, *Handbook of Theoretical Computer Science*, volume A, chapter 2, pp.67–161. Elsevier Science Publishers B. V. (North-Holland), 1990. 43

30. Ph. G. Kolaitis and M. Y. Vardi. Conjunctive-Query Containment and Constraint Satisfaction. *Journal of Computer and System Sciences*, 61:302–332, 2000. 39, 40, 45, 52, 53

31. A. Lustig and O. Shmueli. Acyclic Hypergraph Projections. *J. of Algorithms*, 30:400–422, 1999. 55

32. D. Maier. *The Theory of Relational Databases*, Rochville, Md, Computer Science Press, 1986. 42, 45

33. J. Pearson and P. Jeavons. A survey of tractable constraint satisfaction problems. Technical Report CSD-TR-97-15, Royal Halloway University of London, 1997. 40

34. N. Robertson and P. D. Seymour. Graph Minors II. Algorithmic Aspects of Tree-Width. *J. Algorithms*, 7:309-322, 1986. 44, 53

35. T. J. Schaefer. The Complexity of Satisfiability Problems. In *Proc. STOC'78*. 40

36. P. D. Seymour and R. Thomas. Graph Searching and a Min-Max Theorem for Tree-Width. *J. of Combinatorial Theory, Series B*, 58:22-33, 1993. 51

37. R. E. Tarjan, and M. Yannakakis. Simple linear-time algorithms to test chordality of graphs, test acyclicity of hypergraphs, and selectively reduce acyclic hypergraphs. *SIAM J. Computing*, 13(3):566-579, 1984. 43

38. M. Vardi. Complexity of Relational Query Languages. In *Proc. of 14th ACM STOC*, pp. 137–146, 1982. 43

39. E. Wanke. Bounded Tree-Width and LOGCFL. *Journal of Algorithms*, 16:470–491, 1994. 44

40. A. N. Wilschut, J. Flokstra, and P. M. G. Apers. Parallel evaluation of multi-join queries. In *Proc. of SIGMOD'95*, San Jose, CA USA, pp.115–126, 1995. 43

41. M. Yannakakis. Algorithms for Acyclic Database Schemes. *Proc. VLDB'81*, pp. 82–94, C. Zaniolo and C. Delobel Eds., Cannes, France, 1981. 43

The Strength of Non-size-increasing Computation (Introduction and Summary)

Martin Hofmann

TU Darmstadt, FB 4, Schloßgartenstr. 7, 64289 Darmstadt, Germany
mh@mathematik.tu-darmstadt.de

Abstract. We study the expressive power non-size increasing recursive definitions over lists. This notion of computation is such that the size of all intermediate results will automatically be bounded by the size of the input so that the interpretation in a finite model is sound with respect to the standard semantics. Many well-known algorithms with this property such as the usual sorting algorithms are definable in the system in the natural way. The main result is that a characteristic function is definable if and only if it is computable in time $O(2^{p(n)})$ for some polynomial p. The method used to establish the lower bound on the expressive power also shows that the complexity becomes polynomial time if we allow primitive recursion only. This settles an open question posed in [1,6]. The key tool for establishing upper bounds on the complexity of derivable functions is an interpretation in a finite relational model whose correctness with respect to the standard interpretation is shown using a semantic technique.

Keywords: computational complexity, higher-order functions, finite model, semantics

AMS Classification: 03D15, 03C13, 68Q15, 68Q55

The present document contains an introduction to and a summary of the results presented in the author's invited talk at MFCS'01. A full version of the paper is available as [8]. The author wishes to thank the organisers and the programme committee of MFCS'01 for the invitation to present his results there. He would also like to acknowledge financial support by the Mittag-Leffler institute which funded a research stay at that place during which the bulk of the research presented here was carried out.

Consider the following recursive definition of a function on lists:

$$\begin{aligned}
\texttt{twice}(\mathsf{nil}) &= \mathsf{nil} \\
\texttt{twice}(\mathsf{cons}(x,l)) &= \mathsf{cons}(\mathsf{tt}, \mathsf{cons}(\mathsf{tt}, \texttt{twice}(l)))
\end{aligned} \tag{1}$$

Here nil denotes the empty list, $\mathsf{cons}(x,l)$ denotes the list with first element x and remaining elements l. tt, ff are the members of a type T of truth values. We have that $\texttt{twice}(l)$ is a list of length $2 \cdot |l|$ where $|l|$ is the length of l. Now consider

$$\begin{aligned}
\exp(\mathsf{nil}) &= \mathsf{cons}(\mathsf{tt}, \mathsf{nil}) \\
\exp(\mathsf{cons}(x,l)) &= \texttt{twice}(\exp(l))
\end{aligned} \tag{2}$$

J. Sgall, A. Pultr, and P. Kolman (Eds.): MFCS 2001, LNCS 2136, pp. 58–61, 2001.

We have $|\exp(l)| = 2^{|l|}$ and further iteration leads to elementary growth rates.

This shows that innocuous looking recursive definitions can lead to enormous growth. In order to prevent this from happening it has been suggested in [?,10] to rule out definitions like (2) above, where a recursively defined function, here twice, is applied to the result of a recursive call. Indeed, it has been shown that such discipline restricts the definable functions to the polynomial-time computable ones and moreover every polynomial-time computable *function* admits a definition in this style.

Many naturally occurring *algorithms*, however, do not fit this scheme. Consider, for instance, the definition of insertion sort:

$$\text{insert}(x, \text{nil}) = \text{cons}(x, \text{nil})$$
$$\text{insert}(x, \text{cons}(y, l)) = \text{if } x \leq y \text{ then } \text{cons}(x, \text{cons}(y, l)) \text{ else } \text{cons}(y, \text{insert}(x, l))$$
$$\text{sort}(\text{nil}) = \text{nil}$$
$$\text{sort}(\text{cons}(x, l)) = \text{insert}(x, \text{sort}(l))$$

(3)

Here just as in (2) above we apply a recursively defined function (insert) to the result of a recursive call (sort), yet no exponential growth arises.

It has been argued in [3] and [6] that the culprit is definition (1) because it defines a function that increases the size of its argument and that non size-increasing functions can be arbitrarily iterated without leading to exponential growth.

In [3] a number of partly semantic criteria were offered which allow one to recognise when a function definition is non size-increasing. In [6] we have given syntactic criteria based on linearity (bound variables are used at most once) and a so-called resource type $\diamond$ which counts constructor symbols such as "cons" on the left hand side of an equation.

This means that cons becomes a ternary function taking one argument of type $\diamond$, one argument of some type A (the head) and a third argument of type $L(A)$, the tail. There being no closed terms of type $\diamond$ the only way to apply cons is within a recursive definition; for instance, we can write

$$\text{append}(\text{nil}, l_2) = l_2$$
$$\text{append}(\text{cons}(d, a, l_1), l_2) = \text{cons}(d, a, \text{append}(l_1, l_2))$$

(4)

Alternatively, we may write

$$\text{append}(l_1, l_2) = \text{match } l \text{ with } \text{nil} \Rightarrow l_2 \mid \text{cons}(d, a, l_1') \Rightarrow \text{cons}(d, \text{append}(l_1, l_2))$$ (5)

We notice that the following attempted definition of twice is illegal as it violates linearity (the bound variable d is used twice):

$$\text{twice}(\text{nil}) = \text{nil}$$
$$\text{twice}(\text{cons}(d, x, l)) = \text{cons}(d, \text{tt}, \text{cons}(d, \text{tt}, \text{twice}(l)))$$

(6)

The definition of insert, on the other hand, is in harmony with linearity provided that insert gets an extra argument of type $\diamond$ and, moreover, we assume that the inequality test returns its arguments for subsequent use.

The main result of [6] and [1] was that all functions thus definable by *structural recursion* are polynomial-time computable even when higher-order functions are allowed. In [7] it has been shown that general-recursive first-order definitions admit a translation into a fragment of the programming language C without dynamic memory allocation ("malloc") which on the one hand allows one to automatically construct imperative implementations of algorithms on lists which do not require extra space or garbage collection. More precisely, this translation maps the resource type $\Diamond$ to the C-type void $*$ of pointers. The cons function is translated into the C-function which extends a list by a given value using a provided piece of memory. It is proved that the pointers arising as denotation of terms of type $\Diamond$ always point to free memory space which can thus be safely overwritten.

This translation also demonstrates that all definable functions are computable on a Turing machine with linearly bounded work tape and an unbounded stack (to accommodate general recursion) which by a result of Cook[1] [4] equals the complexity class $DTIME(2^{O(n)})$. It was also shown in [7] that any such function admits a representation.

In the presence of higher-order functions the translation into C breaks down as C does not have higher-order functions. Of course, higher-order functions can be simulated as closures, but this then requires arbitrary amounts of space as closures can grow proportionally to the runtime. In a system based on structural recursion such as [6] this is not a problem as the runtime is polynomially bounded there. The hitherto open question of complexity of general recursion with higher-order functions is settled in this work [8] and shown to require a polynomial amount of space only in spite of the unbounded runtime.

We thus demonstrate that a function is representable with general recursion and higher-order functions iff it is computable in polynomial space and an unbounded stack or equivalently (by Cook's result) in time $O(2^{p(n)})$ for some polynomial p. The lower bound of this result also demonstrates that indeed all characteristic functions of problems in P are definable in the structural recursive system. This settles a question left open in [1,6].

In view of the results presented in the talk (see also [8]), these systems of non size-increasing computation thus provide a very natural connection between complexity theory and functional programming. There is also a connection to finite model theory in that programs admit a sound interpretation in a finite model. This improves upon earlier combinations of finite model theory with functional programming [5] where interpretation in a finite model was achieved in a brute-force way by changing the meaning of constructor symbols, e.g. successor of the largest number N was defined to be N itself. In those systems it is the responsibility of the programmer to account for the possibility of cut-off when reasoning about the correctness of programs. In the systems studied here linearity and the presence of the resource types automatically ensure that cutoff

---

[1] This result asserts that if $L(n) > \log(n)$ then $DTIME(2^{O(L(n))})$ equals the class of functions computable by a Turing machine with an $L(n)$-bounded R/W-tape and an unbounded stack.

never takes place. Formally, it is shown that the standard semantics in an infinite model agrees with the interpretation in a certain finite model for all well-formed programs.

Another piece of related work is Jones' [9] where the expressive power of cons-free higher-order programs is studied. It is shown there that first-order cons-free programs define polynomial time , whereas second-order programs define EXPTIME. This shows that the presence of "cons", tamed by linearity and the resource type changes the complexity-theoretic strength. While loc. cit. also involves Cook's abovementioned result (indeed, this result was brought to the author's attention by Neil Jones) the other parts of the proof are quite different.

References

1. Klaus Aehlig and Helmut Schwichtenberg. A syntactical analysis of non-size-increasing polynomial time computation. In *Proceedings of the Fifteenth IEEE Symposium on Logic in Computer Science (LICS '00), Santa Barbara*, 2000. 58, 60
2. Stephen Bellantoni and Stephen Cook. New recursion-theoretic characterization of the polytime functions. *Computational Complexity*, 2:97–110, 1992.
3. Vuokko-Helena Caseiro. *Equations for Defining Poly-time Functions*. PhD thesis, University of Oslo, 1997. Available by ftp from ftp.ifi.uio.no/pub/vuokko/0adm.ps. 59
4. Stephen A. Cook. Linear-time simulation of deterministic two-way pushdown automata. *Information Processing*, 71:75–80, 1972. 60
5. Andreas Goerdt. Characterizing complexity classes by higher type primitive recursive definitions. *Theoretical Computer Science*, 100:45–66, 1992. 60
6. Martin Hofmann. Linear types and non size-increasing polynomial time computation. To appear in Information and Computation. See www.dcs.ed.ac.uk/home/papers/icc.ps.gz for a draft. An extended abstract has appeared under the same title in Proc. Symp. Logic in Comp. Sci. (LICS) 1999, Trento, 2000. 58, 59, 60
7. Martin Hofmann. A type system for bounded space and functional in-place update. *Nordic Journal of Computing*, 7(4), 2000. An extended abstract has appeared in *Programming Languages and Systems*, G. Smolka, ed., Springer LNCS, 2000. 60
8. Martin Hofmann. The strength of non size-increasing computation. 2001. Presented at the workshop *Implicit Computational Complexity 2001*, Aarhus, 20 May 2001. Submitted for publication. See also www.dcs.ed.ac.uk/home/mxh/icc01_hofmann.ps. 58, 60
9. Neil Jones. The Expressive Power of Higher-Order Types or, Life without CONS. *Journal of Functional Programming*, 2001. to appear. 61
10. Daniel Leivant. Stratified Functional Programs and Computational Complexity. In *Proc. 20th IEEE Symp. on Principles of Programming Languages*, 1993. 59

Introduction to Recent Quantum Algorithms

Peter Høyer*

Department of Computer Science, University of Calgary
2500 University Drive N.W.
Calgary, Alberta, Canada T2N 1N4
hoyer@cpsc.ucalgary.ca

Abstract. We discuss some of the recent progress in quantum algorithmics. We review most of the primary techniques used in proving upper and lower bounds and illustrate how to apply the techniques to a variety of problems, including the threshold function, parity, searching and sorting. We also give a set of open questions and possible future research directions. Our aim is to give a basic overview and we include suggestions to further reading.

1 Introduction

The most famous quantum algorithms are Shor's [38] algorithms for integer factorization and computing discrete logarithms and Grover's [28] algorithm for searching an un-ordered set of elements. In this paper, we discuss and survey many of the quantum algorithms found since the discoveries of Shor and Grover.

The algorithms of Shor and Grover, as well as most other existing quantum algorithms, can be naturally expressed in the so-called black box model. In this model, we are given some function f as input. The function f is given as a black box so that the only knowledge we can gain about f is in asking for its value on points of its domain. We may think of f as an oracle which, when asked some question x, replies by $f(x)$. Our measure of complexity is the number of evaluations of f required to solve the problem of interest.

On a quantum computer, the only two types of operations allowed are unitary operators and measurements. We may without loss of generality assume that all measurements are performed at the end of the computation, and thus any quantum algorithm can be modeled as a sequence of unitary operators followed by a single measurement. Consequently, we need to model our queries to function f so that they are unitary. Let $f : X \to Z$ be any function with $Z = \{0,1\}^m$ for some integer m. Define the unitary operator O_f by

$$|x\rangle|z\rangle|w\rangle \longmapsto |x\rangle|z \oplus f(x)\rangle|w\rangle \qquad (1)$$

for all $x \in X$, $z \in Z$ and $w \in Z$, where $z \oplus f(x)$ denotes the binary exclusive-or of bit-strings z and $f(x)$. Applying the operator O_f twice is equivalent to applying the identity operator and thus O_f is unitary (and reversible) as required.

---

* Supported in part by Canada's NSERC and the Pacific Institute for the Mathematical Sciences.

J. Sgall, A. Pultr, and P. Kolman (Eds.): MFCS 2001, LNCS 2136, pp. 62–74, 2001.

It changes the content of the second register ($|z\rangle$) conditioned on the value of the first register ($|x\rangle$). The purpose of the third register ($|w\rangle$) is simply to allow for extra work-space. We often refer to O_f as the *black box* or as the *oracle*. On a quantum computer, we are given access to operator O_f and our objective is to use the fewest possible applications of O_f to solve the problem at hand. This model of computation is often refered to as the *quantum black box model*, and sometimes also to as *quantum decision trees* [16].

A quantum algorithm Q that uses T applications of O_f (i.e., uses T queries) is a unitary operator of the form [16,4,31]

$$Q = (UO_f)^T U \tag{2}$$

for some unitary operator U. We always apply algorithm Q on the initial state $|0\rangle$. The superposition obtained by applying Q on $|0\rangle$ is $Q|0\rangle$, which is $(UO_f)^T U|0\rangle$. The operators are applied right to left: first apply U, and then iterate UO_f a total number of T times. After applying Q on $|0\rangle$, we always measure the final state in the computational basis. The outcome of the measurement are classical bits, and the rightmost bit(s) of those is the output of the algorithm.

Prior to the discoveries of Shor and Grover, notable quantum algorithms were found by Deutsch [20], Deutsch and Jozsa [21], Bernstein and Vazirani [6] and Simon [39]. Deutsch [20] considered the problem that we are given a function $f : \{0,1\} \to \{0,1\}$ as a black box O_f, and we want to determine whether $f(0) = f(1)$ or $f(0) \neq f(1)$. Deutsch gave a zero-error quantum algorithm that uses only 1 application of O_f and outputs the correct answer with probability $\frac{1}{2}$, and the answer "inconclusive" with complementary probability $\frac{1}{2}$.

The algorithm of Deutsch [20] is generalized and improved by Deutsch and Jozsa [21], Cleve, Ekert, Macchiavello and Mosca [18], Tapp [40] and others. Deutsch's algorithm, as well as the early algorithms by Bernstein and Vazirani [6] and Simon [39], are discussed in many introductions to quantum computing, including the excellent papers by Berthiaume [7], Cleve [17] and Reiffel and Polak [37]. For a thorough introduction to quantum computing, see the recent book by Nielsen and Chuang [34].

The two most widely used tools for constructing quantum algorithms are Fourier transforms and amplitude amplification. Fourier transforms is an intrinsic ingredient in the algorithms of Bernstein and Vazirani [6], Simon [39] and Shor [38]. See for instance Ivanyos, Magniez and Santha [32] and the references therein for applications of Fourier transforms in quantum algorithms. Amplitude amplification was introduced by Brassard and Høyer [9] as a generalization of the underlying subroutine used in Grover's algorithm [28]. We discuss this technique further in Sect. 3 as most of the newer quantum algorithms utilize amplitude amplification.

In the rest of this paper, we review some of the main techniques used in proving upper and lower bounds on quantum algorithms. For the purpose of illustrating the techniques, we also discuss nine concrete problems, each of them hopefully being considered fundamental. Section 2 contains an overview of the nine problems, including the best known upper and lower bounds. We discuss techniques for proving upper bounds in Sect. 3, and lower bounds in Sect. 4. We conclude in Sect. 5 by mentioning some open questions and possible future research directions.

2   Overview of Problems

Below, we give a list of nine problems that have been considered in the quantum black box model. The problems fall in 3 groups. The first group consists of decision problems with solutions depending only on $|f^{-1}(1)|$, the number of elements being mapped to 1. Let $[N] = \{0, 1, \ldots, N-1\}$. We say a problem P is *symmetric* if $P(f) = P(f \circ \pi)$ for all inputs $f : [N] \to Z$ and all permutations $\pi : [N] \to [N]$. All problems in the first group are symmetric. Generally speaking, the quantum complexities of symmetric problems are much more well-understood than those of non-symmetric problems. In the second group, we have put the fundamental computational tasks of searching and sorting, and finally, in the third group, some decision problems related to sorting. Table 1 gives the best known lower and upper bounds for each of these nine problems.

OR   Given Boolean function $f : [N] \to \{0, 1\}$, is $|f^{-1}(1)| > 0$?
THRESHOLD$_S$ Given Boolean function $f : [N] \to \{0, 1\}$, is $|f^{-1}(1)| \geq S$?
MAJORITY   Given Boolean function $f : [N] \to \{0, 1\}$, is $|f^{-1}(1)| > N/2$?
PARITY   Given Boolean function $f : [N] \to \{0, 1\}$, is $|f^{-1}(1)|$ odd?
ORDERED SEARCHING   Given monotone Boolean function $f : [N] \to \{0, 1\}$ and promised that $f(N-1) = 1$, output the smallest index $x \in [N]$ so that $f(x) = 1$.
SORTING   Given function $f : [N] \to Z$, output a permutation $\pi : [N] \to [N]$ so that $f \circ \pi$ is monotone.
COLLISION   Given function $f : [N] \to Z$ and promised that f is either 1-to-1 or 2-to-1, decide which is the case.
ELEMENT DISTINCTNESS   Given a function $f : [N] \to Z$, do there exist distinct elements $x, y \in [N]$ so that $f(x) = f(y)$?
CLAW (monotone case)   Given two monotone functions $f, g : [N] \to Z$, does there exist $(x, y) \in [N]^2$ so that $f(x) = g(y)$?

3   The Quantum Algorithms

For the five problems OR, THRESHOLD$_S$, COLLISION, ELEMENT DISTINCTNESS and CLAW, the quantum complexities listed in Table 1 are asymptotically smaller than the corresponding randomized decision tree complexities. In each case, the speed-up is at most quadratic and is achieved by applying amplitude amplification and estimation. Many newer quantum algorithms rely on these two techniques, and we therefore now give a brief description of them and sketch how to apply them to each of these five problems. For more details on amplitude amplification and estimation, see Brassard, Høyer, Mosca and Tapp [10].

Table 1. The best known lower and upper bounds in the quantum black box model for each of the nine problems defined in Sect. 2. These are the asymptotic bounds for two-sided bounded-error quantum algorithms, with the exception of the bounds for ORDERED SEARCHING which are for exact quantum algorithms. The two rightmost columns contain references to the proofs of each non-trivial lower and upper bound, respectively. We assume the input to each of the four last problems are given as comparison matrices. For the threshold problem, $S^+ = S+1$ and $S^- = S-1$. We use $\log^\star$ to denote the log-star function defined below, and c is some (small) constant

Problem	Lower	Upper	References	
OR	$\sqrt{N}$	$\sqrt{N}$	[5]	[8,28]
THRESHOLDS	$\sqrt{S^+(N-S^-)}$	$\sqrt{S^+(N-S^-)}$	[4]	[4]
MAJORITY	N	N	[4]	
PARITY	N	N	[23,4]	
ORDERED SEARCHING	$0.220\log_2(N)$	$0.526\log_2(N)$	[31]	[25]
SORTING	$N\log(N)$	$N\log(N)$	[31]	
COLLISION		$N^{1/3}\log(N)$		[11]
ELEM. DISTINCTNESS	$\sqrt{N}\log(N)$	$N^{3/4}\log(N)$	[31]	[14]
CLAW (monotone case)	$\sqrt{N}$	$\sqrt{N}c^{\log^\star(N)}$		[14]

Consider we want to solve some problem using some quantum algorithm $\mathcal{A}$. The algorithm $\mathcal{A}$ starts on the initial state $|0\rangle$ and produces some final superposition $|\Psi\rangle = \sum_{i\in\mathbb{Z}}\alpha_i|i\rangle$. We assume $\mathcal{A}$ is a quantum algorithm that uses no measurement during the computation. This assumption is mostly technical and can be safely ignored for the purposes of this paper, so we do not elaborate any further on this. What is crucial, is that we from the output of algorithm $\mathcal{A}$ somehow can deduce if we have solved the problem or not. We formalize this by assuming that we, in addition to $\mathcal{A}$, are given some Boolean function $\chi : \mathbb{Z} \to \{0,1\}$. We say algorithm $\mathcal{A}$ *succeeds* if a measurement of $|\Psi\rangle$ yields an integer i so that $\chi(i) = 1$. Let a denote the success probability of $\mathcal{A}$, that is, let $a = \left|\sum_{i\in\chi^{-1}(1)}\alpha_i|i\rangle\right|^2 = \sum_{i\in\chi^{-1}(1)}|\alpha_i|^2$.

There are (at least) two types of questions one may consider concerning algorithm $\mathcal{A}$ with respect to function χ. Firstly, we may consider the problem of finding an integer i so that $\chi(i) = 1$, i.e., finding a solution i, and secondly, we may ask for the value of a, i.e., what is the success probability of $\mathcal{A}$? These two questions concern searching and estimation, respectively.

Amplitude amplification [10] is a technique that allows fast searching on a quantum computer. On a classical computer, the standard technique to boosting the probability of success is by repetition. By running algorithm $\mathcal{A}$ a total number of j times, the probability of success increases to roughly ja (assuming $ja \ll 1$). Intuitively, we can think of this strategy as each additional run of algorithm $\mathcal{A}$ boosting the probability of success by an additive amount of roughly a. To find an integer i with $\chi(i) = 1$, we require an expected number of $\Theta(\frac{1}{a})$ repetitions. A quantum analogue of boosting the probability of success is to boost the *amplitude* of being in a certain subspace of the Hilbert space. Amplitude amplification is such a technique. It allows us to boost the probability of success to roughly $j^2 a$ using only j applications of algorithm $\mathcal{A}$ and function χ, which implies that we can find an integer i with $\chi(i) = 1$ applying $\mathcal{A}$ and χ an expected number of only $O(\frac{1}{\sqrt{a}})$ times.

Theorem 1 (Amplitude amplification) [10]. *Let $\mathcal{A}$ be any quantum algorithm that uses no measurements, and let $\chi : \mathbb{Z} \to \{0,1\}$ be any Boolean function. Let $\mathcal{A}|0\rangle = \sum_{i \in \mathbb{Z}} \alpha_i |i\rangle$ be the superposition obtained by running $\mathcal{A}$ on the initial state $|0\rangle$. Let $a = |\sum_{i \in \chi^{-1}(1)} \alpha_i |i\rangle|^2$ denote the probability that a measurement of the final state $\mathcal{A}|0\rangle$ yields an integer i so that $\chi(i) = 1$. Provided $a > 0$ we can find an integer i with $\chi(i) = 1$ using an expected number of only $O(\frac{1}{\sqrt{a}})$ applications of $\mathcal{A}$, the inverse $\mathcal{A}^{-1}$ and function χ.*

This theorem is a generalization of Grover's result [28] that a database can be searched for a unique element in $O(\sqrt{N})$ queries. The quantum algorithm [8] for OR can be phrased in these terms: Let $\mathcal{A}$ be any quantum algorithm that maps the initial state $|0\rangle$ to an equally weighted superposition of all possible inputs to function f, $\mathcal{A}|0\rangle = \frac{1}{\sqrt{N}} \sum_{i=0}^{N-1} |i\rangle$. If we measure state $\mathcal{A}|0\rangle$, our probability of seeing an integer i so that $f(i) = 1$ is exactly $a = |f^{-1}(1)|/N$, since every i is equally likely to be measured. Let $\chi = f$. Thus using an expected number of only $O(\frac{1}{\sqrt{a}}) = O(\sqrt{N/|f^{-1}(1)|})$ applications of function f we can find an integer i so that $f(i) = 1$, provided there is such an i. This implies that there is a one-sided error quantum algorithm for OR that, using only $O(\sqrt{N})$ applications of f, outputs "no" with certainty if $|f^{-1}(1)| = 0$, and "yes" with probability at least $\frac{2}{3}$ if $|f^{-1}(1)| > 0$.

The quantum algorithm [11] for COLLISION also uses amplitude amplification. First, pick any subset B of $[N]$ of cardinality $N^{1/3}$ and sort B with respect to its f-values using $O(N^{1/3} \log(N))$ comparisons of the form "Is $f(i) < f(j)$?". Once B is sorted, check that no two consecutive elements in the sorted list map to the same value under f, using an additional $N^{1/3} - 1$ comparisons. If a collision is found, output "2-to-1" and stop. Otherwise, proceed as follows. Define $\chi : [N] \to \{0,1\}$ by $\chi(i) = 1$ if and only if $i \notin B$ and $f(i) = f(j)$ for some $j \in B$. A single evaluation of χ can be implemented using only $O(\log(N))$ comparisons of f-values since B is sorted. If function f is 2-to-1 then $|\chi^{-1}(1)| = N^{1/3}$, and if f is 1-to-1 then $|\chi^{-1}(1)| = 0$. As in the case of OR, there thus is a one-sided error quantum subroutine $\mathcal{A}$ that, using only $O(\sqrt{N/N^{1/3}})$ applications of χ, outputs "no" with certainty if $|\chi^{-1}(1)| = 0$, and "yes" with probability at least $\frac{2}{3}$ if $|\chi^{-1}(1)| = N^{1/3}$. If subroutine $\mathcal{A}$ outputs "yes", then output "2-to-1" and stop, otherwise output "1-to-1" and stop. The total number of comparisons of f-values is $O(N^{1/3} \log(N)) + (N^{1/3} - 1) + O(\sqrt{N/N^{1/3}} \times \log(N))$, which is $O(N^{1/3} \log(N))$ as specified in Table 1.

The quantum algorithm [14] for ELEMENT DISTINCTNESS is similar to the algorithm for COLLISION, except that we now require two nested applications of amplitude amplification. Interestingly, the algorithm for ELEMENT DISTINCTNESS uses only $O(N^{3/4} \log(N))$ comparisons, which is not only sublinear in N, but also much less than the number of comparisons required to sort on a quantum computer. The algorithm [14] for CLAW uses $O(\sqrt{N}c^{\log^\star(N)})$ comparisons for some (small) constant c with $O(\log^\star(N))$ nested applications of amplitude amplification. The log-star function $\log^\star()$ is defined as the minimum number of iterated applications of the logarithm function necessary to obtain a number less than or equal to 1: $\log^\star(M) = \min\{s \geq 0 \mid \log^{(s)}(M) \leq 1\}$, where $\log^{(s)} = \log \circ \log^{(s-1)}$ denotes the sth iterated application of log, and $\log^{(0)}$ is the identity function.

Amplitude estimation [10] is a technique for estimating the success probability a of a quantum algorithm $\mathcal{A}$. This technique is used in the quantum algorithm [4] for THRESHOLD$_S$. On a classical computer, the standard technique to estimating the probability of success is, again, by repetition: If algorithm $\mathcal{A}$ succeeds in j out of k independent runs, we may output $\tilde{a} = j/k$ as an approximation to a. A quantum analogue of estimating the probability of success is to estimate the amount of *amplitude* of being in a certain subspace of the Hilbert space. The following theorem provides a method for doing this.

Theorem 2 (Amplitude estimation) [10]. *Let the setup be as in Theorem 1. There exists a quantum algorithm that given $\mathcal{A}$, function χ and an integer $M > 0$, outputs $\tilde{a}$ $(0 \leq \tilde{a} \leq 1)$ such that*

$$|\tilde{a} - a| \leq 2\pi \frac{\sqrt{a(1-a)}}{M} + \frac{\pi^2}{M^2}$$

with probability at least $\frac{8}{\pi^2}$. The algorithm uses $O(M)$ applications of $\mathcal{A}$, the inverse $\mathcal{A}^{-1}$ and function χ.

A straight-forward classical algorithm for THRESHOLD$_S$ is as follows. Let $\mathcal{A}$ denote the algorithm that outputs a random element $x \in [N]$, taken uniformly. The probability that $\mathcal{A}$ succeeds in outputting an x so that $f(x) = 1$ is exactly $a = |f^{-1}(1)|/N$. Apply $\mathcal{A}$ a total number of k times, and let j denote the number of times $\mathcal{A}$ succeeds. If $\lfloor \frac{j}{k} + \frac{1}{2} \rfloor \geq S$, then output "yes", otherwise output "no". A simple quantum version of this algorithm is as follows. First, we find an estimate $\tilde{a}$ of a by applying the above theorem with $M = 100\sqrt{S^+(N - S^-)}$, where $S^+ = S + 1$ and $S^- = S - 1$. Our estimate $\tilde{a}$ satisfies that $|\tilde{a} - a| \leq 2\pi\sqrt{a(1-a)}/M + \frac{\pi^2}{M^2}$ with probability at least $\frac{8}{\pi^2}$, which implies that if $|f^{-1}(1)| \geq S$ then $\lfloor \tilde{a}N + \frac{1}{2} \rfloor \geq S$ with probability at least $\frac{8}{\pi^2}$, and if $|f^{-1}(1)| \leq S - 1$ then $\lfloor \tilde{a}N + \frac{1}{2} \rfloor \leq S - 1$ with probability at least $\frac{8}{\pi^2}$. Thus, if $\lfloor \tilde{a}N + \frac{1}{2} \rfloor \geq S$ then output "yes", otherwise output "no". The number of evaluations of f used by this algorithm is $O(M)$, which is $O(\sqrt{S^+(N - S^-)})$ as specified in Table 1.

3.1   Other Quantum Algorithms

The quantum complexities of the four symmetric decision problems in the first group in Table 1 are all tight, up to constant factors. The quantum complexities of symmetric decision problems are reasonably well-understood, especially since the seminal work of Beals, Buhrman, Cleve, Mosca and de Wolf [4]. We also have a fair understanding of the quantum complexities of some symmetric non-decision problems that are related to statistics. Four examples of such problems are:

COUNTING   Given Boolean function $f : [N] \rightarrow \{0, 1\}$, compute $|f^{-1}(1)|$.
MINIMUM   Given $f : [N] \rightarrow Z$, find $x \in [N]$ so that $f(x)$ is minimum.
MEAN   Given $f : [N] \rightarrow Z$, compute $\frac{1}{N} \sum_{x \in [N]} f(x)$.
MEDIAN   Given function $f : [N] \rightarrow Z$, find $x \in [N]$ so that $f(x)$ has rank $\lfloor N/2 \rfloor$ in $f([N])$.

All of these problems have efficient quantum algorithms: COUNTING can be solved via amplitude estimation [10], MINIMUM by repeated applications of amplitude amplifications [22], and MEAN and MEDIAN by using both amplitude amplification and estimation [33,27,30].

Often, amplitude amplification and estimation are used in conjunction with techniques from "classical" computing: Novak [35] considers the quantum complexities of integration, Hayes, Kutin and van Melkebeek [29] MAJORITY (see also Alonso, Reingold and Schott [1]), and Ramesh and Vinay [36] string matching.

As mentioned in the introduction, Fourier transforms are also widely used in quantum algorithms, the most famous examples being in Shor's algorithms [38] for factoring and finding discrete logarithms. We refer to [32] and the references therein for many more examples. The quantum algorithm for ordered searching by Farhi, Goldstone, Gutmann and Sipser [25] is one of the few remarkable examples of quantum algorithms based on principles seemingly different from those found in amplitude amplification and Fourier transforms. A generic bounded-error quantum algorithm for solving any problem using at most $N/2 + \sqrt{N}$ applications of f is given by van Dam in [19].

4   The Lower Bounds

Much work has been done on proving lower bounds for the quantum black box model. If trying to categorizing the many approaches, we may consider there being two main methods, the first being by inner products, the second by degrees of polynomials. Very roughly speaking, so far the simplest and tightest lower bounds are proven by the former method for bounded-error quantum algorithms, and by the latter for exact and small-error quantum algorithms.

The primary idea of the latter method is that to any problem P, we can associate a polynomial and the degree of that polynomial yields a lower bound on the number of queries required by any quantum algorithm solving P. The method is introduced by Beals, Buhrman, Cleve, Mosca and de Wolf in [4], and also implicitly used by Fortnow and Rogers in [26]. A beautiful survey of this and related methods for deterministic and randomized decision trees is given by Buhrman and de Wolf in [16]. See also Buhrman, Cleve, de Wolf and Zalka [13] for bounds on the quantum complexity as a function of the required error probability via degrees of polynomials.

The first general method for proving lower bounds for quantum computing was introduced by Bennett, Bernstein, Brassard and Vazirani in their influential paper [5]. Their technique is nicely described in Vazirani's exposition [41], where it is refered to as a "hybrid argument". Recently, Ambainis [3] introduced a very powerful lower bound technique based on entanglement considerations, which he refers to as "quantum arguments". The techniques in [5] and [3] share many properties, and we may view both as being based on inner product arguments. In the rest of this section, we discuss general properties of lower bound techniques based on inner products.

Suppose we are given one out of two possible states. That is, suppose we are given state $|\psi\rangle$ and promised that either $|\psi\rangle = |\psi_0\rangle$ or $|\psi\rangle = |\psi_1\rangle$. We want to find out which is the case. We assume $|\psi_0\rangle$ and $|\psi_1\rangle$ are known states. Then we may ask, what is the best measurement we can perform on $|\psi\rangle$ in attempting correctly guessing whether $|\psi\rangle = |\psi_0\rangle$ or $|\psi\rangle = |\psi_1\rangle$. The answer can be expressed in terms of the inner product $\langle\psi_0|\psi_1\rangle$.

Lemma 1. *Suppose we are given some state $|\psi\rangle$ and promised that either $|\psi\rangle = |\psi_0\rangle$ or $|\psi\rangle = |\psi_1\rangle$ for some known states $|\psi_0\rangle$ and $|\psi_1\rangle$. Then, for all $0 \leq \varepsilon \leq \frac{1}{2}$, the following two statements are equivalent.*

1. *There exists some measurement we can perform on $|\psi\rangle$ that produces one bit b of outcome so that if $|\psi\rangle = |\psi_0\rangle$ then $b = 0$ with probability at least $1 - \varepsilon$, and if $|\psi\rangle = |\psi_1\rangle$ then $b = 1$ with probability at least $1 - \varepsilon$.*
2. *$|\langle\psi_0|\psi_1\rangle| \leq 2\sqrt{\varepsilon(1 - \varepsilon)}$.*

Two states can be distinguished with certainty if and only if their inner product is zero, and they can be distinguished with high probability if and only if their inner product has small absolute value.

With this, we now give the basic idea in the former lower bound method. Our presentation follows that of Høyer, Neerbek and Shi [31]. Consider some decision problem P. Suppose $Q = (UO)^T U$ is some quantum algorithm that solves P with error probability at most ε using T queries to the oracle. Let $\varepsilon' = 2\sqrt{\varepsilon(1 - \varepsilon)}$. Let $R_0 \subseteq \{f : [N] \to Z \mid P(f) = 0\}$ be any non-empty subset of the possible input functions on which the correct answer to problem P is 0. Similarly, let $R_1 \subseteq \{g : [N] \to Z \mid P(g) = 1\}$ be any non-empty subset of the 1-inputs.

Initially, the state of the computer is $|0\rangle$. After j iterations, the state is $|\Psi_f^j\rangle = (\mathsf{UO}_f)^j \mathsf{U}|0\rangle$ if we are given oracle f, and it is $|\Psi_g^j\rangle = (\mathsf{UO}_g)^j \mathsf{U}|0\rangle$ if we are given oracle g. Suppose that $f \in R_0$ and $g \in R_1$. Then we must have that $|\langle \Psi_f^T | \Psi_g^T \rangle| \leq \varepsilon'$ since the presumed algorithm outputs 0 with high probability on every input f in R_0, and it outputs 1 with high probability on every input g in R_1. We thus also have that $|\sum_{f \in R_0} \sum_{g \in R_1} \langle \Psi_f^T | \Psi_g^T \rangle| \leq \varepsilon' |R_0||R_1|$.

For each $j \in \{0, 1, \ldots, T\}$, let

$$W_j = \sum_{f \in R_0} \sum_{g \in R_1} \langle \Psi_f^j | \Psi_g^j \rangle \tag{3}$$

denote the sum of the inner products after j iterations. We may think of W_j is the total "weight" after j iterations. Initially, the total weight $W_0 = |R_0||R_1|$ is large, and after T iterations, the absolute value of the total weight $|W_T| \leq \varepsilon' |R_0||R_1|$ is small. The quantity $|W_j - W_{j+1}|$ is a measure of the "progress" achieved by the jth query.

Theorem 3 ([3]). *If Δ is an upper bound on $|W_j - W_{j+1}|$ for all $0 \leq j < T$, then the algorithm requires at least $(1 - \varepsilon')\frac{W_0}{\Delta}$ queries in computing problem P with error probability at most ε.*

Proof. The initial weight is W_0, and by the above discussion, $|W_T| \leq \varepsilon' W_0$ where $\varepsilon' = 2\sqrt{\varepsilon(1 - \varepsilon)}$. Write $W_0 - W_T = \sum_{j=0}^{T-1}(W_j - W_{j+1})$ as a telescoping sum. Then $|W_0 - W_T| \leq \sum_{j=0}^{T-1} |W_j - W_{j+1}| \leq T\Delta$, and the theorem follows. □

By Theorem 3, we may prove a lower bound on the quantum complexity of some problem P by proving an upper bound on $\max_j |W_j - W_{j+1}|$ that holds for any quantum algorithm for P.

Each of the lower bounds for the five problems OR, THRESHOLD$_S$, MAJORITY, PARITY and CLAW listed in Table 1 can be proven using this inner product method. For instance, for OR, let $R_0 = \{f : [N] \rightarrow \{0,1\} \mid |f^{-1}(1)| = 0\}$ be the singleton set consisting only of the function identical 0, and let $R_1 = \{f : [N] \rightarrow \{0,1\} \mid |f^{-1}(1)| = 1\}$ consists of the N functions mapping a unique element to 1. Then $W_0 = |R_0||R_1| = N$. Using these sets, we can show [3,31] that $|W_j - W_{j+1}| \leq 2\sqrt{N}$ for all $0 \leq j < N$, and thus we require at least $(1 - \varepsilon')\frac{N}{2\sqrt{N}} \in \Omega(\sqrt{N})$ queries to the oracle to solve OR with error probability ε.

The lower bounds for the three problems ORDERED SEARCHING, SORTING and ELEMENT DISTINCTNESS listed in Table 1 can be proven using a generalization proposed in [31]. Re-define the weight W_j to be a (possibly non-uniform) *weighted* sum of the inner products,

$$W_j = \sum_{f \in R_0} \sum_{g \in R_1} \omega(f, g)\langle \Psi_f^j | \Psi_g^j \rangle. \tag{4}$$

where $\omega(f, g) \geq 0$ for all oracles $f \in R_0$ and $g \in R_1$. Allowing non-uniform weights yields lower bounds that are a logarithmic factor better than the corresponding almost-trivial lower bounds [31].

4.1 Other Lower Bounds

Many more good lower bounds on the quantum black box complexities besides the ones we have mentioned so far are known. These include the following. Nayak and Wu [33] prove optimal lower bounds for MEAN and MEDIAN by the polynomial method. Buhrman and de Wolf [15] give the first non-trivial lower bound for ORDERED SEARCHING by a reduction from PARITY. Farhi, Goldstone, Gutmann, and Sipser [23] improve this to $\log_2(N)/2 \log_2 \log_2(N)$, and Ambainis [2] shows the first $\Omega(\log(N))$ lower bound. Buhrman, Cleve and Wigderson [12] prove several lower bounds on quantum black box computing derived from their seminal work on quantum communication complexity. Zalka [42] shows that Grover's original algorithm for finding a unique marked element in a database is optimal also when considering constant factors and low order terms.

5 Conclusion and Open Problems

In this paper, we have tried to present some of main ideas used in quantum algorithmics. The reader interested in studying quantum computing further may benefit from reading some of the many excellent introductions and reviews. Good starting points include [7,16,17,18,34,37,41], all of which can be found at the authors' home pages or on the so-called e-print archive (to which we have given a pointer after the references).

A primary unsolved question in quantum algorithmics is the quantum complexity of COLLISION. How many queries are necessary and sufficient for distinguishing a 1-to-1 function from a 2-to-1 function? Basically all we know, is that it is easy to distinguish the set of 1-to-1 functions from certain highly structured subclasses of the 2-to-1 functions [39,9]. Can this result be extended to for instance subclasses based somehow on pseudo-random number generators?

We also find it interesting to compare the quantum black box model with other models. Let P be any of the nine problems listed in Table 1, and let T b e the quantum complexity of any known quantum algorithm for P. Then the randomized decision tree complexity for P is known to be in $O(T^2)$. We are thus naturally lead to ask for what problems the randomized decision tree complexity is always at most quadratic in the quantum complexity? A related question is to consider time-space tradeoffs—see the conclusions in [14,31].

A challenging and interesting quest is finding new quantum algorithms. Many existing quantum algorithms seem to benefit from symmetries, periodicities, repeated patterns, etc. in the problems under consideration. Maybe other problems that also contain such properties can be solved efficiently on a quantum computer? Does the lack of such properties rule out efficient quantum algorithms? Are there new problems not known to be NP-complete that can be solved in polynomial time on a quantum computer?

Acknowledgements

I am grateful to Richard Cleve for helpful discussions.

References

1. ALONSO, L., REINGOLD, E. M., SCHOTT, R.: Determining the majority. *Information Processing Letters* **47** (1993) 253–255 68

2. AMBAINIS, A.: A better lower bound for quantum algorithms searching an ordered list. *Proc. of 40th IEEE Symposium on Foundations of Computer Science* (1999) 352–357 71

3. AMBAINIS, A.: Quantum lower bounds by quantum arguments. *Journal of Computer and System Sciences* (to appear) 69, 70

4. BEALS, R., BUHRMAN, H., CLEVE, R., MOSCA, M., DE WOLF, R.: Quantum lower bounds by polynomials. *Journal of the ACM* (to appear) 63, 65, 67, 68, 69

5. BENNETT, C. H., BERNSTEIN, E., BRASSARD, G., VAZIRANI, U.: Strengths and weaknesses of quantum computation. *SIAM Journal on Computing* **26** (1997) 1510–1523 65, 69

6. BERNSTEIN, E., VAZIRANI, U.: Quantum complexity theory. *SIAM Journal on Computing* **26** (1997) 1411–1473 63

7. BERTHIAUME, A.: Quantum computation. In: Hemaspaandra, L., Selman, A.L. (eds.): *Complexity Theory Retrospective II.* Springer-Verlag (1997) Chapter 2, 23–50 63, 71

8. BOYER, M., BRASSARD, G., HØYER, P., TAPP, A.: Tight bounds on quantum searching. *Fortschritte Der Physik* **46** (1998) 493–505 65, 66

9. BRASSARD, G., HØYER, P.: An exact quantum polynomial-time algorithm for Simon's problem. *Proc. of 5th Israeli Symposium on Theory of Computing and Systems* (1997) 12–23 63, 71

10. BRASSARD, G., HØYER, P., MOSCA, M., TAPP, A.: Quantum amplitude amplification and estimation. quant-ph/0005055, 2000 64, 66, 67, 68

11. BRASSARD, G., HØYER, P., TAPP, A.: Quantum algorithm for the collision problem. *ACM SIGACT News (Cryptology Column)* **28** (1997) 14–19 65, 66

12. BUHRMAN, H., CLEVE, R., WIGDERSON, A.: Quantum vs. classical communication and computation. *Proc. of 30th ACM Symposium on Theory of Computing* (1998) 63–68 71

13. BUHRMAN, H., CLEVE, R., DE WOLF, R., ZALKA, Ch.: Bounds for small-error and zero-error quantum algorithms. *Proc. of 40th IEEE Symposium on Foundations of Computer Science* (1999) 358–368 69

14. BUHRMAN, H., DÜRR, C., HEILIGMAN, M., HØYER, P., MAGNIEZ, F., SANTHA, M., DE WOLF, R.: Quantum algorithms for element distinctness. *Proc. of 16th IEEE Computational Complexity* (2001) (to appear) 65, 67, 71

15. BUHRMAN, H., DE WOLF, R.: A lower bound for quantum search of an ordered list. *Information Processing Letters* **70** (1999) 205–209 71

16. BUHRMAN, H., DE WOLF, R.: Complexity measures and decision tree complexity: A survey. *Theoretical Computer Science* (to appear) 63, 69, 71

17. CLEVE, R.: An introduction to quantum complexity theory. In: Macchiavello, C., Palma, G. M., Zeilinger, A. (eds.): *Collected Papers on Quantum Computation and Quantum Information Theory.* World Scientific (2000) 103–127 63, 71

18. CLEVE, R., EKERT, A., MACCHIAVELLO, C., MOSCA, M.: Quantum algorithms revisited. *Proceedings of the Royal Society of London* **A454** (1998) 339–354 63, 71

19. VAN DAM, W.: Quantum oracle interrogation: Getting all information for almost half the price. *Proc. of 39th IEEE Symposium on Foundations of Computer Science* (1998) 362–367 68

20. DEUTSCH, D.: Quantum computational networks. *Proceedings of the Royal Society of London* **A425** (1989) 73–90   63
21. DEUTSCH, D., JOZSA, R.: Rapid solutions of problems by quantum computation. *Proceedings of the Royal Society of London* **A439** (1992) 553–558   63
22. DÜRR, Ch., HØYER, P.: A quantum algorithm for finding the minimum. quant-ph/9607014, 1996   68
23. FARHI, E., GOLDSTONE, J., GUTMANN, S., SIPSER, M.: A limit on the speed of quantum computation in determining parity. *Physical Review Letters* **81** (1998) 5442–5444   65, 71
24. FARHI, E., GOLDSTONE, J., GUTMANN, S., SIPSER, M.: A limit on the speed of quantum computation for insertion into an ordered list. quant-ph/9812057, 1998
25. FARHI, E., GOLDSTONE, J., GUTMANN, S., SIPSER, M.: Invariant quantum algorithms for insertion into an ordered list. quant-ph/9901059, 1999   65, 68
26. FORTNOW, L., ROGERS, J.: Complexity limitations on quantum computation. *Journal of Computer and System Sciences* **59** (1999) 240–252   69
27. GROVER, L. K.: A fast quantum mechanical algorithm for estimating the median. quant-ph/9607024, 1996   68
28. GROVER, L. K.: Quantum mechanics helps in searching for a needle in a haystack. *Physical Review Letters* **79** (1997) 325–328   62, 63, 65, 66
29. HAYES, T., KUTIN, S., VAN MELKEBEEK, D.: On the quantum complexity of majority. Technical Report TR-98-11, Department of Computer Science, University of Chicago (1998)   68
30. HØYER, P.: Unpublished, 1998–2000   68
31. HØYER, P., NEERBEK, J., SHI, Y.: Quantum complexities of ordered searching, sorting, and element distinctness. *Proc. of 28th International Colloquium on Automata, Languages and Programming* (2001) (to appear)   63, 65, 69, 70, 71
32. IVANYOS, G., MAGNIEZ, F., SANTHA, M.: Efficient quantum algorithms for some instances of the non-Abelian hidden subgroup problem. *Proc. of 13th ACM Symposium on Parallel Algorithms and Architectures* (2001) (to appear)   63, 68
33. NAYAK, A., WU, F.: The quantum query complexity of approximating the median and related statistics. *Proc. of 31st ACM Symposium on Theory of Computing* (1999) 384–393   68, 71
34. NIELSEN, M. A., CHUANG, I. L.: *Quantum computation and quantum information.* Cambridge University Press (2000)   63, 71
35. NOVAK, E.: Quantum complexity of integration. *Journal of Complexity* **17** (2001) 2–16   68
36. RAMESH, H., VINAY, V.: String matching in $\tilde{O}(\sqrt{n} + \sqrt{m})$ quantum time. quant-ph/0011049, 2000   68
37. REIFFEL, E. G., POLAK, W.: An introduction to quantum computing for non-physicists. *ACM Computing Surveys* **32** (2000) 300–335   63, 71
38. SHOR, P. W.: Polynomial-time algorithms for prime factorization and discrete logarithms on a quantum computer. *SIAM Journal on Computing* **26** (1997) 1484–1509   62, 63, 68
39. SIMON, D. R.: On the power of quantum computation. *SIAM Journal on Computing* **26** (1997) 1474–1483   63, 71
40. TAPP, A., private communication (1997)   63
41. VAZIRANI, U.: On the power of quantum computation. *Philosophical Transactions of the Royal Society of London* **A356** (1998) 1759–1768   69, 71
42. ZALKA, Ch.: Grover's quantum searching algorithm is optimal. *Physical Review A* **60** (1999) 2746–2751   71

Many of the above references can be found at the Los Alamos National Laboratory
e-print archive (http://arXiv.org/archive/quant-ph).

Decomposition Methods and Sampling Circuits in the Cartesian Lattice

Dana Randall*

College of Computing and School of Mathematics, Georgia Institute of Technology
Atlanta, GA 30332-0280
randall@math.gatech.edu

Abstract. Decomposition theorems are useful tools for bounding the convergence rates of Markov chains. The theorems relate the mixing rate of a Markov chain to smaller, derivative Markov chains, defined by a partition of the state space, and can be useful when standard, direct methods fail. Not only does this simplify the chain being analyzed, but it allows a hybrid approach whereby different techniques for bounding convergence rates can be used on different pieces. We demonstrate this approach by giving bounds on the mixing time of a chain on circuits of length $2n$ in $\mathbb{Z}^d$.

1 Introduction

Suppose that you want to sample from a large set of combinatorial objects. A popular method for doing this is to define a Markov chain whose state space Ω consists of the elements of the set, and use it to perform a random walk. We first define a graph H connecting pairs of states that are close under some metric. This underlying graph on the state space representing allowable transitions is known as the Markov kernel.

To define the transition probabilities of the Markov chain, we need to consider the desired stationary distribution π on Ω. A method known as the *Metropolis algorithm* assigns probabilities to the edges of H so that the resulting Markov chain will converge to this distribution. In particular, if Δ is the maximum degree of any vertex in H, and (x, y) is any edge,

$$P(x, y) = \frac{1}{2\Delta} \min\left(1, \frac{\pi(y)}{\pi(x)}\right).$$

We then assign self loops all remaining probability at each vertex, so $P(x, x) \geq 1/2$ for all $x \in \Omega$. If H is connected, π will be the unique stationary distribution of this Markov chain. We can see this by verifying that *detailed balance* is satsified on every edge (x, y), i.e., $\pi(x)P(x, y) = \pi(y)P(y, x)$.

As a result, if we start at any vertex in Ω and perform a random walk according to the transition probabilities defined by P, and we walk long enough,

---

* Supported in part by NSF Grant No. CCR-9703206.

J. Sgall, A. Pultr, and P. Kolman (Eds.): MFCS 2001, LNCS 2136, pp. 74–86, 2001.

we will converge to the desired distribution. For this to be useful, we need that we are converging rapidly to π, so that after a small, polynomial number of steps, our samples will be chosen from a distribution which is provably arbitrarily close to stationarity. A Markov chain with this property is *rapidly mixing*.

Consider, for example, the set of independent sets $\mathcal{I}$ of some graph G. Taking the Hamming metric, we can define H by connecting any two independent sets that differ by the addition or deletion of a single vertex. A popular stationary distribution is the *Gibbs distribution* which assigns weight $\pi(I) = \gamma^{|I|}/Z_\gamma$ to I, where $\gamma > 0$ is an input parameter of the system, $|I|$ is the size of the independent set I, and $Z_\gamma = \sum_{J \in \mathcal{I}} \gamma^{|J|}$ is the normalizing constant known as the *partition function*. In the Metropolis chain, we have $P(I, I') = \frac{1}{2n} \min(1, \gamma)$ if I' is formed by adding a vertex to I, and $P(I, I') = \frac{1}{2n} \min(1, \gamma^{-1})$ if I' is formed by deleting a vertex from I.

Recently there has been great progress in the design and analysis of Markov chains which are provably efficient. One of the most popular proof techniques is *coupling*. Informally, coupling says that if two copies of the Markov chain can be simultaneously simulated so that they end up in the same state very quickly, regardless of the starting states, then the chain is rapidly mixing. In many instances this is not hard to establish, which gives a very easy proof of fast convergence.

Despite the appeal of these simple coupling arguments, a major drawback is that many Markov chains which appear to be rapidly mixing do not seem to admit coupling proofs. In fact, the complexity of typical Markov chains often makes it difficult to use any of the standard techniques, which include bounding the conductance, the log Sobolev constant or the spectral gap, all closely related to the mixing rate.

The decomposition method offers a way to systematically simplify the Markov chain by breaking it into more manageable pieces. The idea is that it should be easier to apply some of these techniques to the simplified Markov chains and then infer a bound on the original Markov chain. In this survey we will concentrate on the *state decomposition theorem* which utilizes some partition of the state space. It says that if the Markov chain is rapidly mixing when restricted to each piece of the partition, and if there is sufficient flow between the pieces (defined by a *"projection"* of the chain), then the original Markov chain must be rapidly mixing as well. The allows us to take a top-down approach to mixing rate analysis, whereby we need only consider the mixing rate of the restrictions and the projection. In many cases it is easier to define good couplings on these simpler Markov chains, or to use one of the other known methods of analysis. We note, however, that using indirect methods such as the decomposition or comparison (defined later) invariably adds orders of magnitude the bounds on the running time of the algorithm. Hence it is wise to use these methods judiciously unless the goal is simply to establish a polynomial bound on the mixing rate.

2   Mixing Machinery

In what follows, we assume that $\mathcal{M}$ is an ergodic (i.e. irreducible and aperiodic), reversible Markov chain with finite state space Ω, transition probability matrix P, and stationary distribution π.

The time a Markov chain takes to converge to its stationary distribution, i.e., the mixing time of the chain, is measured in terms of the distance between the distribution at time t and the stationary distribution. Letting $P^t(x, y)$ denote the t-step probability of going from x to y, the *total variation distance* at time t is

$$\|P^t, \pi\|_{tv} = \max_{x \in \Omega} \frac{1}{2} \sum_{y \in \Omega} |P^t(x, y) - \pi(y)|.$$

For $\varepsilon > 0$, the *mixing time* $\tau(\varepsilon)$ is

$$\tau(\varepsilon) = \min\{t : \|P^{t'}, \pi\|_{tv} \le \varepsilon, \forall t' \ge t\}.$$

We say a Markov chain is *rapidly mixing* if the mixing time is bounded above by a polynomial in n and $\log \frac{1}{\varepsilon}$, where n is the size of each configuration in the state space.

It is well known that the mixing rate is related to the *spectral gap* of the transition matrix. For the transition matrix P, we let $Gap(P) = \lambda_0 - |\lambda_1|$ denote its spectral gap, where $\lambda_0, \lambda_1, \ldots, \lambda_{|\Omega|-1}$ are the eigenvalues of P and $1 = \lambda_0 > |\lambda_1| \ge |\lambda_i|$ for all $i \ge 2$. The following result the spectral gap and mixing times of a chain (see, e.g., [18]).

Theorem 1. *Let* $\pi_* = \min_{x \in \Omega} \pi(x)$. *For all* $\varepsilon > 0$ *we have*

(a) $\tau(\varepsilon) \le \frac{1}{Gap(P)} \log(\frac{1}{\pi_* \varepsilon})$

(b) $\tau(\varepsilon) \ge \frac{|\lambda_1|}{2Gap(P)} \log(\frac{1}{2\varepsilon})$.

Hence, if $1/Gap(P)$ is bounded above by a polynomial, we are guaranteed fast (polynomial time) convergence. For most of what follows we will rely on the spectral gap bound on mixing. Theorem 1 is useful for deriving a bound on the spectral gap from a coupling proof, which provides bounds on the mixing rate.

We now review of some of the main techniques used to bound the mixing rate of a chain, including the decomposition theorem.

2.1   Path Coupling

One of the most popular methods for bounding mixing times has been the coupling method. A *coupling* is a Markov chain on $\Omega \times \Omega$ with the following properties. Instead of updating the pair of configurations independently, the coupling updates them so that i) the two processes will tend to coalesce, or "move together" under some measure of distance, yet ii) each process, viewed in isolation, is performing transitions exactly according to the original Markov chain. A valid coupling ensures that once the pair of configurations coalesce, they agree from that time forward. The mixing time can be bounded by the expected time for configurations to coalesce under any valid coupling.

The method of path coupling simplifies our goal by letting us bound the mixing rate of a Markov chain by considering only a small subset of $\Omega \times \Omega$ [3, 6].

Theorem 2. (Dyer and Greenhill [6]) *Let Φ be an integer valued metric defined on $\Omega \times \Omega$ which takes values in $\{0, \ldots, B\}$. Let U be a subset of $\Omega \times \Omega$ such that for all $(x, y) \in \Omega \times \Omega$ there exists a path $x = z_0, z_1, \ldots, z_r = y$ between x and y such that $(z_i, z_{i+1}) \in U$ for $0 \le i < r$ and*

$$\sum_{i=0}^{r-1} \Phi(z_i, z_{i+1}) = \Phi(x, y).$$

Define a coupling $(x, y) \to (x', y')$ of the Markov chain $\mathcal{M}$ on all pairs $(x, y) \in U$. Suppose that there exists $\alpha < 1$ such that $\mathbf{E}[\Phi(x', y')] \le \alpha \Phi(x, y)$ for all $(x_t, y_t) \in U$, Then the mixing time of $\mathcal{M}$ satisfies

$$\tau(\epsilon) \le \frac{\log(B\epsilon^{-1})}{1 - \alpha}.$$

Useful bounds can also be derived in the case that $\alpha = 1$ in the theorem (see [6]).

2.2   The Disjoint Decomposition Method

Madras and Randall [12] introduced two decomposition theorems which relate the mixing rate of a Markov chain to the mixing rates of related Markov chains. The *state decomposition theorem* allows the state space to be decomposed into overlapping subsets; the mixing rate of the original chain can be bounded by the mixing rates of the *restricted Markov chains*, which are forced to stay within the pieces, and the ergodic flow between these sets. The *density decomposition theorem* is of a similar flavor, but relates a Markov chain to a family of other Markov chains with the same Markov kernel, where the transition probabilities of the original chain can be described as a weighted average of the transition probabilities of the chains in the family.

We will concentrate on the state decomposition theorem, and will present a newer version of the theorem due to Martin and Randall [15] which allows the decomposition of the state to be a partition, rather than requiring that the pieces overlap.

Suppose that the state space is partitioned into m disjoint pieces $\Omega_1, \ldots, \Omega_m$. For each $i = 1, \ldots, m$, define $P_i = P\{\Omega_i\}$ as the restriction of P to Ω_i which rejects moves that leave Ω_i. In particular, the restriction to Ω_i is a Markov chain, $\mathcal{M}_i$, where the transition matrix P_i is defined as follows: If $x \ne y$ and $x, y \in \Omega_i$ then $P_i(x, y) = P(x, y)$; if $x \in \Omega_i$ then $P_i(x, x) = 1 - \sum_{y \in \Omega_i, y \ne x} P_i(x, y)$. Let π_i be the normalized restriction of π to Ω_i, i.e., $\pi_i(A) = \frac{\pi(A \cap \Omega_i)}{\pi(\Omega_i)}$. Notice that if Ω_i is connected then π_i is the stationary distribution of P_i.

Next, define $\overline{P}$ to be the following aggregated transition matrix on the state space $[m]$:

$$\overline{P}(i,j) = \frac{1}{\pi(\Omega_i)} \sum_{\substack{x \in \Omega_i, \\ y \in \Omega_j}} \pi(x)P(x,y).$$

Theorem 3. (Martin and Randall [15]) *Let P_i and $\overline{P}$ be as above. Then the spectral gaps satisfy*

$$Gap(P) \geq \frac{1}{2}Gap(\overline{P}) \min_{i \in [m]} Gap(P_i).$$

A useful corollary allows us to replace $\overline{P}$ in the theorem with the Metropolis chain defined on the same Markov kernel, provided some simple conditions are satisfied. Since the transitions of the Metropolis chain are fully defined by the stationary distribution $\overline{\pi}$, this is often easier to analyze than the true projection.

Define P_M on the set $[m]$, with Metropolis transitions $P_M(i,j) = \min\{1, \frac{\pi(\Omega_j)}{\pi(\Omega_i)}\}$. Let $\partial_i(\Omega_j) = \{y \in \Omega_j : \exists\, x \in \Omega_i \text{ with } P(x,y) > 0\}$.

Corollary 4. [15] *With P_M as above, suppose there exists $\beta > 0$ and $\gamma > 0$ such that*

(a) $P(x,y) \geq \beta$ whenever $P(x,y) > 0$;
(b) $\pi(\partial_i(\Omega_j)) \geq \gamma\pi(\Omega_j)$ whenever $\overline{P}(i,j) > 0$.

Then

$$Gap(P) \geq \frac{1}{2}\beta\gamma\, Gap(P_M) \min_{i=1,\ldots,m} Gap(P_i).$$

2.3   The Comparison Method

When applying the decomposition theorem, we reduce the analysis of a Markov chain to bounding the convergence times of smaller related chains. In many cases it will be much simpler to analyze variants of these auxiliary Markov chains instead of the true restrictions and projections. The comparison method tells us ways in which we can slightly modify one of these Markov chains without qualitatively changing the mixing time. For instance, it allows us to add additional transition edges or to amplify some of the transition probabilities, which can be useful tricks for simplifying the analysis of a chain.

Let $\tilde{P}$ and P be two reversible Markov chains on the same state space Ω with the same stationary distribution π. The comparison method allows us to relate the mixing times of these two chains (see [4] and [17]). In what follows, suppose that $Gap(\tilde{P})$, the spectral gap of $\tilde{P}$, is known (or suitably bounded) and we desire a bound on $Gap(P)$, the spectral gap of P, which is unknown.

Following [4], we let $E(P) = \{(x,y) : P(x,y) > 0\}$ and $E(\tilde{P}) = \{(x,y) : \tilde{P}(x,y) > 0\}$ denote the sets of edges of the two chains, viewed as directed graphs. For each $(x,y) \in E(\tilde{P})$, define a *path* γ_{xy} using a sequence of states

$x = x_0, x_1, \ldots, x_k = y$ with $(x_i, x_{i+1}) \in E(P)$, and let $|\gamma_{xy}|$ denote the length of the path. Let $\Gamma(z, w) = \{(x, y) \in E(\tilde{P}) : (z, w) \in \gamma_{xy}\}$ be the set of paths that use the transition (z, w) of P. Finally, define

$$A = \max_{(z,w) \in E(P)} \left\{ \frac{1}{\pi(z) P(z, w)} \sum_{\Gamma(z,w)} |\gamma_{xy}| \pi(x) \tilde{P}(x, y) \right\}.$$

Theorem 5. (Diaconis and Saloff-Coste [4]) *With the above notation, the spectral gaps satisfy* $Gap(P) \geq \frac{1}{A} Gap(\tilde{P})$

It is worthwhile to note that there are several other comparison theorems which turn out to be useful, especially when applying decomposition techniques. The following lemma helps us reason about a Markov chain by slightly modifying the transition probabilities (see, e.g., [10]). We use this trick in our main application, sampling circuits.

Lemma 6. *Suppose P and P' are Markov chains on the same state space, each reversible with respect to the distribution π. Suppose there are constants c_1 and c_2 such that $c_1 P(x, y) \leq P'(x, y) \leq c_2 P(x, y)$ for all $x \neq y$. Then $c_1 Gap(P) \leq Gap(P') \leq c_2 Gap(P)$.*

3   Sampling Circuits in the Cartesian Lattice

A *circuit* in $\mathbb{Z}^d$ is a walk along lattice edges which starts and ends at the origin. Our goal is to sample from $\mathcal{C}$, the set of circuits of length $2n$. It is useful to represent each walk as a string of $2n$ letters using $\{a_1, \ldots, a_d\}$ and their inverses $\{a_1^{-1}, \ldots, a_d^{-1}\}$, where a_i represents a positive step in the ith direction, and a_i^{-1} represents a negative step. Since these are closed circuits, the number of times a_i appears must equal the number of times a_i^{-1} appears, for all i. We will show how to uniformly sample from the set of all circuits of length $2n$ using an efficient Markov chain. The primary tool will be finding an appropriate decomposition of the state space. We outline the proof here and refer the reader to [16] for complete details.

Using a similar strategy, Martin and Randall showed how to use a Marko chain to sample circuits in regular d-ary trees, i.e., paths of length $2n$ which trace edges of the tree starting and ending at the origin [15]. This problem generalizes to sampling Dyke paths according to a distribution which favors walks that hit the x-axis a large number of times, known in the statistical physics community as "adsorbing staircase walks." Here too the decomposition method was the basis of the analysis. We note that there are other simple algorithms for sampling circuits on trees which do not require Markov chains. In contrast, to our knowledge, the Markov chain based algorithm discussed in this paper is the first efficient method for sampling circuits on $\mathbb{Z}^d$.

3.1   The Markov Chain on Circuits

The Markov chain on C is based on two types of moves: *transpositions* of neighboring letters in the word (which keep the numbers of each letter fixed) and *rotations*, which replace an adjacent (a_i, a_i^{-1}) with (a_j, a_j^{-1}), for some pair of letters a_i and a_j.

We now define the transition probabilities $\mathcal{P}$ of $\mathcal{M}$, where we say $x \in_u X$ to mean that we choose x from set X uniformly. Starting at σ, do the following. With probability $1/2$, pick $i \in_u [n-1]$ and transpose σ_i and σ_{i+1}. With probability $1/2$, pick $i \in_u [n-1]$ and $k \in_u [d]$ and if σ_i and σ_{i+1} are inverses (where σ_i is a step in the positive direction), then replace them with (a_k, a_k^{-1}). Otherwise keep σ unchanged.

The chain is aperiodic, ergodic and reversible, and the transitions are symmetric, so the stationary distribution of this Markov chain is the uniform distribution on C.

3.2   Bounding the Mixing Rate of the Circuits Markov Chain

We bound the mixing rate of $\mathcal{M}$ by appealing to the decomposition theorem. Let $\sigma \in C$ and let x_i equal the number of occurrences of a_i, and hence a_i^{-1} in σ, for all i. Define the *trace* $\mathrm{Tr}(\sigma)$ to be the vector $X = (x_1, ..., x_d)$. This defines a partition of the state space into

$$C = \cup\, C_X,$$

where the union is over all partitions of n into d pieces and C_X is the set of words $\sigma \in C$ such that $\mathrm{Tr}(\sigma) = X = (x_1, ..., x_d)$. The cardinality of the set C_X is $\binom{2n}{x_1, x_1, ..., x_d, x_d}$, the number of distinct words (or permutations) of length $2n$ using the letters with these prescribed multiplicities. The number of sets in the partition of the state space is exactly the number of partitions of n into d pieces, $D = \binom{n+d-1}{d-1}$.

Each restricted Markov chain consists of all the words which have a fixed trace. Hence, transitions in the restricted chains consist of only transpositions, as rotations would change the trace. The projection $\overline{P}$ consists of a simplex containing D vertices, each representing a distinct partition of $2n$. Letting

$$\Phi(X, Y) = \frac{1}{2}||X - Y||_1 ,$$

two points X and Y are connected by an edge of $\overline{P}$ iff $\Phi(X, Y) = 1$, where $||\cdot||_1$ denotes the ℓ_1 metric. In the following we make no attempt to optimize the running time, and instead simply provide polynomial bounds on the convergence rates.

● **Step 1 – The restricted Markov chains:** Consider any of the restricted chains P_X on the set of configurations with trace X. We need to show that this simpler chain, connecting pairs of words differing by a transposition of adjacent letters, converges quickly for any fixed trace.

We can analyze the transposition moves on this set by mapping C_X to the set of linear extensions of a particular partial order. Consider the alphabet $\cup_i\{\{a_{i,1}, ..., a_{i,x_i}\} \cup \{A_{i,1}, ..., A_{i,x_i}\}\}$, and the partial order defined by the relations $a_{i,1} \prec a_{i,2} \prec ... \prec a_{i,x_i}$ and $A_{i,1} \prec A_{i,2} \prec ... \prec A_{i,x_i}$, for all i. It is straightforward to see that there is a bijection between the set of circuits in C_X and the set of linear extensions to this partial order (mapping a^{-1} to A). Furthermore, this bijection preserves transpositions. We appeal to the following theorem due to Bubley and Dyer [3]:

Theorem 7. *The transposition Markov chain on the set of linear extensions to a partial order on n elements has mixing time $O(n^4(\log^2 n + \log \epsilon^{-1}))$.*

Referring to theorem 1, we can derive the following bound.

Corollary 8. *The Markov chain P_X has spectral gap $Gap(P_X) \geq 1/(cn^4 \log^2 n)$ for some constant c.*

• **Step 2 – The projection of the Markov chain:** The states $\overline{\Omega}$ of the projection consist of partitions of n into d pieces, so $|\overline{\Omega}| = D$. The stationary probability of $X = (x_1, ..., x_d)$ is $\overline{\pi}(X) = \binom{2n}{x_1, x_1, ..., x_d, x_d}$, the number of words with these multiplicities.

The Markov kernel is defined by connecting two partitions X and Y if the distance $\Phi(X, Y) = (\|x - y\|_1)/2 = 1$. Before applying corollary 4 we first need to bound the mixing rate of the Markov chain defined by Metropolis probabilities. In particular, if $X = (x_1, ..., x_d)$ and $Y = (x_1, ..., x_i + 1, ..., x_j - 1, ..., x_d)$, then

$$P_M(X, Y) = \frac{1}{2n^2} \cdot \min\left(1, \frac{\overline{\pi}(Y)}{\overline{\pi}(X)}\right)$$

$$= \frac{1}{2n^2} \cdot \min\left(1, \frac{x_j^2}{(x_i + 1)^2}\right).$$

We analyze this Metropolis chain indirectly by first considering a variant P'_M which admits a simpler path coupling proof. Using the same Markov kernel, define the transitions

$$P'_M(X, Y) = \frac{1}{2n^2(x_i + 1)^2}.$$

In particular, the x_i in the denominator is the value which would be increased by the rotation. Notice that detailed balance is satisfied:

$$\frac{\overline{\pi}(X)}{\overline{\pi}(Y)} = \frac{(x_i + 1)^2}{x_j^2} = \frac{P'_M(Y, X)}{P'_M(X, Y)}.$$

This guarantees that P'_M has the same stationary distribution as P_M, namely $\overline{\pi}$.

The mixing rate of this chain can be bounded directly using path coupling. Let $U \subseteq \overline{\Omega} \times \overline{\Omega}$ be pairs of states X and Y such that $\Phi(X, Y) = 1$. We couple by choosing the same pair of indices i and j, and the same bit $b \in \{-1, 1\}$ to update each of X and Y, where the probability for accepting each of these moves is dictated by the transitions of P'_M.

Lemma 9. *For any pair $(X_t, Y_t) \in U$, the expected change in distance after one step of the coupled chain is $E[\Phi(X_{t+1}, Y_{t+1})] \leq (1 - \frac{1}{n^6}) \Phi(X_t, Y_t)$.*

Proof. If $(X_t, Y_t) \in U$, then there exist coordinates k and k' such that $y_k = x_k + 1$ and $y_{k'} = x_{k'} - 1$. Without loss of generality, assume that $k = 1$ and $k' = 2$. We need to determine the expected change in distance after one step of the coupled chain. Suppose that in this move we try to add 1 to x_i and y_i and subtract 1 from x_j and y_j. We consider three cases.

Case 1: If $|\{i, j\} \cap \{1, 2\}| = 0$, then both processes accept the move with the same probability and $\Phi(X_{t+1}, Y_{t+1}) = 1$.

Case 2: If $|\{i, j\} \cap \{1, 2\}| = 1$, then we shall see that the expected change is also zero. Assume without loss of generality that $i = 1$ and $j = 3$, and first consider the case $b = 1$. Then we move from X to $X' = (x_1 + 1, x_2, x_3 - 1, ..., x_d)$ with probability $\frac{1}{2n^2(x_1+1)^2}$ and from Y to $Y' = (x_1 + 2, x_2 - 1, x_3 - 1, ..., x_d)$ with probability $\frac{1}{2n^2(x_1+2)^2}$. Since $P'_M(X, X') > P'_M(Y, Y')$, with probability $P'_M(Y, Y')$ we update both X and Y; with probability $P'_M(X, X') - P'_M(Y, Y')$ we update just X; and with all remaining probability we update neither. In the first case we end up with X' and Y', in the second we end up with X' and Y and in the final case we stay at X and Y. All of these pairs are unit distance apart, so the expected change in distance is zero. If $b = -1$, then $P'_M(X, X') = P'_M(Y, Y') = \frac{1}{2n^2(x_3+1)^2}$ and again the coupling keeps the configurations unit distance apart.

Case 3: If $|\{i, j\} \cap \{1, 2\}| = 2$, then we shall see that the expected change is at most zero. Assume without loss of generality that $i = 1$, $j = 2$ and $b = 1$. The probability of moving from X to $X'' = (x_1 + 1, x_2 - 1, ..., x_d) = Y$ is $P'_M(X, X'') = \frac{1}{2n^2(x_1+1)^2}$. The probability of moving from Y to $Y'' = (x_1 + 2, x_2 - 2, ..., x_d)$ is $P'_M(Y, Y'') = \frac{1}{2n^2(x_1+2)^2}$. So with probability $P'_M(Y, Y'')$ we update both configurations, keeping them unit distance apart, and with probability $P'_M(X, X'') - P'_M(Y, Y'') \geq \frac{1}{2n^6}$ we update just X, decreasing the distance to zero. When $b = -1$ the symmetric argument shows that we again have a small chance of decreasing the distance.

Summing over all of these possibilities yields the lemma.   □

The path coupling theorem implies that the mixing time is bounded by $\tau(\epsilon) \leq O(n^6 \log n)$. Furthermore, we get the following bound on the spectral gap.

Theorem 10. *The Markov chain P'_M on $\overline{\Omega}$ has spectral gap $Gap(P'_M) \geq c'/(n^6 \log n)$ for some constant c'.*

This bounds the spectral gap of the modified Metropolis chain P'_M, but we can readily compare the spectral gaps of P'_M and P_M using lemma 6. Since all the transitions of P_M are at least as large as those of P'_M, we find

Corollary 11. *The Markov chain P_M on $\overline{\Omega}$ has spectral gap $Gap(P_M) \geq c'/(n^6 \log n)$.*

• **Step 3 – Putting the pieces together:** These bounds on the spectral gaps of the restrictions P_i and the Metropolis projection P'_M enable us to apply the decomposition theorem to derive a bound on the spectral gap of P, the original chain.

Theorem 12. *The Markov chain P is rapidly mixing on $\mathcal{C}$ and the spectral gap satisfies $Gap(P) \geq c''/(n^{12}d\log^3 n)$, for some constant c''.*

Proof. To apply corollary 4, we need to bound the parameters β and γ. We find that $\beta \geq \frac{1}{4nd}$, the minimum probability of a transition. To bound γ we need to determine what fraction of the words in C_X are neighbors of a word in C_Y if $\Phi(X,Y) = 1$ (since π is uniform within each of these sets). If $X = (x_1, ..., x_d)$ and $Y = (x_1, ..., x_i + 1, ..., x_j - 1, ..., x_n)$, this fraction is exactly the likelihood that a word in C_X has an a_i followed by an a_i^{-1}, and this is easily determined to be at least $1/n$.

Combining $\beta \geq \frac{1}{4nd}$, $\gamma \geq \frac{1}{n}$ with our bounds from lemmas 8 and 11, corollary 4 gives the claimed lower bound on the spectral gap. □

4 Other Applications of Decomposition

The key step to applying the decomposition theorem is finding an appropriate partition of the state space. In most examples a natural choice seems to be to cluster configurations of equal probability together so that the distribution for each of the restricted chains is uniform, or so that the restrictions share some essential feature which will make it easy to bound the mixing rate.

In the example of section 3, the state space is divided into subsets, each representing a partition of n into d parts. It followed that the vertices of the projection formed a d-dimensional simplex, where the Markov kernel was formed by connecting vertices which are neighbors in the simplex. We briefly outline two other recent applications of the decomposition theorem where we get other natural graphs for the projection. In the first case graph defining the Markov kernel of the projection is one-dimensional and in the second it is a hypercube.

4.1 Independent Sets

Our first example is sampling independent sets of a graph according to the Gibbs measure. Recall that $\pi(I) = \gamma^{|I|}/Z_\gamma$, where $\gamma > 0$ is an input parameter and Z_γ normalizes the distribution. There has been much activity in studying how to sample independent sets for various values of γ using a simple, natural Markov chain based on inserting, deleting or exchanging vertices at each step. Works of Luby and Vigoda [9] and Dyer and Greenhill [5] imply that this chain is rapidly mixing if $\gamma \leq 2/(\Delta - 2)$, where Δ is the maximum number of neighbors of any vertex in G. It was shown by Borgs et al. [1] that this chain is slowly mixing on some graphs for γ sufficiently large.

Alternatively, Madras and Randall [12] showed that this algorithm is fast for *every* value of γ, provided we restrict the state space to independent sets of size at most $n^* = \lfloor |V|/2(\Delta + 1) \rfloor$. This relies heavily on earlier work of Dyer and Greenhill [6] showing that a Markov chain defined by exchanges is rapidly mixing on the set of independent sets of fixed size k, whenever $k \leq n^*$. The decomposition here is quite natural: We partition $\mathcal{I}$, the set of independent sets of G, into pieces $\mathcal{I}_k$ according to their size. The restrictions arising from this partition permit exchanges, but disallow insertions or deletions, as they exit the state space of the restricted Markov chain. These are now exactly the Markov chains proven to be rapidly mixing by Dyer and Greenhill (but with slightly greater self-loop probabilities) and hence can also be seen to be rapidly mixing. Consequently, we need only bound the mixing rate of the projection.

Here the projection is a one-dimensional graph on $\{0, ..., n^*\}$. Further calculation determines that the stationary distribution $\bar{\pi}(k)$ of the projection is unimodal in k, implying that the projection is also rapidly mixing. We refer the reader to [12] for details.

4.2   The Swapping Algorithm

To further demonstrate the versatility and potential of the decomposition method, we review an application of a very different flavor. In recent work, Madras and Zheng [13] show that the *swapping algorithm* is rapidly mixing for the *mean field* Ising model (i.e., the Ising model on the complete graph), as well as for a simpler toy model.

Given a graph $G = (V, E)$, the *ferromagnetic Ising model* consists of a graph G whose vertices represent particles and whose edges represent interactions between particles. A spin configuration is an assignment of *spins*, either $+$ or $-$, to each of the vertices, where adjacent vertices prefer to have the same spin. Let $J_{x,y} > 0$ be the interaction energy between vertices x and y, where $(x, y) \in E$. Let $\sigma \in \Omega = \{+, -\}^{|V|}$ be any assignment of $\{+, -\}$ to each of the vertices. The *Hamiltonian* of σ is

$$H(\sigma) = \sum_{(x,y) \in E} J_{x,y} 1_{\sigma_x \neq \sigma_y},$$

where 1_A is the indicator function which is 1 when the event A is true and 0 otherwise. The probability that the Ising spin state is σ is given by the *Gibbs distribution*:

$$\pi(\sigma) = \frac{e^{-\beta H(\sigma)}}{Z(G)},$$

where β is inverse temperature and

$$Z(G) = \sum_{\sigma} e^{-\beta H(\sigma)}.$$

It is well known that at sufficiently low temperatures the distribution is bimodal (as a function the number of vertices assigned $+$), and any local dynamics will be slowly mixing. The simplest local dynamics, *Glauber dynamics*, is the Markov chain defined by choosing a vertex at random and flipping the spin at that vertex with the appropriate Metropolis probability.

Simulated tempering, which varies the temperature during the runtime of an algorithm, appears to be a useful way to circumvent this difficulty [8, 14]. The chain moves between m close temperatures that interpolate between the temperature of interest and very high temperature, where the local dynamics converges rapidly. The *swapping algorithm* is a variant of tempering, introduced by Geyer [7], where the state space is Ω^m and each configuration $S = (\sigma_1, ..., \sigma_m) \in \Omega^m$ consists of one sample at each temperature. The stationary distribution distribution is $\pi(S) = \Pi_{i=1}^m \pi_i(\sigma_i)$, where π_i is the distribution at temperature i. The transitions of the swapping algorithm consist of two types of moves: with probability $1/2$ choose $i \in [m]$ and perform a local update of σ_i (using Glauber dynamics at this fixed temperature); with probability $1/2$ choose $i \in [m-1]$ and move from $S = (\sigma_1, ..., \sigma_m)$ to $S' = (\sigma_1, ..., \sigma_{i+1}, \sigma_i, ..., \sigma_m)$, i.e., swap configurations i and $i+1$, with the appropriate Metropolis probability.

The idea behind the swapping algorithm, and other versions of tempering, is that, in the long run, the trajectory of each Ising configuration will spend equal time at each temperature, potentially greatly speeding up mixing. Experimentally, this appears to overcome obstacles to sampling at low temperatures.

Madras and Zheng show that the swapping algorithm is rapidly mixing on the mean-field Ising model at all temperatures. Let $\Omega^+ \subset \Omega$ be the set of configurations that are predominantly $+$, and similarly Ω^-. Define the trace of a configuration S to be $\mathrm{Tr}(S) = (v_1, ..., v_m) \in \{+, -\}^m$ where $v_i = +$ if $\sigma_i \in \Omega^+$ and $v_i = -$ if $\sigma_i \in \Omega^-$. The analysis of the swapping chain uses decomposition by partitioning the state space according to the trace.

The projection for this decomposition is the m-dimensional hypercube where each vertex represents a distinct trace. The stationary distribution is uniform on the hypercube because, at each temperature, the likelihood of being in Ω^+ and Ω^- are equal due to symmetry. Relying on the comparison method, it suffices to analyze the following simplification of the projection: Starting at any vertex $V = (v_1, ..., v_m)$ in the hypercube, pick $i \in_u [m]$. If $i = 1$, then with probability $1/2$ flip the first bit; if $i > 1$, then with probability $1/2$ transpose the v_{i-1} and v_i; and with all remaining probability do nothing. This chain is easily seen to be rapidly mixing on the hypercube and can be used to infer a bound on the spectral gap of the projection chain.

To analyze the restrictions, Madras and Zheng first prove that the simple, single flip dynamics on Ω^+ is rapidly mixing at any temperature; this result is analytical, relying on the fact that the underlying graph is complete for the mean-field model. Using simple facts about Markov chains on product spaces, it can be shown that the each of the restricted chains must also be rapidly mixing (even without including any swap moves). Once again decomposition completes the proof of rapid mixing, and we can conclude that the swapping algorithm is efficient on the complete graph.

Acknowledgements

I wish to thank Russell Martin for many useful discussions, especially regarding the technical details of section 3, and Dimitris Achlioptas for improving an earlier draft of this paper.

References

1. C. Borgs, J. T. Chayes, A. Frieze, J. H. Kim, P. Tetali, E. Vigoda, and V. H. Vu. Torpid mixing of some MCMC algorithms in statistical physics. *Proc. 40th IEEE Symposium on Foundations of Computer Science*, 218–229, 1999.
2. R. Bubley and M. Dyer. Faster random generation of linear extensions. *Discrete Mathematics*, **201**:81–88, 1999.
3. R. Bubley and M. Dyer. Path coupling: A technique for proving rapid mixing in Markov chains. *Proc. 38th Annual IEEE Symposium on Foundations of Computer Science* 223–231, 1997.
4. P. Diaconis and L. Saloff-Coste. Comparison theorems for reversible Markov chains. *Annals of Applied Probability*, **3**:696–730, 1993. 78
5. M. Dyer and C. Greenhill. On Markov chains for independent sets. *Journal of Algorithms*, **35**: 17–49, 2000.
6. M. Dyer and C. Greenhill. A more rapidly mixing Markov chain for graph colorings. *Random Structures and Algorithms*, **13**:285–317, 1998.
7. C. J. Geyer Markov Chain Monte Carlo Maximum Likelihood. *Computing Science and Statistics: Proceedings of the 23rd Symposium on the Interface* (E. M. Keramidas, ed.), 156-163. Interface Foundation, Fairfax Sta tion, 1991.
8. C. J. Geyer and E. A. Thompson. Annealing Markov Chain Monte Carlo with Applications to Ancestral Inference. *J. Amer. Statist. Assoc.* **90** 909–920, 1995.
9. M. Luby and E. Vigoda. Fast Convergence of the Glauber dynamics for sampling independent sets. *Random Structures and Algorithms* **15**: 229–241, 1999.
10. N. Madras and M. Piccioni. Importance sampling for families of distributions. *Ann. Appl. Probab.* **9**: 1202–1225, 1999.
11. N. Madras and D. Randall. Factoring graphs to bound mixing rates. *Proc. 37th Annual IEEE Symposium on Foundations of Computer Science*, 194–203, 1996.
12. N. Madras and D. Randall. Markov chain decomposition for convergence rate analysis. *Annals of Applied Probability*, (to appear), 2001.
13. N. Madras and Z. Zheng. On the swapping algorithm. Preprint, 2001.
14. E. Marinari and G. Parisi. Simulated tempering: a new Monte Carlo scheme. *Europhys. Lett.* **19** 451–458, 1992.
15. R. A. Martin and D. Randall Sampling adsorbing staircase walks using a new Markov chain decomposition method. *Proceedings of the 41st Symposium on the Foundations of Computer Science (FOCS 2000)* , 492–502, 2000.
16. R. A. Martin and D. Randall Disjoint decomposition with applications to sampling circuits in some Cayley graphs. Preprint, 2001.
17. D. Randall and P. Tetali. Analyzing Glauber dynamics by comparison of Markov chains. *Journal of Mathematical Physics*, **41**:1598–1615, 2000. 78
18. A. J. Sinclair. *Algorithms for random generation & counting: a Markov chain approach.* Birkhäuser, Boston, 1993. 76

New Algorithms for k-SAT Based on the Local Search Principle

Uwe Schöning

Abteilung Theoretische Informatik, Universität Ulm
89069 Ulm, Germany
schoenin@informatik.uni-ulm.de

Abstract. Recently, several algorithms for the NP-complete problem k-SAT have been proposed and rigorously analyzed. These algorithms are based on the heuristic principle of local search. Their deterministic and their probabilistic versions and variations, have been shown to achieve the best complexity bounds that are known for k-SAT (or the special case 3-SAT). We review these algorithms, their underlying principles and their analyses.

1 Introduction

Consider Boolean formulas in *conjunctive normal form* (CNF), i.e. formulas, which are conjunctions (and's) of disjunction (or's) of variables or negated variables. A variable or negated variable is called a *literal*. We assume here that the variables are $x_1, x_2, \ldots, x_n$, and the complexity of all algorithms is measured as a function of n. A formula in CNF is in k-CNF if all *clauses* (disjunction of literals) contain at most k literals. It is well known that k-SAT is NP-complete provided $k \geq 3$. Here k-SAT is the problem of determining whether a formula in k-CNF is satisfiable. A formula is *satisfiable* if there is an assignment $a : \{x_1, x_2, \ldots, x_n\} \rightarrow \{0, 1\}$ which assigns to each variable a Boolean value such that the entire formula evaluates to true.

Since k-SAT is NP-complete there is a considerable interest in algorithms that run faster than the naive algorithm which just tests all 2^n potential assignments and which therefore takes worst-case time 2^n (up to some polynomial factor). Especially the case of 3-SAT is interesting since k-SAT becomes NP-complete, starting with $k = 3$. A milestone paper in this respect is [10]. For every k an algorithm for k-SAT, based on clever backtracking, is presented. In the case of 3-SAT a worst-case upper bound of 1.619^n is shown in [10]. In the general k-SAT case, the complexity is $(a_k)^n$ where a_k is the solution to the following equation $(a_k)^{k-1} = (a_k)^{k-2} + \ldots + (a_k)^1 + 1$. The method has been improved in [9] to 1.505^n. Starting with the paper [12] probabilistic algorithm came into the scene, where an upper bound of 1.58^n is shown in the case of 3-SAT, and which was later improved in [13] to 1.36^n.

Another probabilistic algorithm, based on local search, was shown to achieve $(4/3)^n 0 (1.3333...)^n$ in the case of 3-SAT [15]. This algorithm and its variations are the theme of this article.

J. Sgall, A. Pultr, and P. Kolman (Eds.): MFCS 2001, LNCS 2136, pp. 87–95, 2001.

The notion "local search" refers to algorithms which wander through the search space $\{0,1\}^n$ of all assignments, at each step altering the actual assignment to some neighbor assignment. Neighborhood is defined in terms of having Hamming-distance $= 1$. The Hamming-distance between two assignments is the number of bits where the assignments differ.

2   A Local Search Procedure for k-SAT

Suppose, input formula F in k-CNF is given. Let F contain n variables $x_1, \ldots, x_n$. An assignment to these variables can be considered as a 0-1-string of length n.

Consider the following recursive procedure local-search.

> **procedure** local-search (a : assignment; m : **integer**) : **boolean**;
> {Returns true iff there is a satisfying assignment for formula F
> within Hamming-distance $\leq m$ from assignment a.}
> **begin**
>   if $F(a) = 1$ **then return true**;
>   if $m = 0$ **then return false**;
>   {Let $C = \{l_1, l_2, \ldots, l_k\}$ be some clause in F with $C(a) = 0$,
>   i.e. all k literals l_i of C are set to 0 under a}
>   **for** $i := 1$ **to** k **do if** local-search$(a|l_i, m-1)$ **then return true**;
>   **return false**;
> **end** local-search

Here $a|l_i$ denotes the assignment a' obtained from a by flipping the value of literal l_i, i.e. changing the value of l_i from 0 to 1. If l_i is the (negated or non-negated) variable x_j, this means that the j-th bit in a is changed from 1 to 0 (or from 0 to 1, respectively). The Hamming distance between a and this new assignment a' is 1.

The correctness of the procedure local-search follows from the following observation. If there is a satisfying assignment a^* of formula F, and the Hamming distance between a^* and the actual assignment a in the procedure evocation is d, then for at least one of the assignments $a|l_i$, $i = 1, \ldots, k$, used in the recursive procedure calls, the Hamming distance to a^* is $d-1$. At least one of the recursive calls of local-search will therefore return the correct result.

It was suggested to modify the algorithm so that it "freezes" the assignment of a variable once it was flipped. This prevents that a variable value is flipped back in a deeper recursive procedure call. This modification is also correct, but will not be considered here. Another heuristic is to choose among the clauses C with $C(a) = 0$ a shortest one.

An evocation of procedure local-search(a, m), if $m > 0$ and $F(a) = 0$, causes up to k evocations of local-search$(\ldots, m-1)$. That is, the recursion tree induced by local-search(a, m) has k^m many leaves. Hence, the complexity of local-search(a, m) is within a polynomial factor of k^m.

Now, here is a simple deterministic algorithm for 3-SAT. We perform two calls of the procedure local-search, one for local-search(0^n, $n/2$) and one for local-search(1^n, $n/2$). It is clear that the entire search space $\{0,1\}^n$ is covered by these two calls since every assignment is within Hamming distance of $n/2$ from either 0^n or 1^n. The complexity of this deterministic 3-SAT algorithm is within a polynomial factor of $3^{n/2} \approx 1.732^n$, a bound which comes close to the Monien-Speckenmeyer bound [10] but is obtained in a much simpler way.

Notice that it is not a good idea to generalize this to k-SAT with $k \geq 4$, since in this case $k^{n/2} \geq 2^n$.

3   Random Initial Assignments

A better idea (introduced in [14]) is to use independent random initial assignments $a_1, a_2, \ldots, a_t \in \{0,1\}^n$, and at the same time use a Hamming radius βn, smaller than $n/2$, so that this new (probabilistic) algorithm for k-SAT looks like this:

> **for** t **times do**
> **begin**
> Choose an assignment $a \in \{0,1\}^n$, uniformly at random;
> **if** local-search(a, βn) **then accept**
> **end**;
> **reject**

The question is how to choose t und β optimally so that the error probability becomes negligible, as well as the overall complexity is minimized. It is clear that the overall (worst-case) complexity is $t \cdot k^{\beta n}$ where t could be a function depending on n.

Regarding the error probability, a single evocation of local-search(a, βn) with randomly chosen $a \in \{0,1\}^n$ finds a satisfying assignment a^* (if it exists) with probability

$$\frac{\sum_{i=0}^{\beta n} \binom{n}{i}}{2^n}$$

Since t random assignments are chosen independently, the (error) probability of missing the satisfying assignment a^* in each evocation of local-search(a, βn) is

$$\left(1 - \frac{\sum_{i=0}^{\beta n} \binom{n}{i}}{2^n}\right)^t \leq e^{-t \cdot \sum_{i=0}^{\beta n} \binom{n}{i} / 2^n}$$

To bring this error probability below some negligible value, like e^{-20}, it suffices to choose

$$t = 20 \cdot 2^n \Big/ \sum_{i=0}^{\beta n} \binom{n}{i}$$

Using Stirling's approximation $n! \asymp (n/e)^n \sqrt{2\pi n}$, it can be seen that, for fixed β, the expression $\sum_{i=0}^{\beta n} \binom{n}{i}$ behaves asymptotically, up to a polynomial factor, like $[(1/\beta)^\beta (1/(1-\beta))^{1-\beta}]^n$ (see [2]). Therefore, it suffices, up to a polynomial factor, to choose t as

$$\left[2 \cdot \beta^\beta \cdot (1-\beta)^{1-\beta}\right]^n$$

The overall complexity of this probabilistic k-SAT algorithm is therefore within a polynomial factor of

$$\left[2 \cdot \beta^\beta \cdot (1-\beta)^{1-\beta} \cdot k^\beta\right]^n$$

By calculating the derivative of the expression in brackets and setting it to zero, it can be seen that the expression in brackets is minimized by the choice $\beta = 1/(k+1)$. Inserting $\beta = 1/(k+1)$ in the expression gives the complexity bound of $[2k/(k+1)]^n$. In the case of 3-SAT this is 1.5^n.

Actually, by analyzing the possible structure of 3 successive recursion levels of local-search by a careful case analysis, one can see that not all 27 possible recursive calls (after 3 levels of local-search) can occur. Therefore, in the case of 3-SAT, the complexity bound can be further reduced, namely to 1.481^n (see [4,5]). (Actually, this is not just a closer analysis but also a modification of the algorithm since the clauses C in the procedure local-search need to be chosen carefully.) This bound also applies in the derandomized version of the algorithm as discussed in the next section.

4   Derandomization Using Covering Codes

The above example $\{0^n, 1^n\}$ of two especially chosen initial assignments (together with an appropriate Hamming distance, which is $n/2$ in this case) is nothing else than a special *covering code* (see [3]). Instead of choosing the initial assignments at random as in the previous section, we are now looking for a systematic, deterministic way of selecting $\{a_1, a_2, \ldots, a_t\}$ such that, together with an Hamming distance βn, as small as possible, we get

$$\{0,1\}^n = \bigcup_{i=1}^{t} H_{\beta n}(a_i)$$

Here $H_m(a) = \{b \in \{0,1\}^n \mid$ the Hamming distance between a and b is at most $m\}$. With such a code, the algorithm of the last section becomes deterministic since we can cycle through all code words (i.e. initial assignments) systematically in a deterministic manner.

In terms of coding theory, we are looking for codes with small covering radius (βn) and a small number of code words (t) such that the enumeration of all code words is possible within time polynomial in the number of code words. In the ideal case of a *perfect code*, the relation between β and t reaches the Hamming bound, i.e. the number of code words t satisfies

$$t = 2^n \Big/ \sum_{i=0}^{\beta n} \binom{n}{i}$$

which is, up to a polynomial factor, $[2 \cdot \beta^\beta \cdot (1 - \beta)^{1-\beta}]^n$. Actually, as seen in the last section, the Hamming bound is reached by a set of random code words, up to a polynomial factor, and up to a very small error probability.

In [4,5] two approaches are suggested to build up a good covering code. The first approach achieves "almost perfect" codes and needs only polynomial space. The second approach achieves perfect codes (up to some polynomial factor), but needs exponential space.

Approach 1: Suppose β is fixed beforehand. Using a probabilistic analysis like in the last section, it can be proved that for every $\varepsilon > 0$ there is a code length n_ε such that a random code $\{a_1, a_2, \ldots, a_{t_\varepsilon}\}$ with this code length will satisfy $\{0,1\}^{n_\varepsilon} = \bigcup_{i=1}^{t_\varepsilon} H_{\beta n_\varepsilon}(a_i)$ with high probability, provided that the number of code words t_ε satisfies $t_\varepsilon = [2 \cdot \beta^\beta \cdot (1 - \beta)^{1-\beta}]^{(1+\varepsilon)n_\varepsilon}$. Now given the actual number of variables n, we assume that n is a multiple of n_ε, i.e. $n = q_\varepsilon \cdot n_\varepsilon$. Then we can design a covering code of code length n as follows: each code word consists of q_ε blocks of length n_ε and each block cycles through all the code words $\{a_1, a_2, \ldots, a_{t_\varepsilon}\}$ independently. That is, this code has $(t_\varepsilon)^{n/n_\varepsilon}$, i.e. $[2 \cdot \beta^\beta \cdot (1 - \beta)^{1-\beta}]^{(1+\varepsilon)n}$ many code words. For every choice of ε the small code of length n_ε can be "hard-wired" in the respective k-SAT algorithm (which is deterministic, although the existence of those constant-size codes is shown by a probabilistic argument), and which has complexity $[2k/(k-1)]^{(1+\varepsilon)n}$. Notice that the space complexity is polynomial.

Approach 2: Fix the constant $c = 6$ and construct a covering code of length n/c in a greedy manner: enumerate all strings of length n/c and always choose among the remaining strings such a string as next code word which covers the most strings not yet covered. The time required by this greedy algorithm is $2^{2n/c}$. Also, exponential space is needed. The number of code words is within a polynomial factor of the Hamming bound (cf. [7]). Then, similar to Approach 1, concatenate c codewords to obtain one code word of length n. Finally, a deterministic k-SAT algorithm is obtained with worst-case time complexity $(2k/(k+1))^n$ and exponential space complexity.

5   Probabilistic Local Search

Now we turn back to a probabilistic procedure (based on [15]). First, we go back to the approach of choosing the initial assignment at random. Second, the deterministic search done by procedure local-search will be substituted by some random searching. Each time a clause C has been selected which is false under the actual assignment a, one of the literals in this clause is selected uniformly at random und its value under the assignment a is flipped. The resulting algorithm looks as follows.

```
for t times do
  begin
    Choose an assignment a ∈ {0,1}ⁿ, uniformly at random;
    for u times do
      begin
        if F(a) = 1 then accept;
        {Let C = {l₁, l₂, ..., lₖ} be some clause in F with C(a) = 0,
        i.e. all k literals lᵢ of C are set to 0 under a}
        Choose one of the k literals lᵢ at random;
        Flip the value of lᵢ in assignment a;
      end;
  end;
reject
```

The algorithm is similar to the one in [11], but there it is only applied to 2-SAT.

We have to determine optimal values for t and u (depending on n and k).

This algorithm can be analyzed using Markov chains. Suppose F is satisfiable. Fix some satisfying assignment a^*. Under assignment a^*, in each clause C of F, at least one literal becomes true. We fix in each clause *exactly one* such literal, and call it the *good literal* of the respective clause.

After guessing the random initial assignment a, we have some Hamming distance to the fixed assignment a^*. Clearly, this Hamming distance is a random variable being symmetrically binomially distributed, i.e. $Pr(\text{Hamming distance} = j) = 2^{-n}\binom{n}{j}$. In each step of the inner for-times-loop the probability of guessing the good literal in clause C is exactly $1/k$. With probability $1 - 1/k$ we choose a "bad" literal (even if it also makes C true). That is, we perform a random walk on an imaginary Markov chain. We move one step closer to our goal (state zero) if we choose the good literal (with probability $1/k$) and otherwise (with probability $1 - 1/k$) we increase the distance to state zero by one. Notice that the state number is an upper bound to the Hamming distance between the actual assignment a and the fixed satisfying assignment a^*.

For this (infinite) Markov chain it is known by standard results (see [6,16]), assuming $k \geq 3$:

- **Pr**(absorbing state 0 is reached | process started in state j) $= (1/(k-1))^j$

- **E**(number of steps until state 0 is reached | process started in state j and the absorbing state 0 is reached) $= jk/(k-2)$

The following facts take the initial probability $\binom{n}{j}2^{-n}$ for state j into account.

- **Pr**(the absorbing state 0 is reached) $= \sum_{j=0}^{n} \binom{n}{j}2^{-n}(1/(k-1))^j = (k/(2(k-1)))^n$

- **E**(number of steps until 0 is reached | the absorbing state 0 is reached) $= n/(k-2)$.

Using Markov's inequality and calculation with conditional probabilities, we obtain finally,

$\mathbf{Pr}$(after at most $2n/(k-2)$ steps the state 0 is reached) $\geq$

$$(1/2) \cdot (k/(2(k-1)))^n$$

Like in section 3, to obtain an error probability of at most e^{-20}, we need to repeat the random experiment $20 \cdot 2 \cdot (2(k-1)/k)^n$ times. Therefore, the optimal choice for u is $u = 2n/(k-2)$ and for t is $t = 40 \cdot (2(k-1)/k)^n$. Thus, the entire probabilistic algorithm has worst-case complexity $(2(k-1)/k)^n$, up to some polynomial factor. In the case of 3-SAT, the algorithm has complexity $(4/3)^n$, up to some polynomial factor.

6   Further Improvement by Independent Clauses

In the following, the bound $(4/3)^n$ in the case of 3-SAT is improved somewhat. The idea (from [17]) is to substitute the "blind" random guessing of an initial assignment $a \in \{0,1\}^n$ by some more clever random process such that "sometimes" the Hamming distance to the fixed satisfying assignment is small (and the subsequent random walk on the Markov chain is more likely to be successful.)

A key notion is that of *independent clauses*. Two clauses are independent if they have no variable in common. The following simple algorithm computes a maximal set $\mathcal{C}$ of independent clauses in a greedy manner. Let the clauses with 3 literals in the input formula F be ordered in some way: $C_1, C_2, \ldots, C_m$.

> $\mathcal{C} := \emptyset$;
> **for** $i := 1$ **to** m **do**
> **if** clause C_i is independent of all clauses in $\mathcal{C}$ **then** $\mathcal{C} := \mathcal{C} \cup \{C_i\}$;

Notice that the obtained set $\mathcal{C}$ has size at most $n/3$. Now either $|\mathcal{C}|$ is "small" ($|\mathcal{C}| \leq \alpha n$ for some constant α to be determined later), or it is "big" (i.e. $|\mathcal{C}| > \alpha n$).

Case 1: $|\mathcal{C}| \leq \alpha n$.
In this case we apply the following algorithm. Notice that each independent clause $C \in \mathcal{C}$ can be satisfied by 7 assignments to the 3 occurring variables. These assignments can be tested independently for the clauses in $\mathcal{C}$. For each of the $7^{|\mathcal{C}|} \leq 7^{\alpha n}$ assignments to the variables occurring in clauses from $\mathcal{C}$ we do the following. This (partial) assignment is applied to the formula. The result is that all clauses from $\mathcal{C}$ are satisfied, and additionally, all remaining clauses either have already 2 literals, or at least one literal is set under the partial assignment. This is because $\mathcal{C}$ was constructed to be a *maximal* set of independent clauses. In other words, the remaining formula is a 2-CNF formula. Satisfiability of 2-CNF formulas can be tested in polynomial time [1]. Therefore, in this case, we have a satisfiability test for the input formula F running in time at most $7^{\alpha n}$, up to some polynomial factor.

Case 2: $|C| > \alpha n$.

In this case we use the random walk algorithm like in the last section, but with a modified initial distribution of the assignments. We choose the initial assignments a at random using the following stochastic process. For each clause $C \in \mathcal{C}$ choose the assignments for the 3 literals in C from the 8 assignments $000, 001, 010, 100, 011, 101, 110, 111$ according to the following probability distribution.

assignment	probability
000	0
001	z
010	z
100	z
011	y
101	y
110	y
111	x

These probabilities x, y, z are then determined by analyzing the complexity of this modified algorithm and solving a set of linear equations. The result is

$$x = 1/7, \quad y = 2/21, \quad z = 4/21.$$

With these values, the success probability becomes

$$(3/4)^{n-3|C|} \cdot (3/7)^{|C|} \geq \left[(3/4)^{(1-3\alpha)} (3/7)^{\alpha} \right]^n$$

The reciprocal value gives the complexity of the algorithm (up to a polynomial factor). The optimal value for α, used to distinguish Cases 1 and 2, can now be determined by solving for

$$7^{\alpha} = (4/3)^{(1-3\alpha)} (7/3)^{\alpha}$$

yielding $\alpha \approx 0.146652$. With these parameters we obtain the total running time 1.3303^n, up to some polynomial factor.

A further minor improvement is possible as follows, cf. [8]. Instead of fixing the values for x, y, z in some way, one can run a finite set of algorithms in parallel which all use different (x, y, z)-values. Even if we don't know beforehand how many and which of the literals in the clauses are satisfied under a^*, at least one of the finitely many algorithms will obtain a somewhat better success probability as the one calculated above. The new bound is 1.3302^n. The details will be presented in the full paper.

References

1. B. Aspvall, M. F. Plass, R. E. Tarjan: A linear time algorithm for testing the truth of certain quantified Boolean formulas, *Information Processing Letters* 8(3) (1979) 121–123. 93
2. R. B. Ash: *Information Theory.* Dover 1965. 90
3. G. Cohen, I. Honkala, S. Litsyn, A. Lobstein: *Covering Codes.* North-Holland 1997. 90
4. E. Dantsin, A. Goerdt, E. A. Hirsch, U. Schöning: Deterministic algorithms for k-SAT based on covering codes and local search. *Proc. 27th International Colloquium on Automata, Languages and Programming 2000.* Springer Lecture Notes in Computer Science, Vol. 1853, pages 236–247, 2000. 90, 91
5. E. Dantsin, A. Goerdt, E. A. Hirsch, R. Kannan, J. Kleinberg, C. Papadimitriou, P. Raghavan, U. Schöning: A deterministic $(2 - \frac{2}{k+1})^n$ algorithm for k-SAT based on local search. To appear in *Theoretical Computer Science.* 90, 91
6. G. R. Grimmett, D. R. Stirzaker: *Probability and Random Processes.* Oxford University Press, 2nd Edition, 1992. 92
7. D. S. Hochbaum (ed.): *Approximation Algorithms for NP-Hard Problems.* PWS Publishing Company, 1997. 91
8. T. Hofmeister, U. Schöning, R. Schuler, O. Watanabe: *A probabilistic 3-SAT algorithm further improved.* submitted. 94
9. O. Kullmann: New methods for 3-SAT decision and worst-case analysis. *Theoretical Computer Science* 223 (1999) 1–72. 87
10. B. Monien, E. Speckenmeyer: Solving satisfiability in less than 2^n steps. *Discrete Applied Mathematics* 10 (1985) 287–295. 87, 89
11. C. H. Papadimitriou: On selecting a satisfying truth assignment. *Proceedings of the 32nd Ann. IEEE Symp. on Foundations of Computer Science*, 163–169, 1991. 92
12. R. Paturi, P. Pudlák, F. Zane: Satisfiability coding lemma. *Proceedings 38th IEEE Symposium on Foundations of Computer Science* 1997, 566–574. 87
13. R. Paturi, P. Pudlák, M. E. Saks, F. Zane: An improved exponential-time algorithm for k-SAT. *Proceedings 39th IEEE Symposium on Foundations of Computer Science* 1998, 628–637. 87
14. U. Schöning: *On The Complexity Of Constraint Satisfaction Problems.* Ulmer Informatik-Berichte, Nr. 99-03. Universität Ulm, Fakultät für Informatik, 1999. 89
15. U. Schöning: A probabilistic algorithm for k-SAT and constraint satisfaction problems. *Proceedings 40th IEEE Symposium on Foundations of Computer Science* 1999, 410–414. 87, 91
16. U. Schöning: A probabilistic algorithm for k-SAT based on limited local search and restart. To appear in *Algorithmica.* 92
17. R. Schuler, U. Schöning, O. Watanabe: *An Improved Randomized Algorithm For 3-SAT.* Techn. Report TR-C146, Dept. of Mathematical and Computing Sciences, Tokyo Institute of Technology, 2001. 93

Linear Temporal Logic and Finite Semigroups

Thomas Wilke

CAU, Kiel, Germany
wilke@ti.informatik.uni-kiel.de

Abstract. This paper gives a gentle introduction to the semigroup-theoretic approach to classifying discrete temporal properties and surveys the most important results.

1 Introduction

Linear temporal logic is a widely used, rigorous formalism for specifying temporal properties conveniently, namely by using formal counterparts of natural-language constructs such as "always", "eventually", "until", "hitherto" and so forth. Developed in the 1960s as an extension of tense (modal) logic, linear temporal logic had been primarily studied in mathematical logic and linguistics [8] before Pnueli [12] and Kröger [9] suggested to use it for specifying temporal properties of nonterminating computations such as the behavior of reactive systems. Since then it has become ever more important in computer science, especially, as it is feasible to determine whether or not a finite-state system (such as a piece of hardware or a communication protocol) satisfies a desirable linear temporal property.

The important feature of linear temporal logic are its temporal operators, which allow one to express relations between the propositional variables that hold true in various points of time. From the most general point of view, a temporal operator is simply a first-order formula with one free first-order variable and free unary predicate symbols [6]. As a consequence, there are infinitely many temporal operators, but only a few of them are in wide-spread use, namely those that are easy to phrase in natural language and, at the same time, powerful enough to express what is necessary.

It is an intriguing question to determine how much each of the temporal operators in use adds to the expressive power of linear temporal logic. Equivalently, one can ask which temporal properties can be expressed if linear temporal logic is restricted to only certain operators. Similarly, it is interesting to determine what can be expressed if the nesting depth of the binary temporal operators (which, for a human, are more difficult to parse) is restricted. More general, it is an interesting question to characterize the various fragments of linear temporal logic.

There are different ways of characterizing fragments of linear temporal logic. One way is to show that the expressive power of a given fragment is the same as the expressive power of a different logic or another formalism. Results of this

J. Sgall, A. Pultr, and P. Kolman (Eds.): MFCS 2001, LNCS 2136, pp. 96–110, 2001.

kind include Kamp's famous result that linear temporal logic and first-order logic are equally expressive (in Dedekind-complete time structures) [8], that linear temporal logic restricted to unary operators is as expressive as first-order logic with two variables [5], and that in linear temporal logic one can define exactly the same formal languages that can be defined by star-free regular expressions [8,13,10]. From a computational point of view, it is more desirable to have an *effective* characterization in the form of an algorithm that given a property determines if it is expressible in the fragment in question and computes, if possible, a respective formula. For example, from the same work mentioned above [8,13,10] it follows that a regular language is expressible in linear temporal logic if and only if its syntactic semigroup contains no subsemigroup which is a non-trivial group. This is an effective description, for from any reasonable representation of a regular language (finite automaton, regular expression, monadic second-order formula) one can effectively obtain the multiplication table of its syntactic semigroup and by an exhaustive search it is easy to check whether it contains a non-trivial group as a subsemigroup.

The above decision procedure follows a pattern which has been used for many effective characterizations of fragments of linear temporal logic and has been most successful in that it has also worked for fairly complicated fragments; the purpose of this paper is to explain this pattern, the "semigroup paradigm".

As described above, the semigroup paradigm is very simple. If one wants to obtain an effective characterization of a certain fragment F (where time is modeled as an initial segment of the natural numbers) one first considers the class L of formal (regular) languages that can be defined by a formula in the fragment. (Since every temporal property defines a regular language, L is a perfect description of the expressive power of F.) Next, one considers the smallest class V of finite semigroups containing all syntactic semigroups of the languages in L and which is closed under boolean operations, homomorphic images, and finite direct products, the pseudovariety generated by the syntactic semigroups of the languages in L. In most cases, V is a perfect description of L in the sense that V contains no syntactic semigroup of a language not in L. In the last step, one uses semigroup theory to show that V is decidable, which also implies that expressibility in F is decidable, for passing from a linear temporal formula to the syntactic semigroup of the language associated with it is effective. In many instances, the most difficult part is the last step, namely to find a decision procedure for V.— The purpose of this paper is to illustrate and explain this approach in more detail.

In most applications, modalities are used that refer only to the present and the future (e. g., "sometime in the future", "until") even though Kamp, when he defined temporal logic, had one future and one past modality in his logic. The restriction to future modalities is no restriction from a theoretical point of view: by a theorem of Pnueli, Shelah, Stavi and Gabbay [7], every linear temporal formula is equivalent to a future formula if one is only interested in defining properties of time structures with a starting point.

As pointed out above, Kröger and Pnueli suggested to use linear temporal logic for specifying properties of non-terminating computations. So time should be modeled as the natural numbers (assuming one is interested in discrete time). In this paper, only finite prefixes of the natural numbers are considered. Most results presented here can be extended to the natural numbers (ω-words), but different techniques are required, which are out of the scope of this paper.

Basic notation. When $\equiv$ is an equivalence relation over a set M and $m \in M$, then $[m]_\equiv$ denotes the equivalence class of m. Accordingly, when $M' \subseteq M$, then $M'/\equiv$ denotes the set of all equivalence classes of elements from M', that is, $M'/\equiv \, = \{[m]_\equiv \mid m \in M'\}$.

Strings are sequences of letters over a finite alphabet. The length of a string u is denoted $|u|$. The i-th letter of a string u is denoted u_i, that is, $u = u_1 \ldots u_{|u|}$. The suffix of a string u starting at position i, the string $u_i \ldots u_{|u|}$, is denoted $\mathrm{suf}_i(u)$ and, similarly, prefixes are denoted $\mathrm{pref}_i(u)$.

As usual, A^* and A^+ denote the set of all strings over A and all non-empty strings over A, respectively.

2   Linear Temporal Logic

In this section, the basics on linear temporal logic are recalled. For more on this, see [4] and [6].

2.1   Syntax

A *future linear temporal logic formula (LTL formula)* over some finite set Σ of *propositional variables* is built from the boolean constant tt and the elements of Σ using boolean connectives and temporal operators: X (neXt), F (eventually in the future, Finally), and U (Until), where only U is binary and the others are unary.

2.2   Semantics

Given an LTL formula φ over some set Σ of propositional variables and a string $u \in (2^\Sigma)^+$, we write $u \models \varphi$ for the fact that u satisfies φ. In particular,

- $u \models$ tt,
- $u \models p$ if $p \in u_1$,
- $u \models \varphi \; U \; \psi$ if there exists j with $1 < j \leq |u|$ such that $\mathrm{suf}_j(u) \models \psi$ and $\mathrm{suf}_i(u) \models \varphi$ for all i with $1 < i < j$.

The semantics of X and F is derived from this:

$$X\varphi = \neg tt \; U \; \varphi \, , \qquad\qquad F\varphi = tt \; U \; \varphi \, . \qquad (1)$$

With every formula φ over Σ, we associate a language over 2^{Σ}:

$$L(\varphi) = \{u \in (2^{\Sigma})^+ \mid u \models \varphi\} \; . \tag{2}$$

We say $L(\varphi)$ is *defined* by φ.

When we want to define languages over arbitrary alphabets (rather than alphabets of the form 2^{Σ}) we will use the following convention. We view A as a subset of some appropriate alphabet 2^{Σ}. More formally, for each alphabet A we choose once and for all a set Σ of propositional variables such that $2^{|\Sigma|-1} < |A| \leq 2^{|\Sigma|}$ and an injective mapping $\iota\colon A \to 2^{\Sigma}$, which we extend to a homomorphism $\iota\colon A^* \to (2^{\Sigma})^*$. For each formula $\varphi \in \mathsf{LTL}_A$ (where the atomic formulas are letters from A), we set

$$L(\varphi) = \{u \in A^+ \mid \iota(u) \models \varphi'\} \; , \tag{3}$$

where φ' is obtained from φ by replacing every occurrence of a letter a by

$$\bigwedge_{p \in \iota(a)} p \wedge \bigwedge_{p \notin \iota(a)} \neg p \; . \tag{4}$$

As above, $L(\varphi)$ is the language *defined* by φ.

With every class of formulas Φ, we associate the class $L(\Phi)$ of all languages defined by formulas in Φ. The alphabet associated with a formula and the alphabet a language is written over will always be implicit. When $L \in L(\Phi)$, we say L is *expressible* in Φ.

Formulas $\varphi, \psi \in \mathsf{LTL}_\Sigma$ are *equivalent*, denoted $\varphi \equiv \psi$, if $L(\varphi) = L(\psi)$. (This assumes that their alphabets are the same.)

2.3   Fragments

A fragment of LTL is determined by which temporal operators are allowed and possible restrictions on the nesting depth on these operators. When $\theta_1, \ldots, \theta_r$ are operators, then $\mathsf{TL}[\theta_1, \ldots, \theta_r]$ is the fragment of LTL where only $\theta_1, \ldots, \theta_r$ are allowed. For instance, $\mathsf{TL}[\mathsf{X}, \mathsf{F}]$ is what Cohen, Pin, and Perrin call *restricted temporal logic*.

Subscripts indicate bounds on nesting depth. For instance, in $\mathsf{TL}[\mathsf{X}_k]$ only X is allowed and the nesting depth in X is at most k. Another example is $\mathsf{TL}[\mathsf{X}, \mathsf{F}, \mathsf{U}_k]$, where X and F are allowed without restriction and the nesting depth in U must not be greater than k. This is also known as the k-th level of the *until hierarchy*.

3   Formal Languages and Semigroups

In this section, the basics on formal languages and finite semigroups are recalled. See also [11], and for semigroup theory see [1].

3.1   Semigroups and Congruences

An equivalence relation $\equiv$ on A^+ is called a *congruence* over A if $uv \equiv u'v'$ whenever $u \equiv u'$ and $v \equiv v'$. A congruence relation has *finite index* if it has a finite number of equivalence classes. A congruence over some alphabet A *saturates* a formal language if it is a union of congruence classes.

Given a language $L \subseteq A^+$, its *syntactic congruence*, denoted $\equiv_L$, is the equivalence relation on A^+ defined by

$$u \equiv_L v \quad \text{iff} \quad \forall x, y \in A^+ (xuy \in L \leftrightarrow xvy \in L) \ ,$$

which is, in fact, an equivalence relation. It is the coarsest congruence saturating L and has a finite index if and only if L is regular. If L is regular and (A, Q, q_I, δ, F) is the minimal DFA recognizing L, then

$$u \equiv_L v \quad \text{iff} \quad \forall q \in Q(\delta^+(q, u) \in F \leftrightarrow \delta^+(q, v) \in F) \ ,$$

where δ^+ is the extended transition function, obtained by extending δ according to $\delta^+(q, xa) = \delta(\delta^+(q, x), a)$.

A semigroup is a set equipped with an associative product. The free semigroup over a set A is the set A^+ with concatenation as product. A homomorphism $h: A^+ \to S$ into a semigroup induces an equivalence relation $\equiv_h$ defined by $u \equiv v$ if $h(u) = h(v)$. Conversely, a congruence relation $\equiv$ over some alphabet A induces a homomorphism $h_\equiv: A^+ \to A^+/\equiv$ into the quotient semigroup $A^+/\equiv$ defined by $h(u) = [u]_\equiv$.

A language L over A is recognized by a homomorphism $h: A^+ \to S$ if L is saturated by $\equiv_h$, that is, when $L = h^{-1}(P)$ for some set $P \subseteq S$. It is recognized by a semigroup S if there exists a homomorphism $h: A^+ \to S$ which recognizes it. The *syntactic semigroup* of a language $L \subseteq A^+$, denoted $S(L)$, is defined by $S(L) = A^+/\equiv_L$. Clearly, L is recognized by $h_{\equiv_L}$ and $S(L)$ and L is regular iff $S(L)$ is finite.

3.2   Pseudovarieties of Semigroups

Often enough, interesting classes of regular languages (such as all LTL expressible languages) are closed under boolean combinations, left and right quotients (if L belongs to the class, then also $a^{-1}L = \{u \mid au \in L\}$ and $La^{-1} = \{u \mid ua \in L\}$), and inverse homomorphic images. Such classes are called *varieties of formal languages*. According to Eilenberg's theorem [3], varieties of formal languages are in a one-to-one correspondence with pseudovarieties of finite semigroups, which are classes of finite semigroups closed under finite direct products, homomorphic images and subsemigroups. More precisely, the following two mappings are inverse to each other: (1) the mapping that associates with every variety $\mathbf{L}$ of formal languages the pseudovariety of semigroups generated by the syntactic semigroups of the languages in $\mathbf{L}$, and (2) the mapping that associates with every pseudovariety $\mathbf{V}$ of finite semigroups the class of formal languages recognized by elements of $\mathbf{V}$ (which is a variety of formal languages).

So studying the corresponding pseudovariety of semigroups is as good as studying a given variety of languages. This is true even when it comes to computational issues, because constructing the syntactic semigroup of a regular language from a representation by a finite automaton or a regular expression is effective. In particular, in order to show that membership to a given variety of formal languages is decidable it is enough to show that membership to the corresponding pseudovariety of semigroups is decidable. And the latter is, often enough, (much) easier.

Decidability for pseudovarieties of semigroups has been studied for quite a while. In certain situations, decidability is obvious, for instance, when a pseudovariety is defined by a finite set of equations. But in other situations it can be very difficult. In fact, there are quite a view pseudovarieties where decidability is an open problem. Fortunately, showing decidability for the pseudovarieties related to LTL is not that difficult.

We will use the following terminology when working with congruences. If the quotient semigroup of a congruence relation $\equiv$ over some alphabet A belongs to a pseudovariety V, we will say that $\equiv$ is a V-congruence.

4 Simple Fragments

We first study fragments of temporal logic which are easy to characterize, but are non-trivial nevertheless. Here, easy means that it is almost straightforward to come up with a finite set of equations (in fact, a single equation) for the corresponding pseudovariety of semigroups.

4.1 Next Only

We first look at the LTL fragment where only X is allowed, that is, we study TL[X]. Observe that although in TL[X] one cannot express complicated properties, even over a fixed alphabet there are infinitely many different properties that can be expressed. So finding an effective characterization is not a trivial issue.

For a start, we consider $TL[X_k]$ for increasing k. The formulas in $TL[X_0]$ are the propositional formulas, and with such a formula we can only speak about the first position of a string. In $TL[X_1]$ we can also speak about the second position. We can, for instance, say that in the second position there is a certain letter (Xa), and we can say that there is no second position $(\neg Xtt)$. In $TL[X_2]$, we can also speak about the third position of a string ... Algebraically, this means that in the syntactic semigroup of a language expressible in $TL[X_k]$ any two products of at least $k + 1$ elements which agree on the first $k + 1$ factors are the same:

Theorem 1. *Let $k \geq 0$ and $L \subseteq A^+$ a regular language. Then the following are equivalent:*

1. *L is expressible in $TL[X_k]$.*
2. *$S(L)$ satisfies*

$$s_1 \ldots s_{k+1} t = s_1 \ldots s_{k+1} \ . \qquad\qquad (X_k)$$

3. L is saturated by the congruence $\equiv_k$ defined by $u \equiv_k v$ if $u = v$ or $|u|, |v| \geq k + 1$ and $\text{pref}_{k+1}(u) = \text{pref}_{k+1}(v)$.

Proof. That 2. and 3. are equivalent is easy to see: the quotient semigroup $A^+/\equiv_k$ satisfies (X_k), and if a semigroup S satisfies (X_k) and $h: S \to A^+$ is a homomorphism, then $\equiv_k \subseteq \equiv_h$.

To complete the proof we show the equivalence between 1. and 3. For the implication from 3. to 1. we only need to show that every equivalence class of $\equiv_k$ is expressible in $\text{TL}[X_k]$. This is simple: if $|u| \leq k$, then $[u]_{\equiv_k}$, which contains just u, is defined by $u_1 \wedge X(u_2 \wedge X(u_3 \wedge \cdots \wedge X(u_{|u|} \wedge \neg Xtt)))$; if $|u| \geq k+1$, then $[u]_{\equiv_k}$ is defined by $u_1 \wedge X(u_2 \wedge X(u_3 \wedge \cdots \wedge Xu_{k+1}))$.

The proof of the implication from 1. to 3. goes by induction on k. The base case, $k = 0$, is simple: a language $L \subseteq A^+$ is expressible in $\text{TL}[X_0]$ if $L = BA^*$ for some $B \subseteq A$, and all these languages are saturated by $\equiv_k$. In the inductive step, assume we are given a formula $\varphi \in \text{TL}[X_{k+1}]$. Then φ is a boolean combination of propositional formulas and formulas of the form $X\psi$ where $\varphi \in \text{TL}[X_k]$. So, by induction hypothesis, $L(\varphi)$ is a boolean combination of languages of the form BA^* and $A[u]_{\equiv_k}$ where $B \subseteq A$ and $u \in A^+$ is a $\equiv_k$-class. Clearly, each language of the first type is saturated by $\equiv_{k+1}$. Moreover, it is easy to see that $A[u]_{\equiv_k} = \bigcup_{a \in A}[au]_{\equiv_{k+1}}$; hence, every language of the second type is also saturated by $\equiv_{k+1}$. □

Example 2. Consider $L = abA^*$ where $A = \{a, b\}$. The syntactic congruence of L has 4 classes: a, aaA^*, bA^*, abA^*. The multiplication table of $S(L)$ is:

	a	b	aa	ab
a	aa	ab	aa	aa
b	b	b	b	b
aa	aa	aa	aa	aa
ab	ab	ab	ab	ab

where the elements are identified with shortest representatives of the corresponding equivalence classes. Clearly, (X_0) is not satisfied, because aa is different from a. On the other hand, it is easy to see that (X_1) is satisfied. So L should be definable by a $\text{TL}[X_1]$ formula. It is simple to come up with such a formula: $a \wedge Xb$. □

Clearly, a characterization of the pseudovariety corresponding to full $\text{TL}[X]$ cannot be given in terms of a finite number of equations. In order to arrive at an effective characterization nevertheless, we allow a new operation to be used in equations—currently, in equations we only use the binary products of the underlying semigroups.

Let S be a finite semigroup. Then, for every $s \in S$, there is a unique element $e \in \{s, s^2, s^3, \ldots\}$ such that $e^2 = e$. That is, in the subsemigroup generated by s there is a unique idempotent element. This will be denoted by s^ω. Clearly, when S and s are given, then s^ω can be determined easily. So if we allow $(.)^\omega$ to be used in equations for defining classes of finite semigroups—this is what we do from now on—, then these equations can still be verified effectively.

The key observation for finding an equation for full TL[X] is that in a finite semigroup idempotent elements represent "long" strings. More precisely, in a finite semigroup S every element which can be written as a product of at least $|S|$ elements can also be written in the form $s'es''$ where e is idempotent (for any product $s_1 \ldots s_n$ consider the partial products s_1, $s_1 s_2$, $\ldots$ and use the pigeon-hole principle). Using this, we obtain:

Lemma 3. *A finite semigroup satisfies one of the equations (X_k) if and only if it satisfies*

$$s^\omega t = s^\omega \ . \tag{X}$$

Proof. Let S be a finite semigroup. If S satisfies (X_k), then S satisfies (X): we have $s^\omega t = (s^\omega)^{k+1} t = (s^\omega)^{k+1} = s^\omega$, where the middle equality is due to (X_k).

Conversely, in a semigroup S with n elements we have $s_1 \ldots s_n y = s'es''y = s'e^\omega s''y = s'e^\omega = s'e^\omega s'' = s_1 \ldots s_n$ for appropriate elements $s', s'' \in S$, e idempotent, and any choice of $s_1, \ldots, s_n \in S$. $\qquad\square$

Example 4. Consider L from Example 2 again. Clearly, b, aa, and ab are the only idempotent elements, and all of them are left zeroes. So (X) is satisfied. $\quad\square$

We conclude with a summary of the results:

Theorem 5. *1. For every k, the fragment TL[X_k] is effectively characterized by (X_k).*
2. The fragment TL[X] is effectively characterized by (X).

4.2   Eventually Only

In this section, we look at a more complicated class of temporal formulas. We want to characterize the languages that are expressible in TL[F].

With nesting depth 0 one can specify that a string starts with a certain letter. With nesting depth 1 one can specify in addition that certain letters occur after the first position (Fa) and others don't ($\neg Fa$). With nesting depth 2 one can, for instance, specify in addition that after the first position a certain letter occurs and after that one certain letters do not occur any more ($F(a \wedge \neg Fa_1 \wedge \ldots \neg Fa_n)$) and other related properties.

We want to phrase algebraically what all the above properties have in common. Assume we are given a set Φ of properties that can be expressed in TL[F], say the set of all properties of nesting depth at most k. Suppose a string of the form u and one of its suffixes, say $\mathrm{suf}_i(u)$ satisfy exactly the same subset Φ' of Φ. Then it should be clear that every string "between" u and $\mathrm{suf}_i(u)$ starting with the same letter as u and v, that is, every $\mathrm{suf}_j(u)$ with $1 \le j \le i$ and $u_1 = u_j$ ($= u_i$) should satisfy exactly Φ'.

This can be phrased very easily in terms of a universal Horn sentence (and in terms of an equation):

Theorem 6. *Let $L \subseteq A^+$ be a regular language. Then the following are equivalent.*

1. *L is expressible in $\mathsf{TL[F]}$.*
2. *$S(L)$ satisfies*

$$stu = u \rightarrow rtu = ru \ . \tag{5}$$

   (Observe that on the right-hand side we could also write $rstu = rtu = ru$.)
3. *$S(L)$ satisfies*

$$rs(ts)^\omega = r(ts)^\omega \ . \tag{6}$$

In the proof, we use the following equivalence relation, which parametrizes the $\mathsf{TL[F]}$ expressible properties. For every k, let $\equiv_k$ be the congruence with $u \equiv_k v$ if u and v satisfy the same $\mathsf{TL[F_k]}$ properties. This equivalence relation is in fact a congruence relation. We will use the following inductive description of $\equiv_k$:

Lemma 7. *Let u and v be arbitrary non-empty strings over A. Then:*

1. *$u \equiv_0 v$ iff $u_1 = v_1$.*
2. *$u \equiv_{k+1} v$ iff*
   (a) *$u_1 = v_1$,*
   (b) *for every i with $1 < i \le |u|$ there exists j with $1 < j \le |v|$ such that $\mathrm{suf}_i(u) \equiv_k \mathrm{suf}_j(v)$, and*
   (c) *for every i with $1 < i \le |v|$ there exists j with $1 < j \le |u|$ such that $\mathrm{suf}_i(v) \equiv_k \mathrm{suf}_j(u)$.* □

This can be proved by a straightforward induction on k; it simply formalizes what was informally described at the beginning of this subsection.

Proof of Theorem 6. It is easy to see that (5) and (6) are equivalent for every finite semigroup. First, assume (5) holds. Let s and t be arbitrary semigroup elements and m such that $(ts)^m = (ts)^\omega$. Then $t(st)^{m-1}s(st)^m = (ts)^m$, so, using (5), $rs(ts)^m = r(ts)^m$, hence $rs(ts)^\omega = r(ts)^\omega$. Conversely, if $stu = u$, then $(st)^m u = u$ for every $m \ge 1$, in particular, $(st)^\omega u = u$, which implies $rt(st)^\omega u = ru$. But $t(st)^\omega u = tu$ anyway, so $rtu = ru$.

To conclude the proof we show that 1. and 2. are equivalent. The proof that 1. implies 2. is by induction on k. It is enough to show that the quotient semigroups $A^+/\equiv_k$ satisfy (5). For $k = 0$, this is trivial. So let $k > 0$ and assume $stu \equiv_k u$. Clearly, rtu and ru start with the same letter, so 2.(a) from Lemma 7. Let $\varphi \in \mathsf{TL[F_{k-1}]}$. First, suppose $\mathrm{suf}_i(ru) \models \varphi$ for some i with $1 < i \le |ru|$. We distinguish two cases. If $i \le |r|$, then $r'u \models \varphi$ for a suffix r' of r, and the induction hypothesis yields $r'tu \models \varphi$, which shows $\mathrm{suf}_i(rtu) \models \varphi$. If $i > |r|$, then $\mathrm{suf}_{i-|r|}(u) \models \varphi$, so $\mathrm{suf}_{i+|t|-|r|}(tu) \models \varphi$, which also means $\mathrm{suf}_{i+|t|}(rtu) \models \varphi$. So, in both cases, $\mathrm{suf}_j(rtu) \models \varphi$ for some j with $1 < j \le |rtu|$, that is, 2.(b) from Lemma 7. In the same fashion, one proves that 2.(c) holds. This shows $rtu \equiv_k ru$.

For the implication from 2. to 1., we again use the above characterization of $\equiv_k$. Assume (5) does not hold for the syntactic semigroup of some language L. Then there exist strings r, s, t, u such that $stu \equiv_L u$ and $rtu \not\equiv_L ru$. As a consequence, we have $rt(st)^k u \not\equiv_L r(st)^k u$ for every k. By definition of $\equiv_L$, this means that for every k there exist $x, y \in A^*$ such that $xrt(st)^k uy \in L$ and $xr(st)^k uy \notin L$, or $xrt(st)^k uy \notin L$ and $xr(st)^k uy \in L$. By a simple induction, one shows $xrt(st)^k uy \equiv_k xr(st)^k uy$. So no formula in $\mathsf{TL}[\mathsf{F}_k]$ can define L. Since this is true for every k, L cannot be expressed in $\mathsf{TL}[\mathsf{F}]$.   □

Example 8. Consider $L = A^* a A^* b$ where $A = \{a, b, c\}$. The classes of $\equiv_L$ are: $A^* a$, $A^* a A^* b$, $(b + c)^* b$, $(b + c)^* c$, $A^* a A^* c$. And the multiplication table of $S(L)$ is:

	a	b	c	ab	ac
a	a	ab	ac	ab	ac
b	a	b	c	ab	ac
c	a	b	c	ab	ac
ab	a	ab	ac	ab	ac
ac	a	ab	ac	ab	ac

The elements a, ab, and ac are right zeros, so it is impossible to violate (6) with $ts \in \{a, ab, ac\}$. But $(ts)^\omega \notin \{a, ab, ac\}$ iff $t, s \in \{b, c\}$, and in the subsemigroup $\{b, c\}$, b and c are right zeros. So $S(L)$ satisfies (6). This is what we expect, because L is defined by $(a \wedge \mathsf{F}b) \vee \mathsf{F}(a \wedge \mathsf{F}b)$.   □

Example 9. Consider the language L from Example 2. The second letter of any word in L is fixed to b. This should not be expressible in $\mathsf{TL}[\mathsf{F}]$. Our decision procedure yields the expected result: in $S(L)$, we have $aa = aa(ba)^\omega \neq a(ba)^\omega = ab$, thus (6) does not hold in $S(L)$.   □

5   Composing Classes

Many fragments of LTL are more difficult to characterize than $\mathsf{TL}[\mathsf{X}]$ and $\mathsf{TL}[\mathsf{F}]$, for instance, $\mathsf{TL}[\mathsf{X}, \mathsf{F}]$, which is known as restricted temporal logic. For fragments like this one, a compositional approach should be taken. The key concepts are substitution on the LTL side and semidirect products on the semigroup side.

5.1   Substitution

It is quite easy to see that every formula in $\mathsf{TL}[\mathsf{X}, \mathsf{F}]$ is equivalent to a formula in that fragment where F does not occur in the scope of X. So X can always be pushed inside. We describe this more formally using substitution.

A function $\sigma \colon \mathsf{LTL}_\Sigma \to \mathsf{LTL}_\Gamma$ is a *substitution* if σ maps each formula $\varphi \in \mathsf{LTL}_\Sigma$ to the formula which is obtained from φ by replacing every occurrence of a propositional variable $p \in \Sigma$ by $\sigma(p)$. Note that a substitution is fully determined by the values for the propositional variables.

When Ψ is a class of LTL formulas, then $\sigma\colon \mathsf{LTL}_\Sigma \to \mathsf{LTL}_\Gamma$ is a Ψ substitution if $\sigma(p) \in \Psi$ for every $p \in \Sigma$. When Φ and Ψ are classes of formulas, then $\Phi \circ \Psi$ is the set of all formulas $\sigma(\varphi)$ where $\varphi \in \Phi$ is a formula over some Σ and $\sigma\colon \mathsf{LTL}_\Sigma \to \mathsf{LTL}_\Gamma$ is a Ψ substitution for some Γ. Clearly, substitution is associative, so it is not necessary to use parentheses.

The above remark about $\mathsf{TL}[\mathsf{X}, \mathsf{F}]$ can now be rephrased as follows. Every formula in $\mathsf{TL}[\mathsf{X}, \mathsf{F}]$ is equivalent to a formula in $\mathsf{TL}[\mathsf{F}] \circ \mathsf{TL}[\mathsf{X}]$.

The most basic lemma about the semantics of LTL is the analogue of the substitution lemma for first-order logic, which is described in what follows. Let $\sigma\colon \mathsf{LTL}_\Sigma \to \mathsf{LTL}_\Gamma$ be a substitution. For each $u \in (2^\Gamma)^+$ of length n, let $\sigma(u)$ be the string $a_1 \ldots a_n$ with

$$a_i = \{p \in \Sigma \mid \mathrm{suf}_i(u) \models \sigma(p)\} \ . \tag{7}$$

So the i-th letter of $\sigma(u)$ encodes which formulas $\sigma(p)$ hold in $\mathrm{suf}_i(u)$ and which don't.

The substitution lemma now splits the problem $u \overset{?}{\models} \sigma(\varphi)$ into two problems:

Lemma 10 (LTL substitution lemma). *Let φ be an LTL formula over Σ and $\sigma\colon \mathsf{LTL}_\Sigma \to \mathsf{LTL}_\Gamma$ a substitution. Then, for every $u \in (2^\Sigma)^+$,*

$$u \models \sigma(\varphi) \quad \textit{iff} \quad \sigma(u) \models \varphi \ . \tag{8}$$

$\square$

The proof of this lemma is a simple induction.

Example 11. Let $\varphi = \mathsf{F}(a \wedge \mathsf{X}b)$ where $A = \{a, b\}$, that is, $L(\varphi) = A^+abA^*$. Identify A with 2^Γ where $\Gamma = \{q\}$, in particular, assume $a = \{q\}$ and $b = \emptyset$. Let $\Sigma = \{p\}$. Then $\varphi = \sigma(\mathsf{F}p)$ with $\sigma\colon p \mapsto a \wedge \mathsf{X}b$, more precisely, $\sigma(p) = q \wedge \mathsf{X}\neg q$. Clearly, $bbbaba \models \varphi$. Using the above lemma, we can verify this as follows. First, $\sigma(bbbaba) = \emptyset\emptyset\emptyset\{p\}\emptyset\emptyset$, and the latter satisfies $\mathsf{F}p$. $\square$

5.2   Semidirect Product/Substitution Principle

Suppose we are given two LTL fragments Φ and Ψ and effective characterizations in terms of finite semigroups, that is, we are given two pseudovarieties V and W such that $L(\Phi) = L(V)$ and $L(\Psi) = L(W)$. We would like to have a characterization of $\Phi \circ \Psi$. To obtain a good picture of what this operation should do, we reconsider the substitution lemma.

There is a straightforward way to model $\sigma(u)$ in an algebraic setting. Assume we are given a congruence $\equiv$ that describes somehow the semantics of the formulas $\sigma(p)$, say it saturates all the languages defined by these formulas. Further, assume u is a non-empty string over A. Then the string $\sigma(u)$ somehow corresponds to the string $u_\equiv$ defined by $u_\equiv = a_1 \ldots a_{|u|}$ where $a_i = [\mathrm{suf}_i(x)]_\equiv$. Note that $u_\equiv$ should strictly be viewed as a string over the semigroup $A^+/\equiv$.

From the above it is clear it would be desirable to have an algebraic operation at hand which given pseudovarieties $\boldsymbol{V}$ and $\boldsymbol{W}$ yields a new pseudovariety generated by all quotients of the form $A^+/\equiv'''$ where $\equiv'''$ is a congruence relation generated by an equivalence relation $\equiv''$ obtained as follows.

(†) For some $\boldsymbol{W}$-congruence $\equiv$ over A and some $\boldsymbol{V}$-congruence $\equiv'$ on $(S^+/\equiv)^+$, the relation $\equiv''$ relates u and v if $u_\equiv \equiv' v_\equiv$.

This would allow us to combine $\boldsymbol{V}$ and $\boldsymbol{W}$ in a way such that the resulting pseudovariety corresponds to $L(\Phi \circ \Psi)$.

Example 12. Reconsider Example 11. Let $\equiv$ denote the syntactic congruence of $L(a \wedge \mathsf{X}b)$, which we already know from Example 2. Suppose $u = aabbba$ and $v = bbbabab$. Then $u_\equiv = [aa][ab][b][b][b][a]$ and $v_\equiv = b\,b\,b\,ab\,b\,ab$. Let $\equiv''$ be the syntactic congruence of the language $S(L)[ab]S(L)^*$. Then $u \equiv'' v$. □

As it turns out, there is an algebraic operation on pseudovarieties that is very close to what we need: the pseudovariety usually denoted $(\boldsymbol{V}^\rho * \boldsymbol{W}^\rho)^\rho$. Here, $(.)^\rho$ means that the multiplication tables are tranposed, and $*$ stands for the semidirect product of pseudovarieties (see below).

The exact description of $(\boldsymbol{V}^\rho * \boldsymbol{W}^\rho)^\rho$ is as follows [16]. It is the pseudovariety generated by semigroups $A^+/\equiv''$ where $\equiv''$ is a congruence obtained as follows.

(‡) For some $\boldsymbol{W}$-congruence $\equiv$ over A and some $\boldsymbol{V}$-congruence $\equiv'$ over $A \times A^+/\equiv$, the relation $\equiv$ relates u and v if
1. $u \equiv v$ and
2. for every $x \in A^*$, $u_\equiv^x \equiv' v_\equiv^x$ with $w_\equiv^x$ defined by $w_\equiv^x = a_1 \ldots a_{|w|}$ where $a_i = (w_i, [wx]_\equiv)$.

(In the above, special care has to be taken when wx is the empty string.)

Observe that (†) and (‡) deviate only little from each other, but the minor difference in the two has some consequences, which will be explained in what follows.

We need some more notation. We write $L'(\boldsymbol{V})$ for the class of all $L \subseteq A^+$ such that L is finite union of languages of the form aL' with $a \in A$ and L' recognized by a monoid homomorphism $A^+ \to S^1$ with $S \in \boldsymbol{V}$. Here, S^1 is S augmented by a neutral element if S contains no neutral element and else S. And we write $\Phi \circ' \Psi$ for the set of all boolean combinations of formulas from $(\Phi \circ \Psi) \cup \Psi$. With this notation, we get:

Theorem 13 (semidirect product substitution principle [17]). *Let Φ and Ψ be classes of LTL formulas and $\boldsymbol{V}$ and $\boldsymbol{W}$ be pseudovarieties of semigroups such that $L(\Phi) = L(\boldsymbol{V})$ and $L(\Psi) = L'(\boldsymbol{W})$. Then*

$$L(\Phi \circ' \Psi) = L(\boldsymbol{V} *^\rho \boldsymbol{W}) \ , \tag{9}$$

where $\boldsymbol{W} *^\rho \boldsymbol{V} = (\boldsymbol{V}^\rho * \boldsymbol{W}^\rho)^\rho$.

One simple application is given in the following example.

Example 14. We want to derive an effective characterization of TL[X, F]. We already know that this amounts to obtain an effective characterization of TL[F] ∘ TL[X]. Clearly, TL[F]∘'TL[X] is at least as expressive as TL[F]∘TL[X] and not more expressive than TL[X, F]. So, by the semidirect product substitution principle, we get

$$L(\mathsf{TL}[\mathsf{X},\mathsf{F}]) = L(\mathbf{L}' *^\rho \mathbf{K}) , \tag{10}$$

where $\mathbf{L}'$ is defined by (6) and $\mathbf{K}$ by (X).   □

It follows from results in semigroup theory [15] that the above pseudovariety is decidable, which means we have an effective characterization of TL[X, F].

A first effective characterization was originally obtained by Cohen, Perrin, and Pin [2] without using the semidirect product substitution principle. They showed

$$L(\mathsf{TL}[\mathsf{X},\mathsf{F}]) = L(\mathbf{LL}) \tag{11}$$

where $\mathbf{LL}$ is the pseudovariety of locally L-trivial semigroups defined by

$$r^\omega s(r^\omega t r^\omega s)^\omega = (r^\omega t r^\omega s)^\omega , \tag{12}$$

which, by Eilenberg's theorem, is the same as $\mathbf{L}' *^\rho \mathbf{K}$.

5.3   Semidirect Product

For the sake of completeness, we conclude this subsection with a formal definition of the semidirect product of two semigroups. Given semigroups S and T and a monoid homomorphism $h\colon T^1 \to \mathrm{End}(S)$ from T^1 to the monoid of endomorphisms of S, the *semidirect product* of S with T (with respect to h), denoted $S *_h T$, is the set $S \times T$ with multiplication defined by $(s_1, t_1)(s_2, t_2) = (s_1 s_2(t_1 h), t_1 t_2)$. The interested reader may want to check how this relates with the constructions in (†) and (‡).

Given pseudovarieties V and W of semigroups, their semidirect product is the pseudovariety of semigroups generated by all semigroups of the form $S *_h T$ where $S \in V$, $T \in W$, and $h\colon T^1 \to \mathrm{End}(S)$ is a monoid homomorphism; it is denoted by $V * W$. This product is associative whereas the semidirect product on individual semigroups is not.

6   Conclusion

Using the semidirect product substitution principle one can easily characterize many other fragments of LTL. For instance, one gets a parametrized version of Theorem 6:

$$L(\mathsf{TL}[\mathsf{F}_k]) = L(\mathbf{Sl} *^\rho \cdots *^\rho \mathbf{Sl}) , \tag{13}$$

where the product has k factors and $\mathbf{Sl}$ is the pseudovariety of semilattices defined by

$$st = ts \; , \qquad\qquad s^2 = s \; . \qquad\qquad (14)$$

The most interesting question is to classify an LTL property according to how deeply U, the only binary operator, needs to be nested to express the property. In other words, one is interested in an effective characterization of the classes

$$L(\mathsf{TL}[\mathsf{X}, \mathsf{F}, \mathsf{U}_0]) \subseteq L(\mathsf{TL}[\mathsf{X}, \mathsf{F}, \mathsf{U}_1]) \subseteq L(\mathsf{TL}[\mathsf{X}, \mathsf{F}, \mathsf{U}_2]) \subseteq \cdots \; , \qquad (15)$$

which is known as the *until hierarchy* of linear temporal logic. In Example 14, we have seen how level 0 can be characterized. In [17] it is shown that

$$L(\mathsf{TL}[\mathsf{X}, \mathsf{F}, \mathsf{U}_k]) = L(\mathbf{L'} *^\rho \mathbf{MD_1}^\rho *^\rho \cdots *^\rho \mathbf{MD_1}^\rho *^\rho \mathbf{K}) \qquad (16)$$

with k factors $\mathbf{MD_1}$ where $\mathbf{MD_1}$ is the pseudovariety generated by the three element monoid $\{1, a, b\}$ determined by $ax = a$ and $bx = b$ for all $x \in \{1, a, b\}$.

All the above characterizations are effective, that is, the respective pseudovarieties are decidable. The decidability proofs are quite involved; they rely, for instance, on results by Straubing [15] and Steinberg [14].

We have seen that decidability of fragments of LTL can be reduced to decidability of iterated semidirect products of pseudovarieties of semigroups. In general, it is not at all clear whether products of decidable pseudovarieties are decidable. For the products that are relevant to characterizing fragments of LTL the known decidability criteria—deep results from semigroup theory—yield positive results. These criteria are, however, quite involved. Especially, they don't allow to prove reasonable upper bounds on the complexity of the decision problems.

References

1. Jorge Almeida. *Finite Semigroups and Universal Algebra*, volume 3 of *Series in Algebra*. World Scientific, Singapore, 1995. 99
2. Joëlle Cohen, Dominique Perrin, and Jean-Eric Pin. On the expressive power of temporal logic. *J. Comput. System Sci.*, 46(3):271–294, 1993. 108
3. Samuel Eilenberg. *Automata, Languages, and Machines*, volume 59-B of *Pure and Applied Mathematics*. Academic Press, New York, 1976. 100
4. Allen E. Emerson. Temporal and modal logic. In Jan van Leeuwen, editor, *Handbook of Theoretical Computer Science*, volume B: Formal Methods and Semantics, pages 995–1072. Elsevier Science Publishers B. V., Amsterdam, 1990. 98
5. Kousha Etessami, Moshe Y. Vardi, and Thomas Wilke. First-order logic with two variables and unary temporal logic. In *Proceedings 12th Annual IEEE Symposium on Logic in Computer Science*, pages 228–235, Warsaw, Poland, 1997. 97
6. Dov M. Gabbay, Ian Hodkinson, and Mark Reynolds. *Temporal Logic: Mathematical Foundations and Computational Aspects*, vol. 1 of *Oxford Logic Guides*. Clarendon Press, Oxford, 1994. 96, 98

7. Dov M. Gabbay, Amir Pnueli, Saharon Shelah, and Jonathan Stavi. On the temporal analysis of fairness. In *Conference Record of the 12th ACM Symposium on Principles of Programming Languages*, pages 163–173, Las Vegas, Nev., 1980. 97

8. Johan Anthony Willem Kamp. *Tense Logic and the Theory of Linear Order*. PhD thesis, University of California, Los Angeles, Calif., 1968. 96, 97

9. Fred Kröger. LAR: A logic of algorithmic reasoning. *Acta Informatica*, 8(3), 1977. 96

10. Robert McNaughton and Seymour Papert. *Counter-Free Automata*. MIT Press, Cambridge, Mass., 1971. 97

11. Jean-Eric Pin. *Varieties of Formal Languages*. Plenum Press, New York, 1986. 99

12. Amir Pnueli. The temporal logic of programs. In *18th Annual Symposium on Foundations of Computer Science*, pages 46–57, Rhode Island, Providence, 1977. 96

13. Marcel P. Schützenberger. On finite monoids having only trivial subgroups. *Inform. and Computation*, 8:190–194, 1965. 97

14. Ben Steinberg. Semidirect products of categories and applications. *Journal of Pure and Applied Algebra*, 142:153–182, 1999. 109

15. Howard Straubing. Finite semigroup varieties of the form $V * D$. *J. Pure Appl. Algebra*, 36:53–94, 1985. 108, 109

16. Denis Thérien and Alex Weiss. Graph congruences and wreath products. *J. Pure Appl. Algebra*, 36:205–215, 1985. 107

17. Denis Thérien and Thomas Wilke. Temporal logic and semidirect products: An effective characterization of the until hierarchy. In *Proceedings of the 37th Annual Symposium on Foundations of Computer Science*, pages 256–263, Burlington, Vermont, 1996. 107, 109

Refined Search Tree Technique for DOMINATING SET on Planar Graphs

Jochen Alber[1] *, Hongbing Fan[2], Michael R. Fellows[2], Henning Fernau[1],
Rolf Niedermeier[1], Fran Rosamond[2], and Ulrike Stege[2]

[1] Universität Tübingen, Wilhelm-Schickard-Institut für Informatik
Sand 13, 72076 Tübingen, Fed. Rep. of Germany
`alber,fernau,niedermr@informatik.uni-tuebingen.de`
[2] Department of Computer Science, University of Victoria
Victoria B.C., Canada V8W 3P6
`hfan,fellows,fran,stege@csr.csc.uvic.ca`

Abstract. We establish refined search tree techniques for the parameterized DOMINATING SET problem on planar graphs. We derive a fixed parameter algorithm with running time $O(8^k n)$, where k is the size of the dominating set and n is the number of vertices in the graph. For our search tree, we firstly provide a set of reduction rules. Secondly, we prove an intricate branching theorem based on the Euler formula. In addition, we give an example graph showing that the bound of the branching theorem is optimal with respect to our reduction rules. Our final algorithm is very easy (to implement); its analysis, however, is involved.

Keywords. dominating set, planar graph, fixed parameter algorithm, search tree

1 Introduction

The parameterized DOMINATING SET problem, where we are given a graph $G = (V, E)$, a parameter k and ask for a set of vertices of size at most k that dominate all other vertices, is known to be $W[2]$-complete for general graphs [8]. The class $W[2]$ formalizes intractability from the point of view of parameterized complexity. It is well-known that the problem restricted to planar graphs is fixed parameter tractable. An algorithm running in time $O(11^k n)$ was claimed in [7,8]. The analysis of the algorithm, however, turned out to be flawed; hence, this paper seems to give the first completely correct analysis of a fixed parameter algorithm for DOMINATING SET on planar graphs with running time $O(c^k n)$ for *small* constant c that even improves the previously claimed constant considerably. We mention in passing that in companion work various approaches that yield algorithms of running time $O(c^{\sqrt{k}} n)$ for PLANAR DOMINATING SET and related problems were considered (see [1,2,3]).[1] Interestingly, very recently

---

* Supported by the Deutsche Forschungsgemeinschaft (DFG), research project PEAL (Parameterized complexity and Exact ALgorithms), NI 369/1-1.

[1] The huge worst case constant c that was derived there is rather of theoretical interest.

J. Sgall, A. Pultr, and P. Kolman (Eds.): MFCS 2001, LNCS 2136, pp. 111–123, 2001.

it was shown that up to the constant c this time bound is optimal unless a very unlikely collapse of a parameterized complexity hierarchy occurs [4,5].

Fixed parameter algorithms based on search trees. A method that has proven to yield easy and powerful fixed parameter algorithms is that of constructing a bounded search tree. Suppose we are given a graph class $\mathcal{G}$ that is closed under taking subgraphs and that guarantees a vertex of degree d for some constant d.[2] Such graph classes are, e.g., given by bounded degree graphs, or by graphs of bounded genus, and, hence, in particular, by planar graphs. More precisely, an easy computation shows that, e.g., the class $\mathcal{G}(S_g)$ of graphs that are embeddable on an orientable surface S_g of genus g guarantees a vertex of degree $d_g := \lceil 2(1 + \sqrt{3g + 1}) \rceil$ for $g > 0$, and $d_0 := 5$.

Consider the k-INDEPENDENT SET problem on $\mathcal{G}$, where, for given $G = (V, E) \in \mathcal{G}$, we seek for an independent set of size at least k. For a vertex u with degree at most d and neighbors $N(u) := \{u_1, \ldots, u_d\}$, we can choose one vertex $w \in N[u] := \{u, u_1, \ldots, u_d\}$ to be in an optimal independent set and continue the search on the graph G' where we deleted $N[w]$. This observation yields a simple $O((d+1)^k n)$ degree-branching search tree algorithm.

In the case of k-DOMINATING SET, the situation seems more intricate. Clearly, again, either u or one of its neighbors can be chosen to be in an optimal dominating set. However, removing u from the graph leaves all its neighbors being already dominated, but still also being suitable candidates for an optimal dominating set. This consideration leads us to formulate our search tree procedure in a more general setting, where there are two kinds of vertices in our graph.

ANNOTATED DOMINATING SET
Input: A *black* and *white* graph $G = (B \uplus W, E)$, and a positive integer k.
Parameter: k
Question: Is there a choice of at most k vertices $V' \subseteq V = B \uplus W$ such that, for every vertex $u \in B$, there is a vertex $u' \in N[u] \cap V'$? In other words, is there a set of k vertices (which may be either black or white) that dominates the set of all black vertices?

In each step of the search tree, we would like to branch according to a low degree black vertex. By our assumptions on the graph class, we can guarantee the existence of a vertex $u \in B \uplus W$ with $\deg(u) \leq d$. However, as long as *not all* vertices have degree bounded by d (as, e.g., the case for graphs of bounded genus g, where only *the existence* of a vertex of degree at most d_g is known), this vertex need not necessarily be black. These considerations show that a direct $O((d+1)^k n)$ search tree algorithm for DOMINATING SET seems out of reach for such graph classes.

Our results. In this paper, we present a fixed parameter algorithm for (ANNOTATED) DOMINATING SET on planar graphs with running time $O(8^k n)$. For that purpose, we provide a set of reduction rules and, then, use a search tree in which we are constantly simplifying the instance according to the reduction rules (see

---

[2] This means that, for each $G = (V, E) \in \mathcal{G}$, there exists a $u \in V$ with $\deg_G(u) \leq d$.

Subsection 3.1). The branching in the search tree will be done with respect to low degree vertices. The analysis of this algorithm will be carried out in a new branching theorem (see Subsection 3.2) which is based on the Euler formula for planar graphs. In addition, we give an example showing that the bound of the branching theorem is optimal (see Section 3.3). Finally, it is worth noting here that the algorithm we present is very simple and easy to implement.

Due to the lack of space, several proof details had to be omitted.

2   Preliminaries

We assume familiarity with basic notions and concepts in graph theory, see, e.g., [6]. For a graph $G = (V, E)$ and a vertex $u \in V$, we use $N(u)$ and $N[u]$, respectively, to denote the open and closed neighborhood of u, respectively. By $\deg_G(u) := |N_G(u)|$, we denote the *degree* of the vertex u in G. A *pendant* vertex is a vertex of degree one. For $V' \subseteq V$, the induced subgraph of V' is denoted by $G[V']$. In particular, we use the abbreviation $G - V' := G[V \setminus V']$. If V' is a singleton, then we omit brackets and simply write $G - v$ for a vertex v. In addition, we write $G - e$ or $G + e$ when we delete or add an edge e to G without changing the vertex set of G.

Let G be a connected planar graph, i.e., a connected graph that admits a crossing-free embedding in the plane. Such an embedding is called a *plane embedding*. A planar graph together with a plane embedding is called a *plane graph*. Note that a plane graph can be seen as a subset of the Euclidean plane $\mathbb{R}^2$. The set $\mathbb{R}^2 \setminus G$ is open; its regions are the *faces* of G. Let $\mathcal{F}$ be the set of faces of a plane graph. The *size of a face* $F \in \mathcal{F}$ is the number of vertices on the boundary of the face. A *triangular face* is a face of size three. If G is a plane graph and $V' \subseteq V$, then $G[V']$ and $G - V'$ can be always considered as plane graphs with an embedding inherited by the embedding of G.

3   The Algorithm and Its Analysis

Our algorithm is based on reduction rules (see Subsection 3.1) and an improved branching theorem (see Subsection 3.2). With respect to our set of reduction rules, we show optimality for the branching theorem (see Subsection 3.3).

3.1   Reduction Rules

We consider the following reduction rules for simplifying the ANNOTATED PLANAR DOMINATING SET problem. In developing the search tree, we will always assume that we are branching from a reduced instance (thus, we are constantly simplifying the instance according to the reduction rules).[3] When a vertex u is

---

[3] The idea of doing so-called *rekernelizations* (i.e., repeated application of reduction rules) while constructing the search tree was already exhibited in [9,10] in a somewhat different context.

placed in the dominating set D by a reduction rule, then the target size k for D is reduced to $k - 1$ and the neighbors of u are whitened.

(R1) Delete edges between white vertices.
(R2) Delete a pendant white vertex.
(R3) If there is a pendant black vertex w with neighbor u (either black or white), then delete w, place u in the dominating set, and lower k to $k - 1$.
(R4) If there is a white vertex u of degree 2, with two black neighbors u_1 and u_2 connected by an edge $\{u_1, u_2\}$, then delete u.
(R5) If there is a white vertex u of degree 2, with black neighbors u_1, u_3, and there is a black vertex u_2 and edges $\{u_1, u_2\}$ and $\{u_2, u_3\}$ in G, then delete u.
(R6) If there is a white vertex u of degree 2, with black neighbors u_1, u_3, and there is a white vertex u_2 and edges $\{u_1, u_2\}$ and $\{u_2, u_3\}$ in G, then delete u.
(R7) If there is a white vertex u of degree 3, with black neighbors u_1, u_2, u_3 for which the edges $\{u_1, u_2\}$ and $\{u_2, u_3\}$ are present in G (and possibly also $\{u_1, u_3\}$), then delete u.

Let us call a set of simplifying reduction rules of a certain problem *sound* if, whenever (G, k) is some problem instance and instance (G', k') is obtained from (G, k) by applying one of the reduction rules, then (G, k) has a solution iff (G', k') has a solution. A simple case analysis shows:

Lemma 1. *The reduction rules are sound.*   □

Suppose that G is a *reduced* graph, that is, none of the above reduction rules can be applied. By using the rules (R1), (R2), (R4) and (R7), we can show:

Lemma 2. *Let $G = (B \uplus W, E)$ be a plane black and white graph. If G is reduced, then the white vertices form an independent set and every triangular face of $G[B]$ is empty.*   □

3.2   A New Branching Theorem

Theorem 3. *If $G = (B \uplus W, E)$ is a planar black and white graph that is reduced, then there exists a black vertex $u \in B$ with $\deg_G(u) \leq 7$.*

The following technical lemma, based on an "Euler argument," will be needed. Note that if there is any counterexample to the theorem, then there is a connected counterexample.

Lemma 4. *Suppose $G = (B \uplus W, E)$ is a connected plane black and white graph with b black vertices, w white vertices, and e edges. Let the subgraph induced by the black vertices be denoted $H = G[B]$. Let c_H denote the number of components of H and let f_H denote the number of faces of H. Let*

$$z = \bigl(3(b + w) - 6\bigr) - e \tag{1}$$

measure the extent to which G fails to be a triangulation of the plane. If the criterion

$$3w - 4b - z + f_H - c_H < 7 \tag{2}$$

is satisfied, then there exists a black vertex $u \in B$ with $\deg_G(u) \leq 7$.

Proof. Let the (total) numbers of vertices, edges and faces of G be denoted v, e, f respectively. Let e_{bw} be the number of edges in G between black and white, and let e_{bb} denote the number of edges between black and black. With this notation, we have the following relationships.

$$v - e + f = 2 \qquad \text{(Euler formula for } G\text{)} \qquad (3)$$
$$v = b + w \qquad\qquad\qquad\qquad (4)$$
$$e = e_{bb} + e_{bw} \qquad\qquad\qquad\qquad (5)$$
$$b - e_{bb} + f_H = 1 + c_H \qquad \text{((extended) Euler formula for } H\text{)} \qquad (6)$$
$$2v - 4 - z = f \qquad \text{(by Eq. (1), (4), and (5))} \qquad (7)$$

If the lemma were false, then we would have, using (5),

$$8b \;\leq\; 2e_{bb} + e_{bw} \;=\; e_{bb} + e. \qquad (8)$$

We will assume this and derive a contradiction. The following inequality holds:

$$
\begin{aligned}
3 + c_H &= v + b - (e_{bb} + e) + f + f_H && \text{(by (3) and (6))}\\
&\leq v + b - 8b + f + f_H && \text{(by (8))}\\
&= 3v - 7b + f_H - 4 - z && \text{(by (7))}\\
&= 3w - 4b + f_H - 4 - z. && \text{(by (4))}
\end{aligned}
$$

This yields a contradiction to 2. □

Proving Theorem 3 by contradiction, it will be helpful to know that a corresponding graph has to be connected and has minimum degree 3.

Lemma 5. *If there is any counterexample to Theorem 3, then there is a connected counterexample where* $\deg_G(u) \geq 3$ *for all* $u \in W$.

Proof. Suppose G is a counterexample to the theorem. Then, G does not have any white vertices of degree 1, else reduction rule (R2) can be applied. Let G' be obtained from G by simultaneously replacing every white vertex u of degree 2 with neighbors x and y by an edge $\{x, y\}$. The neighbors x and y of u are necessarily black, else (R1) can be applied, and in each case the edge $\{x, y\}$ is not already present in G, else rule (R4) would apply. We argue that G' is reduced. If not, then the only possibility is that reduction rule (R7) applies to some white vertex u of degree 3 in G'. If rule (R7) did not apply to u in G, then one of the edges between the neighbors of u must have been created in our derivation of G' from G, i.e., one of these edges replaced a white vertex u' of degree 2. But this implies that reduction rule (R6) could be applied in G to u', contradicting that G is reduced. □

Before giving the proof of Theorem 3, we introduce the following notation:

Notation: Let $G = (B \uplus W, E)$ be a plane black and white graph and let $\mathcal{F}$ be the set of faces of $G[B]$. Then, for each $F \in \mathcal{F}$, we let

- w_F denote the number of white vertices embedded in F,

- z_F denote the number of edges that would have to be added in order to complete a triangulation of that part of the embedding of G contained in F,
- t_F denote the number of edges needed to triangulate F in $G[B]$ (that is, triangulating only between the black vertices on the boundary of F, and noting that the boundary of F may not be connected), and
- c_F denote the number of components of the boundary of F, minus 1.

Observe that the numbers z_F and t_F are indeed well-defined. This can be shown by using [6, Proposition 4.2.6].

Proof (of Theorem 3). We can assume that if there is a counterexample G then G is connected, but the black subgraph $H := G[B]$ might not be connected. Moreover, by Lemma 5 we may assume that $\deg_G(u) \geq 3$ for all $u \in W$. If c_H denotes the number of components of H, by induction on c_H, it is easy to see that $c_H - 1 = \sum_{F \in \mathcal{F}} c_F$. Also, if z is the number of edges needed to triangulate G, we clearly get $z = \sum_{F \in \mathcal{F}} z_F$. The criterion (2) established by Lemma 4 can be rephrased as

$$3 \sum_{F \in \mathcal{F}} w_F - \sum_{F \in \mathcal{F}} z_F - 4b + f_H - c_H < 7,$$

which is equivalent to

$$3 \sum_{F \in \mathcal{F}} (w_F + c_F/3 - z_F/3 + 1/3) - 4b - 2c_H < 6.$$

Now, assume that we can show the inequality

$$w_F + c_F/3 - z_F/3 + 1/3 \leq \alpha t_F + \beta \tag{9}$$

for some constants α and β and for every face F of the subgraph H. Call this our *linear bound* assumption. Then, criterion (2) will hold if

$$3 \sum_{F \in \mathcal{F}} (\alpha t_F + \beta) - 4b - 2c_H = \left(3\alpha \sum_{F \in \mathcal{F}} t_F \right) + \left(3\beta \sum_{F \in \mathcal{F}} 1 \right) - 4b - 2c_H < 6.$$

Noting that $\sum_{F \in \mathcal{F}} t_F$ is the number of edges needed to triangulate H, we have

$$\sum_{F \in \mathcal{F}} t_F = 3b - 6 - e_{bb}.$$

The number of faces of H is $\sum_{F \in \mathcal{F}} 1 = f_H = e_{bb} - b + 1 + c_H$, by Euler's formula (7). Together, these give us the following targeted criterion:

$$3\alpha(3b - 6 - e_{bb}) + 3\beta(e_{bb} - b + 1 + c_H) - 4b - 2c_H < 6.$$

Multiplying out and gathering terms, we need to establish (using the linear bound assumption), that

$$b(9\alpha - 3\beta - 4) + e_{bb}(3\beta - 3\alpha) + 3\beta(1 + c_H) - 18\alpha - 2c_H < 6.$$

This inequality is easily verified for $\alpha = \beta = 2/3$.

To complete the argument, we need to establish that the linear bound assumption (9) with $\alpha = \beta = 2/3$ holds for faces of reduced graphs, i.e., that

$$w_F + c_F/3 - z_F/3 \le 2t_F/3 + 1/3. \tag{10}$$

But this is a consequence of the following Propositions 6 and 8.   □

Proposition 6. *Let $G = (B \uplus W, E)$ be a reduced plane black and white graph and let F be a face of $G[B]$. Then, using the notation above, we have $w_F + c_F \le z_F + 1$.*

Proof. Consider the "face-graph" $G_F := G[B_F \cup W_F]$, where B_F is the set of black vertices forming the boundary of F and W_F is the set of white vertices inside F. Note that G_F may consist of several "black components," connected among themselves through white vertices. Contracting each of these black components into one (black) vertex, we obtain the *bipartite* graph G'_F. Note that both the black and also the white vertices form independent sets in G'_F. Clearly, G'_F is still planar. Since G'_F is a bipartite planar graph, preserving planarity it is easy to show that we can connect the white vertices among themselves by a tree of $w_F - 1$ white white edges and that we can connect the black vertices among themselves by a tree of c_F black black edges. Clearly, this implies that we can also add at least $c_F + w_F - 1$ new edges to G_F without destroying planarity. Hence, we need at least $c_F + w_F - 1$ additional edges to triangulate the interior of F in the graph G.   □

Property 7. If F_1 and F_2 are two faces of $G[B]$ with common boundary edge e, then $t_{F_1} + t_{F_2} + 1$ equals t_F, where we now consider $(G - e)[B]$, and F is the face which results from merging F_1 and F_2 when deleting e.

Proposition 8. *Suppose $G = (B \uplus W, E)$ is a reduced plane black and white graph, with $\deg(u) \ge 3$ for all $u \in W$. Let F be a face of $G[B]$. Then, using the notation above, $w_F \le t_F$.*

Proof. Consider a reduced black and white graph $G = (B \uplus W, E)$ with $\deg(u) \ge 3$ for all $u \in W$. If there is some $u \in W$ with $\deg(u) > 4$, then delete arbitrarily all edges incident with u but four of them. While preserving the black induced subgraph, the resulting graph is still reduced, since no rules apply to white degree-4-vertices. Therefore, we can assume from now on without loss of generality that all white vertices of G have maximum degree of four.

We will now show the claim by induction on the number $\#w^4$ of white vertices of degree four. The hardest part is the induction base, which is deferred to the subsequent Lemma 9. Assume that the claim was shown for each graph with $\#w^4 \le \ell$ and assume now that G has $\ell + 1$ white degree-4-vertices. Choose some arbitrary $u \in W$ with $\deg(u) = 4$. Let $\{b_1, \ldots, b_4\}$ be the clockwisely ordered neighbors of u. Due to planarity, we may assume further that $\{b_1, b_3\} \notin E$ without loss of generality. Consider now $G' = (G - u) + \{b_1, b_3\}$. We prove below that G' (or $G'' = (G - u) + \{b_2, b_4\}$ in one special case) is reduced. This means that the induction hypothesis applies to G'. Hence, $w_F \leq t_F$ for all faces in $G'[B]$. Observe that G' contains all the faces of G except from the face F of G which contains u; F might be replaced by two faces F_1 and F_2 with common boundary edge $\{b_1, b_3\}$. In this case, $w_{F_1} \leq t_{F_1}$, $w_{F_2} \leq t_{F_2}$, $w_{F_1} + w_{F_2} + 1 = w_F$ and, by Property 7, $t_{F_1} + t_{F_2} + 1 = t_F$. Hence, $w_F \leq t_F$ by induction. In the case where face F still exists in G' it is trivial to see that $w_F \leq t_F$.

To complete the proof, we argue why G' has to be reduced. Obviously, this is clear if $\forall b_i \forall v \in N(b_i) \deg(v) = 4$, since no reduction rules apply to degree-4-vertices. We now discuss the case that u has degree-3-vertices as neighbors.

1. If a degree-3-vertex is neighbor of some b_i, but not of b_j, $j \neq i$, then no reduction rule is triggered when constructing G'.
2. Consider the case that a degree-3-vertex is neighbor of two b_i, b_j, $i \neq j$. We can assume that $\{i, j\} \neq \{1, 3\}$, since otherwise $\{b_2, b_4\} \notin E$ and we could consider $G'' = (G - u) + \{b_2, b_4\}$ instead of G' with a argument similar to the case $\{i, j\} = \{1, 3\}$. If $\{i, j\} = \{1, 3\}$, then G' is clearly reduced. If $\{i, j\} = \{1, 2\}$ (or, more generally, $|\{i, j\} \cap \{1, 3\}| = 1$), then no reduction rules are triggered when passing from G to G'.
3. If a degree-3-vertex is neighbor of three b_i, b_j, b_k, then a reasoning similar to the one in the previous point applies.

This concludes the proof of the proposition.   □

The following lemma serves as the induction base in the proof of Proposition 8.

Lemma 9. *Suppose $G = (B \uplus W, E)$ is a reduced plane black and white graph, with $\deg(u) = 3$ for all $u \in W$. Let F be a face of $G[B]$. Then, using the notation above, $w_F \leq t_F$.*

Proof. (Sketch) Let us consider a fixed planar embedding of the graph G, and consider a face F of the black induced subgraph $G[B]$. Let $W_F \subseteq W$ be the set of white vertices in the interior of F, and let $B_F \subseteq B$ denote the black vertices on the boundary of F. We want to find at least $|W_F|$ many black black edges that can be added to $G[B]$ inside F. For that purpose, define the set

$$E^{\text{poss}} := \{e = \{b_1, b_2\} \mid b_1, b_2 \in B_F \wedge e \notin E(G[B])\}$$

of non-existing black black edges.

For a subset $W' \subseteq W_F$ we construct a bipartite graph $H(W') := (W' \cup T(W'), E(W'))$ as follows. In $H(W')$, the first bipartition set is formed by the vertices W' and the second one is given by the set

$$T(W') := \{e = \{b_1, b_2\} \in E^{\text{poss}} \mid \exists u \in W' : e \subset N_G(u)\}.$$

The edges in $H(W')$ are then given by

$$E(W') := \{\{u, e\} \mid u \in W', e \in T(W'), e \subset N_G(u)\}.$$

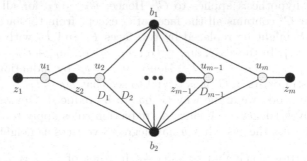

Fig. 1. Illustration of a diamond D generated by a pair vertices $\{b_1, b_2\} \in T(W_F)$

In this way, the set $T(W')$ gives us vertices in $H(W')$ that correspond to pairs $e = \{b_1, b_2\}$ of black vertices in B_F between which we still can draw an edge in $G[B]$. Note that the edge e can even be drawn in the interior of F, since b_1 and b_2 are connected by a white vertex in $W' \subseteq W_F$. This means that

$$|T(W_F)| \leq t_F. \tag{11}$$

Due to reduction rule (R7), for each $u \in W_F$, the neighbors $N(u) \subseteq B_F$ are connected by at most one edge in $G[B]$. By construction of $H(W_F)$, we find:

$$\deg_{H(W_F)}(u) \geq 2 \quad \forall u \in W_F. \tag{12}$$

The degree $\deg_{H(W_F)}(e)$ for an element $e = \{b_1, b_2\} \in T(W_F)$ tells us how many white vertices share the pair $\{b_1, b_2\}$ as common neighbors. We do case analysis according to this degree.

Case 1: Suppose $\deg_{H(W_F)}(e) \leq 2$ for all $e \in T(W_F)$, then $H(W_F)$ is a bipartite graph, in which the first bipartition set has degree at least two (see Eq. (12)) and the second bipartition set has degree at most two. In this way, the second set cannot be smaller, which yields

$$w_F = |W_F| \leq |T(W_F)| \overset{(11)}{\leq} t_F.$$

Case 2: There exist elements $e = \{b_1, b_2\}$ in $T(W_F)$ which are shared as common neighbors by more than 2 white vertices (i.e., $\deg_{H(W_F)}(e) = m > 2$). Suppose we have $u_1, \ldots, u_m \in W_F$ with $N_G(u_i) = \{b_1, b_2, z_i\}$ (i.e., $\{u_i, e\} \in E(W_F)$). We may assume that the vertices are ordered such that the closed region D bounded by $\{b_1, u_1, b_2, u_m\}$ contains all other vertices $u_2, \ldots, u_{m-1}$ (see Figure 1).

We call D the *diamond* generated by $\{b_1, b_2\}$. Note that D consists of $m - 1$ regions, which we call *blocks* in the following; the block D_i is bounded by $\{b_1, u_i, b_2, u_{i+1}\}$ $(i = 1, \ldots, m - 1)$. Let $W_i \subseteq W_F$, and $B_i \subseteq B_F$, respectively, denote the white and black, respectively, vertices that lie in D_i. For the boundary vertices $\{b_1, b_2, u_1, \ldots, u_m\}$ we use the following convention: b_1, b_2 are added to all blocks, i.e., $b_1, b_2 \in B_i$ for all i; and u_i is added to the region where its third neighbor z_i lies in. A block is called *empty* if $B_i = \{b_1, b_2\}$ and, hence, $W_i = \emptyset$. Moreover, let $W_D := \bigcup_{i=1}^{m-1} W_i$ and $B_D := \bigcup_{i=1}^{m-1} B_i$.

We only consider diamonds, where z_1 and z_m are not contained in D (see Figure 1). The other cases can be treated with similar arguments.

Note that each block of a diamond D may contain further diamonds, the blocks of which may contain further diamonds, and so on. Since no diamonds overlap, the topological inclusion forms a natural ordering on the set of diamonds and their blocks.

We can show the following claim by induction on the diamond structure:

Claim: For each diamond D generated by $\{b_1, b_2\}$, we can add t_D (where $t_D \geq |W_D|$) many black black edges to $G[B]$ other than $\{b_1, b_2\}$. All of these additional edges can be drawn inside D so that $\{b_1, b_2\}$ still can be drawn.

Using this claim, we can finish the proof of the induction base of the proposition: Consider all diamonds $D^1, \ldots, D^r$ which are not contained in any further diamond. Suppose D^i has boundary $\{b_1^i, u_1^i, b_2^i, u_{m_i}^i\}$ with $b_1^i, b_2^i \in B_F$ and $u_1^i, u_{m_i}^i \in W_F$. Let

$$W_F' := W_F \setminus \left(\bigcup_{i=1}^{r} W_{D^i} \right).$$

According to the claim we already found $\sum_{i=1}^{r} t_{D^i}$ many black black edges in E^{poss} inside the diamonds D^i. Observe that each pair $e^i = \{b_1^i, b_2^i\}$ is only shared as common neighbors by at most two white vertices (namely, u_1^i and $u_{m_i}^i$) in (sic!) W_F'. Hence, the bipartite graph $H(W_F')$ again has the property that

- $\deg_{H(W_F')}(e) \leq 2$ for all $e \in T(W_F')$ and still

- $\deg_{H(W_F')}(u) \geq 2$ for all $u \in W_F'$.[4]

Similar to "Case 1" this proves that—additionally—we find t' (with $t' \geq |W_F'|$) many edges in E^{poss}. Hence,

$$w_F = |W_F| = |W_F'| + \left| \bigcup_{i=1}^{r} W_{D^i} \right| \leq t' + \left(\sum_{i=1}^{r} t_{D^i} \right) \leq t_F. \qquad \square$$

Using Theorem 3 for the construction of a search tree as elaborated in Section 1, we conclude:

Theorem 10. (ANNOTATED) DOMINATING SET *on planar graphs can be solved in time* $O(8^k n)$. $\qquad \square$

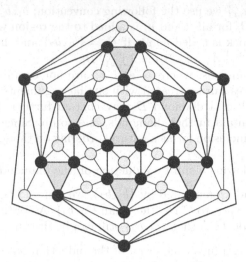

Fig. 2. A reduced graph with all black vertices having degree 7, thus showing the optimality of the bound derived in our branching theorem

3.3   Optimality of the Branching Theorem

We conclude this section by the observation that with respect to the set of reduction rules introduced above the upper bound in our branching theorem is optimal. More precisely, there exists a plane reduced black and white graph with the property that all black vertices have degree at least 7. Such a graph is shown in Figure 2. Moreover, this example can be generalized towards an infinite set of plane graphs with the property that all black vertices have degree at least 7. The given example is the smallest of all graphs in this class. It is an interesting and challenging task to ask for further reduction rules that would yield a provably better constant in the branching theorem. For example, one might think of the following generalization of reduction rule (R6):

(R6') If there are white vertices $u_1, u_2 \in W$ with $N_G(u_1) \subseteq N_G(u_2)$, then delete u_1.

However, the graph in Figure 2 is reduced even with respect to rule (R6'). We leave it as an open question to come up with further reduction rules such that the graph of Figure 2 is no longer reduced.

4   Conclusion and Open Questions

In this paper, we gave the first search tree algorithm proven to be correct for the DOMINATING SET problem on planar graphs. It improves on the original,

---

[4] Note that according to the claim the edges $\{b_1^i, b_2^i\}$ still can be used.

flawed theorem stating an exponential term 11^k, which is now lowered to 8^k. Unfortunately, the proof of correctness has become considerably more involved and fairly technical.

Our work suggests several directions for future research:

- Can we improve the branching theorem by adding further, more involved reduction rules?
- Finding a so-called problem kernel (see [8] for details) of polynomial size $p(k)$ for DOMINATING SET on planar graphs in time $T_K(n,k)$ would improve the running time to $O(8^k + T_K(n,k))$ using the interleaving technique analyzed in [10]. Currently, we even hope for a linear size problem kernel for DOMINATING SET on planar graphs.
- Since our results for the search tree itself are based on the Euler formula, a generalization to the class of graphs $\mathcal{G}(S_g)$ (allowing a crossing-free embedding on an orientable surface S_g of genus g) seems likely.

Acknowledgment. We thank Klaus Reinhardt for discussions on the topic of this work and for pointing to an error in an earlier version of the paper.

References

1. J. Alber, H. L. Bodlaender, H. Fernau, and R. Niedermeier. Fixed parameter algorithms for planar dominating set and related problems. In *7th Scandinavian Workshop on Algorithm Theory SWAT*, volume 1851 of *LNCS*, pages 97–110, Springer-Verlag, 2000. A long version has been accepted for publication in *Algorithmica*. 111
2. J. Alber, H. Fernau, and R. Niedermeier. Graph separators: a parameterized view. Technical Report WSI–2001–8, Universität Tübingen (Fed. Rep. of Germany), Wilhelm-Schickard-Institut für Informatik, 2001. Extended abstract accepted at *COCOON 2001*, to appear in *LNCS*, Springer-Verlag, August 2001. 111
3. J. Alber, H. Fernau, and R. Niedermeier. Parameterized complexity: exponential speedup for planar graph problems. Technical Report TR01–023, ECCC Reports, Trier (Fed. Rep. of Germany), March 2001. Extended abstract accepted at *ICALP 2001*, to appear in *LNCS*, Springer-Verlag, July 2001. 111
4. L. Cai, M. Fellows, D. Juedes, and F. Rosamond. Efficient polynomial-time approximation schemes for problems on planar graph structures: upper and lower bounds. Manuscript, May 2001. 112
5. L. Cai and D. Juedes. Subexponential parameterized algorithms collapse the W-hierarchy. Extended abstract accepted at *ICALP 2001*, to appear in *LNCS*, Springer-Verlag, July 2001. 112
6. R. Diestel. Graph Theory. Springer-Verlag, 1997. 113, 116
7. R. G. Downey and M. R. Fellows. Parameterized computational feasibility. In *Feasible Mathematics II*, pages 219–244. Birkhäuser, 1995. 111
8. R. G. Downey and M. R. Fellows. *Parameterized Complexity*. Springer-Verlag, 1999. 111, 122
9. R. G. Downey, M. R. Fellows, and U. Stege. Parameterized complexity: A framework for systematically confronting computational intractability. *DIMACS Series in Discrete Mathematics and Theoretical Computer Science*, 49:49–99, 1999. 113

10. R. Niedermeier and P. Rossmanith. A general method to speed up fixed-parameter-tractable algorithms. *Information Processing Letters*, 73:125–129, 2000. 113, 122

The Computational Power of a Family of Decision Forests

Kazuyuki Amano, Tsukuru Hirosawa, Yusuke Watanabe, and Akira Maruoka

Graduate School of Information Sciences, Tohoku University
{ama|hiro|yusuke|maruoka}@maruoka.ecei.tohoku.ac.jp

Abstract. In this paper, we consider the help bit problem in the decision tree model proposed by Nisan, Rudich and Saks (FOCS '94). When computing k instances of a Boolean function f, Beigel and Hirst (STOC '98) showed that $\lfloor \log_2 k \rfloor + 1$ help bits are necessary to reduce the complexity of f, for any function f, and exhibit the functions for which $\lfloor \log_2 k \rfloor + 1$ help bits reduce the complexity. Their functions must satisfy the condition that their complexity is greater than or equal to $k - 1$. In this paper, we show new functions satisfying the above conditions whose complexity are only $2\sqrt{k}$. We also investigate the help bit problem when we are only allowed to use decision trees of depth 1. Moreover, we exhibit the close relationship between the help bit problem and the complexity for circuits with a majority gate at the top.

1 Introduction and Definitions

Pick your favorite computational model to compute Boolean function. Let f be a Boolean function and let c be the complexity of f in the model. The help bit problem and the cover problem, which we will mainly deal with in this paper, are informally as follows.

Help Bit Problem Suppose you wish to compute f on two inputs x and y, and are allowed one "help-bit", i.e., an arbitrary function of the two inputs. Is it possible to choose this help-bit function so that, given the help-bit, $f(x)$ and $f(y)$ can each be computed with a computational complexity of less than c, and if so, by how much? How about computing f on k inputs with l help-bits?

Cover Problem Suppose you wish to compute f on two inputs x and y giving two pairs of answers (a_x^1, a_y^1) and (a_x^2, a_y^2) such that at least one pair (a_x^i, a_y^i) computes $(f(x), f(y))$ for $i \in \{1, 2\}$. Does a complexity less than c ever suffice to compute an answer for an input? What if you wish to evaluate f on k inputs and you are allowed to give m answer tuples, generalizing the number of inputs and that of answer tuples?

The notion of a help bit was first introduced by Cai[Cai89] in the context of constant depth circuits. Amir et al.[ABG00] studied general circuit models. Nisan et al.[NRS94] and Beigel and Hirst[BH98] investigated these two problems for the decision tree model. In this paper, with the same spirit, we consider these problems for the decision tree model of computation. There are formal definitions for the cover problem, following Beigel and Hirst[BH98].

J. Sgall, A. Pultr, and P. Kolman (Eds.): MFCS 2001, LNCS 2133, pp. 123–134, 2001.
© Springer-Verlag Berlin Heidelberg 2001

Definition 1. *A* decision tree *over a variable set X is a triple (T, p, a) where T is a rooted ordered binary tree, p is a mapping from the internal nodes of T to the set X, and a is a mapping from the leaves of T to $\{0, 1\}$. For a node v of T, the variable $p(v)$ which is assigned to v is said to be the* label *of v.*

A decision tree T over X computes a Boolean function f as follows: If α is any assignment in $\{0, 1\}^X$, the computation of T on α is the unique path $v_0, v_1, \ldots, v_s$ from the root v_0 of T to some leaf v_s. Inductively, v_{i+1} for $i \geq 0$ is defined to be the $\alpha(p(v_i))$th child of v_i. The output of the computation is given by $a(v_s)$, which is denoted by $T(\alpha)$.

The depth *of a decision tree T is the length of the longest path in T. DEPTH(d) is the class of Boolean functions computable by decision trees having at most depth d.*

Definition 2. *A* decision forest *over a variable set X is an ordered tuple of decision trees over X. The forest $F = (T_1, \ldots, T_k)$ computes the function*

$$F(\alpha_1, \ldots, \alpha_k) = (T_1(\alpha), \ldots, T_k(\alpha)).$$

The depth *of the forest $(T_1, \ldots, T_k)$ is the maximum of the depths of $T_1, \ldots, T_k$.*

Definition 3. *A* family *of decision forests over a variable set X is a set of decision forests over X. The* depth *of a family of decision forests $\{F_1, \ldots, F_k\}$ is the maximum of the depths of $F_1, \ldots, F_k$. The* size *of a family of forests $\mathcal{F}$ is the number of forests in $\mathcal{F}$. The* width *of a family of forests $\mathcal{F}$ is the number of trees in each forest of $\mathcal{F}$. A family $\mathcal{F}$ of forests* covers *a function g if*

$$\forall \alpha \exists F \in \mathcal{F} \quad F(\alpha) = g(\alpha).$$

When $\mathcal{F}$ covers g, we call $\mathcal{F}$ a cover *for g.*

Definition 4. *Let X be the union of mutually disjoint sets $X_1, \ldots, X_k$. For $1 \leq j \leq k$, let f_j be a function over X_j. Let*

$$\mathcal{F} = \{(T_{11}, \ldots, T_{1k}), \ldots, (T_{m1}, \ldots, T_{mk})\}$$

be a cover for $(f_1, \ldots, f_k)$. The cover $\mathcal{F}$ is called pure *if, for any i, j, the nodes of the tree T_{ij} are labeled only with a variable in X_j,* impure *otherwise.*

In this paper, we are mainly interested in covers for a tuple consisting of k isomorphic copies of the same function over the disjoint sets of variables.

Definition 5. *Let f be a Boolean function on n variables. We define*

$$f^{[k]}(\overrightarrow{x_1}, \ldots, \overrightarrow{x_k}) = (f(\overrightarrow{x_1}), \ldots, f(\overrightarrow{x_k})).$$

Let $PCover_d(m, k)$ be the class of Boolean functions f such that $f^{[k]}$ has a pure cover of size m, width k and depth d. Let $Cover_d(m, k)$ be the class of Boolean functions f such that $f^{[k]}$ has a cover of size m, width k and depth d.

By the definition, $\text{PCover}_d(m, k) \subseteq \text{Cover}_d(m, k)$ for any d, m and k. One may think that an impure family of forests cannot do better than a pure family, i.e., $\text{PCover}_d(m, k) = \text{Cover}_d(m, k)$. However this question is open[BH98]. As Beigel and Hirst have pointed out[BH98], we think that this is the most important open problem concerning the cover problem in the decision tree model. The help bit problem and the cover problem are strongly related each other because it is known that a Boolean function f is computable with l help bits iff f has a cover of size 2^l[BH98].

A brute-force construction obviously shows that $\text{PCover}_d(2^k, k)$ contains all the Boolean functions for every d. So it is natural to ask when a nontrivial cover exists. More precisely, for what pair of values m and k does $\text{DEPTH}(d) \subsetneq \text{PCover}_d(m, k)$ or $\text{DEPTH}(d) \subsetneq \text{Cover}_d(m, k)$ hold? Concerning these problems, Beigel and Hirst[BH98] proved the following strong statements.

$$\forall d \; \forall k \quad \text{DEPTH}(d) = \text{PCover}_d(k, k) = \text{Cover}_d(k, k),$$
$$\forall k \geq 2 \; \forall d \geq k - 1 \quad \text{DEPTH}(d) \subsetneq \text{PCover}_d(k + 1, k).$$

The first claim says that $\lfloor \log_2 k \rfloor$ help bits do not help to evaluate a function on k inputs. The second claim says that the first claim is tight in the sense that increasing the size k of a family of forests properly enlarges the set of functions covered, provided that the complexity of the functions is greater that or equal to $k - 1$. In Section 2.2 of the paper, we extend the second result to

$$\forall k \geq 2 \; \forall d \geq 2\sqrt{k} \quad \text{DEPTH}(d) \subsetneq \text{PCover}_d(k + 1, k).$$

In this paper, we also consider the computational power of a family of forests of depth 1. Let $\tau_d^{Pure}(k)$ (respectively, $\tau_d(k)$) be the minimum m such that $\text{DEPTH}(d) \subsetneq \text{PCover}_d(m, k)$ (respectively, $\text{DEPTH}(d) \subsetneq \text{Cover}_d(m, k)$) holds. The values of $\tau_d^{Pure}(k)$ or $\tau_d(k)$ were known only for an elemental case such as $\tau_1^{Pure}(3) = 5$[BH98]. In Section 2.1 of the paper, we determine the asymptotic values of $\tau_1^{Pure}(k)$ and of $\tau_1(k)$ and the exact values of $\tau_1^{Pure}(k)$ for $k \leq 6$.

Finally, in Section 3, we exhibit the close relationship between the cover problem and the complexity for circuits with a majority gate at the top.

2   Cover Problem

In this section, we discuss the computational power of a family of depth-d forests for a relatively small d.

Before going into detail, we show an example of covers. The following example was first suggested by Blum which appeared in [NRS94]. For more examples of covers, see [BH98].

Example 6. The majority function on 3 variables, denoted $\text{MAJ}_3(x, y, z)$, is defined as

$$\text{MAJ}_3(x, y, z) = \begin{cases} 1 & \text{if } x + y + z \geq 2, \\ 0 & \text{otherwise.} \end{cases}$$

Let $X_1 = \{x_1, y_1, z_1\}$, $X_2 = \{x_2, y_2, z_2\}$ and $X = X_1 \cup X_2$. Then the family of forests $\{(x_1, x_2), (y_1, y_2), (z_1, z_2)\}$ covers $\text{MAJ}_3^{[2]}$. This can be easily seen because at most one of the variables x_i, y_i and z_i can be unequal to $\text{MAJ}_3(x_i, y_i, z_i)$. In what follows, when discussing pure covers we will drop the subscript that distinguishes variables from different instances. Accordingly, we represent the cover above as $\{(x, x), (y, y), (z, z)\}$.

2.1   Depth 1

Recall that $\tau_d^{Pure}(k)$ (respectively, $\tau_d(k)$) denotes the minimum m such that $\text{DEPTH}(d) \subsetneq \text{PCover}_d(m, k)$ (respectively, $\text{DEPTH}(d) \subsetneq \text{Cover}_d(m, k)$) holds. One of the most fundamental problems on covers is to determine the values of $\tau_d^{Pure}(k)$ and $\tau_d(k)$ for various values of d and k. These questions seem to be difficult even if we restrict ourselves to $d = 1$. All the results concerning the problem obtained so far are $\tau_1^{Pure}(1) = 2$, $\tau_1^{Pure}(2) = 3$ and $\tau_1^{Pure}(3) = 5$. (See the technical report version of [BH98]). In this subsection, we determine the asymptotic values of $\tau_1^{Pure}(k)$ and $\tau_1(k)$ and the exact values of $\tau_1^{Pure}(k)$ for $k \leq 6$.

For a Boolean function f and for an integer $d \geq 0$, the *agreement probability* between f and $\text{DEPTH}(d)$, denoted $\rho(f, d)$, is defined to be the maximum real number p such that there is a distribution T on $\text{DEPTH}(d)$ that satisfies $\forall \alpha \, \text{Pr}_{T \in \mathcal{T}}[f(\alpha) = T(\alpha)] \geq p$. Nisan et al.[NRS94] obtained the following theorem which gives a relationship between $\rho(d)$ and $\tau_d(k)$.

Theorem 7 ([NRS94]). *For any $d \geq 1$,*

$$\left(\frac{1}{\max_{f \in DEPTH(d)} \rho(f, d)} \right)^k \leq \tau_d(k) \leq \tau_d^{Pure}(k).$$

By virtue of this theorem if we can prove an upper bound on the agreement probability then we get the corresponding lower bound on $\tau_d(k)$.

Theorem 8. *For any $f \notin DEPTH(1)$, $\rho(f, 1) \leq 2/3$.*

Proof. Let f be a function that cannot be computed by a decision tree of depth 1. Let n be the number of input variables of f. Since f is not constant, there are two strings $a \in \{0, 1\}^{i-1}$ and $b \in \{0, 1\}^{n-i}$ such that $f(a0b) \neq f(a1b)$ holds. Assume, in contradiction, that $\rho(f, 1) > 2/3$ holds, i.e., there exists a distribution T on $\text{DEPTH}(1)$ such that, for any $\alpha \in \{0, 1\}^n$, $\text{Pr}_{T \in \mathcal{T}}[T(\alpha) = f(\alpha)] > 2/3$ holds. Without loss of generality we can assume $f(a0b) = 0$ and $f(a1b) = 1$. Then $\text{Pr}_{T \in \mathcal{T}}[T(a0b) = 0] > 2/3$ and $\text{Pr}_{T \in \mathcal{T}}[T(a1b) = 1] > 2/3$ holds. It is obvious that x_i is the only tree of depth 1 such that the output of it is changed from 0 to 1 when we change the input from $a0b$ to $a1b$. It is easily seen that to satisfy the above formulae, the probability of the tree x_i in T must greater than $1/3$. But since the f is not the function x_i, there exists an input β on which f and x_i output different values. Hence $\text{Pr}_{T \in \mathcal{T}}[T(\beta) = f(\beta)] < 1 - 1/3 = 2/3$. This contradicts the assumption and completes the proof. $\square$

Corollary 9. *For any $k \geq 1$, we have $1.5^k \leq \tau_1(k) \leq \tau_1^{Pure}(k)$.*   □

For a pure family of forests, we can obtain a slightly stronger statement.

Corollary 10. *For any $k \geq 1$, we have $\lceil 1.5\tau_1^{Pure}(k-1) \rceil \leq \tau_1^{Pure}(k)$.*   □

For an upper bound on $\tau_1(k)$, we can show the following.

Theorem 11. *For sufficiently large k,*

$$\tau_1(k) \leq \tau_1^{Pure}(k) \leq (1+k)1.5^k.$$

Proof (Sketch). First we observe that $\rho(x \vee y, 1) = 2/3$ because at least 2 of the decision trees 1, x, y are equal to $x \vee y$ for every assignment. Thus as in the construction of a cover for the majority function described in Section 8.2 in [BH98], we can obtain a cover of depth 1 and size $(1+k)(1/\rho(x \vee y, 1))^k$ for $(x \vee y)^{[k]}$ by using Lovasz's theorem on fractional covers[Lov75].   □

In view of Corollary 9 and Theorem 11, it is natural to expect that there is a constant $c(d)$ such that $\lim_{k \to \infty} (\tau_d(k))^{1/k} \to c(d)$ for $d \geq 1$. We succeeded to prove $c(1) = 1.5$. In the next subsection, we will prove that $\rho(x_1 \vee \bar{x}_2 x_3, 2) \geq 4/5$ which implies $c(2) \leq 1.25$. We conjecture that $c(2) = 1.25$, but the exact value of $c(d)$ for $d \geq 2$ remain to be shown.

In the rest of this subsection, we exhibit that the lower bound stated in Corollary 10 is optimal for $k \leq 6$. (The optimum for $k \leq 3$ is shown implicitly in [BH98].) The following table gives exact values of $\tau_1^{Pure}(k)$ for a small k:

$k =$	1	2	3	4	5	6	7	8
$\tau_1^{Pure}(k) =$	2	3	5	8	12	18	$27 \sim 31$	$41 \sim 54$

Theorem 12. *For any $k \leq 6$, $\tau_1^{Pure}(k) = \lceil 1.5\tau_1^{Pure}(k-1) \rceil$.*

Proof. To prove the theorem, it is sufficient to construct pure families of depth 1 with sizes given in the above table that cover a function that cannot be computed by any decision tree of depth 1. We choose the function $x \vee y$ as such a function. Figure 1 gives the pure covers for $(x \vee y)^{[k]}$ where $k = 4, 5$ and 6. We found them after a large number of trials and errors. Unfortunately, we could not generalize our construction to generate an optimal cover for larger values of k.   □

Before closing this subsection, we give another interesting construction of a cover for the function $(x \vee y)^{[k]}$ which implies the upper bounds on $\tau_1^{Pure}(7)$ and $\tau_1^{Pure}(8)$ in the above table.

Let T be a decision tree and $\mathcal{S} = \{(T_{11}, \ldots, T_{1k}), \ldots, (T_{m1}, \ldots, T_{mk})\}$ be a family of forests. Let

$$T \circ \mathcal{S} = \{(T, T_{11}, \ldots, T_{1k}), \ldots, (T, T_{m1}, \ldots, T_{mk})\},$$
$$\mathcal{S}|_T = \{(T, T_{12}, \ldots, T_{1k}), \ldots, (T, T_{m2}, \ldots, T_{mk})\}.$$

$$\left\{\begin{array}{l}(\; x, x, 1, 1) \\ (\; x, y, x, y) \\ (\; x, 1, 1, x) \\ (\; y, x, y, x) \\ (\; y, 1, y, 1) \\ (\; 1, 1, x, x) \\ (\; 1, y, 1, y) \\ (\; 1, x, x, 1) \end{array}\right\} \quad \left\{\begin{array}{l}(\; x, x, y, x, y \;), \\ (\; x, x, 1, 1, x \;), \\ (\; x, y, 1, y, 1 \;), \\ (\; x, 1, x, x, 1 \;), \\ (\; y, x, x, 1, 1 \;), \\ (\; y, y, x, y, x \;), \\ (\; y, 1, 1, x, x \;), \\ (\; y, 1, y, 1, y \;), \\ (\; 1, x, y, y, 1 \;), \\ (\; 1, y, y, 1, x \;), \\ (\; 1, 1, x, y, y \;), \\ (\; 1, y, 1, x, y \;) \end{array}\right\} \quad \left\{\begin{array}{l}(\; x, x, x, y, x, y), \; (\; x, y, 1, 1, x, x), \\ (\; x, x, y, 1, y, 1), \; (\; x, 1, y, y, 1, x), \\ (\; x, y, x, x, 1, 1), \; (\; x, 1, 1, x, y, y), \\ (\; y, x, x, 1, 1, x), \; (\; y, y, 1, y, 1, y), \\ (\; y, x, 1, x, x, 1), \; (\; y, 1, x, y, y, 1), \\ (\; y, y, y, x, y, x), \; (\; y, 1, y, 1, x, y), \\ (\; 1, x, y, x, 1, y), \; (\; 1, y, y, y, x, 1), \\ (\; 1, x, 1, y, y, x), \; (\; 1, 1, x, x, x, x), \\ (\; 1, y, x, 1, y, y), \; (\; 1, 1, 1, 1, 1, 1) \end{array}\right\}$$

Fig. 1. The optimal covers for $(x \vee y)^{[4]}$, $(x \vee y)^{[5]}$ and $(x \vee y)^{[6]}$

We define the family of forests $\mathcal{S}^k$ recursively as follows.

$$\mathcal{S}_a^2 = \{(1,1)\}, \quad \mathcal{S}_b^2 = \{(x,y)\}, \quad \mathcal{S}_c^2 = \{(y,x)\},$$

$$\mathcal{S}_a^{k+1} = 1 \circ \mathcal{S}_b^k, \quad \mathcal{S}_b^{k+1} = x \circ \mathcal{S}_a^k \cup x \circ \mathcal{S}_c^k, \quad \mathcal{S}_c^{k+1} = y \circ \mathcal{S}_c^k|_1 \cup y \circ \mathcal{S}_a^k|_y, \quad \text{(for } k \geq 2\text{)}$$

and $\mathcal{S}^k = \mathcal{S}_a^k \cup \mathcal{S}_b^k \cup \mathcal{S}_c^k$ for $k \geq 2$. It is easy to see that $|\mathcal{S}^k| = F_{k+2}$ where F_n is the Fibonacci number, i.e., $F_1 = F_2 = 1$ and $F_n = F_{n-1} + F_{n-2}(n \geq 3)$. An easy but tedious argument, which we omitted in this extended abstract of the paper, can show that the families $\mathcal{S}^k$ cover $(x \vee y)^{[k]}$ for any $k \geq 2$.

2.2 Depth d

In this subsection, we extend the theorem on the computational power of a family of decision forests of width k and size $k + 1$ by proving the following.

Theorem 13. $\forall k \forall d \geq 2\sqrt{k} \quad DEPTH(d) \subsetneq PCover_d(k + 1, k)$.

The key of the proof is to find a function f such that f cannot be computed by any decision tree of depth d and that the agreement probability between f and $DEPTH(d)$ is high. The agreement probability between OR of $d+1$ variables and $DEPTH(d)$ is $1 - 1/(d+1)$ has been shown[BH98]. Our function, which will be defined below, has a higher agreement probability given by $1 - O(1/d^2)$.

Definition 14. For $d \geq 0$, let ALT_d be the Boolean function on $d + 1$ variables defined as follows.

$$ALT_0(x_1) = x_1,$$

$$ALT_d(x_1, x_2, \ldots, x_{d+1}) = \begin{cases} \bar{x}_1 \cdot ALT_{d-1}(x_2, x_3, \ldots, x_{d+1}), & \text{if } d \text{ is odd;} \\ x_1 \vee ALT_{d-1}(x_2, x_3, \ldots, x_{d+1}), & \text{if } d \text{ is even.} \end{cases}$$

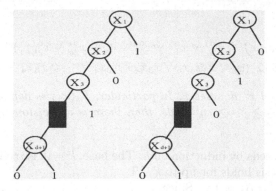

Fig. 2. The function ALT_d (Left:d is odd, Right:d is even)

The decision trees to compute the function ALT_d are shown in Figure 2.2. Since the number of satisfying assignments to ALT_d is odd and the number of satisfying assignments to a function on $d+1$ variables that can be computed by decision trees having depth d or less is even, the function ALT_d is in $\text{DEPTH}(d+1)-\text{DEPTH}(d)$. For $d \geq 0$, we define the set S_d of vectors with length $d+1$ recursively as follows:

$$S_0 = \{0,1\},$$

$$S_d = \begin{cases} 0S_{d-1} \cup 1\{(00)^i(11)^{(d-1)/2-i} \mid i = 0,1,\ldots,(d-1)/2\}1 & \text{if } d \text{ is odd;} \\ 0S_{d-1} \cup 1\{(00)^i(11)^{d/2-i} \mid i = 0,1,\ldots,d/2\} & \text{if } d \text{ is even.} \end{cases}$$

An easy calculation shows the following.

Fact 15. $|S_d| = (d^2 + 4d + 8)/4$ *for even* d *and* $|S_d| = (d^2 + 4d + 7)/4$ *for odd* d. $\qquad\square$

Let f be a Boolean function over X and $v \in \{0,1\}^X$ be an input to f. Let $f \oplus \{v\}$ denote the function such that $(f \oplus \{v\})(v) \neq f(v)$ and $(f \oplus \{v\})(x) = f(x)$ for any $x \neq v$. In what follows, we will prove that, for each $v \in S_d$, the function $\text{ALT}_d \oplus \{v\}$ can be computed by a decision tree of depth d.

Lemma 16. *For any d and for any $v \in S_d$, there exists a decision tree of depth d or less that computes the function $\text{ALT}_d \oplus \{v\}$.*

Before proceeding to the proof of Lemma 16, we show a technical lemma that will be used in the proof of Lemma 16.

Lemma 17. *For $k \geq 0$ and for $\alpha = \alpha_1\alpha_2\cdots\alpha_k \in \{0,1\}^k$, the function F_α on $2k+1$ variables is defined as follows.*

$$F_\alpha(x_1, \ldots, x_{2k+1}) = x_1(x_2^{\alpha_1} \vee x_3^{\alpha_1} \vee x_4^{\alpha_2} \vee x_5^{\alpha_2} \vee \cdots \vee x_{2k}^{\alpha_k} \vee x_{2k+1}^{\alpha_k})$$
$$\vee \, \bar{x}_1 \bar{x}_2(x_3 \vee \bar{x}_3\bar{x}_4 x_5 \vee \bar{x}_3\bar{x}_4\bar{x}_5\bar{x}_6 x_7 \vee \cdots \vee \bar{x}_3\bar{x}_4 \cdots \bar{x}_{2k} x_{2k+1}),$$

where x^1 denotes x and x^0 denotes $\bar{x}$. In particular, $F_\phi(x_1)$ is defined to be the constant 0. If $\alpha_1 \geq \alpha_2 \geq \cdots \geq \alpha_k$ holds, then there is a decision tree of depth $2k$ which computes F_α.

Proof. The proof proceeds by induction on k. The base, $k = 0$, is trivial. Assume the induction hypothesis holds for up to $k - 1$.
Case 1) $\alpha_1 = \alpha_2 = \cdots = \alpha_k = 1$.   Since

$$F_\alpha|_{x_1=0} = \bar{x}_2(x_3 \vee \bar{x}_3\bar{x}_4 x_5 \vee \cdots \vee \bar{x}_3\bar{x}_4 \cdots \bar{x}_{2k} x_{2k+1}),$$

and

$$F_\alpha|_{x_1=1} = x_2 \vee x_3 \vee \cdots \vee x_{2k+1},$$

the decision tree in Figure 3(a) computes $F_\alpha|_{x_1=0}$ and in Figure 3(b) computes $F_\alpha|_{x_1=1}$. Thus the function F_α is computed by the decision tree in Figure 3(c).

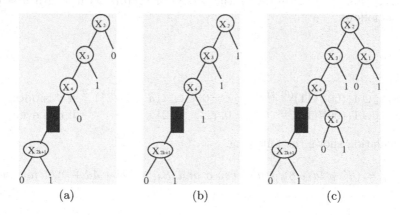

(a)   (b)   (c)

Fig. 3. The construction of the decision trees for F_α.

Case 2) $\alpha_k = 0$.   We have

$$F_\alpha|_{x_{2k+1}=0} = x_1 \vee \bar{x}_1\bar{x}_2(x_3 \vee \bar{x}_3\bar{x}_4 x_5 \vee \cdots \vee \bar{x}_3\bar{x}_4 \cdots \bar{x}_{2k-2} x_{2k-1}),$$
$$F_\alpha|_{x_{2k+1}=1,x_{2k}=0} = x_1 \vee \bar{x}_1\bar{x}_2(x_3 \vee \bar{x}_3\bar{x}_4 x_5 \vee \cdots \vee \bar{x}_3\bar{x}_4 \cdots \bar{x}_{2k-2}),$$
$$F_\alpha|_{x_{2k+1}=1,x_{2k}=1} = x_1(x_2^{\alpha_1} \vee x_3^{\alpha_1} \vee \cdots \vee x_{2k-2}^{\alpha_{k-1}} \vee x_{2k-1}^{\alpha_{k-1}})$$
$$\vee \, \bar{x}_1\bar{x}_2(x_3 \vee \bar{x}_3\bar{x}_4 x_5 \vee \cdots \vee \bar{x}_3\bar{x}_4 \cdots \bar{x}_{2k-2} x_{2k-1})$$
$$= F_{\alpha_1\alpha_2\cdots\alpha_{k-1}}(x_1, x_2, \ldots, x_{2k-1}).$$

Since $F_\alpha|_{x_{2k+1}=0}$ depends on only $2k - 1$ variables, it can be computed by a decision tree of depth $2k - 1$. Similarly, since $F_\alpha|_{x_{2k+1}=1, x_{2k}=0}$ depends on only $2k - 2$ variables, it can be computed by a decision tree of depth $2k - 2$. By the induction hypothesis, $F_\alpha|_{x_{2k+1}=1, x_{2k}=1}$ can be computed by a decision tree of depth $2k - 2$. Hence we can easily construct a decision tree of depth $2k$ by attaching x_{2k+1} to its root, which computes the function F_α. $\square$

Now we proceed to the proof of Lemma 16.

Proof (of Lemma 16). The proof is by induction on d. The base, $d = 0$ is trivial. Assume the induction hypothesis holds for up to $d - 1$.

Case 1) d is even.   Put $d = 2k$. It is sufficient to show that, for every $v \in S_d = 0S_{d-1} \cup 1\{(00)^i(11)^{k-i} \mid i = 0, 1, \ldots, k\}$, we can construct a decision tree of depth $2k$ which computes the function $\mathrm{ALT}_d \oplus \{v\}$. For $v \in 0S_{d-1}$, the induction step follows immediately from the induction hypothesis since $\mathrm{ALT}_d|_{x_1=0} \equiv \mathrm{ALT}_{d-1}$. In what follows, let $v = 1(00)^i(11)^{k-i}$ ($i \in \{0, 1, \ldots, k\}$). Let G_v be the *mask* function defined by $G_v(v) = 0$ and $G_v(x) = 1$ for any $x \neq v$. Since ALT_d outputs 1 on the input v, the function $\mathrm{ALT}_d \oplus \{v\}$ is equivalent to $\mathrm{ALT}_d \cdot G_v$. The function G_v can be expressed as

$$G_v = \bar{x}_1 \vee x_2^{\alpha_1} \vee x_3^{\alpha_1} \vee x_4^{\alpha_2} \vee x_5^{\alpha_2} \vee \cdots \vee x_{2k}^{\alpha_k} \vee x_{2k+1}^{\alpha_k},$$

where $\alpha_1 = \alpha_2 = \cdots = \alpha_i = 1$ and $\alpha_{i+1} = \cdots = \alpha_k = 0$. A simple calculation shows that the function $\mathrm{ALT}_d \cdot G_v$ is equivalent to $F_{\alpha_1 \cdots \alpha_k}$ defined in Lemma 17. Thus, the induction step follows from Lemma 17.

Case 2) d is odd.   Put $d = 2k + 1$. We shall construct a decision tree of depth $2k + 1$ which computes $\mathrm{ALT}_d \oplus \{v\}$ for each $v \in S_d = 0S_{d-1} \cup 1\{(00)^i(11)^{k-i}1 \mid i = 0, 1, \ldots, k\}$. For $v \in 0S_{d-1}$, the induction step follows immediately from the induction hypothesis since $\mathrm{ALT}_d|_{x_1=0} \equiv \mathrm{ALT}_{d-1}$. It is easy to observe that the negation of the function ALT_d can be represented as

$$\overline{\mathrm{ALT}_d} = x_1 \vee \bar{x}_1 \bar{x}_2 x_3 \vee \cdots \vee \bar{x}_1 \bar{x}_2 \cdots \bar{x}_{d-2} x_{d-1} \vee \bar{x}_1 \bar{x}_2 \cdots \bar{x}_d \bar{x}_{d+1}.$$

This function is equal to the function $\mathrm{ALT}_{d+1}(x_1, x_2, \ldots, x_{d+1}, 1)$. By Lemma 17 and by similar arguments to the arguments in the proof of Case 1, we can construct a decision tree of $d + 1$ which computes the function $\mathrm{ALT}_{d+1} \oplus \{v'\}$ where $v' = 1(00)^i(11)^{k-i+1}$. Since the label of the root of the decision tree constructed in the proof of Lemma 17 is x_{d+2}, the right son of the root of this tree, whose depth is d, computes the function $\mathrm{ALT}_d \oplus \{v\}$. $\square$

We are now ready to prove Theorem 13.

Proof (of Theorem 13). To prove the theorem, it is sufficient to show that $\mathrm{ALT}_d^{[k]}(x_1, x_2, \ldots, x_{d+1})$ is in $\mathrm{PCover}_d(k + 1, k)$. By Lemma 16, there are trees $T_1, T_2, \ldots, T_{|S_d|}$ that compute the function $\mathrm{ALT}_d \oplus \{v\}$ for each $v \in S_d$. Since

$d \geq 2\sqrt{k}$, $(d^2+4)/4 \geq k+1$ holds. By Fact 15, we have $|S_d| \geq (d^2+4d+7)/4 \geq (d^2+4)/4 \geq k+1$. Define the family of forests $\mathcal{F}$ as

$$
\mathcal{F} = \left\{
\begin{array}{ccccc}
(\ T_1, & T_1, & T_1, & \cdots, & T_1,\ ), \\
(\ T_2, & T_2, & T_2, & \cdots, & T_2,\ ), \\
\vdots & \vdots & \vdots & \vdots & \vdots \\
(\ T_{k+1}, & T_{k+1}, & T_{k+1}, & \cdots, & T_{k+1}\ )
\end{array}
\right\}
$$

It is easy to see that $\mathcal{F}$ covers $\mathrm{ALT}_d^{[k]}$. This is because, for each column in $\mathcal{F}$, at most one of the decision trees can be unequal to ALT_d. □

3 Relation between Cover and Circuit Complexity

In this section, we reveal that there is a close relation between the cover problem and the complexity for circuits with a majority gate at the top. Let f be a Boolean function whose domain is D and $\mathcal{C}$ be any class of Boolean functions over the same domain. The *correlation* between f and $\mathcal{C}$, denoted $\gamma(f, \mathcal{C})$, is defined as

$$
\gamma(f, \mathcal{C}) \equiv \min_{\mathcal{D}} \max_{h \in \mathcal{C}} \Pr_{\alpha \in \mathcal{D}}[f(\alpha) = h(\alpha)],
$$

where $\mathcal{D}$ denotes a distribution on D. The correlation between f and $\mathcal{C}$ is closely related to the complexity of f for circuits whose top is a majority gate.

Theorem 18 ([Fre95,GHR92]). *Let f be a Boolean function over $\{0,1\}^n$ and $\mathcal{C}$ be a set of functions over the same domain. If $k > (2n \ln 2)(\gamma(f, \mathcal{C}) - 1/2)^{-2}$, then f can be represented as $f(x) = MAJ(h_1(x), \ldots, h_k(x))$ for some $h_i \in \mathcal{C}$. If f can be represented as $f(x) = MAJ(h_1(x), \ldots, h_k(x))$ with $h_i \in \mathcal{C}$ then $(\gamma(f, \mathcal{C}) - 1/2) \geq 1/k$.* □

The proofs of many of our theorems in the last section rely on the analysis of the agreement probability between a target function and a class of decision trees. Now we generalize the definition of the agreement probability to cover other complexity classes. For a Boolean function f and a complexity class $\mathcal{C}$, the *agreement probability* between f and $\mathcal{C}$, denoted $\rho(f, \mathcal{C})$, is defined as

$$
\rho(f, \mathcal{C}) \equiv \max_{\mathcal{T}} \min_{\alpha \in D} \Pr_{h \in \mathcal{T}}[f(\alpha) = h(\alpha)],
$$

where $\mathcal{T}$ denotes a distribution on $\mathcal{C}$. Below we show that the correlation and the agreement probability are identical for any function and any computational class by employing the minimax theorem.

Theorem 19. *For any Boolean function f and any complexity class $\mathcal{C}$, $\rho(f, \mathcal{C}) = \gamma(f, \mathcal{C})$.*

Proof. We have

$$\rho(f,\mathcal{C}) = \max_{\mathcal{T}} \min_{\mathcal{D}} \Pr_{h \in \mathcal{T}, \alpha \in \mathcal{D}}[f(\alpha) = h(\alpha)]$$

$$= \min_{\mathcal{D}} \max_{\mathcal{T}} \Pr_{h \in \mathcal{T}, \alpha \in \mathcal{D}}[f(\alpha) = h(\alpha)] = \gamma(f,\mathcal{C})$$

The first and the third equalities follow from a simple calculation and the second equality follows from the minimax theorem for mixed strategies for two person matrix games.   □

By using Theorems 18 and 19, we can prove the following theorem. This says that if k independent copies of a function f on n variables can be computed by decision forests of depth d using $k-1$ help bits, i.e., if we can save one help bit, then f can be represented as the majority of $O(nk^2)$ functions in DEPTH(d) and an approximate converse to this also holds.

Theorem 20. *Let f be a function on n variables. (i) If $f \in Cover_d(2^k/2, k)$ then f can be expressed as $MAJ(h_1, \ldots, h_{20nk^2})$ for some $h_i \in DEPTH(d)$. (ii) If $f \notin PCover_d(2^k/2, k)$ then f cannot be expressed as $MAJ(h_1, \ldots, h_{2k/(\ln(4nk \ln 2))})$ for any $h_i \in DEPTH(d)$.*

Proof (Sketch). Let f be a function on n variables. Put $\rho(f,d) = 1/2 + 1/z$. (i) Suppose that $f \in Cover_d(2^k/2, k)$. We have $k \geq ((z+2)\ln 2)/2$ since if not, then $1/\rho(f,d)^k > 2^k/2$ which implies $f \notin Cover_d(2^k/2, k)$ by Lemma 4.1 in [NRS94]. Thus $z \leq 2k \ln 2 - 2$. The statement (i) in the theorem immediately follows from Theorems 18 and 19. (ii) Suppose that $f \notin PCover(2^k/2)$. We have $k \leq (z+2)(\ln(4nk \ln 2))/2$ since if not, then $nk \ln 2/\rho(f,d)^k < 2^k/2$ which implies $f \in PCover_d(2^k/2, k)$ by Lemma 4.1 in [NRS94]. Thus $z \geq 2k/(\ln(4nk \ln 2))$. By Theorems 18 and 19, we can easily obtain the statement (ii) in the theorem.   □

Finally, we remark that the above theorem can be applied to other computational models. For example, we may obtain a good lower bound for threshold circuits of depth 3 by investigating covers consisting of threshold circuits of depth 2 instead of decision trees.

Acknowledgments

The authors are grateful to anonymous referees for their constructive comments and suggestions.

References

[ABG00] A. Amir, R. Beigel, W. Gasarch, "Some Connections between Bounded Query Classes and Non-Uniform Complexity (Long Version)", *Electronic Colloquium on Computational Complexity*, Report No. 24, 2000.

[BH98]   R. Beigel and T. Hirst, "One Help Bit Doesn't Help", *Proc. 30th STOC*, pp. 124–129, 1998. (See also "One Help Bit Doesn't", *Technical Report of Yale University, Department of Computer Science*, TR-1118, 1996.)

[Cai89]   J. Cai, "Lower Bounds for Constant Depth Circuits in the Presence of Help Bits", *Proc. 30th FOCS*, pp. 532–537, 1989.

[Fre95]   Y. Freund, "Boosting a Weak Learning Algorithm by Majority", *Information and Computation*, Vol. 121, No. 2, pp. 256–285, 1995.

[GHR92]   M. Goldmann, J. Håstad, A.A. Razborov, "Majority Gates vs. General Weighted Threshold Gates", *Computational Complexity*, Vol. 2, pp. 277–300, 1992.

[Lov75]   L. Lovasz, "On the Ratio of Optimal Integral and Fractional Covers", *Discrete Mathematics*, Vol. 13, pp. 383–390, 1975.

[NRS94]   N. Nisan, S. Rudich and M. Saks, "Products and Help Bits in Decision Trees", *Proc. 35th FOCS*, pp. 318–324, 1994.

Exact Results for Accepting Probabilities of Quantum Automata

Andris Ambainis[1] and Arnolds Ķikusts[2]

[1] Computer Science Division, University of California
Berkeley, CA94720, USA
ambainis@cs.berkeley.edu [***]
[2] Institute of Mathematics and Computer Science, University of Latvia
Raiņa bulv. 29, Rīga, Latvia[†]
sd70053@lanet.lv

Abstract. One of the properties of Kondacs-Watrous model of quantum finite automata (QFA) is that the probability of the correct answer for a QFA cannot be amplified arbitrarily. In this paper, we determine the maximum probabilities achieved by QFAs for several languages. In particular, we show that any language that is not recognized by an RFA (reversible finite automaton) can be recognized by a QFA with probability at most 0.7726....

1 Introduction

A quantum finite automaton (QFA) is a model for a quantum computer with a finite memory. The strength of QFAs is shown by the fact that quantum automata can be exponentially more space efficient than deterministic or probabilistic automata [AF 98].

At least 4 different models of QFAs have been introduced. These models differ in the measurements allowed during the computation. If there is no measurements, the evolution of a quantum system is unitary and this means that no information can be erased. This does not affect unrestricted quantum computation (quantum Turing machines or quantum circuits). However, this severely restricts the power of quantum computations with limited space (like QFAs or space-bounded quantum Turing machines [W 98]).

The most restricted model of QFAs is the "measure-once" model of Crutchfield and Moore [CM 97]. In this model, all transitions of a QFA must be unitary, except for one measurement at the end that is needed to read the result of the computation. The classical counterpart of "measure-once" QFAs are *permutation automata* in which each transition permutes the states of the automaton. It

---

[***] Research supported by Berkeley Fellowship for Graduate Studies, Microsoft Research Fellowship and NSF Grant CCR-9800024.
[†] Research supported by Grant No.01.0354 from the Latvian Council of Science and European Commission, contract IST-1999-11234.

J. Sgall, A. Pultr, and P. Kolman (Eds.): MFCS 2001, LNCS 2136, pp. 135–147, 2001.

turns out that the languages recognized by "measure-once" QFAs are the same as the languages recognized by permutation automata [CM 97,BP 99].

The most general model is QFAs with mixed states [AW 01,C 01,P 99][1]. This model allows arbitrary measurements and it can recognize any regular language.

An intermediate model of QFAs is the "measure-many" model introduced by Kondacs and Watrous [KW 97][2]. This model allows intermediate measurements during the computation but these measurements have to be of a restricted type. (More specifically, they can have 3 outcomes: "accept", "reject", "don't halt" and if one gets "accept" or "reject", the computation ends and this is the result of computation.) The study of this (or other intermediate) models shows to what degree QFAs need measurements to be able to recognize certain languages.

The class of languages recognizable by "measure-many" QFAs has been studied in [KW 97,BP 99,AKV 01]. There are several necessary and sufficient conditions but no complete characterization. Surprisingly, the class of languages recognized by QFAs[3] depends on the probability with which the QFA is required to output the correct answer. This was first discovered by Ambainis and Freivalds [AF 98] who proved that:

- Any language recognizable by a QFA with a probability $7/9 + \epsilon$, $\epsilon > 0$ is recognizable by a reversible finite automaton (RFA)[4].
- The language a^*b^* can be recognized with probability 0.6822.. but cannot be recognized by an RFA.

Thus, the class of languages recognizable with probability 0.6822... is not the same as the class of languages recognizable with probability $7/9 + \epsilon$.

In almost any other computational model, the accepting probability can be increased by repeating the computation in parallel and, usually, this property is considered completely obvious. The above results by Ambainis and Freivalds [AF 98] showed that this is not the case for QFAs. This is caused by the fact that the Kondacs-Watrous model of QFAs combines a reversible (unitary transformation) component with a non-reversible component (measurements).

In this paper, we develop a method for determining the maximum probability with which a QFA can recognize a given language. Our method is based on the quantum counterpart of classification of states of a Markov chain into ergodic and transient states [KS 76]. We use this classification of states to transform the problem of determining the maximum accepting probability of a QFA into a quadratic optimization problem. Then, we solve this problem (analytically in simpler cases, by computer in more difficult cases).

---

[1] Ciamarra[C 01] calls this model "fully quantum finite automata" and Paschen[P 99] calls it "quantum automata with ancilla qubits".

[2] An another model was introduced by Nayak[N 99]. The power of Nayak's model is between "measure-many" QFAs and unrestricted QFAs with mixed states.

[3] For the rest of this paper, we will refer to "measure-many" QFAs as simply QFAs because this is the only model considered in this paper.

[4] Here, "reversible FA" means the corresponding classical model in which the states of the automaton are classical but the automaton can stop the computation similarly to the "accept/reject/don't halt" measurement of a QFA.

Compared to previous work, our new method has two advantages. First, it gives a systematic way of calculating the maximum accepting probabilities. Second, solving the optimization problems usually gives the maximum probability exactly. Most of previous work [AF 98,ABFK 99] used approaches depending on the language and required two different methods: one for bounding the probability from below, another for bounding it from above. Often, using two different approaches gave an upper and a lower bound with a gap between them (like 0.6822... vs. $7/9 + \epsilon$ mentioned above). With the new approach, we are able to close those gaps.

We use our method to calculate the maximum accepting probabilities for a variety of languages (and classes of languages).

First, we construct a quadratic optimization problem for the maximum accepting probability by a QFA of a language that is not recognizable by an RFA. Solving the problem gives the probability $(52 + 4\sqrt{7})/81 = 0.7726...$ This probability can be achieved for the language a^+ in the two-letter alphabet $\{a, b\}$ but no language that is no recognizable by a RFA can be recognized with a higher probability. This improves the $7/9 + \epsilon$ result of [AF 98].

This result can be phrased in a more general way. Namely, we can find the property of a language which makes it impossible to recognize the language by an RFA. This property can be nicely stated in the form of the minimal deterministic automaton containing a fragment of a certain form.

We call such a fragment a "non-reversible construction". It turns out that there are many different "non-reversible constructions" and they have different influence on the accepting probability. The one contained in the a^+ language makes the language not recognizable by an RFA but the language is still recognizable by a QFA with probability 0.7726.... In contrast, some constructions analyzed in [BP 99,AKV 01] make the language not recognizable with probability $1/2 + \epsilon$ for any $\epsilon > 0$.

In the rest of this paper, we look at different "non-reversible constructions" and their effects on the accepting probabilities of QFAs. We consider three constructions: "two cycles in a row", "k cycles in parallel" and a variant of the a^+ construction. The best probabilities with which one can recognize languages containing these constructions are 0.6894..., $k/(2k - 1)$ and 0.7324..., respectively.

The solution of the optimization problem for "two cycles in a row" gives a new QFA for the language a^*b^* that recognizes it with probability 0.6894..., improving the result of [AF 98]. Again, using the solution of the optimization problem gives a better QFA that was previously missed because of disregarding some parameters.

2 Preliminaries

2.1 Quantum Automata

We define the Kondacs-Watrous ("measure-many") model of QFAs [KW 97].

Definition 1. *A QFA is a tuple* $M = (Q; \Sigma; V; q_0; Q_{acc}; Q_{rej})$ *where* Q *is a finite set of states,* Σ *is an input alphabet,* V *is a transition function (explained below),* $q_0 \in Q$ *is a starting state, and* $Q_{acc} \subseteq Q$ *and* $Q_{rej} \subseteq Q$ *are sets of accepting and rejecting states* $(Q_{acc} \cap Q_{rej} = \emptyset)$. *The states in* Q_{acc} *and* Q_{rej}, *are called* halting states *and the states in* $Q_{non} = Q - (Q_{acc} \cup Q_{rej})$ *are called* non halting states.

States of M. The state of M can be any superposition of states in Q (i. e., any linear combination of them with complex coefficients). We use $|q\rangle$ to denote the superposition consisting of state q only. $l_2(Q)$ denotes the linear space consisting of all superpositions, with l_2-distance on this linear space.

Endmarkers. Let κ and $\$$ be symbols that do not belong to Σ. We use κ and $\$$ as the left and the right endmarker, respectively. We call $\Gamma = \Sigma \cup \{\kappa; \$\}$ the *working alphabet* of M.

Transition function. The transition function V is a mapping from $\Gamma \times l_2(Q)$ to $l_2(Q)$ such that, for every $a \in \Gamma$, the function $V_a : l_2(Q) \to l_2(Q)$ defined by $V_a(x) = V(a, x)$ is a unitary transformation (a linear transformation on $l_2(Q)$ that preserves l_2 norm).

Computation. The computation of a QFA starts in the superposition $|q_0\rangle$. Then transformations corresponding to the left endmarker κ, the letters of the input word x and the right endmarker $\$$ are applied. The transformation corresponding to $a \in \Gamma$ consists of two steps.

1. First, V_a is applied. The new superposition ψ' is $V_a(\psi)$ where ψ is the superposition before this step.

2. Then, ψ' is observed with respect to $E_{acc}, E_{rej}, E_{non}$ where $E_{acc} = span\{|q\rangle : q \in Q_{acc}\}$, $E_{rej} = span\{|q\rangle : q \in Q_{rej}\}$, $E_{non} = span\{|q\rangle : q \in Q_{non}\}$. It means that if the system's state before the measurement was

$$\psi' = \sum_{q_i \in Q_{acc}} \alpha_i |q_i\rangle + \sum_{q_j \in Q_{rej}} \beta_j |q_j\rangle + \sum_{q_k \in Q_{non}} \gamma_k |q_k\rangle$$

then the measurement accepts ψ' with probability $p_a = \Sigma \alpha_i^2$, rejects with probability $p_r = \Sigma \beta_j^2$ and continues the computation (applies transformations corresponding to next letters) with probability $p_c = \Sigma \gamma_k^2$ with the system having the (normalized) state $\frac{\psi}{\|\psi\|}$ where $\psi = \Sigma \gamma_k |qk\rangle$.

We regard these two transformations as reading a letter a.

Notation. We use V_a' to denote the transformation consisting of V_a followed by projection to E_{non}. This is the transformation mapping ψ to the non-halting part of $V_a(\psi)$. We use V_w' to denote the product of transformations $V_w' = V_{a_n}' V_{a_{n-1}}' \dots V_{a_2}' V_{a_1}'$, where a_i is the i-th letter of the word w.

Recognition of languages. We will say that an automaton recognizes a language L with probability p $(p > \frac{1}{2})$ if it accepts any word $x \in L$ with probability $\geq p$ and rejects any word $x \notin L$ with probability $\geq p$.

2.2   Useful Lemmas

For classical Markov chains, one can classify the states of a Markov chain into *ergodic* sets and *transient* sets [KS 76]. If the Markov chain is in an ergodic set, it never leaves it. If it is in a transient set, it leaves it with probability $1 - \epsilon$ for an arbitrary $\epsilon > 0$ after sufficiently many steps.

A quantum counterpart of a Markov chain is a quantum system to which we repeatedly apply a transformation that depends on the current state of the system but does not depend on previous states. In particular, it can be a QFA that repeatedly reads the same word x. Then, the state after reading x $k+1$ times depends on the state after reading x k times but not on any of the states before that. The next lemma gives the classification of states for such QFAs.

Lemma 2. *[AF 98] Let $x \in \Sigma^+$. There are subspaces E_1, E_2 such that $E_{non} = E_1 \oplus E_2$ and*

(i) If $\psi \in E_1$, then $V'_x(\psi) \in E_1$ and $\|V'_x(\psi)\| = \|\psi\|$,
(ii) If $\psi \in E_2$, then $\|V'_{x^k}(\psi)\| \to 0$ when $k \to \infty$.

Instead of ergodic and transient sets, we have subspaces E_1 and E_2. The subspace E_1 is a counterpart of an ergodic set: if the quantum process defined by repeated reading of x is in a state $\psi \in E_1$, it stays in E_1. E_2 is a counterpart of a transient set: if the state is $\psi \in E_2$, E_2 is left (for an accepting or rejecting state) with probability arbitrarily close to 1 after sufficiently many x's.

In some of proofs that were omitted due to space constraints, we also use a generalization of Lemma 2 to the case of two (or more) words x and y. This generalization was published in [AKV 01].

3   QFAs vs. RFAs

Ambainis and Freivalds[AF 98] characterized the languages recognized by RFAs as follows.

Theorem 3. *[AF 98] Let L be a language and M be its minimal automaton. L is recognizable by a RFA if and only if there is no q_1, q_2, x such that*

1. $q_1 \neq q_2$,
2. If M starts in the state q_1 and reads x, it passes to q_2,
3. If M starts in the state q_2 and reads x, it passes to q_2, and
4. q_2 is neither "all-accepting" state, nor "all-rejecting" state,

An RFA is a special case of a QFA that outputs the correct answer with probability 1. Thus, any language that does not contain the construction of Theorem 3 can be recognized by a QFA that always outputs the correct answer. Ambainis and Freivalds [AF 98] also showed the reverse of this: any language L with the minimal automaton containing the construction of Theorem 3 cannot be recognized by a QFA with probability $7/9 + \epsilon$.

We consider the question: what is the maximum probability of correct answer than can be achieved by a QFA for a language that cannot be recognized by an RFA? The answer is:

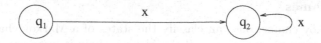

Fig. 1. "The forbidden construction" of Theorem 3

Theorem 4. *Let L be a language and M be its minimal automaton.*

1. *If M contains the construction of Theorem 3, L cannot be recognized by a 1-way QFA with probability more than $p = (52 + 4\sqrt{7})/81 = 0.7726....$*
2. *There is a language L with the minimal automaton M containing the construction of Theorem 3 that can be recognized by a QFA with probability $p = (52 + 4\sqrt{7})/81 = 0.7726....$*

Proof sketch. We consider the following optimization problem.

Optimization problem 1. Find the maximum p such that there is a finite dimensional vector space E_{opt}, subspaces E_a, E_r such that $E_a \perp E_r$, vectors v_1, v_2 such that $v_1 \perp v_2$ and $\|v_1 + v_2\| = 1$ and probabilities p_1, p_2 such that $p_1 + p_2 = \|v_2\|^2$ and

1. $\|P_a(v_1 + v_2)\|^2 \geq p$,
2. $\|P_r(v_1)\|^2 + p_2 \geq p$,
3. $p_2 \leq 1 - p$.

We sketch the relation between a QFA recognizing L and this optimization problem. Let Q be a QFA recognizing L. Let p_{min} be the minimum probability of the correct answer for Q, over all words. We use Q to construct an instance of the optimization problem above with $p \geq p_{min}$.

Namely, we look at Q reading an infinite (or very long finite) sequence of letters x. By Lemma 2, we can decompose the starting state ψ into 2 parts $\psi_1 \in E_1$ and $\psi_2 \in E_2$. Define $v_1 = \psi_1$ and $v_2 = \psi_2$. Let p_1 and p_2 be the probabilities of getting into an accepting (for p_1) or rejecting (for p_2) state while reading an infinite sequence of x's.

Since q_1 and q_2 are different states of the minimal automaton M, there is a word y that is accepted in one of them but not in the other. Without loss of generality, we assume that y is accepted if M is started in q_1 but not if M is started in q_2. Also, since q_2 is not an "all-accepting" state, there must be a word z that is rejected if M is started in the state q_2.

We choose E_a and E_r so that the square of the projection P_a (P_r) of a vector v on E_a (E_r) is equal to the accepting (rejecting) probability of Q if we run Q on the starting state v and input y and the right endmarker $\$$.

Finally, we set p equal to the inf of the set consisting of the probabilities of correct answer of Q on the words y and $x^i y$, $x^i z$ for all $i \in \mathbb{Z}$.

Then, the first requirement of the optimization problem, $\|P_a(v_1 + v_2)\|^2 \geq p$ is true because the word y must be accepted and the accepting probability for it is exactly the square of the projection of the starting state $(v_1 + v_2)$ to P_a.

The second requirement follows from running Q on a word $x^i y$ for some large i. If we pick an *appropriately selected* large i, then reading x^i has the following effect:

1. v_1 gets mapped to a state that is at most ϵ-away (in l_2 norm) from v_1,
2. v_2 gets mapped to an accepting/rejecting state and most ϵ fraction of it stays on the non-halting states.

Together, these two requirements mean that the state of Q after reading x^i is at most 2ϵ-away from v_1. Also, the probabilities of Q accepting and rejecting while reading x^i differ from p_1 and p_2 by at most ϵ.

Since reading y in q_2 leads to a rejection, $x^i y$ must be rejected. The probability of rejection by Q consists of two parts: the probability of rejection during x^i and the probability of rejection during y. The first part differs from p_1 by at most ϵ, the second part differs from $\|P_r(v_1)\|^2$ by at most 4ϵ (because the state of Q when starting to read y differs from v_1 by at most 2ϵ and, by a lemma in [BV 97], if the starting states are close, accepting probabilities are close as well). Therefore, $p_1 + \|P_r(v_1)\|^2$ is at least the accepting probability plus 5ϵ. By appropriately choosing i, we can make this true for any $\epsilon > 0$.

The third requirement is true by considering $x^i z$. This word must be accepted with probability p. Therefore, for any i, Q can only reject during x^i with probability $1 - p$ and $p_2 \leq 1 - p$.

This shows that no QFA can achieve a probability of correct answer more than the solution of optimization problem 1. In the appendix, we show how to solve this optimization problem. The answer is $p = (52 + 4\sqrt{7})/81 = 0.7726....$ This proves the first part of the theorem.

The second part is proven by taking the solution of optimization problem 1 and using it to construct a QFA for the language a^+ in a two-letter alphabet $\{a, b\}$.   □

4   Non-reversible Constructions

We now look at fragments of the minimal automaton that imply that the language cannot be recognized with probability more than p, for some p. We call such fragments "non-reversible constructions". The simplest such construction is the one of Theorem 3. In this section, we present 3 other "non-reversible constructions" that imply that a language can be recognized with probability at most 0.7324..., 0.6894... and $k/(2k - 1)$. This shows that different constructions are "non-reversible" to different extent.

4.1   "Two Cycles in a Row"

The first construction comes from the language $a^* b^*$ considered in Ambainis and Freivalds[AF 98]. This language was the first example of a language that can be recognized by a QFA with some probability (0.6822...) but not with another ($7/9 + \epsilon$). We find the "non-reversible" construction for this language and construct the QFA with the best possible accepting probability.

Theorem 5. *Let L be a language and M its minimal automaton.*

1. *If M contains states q_1, q_2 and q_3 such that, for some words x and y,*
   (a) if M reads x in the state q_1, it passes to q_1,
   (b) if M reads y in the state q_1, it passes to q_2,
   (c) if M reads y in the state q_2, it passes to q_2,
   (d) if M reads x in the state q_2, it passes to q_3,
   (e) if M reads x in the state q_3, it passes to q_3
   then L cannot be recognized by a QFA with probability more than 0.6894....
2. *The language a^*b^* (the minimal automaton of which contains the construc-tion above) can be recognized by a QFA with probability 0.6894....*

The proof (which is omitted due to space constraints) is by a reduction to the following optimization problem.

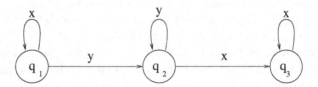

Fig. 2. "The forbidden construction" of Theorem 5

Optimization problem 2. Find the maximum p such that there is a finite-dimensional space E, subspaces E_a, E_r such that $E = E_a \oplus E_r$, vectors v_1, v_2 and v_3 and probabilities p_{a_1}, p_{r_1}, p_{a_2}, p_{r_2} such that

1. $\|v_1 + v_2 + v_3\| = 1$,
2. $v_1 \perp v_2$,
3. $v_1 + v_2 + v_3 \perp v_2$,
4. $v_1 + v_2 \perp v_3$.
5. $\|v_3\|^2 = p_{a_1} + p_{r_1}$;
6. $\|v_2\|^2 = p_{a_2} + p_{r_2}$;
7. $\|P_a(v_1 + v_2 + v_3)\|^2 \geq p$;
8. $\|P_a(v_1 + v_2)\|^2 + p_{a_1} \geq p$;
9. $\|P_a(v_1)\|^2 + p_{a_1} + p_{a_2} \leq 1 - p$.

4.2 k Cycles in Parallel

Theorem 6. *Let $k \geq 2$.*

1. *Let L be a language. If there are words $x_1, x_2, \ldots, x_k$ such that its minimal automaton M contains states $q_0, q_1, \ldots, q_k$ satisfying:*
   (a) if M starts in the state q_0 and reads x_i, it passes to q_i,
   (b) if M starts in the state $q_i (i \geq 1)$ and reads x_j, it passes to q_i,
   Then L cannot be recognized by a QFA with probability greater than $\frac{k}{2k-1}$.

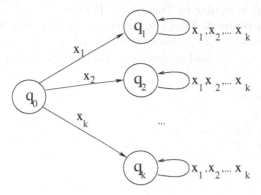

Fig. 3. "The forbidden construction" of Theorem 6

2. *There is a language such that its minimal deterministic automaton contains this construction and the language can be recognized by a QFA with probability $\frac{k}{2k-1}$.*

For $k = 2$, a related construction was considered in [AKV 01]. There is a subtle difference between the two. The "non-reversible construction" in [AKV 01] requires the sets of words accepted from q_2 and q_3 to be incomparable. This extra requirement makes it much harder: no QFA can recognize a language with the "non-reversible construction" of [AKV 01] even with the probability $1/2 + \epsilon$.

A similar effect is true for the constructions in Figures 1 and 4. The construction in Figure 4 is the same as the construction in Figure 1 with an extra requirement that the sets of the words accepted from q_1 and q_2 are incomparable. Again, the construction with this extra requirement is harder: we get the probability 0.7324... instead of 0.7726....

4.3   0.7324... Construction

Theorem 7. *Let L be a language.*

1. *If there are words x, z_1, z_2 such that its minimal automaton M contains states q_1 and q_2 satisfying:*
   (a) if M starts in the state q_1 and reads x, it passes to q_2,
   (b) if M starts in the state q_2 and reads x, it passes to q_2,
   (c) if M starts in the state q_1 and reads z_1, it passes to an accepting state,
   (d) if M starts in the state q_1 and reads z_2, it passes to a rejecting state,
   (e) if M starts in the state q_2 and reads z_1, it passes to a rejecting state,
   (f) if M starts in the state q_2 and reads z_2, it passes to an accepting state.
   Then L cannot be recognized by a QFA with probability greater than $\frac{1}{2} + \frac{3\sqrt{15}}{50} = 0.7324....$
2. *There is a language L with the minimum automaton containing this construction that can be recognized with probability $\frac{1}{2} + \frac{3\sqrt{15}}{50} = 0.7324....$*

The proof (omitted) is similar to Theorem 4. It uses

Optimization problem 3. Find the maximum p such that there is a finite dimensional vector space E_{opt}, subspaces E_a, E_r and vectors v_1, v_2 such that $v_1 \perp v_2$ and $\|v_1 + v_2\| = 1$ and probabilities p_1, p_2 such that $p_1 + p_2 = \|v_2\|^2$ and

1. $\|P_a(v_1 + v_2)\|^2 \geq p$,
2. $\|P_r(v_1 + v_2)\|^2 \geq p$,
3. $1 - \|P_a(v_1)\|^2 - p_1 \geq p$,
4. $1 - \|P_r(v_1)\|^2 - p_2 \geq p$.

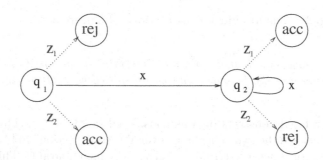

Fig. 4. "The forbidden construction" of Theorem 7

References

ABFK 99. Andris Ambainis, Richard Bonner, Rūsiņš Freivalds, Arnolds Ķikusts. Probabilities to accept languages by quantum finite automata. *Proceedings of COCOON'99*, p. 174-183. Also quant-ph/9904066[5]. 137

AF 98.   Andris Ambainis, Rūsiņš Freivalds. 1-way quantum finite automata: strengths, weaknesses and generalizations. *Proceedings of FOCS'98*, p. 332–341. Also quant-ph/9802062. 135, 136, 137, 139, 141

AKV 01.   Andris Ambainis, Arnolds Ķikusts, Māris Valdats. On the class of languages recognizable by 1-way quantum finite automata. *Proceedings of STACS'01*, p. 75–86. Also quant-ph/0009004. 136, 137, 139, 143

AW 01.   Andris Ambainis, John Watrous. Quantum automata with mixed states. In preparation, 2001. 136

BV 97.   Ethan Bernstein, Umesh Vazirani, Quantum complexity theory. *SIAM Journal on Computing,* 26:1411-1473, 1997. 141

BP 99.   Alex Brodsky, Nicholas Pippenger. Characterizations of 1-way quantum finite automata. quant-ph/9903014. 136, 137

C 01.   M. Pica Ciamarra. Quantum reversibility and a new type of quantum automaton. *Proceedings of FCT'01*, to appear. 136

[5] quant-ph preprints are available at http://www.arxiv.org/abs/quant-ph/preprint-number

G 00.   Jozef Gruska. Descriptional complexity issues in quantum computing. *Journal of Automata, Languages and Combinatorics*, 5:191-218, 2000.

KR 00.   Arnolds Ķikusts, Zigmārs Rasščevskis. On the accepting probabilities of 1-way quantum finite automata. *Proceedings of the workshop on Quantum Computing and Learning*, 2000, p. 72–79.

KS 76.   J. Kemeny, J. Snell. *Finite Markov Chains*. Springer-Verlag, 1976.   136, 139

K 98.   Arnolds Ķikusts. A small 1-way quantum finite automaton. quant-ph/9810065.

KW 97.   Attila Kondacs and John Watrous. On the power of quantum finite state automata. In *Proceedings of FOCS'97*, p. 66–75.   136, 137

CM 97.   C. Moore, J. Crutchfield. Quantum automata and quantum grammars. *Theoretical Computer Science*, 237:275–306, 2000. Also quant-ph/9707031. 135, 136

N 99.   Ashwin Nayak. Optimal lower bounds for quantum automata and random access codes. *Proceedings of FOCS'99*, p. 369-376. Also quant-ph/9904093. 136

P 99.   Katrin Paschen. Quantum finite automata using ancilla qubits. University of Karlsruhe technical report.   136

W 98.   John Watrous. Relationships between quantum and classical space-bounded complexity classes. *Proceedings of Complexity'98*, p. 210-227.   135

Appendix: Solving the Optimization Problem 1

The key idea is to show that it is enough to consider 2-dimensional instances of the problem.

Since $v_1 \perp v_2$, the vectors $v_1, v_2, v_1 + v_2$ form a right-angled triangle. This means that $\|v_1\| = \cos\beta \|v_1 + v_2\| = \cos\beta$, $\|v_2\| = \sin\beta \|v_1 + v_2\| = \sin\beta$ where β is the angle between v_1 and $v_1 + v_2$. Let w_1 and w_2 be the normalized versions of v_1 and v_2: $w_1 = \frac{v_1}{\|v_1\|}$, $w_2 = \frac{v_2}{\|v_2\|}$. Then, $v_1 = \cos\beta w_1$ and $v_2 = \sin\beta w_2$.

Consider the two-dimensional subspace spanned by $P_a(w_1)$ and $P_r(w_1)$. Since the accepting and the rejecting subspaces E_a and E_r are orthogonal, $P_a(w_1)$ and $P_r(w_1)$ are orthogonal. Therefore, the vectors $w_a = P_a(w_1)$ and $w_r = P_r(w_1)$ form an orthonormal basis. We write the vectors w_1, v_1 and $v_1 + v_2$ in this basis. The vector w_1 is $(\cos\alpha, \sin\alpha)$ where α is the angle between w_1 and w_a. The vector $v_1 = \cos\beta w_1$ is equal to $(\cos\beta \cos\alpha, \cos\beta \sin\alpha)$.

Next, we look at the vector $v_1 + v_2$. We fix α, β and v_1 and try to find the v_2 which maximizes p for the fixed α, β and v_1. The only place where v_2 appears in the optimization problem 1 is $\|P_a(v_1 + v_2)\|^2$ on the left hand side of constraint 1. Therefore, we should find v_2 that maximizes $\|P_a(v_1 + v_2)\|^2$. We have two cases:

1. $\alpha \geq \beta$.

   The angle between $v_1 + v_2$ and w_r is at most $\frac{\pi}{2} - \alpha + \beta$ (because the angle between v_1 and w_r is $\frac{\pi}{2} - \alpha$ and the angle between $v_1 + v_2$ and v_1 is β). Therefore, the projection of $v_1 + v_2$ to w_r is at least $\cos(\frac{\pi}{2} - \alpha + \beta) =$

$\sin(\alpha - \beta)$. Since w_r is a part of the rejecting subspace E_r, this means that $\|P_r(v_1 + v_2)\| \geq \sin(\alpha - \beta)$. Since $\|P_a(v_1 + v_2)\|^2 + \|P_r(v_1 + v_2)\|^2 = 1$, this also means that

$$\|P_a(v_1 + v_2)\|^2 \leq 1 - \sin^2(\alpha - \beta) = \cos^2(\alpha - \beta).$$

The maximum $\|P_a(v_1 + v_2)\| = \cos^2(\alpha - \beta)$ can be achieved if we put $v_1 + v_2$ in the plane spanned by w_a and w_r: $v_1 + v_2 = (\cos(\alpha - \beta), \sin(\alpha - \beta))$. Next, we can rewrite constraint 3 of the optimization problem as $1 - p_2 \geq p$. Then, constraints 1-3 together mean that

$$p = \min(\|P_a(v_1 + v_2)\|^2, \|P_r(v_1)\|^2 + p_2, 1 - p_2). \tag{1}$$

To solve the optimization problem, we have to maximize (1) subject to the conditions of the problem. From the expressions for v_1 and $v_1 + v_2$ above, it follows that (1) is equal to

$$p = \min(\cos^2(\alpha - \beta), \sin^2 \alpha \cos^2 \beta + p_2, 1 - p_2) \tag{2}$$

First, we maximize $\min(\sin^2 \alpha \cos^2 \beta + p_2, 1 - p_2)$. The first term is increasing in p_2, the second is decreasing. Therefore, the maximum is achieved when both become equal which happens when $p_2 = \frac{1 - \sin^2 \alpha \cos^2 \beta}{2}$. Then, both $\sin^2 \alpha \cos^2 \beta + p_2$ and $1 - p_2$ are $\frac{1 + \sin^2 \alpha \cos^2 \beta}{2}$. Now, we have to maximize

$$p = \min\left(\cos^2(\alpha - \beta), \frac{1 + \sin^2 \alpha \cos^2 \beta}{2}\right). \tag{3}$$

We first fix $\alpha - \beta$ and try to optimize the second term. Since $\sin \alpha \cos \beta = \frac{\sin(\alpha+\beta)+\sin(\alpha-\beta)}{2}$ (a standard trigonometric identity), it is maximized when $\alpha + \beta = \frac{\pi}{2}$ and $\sin(\alpha + \beta) = 1$. Then, $\beta = \frac{\pi}{2} - \alpha$ and (3) becomes

$$p = \min\left(\sin^2 2\alpha, \frac{1 + \sin^4 \alpha}{2}\right). \tag{4}$$

The first term is increasing in α, the second is decreasing. Therefore, the maximum is achieved when

$$\sin^2 2\alpha = \frac{1 + \sin^4 \alpha}{2}. \tag{5}$$

The left hand side of (5) is equal to $4\sin^2 \alpha \cos^2 \alpha = 4\sin^2 \alpha(1 - \sin^2 \alpha)$. Therefore, if we denote $\sin^2 \alpha$ by y, (5) becomes a quadratic equation in y:

$$4y(1 - y) = \frac{1 + y^2}{2}.$$

Solving this equation gives $y = \frac{4 + \sqrt{7}}{9}$ and $4y(1 - y) = \frac{52 + 4\sqrt{7}}{81} = 0.7726....$

2. $\alpha < \beta$.

We consider $\min(\|P_r(v_1)\|^2 + p_2, 1 - p_2) = \min(\sin^2 \alpha \cos^2 \beta + p_2, 1 - p_2)$. Since the minimum of two quantities is at most their average, this is at most

$$\frac{1 + \sin^2 \alpha \cos^2 \beta}{2}. \tag{6}$$

Since $\alpha < \beta$, we have $\sin \alpha < \sin \beta$ and (6) is at most $\frac{1 + \sin^2 \beta \cos^2 \beta}{2}$. This is maximized by $\sin^2 \beta = 1/2$. Then, we get $\frac{1 + 1/4}{2} = \frac{5}{8}$ which is less than $p = 0.7726...$ which we got in the first case.

Improved Bounds on the Weak Pigeonhole Principle and Infinitely Many Primes from Weaker Axioms

Albert Atserias*

Departament de Llenguatges i Sistemes Informàtics
Universitat Politècnica de Catalunya
Barcelona, Spain
atserias@lsi.upc.es

Abstract. We show that the known bounded-depth proofs of the Weak Pigeonhole Principle PHP_n^{2n} in size $n^{O(\log(n))}$ are not optimal in terms of size. More precisely, we give a size-depth trade-off upper bound: there are proofs of size $n^{O(d(\log(n))^{2/d})}$ and depth $O(d)$. This solves an open problem of Maciel, Pitassi and Woods (2000). Our technique requires formalizing the ideas underlying Nepomnjaščij's Theorem which might be of independent interest. Moreover, our result implies a proof of the unboundedness of primes in $I\Delta_0$ with a provably weaker 'large number assumption' than previously needed.

Keywords. Proof Complexity, Weak Pigeonhole Principle, Bounded Arithmetic.

1 Introduction

The Pigeonhole Principle PHP_n^{n+1} is a fundamental statement about cardinalities of finite sets. It says that $\{0, \ldots, n\}$ cannot be mapped injectively into $\{0, \ldots, n-1\}$. The Pigeonhole Principle is at the heart of many mathematical arguments, although implicitly quite often. As a matter of fact, it implies the Induction Principle, and so its apparent self-evidence makes it even more interesting.

The general form of the Pigeonhole Principle PHP_n^m states that if $m > n$, then $\{0, \ldots, m-1\}$ cannot be mapped injectively into $\{0, \ldots, n-1\}$. When m is substantially larger than n, say $m = 2n$, the principle is called the Weak Pigeonhole Principle. Still, it is quite often the case that this weaker form is enough to carry over many arguments, notably in finite combinatorics and number theory.

The complexity of proving a propositional encoding of the Pigeonhole Principle has been investigated in depth since the problem was proposed by Cook and Reckhow in connection with the $\mathbf{NP} \overset{?}{=} \mathbf{coNP}$ question [5]. Haken proved that PHP_n^{n+1} requires exponential-size proofs in Resolution [10]. On the other hand,

---

* Supported by the CUR, Generalitat de Catalunya, through grant 1999FI 00532. Partially supported by ALCOM-FT, IST-99-14186.

J. Sgall, A. Pultr, and P. Kolman (Eds.): MFCS 2001, LNCS 2136, pp. 148–158, 2001.

Buss proved that PHP_n^{n+1} has polynomial-size proofs in Frege systems [3]. Ajtai proved a superpolynomial lower bound for bounded-depth Frege systems [1], and that was improved to an exponential lower bound by Pitassi, Beame and Impagliazzo [19], and Krajícek, Pudlák and Woods [13] independently. For the Weak Pigeonhole Principle PHP_n^{2n}, the situation is quite different. While it is known that PHP_n^{2n} requires exponential-size proofs in Resolution [4], Paris, Wilkie and Woods [18,12] proved that it has quasipolynomial-size $(n^{O(\log(n))})$ proofs in bounded-depth Frege systems. More recently, Maciel, Pitassi and Woods [15] gave a new quasipolynomial-size proof of optimal depth. Both papers left open, however, whether depth could be traded for size; that is, whether allowing more depth in the proof would allow us reduce the size below $n^{\log(n)}$. We note that such a trade-off is known for the even weaker Pigeonhole Principle $\mathrm{PHP}_n^{n^2}$ [18].

The proofs of Paris, Wilkie and Woods [18] and Maciel, Pitassi and Woods [15] consist in reducing PHP_n^{2n} to $\mathrm{PHP}_n^{n^2}$. In both cases, they build an injective map from $\{0,\dots,n^2-1\}$ to $\{0,\dots,n-1\}$ by repeatedly composing a supposedly injective map from $\{0,\dots,2n-1\}$ to $\{0,\dots,n-1\}$. The difference in their proofs is, essentially, in the proof of $\mathrm{PHP}_n^{n^2}$. Our new contribution is showing that the repeated composition technique can be made more efficient in terms of size. That is, we reduce PHP_n^{2n} to $\mathrm{PHP}_n^{n^2}$ in size $n^{o(\log(n))}$ (notice the small oh). The price we need to pay for that is an increase in depth. More precisely, we show that PHP_n^{2n} reduces to $\mathrm{PHP}_n^{n^2}$ in size $n^{O(d(\log(n))^{2/d})}$ and depth d. This gives us the desired size-depth trade-off upper bound for PHP_n^{2n} since $\mathrm{PHP}_n^{n^2}$ is provable in size $n^{O(\log^{(2)}(n))}$ and depth $O(1)$ [18,12].

The most interesting particular case of our size-depth trade-off is when $d = O(1)$ since it proves that the previously known upper bound in bounded-depth Frege is not optimal. Indeed, $n^{O(d(\log(n))^{2/d})}$ grows slower than $n^{c\log(n)}$ for any constants $d > 2$ and $c > 0$. Thus, any lower bound proof will have to focus on a bound weaker than $n^{\log(n)^\epsilon}$ for any $\epsilon > 0$. We believe this is valuable information. The other interesting particular case is when $d = O(\log\log(n))$. In that case we obtain a proof of size $n^{O(\log\log(n))}$ and depth $O(\log\log(n))$. The bound $n^{O(\log\log(n))}$ is new in the context of PHP_n^{2n}.

The method that we use to reduce the size of the composition technique is inspired from the theory of automata. We observe that checking whether b is the image of a under repeated composition of a function f is a reachability problem in a graph. Therefore, one can use (an analogue of) Savitch's Theorem to efficiently solve the reachability problem. We are more ambitious and we use ideas from an old theorem of Nepomnjaščij that achieves a size-depth trade-off for the same problem [17]. We formalize the ideas in Nepomnjaščij's Theorem into a theorem of Bounded Arithmetic with an automatic translation into propositional Gentzen Calculus. This formalization may be of independent interest. We note that Nepomnjaščij's Theorem has received a renewed deal of attention recently in the context of time-space trade-off lower bounds for the satisfiability problem [6,14,7].

The new bounds on the Weak Pigeonhole Principle that we obtain have some consequences for Feasible Number Theory whose aim is to develop as much number theory as possible without exponentiation. The main open problem of the field is whether the bounded arithmetic theory $I\Delta_0$ can prove that there are unboundedly many primes [18,16,2]. Of course, Euclides' proof cannot be carried over in $I\Delta_0$ since exponentially large numbers are required in the proof. In a major breakthrough, Woods [21] showed that exponentiation can be replaced by a combinatorial argument using the Pigeonhole Principle PHP_n^{n+1}, and Paris, Wilkie and Woods [18] realized that the Weak Pigeonhole Principle PHP_n^{2n} was enough for that proof. As a corollary to their results, they show that $I\Delta_0$ augmented with the statement that $x^{\log(x)}$ exists proves that $(\exists y)(y > x \wedge prime(y))$. Our results improve this to show that, for every standard natural number k, the theory $I\Delta_0$ augmented with the statement that $x^{\log(x)^{1/k}}$ exists proves that $(\exists y)(y > x \wedge prime(y))$. Therefore, the large number assumption "$x^{\log(x)}$ exists" is not optimal. Indeed, for $k > 1$, one can build a model of $I\Delta_0$ with a non-standard element a such that $a^{\log(a)^{1/k}}$ exists in the model but $a^{\log(a)}$ does not.

2   The Proof of the WPHP in Propositional Logic

The propositional form of the Weak Pigeonhole Principle PHP_n^m that we use is formalized by the following sequent:

$$\bigwedge_{i=1}^{m} \bigvee_{j=1}^{n} p_{i,j} \vdash \bigvee_{k=1}^{n} \bigvee_{\substack{i,j=1 \\ i \neq j}}^{m} p_{i,k} \wedge p_{j,k}.$$

We will work with the propositional fragment of the sequent calculus LK. We refer the reader to any standard textbook for a definition [20,12].

Our goal is to prove a size-depth trade-off upper bound for proofs of PHP_n^{2n}. As mentioned in the introduction, our technique consists in reducing PHP_n^{2n} to $\mathrm{PHP}_n^{n^2}$. The difference with previous reductions is that our composition of mappings is done efficiently mimicking the proofs of Savitch's and Nepomnjaščij's Theorems in Complexity Theory. To complete the proof, we will use the fact that $\mathrm{PHP}_n^{n^2}$ has LK proofs of size $n^{O(\log^{(d)}(n))}$ and depth $O(d)$, where $\log^{(d)}(n)$ is the d-wise composition of log with itself [18,12].

For the sake of clarity of exposition, it is more convenient to prove the following extreme case of the trade-off first.

Theorem 1. PHP_n^{2n} has LK proofs of size $n^{O(\log \log(n))}$ and depth $3 \log \log(n)$.

Proof: Let $d = \log \log(n)$. For every $\alpha \in \{0,1\}^{\leq d}$, we define numbers L_α and R_α inductively as follows: Let $L_\lambda = \log(n)$, $R_\lambda = 0$, and $L_{\alpha 0} = L_\alpha$, $R_{\alpha 1} = R_\alpha$, $L_{\alpha 1} = R_{\alpha 0} = \frac{1}{2}(L_\alpha + R_\alpha)$. Now we define sets A_α, B_α and C_α as follows: Let $A_\alpha = \{0, \ldots, 2^{L_\alpha} n - 1\}$, $B_\alpha = \{0, \ldots, 2^{R_\alpha} n - 1\}$ and $C_\alpha = \{0, \ldots, 2^{\frac{1}{2}(L_\alpha + R_\alpha)} n - 1\}$. Observe that $A_\lambda = \{0, \ldots, n^2 - 1\}$ and $B_\lambda = \{0, \ldots, n - 1\}$.

For $\alpha \in \{0,1\}^d$, $x \in A_\alpha$, $y \in B_\alpha$, let R^α_{xy} be defined as follows: If $\lfloor x/2n \rfloor \neq \lfloor y/n \rfloor$, define $R^\alpha_{xy} = 0$; otherwise, define $R^\alpha_{xy} = P_{x'y'}$ where $x' = x \bmod 2n$ and $y' = y \bmod n$. For $\alpha \in \{0,1\}^{<d}$, $x \in A_\alpha$, $y \in B_\alpha$, define $R^\alpha_{xy} = \bigvee_{z \in C_\alpha} R^{\alpha 0}_{xz} R^{\alpha 1}_{zy}$. It is easy to see that the size of R^α_{xy} is bounded by $n^{2(d-|\alpha|)}$ and the depth is bounded by $2(d - |\alpha|)$. In particular, R^λ_{xy} has size bounded by $n^{2 \log \log(n)}$ and depth bounded by $2 \log \log(n)$.

We want to prove the following sequents

$$\bigwedge_{\alpha \in \{0,1\}^k} \bigwedge_x \bigvee_y R^\alpha_{xy} \rightarrow \bigvee_{\alpha \in \{0,1\}^k} \bigvee_y \bigvee_{x_1 \neq x_2} R^\alpha_{x_1 y} R^\alpha_{x_2 y} \tag{1}$$

in size bounded by n^{ck} and depth bounded by $3k$, where c is a sufficiently large constant independent of n ($c = 20$ should work). When $k = d = \log \log(n)$, it is easy to see that sequent (1) is equivalent to PHP^{2n}_n after contraction of repeated formulas, and the theorem will follow. We fix a sufficiently large n, and proceed by induction on k.

Observe that the base case $k = 0$ is precisely the sequent $\mathrm{PHP}^{n^2}_n(R^\lambda)$. Since the sequent $\mathrm{PHP}^{n^2}_n$ has LK proofs of size $n^{O(\log^{(3)}(n))}$ and constant depth, it follows that $\mathrm{PHP}^{n^2}_n(R^\lambda)$ has LK-proofs of size bounded by $n^{c \log \log(n)}$ and depth bounded by $3 \log \log(n)$. Here is where we need n to be sufficiently large.

Suppose next that we have proved sequent (1) for a $k > 0$ in size n^{ck} and depth $3k$. We prove it for $k + 1$. We first prove the following sequents for every $\alpha \in \{0,1\}^k$ and w:

$$\bigwedge_x \bigvee_y R^{\alpha 0}_{xy} \wedge \bigwedge_x \bigvee_y R^{\alpha 1}_{xy} \rightarrow \bigvee_y R^\alpha_{wy}. \tag{2}$$

Recall that R^α_{wy} stands for $\bigvee_z R^{\alpha 0}_{wz} R^{\alpha 1}_{zy}$. Start with the sequents $R^{\alpha 0}_{wz} \rightarrow R^{\alpha 0}_{wz}$ and $\bigwedge_x \bigvee_y R^{\alpha 1}_{xy} \rightarrow \bigvee_y R^{\alpha 1}_{zy}$, and apply right $\wedge$ introduction to obtain

$$R^{\alpha 0}_{wz}, \bigwedge_x \bigvee_y R^{\alpha 1}_{xy} \rightarrow R^{\alpha 0}_{wz} \wedge \bigvee_y R^{\alpha 1}_{zy}. \tag{3}$$

By distributivity we easily get

$$R^{\alpha 0}_{wz}, \bigwedge_x \bigvee_y R^{\alpha 1}_{xy} \rightarrow \bigvee_y R^{\alpha 0}_{wz} R^{\alpha 1}_{zy}. \tag{4}$$

By left $\vee$-introduction, left weakening, left $\wedge$-introduction, right $\vee$-introduction and commutativity of $\vee$, in this order, we get the desired sequent (2).

Next we prove the following sequents for every $\alpha \in \{0,1\}^k$, $w_1 \neq w_2$ and y:

$$R^\alpha_{w_1 y} R^\alpha_{w_2 y} \rightarrow \bigvee_z \bigvee_{x_1 \neq x_2} R^{\alpha 0}_{x_1 z} R^{\alpha 0}_{x_2 z} \vee \bigvee_z \bigvee_{x_1 \neq x_2} R^{\alpha 1}_{x_1 z} R^{\alpha 1}_{x_2 z}. \tag{5}$$

Recall that $R^\alpha_{w_i y}$ stands for $\bigvee_z R^{\alpha 0}_{w_i z} R^{\alpha 1}_{zy}$. Using distributivity, derive the sequent

$$R^\alpha_{w_1 y} R^\alpha_{w_2 y} \rightarrow \bigvee_{z_1, z_2} R^{\alpha 0}_{w_1 z_1} R^{\alpha 1}_{z_1 y} R^{\alpha 0}_{w_2 z_2} R^{\alpha 1}_{z_2 y}. \tag{6}$$

For $z_1 = z_2$, derive the sequent $R^{\alpha 0}_{w_1 z_1} R^{\alpha 1}_{z_1 y} R^{\alpha 0}_{w_2 z_2} R^{\alpha 1}_{z_2 y} \to R^{\alpha 0}_{w_1 z_1} R^{\alpha 0}_{w_2 z_2}$. For $z_1 \neq z_2$, derive the sequent $R^{\alpha 0}_{w_1 z_1} R^{\alpha 1}_{z_1 y} R^{\alpha 0}_{w_2 z_2} R^{\alpha 1}_{z_2 y} \to R^{\alpha 1}_{z_1 y} R^{\alpha 1}_{z_2 y}$. Left $\vee$-introduction and right $\vee$-introduction gives the sequent

$$\bigvee_{z_1,z_2} R^{\alpha 0}_{w_1 z_1} R^{\alpha 1}_{z_1 y} R^{\alpha 0}_{w_2 z_2} R^{\alpha 1}_{z_2 y} \to \bigvee_{z_1=z_2} R^{\alpha 0}_{w_1 z_1} R^{\alpha 0}_{w_2 z_2} \vee \bigvee_{z_1 \neq z_2} R^{\alpha 1}_{z_1 y} R^{\alpha 1}_{z_2 y}. \tag{7}$$

A cut with sequent (6), right weakening, right $\vee$-introduction and commutativity of $\vee$, in this order, give the desired sequent (5).

Now combine all sequents (2) by right $\wedge$-introduction, left $\wedge$-introduction and commutativity of $\wedge$, in this order, to obtain

$$\bigwedge_{\alpha \in \{0,1\}^{k+1}} \bigwedge_x \bigvee_y R^\alpha_{xy} \to \bigwedge_{\alpha \in \{0,1\}^k} \bigwedge_x \bigvee_y R^\alpha_{xy}. \tag{8}$$

Similarly, combine all sequents (5) by left $\vee$-introduction, right $\vee$-introduction and commutativity of $\vee$, in this order, to obtain

$$\bigvee_{\alpha \in \{0,1\}^k} \bigvee_y \bigvee_{x_1 \neq x_2} R^\alpha_{x_1 y} R^\alpha_{x_2 y} \to \bigvee_{\alpha \in \{0,1\}^{k+1}} \bigvee_y \bigvee_{w_1 \neq w_2} R^\alpha_{w_1 y} R^\alpha_{w_2 y}. \tag{9}$$

Finally, two cuts using sequents (1), (8) and (9) give the desired result for $k+1$. If c is sufficiently large, it is easy to check that the size of this proof is bounded by $n^{c(k+1)}$ and its depth is bounded by $3(k+1)$. This completes the induction step. $\square$

An analogous argument mimicking the proof of Nepomnjaščij's Theorem, instead of Savitch's Theorem as above, would give a general size-depth trade-off upper bound. Since the notation in the proof would get fairly tedious, we prefer to state it without proof and get it as a corollary to Theorem 6 below (see the end of Section 4).

Theorem 2. PHP^{2n}_n has LK proofs of size $n^{O(d(\log(n))^{2/d})}$ and depth d.

3   Formalization of Nepomnjaščij's Theorem

Let us briefly recall the proof of a general form of Nepomnjaščij's Theorem. This will be of help later.

Theorem 3. (General Form of Nepomnjaščij's Theorem) *Let $K = K(n)$, $T = T(n)$ and $S = S(n)$ be time-constructible functions such that $K(n) \geq 2$. For every non-deterministic Turing machine running in simultaneous time T and space S, there exists an equivalent alternating Turing machine running in time $O(SK \log(T)/\log(K))$ and $2\log(T)/\log(K)$ alternations.*

Proof: Let M be a non-deterministic Turing machine running in simultaneous time T and space S. The idea is to divide the reachability problem between configurations of M into many equivalent subproblems of smaller size. Hence, configuration C_K is reachable from configuration C_0 in N steps if and only if there exist $K - 1$ intermediate configurations $C_1, \ldots, C_{K-1}$ such that for every $i \in \{0, \ldots, K-1\}$, configuration C_{i+1} is reachable from configuration C_i in N/K steps. An alternating Turing machine can existentially quantify those intermediate configurations, and universally branch to check that every two consecutive configurations are reachable from each other in the appropriate number of steps. Applying this recursively yields an alternating machine that checks whether an accepting configuration is reachable from the initial configuration of M.

The details of the calculations follow. Since M runs in space S, each configuration can be coded by a binary word of length $O(S)$. The outermost reachability problem has length T since M runs in time T. The second level of reachability subproblems have length T/K. In general, the subproblems at level i have length T/K^i. After $\log(T)/\log(K)$ levels, we reached a trivial reachability subproblem. Therefore, the whole computation of the simulating machine takes time $O(SK \log(T)/\log(K))$ and $2 \log(T)/\log(K)$ alternations. $\square$

Our next goal is to formalize the ideas of Theorem 3 into a theorem of the bounded arithmetic theory $I\Delta_0$. We will use the beautiful arithmetization in $I\Delta_0$ of Chapter V, Section 3, in the book of Hájek and Pudlák [9]. In summary, the arithmetization allows us to manipulate sequences provably in $I\Delta_0$. Thus, there are formulas $Seq(s)$ meaning that s is the code of a sequence, $(s)_i = x$ meaning that the i-th element of the sequence s is x, and $s \frown t = p$ meaning that sequence p is the result of appending sequence t to the end of sequence s. The coding is so that if $s = (x)$, the sequence with x as its only element, then $I\Delta_0$ proves $s \leq 9 \cdot x^2$. Morevoer, $I\Delta_0$ proves that for every sequence s, the bound $s \frown (x) \leq 9 \cdot x^2 \cdot s$ holds (see Lemma 3.7 in page 297 of [9]). It follows that $I\Delta_0$ proves that if l^r exists below n and s is a sequence of length r all of whose elements are smaller than l, then $s \leq 9^r \cdot l^{2r}$ (by Δ_0-induction on r). Therefore, the coding is fairly close to its information theoretic bound. Of course, $I\Delta_0$ can prove several other obvious facts about $Seq(s)$, $(s)_i$ and $s \frown t$ (see [9] for details).

Let $\theta(x, y)$ be a Δ_0-formula in a language L extending the usual language of arithmetic $\{+, \times, \leq\}$. Obviously, $\theta(x, y)$ defines a binary relation on any model for the language L that may be interpreted as an infinite directed graph. We define Δ_0-formulas $\Theta_i(x, y)$, with certain parameters, meaning that y is reachable from x under certain conditions that depend on the parameters. More precisely, let $\Theta_0(x, y, t, r, l, n) = \theta(x, y)$ (note that the parameters t, r and l are not used for the moment). Inductively, we define $\Theta_{k+1}(x, y, t, r, l, n)$ as follows:

$$(\exists z \leq n)(Seq(z) \wedge (z)_0 = x \wedge (z)_r = y \wedge (\forall i < r+1)((z)_i \leq l) \wedge$$
$$\wedge (\forall i < r)(\exists c, c' \leq z)((z)_i = c \wedge (z)_{i+1} = c' \wedge \Theta_k(c, c', t, l, l, n))).$$

Informally, the formula $\Theta_k(x, y, t, t, l, n)$ says that y is reachable from x in t^k steps according to the directed graph defined by $\theta(x, y)$ as long as each number in the path is bounded by l. The following theorem states this in the form of recursive equations:

Theorem 4. *Let $M \models I\Delta_0(L)$, and let $x, y, t, r, l, n \in M$ be such that $x, y \leq l$, l^t exists in M, $r < t$, and $9^{t+1} \cdot l^{2(t+1)} \leq n$. Then,*

(i) $\Theta_{k+1}(x, y, t, 0, l, n) \leftrightarrow x = y$,
(ii) $\Theta_{k+1}(x, y, t, r+1, l, n) \leftrightarrow (\exists c \leq l)(\Theta_k(x, c, t, t, l, n) \wedge \Theta_{k+1}(c, y, t, r, l, n))$

hold in M.

Proof: (i) Assume $\Theta_{k+1}(x, y, t, 0, l, n)$ holds. Then, for some $z \leq n$, we have $(z)_0 = x$ and $(z)_0 = y$. Hence, $x = y$. Conversely, if $x = y$, then $z = (x) \leq 9 \cdot x^2 \leq 9 \cdot l^2 \leq n$ is a witness for the existential quantifier in $\Theta_{k+1}(x, y, t, 0, l, n)$. (ii) Assume $\Theta_{k+1}(x, y, t, r+1, l, n)$ holds. Let $z \leq n$ be the witness for its existential quantifier. Let $c = (z)_1$ and note that $(z)_0 = x$. Obviously, $c \leq l$ and $\Theta_k(x, c, t, t, l, n)$ holds. Now let z' be the sequence that results from z when $(z)_0$ is dropped (definable in $I\Delta_0$). It is easily seen that $z' \leq z \leq n$, and $\Theta_{k+1}(c, y, t, r, l, n)$ holds with z' witnessing its existential quantifier. Conversely, if $\Theta_k(x, c, t, t, l, n)$ and $\Theta_{k+1}(c, y, t, r, l, n)$ hold, let $z' \leq n$ be a witness for the existential quantifier in the latter. We can assume that z' codes a sequence of $r + 1$ numbers bounded by l each; if not, just trim z' to the first $r + 1$ numbers (definable in $I\Delta_0$) and the result is still a witness of the existential quantifier. Moreover, $z' \leq 9^{r+1} \cdot l^{2(r+1)}$. Since $x \leq l$ and $r < t$, the sequence $(x) \frown z'$ is coded by some $z \leq 9^{r+2} \cdot l^{2(r+2)} \leq n$. such a z is a witness for the existential quantifier in $\Theta_{k+1}(x, y, t, r+1, l, n)$ and we are done. $\square$

The reader will notice that the recursive equations in Theorem 4 correspond to the inductive definition of the transitive closure of the graph defined by Θ_k. The innermost level of stratification, namely $k = 0$, is the inductive definition of the transitive closure of the graph defined by θ. It is in this sense that we interpret Theorem 4 as a formalization of Nepomnjaščij's Theorem.

4   The Proof of the WPHP in Bounded Arithmetic

The graph of the exponentiation function $x = y^z$ is definable by a Δ_0-formula on the natural numbers. Moreover, Pudlák gave a definition with the basic properties being provable in $I\Delta_0$. Similarly, one can define $y = \lceil \log(x) \rceil$, and $y = \lceil \log^{(k)}(x) \rceil$ in $I\Delta_0$. We make the convention that when expressions such as $\log(a)$ or $(\log(a))^\epsilon$ do not come up integer numbers, the nearest larger integer is assumed unless specified otherwise. Thus, $(\log(a))^\epsilon$ really stands for $\lceil \log(a)^\epsilon \rceil$.

We let L be the usual language of arithmetic $\{+, \times, \leq\}$ extended by a unary function symbol α. We denote $I\Delta_0(L)$ (see the previous section) by $I\Delta_0(\alpha)$. The Weak Pigeonhole Principle PHP_n^m is formalized by the following statement:

$$(\forall x < m)(\alpha(x) < n)) \rightarrow (\exists x, y < m)(x \neq y \wedge \alpha(x) = \alpha(y)).$$

We will abbreviate this statement by $\neg\alpha : m \xrightarrow{1-1} n$. We will make use of the following result:

Theorem 5. [18] *For every* $K > 0$,

$$I\Delta_0(\alpha) \vdash (\exists y)(y = x^{\log^{(K)}(x)}) \to \neg\alpha : x^2 \xrightarrow{1-1} x.$$

Here, $\log^{(0)}(x) = x$ *and* $\log^{(k+1)}(x) = \log(\log^{(k)}(x))$.

Next, we formalize the reduction of PHP_n^{2n} to $\mathrm{PHP}_n^{n^2}$ using the result of Section 3.

Theorem 6. *For every* $K > 0$,

$$I\Delta_0(\alpha) \vdash (\exists y)(y = x^{(\log(x))^{1/K}}) \to \neg\alpha : 2x \xrightarrow{1-1} x.$$

Proof: Let $\epsilon = 1/K$. Let $\mathbf{M} = (M, F)$ be a model of $I\Delta_0(\alpha)$, let $a \in M$ be such that $a^{(\log(a))^\epsilon}$ exists in M, and assume for contradiction that $F : 2a \xrightarrow{1-1} a$. Let $T = (\log(a))^\epsilon$, $L = a^2$, and $N = 9^{T+1} \cdot L^{2(T+1)}$. Observe that N exists in M since $a^{(\log(a))^\epsilon}$ exists, and M is closed under multiplication. Define $\theta(x, y)$ as follows (see the text that follows the formula for the intuition):

$$(\exists r < 2a)(\exists r' < a)(\exists q, q' < a)(x = 2aq + r \wedge y = aq' + r' \wedge \alpha(r) = r' \wedge q = q').$$

Note that this Δ_0-formula could be informally abbreviated by

$$y \bmod a = \alpha(x \bmod 2a) \wedge \lfloor y/a \rfloor = \lfloor x/2a \rfloor$$

when $x, y \in \{0, \ldots, a^2 - 1\}$.

Lemma 7. $\theta(x, y)^{\mathbf{M}} : 2^{i+1}a \xrightarrow{1-1} 2^i a$ *for every* $i < \log(a)$.

Proof: Given $u \in 2^{i+1}a$, let $v = \lfloor u/2a \rfloor \cdot a + \alpha(u \bmod 2a)$. It is not hard to see that $v \in 2^i a$ and $\theta(u, v)$ holds. Moreover, if $w \in 2^i a$ is such that $\theta(u, w)$ holds, then $\lfloor v/a \rfloor = \lfloor u/2a \rfloor = \lfloor w/a \rfloor$ and $v \bmod a = \alpha(u \bmod 2a) = w \bmod a$. Hence, $v = w$. This shows that $\theta(x, y)^{\mathbf{M}}$ is the graph of a function from $2^{i+1}a$ to $2^i a$.

We show next that the function is one-to-one. Let $u, v \in 2^{i+1}a$ and $w \in 2^i a$ be such that $\theta(u, w)$ and $\theta(v, w)$. Then, $\lfloor u/2a \rfloor = \lfloor v/2a \rfloor = \lfloor w/a \rfloor$ and $F(u \bmod 2a) = F(v \bmod 2a) = w \bmod a$. Since F is one-to-one, it must be then that $u \bmod 2a = v \bmod 2a$. Hence, $u = 2a \cdot \lfloor u/2a \rfloor + (u \bmod 2a) = 2a \cdot \lfloor v/2a \rfloor + (v \bmod 2a) = v$. $\square$

Lemma 8. $\Theta_K(x, y, T, T, L, N)^{\mathbf{M}} : a^2 \xrightarrow{1-1} a$.

Proof: We prove that for every $k \leq K$, the formula $\Theta_k(x, y, T, T, L, N)$ defines a one-to-one mapping $\Theta_k : 2^{(i+1)T^k}a \to 2^{iT^k}a$ for every $i < \log(a)/T^k$. The lemma will be proved since $T^K = ((\log(a))^{1/K})^K = \log(a)$ (in fact, $T^K \geq \log(a)$ by our convention on rounding). The proof is by induction on k (this induction is outside M).

Lemma 7 takes care of the base case $k = 0$. We turn to the inductive case $0 < k \leq K$. Fix $i < \log(a)/T^k$. We prove that for every $r \leq T$, the formula

$\Theta_k(x, y, T, r, L, N)$ defines a one-to-one mapping $\Theta_k^r : 2^{(iT+r)T^{k-1}} a \rightarrow 2^{iT^k} a$. That is, we prove that for every $r \leq T$, $x < 2^{(iT+r)T^{k-1}} a$ and $y < 2^{iT^k} a$,

$$\Theta_k(x, z, T, r, L, N) \wedge \Theta_k(y, z, T, r, L, N) \rightarrow x = y$$

holds in $\mathbf{M}$. We use the schema of Δ_0-induction on r in the Δ_0-formula above. The base case $r = 0$ is immediate since $\Theta_k(x, y, T, 0, L, N)$ defines the identity by Theorem 4. Suppose that $0 < r \leq T$, and that $\Theta_k(x, y, T, r-1, L, N)$ defines a one-to-one mapping $\Theta_k^{r-1} : 2^{(iT+r-1)T^{k-1}} a \rightarrow 2^{iT^k} a$. Since $\Theta_{k-1}(x, y, T, T, L, N)$ defines a one-to-one mapping $\Theta_{k-1} : 2^{(iT+r)T^{k-1}} a \rightarrow 2^{(iT+r-1)T^{k-1}} a$ by induction hypothesis on k, and since $\Theta_k(x, y, T, r, L, N)$ defines the composition of Θ_{k-1} and Θ_k^{r-1} by Theorem 4 (observe that $x, y \leq L$, $r-1 < T$ and $9^{T+1} \cdot L^{2(T+1)} \leq N$), it follows that $\Theta_k(x, y, T, r, L, N)$ defines a one-to-one mapping $\Theta_k^r : 2^{(iT+r)T^{k-1}} a \rightarrow 2^{iT^k} a$ as required. $\square$

Since Θ_K is a $\Delta_0(\alpha)$ formula, we have that $(M, \Theta_K(x, y, T, T, L, N)^{\mathbf{M}}) \models I\Delta_0(\alpha)$. Moreover, $a^{\log^{(2)}(a)} < a^{(\log(a))^\epsilon}$ exists in M. It follows from Theorem 5 that $\Theta_K(x, y, T, T, L, N)^{\mathbf{M}}$ is not a one-to-one mapping from $a^2 \rightarrow a$; a contradiction to Lemma 8. $\square$

It is well-known that proofs in $I\Delta_0(\alpha)$ translate into bounded-depth LK proofs of polynomial-size. When statements of the form "$f(x)$ exists" are required as in Theorem 4, the translations come up of size $f(n)^{O(1)}$ (see [12], for example). This gives us Theorem 2 as a corollary.

5   Infinitude of Primes

The existence of infinitely many primes is not guaranteed in weak fragments of arithmetic. For example, it is known that I_{open}, Peano Arithmetic with induction restricted to open formulas, has models with a largest prime [16]. It is an open problem whether $I\Delta_0$ proves the infinitude of primes. It is known, however, that $I\Delta_0$ augmented with the axiom $(\forall x)(\exists y)(y = x^{\log(x)})$ proves it. In addition, a *single* application of this axiom suffices. More precisely,[1]

Theorem 9.   [18] $I\Delta_0 \vdash (\exists y)(y = x^{\log(x)}) \rightarrow (\exists y)(y > x \wedge prime(y))$.

The aim of this section is to show that a weaker axiom suffices, and so the existence of $x^{\log(x)}$ is not the optimal large number assumption. Namely,

Theorem 10. $I\Delta_0 \vdash (\exists y)(y = x^{(\log(x))^{1/K}}) \rightarrow (\exists y)(y > x \wedge prime(y))$ *for every* $K > 0$. *Moreover, there exists a model* $M \models I\Delta_0$ *with a non-standard element* $a \in M$ *such that* $a^{(\log(a))^{1/K}}$ *exists in* M *but* $a^{\log(a)}$ *does not.*

---
[1] This notion of limited use of an axiom also appears in Chapter V, Section 5, Subsection (g) of [9].

Proof: For the second part, let M be a non-standard model of true arithmetic, and let $a \in M$ be its non-standard element. Obviously, $a^{(\log(a))^{1/K}}$ exists in M since the function is total in true arithmetic. Let $N = \{n \in M : (\exists i \in \omega)(M \models n < a^{i(\log(a))^{1/K}})\}$. It is not hard to see that N is a cut of M that is closed under addition and multiplication. It follows that $N \models I\Delta_0$ (see Lemma 5.1.3 in page 64 of [12]). Finally, $a^{(\log(a))^{1/K}}$ still exists in N by absoluteness of the Δ_0-formula expressing the graph of exponentiation. However, $a^{\log(a)}$ does not exist in N because $a^{\log(a)} > a^{i(\log(a))^{1/K}}$ in M for every standard $i \in \omega$.

For the first part, suppose that $a^{(\log(a))^{1/K}}$ exists in $M \models I\Delta_0$. Our goal is to show that no Δ_0-definable function $F : M \to M$ maps $9a \log(a)$ injectively into $8a \log(a)$. The result would follow from Theorem 11 of [18] since then a prime exists in M between a and a^{11}. Let $b = a^2$ and observe that $b^{(\log(b))^{1/K}} = a^{2^{1+1/K}(\log(a))^{1/K}}$ exists in M since it is closed under multiplication. By Theorem 6, no Δ_0-definable function maps $2b$ injectively into b. It follows that no Δ_0-definable functions maps $\frac{9}{8}b$ injectively into b; otherwise we could compose that function with itself a constant number of times to maps $2b$ injectively into b. We conclude that no Δ_0-definable function maps $9a \log(a)$ injectively into $8a \log(a)$; otherwise, we could juxtapose that function with itself to obtain a Δ_0-definable function mapping $\frac{9}{8}b$ injectively into b (break b and $\frac{9}{8}b$ into $a/8 \log(a)$ blocks of size $8a \log(a)$ and $9a \log(a)$ respectively). $\square$

We note that $I\Delta_0$ proves $(\forall x)(\exists y)(y = x^{\log(x)^{\epsilon}}) \to (\forall x)(\exists y)(y = x^{\log(x)})$. However, the second part of Theorem 10 implies that $I\Delta_0$ does not prove $(\exists y)(y = x^{\log(x)^{\epsilon}}) \to (\exists y)(y = x^{\log(x)})$.

6   Discussion and Open Problems

Another major open problem in Feasible Number Theory is whether Fermat's Little Theorem is provable in $I\Delta_0$. Berarducci and Intrigila [2] point out that one important difficulty is that the modular exponentiation relation $x^y \equiv z \pmod{n}$ is not known to be Δ_0-definable. The situation has changed, however. Very recently, Hesse [11] proved that the modular exponentiation relation on numbers of $O(\log(n))$ bits is first-order definable. A well-known translational argument shows then that $x^y \equiv z \pmod{n}$ is Δ_0-definable. The proof of this result, however, seems to rely on Fermat's Little Theorem, and therefore it is not clear whether the basic properties of modular exponentiation are provable in $I\Delta_0$.

Open Problem 1 *Find a Δ_0 definition of the modular exponentiation relation whose basic properties are provable in $I\Delta_0$; namely, $x^y x^z \equiv x^{y+z} \pmod{n}$ and $(x^y)^z \equiv x^{yz} \pmod{n}$.*

We believe that a positive solution to this open problem would help developping the number theory of $I\Delta_0$ in the same way that the Δ_0-definition of the (non-modular) exponentiation relation helped developping the metamathematics of $I\Delta_0$ [8,9].

Acknowledgments. I thank J. L. Balcázar and M. L. Bonet for helpful comments.

References

1. M. Ajtai. The complexity of the pigeonhole principle. In *29th Annual IEEE Symposium on Foundations of Computer Science*, pages 346–355, 1988. 149
2. A. Berarducci and B. Intrigila. Combinatorial principles in elementary number theory. *Annals of Pure and Applied Logic*, 55:35–50, 1991. 150, 157
3. S. R. Buss. Polynomial size proofs of the propositional pigeonhole principle. *Journal of Symbolic Logic*, 52(4):916–927, 1997. 149
4. S. R. Buss and G. Turán. Resolution proofs of generalized pigeonhole principles. *Theoretical Computer Science*, 62:311–317, 1988. 149
5. S. Cook and R. Reckhow. The relative efficiency of propositional proof systems. *Journal of Symbolic Logic*, 44:36–50, 1979. 148
6. L. Fortnow. Time-space tradeoffs for satisfiability. In *12th IEEE Conference in Computational Complexity*, pages 52–60, 1997. To appear in *Journal of Computer and System Sciences*. 149
7. L. Fortnow and D. van Meldebeek. Time-space tradeoffs for non-deterministic computation. In *15th IEEE Conference in Computational Complexity*, 2000. 149
8. H. Gaifman and C. Dimitracopoulos. Fragments of Peano's arithmetic and the MRDP theorem. In *Logic and algorithmic*, number 30 in Monographies de l'Enseignement Mathématique, pages 187–206. Univeristé de Genève, 1982. 157
9. P. Hájek and P. Pudlák. *Metamathematics of First-Order Arithmetic*. Springer, 1993. 153, 156, 157
10. A. Haken. The intractability of resolution. *Theoretical Computer Science*, 39:297–308, 1985. 148
11. W. Hesse. Division is in uniform TC^0. To appear in ICALP'01, 2001. 157
12. J. Krajíček. *Bounded arithmetic, propositional logic, and complexity theory*. Cambridge University Press, 1995. 149, 150, 156, 157
13. J. Krajíček, P. Pudlák, and A. Woods. Exponential lower bound to the size of bounded depth frege proofs of the pigeon hole principle. *Random Structures and Algorithms*, 7(1):15–39, 1995. 149
14. R. J. Lipton and A. Viglas. On the complexity of SAT. In *40th Annual IEEE Symposium on Foundations of Computer Science*, pages 459–464, 1999. 149
15. A. Maciel, T. Pitassi, and A. R. Woods. A new proof of the weak pigeonhole principle. In *32th Annual ACM Symposium on the Theory of Computing*, 2000. 149
16. A. J. Macintyre and D. Marker. Primes and their residue rings in models of open induction. *Annals of Pure and Applied Logic*, 43(1):57–77, 1989. 150, 156
17. V. A. Nepomnjaščij. Rudimentary predicates and Turing calculations. *Soviet Math. Dokl.*, 11:1462–1465, 1970. 149
18. J. B. Paris, A. J. Wilkie, and A. R. Woods. Provability of the pigeonhole principle and the existence of infinitely many primes. *Journal of Symbolic Logic*, 53(4):1235–1244, 1988. 149, 150, 155, 156, 157
19. T. Pitassi, P. Beame, and R. Impagliazzo. Exponential lower bounds for the pigeonhole principle. *Computational Complexity*, 3(2):97–140, 1993. 149
20. G. Takeuti. *Proof Theory*. North-Holland, second edition, 1987. 150
21. A. R. Woods. *Some problems in logic and number theory, and their connections*. PhD thesis, Univerity of Manchester, Department of Mathematics, 1981. 150

Analysis Problems for Sequential Dynamical Systems and Communicating State Machines

Chris Barrett[1], Harry B. Hunt III[2,3], Madhav V. Marathe[1], S. S. Ravi[2,3], Daniel J. Rosenkrantz[2], and Richard E. Stearns[2]

[1] Los Alamos National Laboratory, MS M997
P.O. Box 1663, Los Alamos, NM 87545.
{barrett,marathe}@lanl.gov[†]
[2] Department of Computer Science, University at Albany - SUNY
Albany, NY 12222.
{hunt,ravi,djr,res}@cs.albany.edu[‡]
[3] Part of the work was done while the authors were visiting the Basic and Applied Simulation Sciences Group (TSA-2) of the Los Alamos National Laboratory.

Abstract. Informally, a sequential dynamical system (SDS) consists of an undirected graph where each node v is associated with a state s_v and a transition function f_v. Given the state value s_v and those of the neighbors of v, the function f_v computes the next value of s_v. The node transition functions are evaluated according to a specified total order. Such a computing device is a mathematical abstraction of a simulation system. We address the complexity of some state reachability problems for SDSs. Our main result is a dichotomy between classes of SDSs for which the state reachability problems are computationally intractable and those for which the problems are efficiently solvable. These results also allow us to obtain stronger lower bounds on the complexity of reachability problems for cellular automata and communicating state machines.

1 Introduction and Motivation

We study the computational complexity of some state reachability problems associated with a new class of finite discrete dynamical systems, called **Sequential Dynamical Systems** (SDSs), proposed in [BR99,BMR99]. Informally, an SDS consists of an undirected graph where each node v is associated with a state s_v and a transition function f_v. Given the state value s_v and those of the neighbors of v, the function f_v computes the next value of s_v. The node transition functions are evaluated according to a specified total order. A formal definition of an SDS is given in Section 2.

SDSs are closely related to classical Cellular Automata (CA), a widely studied class of finite discrete dynamical systems used to model problems in physics and

---

[†] The work is supported by the Department of Energy under Contract W-7405-ENG-36

[‡] Supported by a grant from Los Alamos National Laboratory and by NSF Grant CCR-97-34936

J. Sgall, A. Pultr, and P. Kolman (Eds.): MFCS 2001, LNCS 2136, pp. 159–172, 2001.
© Springer-Verlag Berlin Heidelberg 2001

complex systems. Computability aspects of dynamical systems in general and cellular automata in particular have been widely studied in the literature (see for example, [Wo86,Gu89]). Dynamical systems are closely related to networks of communicating automata, transition systems and sequential digital circuits (see for example, [Ra92,AKY99,RH93,Hu73]).

In this paper, we will restrict ourselves to *Simple-SDSs*, that is, SDSs with the following additional restrictions: (i) the state of each node is Boolean and (ii) each local transition function is Boolean and symmetric. Our *hardness results* hold even for such simple-SDSs and thus imply analogous hardness results for more general models.

We study the computational complexity of determining some properties of SDSs. The properties studied include classical questions such as reachability ("Does a given SDS starting from configuration C ever reach configuration C'?") and fixed points ("Does a given SDS have a configuration C such that once C is reached, the SDS stays in C forever?") that are commonly studied by the dynamical systems community. Specifically, we investigate whether such properties can be determined efficiently using computational resources that are polynomial in the size of the SDS representation.

The original motivation to develop a mathematical and computational theory of SDSs was to provide a formal basis for the design and analysis of *large-scale computer simulations*. Because of the widespread use of computer simulations, it is difficult to give a formal definition of a computer simulation that is applicable to all the various settings where it is used. An important characteristic of any computer simulation is the generation of global dynamics by iterated composition of local mappings. Thus, we view simulations as comprised of the following: (i) a collection of entities with state values and local rules for state transitions, (ii) an interaction graph capturing the local dependency of an entity on its neighboring entities and (iii) an update sequence or schedule such that the causality in the system is represented by the composition of local mappings. References [BWO95,BB+99] show how simulations of large-scale transportation systems and biological systems can be modeled using appropriate SDSs.

Following [BPT91], we say that a system is predictable if basic properties such as reachability and fixed point existence can be determined in time that is polynomial in the size of the system specification. Our **PSPACE**-completeness results for predicting the behavior of "very simple" systems essentially imply that the systems are not easily predictable; in fact, our results imply that no prediction method is likely to be more efficient than running the simulation itself. The results here can also be used to show that even simple SDSs are "universal" in that any reasonable model of simulation can be "efficiently locally simulated" by appropriate SDSs that can be constructed in polynomial time. The models investigated include cellular automata, communicating finite state machines, multi-variate difference equations, etc.

We undertake the computational study of SDSs in an attempt to increase our understanding of SDSs in particular and the complex behavior of dynamical systems in general. SDSs are discrete finite analogs of classical dynamical systems, and we aim to obtain a better understanding of "finite discrete computational analogs of chaos". As pointed out in [BPT91,Mo90,Mo91], computational intractability or unpredictability is the closest form of chaotic behavior that such systems can exhibit. Extending the work of [BPT91], we prove a dichotomy result between classes of SDSs whose global behavior is easy to predict and others for which the global behavior is hard to predict. In [Wo86], Wolfram posed the following three general questions in the chapter entitled "Twenty Problems in the Theory of Cellular Automata": (i) **Problem 16:** *How common are computational universality and undecidability* in CA? (ii) **Problem 18:** *How common is computational irreducibility in CA?* (iii) **Problem 19:** *How common are computationally intractable problems about CA?* The results obtained here and in the companion papers [BH+01a,BH+01b,BH+01c] for SDSs (and for CA as direct corollaries) show that the answer to all of the above questions is *"quite common"*. In other words, it is quite common for synchronous as well as sequential dynamical systems to exhibit intractability. In fact, our results show that such intractability is exhibited by *extremely simple* SDSs and CA.

2   Definitions and Problem Formulation

2.1   Formal Definition of an SDS

As stated earlier, we will restrict our selves to *Simple*-SDSs. (Unless otherwise stated, we use "SDS" to mean a simple SDS.) Our definition closely follows the original definition of SDS in [BMR99].

A **Simple Sequential Dynamical System** (SDS) S is a triple $(G, \mathcal{F}, \pi)$, whose components are as follows:

1. $G(V, E)$ is an undirected graph without multi-edges or self loops. G is referred to as the **underlying graph** of S. We use n to denote $|V|$ and m to denote $|E|$. The nodes of G are numbered using the integers 1, 2, ..., n.

2. Each node has one bit of memory, called its **state**. The state of node i, denoted by s_i, takes on a value from $\mathbb{F}_2 = \{0, 1\}$. We use δ_i to denote the degree of node i. Further, we denote by $N(i)$ the neighbors of node i in G, plus node i itself. Each node i is associated with a *symmetric Boolean function* $f_i : \mathbb{F}_2^{\delta_i+1} \to \mathbb{F}_2$, $(1 \leq i \leq n)$. We refer to f_i as a **local transition function**. The inputs to f_i are the state of i and the states of the neighbors of i. By "symmetric" we mean that the function value does not depend on the order in which the input bits are specified; that is, the function value depends only on how many of its inputs are 1. We use $\mathcal{F}$ to denote $\{f_1, f_2, \ldots, f_n\}$.

3. Finally, π is a permutation of $\{1, 2, \ldots, n\}$ specifying the order in which nodes update their states using their local transition functions. Alternatively, π can be envisioned as a total order on the set of nodes.

Computationally, the transition of an SDS from one configuration to another involves the following steps:

---

for $i = 1$ **to** n **do**
    (i) Node $\pi(i)$ evaluates $f_{\pi(i)}$. (This computation uses the *current* values of the state of $\pi(i)$ and those of the neighbors of $\pi(i)$.)
    (ii) Node $\pi(i)$ sets its state $s_{\pi(i)}$ to the Boolean value computed in Step (i).
end-for

Stated another way, the nodes are processed in the *sequential* order specified by permutation π. The "processing" associated with a node consists of computing the value of the node's Boolean function and changing its state to the computed value.

A **configuration** of an SDS is a bit vector $(b_1, b_2, \ldots, b_n)$, where b_i is the state value of node v_i. A configuration $\mathcal{C}$ of an SDS $\mathcal{S} = (G, \mathcal{F}, \pi)$ can also be thought of as a function $\mathcal{C} : V \to \mathbb{F}_2$. The function computed by SDS $\mathcal{S}$, denoted by $F_{\mathcal{S}}$, specifies for each configuration $\mathcal{C}$, the next configuration $\mathcal{C}'$ reached by $\mathcal{S}$ after carrying out the update of node states in the order given by π. Thus, $F_{\mathcal{S}} : \mathbb{F}_2^n \to \mathbb{F}_2^n$ is a global function on the set of configurations. The function $F_{\mathcal{S}}$ can therefore be considered as defining the dynamic behavior of SDS $\mathcal{S}$. We also say that SDS $\mathcal{S}$ moves from a configuration $\mathcal{C}$ at time t to a configuration $F_{\mathcal{S}}(\mathcal{C})$ at time $(t + 1)$. The initial configuration (i.e., the configuration at time $t = 0$) of an SDS $\mathcal{S}$ is denoted by $\mathcal{I}$. Given an SDS $\mathcal{S}$ with initial configuration $\mathcal{I}$, the configuration of $\mathcal{S}$ after t time steps is denoted by $\xi(\mathcal{S}, t)$. We define $\xi(\mathcal{S}, 0) = \mathcal{I}$. For a configuration $\mathcal{C}$, we use $\mathcal{C}(W)$ to denote the states of the nodes in $W \subseteq V$ and $\mathcal{C}(v)$ to denote the state of a particular node $v \in V$.

A **fixed point** of an SDS $\mathcal{S}$ is a configuration $\mathcal{C}$ such that $F_{\mathcal{S}}(\mathcal{C}) = \mathcal{C}$. An SDS $\mathcal{S}$ is said to **cycle** through a sequence of configurations $\langle \mathcal{C}_1, \mathcal{C}_2, \ldots, \mathcal{C}_r \rangle$ if $F_{\mathcal{S}}(\mathcal{C}_1) = \mathcal{C}_2$, $F_{\mathcal{S}}(\mathcal{C}_2) = \mathcal{C}_3$, $\ldots$, $F_{\mathcal{S}}(\mathcal{C}_{r-1}) = \mathcal{C}_r$ and $F_{\mathcal{S}}(\mathcal{C}_r) = \mathcal{C}_1$. A fixed point is a cycle involving only one configuration. Any cycle involving two or more configurations is called a **limit cycle**. The **phase space** $\mathcal{P}_{\mathcal{S}}$ of an SDS $\mathcal{S}$ is a directed graph with one node for each configuration of $\mathcal{S}$. For any pair of configurations $\mathcal{C}$ and $\mathcal{C}'$, there is a directed edge from the node representing $\mathcal{C}$ to that representing $\mathcal{C}'$ if $F_{\mathcal{S}}(\mathcal{C}) = \mathcal{C}'$.

2.2   Problems Considered

Given an SDS $\mathcal{S}$, let $|\mathcal{S}|$ denote the size of the representation of $\mathcal{S}$. In general, this includes the number of nodes, edges and the description of the local transition functions. We assume that evaluating any local transition function given values for its inputs can be done in polynomial time.

The main problems studied in this paper deal with the *analysis* of a given SDS, that is, determining whether a given SDS has a certain property. These problems are formulated below.

Given an SDS $\mathcal{S}$, two configurations $\mathcal{I}$, $\mathcal{B}$, and a positive integer t, the t-REACHABILITY problem is to determine whether $\mathcal{S}$ starting in configuration $\mathcal{I}$ can reach configuration $\mathcal{B}$ in t or fewer time steps. If t is specified in unary, it is easy to solve this problem in polynomial time since we can execute the SDS for t steps and check whether configuration $\mathcal{B}$ is reached at some step. So, we assume that t is specified in binary.

Given an SDS $\mathcal{S}$ and two configurations $\mathcal{I}$, $\mathcal{B}$, the REACHABILITY problem is to determine whether $\mathcal{S}$ starting in configuration $\mathcal{I}$ ever reaches the configuration $\mathcal{B}$. (Note that, for an SDS with n nodes, when $t \geq 2^n$, t-REACHABILITY is equivalent to REACHABILITY.) Given an SDS $\mathcal{S}$ and a configuration $\mathcal{I}$, the FIXED POINT REACHABILITY problem is to determine whether $\mathcal{S}$ starting in state $\mathcal{I}$ reaches a fixed point.

2.3   Extensions of the Basic SDS Model

As defined, the state of each node of an SDS stores a Boolean value and the local transition functions are symmetric Boolean functions. When we allow the state of each node to assume values from a domain $\mathcal{D}$ of a fixed size and allow the local transition functions to have $\mathcal{D}$ as their range, we obtain a **Finite Range SDS** (FR-SDS). If the states may store unbounded values and the local transition functions may also produce unbounded values, we obtain a **Generalized SDS** (Gen-SDS).

Another useful variant is a **Synchronous Dynamical System** (SyDS), an SDS *without* the node permutation. In a SyDS, during each time step, all the nodes *synchronously* compute and update their state values. Thus, SyDSs are similar to classical CA with the difference that the connectivity between cells is specified by an arbitrary graph.

The notion of symmetry can be suitably extended to functions with non-Boolean domains as well. In defining the above models, we did not attempt to relax the symmetry property of local transition functions. Dynamical systems in which the local transition functions are not necessarily symmetric are considered in the companion papers [BH+01a,BH+01b,BH+01c].

3   Summary and Significance of Results

In this paper, we characterize the computational complexity of determining several phase space properties for SDSs and CA. The results obtained are the first such results for SDSs and directly imply corresponding lower bounds on the complexity of similar problems for various classes of CA and communicating finite state machines.

Our main result is a *dichotomy* between easy and hard to predict classes of SDSs. Specifically, we show that t-REACHABILITY, REACHABILITY and FIXED POINT REACHABILITY problems for FR-SDSs are **PSPACE**-complete. Moreover, these results hold even if the local transition functions are identical and the underlying graph is a simple path. We further extend these results to show that the above three problems remain **PSPACE**-complete for SDSs, even when the underlying graph is simultaneously k-regular for some fixed k, bandwidth bounded (and hence pathwidth and treewidth bounded) and the local transition functions are *symmetric*.

In contrast to the above intractability results, we show that these problems are efficiently solvable for SDSs in which each local transition function is *symmetric and monotone*. Specifically, we prove that when each local transition function is a k-simple-threshold function[1] for some $k \geq 0$, these problems can be solved in polynomial time.

As a part of our methodology, we also obtain a number of "simulation" results that show how to simulate one type of SDS (or CA) by another typically more restricted type of SDS (or CA). These simulation results may be of independent interest. For instance, we show (i) how a given FR-SyDS with local transition functions that are not necessarily symmetric can be efficiently simulated by a SyDS, and (ii) how a SyDS can be simulated by an SDS.

The results presented here extend a number of earlier results on the complexity of problems for CA and also have other applications. We briefly discuss these extensions and their significance below.

Reference [RH93] shows the **EXPSPACE**-hardness of local STATE REACH-ABILITY problems for hierarchically-specified linearly inter-connected copies of a single finite automaton. The constructions presented here imply that various local STATE REACHABILITY problems are also **EXPSPACE**-hard, for hierarchically-specified and bandwidth-bounded networks of simple SDSs. Using ideas from [SH+96], this last result implies that determining any simulation equivalence relation or pre-order in the Linear-time/Branching-time hierarchies of [vG90,vG93] is **EXPSPACE**-hard for such hierarchically-specified networks of simple SDSs.

Our reductions are carried out starting from the acceptance problem for deterministic linear space bounded automata (LBAs) and are extremely efficient in terms of time and space requirements. Specifically, these reductions require $O(n)$ space and $O(n \log n)$ time. Thus these results imply tight lower bounds on the deterministic time and space required to solve these problems.

The results in [Su95,Su90] prove the **PSPACE**-completeness of REACHABIL-ITY and FIXED POINT REACHABILITY problems for CA. These papers do not address the effect of restricting the class of local transition functions or restricting the structure of the underlying graph on the complexity of these problems. Our results extend these hardness results to much simpler instances and also provide the first step in proving results that delineate polynomial time solvable and computationally intractable instances.

The results presented here can be contrasted with the work of Buss, Papadimitriou and Tsitsiklis [BPT91] on the complexity of t-REACHABILITY problem for coupled automata. Their identity-independence assumption is similar to our symmetric function assumption, except that they consider first order formulas. In contrast to the polynomial time solvability of the reachability problem for

---

[1] The k-simple-threshold function has the value 1 iff at least k of its inputs are 1. Conventional definition of threshold functions associates a weight with each input [Ko70]. We use the simplified form where all inputs have the same weight.

globally controlled systems of independent automata [BPT91], our results show that a small amount of local interaction suffices to make the reachability problem computationally intractable. Our reduction leads to an interaction graph that is of constant degree, bandwidth bounded and regular.

Other than [BPT91], papers that are most relevant to our work are the following. References [BMR99,BMR00,MR99,Re00,Re00a,LP00] investigate mathematical properties of sequential dynamical systems. Sutner [Su89,Su90,Su95] and Green [Gr87] characterize the complexity of reachability and predecessor existence problems for finite CA. Moore [Mo90,Mo91] makes an important connection between unpredictability of dynamical systems and undecidability of some of their properties. The models considered here are also related to discrete Hopfield networks [FO00,GFP85,BG88].

Alur et al (see for example [AKY99]) consider the complexity of several problems for hierarchically specified communicating finite state machines. SDSs can be viewed as very simple kinds of concurrent state machines. Moreover, the hardness proof obtained here can be extended to obtain **EXPSPACE**-hardness, when we have exponentially many simple automata (vertices in our case) joined to form a bandwidth bounded graph. This result significantly extends a number of known results in the literature concerning concurrent finite state machines by showing that the hardness results hold even for simple classes of individual machines.

Quadratic dynamical systems are a variant of discrete dynamical systems that aim at modeling genetic algorithms. In [ARV94] it is shown that simulating quadratic dynamical systems is **PSPACE**-hard; specifically, it is shown that the t-reachability problem for such systems is **PSPACE**-complete even when t is specified in unary.

4 Complexity of Reachability Problems

4.1 Road Map for the Reductions

In this section we prove our main hardness theorem concerning the t-REACHABILITY, REACHABILITY and FIXED POINT REACHABILITY problems for SDSs.

Theorem 1. (Main hardness theorem) *The t-REACHABILITY, REACHABILITY and* FIXED POINT REACHABILITY *problems for SDSs with symmetric Boolean functions are* **PSPACE**-*hard, even when (i) each node is of constant degree and the graph is regular (i.e. all nodes have the same degree), (ii) the pathwidth and hence the treewidth of the graph is bounded by a constant, and (iii) all the nodes have exactly the same symmetric Boolean function associated with them.*

Overall proof idea: The proof of the above theorem is obtained through a series of local replacement type reductions (steps). The reductions involve building general gadgets that may be of independent interest.

Step 1: First, by a direct reduction from the acceptance problem for a LIN-EAR BOUNDED AUTOMATON (LBA) we can show that the *t*-REACHABILITY, REACHABILITY and FIXED POINT REACHABILITY problems for FR-SyDS (finite CA) and FR-SDS are **PSPACE**-hard even under the following restrictions applied simultaneously. (i) The graph G is a line (which has pathwidth and treewidth of 1). (ii) The number of distinct local transition functions is at most three. (iii) The domain of each function is a small constant, depending only on the size of the LBA encoding. This result is stated as Theorem 2 below; the proof is omitted.

Theorem 2. *(Step 1:) The* *t*-REACHABILITY, REACHABILITY *and* FIXED POINT REACHABILITY *problems for FR-SyDS (Cellular Automata) and FR-SDS are* **PSPACE**-*hard, even when restricted to instances such that: (i) The graph G is a line graph (and thus has pathwidth and treewidth of 1), and (ii) The number of distinct local transition functions is at most three, and (iii) The size of the domain of each function is a small constant.* ∎

Step 2: Next, we show how to transform these problems for FR-SyDS into the corresponding problems for SyDS in which the maximum node degree is bounded. See Section 4.2.

Step 3: Next, we show how a SyDS can be simulated by an SDS where the maximum node degree is bounded. (The underlying graph of the SDS may not be regular and the local transition functions may not be identical.)

Step 4: Finally, we show how to transform the SDS obtained in Step 3 into another SDS whose underlying graph is regular and whose node functions are all identical (same function and same degree).

For reasons of space, we omit the constructions alluded to in Steps 3 and 4.

4.2   SyDS with Symmetric Boolean Functions: Step 2

Definition 3. *Given $k \geq 1$, a **distance-k coloring** of a graph $G(V, E)$ is an assignment of colors $h : V \to \{0, 1, 2, \ldots, |V|-1\}$ to the nodes of G such that for all $u, v \in V$ for which the distance between u and v is at most k, $h(u) \neq h(v)$.*

Proposition 4. *A graph $G(V, E)$ with maximum degree Δ can be distance-2 colored using at most $\Delta^2 + 1$ colors, and such a coloring can be obtained in polynomial time. Thus, for a graph whose node degrees are bounded by a constant, and, in particular, for regular graphs of constant degree, the number of colors used for distance-2 coloring is a constant.* ∎

Theorem 5. *For a given μ and Δ, consider the class of FR-SyDSs where the size of the state domain of each node is at most μ and the degree of each node is at most Δ. There is a polynomial time reduction from a FR-SyDS $\mathcal{S} = (G, \mathcal{F})$ in this class and configurations $\mathcal{I}$ and $\mathcal{B}$ for $\mathcal{S}$ to a usual (having symmetric Boolean functions) SyDS $\mathcal{S}_1 = (G_1, \mathcal{F}_1)$ and configurations $\mathcal{I}_1$ and $\mathcal{B}_1$ for $\mathcal{S}_1$ such that*

1. S starting in configuration $\mathcal{I}$ reaches $\mathcal{B}$ iff S_1 starting in configuration $\mathcal{I}_1$ reaches $\mathcal{B}_1$. Moreover, for each t, S reaches $\mathcal{B}$ in t steps iff S_1 reaches $\mathcal{B}_1$ in t steps.
2. S starting in configuration $\mathcal{I}$ reaches a fixed point iff S_1 starting in $\mathcal{I}_1$ reaches a fixed point.

Proof sketch: Given S, the reduction first constructs a distance-2 coloring h of G, using at most $\Delta^2 + 1$ colors, where the colors are consecutive integers, beginning with zero. The fact that h is a distance-2 coloring is not used in the construction of S_1 from S, but is crucial to the correctness of the reduction.

Next, given graph $G(V, E)$ and coloring h, graph $G_1(V_1, E_1)$ is constructed, as follows. For each node $x_k \in V$, there are $(\mu - 1)\mu^{h(x_k)}$ nodes in V_1. We refer to these nodes as x_{ij}^k, $1 \leq i < \mu$ and $1 \leq j \leq \mu^{h(x_k)}$. Informally, corresponding to a node x_k of S, V_1 contains $\mu - 1$ sets of nodes (which we call *clumps*), each of cardinality $\mu^{h(x_k)}$. For a given node $x_k \in V$, clump $\mathcal{X}_j^k$ refers to the nodes $x_{j,r}^k$ $1 \leq r \leq \mu^{h(x_k)}$. Additionally, we will use $\mathcal{X}^k = \mathcal{X}_1^k \cup \mathcal{X}_2^k \ldots \cup \mathcal{X}_{\mu-1}^k$ to denote the set of all nodes in V_1 corresponding to x_k. E_1 consists of the following two kinds of edges: For each node $x_k \in V$, the nodes in $\mathcal{X}^k$ form a complete graph. Each edge $\{x_k, x_r\} \in E$ is replaced by a complete bipartite graph between the sets of nodes used to replace the nodes x_k and x_r.

Before specifying the construction of the local transition functions of S_1, we define the following mapping ψ_k, for each node $x_k \in V$. Suppose that node $x_k \in V$ has neighbors $y_1, \ldots, y_d$ in G. We define function $\psi_k : \mathbb{N} \to \mu^{d+1}$ (where $\mathbb{N}$ is the set of nonnegative integers) as follows: $\psi_k(w) = \langle c_0, c_1, \ldots, c_d \rangle$, where in the base μ representation of w, c_0 is the coefficient of $\mu^{h(x_k)}$, and c_r, $1 \leq r \leq d$, is the coefficient of $\mu^{h(y_r)}$.

The functions in the set $\mathcal{F}_1$ are defined as follows. We envision the state domain of S to be the integers $0, 1, \ldots, \mu - 1$. Consider a node x_{ij}^k in $\mathcal{X}^k$ and a vector α of Boolean input values for $f_{x_{ij}^k}$. Suppose that in α, exactly w of the input parameters to $f_{x_{ij}^k}$ are equal to 1. Suppose f_{x_k} from $\mathcal{F}$ is the local transition function at node x_k. Then $f_{x_{ij}^k}(\alpha) = 1$ iff $f_{x_k}(\psi_k(w)) \geq i$. In the reduction of S to S_1, function $f_{x_{ij}^k}$ is represented by specifying the subset of counts of input parameters taking value 1 for which the output equals 1.

We now define the following mapping g from the configurations of S to the configurations of S_1, $g : \mu^V \to 2^{V_1}$. Consider a configuration $\mathcal{A}$ of S and node $x_k \in V$. In configuration $g(\mathcal{A})$ of S_1, the nodes in the first $\mathcal{A}(x_k)$ clumps of $\mathcal{X}^k$ have state value 1, and the nodes in the other clumps of $\mathcal{X}^k$ have state value 0. More precisely, for $x_k \in V$, $1 \leq i < \mu$, $1 \leq j \leq \mu^{h(x_k)}$, $g(\mathcal{A})(x_{ij}^k) = 1$ iff $\mathcal{A}(x_k) \geq i$.

The reduction constructs $\mathcal{I}_1$ as $g(\mathcal{I})$, and $\mathcal{B}_1$ as $g(\mathcal{B})$. This completes the construction involved in the reduction. The correctness of this construction is based on showing that the phase space of S is embedded as a subspace of the phase space of S_1, so that S_1 can be used to simulate S. First, we specify which configurations of S_1 are in this subspace.

Define a configuration $\mathcal{A}$ of $\mathcal{S}_1$ to be *proper* if for all k, i, j, p, q,

$$(\mathcal{A}(x_{ij}^k) = 1 \text{ and } i \geq p) \quad \Rightarrow \quad \mathcal{A}(x_{pq}^k) = 1.$$

In other words, a configuration $\mathcal{A}$ of $\mathcal{S}_1$ is *proper* if the value at any node in clump $\mathcal{X}_j^k$ equal to 1 implies that *all* nodes in clumps $\mathcal{X}_1^k, \mathcal{X}_2^k, \ldots, \mathcal{X}_j^k$ are also 1. The following claim provides an important property of the mapping g.

Claim 4.1: Mapping g is a bijection between the configurations of $\mathcal{S}$ and the proper configurations of $\mathcal{S}_1$.

For a mapping $\mathcal{D}$ of a set of states into Boolean values, we let $|\mathcal{D}|$ denote the number of 1's in $\mathcal{D}$. We also take the notational liberty of identifying an assignment of state values to a set of nodes with the vector of values representing the assignment. Also, recall that $N(x)$ denotes the neighbors of node x, plus node x itself. For a configuration $\mathcal{C}$ and set of nodes W, $\mathcal{C}(W)$ denotes the restriction of $\mathcal{C}$ to W.

Claim 4.2: For a configuration $\mathcal{C}$ of $\mathcal{S}$, node x_k of V, and node x_{ij}^k of $\mathcal{X}^k$, $\mathcal{C}(N(x_k)) = \psi_k(|g(\mathcal{C})(N(x_{ij}^k))|)$.

We now state our key claim, which says that $\mathcal{S}_1$ properly simulates $\mathcal{S}$.

Claim 4.3: For every configuration $\mathcal{A}$ of $\mathcal{S}$, $F_{\mathcal{S}_1}(g(\mathcal{A})) = g(F_{\mathcal{S}}(\mathcal{A}))$.

Now, the following claim completes the proof of Theorem 5. The claim can be proven using the above claims and induction on t.

Claim 4.4: Let $\mathcal{S}$ and $\mathcal{S}_1$ be as defined above. Consider $\mathcal{S}$ starting in configuration $\mathcal{I}$ and $\mathcal{S}_1$ starting in configuration $g(\mathcal{I})$. Then (1) $\forall t \geq 0$, $\xi(\mathcal{S}_1, t)$ is proper. (2) $\forall t \geq 0$, $\xi(\mathcal{S}_1, t) = g(\xi(\mathcal{S}, t))$. ∎

5   Polynomial Time Solvable Cases

In this section we prove that t-REACHABILITY, REACHABILITY and FIXED POINT REACHABILITY problems are polynomial time solvable for k-simple-threshold-SDSs, that is, SDSs in which each local transition function is a k-simple-threshold function for some $k \geq 1$. Since each symmetric monotone function is a k-simple-threshold function for some k, these polynomial time results provide the dichotomy between two classes of SDSs: one with symmetric local transition functions and the other with symmetric monotone local transition functions. Some remarks regarding the generality of these polynomial algorithms are provided at the end of this section.

Definition 6. *A k-**simple-threshold**-SDS is an SDS in which the local transition function at each node v_i is a k_i-simple-threshold function, where $1 \leq k_i \leq \min\{k, \delta_i + 1\}$. Here, δ_i is the degree of node v_i.*

Theorem 7. *For any k-simple-threshold-SDS, $1 \leq k \leq n$, the problems t-REACHABILITY, REACHABILITY and FIXED POINT REACHABILITY can be solved by executing at most $3m/2$ steps of the given SDS, where m is the number of edges in the underlying graph.*

Proof sketch: The proof of the theorem is based on a *potential function* argument. A similar approach is used in [GFP85] to establish the convergence rate and to bound the transient length of discrete Hopfield networks with threshold gates.

Given an SDS with underlying graph $G(V, E)$, we assign a potential to each node and each edge in G. For the remainder of the proof, we use k_v to denote the threshold value required for a node v to become 1. For each node v define $T_1(v) = k_v$ and $T(v) = \delta_v + 1$. Recall that s_v denotes the state of node v. Thus $s_v = 1$ iff at least $T_1(v)$ of its inputs are 1; s_v is 0 otherwise. Another interpretation of $T_1(v)$ is that it is the smallest integer such that s_v must be assigned 1 if $T_1(v)$ of v's inputs have value 1. Using this analogy, define $T_0(v)$ to be the smallest integer such that s_v must be assigned 0 if $T_0(v)$ of the inputs to v have value 0. The following observation is an easy consequence of the definitions of k_v-simple-threshold, $T_0(v)$ and $T_1(v)$: For any node $v \in V$, $T_1(v) + T_0(v) = T(v) + 1$. Define the potential $P(v)$ at a node v as follows:

$$P(v) = T_1(v) \text{ if } s_v = 1$$
$$= T_0(v) \text{ if } s_v = 0$$

The following is an easy consequence of the definitions of $T_0(v)$, $T_1(v)$ and the fact that $k_v \geq 1$: For any node $v \in V$, $1 \leq P(v) \leq \delta_v + 1$.

Define the potential $P(e)$ of an edge $e = \{u, v\}$ as follows:

$$P(e) = 1 \quad \text{if } e = \{u, v\} \text{ and } s_u \neq s_v$$
$$= 0 \quad \text{otherwise.}$$

The potential of the entire SDS is given by $P(G) = \sum_{v \in V} P(v) + \sum_{e \in E} P(e)$. It can be seen that the initial potential $P(G)$ (regardless of the initial configuration) is bounded by $3m + n$. Further, for each node $v \in V$, $P(v) \geq 1$; thus, the potential of the SDS at any time is at least n.

If the system has not reached a fixed point, then at least one node v undergoes a state change. Now, fix a global step in the dynamic evolution of the SDS and consider a particular substep in which the state of node v changes from a to b. This state change may modify the potential of v and the potentials of the edges incident on v. It can be shown that each time there is a change in the state of a node, $P(G)$ decreases by at least 2. As argued above, the initial value of $P(G)$ is at most $3m + n$, and the value of $P(G)$ can never be less than n. Thus, the total number of configuration changes is bounded by $[(3m + n) - n]/2 = 3m/2$.

In other words, any k-simple-threshold-SDS reaches a fixed point after at most $3m/2$ steps. Theorem 7 follows.   ∎

The above theorem points out an interesting contrast between CA and SDSs. It is easy to construct instances of k-simple-threshold-CA with limit cycles. In contrast, by Theorem 7, k-simple-threshold-SDSs have fixed points but not limit cycles.

Theorem 7 holds even when each node has a different value of the threshold k. As observed earlier, every symmetric and monotone Boolean function is a k-simple-threshold function for some k. Thus, Theorem 7 implies the polynomial time solvability of reachability problems for SDSs with symmetric monotone local transition functions. Theorem 7 also shows that for k-simple-threshold-SDSs, the length of any transient is at most $3m/2$.

The result of Theorem 7 can also be extended to the case when the local transition function at node v_i is the zero-threshold function (i.e., a function whose output is 1 for all inputs) or the (δ_i+2)-threshold function (i.e., a function whose output is 0 for all inputs). All nodes with such local transition functions will reach their final values during the first step of the SDS. Therefore, in this case, the number of SDS steps needed to reach a fixed point is at most $3m/2+1$.

Acknowledgements: This research has been funded in part by the LDRD-DR project *Foundations of Simulation Science* and LDRD-ER project *Extremal Optimization*. We thank Paul Wollan and Predrag Tosic for reading the manuscript carefully and suggesting changes that substantially improved the readability of the paper. We also thank Gabriel Istrate, Stephen Kopp, Henning Mortveit, Allon Percus and Christian Reidys for fruitful discussions.

References

AKY99.   R. Alur, S. Kannan, and M. Yannakakis. Communicating hierarchical state machines. *Proc. 26th International Colloquium on Automata, Languages and Programming* (ICALP'99), Springer Verlag, 1999. 160, 165

ARV94.   S. Arora, Y. Rabani and U. Vazirani. Simulating quadratic dynamical systems is **PSPACE**-complete. *Proc. 26th Annual ACM Symposium on the Theory of Computing* (STOC'94), Montreal, Canada, May 1994, pp. 459-467. 165

BB+99.   C. Barrett, B. Bush, S. Kopp, H. Mortveit and C. Reidys. Sequential Dynamical Systems and Applications to Simulations. Technical Report, Los Alamos National Laboratory, Sept. 1999. 160

BG88.   J. Bruck and J. Goodman. A generalized convergence theorem for neural networks. *IEEE Trans. on Information Theory*, Vol. 34, 1988, pp. 1089–1092. 165

BH+01a. C. Barrett, H. Hunt III, M. Marathe, S. Ravi, D. Rosenkrantz, R. Stearns and P. Tosic. Gardens of Eden and fixed points in sequential dynamical systems. To appear in *Proc. of the International Conference on Discrete Models - Combinatorics, Computation and Geometry* (DM-CCG), Paris, July 2001. 161, 163

BH+01b. C. Barrett, H. Hunt III, M. Marathe, S. Ravi, D. Rosenkrantz and R. Stearns. Predecessor and permutation existence problems for sequential dynamical systems. Under preparation, May 2001. 161, 163

BH+01c. C. Barrett, H. Hunt III, M. Marathe, S. Ravi, D. Rosenkrantz and R. Stearns. Elements of a theory of computer simulation V: computational complexity and universality. Under preparation, May 2001. 161, 163

BMR99.   C. Barrett, H. Mortveit and C. Reidys. Elements of a theory of simulation II: sequential dynamical systems. *Applied Mathematics and Computation*, 1999, vol 107/2-3, pp. 121–136. 159, 161, 165

BMR00.   C. Barrett, H. Mortveit and C. Reidys. Elements of a theory of computer simulation III: equivalence of SDS. to appear in *Applied Mathematics and Computation*, 2000. 165

BPT91.   S. Buss, C. Papadimitriou and J. Tsitsiklis. On the predictability of coupled automata: an allegory about chaos. *Complex Systems,* 1(5), pp. 525–539, 1991. 160, 161, 164, 165

BR99.   C. Barrett and C. Reidys. Elements of a theory of computer simulation I: sequential CA over random graphs. *Applied Mathematics and Computation,* Vol. 98, pp. 241–259, 1999. 159

BWO95.   C. Barrett, M. Wolinsky and M. Olesen. Emergent local control properties in particle hopping traffic simulations. *Proc. Traffic and Granular Flow,* Julich, Germany, 1995. 160

FO00.   P. Floréen and P. Orponen. Complexity issues in discrete Hopfield networks. *The Computational and Learning Complexity of Neural Networks: Advanced Topics.* Ian Parberry (Editor), 2000. 165

GFP85.   E. Goles F. Fogelman and D. Pellegrin. Decreasing energy functions as a tool for studying threshold networks. *Discrete Applied Mathematics,* vol. 12, 1985, pp. 261–277. 165, 169

Gr87.   F. Green. NP-complete problems in cellular automata. *Complex Systems,* 1(3), pp. 453–474, 1987. 165

Gu89.   H. Gutowitz, Ed. *Cellular Automata: Theory and Experiment.* North Holland, 1989. 160

Hu73.   H. Hunt III. On the Time and Tape Complexity of Languages. Ph.D. Thesis, Cornell University, Ithaca, NY, 1973. 160

Ko70.   Z. Kohavi, *Switching and Finite Automata Theory,* McGraw-Hill Book Company, New York, 1970. 164

LP00.   R. Laubenbacher and B. Pareigis. Finite Dynamical Systems. Technical report, Department of Mathematical Sciences, New Mexico State University, Las Cruces. 165

Mo90.   C. Moore. Unpredictability and undecidability in dynamical systems. *Physical Review Letters,* 64(20), pp. 2354–2357, 1990. 161, 165

Mo91.   C. Moore. Generalized shifts: unpredictability and undecidability in dynamical systems. *Nonlinearity,* 4, pp. 199–230, 1991. 161, 165

MR99.   H. Mortveit and C. Reidys. Discrete sequential dynamical systems. *Discrete Mathematics.* Accepted for publication, 2000. 165

Ra92.   A. Rabinovich. Checking equivalences between concurrent systems of finite state processes. *International Colloquium on Automata Programming and languages* (ICALP'92), LNCS Vol. 623, Springer, 1992, pp. 696–707. 160

Re00.   C. Reidys. On acyclic orientations and SDS. *Advances in Applied Mathematics,* to appear. 165

Re00a.   C. Reidys. Sequential dynamical systems: phase space properties. *Advances in Applied Mathematics,* to appear. 165

RH93.   D. Rosenkrantz and H. Hunt III. The complexity of processing hierarchical specifications. *SIAM Journal on Computing,* 22(3), pp. 627–649, 1993. 160, 164

SH+96.   S. Shukla, H. Hunt III, D. Rosenkrantz and R. Stearns. On the Complexity of Relational Problems for Finite State Processes. *Proc. International Colloquium on Automata, Languages and Programming* (ICALP'96), pp. 466–477. 164

Su89.   K. Sutner. Classifying circular cellular automata. *Physica D,* 45(1-3), pp. 386-395, 1989. 165

Su90.   K. Sutner. De Bruijn graphs and linear cellular automata. *Complex Systems,* 5(1), pp. 19-30, 1990. 164, 165

172   Chris Barrett et al.

Su95.   K. Sutner. On the computational complexity of finite cellular automata. *Journal of Computer and System Sciences,* 50(1), pp. 87–97, February 1995. 164, 165

vG90.   R. van Glabbeek. The linear time-branching time spectrum. Technical Report CS-R9029, Computer Science Department, CWI, Centre for Mathematics and Computer Science, Netherlands, 1990. 164

vG93.   R. van Glabbeek. The linear time-branching time spectrum II (the semantics of sequential systems with silent moves). *LNCS* 715, 1993. 164

Wo86.   S. Wolfram, Ed. *Theory and applications of cellular automata.* World Scientific, 1987. 160, 161

The Complexity of Tensor Circuit Evaluation*
Extended Abstract

Martin Beaudry[1] and Markus Holzer[2]

[1] Département de mathématiques et d'informatique
Université de Sherbrooke, 2500, boul. Université
Sherbrooke (Québec), J1K 2R1 Canada
beaudry@dmi.usherb.ca
[2] Institut für Informatik, Technische Universität München
Arcisstraße 21, D-80290 München, Germany
holzer@informatik.tu-muenchen.de

Abstract. The study of tensor calculus over semirings in terms of complexity theory was initiated by Damm *et al.* in [8]. Here we first look at tensor circuits, a natural generalization of tensor formulas; we show that the problem of asking whether the output of such circuits is non-zero is complete for the class NE = NTIME($2^{O(n)}$) for circuits over the boolean semiring, ⊕E for the field $\mathbb{F}_2$, and analogous results for other semirings. Common sense restrictions such as imposing a logarithmic upper bound on circuit depth are also discussed. Second, we analyze other natural problems concerning tensor formulas and circuits over various semirings, such as asking whether the output matrix is diagonal or a null matrix.

1 Introduction

In a recent paper Damm *et al.* [8] initiated the study of the computational complexity of *multi-linear algebra* expressed by tensors. More precisely, they considered the computational complexity of the evaluation problem for formulas over algebras of matrices defined over a finite, or finitely generated, semiring S, and where the operations are matrix addition, matrix multiplication, and the tensor product—also known as the Kronecker, or outer, or direct product. Central to their work is the analysis of the *non-zero tensor problem*, denoted by $0 \neq \text{val}_S$, which consists in asking whether a given formula yields a 1×1 matrix whose unique entry differs from the zero of S. The complexity of this problem is indicative of where tensor calculus over the specified semiring sits in the hierarchy of complexity classes.

This is one more way of characterizing complexity classes in algebraic terms, which comes after the problem of evaluating formulas and circuits over the Boolean semiring [6,7,11], and the computational models of programs over monoids [2,3], and leaf languages [5], among others. Using tensor calculus in

---

* Supported by the Québec FCAR, by the NSERC of Canada, and by the DFG of Germany.

J. Sgall, A. Pultr, and P. Kolman (Eds.): MFCS 2001, LNCS 2136, pp. 173–185, 2001.
© Springer-Verlag Berlin Heidelberg 2001

this context is especially appealing, because of the many applications matrix algebra finds in various computer science areas, such as the specification of parallel algorithms. Also, since the design of algorithms for quantum computers makes extensive use of unitary matrices [9], tensor formulas and circuits may become useful tools for better understanding the power of this computational model.

In this paper we extend the research from tensor formulas to circuits. We show that for the non-zero problem over unrestricted tensor circuits, the completeness statements transfer smoothly to the exponential time counterparts of the complexity classes encountered in the formula case. However, to achieve this we have to develop a more powerful proof technique which as a byproduct enables us to improve some of the tensor formula completeness results, in that we can write them using polytime many-one instead of polytime Turing reductions (see [8]). More precisely, with notation E for the class $DTIME(2^{O(n)})$, we obtain that:

1. Evaluating a tensor circuit over the natural numbers is #E-complete under polylin many-one reductions, i.e., reductions computable in deterministic polytime producing linear output, and
2. problems $0 \neq val_{\mathbb{B}}$ and $0 \neq val_{\mathbb{F}_2}$ are resp. NE-complete and $\oplus$E-complete under polylin many-one reductions.

We also discuss a number of meaningful sub-cases concerning tensor circuit depth restrictions and consider other natural questions pertaining to tensor circuits based on natural linear-algebraic questions. Since problems of this kind were not addressed in [8], we also solve them for the case of tensor formulas. For instance we investigate the problems of deciding whether a tensor circuit's output is a null matrix or a diagonal matrix. To this end we make heavy use of generalized counting classes on algebraic Turing machines and operator characterizations of complexity classes.

The paper is organized as follows. The next section introduces the definitions and notations needed. In Section 3 we discuss the complexity of the evaluation of tensor circuits in general. Then linear algebraic problems on tensor circuits and formulas are studied in Section 4. Due to space constraints we have to omit most of our proofs. Finally we summarize our results and state some open problems.

2   Definitions, Notations and Basic Techniques

We assume that the reader is familiar with the standard notation from computational complexity (see, e.g., [1]), in particular with the first three items in the inclusion chain $P \subseteq NP \subseteq PSPACE \subseteq EXP \subseteq NEXP$. Here, we use EXP (NEXP, resp.) to denote $DTIME(2^{n^{O(1)}})$ ($NTIME(2^{n^{O(1)}})$, resp.), and we denote by E the class $DTIME(2^{O(n)})$. We refer to the latter as (deterministic) exponential time; its nondeterministic counterpart is referred to as NE. Note that E and NE are not closed under polytime or logspace reductions due to the possibly nonlinear output length of these reductions.

We also assume that the reader is familiar with counterparts to nondeterministic decision classes like NP defined by "counting" or defined as function classes. Here #P is the class of functions f defined with machines M with the same resources as the underlying base class, such that $f(x)$ equals the number of accepting computations of M on x (see, e.g., [16,17]), the parity version $\oplus$P is the class of sets of type $\{\, x \mid f(x) \neq 0 \pmod 2 \,\}$ for some $f \in$ #P, the class MOD_q-P is similarly defined with respect to counting modulo q, and the class $GapP$ equals $\{\, f - g \mid f, g \in$ #P $\,\}$. We also use the class $C_=$P of sets of the type $\{\, x \mid f(x) - g(x) = 0 \,\}$ for some $f, g \in$ #P, the class USP of sets of type $\{\, x \mid f(x) = 1 \,\}$ for some $f \in$ #P (see, e.g., [4]), and DP, the class of those languages which can be written as the difference of two NP languages [13]. Similarly one defines PSPACE, E, and EXP counting classes.

A *semiring* is a tuple $(\mathcal{S}, +, \cdot)$ with $\{0, 1\} \subseteq \mathcal{S}$ and binary operations $+, \cdot :$ $\mathcal{S} \times \mathcal{S} \to \mathcal{S}$ (sum and product), such that $(\mathcal{S}, +, 0)$ is a commutative monoid, $(\mathcal{S}, \cdot, 1)$ is a monoid, multiplication distributes over sum, and $0 \cdot a = a \cdot 0 = 0$ for every a in $\mathcal{S}$ (see, e.g., [12]). A semiring is *commutative* if and only if $a \cdot b = b \cdot a$ for every a and b, it is *finitely generated* if there is a finite set $\mathcal{G} \subseteq \mathcal{S}$ generating all of $\mathcal{S}$ by summation, and is a *ring* if and only if $(S, +, 0)$ is a group. The special choice of $\mathcal{G}$ has no influence on the complexity of problems we study in this paper. Throughout the paper we consider the following semirings: the Booleans $(\mathbb{B}, \vee, \wedge)$, residue class rings $(\mathbb{Z}_q, +, \cdot)$, the naturals $(\mathbb{N}, +, \cdot)$, and the integers $\mathbb{Z} = (\mathbb{Z}, +, \cdot)$.

Let $\mathbf{M}_{\mathcal{S}}$ denote the set of all *matrices* over $\mathcal{S}$, and define $\mathbf{M}_{\mathcal{S}}^{k,\ell} \subseteq \mathbf{M}_{\mathcal{S}}$ to be the set of all *matrices of order* $k \times \ell$. For a matrix A in $\mathbf{M}_{\mathcal{S}}^{k,\ell}$ let $I(A) = [k] \times [\ell]$, where $[k]$ denotes the set $\{1, 2, \ldots, k\}$. The (i, j)th entry of A is denoted by $a_{i,j}$ or $(A)_{i,j}$. Addition and multiplication of matrices in $\mathbf{M}_{\mathcal{S}}$ are defined in the usual way. Additionally we consider the *tensor product* $\otimes : \mathbf{M}_{\mathcal{S}} \times \mathbf{M}_{\mathcal{S}} \to \mathbf{M}_{\mathcal{S}}$ of matrices, also known as Kronecker product, outer product, or direct product, which is defined as follows: for $A \in \mathbf{M}_{\mathcal{S}}^{k,\ell}$ and $B \in \mathbf{M}_{\mathcal{S}}^{m,n}$ let $A \otimes B \in \mathbf{M}_{\mathcal{S}}^{km,\ell n}$ be

$$A \otimes B := \begin{pmatrix} a_{1,1} \cdot B & \ldots & a_{1,\ell} \cdot B \\ \vdots & \ddots & \vdots \\ a_{k,1} \cdot B & \ldots & a_{k,\ell} \cdot B \end{pmatrix}.$$

Hence $(A \otimes B)_{i,j} = (A)_{q,r} \cdot (B)_{s,t}$ where $i = k \cdot (q - 1) + s$ and $j = \ell \cdot (r - 1) + t$.

Proposition 1. *The following hold when the expressions are defined [14]:*

1. $(A \otimes B) \otimes C = A \otimes (B \otimes C)$.
2. $(A + B) \otimes (C + D) = A \otimes C + A \otimes D + B \otimes C + B \otimes D$.
3. $(A \otimes B) \cdot (C \otimes D) = (A \cdot C) \otimes (B \cdot D)$.
4. $(A \cdot B)^{\otimes n} = A^{\otimes n} \cdot B^{\otimes n}$ *for any* $n \geq 1$.
5. $(A + B)^{\mathsf{T}} = A^{\mathsf{T}} + B^{\mathsf{T}}$, $(A \cdot B)^{\mathsf{T}} = B^{\mathsf{T}} \cdot A^{\mathsf{T}}$, *and* $(A \otimes B)^{\mathsf{T}} = A^{\mathsf{T}} \otimes B^{\mathsf{T}}$.

In the proofs of the two main lemmata we make use of special rotation matrices, so called stride permutation matrices [15]. Formally, the *mn-point stride n permutation* $P_n^{mn} \in \mathrm{M}_S^{mn,mn}$ is defined as $P_n^{mn} \left(e_i^m \otimes e_j^n\right)^{\mathsf{T}} = \left(e_j^n \otimes e_i^m\right)^{\mathsf{T}}$, where $e_i^m \in \mathrm{M}_S^{1,m}$ and $e_j^n \in \mathrm{M}_S^{1,n}$ are row unit vectors of appropriate length. In other words, the matrix P_n^{mn} permutes the elements of a vector of length mn with stride distance n. Further we will make use of the following identities on stride permutations, where I_m is the order m identity matrix.

Proposition 2. *The following hold:*

1. $P_n^{mn} = \sum_{i=1}^n \left(e_i^n\right)^{\mathsf{T}} \otimes I_m \otimes \left(e_i^n\right).$
2. $(P_n^{mn})^{-1} = \sum_{i=1}^n \left(e_i^n\right) \otimes I_m \otimes \left(e_i^n\right)^{\mathsf{T}}.$ □

The following notion is borrowed from [8, Definition 11]. An *algebraic Turing machine over a semiring* S is a tuple $M = (Q, \Sigma, \Gamma, \delta, q_0, B, F)$, where Q, $\Sigma \subseteq \Gamma$, Γ, $q_0 \in Q$, $B \in \Gamma$, and $F \subseteq Q$ are defined as for ordinary Turing machines and δ is the *transition relation* taking the form $\delta \subseteq Q \times \Gamma \times Q \times \Gamma \times \{L, S, R\} \times S$. In this machine, a move using the transition $(p, a, q, b, m, s) \in \delta$, is assigned a *weight* s. To a particular computation we associate a weight in the straightforward way, i.e., as the *product* in S of the weights of its successive moves. For completeness, we define the weight of a length-zero computation to be 1.

On input w machine M computes the value $f_M(w)$, a function $f_M : \Sigma^* \to S$, which is defined as the *sum* of the weights of all accepting computations from the initial configuration on input w to an accepting configuration. A language $L_M = \{ w \in \Sigma^* \mid f_M(w) \neq 0 \}$ can also be defined. The definition is adapted in a straightforward way to define algebraic algebraic space or time bounded Turing machines, etc. We define the *generalized counting class* S-#P as the set of all functions $f : \Sigma^* \to S$ such that there is a polytime algebraic Turing machine M over S which computes f. The *generalized language class* S-P is the set of all languages $L \subseteq \Sigma^*$ such that there is a polytime algebraic Turing machine M over S for which $w \in L$ if and only if $f_M(w) \neq 0$.

Let S-#C be a generalized counting class over semiring S which is closed under the pairing operation $\langle \cdot, \cdot \rangle$. Then the "polynomial" counting operator $\#^p$ defines the function class $\#^p \cdot S$-#C as the set of all functions f for which a polynomial p and a function $f_M \in S$-#C can be found, such that

$$f(x) = \sum_{|y| \leq p(|x|)} f_M(x, y),$$

where summation is taken in S. Based on the counting operator $\#^p$ the "polynomial" operators $\exists^p$, $\forall^p$, and $C_=^p$ define the following language classes: $\exists^p \cdot S$-#C ($\forall^p \cdot S$-#C, $\exists!^p \cdot S$-#C, $C_=^p \cdot S$-#C, resp.) is the set of all languages L for which there are some functions f, g in $\#^p \cdot S$-#C such that $x \in L$ if and only if $f(x) \neq 0$, ($f(x) = 0$, $f(x) = 1$, $f(x) = g(x)$, resp.). Obviously, $\#^p \cdot S$-#P $= S$-#P. Table 1 states without proof other significant special cases. The "exponential" operators $\#^e$, $\exists^e$, $C_=^e$, and $\forall^e$ are similarly defined as their polynomial counterparts, by replacing the polynomial p with the function $2^{c \cdot |x|}$ for some constant c.

In the remainder of this section we define tensor circuits over semirings.

Table 1. Polytime classes defined by operators on generalized counting classes. Here χ_L is the characteristic function of the languages L

Class	Semiring S			
	$\mathbb{B}$	$\mathbb{Z}_q$	N	Z
$S\text{-}\#\mathrm{P}$	$\{\chi_L \mid L \in \mathrm{NP}\}$	$\{f \pmod q \mid f \in \#\mathrm{P}\}$	$\#\mathrm{P}$	$Gap\mathrm{P}$
$S\text{-}\mathrm{P}$	NP	$\mathrm{MOD}_q\text{-}\mathrm{P}$	NP	co- $C_=\mathrm{P}$
$\exists^p \cdot S\text{-}\#\mathrm{P}$				
$\forall^p \cdot S\text{-}\#\mathrm{P}$	co-NP	co-$\mathrm{MOD}_q\text{-}\mathrm{P}$	co-NP	$C_=\mathrm{P}$
$\exists!^p \cdot S\text{-}\#\mathrm{P}$	NP		USP	
$C_=^p \cdot S\text{-}\#\mathrm{P}$	co-DP		$C_=\mathrm{P}$	

Definition 3. *A tensor circuit C over a semiring S is a finite directed acyclic graph and its* order *is defined as follows:*

1. *Each node with in-degree zero is labeled with some F from $\mathbb{M}_S^{k \times \ell}$ with entries from S and its order is $k \times \ell$. These nodes are called* inputs.
2. *Non-input nodes, i.e.,* inner *nodes, have in-degree exactly two and are labeled by matrix operations, i.e., addition, multiplication, and tensor product. Let f be an inner node labeled by $\circ$, for $\circ \in \{+, \cdot, \otimes\}$, whose left child g is of order $k \times \ell$ and its right child h is of order $m \times n$. Then we simply say that f is a $\circ$-node, write $f = (g \circ h)$, and its order is defined as:*
   (a) *$k \times \ell$ if node $f = (g + h)$, $k = m$, and $\ell = n$,*
   (b) *$k \times n$ if node $f = (g \cdot h)$ and $\ell = m$, and*
   (c) *$km \times \ell n$ if node $f = (g \otimes h)$.*
3. *There is a unique node with out-degree zero, which is called* output *node. The order of the circuit C is defined to be the order of the output node.*

For a tensor circuit C of order $k \times \ell$ let $I(C) = [k] \times [\ell]$ be its "set of indices." Let $\mathbb{T}_S$ denote the set of all tensor circuits over S, and define $\mathbb{T}_S^{k,\ell} \subseteq \mathbb{T}_S$ to be the set of all tensor circuits of order $k \times \ell$.

A tensor circuit C is called a tensor formula *if its finite directed acyclic graph is actually a binary tree.*

In this paper we consider only finitely generated S, and we assume that entries of the inputs to a tensor circuit are from $\mathcal{G} \cup \{0\}$, where $\mathcal{G}$ is a generating set of S. Hence, labels of input nodes, i.e., matrices, can be string-encoded using list notation such as "[[001][101]]." Tensor circuits can be encoded over the alphabet $\Sigma = \{0\} \cup \mathcal{G} \cup \{[,], (,), \cdot, +, \otimes\}$. Strings over Σ which do not encode valid circuits are deemed to represent the trivial tensor circuit 0 of order 1×1.

Let C be a tensor circuit of order $m \times n$. Its *diameter* is $\max\{m, n\}$, its *size* is the number of nodes and edges, and its *depth* is the maximum number of nodes along a path connecting some input node with the output node (leaf-root path). Besides these usual notions we introduce the *tensor depth* of a circuit as the maximum number of tensor nodes found along a leaf-root path. Concerning tensor circuits we observe that they are much more "powerful" than tensor formulas [8], since they allow to blow up matrix diameter at a rate which is double-exponential. More precisely, if C is a tensor circuit of tensor depth t which has input matrices of diameter at most p, then $|C| \leq p^{2^t}$, and there exists a circuit which outputs a matrix of exactly this diameter.

Proposition 4. *1. Testing whether a string encodes a valid tensor circuit and if so, computing its order, is feasible in deterministic time $(n \cdot 2^{t(n)} \log n)^{O(1)}$, where $t(n)$ denotes the tensor depth of the given tensor circuit.*

2. Verifying that a given tensor circuit meets the requirements to be of depth $d(n)$ (and tensor depth $t(n)$) can be done on a deterministic Turing machine in time $O(n \cdot \log d(n))$. If only tensor depth $t(n)$ has to be checked, the running time is $O(n \cdot \log t(n))$. □

Finally the evaluation of tensor circuits is defined as follows:

Definition 5. *For each semiring S and each k and each ℓ we define $\mathrm{val}_S^{k,\ell}$: $\mathbb{T}_S^{k,\ell} \to \mathbb{M}_S^{k,\ell}$, that is, we associate with node f of order $k \times \ell$ of a tensor circuit C its $k \times \ell$ matrix "value," which is defined as follows:*

1. $\mathrm{val}_S^{k,\ell}(f) = F$ if f is an input node labeled with F,
2. $\mathrm{val}_S^{k,\ell}(f) = \mathrm{val}_S^{k,\ell}(g) + \mathrm{val}_S^{k,\ell}(h)$ if $f = (g+h)$,
3. $\mathrm{val}_S^{k,\ell}(f) = \mathrm{val}_S^{k,m}(g) \cdot \mathrm{val}_S^{m,\ell}(h)$ if $f = (g \cdot h)$ and g is of order $k \times m$, and
4. $\mathrm{val}_S^{k,\ell}(f) = \mathrm{val}_S^{k/m,\ell/n}(g) \otimes \mathrm{val}_S^{m,n}(h)$ if $f = (g \otimes h)$ and h is of order $m \times n$.

The value $\mathrm{val}_S^{k,\ell}(C)$ of a tensor circuit C of order $k \times \ell$ is defined to be the value of the unique output node. Tensor circuits of order 1×1 are called scalar tensor circuits, *and we simply write val_S for the $\mathrm{val}_S^{1,1}$ function.*

The non-zero tensor problem is defined as follows:

Definition 6. *Let S be a semiring. The* non-zero tensor circuit problem *over semiring S is defined to be the set $0 \neq \mathrm{val}_S$ of all scalar tensor circuits C for which $\mathrm{val}_S(C) \neq 0$.*

3   Computational Complexity of the Non-zero Problem

This section contains completeness results on the computational complexity of the tensor circuit evaluation problem over certain semirings. Our main theorem immediately follow from the below given lemmata. For tensor circuit evaluation we obtain the following upper bound. The proof is an adaption of the algorithm on tensor formulas [8, Lemma 16] to the case of circuits.

Lemma 7. *For any finitely generated semiring S, the evaluation problem val_S for scalar tensor circuits of depth $d(n)$ and tensor depth $t(n)$ belongs to the class S-#TISP$(2^{d(n)}(n \cdot 2^{t(n)} \log n)^{O(1)}, (n \cdot 2^{t(n)} \log n)^{O(1)})$.* □

The hardness reads as follows:

Lemma 8. *Let M be an algebraic Turing machine over a finitely generated semiring S running in time $t(n)$ and space $s(n)$, with $s(n) \geq n$, and set $d(n) = \max\{\log t(n), \log s(n)\}$ and $g(n) = \max\{n, d(n)\}$. There is a $(g(n))^{O(1)}$ time and $O(g(n))$ space output computable function f, which on input w, computes a scalar tensor circuit $f(M, w)$ of size $O(g(n))$, depth $O(d(n))$, and tensor depth $O(\log s(n))$ such that $\mathrm{val}_S(f(M, w)) = f_M(w)$.*

Sketch of Proof. It is straightforward to verify that the iterated matrix multiplication at the core of Warshall's algorithm is meaningful over arbitrary semirings; then we are justified to implicitly use this algorithm in the forthcoming reduction. Since $M = (Q, \Sigma, \Gamma, \delta, q_1, B, F)$ is fixed the following assumptions can be imposed without loss of generality:

1. The tape alphabet is $\Gamma = \{0, 1, B\}$; hence input $w \in \Sigma^*$ is binary.
2. Machine M is single-tape and the tape is cyclic with $s(n)$ tape cells, numbered 0 to $s(n) - 1$, so that tape-head motions are from cell i to cell $(i \pm 1)$ (mod $s(n)$). Note that in the case of a machine for a complexity class defined solely in terms of a time bound, these are just enough tape cells to make sure that the machine cannot see the difference between a cyclic tape and one that is infinite in both directions. Moreover, all computations of M have length exactly $t(n)$ and accepting ones are determined solely by the state, regardless of the tape content.
3. A single transition of M can be conceptually split in two phases: First, the machine reads and overwrites the content of the tape cell its read-write head points to, and changes accordingly the state of its finite control. Second, the new state determines whether there will be a head movement, and if so, this is implemented without changing state or tape content by taking a transition assigned the weight 1.

Our second assumption helps us overcome the major hurdle one meets upon building a generic proof, namely finding an efficient way of encoding the tape content and of describing the motions of the read-write head. Here, we keep the head static and move the tape instead; using a cyclic tape enables us to encode this as a rotation matrix, which can be constructed recursively. We thus may assume that the set of all configurations equals $Q \times \Gamma^{s(n)}$, where $\Gamma^{s(n)}$ is a shorthand notation for the set of all strings of length exactly $s(n)$. Let $Q = \{q_1, \ldots, q_m\}$ and $\Gamma = \{a_1, a_2, a_3\}$. Define mappings $c_Q : Q \to M_S^{1,m}$ and $c_\Gamma : \Gamma \to M_S^{1,3}$ as: $c_Q(q_i) = e_i^m$ and $c_\Gamma(a_i) = e_i^3$. With a configuration $C = (q, a_0 \ldots a_{s(n)-1})$ in $Q \times \Gamma^{s(n)}$ we associate the unit vector

$$c(C) = c_Q(q) \otimes \bigotimes_{i=0}^{s(n)-1} c_\Gamma(a_i).$$

Observe, that pre-multiplying $c(C)^{\mathsf{T}}$ by $(I_m \otimes P_{3^{s(n)}}^{3 \cdot 3^{s(n)-1}})$ implements a cyclic tape shift to the left. Similarly cyclic tape shifts to the right are implemented.

The value of the function f_M on input w equals the sum of all labels of accepting paths of length $t(n)$ in the configuration graph G_M with node set $Q \times \Gamma^{s(n)}$. We can associate a matrix A_{G_M} of order $m \cdot 3^{s(n)} \times m \cdot 3^{s(n)}$ with G_M, such that the (i,j)th entry of $A_{G_M}^{t(n)}$ equals the sum of all labels of paths of length $t(n)$ linking configuration C_i with C_j. Let $A = (A_{G_M})^\mathsf{T}$. For an input $w = a_0 \dots a_n$, one has

$$f_M(w) = \mathrm{val}_S \left(V_{accept} \cdot A^{t(n)} \cdot (V_{init})^\mathsf{T} \right),$$

where $V_{accept} = \sum_{C \in F \times \Gamma^{s(n)}} c(C)$ is a row column vectors representing all accepting configurations of M on input w, while $V_{init} = c(C_0)$ and $C_0 = (q_1, a_0 \dots a_{n-1} B^{s(n)-n})$ is the initial configuration of M. The rest of the proof shows how to define an appropriate tensor circuit which meets our requirements on size, depth, and tensor depth.   □

Then we obtain the following completeness result for the evaluation and non-zero problem on scalar tensor circuits.

Theorem 9. *For any finitely generated semiring S, the problem val_S is S-#E-complete and the problem $0 \neq \mathrm{val}_S$ is $\exists^e \cdot S$-#P-complete both under polylin many-one reduction.*   □

In Tables 2 and 3 we list consequences of Theorem 9 on concrete instances of the semiring S and common sense restrictions on the parameters $t(n)$, $d(n)$ and $s(n)$. Not listed in the tables are completeness results under polylin many-one reduction for the case of unrestricted depth and tensor depth; one then captures the class NE and its counting counterparts instead of members of the NEXP family.

Table 2. Complexity of the scalar tensor circuit evaluation problem val_S. Completeness results are meant with respect to polytime many-one reductions

Depth $d(n)$	Tensor depth $t(n)$		Semi-ring S
	$O(\log n)$	unrestricted	
$O(\log n)$	#P		N
	GapP		Z
$O(\log^k n)$	#TISP$(2^{O(\log^k n)}, n^{O(1)})$	#TIME$(2^{O(\log^k n)})$	N
	GapTISP$(2^{O(\log^k n)}, n^{O(1)})$	GapTIME$(2^{O(\log^k n)})$	Z
unrestricted	#PSPACE	#EXP	N
	GapPSPACE	GapEXP	Z

A closer look at Lemma 8 shows that it can be rewritten in terms of tensor formulas instead of circuits. This enables us to significantly improve the results

Table 3. Complexity of the non-zero tensor circuit problem $0 \neq \mathrm{val}_S$. Completeness results are meant with respect to polytime many-one reductions

Depth $d(n)$	Tensor depth $t(n)$		Semi-ring S
	$O(\log n)$	unrestricted	
$O(\log n)$	NP		$\mathbb{B}$
	MOD_q-P		$\mathbb{Z}_q$
	$\oplus$P		$\mathbb{F}_2$
$O(\log^k n)$	$\mathrm{NTISP}(2^{O(\log^k n)}, n^{O(1)})$	$\mathrm{NTIME}(2^{O(\log^k n)})$	$\mathbb{B}$
	$\mathrm{MOD}_q\text{-}\mathrm{TISP}(2^{O(\log^k n)}, n^{O(1)})$	$\mathrm{MOD}_q\text{-}\mathrm{TIME}(2^{O(\log^k n)})$	$\mathbb{Z}_q$
	$\oplus\mathrm{TISP}(2^{O(\log^k n)}, n^{O(1)})$	$\oplus\mathrm{TIME}(2^{O(\log^k n)})$	$\mathbb{F}_2$
unrestricted	PSPACE	NEXP	$\mathbb{B}$
	MOD_q-PSPACE	MOD_q-EXP	$\mathbb{Z}_q$
	$\oplus$PSPACE	$\oplus$EXP	$\mathbb{F}_2$

in [8] on the complexity of the evaluation and non-zero problem on scalar tensor formulas by (1) extending to the case of arbitrary finitely generated semirings, and (2) showing completeness under polytime many-one reduction.

Theorem 10. *Let S denote any finitely generated semiring. For tensor formulas the problem val_S is S-#P-complete and $0 \neq \mathrm{val}_S$ is $\exists^p \cdot S$-#P-complete both under polytime many-one reduction.* $\square$

We observe, that in particular, the evaluation problem for tensor formulas over the naturals is #P-complete with respect to polytime many-one reduction. This statement is not provable with the techniques used in [8].

4   Computational Complexity of Some Related Problems

A large number of decision problems on tensor circuits can be defined, based on natural linear-algebraic questions: Asking whether such a circuit outputs a null matrix, or a diagonal square matrix, etc. These questions were not addressed in [8], so that we look at them also in the case of tensor formulas. We start with a very useful lemma, which is a generalization of a construction given in [8, Lemma 7]. We omit the technical proof, which is based on a generic reduction and makes extensive use of stride permutations.

Lemma 11. *Let M be an algebraic polytime Turing machine with binary input alphabet $\Sigma = \{0, 1\}$ over a finitely generated semiring S and $p(n)$ a polynomial or an exponential function. There is a polylin computable function f, which on input 1^n computes a tensor circuit $C_{M,n} = f(M, 1^n)$ of linear size and of depth $d(n) = O(\log n)$ if $p(n)$ is a polynomial and $d(n) = O(n)$ if $p(n)$ is an exponential* *function, such that for all* $w = a_0 \ldots a_{p(n)-1}$ *with* $a_i \in \Sigma$ *and* $0 \le i \le p(n) - 1$,

$$f_M(w) = \mathrm{val}_S \left(\left(\bigotimes_{i=0}^{p(n)-1} e_{a_i+1}^2 \right) \cdot C_{M,n} \cdot \left(\bigotimes_{i=0}^{p(n)-1} e_{a_i+1}^2 \right)^{\mathsf{T}} \right).$$

In other words the matrix $\mathrm{val}_S^{2^{p(n)}, 2^{p(n)}}(C_{M,n})$ *contains the values of the function* f_M *on all possible inputs of length* $p(n)$ *on the main diagonal, and zero elsewhere.*   □

Our analysis of linear-algebraic problems as mentioned above uses two ancillary problems defined next. Let S be a finitely generated semiring. We define the *zero-value tensor circuit problem* over S, for some $s \in S$, as the set ZERO$_S$ of all triples (C, i, j), where C is a tensor circuit of order $I(C) = k \times \ell$, integers i and j are expressed in binary such that $(i,j) \in [k] \times [\ell]$, and $(\mathrm{val}_S^{k,\ell}(C))_{i,j} = 0$. The *one-value tensor circuit problem* ONE$_S$ over S is defined similarly with the condition $(\mathrm{val}_S^{k,\ell}(C))_{i,j} = 1$. Note that for each problem we define on tensor circuits, a corresponding problem is defined by restricting the domain of the input circuits to the set of all tensor formulas over S; these restrictions are denoted with an initial "F-," e.g., F-ZERO$_S$ for ZERO$_S$, etc. For these cases we find the following situation, which proof is a standard exercise and thus left to the reader.

Lemma 12. *Let* S *be a finitely generated semiring. Then (1) the tensor formula problem* F-ZERO$_S$ *is* $\forall^p \cdot S$-#P-complete *and* F-ONE$_S$ *is* $\exists!^p \cdot S$-#P-complete *both under polytime many-one reduction and (2) the tensor circuit problem* ZERO$_S$ *is* $\forall^e \cdot S$-#P-complete *and* ONE$_S$ *is* $\exists!^e \cdot S$-#P-complete *both under polylin many-one reduction.*   □

Once Lemma 12 is available we apply it to a range of other problems such as, e.g., testing whether the output of a tensor circuit is the null-matrix (NULL$_S$), a diagonal matrix (DIAG$_S$), the identity matrix (ID$_S$), an orthogonal matrix (ORTHO$_S$), a symmetric matrix (SYM$_S$), and verifying whether two tensor circuits received as input are equivalent (EQ$_S$), i.e., whether they output the same matrix. The former two problems are classified as follows:

Theorem 13. *If* S *is a finitely generated semiring, then (1) the tensor formula problems* F-NULL$_S$ *and* F-DIAG$_S$ *are* $\forall^p \cdot (\forall^p \cdot S$-#P)-complete *under polytime many-one reduction and (2) the tensor circuit problems* NULL$_S$ *and* DIAG$_S$ *are* $\forall^e \cdot (\forall^p \cdot S$-#P)-complete *under polylin many-one reduction.*

Proof. We only write the proof for DIAG$_S$. The reasoning for the null tensor circuit problem is identical. First we show containment in $\forall^e \cdot (\forall^p \cdot S$-#P). Let C be a tensor formula instance of DIAG$_S$ and assume that $I(C) = [k] \times [\ell]$. Then $C \in$ DIAG$_S$ if and only if C is a square tensor circuit and

$$\forall (i,j) \in [k] \times [\ell] : \left((i \ne j) \Rightarrow (C, i, j) \in \mathrm{ZERO}_S \right).$$

By Proposition 4, whether a tensor circuit outputs a square matrix can be tested in deterministic exponential time. Next, since ZERO_S is in $\forall^e \cdot S\text{-}\#P$ by Lemma 12, membership in DIAG_S can be checked in $\forall^e \cdot (\forall^p \cdot S\text{-}\#P)$. Observe, that given (exponential-length) binary encodings for indices i and j, an algorithm for ZERO_S will have a computation time polynomial in the length of its input, which consists in i, j, and a description of C.

For the hardness part we argue as follows. Let L be a language in $\forall^e \cdot (\forall^p \cdot S\text{-}\#P)$. Then by definition there is a constant c and an algebraic polynomial time Turing machine M over S such that

$$x \in L \quad \text{if and only if} \quad \forall |y| = 2^{c \cdot |x|} : f_M(x, y) = 0.$$

Define C to be the tensor formula constructed according to Lemma 11 in polytime from M and the constant c such that C evaluates to a matrix of order $2^{2^{c \cdot |x|}} \times 2^{2^{c \cdot |x|}}$ containing all values of f_M on input $\langle x, y \rangle$ for all possible y's of appropriate length on the main diagonal. Then it is easily seen that $x \in L$ if and only if tensor circuit $D_{1,2}^2 \otimes C$, where $D_{i,j}^k$ is the order k dot matrix having one in position (i, j) and zeros elsewhere, evaluates to the all-zero matrix or equivalently to a diagonal matrix. This completes the construction and proves the stated claim.   □

The proofs on the complexity of the remaining decision problems follow similar lines. In Table 4 we summarize the complexity of these decision problems for tensor circuits over specific semirings. There Π_2^e denotes the second level of the weak exponential hierarchy [10,18]. Meanwhile, for the restrictions of these problems to tensor formulas completeness results translate to the polytime counterparts of the complexity classes encountered in the circuit case.

Table 4. The complexity of tensor circuit problems summarized. Completeness results are meant with respect to polylin many-one reductions

Problem	$\mathbb{B}$	$\mathbb{Z}_q$	$\mathbb{F}_2$	$\mathbb{N}$	$\mathbb{Z}$
		Semiring S			
NULL_S	co-NE	$\forall^e \cdot \text{co-MOD}_q\text{-P}$	$\forall^e \cdot \oplus P$	co-NE	$C_=E$
DIAG_S					
ORTHO_S	Π_2^e			$\forall^e \cdot \text{USP}$	
ID_S					
EQ_S				$C_=E$	
SYM_S					

5   Discussion

In their paper on tensor calculus Damm *et al.* [8] showed how the three most important nondeterministic complexity classes NL, LOGCFL, and NP were elegantly captured by a single problem, evaluation of tensor formulas over the boolean semiring. We have shown how tensor circuits extend this nice result to the classes PSPACE, NE, and NEXP. Since tensor formulas and circuits are handy to give compact specifications for possibly very large matrices, we deemed it important to look at a number of other classical problems involving matrices. Other questions remain to be looked at including the ubiquitous one of asking whether a system of linear equations is feasible.

Acknowledgments

The authors thank Carsten Damm for helpful discussions and pointers to useful references. Also thanks to Pierre McKenzie and Jose Manuel Fernandez for fruitful discussions.

References

1. J. L. Balcázar, J. Díaz, and J. Gabarró. *Structural Complexity I*. Texts in Theoretical Computer Science. Springer Verlag, 1995. 174
2. D. A. Mix Barrington. Bounded-width polynomial size branching programs recognize exactly those languages in NC^1. *Journal of Computer and System Sciences*, 38:150–164, 1989. 173
3. D. A. Mix Barrington and D. Thérien. Finite monoids and the fine structure of NC^1. *Journal of the Association of Computing Machinery*, 35:941–952, 1988. 173
4. A. Blass and Y. Gurevich. On the unique satisfiability problem. *Information and Control*, 82:80–88, 1982. 175
5. G. Buntrock, C. Damm, U. Hertrampf, and C. Meinel. Structure and importance of logspace MOD-classes. *Mathematical Systems Theory*, 25:223–237, 1992. 173
6. S. R. Buss. The Boolean formula value problem is in ALOGTIME. In *Proceedings of the 19th Symposium on Theory of Computing*, pages 123–131. ACM Press, 1987. 173
7. S. R. Buss, S. Cook, A. Gupta, and V. Ramachandran. An optimal parallel algorithm for formula evaluation. *SIAM Journal on Computing*, 21(4):755–780, 1992. 173
8. C. Damm, M. Holzer, and P. McKenzie. The complexity of tensor calculus. In *Proceedings of the 15th Conference on Computational Complexity*, pages 70–86. IEEE Computer Society Press, 2000. 173, 174, 176, 178, 181, 183
9. L. Fortnow. One complexity theorist's view of quantum computing. In *Electronic Notes in Theoretical Computer Science*, volume 31. Elsevier, 2000. 174
10. L. Hemachandra. The strong exponential hierarchy collapses. *Journal of Computer and System Sciences*, 39(3):299–322, 1989. 183
11. N. D. Jones and W. T. Laaser. Complete problems for deterministic polynomial time. *Theoretical Computer Science*, 3:105–117, 1976. 173
12. W. Kuich and A. Salomaa. *Semirings, Automata, Languages*, volume 5 of *EATCS Monographs on Theoretical Computer Science*. Springer, 1986. 175
13. C. H. Papadimitriou and M. Yannakakis. The complexity of facets (and some facets of complexity). *Journal of Computer and System Sciences*, 28(2):244–259, 1984. 175
14. W.-H. Steeb. *Kronecker Product of Matrices and Applications*. Wissenschaftsverlag, Mannheim, 1991. 175

15. R. Tolimieri, M. An, and Ch. Lu. *Algorithms for Discrete Fourier Transform and Convolution*. Springer Verlag, 1997. 175

16. L. G. Valiant. The complexity of computing the permanent. *Theoretical Computer Science*, 8:189–201, 1979. 175

17. V. Vinay. Counting auxiliary pushdown automata and semi-unbounded arithmetic circuits. In *Proceedings of the 6th Structure in Complexity Theory*, pages 270–284. IEEE Computer Society Press, 1991. 175

18. C. Wrathall. Complete sets and the polynomial-time hierarchy. *Theoretical Computer Science*, 3:23–33, 1977. 183

Computing Reciprocals of Bivariate Power Series

Markus Bläser

Institut für Theoretische Informatik, Med. Universität zu Lübeck
Wallstr. 40, 23560 Lübeck, Germany
blaeser@tcs.mu-luebeck.de

Abstract. We consider the multiplicative complexity of the inversion and division of bivariate power series modulo the "triangular" ideal generated by all monomials of total degree $n + 1$. For inversion, we obtain a lower bound of $\frac{7}{8}n^2 - O(n)$ opposed to an upper bound of $\frac{7}{3}n^2 + O(n)$. The former bound holds for all fields with characteristic distinct from two while the latter is valid over fields of characteristic zero that contain all roots of unity (like e.g. $\mathbb{C}$). Regarding division, we prove a lower bound of $\frac{5}{4}n^2 - O(n)$ and an upper bound of $3\frac{5}{6}n^2 + O(n)$. Here, the former bound is proven for arbitrary fields whereas the latter bound holds for fields of characteristic zero that contain all roots of unity.
Similar results are obtained for inversion and division modulo the "rectangular" ideal (X^{n+1}, Y^{n+1}).

1 Introduction

The question how many *essential* multiplications and divisions are needed to compute the reciprocal of a power series is an intriguing and challenging problem in algebraic complexity theory. Here, "essential" means that additions, subtractions, and scalar multiplications are free of costs. (We give a formal definition of multiplicative complexity below.) In the literature, mainly the inversion of *univariate* power series is discussed. In the present work, we investigate the multiplicative complexity of inversion of *bivariate* power series. More precisely, given a power series $A = \sum_{i=0}^{\infty} \sum_{j=0}^{\infty} a_{i,j} X^i Y^j$ (with indeterminate coefficients) we want to determine the number of essential operations necessary and sufficient to compute the coefficients $b_{i,j}$ with $i + j \le n$ of $A^{-1} = \sum_{i=0}^{\infty} \sum_{j=0}^{\infty} b_{i,j} X^i Y^j$ from the coefficients $a_{i,j}$ with $i + j \le n$.

The reason why to investigate this problem is threefold: first, the problem of computing reciprocals of bivariate power series is an algebraic modelling of the inversion of univariate power series with arbitrary precision numbers as coefficients. The first indeterminate X refers to the original indeterminate whereas the second indeterminate Y models the precision management. Further details can be found in [8]. Second, the results obtained in the present work may shed a new light on the univariate case, a major open problem in algebraic complexity theory (see e.g. [4, Problem 2.7]). Third, our new results complement nicely the results in [2] on the multiplication of bivariate power series.

What is known in the univariate case? Let C_n denote the number of essential operations necessary and sufficient to compute the first $n+1$ coefficients of $A^{-1} =$

J. Sgall, A. Pultr, and P. Kolman (Eds.): MFCS 2001, LNCS 2136, pp. 186–197, 2001.

$b_0 + b_1 X + b_2 X^2 + \cdots$ from the first $n + 1$ coefficients of the power series $A = a_0 + a_1 X + a_2 X^2 + \cdots$ (where we again consider $a_0, a_1, a_2, \ldots$ as indeterminates over some ground field k). If the characteristic of the ground field k differs from two, then we have

$$n - 2 + \lceil (n-1)/2 \rceil \leq C_n \leq 3\tfrac{3}{4}n. \tag{1}$$

If the characteristic of k is two, then

$$n + 1 \leq C_n \leq 3\tfrac{1}{4}n. \tag{2}$$

Kalorkoti [5] attributes the lower bound in (1) to Hanns-Jörg Stoß. To prove this bound, the problem of computing the lower $n - 1$ coefficients of the square of a polynomial of degree $n-2$ is reduced to the problem of computing $b_0, \ldots, b_n$ using techniques due to Strassen [12]. (The lower bound for computing the lower $d + 1$ coefficients of the square of a polynomial of degree d given in [5] can be improved by one, $d + \lceil (d+1)/2 \rceil$ is the true complexity of this problem.) The upper bound in (1) is due to Kung [6]. Kung actually states the upper bound $4n - \log n$, but a more careful analysis of his algorithm by Schönhage yields the stated upper bound, see again [5] for a discussion.

The lower bound in (2) follows from a standard linear independence argument, the upper bound is again due to Kung. The improvement compared to (1) follows from the fact that computing the square of a polynomial is cheap over fields of characteristic two.

Until today, the bounds in (1) and (2) have withstood any attempts of improvement (including those of the author, see however [3] for an improvement in a restricted setting).

1.1 Computations, Costs and Complexity

In the present section, we briefly introduce the concept of multiplicative complexity. For a detailed description, the reader is referred to [4]. We start with giving a formal definition of computations.

Definition 1. *Let $x_1, \ldots, x_m$ be indeterminates over a field k and $f_1, \ldots, f_n \in K = k(x_1, \ldots, x_m)$.*

1. *A sequence $\beta = (w_1, \ldots, w_\ell)$ of rational functions $w_1, \ldots, w_\ell \in K$ is called a computation over K, if for all $1 \leq \lambda \leq \ell$ there are $i, j < \lambda$ such that $w_\lambda = w_i \circ w_j$ with $\circ \in \{+, -, *, /\}$ and $w_j \neq 0$ if $\circ = /$ or $w_\lambda = \alpha \cdot w_i$ with $\alpha \in k$ or $w_\lambda \in k \cup \{x_1, \ldots, x_m\}$.*
2. *The sequence is called a computation for $f_1, \ldots, f_n$, if in addition $f_1, \ldots, f_n \in \{w_1, \ldots, w_\ell\}$.*

The next step is to define the costs of a computation.

Definition 2. *Let k be a field, $x_1, \ldots, x_m$ indeterminates over k, and $\beta = (w_1, \ldots, w_\ell)$ a computation over $K = k(x_1, \ldots, x_m)$.*

1. *The costs γ_λ in the λth step of β are defined as*

$$
\gamma_\lambda = \begin{cases}
0 & \text{if there are } i, j < \lambda \text{ such that } w_\lambda = w_i \pm w_j, \\
 & \text{or } w_\lambda = \alpha \cdot w_i \text{ with } \alpha \in k, \\
 & \text{or } w_\lambda \in k \cup \{x_1, \ldots, x_m\}, \\
1 & \text{otherwise.}
\end{cases}
$$

2. *The costs $\Gamma(\beta)$ of β are $\Gamma(\beta) = \sum_{\lambda=1}^{\ell} \gamma_\lambda$.*

In the above definition, we count only those steps where the rational function w_λ can solely be expressed as the product or quotient of two elements with smaller index. Such a step is called an *essential* multiplication or division, respectively. This measurement of costs is also called *Ostrowski measure*.

We proceed with defining the complexity of a set of rational functions.

Definition 3. *Let $x_1, \ldots, x_m$ be indeterminates over a field k and $f_1, \ldots, f_n \in K = k(x_1, \ldots, x_m)$. The multiplicative complexity $C(f_1, \ldots, f_n)$ of $f_1, \ldots, f_n$ is defined by $C(f_1, \ldots, f_n) = \min\{\Gamma(\beta) \mid \beta \text{ is a computation for } f_1, \ldots, f_n\}$.*

1.2   Our Results

After settling the model of computation, we are able to formulate our problem precisely: let $A(X, Y) = \sum_{i=0}^{\infty} \sum_{j=0}^{\infty} a_{i,j} X^i Y^j$ be a bivariate power series with indeterminates over some ground field k as coefficients. Let $B(X, Y) = \sum_{i=0}^{\infty} \sum_{j=0}^{\infty} b_{i,j} X^i Y^j$ be the power series defined by $A \cdot B = 1$. (Note that A is invertible, since $a_{0,0}$ is an indeterminate.) The coefficients of B are rational functions in the coefficients of A and only $a_{0,0}$ occurs in the denominators of these rational functions. If we write $A = a_{0,0} - A_1$, then

$$
B = \frac{1}{a_{0,0}} + \frac{1}{a_{0,0}^2} A_1 + \frac{1}{a_{0,0}^3} A_1^2 + \cdots . \tag{3}
$$

Power series are infinite objects but in our model (and in today's computers) we can only deal with a finite number of rational functions. In the univariate case, computing modulo (X^{n+1}) is the canonical choice. In the bivariate case, however, there are more than one meaningful choice. We here mainly consider the case of computing modulo the "triangular" ideal $I_n := (X^{n+1}, X^n Y, X^{n-1} Y^2, \ldots, Y^{n+1})$ generated by all monomials of degree $n + 1$. In this case, we only have to consider terms up to $A_1^n / a_{0,0}^{n+1}$ in (3). Another possibility is to take the "rectangular" ideal (X^{n+1}, Y^{n+1}). We will discuss this case in Section 5 briefly.

In the above language, our problem reads as follows: find good upper and lower bounds for $C(\{b_{i,j} \mid i + j \leq n\})$. Note that the rational functions $b_{i,j}$ with $i + j \leq n$ only depend on the $a_{i,j}$ with $i + j \leq n$. Therefore, inverting power series modulo I_n is equivalent to computing the reciprocal of the generic element $\sum_{i+j\leq n} a_{i,j}X^iY^j$ in the local k-algebra $\mathcal{T}_n := K[X,Y]/I_n$ where $K = k(a_{i,j} \mid i+j \leq n)$. For convenience, we do not always want to choose names for the coefficients of the power series explicitly. Therefore, we assume the following convention about the fields k and K:

The field k denotes the underlying ground field and multiplication with elements from k is for free. On the other hand, the field K is obtained by adjoining to k the coefficients (which are indeterminates over k) of the power series we want to manipulate.

So if for instance we speak about the multiplicative complexity of the multiplication in $K[X,Y]/I_n$, we mean the multiplicative complexity of computing the coefficients of the product AB of two "truncated" power series $A = \sum_{i+j\leq n} a_{i,j}X^iY^j$ and $B = \sum_{i+j\leq n} b_{i,j}X^iY^j$ from the coefficients of A and B. The field K is obtained by adjoining all $a_{i,j}$ and $b_{i,j}$ with $i+j \leq n$ to k. (Note that the coefficients of AB are rational functions in the coefficients of A and B.)

Our main results concerning inversion are subsumed in the following theorems:

Theorem 4. *For fields k of characteristic zero that contain all roots of unity, the multiplicative complexity of inverting in $\mathcal{T}_n$ is at most $\frac{7}{3}n^2 + O(n)$.*

Theorem 5. *The multiplicative complexity of inversion in $\mathcal{T}_n$ has the lower bound $\frac{1}{2}n^2 + \frac{3}{2}n + 1$. Over fields of characteristic distinct from two this can be improved to $\frac{7}{8}n^2 - O(n)$.*

Building up on these results, we obtain the following bounds for the division of power series:

Theorem 6. *Let k be a field of characteristic zero that contains all roots of unity. The multiplicative complexity of division in $\mathcal{T}_n$ is at most $3\frac{5}{6}n^2 + O(n)$.*

Theorem 7. *The multiplicative complexity of division in $\mathcal{T}_n$ is bounded from below by $\frac{5}{4}n^2 - O(n)$.*

For the upper bound in Theorem 4, we use generalizations of the fast multiplication algorithms for $\mathcal{T}_n$ developed in [2]. These algorithms only work over fields of characteristic zero that contain all roots of unity. (More precisely, we only need that the multiplicative complexity in the algebra $K[X]/(X^N - 1)$ equals N for certain values of N. The preceding more restrictive formulation avoids some lengthy case discussions.) Therefore, we make the following convention:

Throughout the remainder of this work, we assume that the underlying ground field k has characteristic zero and contains all roots of unity.

So for instance, $k = \mathbb{C}$ would be an appropriate choice. All lower bounds proven in this work remain valid over fields with characteristic distinct from two, in the case of inversion, and over arbitrary fields, in the case of division. All upper bounds concerning inversion and division can be carried over to arbitrary fields by utilizing the more expensive evaluation–multiplication–interpolation scheme for multiplication in $\mathcal{T}_n$ at the price of larger constants.

2   Multiplying Bivariate Power Series

The inversion of power series is usually done by an iterative process. In the nonscalar model, the costing operations in each iteration are caused by polynomial multiplications of various kinds. In this section, we therefore study the multiplication in T_n extensively.

2.1   Multiplication in $K[X]/(X^N - 1)$

In what follows, we will reduce the multiplication in T_n to the multiplication in the algebra $U_N := K[X]/(X^N - 1)$ for some suitable N. Therefore, we first examine the problem of multiplying two univariate polynomials modulo $X^N - 1$ and state some well known results concerning its complexity (see also [4, Chap. 2.1]). We again assume that the coefficients of the elements we want to multiply are indeterminates over k and that K is obtained from k by adjoining these indeterminates.

Theorem 8 (see [4, Chap. 2.1]). *The multiplicative complexity of the multiplication in U_N equals N.*

Theorem 9. *Suppose we know M coefficients of the product AB mod $X^N - 1$ of two polynomials A and B with indeterminate coefficients in advance. Then the coefficients of AB mod $X^N - 1$ can be computed with $N - M$ essential multiplications.*

A standard linear independence argument show that the number $N - M$ of multiplications is also necessary.

2.2   Multiplication in T_n

Let us turn to the multiplication in T_n. Assume we are given two elements $A, B \in T_n$. One way to multiply A and B is to evaluate both A and B at $2n^2 + 3n + 1$ suitably chosen points, which is free of costs, perform that many pointwise costing multiplications, obtain the coefficients of the (untruncated) product by interpolation, which is again free of costs, and finally throw away all coefficients belonging to monomials of total degree greater than n. By the Alder-Strassen theorem [1], the foregoing method is optimal (w.r.t. multiplicative complexity) if we want to compute the product of two univariate polynomials modulo X^{n+1}. But surprisingly, we can do better in the bivariate case, as is shown in [2]. In the remainder of this section, we present useful variations on and generalizations of this result.

The main idea of [2] is to reduce the multiplication of A and B to the multiplication of two elements A' and B' in U_N for some suitable N. We encode both factors A and B as long univariate polynomials by substituting $Y \mapsto X^{2n+2}$, i.e., $A'(X) = A(X, X^{2n+2})$ and $B'(X) = B(X, X^{2n+2})$. This substitution groups together the coefficients of all monomials with the same degree in Y and inserts some additional zero coefficients between these groups. If we now set

$N = n(2n + 2) + n + 1$, we can read off the coefficients of $R = AB$ from the coefficients of the result $R' = A'B' \mod X^N - 1$. Borrowing the language of [8], we can depict this fact by a pattern

$$\text{d d d d o o o o' d d d o o o o o' d d o o o o o o' d o o o}$$
$$\text{for the factors } A', B'$$

$$\text{d d d d * * * o' d d d * * * o o' d d * * * o o o' d * * * /}$$
$$\text{o o o o * * * o' o o o o * * o o' o o o o *}$$
$$\text{for the result } R',$$

here for $n = 3$, $N = 28$. Lower coefficients are standing on the left, higher ones on the right. An entry "d" (data) marks a coefficient of the original polynomials A, B, and R, respectively, an entry "o" stands for a zero coefficient, and an "*" indicates "garbage", that is, a coefficient that is not needed to recover R. The quotes separate groups of monomials with the same degree in Y. The second line of the result is the "wraparound" produced by the -1 of $X^N - 1$. By inserting the right number of zeros, some of the garbage places are used twice while the coefficients of R are preserved. Moreover, we know $\frac{1}{2}n(n+1)$ coefficient of R' in advance, since they are zero. Thus the upper bound of $N - \frac{1}{2}n(n + 1)$ essential multiplications follows from Theorem 9.

Theorem 10 (see [2, Thm. 6]). *The multiplicative complexity of the multiplication in $\mathcal{T}_n$ is at most $\frac{3}{2}n^2 + \frac{5}{2}n + 1$.*

A lower bound of $\frac{5}{4}n^2 - O(n)$ for the multiplicative complexity of the multiplication in $\mathcal{T}_n$ is proven in [2].

Since the coefficients of R are a subset of the coefficients of R', we immediately obtain the following more general theorem.

Theorem 11. *Suppose we know M coefficients of the product $AB \in \mathcal{T}_n$ of two polynomials A and B with indeterminate coefficients in advance. Then the coefficients of $AB \in \mathcal{T}_n$ can be computed with $\frac{3}{2}n^2 + \frac{5}{2}n + 1 - M$ essential multiplications.*

For any $m \leq n$, we may view $\mathcal{T}_m$ as a subspace of $\mathcal{T}_n$ just by filling the lacking coefficients with zeros. If say A is in $\mathcal{T}_m$ with $m \leq n$, then one moment's reflection shows that we will have an additional amount of $(n - m)(n + 1)$ extra zeros in the result. (Another solution that is more suited for algorithmic purposes is to substitute $Y \mapsto X^{m+n+2}$ instead of $Y \mapsto X^{2n+2}$.) Thus, the below theorem follows.

Theorem 12. *Let $m \leq n$. Suppose we know M coefficients of the product $AB \in \mathcal{T}_n$ of $A \in \mathcal{T}_m$ and $B \in \mathcal{T}_n$ in advance. Then the coefficients of $AB \in \mathcal{T}_n$ can be computed with $\frac{1}{2}n^2 + mn + \frac{3}{2}n + m + 1 - M$ essential multiplications.*

2.3   Squaring in $\mathcal{T}_n$

The proof of the lower bound for the complexity of inversion will be reduced to the proof of a lower bound for squaring. Therefore, we provide a lower bound for the complexity of computing the square of an element in $\mathcal{T}_n$ in this section. Because the computation of squares is interesting on its own, we also provide upper bounds for this problem.

The straight forward method for squaring an element $A \in \mathcal{T}_n$ is of course to use the multiplication algorithm presented in the preceding section. But we can do better: let $h = \lceil (n-1)/2 \rceil$ and write $A = L + U$ where L contains only monomials of total degree less than or equal to h while U consists of all other monomials (of total degree greater than h). We have

$$A^2 = L^2 + 2LU = L(L + 2U) \qquad \text{in } \mathcal{T}_n.$$

Now we may plug in Theorem 12. This yields the following upper bound.

Theorem 13. *Suppose we know M coefficients of the square $A^2 \in \mathcal{T}_n$ of some $A \in \mathcal{T}_n$ with indeterminate coefficients in advance. Then the coefficients of $A^2 \in \mathcal{T}_n$ can be computed with $\frac{1}{2}n^2 + hn + \frac{3}{2}n + h + 1 - M \le n^2 + 2n + 1 - M$ essential multiplications, where $h = \lceil (n-1)/2 \rceil$.*

Compared with the straight forward method, we save about $\frac{1}{2}n^2$ essential multiplications through the above construction.

Next we give a lower bound for the multiplicative complexity of squaring over fields of characteristic distinct from two.

The coefficients $b_{\mu,\nu}$ of A^2 are quadratic forms in the coefficients $a_{\mu,\nu}$ of A. In this case, according to Strassen [12], the fact that the multiplicative complexity of computing the $b_{\mu,\nu}$ from the $a_{\mu,\nu}$ is bounded from above by ℓ is equivalent to the existence of ℓ products $P_\lambda = U_\lambda \cdot V_\lambda$ of linear forms U_λ and V_λ in the $a_{\mu,\nu}$ such that

$$b_{i,j} \in \lin\{P_1, \dots, P_\ell\} \qquad \text{for all } i, j \text{ with } i + j \le n.$$

In other words, we may restrict ourselves to computations that contain only normalized multiplications and no divisions.

Theorem 14. *Over fields of characteristic distinct from two, computing the square of an element from $\mathcal{T}_n$ requires at least $n^2 - \frac{1}{2}h^2 + 3n - \frac{3}{2}h + 1$ essential operations, where $h = \lceil n/2 \rceil$.*

Proof. Consider a computation with ℓ essential operations that computes the coefficients $b_{\mu,\nu}$ of A^2 from the indeterminate coefficients $a_{\mu,\nu}$ of A. By the preceding considerations, we may restrict our attention to computations that have ℓ products of linear forms as its costing operations.

Now we exploit the so-called substitution method (see [7] or [4, Chap. 6]): we substitute one of the coefficients, say $a_{i,j}$, by a linear form in the remaining ones in such a way that one of the products P_λ is trivialized, that is, one of the linear forms U_λ or V_λ becomes zero after this substitution. Thereafter, we know that the multiplicative complexity of the original problem is at least one plus the multiplicative complexity of the quadratic forms obtained through this substitution. We may then repeat this process if desired. Of course, we can only trivialize one of the products by substituting $a_{i,j}$ if there is a product that actually depends on $a_{i,j}$. If one of the computed quadratic forms does, this is certainly the case.

Here, our substitution strategy looks as follows: we define an ordering on the $a_{\mu,\nu}$ by

$$(\mu, \nu) > (\mu', \nu') :\Longleftrightarrow \mu + \nu > \mu' + \nu' \text{ or } \mu + \nu = \mu' + \nu' \text{ and } \mu > \mu'$$

and

$$a_{\mu,\nu} > a_{\mu',\nu'} :\Longleftrightarrow (\mu, \nu) > (\mu', \nu').$$

We then substitute all monomials w.r.t. that ordering starting with $a_{n,0}$ and then stepping down until we reach $a_{0,h+1}$. In other words, we just follow the "diagonals" starting with the outermost and ending with the $(h+2)$th one.

To simplify the notations, we denote by $(s_1, t_1), (s_2, t_2), \ldots, (s_r, t_r)$ the pairs $(n, 0), (n-1, 1), \ldots, (0, h+1)$ (ordered w.r.t. ">"). For technical reasons, we also add an additional pair (s_0, t_0) with the property of being larger than all other of the above pairs. Let $b_{i,j}^{(\rho)}$, $0 \le \rho \le r$, denote the homomorphic image of $b_{i,j}$ after substituting $a_{s_1,t_1}, \ldots, a_{s_\rho,t_\rho}$.

Claim 1. The polynomial $b_{i,j}^{(\rho)}$ is a homogeneous polynomial of degree two in the $a_{\mu,\nu}$ with $(s_\rho, t_\rho) > (\mu, \nu)$.

Claim 2. If $(s_\rho, t_\rho) > (i, j)$ then $b_{i,j}^{(\rho)} = b_{i,j}$.

Claim 3. All monomials $\alpha \cdot a_{u,v} a_{u',v'}$ of $b_{i,j}^{(\rho)}$ fulfil $(i, j) \ge (u + u', v + v')$. (Here, "$\ge$" is the reflexive closure of ">".)

Claim 4. If $\alpha \cdot a_{u,v} a_{u',v'}$ is a monomial of $b_{i,j}^{(\rho)}$ but not of $b_{i,j}$, i.e., $\alpha \cdot a_{u,v} a_{u',v'}$ is the result of some substitution, then even $(i, j) > (u + u', v + v')$.

Claim 5. If $\alpha \cdot a_{p,q} a_{p',q'}$ is a monomial of $b_{i,j}$ with $(s_r, t_r) > (p, q)$ and $(s_r, t_r) > (p', q')$, then $\alpha \cdot a_{p,q} a_{p',q'}$ is a monomial $b_{i,j}^{(\rho)}$.

The first claim is obviously true, since we have substituted each a_{s_ρ,t_ρ} by a linear form in the remaining $a_{\mu,\nu}$ with $a_{s_\rho,t_\rho} > a_{\mu,\nu}$.

For the second claim, note that $b_{i,j}$ does not depend on $a_{\mu,\nu}$ if $(\mu, \nu) > (i, j)$. Therefore, $b_{i,j}$ is unaffected by the substitution of a_{s_ρ,t_ρ} if $(s_\rho, t_\rho) > (i, j)$. Thus, $b_{i,j}^{(\rho)} = b_{i,j}^{(\rho-1)} = \cdots = b_{i,j}^{(0)} = b_{i,j}$ as long as $(s_\rho, t_\rho) > (i, j)$.

For the third, fourth and fifth claim, observe that if $(p, q) > (p', q')$, then also

$$(p + p'', q + q'') > (p' + p'', q' + q'') \qquad \text{for all } (p'', q''). \tag{4}$$

We prove these claims by induction in ρ: each monomial $\alpha \cdot a_{u,v} a_{u',v'}$ of $b_{i,j}^{(0)} = b_{i,j}$ fulfils $(u + u', v + v') = (i, j)$, thus the induction start is clear. Assume that the claims hold for some $\rho < r$. We obtain $b_{i,j}^{(\rho+1)}$ from $b_{i,j}^{(\rho)}$ by replacing $a_{s_{\rho+1},t_{\rho+1}}$ by a linear form in the remaining $a_{\mu,\nu}$ with $a_{s_{\rho+1},t_{\rho+1}} > a_{\mu,\nu}$. Let $\alpha \cdot a_{u,v} a_{s_{\rho+1},t_{\rho+1}}$ be monomial of $b_{i,j}^{(\rho)}$. The $(\rho+1)$th substitution transforms this monomial into a sum of monomials of the form $\alpha' \cdot a_{u,v} a_{\mu,\nu}$. By (4), $(u+s_{\rho+1}, v+t_{\rho+1}) > (u+\mu, v+\nu)$ proving the third and fourth claim. The monomials $\alpha \cdot a_{p,q} a_{p',q'}$ of of $b_{i,j}^{(\rho)}$ with $(s_r, t_r) > (p,q)$ and $(s_r, t_r) > (p',q')$ are not affected by the substitution, since the new monomials $\alpha' \cdot a_{u,v} a_{\mu,\nu}$ introduced by the substitution cannot cancel $\alpha \cdot a_{p,q} a_{p',q'}$, because $(p+p', q+q') = (i,j) > (u+\mu, v+\nu)$. Thus $\alpha \cdot a_{p,q} a_{p',q'}$ is also a monomial of $b_{i,j}^{(\rho+1)}$. This proves the fifth claim.

By the second claim, each of the r substitutions can be done in such a way that one costing product is killed, since $b_{s_{\rho+1},t_{\rho+1}}^{(\rho)} = b_{s_{\rho+1},t_{\rho+1}}$ depends on $a_{s_{\rho+1},t_{\rho+1}}$. Altogether, this kills $\frac{1}{2}n^2 - \frac{1}{2}h^2 + \frac{3}{2}n - \frac{3}{2}h$ products.

Since we only substitute indeterminates on the $(h+2)$th or higher diagonals, each $b_{i,j}^{(r)}$ still contains one of its original monomials after substituting, for instance $\alpha \cdot a_{\lceil i/2 \rceil, \lceil j/2 \rceil} a_{\lfloor i/2 \rfloor, \lfloor j/2 \rfloor}$ with $\alpha \in \{1, 2\}$ by the fifth claim. By the third claim, no other $b_{i',j'}^{(r)}$ with $(i,j) > (i',j')$ can contain this monomial. Thus, the quadratic forms $b_{\mu,\nu}^{(r)}$ are still linearly independent and their multiplicative complexity is a least $\frac{1}{2}n^2 + \frac{3}{2}n + 1$. This proves the theorem. $\square$

Over fields of characteristic two, the situation is quite different: squaring over fields of characteristic two is cheap and we can determine the multiplicative complexity exactly.

Let $A \in \mathcal{T}_n$ with indeterminate coefficients $a_{\mu,\nu}$. Over fields of characteristic two, we have

$$A^2 = \sum_{\mu=0}^{n} \sum_{\nu=0}^{n} \sum_{\substack{p+s=\mu \\ q+t=\nu}} a_{p,q} a_{s,t} X^\mu Y^\nu = \sum_{\mu=0}^{\lfloor n/2 \rfloor} \sum_{\nu=0}^{\lfloor n/2 \rfloor} a_{\mu,\nu}^2 X^{2\mu} Y^{2\nu}.$$

Thus we may compute the coefficients of A^2 just by squaring the appropriate coefficients of A. Since the nonzero coefficients of A^2 are linearly independent, this is optimal.

Theorem 15. *Over fields of characteristic two, the multiplicative complexity of squaring in $\mathcal{T}_n$ is $\frac{1}{2}h^2 + \frac{3}{2}h + 1$, where $h = \lfloor n/2 \rfloor$. If we know M nonzero coefficients in advance, then the multiplicative complexity is $\frac{1}{2}h^2 + \frac{3}{2}h + 1 - M$.*

3   Inverting Bivariate Power Series

In this section, we present a second order iteration scheme (which is superior to third order here) for the inversion of bivariate power series. To avoid lengthy calculations, we state all following bounds as $c \cdot n^2 + O(n)$ with an explicitly given constant c. One should note that in each of these bounds, the constant hidden in the O-notation is rather small. Let A be a bivariate power series with indeterminate coefficients. Given a positive integer n, our task is to compute a polynomial B of total degree at most n such that

$$AB \equiv 1 \mod I_{n+1}.$$

Let $h = \lceil (n-1)/2 \rceil$ and suppose we have already computed a polynomial C of total degree at most h such that

$$AC \equiv 1 \mod I_{h+1}. \tag{5}$$

Squaring yields

$$A \cdot (-AC^2 + 2C) \equiv 1 \mod I_{n+1}.$$

Thus

$$B \equiv -AC^2 + 2C \mod I_{n+1}.$$

So we have to compute the coefficients of $-AC^2 + 2C$ (viewed as an element of $\mathcal{T}_n$). We proceed as follows: we first compute AC. By (5), we know the coefficients of all monomials of AC with total degree h or less. Thus by Theorem 12, the coefficients of AC can be computed with $\frac{7}{8}n^2 + O(n)$ essential multiplications. Next, we compute $C \cdot AC$. Again by (5), we know the coefficients of all monomials of $C \cdot AC$ with total degree h or less, they are simply the coefficients of C. Thus another $\frac{7}{8}n^2 + O(n)$ essential multiplications suffice to compute the coefficients of AC^2. From this, we may obtain the coefficients of B without any further costing operations. Let us analyze the number $E(n)$ of essential operations performed by this algorithm (regarded as a straight-line program after unrolling the recursive calls): we have $E(n) = E(h) + \frac{7}{4}n^2 + O(n)$ with $E(0) = 1$. This yields $E(n) = \frac{7}{3}n^2 + O(n)$ and therefore proves Theorem 4.

What can we say about lower bounds for the multiplicative complexity of inversion in $\mathcal{T}_n$? A standard linear independence argument shows that the number of the computed coefficients, $\frac{1}{2}n^2 + \frac{3}{2}n + 1$, gives a lower bound. Over fields of characteristic two, this is the best we know. Over other fields, we can further improve this. Let $a_{\mu,\nu}$ denote the indeterminate coefficients of the element A we want to invert. The coefficients of the inverse are polynomials in the $a_{\mu,\nu}$ with $(\mu, \nu) \neq (0,0)$ over $k(a_{0,0})$ which we take as our new ground field (i.e., $a_{0,0}$ becomes a scalar). Using the technique by Strassen [12], we can transform a computation for the coefficients of A^{-1} into a computation that computes only terms of degree two and has no more essential operations than the original computation. At quick look at (3) shows that this computation actually computes $k(a_{0,0})$-multiples of the coefficients of A_1^2! Now Theorem 14 implies the bound of Theorem 5. (The missing constant term of A_1^2 reduces the lower bound just by an additive amount of $O(n)$.)

4   Computing Quotients of Bivariate Power Series

Since $A/B = A \cdot B^{-1}$, Theorem 10 and Theorem 4 immediately yield the upper bound stated in Theorem 6.

For the lower bound, we once more exploit Strassen's technique [12]: let $a_{i,j}$ and $b_{i,j}$ denote the coefficients of A and B, respectively. We regard our computation as a computation over the new ground field $k(a_{0,0})$. Now we transform a given computation for the coefficients of B/A into a computation that computes only terms of degree two and has no more essential operations than the original computation. Moreover, this can be done in such a way that all costing operations are products of linear forms in the coefficients of A and B. The resulting computation now computes the coefficients of $B \cdot A_1/a_{0,0}^2$ in $\mathcal{T}_n$. Altogether, we have obtained a computation that merely multiplies two elements of $\mathcal{T}_n$ except for the missing constant term of A_1. Now the result of [2, Thm. 3] for the multiplication in $\mathcal{T}_n$ can be applied. (The missing constant term again reduces the lower bound by an additive amount of $O(n)$). This completes the proof of Theorem 7.

5   The Rectangular Case

In contrast to $\mathcal{T}_n$, in the rectangular case $\mathcal{R}_n := k[X,Y]/(X^{n+1}, Y^{n+1})$ a direct application of the iteration process in Section 3 does not work very satisfactory. The main reason is that only $(X^{n+1}, Y^{n+1})^2 = (X^{2n+2}, X^{n+1}Y^{n+1}, Y^{2n+2})$ holds, while we had $I_n^2 = I_{2n+1}$ before. It is more favourable to embed elements from $\mathcal{R}_n$ into $\mathcal{T}_{2n}$ and then use the techniques of the preceding sections. This yields a upper bound of $9\frac{1}{3}n^2 + O(n)$ for the inversion in $\mathcal{R}_n$. But we can do better: note that when embedding $\mathcal{R}_n$ into $\mathcal{T}_{2n}$ it does not matter how we choose the coefficients of the monomials with X–degree or Y–degree greater than n. So using the substitution lemma from [9, Lemma 2], we may substitute these coefficients (in the same order as proposed in Theorem 14) and we can kill $n^2 + n$ many products yielding the upper bound $8\frac{1}{3}n^2 + O(n)$.

Theorem 16. *For fields k of characteristic zero that contain all roots of unity, the multiplicative complexity of inverting in $\mathcal{R}_n$ has the upper bound $8\frac{1}{3}n^2 + O(n)$.*

For a lower bound, we exploit essentially the same substitution strategy as in the triangular case. However, we substitute the coefficients of all monomials of total degree at least $d = \lceil \frac{2}{3}n \rceil$. This kills $(n+1)^2 - \frac{1}{2}d(d+1)$ many products. In contrast to the triangular case, we can now only conclude that at least $(n+1)^2 - \frac{1}{2}(2n - 2d + 2)(2n - 2d + 3)$ of the coefficients of the square are still linearly independent after substituting. This proves the following lower bound.

Theorem 17. *Over fields of characteristic distinct from two, $\frac{14}{9}n^2 - O(n)$ is a lower bound for the multiplicative complexity of inversion in $\mathcal{R}_n$.*

Building up on these results, we obtain the following bounds for the division by utilizing [2, Thms. 4 and 8].

Theorem 18. *Let k be a field of characteristic zero that contains all roots of unity. The multiplicative complexity of division in $\mathcal{R}_n$ is at most $11\frac{1}{3}n^2 + O(n)$.*

Theorem 19. *The multiplicative complexity of division in $\mathcal{R}_n$ has the lower bound $\frac{7}{3}n^2 - O(n)$.*

Acknowledgements

I would like to thank the referees for pointing out some inaccuracies in the preliminary version.

References

1. A. Alder and V. Strassen. On the algorithmic complexity of associative algebras. *Theoret. Comput. Sci.*, 15:201–211, 1981. 190
2. Markus Bläser. Bivariate polynomial multiplication. In *Proc. 39th Ann. IEEE Symp. on Foundations of Comput. Sci. (FOCS)*, pages 186–191, 1998. 186, 189, 190, 191, 196
3. Markus Bläser. On the number of multiplications needed to invert a monic power series over fields of characteristic two. Technical report, Institut für Informatik II, Universität Bonn, January 2000. 187
4. Peter Bürgisser, Michael Clausen, and M. Amin Shokrollahi. *Algebraic Complexity Theory*. Springer, 1997. 186, 187, 190, 192
5. K. Kalorkoti. Inverting polynomials and formal power series. *SIAM J. Comput.*, 22:552–559, 1993. 187
6. H. T. Kung. On computing reciprocals of power series. *Numer. Math.*, 22:341–348, 1974. 187
7. Victor Ya. Pan. Methods for computing values of polynomials. *Russ. Math. Surv.*, 21:105–136, 1966. 192
8. Arnold Schönhage. Bivariate polynomial multiplication patterns. In *Proc. 11th Applied Algebra and Error Correcting Codes Conf. (AAECC)*, Lecture Notes in Comput. Sci. 948, pages 70–81. Springer, 1995. 186, 191
9. Arnold Schönhage. Multiplicative complexity of Taylor shifts and a new twist of the substitution method. In *Proc. 39th Ann. IEEE Symp. on Foundations of Comput. Sci. (FOCS)*, pages 212–215, 1998. 196
10. Arnold Schönhage. Variations on computing reciprocals of power series. *Inf. Proc. Letters*, 74:41–46, 2000.
11. Malte Sieveking. An algorithm for division of power series. *Computing*, 10:153–156, 1972.
12. Volker Strassen. Vermeidung von Divisionen. *J. Reine Angew. Math.*, 264:184–202, 1973. 187, 192, 195, 196
13. Volker Strassen. Algebraic complexity theory. In J. van Leeuven, editor, *Handbook of Theoretical Computer Science Vol. A*, pages 634–672. Elsevier, 1990.

Automatic Verification of Recursive Procedures with One Integer Parameter

Ahmed Bouajjani, Peter Habermehl, and Richard Mayr

LIAFA - Université Denis Diderot
Case 7014 - 2, place Jussieu, F-75251 Paris Cedex 05. France
{abou,haberm,mayr}@liafa.jussieu.fr
Fax: +33 1 44 27 68 49

Abstract. Context-free processes (BPA) have been used for dataflow-analysis in recursive procedures with applications in optimizing compilers [6]. We introduce a more refined model called BPA($\mathbb{Z}$) that can model not only recursive dependencies, but also the passing of integer parameters to subroutines. Moreover, these parameters can be tested against conditions expressible in Presburger-arithmetic. This new and more expressive model can still be analyzed automatically. We define $\mathbb{Z}$-input 1-CM, a new class of one-counter machines that take integer numbers as input, to describe sets of configurations of BPA($\mathbb{Z}$). We show that the *Post** (the set of successors) of a set of BPA($\mathbb{Z}$)-configurations described by a $\mathbb{Z}$-input 1-CM can be effectively constructed. The *Pre** (set of predecessors) of a regular set can be effectively constructed as well. However, the *Pre** of a set described by a $\mathbb{Z}$-input 1-CM cannot be represented by a $\mathbb{Z}$-input 1-CM in general and has an undecidable membership problem. Then we develop a new temporal logic based on reversal-bounded counter machines that can be used to describe properties of BPA($\mathbb{Z}$) and show that the model-checking problem is decidable.

1 Introduction

Besides their classical use in formal language theory, pushdown automata have recently gained importance as an abstract process model for recursive procedures. Algorithms for model checking pushdown automata have been presented in [3,1,11,4]. Reachability analysis for pushdown automata is particularly useful in formal verification. Polynomial algorithms for reachability analysis have been presented in [1] and further optimized in [5]. For most purposes in formal verification it is sufficient to consider BPA ('Basic Process Algebra'; also called context-free processes), the subclass of pushdown automata without a finite control. BPA have been used for dataflow-analysis in recursive procedures with applications in optimizing compilers [6].

The weakness of BPA is that it is not a very expressive model for recursive procedures. It can model recursive dependencies between procedures, but not the passing of data between procedures or different instances of a procedure with different parameters.

J. Sgall, A. Pultr, and P. Kolman (Eds.): MFCS 2001, LNCS 2136, pp. 198–211, 2001.

Example 1. Consider the following abstract model of recursive procedures P, Q, R, S and F, which take an integer number as argument: ($x|y$ means "x divides y").

$P(x)$: If $x \geq 16$

$\qquad$ If $8|x$ then $Q(x+1)$

$\qquad$ else $P(x-2)$

$\qquad$ else $F(x)$

$Q(x)$: If $2|x$ then $R(x)$

$\qquad$ else $S(x+1)$

If one starts by calling procedure P (with any parameter) then procedure R will never be called, because P never calls Q with an even number as parameter. However, a BPA model for these procedures cannot detect this.

Thus, we define a new more expressive model called BPA($\mathbb{Z}$) that extends BPA with integer parameters. Procedures are now called with an integer parameter that can be tested, modified and passed to subroutines. We limit ourselves to one integer parameter, because two would give the model full Turing power and make all problems undecidable. BPA($\mathbb{Z}$) is a compromise between expressiveness and automatic analysability. On the one hand it is much more expressive than BPA and can model more aspects of full programs. On the other hand it is still simple enough such that most verification problems about BPA($\mathbb{Z}$) stay decidable. For the verification of safety properties, it is particularly useful to have a symbolic representation of sets of configurations and to be able to effectively construct representations of the *Pre** (the set of predecessors) and the *Post** (the set of successors) of a given set of configurations. While finite automata suffice for describing sets of configurations of BPA, a more expressive formalism is needed for BPA($\mathbb{Z}$). We define $\mathbb{Z}$-input 1-CM, a new class of one-counter machines that take integer numbers as input, to describe sets of configurations of BPA($\mathbb{Z}$). We show that the *Post** (the set of successors) of a set described by a $\mathbb{Z}$-input 1-CM can be effectively constructed. The *Pre** (the set of predecessors) of a regular set can be effectively constructed as well. However, the *Pre** of a set described by a $\mathbb{Z}$-input 1-CM cannot be represented by a $\mathbb{Z}$-input 1-CM in general and has an undecidable membership problem.

We develop a new temporal logic based on reversal-bounded counter machines that can be used to describe properties of BPA($\mathbb{Z}$). By combining our result on the constructibility of the *Post** with some results by Ibarra et al. on reversal bounded counter machines [7,8] we show that the model-checking problem is decidable.

2 BPA($\mathbb{Z}$)

We define BPA($\mathbb{Z}$), an extension of BPA, as an abstract model for recursive procedures with integer parameters.

Definition 2. *A n-ary Presburger predicate $P(k_1, \ldots, k_n)$ is an expression in Presburger-arithmetic of type boolean (i.e., the outermost operator is a logical operator or quantifier) that contains exactly n free variables $k_1, \ldots, k_n$ of type integer.*

Definition 3. *We define* integer symbol sequences *(ISS) to describe configurations of processes. ISS are finite sequences of the form* $X_1(k_1)X_2(k_2)\dots X_n(k_n)$, *where the X_i are symbols from a given finite set and the $k_i \in \mathbb{Z}$ are integers. (The brackets are mere 'syntactic sugar' and can be omitted.) Greek letters* $\alpha, \beta, \dots$ *are used to denote ISS. The constant ϵ denotes the empty sequence.*

Definition 4. *Let* $Act = \{\epsilon, a, b, c, \dots\}$ *and* $Const = \{\epsilon, X, Y, Z, \dots\}$ *be disjoint sets of* actions *and* process constants, *respectively. A BPA($\mathbb{Z}$) (α, Δ) is given by an initial configuration α (where α is an ISS) and a finite set Δ of conditional rewrite rules of the form* $X(k) \xrightarrow{a} X_1(e_1)X_2(e_2)\dots X_n(e_n)$, $P(k)$ *where*

- $X \in Const$, $a \in Act$, k *is a free variable of type integer.*
- $\forall i \in \{1, \dots, n\}. X_i \in Const$.
- *For every $i \in \{1, \dots, n\}$ e_i is an expression of one of the following two forms:*
  - $e_i = k_i$ *or* $e_i = k + k_i$ *for some constant $k_i \in \mathbb{Z}$.*
- $P(k)$ *is a unary Presburger predicate.*

Note that n can be 0. In this case the rule has the form $X(k) \xrightarrow{a} \epsilon$, $P(k)$. *We denote the finite set of constants used in Δ by $Const(\Delta)$ and the finite set of actions used in Δ by $Act(\Delta)$. These rewrite rules induce a transition relation on ISS by prefix-rewriting as follows: For any α we have* $X(q)\alpha \xrightarrow{a} X_1(q_1)X_2(q_2)\dots X_n(q_n)\alpha$ *if there is a rewrite rule* $X(k) \xrightarrow{a} X_1(e_1)X_2(e_2)\dots X_n(e_n)$, $P(k)$ *such that the following conditions are satisfied.*

- $P(q)$
- *If $e_i = k_i$ then $q_i = k_i$.*
- *If $e_i = k + k_i$ then $q_i = q + k_i$.*

The Presburger predicates can be used to describe side conditions for the application of rules, e.g., the rule $X(k) \xrightarrow{a} Y(k-7)Z(k+1)$, $3|k \wedge k \geq 8$ can only be applied to ISS starting with $X(q)$ where q is at least 8 and divisible by 3. Furthermore, we can use Presburger predicates to express rules with constants on the left side, e.g., the rule $X(5) \xrightarrow{a} Y(2)Z(17)$ can be expressed by $X(k) \xrightarrow{a} Y(2)Z(17)$, $k = 5$. In the following we sometimes use rules with constants on the left side as a shorthand notation.

Definition 5. *We say that a BPA($\mathbb{Z}$) is in* normal form *if it only contains the following three types of rules:*

$$X(k) \xrightarrow{a} X_1(e_1)X_2(e_2), \quad P(k)$$
$$X(k) \xrightarrow{a} Y(e), \quad P(k)$$
$$X(k) \xrightarrow{a} \epsilon, \quad P(k)$$

where e, e_1, e_2 are expressions and $P(k)$ is a unary Presburger predicate as in Def. 4.

We call the rules of the third type decreasing *and the first two types* nondecreasing.

Remark 6. It is easy to see that general BPA($\mathbb{Z}$) can be simulated by BPA($\mathbb{Z}$) in normal form with the introduction of some auxiliary constants. Long rules are split into several short rules. For example the long rule $X(k) \xrightarrow{a} Y(k+1)$. $Z(k-2).W(k+7)$ is replaced by $X(k) \xrightarrow{a} X'(k).W(k+7)$ and $X'(k) \xrightarrow{\epsilon} Y(k+1).Z(k-2)$. If one is only interested in the set of reachable configurations of the original BPA($\mathbb{Z}$) then one has to filter out the intermediate configurations that contain auxiliary constants. It will turn out in Section 4 that this is possible. We will show that the set of reachable configurations of a BPA($\mathbb{Z}$) can be represented by a $\mathbb{Z}$-input 1-CM (a special type of one-counter machine) that is closed under synchronization with finite automata. Thus, to filter out the intermediate configurations it suffices to synchronize with the finite automaton that accepts exactly all sequences not containing auxiliary constants.

If one extends the model BPA($\mathbb{Z}$) by allowing two integer parameters instead of one, it becomes Turing-powerful, because it can simulate a Minsky 2-counter machine.

It is clear that a BPA($\mathbb{Z}$) can simulate a 1-counter machine. However, the set of reachable configurations of a BPA($\mathbb{Z}$) cannot be described by a normal 1-counter machine.

Example 7. Consider the BPA($\mathbb{Z}$) with just one rule $X(k) \xrightarrow{a} X(k+1)X(k)$ and initial state $X(0)$. The set of reachable configurations are all decreasing sequences of the form $X(n)X(n-1)X(n-2)\ldots X(0)$ for any $n \in \mathbb{N}$. The language consisting of these sequences cannot be accepted by a normal 1-counter machine, no matter how the integer numbers are coded (e.g., in unary coding or in binary as sequences of 0 and 1). The reason is that one cannot test the equality of the counter against the input without losing the content of the counter during the test.

The central problem in this paper is to compute a representation of the set of reachable states of a BPA($\mathbb{Z}$).

Definition 8. Let Δ be the set of rules of a BPA($\mathbb{Z}$) and L a language of ISS (describing configurations of the BPA($\mathbb{Z}$)). By $Post^*_\Delta(L)$ we denote the set of all successors (reachable configurations) of elements of L w.r.t. Δ. $Post^*_\Delta(L) = \{\beta \mid \exists \alpha \in L. \alpha \rightarrow^*_\Delta \beta\}$. By $Pre^*_\Delta(L)$ we denote the set of all predecessors of elements of L w.r.t. Δ. $Pre^*_\Delta(L) = \{\alpha \mid \exists \beta \in L. \alpha \rightarrow^*_\Delta \beta\}$.

3   Automata

Definition 9. An alternating pushdown automaton (APDA for short) is a triple $\mathcal{P} = (P, \Gamma, \Delta_A)$ where P is a finite set of control locations, Γ is a finite stack alphabet, Δ_A is a function that assigns to each element of $P \times \Gamma$ a positive (i.e. negation free) boolean formula over elements of $P \times \Gamma^*$.

An alternating 1-counter machine is a special case of an APDA, because the counter can be simulated by the pushdown stack.

Definition 10. *A pushdown counter automaton (PCA) [7] is a pushdown automaton that is augmented with a finite number of reversal-bounded counters (containing integers). A counter is reversal bounded iff there is a fixed constant k s.t. in any computation the counter can change at most k times between increasing and decreasing.*

Now we define a new class of 1-counter machines with infinite input. These $\mathbb{Z}$-input 1-counter machines consider whole integer numbers as one piece of input and can compare them to constants, or to the internal counter without changing the counter's value. Additionally, they have several other useful features like Presburger-tests on the counter. $\mathbb{Z}$-input 1-counter machines will be used in Section 4 to represent sets of reachable configurations of BPA($\mathbb{Z}$).

Definition 11. *A $\mathbb{Z}$-input 1-counter machine M is described by a finite set of states Q, an initial state $q_0 \in Q$, a final state accept $\in Q$, a non-accepting state fail $\in Q$, and a counter c that contains a (possibly negative) integer value. The initial configuration is given by q_0 and some initial counter value. The machine reads pieces of input of the form $S(i)$ where S is a symbol out of a given finite set and $i \in \mathbb{Z}$ is an integer number. The instructions have the following form:*

1. $(q: \ c := c + 1; \mathsf{goto}\ q')$
2. $(q: \ c := c - 1; \mathsf{goto}\ q')$
3. $(q: \ \mathsf{If}\ c \geq 0\ \mathsf{then\ goto}\ q'\ \mathsf{else\ goto}\ q'')$.
4. $(q: \ \mathsf{If}\ c = 0\ \mathsf{then\ goto}\ q'\ \mathsf{else\ goto}\ q'')$.
5. $(q: \ Read\ input\ S(i).\ \mathsf{If}\ S = X\ \mathsf{and}\ i = K\ \mathsf{then\ goto}\ q'\ \mathsf{else\ goto}\ q'')$.
6. $(q: \ Read\ input\ S(i).\ \mathsf{If}\ S = X\ \mathsf{and}\ i = c\ \mathsf{then\ goto}\ q'\ \mathsf{else\ goto}\ q'')$.
7. $(q: \ \mathsf{If}\ P(c)\ \mathsf{then\ goto}\ q'\ \mathsf{else\ goto}\ q'')$, *where P is a unary Presburger predicate.*

where $X \in Const$ is a symbol constant and $K \in \mathbb{Z}$ is an integer constant.

$\mathbb{Z}$-input 1-counter machines can be nondeterministic, i.e., there can be several instructions at the same control-state. Each transition arc to a new control-state can be labeled with an atomic action. The language $L(M)$ accepted by a machine M is the set of ISS which are read by M in a run from the initial configuration to the accepting state.

In the following we use several shorthand notations for operations which can be encoded by the standard operations above. We use $c := c + j$ (incrementing the counter by a constant j), $c := j$ (setting the counter to a given constant j) and the operation $guess(c)$ (setting the counter to a nondeterministically chosen integer).

It is now easy to see that the set of reachable states of Example 7 can be described by the following $\mathbb{Z}$-input 1-counter machine with initial configuration $(q_0, 0)$:

$q_0 : guess(c);\ \mathsf{goto}\ q_1$

$q_1 :$ Read input $S(i)$. If $S = X$ and $i = c$ then goto q_2 else goto *fail*

$q_2 : c := c - 1; \mathsf{goto}\ q_1$

$q_2 :$ If $c = 0$ then goto *accept* else goto *fail*

While instructions of type 6 (integer input) do increase the expressive power of 1-counter machines, this is not the case for instructions of type 7 (Presburger tests). The following lemma shows that instructions of type 7 can be eliminated if necessary. We use them only as a convenient shorthand notation.

Lemma 12. *For every (alternating) ($\mathbb{Z}$-input) 1-counter machine M with Presburger tests (i.e., instructions of type 7), an equivalent (alternating) ($\mathbb{Z}$-input) 1-counter machine M' without Presburger tests can be effectively constructed. (Equivalent means that it accepts the same input (initial counter value), and the same language.)*

Proof. Any Presburger formula can be written in a normal form that is a boolean combination of linear inequalities and tests of divisibility. As we consider only Presburger formulae with one free variable, it suffices to consider tests of the forms $c \geq k$, $c \leq k$ and $k|c$ for constants $k \in \mathbb{Z}$. Let K be the set of constants k used in these tests. K is finite and depends only on the Presburger predicates used in M. Let $K' = \{k_1, \ldots, k_m\} \subseteq K$ be the finite set of constants used in divisibility tests. For every control-state s of M we define a set of control-states of M' of the form $(s, j_1, \ldots, j_m)$ where $j_i \in \{0, \ldots, k_i - 1\}$ for every $i \in \{1, \ldots, m\}$. Now M' simulates the computation of M in such a way that M' is in a state $(s, j_1, \ldots, j_m)$ iff M is in state s and $j_i = c \bmod k_i$. For example if $K' = \{2, 5\}$ then the step $(s, n) \overset{c:=c+1}{\longrightarrow} (s', n+1)$ of M yields e.g. the step $((s, 1, 2), n) \overset{c:=c+1}{\longrightarrow} ((s', 0, 3), n+1)$ of M'. The divisibility tests thus become trivial in M', because this information is now encoded in the control-states of M'. The linear inequality tests are even easier to eliminate. For example the test $c \geq 5$ can be done by decrementing the counter by 5, testing for ≥ 0 and re-incrementing by 5. Thus, the Presburger tests can be eliminated from M'.   □

It is only a matter of convention if a $\mathbb{Z}$-input 1-CM reads the input from left to right (the normal direction) or from right to left (accepting the mirror image as in the example above). It is often more convenient to read the input from right to left (e.g., in Section 5), but the direction can always be reversed, as shown by the following lemma.

Lemma 13. *Let M be a $\mathbb{Z}$-input 1-CM with initial configuration (q_0, k_0) (with $k_0 \in \mathbb{Z}$) and final state q_f that reads the input from right to left. A $\mathbb{Z}$-input 1-CM M' can be constructed that reads the input from left to right and accepts the same language as M.*

Proof. (sketch) M' has the same control-states as M plus a new initial state q_0' and a new final state q_f'. M' starts in configuration $(q_0', 0)$. It guesses a value for its counter and goes to q_f. Then it does the computation of M in reverse (reading the input from left to right) until it reaches q_0. It tests if the counter has value k_0. If yes, it goes to q_f' and accepts. If no, then it doesn't accept.   □

4   Constructing Post*

Theorem 14. *Let Δ be a set of BPA($\mathbb{Z}$) rules and M a $\mathbb{Z}$-input 1-counter machine. Then a $\mathbb{Z}$-input 1-counter machine M' with $L(M') = Post^*_\Delta(L(M))$ can be effectively constructed.*

To prove this theorem we generalize the proof of a theorem in [1] which shows that the *Post** of a regular set of configurations of a pushdown automaton is regular. This proof uses a saturation method, i.e. adding a finite number of transitions and states to the automata representing configurations.

We cannot directly adapt this proof to BPA($\mathbb{Z}$), because process constants in a configuration can disappear for certain values of the parameter by applying decreasing rules. We show how to calculate a Presburger formula to characterize these values. This allows us to eliminate decreasing rules from Δ. This means that symbols produced by rules in some derivation can not disappear later. Then, we can apply the saturation method.

First, we show how to characterize the set $\{k \mid X(k) \to^*_\Delta \epsilon\}$ by a Presburger formula. We transform the set of rules Δ into an alternating one-counter machine and show that the set of initial values of accepting computations is effectively semilinear. This follows from a corollary of a theorem from [1] which states that the *Pre** of a more general model (alternating pushdown systems) is regular.

Theorem 15. *[1] Given an APDA $\mathcal{P}$ and a regular set of configurations $\mathcal{C}$, $pre^*(\mathcal{C})$ is regular and effectively constructible.*

With an APDA we can easily simulate an alternating one-counter machine with Presburger tests: First, we eliminate the Presburger tests with Lemma 12. Then, with the stack we can easily simulate the counter. We obtain the following:

Corollary 16. *Let M be an (alternating) one-counter machine with Presburger tests. The set of initial counter values for which a computation of M is accepting is effectively Presburger definable.*

Now we can prove the following two lemmas.

Lemma 17. *Let Δ be a set of BPA($\mathbb{Z}$) rules and X a process constant. Then a Presburger formula $P_X(k)$ with $\{k \mid P_X(k)\} = \{k \mid X(k) \to^*_\Delta \epsilon\}$ can be effectively constructed.*

Proof. We construct an alternating one-counter machine M such that: M with initial counter value k has an accepting computation iff $X(k) \to^*_\Delta \epsilon$. Then, we apply Corollary 16.

We construct M as follows: To each process constant Y of the BPA($\mathbb{Z}$) we associate a state q_Y in M. The initial state of the counter machine is q_X and it's final state *accept*. Each non-decreasing rewrite rule $X(k) \xrightarrow{a} X_1(e_1)X_2(e_2)\dots$ $X_n(e_n)$, $P(k)$ is translated into a transition of the counter machine as follows: q_X goes to a conjunction of the states $q_{X_1}, \dots, q_{X_n}$ by testing $P(k)$ and changing the counter according to $e_1, \ldots, e_n$. Each decreasing rule $X(k) \xrightarrow{a} \epsilon$, $P(k)$ is translated into a transition of the counter machine from q_X to *accept* by testing $P(k)$. It is a clear, that a run of M with some initial counter value k is accepting iff $X(k)$ can disappear with rules of Δ.   □

Lemma 18. *Let Δ be a set of BPA($\mathbb{Z}$) rules and M a $\mathbb{Z}$-input 1-counter machine representing a set of configurations. Then, we can effectively construct a set of rules Δ' without decreasing rules and a $\mathbb{Z}$-input 1-counter machine M', such that $Post_\Delta^*(M) = Post_{\Delta'}^*(M')$.*

We use this lemma to prove Theorem 14. To construct a counter machine M' representing $Post_\Delta^*(M)$, given a counter machine M and a set of BPA($\mathbb{Z}$) rules Δ, it suffices to consider non-decreasing rules. We explain the main idea with an example: Suppose we have a rule of the form $X(k) \xrightarrow{a} Y(k+3)Z(k-2)$, $P(k)$ in Δ and the automaton M is of the following form:

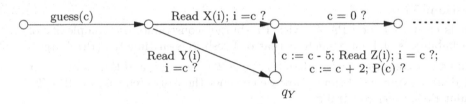

Notice that the counter is not tested before the input instruction. This is not a restriction (see full paper [2]). We add a new state q_Y for Y and transitions to M and obtain:

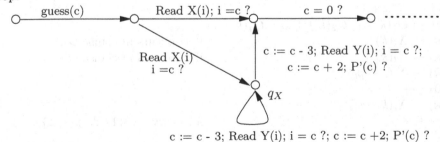

The transition going out of q_Y changes the counter value in such a way that if Y is read with parameter k then Z is read with a parameter $k - 5$, where -5 is the difference between -2 and 3. Then, the transition restores the counter value to the value before application of the rule by adding 2 and tests $P(c)$. Now consider instead a rule $X(k) \xrightarrow{a} X(k+1)Z(k-2)$ $P'(k)$ in Δ. Following the same principle as before we add a state for X and transitions. This will create a loop:

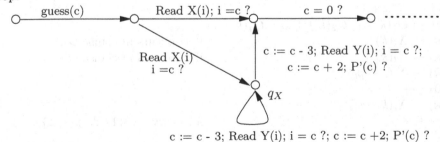

It is clear that in this way we only add a finite number of states and transitions. A lot of cases have to be considered. The details of the proof can be found in the full paper [2].

Remark 19. While the language of reachable states of any BPA($\mathbb{Z}$) can be described by a $\mathbb{Z}$-input 1-CM, the converse is not true. Some $\mathbb{Z}$-input 1-CMs describe languages that cannot be generated by any BPA($\mathbb{Z}$). Consider the language

$$\{X(k)Y(j_1)Y(j_2)\ldots Y(j_n)X(k) \mid k \in \mathbb{N}, n \in \mathbb{N}, j_1, \ldots, j_n \in \mathbb{N}\}$$

It is easy to construct a $\mathbb{Z}$-input 1-CM for this language (it just ignores the values of the j_i). However, no BPA($\mathbb{Z}$) generates this language, since it cannot guess the values of the arbitrarily many j_i without losing the value for k, which it needs again at the end.

The complexity of constructing a representation of Post* must be at least as high as the complexity of the reachability problem for BPA($\mathbb{Z}$). A special case of the reachability problem is the problem if the empty state ϵ is reachable from the initial state.

ϵ-REACHABILITY FOR BPA($\mathbb{Z}$)

Instance: A BPA($\mathbb{Z}$) Δ with initial state $X(0)$.
Question: $X(0) \to^* \epsilon$?

It is clear that for BPA($\mathbb{Z}$) with Presburger constraints the complexity of ϵ-reachability is at least as high as that of Presburger-arithmetic. (Presburger-arithmetic is complete for the class $\bigcup_{k>1} TA[2^{2^{n^k}}, n]$ [10], and thus requires at least doubly exponential time). Now we consider the restricted case of BPA($\mathbb{Z}$) without Presburger constraints.

Theorem 20. *The ϵ-reachability problem for BPA($\mathbb{Z}$) without full Presburger constraints, but with constants on the left sides of rules, is $\mathcal{NP}$-hard.*
Proof. We reduce 3-SAT to ϵ-reachability. Let $Q := Q_1 \wedge \ldots \wedge Q_j$ be a boolean formula in 3-CNF with j clauses over the variables $x_1, \ldots, x_n$. Let p_l be the l-th prime number. We encode an assignment of values to $x_1, \ldots, x_n$ in a natural number x by Gödel coding, i.e., x_i is true iff x is divisible by p_i. The set of rules Δ is defined as follows:

$$
\begin{aligned}
&X(k) \to X(k+1)\\
&X(k) \to Q_1(k+1).Q_2(k+1).\ldots.Q_j(k+1)\\
&Q_i(k) \to X_l(k) &&\text{if } x_l \text{ occurs in clause } Q_i.\\
&Q_i(k) \to \bar{X}_l(k) &&\text{if } \bar{x}_l \text{ occurs in clause } Q_i.\\
&X_l(k) \to X_l(k - p_l)\\
&X_l(0) \to \epsilon\\
&\bar{X}_l(k) \to \bar{X}_l(k - p_l)\\
&\bar{X}_l(r) \to \epsilon &&\text{for every } r \in \{1, \ldots, p_l - 1\}
\end{aligned}
$$

As the l-th prime number is $\mathcal{O}(l \cdot \log l)$, the size of Δ is $\mathcal{O}(j + n^2 \log n)$. It is easy to see that $X(0) \to^* \epsilon$ iff Q is satisfiable. $\qquad\square$

5   The Constructibility of Pre*

In this section we show that the Pre^* of a regular set of configurations (w.r.t. a BPA($\mathbb{Z}$)) is effectively constructible. However, the Pre^* of a set of configurations described by a $\mathbb{Z}$-input 1-CM is not constructible. It is not even representable by a $\mathbb{Z}$-input 1-CM in general. Regular sets are given by finite automata. We define that finite automata ignore all integer input and are only affected by symbols. So, in the context of BPA($\mathbb{Z}$) we interpret the language $(ab)^*$ as $\{a(k_1)b(k_1')\ldots a(k_n)b(k_n') \mid n \in \mathbb{N}_0, \forall i.\, k_i, k_i' \in \mathbb{Z}\}$.

Theorem 21. *Let Δ be a BPA($\mathbb{Z}$) and R a finite automaton. Then a $\mathbb{Z}$-input 1-CM M can be effectively constructed s.t. $M = Pre^*_\Delta(R)$.*

Proof. Every element in $Pre^*_\Delta(R)$ can be written in the form $\alpha X(k)\gamma$ where $\alpha \to^* \epsilon$, $X(k) \to^* \beta$ and $\beta\gamma \in R$. Thus there must exist a state r in R s.t. there is a path from the initial state r_0 of R to r labeled β and a path from r to a final state of R labeled γ. We consider all (finitely many) pairs (X, r) where $X \in Const(\Delta)$ and $r \in states(R)$. Let R_r be the finite automaton that is obtained from R by making r the only final state. We compute the set of integers k for which there exists a β s.t. $X(k) \to^* \beta$ and $\beta \in R_r$. First we compute the $\mathbb{Z}$-input 1-CM M_X that describes $Post^*(X(k))$ as in Theorem 14. Then we compute the product of M_X with R_r, which is again a $\mathbb{Z}$-input 1-CM. The set of initial values k for which $M_X \times R_r$ is nonempty is Presburger and effectively computable. Let $P_{X,r}$ be the corresponding unary Presburger predicate. Let R'_r be the finite automaton that is obtained from R by making r the initial state. We define $M_{X,r}$ to be the $\mathbb{Z}$-input 1-CM that behaves as follows: First it accepts $X(k)$ iff $P_{X,r}(k)$ and then it behaves like R'_r. Let M_ϵ be the $\mathbb{Z}$-input 1-CM that accepts all sequences α s.t. $\alpha \to^* \epsilon$. M_ϵ is effectively constructible, since for every symbol Y the set of k for with $Y(k) \to^* \epsilon$ is Presburger and effectively constructible by Lemma 17. Then finally we get

$$M = M_\epsilon \cdot \bigcup_{X,r} M_{X,r}$$

□

Now we consider the problem of the Pre^* of a set of configurations described by a $\mathbb{Z}$-input 1-CM.

MEMBERSHIP IN Pre^* OF $\mathbb{Z}$-INPUT 1-CM

Instance: A BPA($\mathbb{Z}$) Δ, a $\mathbb{Z}$-input 1-CM M and a state $X(0)$
Question: $X(0) \in Pre^*_\Delta(M)$?

Theorem 22. *Membership in Pre^* of $\mathbb{Z}$-input 1-CM is undecidable.*

The proof is by reduction of the halting problem for Minsky 2-counter machines. It is given in the full paper [2]. Theorem 22 does not automatically imply that the Pre^* of a $\mathbb{Z}$-input 1-CM (w.r.t. Δ) cannot be represented by a $\mathbb{Z}$-input 1-CM. It leaves the possibility that this $\mathbb{Z}$-input 1-CM is just not effectively constructible. (Cases like this occur, e.g., the set of reachable states of a classic lossy counter machine is semilinear, but not effectively constructible [9].) However, the following theorem (proved in the full paper [2]) shows that the Pre^* of a $\mathbb{Z}$-input 1-CM is not a $\mathbb{Z}$-input 1-CM in general.

Theorem 23. *Let Δ be a BPA($\mathbb{Z}$) and M a $\mathbb{Z}$-input 1-CM. The set $Pre_\Delta^*(M)$ cannot be represented by a $\mathbb{Z}$-input 1-CM in general.*

6   The Logic and its Applications

We define a logic called ISL (Integer Sequence Logic) that can be used to verify properties of BPA($\mathbb{Z}$). It is interpreted over ISS (see Def. 3). We define a notion of satisfaction of an ISL formula by a BPA($\mathbb{Z}$) and show that the verification problem is decidable.

Let *const* denote the projection of ISS on sequences of constants obtained by omitting the integers; formally $const(X_1(k_1)X_2(k_2)\ldots X_m(k_m)) = X_1X_2\ldots X_m$. Then, the logic ISL is defined as follows:

Definition 24. *ISL formulae have the following syntax: $F := (A_1,\ldots,A_n,P)$, where $A_1,\ldots,A_n$ are finite automata over an alphabet of process constants, and P is an $(n-1)$-ary Presburger predicate. Formulae are interpreted over sequences w of the form $X_1(k_1)X_2(k_2)\ldots X_m(k_m)$, where the satisfaction relation is defined as follows:*

$w \models F$ *iff there exist words $w_1,\ldots,w_n$, constants $Y_1,\ldots,Y_{n-1}$ and integers $k_1,\ldots,k_{n-1}$ s.t. $w = w_1Y_1(k_1)w_2Y_2(k_2)\ldots w_{n-1}Y_{n-1}(k_{n-1})w_n$ and*

- *$\forall i \in \{1,\ldots,n-1\}$. A_i accepts $const(w_i)Y_i$.*
- *A_n accepts $const(w_n)$ and $P(k_1,\ldots,k_{n-1})$ is true.*

The set of sequences which satisfy a formula F is given by $[\![F]\!] = \{w \mid w \models F\}$.

Intuitively, ISL formulae specify regular patterns (using automata) involving a finite number of integer values which are constrained by a Presburger formula. We use ISL formulae to specify properties on the configurations of the systems and not on their computation sequences, the typical use of specification logics in verification. For instance, when BPA($\mathbb{Z}$)'s are used to model recursive programs with an integer parameter, a natural question that can be asked is whether some procedure X can be called with some value k satisfying a Presburger constraint P. This can be specified by asking whether there is a reachable configuration corresponding to the pattern $Const^*X(k)Const^*$, where $P(k)$ holds. Using ISL formulae, we can specify more complex questions such as whether it is possible that the execution stack of the recursive program can contain two consecutive copies of a procedure with the same calling parameter. This corresponds to the pattern $Const^*X(k_1)(Const - \{X\})^*X(k_2)Const^*$, where $k_1 = k_2$.

The first result we show, is that we can characterize $[\![F]\!]$ by means of reversal bounded counter automata. However, elements of $[\![F]\!]$ are sequences over an infinite alphabet, since they may contain any integer. To characterize over a finite alphabet an element $w \in [\![F]\!]$ we can encode the integers in w *in unary*: a positive (resp. negative) integer k_i is replaced by k_i (resp. $-k_i$) occurrences of a symbol p_i (resp. n_i). Hence, given a set L of ISS, let $\widehat{L}$ denote the set of all sequences in L encoded in this way. We can characterize $\widehat{[\![F]\!]}$ with a reversal bounded counter automaton.

Lemma 25. *We can construct a reversal bounded counter automaton M over a finite alphabet Σ such that $\widehat{[\![F]\!]} = L(M)$.*

Proof. The reversal bounded counter machine M simulates sequentially the automata $A_1, \ldots, A_n$ in order to check if the input is of the correct regular pattern. After reading w_i (A_i has to be in an accepting state), the machine reads a sequence of symbols p_i or n_i and stores their length in corresponding reversal bounded counters. After the input has been completely read, the Presburger formula can be tested by using a finite number of other reversal bounded counters. □

Now, we define a notion of satisfaction between BPA($\mathbb{Z}$)'s and ISL formulae.

Definition 26. *Let (w_0, Δ) be a BPA($\mathbb{Z}$) with initial configuration w_0 and set of rules Δ. Let F be an ISL-formula. We define that (w_0, Δ) satisfies the formula F iff it has a reachable configuration that satisfies F. Formally*

$$(w_0, \Delta) \models F \iff \exists w \in Post^*_\Delta(w_0).\, w \models F$$

To prove the decidability of the verification problem $(w_0, \Delta) \models F$, for a given BPA($\mathbb{Z}$) (w_0, Δ) and a formula F we need the following definition and a lemma.

Definition 27. *Let L be a set of ISS. Then, $L|_k$ is the set of sequences w such that there exists a sequence $w' \in L$ with $k' \geq k$ integers such that w is obtained from w' by removing $k' - k$ integers and encoding the remaining integers in unary.*

Lemma 28. *Let (w_0, Δ) be a BPA($\mathbb{Z}$) with initial configuration w_0 and set of rules Δ. Then we can construct a PCA M such that $L(M) = Post^*_\Delta(w_0)|_k$.*

Proof. First by Theorem 14 we construct a $\mathbb{Z}$-input 1-CM M that accepts $Post^*_\Delta(w_0)$. We construct a PCA from M by (1) using the pushdown store to encode the counter (2) choosing non-deterministically exactly k input values which are compared to the counter. For these comparisons we need k additional reversal bounded counters (to avoid losing the counter value). □

Theorem 29. *Let (w_0, Δ) be a BPA($\mathbb{Z}$) with initial configuration w_0 and set of rules Δ and $F = (A_1, \ldots, A_n, P)$ an ISL-formula. The problem $(w_0, \Delta) \models F$ is decidable.*

Proof. Clearly, we have $Post^*_\Delta(w_0) \cap [\![F]\!] \neq \emptyset$ iff $\widehat{Post^*_\Delta(w_0)} \cap \widehat{[\![F]\!]} \neq \emptyset$, which is also equivalent to $Post^*_\Delta(w_0)|_{n-1} \cap \widehat{[\![F]\!]} \neq \emptyset$ since F cannot constrain more than $n - 1$ integers. Then we show that $\widehat{Post^*_\Delta(w_0)}|_{n-1} \cap \widehat{[\![F]\!]} \neq \emptyset$ is decidable. This follows from Lemma 28, Lemma 25, the fact that the intersection of a CA language with a PCA language is a PCA language (Lemma 5.1 of [7]), and Theorem 5.2 of [7] which states that the emptiness problem of PCA is decidable. □

Finally, we consider another interesting problem concerning the analysis of BPA($\mathbb{Z}$)'s. When used to model recursive procedures, a natural question is to know the set of all the possible values for which a given procedure can be called. More generally, we are interested in knowing all the possible values of the vectors $(k_1, \ldots, k_n)$ such that there is a reachable configuration which satisfies some given ISL formula $F = (A_1, \ldots, A_{n+1}, P)$. We show that this set is effectively semilinear.

Theorem 30. *Let (w_0, Δ) be a BPA($\mathbb{Z}$) with initial configuration w_0 and set of rules Δ, and let F be an ISL formula. Then, the set*

$$\{(k_1, \ldots, k_n) \in \mathbb{Z}^n \mid \exists w = w_1 Y_1(k_1) \ldots w_n Y_n(k_n) w_{n+1} \in Post_\Delta^*(w_0) . w \models F\}$$

is effectively semilinear.

Proof. As in the proof of Theorem 29, we can construct a PCA which recognizes the language $\widehat{Post_\Delta^*(w_0)}|_n \cap \widehat{[\![F]\!]}$. Then, the result follows from the fact that the Parikh image of a PCA language is semilinear (see Theorem 5.1 of [7]).   □

7   Conclusion

We have shown that BPA($\mathbb{Z}$) is a more expressive and more realistic model for recursive procedures than BPA. The price for this increased expressiveness is that a stronger automata theoretic model ($\mathbb{Z}$-input 1-CM) is needed to describe sets of configurations, while simple finite automata suffice for BPA. As a consequence the set of predecessors is no longer effectively constructible for BPA($\mathbb{Z}$) in general. However, the set of successors is still effectively constructible in BPA($\mathbb{Z}$) and thus many verification problems are decidable for BPA($\mathbb{Z}$), e.g., model checking with ISL. Thus, BPA($\mathbb{Z}$) can be used for verification problems like dataflow analysis, when BPA is not expressive enough. We expect that our results can be generalized to more expressive models (e.g., pushdown automata with an integer parameter), but some details of the constructions will become more complex.

References

1. A. Bouajjani, J. Esparza, and O. Maler. Reachability analysis of pushdown automata: application to model checking. In *International Conference on Concurrency Theory (CONCUR'97)*, volume 1243 of *LNCS*. Springer Verlag, 1997. 198, 204
2. A. Bouajjani, P. Habermehl, and R. Mayr. Automatic Verification of Recursive Procedures with one Integer Parameter. Technical Report LIAFA, 2001. available at http://www.informatik.uni-freiburg.de/~mayrri/index.html 205, 206, 207
3. O. Burkart and B. Steffen. Pushdown processes: Parallel composition and model checking. In *CONCUR'94*, volume 836 of *LNCS*, pages 98–113. Springer Verlag, 1994. 198

4. O. Burkart and B. Steffen. Model checking the full modal mu-calculus for infinite sequential processes. In *Proceedings of ICALP'97*, volume 1256 of *LNCS*. Springer Verlag, 1997. 198

5. J. Esparza, D. Hansel, P. Rossmanith, and S. Schwoon. Efficient algorithms for model checking pushdown systems. In *Proc. of CAV 2000*, volume 1855 of *LNCS*. Springer, 2000. 198

6. J. Esparza and J. Knoop. An automata-theoretic approach to interprocedural data-flow analysis. In *Proc. of FoSSaCS'99*, volume 1578 of *LNCS*, pages 14–30. Springer Verlag, 1999. 198

7. O. Ibarra. Reversal-bounded multicounter machines and their decision problems. *Journal of the ACM*, 25:116–133, 1978. 199, 202, 209, 210

8. O. Ibarra, T. Bultan, and J. Su. Reachability analysis for some models of infinite-state transition systems. In *Proc. of CONCUR 2000*, volume 1877 of *LNCS*. Springer, 2000. 199

9. R. Mayr. Undecidable problems in unreliable computations. In *Proc. of LATIN 2000*, volume 1776 of *LNCS*. Springer Verlag, 2000. 207

10. J. van Leeuwen, editor. *Handbook of Theoretical Computer Science: Volume A, Algorithms and Complexity*. Elsevier, 1990. 206

11. I. Walukiewicz. Pushdown processes: games and model checking. In *International Conference on Computer Aided Verification (CAV'96)*, volume 1102 of *LNCS*. Springer, 1996. 198

Graph–Driven Free Parity BDDs: Algorithms and Lower Bounds

Henrik Brosenne, Matthias Homeister*, and Stephan Waack**

Institut für Numerische und Angewandte Mathematik, Georg–August–Universität
Göttingen
Lotzestr. 16–18, 37083 Göttingen, Germany
{homeiste,waack}@math.uni-goettingen.de

Abstract. We investigate graph–driven free parity BDDs, which strictly
generalize the two well–known models parity OBDDs and graph–driven
free BDDs.

The first main result of this paper is a polynomial time algorithm that
minimizes the number of nodes in a graph–driven free parity BDD.

The second main result is an exponential lower bound on the size of
graph–driven free parity BDD for linear code functions.

1 Introduction

Branching Programs or Binary Decision Diagrams are a well established model
for Boolean functions with applications both in complexity theory and in the
theory of data structures for hardware design and verification.

In complexity theory branching programs are a model of sequential space
bounded computations. Upper and lower bounds on the branching program size
for explicitly defined functions are upper and lower bounds on the sequential
space complexity of these functions.

Data structures for Boolean functions have to allow succinct representations
of as many Boolean functions as possible. They have to admit algorithms for the
most important operations. Among others, these are *minimization, synthesis*
and *equivalence test* (for a survey see [15]). Even if a data structure is of more
theoretical rather than of practical interest, minimization and equivalence test
are of structural significance.

A *(nondeterminstic) binary decision diagram (BDD)* $\mathcal{B}$ on the Boolean variables $\{x_1, \ldots, x_n\}$ is a directed acyclic graph with the following properties. Let $\mathcal{N}(\mathcal{B})$ be the set of nodes of $\mathcal{B}$. There are two distinct nodes s and t called the
source and the *target* node. The outdegree of the target and the indegree of the
source are both equal to zero. The source s is joined to each node of its successor set Succ (s) in $\mathcal{B}$ by an unlabeled directed edge. The nodes different from
the source and the target are called *branching nodes*. Each branching node u is
labeled with a Boolean variable var $(u) \in \{x_1, \ldots, x_n\}$, its successor set falls into

---

* Supported by DFG grant Wa766/4-1
** Supported in part by DFG grant Wa766/4-1

J. Sgall, A. Pultr, and P. Kolman (Eds.): MFCS 2001, LNCS 2136, pp. 212–223, 2001.
© Springer-Verlag Berlin Heidelberg 2001

two subsets $\mathrm{Succ}_0(u)$ and $\mathrm{Succ}_1(u)$, where, for $b \in \{0,1\}$, the node u is joined to each $v \in \mathrm{Succ}_b(u)$ by a directed edge labeled with b. For $b \in \{0,1\}$, an element of $\mathrm{Succ}_b(u)$ is called a b-*successor* of the node u. Moreover, we assume $\mathcal{B}$ to be weakly connected in the following sense. For each branching node u, there is a directed path from the source via this node to the target. The *size* of a BDD $\mathcal{B}$, denoted by $\mathrm{SIZE}(\mathcal{B})$, is the number of its nodes. A branching node u of a BDD is called *deterministic*, if $\mathrm{Succ}_b(u) \leq 1$, for all $b \in \{0,1\}$. The source s is called *deterministic*, if $\#\mathrm{Succ}(s) \leq 1$. The BDD $\mathcal{B}$ as a whole is defined to be *deterministic*, if the source and all branching nodes are deterministic.

Let $\mathbb{B}_n$ denote the set of all Boolean functions of n variables. (If we wish to indicate the set of variables on which the functions depend, we write, for example, $\mathbb{B}(x_1, \ldots, x_n)$ or $\mathbb{B}(x_2, x_4, x_6, \ldots, x_{2 \cdot n})$ instead of $\mathbb{B}_n$.) We regard $\mathbb{B}_n$ as an $\mathbb{F}_2$–algebra, where $\mathbb{F}_2$ is the prime field of characteristic 2. The product $f \wedge g$ or fg of two functions $f, g \in \mathbb{B}_n$ is defined by componentwise conjunction. Their sum $f \oplus g$ corresponds to the componentwise exclusive–or. (In line with this notation, "$\oplus$" is also used for the symmetric difference of sets.) There are several ways to let a BDD $\mathcal{B}$ on the set of Boolean variables $\{x_1, \ldots, x_n\}$ represent a Boolean function $f \in \mathbb{B}_n$. Throughout this paper we make use of the *parity acceptance mode*. For each node u of the diagram $\mathcal{B}$, we inductively define its *resulting function* Res_u. The resulting function of the target equals the all–one function. For a branching node u labeled with the variable x, $\mathrm{Res}_u = (x \oplus 1) \wedge \bigoplus_{v \in \mathrm{Succ}_0(u)} \mathrm{Res}_v \oplus x \wedge \bigoplus_{v \in \mathrm{Succ}_1(u)} \mathrm{Res}_v$. If s is the source, then $\mathrm{Res}_s := \bigoplus_{v \in \mathrm{Succ}(s)} \mathrm{Res}_v$. The function $\mathrm{Res}(\mathcal{B}) : \mathbb{F}_2^n \to \mathbb{F}_2$ represented by the whole diagram is defined to be Res_s. Let us denote a BDD equipped with the parity acceptance mode as *parity BDD*. (From now on we speak about a *BDD*, only if it is *deterministic*.)

The variants of decision diagrams tractable in the theory of data structures of Boolean functions as well as in complexity theory restrict the number and the kind of read accesses to the input variables. A very popular model is the following one. A (parity) BDD is defined to be a *free (parity) BDD*, if each variable is tested on each path from the source at most once.

Ordered binary decision diagrams (OBDDs), introduced by Bryant (1986), are the state–of–the–art data structure for Boolean functions in the logical synthesis process, for verification and test pattern generation, and as a part of CAD tools. They are free BDDs with the following additional property. There is a permutation σ, a so–called *variable ordering*, of the set $\{1, 2, \ldots, n\}$ such that if node v labeled with $x_{\sigma(j)}$ is a successor of node u labeled with $x_{\sigma(i)}$, then $i > j$. Perhaps the most important fact about OBDDs for theory is that a size–minimal OBDD for a function f and a fixed variable ordering σ is uniquely determined. It can be efficiently computed by Bryant's minimization algorithm. But many even simple functions have exponential OBDD–size (see [2], [4]). For this reason models less restrictive than OBDDs are studied.

First we mention *parity OBDDs*. They are defined to be free parity BDDs subject to variable orderings in the above sense. Introduced by Gergov and Meinel in [7], they have been intensively studied in [14].

Second free BDDs without any further restriction are considered as a data structure. The problem was to find a counterpart to the variable ordering of OBDDs. It was independently solved by Gergov and Meinel in [6], and by Sieling and Wegener in [12]. Roughly speaking, the characteristic feature of a free BDD in contrast to OBDDs is that we may use different variable orderings for different inputs.

Definition 1. *A graph ordering $\mathcal{B}_0$ on the set of Boolean variables $\{x_1, \ldots, x_n\}$ is a deterministic free BDD which is complete in the following sense. The source s has exactly one successor $\mathrm{succ}(s)$, each branching node u has exactly one 1– successor and exactly one 0–successor, and on each path from the source to the target there is for each variable x_i exactly one node labeled with x_i.*

The relation between a graph ordering and a free BDD guided by it is given in Definition 2. Informally spoken, depending on the Boolean variables tested so far and the corresponding input bits retrieved, the graph ordering predicts the next variable to be tested. For later use in this paper, we let not only free BDDs but also free parity BDDs be guided in that way.

Definition 2. *Let $\mathcal{B}_0$ be a graph ordering on $\{x_1, \ldots, x_n\}$. A free parity BDD $\mathcal{B}$ is defined to be an oracle–driven free parity BDD guided by $\mathcal{B}_0$ if the following condition is satisfied. Let, for any input $b \in \{0,1\}^n$, π_b^0 be the path in $\mathcal{B}_0$ and π_b be an arbitrarily chosen path in $\mathcal{B}$ under the input b. If x_i is tested before x_j when traversing π_b, then this is true when traversing π_b^0, too.*

Note, that OBDDs and parity OBDDs can be regarded as guided by graph orderings, the so–called *line orderings*.

Gergov and Meinel, and Sieling and Wegener were able to show, that oracle–driven free BDDs efficiently support most of the OBDD–operations.

A drawback of oracle–driven free BDDs is, that they do not have "levels" defined by the nodes of the guiding graph ordering such as OBDDs and parity OBDDs have. To enforce the existence of levels in the case of free BDDs, Sieling and Wegener [12] introduced what they called well–structured graph–driven FBDDs. Extending this notion in Definition 3 to the case of free parity BDDs, we get the structure on which this paper is mainly focused.

Definition 3. *An oracle–driven free parity BDD $\mathcal{B}$ guided by $\mathcal{B}_0$ is defined to be* graph–driven, *if there is an additional* level *function $\mathrm{level} : \mathcal{N}(\mathcal{B}) \to \mathcal{N}(\mathcal{B}_0)$ with the following properties.*

- *$\mathrm{level}(s) = s$, $\mathrm{level}(t) = t$;*
- *For each branching node $u \in \mathcal{N}(\mathcal{B})$, $\mathrm{level}(u)$ is a branching node of $\mathcal{B}_0$ that is labeled with the same Boolean variable as u: $\mathrm{var}(u) = \mathrm{var}(\mathrm{level}(u))$.*
- *For $b \in \{0,1\}^n$, let π_b^0 be the path in $\mathcal{B}_0$ and let π_b be an arbitrarily chosen path in $\mathcal{B}$ under the input b. For each node v, if v is contained in π_b, then $\mathrm{level}(v)$ is contained in π_b^0.*

Let $\mathcal{B}$ be a graph–driven free parity BDD guided by $\mathcal{B}_0$, then the set of nodes of $\mathcal{B}$ is partitioned into levels as follows. For each node w of $\mathcal{B}_0$, we define the level $\mathcal{N}_w$ of $\mathcal{B}$ associated with the node w to be the set $\{u \in \mathcal{N}(\mathcal{B}) \mid \mathrm{level}(u) = w\}$.

Note, that graph–driven free parity BDDs have a strictly larger descriptive power than both oracle–driven free BDDs and parity OBDDs. This follows from results due to Sieling. He has proved in [13], that there is an explicitly defined function that has polynomial size parity OBDDs but exponential size free BDDs, whereas another function has polynomial size free BDDs but exponential size parity OBDDs.

The results of this paper can be summarized as follows. An algebraic characterization (see Theorem 7) of the graph–driven free parity BDD complexity serves as basis both for lower and for upper bounds.

Having derived a lower bound criterion (see Corollary 9), we are able to prove exponential lower bounds on the size of graph–driven and of oracle–driven free parity BDDs for linear code functions. This extends an analogous result for parity OBDDs due to Jukna (see [9]).

Moreover, we establish deterministic polynomial time algorithms for solving the minimization problem for the number of nodes (see Theorem 14), and the equivalence test problem (see Corollary 15).

2   Algebraic Characterization and Lower Bounds

Throughout this section let us fix a graph–driven free parity BDD $\mathcal{B}$ on the set of Boolean variables $\{x_1, \ldots, x_n\}$ guided by a graph ordering $\mathcal{B}_0$ that represents a Boolean function $f \in \mathbb{B}_n$.

Our first problem is to determine which nodes of $\mathcal{B}$ can be joined together by a directed edge without violating Definition 3. The following lemma gives the answer. The proof is omitted.

Lemma 4. *Let u be any node of $\mathcal{B}$ joined to a node v by a directed edge.*

If node u is a branching node and the edge is labeled with $b \in \{0, 1\}$, then all paths in the graph ordering $\mathcal{B}_0$ from level (u)'s b–successors to the target node t pass through node level (v).

If node u is the source of $\mathcal{B}$, then all paths in the graph ordering $\mathcal{B}_0$ from the source s to the target node t pass through node level (v).

Lemma 4 provides us with the opportunity to define the *level tree* $\mathcal{T}(\mathcal{B}_0)$ of a graph ordering $\mathcal{B}_0$. To this end, we call a node w'' of $\mathcal{B}_0$ to be an *ascendant* of a branching node w', iff any path from w' to the target node passes through node w''. (Thus each node is an ascendant of itself.) The node w'' is defined to be an *immediate ascendant* of node w', iff w'' is distinct from w', w'' is an ascendant of w', and any ascendant of w' distinct from w' is ascendant of w'', too. Obviously, for each branching node of $\mathcal{B}_0$ there is a uniquely determined immediate ascendent. The set of nodes of the level tree $\mathcal{T}(\mathcal{B}_0)$ is $\mathcal{N}(\mathcal{B}_0) \setminus \{s\}$. Node w'' is joined to node w' by a directed edge, iff w'' is the immediate ascendant of w'. The root of $\mathcal{T}(\mathcal{B}_0)$ is the target node t of $\mathcal{B}_0$. The successor of $\mathcal{B}_0$'s source s is one leaf of $\mathcal{T}(\mathcal{B}_0)$.

It is the aim of this section to characterize the number of nodes of a size–minimal graph–driven free parity BDD in terms of invariants of the level tree $T(\mathcal{B}_0)$ and the function f. For describing what we want, we need a little more notation.

We define the *stalk of the function f at a node w* of the level tree $T(\mathcal{B}_0)$, denoted by $\mathbb{B}_w(f)$, to be the following vector space. If $\mathbb{S}_w(f)$ is the set of all subfunctions $f|_\pi$ of f, where π is a path from the source s to the node w in $\mathcal{B}_0$, then $\mathbb{B}_w(f)$ is spanned by the union of all sets $\mathbb{S}_{w'}(f)$, where w' ranges over all ascendants of w in $T(\mathcal{B}_0)$. Note that if w'' is an ascendant of w', then the stalk at w'' is a subspace of the stalk at w'. The family $\{\mathbb{B}_w(f) \mid w \in T(\mathcal{B}_0)\}$ is called the *sheaf of the function f on the level tree $T(\mathcal{B}_0)$*.

Analogously we define the *sheaf of the free parity BDD $\mathcal{B}$ on the level tree $T(\mathcal{B}_0)$*. For any node w of the level tree, the *stalk $\mathbb{B}_w(\mathcal{B})$ of $\mathcal{B}$ at the node w* is the defined to be the vector space spanned by all functions Res_u, where $\mathrm{level}(u)$ is an ascendant of w.

The next lemma is a direct consequence of the usual path argument.

Lemma 5. *The sheaf of f is a subsheaf of the sheaf of $\mathcal{B}$, i.e. $\mathbb{B}_w(f) \subseteq \mathbb{B}_w(\mathcal{B})$, for all nodes w of the level tree $T(\mathcal{B}_0)$.*

Lemma 6. *Let w' be any branching node of the level tree $T(\mathcal{B}_0)$, and let w'' be its immediate ascendant. Then $\#\mathcal{N}_{w'}(\mathcal{B}) \geq \dim_{\mathbb{F}_2} \mathbb{B}_{w'}(f) - \dim_{\mathbb{F}_2} \mathbb{B}_{w''}(f)$.*

Proof. By Lemma 5 we have a canonical linear mapping

$$\phi : \mathbb{B}_{w'}(f) \to \mathbb{B}_{w'}(\mathcal{B})/\mathbb{B}_{w''}(\mathcal{B})$$
$$g \mapsto g \bmod \mathbb{B}_{w''}(\mathcal{B}).$$

Since $\#\mathcal{N}_{w'}(\mathcal{B}) \geq \dim \mathbb{B}_{w'}(\mathcal{B})/\mathbb{B}_{w''}(\mathcal{B})$, it suffices to show that $\ker \phi = \mathbb{B}_{w''}(f)$.

The inclusion $\ker \phi \supseteq \mathbb{B}_{w''}(f)$ is an immediate consequence of Lemma 5. Let $g \in \ker \phi$. Then $g \in \mathbb{B}_{w''}(\mathcal{B})$. Consequently, g does not essentially depend on any variable tested on any path from w' to w'' in $\mathcal{B}_0$. It follows that $g \in \mathbb{B}_{w''}(f)$.

We are able to formulate and prove Theorem 7 that will prove useful both in designing algorithms and in proving lower bounds.

Theorem 7. *Let $\mathcal{B}$ be a size–minimal graph–driven free parity BDD on $\{x_1, \ldots, x_n\}$ guided by $\mathcal{B}_0$ representing $f \in \mathbb{B}_n$. Let w' be any branching node of $T(\mathcal{B}_0)$, and let w'' its immediate ascendant. Then*

$$\#\mathcal{N}_{w'}(\mathcal{B}) = \dim_{\mathbb{F}_2} \mathbb{B}_{w'}(f) - \dim_{\mathbb{F}_2} \mathbb{B}_{w''}(f).$$

Proof. By Lemma 6, it suffices to construct a graph–driven free parity BDD $\mathcal{B}'$ guided by $\mathcal{B}_0$ representing f such that the asserted equations hold.

First, let us turn to the set of nodes of $\mathcal{B}'$. We define $\mathcal{N}_s(\mathcal{B}')$ to be the source s of $\mathcal{B}'$, $\mathcal{N}_t(\mathcal{B}')$ to be the all-one function, which plays the role of the target node of $\mathcal{B}'$, and $\mathcal{N}_w(\mathcal{B}')$ to be a set of representatives of a basis of the space $\mathbb{B}_w(f)/\mathbb{B}_{\tilde{w}}(f)$, where $\tilde{w}$ is the immediate ascendant of w.

Second, we have to inductively create edges in such a way that $\mathcal{B}'$ represents the Boolean function f. We do that in a bottom–up manner, such that for any node w of the graph ordering $\mathcal{B}_0$, and any $h \in \mathcal{N}_w(\mathcal{B}')$, we have $\text{Res}_h = h$. To this end, we fix a topological ordering of the nodes of $\mathcal{B}_0$ with the least node being the target.

The claim is true for $w = t$. For the induction step, we assume that $w \neq t$, var $(w) = x_l$, for some l, and $h \in \mathcal{N}_w(\mathcal{B}')$. For $b \in \{0,1\}$, let w_b be the b–successor of the node w in the graph ordering $\mathcal{B}_0$. For $b = 0,1$, we conclude by Lemma 4 $h|_{x_l=b} \in \mathbb{B}_{w_b}(f)$. By induction hypothesis, the functions $h|_{x_l=0}$ and $h|_{x_l=1}$ have unique representations as sums of resulting functions of nodes defined so far. We have to "hardwire" these representations in the decision diagram. For $b = 0,1$, the node h is joined to any other node h' by a directed edge labeled b if and only if h' occurs in the sum that represents $h|_{x_l=b}$.

Theorem 7 implicitly contains a lower bound technique. We see that it does not suffice to estimate from below the sum of dimensions of the stalks at a fixed depth level. But if we are able to detect large subspaces of these stalks that have pairwise trivial intersections, we are done.

Taking pattern from [8], we define the following notion.

Definition 8. *A function $f \in \mathbb{B}_n$ is called* linearly k–mixed, *for $k < n$, if for all subsets $V \subset \{x_1, \ldots, x_n\}$ such that $|V| = k$ and all variables $x_l \notin V$ the 2^k assignments of constants to the variables $x_i \in V$ lead to a sequence of subfunctions that is linearly independent modulo the vector space of all functions not depending on x_l.*

Corollary 9 (Lower Bound Criterion). *If f is linearly k–mixed, then any graph–driven free parity BDD representing f has size bounded below by 2^k.*

Proof. Let $\mathcal{B}_0$ be any graph ordering on the set of Boolean variables $\{x_1, \ldots, x_n\}$, let w be any branching node of $\mathcal{B}_0$ at depth level $k+1$, and let w' be the immediate ascendant of w in $\mathcal{T}(\mathcal{B}_0)$. Assume without loss of generality that the node w is labeled with x_{k+1} and that the set of variables tested on any path to w equals $\{x_1, \ldots, x_k\}$. If Π_w is the set of all paths that lead in $\mathcal{B}_0$ to w, then by Theorem 7 its suffices to show that $\dim_{\mathbb{F}_2}(\mathbb{B}_w(f)/\mathbb{B}_{w'}(f)) \geq |\Pi_w|$. Indeed, for each linear combination $\sum_{\pi \in \Pi_w} b_\pi f|_\pi \in \mathbb{B}_{w'}(f) \subseteq \mathbb{B}(x_{k+2}, \ldots, x_n)$ it follows that $b_\pi = 0$, for $\pi \in \Pi_w$, since f is linearly k–mixed. $\qquad \square$

We use Corollary 9 to prove exponential lower bounds on the size of graph–driven free parity BDDs for characteristic functions of linear codes.

A *linear code* C is a linear subspace of $\mathbb{F}_2^n$. We consider the *characteristic function* $f_C : \mathbb{F}_2^n \to \{0,1\}$ defined by $f_C(a) = 1 \iff a \in C$. The *Hamming distance* of two code words $a, b \in C$ is defined to be the number of 1's of $a \oplus b$. The *minimal distance* of a code C is the minimal Hamming distance of two distinct elements of C. The *dual* $C^\perp$ is the set of all vectors b such that $b^T a = 0$, for all elements $a \in C$. (Here $b^T a = b_1 a_1 \oplus \ldots \oplus b_n a_n$ is the standard inner product with respect to $\mathbb{F}_2$.) A set $D \subseteq \mathbb{F}_2^n$ is defined to be k-*universal*, if for any subset of k indicese $I \subseteq \{1, \ldots, n\}$ the projection onto these coordinates restricted to the set D gives the whole space $\mathbb{F}_2^k$.

The next lemma is well–known. See [9] for a proof.

Lemma 10. *If C is a code of minimal distance $k + 1$, then its dual $C^\perp$ is k–universal.*

We shall prove a general lower bound on the size of graph–drive free parity BDDs representing f_C.

Theorem 11. *Let $C \subseteq \mathbb{F}_2^n$ be a linear code of minimal distance d whose dual $C^\perp$ has minimal distance $d^\perp$. Then any graph–driven free parity BDD representing f_C has size greater than or equal to $2^{\min\{d, d^\perp\}-2}$.*

Proof. Let $k = \min\{d, d^\perp\} - 2$. By Corollary 9 is suffices to show that the function f_C is linearly k–mixed. Without loss of generality, we assume the set V of Definition 8 to be $\{x_1, \ldots, x_k\}$ and the additional variable to be x_{k+1}. Let $A = \{a_1, \ldots a_{2^k}\}$ be the set of all assignments of constants to the variables $x_1, \ldots, x_k$. We have to show that the sequence

$$f_C(a_1, x_{k+1}, x_{k+2}, \ldots, x_n), \ldots, f_C(a_{2^k}, x_{k+1}, x_{k+2}, \ldots, x_n)$$

is linearly independent $\mathrm{mod}\,\mathbb{B}(x_{k+2}, \ldots, x_n)$.

Since the distance $d^\perp$ of the dual code $C^\perp$ is greater than k, by Lemma 10 C is k–universal. Consequently, there are assignments $b_1, \ldots, b_{2^k}$ to the variables $x_{k+1}, x_{k+2}, \ldots, x_n$ such that

$$f_C(a_i, b_i) = 1, \text{ for } i = 1, \ldots, 2^k.$$

Since the distance of the code C is greater than or equal to $k + 2$, for $\tilde{b}_i$ being obtained from b_i by skipping the assignment of variable x_{k+1}, we have

$$f_C(a_i, \tilde{b}_j) = 0, \text{ for } i, j = 1, \ldots, 2^k;$$
$$f_C(a_i, b_j) = 0, \text{ for } i, j = 1, \ldots, 2^k, i \neq j.$$

For the sake of deriving a contradiction, let as assume, that there is a nonempty subset $I \subseteq \{1, \ldots, 2^k\}$ such that

$$\bigoplus_{i \in I} f_C(a_i, x_{k+1}, x_{k+2}, \ldots, x_n) = g,$$

where the Boolean function g does not essentially depend on the variable x_{k+1}. Let us fix an index $j \in I$. Then $g(b_j) = 1$, and consequently $g(\tilde{b}_j) = 1$. The latter equation implies, that there is an index $j' \in I$, such that $f_C(a_{j'}, \tilde{b}_j) = 1$. Contradiction.

In order to proof an *explicit lower bound*, recall that the r–th order binary Reed–Muller code $R(r, l)$ of length $n = 2^l$ is the set of graphs of all polynomials in l variables over $\mathbb{F}_2$ of degree at most r. The code $R(r, l)$ is linear and has minimal distance 2^{l-r}. It is known that the dual of $R(r, l)$ is $R(l - r - 1, l)$ (see [10]).

Corollary 12. *Let* $n = 2^l$ *and* $r = \lfloor l/2 \rfloor$. *Then every graph–driven free parity BDD representing the characteristic function of* $R(r, l)$ *has size bounded below by* $2^{\Omega(\sqrt{n})}$.

Proof. Taking the notation of Theorem 11, we have $d = 2^{l-r} = \Omega\left(\sqrt{n}\right)$ and $d^\perp = 2^{r+1} = \Omega\left(\sqrt{n}\right)$. The claim follows.

Corollary 13. *Let* $n = 2^l$ *and* $r = \lfloor l/2 \rfloor$. *Then any oracle–driven free parity BDD guided by* $\mathcal{B}_0$ *representing the characteristic function of* $R(r, l)$ *has size bounded below by* $2^{\Omega(\sqrt{n})}/\mathrm{SIZE}\,(\mathcal{B}_0)$.

Proof. By means of the well–known product construction from automata theory, we can easily prove the following. Let $\mathcal{B}$ be an oracle–driven free parity BDD guided by $\mathcal{B}_0$. Then there is an graph–driven free parity BDD guided by $\mathcal{B}_0$ of size $\mathcal{O}\,(\mathrm{SIZE}\,(\mathcal{B}_0) \cdot \mathrm{SIZE}\,(\mathcal{B}))$ that represents the same Boolean function as $\mathcal{B}$.

Now the claim follows from Corollary 12.

3   Minimizing the Number of Nodes

Let us define *a feasible exponent* ω *of matrix multiplication* over a field k to be a real number such that multiplication of two square matrices of order h may be algorithmically achieved with $\mathcal{O}\,(h^\omega)$ arithmetical operations. It is well–known that matrix multiplication plays a key role in numerical linear algebra. Thus the following problems all have "exponent" ω: matrix inversion, L-R-decomposition, evaluation of the determinant. Up to now, the best known ω is 2.376 (see Coppersmith and Winograd (1990)).

It is the aim of this section to prove the following theorem.

Theorem 14. *Let* ω *be any feasible exponent of matrix multiplication. Let* $\mathcal{B}_0$ *be a fixed graph ordering on the set of Boolean variables* $\{x_1, \ldots, x_n\}$. *Then there is an algorithm that computes taking a graph–driven free parity BDD* $\mathcal{B}$ *guided by* $\mathcal{B}_0$ *as input a size–minimal one representing the same Boolean function as* $\mathcal{B}$ *in time* $\mathcal{O}\,(\mathrm{SIZE}\,(\mathcal{B}_0) \cdot \mathrm{SIZE}\,(\mathcal{B})^\omega)$ *and space* $\mathcal{O}\left(\mathrm{SIZE}\,(\mathcal{B}_0) + \mathrm{SIZE}\,(\mathcal{B})^2\right)$.

Proof. Let $f(x_1, \ldots, x_n)$ be the Boolean function represented by the input BDD $\mathcal{B}$. The algorithm that proves the theorem falls into two phases. The first phase, which we call the *linear reduction phase* insures that the functions represented by nodes of $\mathcal{B}$ are linearly independent. We use a bottom–up approach here, where we refer to the direction of the graph ordering $\mathcal{B}_0$. The second phase, called the *semantic reduction phase*, transforms the input $\mathcal{B}$ in a top–down manner such that afterwards all stalks of $\mathcal{B}$ are subspaces of the corresponding stalks of the function f. Having performed these two phases, the BDD $\mathcal{B}$ thus modified is size–minimal by Theorem 7.

We internally represent $\mathcal{B}$ as two $\mathrm{SIZE}\,(\mathcal{B}) \times \mathrm{SIZE}\,(\mathcal{B})$–adjacency matrices $A^{(0)}$ and $A^{(1)}$ over $\mathbb{F}_2$, where the columns of $A^{(b)}$ represent the b-successor sets of $\mathcal{B}$'s branching nodes, and a column vector R over $\mathbb{F}_2$ of length $\mathrm{SIZE}\,(\mathcal{B})$

that represents the successor set of the source s. (These representations have to be updated again and again.) In line with that, all auxiliary subsets of non–source nodes are represented as $\mathbb{F}_2$–column vectors of length $\text{SIZE}(\mathcal{B})$. For the sake of simplifying notations, we identify the nodes of $\mathcal{B}$ with the functions represented, the nodes of $\mathcal{B}_0$ with the levels indexed.

We describe the linear reduction phase of the algorithm. Assume that we are about to linearly reduce the level w, where all levels that can be reached from w in $\mathcal{B}_0$ by a nontrivial directed path have already been linearly reduced. (More precisely, we are going to reduce the stalk of $\mathcal{B}$ at w, not only the level w.) Let x be the Boolean variable with which w is labeled, for $b \in \{0,1\}$, let w_b be the b–successor of w in $\mathcal{B}_0$, and let w' be the immediate ascendant of $w \in \mathcal{T}(\mathcal{B}_0)$. In order to perform the linear reduction, we use the following two facts.

Fact 1. By Shannon's decomposition we have the following embedding of the stalk of $\mathcal{B}$ at w.

$$\mathbb{B}_w(\mathcal{B}) \to \mathbb{B}_{w_0}(\mathcal{B}) \times \mathbb{B}_{w_1}(\mathcal{B})$$
$$v \mapsto (v|_{x=0}, v|_{x=1})$$

Fact 2. The intersection of the stalk at w_0 and the stalk at w_1 equals the stalk at w', the immediate ascendant of w.

Let $u_1, \ldots, u_\mu$ be the nodes that span the stalk at w', let $u_{\mu+1}, \ldots, u_{\mu+\mu'}$ be the nodes of level w, and let $v_1^0, \ldots, v_{\nu_0}^0$ and $v_1^1, \ldots, v_{\nu_1}^1$ be the nodes that span the stalks at w_1 and w_1, respectively. Moreover, let $\langle .,. \rangle$ denote the inner product in both stalks $\mathbb{B}_{w_0}(\mathcal{B})$ and $\mathbb{B}_{w_1}(\mathcal{B})$ defined by $\langle v_i^b, v_j^b \rangle = \delta_{ij}$, for $i, j = 1, \ldots, \nu_b$, and $b = 0, 1$.

The first step is to set up the matrix $M = (M_{ij})$, where $i = 1, \ldots, \nu_0 + \nu_1$, $j = 1, \ldots, \mu + \mu'$. For $i = 1, \ldots, \nu_0$, $j = 1, \ldots, \mu + \mu'$, define $M_{ij} := \langle v_i^0, u_j|_{x=0} \rangle$, and for $i = 1, \ldots, \nu_1$, define $M_{\nu_0+i,j} := \langle v_i^1, u_j|_{x=1} \rangle$. The columns $M_{.1}, \ldots, M_{.\mu}$ represent the canonical basis of the stalk at w', the columns $M_{.\mu+1}, \ldots, M_{.\mu+\mu'}$ the nodes of level w.

The second step is to find out which of the nodes of level w are superfluous. To this end, we select columns $M_{.\mu+j}$, $j \in J$, such that the columns $M_{.1}, \ldots, M_{.\mu}$, $M_{.\mu+j}$, $j \in J$, form a basis of the space spanned by all columns of M. We then represent the columns not selected in terms of those selected. We assume the result of the second step to be presented as follows.

$$M_{.\mu+l} = \sum_{k=1}^{\mu} \alpha_k M_{.k} + \sum_{j \in J} \alpha_{\mu+j} M_{.\mu+j}, \text{ for } l \notin J,$$
$$U_{\mu+l} := \{u_j \,|\, \alpha_j = 1\}, \text{ for } l \notin J.$$

The third step is to hardwire the results of the second step in the decision diagram $\mathcal{B}$. The problem is, to do so within the desired time bound. Having set up the matrix $\hat{E}$ resulting from the $\text{SIZE}(\mathcal{B}) \times \text{SIZE}(\mathcal{B})$–identity matrix by replacing the columns associated with $u_{\mu+l}$, for $l \notin J$, by the columns $U_{\mu+l}$, we execute the following three instructions.

$$A^{(b)} \leftarrow \hat{E} \cdot A^{(b)}, \text{ for } b = 0, 1,$$

$$R \leftarrow \hat{E} \cdot R.$$

The first and the second one update the branching nodes of $\mathcal{B}$, the third one the source of $\mathcal{B}$. Afterwards the nodes $u_{\mu+l}$, for $l \notin J$, have indegree zero.

In the last step we remove all nodes no longer reachable from a source. This can be done by a depth–first–search traversal.

Now we describe the semantic reduction phase of the algorithm. It suffices to consider the case, that $\mathcal{B}$ has a nonempty level different from s and t. By Theorem 7 the BDD $\mathcal{B}$ has then a uniquely determined nonempty top level w_{top} characterized as follows. The level w_{top} is joined to any other nonempty level by a directed path in $\mathcal{B}_0$. Moreover, w_{top} is an ascendant of the unique successor of $\mathcal{B}_0$'s source.

As in the case of the linear reduction phase, we assume that we are on the point of semantically reducing level w, where all levels that precede w in a topological ordering of the nodes of $\mathcal{B}_0$ have already been reduced.

Case 1. The level w is equal to w_{top}. We have to transform w in such a way that it contains afterwards a single node only. Let $v_1, \ldots, v_\mu$ be the node of level w. We merge these nodes in such a way together that for the resulting node u holds: $u = \bigoplus_{i=1}^{\mu} v_i$. For $b = 0, 1$, the b–successor set of node u is computed by executing the matrix operation $V^{(b)} \leftarrow A^{(b)} \cdot M + L$, where $M = \{v_1, \ldots, v_\mu\}$ and L is the set of the other nodes of of level w.

Case 2. The level w is not equal to w_{top}. *In the first step of this case* for each node u of $\mathcal{B}$ and each Boolean constant b such that there is an b–edge leading from u to a node of level w we do the following. We partition the b–successors of u into the three sets O_u^b, M_u^b, and L_u^b, where M_u^b contains all nodes belonging to level w, L_u^b contains all nodes belonging to levels that are proper ascendants of w, and O_u^b is the remaining set. (By induction hypothesis we have already semantically reduced the levels to which the nodes of O_u^b belong.) Having created a new node $v(u, b)$ of level $\mathcal{N}_a(\mathcal{B})$, we remove all b–edges from u to nodes of $M_u^{(b)} \cup L_u^{(b)}$, and join u to $v(u, b)$ by a directed b–edge. In order to compute the edges outgoing from $v(u, b)$ in such a way that $v(u, b) = \bigoplus_{v \in M_u^{(b)} \cup L_u^{(b)}} v$ holds, we set up the matrices M and L whose columns we have just created. Then we compute for each node $v(u, b)$ its successor sets $\text{Succ}_0(v(u, b))$ and $\text{Succ}_1(v(u, b))$ by means of the matrix operations

$$V^{(b)} \leftarrow A^{(b)} \cdot M + L, \text{ for } b = 0, 1.$$

(Because of the induction hypothesis on the sets O_u^b, we are sure that $v(u, b) \in \mathbb{B}_w(f)$, for all u and b under consideration.)

In the second step we remove all nodes of $\mathcal{B}$ that are no longer reachable from the source.

The new nodes of level w are not necessarily linearly independent from each other and from other nodes of the the the stalk of $\mathcal{B}$ at w. *In the last step* we linearly reduce the stalk of $\mathcal{B}$ at w in the same way as in the first phase.

The runtime of the algorithm is dominated by the $\mathcal{O}(n)$ multiplications of SIZE $(\mathcal{B}) \times$ SIZE $(\mathcal{B})$–matrices. The space demand is obvious.

Let $\mathcal{B}'$ and $\mathcal{B}''$ be two graph–driven free parity BDDs on $\{x_1, \ldots, x_n\}$ guided by $\mathcal{B}_0$. First, using standard techniques, for example the well–known "product construction", and taking pattern from [12], one can easily perform the Boolean synthesis operations in time $\mathcal{O}(\text{SIZE}(\mathcal{B}_0) \cdot (\text{SIZE}(\mathcal{B}') \cdot \text{SIZE}(\mathcal{B}''))^{\omega})$. Second, we have the following.

Corollary 15. *It can be decided in time* $\mathcal{O}(\text{SIZE}(\mathcal{B}_0) \cdot (\text{SIZE}(\mathcal{B}') + \text{SIZE}(\mathcal{B}''))^{\omega})$ *whether or not* $\mathcal{B}'$ *and* $\mathcal{B}''$ *represent the same function.*

4 Open Problems

Problem 1. In contrast to the deterministic case, not all free parity BDDs are oracle–driven or even graph–driven ones.

Seperate the descriptive power of free parity BDDs from that of graph–driven free parity BDDs.

Problem 2. In [11] exponential lower bounds for pointer functions on the size of $(\oplus, k)$–branching programs are proved. (A $(\oplus, k)$–branching program is a free parity BDD with the source being the only nondeterministic node. The parameter k denotes the fan–out of the source.)

Prove that the linear code function of Corollary 12 cannot be represented by an expression $\bigoplus_{i=1}^{k} f_i$, where the functions f_i have graph–driven free parity BDDs (guided by distinct graph orderings) of polynomial size.

Problem 3. Bollig proved in [1] a $2^{\Omega(n/\log n)}$ lower bound on the size of oracle–driven free parity BDDs guided by a tree ordering for the middle bit of the integer multiplication. (A graph ordering $\mathcal{B}_0$ on $\{x_1, \ldots, x_n\}$ is defined to be a tree ordering, if $\mathcal{B}_0$ becomes a tree of size $n^{\mathcal{O}(1)}$ by eliminating the sink and replacing multiedges between nodes by simple edges. By means of Bollig's lower bound arguments it is possible to show that graph–driven free parity BDDs are more powerful than tree–driven ones.)

Extend Bollig's result to graph–driven free parity BDDs.

Acknowledgement

We are indebted to Beate Bollig for her helpful suggestions and comments.

References

1. BOLLIG, B. (2000), Restricted Nondeterministic Read–Once Branching Programs and an Exponential Lower Bound for Integer Multiplication, in "Proceedings, 25th Mathematical Foundations of Computer Science", Lecture Notes in Computer Science **1893**, Springer–Verlag, pp. 222–231. 222

2. BREITBART, Y., HUNT, H. B., AND ROSENKRANTZ, D. (1991) The size of binary decision diagrams representing Boolean functions, Preprint. 213

3. BRYANT, R. E. (1986), Graph–based algorithms for Boolean function manipulation, *IEEE Trans. on Computers* **35**, pp. 677–691.

4. BRYANT, R. E. (1991), On the complexity of VLSI implementations of Boolean functions with applications to integer multiplication, *IEEE Trans. on Computers* **40**, pp. 205–213. 213

5. COPPERSMITH, D., AND WINOGRAD, S. (1990), Matrix multiplication via arithmetic progressions, *J. Symbolic Computation* **9**, pp. 251–280.

6. GERGOV, J., AND MEINEL, CH. (1993), Frontiers of feasible and probabilistic feasible Boolean manipulation with branching programs, in "Proceedings, 10th Symposium on Theoretical Aspects of Computer Science", Lecture Notes in Computer Science **665**, Springer–Verlag, pp. 576–585. 214

7. GERGOV, J., AND MEINEL, CH. (1996), Mod–2–OBDDs — a data structure that generalizes exor-sum-of-products and ordered binary decision diagrams, *Formal Methods in System Design* **8**, pp. 273–282. 213

8. JUKNA, S. (1988), Entropy of contact circuits and lower bounds on their complexity, *Theoretical Computer Science* **57**, pp. 113–129. 217

9. JUKNA, S. (1999), Linear codes are hard for oblivious read-once parity branching programs, *Information Processing Letters* **69**, pp. 267–269. 215, 218

10. MACWILLIAMS, E. J., AND SLOANE, N. J. A. (1977), The theory of errorcorrecting codes, Elsevier, North–Holl. 218

11. SAVICKÝ, P., AND SIELING, D. (2000), A hierarchy result for read–once branching programs with restricted parity nondeterminism, in "Proceedings, 25th Mathematical Foundations of Computer Science", Lecture Notes in Computer Science **1893**, Springer–Verlag, pp. 650–659. 222

12. SIELING, D., AND WEGENER, I. (1995), Graph Driven BDD's — a new Data Structure for Boolean Functions, *Theoretical Computer Science* **141**, pp. 283-310. 214, 222

13. SIELING, D. (1999b), Lower bounds for linear transformed OBDDs and FBDDs, in "Proceedings of Conference on the Foundations of Software Technology and Theoretical Computer Science", Lecture Notes in Computer Science **1738**, Springer–Verlag, pp. 356-368. 215

14. WAACK, ST. (1997), On the descriptive and algorithmic power of parity ordered binary decision diagrams, in "Proceedings, 14th Symposium on Theoretical Aspects of Computer Science", Lecture Notes in Computer Science **1200**, Springer–Verlag, pp. 201–212. 213

15. WEGENER, I. (2000), "Branching programs and binary decision diagrams", SIAM Monographs on Discrete Mathematics and Applications. 212

Computable Versions of Baire's Category Theorem

Vasco Brattka[*]

Theoretische Informatik I, Informatikzentrum
FernUniversität, 58084 Hagen, Germany
`vasco.brattka@fernuni-hagen.de`

Abstract. We study different computable versions of Baire's Category Theorem in computable analysis. Similarly, as in constructive analysis, different logical forms of this theorem lead to different computational interpretations. We demonstrate that, analogously to the classical theorem, one of the computable versions of the theorem can be used to construct interesting counterexamples, such as a computable but nowhere differentiable function.

Keywords: computable analysis, functional analysis.

1 Introduction

Baire's Category Theorem states that a complete metric space X cannot be decomposed into a countable union of nowhere dense closed subsets A_n (cf. [7]). Classically, we can bring this statement into the following two equivalent logical forms:

1. For all sequences $(A_n)_{n \in \mathbb{N}}$ of closed and nowhere dense subsets $A_n \subseteq X$, there exists some point $x \in X \setminus \bigcup_{n=0}^{\infty} A_n$,
2. for all sequences $(A_n)_{n \in \mathbb{N}}$ of closed subsets $A_n \subseteq X$ with $X = \bigcup_{n=0}^{\infty} A_n$, there exists some $k \in \mathbb{N}$ such that A_k is somewhere dense.

Both logical forms of the classical theorem have interesting applications. While the first version is often used to ensure the existence of certain types of counterexamples, the second version is for instance used to prove some important theorems in functional analysis, like the Open Mapping Theorem and the Closed Graph Theorem [7]. However, from the computational point of view the content of both logical forms of the theorem is different. This has already been observed in constructive analysis, where a discussion of the theorem can be found in [6].

We will study the theorem from the point of view of computable analysis, which is the Turing machine based theory of computable real number functions, as it has been developed by Pour-El and Richards [11], Ko [8], Weihrauch [13]

[*] Work supported by DFG Grant BR 1807/4-1

J. Sgall, A. Pultr, and P. Kolman (Eds.): MFCS 2001, LNCS 2136, pp. 224–235, 2001.
© Springer-Verlag Berlin Heidelberg 2001

and others. This line of research is based on classical logic and computability is just considered as another property of classical numbers, functions and sets. In this spirit one version of the Baire Category Theorem has already been proved by Yasugi, Mori and Tsujii [14].

In the representation based approach to computable analysis, which has been developed by Weihrauch and others [13] under the name "Type-2 theory of effectivity", the computational meaning of the Baire Category Theorem can be analysed very easily. Depending on how the sequence $(A_n)_{n \in \mathbb{N}}$ is represented, i.e. how it is "given", we can compute an appropriate point x in case of the first version or compute a suitable index k in case of the second version. Roughly speaking, the second logical version requires stronger information on the sequence of sets than the first version. Unfortunately, this makes the second version of the theorem less applicable than its classical counterpart, since this strong type of information on the sequence $(A_n)_{n \in \mathbb{N}}$ is rarely available.

We close this introduction with a short survey on the organisation of the paper. In the next section we briefly summarize some basic definitions from computable analysis which will be used to formulate and prove our results. In Section 3 we discuss the first version of the computable Baire Category Theorem followed by an example of its application in Section 4, where we construct computable but nowhere differentiable functions. Finally, in Section 5 we discuss the second version of the theorem.

Further applications of the first version of the computable Baire Category Theorem can be found in [3].

2   Preliminaries from Computable Analysis

In this section we briefly summarize some notions from computable analysis. For details the interested reader is refered to [13]. The basic idea of the representation based approach to computable analysis is to represent infinite objects like real numbers, functions or sets, by infinite strings over some alphabet Σ (which should at least contain the symbols 0 and 1). Thus, a *representation* of a set X is a surjective mapping $\delta :\subseteq \Sigma^\omega \to X$ and in this situation we will call (X, δ) a *represented space*. Here the inclusion symbol is used to indicate that the mapping might be partial. If we have two represented spaces (X, δ) and (Y, δ') and a function $f :\subseteq X \to Y$, then f is called (δ, δ')-*computable*, if there exits some computable function $F :\subseteq \Sigma^\omega \to \Sigma^\omega$ such that $\delta' F(p) = f\delta(p)$ for all $p \in \text{dom}(f\delta)$. Of course, we have to define computability of sequence functions $F :\subseteq \Sigma^\omega \to \Sigma^\omega$ to make this definition complete, but this can be done via Turing machines: F is computable if there exists some Turing machine, which computes infinitely long and transforms each sequence p, written on the input tape, into the corresponding sequence $F(p)$, written on the one-way output tape. Later on, we will also need computable multi-valued operations $f :\subseteq X \rightrightarrows Y$, which are defined analogously to computable functions by substituting $\delta' F(p) \in f\delta(p)$ for the equation above. If the represented spaces are fixed or clear from the context, then we will simply call a function or operation f *computable*. A *computable sequence* is a computable function $f : \mathbb{N} \to X$, where we assume that $\mathbb{N}$ is represented by $\delta_{\mathbb{N}}(1^n 0^\omega) := n$ and a point $x \in X$ is called *computable*, if there is a constant computable function with value x.

Given two represented spaces (X, δ) and (Y, δ'), there is a canonical representation $[\delta, \delta']$ of $X \times Y$ and a representation $[\delta \to \delta']$ of certain functions $f : X \to Y$. If δ, δ' are *admissible* representations of T_0–spaces with countable bases (cf. [13]), then $[\delta \to \delta']$ is actually a representation of the set $\mathcal{C}(X, Y)$ of continuous functions $f : X \to Y$. If $Y = \mathbb{R}$, then we write for short $\mathcal{C}(X) := \mathcal{C}(X, \mathbb{R})$. The function space representation can be characterized by the fact that it admits evaluation and type conversion. Evaluation means that $\mathcal{C}(X, Y) \times X \to Y, (f, x) \mapsto f(x)$ is $([[\delta \to \delta'], \delta], \delta')$–computable. Type conversion means that for any represented space (Z, δ'') a function $f : Z \to \mathcal{C}(X, Y)$ is $(\delta'', [\delta \to \delta'])$–computable, if and only if the associated function $\hat{f} : Z \times X \to Y$, defined by $\hat{f}(z, x) := f(z)(x)$, is $([\delta'', \delta], \delta')$–computable. Moreover, the $[\delta \to \delta']$–computable points are just the (δ, δ')–computable functions. Given a represented space (X, δ), we will also use the representation $\delta^{\mathbb{N}} := [\delta_{\mathbb{N}} \to \delta]$ of the set of sequences $X^{\mathbb{N}}$. Finally, we will call a subset $A \subseteq X$ δ–*r.e.*, if there exists some Turing machine that recognizes A in the following sense: whenever an input $p \in \Sigma^\omega$ with $\delta(p) \in A$ is given to the machine, the machine stops after finitely many steps, for all other $p \in \mathrm{dom}(\delta)$ it computes forever.

Many interesting representations can be derived from computable metric spaces and we will also use them to formulate the computable versions of the Baire Category Theorem.

Definition 1 (Computable metric space). A tuple (X, d, α) is called *computable metric space*, if

1. $d : X \times X \to \mathbb{R}$ is a metric on X,
2. $\alpha : \mathbb{N} \to X$ is a sequence which is dense in X,
3. $d \circ (\alpha \times \alpha) : \mathbb{N}^2 \to \mathbb{R}$ is a computable (double) sequence in $\mathbb{R}$.

Here, we tacitly assume that the reader is familiar with the notion of a computable sequence of reals, but we will come back to that point below. Obviously, a computable metric space is especially separable. Given a computable metric space (X, d, α), its *Cauchy representation* $\delta_X : \subseteq \Sigma^\omega \to X$ can be defined by

$$\delta_X(01^{n_0} 01^{n_1} 01^{n_2} ...) := \lim_{i \to \infty} \alpha(n_i)$$

for all n_i such that $d(\alpha(n_i), \alpha(n_j)) \leq 2^{-i}$ for all $j > i$ (and undefined for all other input sequences). In the following we tacitly assume that computable metric spaces are represented by their Cauchy representation. If X is a computable metric space, then it is easy to see that $d : X \times X \to \mathbb{R}$ becomes computable. An important computable metric space is $(\mathbb{R}, d, \alpha_{\mathbb{Q}})$ with the Euclidean metric $d(x, y) := |x - y|$ and some standard numbering of the rational numbers, as $\alpha_{\mathbb{Q}}\langle i, j, k \rangle := (i - j)/(k + 1)$. Here, $\langle i, j \rangle := 1/2(i + j)(i + j + 1) + j$ denotes *Cantor pairs* and this definition is extended inductively to finite tuples. For short we will occasionally write $\overline{k} := \alpha_{\mathbb{Q}}(k)$. In the following we assume that $\mathbb{R}$ is endowed with the Cauchy representation $\delta_{\mathbb{R}}$ induced by the computable metric space given above. This representation of $\mathbb{R}$ can also be defined, if $(\mathbb{R}, d, \alpha_{\mathbb{Q}})$ just fulfills 1. and 2. of the definition above and this leads to a definition of computable real number sequences without circularity.

Other important representations cannot be deduced from computable metric spaces. Especially, we will use representations of the hyperspace of closed subsets $\mathcal{A}(X) := \{A \subseteq X : A \text{ closed}\}$ of a metric space X, which will be defined in the following sections. For a more comprehensive discussion of hyperspace representations, see [4]. Here, we just mention that we will denote the *open balls* of (X, d) by $B(x, \varepsilon) := \{y \in X : d(x, y) < \varepsilon\}$ for all $x \in X$, $\varepsilon > 0$ and correspondingly the *closed balls* by $\overline{B}(x, \varepsilon) := \{y \in X : d(x, y) \leq \varepsilon\}$. Occasionally, we denote complements of sets $A \subseteq X$ by $A^c := X \setminus A$.

3   First Computable Baire Category Theorem

For this section let (X, d, α) be some fixed complete computable metric space, and let $\mathcal{A} := \mathcal{A}(X)$ be the set of closed subsets. We can easily define a representation $\delta_{\mathcal{A}}^{>}$ of $\mathcal{A}$ by

$$\delta_{\mathcal{A}}^{>}(01^{\langle n_0, k_0 \rangle} 01^{\langle n_1, k_1 \rangle} 01^{\langle n_2, k_2 \rangle} \ldots) := X \setminus \bigcup_{i=0}^{\infty} B(\alpha(n_i), \overline{k_i}).$$

We write $\mathcal{A}_>$ to indicate that we use the represented space $(\mathcal{A}, \delta_{\mathcal{A}}^{>})$. The computable points $A \in \mathcal{A}_>$ are the so-called *co-r.e. closed* subsets of X. From results in [4] it directly follows that preimages of $\{0\}$ of computable functions are computable in $\mathcal{A}_>$. We formulate the result a bit more general.

Lemma 2. *The operation* $\mathcal{C}(X) \to \mathcal{A}_>, f \mapsto f^{-1}\{0\}$ *is computable and admits a computable right inverse.*

Using this fact we can immediately conclude that the union operation is computable on $\mathcal{A}_>$.

Proposition 3. *The operation* $\mathcal{A}_> \times \mathcal{A}_> \to \mathcal{A}_>, (A, B) \mapsto A \cup B$ *is computable.*

Proof. Using evaluation and type conversion w.r.t. $[\delta_X \to \delta_{\mathbb{R}}]$, it is straightforward to show that $\mathcal{C}(X) \times \mathcal{C}(X) \to \mathcal{C}(X), (f, g) \mapsto f \cdot g$ is computable, but if $f^{-1}\{0\} = A$ and $g^{-1}\{0\} = B$, then $(f \cdot g)^{-1}\{0\} = A \cup B$. Thus the desired result follows from the previous Lemma 2. $\square$

Since computable functions have the property that they map computable points to computable points, we can deduce that the class of co-r.e. closed sets is closed under intersection.

Corollary 4. *If* $A, B \subseteq X$ *are co-r.e. closed, then* $A \cup B$ *is co-r.e. closed too.*

Moreover, it is obvious that we can compute complements of open balls in the following sense.

Proposition 5. $\left(X \setminus B(\alpha(n), \overline{k})\right)_{\langle n,k\rangle \in \mathbb{N}}$ *is a computable sequence in* $\mathcal{A}_>$.

Using these both observations, we can prove the following first version of the computable Baire Category Theorem just by transferring the classical proof.

Theorem 6 (First computable Baire Category Theorem). *There exists a computable operation* $\Delta :\subseteq \mathcal{A}_>^{\mathbb{N}} \rightrightarrows X^{\mathbb{N}}$ *with the following property: for any sequence* $(A_n)_{n\in\mathbb{N}}$ *of closed nowhere dense subsets of* X*, there exists some sequence* $(x_n)_{n\in\mathbb{N}} \in \Delta(A_n)_{n\in\mathbb{N}}$ *and all such sequences* $(x_n)_{n\in\mathbb{N}}$ *are dense in* $X \setminus \bigcup_{n=0}^{\infty} A_n$.

Proof. Let us fix some $n = \langle n_1, n_2\rangle \in \mathbb{N}$. We construct sequences $(x_{n,k})_{k\in\mathbb{N}}$ in X and $(r_{n,k})_{k\in\mathbb{N}}$ in $\mathbb{Q}$ as follows: let $x_{\langle n_1,n_2\rangle,0} := \alpha(n_1)$, $r_{\langle n_1,n_2\rangle,0} := 2^{-n_2}$. Given $r_{n,i}$ and $x_{n,i}$ we can effectively find some point $x_{n,i+1} \in \text{range}(\alpha) \subseteq X$ and a rational $\varepsilon_{n,i+1}$ with $0 < \varepsilon_{n,i+1} \leq r_{n,i}$ such that

$$B(x_{n,i+1}, \varepsilon_{n,i+1}) \subseteq (X \setminus A_i) \cap B(x_{n,i}, r_{n,i}) = (A_i \cup X \setminus B(x_{n,i}, r_{n,i}))^c.$$

One the one hand, such a point and radius have to exist since A_i is nowhere dense and on the other hand, we can effectively find them, given a $\delta_{\mathcal{A}}^{>\mathbb{N}}$–name of the sequence $(A_n)_{n\in\mathbb{N}}$ and using Propositions 3 and 5. Now let $r_{n,i+1} := \varepsilon_{n,i+1}/2$. Altogether, we obtain a sequence of closed balls

$$\overline{B}(x_{n,i+1}, r_{n,i+1}) \subseteq \overline{B}(x_{n,i}, r_{n,i}) \subseteq ... \subseteq \overline{B}(x_{n,0}, r_{n,0})$$

with $r_{n,i} \leq 2^{-i}$ and thus $x_n := \lim_{i\to\infty} x_{n,i}$ exists since X is complete and the sequence $(x_{n,i})_{i\in\mathbb{N}}$ is even rapidly converging. Finally, the sequence $(x_n)_{n\in\mathbb{N}}$ is dense in $X \setminus \bigcup_{n=0}^{\infty} A_n$, since for any pair (n_1, n_2) we obtain by definition $x_{\langle n_1,n_2\rangle} \in B(\alpha(n_1), 2^{-n_2})$. Altogether, the construction shows how a Turing machine can transform each $\delta_{\mathcal{A}}^{>\mathbb{N}}$–name of a sequence $(A_n)_{n\in\mathbb{N}}$ into a δ_X–name of a suitable sequence $(x_n)_{n\in\mathbb{N}}$. $\qquad\square$

As a direct corollary of this uniformly computable version of the Baire Category Theorem we can conclude the following weak version.

Corollary 7. *For any computable sequence* $(A_n)_{n\in\mathbb{N}}$ *of co-r.e. closed nowhere dense subsets* $A_n \subseteq X$*, there exists some computable sequence* $(x_n)_{n\in\mathbb{N}}$ *which is dense in* $X \setminus \bigcup_{n=0}^{\infty} A_n$.

Since any computable sequence $(A_n)_{n\in\mathbb{N}}$ of co-r.e. closed nowhere dense subsets $A_n \subseteq X$ is "sequentially effectively nowhere dense" in the sense of Yasugi, Mori and Tsujii, we can conclude the previous corollary also from their effective Baire Category Theorem [14].

It is a well-known fact that the set of computable real numbers $\mathbb{R}_c$ cannot be enumerated by a computable sequence [13]. We obtain a new proof for this fact and a generalization for computable complete metric spaces without isolated points. First we prove the following simple proposition.

Proposition 8. *The operation* $X \to \mathcal{A}_>, x \mapsto \{x\}$ *is computable.*

Proof. This follows directly from the fact that $d : X \times X \to \mathbb{R}$ is computable and $\{x\} = X \setminus \bigcup \{B(\alpha(n), \overline{k}) : d(\alpha(n), x) > \overline{k}$ and $n, k \in \mathbb{N}\}$.   □

If X is a metric space without isolated points, then all singleton sets $\{x\}$ are nowhere dense closed subsets. This allows to combine the previous proposition with the computable Baire Category Theorem 7.

Corollary 9. *If X is a computable complete metric space without isolated points, then for any computable sequence $(y_n)_{n \in \mathbb{N}}$ in X, there exists a computable sequence $(x_n)_{n \in \mathbb{N}}$ in X such that $(x_n)_{n \in \mathbb{N}}$ is dense in $X \setminus \{y_n : n \in \mathbb{N}\}$.*

Using Theorem 6 it is straightforward to derive even a uniform version of this theorem which states that we can effectively find a corresponding sequence $(x_n)_{n \in \mathbb{N}}$ for any given sequence $(y_n)_{n \in \mathbb{N}}$. Instead of formulating this uniform version, we include the following corollary which generalizes the statement that $\mathbb{R}_c$ cannot be enumerated by a computable sequence.

Corollary 10. *If X is a computable complete metric space without isolated points, then there exists no computable sequence $(y_n)_{n \in \mathbb{N}}$ such that $\{y_n : n \in \mathbb{N}\}$ is the set of computable points of X.*

4   Computable but Nowhere Differentiable Functions

In this section we want to effectivize the standard example of an application of the Baire Category Theorem. We will show that there exists a computable but nowhere differentiable function $f : [0, 1] \to \mathbb{R}$. It is not to difficult to construct an example of such a function directly and actually, some typical examples of continuous nowhere differentiable functions, like *van der Waerden's function* $f : [0, 1] \to \mathbb{R}$ or *Riemann's function* $g : [0, 1] \to \mathbb{R}$ (cf. [9]), defined by

$$f(x) := \sum_{n=0}^{\infty} \frac{\langle 4^n x \rangle}{4^n} \quad \text{and} \quad g(x) := \sum_{n=0}^{\infty} \frac{\sin(n^2 \pi x)}{n^2},$$

where $\langle x \rangle := \min\{x - [x], 1 + [x] - x\}$ denotes the distance of x to the nearest integer, can easily be seen to be computable. The purpose of this section is rather to demonstrate that the computable version of the Baire Category Theorem can be applied in similar situations as the classical one.

In this section we will use the computable metric space of continuous functions $(\mathcal{C}[0, 1], d_{\mathcal{C}}, \alpha_{\mathbb{Q}[x]})$, where $d_{\mathcal{C}}$ denotes the *supremum metric*, which can be defined by $d_{\mathcal{C}}(f, g) := \max_{x \in [0,1]} |f(x) - g(x)|$ and $\alpha_{\mathbb{Q}[x]}$ denotes some standard numbering of the set $\mathbb{Q}[x]$ of rational polynomials $p : [0, 1] \to \mathbb{R}$. By $\delta_{\mathcal{C}}$ we denote the Cauchy representation of this space and in the following we tacitly assume that $\mathcal{C}[0, 1]$ is endowed with this representation. For technical simplicity we assume that functions $f : [0, 1] \to \mathbb{R}$ are actually functions $f : \mathbb{R} \to \mathbb{R}$ extended constantly, i.e. $f(x) = f(0)$ for $x \leq 0$ and $f(x) = f(1)$ for $x \geq 1$. It is well-known that a function $f : [0, 1] \to \mathbb{R}$ is $\delta_{\mathcal{C}}$–computable, if it is computable considered as a function $f : \mathbb{R} \to \mathbb{R}$ and we can actually replace δ_C by the restriction of $[\delta_{\mathbb{R}} \to \delta_{\mathbb{R}}]$ to $C[0,1]$ whenever it is helpful [13].

We will consider differentiability for functions $f : [0,1] \to \mathbb{R}$ only within $[0,1]$. If a function $f : [0,1] \to \mathbb{R}$ is differentiable at some point $t \in [0,1]$, then the quotient $\left| \frac{f(t+h)-f(t)}{h} \right|$ is bounded for all $h \neq 0$. Thus f belongs to the set

$$D_n := \left\{ f \in C[0,1] : (\exists t \in [0,1])(\forall h \in \mathbb{R} \setminus \{0\}) \left| \frac{f(t+h)-f(t)}{h} \right| \leq n \right\}$$

for some $n \in \mathbb{N}$. Because of continuity of the functions f, it suffices if the universal quantification over h ranges over some dense subset of $\mathbb{R} \setminus \{0\}$ such as $\mathbb{Q} + \pi$ in order to obtain the same set D_n.

It is well-known, that all sets D_n are closed and nowhere dense [7]. Thus, by the classical Baire Category Theorem, the set $C[0,1] \setminus \bigcup_{n=0}^{\infty} D_n$ is non-empty and there exists some continuous but nowhere differentiable function $f : [0,1] \to \mathbb{R}$. Our aim is to prove that $(D_n)_{n \in \mathbb{N}}$ is a computable sequence of co-r.e. closed nowhere dense subsets of $C[0,1]$, i.e. a computable sequence in $\mathcal{A}_>(C[0,1])$. Then we can apply the computable Baire Category Theorem 6 to ensure the existence of a computable but nowhere differentiable function $f : [0,1] \to \mathbb{R}$.

The crucial point is to get rid of the existential quantification of t over $[0,1]$ since arbitrary unions of co-r.e. closed sets need not to be (co-r.e.) closed again. The main tool will be the following Proposition which roughly speaking states that co-r.e. closed subsets are closed under parametrized countable and computable intersection and compact computable union.

Proposition 11. *Let (X, δ) be some represented space and let (Y, d, α) be some computable metric space.*

1. *If the function $A : X \times \mathbb{N} \to \mathcal{A}_>(Y)$ is computable, then the countable intersection $\cap A : X \to \mathcal{A}_>(Y), x \mapsto \bigcap_{n=0}^{\infty} A(x,n)$ is computable too.*

2. *If the function $U : X \times \mathbb{R} \to \mathcal{A}_>(Y)$ is computable, then the compact union $\cup U : X \to \mathcal{A}_>(Y), x \mapsto \bigcup_{t \in [0,1]} U(x,t)$ is computable too.*

Proof. 1. Let $A : X \times \mathbb{N} \to \mathcal{A}_>(Y)$ be computable. If for some fixed $x \in X$ we have $A(x,n) = Y \setminus \bigcup_{k=0}^{\infty} B(\alpha(i_{nk}), \overline{j_{nk}})$ with $i_{nk}, j_{nk} \in \mathbb{N}$ for all $n, k \in \mathbb{N}$, then

$$\bigcap_{n=0}^{\infty} A(x,n) = \bigcap_{n=0}^{\infty} \left(Y \setminus \bigcup_{k=0}^{\infty} B(\alpha(i_{nk}), \overline{j_{nk}}) \right) = Y \setminus \left(\bigcup_{\langle n,k \rangle=0}^{\infty} B(\alpha(i_{nk}), \overline{j_{nk}}) \right).$$

Thus, it is straightforward to show that $\cap A : X \to \mathcal{A}_>(Y)$ is computable too.

2. Now let $U : X \times \mathbb{R} \to \mathcal{A}_>(Y)$ be computable. Let $\delta_{[0,1]} :\subseteq \Sigma^{\omega} \to [0,1]$ be the *signed digit representation* of the unit interval, where $\Sigma = \{0, 1, -1\}$ and $\delta_{[0,1]}$

is defined in all possible cases by

$$\delta_{[0,1]}(p) := \sum_{i=0}^{\infty} p(i)2^{-i}.$$

It is known that $\mathrm{dom}(\delta_{[0,1]})$ is compact and $\delta_{[0,1]}$ is computably equivalent to the Cauchy representation $\delta_{\mathbb{R}}$, restricted to $[0,1]$ (cf. [13]). Thus, U restricted to $X \times [0,1]$ is $([\delta, \delta_{[0,1]}], \delta_{\mathcal{A}}^{>})$–computable. Then there exists some Turing machine M which computes a function $F :\subseteq \Sigma^{\omega} \to \Sigma^{\omega}$ which is a $([\delta, \delta_{[0,1]}], \delta_{\mathcal{A}}^{>})$–realization of $U : X \times \mathbb{R} \to \mathcal{A}_{>}(Y)$. Thus, for each given input sequence $\langle p, q \rangle \in \Sigma^{\omega}$ with $x := \delta(p)$ and $t := \delta_{[0,1]}(q)$ the machine M produces some output sequence $01^{\langle n_{q0}, k_{q0} \rangle} 01^{\langle n_{q1}, k_{q1} \rangle} 01^{\langle n_{q2}, k_{q2} \rangle}...$ such that

$$U(x,t) = Y \setminus \bigcup_{i=0}^{\infty} B(\alpha(n_{qi}), \overline{k_{qi}}).$$

Since we will only consider a fixed p, we do not mention the corresponding dependence in the indices of the values n_{qi}, k_{qi}. It is easy to prove that the set $W := \{w \in \Sigma^* : (\exists q \in \mathrm{dom}(\delta_{[0,1]}))\ w$ is a prefix of $q\}$ is recursive.

We will sketch the construction of a machine M' which computes the operation $\bigcup U : X \to \mathcal{A}_{>}(Y)$. On input p the machine M' works in parallel phases $\langle i, j, k \rangle = 0, 1, 2, ...$ and produces an output r. In phase $\langle i, j, k \rangle$ it simulates M on input $\langle p, w0^{\omega} \rangle$ for all words $w \in \Sigma^k \cap W$ and exactly k steps. Let $01^{\langle n_{w0}, k_{w0} \rangle} 01^{\langle n_{w1}, k_{w1} \rangle}...01^{\langle n_{wl_w}, k_{wl_w} \rangle}0$ be the corresponding output of M (more precisely: the longest prefix of the output which ends with 0). Then the machine M' checks whether for all $w \in \Sigma^k \cap W$ there is some $\iota_w = 0, ..., l_w$ such that $d(\alpha(i), \alpha(n_{wl_w})) + \overline{j} < \overline{k_{wl_w}}$ holds, which especially implies

$$B(\alpha(i), \overline{j}) \subseteq \bigcap_{w \in \Sigma^k \cap W} B(\alpha(n_{wl_w}), \overline{k_{wl_w}}) \subseteq \bigcap_{t \in [0,1]} Y \setminus U(x,t) = Y \setminus \bigcup U(x).$$

The verification is possible since (X, d, α) is a computable metric space. As soon as corresponding values ι_w are found for all $w \in \Sigma^k \cap W$, phase $\langle i, j, k \rangle$ is finished with extending the output by $01^{\langle i, j \rangle}$. Otherwise it might happen that the phase never stops, but other phases may run in parallel.

We claim that this machine M' actually computes $\bigcup U$. On the one hand, it is clear that $B(\alpha(i), \overline{j}) \subseteq Y \setminus \bigcup U(x)$ whenever $01^{\langle i, j \rangle}$ is written on the output tape by M'. Thus, if M' actually produces an infinite output r, then we obtain immediately $\delta_{\mathcal{A}}^{>}(r) \subseteq \bigcup U(\delta(p))$. On the other hand, let $y \in Y \setminus \bigcup U(\delta(p))$. Then for any $q \in \mathrm{dom}(\delta_{[0,1]})$ the machine M produces some output sequence $01^{\langle n_{q0}, k_{q0} \rangle} 01^{\langle n_{q1}, k_{q1} \rangle} 01^{\langle n_{q2}, k_{q2} \rangle}...$ and there has to be some l_q such that $y \in B(\alpha(n_{ql_q}), \overline{k_{ql_q}})$ and a finite number k of steps such that M produces $01^{\langle n_{ql_q}, k_{ql_q} \rangle}0$ on the output tape. Since $\mathrm{dom}(\delta_{[0,1]})$ is compact, there is even a common such k for all $q \in \mathrm{dom}(\delta_{[0,1]})$. Let $w' := w0^{\omega}$ for all $w \in \Sigma^*$. Then there exist $i, j \in \mathbb{N}$ such that

$$y \in B(\alpha(i), \overline{j}) \subseteq \bigcap_{w \in \Sigma^k \cap W} B(\alpha(n_{w'l_{w'}}), \overline{k_{w'l_{w'}}})$$

and $d(\alpha(i), \alpha(n_{w'l_{w'}})) + \bar{j} < \overline{k_{w'l_{w'}}}$. Thus M' will produce $01^{\langle i,j\rangle}$ on the output tape in phase $\langle i, j, k\rangle$. Altogether, this proves $\delta_{\mathcal{A}}^{>}(r) = \cup U(\delta(p))$ and thus the operation $\cup U : X \to \mathcal{A}_{>}(Y)$ is computable. $\qquad\square$

Now using this proposition, we can directly prove the desired result.

Theorem 12. *There exists a computable sequence $(f_n)_{n\in\mathbb{N}}$ of computable but nowhere differentiable functions $f_n : [0, 1] \to \mathbb{R}$ such that $\{f_n : n \in \mathbb{N}\}$ is dense in $\mathcal{C}[0, 1]$.*

Proof. If we can prove that $(D_n)_{n\in\mathbb{N}}$ is a computable sequence of co-r.e. nowhere dense closed sets, then Corollary 7 implies the existence of a computable sequence of computable functions f_n in $\mathcal{C}[0, 1] \setminus \bigcup_{n=0}^{\infty} D_n$. Since all somewhere differentiable functions are included in some D_n, it follows that all f_n are nowhere differentiable. Since it is well-known that all D_n are nowhere dense, it suffices to prove the computability property. We recall that it suffice to consider values $h \in \mathbb{Q} + \pi$ in the definition of D_n because of continuity of the functions f. We define a function $F : \mathbb{N} \times \mathbb{R} \times \mathbb{N} \times \mathcal{C}[0, 1] \to \mathbb{R}$ by

$$F(n, t, k, f) := \max\left\{\left|\frac{f(t + \bar{k} + \pi) - f(t)}{\bar{k} + \pi}\right| - n, 0\right\}.$$

Then using the evaluation property of $[\delta_{\mathbb{R}} \to \delta_{\mathbb{R}}]$, one can prove that F is computable. Using type conversion w.r.t. $[\delta_{\mathcal{C}} \to \delta_{\mathbb{R}}]$ one obtains computability of $\hat{F} : \mathbb{N} \times \mathbb{R} \times \mathbb{N} \to \mathcal{C}(\mathcal{C}[0, 1])$, defined by $\hat{F}(n, t, k)(f) := F(n, t, k, f)$. Using Lemma 2 we can conclude that the mapping $A : \mathbb{N} \times \mathbb{R} \times \mathbb{N} \to \mathcal{A}_{>}(\mathcal{C}[0, 1])$ with $A(n, t, k) := (\hat{F}(n, t, k))^{-1}\{0\}$ is computable. Thus by the previous proposition $\cap A : \mathbb{N} \times \mathbb{R} \to \mathcal{A}_{>}(\mathcal{C}[0, 1])$ is also computable and thus $\cup \cap A : \mathbb{N} \to \mathcal{A}_{>}(\mathcal{C}[0, 1])$ too. Now we obtain

$$\cup \cap A(n) = \bigcup_{t\in[0,1]} \bigcap_{k=0}^{\infty} \left\{f \in \mathcal{C}[0, 1] : \left|\frac{f(t + \bar{k} + \pi) - f(t)}{\bar{k} + \pi}\right| \le n\right\} = D_n.$$

Thus, $(D_n)_{n\in\mathbb{N}}$ is a computable sequence of co-r.e. closed subsets of $\mathcal{C}[0, 1]$. $\qquad\square$

5   Second Computable Baire Category Theorem

While the first version of the computable Baire Category Theorem has been proved by a direct adaptation of the classical proof, the second version will even be a consequence of the classical version. Whenever a classical theorem for complete computable metric spaces X, Y has the form

$$(\forall x)(\exists y)R(x, y)$$

with a predicate $R \subseteq X \times Y$ which can be proven to be r.e. open, then the theorem admits a computable multi-valued realization $F : X \rightrightarrows Y$ such that $R(x, y)$ holds for all $y \in F(x)$ (cf. the Uniformization Theorem 3.2.40 in [2]).

Actually, a computable version of the second formulation of the Baire Category Theorem, given in the Introduction, can be derived as such a direct corollary of the classical version.

Given a co-r.e. set $A \subseteq X$, the closure of its complement $\overline{A^c}$ needs not to be co-r.e. again (cf. Proposition 5.4 in [1]). Thus, the operation $\mathcal{A}_> \to \mathcal{A}_>, A \mapsto \overline{A^c}$ cannot be computable (and actually it is not even continuous in the corresponding way). In order to overcome this deficiency, we can simply include the information on $\overline{A^c}$ into a representing sequence of A. This is a usual trick in topology and computable analysis to make functions continuous or computable, respectively. So, if δ is an arbitrary representation of $\mathcal{A}$, then the representation δ^+ of $\mathcal{A}$, defined by

$$\delta^+ \langle p, q \rangle := A :\Longleftrightarrow \delta(p) = A \text{ and } \delta_{\mathcal{A}}^>(q) = \overline{A^c},$$

has automatically the property that $\mathcal{A} \to \mathcal{A}, A \mapsto \overline{A^c}$ becomes $(\delta^+, \delta_{\mathcal{A}}^>)$–computable. Here $\langle \ \rangle : \Sigma^\omega \times \Sigma^\omega \to \Sigma^\omega$ denotes some appropriate computable pairing function [13]. We can especially apply this procedure to $\delta := \delta_{\mathcal{A}}^>$. The corresponding $\delta_{\mathcal{A}}^{>+}$–computable sets $A \subseteq X$ are called *bi-co-r.e.* closed sets. In this case we write $\mathcal{A}_{>+}$ to denote the represented space $(\mathcal{A}, \delta_{\mathcal{A}}^{>+})$. Now we can directly conclude that the property "somewhere dense" is r.e.

Proposition 13. *The set $\{A \in \mathcal{A} : A \text{ is somewhere dense}\}$ is r.e. in $\mathcal{A}_{>+}$.*

The proof follows directly from the fact that a closed set $A \subseteq X$ is somewhere dense, if and only if there exist $n, k \in \mathbb{N}$ such that $B(\alpha(n), \overline{k}) \subseteq A^\circ = \overline{A^c}^{\;c}$. We can now directly conclude the second computable version of the Baire Category Theorem as a consequence of the classical version (and thus especially as a consequence of the first computable Baire Category Theorem 6).

Theorem 14 (Second computable Baire Category Theorem). *There exists a computable operation $\Sigma :\subseteq \mathcal{A}_{>+}^{\mathbb{N}} \rightrightarrows \mathbb{N}$ with the following property: for any sequence $(A_n)_{n \in \mathbb{N}}$ of closed subsets of X with $X = \bigcup_{n=0}^\infty A_n$, there exists some $\langle i, j, k \rangle \in \Sigma(A_n)_{n \in \mathbb{N}}$ and for all such $\langle i, j, k \rangle$ we obtain $B(\alpha(i), \overline{j}) \subseteq A_k$.*

Of course, if we replace $\mathcal{A}_{>+}$ by $(\mathcal{A}, \delta^+)$ with any other underlying representation δ instead of $\delta_{\mathcal{A}}^>$, then the theorem would also hold true. We mention that the corresponding constructive version of the theorem (Theorem 2.5 in [6]), if directly translated into a computable version, leads to a weaker statement than Theorem 14: if the sequence $(A_n)_{n \in \mathbb{N}}$ would be effectively given by the sequences of distance functions of A and A^c, this would constitute a stronger input information than it is the case if it is given by $\delta_{\mathcal{A}}^{>+}$. Now we can formulate a weak version of the second Baire Category Theorem.

Corollary 15. *For any computable sequence $(A_n)_{n \in \mathbb{N}}$ of bi-co-r.e. closed subsets $A_n \subseteq X$ with $X = \bigcup_{j=0}^\infty A_{\langle i,j \rangle}$ for all $i \in \mathbb{N}$, there exists a total computable function $f : \mathbb{N} \to \mathbb{N}$ such that $A_{\langle i, f(i) \rangle}$ is somewhere dense for all $i \in \mathbb{N}$.*

By applying some techniques from recursion theory [12,10], we can prove that the previous theorem and its corollary do not hold true with $\mathcal{A}_>$ instead of $\mathcal{A}_{>+}$. For this result we use as metric space the Euclidean space $X = \mathbb{R}$.

Theorem 16. *There exists a computable sequence $(A_n)_{n\in\mathbb{N}}$ of co-r.e. closed subsets $A_n \subseteq [0,1]$ with $[0,1] = \bigcup_{j=0}^{\infty} A_{\langle i,j\rangle}$ for all $i \in \mathbb{N}$ such that for every computable $f : \mathbb{N} \to \mathbb{N}$ there is some $i \in \mathbb{N}$ such that $A_{\langle i,f(i)\rangle}$ is nowhere dense.*

Proof. We use some total Gödel numbering $\varphi : \mathbb{N} \to P$ of the set of partial recursive functions $P := \{f :\subseteq \mathbb{N} \to \mathbb{N} : f \text{ computable}\}$ to define sets

$$A'_{\langle i,j\rangle} := \bigcup_{k=0}^{\min \varphi_i^{-1}\{j\}} \left\{\frac{m}{2^k} : m = 0, ..., 2^k\right\}.$$

For this definition we assume $\min \emptyset = \infty$. Whenever $i \in \mathbb{N}$ is the index of some total recursive function $\varphi_i : \mathbb{N} \to \mathbb{N}$ such that $\text{range}(\varphi_i) \neq \mathbb{N}$, then we obtain $\bigcup_{j=0}^{\infty} A'_{\langle i,j\rangle} = [0,1]$ and $A'_{\langle i,j\rangle}$ is somewhere dense, if and only if $j \notin \text{range}(\varphi_i)$. Using the smn-Theorem one can inductively prove that there is a total recursive function $r : \mathbb{N} \to \mathbb{N}$ such that $\varphi_{r\langle i,j\rangle}$ is total if φ_i is and

$$\text{range}(\varphi_{r\langle i,\langle k,\langle n_0,...,n_k\rangle\rangle\rangle}) = \text{range}(\varphi_i) \cup \{n_0, ..., n_k\}.$$

Let i_0 be the index of some total recursive function which enumerates some simple set $S := \text{range}(\varphi_{i_0})$ and define $A_{\langle i,j\rangle} := A'_{\langle r\langle i_0,i\rangle,j\rangle}$. Then $(A_n)_{n\in\mathbb{N}}$ is a computable sequence of co-r.e. closed subsets $A_n \subseteq [0,1]$. Let us assume that there exists a total recursive function $f : \mathbb{N} \to \mathbb{N}$ with the property that $A_{\langle i,f(i)\rangle}$ is somewhere dense for all $i \in \mathbb{N}$. Let $j_0 \in \mathbb{N} \setminus S$ and define a function $g : \mathbb{N} \to \mathbb{N}$ inductively by $g(0) := j_0$ and $g(n+1) := f(r\langle i_0, \langle n, \langle g(0), ...g(n)\rangle\rangle\rangle)$. Then g is computable and $\text{range}(g)$ is some infinite r.e. subset of the immune set $\mathbb{N} \setminus S$. Contradiction! □

The reader might notice that the constructed sequence $(A_n)_{n\in\mathbb{N}}$ is even a computable sequence of recursive closed sets (cf. [5,13]). Even a simpler variant of the same idea can be used to prove that in a well-defined sense there exists no continuous multi-valued operation $\Sigma :\subseteq \mathcal{A}_>^{\mathbb{N}} \rightrightarrows \mathbb{N}$ which meets the conditions of Theorem 14.

Unfortunately, the simplicity of the proof of the second computable Baire Category Theorem 14 corresponds to its uselessness. The type of information that one could hope to gain from an application of the theorem has already to be fed in by the input information. However, Theorem 16 shows that a substantial improvement of Theorem 14 seems to be impossible.

Acknowledgement

The author would like to thank Hajime Ishihara for an insightful discussion on the rôle of the Baire Category Theorem in constructive analysis and for pointing him to the corresponding section in [6].

References

1. V. Brattka. Computable invariance. *Theoretical Computer Science*, 210:3–20, 1999. 233
2. V. Brattka. Recursive and computable operations over topological structures. Informatik Berichte 255, FernUniversität Hagen, July 1999. Dissertation. 232
3. V. Brattka. Computability of Banach Space Principles. Informatik Berichte, FernUniversität Hagen, June 2001. 225
4. V. Brattka and G. Presser. Computability on subsets of metric spaces. submitted, 2000. 227
5. V. Brattka and K. Weihrauch. Computability on subsets of Euclidean space I: Closed and compact subsets. *Theoretical Computer Science*, 219:65–93, 1999. 234
6. D. Bridges and F. Richman. *Varieties of Constructive Mathematics*. Cambridge University Press, Cambridge, 1987. 224, 233, 234
7. C. Goffman and G. Pedrick. *First Course in Functional Analysis*. Prentince-Hall, Englewood Cliffs, 1965. 224, 230
8. K.-I. Ko. *Complexity Theory of Real Functions*. Birkhäuser, Boston, 1991. 224
9. S. S. Kutateladze. *Fundamentals of Functional Analysis*. Kluwer Academic Publishers, Dordrecht, 1996. 229
10. P. Odifreddi. *Classical Recursion Theory*. North-Holland, Amsterdam, 1989. 234
11. M. B. Pour-El and J. I. Richards. *Computability in Analysis and Physics*. Springer, Berlin, 1989. 224
12. K. Weihrauch. *Computability*. Springer, Berlin, 1987. 234
13. K. Weihrauch. *Computable Analysis*. Springer, Berlin, 2000. 224, 225, 226, 228, 230, 231, 233, 234
14. M. Yasugi, T. Mori, and Y. Tsujii. Effective properties of sets and functions in metric spaces with computability structure. *Theoretical Computer Science*, 219:467–486, 1999. 225, 228

Automata on Linear Orderings

Véronique Bruyère[1] and Olivier Carton[2]

[1] Institut de Mathématique et d'Informatique
Université de Mons-Hainaut, Le Pentagone,
6 avenue du Champ de Mars, B-7000 Mons, Belgium
Veronique.Bruyere@umh.ac.be
http://sun1.umh.ac.be/~vero/
[2] Institut Gaspard Monge, Université de Marne-la-Vallée
5 boulevard Descartes, F-77454 Marne-la-Vallée Cedex 2, France
Olivier.Carton@univ-mlv.fr
http://www-igm.univ-mlv.fr/~carton/

Abstract. We consider words indexed by linear orderings. These extend finite, (bi-)infinite words and words on ordinals. We introduce automata and rational expressions for words on linear orderings. We prove that for countable scattered linear orderings they are equivalent. This result extends Kleene's theorem. The proofs are effective.

1 Introduction

The theory of automata finds its origin in the paper of S. C. Kleene of 1956 where the basic theorem, known as Kleene's theorem, is proved for finite words [14]. Since then, automata working on infinite words, trees, traces ... have been proposed and this theory is a branch of theoretical computer science that has developed into many directions [16,20].

In this paper we focus on automata accepting objects which can be linearly ordered. Examples are finite, infinite, bi-infinite and ordinal words, where the associated linear ordering is a finite ordering, the ordering of the integers, the ordering of the relative integers and the ordering of an ordinal number respectively. Each such class of words has its automata and a Kleene-like theorem exists. Büchi introduced the so-called Büchi automata to show the decidability of the monadic second order theory of $\langle \mathbb{N}, < \rangle$ [7]. He later extended the method to countable ordinals thanks to appropriate automata [8]. Büchi already introduced ω-rational operations in [8]. Rational operations and a Kleene's theorem are proposed in [9] for words indexed by an ordinal less than ω^ω, they are extended to any countable ordinal in [22]. The case of bi-infinite words is solved in [15,11].

Our goal is a unified approach through the study of words indexed by any countable linear ordering. We propose a new notion of automaton which is simple, natural and includes previous automata. We also propose rational expressions and the related Kleene's theorem. Our constructions are effective.

Words indexed by a countable linear ordering were introduced in [10]. They are exactly the frontier of a labeled binary tree read from left to right. Some

J. Sgall, A. Pultr, and P. Kolman (Eds.): MFCS 2001, LNCS 2136, pp. 236–247, 2001.
© Springer-Verlag Berlin Heidelberg 2001

rational expressions are already studied in [10,13,19] with the nice property that they are total functions. For instance, an ω-power can be concatenated with a finite word and the resulting word can be iterated thanks to a reversed ω-power. These operations lead to a characterization of the words which are frontier of regular trees.

The rational operations that we propose include the usual union, concatenation, finite iteration and omega iteration, as well as the ordinal iteration introduced in [22] for ordinal words. There are also the reverse omega iteration and the reverse ordinal iteration to capture the left-infinite ordinal words and the bi-infinite words. A last operation is necessary which is the iteration for all linear orderings. This binary operation is subtle since it has to take into account the cuts of a linear ordering as defined in [17].

Our automata work as follows. Consider the case of a finite word w of length n. The underlying ordering is $1 < 2 < \cdots < n$. The $n+1$ states of a path for w are inserted between the letters of w, i.e., at the cuts of the ordering. For a word w indexed by a linear ordering, the associated path has its states indexed by the cuts of the ordering. The automaton has three types of transitions: the usual successor transitions, left limit and right limit transitions. For two states consecutive on the path, there must be a successor transition labeled by the letter in between. For a state q which has no predecessor on the path, the left limit (with respect to q) set P of states is computed and there must be a left limit transition between P and q. Right limit transitions are used when a state has no successor on the path. For a Muller automaton, such a left limit set P is nothing else than the states appearing infinitely often along the path. In our case, the path then ends with an additional left limit transition to a state q which is final.

Recently ordinal words (called Zeno words) were considered as modeling infinite sequences of actions which occur in a finite interval of time [12,4]. While the intervals of time are finite, infinite sequences of actions can be concatenated. A Kleene's theorem already exists for classical timed automata (where infinite sequences of actions are supposed to generate divergent sequences of times) [2,1,5]. In [4], automata of Choueka and Wojciechowski are adapted to Zeno words. A kind of Kleene's theorem is proved, that is, the class of Zeno languages is the closure under an operation called refinement of the class of languages accepted by classical timed automata.

The paper is organized as follows. We recall the basic notions on linear orderings in Section 2. Words indexed by linear orderings are introduced in Section 3 and the rational expressions in Section 4. The related automata are defined in Section 5. The main theorem is stated in Section 6. In the last section, we mention some open problems.

2   Linear Orderings

In this section, we recall the definitions and fix the terminology [17].

A *linear ordering* J is a set equipped with an ordering $<$ which is total, that is, for any $j \neq k$ in J, either $j < k$ or $k < j$ holds. The ordering of the integers, of the relative integers and of the rational numbers are linear orderings, respectively denoted by ω, ζ and η. We recall that an *ordinal* is a linear ordering which is well-ordered, that is, any nonempty subset has a least element. Notation $\mathcal{N}$ and $\mathcal{O}$ is used for the class of finite linear orderings and the class of countable ordinals. In the sequel, we freely say that two orderings are equal if they are actually isomorphic.

Let J and K be two linear orderings. We denote by $-J$ the backwards linear ordering obtained by reversing the ordering relation. The linear ordering $J + K$ is the ordering on the disjoint union $J \cup K$ extended with $j < k$ for any $j \in J$ and any $k \in K$. More generally, let K_j be a linear ordering for any $j \in J$. The linear ordering $\sum_{j \in J} K_j$ is the set of pairs (k, j) such that $k \in K_j$. The relation $(k_1, j_1) < (k_2, j_2)$ holds iff $j_1 < j_2$ or $j_1 = j_2$ and $k_1 < k_2$ in K_{j_1}.

Example 1. The ordering $-\omega + \omega$ is equal to the ordering ζ. The ordering ζ^2 is equal to the sum $\sum_{j \in \zeta} \zeta$. It is made of ζ copies of the ordering ζ.

A *Dedekind cut* or simply a *cut* of a linear ordering J is a pair (K, L) of intervals such that J is the disjoint union $K \cup L$ and for any $k \in K$ and $l \in L$, $k < l$. The set of all cuts of the ordering J is denoted by $\hat{J}$. We also denote by $\hat{J}^*$ the set $\hat{J} \setminus \{(\varnothing, J), (J, \varnothing)\}$. The set $\hat{J}$ can be linearly ordered as follows. For any cuts $c_1 = (K_1, L_1)$ and $c_2 = (K_2, L_2)$, we define the relation $c_1 < c_2$ iff $K_1 \subsetneq K_2$.

Example 2. Let J be the ordinal ω. The set $\hat{J}$ contains the cut $(\{0, \ldots, n - 1\}, \{n, n + 1, \ldots\})$ for any integer n and the last cut $(\omega, \varnothing)$. The ordering $\hat{J}$ is thus the ordinal $\omega + 1$. The ordering $\hat{J}$ for $J = \eta$ is not countable since it contains the usual ordering on the set of real numbers.

2.1 Scattered Linear Orderings

A linear ordering J is said to be *dense* if for any $i < k$ in J, there is $j \in J$ such that $i < j < k$. It is *scattered* if it contains no dense subordering. The following characterization of countable scattered linear orderings is due to Hausdorff.

Theorem 3 (Hausdorff). *A countable linear ordering J is scattered iff J belongs to $\bigcup_{\alpha \in \mathcal{O}} V_\alpha$ where the classes V_α are inductively defined by*

1. $V_0 = \{0, 1\}$;
2. $V_\alpha = \{\sum_{j \in J} K_j \mid J \in \mathcal{N} \cup \{\omega, -\omega, \zeta\}$ and $K_j \in \bigcup_{\beta < \alpha} V_\beta\}$.

The set of all countable scattered linear orderings is denoted by $\mathcal{S}$. The ordering η is not scattered. The ordering ζ^2 is scattered as it belongs to the class V_2. It can be proved that J is a countable scattered linear ordering iff $\hat{J}$ is a countable scattered linear ordering.

2.2   The Ordering $J \cup \hat{J}$

The orderings of J and $\hat{J}$ can be extended to an ordering on the disjoint union $J \cup \hat{J}$ as follows. For $j \in J$ and a cut $c = (K, L)$, define the relations $j < c$ and $c < j$ by respectively $j \in K$ and $j \in L$. Note that exactly one of these two relations holds since (K, L) is a partition of J. These relations together with the orderings of J and $\hat{J}$ endows $J \cup \hat{J}$ with a linear ordering.

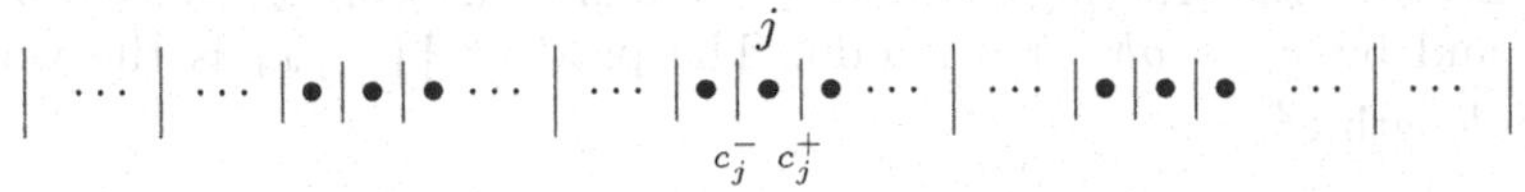

$$
\left| \cdots \left| \cdots \right| \bullet \left| \bullet \right| \bullet \cdots \left| \cdots \right| \bullet \left| \bullet \right| \bullet \bullet \cdots \right| \cdots \left| \bullet \right| \bullet \left| \bullet \right| \cdots \left| \cdots \right|
$$
$$
\underset{c_j^- \; c_j^+}{\overset{j}{}}
$$

Fig. 1. Ordering $J \cup \hat{J}$ for $J = \zeta^2$

Example 4. For the ordering $J = \zeta^2$, there is a cut between each consecutive elements in each copy of ζ, but there is also a cut between consecutive copies of ζ. There are also the first and the last cuts. The ordering $J \cup \hat{J}$ is pictured in Figure 1 where each element of J is represented by a bullet and each cut by a vertical bar.

For any element j, there are two consecutive cuts c_j^- and c_j^+ such that $c_j^- < j < c_j^+$. They are defined as $c_j^- = (K, \{j\} \cup L)$ and $c_j^+ = (K \cup \{j\}, L)$ with $K = \{k \mid k < j\}$ and $L = \{k \mid j < k\}$. We denote by $J \cup \hat{J}^*$ the set $J \cup \hat{J} \setminus \{(\varnothing, J), (J, \varnothing)\}$. It follows from Theorem 3 that if J is a countable scattered ordering, then both orderings $J \cup \hat{J}$ and $J \cup \hat{J}^*$ are also countable and scattered.

In the sequel, we only consider linear orderings which are *countable* and *scattered*. This hypothesis is needed in the proof of the main result (Theorem 14).

3   Words on Linear Orderings

Let A be a finite alphabet whose elements are called letters. For a linear ordering J, a *word* of *length* J over A is a function from J to A which maps any element j of J to a letter a_j of A. A word $(a_j)_{j \in J}$ on a linear ordering J can be seen as a labeled ordering where each point of J has been decorated by a letter. The word whose length is the empty set is called the *empty word* and it is denoted by ε. The set of all words is denoted by $A^\diamond$.

The notion of word we have introduced generalizes the notions of word already considered in the literature. If the ordering J is finite with n elements, a words of length J is a finite sequence $a_1 \ldots a_n$ of letters [16]. A word of length ω is an ω-sequence $a_0 a_1 a_2 \ldots$ of letters which is usually called an ω-word or an infinite word [20]. A word of length ζ is a sequence $\ldots a_{-2} a_{-1} a_0 a_1 a_2 \ldots$ of letters which is usually called a bi-infinite word. An ordinal word is a word indexed by a countable ordinal.

Example 5. The word $x = b^{-\omega} a b^\omega$ is the word of length $J = \zeta$ defined by $x_j = a$ if $j = 0$ and by $y_j = b$ otherwise.

Let $x = (a_j)_{j \in J}$ and $y = (b_k)_{k \in K}$ be two words of length J and K. The *product* xy (or the *concatenation*) of x and y is the word $z = (c_i)_{i \in J+K}$ of length $J + K$ such that $c_i = a_i$ if $i \in J$ and $c_i = b_i$ if $i \in K$. More generally, let J be a linear ordering and for each $j \in J$, let x_j be a word of length K_j. The *product* $\prod_{j \in J} x_j$ is the word z of length $K = \sum_{j \in J} K_j$ defined as follows. Suppose that each word x_j is equal to $(a_{k,j})_{k \in K_j}$ and recall that K is the set of all pairs (k, j) such that $k \in K_j$. The product z is then equal to $(a_{k,j})_{(k,j) \in K}$.

Example 6. Let J be the ordering ζ and for $j \in J$ define the word x_j by $x_j = b^{-\omega}$ if j is even and by $x_j = ab^\omega$ if j is odd. The product $\prod_{j \in J} x_j$ is the word $(b^{-\omega}ab^\omega)^\zeta$ of length ζ^2.

Two words $x = (a_j)_{j \in J}$ and $y = (b_k)_{k \in K}$ of length J and K are *isomorphic* if there is an ordering isomorphism f from J into K such that $a_j = b_{f(j)}$ for any j in J. This obviously defines an equivalence relation on words. In this paper, we identify isomorphic words and a word is actually a class of isomorphic words. It makes sense to identify isomorphic words since automata and rational expressions that we introduce do not distinguish isomorphic words.

Note that some orderings like ζ have non trivial internal isomorphisms. For instance, let x be the word of Example 5 and y be the word defined by $y_j = a$ if $j = 1$ and by $y_j = b$ otherwise. The two words x and y are isomorphic since the function f given by $f(x) = x + 1$ is an isomorphism from ζ to ζ.

4   Rational Expressions

In this section, the rational operations used to define rational sets of words on linear orderings are introduced. These operations include of course the usual Kleene operations for finite words which are the union $+$, the concatenation $\cdot$ and the star operation $*$. They also include the omega iteration ω usually used to construct ω-words and the ordinal iteration $\sharp$ introduced by Wojciechowski [22] for ordinal words. Three new operations are also needed: the backwards omega iteration $-\omega$, the backwards ordinal iteration $-\sharp$ and a last binary operation denoted $\diamond$ which is a kind of iteration for all linear orderings.

We first define the iterations in a unified framework. The general iteration $X^{\mathcal{J}}$ with respect to a class $\mathcal{J}$ of linear orderings can be defined by

$$X^{\mathcal{J}} = \{\prod_{j \in J} x_j \mid J \in \mathcal{J} \text{ and } x_j \in X\}.$$

The sets X^*, X^ω, $X^{-\omega}$, $X^\sharp$ and $X^{-\sharp}$ are then respectively equal to $X^{\mathcal{J}}$ for $\mathcal{J}$ equal to the class $\mathcal{N}$ of all finite linear orderings, the class $\{\omega\}$ which only contains the ordering ω, the class $\{-\omega\}$, the class $\mathcal{O}$ of all countable ordinals and the class $-\mathcal{O} = \{-\alpha \mid \alpha \in \mathcal{O}\}$.

We now define the binary operations. Let X and Y be two sets of words. The sets $X + Y$, $X \cdot Y$ and $X \diamond Y$ are defined by

$$X + Y = X \cup Y \quad \text{and} \quad X \cdot Y = \{xy \mid x \in X \text{ and } y \in Y\},$$

$$X \diamond Y = \{ \prod_{j \in J \cup \hat{J}^*} z_j \mid J \in \mathcal{S} \text{ and } z_j \in X \text{ if } j \in J \text{ and } z_j \in Y \text{ if } j \in \hat{J}^*\}.$$

A word x belongs to $X \diamond Y$ iff there is a countable scattered linear ordering J such that x is the product of a sequence of length $J \cup \hat{J}^*$ of words where each word indexed by an element of J belongs to X and each word indexed by a cut in $\hat{J}^*$ belongs to Y.

An abstract *rational expression* is a well-formed term of the free algebra over $A \cup \{\varepsilon\}$ with the symbols denoting the rational operations as function symbols. Each rational expression denotes a set of words which is inductively defined by the above definitions of the rational operations. A set of words is *rational* if it can be denoted by a rational expression. We use the abbreviation X^ς for $X^{-\omega} X^\omega$ and $X^\diamond$ for $X \diamond \varepsilon$.

Example 7. The expressions $A^\diamond$, $A^\diamond a A^\diamond$ respectively denote the set of all words and the set of words having an occurrence of the letter a. The expression $(A^\omega)^{-\omega}$ denote the set of words which are sequences of length $-\omega$ of ω-words. A word x belongs to the set denoted by $a^\varsigma \diamond b$ iff, for some linear ordering J, x is a sequence of length $J \cup \hat{J}^*$ of words x_j such that $x_j = a^\varsigma$ for any $j \in J$ and $x_j = b$ for any $j \in \hat{J}^*$.

Example 8. The expression AA^* denotes the set of nonempty finite words and the expression $(A^\diamond)^\omega A^\diamond + A^\diamond (A^\diamond)^{-\omega}$ denotes its complement. Indeed, a linear ordering $J \neq \varnothing$ is not finite if it has at least a cut (K, L) such that either K does not have a greatest element or L does not have a least element. The rational expression $(A^\diamond)^\omega$ denotes the set of words whose length does not have a last element. Therefore, the expression $(A^\diamond)^\omega A^\diamond$ denotes the set whose length has a cut (K, L) such that K does not have a greatest element. Symmetrically, the expression $A^\diamond (A^\diamond)^{-\omega}$ denotes the set whose length has a cut (K, L) such that L does not have a least element.

5   Automata

In this section, automata on words on linear orderings are defined. They are a natural generalization of Muller automata on ω-words and of automata introduced by Büchi [8] on ordinal words. The latter automata are usual (Kleene) automata with additional limit transitions of the form $P \to p$ used for limit ordinals. The automata that we introduce have limit transitions of the form $P \to p$ as well as of the form $p \to P$. We point out that these automata make sense for all linear orderings.

Definition 9. *Let A be a finite alphabet. An automaton $\mathcal{A}$ over A is a 4-tuple (Q, E, I, F) where Q is a finite set of states, $E \subseteq (Q \times A \times Q) \cup (\mathcal{P}(Q) \times Q) \cup (Q \times \mathcal{P}(Q))$ is the set of transitions, $I \subseteq Q$ is the set of initial states and $F \subseteq Q$ is the set of final states.*

Since the alphabet and the set of states are finite, the set of transitions is also finite. Transitions are either *successor* transitions of the form $p \xrightarrow{a} q$, or *left limit* transitions of the form $P \to q$, or *right limit* transitions of the form $q \to P$, where P is a subset of Q.

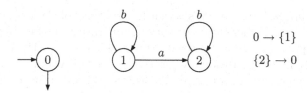

$$0 \to \{1\}$$
$$\{2\} \to 0$$

Fig. 2. Automaton of example 10

Example 10. The automaton pictured in Figure 2 has 3 successor transitions which are pictured like a labeled graph. It has also a left limit transition $\{2\} \to 0$ and a right limit transition $0 \to \{1\}$. The state 0 is the only initial state and the only final state.

In order to define the notion of path in such an automaton, the following notion of limits is needed. We define it for an arbitrary linear ordering J but we use it when the considered ordering is actually the ordering $\hat{J}$ of cuts of a given ordering J. Let Q be a finite set, let J a linear ordering and let $\gamma = (q_j)_{j \in J}$ be a word over Q. Let j be a fixed element of J. The *left* and *right limit set* of γ at j are the two subsets $\lim_{j^-} \gamma$ and $\lim_{j^+} \gamma$ of Q defined as follows.

$$\lim_{j^-} \gamma = \{q \in Q \mid \forall k < j \; \exists i \quad k < i < j \text{ and } q = q_i\},$$

$$\lim_{j^+} \gamma = \{q \in Q \mid \forall j < k \; \exists i \quad j < i < k \text{ and } q = q_i\}.$$

Note that if the element j has a predecessor, the limit set $\lim_{j^-} \gamma$ is empty. Conversely, if the element j is not the least element of J and if it has no predecessor, the limit set $\lim_{j^-} \gamma$ is nonempty since the set Q is finite. Similar results hold for right limit sets.

We now come to the definition of a path in an automaton on linear orderings. Let x be a word of length J. Roughly speaking, a path associated with x is a labeling of each cut of J by a state of the automaton such that local properties are satisfied. If two cuts are consecutive there must be a successor transition labeled by the letter of x in between the two cuts. If a cut does not have a predecessor or a successor, there must be a limit transition between the left or right limit set of states and the state of the cut.

Definition 11. *Let $\mathcal{A}$ be an automaton and let $x = (a_j)_{j \in J}$ be a word of length J. A path γ labeled by x is a sequence of states $\gamma = (q_c)_{c \in \hat{J}}$ of length $\hat{J}$ such that*

- *For any consecutive cuts c_j^- and c_j^+, $q_{c_j^-} \xrightarrow{a_j} q_{c_j^+}$ is a successor transition.*
- *For any cut c which is not the first cut and which has no predecessor, $\lim_{c^-} \gamma \to q_c$ is a left limit transition.*
- *For any cut c which is not the last cut and which has no successor, $q_c \to \lim_{c^+} \gamma$ is a right limit transition.*

By the previous definition, there is a transition entering the state q_c for any cut c which is not the first cut. This transition is a successor transition if the cut c has a predecessor in $\hat{J}$ and it is a left limit transition otherwise. Similarly, there is a transition leaving q_c for any cut c which is not the last cut.

Since the ordering $\hat{J}$ has a least and a greatest element, a path always has a first and a last state which are indexed by the first and the last cut. A path is *successful* iff its first state is initial and its last state is final. A word is *accepted* by the automaton iff it is the label of a successful path. A set of words is *recognizable* if it is the set of words accepted by some automaton.

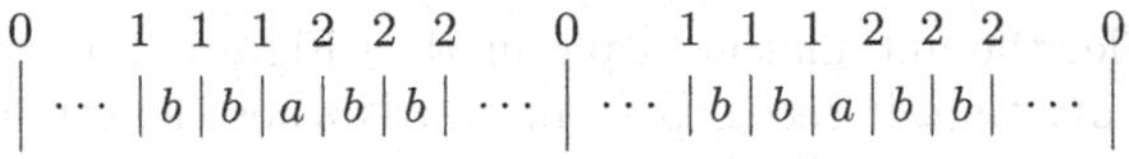

$$
\begin{array}{ccccccccccccccc}
0 & & 1 & 1 & 1 & 2 & 2 & 2 & 0 & & 1 & 1 & 1 & 2 & 2 & 2 & & 0
\end{array}
$$

$$
\Big| \; \cdots \; \big| b \big| b \big| a \big| b \big| b \big| \; \cdots \; \Big| \; \cdots \; \big| b \big| b \big| a \big| b \big| b \big| \; \cdots \; \Big|
$$

Fig. 3. A path labeled by $(b^{-\omega} a b^{\omega})^2$

Example 12. Consider the automaton $\mathcal{A}$ of Figure 2 and let x be the word $(b^{-\omega} a b^{\omega})^2$ of length $\zeta + \zeta$. A successful path γ labeled by x is pictured in Figure 3. This path is made of two copies of the path $01^{-\omega} 2^{\omega} 0$. A path for the word $(b^{-\omega} a b^{\omega})^\zeta$ cannot be made by ζ copies of $01^{-\omega} 2^{\omega} 0$ because the right limit set of the first state would be $\{0, 1, 2\}$ and the automaton has no transition of the form $q \to \{0, 1, 2\}$. The automaton $\mathcal{A}$ thus recognizes the set $(b^{-\omega} a b^{\omega})^*$.

The notion of path we have introduced for words on orderings coincide with the usual notion of paths considered in the literature for finite words, ω-words and ordinal words. Let x be a finite word $a_1 \ldots a_n$. The set of cuts of the finite ordering $\{1, \ldots, n\}$ can be identified with $\{1, \ldots, n+1\}$. In our setting, a path labeled by x is then a finite sequence $q_1, \ldots, q_{n+1}$ of states such that $q_j \xrightarrow{a_j} q_{j+1}$ is a successor transition for any j in $\{1, \ldots, n\}$. This matches the usual definition of a path in an automaton [16, p. 5].

Let $x = a_0 a_1 a_2 \ldots$ be an ω-word. The set of cuts of the ordering $J = \omega$ is the ordinal $\omega + 1 = \{0, 1, 2, \ldots, \omega\}$ (see Example 2). The pairs of consecutive cuts are the pairs $(j, j+1)$ for $j < \omega$ whereas the cut $c = \omega$ has no predecessor. In our setting, a path γ labeled by x is a sequence $q_0, q_1, q_2, \ldots, q_\omega$ of states such that $q_j \xrightarrow{a_j} q_{j+1}$ is a successor transition for any $j < \omega$ and such that $\lim_{\omega^-} \gamma \to q_\omega$ is a left limit transition. Note that $\lim_{\omega^-} \gamma$ is the set of states which occur infinitely many times in γ. This path is successful iff q_0 is initial and q_ω is final. Define the family $\mathcal{T}$ of subsets of states by

$$\mathcal{T} = \{P \mid \exists q \in F \text{ such that } P \to q \in E\}.$$

The path γ is then successful iff q_0 is initial and if the set $\lim_{\omega^-} \gamma$ of states belongs to the family $\mathcal{T}$. This matches the definition of a successful path in a Muller automaton [20, p. 148].

The set of cuts of an ordinal α is the ordinal $\alpha + 1$. Therefore, the notion of path we have introduced coincide for ordinal words with the notion of path considered in [3].

$$\{0\} \to 0$$

Fig. 4. Automaton of example 13

Example 13. Consider the automaton $\mathcal{A}$ pictured in Figure 4. This automaton has no right limit transition. It recognizes the words whose length is an ordinal since a linear ordering J is an ordinal iff any of its cuts (except the last one) has a successor in $\hat{J}$. The automaton obtained by suppressing the left limit transition of $\mathcal{A}$ recognizes the set of finite words since a linear ordering J is finite iff any of its cuts (except the first or the last one) has a successor and a predecessor in $\hat{J}$.

6   Rational Expressions vs Automata

In this section, we state that rational expressions and automata are equivalent. This result extends the well-known Kleene's theorem on finite words. It has been first extended to words of length ω by Büchi [6] and to words of ordinal length by Wojciechowski [22]. The proof that we propose is effective.

Theorem 14. *A set of words on countable scattered linear orderings is rational iff it is recognizable.*

The proof that any rational set of words is recognized by an automaton is by induction on the rational expression denoting the set. For each rational operation, we describe a corresponding construction for the automata. The constructions for the union, the concatenation and the finite iteration are very similar to the classical ones for automata on finite words [16, p. 15].

The proof that any set of words recognized by an automaton is rational is by induction on the number of states of the automaton. It is a generalization of McNaughton and Yamada algorithm and it is the most difficult part.

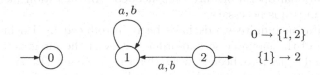

$$0 \to \{1, 2\}$$
$$\{1\} \to 2$$

Fig. 5. Automaton recognizing $(A^\omega)^{-\omega}$

Example 15. The automaton pictured in Figure 5 recognizes the set denoted by the rational expression $(A^\omega)^{-\omega}$. The part of the automaton given by state 1 and the left limit transition $\{1\} \to 2$ accepts the set A^ω. The successor transition from state 2 to state 1 allows to concatenate two words of A^ω. The right limit transition $0 \to \{1, 2\}$ leads to a sequence of length $-\omega$ of words of A^ω.

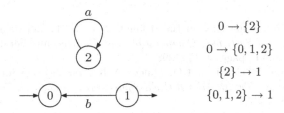

$$0 \to \{2\}$$
$$0 \to \{0, 1, 2\}$$
$$\{2\} \to 1$$
$$\{0, 1, 2\} \to 1$$

Fig. 6. Automaton recognizing $a^\varsigma \diamond b$

Example 16. The automaton pictured in Figure 6 recognizes the set $a^\varsigma \diamond b$. The part of the automaton given by state 2 and the two limit transitions $0 \to \{2\}$ and $\{2\} \to 1$ accepts the word a^ς whereas the part given by the successor transition from state 1 to state 0 accepts the word b. Any occurrence of a^ς is preceded and followed by an occurrence of b in the automaton. More generally, thanks to the limit transitions $0 \to \{0, 1, 2\}$ and $\{0, 1, 2\} \to 1$, the occurrences of a^ς are indexed by a linear ordering J, the occurrences of b are indexed by the ordering $\hat{J}^*$ and they are interleaved according to the ordering $J \cup \hat{J}^*$.

Conclusion

In this paper, we have introduced automata and rational expressions for words on linear orderings. For words on countable and scattered linear orderings, these two notions are equivalent. This result extends the usual Kleene's theorem for finite words.

We mention some open problems. A natural generalization of the result would be to remove the restrictions on the orderings and to first consider words on countable linear orderings and then words on all linear orderings. Automata that we have introduced are suitable for all linear orderings. It seems however that new rational operations are then needed. An operation like the η-shuffle introduced in [13] is necessary.

Automata on infinite words have been introduced by Büchi to prove the decidability of the monadic second-order theory of the integers [6]. Since then, automata and logics have been shown to have strong connections [20]. It is known that the monadic second-order theory of all linear orderings is decidable [18] but the proof is based on a model theoretic approach, called the composition method, avoiding the use of automata. We refer to [21] for a review of the composition method and the Shelah's proof of Büchi's decidability result. Therefore, the next step is to investigate the connections between logics and the automata that we have introduced. Such a study has to begin with the closure of the class of recognizable sets under the boolean operations. We do not know whether this class is closed under the complementation.

References

1. E. Asarin. Equations on timed languages. In T. Henzinger and S. Sastry, editors, *Hybrid Systems : Computation and Control*, number 1386 in Lect. Notes in Comput. Sci., pages 1–12, 1998. 237

2. E. Asarin, P. Caspi, and O. Maler. A Kleene theorem for timed automata. In *Proceedings, Twelth Annual IEEE Symposium on Logic in Computer Science*, pages 160–171, 1997. 237

3. N. Bedon and O. Carton. An Eilenberg theorem for words on countable ordinals. In Cláudio L. Lucchesi and Arnaldo V. Moura, editors, *Latin'98: Theoretical Informatics*, volume 1380 of *Lect. Notes in Comput. Sci.*, pages 53–64. Springer-Verlag, 1998. 244

4. B. Bérard and C. Picaronny. Accepting Zeno words without making time stand still. In *Mathematical Foundations of Computer Science 1997*, volume 1295 of *Lect. Notes in Comput. Sci.*, pages 149–158, 1997. 237

5. P. Bouyer and A. Petit. A Kleene/Büchi-like theorem for clock languages. *J. of Automata, Languages and Combinatorics*, 2001. To appear. 237

6. J. R. Büchi. Weak second-order arithmetic and finite automata. *Z. Math. Logik und grundl. Math.*, 6:66–92, 1960. 244, 246

7. J. R. Büchi. On a decision method in the restricted second-order arithmetic. In *Proc. Int. Congress Logic, Methodology and Philosophy of science, Berkeley 1960*, pages 1–11. Stanford University Press, 1962. 236

8. J. R. Büchi. Transfinite automata recursions and weak second order theory of ordinals. In *Proc. Int. Congress Logic, Methodology, and Philosophy of Science, Jerusalem 1964*, pages 2–23. North Holland, 1965. 236, 241

9. Y. Choueka. Finite automata, definable sets, and regular expressions over ω^n-tapes. *J. Comput. System Sci.*, 17(1):81–97, 1978. 236

10. B. Courcelle. Frontiers of infinite trees. *RAIRO Theoretical Informatics*, 12(4):319–337, 1978. 236, 237

11. D. Girault-Beauquier. Bilimites de langages reconnaissables. *Theoret. Comput. Sci.*, 33(2–3):335–342, 1984. 236

12. M. R. Hansen, P. K. Pandya, and Z. Chaochen. Finite divergence. *Theoret. Comput. Sci.*, 138(1):113–139, 1995. 237

13. S. Heilbrunner. An algorithm for the solution of fixed-point equations for infinite words. *RAIRO Theoretical Informatics*, 14(2):131–141, 1980. 237, 246
14. S. C. Kleene. Representation of events in nerve nets and finite automata. In C. E. Shannon, editor, *Automata studies*, pages 3–41. Princeton University Press, Princeton, 1956. 236
15. M. Nivat and D. Perrin. Ensembles reconnaissables de mots bi-infinis. In *Proceedings of the Fourteenth Annual ACM Symposium on Theory of Computing*, pages 47–59, 1982. 236
16. D. Perrin. Finite automata. In J. van Leeuwen, editor, *Handbook of Theoretical Computer Science*, volume B, chapter 1, pages 1–57. Elsevier, 1990. 236, 239, 243, 244
17. J. G. Rosenstein. *Linear ordering*. Academic Press, New York, 1982. 237
18. S. Shelah. The monadic theory of order. *Annals of Mathematics*, 102:379–419, 1975. 246
19. W. Thomas. On frontiers of regular sets. *RAIRO Theoretical Informatics*, 20:371–381, 1986. 237
20. W. Thomas. Automata on infinite objects. In J. van Leeuwen, editor, *Handbook of Theoretical Computer Science*, volume B, chapter 4, pages 133–191. Elsevier, 1990. 236, 239, 244, 246
21. W. Thomas. Ehrenfeucht games, the composition method, and the monadic theory of ordinal words. In *Structures in Logic and Computer Science, A Selection of Essays in Honor of A. Ehrenfeucht*, number 1261 in Lect. Notes in Comput. Sci., pages 118–143. Springer-Verlag, 1997. 246
22. J. Wojciechowski. Finite automata on transfinite sequences and regular expressions. *Fundamenta Informaticæ*, 8(3-4):379–396, 1985. 236, 237, 240, 244

Algorithmic Information Theory and Cellular Automata Dynamics

Julien Cervelle, Bruno Durand, and Enrico Formenti

LIM - CMI
39 rue Joliot-Curie, 13453 Marseille Cedex 13, France
{cervelle,bdurand,eforment}@cmi.univ-mrs.fr

Abstract. We study the ability of discrete dynamical systems to transform/generate randomness in cellular spaces. Thus, we endow the space of bi-infinite sequences by a metric inspired by information distance (defined in the context of Kolmogorov complexity or algorithmic information theory). We prove structural properties of this space (non-separability, completeness, perfectness and infinite topological dimension), which turn out to be useful to understand the transformation of information performed by dynamical systems evolving on it. Finally, we focus on cellular automata and prove a dichotomy theorem: continuous cellular automata are either equivalent to the identity or to a constant one. This means that they cannot produce any amount of randomness.

Keywords: Kolmogorov complexity, topology, cellular automata, discrete dynamical systems

1 Introduction

Cellular automata are formal models for complex systems. They were introduced by J. von Neumann for modeling cellular growth and self-replication. Afterwards, they have been successfully applied in a large number of scientific disciplines ranging from mathematics to computer science, from physics to chemistry and geology.

They are essentially a massive parallel model made of a lattice of elementary identical finite state machines (automata), usually called *cells*. Cells are updated synchronously according to a finite set of local rules that compute the new state of the cell from the states of a finite set of neighboring cells (eventually including the cell itself). A snapshot of the states of the cells is called a *configuration*. A stack of configurations in which each one is obtained from the preceding, applying the updating rule, is called an *evolution*.

The success of the model relies essentially on its simple definition coupled with the rich variety of different evolutions. Many evolutions display a complex or random behavior. It is a recurring question in the community to understand if the complexity or randomness in such evolutions is an intrinsic property of the model or if it is a side effect due to inappropriate topological context, or,

J. Sgall, A. Pultr, and P. Kolman (Eds.): MFCS 2001, LNCS 2136, pp. 248–260, 2001.

which is worst, due to unsuitable complexity definitions. In this paper, we study the complexity of cellular automata evolutions in the general context of discrete dynamical systems.

Dynamical systems are mathematical representations of dynamical processes appearing in many scientific fields. A large quantity of literature on the subject has been published in recent years. The contribution of computer science to this discipline is mainly devoted to the domain of validation and prediction. In this context, a fundamental issue is to understand the "complexity" of the system. Complexity may have different definitions according to properties investigated. In our context, "complex" is defined via algorithmic information theory since we are interested in the ability of systems to transform information.

The idea to use Kolmogorov complexity to study dynamical systems is not new (see [5,7,25,10]). Consider a finite open partition of a compact space X ($P_0 \cup P_1 = X$ for simplicity) and a dynamical system f on X (*i.e.* a continuous function from X to X). The orbit of a point $x_0 \in X$ is associated with a sequence $c_{x_0} \in \{0,1\}^N$ as follows: $c_{x_0}(i) = \alpha$ if $f^i(x_0) \in P_\alpha$, where $\alpha \in \{0,1\}$ and f^i is the i-fold composition of f with itself. We can define a *labeling function* $\ell : X \to \{0,1\}^N$ as $\ell(x) = c_x$ (see Fig. 1). It is natural to expect that the more complex the system, the more complex the labeling.

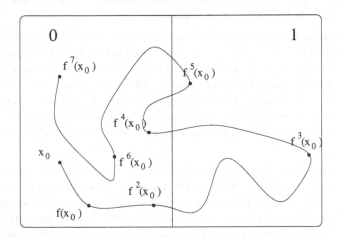

Fig. 1. An example of orbit labeling. Here $\ell(x_0) = 00010100\ldots$.

In [5], for $x \in X$, Brudno defines the complexity of the orbit O_x as follows:

$$K(O_x) = \limsup_{n \to \infty} \frac{K(\ell(x)_{0:n})}{n+1}.$$

where $\ell(x)_{0:n}$ are the first $n+1$ bits of $\ell(x)$, and $K(x)$ stands for the Kolmogorov Complexity of x, which is defined section 2.

It is well-known (see [5]) that, in a measure-theoretic sense, almost every orbit of a homeomorphism has maximal complexity. In particular, the *shift* map has maximal complexity according to this approach. This fact is surprising since for intuition the shift map is a very "simple" cellular automaton. A rough explanation is that all the complexity is contained in the initial configuration and the shift map preserves it. We would like to conclude that the complexity observed is not an intrinsic property of this system, but it is due to the structure of the topological space in which the system is embedded. For this reason, we perform a deep change of the topological space where the dynamic and the complexity of cellular automata is studied. In this way we hope that the intrinsic complexity of the cellular automata is not biased by the complexity of the initial configuration. This deep change undergoes several steps. First, we define a new topology and show that it is suitable for the study of dynamical systems on it. As a second step, we give a natural definition of complexity. Finally, we prove that cellular automata are not able to continuously modify this complexity.

In the case of cellular automata, two completely different approaches are possible. One is exposed by Calude *et al.* in [7], the latter in this paper. Roughly speaking both draw the same conclusions:

  i) randomness is preserved by surjective CA;
 ii) non-surjective CA destroy randomness;
iii) all CA preserve non-randomness (this result remains open for dimensions greater than one in Calude *et al.*'s approach).

These results confirm the importance of the surjectivity for CA and give a new perspective to previous work [20,21,19,13,11]. Note that Calude *et al.* approach and ours are rather different. In particular, in [7], the third question (see above) is open for CA in dimensions greater than 2, while our proofs can be easily extended to higher dimensions. Moreover, their approach seems to work only in separable spaces. Note also that many recent works on CA dynamics have begun to study CA on unseparable topological spaces (Weyl and Besicovitch topologies) [16,1,8,4].

The paper is structured as follows. Section 2 introduces basic notions of algorithmic complexity and the fundamental definition of incompressible strings. Section 3 is essentially devoted to the study of a new metric space whose distance is based on *information distances*. In this section, we intend, by topological results, to prove that this space is neither too simple (countable, separable, finite dimensional) nor too complex (not connected, not perfect). Finally, Section 4 is devoted to cellular automata and uses the results of Section 3 to prove our main theorems.

Some more ambitious use of our work would be to apply it for general dynamical systems, or to study the relationships with classical ergodic theory.

2   Algorithmic Complexity

Measuring complexity of mathematical objects is not easy but very fruitful for new ideas and applications.

In this paper we are interested in the complexity of *configurations* i.e. bi-infinite sequences of symbols chosen from a finite alphabet (usually $\{0,1\}$). In this context complexity is measured using the approach outlined by Kolmogorov [14] and then Chaitin [9]. A more modern presentation and a state of art on the subject are given in [15,6].

We will note X the set of all the configurations over the alphabet $\{0,1\}$. $\underline{0}$ is the configuration made only of 0s. Similarly, $\underline{1}$ is the configuration made only of 1s. For any bi-infinite sequence, or any finite word u, u_a is the letter of u at position a and $u_{a:b}$, $a, b \in \mathbb{Z}$ is the subword $u_{a:b} = u_a u_{a+1} \dots u_{b-1} u_b$.

Given a word w of symbols from a finite alphabet S (that we call the set of *states*), its complexity is the shortest program p which "computes" w. We say that a program p "computes" a word w on input z, if $\varphi(\langle p, z \rangle) = w$, where φ is a fixed computable function (total or partial), and $x, y \mapsto \langle x, y \rangle$ is any fixed computable bijection pairing. In other words, p is a *description* for w, when z is given, according to the "description mode" φ (see [24]).

Definition 1. *Let x, y and p be finite words over some fixed finite alphabet, and φ be a computable function. The Kolmogorov complexity K_φ of x conditional to y w.r.t. the description mode φ is*

$$K_\varphi(x|y) = min\{|p|, \varphi(\langle y, p \rangle) = x\},$$

and $K_\varphi(x|y) = \infty$ if there is no such p. $x, y \mapsto \langle x, y \rangle$ is any fixed computable bijection pairing.

The previous definition depends on the choice of the reference description mode. The following fundamental result overcomes this inconvenience, since it proves the existence of an "optimal" description mode.

Theorem 2 (Optimality theorem). *There is a description mode φ_0 such that for all computable function φ, there is a constant c $\forall x, y$, $K_{\varphi_0}(x|y) \leqslant K_\varphi(x|y) + c$. Such a mode φ_0 is said to be* additively optimal.

Definition 3. *Fix an additively optimal mode φ_0 as reference for K. We define conditional Kolmogorov complexity as $K(x|y) = K_{\varphi_0}(x|y)$. The unconditional Kolmogorov complexity K is simply $K(x) = K(x|\varepsilon)$ where ε denotes the empty word.*

A description of a word can be interpreted as a compressed version of itself. Thus Kolmogorov complexity and the notion of compression are deeply related. Kolmogorov proposed to define that a word w is *c-incompressible* if $|w| - K(w)$ is less than c. Martin-Löf proved in [17] that a word is *c-random* if it cannot be selected by any possible computable tests at level of precision c (see [15] page 129 for definitions).

Working with infinite sequences is more satisfactory since the constant c disappears when taking limits. For infinite sequences, the very important Levin-Schnorr theorem proves the equivalence between the class of Martin-Löf random words and the class of incompressible ones. This theorem has many applications in combinatorial complexity, and we refer to the books [15,6] and to the review paper [23] for further details. The results of this theory are central for several of our proofs.

Definition 4. *A configuration or a sequence x is called* random *if there is a constant c such that $\forall n$, $K(x_{-n:n}) \geqslant 2n - c$.*

Definition 5. *A configuration u is called* weakly-incompressible *if*

$$\limsup_{n \to \infty} \frac{K(u_{-n:n})}{2n + 1} = 1.$$

Remark 6. The set of weakly-incompressible configurations strictly contains the set of random sequences [15] with respect to uniform Bernoulli measure μ. Thus, μ-almost all configurations are weakly-incompressible. □

3   Metrics on Space of Configurations

A minimal property that a complexity distance d must satisfy is that

$$d(u, \sigma(u)) = 0 \tag{1}$$

where σ is the usual shift operation on configurations. The point is that shifting a configuration must not change its informational content.

In [3] information distances over finite words have been introduced and studied. A fundamental result of [3] is that all such distances are equal up to a logarithmic term (which is negligible in the infinite case) or do not satisfy property 1. As a consequence, we consider only one of those in [3]: $\forall x, y \in X$,

$$d(x, y) = \limsup_{n \to \infty} \frac{K(x_{-n:n}|y_{-n:n}) + K(y_{-n:n}|x_{-n:n})}{4n + 1}.$$

As defined, d is not a distance over X since there are distinct configurations x and y with $d(x, y) = 0$. Define the equivalence relation $\doteq$ by $x \doteq y$ if and only if $d(x, y) = 0$. We will work in the space $\dot{X}$, which is the quotient $X/\doteq$.

Remark 7. The definition of d is robust since it is independent from the additively optimal mode used to define K and of the enumeration of machines.

Topological properties of a space may deeply condition the behavior of dynamical systems defined on it (see [4] for some examples). For this reason we carefully study topological properties of $\dot{X}$.

Notation. In the sequel, the expression $\Bumpeq^x$ stands for 2^{2^x}. Moreover any topological consideration, like continuity or openness, is to be understood with respect to the new topology.

Proposition 8. *The space $\dot{X}$ is not separable i.e. there is no countable dense set.*

Proof. Let A be a countable subset of X. We prove that there exists an infinite sequence of points which is at non-zero distance from all points in A.

Let $A = \{a^0, a^1, \ldots, a^n, \ldots\}$ and b^n the sequences defined as follows:

$$\forall k, n \in \mathbb{N}, \; b^{2^{k+1}n + 2^k - 1} = a^k .$$

Note that the $(b^n)_{n \in \mathbb{N}}$ sequence contains infinitely many occurrences of each a^i.

For $i > 0$, let $\alpha^i = b^i_{\ell^{i-1}+1:\ell^i}$, and let β^i be a word of length $\ell^i - \ell^{i-1}$ such that $K(\beta^i|\alpha^i) \geqslant \ell^i - \ell^{i-1}$. Let ω be the bi-infinite sequence:

$$\omega = \ldots 000000000\,0\,00\beta^1 \ldots \beta^{n-1}\beta^n \ldots$$

where the last 0 is at index 2 in ω. Thus $\omega_{\ell^{i-1}+1:\ell^i} = \beta^i$.

We claim that for all k, $d(a^k, \omega) \geqslant \frac{1}{4}$. For all $n \in \mathbb{N}$, let $u_n = 2^{k+1}n + 2^k - 1$. Since for all $n \in \mathbb{N}$ we have

$$\omega_{\ell^{u_n-1}+1:\ell^{u_n}} = \beta^{u_n},$$

and

$$a^k_{\ell^{u_n-1}+1:\ell^{u_n}} = b^{u_n}_{\ell^{u_n-1}+1:\ell^{u_n}} = \alpha^{u_n} ,$$

we obtain

$$K(\omega_{-\ell^{u_n}:\ell^{u_n}} | a^k_{-\ell^{u_n}:\ell^{u_n}})$$
$$\vee$$
$$K(\omega_{\ell^{u_n-1}+1:\ell^{u_n}} | a^k_{\ell^{u_n-1}+1:\ell^{u_n}}) + o(\ell^{u_n})$$
$$\vee$$
$$K(\beta^{u_n} | \alpha^{u_n}) + o(\ell^{u_n})$$
$$\vee$$
$$\ell^{u_n} - \ell^{u_n-1} + o(\ell^{u_n})$$

and hence

$$d(a^k, \omega) \geqslant \limsup_{n \to \infty} \frac{K(\omega_{-\ell^{u_n}:\ell^{u_n}} | a^k_{-\ell^{u_n}:\ell^{u_n}})}{4 \cdot \ell^{u_n} - 1}$$
$$\vee$$
$$\limsup_{n \to \infty} \frac{\ell^{u_n} - \ell^{u_n-1} + o(\ell^{u_n})}{4 \cdot \ell^{u_n} - 1} \geqslant \frac{1}{4}$$

□

Proposition 9. *The space $\dot{X}$ is not locally compact.*

Proof. Let u be a random configuration. Consider the configurations c^n defined as follows for any fixed k: $c^n_{ki} = u_{2^n(2i+1)}$ and 0 elsewhere. By construction $\forall n$, $d(c^n, \underline{0}) \leqslant 1/k$. As, for all integers p and q, the subsequences of u, $(u_{2^p(2i+1)})_{i \in \mathbb{N}}$ and $(u_{2^q(2i+1)})_{i \in \mathbb{N}}$ are extracted recursively, independently, and do not overlap, they are mutually independent since u is random. Hence $d(c^p, c^q) \geqslant 1/k$.

□

Proposition 10. *The space $\dot{X}$ is perfect, pathwise connected and infinite dimensional.*

Proof. For each real $\alpha \in [0,1]$, choose a sequence of integers $(p_i^\alpha)_{i \in \mathbb{N}}$ such that $\lim\limits_{i \to \infty} \dfrac{p_i^\alpha}{\not{k}^{i+1} - \not{k}^i} = \alpha$. Let us fix a random configuration u. Starting from u, we progressively erase bits of u continuously, until we reach $\underline{0}$. We define $f \colon \mathbb{R} \to \dot{X}$ by $\forall n \in \mathbb{Z}$, $\forall \not{k}^n < i \le \not{k}^n + p_{|n|}^\alpha$, $f(\alpha)_i = u_i$, and 0 otherwise.

Continuity of f. Let $\alpha < \beta$ be two reals: $\exists N$, $n > N \implies p_n^\alpha < p_n^\beta$. Since we have less bits equal to 1 in $f(\alpha)$ than in $f(\beta)$, and since we can compute which bits are erased only knowing the sequence p^α, we have

$$\limsup_{n \to \infty} \frac{K(f(\alpha)_{-n:n} | f(\beta)_{-n:n})}{4n + 1} = 0 \ .$$

Let $m > N$ be an integer and and n an integer such that $\not{k}^m \le n < \not{k}^{m+1}$. We know that $K(f(\beta)_{-n:n} | f(\alpha)_{-n:n})$ is equal, up to a logarithmic term, to the complexity of the bits taken from u of $f(\alpha)$ knowing the bits taken from u of $f(\beta)$. The bits which are in $f(\beta)_{-n:n}$ and not in $f(\alpha)_{-n:n}$ and the first terms of p^β are sufficient to compute the former from the latter. Thus the complexity of $f(\beta)_{-n:n}$ knowing $f(\alpha)_{-n:n}$ is less than the number of those bits, up to an additive constant. This number is $\sum\limits_{i=0}^{m-1} (p_i^\beta - p_i^\alpha) + \max(\min(p_m^\beta, n) - p_m^\alpha, 0)$.

Therefore we have

$$\frac{K(f(\alpha)_{-n:n} | f(\beta)_{-n:n})}{4n + 1}$$
$$\wedge$$
$$\frac{\sum_{i=0}^{m-1} p_i^\beta - p_i^\alpha + \max(\min(p_m^\beta, n) - p_m^\alpha, 0) + 4\log_2 n + c}{4n + 1}$$
$$\wedge$$
$$2 \frac{p_m^\beta - p_m^\alpha + \log_2 \not{k}^m}{\not{k}^m}$$

which implies that

$$\limsup_{n \to \infty} \frac{K(f(\alpha)_{-n:n} | f(\beta)_{-n:n})}{4n + 1} \le 4(\beta - \alpha).$$

We conclude that

$$d(f(\beta), f(\alpha)) \le 4(\beta - \alpha)$$

and, using f, that $\dot{X}$ is pathwise connected.

Perfectness. The sequence $f(\frac{1}{n})$ has limit u.

Injectivity of f. For all $m > N$,

$$K(f(\beta)_{-\not{k}^m : \not{k}^m} | f(\alpha)_{-\not{k}^m : \not{k}^m})$$
$$\vee$$
$$K(u_{\not{k}^m + 1 : \not{k}^m + p_m^\beta} | u_{\not{k}^m + 1 : \not{k}^m + p_m^\alpha}) - \mathcal{O}(\log_2(\not{k}^m)).$$

As $K(x|y) \geqslant K(x) - K(y) - \mathcal{O}(\log_2(|x| + |y|))$ and by incompressibility of u
$K(u_{\not\kappa m + 1 : \not\kappa m + p_m^\alpha}) = p_n^\alpha + \mathcal{O}(\log_2(\not\kappa^m))$, we get

$$K(f(\beta)_{-\not\kappa m : \not\kappa m} | f(\alpha)_{-\not\kappa m : \not\kappa m}) \geqslant p_n^\beta - p_n^\alpha - \mathcal{O}(\log_2(\not\kappa^m))$$

hence

$$d(f(\alpha), f(\beta)) \geqslant \beta - \alpha > 0.$$

Infinite dimension of $\dot{X}$. The continuity and injectivity of f prove that $\dot{X}$ is at least one-dimensional. Let n be positive integer. Define a function h as follows:

$$h : \begin{cases} \left(\dot{X}\right)^n \to \dot{X} \\ (c^i)_{1 \leqslant i \leqslant n} \mapsto f\left((c^i)_{1 \leqslant i \leqslant n}\right) \end{cases}$$

with

$$f\left((c^i)_{1 \leqslant i \leqslant n}\right) = \dots c^1_{-k} c^2_{-k} \dots c^n_{-k} c^1_{-k+1} \dots c^n_{-k+1} \dots c^1_k c^2_k \dots c^n_k \dots .$$

The function h is continuous: $\forall x, y \in \left(\dot{X}\right)^n$,

$$d(x^i, y^i) \leqslant \delta \implies d(h(x), h(y)) \leqslant \delta$$

since

$$K(h(x)_{-k:k} | h(y)_{-n:n} k) = \sum_{i=1}^{n} K(x^i_{-\lceil \frac{k}{n} \rceil : \lceil \frac{k}{n} \rceil} | y^i_{-\lceil \frac{k}{n} \rceil : \lceil \frac{k}{n} \rceil}) + n \log_2 \left\lceil \frac{k}{n} \right\rceil .$$

h is injective (similar proof as for f). Let f_n be defined as follows

$$f_n : \begin{cases} [0,1]^n \longrightarrow \dot{X} \\ (\alpha^i)_{1 \leqslant i \leqslant n} \longmapsto h(f(\alpha_1), f(\alpha_2), \dots, f(\alpha_n)) . \end{cases}$$

Since f_n is the composition of f and h, it is continuous and injective.   □

4   Cellular Automata

A *cellular automaton* (CA for short) is a map $f: S^{\mathbb{Z}} \to S^{\mathbb{Z}}$ defined for any $x \in S^{\mathbb{Z}}$ and $i \in \mathbb{N}$ by

$$f(x)_i = \lambda(x_{i-r} \dots x_i \dots x_{i+r})$$

where $\lambda: S^{2r+1} \to S$ is the *local rule* and r is the (neighborhood) *radius*. S is a finite set, usually called the set of *states* of the CA. f is called the *global rule* of the CA. We will note 0 one state of S. A *finite configuration* is a configuration with a finite number of states which differ from 0. It can be noted $0^\infty w 0^\infty$, where w is a word made of letters from S.

A CA is *compatible* w.r.t. $\doteq$ if and only if $\forall x, y \in S^{\mathbb{Z}}$, $x \doteq y \Rightarrow f(x) \doteq f(y)$.

A CA is *surjective* if f is surjective. A CA is *pre-injective* if f is injective on finite configurations. By Moore-Myhill's theorem [20,21], a CA f such that $f(\underline{0}) = \underline{0}$ is surjective if and only if it is pre-injective.

Remark 11. For any CA of global rule f, any radius r and any alphabet, it holds that

$$\forall x \in X, \limsup_{n \to \infty} \frac{K(f(x)_{-n:n}|x_{-n:n})}{2n+1} = 0 \ ,$$

since the knowledge of the local rule and the $2r$ letters of $x_{-n-r:-n-1}$ and $x_{n+1:n+r}$, whose complexity is bounded by a constant, is enough to compute $f(x)_{-n:n}$ from $x_{-n:n}$.

Furthermore, all recursive infinite sequences are in the same class as $\underline{0}$.   □

Lemma 12. *Surjective CA are the identity map of $\dot{X}$.*

Proof. Let f be a surjective CA.

In [18] is proved that, as f is surjective, f is balanced, that is to say that all words of the same length k have the same number of pre-images (of length $k + 2r$) by f. Since the number of words of length k is 2^k, there are exactly 2^{2r} pre-images of a given word of length k.

Consider a program p which on input $\langle y, i \rangle$ acts as follows:

- create the set P of the 2^{2r} pre-images of y;
- outputs the ith word of P, following the lexicographic order, truncated to its $|y|$ central letters.

when $i \leqslant 2^{2r}$ and never halts else.

Thus, the only thing we need in order to compute $x_{-n:n}$ from $f(x)_{-n:n}$ in the description mode p is the number i of $x_{-n:n}$ in the set P of the pre-images of $f(x)_{-n:n}$, which is bounded by a constant. We conclude that

$$\forall n \in \mathbb{N}, K(x_{-n:n}|f(x)_{-n:n}) = O(1) \ .$$

Using remark 11, we conclude that: $d(x, f(x)) = 0$.   □

Since the identity map is continuous, we have the following corollary.

Corollary 13. *Surjective CA are continuous.*

This result can be strengthened as follows:

Proposition 14. *Continuous CA are either surjective or constant. Non-continuous CA are not compatible.*

Proof. Let g be the global function of a non-surjective, non-constant CA of radius r. Let s be the state obtained applying the local rule to 0^{2r+1}. We define h, a state-to-state function on configurations which maps 0 to s, s to 0 and all other states onto themselves. We define $f = h \circ g$. Since h is a involution, f is also non-surjective and non-constant.

Using Moore-Myhill's theorem, f is not injective when restricted to finite configurations, and thus there exist two finite configurations such that

$$0^\infty w_0 0^\infty \to 0^\infty z_0 0^\infty \text{ and } 0^\infty w_2 0^\infty \to 0^\infty z_0 0^\infty \ ,$$

for suitable finite words w_0, w_2, z_0. Moreover, since the CA is not constant, there exists at least one finite configuration $0^\infty w_1 0^\infty$ whose image is $0^\infty z_1 0^\infty$ with $z_1 \neq z_0$. Without loss of generality one can choose $|w_0| = |w_1| = |w_2|$ and $|z_1| = |z_0| = |w_0| + 2r$. Let x be a random sequence of 0s and 1s. Define two configurations α and β such that

$$\alpha = \ldots 0^r w_{x_i} 0^r w_{x_{i+1}} 0^r \ldots$$

and

$$\beta = \ldots 0^r w_{2x_i} 0^r w_{2x_{i+1}} 0^r \ldots$$

where r is the radius of the neighborhood. Note that $\alpha \doteq \beta$. We have

$$f(\alpha) = \ldots z_{x_i} z_{x_{i+1}} \ldots \quad \text{and} \quad f(\beta) = z_0^\infty$$

and hence $d(f(\alpha), f(\beta)) > 0$. Since $g = h \circ f$, and h is a just renaming the states, $d(g(\alpha), g(\beta)) > 0$. Hence g is not compatible, and therefore not continuous.

Conversely, if g is constant, it is continuous, and with lemma 13, if g is surjective, it is continuous.   □

Proposition 14 precisely characterizes the set of continuous CA on this space. In dimension one, the class of continuous CA is decidable since surjectivity is decidable [2,22]. Although continuity fails for non-surjective CA the following proposition says that a small amount of continuity still persists.

Proposition 15. *For all CA f,*

i) $f(\underline{0}) \doteq \underline{0}$, i.e. $\underline{0}$ is a fix point of f in $\dot{X}$;
ii) $\underline{0}$ is an equicontinuity point of f.

Proof. i) The first part is straightforward since the image of $\underline{0}$ is a constant configuration (made of only one state) and all of them are in the same class (see Remark 11).
ii) Remark that $\forall x \in X, \exists c \in \mathbb{N}, \forall t \in \mathbb{N}$,

$$d(f^t(x), f^t(\underline{0})) = \limsup_{n \to \infty} \frac{K(f^t(x)_{-n:n}|0^{2n+1}) + K(0^{2n+1}|f^t(x)_{-n:n})}{2n+1}$$

$$\wedge$$

$$\limsup_{n \to \infty} \frac{K(x_{-n:n}|0^{2n+1}) + c}{2n+1} = d(x, \underline{0}) \ .$$

□

Often, it is not interesting to consider the behavior of the CA over the whole set of configurations. For instance, configurations which are destroyed by the automaton after one step (*i.e.* any configuration which has no pre-image through the CA) are not relevant when studying the evolution of the CA. In this way, the notion of limit set have been introduced.

Definition 16 (limit set). *Let f be the global rule of a CA. The limit set of this CA is the set*

$$\Omega = \bigcap_{n \in \mathbb{N}} f^n(X)$$

where X is the set of the configurations.

The limit set is the set of all the configurations having infinitely many pre-images through f. We note Ω_n the set of all words of size n which are subwords of a configuration of Ω.

It is a natural question to wonder whether non-surjective CA are continuous at least on their limit sets (where they act surjectively). The following proposition is a partial answer to this question.

Theorem 17. *Let $g \colon \mathbb{N} \to \mathbb{N}$ be a computable function such that $g(n) = o(n)$. Let f be the global function of a CA. Let r be the radius of the CA. If the following condition (that limits the growth of the number of pre-images) is verified*

$$\forall x \in X, \ \forall n \in \mathbb{N}, \ \log_2 |f^{-1}(x_{-n:n}) \cap \Omega_{n+2r}| \leqslant g(n) \tag{2}$$

then $f|_\Omega$ is equivalent to the identity map on Ω.

Proof. Using Remark 11,

$$\limsup_{n \to \infty} \frac{K(f(x)_{-n:n}|x_{-n:n})}{n} = 0.$$

It remains to prove that, for all x in Ω,

$$\limsup_{n \to \infty} \frac{K(x_{-n:n}|f(x)_{-n:n})}{n} = 0.$$

Consider a program p which on input $\langle y, i \rangle$ acts as follows:

- creates the set P of all the pre-images of y;
- enumerates the set of forbidden words F
- erases from P all the words containing words in F, until P contains less than $2^{g(|y|)}$ words;
- outputs the ith word of P, following the lexicographic order.

This program p is well-defined since F is recursively enumerable. Since condition (2) holds, p always halts. Hence $K(x_{-n-2r:n+2r}|f(x_{-n:n})) = g(n)$. ☐

Theorem 18. *In $\dot{X}$ the following statements hold:*

i) randomness and weak-incompressibility is preserved by surjective CA;
ii) non-surjective CA destroy both randomness and weak-incompressibility;
iii) all CA preserve non weak-incompressibility.

Proof. i) Lemma 12;
ii) Non-surjective CA erase at least a finite pattern (see for example [12]). Thus a random sequence is mapped to a non-random-one according to both definitions.
iii) Proposition 15. ☐

References

1. P. Kůrka A. Maass. Stability of subshifts in cellular automata. *to appaer in Fondamenta informaticæ*, 2000. 250
2. S. Amoroso and Y. N. Patt. Decision procedures for surjectivity and injectivity of parallel maps for tesselation structures. *J. Comp. Syst. Sci.*, 6:448–464, 1972. 257
3. C. H. Bennet, P. Gács, M. Li, P. M. B. Vitányi, and W. H. Zurek. Information distance. *IEEE Trans. Inform. Theory*, 44(4):1407–1423, 1998. 252
4. F. Blanchard, E. Formenti, and P. Kůrka. Cellular automata in Cantor, Besicovitch and Weyl topological spaces. *Complex Systems*, 11-2, 1999. 250, 252
5. A. A. Brudno. Entropy and complexity of the trajectories of a dynamical system. *Trans. Moscow Math. Soc.*, 2:127–151, 1983. 249, 250
6. C. Calude. *Information and Randomness*. Springer-Verlag, 1994. 251, 252
7. C. Calude, P. Hertling, H. Jürgensen, and K. Weihrauch. Randomness on full shift spaces. *Chaos, Solitons & Fractals*, 1:1–13, 2000. 249, 250
8. G. Cattaneo, E. Formenti, G. Manzini, and L. Margara. Ergodicity and regularity for cellular automata over Z_m. *Theoretical Computer Science*, 233(1-2):147–164, 1999. 250
9. G. J. Chaitin. On the length of programs for computing finite binary sequences. *J. of ACM*, 13:547–569, 1966. 251
10. J.-C. Dubacq, B. Durand, and E. Formenti. Kolmogorov complexity and cellular automata classification. *Theor. Comp. Sci.*, 259(1-2):271–285, 2001. 249
11. B. Durand. The surjectivity problem for 2D cellular automata. *Journal of Computer and Systems Science*, 49(3):718–725, 1994. 250
12. B. Durand. Global properties of cellular automata. In E. Goles and S. Martinez, editors, *Cellular Automata and Complex Systems*. Kluwer, 1998. 258
13. J. Kari. Reversibility and surjectivity problems of cellular automata. *Journal of Computer and System Sciences*, 48:149–182, 1994. 250
14. A. N. Kolmogorov. Three approaches to the definition of the concept "quantity of information". *Problems of information transmission*, 1:3–11, 1965. 251
15. M. Li and P. Vitányi. *An Introduction to Kolmogorov complexity and its applications*. Springer-Verlag, second edition, 1997. 251, 252
16. B. Martin. Apparent entropy of cellular automata. *Complex Systems*, 12, 2000. 250
17. P. Martin-Löf. The definition of a random sequence. *Information & Control*, 9:602–619, 1966. 251
18. A. Maruoka and M. Kimura. Conditions for injectivity of global maps for tessellation automata. *Information & control*, 32:158–162, 1976. 256
19. A. Maruoka and M. Kimura. Injectivity and surjectivity of parallel maps for cellular automata. *Journal of Computer and System Sciences*, 18:158–162, 1979. 250
20. E. F. Moore. Machine models of self-reproduction. *Proc. Symp. Appl. Math., AMS Rep.*, 14:17–34, 1963. 250, 255
21. J. Myhill. The converse of Moore's Garden of Eden theorem. *Proc. Amer. Math. Soc.*, 14:685–686, 1963. 250, 255
22. K. Sutner. De Bruijn graphs and linear cellular automata. *Complex Systems*, 5:19–30, 1991. 257
23. V. A. Uspensky, A. L. Semenov, and A. Kh. Shen. Can individual sequences of zeros and ones be random? *Russ. Math. Surveys*, 45:121–189, 1990. 252
24. V. A. Uspensky and A. Kh. Shen. Relations between varieties of Kolmogorov complexities. *Math. Syst. Theory*, 29(3):270–291, 1996. 251

25. H. S. White. Algorithmic complexity of points in dynamical systems. *Ergod. Th. & Dynam. Sys.*, 13:807–830, 1993. 249

The k-Median Problem for Directed Trees*

Extended Abstract

Marek Chrobak[1], Lawrence L. Larmore[2], and Wojciech Rytter[3,4]

[1] Department of Computer Science, University of California
Riverside, CA 92521
[2] Department of Computer Science, University of Nevada
Las Vegas, NV 89154-4019
[3] Instytut Informatyki, Uniwersytet Warszawski
Banacha 2, 02–097, Warszawa, Poland
[4] Department of Computer Science, University of Liverpool
Liverpool L69 7ZF, UK

Abstract. The k-median problem is a classical facility location problem. We consider the k-median problem for *directed trees*, motivated by the problem of locating proxies on the World Wide Web. The two main results of the paper are an $O(n \log n)$ time algorithm for $k=2$ and an $O(n \log^2 n)$ time algorithm for $k=3$. The previously known upper bounds for these two cases were $O(n^2)$.

1 Introduction

In the *k-median problem* we are given a graph G in which each node u has a non-negative weight w_u and each edge (u, v) has a non-negative length d_{uv}. Our goal is to find a set F of k vertices that minimizes $Cost(F) = \sum_{x \in G} \min_{u \in F} w_x d_{xu}$. We think of the nodes of G as customers, with each customer x having a demand w_x for some service. F is a set of k *facilities* that provide this service. It costs d_{xu} to provide a unit of service to customer at x from a facility located at u. We wish to place k facilities in G so that the overall service cost is minimized. This optimal set is called the *k-median* of the weighted graph G. The k-median problem is a classical facility location problem which has been a subject of study for several decades. The general case is known to be NP-complete (see Problem ND51 in [8]), and most of the past work focussed on efficient heuristics (see, for example, [14]) and approximation algorithms [2,6,5].

Trees. In this paper we focus on the case of the k-median problem when G is a tree. In 1979 Kariv and Hakimi [13] presented a dynamic programming algorithm for this problem that runs in time $O(k^2 n^2)$. Hsu [12] gave an algorithm with running time $O(kn^3)$. More recently, Tamir [18] showed that the Kariv-Hakimi

---

* Research supported by grant EPSRC GR/N09077 (UK). M.Chrobak was also partially supported by by NSF grant CCR-9988360. W.Rytter was also partially supported by grant KBN 8T11C03915.

J. Sgall, A. Pultr, and P. Kolman (Eds.): MFCS 2001, LNCS 2136, pp. 260–271, 2001.

algorithm is faster than previously thought, namely that its running time is only $O(kn^2)$. This is the best upper bound known to this date.

Some other special cases of the problem have been considered. Hassin and Tamir [11] gave an $O(kn)$ time algorithm when the tree is a path. Another $O(kn)$-time algorithm for this problem was later given by Auletta et al. [4]. The case $k = 2$ was considered by Gavish and Sridhar [9] who presented an algorithm with running time $O(n \log n)$. The dynamic version of the case $k = 2$ was studied by Auletta et al. in [3][1].

Directed trees. We consider the k-median problem for *rooted directed trees*. Let T be a directed tree with root r. If y is the father of x, denote by d_{xy} the length of edge (x, y). The length function extends in a natural way to any pair x, y where y is an ancestor of x. As in the undirected case, each node x is assigned a non-negative weight w_x. Our goal is to find a set of nodes F of size k that minimizes the quantity:

$$Cost(F) = \sum_{x \in T} \min_{u \in F + r} w_x d_{xu}.$$

We use the same terminology as in the undirected case, that is, the nodes in F are called *facilities*, and the optimal set F is called the *k-median* of T. Note, however, that in our case we actually have $k + 1$ facilities in the tree, since the default "root facility" at r is fixed and not counted in F. For specific values of k this distinction makes the undirected case quite different from the undirected one. For example, for $k = 2$, in the undirected case we need to partition the tree into two subtrees that form the domains of the two facilities. This task can be accomplished by searching the tree for an edge that separates the tree into two appropriately balanced subtrees [9,3]. In the directed case, however, we need to partition T into *three* subtrees.

The k-median problem for directed trees was introduced by Li et al. [16] as a mathematical model for optimizing the placement of web proxies to minimize average latency. In [16], the authors present a $O(k^2 n^3)$-time algorithm for this problem. This was later improved to $O(k^2 n^2)$ by Vigneron et al. [19]. In [19] the authors also noted that the Kariv-Hakimi algorithm can be modified to find k-medians in directed trees in time $O(k^2 n^2)$. Tamir (private communication) observed that his method from [18] can be adapted to reduce the time complexity to $O(kn^2)$. For the line, Li et al. [15] give an algorithm with running time $O(kn^2)$. This was just recently improved by Woeginger [20] to $O(kn)$.

Our results. We present several algorithms for the k-median problem on directed trees. We start by describing, in Section 2, the basic dynamic programming algorithm, Algorithm AncDP, which is an adaptation of the Kariv-Hakimi algorithm to directed trees (see also [19]). Then, in Section 3, we propose an alternative dynamic programming algorithm, Algorithm DepthDP. For $k = 2$ this algorithm can be implemented in time $O(n \log^2 n)$. The problem of estimating

---

[1] In [3], the authors also claimed a linear-time algorithm; however, their running time analysis is flawed.

the worst-case running time of Algorithm DepthDP reduces to an interesting combinatorial problem of estimating the number of line segments in a certain piece-wise linear function. We show that for $k = 2$ this number is $O(n)$ and we conjecture that it is also linear for any fixed k. This conjecture would imply that, for each fixed k, Algorithm DepthDP has time complexity $O(n^2)$. We believe that Algorithm DepthDP is faster than Algorithm AncDP on realistic data. Both algorithms are being implemented and our experimental results will be reported in the full version of this paper.

In Section 4 we show that with an additional requirement that the facilities are independent (no facility is an ancestor of another) the problem can be solved in time $O(k^2 n)$.

Our main results are given in Sections 5 and 6. In Section 5 we improve the asymptotic running time for $k = 2$ to $O(n \log n)$. By the result from Section 4, it is sufficient to consider the case when one facility is an ancestor of the other. For this case we show that the objective function satisfies the Monge property. Our algorithm Fast2Median uses the divide-and-conquer approach. A given tree T is split into two smaller trees, for which the solutions are computed recursively. Yet another tree is constructed, on which we apply the SMAWK algorithm [1] for computing maxima in Monge matrices. From the solutions for these three instances we obtain the solution for T.

In Section 6, we deal with the case $k = 3$. For this case we present Algorithm Fast3Median that runs in time $O(n \log^2 n)$. The algorithm breaks the problem into cases depending on various genealogical patterns among facilities, and solves each case independently. The cases are solved by a combination of dynamic programming and divide-and-conquer methods.

2   Ancestor-Based Dynamic Programming

Throughout the paper, we will assume that T is a binary tree with n nodes in which each non-leaf node has two sons. Any tree can be transformed into such a binary tree by adding a linear number of nodes with zero weight. By T_u we denote the subtree of T rooted at u, and $n_u = |T_u|$ is the number of descendants of u. By $d_u = d_{ur}$ we denote the *depth* of u, and by ω_u the total weight of T_u, that is $\omega_u = \sum_{x \in T_u} w_x$. We often refer to ω_u as the *tree weight* of u.

We now show an adaptation of the Kariv-Hakimi algorithm to directed trees. Let $Cost_x^j(u)$ be the optimal cost of placing j facilities within T_x given that the closest facility above x is u. Formally, define T_x^u to be the tree that consists of T_x with an additional root node r_u whose only son is x, and with edge (x, r_u) having length d_{xu}. Then $Cost_x^j(u)$ is the minimum cost of placing j facilities in T_x^u. Note that we allow $u = x$, in which case $d_{xu} = 0$, and $Cost_x^j(x)$ will be equal to the minimum cost of placing j facilities in T_x (not counting the default facility at x).

Algorithm AncDP. We only show how to compute optimal cost. The algorithm can be easily modified to determine the k-median. We compute all quantities $Cost_x^j(u)$ bottom-up. If x is a leaf, set $Cost_x^0(u) = w_x d_{xu}$ and $Cost_x^j(u) = 0$ for

$j \geq 1$. Suppose that x is not a leaf and y, z are the sons of x. For each u that is an ancestor of x and $j = 0, \ldots, k$ we then compute:

$$Cost_x^j(u) = \min \begin{cases} \min_{p+q=j} \left\{ Cost_y^p(u) + Cost_z^q(u) \right\} + w_x d_{xu} \\ Cost_x^{j-1}(x) \end{cases}$$

For $j = 0$, assume that $Cost_x^{j-1}(x) = \infty$. In the first case we have p facilities in T_y, q facilities in T_z and no facility at x. In the second case we have one facility at x. When we are done, return $Cost_r^k(r)$.

Running time. Algorithm AncDP clearly runs in time $O(k^2 n^2)$. As noted by Tamir (private communication), the bound can be improved to $O(kn^2)$ using the method from [10,18].

3   Depth-Based Dynamic Programming

In this section we give a different algorithm called DepthDP. The idea is that instead of computing $Cost_x^j(u)$ for the ancestors of u, we will compute a real-argument function $Cost_x^j(\alpha)$, for an appropriate range of α, which represents the distance d_{xu}. For $k = 2$ we show an implementation of this algorithm called DepthDP2 that runs in time $O(n \log^2 n)$. Although later in Section 5 we improve the upper bound further to $O(n \log n)$, we include DepthDP2 in the paper, since it is considerably simpler and easier to implement. Also, the main idea of DepthDP will play a major role in Section 6, where we improve the upper bound for $k = 3$ to $O(n \log^2 n)$.

Let $Cost_x^j(\alpha)$ be the optimal cost of placing j facilities within T_x given that the closest facility above x is at distance α. Formally, define T_u^α to be the tree that consists of T_x and an extra root node r_α whose only son is x and with edge (x, r_α) having length α. Then $Cost_x^j(\alpha) = Cost_x^j(r_\alpha)$, the minimum cost of placing j facilities in T_x^α. We allow $\alpha = 0$, in which case $Cost_x^j(0) = Cost_x^j(x)$.

Algorithm DepthDP. As before, we compute all functions $Cost_x^j(\alpha)$ bottom-up. If x is a leaf, set $Cost_x^0(\alpha) = w_x \alpha$ and $Cost_x^j(\alpha) = 0$ for $j \geq 1$. Suppose that x is not a leaf and y, z are the sons of x. Similarly as in Algorithm AncDP, we have

$$Cost_x^j(\alpha) = \min \begin{cases} \min_{p+q=j} \left\{ Cost_y^p(\alpha + d_{yx}) + Cost_z^q(\alpha + d_{zx}) \right\} + w_x \alpha \\ Cost_x^{j-1}(0) \end{cases}$$

where $Cost_x^{-1}(0) = \infty$. When we are done, return $Cost_r^k(0)$.

Functions $Cost_x^j(\alpha)$. We still need to explain how to store and maintain the functions $Cost_x^j(\alpha)$ throughout the computation. Given a set $F \subseteq T_x^\alpha$ of size j, denote by $Cost_x^F(\alpha)$ the cost of having j facilities in F. For a fixed F, $Cost_x^F(\alpha)$ is a linear function, namely $Cost_x^F(\alpha) = a_F \alpha + b_F$ for some constants a_F, b_F. Thus $Cost_x^j(\alpha)$ is a concave, piece-wise linear-function.

The crucial questions are: (1) What is the complexity (number of segments) of $Cost_x^j(\alpha)$? (2) How to store and update the functions $Cost_x^j(\cdot)$ efficiently?

Algorithm DepthDP2 for 2 medians. We now consider the case $k = 2$. For $j = 0$, $Cost_x^0(\cdot)$ is a linear function and is easy to maintain. For $j = 2$, we only need $Cost_x^2 = Cost_x^2(d_x)$. The hardest case is $j = 1$. In this case, the problem reduces to keeping track of a lower envelope of a number of line segments, which can be done in time $O(\log n)$ per operation (details in the full paper).

Theorem 1. *Algorithm* DepthDP2 *computes 2-medians in directed trees in time* $O(n \log^2 n)$.

Minimizers. In order to estimate the running time of Algorithm DepthDP, one needs to bound the complexity (that is, the number of segments) of functions $Cost_x^j(\cdot)$. We now consider this issue. Without loss of generality we can restrict our attention to the case when x is the root of T and $j = k$. For simplicity, we will write $Cost(\alpha)$ for $Cost_x^k(\alpha)$.

For a set F of k vertices in T, let $Cost^F(\alpha)$ be the cost of placing k facilities on F in T^α. We say that F is a *minimizer of* $Cost(\cdot)$ *at* α if F minimizes $Cost^F(\alpha)$ among all k-element sets. Each segment of $Cost(\cdot)$ belongs to some line $Cost^F(\cdot)$. Thus we are interested in the following question: How many different minimizers we may have? Clearly, this number cannot exceed $\binom{n}{k}$. We conjecture that this number is substantially smaller.

Conjecture 2. For each fixed k, the number of minimizers is $O(n)$.

The above linearity conjecture immediately implies that, for any fixed k, Algorithm DepthDP runs in time $O(n^2)$. For varying k, the complexity of the algorithm depends on how fast the bound on number of maximizers grows with k.

The linearity conjecture is trivially true for $k = 1$. We show that it also holds for $k = 2$. The proof is based on amortized analysis and will appear in the full version of this paper.

Theorem 3. *For* $k = 2$, *function* $Cost(\cdot)$ *has* $O(n)$ *minimizers.*

4   Independent Facilities

Suppose that instead of any set of k facilities, we wish to find a set of k *independent* facilities that minimizes cost. We show that this problem can be solved in linear time for any fixed k.

Algorithm IndDP. The algorithm works by simple bottom-up dynamic programming. For each node x and each j we compute $IndGain_x^j$, the maximum gain resulting from placing up to j independent facilities in T_x. If x is a leaf, set $IndGain_x^0 = 0$ and $IndGain_x^j = w_x d_x$ for $j > 0$. Suppose that x is not a leaf and y, z are the sons of x. For other nodes:

$$IndGain_x^j = \max \begin{cases} \max_{p+q=j} \left\{ IndGain_y^p + IndGain_z^q \right\} \\ w_x d_x \end{cases}$$

Algorithm IndDP clearly runs in time $O(k^2 n)$.

5 A Faster Algorithm for $k = 2$

We now improve the upper bound for $k = 2$ to $O(n \log n)$. We have two subcases: the independent case, when the two facilities are independent, and the ancestral case, when one facility is an ancestor of the other. The independent case has been solved in Section 4.

The ancestral case. In the ancestral case, our task is to find u and $v \in T_u$ that maximize the gain resulting from placing two facilities at u, v:

$$Gain(u, v) = d_u \omega_u + (d_v - d_u)\omega_v. \tag{1}$$

The algorithm for the ancestral case uses the divide-and-conquer approach. At each step we make a recursive call to some tree S to compute an ancestral pair u, v that maximizes (1) among all pairs in S. Let x be the root of S. It is well-known – and easy to see – that S always has a node y such that $\frac{1}{3} \cdot |S| \le |S_y| \le \frac{2}{3} \cdot |S|$. We call this node y a *centroid* of S. Using y, we can partition S into disjoint subtrees S_y and $S - S_y$. This gives rise to three possible locations of the optimal pair u, v: (i) $u, v \in S_y$, (ii) $u, v \in S - S_y$, or (iii) $u \in path(x, y)$ and $v \in S_y$, where $path(x, y)$ denotes the path from x to y in S (See Figure 1), excluding y. Recursively, we find optimal pairs in S_y and $S - S_y$.

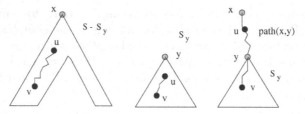

Fig. 1. Three possible location of an optimal ancestral pair in a partitioned tree S

We now show how to find an optimal pair that is split between $path(x, y)$ and S_y. Denote by $P = path(x, y)$ and $Q = S_y$ the lists of all ancestors (excluding y) and descendants of y, ordered by decreasing tree weight ω. By routine calculations we get the following lemma.

Lemma 4. *Let $P = (p_1, \ldots, p_a)$ be a list of nodes ordered by increasing depth and $Q = (q_1, \ldots, q_b)$ a list of nodes ordered by decreasing tree weight. Then function $Gain(\cdot, \cdot)$ has the Monge property on $P \times Q$, that is, for all $i < j$ and $g < h$ we have*

$$Gain(p_i, q_g) + Gain(p_j, q_h) \ge Gain(p_i, q_h) + Gain(p_j, q_g)$$

Procedure AncSplitPair(P, Q). The procedure receives on input two lists of vertices P and Q that satisfy the conditions of Lemma 4. It outputs a pair

$\{u, v\}$, for $u \in P$ and $v \in Q$, that maximizes $Gain(u, v)$. To find u and v, we simply apply the linear-time algorithm called SMAWK (see [1]) that computes maxima of functions with the Monge property.

Procedure AncPair(S). The input S is a tree with root x. The procedure produces two outputs: the optimal ancestral pair from S, and the list of all nodes of S sorted by decreasing tree weight. If $S = \{x\}$, return $\{x, x\}$ and a one-element list (x). If $|S| > 1$, we proceed as follows:

0. Find a centroid y of S.
1. Call AncPair(S_y) to compute the best pair $\{u_1, v_1\}$ in S_y and the list Q_1 of the nodes in S_y sorted by decreasing tree weight.
2. Call AncPair($S - S_y$) to compute the best pair $\{u_2, v_2\}$ in $S - S_y$ and the list Q_2 of the nodes in $S - S_y$ sorted by decreasing tree weight.
3. Call AncSplitPair(P, Q_1), where $P = path(x, y)$, to compute the best pair $\{u_3, v_3\}$, where $u_3 \in path(x, y)$ and $v_3 \in S_y$.
4. Return the best of the three pairs $\{u_i, v_i\}$, $i = 1, 2, 3$.
5. Merge lists Q_1 and Q_2 and return the resulting sorted list.

Algorithm Fast2Median. Given a weighted tree T, we use Algorithm IndDP with $k = 2$ to find the best independent pair. Then we use Procedure AncPair to find the best ancestral pair. We output the better of these two.

Theorem 5. *Algorithm* Fast2Median *computes 2-medians in directed trees in time* $O(n \log n)$.

Proof. Consider one recursive call of AncPair(S), with $|S| = m$ and $|S_y| = m'$. The running time of AncPair(S) satisfies the recurrence $t(m) = t(m') + t(m - m') + O(m)$, where $m/3 \le m \le 2m/3$, since computing y, merging sorted sequences, and the call to AncSplitPair all take time $O(m)$. This immediately implies that the running time of AncPair is $t(n) = O(n \log n)$.

6   An Algorithm for $k = 3$

In this section we improve the time complexity for $k = 3$ to $O(n \log^2 n)$. We start by describing some auxiliary procedures.

Best-descendant function. First, we solve the following task. For each u, we want to compute $v \in T_u$ that maximizes $Gain(u, v) = d_u \omega_u + (d_v - d_u)\omega_v$. We refer to this v as the *best descendant* of u and denote it $BestDesc(u)$.

We compute the function $BestDesc(\cdot)$ by essentially the same algorithm as the one for solving the ancestral case of the 2-median problem, AncPair. Referring to the description of AncPair in Section 5, the following changes need to be made. The recursive call to AncPair(S), instead of outputting just one optimal pair, computes $BestDesc(u)$ for all $u \in S$. As before, it also produces a list of nodes in S sorted according to decreasing tree weight. After the two recursive calls to S_y and $S - S_y$, we call a modified version of AncSplitPair(P, Q) that, for each $u \in P$ computes an optimal $v \in Q$ that maximizes $Gain(u, v)$. This, again, is done using the SMAWK algorithm. For $u \in S_x - path(x, y)$ this gives us the desired values of $BestDesc(u)$. For $u \in path(x, y)$, $BestDesc(u)$ is the better of the nodes produced by the recursive call to $\mathsf{AncPair}(S - S_y)$ and by $\mathsf{AncSplitPair}(P, Q_1)$. The running time of this modification of $\mathsf{AncPair}$ is also $O(n \log n)$.

Forbidden path. We need another, rather technical procedure, similar to the one given above for computing $BestDesc(\cdot)$. In this case, the input is some tree U with root x. We are also given a node $z \in U$. The goal is to compute, for each $u \in path(x, z)$ the best descendant v of u in $U - path(x, z)$ (the one that maximizes $Gain(u, v)$). We call $path(x, z)$ the *forbidden path*. We accomplish this task with a recursive procedure $\mathsf{PathBestDesc}(U, z)$ based on the centroid partitions. We find a centroid y of U and we partition U into U_y and $U - U_y$. We now have two cases (see Figure 2).

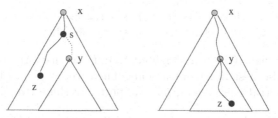

Fig. 2. Two cases in computing descendants with a forbidden path

Suppose that y and z are independent, and denote by s their lowest common ancestor. We first make a recursive call to $\mathsf{PathBestDesc}(U - U_y, z)$. This will give us the correct descendant for each $u \in path(x, z) - path(x, s)$. Then for all $u \in path(x, s)$ we compute the best descendant in U_y using the SMAWK algorithm, and we pick the better of the two choices.

We now examine the other case, when $z \in U_y$. We first make a recursive call to $\mathsf{PathBestDesc}(U_y, z)$. This will give us the correct descendant for each $u \in path(y, z)$. Next, we make a recursive call to $\mathsf{PathBestDesc}(U - U_y, y')$, where y' is the father of y. Then, for all $u \in path(x, y')$ we compute the best ancestor in $U_y - path(y, z)$ using the SMAWK algorithm. (Removing $path(y, z)$ does not affect the algorithm, we just need to change the second input list). This gives us up to two candidate descendants for all $u \in path(x, y')$ and we pick the better one.

Overall, if $|U| = m$, the above procedure in time $O(m \log m)$ for each $u \in path(x, z)$ will compute its best descendant that is not on $path(x, z)$.

The algorithm. We consider separately four cases corresponding to different genealogical relations between the facilities and choose the optimal solution. These four cases are:

Independent: All three facilities are independent.

Partially independent: A pair of nodes u, v, with u being an ancestor of v, and one node t independent of u and v.

Ancestral: Three nodes u, v, t on a path from root to a leaf.

Partially ancestral: A node u with two independent descendants v, t.

To solve the independent case we can simply apply Algorithm IndDP from Section 4 with $k = 3$. We now describe how to solve the remaining three cases.

Partially independent case. In this case, we use the function $BestDesc(\cdot)$. We perform dynamic programming in a bottom-up order, similarly to the independent case from Section 4. For each node x we compute three values: an optimal singleton in T_x, and optimal ancestral pair in T_x, and an optimal partially independent triple in T_x. Suppose we are now at x. The optimal singleton in T_x is either x, or the optimal singleton in T_y, or the optimal singleton in T_z. The optimal ancestral pair is either $\{x, v\}$, for $v = BestDesc(x)$, or the optimal pair from T_y, or the optimal from T_z. The optimal partially independent triple is obtained by combining the optimal pair from one subtree with the optimal singleton from the other and adding up their gain values. This can be done in two ways, and we choose the better of the two triples.

Ancestral case. The totally ancestral case is solved, in essence, by applying the ancestral case of the 2-median algorithm twice. We first compute function $BestDesc(\cdot)$. With each u and $v \in T_u$ we can associate a modified gain value $Gain^*(u, v) = d_u \omega_u + (d_v - d_u)\omega_v + G(v)$, where $G(v) = (d_t - d_v)\omega_t$, for $t = BestDesc(v)$. Calculations similar to the proof of Lemma 4 show that $Gain^*(u, v)$ has the Monge property. The maximum of $Gain^*(u, v)$ is equal to the maximum gain of an ancestral triple $\{u, v, t\}$. Thus we can solve this case by running the procedure AncPair(T) with the modified gain function $Gain^*(u, v)$ instead of $Gain(u, v)$.

Partially ancestral case. This is the most difficult case. The algorithm is again based on divide-and-conquer using the centroid tree partition. As before, a given tree S is partitioned into two subtrees S_y and $S - S_y$. There are four possibilities for the location of the optimal partially ancestral triple, see Figure 3. The first two cases are solved recursively. The third case is solved using the depth-based dynamic programming approach from Section 3. The last case is hardest. In this case $u \in path(x, y)$, one of its descendants v is in $S - S_y - path(x, y)$ and the other descendant t is in S_y. The idea is that we can compute v and t independently of each other. We compute t using the SMAWK algorithm. To compute v, we call the procedure PathBestDesc$(S - S_y, y)$ that finds the best descendant that does not lie on $path(x, y)$.

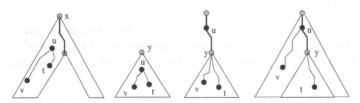

Fig. 3. Four possible location of an optimal partially ancestral triple in a partitioned tree S

We now describe the whole procedure in more detail. For a tree S with root x, let $Gain_S^1(\alpha)$ denote the optimal gain of having one facility in S given that there is another facility located at distance α above x. Similarly, let $IndGain_S^2(\alpha)$ denote the optimal gain of having two independent facilities in S given that there is another facility located at distance α above x. (We slightly overload the notation introduced in Section 3 to avoid clutter).

Procedure PartAncTriple(S). The input S is a tree with root x. The procedure produces three outputs. The first output is an optimal partially ancestral triple $\{u, v, t\}$ in S (or nil if no such triple exists). The second output is the function $Gain_S^1(\cdot)$, and the third output is the function $IndGain_S^2(\cdot)$. Both functions are piece-wise linear functions, as explained in Section 3. For both functions, with each segment, we store the node (or the pair of nodes) that corresponds to this segment.

If $|S| = \{x\}$, return nil. If $|S| > 1$, we proceed as follows:

0. Find a centroid y of S. Let $S' = S - S_y$.
1. Call PartAncTriple(S_y). Let $\{u_1, v_1, t_1\}$ be the optimal triple in S_y.
2. Call PartAncTriple(S'). Let $\{u_2, v_2, t_2\}$ be the optimal triple in S'.
3. We divide this step into three sub-steps. (3a) Call PathBestDesc(S', y) to compute, for each $u \in path(x, y)$ the best descendant v of u in $S' - path(x, y)$. (3b) For each $u \in path(x, y)$ compute its best descendant $t \in S_y$. This again can be done using the SMAWK algorithm. (3c) For each $u \in path(x, y)$ form a partially ancestral triple $\{u, v, t\}$ by using nodes v and t from the previous two steps. Among these triples, let $\{u_3, v_3, t_3\}$ be the triple that has the maximum gain value.
4. For each $u \in path(x, y)$ compute $IndGain_{S_y}(d_y - d_u)$. Let u_4 be the node that achieves this maximum, and v_4, t_4 the independent nodes in S_y that correspond to the line segment that contains $\alpha = d_y - d_u$. Form a partially ancestral triple $\{u_4, v_4, t_4\}$.
5. Return the best of the triples $\{u_i, v_i, t_i\}$, $i = 1, 2, 3, 4$.
6. Compute $Gain_S^1(\cdot)$. To determine $Gain_S^1(\cdot)$, we use the gain functions computed by the recursive calls in Step 1 and 2. We take $Gain_S^1(\alpha)$ to be the maximum of $Gain_{S_y}^1(\alpha + d_{yx})$ and $Gain_{S'}^1(\alpha)$.
7. Compute $IndGain_S^2(\cdot)$. To determine $IndGain_S^2(\alpha)$, we compute the maximum of three functions that correspond to three possible locations of the independent facilities v, t in S: $v, t \in S_y$, $v, t \in S'$, and $v \in S_y$ and $t \in S'$. Thus the first function is $IndGain_{S_y}^2(\alpha + d_{yx})$, the second function is $IndGain_{S'}^2(\alpha + d_{yx})$, and the third function is $Gain_{S_y}^1(\alpha + d_{yx}) + Gain_{S'}^1(\alpha)$.

Algorithm Fast3Median. Given a weighted tree T, we compute separately the optimal triples for each of the four genealogical patterns. The independent case was solved in Section 4. The algorithms for the other three cases were described above.

Theorem 6. *Algorithm* Fast3Median *computes 3-medians in directed trees in time* $O(n \log^2 n)$.

Proof. It is sufficient to estimate the running time of Procedure PartAncTriple. Consider a recursive call to a tree S. Let $m = |S|$ and $m' = |S_y|$. The recursive calls take time $t(m')$ and $t(m - m')$. The invocation of PathBestDesc(S', y) in Step 3 executes in time $O(m \log m)$. The rest of the computation in this step requires only time $O(m)$. Steps 4-7 take time $O(m)$. So, overall, the running time satisfies the recurrence: $t(m) = t(m') + t(m - m') + O(m \log m)$, where $m/3 \leq m \leq 2m/3$. This immediately implies that the running time of PartAncTriple is $t(n) = O(n \log^2 n)$.

7 Final Comments

We presented an $O(n \log n)$-time algorithm for 2-medians and an $O(n \log^2 n)$-time algorithm for 3-medians in directed trees, improving the $O(n^2)$ upper bound for these cases. Many open problems remain.

Can the 2-median problem be solved in linear time? The answer is positive if the tree is a path. However, for general trees, we remain skeptical. The hard case appears to be a tree with a long path from the root r to a node y and a "bushy" subtree T_y. If T_y is sorted by decreasing tree weight, we can apply the SMAWK algorithm to get a solution in linear time. So the problem appears to reduce to the question of whether sorting is necessary.

What is the complexity of the problem for general k? We conjecture that it can be solved in time $O(c_k n \log^a n)$, where a is a small constant and c_k is a constant that depends only on k. An analogous question can be asked for the undirected case. We believe that our approach from Section 6 can be used to get an $O(n \log^2 n)$-time algorithm for this problem in case $k = 3$.

The linearity conjecture remains open for $k \geq 3$. Getting good estimates on the number of minimizers of the cost function would help to get better upper bounds on the complexity of the k-median problem. The linearity conjecture is closely related to some problems in combinatorial geometry. For example, if T is a star, that is, a tree consisting only of a root with $n - 1$ sons, then the problem is equivalent to counting the number of sets consisting of k top lines in an arrangement of $n - 1$ lines, which is known to be $O(kn)$. We suspect that nearly linear upper bounds can be obtained using Davenport-Schinzel sequences [7,17], but so far we were able to apply this method only for $k = 2$.

References

1. A. Aggarwal, M. M. Klawe, S. Moran, and R. Wilber. Geometric applications of a matrix-searching algorithm. *Algorithmica*, 2:195–208, 1987. 262, 266
2. S. Arora, P. Raghavan, and S. Rao. Approximation schemes for euclidean k-medians and related problems. In *Proc. 30th Annual ACM Symposium on Theory of Computing (STOC'98)*, pages 106–113, 1998. 260
3. V. Auletta, D. Parente, and G. Persiano. Dynamic and static algorithms for optimal placement of resources in a tree. *Theoretical Computer Science*, 165:441–461, 1996. 261

4. V. Auletta, D. Parente, and G. Persiano. Placing resources on a growing line. *Journal of Algorithms*, 26:87–100, 1998. 261
5. M. Charikar and S. Guha. Improved combinatorial algorithms for facility location and k-median problems. In *Proc. 40th Symposium on Foundations of Computer Science (FOCS'99)*, pages 378–388, 1999. 260
6. M. Charikar, S. Guha, E. Tardos, and D. Shmoys. A constant-factor approximation algorithm for the k-median problem. In *Proc. 31st Annual ACM Symposium on Theory of Computing (STOC'99)*, 1999. 260
7. H. Davenport and A. Schinzel. A combinatorial problem connected with differential equations. *American J. Math.*, 87:684–694, 1965. 270
8. M. R. Garey and D. S. Johnson. *Computers and Intractability: a Guide to the Theory of NP-completeness*. W. H. Freeman and Co., 1979. 260
9. R. Gavish and S. Sridhar. Computing the 2-median on tree networks in $O(n \log n)$ time. *Networks*, 26:305–317, 1995. 261
10. M. H. Halldorson, K. Iwano, N. Katoh, and T. Tokuyama. Finding subsets maximizing minimum structures. In *Proc. 6th Annual Symposium on Discrete Algorithms (SODA'95)*, pages 150–157, 1995. 263
11. R. Hassin and A. Tamir. Improved complexity bounds for location problems on the real line. *Operation Research Letters*, 10:395–402, 1991. 261
12. W. L. Hsu. The distance-domination numbers of trees. *Operation Research Letters*, 1:96–100, 1982. 260
13. O. Kariv and S. L. Hakimi. An algorithmic approach to network location problems II: The p-medians. *SIAM Journal on Applied Mathematics*, 37:539–560, 1979. 260
14. M. R. Korupolu, C. G. Plaxton, and R. Rajaraman. Analysis of a local search heuristic for facility location problems. *Journal of Algorithms*, 37:146–188, 2000. 260
15. B. Li, X. Deng, M. Golin, and K. Sohraby. On the optimal placement of web proxies on the Internet: linear topology. In *Proc. 8th IFIP Conference on High Peformance Netwworking (HPN'98)*, pages 00–00, 1998. 261
16. B. Li, M. J. Golin, G. F. Italiano, X. Deng, and K. Sohraby. On the optimal placement of web proxies in the Internet. In *IEEE InfoComm'99*, pages 1282–1290, 1999. 261
17. M. Sharir and P. K. Agarwal. *Davenport-Schinzel sequences and their geometric applications*. Cambridge University Press, 1995. 270
18. A. Tamir. An $O(pn^2)$ algorithm for the p-median and related problems on tree graphs. *Operations Research Letters*, 19:59–64, 1996. 260, 261, 263
19. A. Vigneron, L. Gao, M. Golin, G. Italiano, and B. Li. An algorithm for finding a k-median in a directed tree. *Information Processing Letters*, 74:81–88, 2000. 261
20. G. Woeginger. Monge strikes again: optimal placement of web proxies in the Internet. *Operations Research Letters*, 27:93–96, 2000. 261

On Pseudorandom Generators in NC^0⋆

Mary Cryan and Peter Bro Miltersen

BRICS, Basic Research in Computer Science
Centre of the Danish National Research Foundation
Department of Computer Science, University of Aarhus
{maryc,bromille}@brics.dk

Abstract. In this paper we consider the question of whether NC^0 circuits can generate pseudorandom distributions. While we leave the general question unanswered, we show

- Generators computed by NC^0 circuits where each output bit depends on at most 3 input bits (i.e, NC^0_3 circuits) and with stretch factor greater than 4 are not pseudorandom.
- A large class of "non-problematic" NC^0 generators with superlinear stretch (including all NC^0_3 generators with superlinear stretch) are broken by a statistical test based on a linear dependency test combined with a pairwise independence test.
- There is an NC^0_4 generator with a super-linear stretch that passes the linear dependency test as well as k-wise independence tests, for any constant k.

1 Introduction

The notion of deterministically expanding a short seed into a long string that looks random to efficient observers, i.e., the notion of a *pseudorandom generator*, has been a fundamental idea in complexity as well as cryptography. Nevertheless, the question of whether strong pseudorandom generators actually exist is a huge open problem, as their existence (at least with cryptographic parameters) implies that $P \neq NP$: One can prove that a generator G is not pseudorandom by presenting a polynomial-time algorithm to decide range(G). Therefore if $P = NP$, strong pseudorandom generators do not exist.

On the other hand, that G has an NP-hard range does by no means guarantee or even suggest that G is a strong generator, as there may be statistical regularities in the output of the generator. For this reason, existing pseudorandom generators are proven to be strong under stronger hardness assumptions than $P \neq NP$. Some of the most important research in this area was presented in a series of papers by Blum and Micali [2], Yao [18] and Goldreich and Levin [6]. Håstad, Impagliazzo, Levin and Luby [9,8], building on this research, finally showed that the existence of a cryptographic one-way function is sufficient (and

---

⋆ Full version at http://www.brics.dk/~maryc/publ/psu.ps. Partially supported by the IST Programme of the EU, contract number IST-1999-14186 (ALCOM-FT).

J. Sgall, A. Pultr, and P. Kolman (Eds.): MFCS 2001, LNCS 2136, pp. 272–284, 2001.

trivially necessary) for the existence of a cryptographic pseudorandom generator. This general construction only depends on a generic one-way function, but the resulting generator is quite involved and seems to need the full power of the complexity class **P**.

The general construction left open the precise computational power needed to produce pseudorandomness. Kharitonov [11] showed, under a more specific hardness assumption, that there is a secure pseudorandom generator in $\mathbf{NC^1}$ for any polynomial stretch function. Impagliazzo and Naor [10] showed how to construct secure pseudorandom generators based on the assumed intractability of the subset sum problem. In particular, they showed how to construct a generator with a non-trivial stretch function (expanding n bits to $n + \Theta(\log(n))$ bits) in the complexity class $\mathbf{AC^0}$. This suggests that even rudimentary computational resources are sufficient for producing pseudorandomness (it is worth noting that Linial, Mansour and Nisan [14] proved that there are no pseudorandom *function* generators in $\mathbf{AC^0}$ with very good security parameters).

From both a theoretical and a practical point of view it seems worth finding out *how* rudimentary pseudorandom generators can be. There are different interpretations of this question, depending on how we formalise "rudimentary" or "simple". In previous work Kharitonov et al. [12] and Yu and Yung [19] proved strong negative results about the ability of various automata and other severely space-restricted devices to produce pseudorandomness. In this paper, we interpret "simple" in terms of circuit complexity and ask: *Are there pseudorandom generators in* $\mathbf{NC^0}$?, that is, are there pseudorandom generators where each output bit depends on only a constant number of input bits? Such generators would be appealing in practice as they can be evaluated in constant time using only bounded fan-in hardware. However, it seems that with such a severe constraint on the generator, there would certainly be statistical regularities in the output, so that we could construct a sequence of circuits to distinguish between the uniform distribution and output from the generator. This is a tempting conjecture, but we have not been able to prove it without further restrictions.

Our main results are:

Theorem 11 *There is no strong pseudorandom generator with a stretch factor greater than 4 for which every output bit depends on at most 3 input bits (that is, no* $\mathbf{NC_3^0}$*-generator).*

We can actually prove a weaker version of the theorem above for a more general class of "non-problematic" generators of which $\mathbf{NC_3^0}$ generators are a special case, and we do this in Theorem 9. It is interesting to note that deciding the *range* of a $\mathbf{NC_3^0}$ circuit is, in general, **NP**-hard (see Proposition 1), so we cannot use the approach of inverting the generator's output to prove either of these two theorems. Instead, in Theorem 9 we show that we can break any such generator with a statistical test consisting of a linear dependency test and a test for pairwise independence of the output bits. On the other hand, we show

Theorem 14 *There is a generator with a superlinear stretch for which every output bit depends on at most 4 input bits (i.e., an* $\mathbf{NC_4^0}$*-generator) so that*

that there are no linear dependencies among the output bits and so that for any constant k and for sufficiently large input size, all bits of the output are k-wise independent.

The question of whether there are true pseudorandom generators in $\mathbf{NC}_4^0$ is still open. We have no construction that we believe is truly pseudorandom, nor do we have a general scheme for breaking such generators. We have been able to reduce the number of boolean functions on 4 variables that could possibly serve as basis for such a generator to 4, up to isomorphism.

2   Definitions and Background

Definition 1. A circuit $C : \{0,1\}^n \to \{0,1\}^N$ is in $\mathbf{NC}_c^0$ if every output bit of C is a function of at most c input bits. The circuit C is associated with an induced distribution on $\{0,1\}^N$, where the probability of $y \in \{0,1\}^N$ is the probability that C outputs y when the input to C is chosen from the uniform distribution U_n on $\{0,1\}^n$. We will be flexible about notation and use C to denote this induced distribution as well as the circuit itself. We say a *sequence* of circuits $\{C_n\}_n : \{0,1\}^n \to \{0,1\}^{\ell(n)}$ is in $\mathbf{NC}_c^0$ if C_n is in $\mathbf{NC}_c^0$ for all n. The function $\ell(n)$ is known as the *stretch function* of the sequence and the *stretch factor* is $\ell(n)/n$.

Definition 2. A function $f : \mathbf{N} \to [0,1]$ is said to be *negligible* if for every $c > 0$, there is some constant n_c such that $f(n) < n^{-c}$ for every $n \geq n_c$. It is said to be *overwhelming* if $1 - f$ is negligible.

Definition 3 (from [10]). A function $G : \{0,1\}^n \to \{0,1\}^{\ell(n)}$ is a *pseudorandom generator* if every non-uniform polynomial-time algorithm A has a *negligible* probability of distinguishing between outputs of G and truly random sequences; that is, for every algorithm A,

$$|\Pr[A(G(x)) = 1] - \Pr[A(y) = 1]| \quad \text{is negligible,}$$

where x and y are chosen from U_n and $U_{\ell}(n)$ respectively.

Deviating slightly from the above definition, we prove, for convenience, our non-pseudorandomness results for sequences of $\mathbf{NC}_c^0$ distributions by using a *statistical test* to distinguish between *several* samples of the uniform distribution and several samples of the $\mathbf{NC}_c^0$ distributions, rather than one sample.

Definition 4. An efficient *statistical test* is a procedure which takes as input a parameter N and $m = m(N)$ *samples* (for some $m(N) = N^{O(1)}$) $y_1, \ldots, y_m$ from $\{0,1\}^N$, runs in polynomial time, and either ACCEPTs or REJECTs. Furthermore, if $y_1, \ldots, y_m$ are chosen from the uniform distribution U_N, the probability of acceptance is overwhelming (that is, the probability of rejection is negligible).

Definition 5. Let A be a statistical test. An ensemble $\{D_n\}_{n\in N}$ of probability distributions on $\{0,1\}^{\ell(n)}$, $\ell : \mathbf{N} \to \mathbf{N}$ is said to *fail* A if the probability that A rejects $y_1, \ldots, y_{m(\ell(n))}$ chosen independently from D_n is *not* negligible. Otherwise the ensemble is said to *pass* A.

Although Definition 3 is the traditional definition of pseudorandomness, it is well known that an ensemble of distributions generated by a generator fails *some* (possibly non-uniform) statistical test in the sense of Definition 5 if and only if the generator is *not* pseudorandom in the sense of Definition 3 (see Goldreich [5], page 81] for details).

Some of the results in Sections 3 and 4 refer to special $\mathbf{NC}_c^0$ circuits that we refer to as *non-problematic* circuits. We need the following definition:

Definition 6. A function $f : \{0,1\}^c \to \{0,1\}$ is *affine* if $f(x_1, \ldots, x_c)$ can be written in the form $\alpha_1 x_1 + \ldots \alpha_c x_c + \alpha_{c+1}$ (mod 2), for $\alpha_i \in \{0,1\}$. We say that f is *statistically dependent* on the input variable x_i if either $\Pr[f(x) = 1 \mid x_i = 1] \neq 1/2$ or $\Pr[f(x) = 1 \mid x_i = 0] \neq 1/2$. An $\mathbf{NC}_c^0$ circuit is *non-problematic* if for every output gate $y_j = f_j(x_{j,1}, \ldots, x_{j,c})$, the function f_j is either an affine function or depends statistically on one of its variables.

Before we present our results, we discuss previous negative results on generation of pseudorandomness. Lower bounds for pseudorandom *function* generators were studied in [14,16,13]. The only papers we are aware of that give impossibility results or lower bounds for "plain" pseudorandom generators as defined above, are the papers by Kharitonov et al. [12] and Yu and Yung [19].

Kharitonov et al. [12] proved that no generator that is either a one-way logspace machine or a one-way pushdown machine is a strong generator, even when the generator is only required to extend the input by a single bit. They also considered the question of whether sublinear-space generators exist and related this issue to the $\mathbf{L}$ vs $\mathbf{P}$ question. It is worth noting that the proofs of non-pseudorandomness that Kharitonov et al. obtain for one-way logspace machines and one-way pushdown machines (and for finite reversal logspace machines) depend on showing that the range of these generators can be recognized in polynomial time.

Yu and Yung [19] considered bidirectional finite state automata and bidirectional machines with sublogarithmic space ($o(\log n)$ space) as generators. They proved that bidirectional finite state automata are not strong pseudorandom generators, even when the generator only extends the input by one bit. For these generators, they also showed that the tester for distinguishing between the generator's output and the uniform distribution can be assumed to lie in $\mathbf{L}^2$. For bidirectional machines with $o(\log n)$-space, they showed that any generator in this class with superlinear stretch is not pseudorandom. All the impossibility results of Yu and Yung are again obtained by showing that the range of the generator can be recognized in polynomial time.

We can use a similar strategy to show that pseudorandom generators in $\mathbf{NC}_2^0$ do not exist: Note that for all uniform sequences $\{C_n\}_n$ of circuits in $\mathbf{NC}_2^0$, range(C_n) is in $\mathbf{P}$. This follows because the problem is a special case of 2-SAT.

However, this is a strategy we cannot use (in general) for distributions generated by $\mathbf{NC^0}$-circuits, because we now show that there are sequences $\{C_n\}_n$ of circuits in $\mathbf{NC^0_3}$ such that deciding $\text{range}(C_n)$ is $\mathbf{NP}$-complete. The proposition below is an improvement of a result due to Agrawal et al. [1, Proposition 1], who showed that there is a function f in $\mathbf{NC^0_4}$ such that that is $\mathbf{NP}$-complete to invert f. It also improves an unpublished theorem of Valiant (Garey and Johnson [4], page 251), stating that inverting multinomials over $GF(2)$ is $\mathbf{NP}$-hard: we prove that this is the case even for a specific sequence of multinomials.

Proposition 7. *There is a uniform sequence $\{C_n\}_{n \in \mathbf{N}}$ of $\mathbf{NC^0_3}$ circuits such that $\text{range}(C_n)$ is $\mathbf{NP}$-complete. Also, every output gate in every circuit in the sequence is a degree-2 multinomial over $GF(2)$.*

Proof. The reduction is from the $\mathbf{NP}$-complete problem 3-SAT. For every $n \in \mathbf{N}$ we construct a $\mathbf{NC^0_3}$ circuit to model 3-SAT problem on n variables. The circuit has an input gate x_i for $1 \le i \le n$ for each logical variable, and three helper inputs $h_{j,1}, h_{j,2}, h_{j,3}$ for every possible three-literal clause c_j (there are $8\binom{n}{3}$ c_js). The circuit has four output gates $y_{j,0}, y_{j,1}, y_{j,2}, y_{j,3}$ for every c_j. Let $N = \frac{16}{3}n(n-1)(n-2)$ denote the total number of output gates.

The gates are connected to form a $\mathbf{NC^0_3}$ circuit as follows: for every possible clause c_j, we connect $h_{j,1}, h_{j,2}$ and $h_{j,3}$ (the "helper" inputs) to $y_{j,0}$ as follows:

$$y_{j,0} =_{def} (1 + h_{j,1} + h_{j,2} + h_{j,3}) \pmod 2 \tag{1}$$

It is easy to check that $y_{j,0} = 1$ iff an even number (0 or 2) of the helper inputs $\{h_{j,1}, h_{j,2}, h_{j,3}\}$ are switched on. Let the literals in clause c_j be $\ell_{j,1}, \ell_{j,2}$ and $\ell_{j,3}$. Each of the output gates $\{y_{j,1}, y_{j,2}, y_{j,3}\}$ is the disjunction of one helper input and one of the literals in c_j:

$$y_{j,i} =_{def} 1 - (1 - \ell_{j,i})(1 - h_{j,i}) \pmod 2 \quad i = 1, 2, 3 \tag{2}$$

Two facts:

(i) If all of the helper inputs for c_j are 1 then $y_{j,0} = 0$ and $y_{j,1}y_{j,2}y_{j,3} = 111$ both hold.
(ii) Suppose the x_i inputs are fixed but the helper inputs are not. Then we can set values for $h_{j,1}, h_{j,2}, h_{j,3}$ to give $y_{j,0}y_{j,1}y_{j,2}y_{j,3} = 1111$ iff at least one of the literals in c_j is true under the truth assignment given by the x_i input variables.

For any instance (X, C) of 3-SAT with $n = |X|$ variables, define $a \in \{0,1\}^N$ by setting $a_{j,0}a_{j,1}a_{j,2}a_{j,3} = 1111$ for every $c_j \in C$ and $a_{j,0}a_{j,1}a_{j,2}a_{j,3} = 0111$ for every $c_j \notin C$. By fact (i), we can imagine that all the helper inputs for clauses outside C are set to 1. Then by fact (ii), $a \in range(C_n)$ iff there is some truth assignment for the x_i variables that satisfies some literal of every clause in C.

Note that the functions defined for the output gates (Equations 1 and 2) are all degree 2 multinomials.

3 Lower Bounds

In this section we prove that sequences of non-problematic $\mathbf{NC}_c^0$ circuits with a "large enough" (constant) stretch factor are not pseudorandom generators.

The particular statistical test that we use to detect non-pseudorandomness is the LINPAIR test, which tests for non-trivial linear dependencies in the output (Step (i)) and also tests that the distribution on every pair of indices is close to uniform. We also use a generalization of the LINPAIR test called the LIN(k) test, which instead of merely checking that the distribution on every *pair* of outputs is close to uniform, checks that the distribution on every group of k indices is close to uniform (Step (ii)). Note that Step (i) of LIN(k) can be computed in polynomial-time by Gaussian elimination. Also for any constant $k \in \mathbf{N}$, Step (ii) of LIN(k)[m, N] is polynomial in m and N, because there are only $2^k \binom{N}{k}$ different $F_{j_1,\ldots,j_k}[b_1,\ldots,b_k]$ values to calculate and test.

Throughout the paper we will use LINPAIR[m, N] to denote LIN(2)[m, N].

Algorithm 1 LIN(k)[m, N]

---

input: m samples $a_1,\ldots,a_m \in \{0,1\}^N$. Each a_i is written as $a_{i,1}\ldots a_{i,N}$.
output: ACCEPT or REJECT

(i) Check the linear system $\sum_j z_j a_{i,j} = z_{N+1}(\bmod\ 2)$ ($1 \leq i \leq m$) for a non-trivial solution;
If *there is a non-trivial solution* **then** REJECT;
else

    (ii) For every set of k indices $j_1,\ldots,j_k$ ($1 \leq j_1 < \ldots j_k \leq N$), calculate

        $F_{j_1,\ldots,j_k}[b_1,\ldots,b_k] =_{def} \frac{1}{m}|1 \leq i \leq m : a_{i,j_1}\ldots a_{i,j_k} = b_1\ldots b_k]$,
        for every tuple $b_1\ldots b_k \in \{0,1\}^k$;
    if $|F_{j_1,\ldots,j_k}[b_1,\ldots,b_k] - (1/2^k)| > \frac{1}{N}$ for any $j_1,\ldots,j_k$ and $b_1,\ldots,b_k$
    then
        REJECT;
    else
        ACCEPT;
    end
end

Now we show that LIN(k) is a statistical test for any k (it accepts the uniform distribution):

Lemma 8. *Let* $m : \mathbf{N} \to \mathbf{N}$ *be any function satisfying* $m(N) \geq N^2 \log^2 N$. *Then if the* LIN(k) *algorithm is run with* m *samples from* U_N, *the probability that* LIN(k) *accepts is overwhelming.*

Proof. First consider Step (i). First note that since we are working with binary arithmetic modulo 2, any non-trivial solution to the linear system $\sum_{j=1}^N z_j a_{i,j} =$

z_{N+1}(mod 2) corresponds to a non-empty subset $S \subseteq \{1, \ldots, N\}$ such that either (a) $\sum_{j \in S} a_{i,j} = 0$ holds for all $1 \leq i \leq m$ or (b) $\sum_{j \in S} a_{i,j} = 1$ holds for all i. For any fixed set S of indices, the probability that $\sum_{j \in S} a_{i,j} = 0$ holds for a single a_i chosen from the uniform distribution is $1/2$. The probability that (a) holds for all $1 \leq i \leq m$ is $1/2^m$, and the probability that either (a) holds or (b) holds is $1/2^{m-1}$. There are only 2^N subsets of $\{1, \ldots, N\}$, so we can bound the probability of finding any non-trivial solution to $\sum_{j=1}^{N} z_j a_{i,j} = z_{N+1}$(mod 2) by $2^N/2^{m-1}$, which is at most 2^{-N} when $m \geq N^2 \log^2 N$.

Next we bound the probability that Step (ii) rejects. For any k indices $j_1, \ldots, j_k$, and any k-tuple of bits $b_1, \ldots b_k \in \{0,1\}^k$, the probability that $j_1, \ldots, j_k$ reads $b_1 \ldots b_k$ in a sample from U_N is $1/2^k$. By Hoeffding's Inequality (see McDiarmid [15, Corollaries (5.5) and (5.6)]), if we take m samples from U_N, then

$$\Pr\left[|F_{j_1,\ldots,j_k}[b_1, \ldots, b_k]| - 1/2^k| \geq 1/N\right] \leq 2\exp[-1/3(2^k/N)^2 m(1/2^k)]$$
$$\leq 2\exp[-2^k/3(\log N)^2]$$

which is at most $2N^{-(\log N)2^{k-2}}$. There are $2^k N^k$ different tests in Step (ii), so the probability that any one of these fails is at most $2^{k+1} N^{k-(\log N)2^{k-2}}$, which is at most $N^{-\log N/2}$ for large N.

So LIN(k) succeeds with probability at least $1 - 2^{-N} - N^{-\log N/2}$.

Theorem 9. *For any constant $c \geq 2$ there is a constant $d \in \mathbf{N}$ so that the following holds: If $\{C_n\}_{n \in \mathbf{N}}$ is a family of non-problematic $\mathbf{NC}_c^0$ circuits whose stretch function satisfies $\ell(n) \geq dn$, then $\{C_n\}_{n \in \mathbf{N}}$ fails LINPAIR$[m, \ell(n)]$ for any function m with $m(\ell(n)) \geq \log^2 \ell(n)$. Hence $\{C_n\}_{n \in \mathbf{N}}$ is not pseudorandom.*

Proof. The constant $d = (2^{(c-1)2^c}(c-1)!) + 1$ is large enough for this theorem. We will show that for all such circuits, LINPAIR$[m, \ell(n)]$ fails for all $m \geq \log^2 \ell(n)$ (for large enough n).

There are two cases. The circuits are non-problematic, so every output bit of C_n is either an affine bit or is statistically dependent on one of its input variables.

Case (i): First we prove the theorem for the case when C_n has at least $n+1$ affine outputs. Without loss of generality, let the affine outputs be $y_1, \ldots, y_{n+1}$. Since there are only n input variables, and each output y_i is an affine combination of some input variables, there is a non-trivial affine dependence among the y_i's. That is, we can find constants $\alpha_1, \ldots, \alpha_{n+1} \in \{0,1\}$ such that not all of the α_i are 0 and such that either $\sum_{i=1}^{n} \alpha_i y_i = 0$(mod 2) or $\sum_{i=1}^{n} \alpha_i y_i = 1$(mod 2) holds for all $x_1, \ldots, x_n \in \{0,1\}$. Therefore, Step (i) of the LINPAIR algorithm will always find a non-trivial solution (regardless of the number m of samples), and the algorithm will reject.

Case (ii): If there are less than $n+1$ affine output bits, then there are at least $n(d-1)$ output bits that are statistically dependent on at least one of their input variables. Therefore there is at least one input bit x such that $(d-1)$ different output bits are all statistically dependent on x. Assume wlog that these output bits are $y_1, \ldots, y_{d-1}$. We will prove that the distribution on some pair of these bits deviates from the uniform distribution U_2, and therefore will fail Step (ii) of LINPAIR.

We use the Erdős-Rado sunflower lemma (see Papadimitriou[17, page 345]), which gives conditions that are sufficient to ensure that a family $\mathcal{F}$ of sets contains a sunflower of size p, where a sunflower is a subfamily $\mathcal{F}' \subset \mathcal{F}$ of size p, where every pair of subsets in $\mathcal{F}'$ has the same intersection. The family $\mathcal{F}$ that we consider contains a set I_i for every $1 \le i \le d - 1$, where I_i contains all the input variables feeding y_i except x. By the sunflower lemma, our definition of d ensures that there is a sunflower of size $d' = 2^{2^c} + 1$ in $\mathcal{F}$. That is, we have d' special output bits $y_1, \ldots, y_{d'}$, such that each y_i depends on the bit x, on some additional input bits $u = (u_1, \ldots, u_r)$ common to $y_1, \ldots, y_{d'}$, and on some extra input bits z_i not shared by any two outputs in $y_1, \ldots, y_{d'}$.

There are only 2^{2^c} different functions on c variables, so if we take the most common function f among the output bits $y_1, \ldots, y_{d'}$ (with the input bits ordered as x first, then u, then the private input bits), then we can find at least 2 output bits with the same function. Assume these are y_1 and y_2. So we have

$$y_1 = f(x, u_1, \ldots, u_r, z_1) \quad \text{and} \quad y_2 = f(x, u_1, \ldots, u_r, z_2)$$

where z_1 and z_2 are vectors of input bits, and z_1 and z_2 are disjoint and of the same length. We can now show that the distribution on (y_1, y_2) is not U_2. Calculating, we have

$$\Pr[y_1 y_2 = 11] = \frac{1}{2}(\Pr[y_1 y_2 = 11 | x = 1] + \Pr[y_1 y_2 = 11 | x = 0])$$

$$= \frac{1}{2}(\text{Avg}_j(\Pr[y_1 y_2 = 11 | xu = 1j]) + \text{Avg}_j(\Pr[y_1 y_2 = 11 | xu = 0j]))$$

$$= \frac{1}{2}(\text{Avg}_j(\Pr[y_1 = 1 | xu = 1j]^2) + \text{Avg}_j(\Pr[y_1 = 1 | xu = 0j]^2))$$

$$= \frac{1}{2}(\text{Avg}_j(\beta_j)^2 + \text{Avg}_j(\gamma_j)^2)$$

where Avg_j denotes the average of its argument over all $j \in \{0,1\}^r$, and $\beta_j = \Pr[y_1 = 1 | xu = 1j]$ and $\gamma_j = \Pr[y_1 = 1 | xu = 0j]$. We know that the output y_1 is statistically dependent on x, so we assume wlog that $\Pr[y_1 = 1 | x = 1] = 1/2 + \epsilon$, where $\epsilon \ge 1/2^c$ (since y_1 is a function on c inputs). Then, assuming $\Pr[y_1 = 1] = 1/2$, we have $\Pr[y_1 = 1 | x = 0] = 1/2 - \epsilon$ (if we have $\Pr[y_1 = 1] \ne 1/2$, then we are already finished).

For any sequence of numbers $\{t_j\}$, $\text{Avg}_j(t_j)^2 \ge (\text{Avg}_j t_j)^2$. Also, $\Pr[y_1 = 1 | x = 1] = \text{Avg}_j \beta_j$ and $\Pr[y_1 = 1 | x = 0] = \text{Avg}_j \gamma_j$. Therefore

$$\Pr[y_1 y_2 = 11] \ge \frac{1}{2}(\Pr[y_1 = 1 | x = 1]^2 + \Pr[y_1 = 1 | x = 0]^2)$$

$$= \frac{1}{2}((1/2 + \epsilon)^2 + (1/2 - \epsilon)^2)$$

$$= \frac{1}{2}(1/2 + 2\epsilon^2)$$

Now consider the calculation of $F_{1,2}[1,1]$ in Step (ii). The expected value of $F_{1,2}[1,1]$ is at least $1/4 + 1/2^{2c}$. Therefore, if the deviation of $F_{1,2}[1,1]$ from its expectation is bounded by $1/2^{2c+1}$, we have $|F_{1,2}[1,1] - 0.25| \geq 1/2^{2c+1}$. Using Chernoff Bounds (see McDiarmid [15]), the probability that $F_{1,2}[1,1]$ deviates from its expectation by $1/2^{2c+1}$ is at most $2\exp[-2m(\ell(n))/2^{2(2c+1)}]$. Thus for $m(\ell(n)) \geq \log^2 \ell(n)$ and large enough n, we have $|F_{1,2}[1,1] - 0.25| \geq 1/2^{2c+1} \geq 1/\ell(n)$ with probability at least $1 - 2\exp[-2(\log \ell(n))^{3/2}]$. Therefore for sufficiently large $\ell(n)$, Step (ii) fails and LINPAIR$[m, \ell(n)]$ rejects with probability at least $1 - 2\ell(n)^{-2(\log^{1/2}\ell(n))}$.

One of our referees has pointed out that the bound on d in Theorem 9 can be improved by testing the variance of sums of output variables. For any discrete random variable X, the variance of X, denoted var(X), is defined as the expectation of $(X - \mathbf{E}[X])^2$. In case (ii) of Theorem 9, we find a set of output variables $y_1, \ldots, y_{d-1}$ such that each y_i is statistically dependent on the same variable x. Suppose $d \geq 2^{2c} + 1$. Then var$(\sum_{i=1}^{d-1} z_i)$ is at least $d - 1$ for some choice of $z_i \in \{y_i, \bar{y}_i\} : i = 1, \ldots d - 1$, whereas the variance is $(d-1)/4$ when the z_i are independent uniform random bits (see Grimmett and Stirzaker [7]). The following statistical test distinguishes between the $\mathbf{NC}_c^0$ circuits of case (ii) and the uniform distribution: for every $z_{i_1}, \ldots, z_{i_{d-1}}$ where $z_{i_j} \in \{y_i, \bar{y}_i\}$ and the i_j are distinct indices, calculate the average of $(\sum_{j=1}^{d-1} z_{i_j} - (d-1)/2)^2$ over the set of samples. If any of these estimates is greater than $(d-1)/2$, then reject.

The following observation can be easily verified by computer.

Observation 10 *Every $\mathbf{NC}_3^0$ circuit is non-problematic.*

By Observation 10, Theorem 9 holds for *all* sequences of $\mathbf{NC}_3^0$ circuits. By a more careful analysis of $\mathbf{NC}_3^0$ circuits, we can prove a stronger result:

Theorem 11. *Let $\{C_n\}_{n\in\mathbf{N}}$ be any ensemble of $\mathbf{NC}_3^0$ circuits which has a stretch function $\ell(n)$ with $\ell(n) \geq 4n + 1$. Then $\{C_n\}_{n\in\mathbf{N}}$ fails LIN$(4)[m, \ell(n)]$ for any m satisfying $m(\ell(n)) \geq \log^2 \ell(n)$ and therefore is not pseudorandom.*

4   A Generator That Passes LIN(k)

Our original goal was to prove that for all $c \in \mathbf{N}$ there exists some constant $d \in \mathbf{N}$ such that any sequence of $\mathbf{NC}_c^0$ circuits with stretch factor at least d is not pseudorandom. So far we have only been able to prove this for sequences of non-problematic circuits. For the set of $\mathbf{NC}_4^0$ circuits, we can classify the number of different problematic output gate functions in the following way:

Observation 12 *Consider the set of all problematic functions $f : \{0,1\}^4 \to \{0,1\}$ on four variables. Define an equivalence relation on this set by saying $f \equiv g$ iff*

$$f(x_1, x_2, x_3, x_4) = b_5 + g(\pi(x_1 + b_1, \ldots, x_4 + b_4))(\mathrm{mod}\ 2)$$

for some permutation π and some five boolean values $b_1, \ldots, b_5 \in \{0, 1\}$. That is, two functions are equivalent if one can be obtained from the other by permuting and possibly negating input variables and possibly negating the output. Then, there are only four problematic non-equivalent functions on 4 inputs, namely

$$f_1(x_1, \ldots, x_4) = x_1 + x_2 + x_3 x_4 \pmod{2}$$
$$f_2(x_1, \ldots, x_4) = x_1 + x_2 x_3 + x_3 x_4 + x_2 x_4 \pmod{2}$$
$$f_3(x_1, \ldots, x_4) = x_1 + x_2 + x_4(x_2 + x_3) \pmod{2}$$
$$f_4(x_1, \ldots, x_4) = x_1 + x_2 + (x_1 + x_4)(x_2 + x_3) \pmod{2}$$

Proof. Case analysis (by computer).

Any $\mathbf{NC}_4^0$ circuit with a large enough stretch factor is guaranteed to either (a) contain enough non-problematic output gates to allow us to use Theorem 9 to prove non-pseudorandomness, or (b) contain a large number of output gates of the form f_i, for one of the $f_1, \ldots, f_4$ functions. So when we consider candidates for pseudorandom generators in $\mathbf{NC}_4^0$ with superlinear stretch, we only have to consider generators using gates of the form of *one* of f_1, f_2, f_3, f_4.

We show that Theorem 9 cannot be extended to non-problematic functions by exhibiting an $\mathbf{NC}_4^0$ generator using only output gates of the form f_1 and passing $\mathrm{LIN}(k)$ for all constants k. We need the following lemma (a variation of the well-known Schwartz-Zippel lemma):

Lemma 13. *For any r-variable multilinear polynomial f of degree at most 2 over $GF(2)$ which is not a constant function, $\mathrm{Pr}_x[f(x) = 0] \in [1/4, 3/4]$ when x is chosen uniformly at random from $\{0, 1\}^r$.*

Theorem 14. *Let $k : \mathbf{N} \to \mathbf{N}$ be any function satisfying $2 \le k(n) < \log n$. Then there is a generator in $\mathbf{NC}_4^0$ with stretch function $\ell(n) = n^{1+\Theta(1/k)}$, such that for any function m satisfying $m \ge \ell(n)^2 (\log^2 \ell(n))$, the generator passes $\mathrm{LIN}(k(n))[m, \ell(n)]$ with overwhelming probability.*

Proof. In this proof we will informally use k to denote $k(n)$ and ℓ to denote $\ell(n)$. Our construction uses the following result from extremal graph theory (see Bollabás [3, page 104]): For all n there exists a graph G on $n/2$ nodes with ℓ edges, such that the girth of G (i.e., the length of the shortest cycle of G) is at least k. We use this graph to construct a circuit, with an output gate for every edge in G: the input bits of the generator are split into two sets, the set $\{z_1, \ldots, z_{n/2}\}$ which will represent the nodes of G, and another set $\{x_1, \ldots, x_{n/2}\}$. We assume some arbitrary enumeration $\{(u_i, v_i) : i = 1, \ldots, \ell\}$ of the edges of the graph, and we also choose any ℓ different ordered pairs $\{(i_1, i_2) : i = 1, \ldots, \ell\}$ from $\{1, \ldots, n/2\}^2$. Then, for every output bit y_i, we define

$$y_i =_{def} x_{i_1} x_{i_2} + z_{u_i} + z_{v_i} \pmod{2},$$

where z_{u_i} and z_{v_i} are the inputs from $\{z_1, \ldots, z_{n/2}\}$ representing u_i and v_i respectively. Note that $\mathrm{Pr}[y_i = 1] = 1/2$ for every i.

We show that the generator passes the linear dependency test of $\mathrm{LIN}(k)$ with high probability and also that the output bits are k-wise independent, and therefore pass Step (ii) of $\mathrm{LIN}(k)$.

First consider the test for linear dependence among the outputs $y_1, \ldots, y_\ell$ (Step (i)). By definition of the y_i functions, no two y_i share the same $x_{i_1} x_{i_2}$ term. Then for every sequence $\alpha_1, \ldots, \alpha_\ell \in \{0,1\}$ containing some non-zero term, $\sum_{i=1}^\ell \alpha_i y_i$ is not a constant function. Step (i) rejects (given m samples $a_1, \ldots, a_m \in \{0,1\}^\ell$) iff there exist $\alpha_1, \ldots, \alpha_\ell \in \{0,1\}$ not all zero such that

$$\sum_{i=1}^\ell \alpha_i a_{j,i} \quad \text{is constant for all } 1 \leq j \leq m$$

For any particular sequence $\alpha_1, \ldots, \alpha_\ell \in \{0,1\}$, Lemma 13 implies that the probability that m samples satisfy the equation above is at most $2(3/4)^m$. There are 2^ℓ different α-sequences, so the total probability that Step (i) rejects is at most $2^\ell 2(3/4)^m = 2^\ell 2(3/4)^{\ell^2 \log^2 \ell(n)} \leq (3/4)^{\ell^2/2}$, which is negligible.

For Step (ii), we show that any k outputs are mutually independent. Let k' be the minimum k' for which k' output bits are mutually dependent. Assume wlog that these output bits are $y_1, \ldots, y_{k'}$. Now, suppose there is some y_i with $1 \leq i \leq k'$ such that one of z_{u_i}, z_{v_i} does not appear in any other output function for $y_1, \ldots, y_{k'}$. If this is the case, then regardless of the values along the output gates $y_1 \ldots y_{i-1} y_{i+1} \ldots y_{k'}$, the probability that $y_i = 1$ is always $1/2$ (y_i is mutually independent of all the other y_i's). Therefore the set of outputs $y_1, \ldots y_{i-1}, y_{i+1} \ldots y_{k'}$ must be mutually dependent and we obtain a contradiction. Therefore if k' is the minimum value for which k' outputs are mutually dependent, then for every y_i, z_{u_i} and z_{v_i} appear in at least one other output from $y_1, \ldots, y_{k'}$. In terms of the original graph that we used to construct our circuit, we find that in the subgraph consisting of the set of edges for $y_1, \ldots, y_{k'}$, each vertex has degree at least 2. Then this subgraph contains a cycle of length at most k', so $k' \geq k$, as required.

Then an argument similar to the argument for $\mathrm{LIN}(k)$ in Lemma 8 shows that Step (ii) rejects with negligible probability.

5   Open Problems

The main open problem is of course whether $\mathbf{NC^0}$ circuits in general can be pseudorandom generators. We believe this may turn out to be a difficult question. Some, perhaps, easier subquestions are the following:

We have been able to show that there is no pseudorandom generator in $\mathbf{NC^0_3}$ that expands n bits to $4n+1$ bits. It would be interesting to optimise this to show the non-existence of generators in $\mathbf{NC^0_3}$ expanding n bits to $n+1$ bits. . Another goal that may be within reach is to prove that $\mathbf{NC^0_4}$ generators with superlinear stretch cannot be pseudorandom generators. There is no reason to believe that the construction of Theorem 14 is unbreakable. Indeed, note that the generator is not specified completely as the graph and the exact enumeration of pairs is unspecified. It is easy to give examples of specific graphs and enumerations where the resulting generator can be easily broken by testing whether a particular linear combination of the output bits yields an unbiased random variable. If one believes that $\mathbf{NC_4^0}$ generators in general are breakable (as we tend to do), Observation 12 suggests that a limited number of *ad hoc* tests may be sufficient to deal with all cases. For instance, it is conceivable that every such generator is broken by testing whether a particular linear combination of the output bits yields an unbiased random variable.

References

1. M. Agrawal, E. Allender and S. Rudich, "Reductions in Circuit Complexity: An Isomorphism Theorem and a Gap Theorem"; *Journal of Computer and System Sciences*, Vol **57**(2): pages 127-143, 1998. 276
2. M. Blum and S. Micali, "How to Generate Cryptographically Strong Sequences of Pseudo-Random Bits"; *SIAM Journal on Computing*, Vol **13**: pages 850-864, 1984. 272
3. B. Bollabás, *Extremal Graph Theory*, Academic Press Inc (London), 1978. 281
4. M. R. Garey and D. S. Johnson, *Computers and Intractability: A Guide to the Theory of NP-completeness*, W. H. Freeman and Company (1979). 276
5. O. Goldreich, *Modern Cryptography, Probabilistic Proofs and Pseudo-randomness*; Vol **17** of series on Algorithms and Combinatorics, Springer-Verlag, 1999. 275
6. O. Goldreich and L. A. Levin, "Hard-Core Predicates for any One-Way Function"; *Proceedings of the 21st Annual ACM Symposium on Theory of Computing*, pages 25-32, 1989. 272
7. G. R. Grimmett and D. R. Stirzaker, *Probability and Random Processes*, Oxford University Press, 1992. 280
8. J. Håstad, R. Impagliazzo, L. A. Levin and M. Luby, "A Pseudorandom Generator from any One-way Function"; *SIAM Journal on Computing*, Vol **28**(4): pages 1364-1396, 1999. 272
9. R. Impagliazzo, L. A. Levin and M. Luby, "Pseudo-random Generation from One-way Functions"; *Proceedings of the 21st Annual ACM Symposium on Theory of Computing*, pages 12-24, 1989. 272
10. R. Impagliazzo, M. Naor, "Efficient Cryptographic Schemes Provably as Secure as Subset Sum"; *Journal of Cryptology*, **9**(4): pages 199-216, 1996. 273, 274
11. M. Kharitonov, "Cryptographic Hardness of Distribution-specific Learning"; *Proceedings of the 25th Annual ACM Symposium on Theory of Computing*, pages 372-381, 1993. 273
12. M. Kharitonov, A. V. Goldberg and M. Yung, "Lower Bounds for Pseudorandom Number Generators"; *Proceedings of the 30th Annual Symposium on Foundations of Computer Science*, pages 242-247, 1989. 273, 275
13. M. Krause and S. Lucks, "On the minimal Hardware Complexity of Pseudorandom Function Generators", *Proceedings of the 18th Symposium on Theoretical Aspects of Computer Science*, 2001. 275
14. N. Linial, Y. Mansour and N. Nisan, "Constant Depth Circuits, Fourier Transform, and Learnability", *Journal of the ACM*, Vol **40**(3): pages 607-620, 1993. 273, 275
15. C. McDiarmid, "On the method of bounded differences", *London Mathematical Society Lecture Note Series* **141**, Cambridge University Press, 1989, 148–188. 278, 280

16. M. Naor and O. Reingold, "Synthesizers and Their Application to the Parallel Construction of Pseudorandom Functions", *Journal of Computer and Systems Sciences*, **58**(2): pages 336-375, 1999. 275
17. C. H. Papadimitriou, *Computational Complexity*, Addison-Wesley, 1994. 279
18. A.C.C. Yao, "Theory and Applications of Trapdoor Functions"; *Proceedings of the 23rd Annual Symposium on Foundations of Computer Science*, pages 80-91, 1982. 272
19. X. Yu and M. Yung, "Space Lower-Bounds for Pseudorandom-Generators"; *Proceedings of the Ninth Annual Structure in Complexity Theory Conference*, pages 186-197, 1994. 273, 275

There Are No Sparse NP_W-Hard Sets

Felipe Cucker[1] * and Dima Grigoriev[2]

[1] Department of Mathematics, City University of Hong Kong
83 Tat Chee Avenue, Hong Kong, P. R. of China
macucker@math.cityu.edu.hk
[2] IMR, Université de Rennes I, Campus de Beaulieu, Rennes 35042, France
dima@maths.univ-rennes1.fr

Abstract. In this paper we prove that, in the context of weak machines over $\mathbb{R}$, there are no sparse NP-hard sets.

1 Introduction

In [2] Berman and Hartmanis conjectured that all NP-complete sets are polynomially isomorphic. That is, that for all NP-complete sets A and B, there exists a bijection $\varphi : \Sigma^* \to \Sigma^*$ such that $x \in A$ if and only if $\varphi(x) \in B$. In addition both φ and its inverse are computable in polynomial time. Here Σ denotes the set $\{0, 1\}$ and Σ^* the set of all finite sequences of elements in Σ.

Should this conjecture be proved, we would have as a consequence that no "small" NP-complete set exists in a precise sense of the word "small". A set $S \subseteq \Sigma^*$ is said to be *sparse* when there is a polynomial p such that for all $n \in \mathbb{N}$ the subset S_n of all elements in S having size n has cardinality at most $p(n)$. If the Berman-Hartmanis conjecture is true, then there are no sparse NP-complete sets.

In 1982 Mahaney ([11]) proved this weaker conjecture showing that there exist sparse NP-hard sets if and only if P = NP. After this, a whole stream of reserach developed around the issue of reductions to "small" sets (see [1]).

In a different line of research, Blum, Shub and Smale introduced in [4] a theory of computability and complexity over the real numbers with the aim of modelling the kind of computations performed in numerical analysis. The computational model defined in that paper deals with real numbers as basic entities, and performs arithmetic operations on them as well as sign tests. Inputs and outputs are vectors in $\mathbb{R}^n$ and decision problems are subsets of $\mathbb{R}^\infty$, the disjoint union of $\mathbb{R}^n$ for all $n \geq 1$. The classes $P_{\mathbb{R}}$ and $NP_{\mathbb{R}}$ —which are analogous to the well known classes P and NP— are then defined and one of the main results in [4] is the existence of natural $NP_{\mathbb{R}}$-complete problems.

Clearly, the sparseness notion defined above for sets over $\{0, 1\}$ will not define any meaningful class over $\mathbb{R}$ since now the set of inputs of size n is $\mathbb{R}^n$ and this is an infinite set. A notion of sparseness over $\mathbb{R}$ capturing the main features of the discrete one (independence of any kind of computability notion and capture

---

* Partially supported by CERG grant 9040393.

J. Sgall, A. Pultr, and P. Kolman (Eds.): MFCS 2001, LNCS 2136, pp. 285–291, 2001.
© Springer-Verlag Berlin Heidelberg 2001

of a notion of "smallness"), however, was proposed in [6]. Let $S \subseteq \mathbb{R}^\infty$. We say that S is *sparse* if, for all $n \geq 1$, the set

$$S_n = \{x \in S \mid x \in \mathbb{R}^n\}$$

has dimension at most $\log^q n$ for some fixed q. Here dimension is the dimension, in the sense of algebraic geometry, of the Zariski closure of S_n. Note that this notion of sparseness parallels the discrete one in a very precise way. For a subset $S_n \subseteq \{0, 1\}^n$ its cardinality gives a measure of its size and a sparse set is one for which, for all n, this cardinality is polylogarithmic in the largest possible (i.e. 2^n the cardinality of $\{0, 1\}^n$). For a subset $S_n \subseteq \mathbb{R}^n$, we take the dimension to measure the size of S_n, and again define sparseness by the property of having this measure polylogarithmic in the largest possible (which is now n, the dimension of $\mathbb{R}^n$).

Using this definition of sparseness for subsets of $\mathbb{R}^\infty$ the main result of [6] proves that there are no sparse NP-complete sets in the context of machines over $\mathbb{R}$ which do not perform multiplications or divisions and branch on equality tests only. Note that this result is not conditioned to the inequality P $\neq$ NP since this inequality is known to be true in this setting (cf. [12]).

A variation on the BSS model attempting to get closer to the Turing machine (in the sense that iterated multiplication is somehow penalized) was introduced by Koiran in [10]. This model, which Koiran called *weak*, takes inputs from $\mathbb{R}^\infty$ but no longer measures the cost of the computation as the number of arithmetic operations performed by the machine. Instead, the cost of each individual operation $x \circ y$ depends on the sequences of operations which lead to the terms x and y from the input data and the machine constants.

In this paper we extend Mahaney's theorem to machines over $\mathbb{R}$ endowed with the weak cost. Again, this is not a conditional result since it is known that P $\neq$ NP in this context too (cf. [7]). If NP_W denotes the class of sets decided in non-deterministic polynomial cost, our main result is the following.

Theorem 1. *There are no sparse* NP_W*-hard sets.*

2   The Weak Cost

Let M be a machine over $\mathbb{R}$, let $\alpha_1, \ldots, \alpha_s$ be its constants and $a = (\alpha_1, \ldots, \alpha_s) \in \mathbb{R}^s$. Let $x = (x_1, \ldots, x_n) \in \mathbb{R}^\infty$. At any step ν of the computation of M with input x, the intermediate value $z \in \mathbb{R}$ produced in this step can be written as a rational function of a and x, $z = \varphi(a, x)$. This rational function only depends on the computation path followed by M up to ν (i.e. on the sequence steps previously performed by M) and is actually a coordinate of the composition of the arithmetic operations performed along this path (see [3] for details). Let $\varphi = \frac{g_\nu}{h_\nu}$ be the representation of φ obtained by retaining numerators and denominators in this composition. For example, the representation of the product $\frac{g}{h} \cdot \frac{r}{s}$ is always $\frac{gr}{hs}$ and the one of the addition $\frac{g}{h} + \frac{r}{s}$ always $\frac{gs+hr}{hs}$. We will now use g_ν and h_ν to define weak cost.

Definition 1. The *weak cost* of any step ν is defined to be the maximum of $\deg(g_\nu)$, $\deg(h_\nu)$, and the maximum bit size of the coefficients of g_ν and h_ν. For any $x \in \mathbb{R}^\infty$ the *weak cost of M on x* is defined to be the sum of the costs of the steps performed by M with input x.

The class P$_W$ of sets decided within *weak polynomial cost* is now defined by requiring that for each input of size n the weak cost of its computation is bounded by a polynomial in n. A set S is decided in *weak nondeterministic polynomial cost* (we write $S \in$ NP$_W$) if there is a machine M working within weak polynomial cost satisfying the following: for each $x \in \mathbb{R}^\infty$, $x \in S$ if and only if there is $y \in \mathbb{R}^\infty$ with size polynomial in n such that M accepts the pair (x, y).

Remark 1. The definitions above do not fully coincide with those given in [10] since this reference requires the representation of the rational functions φ above to be relatively prime. The definitions we give here, which are taken from [3], are essentially equivalent. For, if a set is in P$_W$ with the definition above, it is clearly in P$_W$ with Koiran's. The converse is more involved to prove. Roughly speaking, any machine can be simulated by another which keeps "programs" instead of performing the arithmetic operations at the computation nodes. When the computation reaches a branch node the program for the register whose value is tested for positivity is evaluated at the pair (a, x) to decide such positivity. Now note that one can use algorithms of symbolic computation to make the numerator and denominator of the rational function computed by the program relatively prime before evaluating.

3   Proof of the Main Result

Let $n \geq 1$. Consider the polynomial

$$f_n = x_1^{2^n} + \ldots + x_n^{2^n} - 1$$

and let $C_n = \{x \in \mathbb{R}^n \mid f_n(x) = 0\}$. The polynomial f_n is irreducible and the dimension of C_n is $n-1$. Let $C \subset \mathbb{R}^\infty$ be given by $C = \cup C_n$. We know (cf. [7]) that $C \in$ NP$_W$ but $C \notin$ P$_W$.

Let $S \subset \mathbb{R}^\infty$ be a NP$_W$-hard set. Then C reduces to S. That is, there exists a function $\varphi : \mathbb{R}^\infty \to \mathbb{R}^\infty$ computable with polynomial cost such that, for all $x \in \mathbb{R}^\infty$, $x \in C \iff \varphi(x) \in S$. For each $n \geq 1$, the restriction of φ to $\mathbb{R}^n$ is a piecewise rational function. Our first result, Proposition 2, gives some properties of this function. It uses the following simple fact in real algebraic geometry whose proof can be found in Chapter 19 of [3].

Proposition 1. *Let $f \in \mathbb{R}[x_1, \ldots, x_n]$ be an irreducible polynomial such that the dimension of its zero set $\mathcal{Z}(f) \subseteq \mathbb{R}^n$ is $n-1$. Then, for any polynomial $g \in \mathbb{R}[x_1, \ldots, x_n]$, g vanishes on $\mathcal{Z}(f)$ if and only if g is a multiple of f.* □

Proposition 2. *Let n be sufficiently large. There exist $x \in C_n$ and $U \subset \mathbb{R}^n$ an open ball centered at x such that the restriction of φ to U is a rational map $h : U \to \mathbb{R}^m$ for some m bounded by a polynomial in n. In addition, if $h_1, \ldots, h_m$ are the coordinates of h, then the degrees of the numerator and denominator of h_i are also bounded by a polynomial in n for $i = 1, \ldots, m$.*

Proof. Let M be a machine computing φ within weak polynomial cost. By unwinding the computation of M in a standard manner we obtain an algebraic computation tree of depth polynomial in n. To each branch η in this tree one associates a set $D_\eta \subseteq \mathbb{R}^n$ such that the D_η partition $\mathbb{R}^n$ (i.e. $\cup D_\eta = \mathbb{R}^n$ and $D_\eta \cap D_\gamma = \emptyset$ for $\eta \neq \gamma$). In addition, each branch η computes a rational map h_η and $\varphi_{|D_\eta} = h_\eta$. The set D_η is the set of points in $\mathbb{R}^n$ satisfying a system

$$\bigwedge_{i=1}^{s_\eta} q_i(x_1, \ldots, x_n) \geq 0 \wedge \bigwedge_{i=s_\eta+1}^{t_\eta} q_i(x_1, \ldots, x_n) < 0 \tag{1}$$

where the $q_i(X_1, \ldots, X_n)$ are the rational functions tested along the branch. Since M works within weak cost, the numerators and denominators of the q_i as well as those of h_η have degrees bounded by a polynomial in n.

Everything we need now to see is that for some branch η, D_η contains an open neighbourhood of a point $x \in C_n$.

To do so first notice that, by replacing each q_i by the product of its numerator and denominator, we can assume that the q_i are polynomials. Also, by writing $q_i \geq 0$ as $q_i = 0 \vee q_i > 0$ and distributing the disjunctions in (1) we can express D_η as a finite union of sets satisfying a system

$$\bigwedge_{i=1}^{s} q_i(x_1, \ldots, x_n) = 0 \wedge \bigwedge_{i=s+1}^{t} q_i(x_1, \ldots, x_n) < 0. \tag{2}$$

We have thus described $\mathbb{R}^n$ as a union of sets which are solutions of systems like (2). Since this union is finite there exists one such set D containing a subset H of C_n of dimension $n - 1$. Let D be the solution of a system like (2). We claim that there are no equalities in such system. Assume the contrary. Then there is a polynomial q such that $H \subset \mathcal{Z}(q)$. Since $\dim H = n - 1$ and C_n is irreducible this implies that $q(C_n) = 0$ and, by Proposition 1, that q is a multiple of f_n. Since $\deg f_n = 2^n$ this is not possible for sufficiently large n.

The above implies that D is an open set from which the statement follows.

For the next result we keep the notation of the statement of Proposition 2.

Proposition 3. *Let $k = \dim h(U)$.*

(i) *There exist indices $i_1, \ldots, i_k \in \{1, \ldots, m\}$, a polynomial $g \in \mathbb{R}[y_1, \ldots, y_k]$ and a rational function $q \in \mathbb{R}(x_1, \ldots, x_n)$ with both numerator and denominator relatively prime with f_n such that*

$$g(h_{i_1}, \ldots, h_{i_k}) = f_n^\ell q$$

for some $\ell > 0$.

(ii) *Let n be sufficiently large. Then $k \geq n$.*

Proof. For part (i), first, notice that, since $\dim(h(U)) = k$, there exist $i_1, \ldots, i_k \in \{1, \ldots, m\}$ such that the functions $h_{i_1}, \ldots, h_{i_k}$ are algebraically independent. We want to show that $\dim(U \cap C_n) < k$. To do so let $X = h(U)$, $Y = h(U - C_n)$ and $Z = h(U \cap C_n)$. We have that all X, Y and Z are semialgebraic subsets of $\mathbb{R}^m$. In addition, Z is contained in the closure of Y with respect to the Euclidean topology relative to X since h is continuous and $Y \cap Z = \emptyset$ since h is the restriction of φ to U and φ is a reduction.

From here it follows that Z is included in the boundary of Y relative to X. Hence, $\dim Z < \dim Y = \dim X$ (see e.g. Proposition 2.8.12 of [5]).

The above shows that $\dim h(U \cap C_n) < k$. Therefore, there exists $g \in \mathbb{R}[y_1, \ldots, y_k]$ such that, for all $x \in U \cap C_n$, $g(h_{i_1}(x), \ldots, h_{i_k}(x)) = 0$. Write this as a rational function $g(h) = a/b$ with $a, b \in \mathbb{R}[x_1, \ldots, x_n]$ relatively prime. Then $a(C_n) = 0$ and $a \neq 0$ (since $h_{i_1}, \ldots, h_{i_k}$ are algebraically independent). By Proposition 1 this implies that there exists $r \in \mathbb{R}[x_1, \ldots, x_n]$ such that $a = r f_n$. If ℓ is the largest power of f_n dividing a then the result follows by taking $q = \frac{r'}{b}$ where r' is the quotient of r divided by $f_n^{\ell-1}$.

We now proceed to part (ii). To simplify notation, assume that $i_j = j$ for $j = 1, \ldots, k$. Also, let d be a bound for the degrees of the numerators and denominators of the h_j. Recall from Proposition 2 that d is bounded by a polynomial in n.

By part (i) there exists $q \in \mathbb{R}(x_1, \ldots, x_n)$ relatively prime with f_n such that

$$f_n^\ell q = g(h_1, \ldots, h_k)$$

for a certain $\ell \geq 1$. Taking derivatives on both sides we obtain that, for all $x \in \mathbb{R}^n$,

$$\nabla(f_n^\ell q)(x) = \nabla(g)(h(x)) \circ Dh(x) \tag{3}$$

where ∇ denotes the gradient and $Dh(x)$ is the Jacobian matrix of h at x.

Assume that $k < n$. Transposing (3) one sees that $\nabla(f^\ell q)(x)$ is the image of a vector of dimension k. Thus, there exists a linear dependency among the first $k + 1$ coordinates of $\nabla(f^\ell q)(x)$

$$\sum_{i=1}^{k+1} \lambda_i \frac{\partial f_n^\ell q}{\partial x_i} = 0 \tag{4}$$

and the coefficients λ_i of this linear dependency are the determinants of some minors of $Dh(x)$. Thus, for $i = 1, \ldots, k + 1$, λ_i is a rational function of x whose numerator and denominator have degrees bounded by kd. Since the submatrix of $Dh(x)$ obtained by keeping its first $k+1$ rows contains at most $k(k+1)$ different denominators, multiplying equation (4) by the product of all of all of them allows one to assume that the λ_i are polynomials with degree at most $kd(k+1)$.

By the product rule we get

$$\sum_{i=1}^{k+1} \lambda_i \left(\ell f_n^{\ell-1} q \frac{\partial f_n}{\partial x_i} + f_n^\ell \frac{\partial q}{\partial x_i} \right) = 0$$

i.e.

$$\ell f_n^{\ell-1} q \sum_{i=1}^{k+1} \lambda_i \frac{\partial f_n}{\partial x_i} + f_n^\ell \sum_{i=1}^{k+1} \lambda_i \frac{\partial q}{\partial x_i} = 0.$$

Since f_n^ℓ divides the second term above it must also divide the first from which, using that f_n and q are relatively prime, it follows that f_n divides $\sum_{i=1}^{k+1} \lambda_i \frac{\partial f_n}{\partial x_i}$. That is, there exists a polynomial p such that

$$f_n p = \sum_{i=1}^{k+1} \lambda_i \frac{\partial f_n}{\partial x_i}$$

i.e.

$$p \left(\sum_{i=1}^n x_i^{2^n} - 1 \right) = 2^n \sum_{i=1}^{k+1} \lambda_i x_i^{2^n - 1}.$$

Now, for n large enough, the degrees of the λ_i are smaller than $2^n - 1$ since $kd(k+1)$ is polynomial in n. This implies that the degree of p must also be bounded by $kd(k+1)$. But then, for each $i \le k+1$, $px_i^{2^n} = \lambda_i x_i^{2^n - 1}$, i.e., $px_i = \lambda_i$. And from here it follows that $-p = 0$, a contradiction.

Theorem 1 now readily follows. For all $n \in \mathbb{N}$, Proposition 2 ensures the existence of an open ball $U \subset \mathbb{R}^n$ whose image by the reduction φ is included in $\mathbb{R}^m$ with m polynomially bounded on n. But for all n sufficiently large this image, by Proposition 3 (ii), has dimension at least n and therefore it can not be polylogarithmic on m.

Remark 2. The result of Theorem 1, together with that in [6], supports the conjecture that there are no sparse NP-hard sets over the reals unless P = NP. There are two main settings where this remains to be proved. On the one hand, machines which do not multiply nor divide but which branch over sign tests. On the other hand, the unrestricted case in which the machine can multiply or divide (and branch over sign tests) with unit cost. In these two cases, the result seems harder since there is no proof that P $\ne$ NP. In the first case, we would like to remark that, if many-one reductions are replaced by Turing reductions and we assume that P $\ne$ NP then Mahaney's conjecture is false. This is due to a result of Fournier and Koiran [9] proving that any NP-complete set in the Boolean setting (i.e. over $\{0,1\}$) is NP-complete over the reals with addition and order for Turing reductions. Since the subsets of elements of size n of any such set S have dimension 0 the sparseness of S is immediate. For more on this see [8].

References

1. V. Arvind, Y. Han, L. Hemachandra, J. Köbler, A. Lozano, M. Mundhenk, M. Ogiwara, U. Schöning, R. Silvestri, and T. Thierauf. Reductions to sets of low information content. In K. Ambos-Spies, S. Homer, and U. Schöning, editors, *Complexity Theory: current research*, pages 1–45. Cambridge University Press, 1993. 285

2. L. Berman and J. Hartmanis. On isomorphism and density of NP and other complete sets. *SIAM Journal on Computing*, 6:305–322, 1977. 285

3. L. Blum, F. Cucker, M. Shub, and S. Smale. *Complexity and Real Computation*. Springer-Verlag, 1998. 286, 287

4. L. Blum, M. Shub, and S. Smale. On a theory of computation and complexity over the real numbers: NP-completeness, recursive functions and universal machines. *Bulletin of the Amer. Math. Soc.*, 21:1–46, 1989. 285

5. J. Bochnak, M. Coste, and M.-F. Roy. *Géométrie algébrique réelle*. Springer-Verlag, 1987. 289

6. F. Cucker, P. Koiran, and M. Matamala. Complexity and dimension. *Information Processing Letters*, 62:209–212, 1997. 286, 290

7. F. Cucker, M. Shub, and S. Smale. Complexity separations in Koiran's weak model. *Theoretical Computer Science*, 133:3–14, 1994. 286, 287

8. H. Fournier. Sparse NP-complete problems over the reals with addition. *Theoretical Computer Science*, 255:607–610, 2001. 290

9. H. Fournier and P. Koiran. Lower bounds are not easier over the reals: Inside PH. In *28th International Colloquium on Automata, Languages and Programming*, volume 1853 of *Lect. Notes in Comp. Sci.*, pages 832–843. Springer-Verlag, 2000. 290

10. P. Koiran. A weak version of the Blum, Shub & Smale model. *J. Comput. System Sci.*, 54:177–189, 1997. A preliminary version appeared in *34th annual IEEE Symp. on Foundations of Computer Science*, pp. 486–495, 1993. 286, 287

11. S. R. Mahaney. Sparse complete sets for NP: Solution of a conjecture by Berman and Hartmanis. *J. Comput. System Sci.*, 25:130–143, 1982. 285

12. K. Meer. A note on a P $\neq$ NP result for a restricted class of real machines. *Journal of Complexity*, 8:451–453, 1992. 286

Sharing One Secret vs. Sharing Many Secrets: Tight Bounds for the Max Improvement Ratio*

Giovanni Di Crescenzo

Telcordia Technologies
445 South Street, Morristown, NJ, 07960-6438, USA
giovanni@research.telcordia.com

Abstract. A secret sharing scheme is a method for distributing a secret among several parties in such a way that only qualified subsets of the parties can reconstruct it and unqualified subsets receive no information about the secret. A multi secret sharing scheme is the natural extension of a secret sharing scheme to the case in which many secrets need to be shared, each with respect to possibly different subsets of qualified parties. A multi secret sharing scheme can be trivially realized by realizing a secret sharing scheme for each of the secrets. A natural question in the area is whether this simple construction is the most efficient as well, and, if not, how much improvement is possible over it.

In this paper we address and answer this question, with respect to the most widely used efficiency measure, that is, the *maximum* piece of information distributed among all the parties. Although no improvement is possible for several instances of multi secret sharing, we present the first instance for which some improvement is possible, and, in fact, we show that for this instance an improvement factor equal to the number of secrets over the above simple construction is possible. The given improvement is also proved to be the best possible, thus showing that the achieved bound is tight.

Keywords: Cryptography, Secret Sharing, Multi-Secret Sharing

1 Introduction

A secret sharing scheme is a pair of efficient algorithms: a distribution algorithm and a reconstruction algorithm, run by a dealer and some parties. The distribution algorithm is executed by a dealer who, given a secret, computes some shares of it and gives them to the parties. The reconstruction algorithm is executed by a qualified subset of parties who, by putting together their own shares, can therefore reconstruct the secret. A secret sharing scheme satisfies the additional property that any non-qualified subset of participants does not obtain any information about the secret. The notion of secret sharing was introduced by Blackley [2] and Shamir [21], who considered the important case in which the

---

J. Sgall, A. Pultr, and P. Kolman (Eds.): MFCS 2001, LNCS 2136, pp. 292–304, 2001.
© Springer-Verlag Berlin Heidelberg 2001

set of qualified subsets of participants is the set of all subsets of size at least k, for some integer k.

Since their introduction, secret sharing schemes have been widely employed in the construction of more elaborated cryptographic primitives and several types of cryptographic protocols. Being so often employed, central research questions in this area are both the construction of efficient algorithms for this task, and finding bounds on the possible efficiency that such algorithms can achieve, where the efficiency measure mostly studied in the literature, and the one that we will also consider in this paper, is related to the size of the largest distributed share (typically called "information rate", for its analogy with a so-called coding theory notion). Several efficient algorithms have been given in the literature (see, e.g., [21,1,15]) and many lower bounds on the size of the shares distributed to participants have been presented (see, e.g., [8,5,23,9,25]).

A natural extension of secret sharing schemes, motivated by several application scenarios, is to consider the case in which many secrets need to be shared, each requiring a possibly different set of qualified subsets of parties. These schemes are called multi-secret sharing schemes and were considered, for instance, in [4,16,17,7].

Note that the scheme obtained by composing a (single) secret sharing scheme for each of the secrets results in a multi secret sharing scheme. From an efficiency point of view, however, it is interesting to ask whether this is the best possible construction in general (that is, for all possible sets of qualified subsets of parties) or some improvement is possible, using some different construction. Moreover it is of interest to ask how much improvement, if any at all, can be achieved. This paper addresses and completely answers both these questions.

Previous results. First of all notice that it is not hard to come up with elementary classes of sets of qualified parties for which no improvement is possible (for instance, just think of each secret being recovered by subsets of a new group of parties, or each secret being recovered by exactly the same list of subsets of parties). Moreover, previous results in [4] imply that no improvement is possible for a large class of sets of qualified parties (namely, the sets of parties of size larger than a given threshold). On a different note, results in [4] imply that some improvement is possible for the case in which 2 or 3 secrets are shared, if one considers another efficiency measure which has sometimes been studied in the literature; namely, a measure related to the sum of the shares distributed to the parties (also called "average information rate"). The study of the analogue questions of this paper for this measure was done in [11], where it was shown that if m secrets are to be shared, there exists an instance for which an improvement by a factor of $1/m + \epsilon$, for any $\epsilon > 0$, is possible. None of the techniques in [4,11] naturally extends to the most used (and more difficult to treat) measure of the maximum size of the shares, that is considered in this paper. In fact, neither the schemes presented in [4,11] nor other schemes in the literature give any improvement at all on the maximum size of the shares, thus leaving open the questions of whether any improvement is possible; and, if so, how much.

Our results. In this paper we solve the above open questions.

First of all we observe that the maximum size of the shares of any multi secret sharing scheme will always be at least $1/m$ times the maximum size of the shares of the basic multi secret sharing scheme, if m is the number of secrets. This observation directly follows from the fact that a multi secret sharing scheme is a (single) secret sharing scheme with respect to all secrets.

Our main result is the following construction. Specifically, we show the first instance of multi secret sharing for which some improvement is achievable. Even more interestingly, for this construction -the best possible- improvement is achievable, therefore showing that the previous bound is tight. In other words, there exists an instance of multi secret sharing and a multi secret sharing scheme for it such that the maximum size of the shares distributed according to this scheme is equal to $1/m$ times the maximum size of the shares distributed according to the basic multi secret sharing scheme. Establishing this result requires understanding the combinatorial structure that such an instance should satisfy, finding such an instance and finding an especially efficient multi secret sharing scheme for it. We believe most of the techniques used can find applications in the design of efficient multi secret sharing schemes.

Our investigation could also be considered as the study of the "direct product problem" for secret sharing schemes. This type of problem has received a lot of attention in several areas of theoretical computer science (see Appendix A).

Organization of the paper. In Section 2 we present all definitions of interest for the rest of the paper. In Section 3 we present our observation on the best possible improvement in the maximum share size when designing multi secret sharing schemes. Our main result is presented in Section 4. Given the space constraints, we omit several definitions and proofs.

2   Definitions

In this section we recall the definitions of secret sharing schemes and multi-secret sharing schemes, and define quantities that will be of interest in this paper, as max share size, optimal max share size and max improvement ratios.

SECRET SHARING SCHEMES. Informally, a secret sharing scheme is a pair of efficient algorithms (the *distribution* and the *reconstruction* algorithm) which allow a dealer to divide a certain secret s into n pieces of information, called *shares*, in such a way that the following two requirements hold: qualified subsets of the n pieces of information allow to compute s (this is the *correctness* requirement) but given any non-qualified subset of such pieces, no algorithm can compute any information about s (this is the *privacy* requirement). Whether a certain subset is qualified or not is determined by a fixed, so-called, access structure.

More formally, let $\mathcal{P} = \{P_1, P_2, \ldots, P_n\}$ be a set of n participants, and define an *access structure* $\mathcal{A}$ over $\mathcal{P}$ as a set of subsets of $\mathcal{P}$. We say that an access structure $\mathcal{A}$ is *monotone* if $A \in \mathcal{A}$ implies $B \in \mathcal{A}$, for any B such that $A \subseteq B$. Any monotone function f over n boolean variables $x_1, \ldots, x_n$ implicitly defines an access structure in the standard way: if $f(x_1, \ldots, x_n) = 1$ then subset A

belongs to $\mathcal{A}$, where $P_i \in A$ if and only if $x_i = 1$. In this paper, as usually done in secret sharing, we will only consider monotone access structures. Then the correctness requirement is formalized by saying that, for any subset $A \in \mathcal{A}$, given the shares returned by the distribution algorithm and corresponding to parties in A, the reconstruction algorithm returns a value equal to the secret. Moreover, the privacy requirement is formalized by saying that, for any subset $A \notin \mathcal{A}$, the value of the secret is independent from the value of the shares returned by the distribution algorithm and corresponding to parties in A.

MAX SHARE SIZE. We denote by rS the random variable denoting the secret and taking value $s \in S$ and by $rSH_1, \ldots, rSH_n$ the random variables denoting the shares, each taking value $sh_i \in Sh_i$ (all variables being determined by the distribution from which the secret s is selected and by an execution of the distribution algorithm $\mathcal{D}$ on input s). We define the *max share size for access structure $\mathcal{A}$ with respect to secret sharing scheme $(\mathcal{D}, \mathcal{R})$* as the value

$$\mathsf{MaxShSize}(\mathcal{A}, (\mathcal{D}, \mathcal{R})) = \max_{i=1}^{n} \mathcal{H}(rSH_i),$$

where by $\mathcal{H}$ we denote the binary entropy version (see, e.g. [10]), and the *optimal max share size for access structure $\mathcal{A}$* as the value

$$\mathsf{OpMaxShSize}(\mathcal{A}) = \min_{(\mathcal{D}, \mathcal{R})} \mathsf{MaxShSize}(\mathcal{A}, (\mathcal{D}, \mathcal{R})).$$

2.1   Multi-secret Sharing Schemes

Informally, a multi secret sharing scheme is a pair of efficient algorithms (the *distribution* and the *reconstruction* algorithm) which allow a dealer to divide m secrets $s_1, \ldots, s_m$ into n shares in such a way that the following two requirements hold for each $i = 1, \ldots, n$: qualified subsets of the n pieces of information allow to compute s_i (this is the *correctness* requirement), but given any non-qualified subset of such pieces, no algorithm can compute any information about the value of s_i other than the information that is given by the secrets determined by such pieces; moreover, this holds even if the value of all other secrets is known (this is the *privacy* requirement). For each $i = 1, \ldots, m$, whether a certain subset is qualified or not for the computation of the i-th secret s_i is determined by a fixed access structure $\mathcal{A}_i$. An important observation is that the m access structures, each associated with secret s_i, for $i = 1, \ldots, m$, are not, in general, equal. More formally, the correctness requirement is defined by saying that for any $i = 1, \ldots, m$, and for any subset $A \in \mathcal{A}_i$, given the shares returned by the distribution algorithm and corresponding to parties in A and value i, the reconstruction algorithm returns a value equal to secret s_i. Moreover, the privacy requirement is formalized by saying that, for any $i = 1, \ldots, m$, and for any subset $A \notin \mathcal{A}$, the value of the i-th secret s_i is independent from the value of the shares returned by the distribution algorithm and corresponding to parties in A, even given the value of all other secrets.

MAX SHARE SIZE. In the rest of the paper we will assume, for simplicity, that all secrets have the same size, i.e., $|s_1| = \cdots = |s_m| = l$. Similarly as for the case of single secret sharing, we will denote by rS^m the random variable denoting the m-tuple of secrets and taking value $s_1, \ldots, s_m \in S^m$ and by $rSH_1, \ldots, rSH_n$ the random variables denoting the shares returned by the distribution algorithm, each taking value $sh_i \in SH_i$. We define the *max share size of access structures* $\mathcal{A}_1, \ldots, \mathcal{A}_m$ *with respect to multi secret sharing scheme* $(\mathcal{D}, \mathcal{R})$ as the value

$$\mathsf{MaxShSize}(\mathcal{A}_1, \ldots, \mathcal{A}_m, (\mathcal{D}, \mathcal{R})) = \max_{i=1}^{n} \mathcal{H}(rSH_i),$$

and the *optimal max share size of access structures* $\mathcal{A}_1, \ldots, \mathcal{A}_m$ as the value

$$\mathsf{OpMaxShSize}(\mathcal{A}_1, \ldots, \mathcal{A}_m) = \min_{(\mathcal{D}, \mathcal{R})} \mathsf{MaxShSize}(\mathcal{A}_1, \ldots, \mathcal{A}_m, (\mathcal{D}, \mathcal{R})).$$

2.2   Improvement Ratios

THE BASIC MULTI SECRET SHARING SCHEME. The first approach to consider in constructing a multi secret sharing scheme for access structures $\mathcal{A}_1, \ldots, \mathcal{A}_m$ is certainly to combine m (single-secret) sharing schemes, each for access structure $\mathcal{A}_i$, for $i = 1, \ldots, m$. Specifically, given secret sharing scheme $(\mathcal{D}_i, \mathcal{R}_i)$ for access structure $\mathcal{A}_i$, for $i = 1, \ldots, m$, define the $((\mathcal{D}_1, \mathcal{R}_1), \ldots, (\mathcal{D}_m, \mathcal{R}_m))$-*composed scheme* as the following multi secret sharing scheme (c-$\mathcal{D}$, c-$\mathcal{R}$): algorithm c-$\mathcal{D}$ sequentially runs algorithms $\mathcal{D}_1, \ldots, \mathcal{D}_m$, using each time independently chosen random bits; algorithm c-$\mathcal{R}$ takes as additional input an index i (meaning that it is trying to recover the i-th secret) and runs algorithm $\mathcal{R}_i$ using as additional input the output of $\mathcal{D}_i$. We can now define the *basic multi secret sharing scheme* for access structures $\mathcal{A}_1, \ldots, \mathcal{A}_m$, that we denote as (b-$\mathcal{D}$, b-$\mathcal{R}$), as the $((\mathcal{D}_1, \mathcal{R}_1), \ldots, (\mathcal{D}_m, \mathcal{R}_m))$-composed scheme, for the same structures, where, for $i = 1, \ldots, m$, each $(\mathcal{D}_i, \mathcal{R}_i)$ is chosen so that it minimizes $\mathsf{OpMaxShSize}(\mathcal{A}_i)$; moreover, if for some i there are many schemes $(\mathcal{D}_i, \mathcal{R}_i)$ which minimize $\mathsf{OpMaxShSize}(\mathcal{A}_i)$, we choose the one that results in the minimum value for $\mathsf{MaxShSize}(\mathcal{A}_1, \ldots, \mathcal{A}_m, (\text{c-}\mathcal{D}, \text{c-}\mathcal{R}))$.

IMPROVEMENT RATIOS. In order to study all other possible approaches for constructing a multi secret sharing scheme, we define quantities that measure how well these approaches perform, when compared with the basic multi secret sharing scheme for the same access structures. Specifically, we define the *max-improvement ratio for access structures* $\mathcal{A}_1, \ldots, \mathcal{A}_m$ *with respect to multi secret sharing scheme* $(\mathcal{D}, \mathcal{R})$ as the value

$$\mathsf{MaxIR}(\mathcal{A}_1, \ldots, \mathcal{A}_m, (\mathcal{D}, \mathcal{R})) = \frac{\mathsf{MaxShSize}(\mathcal{A}_1, \ldots, \mathcal{A}_m, (\mathcal{D}, \mathcal{R}))}{\mathsf{MaxShSize}(\mathcal{A}_1, \ldots, \mathcal{A}_m, (\text{b-}\mathcal{D}, \text{b-}\mathcal{R}))},$$

and the *optimal max-improvement ratio for access structures* $\mathcal{A}_1, \ldots, \mathcal{A}_m$ as the value

$$\mathsf{OpMaxIR}(\mathcal{A}_1, \ldots, \mathcal{A}_m) = \min_{(\mathcal{D}, \mathcal{R})} \mathsf{MaxIR}(\mathcal{A}_1, \ldots, \mathcal{A}_m, (\mathcal{D}, \mathcal{R})).$$

3   A Lower Bound on the Optimal Max-improvement Ratio

In this section we present a lower bound on the optimal max improvement ratio for any tuple of access structures. Informally, this result gives a limit on the possibility of designing non-trivial algorithms for multi secret sharing schemes versus the trivial approach of the above defined basic multi secret sharing scheme. The limit consists of the fact that the largest share of any multi secret sharing scheme can be smaller than the largest share of the basic multi secret sharing scheme for the same access structures by a factor of at most the number of secrets. Formally, we have the following

Theorem 1. *Let* m, n *be positive integers and let* $\mathcal{A}_1, \ldots, \mathcal{A}_m$ *be access structures over a set of size* n. *It holds that* $\mathsf{OpMaxIR}(\mathcal{A}_1, \ldots, \mathcal{A}_m) \geq 1/m$.

The proof of the above theorem directly follows from the fact that any multi-secret sharing scheme is also a single secret sharing scheme for each of the secrets and can be extended so that it applies to a large class of practical efficiency measures.

4   An Upper Bound on the Max-improvement Ratio

In this section we present an upper bound on the optimal max improvement ratio. Our bound is obtained by exhibiting a specific tuple of access structures and a multi secret sharing schemes for it. This construction gives the first example of a multi secret sharing scheme that is more efficient than the basic multi secret sharing scheme for the same access structures. Moreover, the improvement in efficiency is the best possible; in other words, the largest share distributed according to our multi secret sharing scheme is smaller than the largest share distributed according to the basic multi secret sharing scheme by a factor equal to m, the number of secrets. Formally, we obtain the following

Theorem 2. *Let* m *be an integer. There exists an integer* n *and access structures* $\mathcal{A}_1, \ldots, \mathcal{A}_m$ *over a set of size* n *such that* $\mathsf{OpMaxIR}(\mathcal{A}_1, \ldots, \mathcal{A}_m) = 1/m$.

REMARKS. Intuitively, this result guarantees the possibility of designing non-trivial multi secret sharing schemes which have much better max share size than using the basic multi secret sharing scheme. The fact that, as we will see later, the proofs only uses graph-based access structures (a relatively elementary class of access structures) should be viewed as evidence that the class of access structures for which the improvement ratio is strictly smaller than 1 is indeed quite large. Most importantly, our construction matches the bound of Theorem 1, and, therefore, it shows that such bound exactly quantifies the possible improvement in the design of a multi secret sharing scheme.

SUBTLETIES TOWARDS PROVING THE THEOREM. We note that no construction of a multi secret sharing scheme was given in the literature which gives *any* saving in terms of max share size with respect to the basic multi secret sharing scheme construction. (Even the construction of a tuple of access structures presented in [11], which does give some saving with respect to the *average* share size, gives no saving with respect to the max share size.) Therefore we have to follow a significantly different approach in our construction. Finding a tuple of access structure and a scheme for it that is better than the basic multi secret sharing scheme requires the following steps. First, a tuple of access structures needs to be carefully selected, in such a way that the following two steps can be successfully performed. Second, a single secret sharing scheme with optimal share size needs to be presented for each of these access structures, so that a basic multi secret sharing scheme can be constructed. (Here note that the number of access structures that are known in the literature to have optimal constructions for single secret sharing is very small.) Third, an efficient multi-secret sharing scheme needs to be presented for the access structures, so that the max share size of this scheme is smaller than that of the basic multi secret sharing scheme. (Here, note that no such construction had been previously given in the literature.)

INFORMAL DESCRIPTION OF THE PROOF. The rest of this section is devoted to the proof of the above theorem. We start by presenting an m-tuple of access structure and a multi secret sharing scheme for it based on a recursive approach. This scheme achieves max improvement ratio equal to $(1 + \log m)/m$. Although this construction already gives a significant improvement over the basic multi secret sharing scheme, it is still far from the optimal by a logarithmic factor. In order to gain such factor, we take a single graph access structure which has quite large max share size (such constructions exist in the literature) and carefully compose it with the m-tuple of access structure constructed so far. Specifically, we purposely increase the max share size of each single access structure by a logarithmic (in m) factor. As a result, the basic multi secret sharing scheme achieves a max share size of $m \log m + m$. Finally, because of the careful insertion of the latter graph access structure in the previously constructed m-tuple, we can use the previously constructed multi secret sharing scheme and achieve max improvement ratio equal to $(1 + \log m)/m(1 + \log m) = 1/m$.

4.1   A first m-Tuple of Access Structures

We present a tuple of access structures, we analyze the performance of the basic multi secret sharing scheme and present an improved multi secret sharing scheme for it. These structures achieve a max improvement ratio equal to $(1+\log m)/m$, where m is the number of secrets.

THE ACCESS STRUCTURES. Let m be a positive integer; for simplicity, let us assume m can be written as 2^l, for some positive integer l. Let $n = m + 1$, and define the set $\mathcal{P}$ of participants as $\mathcal{P} = \{X_1, \ldots, X_m, Y\}$. For $i = 1, \ldots, m$, define access structure $\mathcal{A}_i$ as the set of subsets Z of $\mathcal{P}$ such that $X_i \in Z$ and $U \in Z$, for any $U \in \mathcal{P} \setminus \{X_i\}$.

THE BASIC MULTI SECRET SHARING SCHEME FOR $\mathcal{A}_1, \ldots, \mathcal{A}_m$. For $i = 1, \ldots, m$, a single secret sharing scheme $(\mathcal{D}_i, \mathcal{R}_i)$ for access structure $\mathcal{A}_i$ is obtained as follows. On input a secret s, algorithm D_i uniformly chooses a string a of the same length as s and distributes string a to participants $X_1, \ldots, X_{i-1}, X_{i+1}, \ldots,$ X_m, Y, and string $a \oplus s$ to participant X_i. The algorithm R_i is straightforward: first note that Y and X_i can compute the logical xor of $a \oplus s$ and s and therefore recover s; then note that participants X_i and X_j, for any $j \in \{1, \ldots, n\} \setminus \{i\}$, can compute s in the same way. Note that this scheme is optimal with respect to the max share size measure since each party obtains a share of the same size of the secret, which is the minimum necessary, as proved in several papers. Moreover, any other scheme optimal for A_i distributes the same amount of information to all parties. Therefore the basic multi secret sharing scheme (b-D, b-R) for $A_1, \ldots, A_m$ is obtained by composing the above defined schemes (D_i, R_i) according to the definition of basic multi secret sharing scheme. We observe that the optimal max share size $\mathsf{MaxShSize}(A_i, (D_i, R_i))$ of the above scheme (D_i, R_i) is equal to $\mathcal{H}(rS)$, and therefore the max share size $\mathsf{MaxShSize}(A_1, \ldots, A_m, (\text{b-}D, \text{b-}R))$ of the scheme (b-D, b-R) is equal to $m \cdot \mathcal{H}(rS)$.

AN EFFICIENT MULTI SECRET SHARING SCHEME FOR $A_1, \ldots, A_m$. We now describe a more efficient multi secret sharing scheme (D, R) for the above access structures $A_1, \ldots, A_m$.

Instructions for algorithm D: On input a k-bit secret s, do the following:
1. Uniformly and independently choose k-bit strings $a_1, \ldots, a_m$.
2. Give $a_1 \oplus \cdots \oplus a_m$ to Y.
3. Give $a_i \oplus s_i$ to X_i, for $i = 1, \ldots, m$.
4. Run procedure $Proc1(1, m, aa)$, where $aa = (a_1, \ldots, a_m)$.

Instructions for Procedure $Proc1$: On input min, max, aa, do the following:
1. Let $med = (min + max + 1)/2$.
2. Let $S_0 = \{min, \ldots, med - 1\}$ and $S_1 = \{med, \ldots, max\}$.
3. Give $\oplus_{i \in S_0} a_i$ to X_j, for all $j \in S_1$.
4. Give $\oplus_{i \in S_1} a_i$ to X_j, for all $j \in S_0$.
5. Run procedures $Proc1(min, med - 1, aa)$ and $Proc1(med, max, aa)$.
6. Return.

Instructions for algorithm R: On input $i \in \{1, \ldots, m\}$, and $A \in A_i$, do the following
1. If $A = \{Y, X_i\}$ then
   let t be the share given to Y;
   let $t_1, \ldots, t_{\log m + 1}$ be the shares given to X_i;
   output: $s_i = t_1 \oplus \cdots \oplus t_{\log m + 1} \oplus t$.
2. If $A = \{X_i, X_j\}$ then run procedure $Proc2(1, m, aa)$.

Instructions for Procedure $Proc2$: On input min, max, aa, do the following:
1. Let $med = (min + max + 1)/2$.
2. Let $S_0 = \{min, \ldots, med - 1\}$ and $S_1 = \{med, \ldots, max\}$.
3. If $X_i, X_j \in S_0$ then run procedure $Proc2(min, med - 1, aa)$ and return.
4. If $X_i, X_j \in S_1$ then run procedure $Proc2(med, max, aa)$ and return.
5. If $X_i \in S_0$ and $X_j \in S_1$ then
   let t be the share given to X_j equal to $a_{min} \oplus \cdots \oplus a_{med-1}$;
   let $t_1, \ldots, t_l$ be the shares given to X_i using any of the values $a_{min}, \ldots, a_{med-1}$
   in their computation;
   output: $s_i = t_1 \oplus \cdots \oplus t_l \oplus t$ and return.

6. If $X_j \in S_0$ and $X_i \in S_1$ then

    let t be the share given to X_j equal to $a_{med} \oplus \cdots \oplus a_{max}$;

    let $t_1, \ldots, t_l$ be the shares given to X_i using any of the values $a_{med}, \ldots, a_{max}$ in their computation;

    output: $s_i = t_1 \oplus \cdots \oplus t_l \oplus t$ and return.

We describe an example for how the above construction works for the case $m = 8$. Recall that the set of participants is $\mathcal{P} = \{X_1, \ldots, X_8, Y\}$; for $i = 1, \ldots, 8$, access structure $\mathcal{A}_i$ includes all subsets of $\mathcal{P}$ that include the subsets $\{X_j, X_i\}$, for all $j \in \{1, \ldots, 8\} \setminus \{i\}$, and subset $\{Y, X_i\}$. Now, consider scheme $(\mathcal{D}, \mathcal{R})$; let us call $s_1, \ldots, s_8$ the 8 secrets, and $a_1, \ldots, a_8$ the random values chosen during the execution of algorithm $\mathcal{D}$; then the shares distributed to the participants are of the following form:

- $a_1 \oplus \cdots \oplus a_8$ to participant Y;
- $a_5 \oplus a_6 \oplus a_7 \oplus a_8, a_3 \oplus a_4, a_2, a_1 \oplus s_1$ to participant X_1;
- $a_5 \oplus a_6 \oplus a_7 \oplus a_8, a_3 \oplus a_4, a_1, a_2 \oplus s_2$ to participant X_2;
- $a_5 \oplus a_6 \oplus a_7 \oplus a_8, a_1 \oplus a_2, a_4, a_3 \oplus s_3$ to participant X_3;
- $a_5 \oplus a_6 \oplus a_7 \oplus a_8, a_1 \oplus a_2, a_3, a_4 \oplus s_4$ to participant X_4;
- $a_1 \oplus a_2 \oplus a_3 \oplus a_4, a_7 \oplus a_8, a_6, a_5 \oplus s_5$ to participant X_5;
- $a_1 \oplus a_2 \oplus a_3 \oplus a_4, a_7 \oplus a_8, a_5, a_6 \oplus s_6$ to participant X_6;
- $a_1 \oplus a_2 \oplus a_3 \oplus a_4, a_5 \oplus a_6, a_8, a_7 \oplus s_7$ to participant X_7;
- $a_1 \oplus a_2 \oplus a_3 \oplus a_4, a_5 \oplus a_6, a_7, a_8 \oplus s_8$ to participant X_8.

We note that the max share size $\mathsf{MaxShSize}(\mathcal{A}_1, \ldots, \mathcal{A}_m, (\mathcal{D}, \mathcal{R}))$ of scheme $(\mathcal{D}, \mathcal{R})$ is equal to $(1 + \log m) \cdot \mathcal{H}(rS)$ (since all parties X_i receive $1 + \log m$ shares of the same size as the secret, which is assumed wlog to be equal to its entropy). Instead, scheme $(\text{b-}\mathcal{D}, \text{b-}\mathcal{R})$ has max share size $\mathsf{MaxShSize}(\mathcal{A}_1, \ldots, \mathcal{A}_m, (\text{b-}\mathcal{D}, \text{b-}\mathcal{R})) = m \cdot \mathcal{H}(rS)$. We then have that $\mathsf{MaxIR}(\mathcal{A}_1, \ldots, \mathcal{A}_m, (\mathcal{D}, \mathcal{R}))$ is equal to

$$\frac{\mathsf{MaxShSize}(\mathcal{A}_1, \ldots, \mathcal{A}_m, (\mathcal{D}, \mathcal{R}))}{\mathsf{MaxShSize}(\mathcal{A}_1, \ldots, \mathcal{A}_m, (\text{b-}\mathcal{D}, \text{b-}\mathcal{R}))} = \frac{(1 + \log m) \cdot \mathcal{H}(rS)}{m \cdot \mathcal{H}(rS)} = \frac{1 + \log m}{m}.$$

4.2 A Specific Graph-Based Access Structure

In this section we combine some results in the literature to obtain an access structure with certain specific properties. Oddly enough, a first property we require from such access structure is that the optimal max share size of it has to be much larger than the size of the secret; specifically, we would like the ratio of the size of the secret to the optimal max share size to go to 0 as the number of secrets grows. A second property we require is that there exists a construction of a secret sharing scheme that exactly achieves the optimal max share size. One construction in the literature that satisfies the above two property is presented in [3], based on an access structure presented in [25]. We now recall some necessary definitions and results. First of all, we say that an access structure $\mathcal{A}$ is *graph-based* if for each subset $A \in \mathcal{A}$, there exists a subset B such that $B \in \mathcal{A}$, $B \subseteq A$, and $|B| = 2$. A graph-based access structure $\mathcal{A}$ can be described by a graph $G_{\mathcal{A}}$, which is called the *graph associated to* $\mathcal{A}$, and is defined as follows: the set of vertices is the set of participants, and the set of edges is defined by all subsets B in $\mathcal{A}$ such that $|B| = 2$.

Fact 1 [3,25] *For any even positive integer $q, p \geq 6$, and any positive integer $d \geq 2$, there exists a graph-based access structure $\mathcal{A}$ such that the graph associated to $\mathcal{A}$ is d-regular and has qp^{d-2} nodes, and such that the optimal max share size of $\mathcal{A}$ is greater or equal to $(d+1)\mathcal{H}(rS)/2$.*

Fact 2 [22] *For any positive integer d, and any graph-based access structure $\mathcal{A}$, such that the graph associated to $\mathcal{A}$ has max degree d, it is possible to construct a secret sharing scheme $(\mathcal{D}, \mathcal{R})$ for $\mathcal{A}$ such that the max share size of $\mathcal{A}$ with respect to $(\mathcal{D}, \mathcal{R})$ is equal to $(d+1)\mathcal{H}(rS)/2$.*

By combining the above two facts, we obtain the following

Fact 3 [22,3,25] *For any even positive integer $q, p \geq 6$, and any positive integer $d \geq 2$, there exists a graph-based access structure $\mathcal{A}$, such that the graph associated to $\mathcal{A}$ has max degree d and qp^{d-2} nodes, and such that (a) the optimal max share size of $\mathcal{A}$ is greater or equal to $(d+1)\mathcal{H}(rS)/2$, and (b) it is possible to construct a secret sharing scheme $(\mathcal{D}, \mathcal{R})$ for $\mathcal{A}$ such that the max share size of $\mathcal{A}$ with respect to $(\mathcal{D}, \mathcal{R})$ is equal to $(d+1)\mathcal{H}(rS)/2$.*

4.3   A Substitution Based Construction

We now conclude the description of our construction by describing a final step. Informally, we would like to replace the participant Y in the access structures $\mathcal{A}_1, \ldots, \mathcal{A}_m$ with any qualified subset of the access structure $\mathcal{A}$ from Fact 3.

THE ACCESS STRUCTURES. Let m be a positive integer, which we assume for simplicity to be equal to 2^l, for some positive integer l; let $q = p = 6$, $d = 2(1 + \log m) - 1$, $k = qp^{d-2}$ and $n = m + k$. Define the set $\mathcal{P}$ of participants as $\mathcal{P} = \{X_1, \ldots, X_m, U_1, \ldots, U_k\}$. Let $\mathcal{A}$ be the access structure over $\{U_1, \ldots, U_k\}$ guaranteed by Fact 3. For $i = 1, \ldots, m$, define access structure $\mathcal{A}_i$ as the set of subsets Z of $\mathcal{P}$ such that either: (a) $X_i \in Z$ and $X_j \in Z$, for any $j \neq i$, or (b) $X_i \in Z$ and $\exists Y \subseteq \{U_1, \ldots, U_k\}$ such that $Y \subseteq Z$ and $Y \in \mathcal{A}$.

THE MAX-BASIC MULTI SECRET SHARING SCHEME FOR $\mathcal{A}_1, \ldots, \mathcal{A}_m$. A crucial fact to observe is that any secret sharing scheme for access structure $\mathcal{A}_i$ has max share size at least $(1 + \log m) \cdot \mathcal{H}(rS)$. Assume for the sake of contradiction that this is not the case; then, observe that $1 + \log m = (d+1)/2$, and therefore there exists a secret sharing scheme for $\mathcal{A}_i$ that has max share size smaller than $(d+1)\mathcal{H}(rS)/2$. This scheme can be used, for instance as done in the proof of Lemma 4.1 in [6], in order to construct a secret sharing scheme for access structure $\mathcal{A}$ having max share size smaller than $(d+1)\mathcal{H}(rS)/2$. This contradicts Fact 3. Finally, we derive that the max-basic multi secret sharing scheme for $\mathcal{A}_1, \ldots, \mathcal{A}_m$ has max share size at least $m(1 + \log m) \cdot \mathcal{H}(rS)$ (in fact, this inequality can be made tight by using Fact 3).

AN EFFICIENT MULTI SECRET SHARING SCHEME FOR $\mathcal{A}_1, \ldots, \mathcal{A}_m$. A multi secret sharing scheme $(\mathcal{D}, \mathcal{R})$ for $\mathcal{A}_1, \ldots, \mathcal{A}_m$ can then simply be obtained from the multi secret sharing scheme given in Section 4.1, call it $(\mathcal{D}_0, \mathcal{R}_0)$, and the single secret sharing scheme for access structure $\mathcal{A}$ guaranteed by Fact 3, call it

$(\mathcal{D}_1, \mathcal{R}_1)$, as follows. Algorithm $\mathcal{D}$ runs algorithm $\mathcal{D}_0$ with the following modification: when algorithm $\mathcal{D}_0$ sends $a_1 \oplus \cdots \oplus a_m$ to Y, algorithm $\mathcal{D}$ shares $a_1 \oplus \cdots \oplus a_m$ among the participants $U_1, \ldots, U_k$, and according to algorithm $\mathcal{D}_1$. Algorithm $\mathcal{R}$ runs algorithm $\mathcal{R}_0$ with the following modification: when algorithm $\mathcal{R}_0$ require participant Y to provide the share equal to $a_1 \oplus \cdots \oplus a_m$, algorithm $\mathcal{R}$ requires participants $U_1, \ldots, U_k$ to run algorithm $\mathcal{R}_1$ to recover the value $a_1 \oplus \cdots \oplus a_m$ and to provide such value for the remaining computation made by $\mathcal{R}$.

PROPERTIES AND MAX IMPROVEMENT RATIO OF OUR SCHEME. The correctness property of our scheme $(\mathcal{D}, \mathcal{R})$ follows directly from the analogue property of schemes $(\mathcal{D}_0, \mathcal{R}_0)$ and $(\mathcal{D}_1, \mathcal{R}_1)$. The privacy property of $(\mathcal{D}, \mathcal{R})$ follows the same reasoning done in order to prove the analogue property of scheme $(\mathcal{D}_0, \mathcal{R}_0)$ after replacing participant Y with the participants $U_1, \ldots, U_k$. We note that in our scheme $(\mathcal{D}, \mathcal{R})$ all parties X_i receive $1 + \log m$ shares of the same size as the secret, as can be seen by the construction of algorithm $\mathcal{D}_0$, and all parties U_i also receive $1 + \log m$ shares of the same size as the secret, because of Fact 3. Therefore the max share size $\mathsf{MaxShSize}(\mathcal{A}_1, \ldots, \mathcal{A}_m, (\mathcal{D}, \mathcal{R}))$ of scheme $(\mathcal{D}, \mathcal{R})$ is equal to $(1 + \log m) \cdot \mathcal{H}(rS)$. Instead, scheme (b-$\mathcal{D}$, b-$\mathcal{R}$) has max share size $\mathsf{MaxShSize}(\mathcal{A}_1, \ldots, \mathcal{A}_m, (\text{b-}\mathcal{D}, \text{b-}\mathcal{R})) = m(1 + \log m) \cdot \mathcal{H}(rS)$. We conclude the proof of Theorem 2 by observing that $\mathsf{MaxIR}(\mathcal{A}_1, \ldots, \mathcal{A}_m, (\mathcal{D}, \mathcal{R}))$ is equal to

$$\frac{\mathsf{MaxShSize}(\mathcal{A}_1, \ldots, \mathcal{A}_m, (\mathcal{D}, \mathcal{R}))}{\mathsf{MaxShSize}(\mathcal{A}_1, \ldots, \mathcal{A}_m, (\text{b-}\mathcal{D}, \text{b-}\mathcal{R}))} = \frac{(1 + \log m) \cdot \mathcal{H}(rS)}{m(1 + \log m) \cdot \mathcal{H}(rS)} = \frac{1}{m}.$$

References

1. J. C. Benaloh, J. Leichter, *Generalized Secret Sharing and Monotone Functions*, Proc. of CRYPTO 88. 293
2. G. R. Blakley, *Safeguarding cryptographic keys*, In *Proc. Nat. Computer Conf. AFIPS Conf. Proc.*, pp. 313–317, 1979, vol.48. 292
3. C. Blundo, A. De Santis, R. De Simone, and U. Vaccaro, *Tight Bounds on the Information Rate of Secret Sharing Schemes*, Design, Codes, and Cryptography, vol. 11, 1997, pp. 107–122. 300, 301
4. C. Blundo, A. De Santis, G. Di Crescenzo, A. Giorgio Gaggia and U. Vaccaro, *Multi-Secret Sharing Schemes,* in "Advances in Cryptology – CRYPTO 94", Lecture Notes in Computer Science, vol. 839, Springer Verlag, pp. 150–163. 293
5. C. Blundo, A. De Santis, L. Gargano and U. Vaccaro, *On the Information Rate of Secret Sharing Schemes,* in Theoretical Computer Science, vol. 154, pp. 283–306, 1996 (previous version in CRYPTO 92). 293
6. C. Blundo, A. De Santis, A. Giorgio Gaggia, and U. Vaccaro, *New Bounds on the Information Rate of Secret Sharing Schemes,* in IEEE Transactions on Information Theory, vol. IT-41, n. 2, march 1955. 301
7. C. Blundo, A. De Santis, and U. Vaccaro, *Efficient Sharing of Many Secrets,* in Proceedings of STACS 93. 293
8. R. M. Capocelli, A. De Santis, L. Gargano, and U. Vaccaro, *On the Size of Shares for Secret Sharing Schemes,* Journal of Cryptology, Vol. **6**, pp. 57–167, 1993. 293
9. L. Csirmaz, *The Size of a Share Must be Large,* Journal of Cryptology, Vol. **10**, n. 4, pp. 223–231, 1997. 293

10. T. Cover and J. Thomas, ELEMENTS OF INFORMATION THEORY, John Wiley and Sons, 1991. 295
11. G. Di Crescenzo, *Sharing One Secret vs. Sharing Many Secrets: Tight Bounds on The Average Improvement Ratio,* in Proceedings of SODA 2000. 293, 298
12. G. Di Crescenzo and R. Impagliazzo, *Proofs of Membership vs. Proofs of Knowledge,* in Proceedings of Computational Complexity 1998. 304
13. T. Feder, E. Kushilevitz, and M. Naor, *Amortized Communication Complexity,* in Proceedings of FOCS 91, 1991, pp. 239–248. 303
14. R. Impagliazzo, R. Raz and A. Wigderson, *A direct product theorem,* in Proceedings of 11th Annual IEEE Conference on Structure in Complexity Theory, 1994. 303
15. M. Ito, A. Saito, and T. Nishizeki, *Secret Sharing Scheme Realizing General Access Structure,* Proceedings of IEEE Global Telecommunications Conference, Globecom 87, Tokyo, Japan, pp. 99–102, 1987. 293
16. W. Jackson, K. Martin, and C. O'Keefe, *Multi-Secret Threshold Schemes,* in Design, Codes and Cryptography, vol. 9, n. 3, pp. 287–303, 1996. 293
17. W. Jackson, K. Martin, and C. O'Keefe, *Ideal Secret Sharing Schemes with Multiple Secrets,* in Journal of Cryptology, Vol. **9**, pp. 233–250, 1996. 293
18. M. Karchmer, and A. Wigderson, *On Span Programs,* in Proceedings of 8th Annual IEEE Conference on Structure in Complexity Theory, 1993.
19. N. Nisan, S. Rudich and M. Saks, *Products and Help Bits in Decision Trees,* in Proceedings of FOCS 94, 1994, pp. 318–329. 304
20. R. Raz, *A parallel repetition theorem,* in Proceedings of STOC 95. 304
21. A. Shamir, *How to Share a Secret,* Communication of the ACM, vol. 22, n. 11, November 1979, pp. 612–613. 292, 293
22. D. R. Stinson, *Decomposition Constructions for Secret Sharing Schemes,* Design, Codes and Cryptography, Vol. **2**, pp. 357–390, 1992. 301
23. D. R. Stinson, *An Explication of Secret Sharing Schemes,* Design, Codes and Cryptography, Vol. **2**, pp. 357–390, 1992. 293
24. D. R. Stinson, *Bibliography on Secret Sharing Schemes,* http://bibd.unl.edu/ stinson/ssbib.html, October 2, 1997.
25. M. van Dijk, *On the Information Rate of Perfect Secret Sharing Schemes,* in Design, Codes and Cryptography, vol. 6, pp. 143–169, 1995. 293, 300, 301
26. A. Yao, *Coherent Functions and Program Checkers,* in Proceedings of STOC 90. 303

A   Direct Product Problems

A direct product problem addresses the following question. Assume we are given a certain model of computation, a certain problem P, an instance x for problem P, and an algorithm A that is supposed to solve the instance x for problem P; moreover, let us associate a complexity value to algorithm A. Then the questions asks whether the complexity of an algorithm solving k instances $x_1, \ldots, x_k$ of problem P can be smaller than the complexity of the algorithm which solves each of the instances separately. In case the answer to this question is affirmative, it is of interest to quantify what type of improvement is possible, and to exhibit examples which achieve as large as possible an improvement.

Direct product problems have been studied in the literature in the contexts of communication complexity (see [13,14]), and computational complexity (see [26]

for boolean circuits, [19] for boolean decision trees, [20] for 2-Prover interactive
proofs, and [12] for interactive proofs of knowledge and computational ability).

(H,C,K)-Coloring: Fast, Easy, and Hard Cases[*]

Josep Díaz, Maria Serna, and Dimitrios M. Thilikos

Departament de Llenguatges i Sistemes Informàtics
Universitat Politècnica de Catalunya, Campus Nord – Mòdul C5
c/Jordi Girona Salgado, 1-3. E-08034, Barcelona, Spain
{diaz,mjserna,sedthilk}@lsi.upc.es

Abstract. We define a variant of the H-coloring problem by fixing the number of preimages of a subset C of the vertices of H, thus allowing parameterization. We provide sufficient conditions to guarantee that the problem can be solved in $O(kn + f(k, H))$ steps where f is a function depending only on the number k of fixed preimages and the graph H, and in $O(n^{k+c})$ steps where c is a constant independent of k. Finally, we prove that whenever the non parameterized vertices induce in G a graph that is bipartite and loopless the problem is NP-complete.

1 Introduction

For a given input graph G, let us consider the set of labelings that are given as homomorphisms to a fixed graph H. Recall that given graphs $H = (V(H), E(H))$ and $G = (V(G), E(G))$, an homomorphism is a mapping $\theta : V(G) \to V(H)$ such that $\{v, u\} \in E(G)$ if $\{\theta(v), \theta(u)\} \in E(H)$.

For a fixed graph H, the *H-coloring problem* checks whether, there exists an homomorphism from an input graph G to H. The name of H coloring comes from the fact that each vertex in H can be thought as a "color" that can be assigned to v. In fact, if H is the complete graph on c nodes, the set of H-coloring of G coincides with the set of proper c coloring of G. An interesting generalization of the H-coloring problem is the *list H-coloring* problem where each vertex of G is given together with a list of vertices of H, the vertices where it can be mapped. For results related to the complexity of the H-coloring, see [HN90], and for the list H-coloring problem, see [FH98, FHH99].

In recent times, the H-coloring problem has received a lot of attention, from the structural combinatorics point of view as well as from the algorithmic point of view. For instance, when H is bipartite or it has a loop the H-coloring problem can be trivially solved in polynomial time, but in the case that H is loopless and not-bipartite the problem is known to be NP-complete [HN90] (see also [GHN00] for the complexity of the H-coloring problem for bounded degree graphs). The

---

[*] Research supported by the EU project ALCOM-FT (IST-99-14186) and by the Spanish CYCIT TIC-2000-1970-CE. The research of the 3rd author was supported by the Ministry of Education and Culture of Spain, Grant number MEC-DGES SB98 0K148809.

J. Sgall, A. Pultr, and P. Kolman (Eds.): MFCS 2001, LNCS 2136, pp. 304–315, 2001.
© Springer-Verlag Berlin Heidelberg 2001

classical approach to cope with the hardness of the H-coloring problem is to consider subproblems in the case where G is restricted. The most general result in this direction was given recently in [DST01b] where a $O(nh^{k+1}\min\{k, h\})$ time algorithm is given for the case where when G is a partial k-tree, where $n = |V(G)|$ and $h = |V(H)|$ (see also [TP97]).

In this paper, we define a natural version of the H-coloring problem that allows parameterization. In particular, we set up a weighting K of a subset C of $V(G)$ with non negative integers and we say that an input graph G has a (H, C, K)-coloring if there exists an H-homomorphism $\chi : V(G) \to V(H)$ such that the number of the preimages of any weighted vertex is equal to its weight. We call the triple (H, C, K) partial weighted assignment. If we additionally assign to each vertex of G a list of permissible images, we have a more general version of the (H, C, K)-coloring problem, the *list (H, C, K)-coloring* problem.

A well known and popular interpretation of the various versions of the H-coloring problem is based on a model where the vertices of H are processors and the edges of H represent communication links between them. An instance of the H-coloring problem represent the problem where we have an input graph G where each vertex represents a job to be processed and each edge represents a communication demands between two jobs. A solution to the H-coloring problem is an assignment of jobs to the processors satisfying all the communication demands, see [FH98]. If now, for some subset C of processors, there is a set of restrictions on the number of jobs to be assigned represented by a tuple K then we have an instance of the (H, C, K)-problem introduced and studied in this paper. If, additionally, we demand that each job can be processed by only a list of processors, we have an instance of the list H-coloring and if again we bound the number of jobs that may be processed by some of the processors we have an instance of the list (H, C, K)-coloring problem.

In the (H, C, K)-coloring problem, we can consider the integers in K to be fixed constants and this constitutes a parameterization of the classic H-coloring problem and initiates an alternative approach to study the complexity of the problem. The independent set problem and the vertex cover problem are examples of parameterized problems expressed that way (See Figure 1).

In general, many NP-complete problems can be associated with one or more parameters. In many cases, when the parameter is considered to be a fixed constant, the problem becomes polynomially solvable and the question then is whether it is possible to find an algorithm of complexity $O(f(k)n^\alpha)$, where α is a constant. In such a case, the problem is called *Fixed Parameter Tractable* (FPT). On the other hand, not all the parameterizations yield an FTP or even a polynomial time solution. A characteristic example of this behaviour is the coloring problem, where the parameter k is the number of the used colors: It is in P when $k \leq 2$ and it is NP-complete when $k \geq 3$. In this paper we will attempt a classification of the partial weighted assignments (H, C, K), where the integers in K are fixed constants. We will give properties for (H, C, K) that guarantee that the corresponding problem is in FPT, in P, or NP-complete.

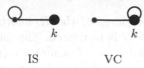

k k

IS   VC

Fig. 1. The graphs (H, C, K) for the parameterized independent set and vertex cover as parameterized colorings (the big vertices represent the labeled vertices of H)

Our main results are three theorems. We first prove that there exists a $O(kn + f(k, h))$ step algorithm for solving the (H, C, K)-coloring problem when the non-weighted vertices of H form an independent set, where k is the sum of the integers in K. This algorithm is the main subroutine for our next result that is a sufficient condition for (H, C, K)-coloring problem to be in FPT. This condition is a set of properties to be satisfied by the connected components of H and the algorithm involved runs in $O((k + h)n + \gamma(G)f(k, h))$ steps, where $\gamma(G)$ is the number of connected components of G. Our second result is a condition for the (H, C, K)-coloring problem to have a polynomial time algorithm. In particular, we provide a $O(n^{k+c})$ time algorithm for the case where the non parameterized vertices induce a graph for which the list H-coloring is in P. Our third result is a sufficient condition for the parameterized H-coloring problem to remain an NP-complete problem. We prove that this is the case when the non parameterized vertices induce a loopless and non-bipartite graph. All those results depend only on properties of the graph induced by the non-weighted vertices.

We also show that for a any graph F for which the F-coloring is in P but the list F-coloring is NP-complete, it is possible to connect additional bounded vertices in such a way that the resulting (H, C, K)-coloring problem is in P. Moreover, we provide some examples of NP-complete (H, C, K)-coloring problems verifying that the list $(H - C)$-coloring is NP-complete and the (H, C, K)-coloring is also NP-complete.

Finally we prove that a very similar classification exists for the list (H, C, K)-coloring problem. In particular, we show that the problem is in FPT when the non-weighted vertices of H form an independent set. Moreover, we prove that the list (H, C, K)-coloring problem is NP-complete if the list $H - C$-coloring problem is NP-complete. Otherwise the problem is in P.

In Section 2, we give some basic definitions as well as a formal description of the problems introduced in this paper. In Section 3 we present the algorithm showing the fixed parameter tractability, as well as the theorem supporting its correctness. In Section 4 we present the cases where our parameterization accepts a polynomial solution and the NP-completeness proof is presented in Section 5. Section 6 presents the extensions to list H-colorings. Finally, in Section 7 we conclude with some remarks and open problems.

2   Definitions

All the graphs in this paper are undirected, have no multiedges, but can have loops. Following the terminology of [FH98, FHH99], we call *reflexive* (*irreflexive*) any graph where any (none) of its vertices have a loop. The vertex (edge) set of a graph G is denoted as $V(G)$ ($E(G)$). If $S \subseteq V(G)$, we call the graph $(S, E(G) \cap \{\{x, y\} \mid x, y \in S\})$ the *subgraph of G induced by S* and we denote it by $G[S]$. We denote by $\gamma(G)$ the number of connected components of G. We also use the notation $G - S$ for the graph $G[V(G) - S]$. The neighborhood of a vertex v in graph G is the set of vertices in G that are adjacent to v in G and we denote it as $N_G(v)$ (clearly, if v has a loop then it is a neighbor of itself). A *reflexive clique* is a clique where all the vertices have loops. A graph G is bipartite when $V(G)$ can be partitioned to two parts such that all the non-loop edges of G have endpoints in both parts. Unless otherwise mentioned, we assume that all the bipartite graphs in this paper are irreflexive. For any function $\varphi : A \to B$ and any subset C of A we define *the restriction of φ to C* as $\varphi|_C = \{(a, b) \in \varphi \mid a \in C\}$.

Let $C = \{a_1, \ldots, a_r\}$ be a set of r vertices of H and let $K = (k_1, \ldots, k_r)$ and $K' = (k'_1, \ldots, k'_r)$ be two r-tuple of non negative integers. We say that $K' \leq K$ when for all i, $1 \leq i \leq r$, we have $k'_i \leq k_i$. Moreover, we define $K + K' = (k_1 + k'_1, \ldots, k_r + k'_r)$. We call the triple (H, C, K) *partial weighted assignment* on H. We say that a mapping $\chi : V(G) \to V(H)$ is an (H, C, K)-*coloring* of G if χ is an H-coloring of G such that for all i, $1 \leq i \leq r$, we have $|\chi^{-1}(a_i)| = k_i$. If in the above definition we replace "=" with "$\leq$" we say that χ is a $(H, C, \leq K)$-*coloring* of G. We refer to the vertices in C as the *weighted vertices* in H. We say that two partial weighted assignments (H, C, K) and (H', C', K') are *equivalent*, $(H, C, K) \sim (H', C', K')$, when for any graph G, G has a (H, C, K)-coloring iff G has a (H', C', K')-coloring. We also call them $\leq$-*equivalent* and we denote it as $(H, C, K) \sim_\leq (H', C', K')$ if for any graph G, G has a $(H, C, \leq K)$-coloring iff G has a $(H', C', \leq K')$-coloring. In Figure 1, we show two examples where the (H, C, K)-coloring can express the independent set and the vertex cover problem.

For each parameter assignment (H, C, K) we will use the notation $h = |V(H)|$ and $k = \sum_{i=1,\ldots,r} k_i$. Finally, for any $i, 1 \leq i \leq r$, $(H, C, K)[-i]$ denotes the partial weighted assignment (H', C', K') where $H' = H - \{a_i\}$, $C' = C - \{a_i\}$, and $K' = (k_1, \ldots, k_{i-1}, k_{i+1}, \ldots, k_r)$. In the obvious way, we can define the extension $(H, C, K)[-S]$ for $S \subset \{1, \ldots, r\}$. Similarly we define the notions of $(H, C, \leq K)[-i]$ and $(H, C, \leq K)[-S]$. We call a (H, C, K) *positive* if all the integers in K are positive. A partial weighted assignment (H, C, K) is a *weighted extension* of a graph F when $H - C = F$.

For fixed H, C, and K, the (H, C, K)-*coloring* problem asks whether there exists an (H, C, K)-coloring of an input graph G. We will also consider the $(H, C, \leq K)$-*coloring* problem defined in the same way for $(H, C, \leq K)$-colorings. Notice that, for any G, the existence of a (H, C, K)-coloring implies the existence of a $(H, C, \leq K)$-coloring. However, the existence of an $(H, C, \leq K)$-coloring of G does not necessarily imply the existence of a (H, C, K)-coloring of G. In whatever concerns the parameterized complexity of the (H, C, K)-coloring, its parameters will be the integers in K.

For a fixed graph H, given a graph G and given a subset $L(u) \subseteq V(H)$, for each vertex $u \in V(G)$, the *list* H-coloring problem asks whether, there is an H-coloring χ of G so that for every $u \in V(G)$ we have $\chi(u) \in L(u)$. This problem can be parameterized in the same way as the H-coloring, to define the *list* (H, C, K)-coloring problem.

3   Cases in FPT

In this section we will give a sufficient condition for the parameterized (H, C, K)-coloring problem to be in FPT. The basic component of the general algorithm will be a procedure that outputs, if it exists, an (H, C, K)-coloring of an input graph G, when C is a vertex cover of H, in $O(kn + f(k, h))$ steps,. The main idea of the proof is to create an equivalent problem whose size does not depend in the size of G (steps 1–8 of the algorithm Find Coloring).

Proposition 1. *If (H, C, K) is a partial weighted assignment where $E(H-C) = \emptyset$, then the algorithm Find-Coloring in Figure 2 checks in $O(kn + h^{k^2 + 2k + (k+1) \cdot 2^k})$ steps whether an input graph G with n vertices has an (H, C, K)-coloring and, if yes, outputs the corresponding mapping.*

Given a partial weighted assignment (H, C, K), we say that it is *compact* if each connected component H_i of H satisfies one of the following conditions:

1. $E(H_i - C) = \emptyset$,
2. $H_i[C]$ is a non-empty reflexive clique with all its vertices adjacent with one looped vertex of $H_i - C$, or
3. $V(H_i) \cap C = \emptyset$ and H_i contains at least one looped vertex.

Notice that any connected component of H satisfies exactly one of the three conditions in the above definition. Let (H, C, K) be a compact partial weighted assignment. We call a connected component of H that satisfies (1) *1-component* of H. Analogously, we define the *2-components* and the *3-components* of H. We call a 2-component H_i of H *compressed* if $|V(H_i) \cap C| = 1$ and we call it *small* if it is compressed and $|V(H_i) - C| = 1$. We call a compact partial weighted assignment (H, C, K) *compressed (small)* if it has no 2-components or if all the 2-components of H are compressed (small). We call a 3-component of H *tiny* when it consists only of a looped non-weighted vertex. We call a compact parameter assignment (H, C, K) *tiny* if it has not 3-components or of all its 3-components are tiny. For an example of these notions see Figure 3.

We can now state the following theorem.

Theorem 2. *If (H, C, K) is compact, then the (H, C, K)-coloring problem, parameterized by K, is in FPT.*

For lack of space we explain briefly the main ideas of the algorithm used to prove Theorem 2. We first prove that, for any compact partial weighted assignment (H, C, K), there exists an equivalent weighted assignment (H', C', K') that

Algorithm Find-Coloring(G, H, C, K).

Input: Two graphs G, H, and a partial weighted assignment (H, C, K) on H
such that $E(H - C) = \emptyset$.
Output: If exists, an (H, C, K)-coloring of G and,
if not, a report that ``no (H, C, K)-coloring of G exists''.

1: Let R_1 be the set of vertices in G of degree $> k$.
2: If $|R_1| > k$ or $|E(G)| > kn$ then return ``no (H, C, K)-coloring of G
exists''.
3: Set $G' = G[V(G) - R_1]$.
4: Let R_2 be the non isolated vertices in G' and let R_3 be the isolated vertices
in G'.
5: If $|R_2| > k^2 + k$ then return ``no (H, C, K)-coloring of G exists''
and stop.
6: Set up a partition $\mathcal{R} = (P_1, \ldots, P_q)$ of R_3 where
(q is the number of different neighborhoods of vertices in R_3 and
for all $v, u \in R_3$ ($\exists_{1 \le i \le q}$ $\{v, u\} \in P_i \Leftrightarrow N_G(v) = N_G(u)$).
7: Let $Q = \emptyset$.
8: For $i = 1, \ldots, q$,
if $|P_i| \le k + 1$, then set $F_i = P_i$,
otherwise let F_i be any subset of P_i where $|F_i| = k + 1$.
Set $Q = Q \cup F_i$ and $P_i = P_i - F_i$.
9: Let $\mathcal{H}$ be the set of all the (H, C, K)-colorings of $G[R_1 \cup R_2 \cup Q]$.
10: If $|\mathcal{H}| = \emptyset$ then output ``no (H, C, K)-coloring of G exists'' and stop.
11: Let χ be an (H, C, K)-coloring in $\mathcal{H}$.
12: For any $i = 1, \ldots, q$ such that $P_i \ne \emptyset$,
Let u be a vertex of F_i such that $\chi(u) \in V(H) - C$.
For any $w \in P_i$, set $\chi'(w) = \chi(u)$.
13: Output $\chi \cup \chi'$ as an (H, C, K)-coloring of G.
14: End.

Fig. 2. The algorithm Find-Coloring

is compact, positive, small, tiny, and such that H' contains only 2-components
or only 3-components but not both. The construction of this new weighted assignment can be done in $O(rh)$ steps.

We observe that if $\chi : V(G) \to V(H)$ is a (H', C', K')-coloring of G, then
the total size of the connected components of G that are mapped to the vertices
of some 2-component H'_i, is at least k'_i, where a'_i is the unique weighted vertex
in H'_i. This observation is the main necessary condition that satisfy the vertices
of G that are mapped to 2-components in a (H', C', K')-coloring of G. There
are two main cases; the first case is when G has "many" connected components.
Then we use the Find-Coloring algorithm to check whether a "small" number of
them are mapped to the 1-components of H' and after we are left with "many"
components that will allow us to extend the mapping. On the other hand, if G
has only "few" connected components then the number the components of G
that can be "mapped" to the components of H does not depend on the size

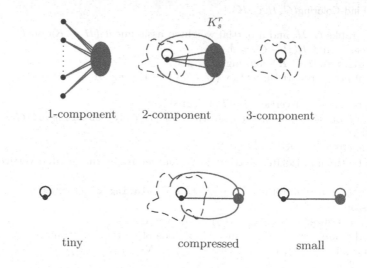

K_s^r

1-component   2-component   3-component

tiny   compressed   small

Fig. 3. Components for a compact partial weighted assignment

of G and the existence of an (H', C', K')-coloring is checked by a combination of exhaustive search and application of the Find-Coloring algorithm.

Observe that if there exists an FPT algorithm for the (H, C, K)-coloring with K as parameter, then by running it $O(k^r)$ times we have an algorithm for the parameterized $(H, C, \leq K)$-coloring problem.

Corollary 3. *If (H, C, K) is compact, then the $(H, C, \leq K)$-coloring problem, parameterized by K, is in FPT.*

4   Cases in P

In this section we present some cases of the (H, C, K)-coloring problem that can be solved by a polynomial time algorithm within time bound $O(n^{O(k)})$.

Our first result relates the decision problem with the *list H-coloring* problem. In particular we provide a polynomial time algorithm if the list coloring can be solved in the graph induced by the non-weighted vertices.

Theorem 4. *For any K, the (H, C, K)-coloring problem can be solved in $O(n^{k+c})$ steps whenever the list $(H - C)$-coloring can be solved in $O(n^c)$ steps.*

Proof. The basis of the algorithm is a reduction to the list $(H - C)$-coloring problem.

Assume that we have a $H[C]$-coloring χ for some $D \subseteq V(G)$, with k vertices. We consider the graph $G' = G[V(G) - D]$ and associate to each vertex $v \in V(G')$

the list

$$L(v) = \begin{cases} \bigcap_{N_G(v) \cap D}(N_H(\varphi(u)) - C), & \text{if } N_G(v) \cap D \neq \emptyset \\ V(H - C) & \text{otherwise} \end{cases}$$

Notice that χ can be completed to an H-coloring if and only if (G', L) has a list $(H - C)$-coloring. As the number of $H[C]$-colorings of subgraphs with k vertices is at most n^k we obtain the desired result.

The previous theorem includes the case when $H - C$ is a reflexive interval graph [FH98] and the case when $H - C$ is an irreflexive graph whose complement is a circular arc graph of clique covering number two [FHH99].

The above proof can be extended to show tractability of a weighted extension of any graph F for which the list H-coloring is still NP-complete, but the F-coloring is in P.

Theorem 5. *Let F_1 be either a bipartite graph with or without loops or a graph with at least one loop, and let F_2 be any graph. Then, for any K, the $(F_1 \oplus F_2, V(F_2), K)$-coloring can be solved in polynomial time for any partial weighted assignment K.*

In the previous theorem, $F_1 \oplus F_2$ denotes the graph obtained from F_1 and F_2 by adding all the edges between vertices in different graphs.

Notice that as we have all to all connections between F_1 and F_2, the lists associated to each vertex are just all the vertices in F_1. Therefore we just have to solve a F_1-coloring problem that indeed is in P. Finally we mention that any weighted extension of a graph F for which the F-coloring problem is in P, is also polynomially solvable in the trivial case when $K = (0, \ldots, 0)$.

5   NP-Complete Cases

In this section first we will prove that if $H - C$ is not bipartite and it does not contain a loop then the (H, C, K)-coloring problem is NP-complete. After we show that some weighted extensions of bipartite graphs, with or without, looped vertices, are also NP-complete. We will consider only the essential cases where (H, C, K) is positive.

Let (H, C, K) be a partial weighted assignment and let a_i be a vertex in C. We say that (H, C, K) is *i-reducible* if H is H'-colorable, where $H' = H - \{a_i\}$. We call the vertex a_i a *reducible* vertex of (H, C, K). We say that (H, C, K) is *reducible* if it is *i-reducible*, for some $1 \leq i \leq r$.

The following lemma indicates that for any (H, C, K)-coloring problem in which (H, C, K) is *reducible*, there exists an equivalent (H', C', K')-coloring problem where (H', C', K') is not reducible.

Lemma 6. *If $(H, C, \leq K)$ is i-reducible then $(H, C, \leq K) \sim_\leq (H, C, \leq K)[-i]$.*

Therefore, by removing all the reducible weighted vertices we obtain an equivalent problem, and we can prove the main result in this section.

Lemma 7. *For any (H,C,K), where the $(H-C)$-coloring problem is NP-complete, the $(H,C,\leq K)$-problem is also NP-complete.*

Proof. From Lemma 6 we can assume that (H,C,K) is a non reducible partial weighted assignment. We set $H' = H - C$ and, in what follows, we will reduce the H'-coloring problem to the $(H,C,\leq K)$-coloring problem. Let G be an instance of the H'-coloring problem. We will construct an instance G' of the $(H,C,\leq K)$-coloring problem as follows: Take as G' the disjoint union of G and a graph $\tilde{H}$ constructed from H' such that each vertex $a_i \in C$ is replaced by a set of k_i vertices $V_i = \{v_i^1, \ldots, v_i^{k_i}\}$, and each vertex $b \in V(H) - C$ is maintained. Any edge connecting two vertices in $V(H) - C$ is copied, an edge connecting a vertex in $V(H) - C$ and a vertex $a_i \in C$, is replicated for each vertex in V_i, finally an edge connecting a vertex v_i^j and a vertex $v_{i'}^{j'}$ is added if $\{a_i, a_{i'}\} \in E(H)$. We set $C' = \cup_{1 \leq i \leq |C|} V_i$.

We claim that G has an H'-coloring iff G' has a $(H,C,\leq K)$-coloring. Making $\chi(a) = a$ for $a \in V(\tilde{H}) - C' = V(H) - C$ and $\chi(v_i^j) = v_i$ for $v_i^j \in C'$ we see that $\tilde{H}$ is $(H,C,\leq K)$-colorable. Notice now that if G is H'-colorable, then G' is also $(H,C,\leq K)$-colorable. It remains to show that if G' is $(H,C,\leq K)$-colorable then also G is H'-colorable.

Let θ be an $(H,C,\leq K)$-coloring of G'. In the case where for some $v_i^j \in V(\tilde{H})$ $\theta(v_i^j) \in V(H) - C$, we set $\tilde{V} = V(\tilde{H}) - C' \cup \{v_1^j, \ldots, v_{|C|}^j\}$. Otherwise, we have that $\tilde{\theta} = \theta|_{\tilde{V}}$ is an $(H - \{v_i\})$-coloring of H and (H,C,K) is reducible, a contradiction. Therefore, $\theta(C') \subseteq C$ and as $|C'| = \sum_{1 \leq i \leq} k_i$, the requirement $|\theta^{-1}(v_i)| \leq k_i$ forces $\theta^{-1}(C) = C'$, which implies that $\theta^{-1}(C) \cap V(G) = \emptyset$ and $\theta(V(G)) \subseteq V(H - C) = V(H')$. This means that $\theta|_{V(G)}$ is a H'-coloring of G.

Using now Lemma 7 and the result of P. Hell and J. Nešetřil, on the NP-completeness of H-coloring, we conclude to the main result in this section.

Theorem 8. *The $(H,C,\leq K)$-coloring problem is NP-complete if $H - C$ is not bipartite and does not contain a loop.*

Notice that the existence of a polynomial algorithm that solves the (H,C,K)-coloring problem implies a polynomial algorithm for the $(H,C,\leq K)$-coloring problem. Therefore, we can state the following corollary.

Corollary 9. *The (H,C,K)-coloring problem is NP-complete if $H - C$ is not bipartite and does not contain a loop.*

The condition of Theorem 8 is not a necessary condition. In the next theorem we prove that the (H,C,K)-coloring remains NP-compete for certain weighted extensions of graphs that are either bipartite or that contain at least one looped vertex. We omit the proof that basically adapts the reductions given in the proofs of NP-completeness for the corresponding list $(H-C)$-coloring problems. Notice that in the case of looped vertices the $(H,C,\leq K)$-coloring is trivially in P.

Theorem 10. *The (H,C,K)-coloring problem is NP-complete for the partial weighted assignments depicted in Figure 4 provided that (H,C,K) is positive.*

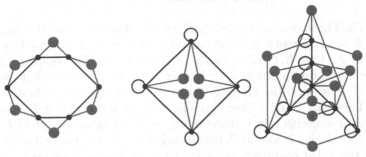

Fig. 4. Some hard partial weighted assignments

Observe that, for the irreflexive graphs in Figure 4, the $(H, C, \leq K)$-coloring problem remains NP-complete if K is positive, but it is trivially in P for the other two.

Finally notice that, by Theorem 5, the subgraph induced by the non parameterized vertices in each of the three problem in Figure 4 has another weighted extension so that the corresponding parameterized coloring problem is in P.

6   Results on List (H, C, K)-Coloring

Proposition 1, and Theorems 4 and 8 can be generalized for the list (H, C, K)-coloring as follows.

Theorem 11. *If (H, C, K) is a partial weighted assignment where $E(H - C) = \emptyset$, then there exists an $O(kn + h^{k^2 + 2k + (k+1) \cdot 2^k})$ time algorithm checking whether an input graph G with n vertices has a list (H, C, K)-coloring and, if yes, outputs the corresponding mapping.*

The algorithm involved in the above theorem is an extension of the Find Coloring algorithm.

Theorem 12. *For any K, the list (H, C, K)-coloring problem can be solved in n^{k+c} steps whenever the list $(H - C)$-coloring can be solved in $O(n^c)$ steps.*

The proof of this theorem is a direct extension of the proof of Theorem 4.

Theorem 13. *The list $(H, C, \leq K)$-coloring and the list (H, C, K)-coloring problems are NP-complete if the list $(H - C)$-coloring is NP-complete.*

The proof of the above theorem is based on a simpler version of the reduction in Theorem 8 that does not require an analogue of Lemma 6.

7   Remarks and Open Problems

The results in Theorems 4 and 8 are sharp in the following sense: If F is a graph where the F-coloring problem is NP-complete then, for any weighted extension (H, C, K) of F, the (H, C, K)-coloring problem is also NP-complete. Moreoever, if the F-coloring problem is in P, then *there exists* a weighted extension (H, C, K) of F so that the (H, C, K)-coloring problem is in P. Moreover the results in Theorems 4 and 10 are also sharp in the sense that if F is a graph where the list F-coloring is in P then for any weighted extension (H, C, K) of F, the (H, C, K)-coloring is also in P. On the other hand, we provided some examples where the F-list coloring is NP-complete and F has a weighted extension (H, C, K) such that (H, C, K)-coloring is also NP-complete.

The above observations argue that there are graphs that for different weighted extensions produce parameterized coloring problems with different complexities. It seems to be a hard problem to achieve a dichotomy distinguishing those parameterized assignments (H, C, K) of a given graph F for which the (H, C, K)-coloring problem is either NP-complete or it is in P. However we conjecture the following.

Conjecture 14. For any graph F such that the list F-coloring is NP-complete, there is a weighted extension (H, C, K) of F such that the (H, C, K)-coloring is also NP-complete.

All of our results on NP-completeness indicate that this frontier depends not only on $H - C$ but also on the structure imposed by the weighted vertices. However, we observe that in all our complexity results – positive or negative – this dichotomy does not depend on the choice of the numbers in a positive K.

We now fix our attention on whether there are cases more general than those in Theorem 2, where the (H, C, K)-coloring problem is in FPT. In particular we conjecture that the condition of Theorem 2 is also necessary.

Conjecture 15. For any partial weighted assignment (H, C, K), if (H, C, K) is compact then the (H, C, K)-coloring problem, with the numbers in K as parameters, is in FPT, otherwise it is W[1]-hard.

In support of our dichotomy conjecture we recall that the parameterized independent set problem, that is known to be W[1]-complete [DF99] falls in this area.

We mention that our second conjecture is quite general to express several open problems in parameterized complexity such as

1. (Parameter: k) Does G contain an independent set S, where $|S| = k$, and such that $G[V(G) - S]$ is bipartite? (This problem can be seen as a parameterization of 3-coloring where some color should be used exactly k times.)
2. (Parameter: k) Is there a set $S \subseteq V(G)$, where $|S| = k$ and $G[V(G) - S]$ is bipartite?
3. (Parameters: k, l) Does G contain $K_{k,l}$ as subgraph?

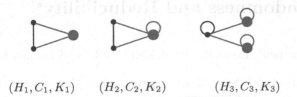

(H_1, C_1, K_1) (H_2, C_2, K_2) (H_3, C_3, K_3)

Fig. 5. Three weighted extensions conjectured to be W[1]-hard when K is parameterized

This problems correspond to the the parameterized colorings given in Figure 5. The (H_3, C_2, K_3)-problem is equivalent to ask if the complement of G contains K_{k_1, k_2} as a subgraph.

For results on counting (H, C, K)-colorings and list (H, C, K)-colorings, see [DST01a].

References

[DF99]   R. G. Downey and M. R. Fellows. *Parameterized complexity.* Springer-Verlag, New York, 1999. 314

[DST01a]   Josep Díaz, Maria Serna, and Dimitrios M. Thilikos. Counting list *H*-colorings and variants. Technical Report LSI-01-27-R, Departament de Llenguatges i Sistemes Informàtics, Universitat Politècnica de Catalunya, Barcelona, Spain, 2001. 315

[DST01b]   Josep Díaz, Maria Serna, and Dimitrios M. Thilikos. Evaluating H-colorings of graphs with bounded decomposability. In *Proceedings of The 7th Annual International Computing and Combinatorics Conference, COCOON 2001,* 2001. 305

[FH98]   Tomas Feder and Pavol Hell. List homomorphisms to reflexive graphs. *Journal of Combinatorial Theory (series B),* 72(2):236–250, 1998. 304, 305, 307, 311

[FHH99]   Tomas Feder, Pavol Hell, and Jing Huang. List homomorphisms and circular arc graphs. *Combinatorica,* 19(4):487–505, 1999. 304, 307, 311

[GHN00]   Anna Galluccio, Pavol Hell, and Jaroslav Nešetřil. The complexity of *H*-colouring of bounded degree graphs. *Discrete Mathematics,* 222(1-3):101–109, 2000. 304

[HN90]   P. Hell and J. Nešetřil. On the complexity of *H*-coloring. *Journal of Combinatorial Theory (series B),* 48:92–110, 1990. 304

[TP97]   Jan Arne Telle and Andrzej Proskurowski. Algorithms for vertex partitioning problems on partial *k*-trees. *SIAM Journal on Discrete Mathematics,* 10(4):529–550, 1997. 305

Randomness and Reducibility[*]

Rod G. Downey[1], Denis R. Hirschfeldt[2], and Geoff LaForte[3]

[1] School of Mathematical and Computing Sciences, Victoria University of Wellington
Rod.Downey@mcs.vuw.ac.nz
[2] Department of Mathematics, The University of Chicago
drh@math.uchicago.edu
[3] Department of Computer Science, University of West Florida
glaforte@uwf.edu

1 Introduction

How random is a real? Given two reals, which is more random? If we partition reals into equivalence classes of reals of the "same degrees of randomness", what does the resulting structure look like? The goal of this paper is to look at questions like these, specifically by studying the properties of reducibilities that act as measures of relative randomness, as embodied in the concept of initial-segment complexity. One such reducibility, called domination or Solovay reducibility, was introduced by Solovay [34], and has been studied by Calude, Hertling, Khoussainov, and Wang [8], Calude [3], Kučera and Slaman [22], and Downey, Hirschfeldt, and Nies [15], among others. Solovay reducibility has proved to be a powerful tool in the study of randomness of effectively presented reals. Motivated by certain shortcomings of Solovay reducibility, which we will discuss below, we introduce two new reducibilities and study, among other things, the relationships between these various measures of relative randomness.

We work with reals between 0 and 1, identifying a real with its binary expansion, and hence with the set of natural numbers whose characteristic function is the same as that expansion. (We also identify finite binary strings with rationals.) Our main concern will be reals that are limits of computable increasing sequences of rationals. We call such reals *computably enumerable* (c.e.), though they have also been called *recursively enumerable, left computable* (by Ambos-Spies, Weihrauch, and Zheng [2]), and *left semicomputable*. If, in addition to the existence of a computable increasing sequence $q_0, q_1, \ldots$ of rationals with limit α, there is a total computable function f such that $\alpha - q_{f(n)} < 2^{-n}$ for all n, then α is called *computable*. These and related concepts have been widely studied. In addition to the papers and books mentioned elsewhere in this introduction, we may cite, among others, early work of Rice [28], Lachlan [23], Soare [30], and Ceĭtin [10], and more recent papers by Ko [19,20], Calude, Coles, Hertling, and Khoussainov [7], Ho [18], and Downey and LaForte [16].

An alternate definition of c.e. reals can be given by using the following definition.

---

[*] The authors' research was supported by the Marsden Fund for Basic Science.

J. Sgall, A. Pultr, and P. Kolman (Eds.): MFCS 2001, LNCS 2136, pp. 316–327, 2001.
© Springer-Verlag Berlin Heidelberg 2001

Definition 1. *A set* $A \subseteq \mathbb{N}$ *is* nearly computably enumerable *if there is a computable approximation* $\{A_s\}_{s \in \omega}$ *such that* $A(x) = \lim_s A_s(x)$ *for all* x *and* $A_s(x) > A_{s+1}(x) \Rightarrow \exists y < x(A_s(y) < A_{s+1}(y))$.

As shown by Calude, Coles, Hertling, and Khoussainov [7], a real $0.\chi_A$ is c.e. if and only if A is nearly c.e.. An interesting subclass of the class of c.e. reals is the class of strongly c.e. reals. A real $0.\chi_A$ is said to be *strongly c.e.* if A is c.e.. Soare [31] noted that there are c.e. reals that are not strongly c.e..

A computer M is *self-delimiting* if, for all finite binary strings σ and $\tau \subsetneq \tau'$, we have $M^\sigma(\tau) \downarrow \; \Rightarrow M^\sigma(\tau') \uparrow$, where M^σ means that M uses σ as an oracle. It is *universal* if for each self-delimiting computer N there is a constant c such that, for all binary strings σ and τ, if $N^\sigma(\tau) \downarrow$ then $M^\sigma(\mu) \downarrow = N^\sigma(\tau)$ for some μ with $|\mu| \leqslant |\tau| + c$. We call c the *coding constant* of N.

Fix a self-delimiting universal computer M. We can define Chaitin's number $\Omega = \Omega_M$ via $\Omega = \sum_{M(\sigma)\downarrow} 2^{-|\sigma|}$, which is the *halting probability* of the computer M. The properties of Ω relevant to this paper are independent of the choice of M. A c.e. real is an *Ω-number* if it is Ω_M for some self-delimiting universal computer M.

The c.e. real Ω is random in the canonical Martin-Löf sense [27] of c.e. randomness. There are many equivalent formulations of c.e. randomness. The one that is most relevant to us here is Chaitin randomness, which we define below. The history of effective randomness is quite rich; references include van Lambalgen [36], Calude [4], Li and Vitanyi [26], and Ambos-Spies and Kučera [1].

Recall that the prefix-free complexity $H(\tau)$ of a binary string τ is the length of the shortest binary string σ such that $M(\sigma) \downarrow = \tau$. (The choice of self-delimiting universal computer M does not affect the prefix-free complexity, up to a constant additive factor.) Most of the statements about $H(\tau)$ made below also hold for the standard Kolmogorov complexity $K(\tau)$. For more on the the definitions and basic properties of $H(\tau)$ and $K(\tau)$, see Chaitin [14], Calude [4], Li and Vitanyi [26], and Fortnow [17]. Among the many works dealing with these and related topics, and in addition to those mentioned elsewhere in this paper, we may cite Solomonoff [32,33], Kolmogorov [21], Levin [24,25], Schnorr [29], Chaitin [11], and the expository article Calude and Chaitin [5].

A real α is *random*, or more precisely, 1-random, if there is a constant c such that $\forall n(H(\alpha \upharpoonright n) \geqslant n - c)$.

Many authors have studied Ω and its properties, notably Chaitin [12,13,14] and Martin-Löf [27]. In the very long and widely circulated manuscript [34] (a fragment of which appeared in [35]), Solovay carefully investigated relationships between Martin-Löf-Chaitin prefix-free complexity, Kolmogorov complexity, and properties of random languages and reals. See Chaitin [12] for an account of some of the results in this manuscript.

Solovay discovered that several important properties of Ω (whose definition is model-dependent) are shared by another class of reals he called Ω-like, whose definition is model-independent. To define this class, he introduced the following reducibility relation between c.e. reals.

Definition 2. *Let α and β be c.e. reals. We say that α dominates β and that β is Solovay reducible (S-reducible) to α, and write $\beta \leqslant_s \alpha$, if there are a constant c and a partial computable function $\varphi : \mathbb{Q} \to \mathbb{Q}$ such that for each rational $q < \alpha$ we have $\varphi(q) \downarrow < \beta$ and $\beta - \varphi(q) \leqslant c(\alpha - q)$. We write $\alpha \equiv_s \beta$ if $\alpha \leqslant_s \beta$ and $\beta \leqslant_s \alpha$.*

Solovay reducibility is reflexive and transitive, and hence $\equiv_s$ is an equivalence relation on the c.e. reals. Thus we can define the *Solovay degree* $\deg_s(\alpha)$ of a c.e. real α to be its $\equiv_s$ equivalence class.

Solovay reducibility is naturally associated with randomness due to the following fact.

Theorem 3 (Solovay [34]). *Let $\beta \leqslant_s \alpha$ be c.e. reals. There is a constant c such that $H(\beta \upharpoonright n) \leqslant H(\alpha \upharpoonright n) + c$ for all n.*

It is this property of Solovay reducibility (which we will call the *Solovay property*), which makes it a measure of relative randomness. This is in contrast with Turing reducibility, for example, which does not have the Solovay property, since the complete c.e. Turing degree contains both random and nonrandom reals.

Solovay observed that Ω dominates all c.e. reals, and Theorem 3 implies that if a c.e. real dominates all c.e. reals then it must be random. This led him to define a c.e. real to be Ω-*like* if it dominates all c.e. reals (that is, if it is S-complete). The point is that the definition of Ω-like seems quite model-independent (in the sense that it does not require a choice of self-delimiting universal computer), as opposed to the model-dependent definition of Ω. However, Calude, Hertling, Khoussainov, and Wang [8] showed that the two notions coincide. This circle of ideas was completed recently by Kučera and Slaman [22], who showed that all random c.e. reals are Ω-like.

For more on c.e. reals and S-reducibility, see for instance Chaitin [12,13,14], Calude, Hertling, Khoussainov, and Wang [8], Calude and Nies [9], Calude [3], Kučera and Slaman [22], and Downey, Hirschfeldt, and Nies [15].

Solovay reducibility is an excellent tool in the study of the relative randomness of reals, but it has several shortcomings. One such shortcoming is that S-reducibility is quite unnatural outside the c.e. reals. It is not very hard to construct a noncomputable real that is not S-above the computable reals (in fact, this real can be chosen to be d.c.e., that is, of the form $\alpha - \beta$ where α and β are c.e.). This and similar facts show that S-reducibility is very unnatural when applied to non-c.e. reals. Another problem with S-reducibility is that it is uniform in a way that relative initial-segment complexity is not. This makes it too strong, in a sense, and appears to preclude its having a natural characterization in terms of initial-segment complexity. In particular, Calude and Coles [6] answered a question of Solovay by showing that the converse of Theorem 3 does not hold (see below for an easy proof of this fact). One consequence of the uniformity of S-reducibility is that there does not appear to be a natural characterization of S-reducibility in terms of initial-segment complexity. Thus, if our goal is to study relative initial segment complexity of reals, it behooves us to look beyond S-reducibility.

In this paper, we introduce two new measures of relative randomness that provide additional tools for the study of the relative randomness of reals, and study their properties and the relationships between them and S-reducibility. We begin with sw-reducibility, which has some nice features but also some shortcomings, and then move on to the more interesting rH-reducibility, which shares many of the best features of S-reducibility while not being restricted to the c.e. reals, and has a very nice characterization, in terms of relative initial-segment complexity, which can be seen as a partial converse to the Solovay property. (Indeed, rH stands for "relative H".)

2   Strong Weak Truth Table Reducibility

Solovay reducibility has many attractive features, but it is not the only interesting measure of relative randomness. In this section, we introduce another such measure, sw-reducibility, which is more explicitly derived from the idea of initial segment complexity, and which is in some ways nicer than S-reducibility. In particular, sw-reducibility is much better adapted to dealing with non-c.e. reals. Furthermore, sw-reducibility is also helpful in the study of S-reducibility, as we will indicate below, and provides a motivation for the definition of rH-reducibility in the next section, since rH-reducibility is a kind of "sw-reducibility with advice". In the interest of space, we omit most proofs in this section.

Recall that a Turing reduction $\Gamma^A = B$ is called a weak truth table (wtt) reduction if there is a computable function φ such that the use function $\gamma(x)$ is bounded by $\varphi(x)$.

Definition 4. *Let $A, B \subseteq \mathbb{N}$. We say that B is* strongly weak truth table reducible *(sw-reducible) to A, and write $B \leqslant_{sw} A$, if there are a constant c and a wtt reduction Γ such that $B = \Gamma^A$ and $\forall x(\gamma(x) \leqslant x + c)$. For reals $\alpha = 0.\chi_A$ and $\beta = 0.\chi_B$, we say that β is sw-reducible to α, and write $\beta \leqslant_{sw} \alpha$, if $B \leqslant_{sw} A$.*

Since sw-reducibility is reflexive and transitive, we can define the *sw-degree* $\deg_{sw}(\alpha)$ of a real α to be its sw-equivalence class.

Solovay [34] noted that for each k there is a constant c such that for all $n \geqslant 1$ and all binary strings σ, τ of length n, if $|0.\sigma - 0.\tau| < k2^{-n}$ then $|H(\tau) - H(\sigma)| \leqslant c$. Using this result, it is easy to check that sw-reducibility has the Solovay property.

Theorem 5. *Let $\beta \leqslant_{sw} \alpha$ be c.e. reals. There is a constant c such that $H(\beta \restriction n) \leqslant H(\alpha \restriction n) + c$ for all $n \in \omega$.*

Theorem 7 below shows that the converse of Theorem 5 does not hold even for c.e. reals.

We now explore the relationship between S-reducibility and sw-reducibility on the c.e. and strongly c.e. reals. In general, neither of the reducibilities under consideration implies the other.

Theorem 6. *There exist c.e. reals $\alpha \leqslant_{sw} \beta$ such that $\alpha \nleqslant_s \beta$. Moreover, α can be chosen to be strongly c.e..*

The proof of this theorem is a finite injury argument, in which we meet requirements of the form $\exists q \in \mathbb{Q}(c(\beta - q) \neq \alpha - \Phi_e(q))$, where Φ_e is the eth partial computable function.

We note that, since sw-reducibility has the Solovay property, the previous result gives a quick proof of the theorem, due to Calude and Coles [6], that the converse of Theorem 3 does not hold. This is one example of the usefulness of sw-reducibility in the study of S-reducibility.

Theorem 7. *There exist c.e. reals $\alpha \leqslant_s \beta$ such that $\alpha \nleqslant_{sw} \beta$ (in fact, even $\alpha \nleqslant_{wtt} \beta$). Moreover, β can be chosen to be strongly c.e..*

The proof of this theorem is a diagonalization argument similar to the proof of the previous theorem.

The counterexamples above can be jazzed up with relatively standard degree control techniques to prove the following result.

Theorem 8. *Let* **a** *be a nonzero c.e. Turing degree. There exist c.e. reals α and β of degree* **a** *such that α is strongly c.e., $\alpha \leqslant_{sw} \beta$, and $\alpha \nleqslant_s \beta$. There also exist c.e. reals γ and δ of degree* **a** *such that δ is strongly c.e., $\gamma \leqslant_s \delta$, and $\gamma \nleqslant_{sw} \delta$.*

On the strongly c.e. reals, however, S-reducibility and sw-reducibility coincide. Since sw-reducibility is sometimes easier to deal with than S-reducibility, this fact makes sw-reducibility a useful tool in the study of S-reducibility on strongly c.e. reals. An example of this phenomenon is Theorem 12 below, which is most easily proved using sw-reducibility, as the proof sketch included below illustrates.

Theorem 9. *If β is strongly c.e. and α is c.e. then $\alpha \leqslant_{sw} \beta$ implies $\alpha \leqslant_s \beta$.*

Theorem 10. *If α is strongly c.e. and β is c.e. then $\alpha \leqslant_s \beta$ implies $\alpha \leqslant_{sw} \beta$.*

Corollary 11. *If α and β are strongly c.e. then $\alpha \leqslant_s \beta$ if and only if $\alpha \leqslant_{sw} \beta$.*

There is a greatest S-degree of c.e. reals, namely that of Ω, but the situation is different for strongly c.e. reals.

Theorem 12. *Let α be strongly c.e.. There is a strongly c.e. real that is not sw-below α, and hence not S-below α.*

Proof. The argument is nonuniform, but is still finite injury. Since sw-reducibility and S-reducibility coincide for strongly c.e. reals, it is enough to build a strongly c.e. real that is not sw-below α. Let A be such that $\alpha = 0.\chi_A$. We build c.e. sets B and C to satisfy the following requirements.

$$\mathcal{R}_{e,i} : \Gamma_e^A \neq B \vee \Gamma_i^A \neq C \ ,$$

where Γ_e is the eth wtt reduction with use less than $x + e$. It will then follow that either $0.\chi_B \not\leqslant_{sw} \alpha$ or $0.\chi_C \not\leqslant_{sw} \alpha$.

The idea for satisfying a single requirement $\mathcal{R}_{e,i}$ is simple. Let $l(e,i,s) = \max\{x \mid \forall y \leqslant x(\Gamma_{e,s}^{A_s}(y) = B_s(y) \wedge \Gamma_{i,s}^{A_s} = C_s(y))\}$. Pick a large number $k >> e, i$ and let $\mathcal{R}_{e,i}$ assert control over the interval $[k, 3k]$ in both B and C, waiting until a stage s such that $l(e,i,s) > 3k$.

First work with C. Put $3k$ into C, and wait for the next stage s' where $l(e,i,s') > 3k$. Note that some number must enter $A_{s'} - A_s$ below $3k + i$. Now repeat with $3k - 1$, then $3k - 2, \ldots, k$. In this way, $2k$ numbers are made to enter A below $3k + i$. Now we can win using B, by repeating the process and noticing that, by the choice of the parameter k, A cannot respond another $2k$ times below $3k + e$.

The theorem now follows by a standard application of the finite injury method. □

Some structural properties are much easier to prove for sw-reducibility than for S-reducibility. One example is the fact that there are no minimal sw-degrees of c.e. reals, that is, that for any noncomputable c.e. real α there is a c.e. real strictly sw-between α and the computable reals. The analogous property for S-reducibility was proved by Downey, Hirschfeldt, and Nies [15] with a fairly involved priority argument.

Definition 13. *Let A be a nearly c.e. set. The* sw-canonical c.e. set A^* *associated with A is defined as follows. Begin with $A_0^* = \emptyset$. For all x and s, if either $x \notin A_s$ and $x \in A_{s+1}$, or $x \in A_s$ and $x \notin A_{s+1}$, then for the least j with $\langle x, j \rangle \notin A_s^*$, put $\langle x, j \rangle$ into A_{s+1}^*.*

Lemma 14. $A^* \leqslant_{sw} A$ and $A \leqslant_{tt} A^*$.

Proof. Since A is nearly c.e., $\langle x, j \rangle$ enters A^* at a given stage only if some $y \leqslant x$ enters A at that stage. Such a y will also be below $\langle x, j \rangle$. Hence $A^* \leqslant_{sw} A$ with use x. Clearly, $x \in A$ if and only if A^* has an odd number of entries in row x, and furthermore, since A is nearly c.e., the number of entries in this row is bounded by x. Hence $A \leqslant_{tt} A^*$. □

Corollary 15. *If A is nearly c.e. and noncomputable then there is a noncomputable c.e. set $A^* \leqslant_{sw} A$.*

Corollary 16. *There are no minimal sw-degrees of c.e. reals.*

Proof. Let A be nearly c.e. and noncomputable. Then $A^* \leqslant_{sw} A$ is noncomputable, and we can c.e. Sacks split A^* into two disjoint c.e. sets A_1^* and A_2^* of incomparable Turing degree. Note that $A_i^* \leqslant_{sw} A^*$. (To decide whether $x \in A_i^*$, ask whether $x \in A^*$ and, if the answer is yes, then run the enumerations of A_1^* and A_2^* to see which set x enters.) So $\emptyset <_{sw} A_1^* <_{sw} A^* \leqslant_{sw} A$. □

Actually, while the above proof yields more than just nonminimality, there is an easier proof that the sw-degrees of c.e. reals have no minimal members.

Given a c.e. real $A = 0.a_1a_2\ldots$, consider the c.e. real $B = 0.a_10a_200a_3000a_4\ldots$. It is easy to prove that if A is noncomputable then so is B. But it is also easy to see that $B \leqslant_{sw} A$, and that if it were the case that $A \leqslant_{sw} B$ then A would be computable. Hence $\emptyset <_{sw} B <_{sw} A$.

One thing we can get out of the proof of Corollary 16 is that every c.e. real has a noncomputable strongly c.e. real sw-below it. The same is not true for S-reducibility.

Theorem 17. *There is a noncomputable c.e. real α such that all strongly c.e. reals dominated by α are computable.*

The proof of this theorem uses a result of Downey and LaForte [16]. A c.e. set $A \subseteq \{0,1\}^*$ *presents* a c.e. real α if A is prefix-free and $\alpha = \sum_{\sigma \in A} 2^{-|\sigma|}$. Downey and LaForte constructed a noncomputable c.e. real α such that if A presents α then A is computable. Given a strongly c.e. real $\beta \leqslant_s \alpha$, it is possible to build a presentation A of α that "encodes" β, in the sense that, for some constant k, by knowing how many strings of length $n + k$ are in A, we can tell whether the nth bit of β is 1. Since A must be computable, this allows us to compute β.

As we have seen, in some ways the sw-degrees are nicer than the S-degrees. Unfortunately, the theorem below shows that this is not always the case. There is a simple join operator, arithmetic addition, which induces a join operation on the S-degrees. No such operation exists for the sw-degrees.

Theorem 18. *There exist nearly c.e. sets A and B such that for all nearly c.e. $W \geqslant_{sw} A, B$ there is a nearly c.e. Q with $A, B \leqslant_{sw} Q$ but $W \not\leqslant_{sw} Q$. Thus the sw-degrees of c.e. reals do not form an uppersemilattice.*

The idea of the proof of this theorem is that if $W \geqslant_{sw} B$ then, by changing B very often, we can cause W to change very often, and hence force W to contain large blocks of 1's. We can then use A to force a change in W somewhere within such a block, which, because W is nearly c.e., forces W to change below the block. But if the block is large enough then we can use this W change to destroy a potential sw-reduction from W to Q, while still allowing Q to be sw-above A and B. The full details involve a finite injury priority argument which is nonuniform in the sense that we prevent a given $W \geqslant_{sw} A, B$ from being a join by constructing infinitely many c.e. reals $Q_i \geqslant_{sw} A, B$ and using the argument outlined above to show that $W \not\leqslant_{sw} Q_i$ for at least one of the Q_i.

The lack of a join operation leads to difficulties in exploring the structure of the sw-degrees beyond what is done here, and is one of the motivations for the introduction of rH-reducibility in the following section.

3　Relative H Reducibility

Both S-reducibility and sw-reducibility are uniform in a way that relative initial-segment complexity is not. This makes them too strong, in a sense, and it is natural to wish to investigate nonuniform versions of these reducibilities. Motivated by this consideration, as well as by the problems with sw-reducibility, we introduce another measure of relative randomness, called relative H reducibility, which can be seen as a nonuniform version of both S-reducibility and sw-reducibility, and which combines many of the best features of these reducibilities. Its name derives from a characterization, discussed below, which shows that there is a very natural sense in which it is an *exact* measure of relative randomness.

Definition 19. *Let α and β be reals. We say that β is relative H reducible (rH-reducible) to α, and write $\beta \leqslant_{rH} \alpha$, if there are a constant k and a partial computable binary function f such that for each n there is a $j \leqslant k$ for which $f(\alpha \upharpoonright n, j) \downarrow = \beta \upharpoonright n$.*

Since rH-reducibility is reflexive and transitive, we can define the *rH-degree* $\deg_{rH}(\alpha)$ of a real α to be its rH-equivalence class.

There are several characterizations of rH-reducibility, each revealing a different facet of the concept. We mention three, beginning with a "relative entropy" characterization whose proof is quite straightforward. For a c.e. real β and a fixed computable approximation $\beta_0, \beta_1, \ldots$ of β, we will let the mind-change function $m(\beta, n, s, t)$ be the cardinality of $\{u \in [s, t] \mid \beta_u \upharpoonright n \neq \beta_{u+1} \upharpoonright n\}$.

Proposition 20. *Let α and β be c.e. reals. The following condition holds if and only if $\beta \leqslant_{rH} \alpha$. There are a constant k and computable approximations $\alpha_0, \alpha_1, \ldots$ and $\beta_0, \beta_1, \ldots$ of α and β, respectively, such that for all n and $t > s$, if $\alpha_t \upharpoonright n = \alpha_s \upharpoonright n$ then $m(\beta, n, s, t) \leqslant k$.*

The following is a more analytic characterization of rH-reducibility, which clarifies its nature as a nonuniform version of both S-reducibility and sw-reducibility.

Proposition 21. *For any reals α and β, the following condition holds if and only if $\beta \leqslant_{rH} \alpha$. There are a constant c and a partial computable function φ such that for each n there is a τ of length $n + c$ with $|\alpha - \tau| \leqslant 2^{-n}$ for which $\varphi(\tau) \downarrow$ and $|\beta - \varphi(\tau)| \leqslant 2^{-n}$.*

Proof. First suppose that $\beta \leqslant_{rH} \alpha$ and let f and k be as in Definition 19. Let c be such that $2^c \geqslant k$ and define the partial computable function φ as follows. Given a string σ of length n, whenever $f(\sigma, j) \downarrow$ for some new $j \leqslant k$, choose a new $\tau \supseteq \sigma$ of length $n + c$ and define $\varphi(\tau) = f(\sigma, j)$. Then for each n there is a $\tau \supseteq \alpha \upharpoonright n$ such that $\varphi(\tau) \downarrow = \beta \upharpoonright n$. Since $|\alpha - \tau| \leqslant |\alpha - \alpha \upharpoonright n| \leqslant 2^{-n}$ and $|\beta - \beta \upharpoonright n| \leqslant 2^{-n}$, the condition holds.

Now suppose that the condition holds. For a string σ of length n, let S_σ be the set of all μ for which there is a τ of length $n + c$ with $|\sigma - \tau| \leqslant 2^{-n+1}$ and $|\mu - \varphi(\tau)| \leqslant 2^{-n+1}$. It is easy to check that there is a k such that $|S_\sigma| \leqslant k$ for all σ. So there is a partial computable binary function f such that for each σ and each $\mu \in S_\sigma$ there is a $j \leqslant k$ with $f(\sigma, j) \downarrow = \mu$. But, since for any real γ and any n we have $|\gamma - \gamma \upharpoonright n| \leqslant 2^{-n}$, it follows that for each n we have $\beta \upharpoonright n \in S_{\alpha \upharpoonright n}$. Thus f and k witness the fact that $\beta \leqslant_{rH} \alpha$. $\qquad \Box$

The most interesting characterization of rH-reducibility (and the reason for its name) is given by the following result, which shows that there is a very natural sense in which rH-reducibility is an exact measure of relative randomness. Recall that the prefix-free complexity $H(\tau \mid \sigma)$ of τ relative to σ is the length of the shortest string μ such that $M^\sigma(\mu) \downarrow = \tau$, where M is a fixed self-delimiting universal computer.

Theorem 22. *Let α and β be reals. Then $\beta \leqslant_{\mathrm{rH}} \alpha$ if and only if there is a constant c such that $H(\beta \restriction n \mid \alpha \restriction n) \leqslant c$ for all n.*

Proof. First suppose that $\beta \leqslant_{\mathrm{rH}} \alpha$ and let f and k be as in Definition 19. Let m be such that $2^m \geqslant k$ and let $\tau_0, \ldots, \tau_{2^m-1}$ be the strings of length m. Define the prefix-free machine N to act as follows with σ as an oracle. For all strings μ of length not equal to m, let $N^\sigma(\mu) \uparrow$. For each $i < 2^m$, if $f(\sigma, i) \downarrow$ then let $N^\sigma(\tau_i) \downarrow = f(\sigma, i)$, and otherwise let $N^\sigma(\tau_i) \uparrow$. Let e be the coding constant of N and let $c = e + m$. Given n, there exists a $j \leqslant k$ for which $f(\alpha \restriction n, j) \downarrow = \beta \restriction n$. For this j we have $N^{\alpha \restriction n}(\tau_j) \downarrow = \beta \restriction n$, which implies that $H(\beta \restriction n \mid \alpha \restriction n) \leqslant |\tau_j| + e \leqslant c$.

Now suppose that $H(\beta \restriction n \mid \alpha \restriction n) \leqslant c$ for all n. Let $\tau_0, \ldots, \tau_k$ be a list of all strings of length less than or equal to c and define f as follows. For a string σ and a $j \leqslant k$, if $M^\sigma(\tau_j) \downarrow$ then $f(\sigma, j) \downarrow = M^\sigma(\tau_j)$, and otherwise $f(\sigma, j) \uparrow$. Given n, since $H(\beta \restriction n \mid \alpha \restriction n) \leqslant c$, it must be the case that $M^{\alpha \restriction n}(\tau_j) \downarrow = \beta \restriction n$ for some $j \leqslant k$. For this j we have $f(\alpha \restriction n, j) \downarrow = \beta \restriction n$. Thus $\beta \leqslant_{\mathrm{rH}} \alpha$. $\qquad\square$

An immediate consequence of this result is that rH-reducibility satisfies the Solovay property.

Corollary 23. *If $\beta \leqslant_{\mathrm{rH}} \alpha$ then there is a constant c such that $H(\beta \restriction n) \leqslant H(\alpha \restriction n) + c$ for all n.*

On the other hand, the converse of this corollary is not true even for strongly c.e. reals. This follows from Theorem 28 below and a result of Zambella [37], who showed, using a technique due to Solovay [34], that there is a noncomputable strongly c.e. real β such that $H(\beta \restriction n) \leqslant H(1^n) + O(1)$.

The next two theorems, which show that rH-reducibility is a common weakening of S-reducibility and sw-reducibility, follow easily from Proposition 21.

Theorem 24. *Let α and β be c.e. reals. If $\beta \leqslant_{\mathrm{s}} \alpha$ then $\beta \leqslant_{\mathrm{rH}} \alpha$.*

Corollary 25. *A c.e. real α is rH-complete if and only if it is random.*

Theorem 26. *If $\beta \leqslant_{\mathrm{sw}} \alpha$ then $\beta \leqslant_{\mathrm{rH}} \alpha$.*

Theorems 6 and 7 show that the converses of Theorems 24 and 26 do not hold, but even among strongly c.e. reals, where S-reducibility and sw-reducibility agree, rH-reducibility is not equivalent to its stronger counterparts.

Theorem 27. *There exist strongly c.e. reals α and β such that $\beta \leqslant_{\mathrm{rH}} \alpha$ but $\beta \not\leqslant_{\mathrm{sw}} \alpha$ (equivalently, $\beta \not\leqslant_{\mathrm{s}} \alpha$).*

The proof of this theorem is a straightforward finite injury argument.

It is interesting to note that, despite the nonuniform nature of its definition, rH-reducibility implies Turing reducibility. Since any computable real is obviously rH-reducible to any other real, this implies that the computable reals form the least rH-degree.

Theorem 28. *If $\beta \leqslant_{rH} \alpha$ then $\beta \leqslant_T \alpha$.*

Proof. Let k be the least number for which there exists a partial computable binary function f such that for each n there is a $j \leqslant k$ with $f(\alpha \upharpoonright n, j)\!\downarrow = \beta \upharpoonright n$. There must be infinitely many n for which $f(\alpha \upharpoonright n, j) \downarrow$ for all $j \leqslant k$, since otherwise we could change finitely much of f to contradict the minimality of k. Let $n_0 < n_1 < \cdots$ be an α-computable sequence of such n. Let T be the α-computable subtree of 2^ω obtained by pruning, for each i, all the strings of length n_i except for the values of $f(\alpha \upharpoonright n_i, j)$ for $j \leqslant k$.

If γ is a path through T then for all i there is a $j \leqslant k$ such that γ extends $f(\alpha \upharpoonright n_i, j)$. Thus there are at most k many paths through T, and hence each path through T is α-computable. But β is a path through T, so $\beta \leqslant_T \alpha$. $\square$

On the other hand, by Theorem 7, S-reducibility does not imply wtt-reducibility, even among c.e. reals, and hence rH-reducibility does not imply wtt-reducibility.

Structurally, the rH-degrees of c.e. reals are nicer than the sw-degrees of c.e. reals.

Theorem 29. *The rH-degrees of c.e. reals form an uppersemilattice with least degree that of the computable sets and highest degree that of Ω. The join of the rH-degrees of the c.e. reals α and β is the rH-degree of $\alpha + \beta$.*

Proof. All that is left to show is that addition is a join. Since $\alpha, \beta \leqslant_S \alpha + \beta$, it follows that $\alpha, \beta \leqslant_{rH} \alpha + \beta$. Let γ be a c.e. real such that $\alpha, \beta \leqslant_{rH} \gamma$. Then Proposition 20 implies that $\alpha + \beta \leqslant_{rH} \gamma$, since for any n and $s < t$ we have $m(\alpha + \beta, n, s, t) \leqslant 2(m(\alpha, n, s, t) + m(\beta, n, s, t)) + 1$. $\square$

In [15], Downey, Hirschfeldt, and Nies studied the structure of the S-degrees of c.e. reals. They showed that the S-degrees of c.e. reals are dense. They also showed that every incomplete S-degree splits over any lesser degree, while the complete S-degree does not split at all. The methods of that paper can be adapted to prove the analogous results for rH-degrees of c.e. reals.

Theorem 30. *For any rH-degrees $\mathbf{a} < \mathbf{b}$ of c.e. reals there is an rH-degree $\mathbf{c}$ of c.e. reals such that $\mathbf{a} < \mathbf{c} < \mathbf{b}$.*

For any rH-degrees $\mathbf{a} < \mathbf{b} < \deg_{rH}(\Omega)$ of c.e. reals, there are rH-degrees $\mathbf{c_0}$ and $\mathbf{c_1}$ of c.e. reals such that $\mathbf{a} < \mathbf{c_0}, \mathbf{c_1} < \mathbf{b}$ and $\mathbf{c_0} \vee \mathbf{c_1} = \mathbf{b}$.

For any rH-degrees $\mathbf{a}, \mathbf{b} < \deg_{rH}(\Omega)$ of c.e. reals, $\mathbf{a} \vee \mathbf{b} < \deg_{rH}(\Omega)$.

Thus we see that rH-reducibility shares many of the nice structural properties of S-reducibility on the c.e. reals, while still being a reasonable reducibility on non-c.e. reals. Together with its various characterizations, especially the one in terms of relative H-complexity of initial segments, this makes rH-reducibility a tool with great potential in the study of the relative randomness of reals.

References

1. K. Ambos-Spies and A. Kučera, Randomness in computability theory, in P. A. Cholak, S. Lempp, M. Lerman, and R. A. Shore (eds.), Computability Theory and Its Applications (Boulder, CO, 1999), vol. 257 of Contemp. Math. (Amer. Math. Soc., Providence, RI, 2000) 1–14. 317

2. K. Ambos-Spies, K. Weihrauch, and X. Zheng, Weakly computable real numbers, J. Complexity 16 (2000) 676–690. 316

3. C. S. Calude, A characterization of c.e. random reals, to appear in Theoret. Comput. Sci. 316, 318

4. C. S. Calude, Information and Randomness, an Algorithmic Perspective, Monographs in Theoretical Computer Science (Springer–Verlag, Berlin, 1994). 317

5. C. S. Calude and G. Chaitin, Randomness everywhere, Nature 400 (1999) 319–320. 317

6. C. S. Calude and R. J. Coles, Program-size complexity of initial segments and domination reducibility, in J. Karhumäki, H. Maurer, G. Păun, and G. Rozenberg (eds.), Jewels Are Forever (Springer-Verlag, Berlin, 1999) 225–237. 318, 320

7. C. S. Calude, R. J. Coles, P. H. Hertling, and B. Khoussainov, Degree-theoretic aspects of computably enumerable reals, in S. B. Cooper and J. K. Truss (eds.), Models and Computability (Leeds, 1997), vol. 259 of London Math. Soc. Lecture Note Ser. (Cambridge Univ. Press, Cambridge, 1999) 23–39. 316, 317

8. C. S. Calude, P. H. Hertling, B. Khoussainov, and Y. Wang, Recursively enumerable reals and Chaitin Ω numbers, Centre for Discrete Mathematics and Theoretical Computer Science Research Report Series 59, University of Auckland (October 1997), extended abstract in STACS 98, Lecture Notes in Comput. Sci. 1373, Springer, Berlin, 1998, 596–606. 316, 318

9. C. S. Calude and A. Nies, Chaitin Ω numbers and strong reducibilities, J. Univ. Comp. Sci. 3 (1998) 1162–1166. 318

10. G. S. Ceĭtin, A pseudofundamental sequence that is not equivalent to a monotone one, Zap. Naučn. Sem. Leningrad. Otdel. Mat. Inst. Steklov. (LOMI) 20 (1971) 263–271, 290, in Russian with English summary. 316

11. G. J. Chaitin, A theory of program size formally identical to information theory, J. Assoc. Comput. Mach. 22 (1975) 329–340, reprinted in [14]. 317

12. G. J. Chaitin, Algorithmic information theory, IBM J. Res. Develop. 21 (1977) 350–359, 496, reprinted in [14]. 317, 318

13. G. J. Chaitin, Incompleteness theorems for random reals, Adv. in Appl. Math. 8 (1987) 119–146, reprinted in [14]. 317, 318

14. G. J. Chaitin, Information, Randomness & Incompleteness, vol. 8 of Series in Computer Science, 2nd ed. (World Scientific, River Edge, NJ, 1990). 317, 318, 326

15. R. Downey, D. R. Hirschfeldt, and A. Nies, Randomness, computability, and density, to appear in SIAM J. Comp. (extended abstract in A. Ferreira and H. Reichel (eds.), STACS 2001 Proceedings, Lecture Notes in Computer Science 2010 (Springer, 2001), 195–205). 316, 318, 321, 325

16. R. Downey and G. LaForte, Presentations of computably enumerable reals, to appear in Theoret. Comput. Sci. 316, 322

17. L. Fortnow, Kolmogorov complexity, to appear in R. G. Downey and D. R. Hirschfeldt (eds.), Aspects of Complexity, Minicourses in Algorithmics, Complexity, and Computational Algebra, NZMRI Mathematics Summer Meeting, Kaikoura, New Zealand, January 7–15, 2000. 317

18. C. Ho, Relatively recursive reals and real functions, Theoret. Comput. Sci. 210 (1999) 99–120. 316
19. K.-I. Ko, On the definition of some complexity classes of real numbers, Math. Systems Theory 16 (1983) 95–100. 316
20. K.-I. Ko, On the continued fraction representation of computable real numbers, Theoret. Comput. Sci. 47 (1986) 299–313. 316
21. A. N. Kolmogorov, Three approaches to the quantitative definition of information, Internat. J. Comput. Math. 2 (1968) 157–168. 317
22. A. Kučera and T. Slaman, Randomness and recursive enumerability, to appear in SIAM J. Comput. 316, 318
23. A. H. Lachlan, Recursive real numbers, J. Symbolic Logic 28 (1963) 1–16. 316
24. L. Levin, On the notion of a random sequence, Soviet Math. Dokl. 14 (1973) 1413–1416. 317
25. L. Levin, The various measures of the complexity of finite objects (an axiomatic description), Soviet Math. Dokl. 17 (1976) 522–526. 317
26. M. Li and P. Vitanyi, An Introduction to Kolmogorov Complexity and its Applications, 2nd ed. (Springer–Verlag, New York, 1997). 317
27. P. Martin-Löf, The definition of random sequences, Inform. and Control 9 (1966) 602–619. 317
28. H. Rice, Recursive real numbers, Proc. Amer. Math. Soc. 5 (1954) 784–791. 316
29. C.-P. Schnorr, Process complexity and effective random tests, J. Comput. System Sci. 7 (1973) 376–388. 317
30. R. Soare, Cohesive sets and recursively enumerable Dedekind cuts, Pacific J. Math. 31 (1969) 215–231. 316
31. R. Soare, Recursion theory and Dedekind cuts, Trans. Amer. Math. Soc. 140 (1969) 271–294. 317
32. R. J. Solomonoff, A formal theory of inductive inference I, Inform. and Control 7 (1964) 1–22. 317
33. R. J. Solomonoff, A formal theory of inductive inference II, Inform. and Control 7 (1964) 224–254. 317
34. R. M. Solovay, Draft of a paper (or series of papers) on Chaitin's work . . . done for the most part during the period of Sept.–Dec. 1974 (May 1975), unpublished manuscript, IBM Thomas J. Watson Research Center, Yorktown Heights, NY, 215 pages. 316, 317, 318, 319, 324
35. R. M. Solovay, On random r.e. sets, in A. Arruda, N. Da Costa, and R. Chuaqui (eds.), Non-Classical Logics, Model Theory, and Computability. Proceedings of the Third Latin-American Symposium on Mathematical Logic, Campinas, July 11–17, 1976, vol. 89 of Studies in Logic and the Foundations of Mathematics (North-Holland, Amsterdam, 1977) 283–307. 317
36. M. van Lambalgen, Random Sequences, PhD Thesis, University of Amsterdam (1987). 317
37. D. Zambella, On sequences with simple initial segments, Tech. rep., Institute for Language, Logic and Information, University of Amsterdam (1990). 324

On the Computational Complexity of Infinite Words[*]

Pavol Ďuriš[1] and Ján Maňuch[2]

[1] Faculty of Mathematics and Physics, Comenius University
Bratislava, Slovakia
duris@dcs.fmph.uniba.sk
[2] TUCS and Dep. of Math., Univ. of Turku
Datacity, Lemminkäisenkatu 14A, Turku, Finland
jamanu@utu.fi

Abstract. This paper contains answers to several problems in the theory of the computational complexity of infinite words. We show that the problem whether all infinite words generated by iterating dgsm's have logarithmic space complexity is equivalent to the open problem asking whether the unary classes of languages in P and in DLOG are equivalent. Similarly, the problem to find a concrete infinite word which cannot be generated in logarithmic space is equivalent to the problem to find a concrete language which does not belong to DSPACE(n). Finally, we separate classes of infinite words generated by double and triple D0L TAG systems.

Introduction

The study of infinite words is an important research topic in Combinatorics on words. The main stream of research focused on combinatorial properties, cf. [Lo], and the descriptional complexity of infinite words, i.e., the measure how complicated simple mechanisms are needed to generate particular infinite words. An extensive study of different machineries for infinite word generation can be found in [CK].

In [HKL] a new area of investigation was introduced, the computational complexity of words, i.e., the measure how much resources (such as time and space) are needed to generate a certain infinite word. The paper concentrates on relations between the two mentioned complexities: descriptional and computational. Further results in this area can be found in [HK].

In [CK], [HKL] and [HK] several interesting problems are proposed. In this paper we will show that even some of the simplest problems proposed are equivalent to well-known hard open problems in complexity theory of Turing machines.

The paper is organized as follows:

---

[*] Research supported by DAAD and Academy of Finland under a common grant No. 864524.

J. Sgall, A. Pultr, and P. Kolman (Eds.): MFCS 2001, LNCS 2136, pp. 328–337, 2001.

In Section 1 we fix our terminology. In Section 1.1 we recall the definition of the computational complexity, while in Section 1.2 we define several simple methods for generating infinite words:

- iterating morphism, the most commonly used method introduced already in [Th];
- iterating a deterministic generalized sequential machine (a dgsm), i.e. a deterministic finite state transducer;
- double and triple D0L TAG systems.

In Section 2.1 we consider the open problem, proposed in [HKL], namely whether all infinite words generated by iterating dgsm's have logarithmic space complexity. This problem was attacked in [Le], claiming that the answer to the problem is affirmative. On the other hand, here we show that the problem is equivalent to an other hard open problem asking whether unary classes of languages P and DLOG (denoted u-P and u-DLOG, respectively) are equivalent. One can easily observe that u-P = u-DLOG if and only if $\cup_{c>0} \text{DTIME}(c^n) = \text{DSPACE}(n)$.

Section 2.1 contains two other small results concerning the growth of dgsm's, It was shown in [HKL] that the greatest possible growth of a dgsm is exponential and that infinite words generated by such dgsm's have logarithmic space complexity. We show that the smallest non-trivial growth is $\Theta(n \log n)$ and that, similarly, dgsm's with such growth generate infinite words with logarithmic space complexity.

In [HKL] an another interesting problem is proposed: to find a concrete infinite word which cannot be generated in the logarithmic space. It is mentioned already in [HK] that this problem is at least as hard as to prove $L \notin \text{DLOG}$ for some $L \in \text{NP}$. In Section 2.2 we show that it is exactly as hard as the problem to find a concrete language, which does not belong to $\text{DSPACE}(n)$. Note that even the problem to find a concrete language, which does not belong to $\text{DSPACE}(\log n) = \text{DLOG}$ is a hard open problem.

Finally, in Section 2.3 we separate the classes of infinite words generated by double and triple D0L TAG systems as it was conjectured in [CK].

1   Preliminaries

In this section we fix our terminology. Let Σ be a finite alphabet. The sets of all finite and infinite words over Σ are denoted Σ^* and $\Sigma^{\mathbb{N}}$, respectively. In the following sections we define the computational complexity of infinite words and describe several iterative devices generating infinite words.

1.1   The Computational Complexity of Infinite Words

The best way how to define the computational complexity of an object is to describe it in the terms of Turing machines. For example, the Kolmogorov complexity of a finite word is the size of the smallest Turing machine generating the word, cf. [Ko]. In the case of infinite words we will use the model of computation based on *k-tape Turing machine*, which consists of

1. a finite state control;
2. k infinite one-way working tapes each containing one two-way read/write head;
3. one infinite output tape containing one one-way write-only head.

We assume that k-tape TM starts in the initial state with all tapes empty and behaves as normal Turing machine. We say that k-tape TM generates an infinite word $w \in \Sigma^{\mathbb{N}}$ if

1. in each step of computation, the content of the output tape is a prefix of w;
2. for each u, prefix of w, there is an integer n such that u is a prefix of the content of the output tape after n steps of computation.

Let M be a k-tape TM generating a word w. The *time* and *space complexities* of M are functions $T_M : \mathbb{N} \to \mathbb{N}$ and $S_M : \mathbb{N} \to \mathbb{N}$ defined as follows:

- $T_M(n)$ is the smallest number of steps of the computation of M when the prefix of w of length n is already written on the output tape;
- $S_M(n)$ is the space complexity of working tapes during first $T_M(n)$ steps of the computation, i.e., the maximum of lengths of words written on working tapes.

Finally, for any integer function $s : \mathbb{N} \to \mathbb{N}$ we define the following complexity classes:

- GTIME$(s) = \{w \in \Sigma^{\mathbb{N}};$ there exists a k-tape TM M generating w and $T_M(n) \leq s(n)$ for all $n \geq 1\}$;
- GSPACE$(s) = \{w \in \Sigma^{\mathbb{N}};$ there exists a k-tape TM M generating w and $S_M(n) \leq s(n)$ for all $n \geq 1\}$;

It follows from the speed-up argument, as in ordinary complexity theory, that functions $s(n)$ and $c.s(n)$, where c is a constant, have the same space computational complexities.

1.2   Iterative Devices Generating Infinite Words

In this section we define several simple methods used for generating infinite words. The simplest and most commonly used method is to iterate a morphism $h : \Sigma^* \to \Sigma^*$: if h is nonerasing and for a letter $a \in \Sigma$, a is a prefix of $h(a)$, then there exists the limit

$$w = \lim_{n \to \infty} h^n(a) .$$

A natural generalization of this method is to use more powerful mapping in the iteration, a deterministic gsm. As a convention we assume throughout that all dgsm's are non-erasing.

The further generalization leads to *double D0L TAG system* which consist of two infinite one-way tapes each of which containing a one-way read-only head and a one-way write-only head. In each step of the generation both read-only heads read a symbol and move right to the next square while the write-only heads write the corresponding outputs to the first empty squares of these tapes. We assume that the infinite word generated by a double D0L TAG system is written on the first tape. A double D0L TAG system can be specified in the terms of the rewriting rules of the form:

$$\begin{pmatrix} a \\ b \end{pmatrix} \rightarrow \begin{pmatrix} \alpha \\ \beta \end{pmatrix}, \text{ where } a, b \in \Sigma,\ \alpha, \beta \in \Sigma^+.$$

Assuming that in each rewriting rule, $|\beta| = 1$, we get a mechanism which iterates a dgsm. Finally, we can define *triple D0L TAG system* by extending the number of tapes to three.

2   The Results

2.1   Dgsm's

In [HKL] the following problem is proposed: are all words generated by dgsm's in GSPACE($\log n$)? We prove here that this problem is equivalent to the problem whether classes u-DLOG and u-P are equivalent. Note that this in some sense contradicts the result of [Le] that all infinite words generated by iterating dgsm's have logarithmic space complexity: if the result in [Le] is correct, then, together with the following theorem, we have D-EXPTIME = DSPACE(n), which is unlikely. [Le] gives only the sketch of the proof of the result and, hence, we are unable to check if it is correct.

Theorem 1. *All infinite words generated by iterating dgsm's have logarithmic space complexity if and only if* u-P = u-DLOG.

Proof. First, let us assume that u-P = u-DLOG. Take a dgsm τ over finite alphabet Σ generating an infinite word $w = w_1 w_2 \dots$. We prove that the space complexity of w is $\mathcal{O}(\log n)$. It is obvious that there is a 1-tape Turing machine M generating the word w in quadratic time. Consider the languages $L_c = \{0^n; \ n \geq 1,\ w_n = c\}$ for all $c \in \Sigma$. Note that L_c is an unary language. We can easily construct a Turing machine recognizing L_c in quadratic time using Turing machine M. By the assumption there exist Turing machines M_c recognizing the languages L_c in logarithmic space. Now, consider 3-tape TM, which runs M_c's to generate n-th letter of w by using the third tape as working tape. It stores the binary representation of n on the first tape and the position of the head of TM M_c on the input tape on the second tape. Before each run of any M_c it erases the working tape and writes on the second tape position 1. It runs M_c for each letter c of the alphabet of τ until some M_c accepts and then it writes the letter c on the output tape. In each step of the simulation of any M_c it checks whether the position represented on the second tape is the last one. Clearly such machine generates the word w in logarithmic space.

Second, assume that all words generated by iterating a dgsm have logarithmic space complexity. Take a Turing machine M working in polynomial time, i.e., $T(M) = \mathcal{O}(n^k)$, recognizing language $L \subseteq 0^*$. We construct 1-tape TM M' for generating an infinite word with the tape divided into three layers. On the first layer it generates unary inputs in increasing order, on the second layer it simulates computations of M on the input stored on the first layer, and on the third layer it writes 1, if computation ends in an accepting state, or 0, if it ends in a rejecting state. Before each simulation it erases the second and third layer of the tape.

Now, consider a dgsm τ which carries out the computations

$$C_i \rightarrow C_{i+1}, \text{ for } i \geq 0,$$

where C_i corresponds to the i-th configuration of M'. It also maps the starting letter $ into the starting configuration of M'. Clearly, the iteration of τ will generate infinite word, the sequence of all configurations of TM M': $W = \$C_0C_1C_2\ldots$. By the assumption the infinite word W has logarithmic space complexity, i.e., there exists a TM M'' generating W in logarithmic space. Finally, we define TM M''' recognizing L, which on the input 0^n runs M'', but instead of writing the bits of W to the output tape, it compares its input with the input on the first layer of each generated configuration, and moreover, it checks the first letter on the third layer of each generated configuration. When the compared inputs coincide and the first letter on the third layer is 0 or 1 then TM M''' halts in the rejecting or in the accepting state, respectively. Otherwise, it continues in generating bits of the next configuration. Clearly, M''' recognize the language L.

Now, it suffices to show that M''' works in logarithmic space. Let C_{i_n} be the configuration in which M' writes 0 or 1 on the first place of the third layer, when the first layer contains 0^n. Hence in the block of configurations $B_n = C_{i_{n-1}+1} \ldots C_{i_n}$, M' erases the second and the third layer of the tape, changes the input on the first layer to 0^n and runs TM M on this input. Since M works in time $\mathcal{O}(n^k)$ the length of any configuration in block B_n is at most $\mathcal{O}(n^k)$ and the number of configurations in B_n is at most $\mathcal{O}(n^k)$. Hence the length of the block B_n is at most $\mathcal{O}(n^{2k})$. Turing machine M''' on the input 0^x generates first x blocks of configurations until it halts. Hence it generates the prefix of the infinite word W of length

$$\sum_{n=0}^{x} |B_n| = \mathcal{O}(x^{2k+1}).$$

Since M'' works in logarithmic space, M''' will use space $\mathcal{O}(\log x)$ to carry out the computation on the input 0^x.

Next, we will show two small results concerning the growth of a dgsm τ, i.e., the integer function $g : \mathbb{N} \rightarrow \mathbb{N}$, where $g(n)$ is the length of $\tau^n(a)$. In [HKL] it was proved that the infinite word generated by a dgsm with the exponential growth has logarithmic space complexity. Note that the exponential growth is the greatest possible growth. Next, we show that the smallest non-trivial growth of a dgsm is $\Theta(n \log n)$ and also that dgsm's with the growth $\Theta(n \log n)$ generate infinite words with logarithmic space complexity. More precisely:

Lemma 2. *If a dgsm has the growth* $o(n \log n)$*, then it generates an ultimately periodic infinite word. Such word can be generated in constant space.*

Proof. Let τ be a dgsm with the growth $o(n \log n)$ generating an infinite word $w = w_1 w_2 \ldots$, with $w_i \in \Sigma$. Let $\tau_q(z)$ $[\sigma_q(z)]$ be the output [the last state] of dgsm τ when reading the input z and starting in the state q.

We define the two sequences of words and one sequence of states of dgsm τ. Let a be the starting symbol, q_0 be the initial state and let $\tau(a) = \tau_{q_0}(a) = av$. Then put:

$$u_1 = a, \qquad\qquad v_1 = v, \qquad\qquad q_1 = \sigma_{q_0}(a)$$
$$u_n = u_{n-1}v_{n-1}, \qquad v_n = \tau_{q_{n-1}}(v_{n-1}), \qquad q_n = \sigma_{q_{n-1}}(v_{n-1}).$$

Observe that $\tau^n(a) = u_{n+1} = u_n v_n = \cdots = u_1 v_1 v_2 \ldots v_n$. This implies $1 + \sum_{j=1}^{n} |v_j| = |\tau^n(a)| = o(n \log n)$. Next, we try to estimate the length of the increment $|v_n|$. Since dgsm τ is non-erasing, for all $i < n$ we have $|v_i| < |v_n|$. We have:

$$n|v_n| \leq \sum_{j=n+1}^{2n} |v_j| \leq \sum_{j=1}^{2n} |v_j| = o(2n \log 2n) = o(n \log n).$$

Hence, $|v_n| = o(\log n)$ and for any constant c there exists n such that $c^{|v_n|} < n$. If we take $c = |\Sigma| + 1$, there must be a repetition among the words $v_1, \ldots, v_n$, let say $v_i = v_{i+k}$ for some integers $i, i+k \leq n$ and $k > 0$. Then $v_j = v_{j+k}$ for all $j \geq i$, hence the infinite word $w = u_1 v_1 v_2 v_3 \ldots$ is ultimately periodic and it can be then generated in constant space, cf. [HKL], Lemma 3.1.

Lemma 3. *An infinite word generated by a dgsm with the growth* $\Theta(n \log n)$ *has logarithmic space complexity.*

Proof. Let τ be a dgsm with the growth $\Theta(n \log n)$ generating an infinite word $w = w_1 w_2 \ldots$. Consider the sequences $\{u_n\}_{n\geq 0}$, $\{v_n\}_{n\geq 0}$, $\{q_n\}_{n\geq 0}$ from the previous proof.

We construct TM M generating w as follows. In the first step it writes u_1 on the output tape, v_1 on the first tape and it sets to the state q_1. In each following step n it simulates dgsm τ on input written on the first tape starting in the state q_{n-1}. The output of the simulation of τ is written, at the same time, to the output tape and to some temporary tape, so that after the simulation it can be copied back to the first tape. Hence, in the end of step n the first tape contains word v_n. The last state of the simulation of τ is the state q_n, from which the next simulation will start.

The space needed to generate the m-th letter of w is at most $|v_n|$, where n is an integer such that v_n contains the letter w_m, i.e.,

$$\sum_{i=1}^{n-1} |v_i| < m \le \sum_{i=1}^{n} |v_i|.$$

One can show in the same way as in in the proof of Lemma 2 that $|v_n| = \mathcal{O}(\log n)$. Moreover, since $n \le 1 + \sum_{i=1}^{n-1} |v_i| \le m$, we have $|v_n| = \mathcal{O}(\log n) = \mathcal{O}(\log m)$. Hence, TM M works in logarithmic space.

2.2   Logarithmic Space Complexity

The second part of Problem 5.2 in [HK] asks to find a specific infinite word which cannot be generated in logarithmic space. We show that this problem is as hard as the problem to find a specific language which does not belong to DSPACE(n), and this is a hard open problem.

Notation 1. Denote the n-th binary word in the lexicographical order by lex(n). Note that for $n \ge 1$, bin(n) = 1 lex(n).

Definition 4. *Let w be an infinite binary word and $L \subseteq \{0,1\}^*$ an binary language. We say that w determines language L if for every positive integer n: the n-th letter of w is 1 if and only if lex(n) belongs to L.*

Theorem 5. *Let w be an infinite binary word and L the language determined by w. Then, the word w is in GSPACE($\log n$) if and only if L belongs to DSPACE(n).*

Proof. First, assume that w has logarithmic space complexity. Let M be a Turing machine generating w in logarithmic space. We construct Turing machine M' recognizing language L. Let lex(n) be the word on the input tape of M', where n is a positive integer. The length of input is $\Theta(\log n)$. M' simulates M in the following way: it remembers only the last letter generated by M and counts the number of them on a special working tape. When this number is equal to n, it stops and accepts the input if and only if the last output letter was 1.

Since M uses space only $\Theta(\log n)$ to generate first n output letters and the same space is needed for counting of the number of output letters, M' works in space $\Theta(\log n)$, which is linear to the length of input. Hence, $L \in$ DSPACE(n).

Next, assume that $L \in$ DSPACE(n). So we have a Turing machine M recognizing L in linear space. Let M' be a Turing machine such that it generates words in lexicographical order on the first working tape and runs M on each generated word. Depending on if the word was accepted or rejected it writes 1 or 0 on the output tape. Clearly, the length of the n-th word on the first working tape is $\Theta(\log n)$. M works in linear space, hence in space $\Theta(\log n)$. Therefore, M' uses logarithmic space to generate n-th letter: $w \in$ GSPACE($\log n$).

As a consequence we have that if we would be able to show about a specific infinite word that it does not belong to GSPACE($\log n$), then we would have also a specific language which does not belong to DSPACE(n), and vice versa. Note, that even the problem to show that a specific language does not belong to DSPACE($\log n$) is open.

2.3   Separation of Double and Triple D0L TAG Systems

In the Section 6 of [HKL] is mentioned that the generation of infinite words by double D0L TAG systems is very powerful mechanism, and that it is not known any concrete example of an infinite word which cannot be generated by this mechanism, although by a diagonalization argument such words clearly exist. In [CK] (Conjecture 4) is conjectured that there exists an infinite word that can be generated by a triple D0L TAG system, but not by any double D0L TAG system. In what follows we are going to give the whole class of infinite words which cannot be generated by any double D0L TAG system. Combining this result with some results in [CK] we can also give an affirmative answer to Conjecture 4 of [CK]. First, let us fix some notation.

Notation 2. Let $w = c_1 \ldots c_n$ be a word with $c_i \in \Sigma$. Then $\mathrm{symb}(w) = \{c_i, 1 \le i \le n\}$ denotes the set of all symbols occurring in the word.

Theorem 6. *Let $s : \mathbb{N} \to \mathbb{N}$ be an integer function such that $s(i) \in 2^{\omega(i)}$, i.e., $s(i)$ grows faster than exponentially. Then the infinite word*

$$w = 10^{s(1)}10^{s(2)}10^{s(3)} \ldots$$

cannot be generated by any double D0L TAG system.

Proof. Assume that w can be generated by double D0L TAG system, and let τ be such a system. Let M be the set of all symbols occurring on the second tape of τ and B the maximal number of symbols written on any tape in one step, i.e.,

$$B = \max\{\max(|\alpha|, |\beta|), \text{ where } \binom{a}{b} \to \binom{\alpha}{\beta} \text{ is a rewriting rule of } \tau\}.$$

Let $w_i = 10^{s(i)}$. First, we show that for any constant $k \ge 1$ there is a positive integer j such that:

$$s(j) + 1 = |w_j| > k|w_1 \ldots w_{j-1}1| + 1. \tag{1}$$

Since $s(i) \in 2^{\omega(i)}$, for any number $c > 1$ there is an integer j such that:

$$|w_j| > c^j,$$
$$|w_i| \le c^i \text{ for all } 1 \le i \le j-1.$$

Then we have

$$k|w_1 \ldots w_{j-1}1| \le k \sum_{i=0}^{j-1} c^i \le k \cdot \frac{c^j - 1}{c - 1}.$$

Taking $c = k+1$, we get $k|w_1 \ldots w_{j-1}1| + 1 \le c^j < |w_j|$.

Consider that we are reading the first 0 of $w_j = 10^{s(j)}$ in the word w. Let u_1 (v_1) be the word written on the first (second) tape between the reading and the writing head. And let, recursively, u_l (resp. v_l) be the word added on the first (resp. second) tape after reading v_{l-1}. Note that for all $l \geq 1$, we have $|v_1 \ldots v_{l-1}| < |u_1 \ldots u_l|$.

We define also set $M_0 \subseteq M$ as follows. For $x \in M$, let $\binom{0}{x} \to \binom{\alpha}{\beta}$ be a rule of τ. Then, $x \in M_0$ if and only if $\alpha \in 0^+$. Hence, M_0 is the set of all symbols such that when reading 0 on the first tape and such symbol on the second tape, it writes only 0's on the first tape.

Assume that for some $i \geq 1$:

$$\sum_{l=1}^{i} |u_l| \leq s(j) . \tag{2}$$

By (2), words $u_1, \ldots, u_i$ contain only 0's, hence when reading words $v_1, \ldots, v_{i-1}$, only 0's are written on the first tape. Since $|v_1 \ldots v_{l-1}| < |u_1 \ldots u_l|$, we have also that while reading words $v_1, \ldots, v_{i-1}$, only 0's are read from the first tape.

Consider the following oriented graph. The vertices are elements of M. There is an arc $x \to y$, if for rule $\binom{\alpha}{x} \to \binom{\alpha}{\beta}$, $y \in \mathrm{symb}(\beta)$ (see Notation 2). For $X \subseteq M$, let $\mathrm{clos}(X)$ be the set of all vertices of the graph to which we can get from any vertex of X following the arcs. If $\mathrm{clos}(\mathrm{symb}(v_1)) \subseteq M_0$, then since $u_1 \in 0^+$, we get by induction that $u_l \in 0^+$ and $\mathrm{symb}(v_l) \subseteq \mathrm{clos}(\mathrm{symb}(v_1)) \subseteq M_0$ for all $l \geq 1$. This is a contradiction since, there must be u_l containing 1.

Hence, there must be an oriented path starting in a vertex of $\mathrm{symb}(v_1)$ and ending in a vertex of $M - M_0$ of length $i_0 \leq |M_0|$. Let i_0 be the shortest such path. If we prove that (2) holds true for $i = |M| + 1 \geq |M_0| + 2 \geq i_0 + 2$, then since during reading $v_1, \ldots, v_{i_0+1}$ only 0's are read and written on the first tape, we have $\mathrm{symb}(v_1), \ldots, \mathrm{symb}(v_{i_0}) \subseteq M_0$ and $\mathrm{symb}(v_{i_0+1}) \cap (M - M_0) \neq \emptyset$. This implies that $1 \in \mathrm{symb}(u_{i_0+2})$, which is a contradiction with (2).

Now it suffices to prove that Equation (2) holds for $i = |M| + 1$. After reading one symbol, system τ can write on any tape at most B symbols. Hence, we have for all $l \geq 1$, $|u_{l+1}|, |v_{l+1}| \leq B|v_l|$. We estimate the left hand side of (2):

$$\sum_{l=1}^{|M|+1} |u_l| \leq |u_1| + \sum_{l=2}^{|M|+1} B^{l-1}|v_1| \leq \max(|u_1|, |v_1|) . \sum_{l=0}^{|M|} B^l . \tag{3}$$

Notice that $T = \sum_{l=0}^{|M|} B^l$ is a constant for τ.

Next, we try to estimate $|u_1|$ and $|v_1|$. Consider the situation when the reading heads are on the $(n+1)$-th symbols of both tapes. Then n symbols have been already read and hence at most Bn symbols written. So, there is at most $(B-1)n$ symbols between writing and reading head on any tape. This implies:

$$\max(|u_1|, |v_1|) \leq (B-1)|w_1 \ldots w_{j-1}1| . \tag{4}$$

Taking $k = T(B-1)$ we get

$$\sum_{l=1}^{|M|+1} |u_l| \overset{(3)}{\leq} T . \max(|u_1|, |v_1|) \overset{(4)}{\leq} T(B-1)|w_1 \ldots w_{j-1}1| \overset{(1)}{<} s(j)$$

as desired.

Let us recall one lemma proved in Examples 11 and 13 of [CK]:

Lemma 7. *Let* $s : \mathbb{N} \to \mathbb{N}$ *be an integer function which is computable, i.e., can be computed by a TM. Then there exists* $t : \mathbb{N} \to \mathbb{N}$ *such that* $t(n) \geq s(n)$ *for all* $n \geq 1$ *and the word*

$$w = 10^{t(1)}10^{t(2)}10^{t(3)} \ldots$$

can be generated by a triple D0L TAG system. Moreover, such integer function t *can be effectively computed.*

The proof of the lemma is based on the following idea. Let M be a TM computing unary strings $1^{s(1)}, 1^{s(2)}, \ldots$ and let τ be a dgsm generating the sequence of configurations of computation of M. We can easily extend the dgsm τ to a triple D0L TAG system by coding all letters generated by τ to 0, except for the last letters of the strings in the sequence $1^{s(1)}, 1^{s(2)}, \ldots$, which are coded to 1. Together with our result we have the following corollary:

Corollary 8. *There exists an infinite word which can be generated by a triple D0L TAG system, but not by any double D0L TAG system.*

Hence, the inclusion "double D0L $\subseteq$ triple D0L" is proper, as conjectured in Conjecture 4 in [CK].

References

[HK]   Hromkovič, J., Karhumäki, J., *Two lower bounds on computational complexity of infinite words*, New trends in formal languages, LNCS 1218, 366–376, 1997. 328, 329, 334

[HKL]   Hromkovič, J., Karhumäki, J., Lepistö, A., *Comparing descriptional and computational complexity of infinite words*, Results and trends in theoretical computer science, LNCS 812, 169–182, 1994. 328, 329, 331, 332, 333, 335

[CK]   Culik, K., II, Karhumäki, J., *Iterative devices generating infinite words*, Int. J. Found. Comput. Sci., Vol. 5 No. 1, 69-97, 1994. 328, 329, 335, 337

[Le]   Lepistö, A., *On the computational complexity of infinite words*, In: Developments in Language Theory II (eds.: J. Dassow, G.Rozenberg, A. Salomaa), World-Scientific, Singapore, 350–359, 1996. 329, 331

[Ko]   Kolmogorov, A. N., *Three approaches to the quantitative definition of information*, Problems Inform. Transmission 1, 662–664, 1968. 330

[Lo]   Lothaire, M., *Combinatorics on words*, Addison-Wesley, Reading, Massachusetts, 1981. 328

[Th]   Thue, A., *Über unendliche Zeichenreihen*, Norske Vid. Selsk. Skr., I Mat. Nat. KI., Kristiania 7, 1–22, 1906. 329

Lower Bounds for On-Line Single-Machine Scheduling

Leah Epstein[1] and Rob van Stee[2],[*]

[1] School of Computer and Media Sciences, The Interdisciplinary Center
Herzliya, Israel
Epstein.Leah@idc.ac.il
[2] Centre for Mathematics and Computer Science (CWI)
Amsterdam, The Netherlands
Rob.van.Stee@cwi.nl

Abstract. The problem of scheduling jobs that arrive over time on a single machine is well-studied. We study the preemptive model and the model with restarts. We provide lower bounds for deterministic and randomized algorithms for several optimality criteria: weighted and unweighted total completion time, and weighted and unweighted total flow time. By using new techniques, we provide the first lower bounds for several of these problems, and we significantly improve the bounds that were known.

1 Introduction

We consider on-line scheduling of n jobs on a single machine. The jobs arrive over time. Job J_j with processing time (or size) p_j is released (or arrives) at time r_j. This is also the time when it is revealed to the algorithm. The algorithm is required to assign each job to a machine. A job can be assigned at its arrival time or later. The algorithm may run at most one job on each machine at any time. In weighted problems, each job J_j is also given a positive weight w_j which represents its importance. We consider both deterministic and randomized algorithms.

Scheduling on a single machine simulates (e.g.) processing tasks on a serial computer. This important problem has been widely studied both on-line and off-line, considering various optimality criteria [1,4,5,8,9,17]. However, until now only relatively weak lower bounds were known, especially for the weighted problems, and in some cases no bounds were known at all. We make significant progress in this area by providing strong (or stronger) lower bounds for several optimality criteria.

In the standard scheduling model, a job which was assigned to a machine must be processed continuously to its completion. The preemptive scheduling model allows the algorithm to stop a running job and resume it later. A third model does not allow preemptions but allows restarts. In this case a running job

---

[*] Research supported by the Netherlands Organization for Scientific Research (NWO), project number SION 612-30-002.

J. Sgall, A. Pultr, and P. Kolman (Eds.): MFCS 2001, LNCS 2136, pp. 338–350, 2001.

may be stopped, but it has to be started from scratch when it is scheduled again. In this paper we focus on preemptive algorithms, and algorithms with restarts.

We consider two optimality criteria. Each one is considered both in the weighted case and in the unweighted case. Let C_j be the completion time of J_j. The flow time F_j is the total time J_j exists in the system, i.e. $F_j = C_j - r_j$. This gives the following four criteria:

1. Minimizing the total completion time ($\sum C_j$).
2. Minimizing the total weighted completion time ($\sum w_j C_j$).
3. Minimizing the total flow time ($\sum F_j$).
4. Minimizing the total weighted flow time which is $\sum w_j F_j$.

The flow time measure is used in applications where it is important to finish tasks fast, relative to their release time. On the other hand, the completion time measure is used when tasks need to be finished as fast as possible, relative to a starting time of the computer, with no connection to their arrival time. The weighted versions of the problems model cases where different jobs have different importance.

We study these problems in terms of competitive analysis. Thus we compare an (on-line) algorithm to an optimal off-line algorithm OPT that knows all jobs in advance, but cannot assign a job before its release time. If the on-line algorithm is allowed to preempt jobs, we assume that OPT can preempt as well. In the other models we only consider non-preemptive off-line schedules. Let T_B be the cost of algorithm B. An algorithm A is R-competitive if for every sequence $T_A \leq R \cdot T_{OPT}$. The competitive ratio of an algorithm is the infimum value of R such that the algorithm is R-competitive.

Known results There are several cases where the optimal schedule has a simple structure.

The optimal schedule is the same both for (weighted) flow time and for (weighted) completion time, since the optimal costs differ by the constant $\sum r_i w_i$. For the weighted case, if all release times are zero (all jobs are released at the same time, hence the problem is always off-line), then an optimal off-line schedule is achieved by sorting the jobs by their ratios of size to weight (p_j/w_j), and processing them in non-decreasing order [14]. For the unweighted case, an optimal preemptive schedule can be built on-line by applying the SRPT algorithm. At all times, this algorithm processes the job with the smallest remaining processing time [10].

In an off-line environment, the simple structure makes the complexity of those problems polynomial. However, all weighted versions of the problem with general release times are strongly NP-hard [9], and so are the unweighted non-preemptive problems. (Naturally, in an off-line environment nothing changes if restarts are allowed, since all jobs are known in advance). Moreover, it is NP-hard to approximate the non-preemptive problem of minimizing total flow time to a factor of $O(n^{1/2-\varepsilon})$ [8]. This paper gives an off-line approximation of performance ratio $\Theta(\sqrt{n})$ for that problem. Polynomial time approximation schemes for preemptive and non-preemptive weighted completion time, and for non-preemptive total completion time, were given recently by [1].

Table 1. Known results and new lower bounds

			Old LB		New LB	Upper bound	
$\sum C_j$	restarts	det.	1.112	[17]	1.2108	2	[7,11,15]
		rand.	1		1.1068	$e/(e-1)$	[4,16]
$\sum F_j$	restarts	det.	$\Omega(n^{1/4})$	[17]	$\Omega(\sqrt{n})$	$\Theta(n)$	
		rand.	1		$\Omega(\sqrt{n})$	$\Theta(n)$	
$\sum w_j C_j$	restarts	det.	31/30	[13]	1.2232	2.415	[6]
		rand.	113/111	[13]	1.1161	1.686	[5]
	preemptions	det.	31/30	[13]	1.0730	2	[6,12]
		rand.	113/111	[13]	1.0389	4/3	[12]
$\sum w_j F_j$	restarts	det.	$\Omega(n^{1/4})$	[17]	$\Omega(n)$	-	
		rand.	1		$\Omega(n)$	-	
	preemptions	det.	1		2	-	
		rand.	1		4/3 [3]	-	

The known results and our new lower bounds for on-line algorithms are summarized in Table 1. In this table, the columns 'Old LB' and 'New LB' contain the best known lower bounds so far and our new lower bounds, respectively. As mentioned earlier, it is possible to get an optimal algorithm (i.e. achieve the competitive ratio 1) for minimizing the total completion time or the total flow time in the preemptive model. We find that this is not the case for the weighted versions of these criteria.

Stougie and Vestjens [16] gave a lower bound of $\Omega(\sqrt{n})$ for randomized algorithms to minimize the total flow time in the standard model (no restarts or preemptions). It is easy to see that there can be no competitive (deterministic or randomized) algorithm to minimize the total weighted flow time in the standard model. No other results for weighted flow time were known until recently. Chekuri, Khanna and Zhu [3] studied this problem in both off-line an on-line environments. Their work was done independently and in parallel to our work. They give a semi-online algorithm with competitive ratio $O(\log^2 P)$, where P is the ratio between maximum and minimum job sizes. They also give lower bounds of 1.618 for deterministic on-line algorithms and 4/3 for randomized algorithms.

Our results We give some new lower bounds, and improve some previously known lower bounds. Specifically, we improve the lower bounds of [17] for scheduling with restarts, both for total flow time and total completion time. We also improve the bounds of [13] for preemptive (deterministic and randomized) scheduling with the goal of minimizing the total weighted flow time. The existing lower bounds for these problems were very close to 1. The substantial improvements we show are due to new techniques we are using.

We begin by discussing several useful lower bounding methods, that are used in more than one proof, in Section 2. Section 3 contains our results on total (weighted) completion time, and Section 4 discusses the total (weighted) flow time measure.

2   Methods

To prove lower bounds for *randomized algorithms* we use the adaptation of Yao's theorem [18]. It states that a lower bound for the competitive ratio of deterministic algorithms on a fixed distribution on the input is also a lower bound for randomized algorithms and is given by $E(T_A/T_{\mathrm{OPT}})$.

A useful method for *weighted problems* is as follows. Assume that at time t, the on-line algorithm is left with one job of size $a \neq 0$ and weight b, and OPT has either completed all the jobs or it is left with a job of a smaller ratio of weight to size. We let k jobs of size ε arrive at times $t + (i-1)\varepsilon$ for $i = 1, \ldots, k$. Each such job has weight $\frac{b}{a}\varepsilon$. Hence it does not matter for the total completion time or the total flow time in which order the on-line algorithm completes the jobs, and all the new jobs are interchangeable with the job of size a. Let $c = k\varepsilon$ and let ε tend to 0, keeping c constant. If OPT has no jobs left, and we are considering the total weighted completion time, then the extra cost of OPT is $tcb/a + c^2b/(2a)$ and the extra cost of the on-line algorithm is $tcb/a + cb + c^2b/(2a)$. The extra cost for other cases can be calculated similarly.

For algorithms that are allowed to *restart* jobs, it can be useful to let jobs of size 0 arrive at such a time that the on-line algorithm is forced to restart the job it is running, whereas OPT can run the jobs immediately due to its different schedule of the other jobs, or possibly delay them (in the case that more jobs arrive). This can be combined with a sequence of jobs with exponentially increasing sizes. By timing the arrival of the jobs, it is possible to force the on-line algorithm to restart every job in such a sequence (if it does not restart, we stop the sequence at that point).

3   Total Completion Time

3.1   Lower Bounds for Algorithms with Restarts

We begin by showing bounds for the problem where all jobs have the same weight, first for deterministic algorithms and then for randomized algorithms.

Theorem 1. *Any deterministic algorithm for minimizing the total completion time on a single machine which is allowed to restart jobs, has a competitive ratio of at least* $\mathcal{R}_1 = 1.2102$.

Proof. Assume there is an algorithm $\mathcal{A}$ that has a competitive ratio of $\mathcal{R}_1 = 1.2102009$. A job of size 1 arrives at time 0. Since restarts are allowed, we may assume $\mathcal{A}$ starts it immediately. A sequence of jobs will now arrive in steps. In each step the online algorithm must restart. If it does not, the sequence stops at that point. Otherwise, the next item in the sequence arrives.

1. A job of size 0 arrives at time $x = 1/\mathcal{R}_1 - 1/2 \approx 0.326309$.
2. A job of size 0 arrives at time $y = 3/(2\mathcal{R}_1^2) - 1/(4\mathcal{R}_1) - 1/4 \approx 0.567603$.
3. Three jobs of size 0 arrive at time 1. If $\mathcal{A}$ does not restart, the implied competitive ratio is $(x + 5y + 4)/(x + y + 5) > \mathcal{R}_1$.

If $\mathcal{A}$ has restarted three times so far, we repeat the following for $i = 1, \ldots, 5$ or as long as $\mathcal{A}$ keeps restarting in step 5. OPT will complete the first six jobs by time 1 and pay 6 for them. Denote the first job that arrived (with size $x_0 = 1$) by J_0.

4. A job of size x_i arrives at the time OPT finishes J_{i-1}.
5. a_i jobs of size 0 arrive at the time OPT finishes J_i. ($\mathcal{A}$ is still executing J_i at this moment.)

If we fix $a_1, \ldots, a_5$ we can determine $x_1, \ldots, x_5$ so that if $\mathcal{A}$ does not restart on arrival of the a_i jobs of size 0, it pays exactly $\mathcal{R}_1$ times the optimal cost. Note that if $\mathcal{A}$ runs any J_i before J_{i-1}, it pays more than $\mathcal{R}_1$ times the optimal cost, and the sequence stops immediately without the arrival of a_i jobs of step 5 (when J_i arrives, the only job which is still not completed in the schedule of $\mathcal{A}$ is J_{i-1}).

i	1	2	3	4	5
a_i	3	2	2	2	1
x_i	2.13118	4.04404	8.33794	18.1366	36.2732

By fixing a_i ($i = 1, \ldots, 5$) as in this table, we can ensure that $\mathcal{A}$ pays more than $\mathcal{R}_1$ times the optimal cost for the entire sequence when the last job arrives. Since $x_5 = 2x_4$, $\mathcal{A}$'s costs are the same if it restarts the job of size x_5 for the last job and if it does not.   □

Using a computer, we have been able to improve this bound slightly using $a_1 = 3$, $a_2 = \ldots = a_{45} = 2$, giving $\mathcal{R}_2 = 1.210883$. After J_{45} arrives, $\mathcal{A}$ has a cost of at least $\mathcal{R}_2$ times the optimal cost whether it restarts or not on arrival of the last 2 jobs of size 0.

Theorem 2. *Any randomized algorithm for minimizing the total completion time on a single machine which is allowed to restart jobs, has a competitive ratio of at least $\mathcal{R}_3 = 114/103 \approx 1.1068$.*

Proof. We use Yao's minimax principle [18] and consider a randomized adversary against a deterministic algorithm. Assume there exists an on-line algorithm $\mathcal{A}$ with a competitive ratio of $\mathcal{R}_3$. At time 0, a job of size 1 arrives. $\mathcal{A}$ will certainly start this job immediately since it is allowed to restart. At time $1/3$, two jobs of size 0 arrive. With probability p, 10 more jobs of size 0 arrive at time 1, followed by 4 jobs of size 1 (either all these jobs arrive, or none of them).

If $\mathcal{A}$ restarts at time $1/3$ and the jobs at time 1 do arrive, it has cost $30\frac{2}{3}$ independent of whether it restarts again. The optimal cost in case all jobs arrive is 27.

This implies that if $\mathcal{A}$ restarts at time $1/3$, it has competitive ratio of at least $30\frac{2}{3}p/27 + (1-p)$; otherwise, it has competitive ratio $p + 3(1-p)/2$. These ratios are equal for $p = 81/103$, and are then $114/103$. This implies a competitive ratio of $\mathcal{R}_3$. $\qquad\square$

The methods in these proofs can be adapted for the weighted problem to give somewhat higher bounds.

Theorem 3. *Any deterministic algorithm for minimizing the total weighted completion time on a single machine which is allowed to restart jobs, has a competitive ratio of at least $\mathcal{R}_4 = 1.2232$.*

Proof. Assume there is an algorithm $\mathcal{A}$ that has a competitive ratio of $\mathcal{R}_4 = 1.2232$. We use a somewhat similar structure as in Theorem 1. A job of size 1 and weight 1 arrives at time 0. Again we assume $\mathcal{A}$ starts it immediately. A sequence of jobs will now arrive in steps. In each step the online algorithm must restart. If it does not, the sequence stops at that point. Otherwise, the next item in the sequence arrives. We fix two weights $W = 0.79$ and $W' = 1.283$ to be used for the first part of the sequence.

1. A job of size 0 and weight W arrives at time $x = 1/\mathcal{R}_4 - 1/(W+1) \approx$ 0.258869.
2. A job of size 0 and weight W' arrives at time $y = (\frac{xW + (1+x)(1+W')}{\mathcal{R}_4} - xW - 1)/(W'+1) \approx 0.574794$.
3. A job of size 0 and weight $Z = (xW\mathcal{R}_4 + yW'\mathcal{R}_4 + 2\mathcal{R}_4 - xW - yW' - y - 1)/(y+1-\mathcal{R}_4) \approx 3.07699$ arrives at time 1.

In all three cases, if $\mathcal{A}$ does not restart, the implied competitive ratio is $\mathcal{R}_4$. If $\mathcal{A}$ has restarted three times so far, we follow the procedure described below. OPT will complete the first four jobs by time 1 and will pay 6.14999 for them. The on-line cost for these jobs is $T_{ONL} = 6.01896$. Denote the first job that arrived (with size and weight 1) by J_0. Put $i = 1$. Let $x_0 = 1$ (denotes its size) $z_0 = 1$ (denotes its weight).

4. A job J_i of size $x_i = 2^i$ and weight z_i arrives at the time OPT finishes J_{i-1}, i.e. at time $2^i - 1$. If $\mathcal{A}$ completes J_{i-1} before J_i, go to step 5, otherwise go to step 6.
5. A job of size 0 and weight $w_i - z_i$ arrives at the time OPT finishes J_i (time $2^{i+1} - 1$). $\mathcal{A}$ is still executing J_i at this moment. If $\mathcal{A}$ does not restart, or $i = 5$, stop the sequence. Otherwise, increase i by 1 and go to Step 4.
6. k jobs of size ε and weight $\varepsilon z_{i-1}/x_{i-1}$ arrive at time $2^{i+1} - 1$, where $k\varepsilon = c_i$. ($\mathcal{A}$ can complete J_i no earlier than this.) The sequence stops.

In step 5, if it is possible to force a restart of J_i, then the cost of OPT will grow by $(2^{i+1} - 1)w_i$ whereas the on-line cost will grow by $(2^{i+1} - 1)(w_i - z_i) + (2^{i+1} + 2^i - 1)z_i$, hence the value z_i should be as large as possible. On the other hand, z_i should be small enough so that $\mathcal{A}$ has a competitive ratio of at least $\mathcal{R}_4$ if it runs J_i before J_{i-1} (as in Theorem 1, all smaller jobs are already completed by $\mathcal{A}$ when J_i arrives). We determine c_i in such a way that z_i is maximized, i.e. $c_i = (2^{i+1} + 2^{i-1} - 1 - \mathcal{R}_4(2^{i+1} - 1))/(\mathcal{R}_4 - 1)$. Now that we know c_i, we can calculate z_i and w_i to force a competitive ratio of $\mathcal{R}_4$ if $\mathcal{A}$ does not restart in step 5 or if it uses the wrong order for the jobs (step 6). We give the results in the following table.

i	1	2	3	4	5
z_i	1.10638	1.48772	2.24592	3.69664	6.91845
w_i	4.55118	5.34374	7.26472	10.2624	11.8410

In the last step, the competitive ratio of $\mathcal{A}$ is at least $\mathcal{R}_4$, independent of $\mathcal{A}$'s schedule.   □

Using a computer, we have been able to improve this bound very slightly using 11 phases instead of 5. Fixing $W = 0.79$ and $W' = 1.285$ we can achieve a lower bound of 1.22324655.

Theorem 4. *Any randomized algorithm for minimizing the total weighted completion time on a single machine which is allowed to restart jobs, has a competitive ratio of at least $\mathcal{R}_5 = 1.1161$.*

Proof. We use Yao's minimax principle and consider a randomized adversary against a deterministic algorithm. We use the following job sequence.

time	size	weight	number
0	1	1	1
0.379739	0	1.88288	1
1	0	7.03995	1
1	ε	ε	k

where $k\varepsilon = c = 3.31003$ and the jobs at time 1 arrive with probability $p = 0.691404$ (either they all arrive, or none of them).

Suppose the jobs at time 1 do arrive, then if the online algorithm $\mathcal{A}$ restarts at time $t = 0.37978$, it can choose to restart again at time 1. If it does, it has costs $1.88288 \cdot t + 7.03995 + 2 + \varepsilon \sum_{i=1}^{k}(2 + i\varepsilon) = 9.75495 + 2c + \varepsilon^2 k(k+1)/2$. For $\varepsilon \to 0$, this tends to $16.3750 + c^2/2 = 21.8532$. If $\mathcal{A}$ does not restart again, it has costs $1.88288 \cdot t + (7.03995 + 1 + c)(t + 1) + \varepsilon^2 k(k+1)/2$ which tends to the same limit.

The optimal costs in this case are $2.88288 + 7.03995 + c + \varepsilon^2 k(k+1)/2 \to 18.7110$.

This implies that if $\mathcal{A}$ restarts at time t, it has a competitive ratio of at least $p \cdot 21.8532/18.7110 + 1 - p$, and otherwise, it has a competitive ratio of at least $p + (1 - p) \cdot 2.88288/(2.88288 \cdot t + 1)$. These ratios are equal for $p = 0.691404$, and are then 1.11610796.   □

3.2   Lower Bounds for Preemptive Algorithms

Since the unweighted problem can be solved to optimality, we only consider the weighted problem in this section. We show this problem cannot be solved optimally. In the unweighted problem, $SRPT$ is optimal. However, in the case that jobs have weights, it is possible that when a new job arrives, the optimal schedule before that time is different compared to the situation where the new job does not arrive. This cannot occur in the unweighted version of the problem. We use this idea to show the following lower bounds.

Theorem 5. *Any deterministic preemptive on-line algorithm for minimizing the total weighted completion time, has a competitive ratio of at least* $\mathcal{R}_6 = 1.0730$.

Proof. The sequence starts with two jobs arriving at time zero. One job of size 1 and weight 1, and the other of size α and weight β, where $1 < \beta < \alpha$. Consider an on-line algorithm $\mathcal{A}$ at time α. If the smaller job is completed by then, k very small jobs of size ε and weight $\beta\varepsilon$, of total length c, arrive ($c = k\varepsilon$). Otherwise, no more jobs arrive. In the first case, OPT runs the larger job, then the small jobs and then the unit job. For $\varepsilon \to 0$, the cost is $T_{OPT} = \alpha\beta + c\alpha\beta + \beta k(k+1)\varepsilon^2/2 + \alpha + c + 1 = (c+1)(\alpha\beta+1) + \alpha + c^2\beta/2$. $\mathcal{A}$ is left with a piece of size 1 of the larger job, hence it does not matter in which order it completes the remaining jobs. We can assume that it runs the unit job first, then the larger job, and then the small jobs. Its cost is at least $T_{\mathcal{A}} \geq 1 + \beta(\alpha+1) + (c+1)\alpha\beta + \alpha + c^2\beta/2$. In the second case, OPT runs the unit job first, and hence $T_{OPT} = 1 + (\alpha+1)\beta$, whereas $\mathcal{A}$ can either finish the unit job first, but no earlier than time α (and pay $\alpha + \beta(\alpha+1)$), or finish the larger job first (and pay $\alpha\beta+\alpha+1$). The second cost is always smaller since $\beta > 1$. Using a computer to search for good values for α, β and c, such that the competitive ratio in both cases is high, we get that for $\alpha = 3.4141$, $\beta = 2.5274$, and $c = 4.4580$, the competitive ratio is at least 1.073042. $\square$

Theorem 6. *Any randomized preemptive on-line algorithm for minimizing the total weighted completion time, has a competitive ratio of at least* $\mathcal{R}_7 = 1.0388$.

Proof. We use Yao's minimax principle and consider a randomized adversary against a deterministic algorithm. We use the sequence from Theorem 5. The small jobs arrive at time α with probability p. Consider a deterministic algorithm $\mathcal{A}$. Let $\mathcal{R}_8$ be the competitive ratio in the case $\mathcal{A}$ completes the smaller job by time α, and $\mathcal{R}_9$ be the competitive ratio if it does not. Then in the first case $E(T_{\mathcal{A}}/T_{OPT}) \geq \mathcal{R}_8 p + (1-p)$, and in the second case $E(T_{\mathcal{A}}/T_{OPT}) \geq \mathcal{R}_9(1-p) + p$. The best value of p for given $\mathcal{R}_8$ and $\mathcal{R}_9$ can be calculated by making the two expected competitive ratios equal. Using a computer to search for good values for α, β and c, such that the competitive ratio is high, we get that for $\alpha = 3.7299$, $\beta = 2.4036$, and $c = 5.4309$ (and $p = 0.36251$), the expected competitive ratio is at least 1.038872. $\square$

4   Total Flow Time

For the standard problem without weights, it is known that the competitive ratio is $\Theta(n)$. It is easy to see that there cannot be a competitive algorithm for the standard weighted problem.

Lemma 7. *Any (deterministic or randomized) algorithm for minimizing the total weighted flow time on a single machine that is not allowed to restart or preempt jobs, has an unbounded competitive ratio.*

Proof. We use Yao's minimax principle and consider a randomized adversary against a deterministic algorithm. The adversary works as follows: at time 0, a job of size and weight 1 arrives. At some time t, uniformly distributed over $(0, N)$, where $N > 1$ is some constant, a second job arrives of size 0 and weight N^2. For all t, the optimal total flow time is bounded by 2. We will show the competitive ratio of any algorithm is bounded by $\Omega(N)$.

Suppose the on-line algorithm starts the first job at time S. If $S \geq N/2$, its expected cost is at least $N/2$ and we are done.

Otherwise, there is a probability of $1/(2N)$ that the second job arrives in the interval $(S, S + 1/2)$, in which case the algorithm has a cost of at least $N^2/2$. This implies its expected cost is at least $\frac{1}{2N} \cdot \frac{N^2}{2} = \Omega(N)$.

Since we can choose $N > 1$ arbitrarily high, the lemma follows. $\qquad\square$

We therefore turn to models where restarts or preemptions are allowed.

4.1   Lower Bounds for Algorithms with Restarts

Theorem 8. *Any (deterministic or randomized) on-line algorithm for minimizing the total flow time, which is allowed to restart jobs, has a competitive ratio of $\Omega(\sqrt{n})$.*

Proof. Consider an on-line algorithm $\mathcal{A}$. We use a job sequence consisting of $n - 2$ jobs of size 0, one job of size 3 and one job of size 2. Let $q = \lfloor\sqrt{n-2}\rfloor$. The two large jobs become available at time 0. Also $n - 2 - q^2$ jobs of size 0 arrive at time 0, to make sure we have n jobs in total. There are two cases to consider.

Case 1. If $\mathcal{A}$ completes the job of size 2 strictly before time 3, we continue as follows: at each time $3 + 2i$ (for $i = 0, 1 \ldots, q - 1$) , q short jobs arrive. If $\mathcal{A}$ does not delay the process of any small job, then it can start the job of size 3 only at time $1 + 2q$, and $T_{\mathcal{A}} \geq 2q$.

If $\mathcal{A}$ runs the job of size 3 earlier than that, then at least one set of small jobs is delayed by at least one unit of time and $T_{\mathcal{A}} \geq q$. OPT assigns the longest job first, and the job of size 2 at time 3, hence no short jobs are delayed and $T_{\text{OPT}} = 8$.

Case 2. Otherwise, if at time 2, $\mathcal{A}$ is not in a mode where it can complete the job of size 2 strictly before time 3, then q jobs of size 0 arrive at time 2. If $\mathcal{A}$ is running some job at that point, and does not stop it then all small jobs will be delayed till time 3 (this is true for any non-zero job) and $T_{\mathcal{A}} \geq q$. All other jobs arrive at time 5 (or any time later). OPT assigns the job of size 2 at time 0 and the other big job at time 2 and $T_{OPT} = 7$. Otherwise an additional $q - 1$ sets of q small jobs each, arrive at times $5 + i$, for $i = 0, \ldots, q - 2$. $\mathcal{A}$ can only complete one big job till time 5. The other big job is either postponed till time $1 + q$, or processed later, and then at least one set of short jobs is delayed by at least one unit of time, hence $T_{\mathcal{A}} \geq q$. In both cases OPT completes both big jobs at time 5 and $T_{OPT} = 7$.

In all cases $T_{OPT} \leq 8$ and $T_{\mathcal{A}} \geq q$, hence the competitive ratio is $\Omega(\sqrt{n})$.

The proof can be extended for randomized algorithms with restarts. We use Yao's minimax principle and consider a randomized adversary against a deterministic algorithm. In this case, we use the following distribution on the input: choose with equal probability the first or the second sequence from the proof above. This gives the lower bound of $\Omega(\sqrt{n})$. □

If the jobs can have different weights, the competitive ratio increases to n.

Theorem 9. *Any (deterministic or randomized) on-line algorithm for minimizing the total weighted flow time, which is allowed to restart jobs, has a competitive ratio of $\Omega(n)$.*

Proof. The proof is very similar to that of the previous theorem. We make the following changes:

- let $q = n$,
- at time 0, one job of size 2 and weight 1 arrives and one job of size 3 and weight 1 (and no other jobs),
- in all places where q jobs used to arrive in that proof, we now let one job arrive of size 0 and weight $q = n$.

It is easy to see that still the competitive ratio is $\Omega(q)$, proving the theorem. □

4.2   Lower Bounds for Preemptive Algorithms

In this section we again consider only weighted flow time, since SRPT clearly gives an optimal solution for total flow time. We can show the following lower bound for deterministic algorithms.

Theorem 10. *Any preemptive deterministic on-line algorithm for minimizing the total weighted flow time, has a competitive ratio of at least $\mathcal{R}_{10} = 2$.*

Proof. Consider the following sequence. At time 0 a job of size α and weight β arrives (we call this job the *large* job), such that $1 < \beta < \alpha$. For $i = 1 \ldots q$ ($q < \alpha$), a job of size and weight 1 arrives at time $i - 1$ (*medium* jobs). Consider the on-line algorithm $\mathcal{A}$ at time α. Let V be the total length of medium jobs that $\mathcal{A}$ processed till that time.

If $V < 1$, no more jobs arrive. In this case $\mathcal{A}$ is left with less than size 1 of the large job. Since running pieces of different medium jobs only increases the cost, we assume that $\mathcal{A}$ completed a size V of one of the medium jobs. Since no other jobs arrive, there are two cases to consider. It is either best to complete the large job and then all medium jobs, or to complete one medium job, then the large job and then the rest of the medium jobs. In both cases, $T_A \geq \alpha\beta + q(\alpha + 1)$. OPT will run all medium jobs before the large job and $T_{OPT} = q + \beta(\alpha + q)$.

If $V \geq 1$, A is left with at least 1 unit of the large job at time α. Let $\mu \geq \alpha$ be the time where A is left with exactly 1 unit of the large job. At each time $\mu + (i-1)\varepsilon$, for $i = 1, \ldots, k$, a job of size ε and weight $\beta\varepsilon$ is released (*small* jobs). Again we may assume that A does not start a medium job before completing the previous one. Then let $j = \lfloor \mu - \alpha + 1 \rfloor$ be the number of medium jobs completed by A before time μ. Let $V' = \mu - \alpha + 1 - j$, this is the part of a medium job that started its process by A but was not completed by time μ. Since no more jobs arrive, A decides whether it should complete this job before the small jobs. The rest of the medium jobs clearly run after the small jobs. Let j' be the number of medium jobs that A runs before the small jobs ($j' \in \{j, j+1\}$). Hence $T_A \geq j' + (\alpha + j')\beta + k(\varepsilon + 1)\beta\varepsilon + (q - j')(\alpha + k\varepsilon + 1)$. OPT runs only $j' - 1$ medium jobs before time μ and completes the large job at time μ. At time $j' + \alpha - 1$ OPT is left with all small jobs and $q - j' + 1$ medium jobs which have lower priority. Hence $T_{OPT} = j' - 1 + (\alpha + j' - 1)\beta + k\varepsilon\beta\varepsilon + (q - j' + 1)(\alpha + k\varepsilon + 1)$. Taking q to be large enough, $\beta = q$, $\alpha = q^2$ and $k\varepsilon = q^3$ where ε tends to zero, we get that the competitive ratio in both cases tends to 2.   □

We use a similar method for randomized algorithms.

Theorem 11. *Any preemptive randomized on-line algorithm for minimizing the total weighted flow time, has a competitive ratio of at least $R_{11} = 4/3$.*

Proof. We use Yao's minimax principle and consider a randomized adversary against a deterministic algorithm. Consider the sequence introduced in Theorem 10. We use the same sequence for $q = 1$ (this sequence is similar to the one given in the proofs for completion time, except that now the small jobs arrive with intervals of ε apart). Since $V \leq 1$, the second case must satisfy $\mu = \alpha$. With probability p, the small jobs arrive starting at time α as in Theorem 10 (they all arrive, or none of them arrives).

We fix $\alpha \gg \beta$, and $k\varepsilon = \alpha^2$. Then the competitive ratio in the first case is $(\beta+1)/\beta$ if the small jobs do not arrive (and 1 otherwise). In the second case, $j' = 1$ and the competitive ratio is β. The best choice for p is $(\beta^2 - \beta)/(\beta^2 - \beta + 1)$. Then the expected competitive ratio is $\beta^2/(\beta^2 - \beta + 1)$. Maximizing this expression we get $\beta = 2$ and expected competitive ratio of at least 4/3.   □

5   Conclusions and Open Questions

An interesting general question is what the difference is between minimizing the weighted and unweighted total completion time, in terms of the competitive ratios that can be achieved.

Based on the results in this paper, we know that the preemptive versions of the weighted completion and flow time problems are different from the unweighted versions, since if the jobs all have the same weight it is possible to schedule the jobs optimally and have a 1-competitive algorithm, both for completion times and for flow times.

It is possible however, that minimizing the total completion time in the standard model and in the model with restarts is as hard in the weighted problem as it is in the unweighted problem. These problems are still open.

References

1. F. Afrati, E. Bampis, C. Chekuri, D. Karger, C. Kenyon, S. Khanna, I. Milis, M. Queyranne, M. Skutella, C. Stein, and M. Sviridenko. Approximation schemes for minimizing average weighted completion time with release dates. In *Proceedings of the 40th Annual IEEE Symposium on Foundations of Computer Science*, pages 32–43, October 1999. 338, 340
2. A. Borodin and R. El-Yaniv. *Online Computation and Competitive Analysis*. Cambridge University Press, 1998.
3. C. Chekuri, S. Khanna, and A. Zhu. Algorithms for minimizing weighted flow time. In *Proc. of the 33th Symp. on Theory of Computing (STOC)*, 2001. 340
4. C. Chekuri, R. Motwani, B. Natarajan, and C. Stein. Approximation techniques for average completion time scheduling. In *Proceedings of the Eighth Annual ACM-SIAM Symposium on Discrete Algorithms (SODA'97)*, pages 609–618. SIAM, Philadelphia, PA, 1997. 338, 340
5. M. X. Goemans, M. Queyranne, A. S. Schulz, M. Skutella, and Y. Wang. Single machine scheduling with release dates. manuscript, 1999. 338, 340
6. Michel X. Goemans. Improved approximation algorithms for scheduling with release dates. In *Proceedings of the 8th Annual ACM-SIAM Symposium on Discrete Algorithms*, pages 591–598, New York / Philadelphia, 1997. ACM / SIAM. 340
7. J. A. Hoogeveen and A. P. A. Vestjens. Optimal on-line algorithms for single-machine scheduling. In *Proc. 5th Int. Conf. Integer Programming and COmbinatorial Optimization*, LNCS, pages 404–414. Springer, 1996. 340
8. H. Kellerer, T. Tautenhahn, and G. J. Woeginger. Approximability and nonapproximability results for minimizing total flow time on a single machine. In *Proceedings of the Twenty-Eighth Annual ACM Symposium on the Theory of Computing*, pages 418–426, Philadelphia, Pennsylvania, 1996. 338, 339
9. J. Labetoulle, E. L. Lawler, J. K. Lenstra, and A. H. G. Rinnooy Kan. Preemptive scheduling of uniform machines subject to release dates. *Progress in Combinatorial Optimization*, pages 245–261, 1984. 338, 339
10. E. L. Lawler, J. K. Lenstra, A. H. G. Rinnooy Kan, and D. B. Shmoys. Sequencing and scheduling: algorithms and complexity. In *Handbooks in operations research and management science*, volume 4, pages 445–522. North Holland, 1993. 339
11. C. A. Phillips, C. Stein, and J. Wein. Scheduling jobs that arrive over time. In *Proceedings of the 4th Workshop on Algorithms and Data Structures (WADS'95)*, volume 955 of *Lecture Notes in Computer Science*, pages 86–97. Springer, 1995. 340
12. A. S. Schulz and M. Skutella. The power of alpha-points in preemptive single machine scheduling. manuscript, 1999. 340
13. M. Skutella. personal communication, 2000. 340
14. W. E. Smith. Various optimizers for single-stage production. *Naval Research and Logistics Quarterly*, 3:59–66, 1956. 339
15. L. Stougie. Unpublished manuscript, 1995. 340

16. L. Stougie and A. P. A. Vestjens. Randomized on-line scheduling: How low can't you go? Unpublished manuscript, 1997. 340
17. A. P. A. Vestjens. On-line machine scheduling. Technical report, Ph.D. thesis, Eindhoven University of Technology, The Netherlands, 1997. 338, 340
18. A. C. Yao. Probabilistic computations: Towards a unified measure of complexity. In *Proc. 18th Annual Symposium on Foundations of Computer Science*, pages 222–227. IEEE, 1977. 341, 342

Approximation Algorithms and Complexity Results for Path Problems in Trees of Rings

Thomas Erlebach

Computer Engineering and Networks Laboratory
ETH Zürich, CH-8092 Zürich, Switzerland
erlebach@tik.ee.ethz.ch

Abstract. A tree of rings is a network that is obtained by interconnecting rings in a tree structure such that any two rings share at most one node. A connection request (call) in a tree of rings is given by its two endpoints and, in the case of prespecified paths, a path connecting these two endpoints. We study undirected trees of rings as well as bidirected trees of rings. In both cases, we show that the path packing problem (assigning paths to calls so as to minimize the maximum load) can be solved in polynomial time, that the path coloring problem with prespecified paths can be approximated within a constant factor, and that the maximum (weight) edge-disjoint paths problem is $\mathcal{NP}$-hard and can be approximated within a constant factor (no matter whether the paths are prespecified or can be determined by the algorithm). We also consider fault-tolerance in trees of rings: If a set of calls has been established along edge-disjoint paths and if an arbitrary link fails in every ring of the tree of rings, we show that at least one third of the calls can be recovered if rerouting is allowed. Furthermore, computing the optimal number of calls that can be recovered is shown to be polynomial in undirected trees of rings and $\mathcal{NP}$-hard in bidirected trees of rings.

1 Introduction

A *ring* is a graph that consists of a single cycle of length at least three. The *trees of rings* are the class of graphs that can be defined inductively as follows:

1. A single ring is a tree of rings.
2. If T is a tree of rings, then the graph obtained by adding a node-disjoint ring R to T and then identifying one node of R with one node of T is also a tree of rings.
3. No other graphs are trees of rings.

Equivalently, a tree of rings is a connected graph T whose edges can be partitioned into rings such that any two rings have at most one node in common and such that, for all pairs (u, v) of nodes in T, all simple paths from u to v touch precisely the same rings. (We say that a path touches a ring if it contains at least one edge of that ring. Furthermore, a path touches a node if it starts at that node, ends at that node, or passes through that node. Two paths *intersect*

J. Sgall, A. Pultr, and P. Kolman (Eds.): MFCS 2001, LNCS 2136, pp. 351–362, 2001.

if they share an edge.) A third equivalent definition is that a tree of rings is a connected graph in which all biconnected components are rings.

Trees of rings are an interesting topology for the construction of communication networks. On the one hand, they are not expensive to build and require few additional links as compared to a tree topology. On the other hand, a tree of rings remains connected even if an arbitrary link fails in each ring, thus achieving much better fault tolerance than a tree network. Trees of rings are also a natural topology for all-optical networks: fiber rings have been employed for a long time (e.g., SONET rings), and it is natural to interconnect different rings in a tree structure.

In modern communication networks, establishing a connection between two nodes often requires allocating resources on all links along a path between the two nodes. For example, this is the case in ATM networks (where bandwidth is reserved along the path from sender to receiver) or in all-optical networks with wavelength-division multiplexing (where a wavelength is reserved along the path from sender to receiver). Resource allocation and call admission control in such networks lead to optimization problems involving edge-disjoint paths. (See [11] for more about the background of such problems.) In this paper, we investigate the complexity and approximability of these problems in networks with tree-of-rings topology.

1.1 Problem Definitions

Each of the following problems can be studied for undirected paths in undirected graphs and for directed paths in bidirected graphs. (A bidirected graph is the graph obtained from an undirected graph by replacing each undirected edge by two directed edges with opposite directions. We use the term "link" to refer to an undirected edge in an undirected graph or to a pair of directed edges with opposite directions in a bidirected graph.) In the undirected case, a connection request (call) is given by its two endpoints, and the call is established along an undirected path between these two endpoints. In the bidirected case, a call specifies sender and receiver, and it is established along a directed path from sender to receiver.

Given a set of paths in a graph, the *load* $L(e)$ of a (directed or undirected) edge e is the number of paths containing that edge. The *maximum load* is the maximum of the values $L(e)$, taken over all edges e of the graph. In this paper, the maximum load is denoted by L. The set of paths that creates this maximum load will always be clear from the context.

Now we define the relevant optimization problems.

Path Coloring (PC): Given a set of calls, assign paths and colors to the calls such that calls receive different colors if their paths intersect. Minimize the number of colors. (Application: Minimize the number of wavelengths in an all-optical WDM network.)

PC with Prespecified Paths (PCwPP): Same as PC, but the paths are specified as part of the input.

Path Packing (PP): Given a set of calls, assign paths to the calls such that the maximum load is minimized. (Application: Minimize the required link capacity if all calls request the same bandwidth.)

Maximum Edge-Disjoint Paths (MEDP): Given a set of calls, select a subset of the calls and assign edge-disjoint paths to the calls in that subset. Maximize the cardinality of the subset. (Application: Maximize the number of established calls.)

Maximum Weight Edge-Disjoint Paths (MWEDP): Given a set of calls with positive weights, select a subset of the calls and assign edge-disjoint paths to the calls in that subset. Maximize the total weight of the subset.

MEDP with Prespecified Paths (MEDPwPP): Same as MEDP, but the paths are specified as part of the input.

MWEDP with Prespecified Paths (MWEDPwPP): Same as MWEDP, but the paths are specified as part of the input.

Optimal Recovery (OR): Given a set C of calls that have been assigned edge-disjoint paths in a graph $T = (V, E)$, and a set $F \subseteq E$ of faulty links, select a subset $C' \subseteq C$ and assign edge-disjoint paths in $T \setminus F$ to the calls in C'. Maximize the cardinality of C'. (Application: Maximize the number of calls that can remain active in spite of the link failures.)

Optimal Weighted Recovery (OWR): Same as OR, but the calls are assigned positive weights and the goal is to maximize the total weight of the calls in C'.

For maximization problems, an algorithm is a ρ-approximation algorithm if it runs in polynomial time and always computes a solution whose objective value is at least a $\frac{1}{\rho}$-fraction of the optimum. For minimization problems, an algorithm is a ρ-approximation algorithm if it runs in polynomial time and always computes a solution whose objective value is at most ρ times the optimum.

MEDP and OR as well as their variants can also be studied for the case that the network is an all-optical network with W wavelengths. In the case of MEDP, for example, this means that the goal is to compute a subset of the given calls and an assignment of paths and at most W different colors to the calls in that subset such that calls with the same color are routed along edge-disjoint paths. Approximation algorithms for MEDP and its variants in the case of W wavelengths can be obtained by calling an algorithm for one wavelength W times. The resulting approximation ratio is only slightly worse than for one wavelength, i.e., at most $1/(1 - e^{-1/\rho}) \leq \rho + 1$ if the approximation ratio for one wavelength is ρ [15].

1.2   Previous Work on Rings and Trees

Most of the problems defined in the previous section have been studied intensively for rings and for trees (and, of course, for many other topologies). For undirected and bidirected ring networks, it is known that PC is $\mathcal{NP}$-hard no matter whether the paths are prespecified [9] or not [7,16], and that PP can be solved optimally in polynomial time [16]. MEDP, MWEDP, MEDPwPP, and MWEDPwPP are polynomial for undirected and bidirected rings.

In undirected and bidirected tree networks, PC is also $\mathcal{NP}$-hard [7,12]. A given set of paths with maximum load L can be efficiently colored using at most $3L/2$ colors in the undirected case [14] and at most $\min\{2L - 1, \lceil 5L/3 \rceil\}$ colors in the bidirected case [8]. MEDP and MWEDP are polynomial in undirected trees of arbitrary degree and in bidirected trees of constant degree, but MAX SNP-hard in bidirected trees of arbitrary degree [5]. For every positive constant ε, there is a $(\frac{5}{3} + \varepsilon)$-approximation algorithm for MEDP and MWEDP in bidirected trees of arbitrary degree [5,6].

1.3   Previous Work on Trees of Rings

It is known that an algorithm for PC in trees that uses at most αL colors can be used to obtain a 2α-approximation algorithm for PC in trees of rings, both in the undirected [14] and in the bidirected case [13]: It is sufficient to remove an arbitrary link from each ring in the tree of rings (the "cut-one-link" heuristic) and to use the tree algorithm in the resulting tree; the maximum load of the obtained paths is at most twice the load of the paths in the optimal solution, which in turn is a lower bound on the optimal number of colors. In this way, a 3-approximation algorithm is obtained in the undirected case and a $\frac{10}{3}$-approximation algorithm in the bidirected case. For undirected trees of rings, a 2-approximation algorithm for PCwPP was given in [3] for the special case in which each node is contained in at most two rings (i.e., in trees of rings with maximum vertex degree equal to four).

The all-to-all instance (the set of calls containing one call for every ordered pair of nodes) in bidirected trees of rings was studied in [1]. It was shown that a routing that minimizes the maximum load L can be computed in polynomial time, and that the resulting paths can be colored optimally with L colors.

1.4   Summary of Results

In this paper, we show that PP is polynomial for trees of rings (Sect. 2) and give practical approximation algorithms with constant approximation ratios for PCwPP (Sect. 3) and for MEDP, MWEDP, MEDPwPP, and MWEDPwPP (Sect. 5). In Sect. 4, we show that at least one third of the active calls in a tree of rings can be recovered after an arbitrary link fails in every ring of the tree of rings, provided that rerouting is allowed. OR and OWR are proved to be polynomial for undirected trees of rings, while OR is shown to be $\mathcal{NP}$-hard for bidirected trees of rings even if $T \setminus F$ is a tree.

2   Path Packing

Path packing in trees of rings can be reduced to path packing in rings. Let c be a call and let p be a path connecting the endpoints of c. For each ring r in the given tree of rings that is touched by p, the node $in(r, c)$ at which p enters the ring (or begins) and the node $out(r, c)$ at which p leaves the ring (or terminates) are uniquely determined by the endpoints of c. Therefore, PP can be tackled by considering each ring separately. For a ring r, simply view each call c that touches r as a call from $in(r, c)$ to $out(r, c)$ and compute a routing that minimizes the maximum load using the known polynomial algorithms for PP in ring networks. The resulting routings can be combined to obtain an optimal routing for the tree of rings.

Theorem 1. *PP can be solved optimally in polynomial time for undirected and bidirected trees of rings.*

Note that Theorem 1 implies that one can decide in polynomial time whether *all* given calls can be established along edge-disjoint paths. Hence, this decision version of the edge-disjoint paths problem is polynomial for trees of rings.

3   Path Coloring with Prespecified Paths

As PCwPP is $\mathcal{NP}$-hard in rings, PCwPP is also $\mathcal{NP}$-hard in trees of rings. Given a set P of paths in a tree of rings $T = (V, E)$, we propose the following greedy approximation algorithm. It is a generalization of the simple greedy algorithm that uses at most $2L - 1$ colors for PC in trees.

algorithm GreedyColoring(T, P):

1. Initially, all paths are uncolored.
2. Choose an arbitrary node $s \in V$ and perform a depth-first search (DFS) of T starting at s. When the DFS reaches a node v, consider all uncolored paths $p \in P$ that touch v, in arbitrary order, and assign to each of them the smallest available color (i.e., the color with smallest index such that no path intersecting p has already been assigned that color).

In order to derive an upper bound on the number of colors used by GreedyColoring, we consider an arbitrary path p at the time it is assigned its color and show that it can intersect only a bounded number of paths that have already been assigned a color prior to p.

See Fig. 1. Assume that the dark nodes have been visited by the DFS already, and that the DFS now reaches v for the first time. Let r denote the ring containing v and a node adjacent to v that has been visited already (there is only one such ring). The uncolored paths touching v do not touch any dark node. They can be classified into two basic types. The first type does not use any edge in r; for the arguments given below, take a path connecting nodes a and b in the figure as a representative path for the first type. The second type uses at least one edge in r; take a path connecting b and c as a representative path.

First, consider the case of undirected paths. If p belongs to the first type, all previously colored paths that intersect p must also touch v and touch a ring containing v that is touched by p. But there are at most four edges incident to v that belong to rings touched by p (in the figure, these are the edges labeled 1, 2, 3, 4), and each conflicting path must use one of these edges. As the maximum load is L, there are less than $4L$ colored paths intersecting p. If p

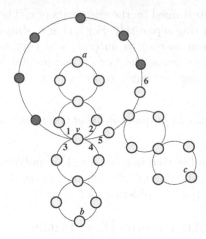

Fig. 1. Illustration of the argument used to prove Theorem 2

belongs to the second type, the number of colored paths that can intersect p is also less than $4L$: if we denote the edges as shown in Fig. 1, there can be at most $L - 1$ colored paths using edge 5, at most L colored paths using edge 6, and at most $2L - 1$ colored paths using edges 3 and 4. Thus, GreedyColoring uses at most $4L$ colors in the undirected case.

The same argument allows us to bound the number of conflicting paths by $8L - 1$ in the bidirected case. We lose a factor of two because now there can be $2L$ paths using the link between two adjacent nodes, L in each direction.

Theorem 2. *For PCwPP in trees of rings, GreedyColoring is a polynomial-time algorithm that uses at most $4L$ colors in the undirected case and at most $8L$ colors in the bidirected case.*

Note that a PC algorithm can be derived from a PCwPP algorithm as follows. First, compute a routing that minimizes the maximum load L (Theorem 1); then, use a PCwPP algorithm to assign colors to the resulting paths. If the PCwPP algorithm uses at most αL colors, an α-approximation algorithm for PC is obtained in this way. Using Theorem 2, we obtain a PC algorithm with approximation ratio at most 4 in the undirected case and at most 8 in the bidirected case. These ratios are worse than the ratios achieved by the "cut-one-link" heuristic (see Sect. 1.3), but it is not clear whether our upper bounds are tight, and it is conceivable that our new algorithms perform better in practice.

4   Recovery after Link Failures

In this section we study the problems OR and OWR in trees of rings. For bidirected trees of rings, we assume that a failure always affects both edges in a pair of directed edges with opposite directions between two adjacent nodes.

Let $T = (V, E)$ be a tree of rings and let $F \subset E$ be the set of faulty links. The worst case among all failure patterns that leave the remaining network connected is the case that $T \setminus F$ is a tree.

Assume that a set C of calls has been active at the time of the failure, i.e., the calls in C are assigned edge-disjoint paths in T. In the worst case, every call in C used a link in F, and all calls are affected by the link failures. Therefore, it is important to allow *rerouting*, i.e., changing the paths assigned to the calls.

If $T \setminus F$ is a tree, there is only one possible routing in $T \setminus F$ for each call $c \in C$. Let P be the set of paths obtained from C by routing each call in $T \setminus F$. The maximum load L created by the paths in P is at most two. (For any edge $e \in E \setminus F$, the paths in P that contain e either contained e already before the link failures, or they contained the faulty link in the ring to which e belongs.) In this case, solving the OR (OWR) problem reduces to solving the MEDP (MWEDP) problem for the set of paths P in the tree $T \setminus F$. As MWEDP is polynomial in undirected trees, this shows that OR and OWR are polynomial for undirected trees of rings (under the assumption that $T \setminus F$ is a tree). For bidirected trees of arbitrary degree, it was not known previously whether MEDP is already $\mathcal{NP}$-hard if the input is a set of paths with maximum load two. However, we have obtained the following theorem.

Theorem 3. *MEDP in bidirected trees is $\mathcal{NP}$-hard even if the given set of paths has maximum load $L = 2$.*

We give a brief outline of the proof. As a first step, we prove by a reduction from 3SAT that the maximum 3-dimensional matching problem (MAX-3DM) is $\mathcal{NP}$-hard even if each element occurs in at most two triples. Then we reduce MAX-3DM to MEDP in bidirected trees by adapting the reduction given by Garg et al. for proving $\mathcal{NP}$-hardness of integral multicommodity flow in undirected trees with edge capacities one or two [10]. The maximum load of the resulting set of paths equals the maximum number of occurrences of an element in the triples of the MAX-3DM instance. The full proof of Theorem 3 is given in [4].

For any set P of paths in a bidirected tree T_1 with maximum load $L = 2$, one can construct a set C of calls that can be routed along edge-disjoint paths in a bidirected tree of rings T_2 and a set F of faulty links in T_2 such that the conflict graph of the paths obtained by routing the calls in C in $T_2 \setminus F$ is identical to the conflict graph of P in T_1. Therefore, Theorem 3 implies that OR and OWR are $\mathcal{NP}$-hard for bidirected trees of rings T even if $T \setminus F$ is a tree. The $(\frac{5}{3} + \varepsilon)$-approximation for MWEDP in bidirected trees from [6] gives a $(\frac{5}{3} + \varepsilon)$-approximation for OR and OWR in bidirected trees of rings in this case.

Furthermore, we can extend this approach to arbitrary sets F of faulty links (i.e., $T \setminus F$ could be disconnected or still contain some complete rings). In general, $T \setminus F$ consists of several connected components. All calls in C whose endpoints are in different components cannot be recovered. Consider one particular connected component H of $T \setminus F$. Intuitively, only the "tree part" of H is relevant for OR and OWR. Each ring in H that does not contain a faulty link can be shrunk into a single node. The resulting network is a tree T'. Then the MWEDP algorithm

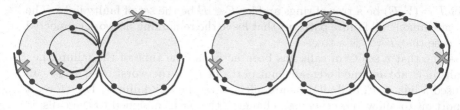

Fig. 2. Examples of bad link failures

for trees is applied to T'. The resulting set of edge-disjoint paths in T' gives a set of edge-disjoint paths in H by using the original routing in the non-faulty rings. Calls that touch only non-faulty rings are always recovered. Analyzing this approach in detail, we obtain the following theorem.

Theorem 4. *For undirected trees of rings, OR and OWR can be solved optimally in polynomial time. For bidirected trees of rings, OR and OWR are $\mathcal{NP}$-hard even if $T \setminus F$ is a tree, and there is a $(\frac{5}{3} + \varepsilon)$-approximation for arbitrary sets F of faulty links.*

A set P of paths in a tree with $L = 2$ can always be colored (efficiently and optimally) with at most 3 colors, both in the undirected case and in the bidirected case. The paths assigned the same color (a color class) are edge-disjoint. Therefore, there is a set $S \subseteq P$ of edge-disjoint paths such that $|S| \geq |P|/3$. In the weighted case we can also infer that there is a set $S' \subseteq P$ of edge-disjoint paths whose total weight is at least one third of the total weight of P. The set S resp. S' can be computed efficiently by coloring P and taking the best of the color classes. If $T \setminus F$ still contains some rings, we can again shrink the rings into single nodes before applying the path coloring algorithm.

Theorem 5. *Assume that a set C of calls has been established along edge-disjoint paths in an undirected or bidirected tree of rings $T = (V, E)$. For any set F of faulty links such that F contains at most one link from every ring in T, it is possible to recover (efficiently) at least $|C|/3$ calls (or calls whose total weight is at least one third of the total weight of C), provided that rerouting is allowed.*

For undirected trees of rings, the bound given in Theorem 5 is tight, i.e., there exists a set of edge-disjoint paths in a tree of rings and a set of link failures such that only one third of the paths can be recovered. See the left-hand side of Fig. 2 for an example. For bidirected trees of rings, we can construct an example where at most $2|C|/5$ calls can be recovered after failure, even if an optimal rerouting is used (right-hand side of Fig. 2). So we know that the fraction of calls that can be recovered after such link failures (where $T \setminus F$ is a tree) in the worst case is at least $\frac{1}{3}$ and at most $\frac{2}{5}$ for bidirected trees of rings.

Now we consider the effect of link failures in an all-optical network with wavelength-division multiplexing (WDM). If W wavelengths (colors) are available, a set C of calls can be established simultaneously if the calls are assigned paths and colors such that intersecting paths receive different colors and such that at most W different colors are used altogether. Corollary 6 follows from Theorem 5 by considering the calls with each wavelength separately.

Corollary 6. *Assume that a set C of calls has been established in an undirected or bidirected tree of rings $T = (V, E)$ using W wavelengths. For any set F of faulty links such that F contains at most one link from every ring in T, it is possible to recover at least $|C|/3$ calls (or calls whose total weight is at least one third of the total weight of C) without changing the wavelength of a call, provided that rerouting is allowed.*

5   Maximum (Weight) Edge-Disjoint Paths

While MEDP and MWEDP are polynomial for rings and for undirected trees, we can show that MEDP and its variants are all $\mathcal{NP}$-hard (even MAX SNP-hard) in undirected and bidirected trees of rings.

Theorem 7. *MEDP, MWEDP, MEDPwPP, and MWEDPwPP are MAX SNP-hard for trees of rings, both in the undirected case and in the bidirected case.*

The full proof of Theorem 7 can be found in [4]. Here we sketch the main ideas. For MEDPwPP in bidirected trees of rings, MAX SNP-hardness follows directly from MAX SNP-hardness of MEDP in bidirected trees, because any set of paths in a bidirected tree can also be viewed as a set of paths in a bidirected tree of rings that contains the tree as a subgraph. Furthermore, MEDP in bidirected trees can be reduced to MEDPwPP in undirected trees of rings by replacing each pair of directed edges with opposite directions by a ring of four nodes and routing the paths with different directions through opposite sides of that ring. Thus, MEDPwPP is MAX SNP-hard for undirected trees of rings. For MEDP, we use *short calls* (calls between two adjacent nodes) in order to make certain edges unusable for the remaining calls. This works because any solution that does not accept all short calls and route them along the direct edges can be transformed into a solution of this form that is at least as good. With the short calls, the MAX SNP-hardness proof for MEDP in bidirected trees can be adapted to MEDP in trees of rings with minor modifications. It is only necessary to show that the number of short calls is small enough so that the reduction is still approximation preserving.

In view of the hardness result of Theorem 7, we are interested in obtaining efficient approximation algorithms.

5.1   Arbitrary Paths

Consider the "cut-one-link" heuristic for MEDP and MWEDP: remove an arbitrary link from each ring and then use an algorithm for MEDP or MWEDP in trees. Theorem 5 implies that the optimal solution in the resulting tree has an objective value that is at least one third of the optimal solution in the full tree of rings.

Theorem 8. *Using the "cut-one-link" heuristic and an α-approximation algorithm for MEDP (for MWEDP) in trees gives a 3α-approximation algorithm for MEDP (for MWEDP) in trees of rings, both in the undirected case and in the bidirected case.*

Noting that MWEDP is polynomial (i.e., admits a 1-approximation algorithm) in undirected trees and that there is a $(\frac{5}{3}+\varepsilon)$-approximation for MWEDP in bidirected trees, we obtain the following corollary.

Corollary 9. *There is a 3-approximation algorithm for MWEDP in undirected trees of rings and a $(5+\varepsilon)$-approximation algorithm for MWEDP in bidirected trees of rings, where ε is an arbitrary positive constant.*

5.2   Prespecified Paths

Given a set P of paths in a tree of rings $T = (V, E)$, we propose the following greedy algorithm for MEDPwPP.

algorithm GreedySelection(T, P):

1. Maintain a set $S \subseteq P$ of selected paths. Initially, $S = \emptyset$.
2. Choose an arbitrary node $s \in V$ and perform a DFS of T starting at s. When the DFS is about to retreat from a node v (i.e., when all neighbors of v have been visited and the DFS moves back to the parent of v in the DFS tree), consider all paths $p \in P$ that touch v and that do not touch any node from which the DFS has not yet retreated, in arbitrary order. If p is edge-disjoint from all paths in S, insert p into S, otherwise discard p.
3. Output S.

Intuitively, the algorithm processes the DFS-tree of T in a bottom-up fashion and greedily selects paths that touch only nodes in subtrees that have already been processed. For the analysis of the approximation ratio of GreedySelection, please refer again to Fig. 1. Now the dark nodes represent nodes from which the DFS has not yet retreated. If the DFS is about to retreat from v, the paths that are considered can be classified into the same two types as in Sect. 3. If GreedySelection selects a path p that is not in the optimal solution, we can bound the number of later paths (paths that are considered after p by GreedySelection) that are in the optimal solution and that are intersected by p using the same arguments as in Sect. 3. In this way we obtain that there can be at most 4 such paths in the undirected case and at most 8 such paths in the bidirected case. (For example, in the undirected case, all later paths that are blocked by a path p from a to b in Fig. 1 must use one of the four edges labeled 1 to 4.) Therefore, each path selected by GreedySelection blocks at most 4 resp. 8 later paths that could be selected in an optimal solution instead. This reasoning can easily be turned into a formal proof that establishes the following theorem.

Theorem 10. *GreedySelection has approximation ratio at most 4 for MEDPwPP in undirected trees of rings and approximation ratio at most 8 for MEDPwPP in bidirected trees of rings.*

GreedySelection can be converted into an approximation algorithm for the problem MWEDPwPP by adopting a two-phase approach similar to the one used by Berman et al. for an interval selection problem [2]. The algorithm maintains a stack S of paths. When a path is pushed on the stack, it is assigned a *value* that may be different from its weight. In the end, the paths are popped from the stack and selected in a greedy way.

algorithm TwoPhaseSelection(T, P):

1. Maintain a stack S of paths with values. Initially, S is the empty stack.
2. Phase One: Choose an arbitrary node $s \in V$ and perform a DFS of T starting at s. When the DFS is about to retreat from a node v, consider all paths $p \in P$ that touch v and that do not touch any node from which the DFS has not yet retreated, in arbitrary order. Compute $total(p, S)$, the sum of the values of the paths in S that intersect p. If the weight $w(p)$ of p is strictly larger than $total(p, S)$, then push p onto the stack and assign it the value $w(p) - total(p, S)$. Otherwise, discard p.
3. Phase Two: After the DFS is complete, pop the paths from the stack and select each path if it does not intersect any previously selected path.
4. Output the set of paths selected in Phase Two.

We can prove the same upper bound on the approximation ratio of TwoPhase-Selection for MWEDPwPP as for GreedySelection for MEDPwPP.

Theorem 11. *Algorithm TwoPhaseSelection is a 4-approximation algorithm for MWEDPwPP in undirected trees of rings and an 8-approximation algorithm for MWEDPwPP in bidirected trees of rings.*

Proof. Let $\overline{S}$ denote the contents of the stack at the end of Phase One, and let t denote the sum of the values of the paths in $\overline{S}$. Fix an arbitrary optimal solution P^*. The proof consists of two parts.

First, we show that the total weight of the paths in P^* is at most $4t$ (at most $8t$) for undirected (bidirected) trees of rings. To see this, let $val(p, \overline{S})$ denote the sum of the values of paths in $\overline{S}$ that intersect p and that were processed in Phase One not after p. (If $p \in \overline{S}$, then the value of p also contributes to $val(p, \overline{S})$.) By the definition of the algorithm, $val(p, \overline{S}) \geq w(p)$ for all paths p. Now the key observation is that the value of every path $q \in \overline{S}$ is counted at most four (eight) times in the sum $\sum_{p \in P^*} val(p, \overline{S}) \geq w(P^*)$. This can again be seen by referring to Fig. 1 and considering the path q to be either of the first type (from a to b) or of the second type (from b to c). In the undirected case, there can be at most four paths in P^* that are processed after q in Phase One and that intersect q. In the bidirected case, there can be at most eight such paths. Therefore, $w(P^*) \leq 4t$ in the undirected case (resp. $w(P^*) \leq 8t$ in the bidirected case).

Second, the solution output by the algorithm has weight at least t: when a path p is selected in Phase Two, the weight of the solution increases by $w(p)$, while the sum of the values of the paths that are still on the stack and that are not intersected by p decreases by at most $w(p)$. □

References

1. B. Beauquier, S. Pérennes, and D. Tóth. All-to-all routing and coloring in weighted trees of rings. In *Proceedings of the 11th Annual ACM Symposium on Parallel Algorithms and Architectures SPAA'99*, pages 185–190, 1999. 354
2. P. Berman and B. DasGupta. Multi-phase algorithms for throughput maximization for real-time scheduling. *Journal of Combinatorial Optimization*, 4(3):307–323, 2000. 361
3. X. Deng, G. Li, W. Zang, and Y. Zhou. A 2-approximation algorithm for path coloring on trees of rings. In *Proceedings of the 11th Annual International Symposium on Algorithms and Computation ISAAC 2000*, LNCS 1969, pages 144–155, 2000. 354
4. T. Erlebach. Approximation algorithms and complexity results for path problems in trees of rings. TIK-Report 109, Computer Engineering and Networks Laboratory (TIK), ETH Zürich, June 2001. Available electronically at ftp://ftp.tik.ee.ethz.ch/pub/publications/TIK-Report109.pdf. 357, 359
5. T. Erlebach and K. Jansen. Maximizing the number of connections in optical tree networks. In *Proceedings of the 9th Annual International Symposium on Algorithms and Computation ISAAC'98*, LNCS 1533, pages 179–188, 1998. 354
6. T. Erlebach and K. Jansen. Conversion of coloring algorithms into maximum weight independent set algorithms. In *ICALP Workshops 2000*, Proceedings in Informatics 8, pages 135–145. Carleton Scientific, 2000. 354, 357
7. T. Erlebach and K. Jansen. The complexity of path coloring and call scheduling. *Theoretical Computer Science*, 255(1–2):33–50, 2001. 353, 354
8. T. Erlebach, K. Jansen, C. Kaklamanis, M. Mihail, and P. Persiano. Optimal wavelength routing on directed fiber trees. *Theoretical Computer Science*, 221:119–137, 1999. Special issue of ICALP'97. 354
9. M. R. Garey, D. S. Johnson, G. L. Miller, and C. H. Papadimitriou. The complexity of coloring circular arcs and chords. *SIAM J. Algebraic Discrete Methods*, 1(2):216–227, 1980. 353
10. N. Garg, V. V. Vazirani, and M. Yannakakis. Primal-dual approximation algorithms for integral flow and multicut in trees. *Algorithmica*, 18:3–20, 1997. 357
11. J. Kleinberg. *Approximation algorithms for disjoint paths problems*. PhD thesis, MIT, 1996. 352
12. S. R. Kumar, R. Panigrahy, A. Russel, and R. Sundaram. A note on optical routing on trees. *Inf. Process. Lett.*, 62:295–300, 1997. 354
13. M. Mihail, C. Kaklamanis, and S. Rao. Efficient access to optical bandwidth. In *Proceedings of the 36th Annual Symposium on Foundations of Computer Science FOCS'95*, pages 548–557, 1995. 354
14. P. Raghavan and E. Upfal. Efficient routing in all-optical networks. In *Proceedings of the 26th Annual ACM Symposium on Theory of Computing STOC'94*, pages 134–143, 1994. 354
15. P.-J. Wan and L. Liu. Maximal throughput in wavelength-routed optical networks. In *Multichannel Optical Networks: Theory and Practice*, volume 46 of *DIMACS Series in Discrete Mathematics and Theoretical Computer Science*, pages 15–26. AMS, 1998. 353
16. G. Wilfong and P. Winkler. Ring routing and wavelength translation. In *Proceedings of the Ninth Annual ACM-SIAM Symposium on Discrete Algorithms SODA'98*, pages 333–341, 1998. 353

A 3-Approximation Algorithm for Movement Minimization in Conveyor Flow Shop Processing

Wolfgang Espelage* and Egon Wanke

Department of Computer Science, D-40225 Düsseldorf, Germany
{espelage,wanke}@cs.uni-dueseldorf.de

Abstract. We consider the movement minimization problem in a conveyor flow shop processing controlled by one worker for all machines. A machine can only execute tasks if the worker is present. Each machine can serve as a buffer for exactly one job. The worker has to cover a certain distance to move from one machine to the next or previous one. The objective is to minimize the total distance the worker has to cover for the processing of all jobs. We introduce the first polynomial time approximation algorithm for this problem with a performance bounded by some fixed factor.

1 Introduction

A *conveyor flow shop* consists of machines $P_1, \ldots, P_m$ and jobs $J_1, \ldots, J_n$. Each job passes the machines in the order $P_1, \ldots, P_m$. At every machine the jobs arrive in the order $J_1, \ldots, J_n$. The jobs usually have to be executed only at some of the machines and not at all of them. All machines are controlled by only one worker operating in front of them. A machine can only execute a task of a job if the worker is present. Thus the worker has to change his working position such that the processing continues. The objective is to minimize the movement of the worker along the flow shop.

In the conveyor flow shop processing which we consider there are no buffers in the usual sense, but each machine itself can be used to buffer one job. For example, if job J_i has finished its execution at machine P_j while the preceding job J_{i-1} still occupies the succeeding machine P_{j+1} then J_i resides at P_j such that P_j can not be used for the execution of the succeeding jobs. This blocking mechanism arises from characteristics of the processing technology.

Movement minimization is especially interesting for *pick-to-belt orderpicking systems* [4, 8], where each task merely consists of picking up certain material from a shelf rack and putting it into a box or a bin. Since the bins are moving on a conveyor they can not bypass each other. Changing the position of the worker takes a considerable amount of time compared to the execution time of a task. For example, in today's order picking systems it is quite usual that the worker

---

* The work of the first author was supported by the German Research Association (DFG) grant WA 674/8-2.

J. Sgall, A. Pultr, and P. Kolman (Eds.): MFCS 2001, LNCS 2133, pp. 363–374, 2001.

spends 50% of his time by moving between the machines. This deplorable state can usually not be redressed by additional workers, because then idle times have to be taken into account.

We should emphasize that movement minimization in conveyor flow shop processing is quit different from *flow shop scheduling*. See [1] for an overview of scheduling in computer and manufacturing systems, see [7] for efficiently solvable cases of permutation flow shop scheduling, and see [3] for now-wait and blocking scheduling problems. The main difference between movement minimization and scheduling is the objective. Movement minimization is totally independent of the exact processing times of the jobs at the machines. The total processing time is always the sum of all processing times of the jobs at the machines, because we consider one worker and, thus, have no parallelism. We only need to distinguish between zero and non-zero processing times. Such processing times are also called unit execution times. All the objectives considered for scheduling of jobs with unit execution times, as for example in [5], are completely insignificant for movement minimization.

In this paper, we concentrate on movement minimization with respect to *additive* distances $d_{j,j'} \geq 0$ between two machines P_j and $P_{j'}$, that is, for all $1 \leq j_1 \leq j_2 \leq j_3 \leq m$ the distances satisfy $d_{j_1,j_2} + d_{j_2,j_3} = d_{j_1,j_3}$. The complexity of movement minimization with respect to additive distances is still open, i.e., not known to be NP-hard or solvable in polynomial time. If the distances only satisfy the triangle inequalities $d_{j_1,j_2} + d_{j_2,j_3} \leq d_{j_1,j_3}$ or the job order can be chosen then the problem becomes NP-hard [2]. There is also a simple $(2 \cdot \log(m+2) - 1)$-competitive online algorithm for movement minimization with additive distances if all distances between consecutive machines are equal, see [2]. If we consider k workers instead of one, then we get the k server problem on a line as a very special case [6]. In this paper, we show that there is a polynomial time algorithm that computes a processing for one worker whose total distance is at most three times the total distance of a minimal processing of a given task matrix R. This is the first approximation algorithm for movement minimization with additive distances whose performance is bounded by some fixed factor.

2   Preliminaries

An instance of the *conveyor flow shop problem* consists of a 0/1 *task matrix* $R = (r_{i,j})_{1 \leq i \leq n, 1 \leq j \leq m}$, $r_{i,j} \in \{0,1\}$. We say, there are n jobs $J_1, \ldots, J_n$ and m *machines* $P_1, \ldots, P_m$. Each job J_i consists of m *tasks* $T_{i,1}, \ldots, T_{i,m}$. Task $T_{i,j}$ is called a *zero task* if $r_{i,j} = 0$ and a *non-zero task* otherwise. Only non-zero tasks have to be executed.

A *processing* of the jobs $J_1, \ldots, J_n$ consists of a sequence of *processing steps*. The *state* of the system between two processing steps is defined by a so-called *state function* $h : \{1, \ldots, n\} \rightarrow \{1, \ldots, m+1\}$. If $h(i) < m+1$ for some job J_i then $T_{i,h(i)}$ is the non-zero task of job J_i that has to be executed next. All non-zero tasks $T_{i,j}$ with $j < h(i)$ have already been executed, while all non-zero tasks $T_{i,j}$ with $j \geq h(i)$ have not been executed yet. If $h(i) = m + 1$ then all non-zero tasks of job J_i are already executed.

We assume that all jobs are moving as far as possible to the machines where they have to be processed next. Thus, we define the *position* $\delta_h(i)$ *of job* J_i *in state h* recursively as follows: $\delta_h(1) := h(1)$ and for $i > 1$

$$\delta_h(i) := \begin{cases} h(i), & \text{if } \delta_h(i-1) = m+1, \\ \min\{h(i), \delta_h(i-1) - 1\}, & \text{if } \delta_h(i-1) < m+1. \end{cases}$$

The least position of a job is $2 - n$, the greatest position is $m + 1$. The position $m + 1$ indicates that the job has already passed the last machine P_m and a position less than 1 indicates that the job has not yet reached the first machine P_1. It is easy to see that the position $\delta_h(i)$ can also be expressed by the following formula

$$\delta_h(i) = \min \left\{ \begin{array}{l} h(i), \\ \min\{h(i') - (i - i') \mid 1 \leq i' \leq i - 1, \ h(i') < m + 1\} \end{array} \right\}. \quad (1)$$

We say, a non-zero task $T_{i,j}$ is *waiting for execution* in state h if $h(i) = j$ and $\delta_h(i) = j$.

State h_{init} defined by $h_{\text{init}}(i) := \min\left(\{j \mid 1 \leq j \leq m \wedge r_{i,j} = 1\} \cup \{m+1\}\right)$ is called the *initial state*. It defines for every $i, 1 \leq i \leq n$, the machine index of the first non-zero task of job J_i, or $m + 1$ if no such machine exists. State h_{final} defined by $h_{\text{final}}(i) := m + 1$ for $1 \leq i \leq n$ is called the *final state*. A state h' is called a *successor state* of state h with respect to *working position* j, $1 \leq j \leq m$, written by $h \rightarrow_j h'$, if there is a task $T_{i,j}$ waiting for execution in state h, i.e., $\delta_h(i) = j$ and $h(i) = j$, and

$$h'(k) := \begin{cases} \min\left(\{j' \mid j < j' \leq m \wedge r_{k,j} = 1\} \cup \{m+1\}\right) & \text{if } k = i, \\ h(k) & \text{otherwise.} \end{cases}$$

We say, that the non-zero task $T_{i,j}$ *has been executed* between the two states h, h'.

A *processing* of a task matrix R is a sequence

$$h_0 \rightarrow_{j_1} h_1 \rightarrow_{j_2} \cdots \rightarrow_{j_k} h_k$$

of states, where h_0 is the initial state, h_l, $1 \leq l \leq k$, is a successor state of h_{l-1} with respect to working position j_l, and h_k is the final state.

To keep the proofs as simple as possible, we define the *total distance* of a processing by

$$(j_1 - 1) + \left(\sum_{s=1}^{k-1} |j_s - j_{s+1}|\right) + (j_k - 1).$$

This corresponds to the assumption that the starting and the final position of the worker is the first machine P_1 and that the distance between P_j and $P_{j'}$ is $|j - j'|$. Figure 1 shows an example of a processing. Note that all the results

$$R = \begin{pmatrix} & P_1 & P_2 & P_3 & P_4 & P_5 \\ \hline J_1 & 0 & 1 & 1 & 0 & 0 \\ J_2 & 0 & 0 & 0 & 0 & 1 \\ J_3 & 1 & 1 & 0 & 0 & 1 \\ J_4 & 1 & 0 & 0 & 1 & 0 \end{pmatrix}$$

i	1	2	3	4
$h_4(i)$	6	5	5	1

i	1	2	3	4
$\delta_{h_4}(i)$	6	5	4	1

k	Input	P_1	P_2	P_3	P_4	P_5	Output	$h_k(1)$	$h_k(2)$	$h_k(3)$	$h_k(4)$
0	$J_4\,J_3\,J_2\,J_1$							2	5	1	1
1	J_4	$J_3\,J_2\,J_1$						3	5	1	1
2		J_4	$J_3\,J_2\,J_1$					3	5	2	1
3		$J_4\,J_3$			J_2		J_1	6	5	2	1
4		J_4		$J_3\,J_2$			J_1	6	5	5	1
5		J_4			J_3		$J_2\,J_1$	6	6	5	1
6		J_4				$J_3\,J_2\,J_1$		6	6	6	1
7				J_4		$J_3\,J_2\,J_1$		6	6	6	4
8							$J_4\,J_3\,J_2\,J_1$	6	6	6	6

Fig. 1. A task matrix R in that the columns and rows are named by the machines and jobs, respectively, and a processing of R with total distance $(2 - 1) + |2 - 1| + |1 - 3| + |3 - 2| + |2 - 5| + |5 - 5| + |5 - 1| + |1 - 4| + (4 - 1) = 18$. This is not a minimal processing, because there is some processing of R with total distance 12. Each row represents a state. The lower left part illustrates the position functions. The input column contains the jobs at a position less than 1. The output column contains the jobs at position $m + 1 = 6$. The tasks in the dark shaded squares are those waiting for execution. The working position is indicated by a '•' between the states. In state h_4 the tasks $T_{1,2}, T_{1,3}, T_{3,1}, T_{3,2}$ have already been executed, $T_{2,5}, T_{3,5}, T_{4,1}, T_{4,4}$ have not been been executed yet, and $T_{2,5}$ and $T_{4,1}$ are waiting for execution.

in this paper hold for arbitrary additive distances. Our objective is to find a processing with a minimal total distance. Such a processing is called a *minimal processing* of R.

It is easy to see that every state h of a processing satisfies the following property:

$$\forall i, j, 1 \le i \le n, 1 \le j \le m: \quad \text{if } \delta_h(i) < j < h(i) \text{ then } r_{i,j} = 0. \tag{2}$$

Intuitively speaking, there is no non-zero task $T_{i,j}$ of job J_i between the current position $\delta_h(i)$ of J_i in state h and the position $h(i)$ at which J_i has to be processed next. Note that one can define a state h that does not satisfy property 2, but such a state can not appear during a processing that starts with the initial state h_{init}.

Every processing of a task matrix defines an arrangement of the non-zero tasks by the order in which they are executed during the processing. However, not every arrangement of the non-zero tasks defines a processing. We call an arrangement of the non-zero tasks *executable*, if there is a processing that executes the non-zero tasks in the order given by the arrangement.

Definition 1. *For two non-zero tasks $T_{i,j}$, $T_{i',j'}$ the successor relation '$\prec$' is defined by*

$$T_{i,j} \prec T_{i',j'} \text{ if } i \leq i' \text{ and } i + j < i' + j'.$$

If $T_{i,j} \prec T_{i',j'}$ then $T_{i',j'}$ is called a successor of $T_{i,j}$, and $T_{i,j}$ is called a predecessor of $T_{i',j'}$.

	P_1	P_2	P_3	P_4	P_5	P_6
J_1	0	0	0	0	1	1
J_2	1	1	1	0	0	1
J_3	1	1	1	1	0	1
J_4	0	0	0	1	0	1
J_5	0	0	1	0	1	1
J_6	1	1	1	0	0	1
J_7	1	1	0	1	0	1

	P_1	P_2	P_3	P_4	P_5	P_6
J_1	0	0	0	0	1	1
J_2	1	1	1	0	0	1
J_3	1	1	1	1	0	1
J_4	0	0	0	1	0	1
J_5	0	0	1	0	1	1
J_6	1	1	1	0	0	1
J_7	1	1	0	1	0	1

Fig. 2. The shaded regions cover all tasks $T_{i',j'}$ with $T_{3,3} \prec T_{i',j'}$ or $T_{i',j'} \prec T_{5,3}$, respectively.

The following fact is easy to see. If $T_{i,j}$ is a predecessor of $T_{i',j'}$ then in any processing, task $T_{i,j}$ has to be executed before task $T_{i',j'}$. If $i = i'$ this is obvious, because then $j < j'$. Otherwise we have $i < i'$. Consider some state h in a processing in that task $T_{i,j}$ is executed next. Then we have $\delta_h(i) = h(i) = j$. Equation (1) for δ_h and $T_{i,j} \prec T_{i',j'}$ now imply

$$\delta_h(i') \leq h(i) - (i' - i) = j - (i' - i) < j'.$$

Thus, $T_{i',j'}$ has not been executed yet.

We say, a sequence of non-zero tasks $T_{i_1,j_1}, \ldots, T_{i_k,j_k}$ *respects the successor relation* $\prec$ if there is no pair l, r, $1 \leq l < r \leq k$, such that $T_{i_r,j_r} \prec T_{i_l,j_l}$.

Lemma 2. *An arrangement $T_{i_1,j_1}, \ldots, T_{i_k,j_k}$ of the non-zero tasks of task matrix R is executable if and only if the arrangement respects the successor relation $\prec$.*

Proof. "⇒" This is already shown above.

"⇐" Assume the arrangement respects the successor relation. Consider a state h of a processing and a non-zero task $T_{i,j}$ such that $T_{i,j}$ has not been executed yet but all non-zero tasks T with $T \prec T_{i,j}$ have already been executed. We claim that in each such state the non-zero task $T_{i,j}$ is waiting for execution and thus can be processed next. A simple induction then shows that all the non-zero tasks can be executed in the order $T_{i_1,j_1}, \ldots, T_{i_k,j_k}$.

To prove the claim, we first remark that all non-zero tasks $T_{i,j'}$, $j' < j$, are predecessors of $T_{i,j}$ and thus we have $h(i) = j$. Assume $T_{i,j}$ is not waiting for execution, that is $\delta_h(i) < h(i) = j$. Then, by the definition of δ_h, there is some $i' < i$ such that $h(i') \leq m$ and $\delta_h(i) = h(i') - (i - i')$. This implies that $T_{i',h(i')}$ is a non-zero task that has not been executed yet. Since $i' < i$ and

$$i + j > i + \delta_h(i) = i' + h(i')$$

we get $T_{i',h(i')} \prec T_{i,j}$. This contradicts the assumption that all non-zero tasks T with $T \prec T_{i,j}$ have already been executed.

Lemma 2 shows that finding a processing with minimal total distance corresponds to finding an arrangement of the non-zero tasks with minimal total distance that respects our successor relation. Such an arrangement of the non-zero tasks is also called a *processing* of R.

For the rest of this paper, we assume that $T_{1,1}$ and $T_{n,1}$ are non-zero tasks and all tasks $T_{i,j}$ with $j > 1$ and $i + j > n$ are zero tasks. This can be obtained by a simple modification of R as follows. If $T_{1,1}$ is a zero tasks we simply set $r_{1,1} = 1$. If not all $T_{i,j}$ with $j > 1$ and $i + j > n$ are zero tasks then we extend R by $t \leq m$ additional jobs which have only zero tasks such that the condition above is satisfied. After that we set $r_{n,1} = 1$, where n is the number of jobs of the new task matrix. If R satisfies our assumption then $T_{1,1}$ is always the first task and $T_{n,1}$ is always the last task in every processing of R, because all non-zero tasks of R (except $T_{1,1}$) are successor tasks of $T_{1,1}$ and all non-zero tasks of R (except $T_{n,1}$) are predecessor tasks of $T_{n,1}$.

Since every processing starts and ends at the first machine, the simple modifications to get a task matrix which satisfies our assumption have no influence on the total distance of a minimal processing, but simplify the technical details in the proofs of the forthcoming lemmas. If $T_{i_1,j_1}, \ldots, T_{i_k,j_k}$ is an arrangement of all non-zero tasks defined by some processing of R then now $T_{i_1,j_1} = T_{1,1}$, $T_{i_k,j_k} = T_{n,1}$, and the total distance of the processing is

$$\sum_{s=1}^{k-1} |j_s - j_{s+1}|.$$

3   An Approximation Algorithm

Definition 3. *Let s, $1 \leq s \leq m-1$, be a machine index and T_{i_1,j_1} be a non-zero task.*

1. T_{i_1,j_1} *is called an* R_s-*task if* $j_1 > s$ *and all tasks* $T_{i,j}$ *with* $j > s$, $i \geq i_1$ *and* $i + j < i_1 + j_1$ *are zero tasks*.
2. T_{i_1,j_1} *is called an* L_s-*task if* $j_1 \leq s$ *and all tasks* $T_{i,j}$ *with* $j \leq s$, $i < i_1$ *and* $i + j \geq i_1 + j_1$ *are zero tasks*.

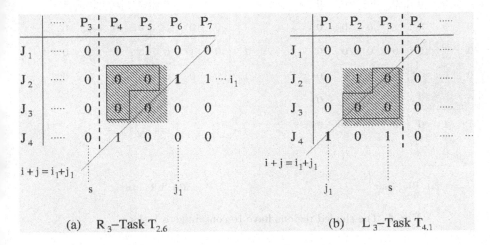

(a) R_3–Task $T_{2.6}$ (b) L_3–Task $T_{4.1}$

Fig. 3. The shaded regions have to contain zero tasks.

Definition 4. *Let* s, $1 \leq s \leq m - 1$, *be a machine index and* $(T_{i_1,j_1}, T_{i_2,j_2})$ *be a pair of two non-zero tasks*.

1. $(T_{i_1,j_1}, T_{i_2,j_2})$ *is called an* RL_s-*pair if the following two conditions hold:*
   (a) T_{i_1,j_1} *is an* R_s-*task,* T_{i_2,j_2} *is an* L_s-*task, and* $T_{i_1,j_1} \prec T_{i_2,j_2}$.
   (b) *All tasks* $T_{i,j}$ *with* $j \leq s$, $i < i_2$, *and* $i + j > i_1 + j_1$ *are zero tasks*.
2. $(T_{i_1,j_1}, T_{i_2,j_2})$ *is called a* LR_s-*pair if the following two conditions hold:*
   (a) T_{i_1,j_1} *is an* L_s-*task,* T_{i_2,j_2} *is an* R_s-*task and* $T_{i_1,j_1} \prec T_{i_2,j_2}$.
   (b) *All tasks* $T_{i,j}$ *with* $j > s$, $i \geq i_1$ *and* $i + j < i_2 + j_2$ *are zero tasks*.

The following lemma follows immediately from Definition 1, 3, 4, and Lemma 2.

Lemma 5.

1. *Let* $(T_{i_1,j_1}, T_{i_2,j_2})$ *be an* RL_s-*pair and let* T_{i_p,j_p} *be a successor task of* T_{i_1,j_1} *with* $j_p \leq s$. *Then* $i_p \geq i_2$ *and in every processing of R all non-zero tasks* $T_{i,j}$ *with* $i < i_2$ *and* $j \leq s$ *have already been executed when* T_{i_p,j_p} *is executed.*
2. *Let* $(T_{i_1,j_1}, T_{i_2,j_2})$ *be an* LR_s-*pair and let* T_{i_p,j_p} *be successor task of* T_{i_1,j_1} *with* $j_p > s$. *Then* $i_p + j_p \geq i_2 + j_2$ *and in every processing of R all non-zero tasks* $T_{i,j}$ *with* $j > s$ *and* $i + j < i_2 + j_2$ *have already been executed when* T_{i_p,j_p} *is executed.*

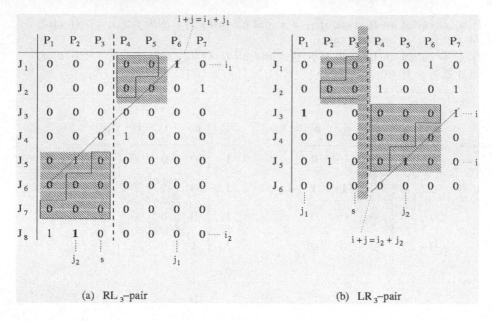

(a) RL₃-pair   (b) LR₃-pair

Fig. 4. The shaded regions have to contain zero tasks.

Definition 6. *Let* s, $1 \leq s \leq m-1$, *be a machine index.*

A triple $(T_{i_1,j_1}, T_{i_2,j_2}, T_{i_3,j_3})$ *is called an* RLR_s-*triple if* $(T_{i_1,j_1}, T_{i_2,j_2})$ *is an* RL_s-*pair and* $(T_{i_2,j_2}, T_{i_3,j_3})$ *is an* LR_s-*pair. It is called an* LRL_s-*triple if* $(T_{i_1,j_1}, T_{i_2,j_2})$ *is an* LR_s-*pair and* $(T_{i_2,j_2}, T_{i_3,j_3})$ *is an* RL_s-*pair.*

A sequence of non-zero tasks $T_{i_1,j_1}, \ldots, T_{i_z,j_z}$ *is called an* s-*chain if* $(T_{i_1,j_1}, T_{i_2,j_2})$ *is either an* LR_s-*pair or an* RL_s-*pair and for* $k = 1, \ldots, z-2$, *if* $(T_{i_k,j_k}, T_{i_{k+1},j_{k+1}})$ *is an* LR_s-*pair then* $(T_{i_{k+1},j_{k+1}}, T_{i_{k+2},j_{k+2}})$ *is an* RL_s-*pair, and if* $(T_{i_k,j_k}, T_{i_{k+1},j_{k+1}})$ *is an* RL_s-*pair then* $(T_{i_{k+1},j_{k+1}}, T_{i_{k+2},j_{k+2}})$ *is an* LR_s-*pair.*

We call a pair of two consecutive tasks $(T_{i,j}, T_{i',j'})$ of a processing a *backward-s-move* if $j > s$ and $j' \leq s$ and a *forward-s-move* if $j \leq s$ and $j' > s$.

Let $\#_R(s)$ be the maximum number of R_s-tasks that can occur in an s-chain. Since the number of R_s-tasks of every s-chain is a lower bound on the number of forward-s-moves of any processing, and since in every processing the number of backward-s-moves is equal to the number of forward-s-moves, we know that

$$\text{LB} := 2 \sum_{s=0}^{m-1} \#_R(s) \tag{3}$$

is a lower bound on the total distance of any processing.

Lemma 7. *Let* s, $1 \leq s \leq m-1$, *be a machine index and let* $\Lambda = T_{i_1,j_1}, T_{i_2,j_2}, \ldots, T_{i_z,j_z}$ *be a minimal processing of* R. *Let* $(T_{i_a,j_a}, T_{i_b,j_b}, T_{i_c,j_c})$, $1 \leq a < b <$

$c \leq z$, be an RLR_s-triple and let

$$p := \min\left\{k \mid a < k \leq b, j_p \leq s, \text{ and } T_{i_a,j_a} \prec T_{i_p,j_p}\right\}$$

and

$$q := \min\left\{k \mid b < k \leq c, j_q > s, \text{ and } T_{i_b,j_b} \prec T_{i_q,j_q}\right\}.$$

(Since $T_{i_a,j_a} \prec T_{i_b,j_b}$ and $T_{i_b,j_b} \prec T_{i_c,j_c}$, p and q are well defined.)
  Then there are at most two forward-s-moves between T_{i_p,j_p} and T_{i_q,j_q} in Λ.

Proof. Let $(T_{i_a,j_a}, T_{i_b,j_b}, T_{i_c,j_c})$ be an RLR_s-triple as shown in the left part of figure 5. Divide the sequence $T_{i_p,j_p}, \ldots, T_{i_q,j_q}$ into maximal nonempty subsequences

$$L_1, R_1, L_2, R_2, L_3, R_3, \ldots, L_t, R_t \tag{4}$$

such that all L_r, $1 \leq r \leq t$, contain only tasks on machines $\leq s$ and all R_r, $1 \leq r \leq t$, contain only tasks on machines $> s$. We show that $t \leq 2$.

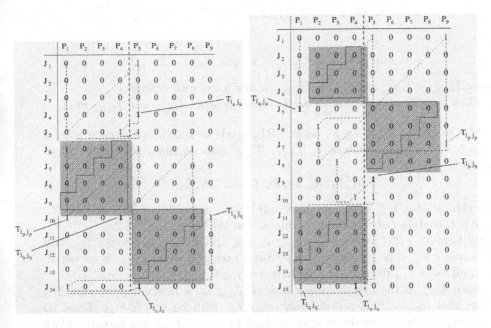

Fig. 5. An RLR_s-triple and an LRL_s-triple. The dashed lines show the movement of the worker. A vertical line indicates that the worker stays at the machine while the jobs move on as far as possible. A diagonal line through task $T_{i,j}$ means that the worker passes machine P_j which is occupied by job J_i. A horizontal line between the tasks $T_{i,j}$ and $T_{i+1,j}$ indicates that the worker passes machine P_j while job J_i has already passed P_j and job J_{i+1} has not yet reached P_j.

By Lemma 5 (1.) we have $i_p \geq i_b$ and all non-zero tasks $T_{i,j}$ with $i < i_b$ and $j \leq s$ have already been executed when T_{i_p,j_p} is executed. Thus, for all tasks $T_{i,j}$ from $L_2, \ldots, L_t$ we have $i \geq i_b$.

Since T_{i_q,j_q} is the first successor task of T_{i_b,j_b} on a machine $j_q > s$ in Λ it follows that $R_1, \ldots, R_{t-1}$ contain only tasks $T_{i,j}$ with $i < i_b$. Thus, no task in $L_2, \ldots, L_t$ is a predecessor of any task in $R_1, \ldots, R_{t-1}$ and we can replace the sequence (4) by the sequence

$$L_1, R_1, R_2, R_3, \ldots, R_{t-1}, L_2, L_3, \ldots, L_t, R_t$$

which yields by Lemma 2 a valid processing whose total distance is less than the total distance of Λ if $t \geq 3$. Since Λ is minimal $t \leq 2$ follows.

Lemma 8. *Let s, $1 \leq s \leq m-1$, be a machine index and let $\Lambda = T_{i_1,j_1}, T_{i_2,j_2}, \ldots, T_{i_z,j_z}$ be a minimal processing of R. Let $(T_{i_a,j_a}, T_{i_b,j_b}, T_{i_c,j_c})$, $1 \leq a < b < c \leq z$, be an LRL_s-triple and let*

$$p := \min\left\{k \mid a < k \leq b, j_p > s, \text{ and } T_{i_a,j_a} \prec T_{i_p,j_p}\right\}$$

and

$$q := \min\left\{k \mid b < k \leq c, j_q \leq s, \text{ and } T_{i_b,j_b} \prec T_{i_q,j_q}\right\}.$$

(Since $T_{i_a,j_a} \prec T_{i_b,j_b}$ and $T_{i_b,j_b} \prec T_{i_c,j_c}$, p and q are well defined.)
Then there is at most one forward-s-move between T_{i_p,j_p} and T_{i_q,j_q} in Λ.

Proof. Let $(T_{i_a,j_a}, T_{i_b,j_b}, T_{i_c,j_c})$ be an LRL_s-triple as shown in the right part of figure 5. Divide the sequence $T_{i_p,j_p}, \ldots, T_{i_q,j_q}$ into maximal nonempty subsequences

$$R_1, L_1, R_2, L_2, R_3, L_3, \ldots, R_t, L_t \qquad (5)$$

such that all L_r, $1 \leq r \leq t$, contain only tasks on machines $\leq s$ and all R_r, $1 \leq r \leq t$, contain only tasks on machines $> s$. We show that $t \leq 2$.

By Lemma 5 (2.) we have $i_p + j_p \geq i_b + j_b$ and all non-zero tasks $T_{i,j}$ with $j > s$ and $i + j < i_b + j_b$ have already been executed when T_{i_p,j_p} is executed. Thus, all tasks $T_{i,j}$ from $R_2, \ldots, R_t$ have $i + j \geq i_b + j_b$.

Since T_{i_q,j_q} is the first successor task of T_{i_b,j_b} on a machine $j_q \leq s$ in Λ it follows that $L_1, \ldots, L_{t-1}$ contain only tasks $T_{i,j}$ with $i + j \leq i_b + j_b$. Thus, no task in $R_2, \ldots, R_t$ is a predecessor of any task in $L_1, \ldots, L_{t-1}$ and we can replace the sequence (5) by the sequence

$$R_1, L_1, L_2, L_3, \ldots, L_{t-1}, R_2, R_3, \ldots, R_t, L_t$$

which yields by Lemma 2 a valid processing whose total distance is less than the total distance of Λ if $t \geq 3$. Since Λ is minimal $t \leq 2$ follows.

Theorem 9. *The total distance of a minimal processing is at most $3LB$ where LB is the lower bound given in equation (3).*

Proof. Let $\Lambda = T_{i_1,j_1}, \ldots, T_{i_z,j_z}$ be a minimal processing for a given task matrix R and let s, $1 \leq s \leq m-1$, be an arbitrary machine index. Let

$$\Gamma = T_{i_{k_1},j_{k_1}}, T_{i_{k_2},j_{k_2}}, \ldots, T_{i_{k_y},j_{k_y}},$$

be an s-chain that starts with L_s-task $T_{1,1}$, i.e., $T_{i_{k_1},j_{k_1}} = T_{1,1}$, and that cannot be extended at the end by further tasks. Since $T_{n,1}$ is a successor task of all other non-zero tasks, it is easy to see that $j_{k_y} \leq s$ and there is no successor task $T_{i,j}$ of $T_{i_{k_y},j_{k_y}}$ with $j > s$. Therefore Γ has always an odd number $y = 2x+1$ of tasks.

For $r = 1, \ldots, 2x$ we define q_r to be the smallest index such that $T_{i_{q_r},j_{q_r}}$ is a successor of $T_{i_{k_r},j_{k_r}}$ with $j_{q_r} > s$ if r is odd and $j_{q_r} \leq s$ if r is even. Obviously, the values for q_r are well defined and $q_r \leq k_{r+1}$. It is sufficient to show that there are at most $3x$ forward-s-moves in Λ.

By Lemma 7 and Lemma 8, we know that for $1 \leq r \leq 2x-1$ there is at most one forward-s-move in Λ between $T_{i_{q_r},j_{q_r}}$ and $T_{i_{q_{r+1}},j_{q_{r+1}}}$ if r is odd, and there are at most two forward-s-moves if r is even. Thus, there are at most $3(x-1)+1$ forward-s-moves in Λ between the execution of $T_{i_{q_1},j_{q_1}}$ and $T_{i_{q_{2x}},j_{q_{2x}}}$.

Since $T_{i_{q_1},j_{q_1}}$ is the first task in Λ on a machine $> s$ there is exactly one forward-s-move before $T_{i_{q_1},j_{q_1}}$. We will now prove the theorem by showing that there is at most one forward-s-move after the execution of $T_{i_{q_{2x}},j_{q_{2x}}}$. To see this, divide the sequence $T_{i_{q_{2x}},j_{q_{2x}}}, \ldots, T_{i_z,j_z}$ into maximal nonempty subsequences

$$L_1, R_1, L_2, R_2, L_3, R_3, \ldots, L_t, R_t, L_{t+1} \tag{6}$$

such that all L_r, $1 \leq r \leq t+1$, contain only tasks on machines $\leq s$ and all R_r, $1 \leq r \leq t$, contain only tasks on machines $> s$.

By Lemma 5 (1.), we know that $i_{q_{2x}} \geq i_{k_{2x+1}}$ and all non-zero tasks $T_{k,l}$ with $k < i_{k_{2x+1}}$ and $l \leq s$ are already executed when $T_{i_{q_{2x}},j_{q_{2x}}}$ is executed. Since the s-chain Γ cannot be extended by an R_s-task there are no non-zero tasks $T_{k,l}$ with $k \geq i_{k_{2x+1}}$ and $l > s$. Thus, none of the subsequences $L_2, \ldots, L_t$ contains a predecessor of any task from $R_1, \ldots, R_t$. Therefore we can replace the task sequence (6) by the sequence

$$L_1, R_1, R_2, \ldots, R_t, L_2, L_3, \ldots, L_t, L_{t+1},$$

which yields a processing with a smaller distance than Λ if $t > 1$. Since Λ is a minimal processing it follows that $t \leq 1$. That is, there is at most one forward-s-move after $T_{i_{q_{2x}},j_{q_{2x}}}$.

The rearrangements of the subsequences as in the proof of Lemma 7 immediately lead to a simple polynomial time algorithm that computes a processing whose total distance is at most three times the total distance of a minimal processing. The algorithm starts, for example, with the trivial processing which corresponds to the arrangement $T_{i_1,j_1}, \ldots, T_{i_k,j_k}$ of all non-zero tasks such that for $l = 1, \ldots, k-1$, $i_l \leq i_{l+1}$ and if $i_l = i_{l+1}$ then $j_l < j_{l+1}$. Then we improve the processing step by step as follows. For each s, we divide the current processing into maximal nonempty subsequences $L_1, R_1, L_2, R_2, \ldots, L_t, R_t, L_{t+1}$ such that all L_i contain only tasks on machines $\leq s$ and all R_i contain only tasks on machines $> s$. Now we exchange a pair (R_i, L_{i+i}) or a pair (L_i, R_{i+1}) if this does not violate the successor relation. Such a step decreases the number of forward s-moves, but does not increase the number of forward s'-moves for $s' \neq s$. From the proofs of lemma 7, lemma 8, and theorem 9 it follows that for each $s \in \{1, \ldots, m-1\}$ the number of forward-s-moves of the resulting processing is at most three times the number of forward-s-moves of a minimal processing. Thus we get the following corollary.

Corollary 10. *There is a polynomial time algorithm that computes for a given task matrix R a processing with a total distance of at most three times the total distance of a minimal processing of R.*

4   Conclusion

It is not possible to prove an upper bound better than 3 by separately comparing the number of forward-s-moves of a minimal processing with our lower bound on forward-s-moves for every s. This can be shown by a simple example for 6 machines and some arbitrary large k such that there is a minimal processing with $3k$ forward-3-moves, but any 3-chain has at most $k+1$ R_3-tasks. This example shows that the number of forward-s-moves of a minimal processing can be arbitrary close to three times the lower bound for some s. However, we do not know whether this can happen simultaneously for all s. On the other hand, there are examples showing that the upper bound on the ratio of a minimal processing and such a lower bound can not be below $\frac{5}{3}$.

References

1. J. Błażewicz, K.H. Ecker, E. Pesch, G. Schmidt, and J. Węglarz. *Scheduling Computer and Manufacturing Processes.* Springer, Berlin, Heidelberg, New York, 1996.
2. W. Espelage and E. Wanke. Movement Minimization in Conveyor Flow Shop Processing with one Worker. *Technical report*, University of Duesseldorf, Germany, 2000.
3. N.G. Hall and C. Sriskandarajah. S survey of machine scheduling problems with blocking and no-wait in process. *Operations Research*, 44(3):510–525, 1996.
4. R. de Koster. Performance approximation of pick-to-belt orderpicking systems. *European Journal of Operational Research*, 92:558–573, 1994.
5. T. Gonzalez. Unit execution time shop problems. *Mathematics of Operations Research*, 7:57–66, 1982.
6. E. Koutsoupias and C. Papadimitriou. On the k server conjecture. *Jornal of the ACM*, 42:971–983, 1995.
7. C.L. Monma and A.H.G. Rinnooy Kan. A concise survey of efficiently solvable special cases of the permutation flow-shop problem. *RAIRO Rech. Opér*, 17:105–119, 1983.
8. J. Rethmann and E. Wanke. Storage controlled pile-up systems, theoretical foundations. *European Journal of Operational Research*, 103(3):515–530, 1997.

Quantifier Rank for Parity of Embedded Finite Models

Hervé Fournier

Laboratoire de l'Informatique du Parallélisme
Ecole Normale Supérieure de Lyon
46, allée d'Italie, 69364 Lyon Cedex 07, France
Herve.Fournier@ens-lyon.fr

Abstract. We are interested in the quantifier rank necessary to express the parity of an embedded set of cardinal smaller than a given bound. We consider several embedding structures like the reals with addition and order, or the field of the complex numbers. We provide both lower and upper bounds. We obtain from these results some bounds on the quantifier rank needed to express the connectivity of an embedded graph, when a bound on its number of vertices is given.

Keywords: Parity, Connectivity, Reachability, Ehrenfeucht-Fraïssé games, Uniform Quantifier Elimination, Constraint databases.

1 Introduction

There are numerous works about the expressiveness obtained by embedding a finite structure into an infinite one M. These studies have been carried out because of their fundamental role in the constraint database model. Among these results, the generic collapse results are of great importance. They state that embedding a finite model into some infinite structures does not help to express a large class of queries, called generic. These results hold for structures M having some good model-theoretic properties: the stronger result deals with structures without the independence property [1]. One of these generic queries is parity, which asks if the cardinal of a finite set $\mathcal{I}$ is even. As a special case of some general collapse theorems, we obtain, for some structures M, that there is no first-order sentence defining parity [3,1,4,5] – for more references see the book [12]. However, when restricting to the case where $|\mathcal{I}|$ is smaller than a given bound, such a formula exists and one can wonder which is the minimal quantifier rank possible. Can we do better than in the case where the finite set stands alone?

Main results are summarized in figure 1. The first column presents the quantifier rank needed for a formula to express that a given set of cardinal bounded by n is even, when this set is embedded into a structure to be read on the left. The last column gives some results concerning connectivity of an embedded graph, where the bound n is on the number of vertices. The first two lines are well-known bounds from finite model theory. Comparing them

J. Sgall, A. Pultr, and P. Kolman (Eds.): MFCS 2001, LNCS 2136, pp. 375–386, 2001.

to the next lines allows oneself to understand how addition and product can (or not) improve these bounds. For example, concerning parity, adding the addition to $(\mathbb{C}, =)$ make the quantifier rank decrease exponentially (from n to $\lceil \log n \rceil + 1$), but adding further the product allows no gain at all. On the reals with order, the addition allows to decrease the quantifier rank for parity from $\log n + \Theta(1)$ to $\Theta(\sqrt{\log n})$. It would be interesting to have precise bounds when the embedding structure is a real closed field, for example $(\mathbb{R}, +, -, \times, \leqslant)$ – see question 1 about this. At last, it was pointed out in [9] that parity and connectivity are expressible over $(\mathbb{Q}, +, -, \times, =)$: this comes from the definability of the integers over this structure [14,7], allowing the power of arithmetic.

	even	connected
$(\mathbb{Q}, =), (\mathbb{R}, =), (\mathbb{C}, =)$	n	$\Theta(\log n)$
$(\mathbb{Q}, <), (\mathbb{R}, <)$	$\log n + \Theta(1)$	$\Theta(\log n)$
$(\mathbb{Q}, +, -, =), (\mathbb{R}, +, -, =), (\mathbb{C}, +, -, =)$	$\lceil \log n \rceil + 1$	$\Theta(\sqrt{\log n})$
$(\mathbb{Q}, +, -, \times, =), (\mathbb{C}, +, -, \times, =)$	$\lceil \log n \rceil + 1$	$\Theta(\sqrt{\log n})$
$(\mathbb{Q}, +, -, <), (\mathbb{R}, +, -, <)$	$\Theta(\sqrt{\log n})$	$\Theta(\sqrt{\log n})$
$(\mathbb{R}, +, -, \times, <)$	$\Omega(\log \log n),$ $O(\sqrt{\log n})$	$\Omega(\log \log n),$ $O(\sqrt{\log n})$
$(\mathbb{Q}, +, -, \times, =), (\mathbb{Q}, +, -, \times, <)$	$\Theta(1)$	$\Theta(1)$

Fig. 1. Quantifier rank for parity and connectivity

Organization. The question we deal with is formally defined in section 2, and first remarks are made there. In section 3, we show some upper and lower bounds for parity in zero characteristic algebraically closed fields and $\mathbb{Q}$-vector spaces (for instance, the reals with addition and equality). Section 4 deals with ordered $\mathbb{Q}$-vector spaces (for instance, the reals with addition and order). We derive from these results some bounds on connectivity of embedded graphs in section 5. In the last section, we relate these bounds with the notion of active-natural collapse.

2   Notations and First Remarks

We are interested in the following problem. We embed a finite set $\mathcal{I}$ in either an algebraically closed field or an ordered $\mathbb{Q}$-vector space: we shall call M this structure. Thus, besides the signature of M, we consider a new predicate I that is interpreted as $\mathcal{I}$. We shall be interested in the query Even, asking if $|\mathcal{I}|$ is even, and Card_m which asks if $|\mathcal{I}| \geqslant m$. For a query Q, we note $\mathrm{QR}_M(Q, n)$ the smallest possible quantifier rank of a first-order formula expressing the query Q when it is known that $|\mathcal{I}| \leqslant n$. Our aim is to find some bounds on $\mathrm{QR}_M(\mathrm{Even}, n)$.

We recall that the quantifier rank $qr(\phi)$ of a formula ϕ is defined by induction on its structure. If ϕ is an atomic formula, $qr(\phi) = 0$. Otherwise $qr(\phi \vee \psi) = qr(\phi \wedge \psi) = \max(qr(\phi), qr(\psi))$ and $qr(\exists x \phi) = qr(\forall x \phi) = 1 + qr(\phi)$. Now let us make the following remark.

Lemma 1 *If two structures M and M' are elementarily equivalent, then for all n_0, we have $QR_M(Even, n_0) = QR_{M'}(Even, n_0)$.*

Proof. Indeed, let n_0 be fixed and suppose we have a first-order formula ϕ such that if $|\mathcal{I}| \leq n_0$, $(M, \mathcal{I}) \models \phi$ if and only if $|\mathcal{I}|$ is even. Let M' be a structure elementarily equivalent to M, and $n \leq n_0$. Let $\tilde{\phi}(x_1, \ldots, x_n)$ be the formula ϕ where $I(x)$ is replaced with $\bigvee_{i=1}^{n} x = x_i$, and $\psi_n = \forall x_1, \ldots, x_n \bigwedge_{i<j} x_i \neq x_j \rightarrow \tilde{\phi}(x_1, \ldots, x_n)$. If n is even, $M' \models \psi_n$ since $M \models \psi_n$; and if n is odd, $M' \models \neg\psi_n$. Thus if $\mathcal{I}' \subset M'$ with $|\mathcal{I}'| \leq n_0$, $(M', \mathcal{I}') \models \phi$ if $|\mathcal{I}'|$ is even and $(M', \mathcal{I}') \models \neg\phi$ if $|\mathcal{I}'|$ is odd. Hence $QR_{M'}(Even, n_0) \leq QR_M(Even, n_0)$ and by symmetry $QR_{M'}(Even, n_0) = QR_M(Even, n_0)$. $\square$

Of course the previous remark also applies to the queries $Card_m$. This justifies the notation $QR_T(Even, n)$ and $QR_T(Card_m, n)$ for a complete theory T. Let us introduce some notations for the theories we shall be interested in, and give some examples of models of these theories.

- zero characteristic algebraically closed field: ACF_0
  ex: $(\overline{\mathbb{Q}}, +, -, \times, =)$, $(\mathbb{C}, +, -, \times, =)$.
- $\mathbb{Q}$-vector space: Qvs
  ex: $(\mathbb{Q}, +, -, =)$, $(\mathbb{R}, +, -, =)$, $(\mathbb{C}, +, -, =)$.
- ordered $\mathbb{Q}$-vector space: Ovs
  ex: $(\mathbb{Q}, +, -, <)$, $(\mathbb{R}, +, -, <)$.

Our main tool will be the back-and-forth games defined by Ehrenfeucht and Fraïssé [6,11,10]. Consider two $\mathcal{L}$-structures M and N. A game of length n between the two structures M and N proceeds as follows. At the i-th step, the first player chooses a point either in M or N; then the second player must choose an element in the other structure. Let us call a_i the point chosen in M and b_i the point chosen in N. After n moves, the game ends, and the second player wins iff the same atomic formulas are true in the structures $(M, a_1, \ldots, a_n)$ and $(N, b_1, \ldots, b_n)$. The second player is said to have a strategy to win the game of length n between M and N if he can win no matter what the first player plays. We shall use the following fundamental property of Ehrenfeucht-Fraïssé games.

Fact 1 *If the second player has a strategy to win back-and-forth games of length n between two structures M and N, then the same formulas of quantifier rank at most n are true in M and N.*

We shall use this fact to establish lower bounds. For example, if we want to show that $QR_T(Even, n) > B$ for a given complete theory T of signature

$\mathcal{L}$, we may proceed as follows. We choose two models M and N of T, and two finite sets $\mathcal{I} \subset M$ and $\mathcal{J} \subset N$ of cardinal at most n, with $|\mathcal{I}|$ odd and $|\mathcal{J}|$ even. Now if we show that the second player has a strategy to win the game of length B between the two $\mathcal{L} \cup \{I\}$-structures $(M, \mathcal{I})$ and $(N, \mathcal{J})$, then of course no first-order formula over $\mathcal{L} \cup \{I\}$ of quantifier rank at most B can express the restricted parity we are looking for.

Now let us examine some bounds on the quantifier rank when the finite structure stands alone. When no order is available, we shall write $\mathrm{QR}_=(\mathrm{Even}, n)$ the minimal quantifier rank of a first-order formula expressing parity. In the same way, we write it $\mathrm{QR}_<(\mathrm{Even}, n)$ when the universe is totally ordered. By some usual back-and-forth games [6,11], we have $\mathrm{QR}_=(\mathrm{Even}, n) = n$ and $\mathrm{QR}_<(\mathrm{Even}, n) = \log n + O(1)$.

3   Parity on Unordered Structures

In this section, we show results concerning parity in 0-characteristic algebraically closed fields and $\mathbb{Q}$-vector spaces. Because a 0-characteristic algebraically closed field is a $\mathbb{Q}$-vector space, $\mathrm{QR}_{ACF_0}(\mathrm{Even}, n) \leq \mathrm{QR}_{\mathbb{Q}vs}(\mathrm{Even}, n)$. Thus this section is divided in two parts: on one hand we show a lower bound on $\mathrm{QR}_{ACF_0}(\mathrm{Even}, n)$, on the other hand we show an upper bound on $\mathrm{QR}_{\mathbb{Q}vs}(\mathrm{Even}, n)$.

3.1   Lower Bound in an Algebraically Closed Field

We shall note $\overline{A}$ the algebraic closure of $A \subseteq \mathbb{C}$. In this section, we shall prove the following lower bound.

Theorem 1 $\mathrm{QR}_{ACF_0}(\mathrm{Even}, n) \geq \lceil \log n \rceil + 1$.

Proof. Thanks to Lemma 1, it is enough to show this lower bound in a given algebraically closed field of characteristic 0. We shall work in the field of the complex numbers $\mathbb{C}$. Let $\mathcal{M}$ be a set of 2^{n-1} algebraically independent elements of $\mathbb{C}$, and $\mathcal{N}$ a set of $2^{n-1} + 1$ algebraically independent elements of $\mathbb{C}$. We are going to prove that the second player can win the back-and-forth game of length n between $(\mathbb{C}, \mathcal{M})$ and $(\mathbb{C}, \mathcal{N})$.

At the beginning, let $E_n = F_n = \overline{\mathbb{Q}}$ and $\varphi = \mathrm{Id}_{\overline{\mathbb{Q}}}$. In the following, φ is a partial application from the first structure $(\mathbb{C}, \mathcal{M})$ to the second one $(\mathbb{C}, \mathcal{N})$, its domain (resp. image) containing the points chosen by the two players in the first (resp. second) structure. At each step we shall extend φ such that its domain and image contain the points newly chosen. When it remains j steps to do, we note E_j the field where φ is defined and $F_j = \varphi(E_j)$. At each step, φ is an isomorphism of algebraically closed field "with points" from E_j onto F_j: this means that for all x in E_j, $x \in \mathcal{M}$ iff $\varphi(x) \in \mathcal{N}$. We also maintain the following property $\mathcal{P}_j$.

First $|\mathcal{M} \setminus E_j|, |\mathcal{N} \setminus F_j| \geqslant 2^{j-1}$. *Moreover, if there exists* $a \in \mathcal{M} \setminus E_j$ *and* $A \subset \mathcal{M} \setminus (E_j \cup \{a\})$ *such that* $a \in \overline{E_j \cup A}$, *then* $|A| \geqslant 2^{j-1}$. *And the corresponding property in* $(\mathbb{C}, \mathcal{N})$: *if there exists* $a \in \mathcal{N} \setminus F_j$ *and* $A \subset \mathcal{N} \setminus (F_j \cup \{a\})$ *such that* $a \in \overline{F_j \cup A}$, *then* $|A| \geqslant 2^{j-1}$.

First let us check that $\mathcal{P}_n$ is verified. We have $|\mathcal{M} \setminus E_n| = |\mathcal{M}| \geqslant 2^{n-1}$. Moreover, there is no $a \in \mathcal{M}$ with $a \in \overline{\mathbb{Q} \cup A}$ such that $A \subseteq \mathcal{M}$ and $a \notin A$ because elements of $\mathcal{M}$ are algebraically independent over $\mathbb{Q}$. And the same in $(\mathbb{C}, \mathcal{N})$.

Let us suppose that $n - j - 1$ steps have been done. The isomorphism φ is defined on E_{j+1} and it remains $j + 1$ steps to do. Property $\mathcal{P}_{j+1}$ is verified by induction hypothesis. By symmetry, we can assume that the point is chosen in $(\mathbb{C}, \mathcal{M})$. Let us note v this point. We can also assume $v \notin E_{j+1}$. There are two cases. First case: $v \in \overline{E_{j+1} \cup \{a_1, \ldots, a_r\}}$ with $a_i \in \mathcal{M} \setminus E_{j+1}$ distinct and $r \leqslant 2^{j-1}$. Then we choose some distinct elements $b_1, \ldots, b_r$ in $\mathcal{N} \setminus F_{j+1}$ and we define $\varphi(a_i) = b_i$. Thus $E_j = \overline{E_{j+1} \cup \{a_1, \ldots, a_r\}}$. Let us show that $\mathcal{P}_j$ is verified. If there exists $d \in \mathcal{M} \setminus E_j$, with $d \in \overline{E_j \cup \{c_1, \ldots, c_l\}}$, $c_i \in \mathcal{M} \setminus (E_j \cup \{d\})$ and $l \leqslant 2^{j-1} - 1$, then $d \in \overline{E_{j+1} \cup \{a_1, \ldots, a_r, c_1, \ldots, c_l\}}$. But $r + l \leqslant 2^{j-1} + 2^{j-1} - 1 = 2^j - 1$. Therefore we should have $d \in E_{j+1}$ by property $\mathcal{P}_{j+1}$, this is absurd. We have the same property in $(\mathbb{C}, \mathcal{N})$. Moreover, $|\mathcal{M} \setminus E_j|, |\mathcal{N} \setminus F_j| \geqslant 2^j - 2^{j-1} = 2^{j-1}$ so $\mathcal{P}_j$ is verified. Exactly in the same way, we show that there are no other points from $\mathcal{M} \setminus E_{j+1}$ in E_j besides the a_i: if $d \in (\mathcal{M} \cap E_j) \setminus (E_{j+1} \cup \{a_1, \ldots, a_r\})$, then $d \in \overline{E_{j+1} \cup \{a_1, \ldots, a_r\}}$ and we conclude with $\mathcal{P}_{j+1}$. This also holds in $(\mathbb{C}, \mathcal{N})$, and it shows that φ is an isomorphism. Second case: let $f \notin \overline{F_{j+1} \cup \mathcal{N}}$. Let $\varphi(v) = f$. We set $E_j = \overline{E_{j+1} \cup \{v\}}$. Let us show that $\mathcal{P}_j$ is verified. Let $a \in \mathcal{M} \setminus E_j$ such that $a \in \overline{E_j \cup A}$ for $A \subseteq \mathcal{M} \setminus (E_j \cup \{a\})$ with $|A| < 2^{j-1}$. Thus $a \in \overline{E_{j+1} \cup \{v\} \cup A}$. This shows $v \in \overline{E_{j+1} \cup \{a\} \cup A}$, because $a \in \overline{E_{j+1} \cup A}$ is impossible by $\mathcal{P}_{j+1}$. But we should be in the first case since $|A \cup \{a\}| \leqslant 2^{j-1}$. This also holds in $(\mathbb{C}, \mathcal{N})$ by the choice of f. Moreover, there is no point of $\mathcal{M}$ in $E_j \setminus E_{j+1}$ because if $a \in \mathcal{M} \cap E_j \setminus E_{j+1}$, then $a \in \overline{E_{j+1} \cup \{v\}}$ and as $a \notin E_{j+1}$ we would have $v \in \overline{E_{j+1} \cup \{a\}}$ which is absurd. This also holds in $(L, \mathcal{N})$ thanks to the choice of f, thus φ remains an isomorphism. Moreover, $|\mathcal{M} \setminus E_j| = |\mathcal{M} \setminus E_{j+1}| \geqslant 2^{j-1}$ which ends to show $\mathcal{P}_j$. This ends the back-and-forth game. Thus we have shown $\mathrm{QR}_{ACF_0}(\text{Even}, 2^{n-1} + 1) > n$. As $\mathrm{QR}_{ACF_0}(\text{Even}, \cdot)$ is an increasing function, we obtain $\mathrm{QR}_{ACF_0}(\text{Even}, n) \geqslant \lceil \log n \rceil + 1$. $\square$

3.2   Upper Bound in a $\mathbb{Q}$-Vector Space

The proof will proceed in three steps.

- First we show that it is possible to express $|\mathcal{I}| \geqslant m$ with a formula of quantifier rank $\lceil \log m \rceil + 2$ in the special case where the elements of $\mathcal{I}$ are known to be linearly independent over $\mathbb{Q}$.
- Then we generalize this bound to the general case.

– At last, we show how to decrease the quantifier rank of these formulas by 1.

We need to build a family of formulas $S_{(\alpha_1,...,\alpha_p)}(x)$ for $p \geqslant 1$ and $(\alpha_1, \ldots, \alpha_p) \in \mathbb{N}^p$. These formulas should satisfy the following:

$$(M,\mathcal{I}) \models S_{(\alpha_1,...,\alpha_p)}(x) \iff \exists x_1, \ldots, x_p \in \mathcal{I} \; x = \sum_{i=1}^{p} \alpha_i x_i \, .$$

Moreover, we shall design such formulas with small quantifier rank. We define them as follows: we take $S_{(1)}(x) := I(x)$, and $S_{(\alpha_1)}(x) := \exists y I(y) \wedge x = y + \ldots + y$ for $\alpha_1 \neq 1$. At last, we define

$$S_{(\alpha_1,...,\alpha_p)}(x) := \exists y \; S_{(\alpha_1,...,\alpha_{\lfloor p/2 \rfloor})}(y) \wedge S_{(\alpha_{\lceil p/2 \rceil},...,\alpha_p)}(x - y) \, .$$

One can check that the quantifier rank of $S_{(\alpha_1,...,\alpha_p)}(x)$ is bounded above by $\lceil \log p \rceil + 1$.

Proposition 1 *In a $\mathbb{Q}$-vector space, if we restrict ourselves to the case where the elements of $\mathcal{I}$ are linearly independent over $\mathbb{Q}$, we can express that $|\mathcal{I}| \geqslant m$ with a formula of quantifier rank $\lceil \log m \rceil + 2$.*

Proof. Let $\bar{\alpha}_m$ be $(1, 1, \ldots, 1) \in \mathbb{N}^m$ and $\bar{\beta}_m$ be $(2, 1, 1, \ldots, 1) \in \mathbb{N}^{m-1}$. Let us define $F_m = \exists x \; S_{\bar{\alpha}_m}(x) \wedge \neg S_{\bar{\beta}_m}(x)$. Note that $\mathrm{qr}(F_m) \leqslant \lceil \log m \rceil + 2$. We claim that F_m expresses $|\mathcal{I}| \geqslant m$. Indeed if F_m is true, that means that there exists x which is a sum of m *different* elements of $\mathcal{I}$: these elements must be different because the second part of F_m ensures that x is not a linear combination of $m-1$ elements of $\mathcal{I}$ with coefficients $(2, 1, \ldots, 1)$. Conversely, if $|\mathcal{I}| \geqslant m$, take s to be the sum of m different elements of $\mathcal{I}$. The formula $S_{\bar{\alpha}_m}(s) \wedge \neg S_{\bar{\beta}_m}(s)$ will be true because the elements of $\mathcal{I}$ are linearly in-Dependant, thus F_m will be true. $\square$

Remark 1 *Both theorem 1 and proposition 1 hold in positive characteristic.*

Proposition 2 $\mathrm{QR}_{\mathbb{Q}vs}(\mathrm{Card}_m, n) \leqslant \lceil \log m \rceil + 2.$

Proof. We shall work in $\mathbb{Q}$, and assume $|\mathcal{I}| \leqslant n$. Let us notice that if the formula described in the previous proof is true, then $|\mathcal{I}| \geqslant m$. And if it is false, then we don't know – because we do not have the hypothesis of linear independence anymore. To get rid of this hypothesis, the trick is to weigh the sum in the previous proof by some integer coefficients.

Let us notice that $S_{\bar{\alpha}}(x)$ is equivalent to $S_{\bar{\alpha}'}(x)$ where $\bar{\alpha}'$ is obtained from $\bar{\alpha}$ by permuting the elements. That is why we consider in the following only non-decreasing tuples. For a tuple $\bar{\alpha} = (\alpha_1, \ldots, \alpha_p) \in \mathbb{N}^p$, let us define $s(\bar{\alpha})$ to be the set of increasing tuples of $\mathbb{N}^{p-1}$ obtained by replacing in $\bar{\alpha}$ any two elements α_i and α_j by their sum. For example, $s((1, 4, 7)) = \{(5, 7), (4, 8), (1, 11)\}$ and $s((1, 2, 2, 3)) = \{(2, 3, 3), (2, 2, 4), (1, 3, 4), (1, 2, 5)\}$. Let us define the following formula

$$J_{\bar{\alpha}} = \exists x S_{\bar{\alpha}}(x) \wedge \bigwedge_{\bar{\beta} \in s(\bar{\alpha})} \neg S_{\bar{\beta}}(x) \, .$$

For the reason mentioned above, if $J_{(\alpha_1,\ldots,\alpha_p)}$ is true then $|\mathcal{I}| \geqslant p$.

We are ready to build a formula expressing $|\mathcal{I}| \geqslant m$, under the hypothesis $|\mathcal{I}| \leqslant n$. Let $N = m!n^{m-1}$. Let $\mathcal{A}$ be a set of $Nm + 1$ elements of $\mathbb{N}^m$ in general position: by this we mean that no $m+1$ of these elements lie on a same hyperplane of $\mathbb{R}^m$. We claim that $H_m = \bigvee_{\bar{\alpha} \in \mathcal{A}} J_{\bar{\alpha}}$ is true if and only if $|\mathcal{I}| \geqslant m$.

Proof of the claim: if H_m is true, then for a $\bar{\alpha} \in \mathcal{A}$ the formula $J_{\bar{\alpha}}$ is true and this implies that $|\mathcal{I}| \geqslant m$. For the converse, let us suppose $|\mathcal{I}| \geqslant m$. Then $\mathcal{I} = \{x_1, \ldots, x_l\}$, where the x_i are all different, and $m \leqslant l \leqslant n$. Let us consider all the equations

$$\sum_{i=1}^{m} A_i x_i = \sum_{i=1}^{m-2} A_{\sigma(i)} x_{t(i)} + (A_{\sigma(m-1)} + A_{\sigma(m)}) x_{t(m-1)}$$

where σ runs over all the permutations of $\{1, \ldots, m\}$ and t runs over all the applications from $1, \ldots, m-1$ to $\{1, \ldots, l\}$. These equations define a family $\mathcal{H}_{\bar{x}}$ of hyperplanes of $\mathbb{R}^m$ in $A_1, \ldots, A_m$ parameterized by $(x_1, \ldots, x_l)$. First these are all *true* hyperplanes. Let us also remark that $|\mathcal{H}_{\bar{x}}| \leqslant N$. On each hyperplane of $\mathcal{H}_{\bar{x}}$ there are at most m elements of $\mathcal{A}$ since they are in general position. As $|\mathcal{A}| > |\mathcal{H}_{\bar{x}}| m$, there must be at least one $\bar{\alpha} \in \mathcal{A}$ which is not on any hyperplane of $\mathcal{H}_{\bar{x}}$. For such a $\bar{\alpha}$ the formula $J_{\bar{\alpha}}$ is true (take $x = \sum_{i=1}^{m} \alpha_i x_i$ for the first existential quantifier). Thus H_m is true, and the claim is proved. $\square$

Theorem 2 $QR_{Qvs}(Card_m, n) \leqslant \lceil \log m \rceil + 1$.

Proof. Once again we assume $|\mathcal{I}| \leqslant n$. We just have to use one quantifier less than in the previous proposition. We can extend the definition of $S_{(\alpha_1,\ldots,\alpha_m)}(x)$ to the case where the α_i are rationals. We define $S_{(1/q)}(x) := I(qx)$, where qx is $x + \ldots + x$ (q times). We also define $S_{(p/q)}(x) := \exists y\, I(qy) \wedge x = py$ when $p \neq 1$. As previously, we take $S_{(\alpha_1,\ldots,\alpha_m)}(x) := \exists y\, S_{(\alpha_1,\ldots,\alpha_{\lfloor m/2 \rfloor})}(y) \wedge S_{(\alpha_{\lceil m/2 \rceil},\ldots,\alpha_m)}$ (x-y). Now, we take $\mathcal{A}$ in the previous proof to be a set of $Nm + 1$ elements of $(1/(\mathbb{N} \setminus \{0\}))^m$. For $\bar{\alpha} = (\alpha_1, \ldots, \alpha_m) \in \mathcal{A}$, what is the quantifier rank of $J_{\bar{\alpha}}$? The quantifier rank of $S_{\bar{\alpha}}(x)$ is $\lceil \log m \rceil$. Moreover, we claim that for each $\bar{\beta}$, we can permute the β_i such that $qr(S_{\bar{\beta}}(x)) = \lceil \log m \rceil$. Notice that, in a $\bar{\beta}$, all coefficients have numerator 1, except maybe the one which is of the form $\alpha_i + \alpha_j$. Two cases happen. If $m - 1$ is a power of 2, then $qr(J_{\bar{\beta}}(x)) = \lceil \log(m-1) \rceil + 1$, which equals to $\lceil \log m \rceil$. If $m - 1$ is not a power of 2, then we can permute the β_i such that $qr(J_{\bar{\alpha}}(x)) = \lceil \log(m-1) \rceil$: all we have to do is to put the coefficient p/q with $p \neq 1$ in a part of the tree that does not reach the bottom. In all cases, $qr(S_{(\alpha_1,\ldots,\alpha_m)}) = \lceil \log m \rceil + 1$. $\square$

Corollary 1 $QR_{Qvs}(Even, n) = QR_{ACF_0}(Even, n) = \lceil \log n \rceil + 1$.

Proof. Let n be fixed. By theorem 2, for any $m \leqslant n$, there is a formula F_m expressing $|\mathcal{I}| \geqslant m$, with $qr(F_m) = \lceil \log m \rceil + 1 \leqslant \lceil \log n \rceil + 1$. Of course $F_m \wedge \neg F_{m+1}$ expresses that $|\mathcal{I}| = m$. Now if we know that $|\mathcal{I}| \leqslant n$, $|\mathcal{I}|$ is even if and only if $\bigvee_{2k \leqslant n} |\mathcal{I}| = 2k$. Remark that $|\mathcal{I}| \geqslant n$ is equivalent to $|\mathcal{I}| = n$ since we know that $|\mathcal{I}| \leqslant n$. Thus our formula expressing parity will be $\bigvee_{2k \leqslant n} F_{2k} \wedge$

$\neg F_{2k+1}$ when n is odd, and $\bigvee_{2k<n}(F_{2k} \wedge \neg F_{2k+1}) \vee F_n$ when n is even. That allows to obtain the desired upper bound. The lower bound was established in theorem 1. $\square$

4 Parity on Ordered Structures

We recall that $\mathcal{O}vs$ is the notation used for the theory of $\mathbb{Q}$-ordered vector spaces. Using back-and-forth games again, we can establish the following lower bound (proofs of this section have been omitted).

Theorem 3 $QR_{\mathcal{O}vs}(\text{Even}, n) = \Omega(\sqrt{\log n})$.

As in the previous section, we can build some formulas whose quantifier rank match (up to a constant factor) the one obtained above.

Corollary 2 $QR_{\mathcal{O}vs}(\text{Even}, n) = \Theta(\sqrt{\log n})$.

In fact, we have the following stronger statement.

Proposition 3 *In an ordered $\mathbb{Q}$-vector space, we can express that $|\mathcal{I}| \geqslant m$ with a formula of quantifier rank $O(\sqrt{\log m})$.*

Remark 2 *Can we also remove the hypothesis about the bound on $|\mathcal{I}|$ on Proposition 2?*

5 Connectivity of Embedded Graphs

In this section we consider a finite graph G embedded in an infinite structure M. Thus we shall add two predicates to the signature of M: a unary predicate V which interprets the vertices $\mathcal{V}$ of the embedded graph, and a binary one E for the edges. We shall use $d(\cdot, \cdot)$ for the distance in the graph.

We are interested in connectivity and reachability. The query Connected asks if the graph G is connected. The query Reach_m will have two free variables a and b and will be true if $a, b \in \mathcal{V}$ and $d(a,b) \leqslant m$. The query Reach is defined in the same way, except there is no limit on the length of the path anymore. Once again we shall consider restriction of these queries to the case where we have a bound on $|\mathcal{V}|$. Thus $QR_M(\text{Connected}, n)$ is the smallest quantifier rank possible for a formula expressing that a graph embedded in M is connected, assuming that is has at most n vertices. Of course the result of Lemma 1 holds for these queries too: if two structures M and M' are elementarily equivalent, then $QR_M(\text{Connected}, n) = QR_{M'}(\text{Connected}, n)$ and $QR_M(\text{Reach}_m, n) = QR_{M'}(\text{Reach}_m, n)$. Let us notice that $QR_M(\text{Reach}, n) = QR_M(\text{Reach}_{n-1}, n)$. Another remark is that $\forall a, b\, V(a) \wedge V(b) \rightarrow \text{Reach}(a, b)$ expresses the connectivity, so $QR_M(\text{Connected}, n) \leqslant QR_M(\text{Reach}, n)+2$. We can obtain a result similar to theorem 3 for a finite graph embedded in an ordered $\mathbb{Q}$-vector space.

Corollary 3 $QR_{Ovs}(\text{Connected}, n) = \Omega(\sqrt{\log n})$.

Proof. We use a usual first-order reduction from parity to connectivity. Let $\mathcal{I}$ be a set composed of the elements $v_1 < v_2 < \ldots < v_n$. We consider the graph $G_n = (V, E)$ over $V = \{v_1, \ldots, v_n\}$ where $E(v_i, v_j)$ holds iff $|i - j| = 2$ or $\{i, j\} = \{1, n\}$. For $n \geqslant 2$, G_n is connected iff $n = |\mathcal{I}|$ is even. As we can express E with a formula of quantifier rank 2 (with the help of predicate I interpreting $\mathcal{I}$) in any ordered structure M, we obtain $QR_M(\text{Even}, n) \leqslant QR_M(\text{Connected}, n) + 2$. It remains to apply this to the theory Ovs. $\square$

Using techniques similar to the previous ones, we can establish a lower bound for algebraically closed fields (remaining proofs of this section have been omitted).

Proposition 4 $QR_{ACF_0}(\text{Connected}, n) = \Omega(\sqrt{\log n})$.

We can establish an upper bound for reachability in $\mathbb{Q}$-vector spaces – which provides the same bound for theories Ovs and ACF_0.

Proposition 5 $QR_{Qvs}(\text{Reach}_m, n) = O(\sqrt{\log m})$.

Corollary 4 *For* $T \in \{Qvs, ACF_0, Ovs\}$,
$$\begin{aligned} & QR_T(\text{Reach}_m, n) = \Theta(\sqrt{\log m}) \\ & QR_T(\text{Reach}, n) = \Theta(\sqrt{\log n}) \\ & QR_T(\text{Connected}, n) = \Theta(\sqrt{\log n}) \end{aligned}$$

6   Relationship with Active-Natural Collapse

We consider a relational database of signature $S = \{R_1, R_2, \ldots, R_f\}$ with R_i of arity r_i, embedded in an infinite L-structure M. An interpretation of the symbols of S is called a database. We shall deal here with finite databases only: each R_i interprets a finite set of M^{r_i}. Given a finite database D, we shall note A the active domain of D, that is to say the set of the coordinates of all points in the database. We recall that in an active semantics formula, quantifiers are of the type $\exists x \in A$ and $\forall x \in A$; quantifiers ranging over the whole universe are not allowed. In the following, we shall write $\exists^a$ and $\forall^a$ when working with active quantifiers.

Definition 1 *An L-structure M is said to have the active-natural collapse if, for any relational signature S and any first-order formula ϕ over $S \cup L$, there exists a first-order active formula ψ over $S \cup L$ such that for any finite database D, we have $(M, D) \models \phi \leftrightarrow \psi$.*

On a structure M with active-natural collapse, there is a relationship between how the quantifier rank grows when transforming a natural semantics formula into an equivalent one in active semantics, and the quantifier rank needed to express parity. This is detailed in the next two remarks.

Proposition 6 *Let M be an o-minimal structure that admits quantifier elimination. Then $\mathrm{QR}_M(\mathrm{Even}, n) \geqslant \log\log n + O(1)$.*

Proof. Let ψ be a formula expressing parity of $|\mathcal{I}|$ for $|\mathcal{I}| \leqslant n$. We consider a structure M', elementarily equivalent to M, that contains a sequence of indiscernibles $E = \{e_i,\ i \in \mathbb{N}\}$. The same formula ψ still works in M'. Let ψ' be the formula obtained by replacing $I(t)$ by $\forall z\ (z = t \to I(z))$ in ψ, where z is a new variable. Remark that $\mathrm{qr}(\psi') \leqslant \mathrm{qr}(\psi) + 1$. Here we can apply to ψ' the algorithm of [4] to obtain an active semantics equivalent formula ψ_{act} (as mentioned there, there is no need for ψ to be in prenex form to apply this algorithm). One can check that $\mathrm{qr}(\psi_{act}) \leqslant 2^{\mathrm{qr}(\psi') + O(1)}$. Now when we restrict ourselves to the case where $\mathcal{I} \subset E$, the formula ψ_{act} is equivalent to a pure order formula ψ_o with $\mathrm{qr}(\psi_o) = \mathrm{qr}(\psi_{act})$. Applying the bound about the pure ordered case recalled in the introduction, we obtain $\mathrm{qr}(\psi_o) \geqslant \log n + O(1)$. Thus $\mathrm{qr}(\psi) \geqslant \log\log n + O(1)$. $\square$

Question 1 *We have proved $\Omega(\log\log n) \leqslant \mathrm{QR}_{\mathbb{R}}(\mathrm{Even}, n) \leqslant O(\sqrt{\log n})$, where $\mathbb{R}$ stands for $(\mathbb{R}, +, -, \times, <)$. Is it possible to make some back-and-forth games in a real closed field to improve this lower bound?*

Proposition 7 *One cannot avoid an exponential growth of the quantifier rank when transforming natural semantics formulas of $\mathbb{Q}vs$ into equivalent ones in active semantics.*

Proof. For a first-order formula ϕ of $\mathbb{Q}vs$ with an extra unary predicate I, let $a(\phi)$ be the minimum quantifier rank of an equivalent active semantics formula. Let $\alpha(r) = \max\{a(\phi),\ \mathrm{qr}(\phi) = r\}$. We want to show that $\alpha(p) \geqslant 2^p$. Consider the formula ϕ_n expressing that $\mathcal{I}$ has at least n elements – see Theorem 2. Let ϕ_n^a be an equivalent active semantics formula. When restricting to the case where $\mathcal{I} \subset E$ with $E = \{e_i,\ i \in \mathbb{N}\}$ a set of indiscernibles, we obtain $\tilde{\phi}_n^a$ a pure equality formula expressing that $\mathcal{I}$ has at least n elements. As $\mathrm{qr}(\phi_n^a) = \mathrm{qr}(\tilde{\phi}_n^a) \geqslant n$ and $\mathrm{qr}(\phi_n) = \lceil \log n \rceil$, taking $n = 2^p$ gives the result. $\square$

We are now interested in a notion introduced by Basu: uniform quantifier elimination [2]. A uniform family of formulas $(\phi_n)_{n \in \mathbb{N}}$ over the structure M is of the form

$$\phi_n(\bar{x}, \bar{y}) = Q^1_{1 \leqslant i_1 \leqslant n} \cdots Q^m_{1 \leqslant i_m \leqslant n} \phi(\bar{x}, y_{i_1}, \ldots, y_{i_m})$$

where $Q_i \in \{\bigvee, \bigwedge\}$, and ϕ is a first order formula over M. A structure M is said to have uniform quantifier elimination if given a uniform family $(\phi_n(z, \bar{x}, \bar{y}))_{n \in \mathbb{N}}$ over M, there exists a uniform family $(\psi_n(\bar{x}, \bar{y}))_{n \in \mathbb{N}}$ over M such that for all n, $\psi_n(\bar{x}, \bar{y})$ is a quantifier-free formula equivalent to $\exists z \phi_n(z, \bar{x}, \bar{y})$. Let us make a remark.

Proposition 8 *A structure has uniform quantifier elimination if and only if it has the active natural collapse for a single unary relation.*

Proof. First of all, let us remark that in the case where there is a single unary relation I, any active formula can be written without using I: just replace $I(t(\bar{w}))$ with $\exists^a v\ v = t(\bar{w})$. That is what we are going to do in the following.

Now to each active formula we associate a uniform family of formulas, and conversely. We shall do it for prenex ones – but this is not imperative. Let $\Phi(\bar{x}, \bar{y})$ be the uniform family

$$\phi_n(\bar{x}, \bar{y}) = Q^1_{1 \leqslant i_1 \leqslant n} \cdots Q^m_{1 \leqslant i_m \leqslant n} \phi(\bar{x}, y_{i_1}, \ldots, y_{i_m})$$

where $Q_i \in \{\bigvee, \bigwedge\}$. To this family will correspond the following active formula

$$\psi(\bar{x}) := \mathcal{Q}^1 t_1 \ldots \mathcal{Q}^m t_m\ \phi(\bar{x}, t_1, \ldots, t_m)$$

with $\mathcal{Q}^i = \exists^a$ (resp. $\forall^a$) if $Q^i = \bigvee$ (resp. $\bigwedge$). Of course one can make the correspondence the other way. Moreover $\Phi(\bar{x}, \bar{y})$ and $\psi(\bar{x})$ are related in this way: $(M, \mathcal{I}) \models \psi(\bar{x})$ if and only if $M \models \phi_n(\bar{x}, \bar{y}_{\mathcal{I}})$ where $n = |\mathcal{I}|$ and $\bar{y}_{\mathcal{I}}$ lists the elements of $\mathcal{I}$. A structure has uniform quantifier elimination iff for any uniform family $\Phi(\bar{x}, \bar{y}, z)$, there exists a uniform family $\tilde{\Phi}(\bar{x}, \bar{y})$ such that $\forall \bar{x} \in M\ \forall n \in \mathbb{N}\ (\exists z \in M\ \phi_n(\bar{x}, \bar{y}, z) \leftrightarrow \tilde{\phi}_n(\bar{x}, \bar{y}))$. In the same way, a structure has the active-natural collapse for a unary relation iff for any active formula $\psi(\bar{x}, z)$, there exists an active formula $\tilde{\psi}(\bar{x})$ such that $\forall \bar{x} \in M(\exists z \in M\ \psi(\bar{x}, z) \leftrightarrow \tilde{\psi}(\bar{x}))$. Thus it is clear that these two notions are equivalent. $\square$

In [2] the question whether the theories ACF_p (p prime or zero) and DCF (the theory of zero-characteristic differentially closed fields) have uniform quantifier elimination was raised. The answer to this question is positive. It is indeed proved in [8] that the No Finite Cover Property (NFCP) is a sufficient condition for a structure to have the active-natural collapse, and both theories mentioned above have NFCP – see [13] for differentially closed fields. In particular, they have the active-natural collapse in the case where the database signature is composed of one unary relation, which is equivalent to uniform quantifier elimination by Proposition 8.

We have the following: active-natural collapse implies uniform quantifier elimination, which in turn implies quantifier elimination. Moreover, there exists a structure that eliminates quantifiers but not uniformly: the ternary random structure [4]. Is there a structure that admits uniform quantifier elimination but not the (full) active-natural collapse? We point out that no stable structure can satisfy this – this is a consequence of [8]. One can also wonder about all intermediate questions: if $\bar{\alpha} = (\alpha_1, \ldots, \alpha_p)$ is smaller than $\bar{\beta} = (\beta_1, \ldots, \beta_q)$ for the lexicographic ordering, then a structure having the active-natural collapse for database signature $\bar{\alpha}$ also has it for database signature $\bar{\beta}$. Does the converse hold? If not, there may be an infinite hierarchy between quantifier elimination and full active-natural collapse.

Acknowledgments

I would like to thank Luc Ségoufin and Pascal Koiran for useful comments and careful reading of the paper.

References

1. J. T. Baldwin and M. Benedikt. Stability theory, permutations of indiscernibles, and embedded finite models. *Trans. Amer. Math. Soc.*, 352(11):4937–4969, 2000. 375
2. S. Basu. New results on quantifier elimination over real closed fields and applications to constraint databases. *Journal of the ACM*, 46(4):537–555, july 1999. 384, 385
3. M. Benedikt, G. Dong, L. Libkin, and L. Wong. Relational expressive power of constraint query languages. *Journal of the ACM*, pages 1–34, 1998. 375
4. M. Benedikt and L. Libkin. Relational queries over interpreted structures. *Journal of the ACM*, 47(4):644–680, 2000. 375, 384, 385
5. O. Chapuis and P. Koiran. Definability of geometric properties in algebraically closed fields. *Mathematical Logic Quarterly*, 45(4):533–550, 1999. 375
6. H.-D. Ebbinghaus and J. Flum. *Finite Model Theory*. Springer-Verlag, 1995. 377, 378
7. D. Flath and S. Wagon. How to pick out the integers in the rationals: An application of number theory to logic. *American Mathematical Monthly*, 98:812–823, 1991. 376
8. J. Flum and Ziegler M. Pseudo-finite homogeneity and saturation. *Journal of Symbolic Logic*, 64(4):1689–1699, 1999. 385
9. S. Grumbach and J. Su. Queries with arithmetical constraints. *Theoretical Computer Science*, 173:151–181, 1997. 376
10. W. Hodges. *A Shorter Model Theory*. Cambridge University Press, 1997. 377
11. N. Immerman. *Descriptive Complexity*. Graduate Texts in Computer Science. Springer, 1998. 377, 378
12. G. Kuper, L. Libkin, and J. Paredaens, editors. *Constraint Databases*. Springer-Verlag, 2000. 375
13. D Marker. Model theory of differential fields. In *Model theory of fields*, number 5 in Lecture notes in logic. Springer, 1996. 385
14. J. Robinson. Definability and decision problems in arithmetic. *Journal of Symbolic Logic*, 14:98–114, 1949. 376

Space Hierarchy Theorem Revised*

Viliam Geffert

Department of Computer Science, P. J. Šafárik University
Jesenná 5, 04154 Košice, Slovakia
geffert@kosice.upjs.sk

Abstract. We show that, for an arbitrary function $h(n)$ and each recursive function $\ell(n)$, that are separated by a nondeterministically fully space constructible $g(n)$, such that $h(n) \in \Omega(g(n))$ but $\ell(n) \notin \Omega(g(n))$, there exists a unary language $\mathcal{L}$ in $\mathrm{NSPACE}(h(n)) - \mathrm{NSPACE}(\ell(n))$. The same holds for the deterministic case.

The main contribution to the well-known *Space Hierarchy Theorem* is that *(i)* the language $\mathcal{L}$ separating the two space classes is unary (tally), *(ii)* the hierarchy is independent of whether $h(n)$ or $\ell(n)$ are in $\Omega(\log n)$ or in $o(\log n)$, *(iii)* the functions $h(n)$ or $\ell(n)$ themselves need not be space constructible nor monotone increasing. This allows us, using diagonalization, to present unary languages in such complexity classes as, for example, $\mathrm{NSPACE}(\log \log n \cdot \log^* n) - \mathrm{NSPACE}(\log \log n)$.

Keywords. Computational complexity; Space complexity.

1 Introduction

The first applications of diagonalization in complexity theory, by Hartmanis, Lewis, and Stearns in 1965 [8], gave deterministic time and space hierarchies. That is, with a small increase in space, we can solve new problems that could not be solved before: For each fully space constructible $h(n) \in \Omega(\log n)$ and each $\ell(n) \notin \Omega(h(n))$, there exists a language $\mathcal{L}$ in $\mathrm{DSPACE}(h(n)) - \mathrm{DSPACE}(\ell(n))$. The separating language is

$$\mathcal{L} = \{w \in \{0,1\}^*; \ M_{\pi(w)} \text{ does not accept } w, \text{ or uses space above } h(|w|)\}.$$

Here $M_0, M_1, \ldots$ denotes a standard enumeration of deterministic Turing machines, equipped with a finite state control, a two-way read-only binary input tape with input enclosed between two end markers, and a semi-infinite two-way read-write work tape that is also binary. (For nondeterministic space hierarchy, we shall consider the corresponding enumeration of *nondeterministic* Turing machines). By some syntax convention, the list of instructions for the machine M_k is decoded from the binary representation of the number k. The function $\pi: \{0,1\}^* \to \mathcal{N}$ must fulfil the following requirements:

* This work was supported by the Slovak Grant Agency for Science (VEGA) under contract #1/7465/20 "Combinatorial Structures and Complexity of Algorithms."

J. Sgall, A. Pultr, and P. Kolman (Eds.): MFCS 2001, LNCS 2136, pp. 387–397, 2001.
© Springer-Verlag Berlin Heidelberg 2001

(i)   The value of $\pi(w)$ must be computed and stored within $O(h(|w|))$ space.
(ii) For each $k \in \mathcal{N}$, there must exist an infinite number of w_n's in $\{0,1\}^*$ satisfying $k = \pi(w_n)$, such that $\lim_{n \in \mathcal{N}} \ell(|w_n|)/h(|w_n|) = 0$.

The condition (i) keeps $\mathcal{L}$ in DSPACE($h(n)$), while (ii) is necessary to prove that $\mathcal{L}$ is not in DSPACE($\ell(n)$): We have to show, for each $k, c \in \mathcal{N}$, that M_k accepting $\mathcal{L}$ uses more than $c \cdot \ell(n)$ work tape cells for some input. The existence of at least one infinite sequence of integers $\mathcal{P} \subseteq \mathcal{N}$ satisfying $\lim_{n \in \mathcal{P}} \ell(n)/h(n) = 0$ follows from $\ell(n) \notin \Omega(h(n))$. A simple way of defining $\pi(w)$ is:

- If $w = 1^k 01^{n-k-1}$, for some $k \in \mathcal{N}$, then $\pi(w) = k$,
- if $w \notin 1^* 01^*$, then $\pi(w) = 0$.

For this π, the condition (i) requires $h(n) \in \Omega(\log n)$. Another reason for the assumption $h(n) \in \Omega(\log n)$ came from the fact that we have to decide whether the machine $M_{\pi(w)}$ *rejects* the input w. It was not known whether the sublogarithmic space is closed under complement. For $h(n)$ below $\log n$, it was not so easy to detect that a machine rejects by going into an infinite cycle. In 1980, Sipser removed this problem by showing that the assumption $h(n) \in \Omega(\log n)$ is not necessary to detect loops [14], but the problem reappeared for the nondeterministic space hierarchy.

For a long time it was believed that nondeterministic space is not closed under complement. Nevertheless, Ibarra [10] obtained a pretty good nondeterministic space hierarchy. The method of inductive counting [11,15] showed that NSPACE($s(n)$) *is closed* under complement for $s(n) \in \Omega(\log n)$, which gave us a simple proof for a tight hierarchy: For each fully space constructible $h(n) \in \Omega(\log n)$ and each $\ell(n) \notin \Omega(h(n))$, there exists a language $\mathcal{L}$ in NSPACE($h(n)$) $-$ NSPACE($\ell(n)$).

Inductive counting does not seem, in general, to work below $\log n$. However, we know since 1993 that inductive counting can be used below $\log n$ for languages with low information content, like, e.g., unary or bounded sets [5].

It should be pointed out that diagonalization itself may work even without a proven closure under complement for the complexity classes under consideration. For example, Seiferas, Fischer, and Meyer proved a tight hierarchy for nondeterministic time classes [2]. In 1983, Žák simplified the proof [17]. However, these methods did not seem to be applicable for unary sets in sublogarithmic space.

To establish a space hierarchy below $\log n$, a different approach was used [9]. Roughly speaking, a language $\mathcal{L}$ separating DSPACE($h(n)$) from NSPACE($\ell(n)$) consists of words w with the prefix of the first $2^{h(n)}$ bits equal to the suffix of the last $2^{h(n)}$ bits. The lower bound is proved by a crossing sequence argument. Thus, for each space constructible $h(n) < \log \frac{n}{2}$ and each $\ell(n) \notin \Omega(h(n))$, there exists a language $\mathcal{L}$ in DSPACE($h(n)$) $-$ NSPACE($\ell(n)$).

Note that the combination of the above statements does not cover all cases; there are functions with values above $\log n$ along one infinite sequence of n's and, at the same time, satisfying $\lim_{n \in \mathcal{P}} h(n)/\log n = 0$ for another sequence of n's.

Moreover, the crossing sequence argument does not present unary (nor even bounded) separating languages. Unary (tally) languages play an important role in complexity theory; they allow a "downward" translation. By compressing unary strings $1^n \in 1^*$ into their "binary counterparts," we get, for example, that NSPACE($\log n$) is equal to DSPACE($\log n$) if and only if NSPACE($\log \log n$) and DSPACE($\log \log n$) are equal for unary languages [7]. Similar relations hold for higher complexity classes [13].

Using a diagonalization scheme, we get the following unary language.

$$\mathcal{L} = \{1^n; \ M_{\pi(n)} \text{ does not accept } 1^n, \text{ or uses more work tape cells than } h(n)\}.$$

Note that here we have $\pi: \mathcal{N} \to \mathcal{N}$. This complicates the choice of a π satisfying the condition (ii) above. Unlike in the binary case, each $k \in \mathcal{N}$ must have its own infinite sequence of n's with $k = \pi(n)$, and therefore there must necessarily exist infinitely many n's with $\pi(n) \neq k$. But then we cannot guarantee that $\mathcal{P}$, the sequence along which $\lim_{n \in \mathcal{P}} \ell(n)/h(n) = 0$, contains a sufficiently large n satisfying $k = \pi(n)$. A usual solution of this problem in the literature is to replace the assumption $\ell(n) \notin \Omega(h(n))$ by $\ell(n) \in o(h(n))$. Then $\lim_{n \in \mathcal{P}} \ell(n)/h(n) = 0$ along $arbitrary$ infinite $\mathcal{P} \subseteq \mathcal{N}$. This allows us to define π, for example, as follows [16, Thm. 11.1.1]:

- If $n = 2^k \cdot (2m+1)$, for some $k, m \in \mathcal{N}$, then $\pi(n) = k$.

Note that k, m are unique for each n. This gives that for each fully space constructible function $h(n) \in \Omega(\log n)$ and each $\ell(n) \in o(h(n))$, there exists a $unary$ language $\mathcal{L}$ in NSPACE($h(n)$) $-$ NSPACE($\ell(n)$).

At the first glance, the assumption $h(n) \in \Omega(\log n)$ is included only to satisfy the condition (i) for π. However, even using a more sophisticated definition of π does not help here: Suppose that $h(n) \notin \Omega(\log n)$. Then $\lim_{n \in \mathcal{P}} h(n)/\log n = 0$ for some infinite $\mathcal{P} \subseteq \mathcal{N}$, and hence, for some sufficiently large $\tilde{n} \in \mathcal{P}$, we get $h(\tilde{n}) = h(\tilde{n}+i \cdot \tilde{n}!)$, for each $i \in \mathcal{N}$. This is because the machine constructing the value of $h(\tilde{n})$ does not have enough space to traverse the input tape without going into a cycle, i.e., repeating the same work tape configuration, and hence it cannot distinguish between $1^{\tilde{n}}$ and $1^{\tilde{n}+i \cdot \tilde{n}!}$ [3,8]. The same holds even if $h(\tilde{n})$ is constructed by a nondeterministic machine [4]. Thus, if $\ell(n) \geq 1$ and $h(n) \notin \Omega(\log n)$, we get $\ell(\tilde{n}+i \cdot \tilde{n}!)/h(\tilde{n}+i \cdot \tilde{n}!) \geq 1/h(\tilde{n}+i \cdot \tilde{n}!) = 1/h(\tilde{n}) > 0$, for each $i \in \mathcal{N}$. This contradicts $\ell(n) \in o(h(n))$.

To overcome the above obstacles and extend the space hierarchy for unary languages below $\log n$, we do not insist on any space constructibility of $h(n)$ or $\ell(n)$. We only assume they are separated by a (non)deterministically fully space constructible $g(n)$, such that $h(n) \in \Omega(g(n))$ but $\ell(n) \notin \Omega(g(n))$. This assumption about the growth rates of $h(n)$ and $\ell(n)$ is independent of whether they are above $\log n$, space constructible, or monotone increasing. We also assume that $\ell(n)$ is recursive, without a limit for the space used to compute $\ell(n)$.

For example, this allows us to present unary languages in such complexity classes as NSPACE($\log \log n \cdot \log^* n$) $-$ NSPACE($\log \log n$), despite the fact that neither of these two space bounds is fully space constructible. The separation is based on diagonalization, using a variant of inductive counting that, for bounded or tally inputs, does not depend on the $\log n$ space bound or constructibility [5].

Before doing so, we need to find, for any nondeterministic machine constructing $g(n)$ and any machine computing $\ell(n)$, a function $\pi(n)$ producing each integer infinitely many times along a sequence of n's, for which $\lim_{n\in\mathcal{P}} \ell(n)/g(n)=0$. Moreover, the values of $\pi(n)$ must be computed within $g(n)$ space, hence, they depend on the (not necessarily monotone) growth rates of $g(n)$ and $\ell(n)$.

2 Preliminaries

We first briefly recall some fundamental notions that are used throughout.

A (non)deterministic machine is *(a) strongly $s(n)$ space bounded*, if, for each input of length n, no computation path uses more than $s(n)$ space, *(b) weakly $s(n)$ space bounded*, if, for each accepted input of length n, there exists at least one accepting path using at most $s(n)$ space.

The classes of languages accepted by strongly and weakly $O(s(n))$ space bounded nondeterministic machines are denoted by strong–NSPACE($s(n)$) and weak–NSPACE($s(n)$), while strong–DSPACE($s(n)$) and weak–DSPACE($s(n)$) denote their deterministic variants, respectively. To keep the notation simple, we omit the prefixes "strong–" and "weak–" if the difference does not matter.

A function $s(n)$ is called *(non)deterministically fully space constructible*, if there exists a (non)deterministic machine such that, for each input of length n, no computation path uses more work tape cells than $s(n)$, but at least one computation path uses exactly $s(n)$ cells.

A function $f(n)$ is *(non)deterministically computable* in $s(n)$ space, if there exists a (non)deterministic machine such that, for the input 1^n, no computation path uses more work tape cells than $s(n)$, at least one computation path halts in an accepting state, and each path halting in an accepting state writes a binary representation of $f(n)$ on the work tape.

We need the following partial extension of the classical results below $\log n$.

Lemma 1 ([14,5]). *Let $s(n)$ be an arbitrary function, and let $\mathcal{L}$ be a binary language in* strong–NSPACE($s(n)$). *If $\mathcal{L}$ is k-bounded, i.e., for some constant $k \in \mathcal{N}$, the number of zeros in each $w \in \mathcal{L}$ is bounded by k, then $\mathcal{L}^c$, the complement of $\mathcal{L}$, is also in* strong–NSPACE($s(n)$).

The same holds for deterministic space complexity classes.

We shall use only 0- or 1-bounded sets, i.e., $\mathcal{L} \subseteq 1^*$ or $\mathcal{L} \subseteq 1^*01^*$, but the statement holds even if the number of zeros is bounded by $c^{s(n)}$, for any constant $c>1$ [5,7].

Theorem 2. *A function $s(n)$ is (non)deterministically fully space constructible if and only if $s(n)$ is (non)deterministically computable in $s(n)$ space.*

Proof. We shall present the argument for the nondeterministic case only. The corresponding argument for deterministic machines is simpler.

The "$\Leftarrow$" part is obvious: For any input of length n, simulate the nondeterministic machine C that computes $s(n)$, imitating the input 1^n. If C accepts, a certified value of $s(n)$ has been written, in binary, on the work tape. This allows us to allocate $s(n)$ work tape cells. If C rejects, the simulator simply halts.

Conversely, let A be a nondeterministic machine constructing $s(n)$. Before presenting the machine C computing $s(n)$, consider the following language.

$$\mathcal{L} = \{1^r 01^{n-r-1}; \; r < s(n)\}.$$

Clearly, $\mathcal{L} \in$ strong–NSPACE$(s(n))$. First, simulate A. After each step, nondeterministically decide whether to carry on the simulation or to abort A. This marks off $\tilde{s}$ work tape cells, for some $\tilde{s} \le s(n)$. Then compare $\tilde{s}$ with r and, if $r < \tilde{s}$, accept. Clearly, if $r < s(n)$, then at least one computation path of A will allocate $\tilde{s} = s(n) > r$ space, and hence the input is accepted. On the other hand, if $r \ge s(n)$, then no path can allocate enough space and the input is rejected.

But then $\mathcal{L}' = \mathcal{L}^c \cap 1^*01^* = \{1^r01^{n-r-1}; \; r \ge s(n)\} \in$ strong–NSPACE$(s(n))$, by Lem. 1, since $\mathcal{L}$ is 1-bounded. Thus, we have a nondeterministic machine A' verifying if $r \ge s(n)$ and never using space above $s(n)$.

We are now ready to present C. First, simulate A on the input 1^n, aborting simulation nondeterministically at any time. This allocates r work tape cells, for some $r \le s(n)$. Then verify if $r \ge s(n)$. Thus, simulate A', pretending that 1^r01^{n-r-1} is present on the input tape. Since the real input is 1^n, this only requires to interpret the symbol at the position $r+1$ as zero.

If A fails to allocate enough space in the first phase, all computation paths of A' will reject. But, for the right sequence of nondeterministic guesses, A will allocate $r = s(n)$ cells, and hence, for the right sequence of nondeterministic guesses, A' will accept. When this happens, C converts the number of the used work tape cells into a number in binary notation and accepts. $\square$

3   Space Hierarchy

Definition 3. *Let $g(n)$ and $\ell(n)$ be two functions. A* prominent sequence *for g/ℓ is the set $\mathcal{P}$ of all n's in $\mathcal{N}$ satisfying*

$$\lfloor g(n)/\ell(n) \rfloor > \max\{\lfloor g(1)/\ell(1) \rfloor, \lfloor g(2)/\ell(2) \rfloor, \ldots, \lfloor g(n-1)/\ell(n-1) \rfloor\}.$$

By definition, $1 \in \mathcal{P}$. An $n \in \mathcal{P}$ is called a prominent point.

Lemma 4. *Let $g(n)$ and $\ell(n)$ be two functions, $\ell(n) \notin \Omega(g(n))$. Then the prominent sequence $\mathcal{P}$ for g/ℓ is infinite. Moreover, for each $c > 0$, there exists an $n_0 \in \mathcal{N}$ such that, for each $n \in \mathcal{P}$, $n \ge n_0$, we have $g(n) > c \cdot \ell(n)$.*

The proof uses the fact that $\ell(n) \notin \Omega(g(n))$ if and only if $\inf_{n \in \mathcal{N}} \ell(n)/g(n) = 0$.

Now we can show how to generate each $k \in \mathcal{N}$ along any $\mathcal{P} \subseteq \mathcal{N}$, effectively given in a form of a prominent sequence, keeping the given space bound $g(n)$.

Lemma 5 (Main). *Let $g(n)$ be (non)deterministically fully space constructible, and $\ell(n)$ be an arbitrary recursive function, $\ell(n) \notin \Omega(g(n))$. Then there exists a function $\tilde{\pi}(n)$, (non)deterministically computable in $O(g(n))$ space, such that, for each $k \in \mathcal{N}$, there exists an $n \in \mathcal{P}$ satisfying $k = \tilde{\pi}(n)$. Here $\mathcal{P}$ denotes the prominent sequence for g/ℓ.*

Proof. Since $\ell(n)$ is recursive, we have a deterministic machine C_ℓ which, for any input 1^n, halts with a binary representation of $\ell(n)$ written on the work tape. Let $s_\ell(n)$ denote the space used by C_ℓ on the input 1^n. Similarly, by Thm. 2, we have a (non)deterministic machine C_g halting in an accepting state with a certified value of $g(n)$ written on the work tape. Note that C_g may also have some rejecting computation paths, or paths that never halt. But, by assumption, C_g never uses more space than $g(n)$.

We shall present a machine $C_{\tilde{\pi}}$ computing the values of some $\tilde{\pi}$. We must also keep track of $s_{\tilde{\pi}}(n)$, the space used by $C_{\tilde{\pi}}$ on the input 1^n.

Let us describe how the values of $\tilde{\pi}(n)$ are defined. If $n = 1$, then, by definition, $\tilde{\pi}(n) = 1$. For $n > 1$, let $B = g(n)$. If $B = 1$, then, by definition, $\tilde{\pi}(n) = 1$. Consider now the general case, i.e., $n > 1$ and $B = g(n) > 1$. Let A be the maximal integer satisfying

(a) $A \le 2^B - 1$,

(b) for each $i \le A$, $s_\ell(i) \le B = g(n)$, and $g(i) \le B/2 = g(n)/2$.

Now take the minimal $k \in \mathcal{N}$, such that

(c) $k \notin \{\tilde{\pi}(i);\ i \in \mathcal{P} \cap \{1, \ldots, A\}\}$.

Then, by definition, $\tilde{\pi}(n) = k$. If already $A = 1$ does not satisfy the condition (b), then $A = 0$, $\{1, \ldots, A\} = \varnothing$, and $\tilde{\pi}(n) = 1$.

To compute the value of $\tilde{\pi}(n)$, we compute, recursively, the values of $\tilde{\pi}(i)$ for prominent points satisfying $i \le A$. To prove the correctness of this recursive definition of $\tilde{\pi}(n)$, it is sufficient to verify that

(d) $A < n$.

This is easy, since $B = g(n) > 1$ and, by (b), we have $g(i) \le B/2$, for each $i \le A$. Thus, if $n \le A$, we would get $g(n) \le B/2 < B = g(n)$, which is a contradiction.

Claim 1. *The machine $C_{\tilde{\pi}}$ computing $\tilde{\pi}(n)$ does not use more than $O(g(n))$ space.*

For $n = 1$ or $B = g(n) = 1$, we have $s_{\tilde{\pi}}(n) = 1$: A certified value of $B = g(n)$ is computed by simulation of C_g. This does not take more space than $g(n)$. If the simulation of C_g fails, due to a wrong sequence of nondeterministic guesses, then $C_{\tilde{\pi}}$ aborts, returning no value of $\tilde{\pi}(n)$ at all. However, for at least one computation path, C_g halts with a certified value of $g(n)$ on the work tape.

The value of A is found by checking $i = 1, 2, 3, \ldots$, until we get the first i such that (a) $i = 2^B$, or (b) either $s_\ell(i) > B$, or $g(i) > B/2$. The space required for the counter representing i is bounded by $B = g(n)$, since $i \le 2^B$.

Checking if $s_\ell(i) > B$ can be performed in $O(g(n))$ space; we simply simulate the machine C_ℓ computing $\ell(i)$ on the input 1^i. The simulation is aborted as soon as C_ℓ tries to use space above B, hence, $B = g(n)$ space is sufficient. In addition, we need $\log i$ bits to represent the position of the input tape head for C_ℓ on 1^i, but this amount of space is also bounded by $O(g(n))$, since $i \leq 2^B$.

We check also if $g(i) > B/2$. Here we simulate C_g computing $g(i)$ on the input 1^i, along a nondeterministically chosen path. There are the following cases. If the chosen path tries to use the space above $B/2$, the simulation of C_g is aborted; this means that $g(i) > B/2$ since no path of C_g uses space above $g(i)$. If the chosen path returns a certified value of $g(i)$ not using space above $B/2$, we can correctly decide if $g(i) > B/2$ in $B = g(n)$ space. Finally, the chosen path may also reject or loop forever, not using space above $B/2$. In this case $C_{\tilde\pi}$ returns no value of $\tilde\pi(n)$ at all. However, at least one computation path of C_g returns a certified value of $g(i)$, be the space used below $B/2$ or not. Summing up, for each $i \leq 2^B$, $C_{\tilde\pi}$ can correctly decide if $g(i) > B/2$ along at least one computation path. In any case, it does not use space above $B = g(n)$. The space required to imitate the input head movement on 1^i is the same as for C_ℓ, i.e., $\log i$ bits.

Before showing how the machine finds $k \in \mathcal{N}$ satisfying (c), let us analyze how much space it needs to decide if $i \in \mathcal{P}$. Once the value of A has been determined, we know that $\ell(i)$ and $g(i)$ must be computed within $O(g(n))$ space, since $s_\ell(i) \leq B$ and $g(i) \leq B/2$, for each $i \leq A$. But then $f_i = \lfloor g(i)/\ell(i) \rfloor$ can also be computed in $O(g(n))$ space, since $f_i \leq g(i)/\ell(i) \leq g(i)$. Finally, to decide if $i \in \mathcal{P}$, we only have to compute, one after another, the values $f_1, f_2, \ldots, f_{i-1}$, and to compare each of these with f_i (see Def. 3). Again, such computation may fail due to a wrong sequence of nondeterministic guesses, while simulating C_g.

Now we are ready to analyze the search for the value of $\tilde\pi(n)$. This is found by checking $k = 1, 2, 3, \ldots,$ until we find the first k such that $k \notin \{\tilde\pi(i); i \in \mathcal{P} \cap \{1, \ldots, A\}\}$. Note that, by (a), there are at most $2^B - 1$ prominent points $i \leq A$, and hence at most $2^B - 1$ different values of $\tilde\pi(i)$ in the set. Therefore, the k with the desired properties must be found among the first 2^B numbers. That is, $k \leq 2^B$, and hence it can be stored in $g(n)$ space.

For each k, we run a nested loop for $i = 1, \ldots, A$, and check if $i \in \mathcal{P}$. For each prominent point, we compare $\tilde\pi(i)$ with k. By the argument above, checking if $i \in \mathcal{P}$ does not require more space than $O(g(n))$. The values of $\tilde\pi(i)$, for $i \in \mathcal{P}$, $i \leq A$, can be computed recursively. That is, the machine $C_{\tilde\pi}$ simulates itself on the input 1^i. This requires $\log i \leq \log A \leq g(n)$ bits to imitate the input head movement on 1^i, plus $s_{\tilde\pi}(i)$ space for the work tape of the nested version of $C_{\tilde\pi}$. Note that here we have $g(i) \leq B/2 = g(n)/2$, by (b), and $i \leq A < n$, by (d).

Summing up, there exists $c > 0$, such that the space used is bounded by

$$s_{\tilde\pi}(n) \leq c \cdot g(n) + \max\{s_{\tilde\pi}(i); i \in \mathcal{P} \cap \{1, \ldots, A\}\}.$$

Let $i_1 \leq A$ be the prominent point with the maximal value of $s_{\tilde\pi}(i)$. Then

$$s_{\tilde\pi}(n) \leq c \cdot g(n) + s_{\tilde\pi}(i_1), \quad \text{for some } i_1 < n \text{ with } g(i_1) \leq g(n)/2.$$

The conditions for i_1 follow from (b) and (d). By induction for $i_1 < n$, we have

$$s_{\tilde{\pi}}(n) \leq c \cdot g(n) + c \cdot g(i_1) + s_{\tilde{\pi}}(i_2), \quad \text{for some } i_2 < i_1 \text{ with } g(i_2) \leq g(i_1)/2.$$

Repeating this process $r+1$ times, we get

$$s_{\tilde{\pi}}(n) \leq c \cdot g(n) + c \cdot g(i_1) + \cdots + c \cdot g(i_r) + s_{\tilde{\pi}}(i_{r+1}),$$

for some $i_{r+1} < i_r < \ldots < i_1 < n$, with $g(i_{j+1}) \leq g(i_j)/2$, for each $j = 1, \ldots, r$.

But then we must get, sooner or later, either $i_{r+1} = 1$ or $g(i_{r+1}) = 1$. In both cases, $s_{\tilde{\pi}}(i_{r+1}) = 1$. It is obvious that $g(i_j) \leq g(n)/2^j$, for each $j = 1, \ldots, r$. Then

$$s_{\tilde{\pi}}(n) \leq c \cdot g(n) \cdot (1 + \tfrac{1}{2} + \tfrac{1}{4} + \cdots + \tfrac{1}{2^r}) + 1 \leq 2c \cdot g(n) + 1 \in O(g(n)).$$

Claim 2. *For each $k \in \mathcal{N}$, there exists an $n \in \mathcal{P}$, such that $k = \tilde{\pi}(n)$.*

For $k = 1$, we have $1 = \tilde{\pi}(1)$. By Def. 3, $1 \in \mathcal{P}$. Assume now, inductively, that we have already found some prominent points $n_1, n_2, \ldots, n_k$, such that $j = \tilde{\pi}(n_j)$, for each $j = 1, \ldots, k$. Define

$$A' = \max\{n_1, n_2, \ldots, n_k\},$$
$$B' = \max\{2, \log(A'+1), s_\ell(1), \ldots, s_\ell(A'), 2g(1), \ldots, 2g(A')\}.$$

Note that

(a') $A' \leq 2^{B'} - 1$,
(b') for each $i \leq A'$, $s_\ell(i) \leq B'$, and $g(i) \leq B'/2$.

By Lem. 4, there exists an $\tilde{n} \in \mathcal{P}$, $\tilde{n} > 1$, such that $g(\tilde{n}) = B > B'\ell(\tilde{n}) \geq B' \geq 2$. Consider now how $\tilde{\pi}(\tilde{n})$ is computed. Since $\tilde{n} > 1$ and $g(\tilde{n}) = B > 1$, we first find the maximal A satisfying (a) and (b). Using $B' < B$, (a'), and (b'), we get that $A \geq A'$. Note that $A \geq A' = \max\{n_1, n_2, \ldots, n_k\}$ and that $n_1, n_2, \ldots, n_k$ are the prominent points satisfying $j = \tilde{\pi}(n_j)$, for each $j = 1, \ldots, k$. Therefore,

$$\{1, 2, \ldots, k\} \subseteq \{\tilde{\pi}(i); \, i \in \mathcal{P} \cap \{1, \ldots, A\}\}.$$

But then the search for the first $\tilde{k}$ satisfying (c) does not stop sooner than it reaches some $\tilde{k} \geq k+1$. There are now two cases to consider. If $\tilde{k} = k+1$, then we are done; we have found a prominent point with $\tilde{\pi}(\tilde{n}) = k+1$. If the search does not stop at $k+1$, that is, $\tilde{k} > k+1$, then $k+1 \in \{\tilde{\pi}(i); \, i \in \mathcal{P} \cap \{1, \ldots, A\}\}$. Thus, we have some other prominent point $i \in \mathcal{P}$, for which $\tilde{\pi}(i) = k+1$. $\quad\square$

Now, let $\pi(n)$ be defined as the number of ones in the binary notation of the number $\tilde{\pi}(n)$. This gives:

Theorem 6. *Let $g(n)$ be (non)deterministically fully space constructible, and $\ell(n)$ be an arbitrary recursive function, $\ell(n) \notin \Omega(g(n))$. Then there exists a function $\pi(n)$, (non)deterministically computable in $O(g(n))$ space, such that, for each $k \in \mathcal{N}$, there exists an infinite number of n's in $\mathcal{P}$ satisfying $k = \pi(n)$. Here $\mathcal{P}$ denotes the prominent sequence for g/ℓ.*

We are now ready to present a space hierarchy.

Theorem 7. *For each $h(n)$ and each recursive $\ell(n)$, separated by a nondeterministically fully space constructible $g(n)$, with $h(n) \in \Omega(g(n))$ but $\ell(n) \notin \Omega(g(n))$, there exists a unary language $\mathcal{L}$ in $\mathrm{NSPACE}(h(n)) - \mathrm{NSPACE}(\ell(n))$.*

More precisely, $\mathcal{L} \in \mathrm{strong\text{-}NSPACE}(h(n)) - \mathrm{weak\text{-}NSPACE}(\ell(n))$, that is, the space hierarchy is established both for strong and weak space classes.

The same holds for the deterministic case.

Proof. By Thm. 2 and 6, we have two machines C_g and C_π, computing certified values of functions $g(n)$ and $\pi(n)$, respectively. Recall that $\pi(n)$ produces each $k \in \mathcal{N}$ infinitely many times, along the prominent sequence $\mathcal{P}$ for g/ℓ. Let

$$\mathcal{L}' = \{1^n; \ M_{\pi(n)} \text{ accepts } 1^n, \text{ not using more work tape cells than } g(n)\}.$$

We first show that $\mathcal{L}'$ is in $\mathrm{strong\text{-}NSPACE}(g(n))$. The machine A' accepting $\mathcal{L}'$ first simulates C_g and C_π, to obtain the values of $g(n)$ and $\pi(n)$, respectively. If any of them does not halt in an accepting state, A' rejects. Otherwise, the certified values of $g(n)$ and $\pi(n)$ have been written on the work tape. Having loaded the binary code of $M_{\pi(n)}$ on the work tape, the machine A' can simulate $M_{\pi(n)}$ along a nondeterministically chosen computation path on the input 1^n. If the chosen path accepts, not using more than $g(n)$ binary work tape cells, A' accepts. If the chosen path tries to exceed this space limit, A' aborts and rejects 1^n. This gives that $\mathcal{L}'$ is in $\mathrm{strong\text{-}NSPACE}(g(n))$. Now, let

$$\mathcal{L} = \{1^n; \ M_{\pi(n)} \text{ does not accept } 1^n, \text{ or uses more work tape cells than } g(n)\}.$$

That is, $\mathcal{L}$ is the complement of $\mathcal{L}'$. By Lem. 1, $\mathcal{L}$ is in $\mathrm{strong\text{-}NSPACE}(g(n))$, since $\mathcal{L}'$ is unary, i.e., a 0-bounded language. But then $\mathcal{L}$ is also in $\mathrm{strong\text{-}NSPACE}(h(n))$, using $h(n) \in \Omega(g(n))$.

On the other hand, let M_k be an arbitrary machine, equipped with a binary work tape and weakly space bounded by $c \cdot \ell(n)$, for some $c > 0$. By Lem. 4, there exists an $n_0 \in \mathcal{N}$ such that $g(n) > c \cdot \ell(n)$, for each $n \in \mathcal{P}$ satisfying $n \geq n_0$. By Thm. 6, there exists an $\tilde{n} \geq n_0$ in $\mathcal{P}$ satisfying $k = \pi(\tilde{n})$. This gives

$$k = \pi(\tilde{n}), \quad c \cdot \ell(\tilde{n}) < g(\tilde{n}).$$

Thus, $1^{\tilde{n}}$ is in $L(M_k) = L(M_{\pi(\tilde{n})})$ if and only if there exists an accepting path of $M_{\pi(\tilde{n})}$ on the input $1^{\tilde{n}}$ not using more work tape cells than $c \cdot \ell(\tilde{n}) < g(\tilde{n})$. On the other hand, $1^{\tilde{n}}$ is in $\mathcal{L}$ if and only if each computation path of $M_{\pi(\tilde{n})}$ on the input $1^{\tilde{n}}$ either rejects, loops, or uses more work tape cells than $g(\tilde{n})$. From this we have that $M_k = M_{\pi(\tilde{n})}$ does not recognize $\mathcal{L}$. This implies that $\mathcal{L}$ is not in $\mathrm{weak\text{-}NSPACE}(\ell(n))$.   □

Thus, a machine not using space above $h(n)$ along any computation path (worst case cost) can diagonalize against any machine working in $\ell(n)$ space, even if the $\ell(n)$ space restriction concerns only accepting computations using minimal space (best cost of acceptance). As an example:

Corollary 8. *There exists a unary language* $\mathcal{L}$ *in the set*

$$\text{strong--NSPACE}(\log \log n \cdot \log^* n) - \text{weak--NSPACE}(\log \log n).$$

The same holds for the corresponding deterministic space complexity classes.

The argument uses the function $g(n) = \log f(n) \cdot \log^* f(n)$, where $f(n)$ is the first number not dividing n. The function $g(n)$ satisfies the conditions of Thm. 7 for functions $\ell(n) = \log \log n$ and $h(n) = \log \log n \cdot \log^* n$, since $\log f(n)$ is fully space constructible, $\log f(n) \in O(\log \log n)$, but $\log f(n) \notin o(\log \log n)$ [7,16].

If the separating language need not necessarily be unary, the separation can be simplified significantly. In addition, the recursivity assumption for $\ell(n)$ can be discarded, even if only a single zero on the input is allowed.

An open problem is a tight *space* hierarchy for classes of the *alternating* hierarchy. By inductive counting [11,15], the hierarchy of $s(n)$ space bounded machines with a constant number of alternations collapses to the first level for $s(n) \in \Omega(\log n)$. Hence, the space hierarchy for such machines coincides with the space hierarchy of nondeterministic machines. But sublogarithmic space classes Σ_k- and $\Pi_k-\text{SPACE}(s(n))$ are provably not closed under complement [6,1,12]. The first step in this direction was made in [18], where a tight hierarchy is established for the above space classes. However, the result is proved for a different computational model, the so-called demon machines, having $s(n)$ space marked off automatically, at the very beginning of the computation.

Finally, as a side note, we point out that the technique presented in Thm. 6 allows us to compute an unpairing function $\pi(n)$ along any recursive set $\mathcal{P}$, bounding the space used by any (arbitrarily small) function $g(n)$:

Corollary 9. *For any (non)deterministically fully space constructible* $g(n)$, *and any recursive* $\mathcal{P} \subseteq \mathcal{N}$, *such that* $\sup_{n \in \mathcal{P}} g(n) = +\infty$, *there exists a function* $\pi(n)$, *(non)deterministically computable in* $O(g(n))$ *space, such that, for each* $k \in \mathcal{N}$, *there exists an infinite number of* n's *in* $\mathcal{P}$ *satisfying* $k = \pi(n)$.

References

1. Braunmühl, B. von, Gengler, R., Rettinger, R.: The alternation hierarchy for sublogarithmic space is infinite. *Comput. Complexity* **3** (1993) 207–30. 396
2. Fischer, M., Meyer, A., Seiferas, J.: Separating nondeterministic time complexity classes. *J. Assoc. Comput. Mach.* **25** (1978) 146–67. 388
3. Freedman, A. R., Ladner, R. E.: Space bounds for processing contentless inputs. *J. Comput. System Sci.* **11** (1975) 118–28. 389
4. Geffert, V.: Nondeterministic computations in sublogarithmic space and space constructibility. *SIAM J. Comput.* **20** (1991) 484–98. 389
5. Geffert, V.: Tally versions of the Savitch and Immerman-Szelepcsényi theorems for sublogarithmic space. *SIAM J. Comput.* **22** (1993) 102–13. 388, 390
6. Geffert, V.: A hierarchy that does not collapse: Alternations in low level space. *RAIRO Inform. Théor.* **28** (1994) 465–512. 396
7. Geffert, V.: Bridging across the log(n) space frontier. *Inform. & Comput.* **142** (1998) 127–58. 389, 390, 396

8. Hartmanis, J., Lewis II, P. M., Stearns, R. E.: Hierarchies of memory limited computations. In *IEEE Conf. Record on Switching Circuit Theory and Logical Design* (1965) 179–90. 387, 389

9. Hopcroft, J. E., Ullman, J. D.: Some results on tape-bounded Turing machines. *J. Assoc. Comput. Mach.* **16** (1969) 168–77. 388

10. Ibarra, O. H.: A note concerning nondeterministic tape complexities. *J. Assoc. Comput. Mach.* **19** (1972) 608–12. 388

11. Immerman, N.: Nondeterministic space is closed under complementation. *SIAM J. Comput.* **17** (1988) 935–38. 388, 396

12. Liśkiewicz, M., Reischuk, R.: The sublogarithmic alternating space world. *SIAM J. Comput.* **25** (1996) 828–61. 396

13. Savitch, W. J.: Relationships between nondeterministic and deterministic tape complexities. *J. Comput. System Sci.* **4** (1970) 177–92. 389

14. Sipser, M.: Halting space bounded computations. *Theoret. Comput. Sci.* **10** (1980) 335–38. 388, 390

15. Szelepcsényi, R.: The method of forced enumeration for nondeterministic automata. *Acta Inform.* **26** (1988) 279–84. 388, 396

16. Szepietowski, A.: *Turing Machines with Sublogarithmic Space*. Springer-Verlag, *Lect. Notes Comput. Sci.* **843** (1994). 389, 396

17. Žák, S.: A Turing machine time hierarchy. *Theoret. Comput. Sci.* **26** (1983) 327–33. 388

18. Žák, S.: A sharp separation below log(n). Tech. Rep. 655/1995, Inst. Comput. Sci., Czech Academy of Sciences (1995). 396

Converting Two–Way Nondeterministic Unary Automata into Simpler Automata

Viliam Geffert[1,*], Carlo Mereghetti[2], and Giovanni Pighizzini[3]

[1] Department of Computer Science, P. J. Šafárik University
Jesenná 5, 04154 Košice, Slovakia
geffert@kosice.upjs.sk
[2] Dipartimento di Inf., Sist. e Com., Università degli Studi di Milano – Bicocca
via Bicocca degli Arcimboldi 8, 20126 Milano, Italy
mereghetti@disco.unimib.it
[3] Dipartimento di Scienze dell'Informazione, Università degli Studi di Milano
via Comelico 39, 20135 Milano, Italy
pighizzi@dsi.unimi.it

Abstract. We show that, on inputs of length exceeding $5n^2$, any n–state unary 2nfa (two–way nondeterministic finite automaton) can be simulated by a $(2n+2)$–state quasi sweeping 2nfa. Such a result, besides providing a "normal form" for 2nfa's, enables us to get a subexponential simulation of unary 2nfa's by 2dfa's (two–way deterministic finite automata). In fact, we prove that any n–state unary 2nfa can be simulated by a 2dfa with $O(n^{\log n + 4})$ states.

Keywords: formal languages; finite state automata; unary languages

1 Introduction

The problem of evaluating the costs — in terms of states — of the simulations between different kinds of finite state automata has been widely investigated in the literature. In particular, several contributions deal with two-way automaton simulations (see, e.g., [7], this Volume). The interest is also due to its relationships with some important topics in complexity theory.

The main open question in this field is certainly that posed by Sakoda and Sipser in 1978 [8], which asks for *the cost of turning a two–way nondeterministic or a one–way nondeterministic n–state finite state automaton (2nfa, 1nfa, resp.) into a two–way deterministic finite state automaton (2dfa)*. They conjecture such a cost to be exponential, and Sipser [9] proves that this is exactly the case when 2dfa's are required to be *sweeping*, i.e., to have head reversals only at the ends of the input tape. Berman and Lingas [1] state a lower bound of $\Omega(n^2/\log n)$ for the general problem, and provide an interesting connection with the celebrated open problem $\text{DLOGSPACE} \overset{?}{=} \text{NLOGSPACE}$. In [2], Chrobak rises the $\Omega(n^2/\log n)$ lower bound to $\Omega(n^2)$.

---

* Supported by the Slovak Grant Agency for Science (VEGA) under contract #1/7465/20 "Combinatorial Structures and Complexity of Algorithms."

J. Sgall, A. Pultr, and P. Kolman (Eds.): MFCS 2001, LNCS 2136, pp. 398–407, 2001.

Given its difficulty in the general case, restricted versions of the open question of Sakoda–Sipser have been tackled. In this regard, one of the most promising lines of research goes through the study of optimal simulations between *unary automata*, i.e., automata working with a single letter input alphabet. The problem of evaluating the costs of unary automata simulations was first settled in [9], and has lead to emphasize some relevant differences with the general case. For instance, we know that $O(e^{\sqrt{n \ln n}})$ states suffice to simulate a unary n–state 1nfa or 2dfa by a one–way deterministic finite state automaton (1dfa). Furthermore, a unary n–state 1nfa can be simulated by a 2dfa having $O(n^2)$ states, and this closes the open problem of Sakoda–Sipser about 1nfa's vs. 2dfa's, at least in the unary case. All these results and their optimality are proved in [2].

In [6], the authors prove that $O(e^{\sqrt{n \ln n}})$ is the optimal cost of simulating unary 2nfa's by 1dfa's. Moreover, just paying by a *quadratic* increase in the number of states, unary 2nfa's can always be regarded to as being *quasi sweeping*, namely, having both reversals and nondeterministic choices *only* at the ends of the input. This may be seen as another step toward a "simplification" of the open question of Sakoda–Sipser.

This work aims to give further contributions that could be helpful shedding some light on the Sakoda–Sipser open problem. Our first result, in Section 3, improves [6], which says that for any n–state unary 2nfa A there exists an equivalent quasi sweeping $O(n^2)$–state 2nfa A'. Here we show that the number of states in A' can be reduced to $2n+2$, provided that the resulting automaton is only *almost equivalent* to the original machine, that is, A and A' are allowed to disagree on a finite number of inputs. This result can be regarded to as providing a sort of *normal form* for 2nfa's, which is a two–way counterpart of the well–known Chrobak Normal Form for 1nfa's [2].

Our almost equivalent quasi sweeping simulation becomes a useful tool for the unary version of 2nfa's vs. 2dfa's question in Section 4. Using a divide–and–conquer technique, we first show that *any n–state quasi sweeping unary 2nfa can be simulated by a 2dfa with no more than $n^{\log n +4}$ states*. Then, by combining this with the above linear almost equivalence between 2nfa's and quasi sweeping 2nfa's, we prove a subexponential simulation of unary 2nfa's by 2dfa's: *for each n–state unary 2nfa, there exists an equivalent 2dfa with $O(n^{\log n+4})$ states*. To the best of authors' knowledge, the previously known best simulation of unary 2nfa's by 2dfa's uses $O(e^{\sqrt{n \ln n}})$ states [6].

2   Finite State Automata

Here, we briefly recall some basic definitions on finite state automata. For a detailed exposition, we refer the reader to [4]. Given a set S, $|S|$ denotes its cardinality and 2^S the family of all its subsets.

A *two–way nondeterministic finite automaton* (2nfa, for short) is a quintuple $A = (Q, \Sigma, \delta, q_0, F)$ in which Q is the finite set of states, Σ is the finite input alphabet, $\delta : Q \times (\Sigma \cup \{\vdash, \dashv\}) \rightarrow 2^{Q \times \{-1,0,+1\}}$ is the transition function, $\vdash, \dashv \notin \Sigma$ are two special symbols, called the left and the right endmarker, respectively, $q_0 \in$

Q is the initial state, and $F \subseteq Q$ is the set of final states. Input is stored onto the input tape surrounded by the two endmarkers, the left endmarker being at the position zero. In a move, A reads an input symbol, changes its state, and moves the input head one position forward, backward, or keeps it stationary depending on whether δ returns $+1$, -1, or 0, respectively. Without loss of generality, we can assume the machine starts and accepts with input head at the left endmarker. The language accepted by A is denoted by $L(A)$. A is a two–way *deterministic* finite state automaton (2dfa) whenever $|\delta(q, \sigma)| = 1$, for any $q \in Q$ and $\sigma \in \Sigma \cup \{\vdash, \dashv\}$.

A 2nfa is called *quasi sweeping* [5] if and only if it presents *both input head reversals and nondeterministic choices at the endmarkers only*. We call *unary* any automaton that works with a single letter input alphabet. We say that an automaton A is *almost equivalent* to an automaton A' if and only if the languages accepted by A and A' coincide, with the exception of a finite number of strings.

3   Linear, Almost Equivalent, Sweeping Simulation

In this section, we show how to get from a unary n–state 2nfa, an almost equivalent quasi sweeping 2nfa, with no more than $2n+2$ states. From now on, we will always refer to a *unary 2nfa A with n states*.

First of all, we recall some results, mainly from [3,6], concerning the "form" of accepting computations of A. By a *loop of length ℓ*, we mean a computation path of A beginning in a state p with the input head at a position i, and ending in the same state with the input head at the position $i + \ell$.

Consider an accepting computation of A on input 1^m. Let $r_0, r_1, \ldots, r_p$ be the sequence of all states in which the input head scans either of the endmarkers. For $1 \leq j \leq p$, the following two possibilities arise:

- In both r_{j-1} and r_j, the input head scans the same endmarker. This segment of computation is called a *U–turn*.
- In r_{j-1} the input head scans one of the two endmarkers, while in r_j it scans the other. This segment of computation is called a *(left to right or right to left) traversal* (notice that, within a traversal, the endmarkers are never touched).

Lemma 1. *Given two states q_1, q_2 of A, and an input 1^m:*

(i) for each U–turn from q_1 to q_2, there is another U–turn from q_1 to q_2 in which the input head is never moved farther than n^2 positions from the endmarker;

(ii) for each traversal from q_1 to q_2, there is a traversal from q_1 to q_2 where A:

  (a) having traversed the starting endmarker and s_1 positions,
  (b) gets into a loop (called dominant loop) of length ℓ, which starts from a state p and is repeated λ times,
  (c) then traverses the remaining s_2 input squares, and finally gets the other endmarker,

  for some p, and λ, s_1, s_2, ℓ satisfying $0 \leq \lambda$, $1 \leq |\ell| \leq n$, and $s_1 + s_2 \leq 3n^2$.

To study possible loop lengths, it is useful to consider the weighted digraph $\mathcal{A}$ representing the transition diagram of our 2nfa A after removing transitions on the endmarkers, and in which we set weights $+1$, -1, or 0 to arcs depending on whether they represent transitions where the input head is moved right, left, or kept stationary, respectively. It is straightforward that *any cycle of weight ℓ in $\mathcal{A}$ represents a computation loop of length ℓ in the automaton A*. Let us partition the digraph $\mathcal{A}$ into strongly connected components $\mathcal{C}_1, \mathcal{C}_2, \ldots, \mathcal{C}_r$. Let $\ell_i > 0$ be the greatest common divisor of cycle weights in $\mathcal{C}_i$, for $1 \leq i \leq r$. It is easy to see that $\ell_1 + \ell_2 + \cdots + \ell_r \leq n$.

By Lemma 1(ii), it is possible to prove that input traversals can be both expanded and compressed. More precisely, it can be shown that [6]:

Lemma 2. *If there exists a traversal on 1^m from a state q_1 to a state q_2, whose dominant loop uses states belonging to the component $\mathcal{C}_i$, then there also exists another traversal from q_1 to q_2 on $1^{m+\mu\ell_i}$, for any integer $\mu > (m - 5n^2)/\ell_i$.*

Using these results, a quasi sweeping automata A' equivalent to A with $O(n^2)$ states is built in [6]. We now show how to improve this construction to get a linear instead of quadratic simulation on the input strings of length exceeding $5n^2$.

The state set Q' of A' will be the union of the state set Q of A with two new sets of states Q^+ and Q^-. The set Q will be used to simulate A on the endmarkers. In particular, U–turns will be precomputed and simulated with stationary moves. The set Q^+ (Q^-, resp.) will be used to simulate left to right (right to left, resp.) traversals. For instance, a cycle of ℓ_i states belonging to Q^+ will be used to simulate "dominant loops" involving states of the strongly connected component $\mathcal{C}_i$ in a left to right traversal. More precisely, for $1 \leq i \leq r$, we define $Q_i^+ = \{q_{i,0}^+, q_{i,1}^+, \ldots, q_{i,(\ell_i-1)}^+\}$ and $Q_i^- = \{q_{i,0}^-, q_{i,1}^-, \ldots, q_{i,(\ell_i-1)}^-\}$, with the following deterministic transitions:

$$\delta'(q_{i,k}^+, 1) = \{(q_{i,(k+1) \text{ MOD } \ell_i}^+, +1)\} \text{ and } \delta'(q_{i,k}^-, 1) = \{(q_{i,(k+1) \text{ MOD } \ell_i}^-, -1)\}.$$

The state set of A' is $Q' = Q \cup Q^+ \cup Q^-$, where $Q^+ = \bigcup_{i=1}^r Q_i^+$, and $Q^- = \bigcup_{i=1}^r Q_i^-$. Since $\ell_1 + \ell_2 + \cdots + \ell_r \leq n$, it is obvious that there are at most $3n$ states in Q'.

Let us now see how to map the states of A into the states of A'. For the sake of brevity, we explain how this mapping works on states involved in left to right traversals. Its extension to encompass traversals in the opposite way can be argued easily. Before defining $\varphi : Q \to Q^+$, we need some notations. Given two states $p, q \in Q$, we write $p \triangleright^{+x} q$ if there exists a computation path of A which starts from p with the input head at a position d, ends in q at the position $d + x$, and which does not visit the endmarkers. By Lemma 1(i), such path does not depend on the input head position, provided that both d and $d + x$ are at least n^2 positions away from either endmarker. A component $\mathcal{C}_i$ is said to be *positive* (*negative*, resp.) if it contains at least one cycle of positive (negative) weight. Note that a component can be, at the same time, positive and negative. Furthermore, all the states used in the dominant loop of a left to right traversal (see Lemma 1) belongs to the same positive component. Given a *positive* component C_i, we designate some state $q_c \in C_i$ as the *center* of the component and, for any state $p \in C_i$, we define

$$\kappa(p) = \min\{x \in \mathbf{N} : q_c \triangleright^{+x} p\}, \quad \text{and} \quad \varphi(p) = q_{i,\kappa(p) \bmod \ell_i}^+.$$

This mapping is well defined since $C_1, C_2, \ldots, C_r$ are disjoint sets. Note also that $\varphi(q_c) = q_{i,0}^+$, and that $q_c \triangleright^{+\kappa(p)} p$, for each $p \in C_i$. The mapping φ defines a partition of the set of states in C_i; states in the same class have the same "distance" modulo ℓ_i from the center and, for sufficiently large inputs, they can be considered equivalent:

Lemma 3. *Let $q, p \in Q$ be in the same positive component C_i. Then $q \triangleright^{+m} p$ in A if and only if $\varphi(q) \triangleright^{+m} \varphi(p)$ in A', for each $m \geq 2n^2 + n$.*

Proof. First, for each two states q', p' within the same component C_i, and each two integers x_1, x_2, $q' \triangleright^{+x_1} p'$ and $q' \triangleright^{+x_2} p'$ implies that $x_1 \bmod \ell_i = x_2 \bmod \ell_i$. The strongly connected C_i must have a path $p' \triangleright^{+h} q'$, for some integer h, and hence also two loops, namely $q' \triangleright^{+(x_1+h)} q'$ and $q' \triangleright^{+(x_2+h)} q'$. But then both $x_1 + h$ and $x_2 + h$ are integer multiples of ℓ_i, i.e., $x_1 \bmod \ell_i = x_2 \bmod \ell_i$.

Thus, if we replace the path $q_c \triangleright^{+\kappa(p)} p$ by a path $q_c \triangleright^{+\kappa(q)} q \triangleright^{+m} p$, we get $\kappa(p) \bmod \ell_i = (\kappa(q) + m) \bmod \ell_i$. Therefore, $q \triangleright^{+m} p$ implies that $m \bmod \ell_i = (\kappa(p) - \kappa(q)) \bmod \ell_i$.

Conversely, let $m \bmod \ell_i = (\kappa(p) - \kappa(q)) \bmod \ell_i$, for some $m \geq 2n^2 + n$. Since C_i is strongly connected, we can find a path $q \triangleright^{+m'} p$, for some $m' < n$. By the argument above, m' must satisfy $m' \bmod \ell_i = (\kappa(p) - \kappa(q)) \bmod \ell_i = m \bmod \ell_i$. Thus, to get a path $q \triangleright^{+m} p$ of weight m, we only have to insert a suitable number of cycles beginning and ending in q, into the path $q \triangleright^{+m'} p$. This is possible, since the set of integers $z > 2n^2$ such that A, starting from the state q with the input head at the position i, reaches the $(i+z)$th input square in the same state, not visiting the endmarkers, coincides with the set of integer multiples of ℓ_i greater than $2n^2$. For details, see [6, Lemma 3.4]. Thus, $q \triangleright^{+m} p$ if and only if $m \bmod \ell_i = (\kappa(p) - \kappa(q)) \bmod \ell_i$, for each $m \geq 2n^2 + n$.

On the other hand, from the definition of δ' and φ, the machine A' has a path $\varphi(q) \triangleright^{+m} \varphi(p)$ if and only if $m \bmod \ell_i = (\kappa(p) - \kappa(q)) \bmod \ell_i$, for each $m > 0$. $\square$

Let us now complete the definition of the transition function δ' of A' by stating its behavior on the endmarkers. We recall that, as previously observed, a computation of A can be decomposed in U–turns and traversals. First, to simulate U–turns, for each $p, q \in Q$, we set

- $(q, 0) \in \delta'(p, \vdash)$ (similarly, $(q, 0) \in \delta'(p, \dashv)$) if and only if there exists a U–turn on the left (right) endmarker starting in the state p and ending in q.

To simulate traversals, we must introduce further moves on the endmarkers. Again, for the sake of brevity, we concentrate on left to right traversals. Using the form recalled in Lemma 1(ii), we substitute such a traversal with a deterministic computation executing one of the simple cycles we have introduced. The initial and the final part of the traversal are precomputed. We define such moves by considering a very large input. In what follows, we let

$$M = \min\{k\rho : k \in \mathbf{N}, \, \rho = \operatorname{lcm}(\ell_1, \ldots, \ell_r) \text{ and } k\rho > 3n^2\}.$$

Suppose, for instance, the input length is $2M$. After reading the left end-marker and the first M input symbols, a state p is reached, which belongs to a positive component $\mathcal{C}_i$. It is easy to see that p belongs to the dominant loop. We want the simulating automaton A' to reach the state $\varphi(p)$ after reading this portion of the input. To this aim, since ℓ_i divides M, it suffices to start the traversal directly in the state $\varphi(p)$. So, we simulate this segment of computation by a single move on the left endmarker. A similar argument can be used for the final part of the traversal. Thus, for $q \in Q$ and $\tilde{p} \in Q^+$, we set

- $(\tilde{p}, 1) \in \delta'(q, \vdash)$ if and only if there exists $r \in Q$ and $p \in \varphi^{-1}(\tilde{p})$ such that $(r, 1) \in \delta(q, \vdash)$ and $r \triangleright^{+M} p$ in A;
- $(q, 0) \in \delta'(\tilde{p}, \dashv)$ if and only if there exists $p \in \varphi^{-1}(\tilde{p})$ with $p \triangleright^{+M} q$ in A.

Similar moves can be defined to simulate right to left traversals. At this point, our simulating automaton is completely defined as $A' = (Q', \{1\}, \delta', q_0, F)$, where Q' and δ' have been presented so far, while q_0 and F are, respectively, the initial state and the set of final states of A. Now we prove the equivalence for left to right traversals. Its validity for symmetrical traversals comes straightforwardly.

Theorem 4. *Let $m > 5n^2$ and $q_1, q_2 \in Q$. There exists a left to right traversal of A on the input 1^m from q_1 to q_2 if and only if there exists a left to right traversal of A' from q_1 which is followed by a stationary move to reach q_2.*

Proof. Since each loop length ℓ_i divides M, by Lemma 2, it is not hard to prove that, for $m > 5n^2$, there exists a left to right traversal of A on the input 1^m from q_1 to q_2 if and only if the same holds true on the input 1^{m+2M}. So, for A, consider the input 1^{m+2M} instead of 1^m.

Only if–part. We subdivide a left to right traversal of A on 1^{m+2M} into the following three phases:

Phase 1. *A leaves the left endmarker from the state q_1 and enters a state $r \in Q$, in a single step. From r, it starts consuming the next M input symbols till it reaches a state p.*

Phase 2. *From p, which belongs to a certain positive component $\mathcal{C}_i$, a loop of length ℓ_i begins, which is repeated till reaching a state q with the input head M positions far from the right endmarker. In this phase, m input symbols are consumed.*

Phase 3. *From q, the last M symbols are consumed, and the right endmarker is finally reached in the state q_2.*

Let us see how these phases are reproduced by A' on the input 1^m. By definition of δ', Phase 1 implies that $(\varphi(p), 1) \in \delta'(q_1, \vdash)$, i.e., A' reaches the state

$\varphi(p)$ by consuming the left endmarker. Next, by Phase 2 and Lemma 3, we get that $\varphi(p) \triangleright^{+m} \varphi(q)$, i.e., A' reaches the right endmarker in the state $\varphi(q)$. Finally, Phase 3 and the definition of δ' ensure that $(q_2, 0) \in \delta'(\varphi(q), \dashv)$, i.e., A' enters the state q_2 with a stationary move on the right endmarker.

If–part. A left to right traversal of A' on 1^m from q_1 to q_2 can be decomposed into a first move that takes A' from q_1 to some state $\tilde{p} \in Q^+$ with the input head on the first '1'. By definition of δ', we know that there exists $p \in \varphi^{-1}(\tilde{p})$ that is reached from q_1 on A after consuming the left endmarker and M symbols. Next, A' enters a loop that takes it to a state $\tilde{q}$ upon consuming the m symbols of the input. Lemma 3 ensures that there exists a computation of A from p to some $q \in \varphi^{-1}(\tilde{q})$, traversing m input tape positions. Finally, A' has a stationary move on the right endmarker that takes it from $\tilde{q}$ to q_2; by definition of δ', there exists a computation of A which, starting in q, upon reading M input symbols, reaches the right endmarker in q_2. In conclusion, we get the existence of a left to right traversal of A from q_1 to q_2 on the input 1^{m+2M}. □

We are now ready to prove the main result of this section.

Lemma 5. *For each n–state unary 2nfa A, there exists an almost equivalent quasi sweeping 2nfa A' with no more than $3n$ states. Moreover, $L(A)$ and $L(A')$ coincide on strings of length greater than $5n^2$.*

Proof. The argument is a straightforward induction on the number of times the input head visits the endmarkers, using Theorem 4, its right–to–left counterpart, and the fact that U–turns are precomputed in A'. □

Theorem 6. *For each n–state unary 2nfa A, there exists an almost equivalent quasi sweeping 2nfa A'' with no more than $2n + 2$ states. Moreover, $L(A)$ and $L(A'')$ coincide on strings of length greater than $5n^2$.*

Proof. We replace A' of Lemma 5 by a machine A'' using only the states in Q^+ and Q^-, i.e., the original states are removed: Each sequence of stationary moves, at either endmarker, can be replaced by a single computation step, switching from Q^+ to Q^-, or vice versa. We need also a new initial and final state. □

In [2], a unary n–state one–way nondeterministic finite automaton (1nfa) is turned into an equivalent 1nfa consisting of an initial path of $O(n^2)$ states ending in a state where a nondeterministic choice is taken. Such a choice leads into one among a certain number of disjoint cycles, and the rest of the computation is deterministic. The total number of states included in the cycles does not exceed n. This structure is usually known as Chrobak Normal Form for 1nfa's. In our framework, this result can be reformulated by saying that there exists an *almost equivalent* n–state 1nfa in which the only nondeterministic decision is taken at the beginning of the computation. From this point of view, Theorem 6 can be regarded to as an extension of this result to the two–way machines, and might suggest a sort of normal form for 2nfa's.

Finally, we observe that if the language accepted by the given 2nfa A is cyclic — i.e., for some $\lambda > 0$, $1^m \in L(A)$ if and only if $1^{m+\lambda} \in L(A)$ — the automaton A'' yielded by our construction is fully equivalent to A.

4   Subexponential Deterministic Simulation

In this section, we show how to simulate an n–state unary 2nfa by a deterministic 2dfa with only $O(n^{\log n+4})$ states, thus improving previous bounds in the literature. To this purpose, by using a divide–and–conquer technique, we first prove that if the 2nfa is quasi sweeping, then the simulation requires no more than $n^{\log n+4}$ states.

Let Q be the set of n states of a given quasi sweeping 2nfa A, and let $q_0 \in Q$ be the initial state. Without loss of generality, assume that A accepts by entering a unique state q_f with the input head scanning the left endmarker. The core of our simulation technique is the implementation of the predicate *reachable* which is defined as follows. Fix an input length m, the states $q, p \in Q$, and an integer $k \geq 1$. Then *reachable*(q, p, k) is true if and only if there exists a computation path of A which starts and ends with the input head scanning the left endmarker in the state q and p, respectively, and visits that endmarker at most $k+1$ times. It is easy to see that if *reachable*(q, p, k) holds true, then there exists a witness computation path where the states encountered when the input head scans the left endmarker are all different. This implies that 1^m *is accepted by A if and only if* reachable(q_0, q_f, n) *holds true*.

To evaluate the predicate *reachable*, it is useful to concentrate first on computing its "simplest" case *reach1*(q, p) = *reachable*$(q, p, 1)$, for any $q, p \in Q$. There are two possibilities: either (i) $(p, 0) \in \delta(q, \vdash)$, or (ii) the path is a left to right traversal of the input followed by a traversal in the opposite way. Case (i) can be immediately tested. For case (ii) (see Theorem 6), observe that all the moves involved are deterministic with the following possible exceptions:

- the first move from the left endmarker, in the state q,
- the first move from the right endmarker, at the beginning of the right to left traversal.

So, case (ii) can be tested by directly simulating the computation of A and trying all the possible nondeterministic choices above listed. In particular, we have no more than n nondeterministic possibilities on each endmarker. One can easily argue that *reach1* can be realized on a 2dfa with n^3 states.

Now, we use *reach1* as a subroutine to compute the predicate *reachable*:

```
function reachable(q, p, k)
if k = 1 then return reach1(q, p)
else begin
    for each state r ∈ Q do
        if reachable(q, r, ⌊k/2⌋) then
            if reachable(r, p, ⌈k/2⌉) then
                return TRUE
    return FALSE
end
```

Such a function can be implemented by using a pushdown store in which, at each position, a pair of states and an integer, corresponding to one activation of the function, are kept. The maximal pushdown height is $\lceil \log_2 n \rceil + 1$.

To reduce the number of possible configurations, we substitute the pushdown with a stack, i.e., we utilize not only the element at the top position, but all information that is currently stored. When a state $r \in Q$ is considered in the **for**–loop of the function, the call $reachable(q, r, \lfloor k/2 \rfloor)$ is simulated by adding the state r on top of the stack, while the call $reachable(r, p, \lceil k/2 \rceil)$ is indicated by replacing r with a marked copy $\hat{r}$ on top of the stack. We now informally explain how, from this stack, it is possible to recover the original pushdown store. For the sake of simplicity, we give an example of the computation of $reachable$ on a quasi sweeping 2nfa with $n = 32$ states (with some technicalities, the argument can be easily extended to any n). Suppose that the current stack content, from the bottom to the top, is $\hat{q}_3\, q_2\, q_5\, \hat{q}_1$. By also representing, implicitly, q_f and $\hat{q}_0$ at the bottom, we have the following situation:

$$\text{STACK} \implies \frac{position}{content} \begin{array}{|cccccc} -1 & 0 & 1 & 2 & 3 & 4 \\ q_f & \hat{q}_0 & \hat{q}_3 & q_2 & q_5 & \hat{q}_1 \end{array}$$

Each stack position s containing a marked state corresponds to an activation of $reachable$ in the inner **if**–statement. The other state of this activation is the first nonmarked state to the left of position s. The third parameter of this activation is exactly $k = n/2^s$ (if n is a power of 2, otherwise k can be easily computed as well). For example, the state $\hat{q}_3$ at position 1 represents the function call $reachable(q_3, q_f, 16)$, while the state $\hat{q}_1$ at position 4 represents $reachable(q_1, q_5, 2)$. The state at position 0 corresponds to the main activation of $reachable$, namely that with parameters $(q_0, q_f, 32)$. In a similar way, we can recover activations corresponding to nonmarked states in the stack. In the example, the sequence of current activations of $reachable$ is

$$(q_0,\, q_f,\, 32),\ (q_3,\, q_f,\, 16),\ (q_3,\, q_2,\, 8),\ (q_3,\, q_5,\, 4),\ (q_1,\, q_5,\, 2).$$

The computation of $reachable(q_0, q_f, n)$ can be implemented by a 2dfa A' which keeps in its state a stack configuration corresponding to the bottom of the recursion of $reachable$. Starting with the head on the left endmarker, A' simulates the execution of the function $reach1$, and then, depending on the outcome of this simulation, it selects another stack configuration. The number of stack configurations corresponding to the bottom of the recursion is bounded[1] by $n^{\log n} 2^{\log n}$. In conclusion, by also considering the cost n^3 of implementing the function $reach1$, we get the following

Theorem 7. *Each n–state quasi sweeping unary 2nfa can be simulated by a 2dfa with $n^{\log n + 4}$ states.*

---

[1] Notice that for any accepting computation of A, there actually exists an accepting computation which does not visit the left endmarker in the same state twice. Thus, we can implement the **for**–loop of the function $reachable$, by considering only states that are not already in the stack. This enables us to lower the total number of stack configurations to $(n-1) \cdot (n-2) \cdot \ldots \cdot (n - \log n) \cdot 2^{\log n}$, i.e., $n \cdot (n-1) \cdot \ldots \cdot (n - \log n)$.

By combining this with results in Section 3, we get

Theorem 8. *For each n–state unary 2nfa, there exists an equivalent 2dfa with* $O(n^{\log n+4})$ *states.*

Proof. (outline) From a given n–state 2nfa A, according to Theorem 6, we can build an almost equivalent quasi sweeping 2nfa A'' with no more than $2n+2$ states. In turn, this automaton can be simulated by a 2dfa B, as shown above. By considering also the structure of A'', one can prove that B has no more than $2n^{\log n+4}$ states. The final 2dfa B' equivalent to A works as follows:

- in a first phase, it simulates an automaton with $5n^2+1$ states accepting the strings in $L(A)$ of length not exceeding $5n^2$;
- if the input length exceeds $5n^2$, it simulates B.

It is easy to show that the total number of states of B' is $O(n^{\log n+4})$.   □

We remark that, as far as the authors know, the subexponential simulation cost contained in Theorem 8 represents an improvement of the best unary simulation of 2nfa's by 2dfa's known in the literature which used $O(e^{\sqrt{n \ln n}})$ states [6].

References

1. Berman P., Lingas A.: On the complexity of regular languages in terms of finite automata. Tech. Report 304, Polish Academy of Sciences, 1977. 398
2. Chrobak M.: Finite automata and unary languages. Theoretical Computer Science, **47** (1986) 149–58. 398, 399, 404
3. Geffert V.: Nondeterministic computations in sublogarithmic space and space constructibility. SIAM J. Computing, **20** (1991) 484–98. 400
4. Hopcroft J., Ullman J.: *Introduction to automata theory, languages, and computation.* Addison–Wesley, Reading, MA, 1979. 399
5. Mereghetti C., Pighizzini G.: Two–way automata simulations and unary languages. J. Aut., Lang. Comb., **5** (2000) 287–300. 400
6. Mereghetti C., Pighizzini G.: Optimal simulations between unary automata. SIAM J. Computing, **30** (2001) 1976–92. 399, 400, 401, 402, 407
7. Piterman N., Vardi M. Y.: From bidirectionality to alternation. This Volume. 398
8. Sakoda W., Sipser M.: Nondeterminism and the size of two–way finite automata. In Proc. 10th ACM Symp. Theory of Computing, 1978, pp. 275–86. 398
9. Sipser M.: Lower bounds on the size of sweeping automata. J. Computer and System Science, **21** (1980) 195–202. 398, 399

The Complexity of the Minimal Polynomial*

Thanh Minh Hoang and Thomas Thierauf

Abt. Theoretische Informatik, Universität Ulm
89069 Ulm, Germany
{hoang,thierauf}@informatik.uni-ulm.de

Abstract. We investigate the computational complexity of the minimal polynomial of an integer matrix.
We show that the computation of the minimal polynomial is in $\mathbf{AC}^0(\mathbf{GapL})$, the $\mathbf{AC}^0$-closure of the logspace counting class $\mathbf{GapL}$, which is contained in $\mathbf{NC}^2$. Our main result is that the problem is hard for $\mathbf{GapL}$ (under $\mathbf{AC}^0$ many-one reductions). The result extends to the verification of all invariant factors of an integer matrix.
Furthermore, we consider the complexity to check whether an integer matrix is diagonalizable. We show that this problem lies in $\mathbf{AC}^0(\mathbf{GapL})$ and is hard for $\mathbf{AC}^0(\mathbf{C_=L})$ (under $\mathbf{AC}^0$ many-one reductions).

1 Introduction

The motivation for our work is twofold: 1) we want to understand the computational complexity of some classical problems in linear algebra, 2) by locating such problems in small space complexity classes we want to clarify the inclusion relationship of such classes.

The *minimal polynomial* of a matrix plays an important role in the theory of matrices. Algorithms to compute the minimal polynomial of a matrix have been studied for a long time. The best known deterministic algorithm to compute the minimal polynomial of an $n \times n$ matrix makes $O(n^3)$ field operations [Sto98]. The Smith normal form of a polynomial matrix can be computed by a randomized $\mathbf{NC}^2$-circuit, i.e., in $\mathbf{RNC}^2$. Therefore the rational canonical form of a matrix and the minimal polynomial of a matrix can be computed in $\mathbf{RNC}^2$ as well (see [KS87, vzGG99] for details). In the case of integer matrices there are even $\mathbf{NC}^2$-algorithms [Vil97].

We take a different approach to compute the minimal polynomial of an integer matrix: we show that the problem can be reduced to matrix powering and solving systems of linear equations. Therefore it is in the class $\mathbf{AC}^0(\mathbf{GapL})$, a subclass of $\mathbf{NC}^2$. Our main result is with respect to the hardness of the problem: we show that the computation of the determinant of a matrix can be reduced to the computation of the minimal polynomial of a matrix. Therefore the problem is hard for $\mathbf{GapL}$.

The minimal polynomial is the first polynomial of the system of all *invariant factors* of a matrix. This system completely determines the structure of the

---

* This work was supported by the Deutsche Forschungsgemeinschaft

J. Sgall, A. Pultr, and P. Kolman (Eds.): MFCS 2001, LNCS 2136, pp. 408–420, 2001.

matrix. Its computation is known to be in $\mathbf{NC}^2$ [Vil97] for integer matrices. We extend our results and techniques to the *verification* of all the invariant factors of a given integer matrix: it is in $\mathbf{AC}^0(\mathbf{C_=L})$ and hard for $\mathbf{C_=L}$.

Using the results about the minimal polynomial, we can classify some more classical problems in linear algebra: a matrix is diagonalizable if it is similar to a diagonal matrix. Testing similarity of two matrices is known to be in $\mathbf{AC}^0(\mathbf{C_=L})$ [HT00]. We show that the problem to decide whether a given integer matrix is diagonalizable is in $\mathbf{AC}^0(\mathbf{GapL})$ and hard for $\mathbf{AC}^0(\mathbf{C_=L})$.

To obtain the latter result, we have to solve a problem that is interesting in its own: decide, whether all eigenvalues of a given integer matrix are pairwise different. This can be done in $\mathbf{AC}^0(\mathbf{C_=L})$.

2   Preliminaries

For a nondeterministic logspace bounded Turing machine M, we denote the number of accepting paths on input x by $acc_M(x)$, and by $rej_M(x)$ the number of rejecting paths. The difference of these two numbers is $gap_M(x) = acc_M(x) - rej_M(x)$.

For the counting classes, we have #$\mathbf{L}$, the class of functions $acc_M(x)$ for some nondeterministic logspace bounded Turing machine M, and $\mathbf{GapL}$ based analogously on functions gap_M. Based on counting, we consider the language class $\mathbf{C_=L}$: a set A is in $\mathbf{C_=L}$, if there exists a $f \in \mathbf{GapL}$ such that for all x: $x \in A \Longleftrightarrow f(x) = 0$.

For sets A and B, A is $\mathbf{AC}^0$-*reducible to* B, if there is a logspace uniform circuit family of polynomial size and constant depth that computes A with unbounded fan-in and-, or-gates and oracle gates for B. In particular, we consider the classes $\mathbf{AC}^0(\mathbf{C_=L})$ and $\mathbf{AC}^0(\mathbf{GapL})$ of sets that are $\mathbf{AC}^0$-reducible to a set in $\mathbf{C_=L}$, respectively a function in $\mathbf{GapL}$.

A is $\mathbf{AC}^0$ *many-one reducible to* B, in symbols: $A \leq_m^{AC^0} B$, if there is a function $f \in \mathbf{AC}^0$ such that for all x we have $x \in A \Longleftrightarrow f(x) \in B$. All reductions in this paper are $\mathbf{AC}^0$ many-one reductions.

Let $A \in \mathbf{F}^{n \times n}$ be a matrix over the field $\mathbf{F}$. A nonzero polynomial $p(x)$ over $\mathbf{F}$ is called an *annihilating polynomial* of A if $p(A) = 0$. The Cayley-Hamilton Theorem states that the characteristic polynomial $\chi_A(x)$ of A is an annihilating polynomial. The characteristic polynomial is a *monic polynomial*: its highest coefficient is one. The *minimal polynomial* of A, denoted $\mu_A(x)$, is the unique monic annihilating polynomial of A with minimal degree.

Let polynomial $d_k(x)$ be the greatest common divisor of all sub-determinants of $(xI - A)$ of order k. For example $d_n(x) = \chi_A(x)$. We see that d_k divides d_{k+1} for each index $0 \leq k \leq n$. Define $d_0(x) \equiv 1$. The *invariant factors* of $(xI - A)$ (or A, for short) are defined as the following (monic) polynomials:

$$i_1(x) = \frac{d_n(x)}{d_{n-1}(x)}, \quad i_2(x) = \frac{d_{n-1}(x)}{d_{n-2}(x)}, \quad \ldots, \quad i_n(x) = \frac{d_1(x)}{d_0(x)}.$$

The characteristic polynomial of A is the product of all the invariant factors: $\chi_A(x) = i_1(x) \cdots i_n(x)$. The $n \times n$ polynomial matrix that has the invariant factors of A as its diagonal entries (starting with $i_n(x)$) and zero elsewhere is the *Smith normal form* of $xI - A$, denoted by $diag\{i_n(x), \ldots, i_1(x)\}$.

We decompose the invariant factors into irreducible divisors over the given number field $\mathbf{F}$:

$$i_1(x) = [e_1(x)]^{c_1} [e_2(x)]^{c_2} \cdots [e_s(x)]^{c_s},$$
$$i_2(x) = [e_1(x)]^{d_1} [e_2(x)]^{d_2} \cdots [e_s(x)]^{d_s},$$
$$\vdots \qquad\qquad \vdots \qquad\qquad (0 \le l_k \le \cdots \le d_k \le c_k; \; k = 1, 2, \ldots, s).$$
$$i_n(x) = [e_1(x)]^{l_1} [e_2(x)]^{l_2} \cdots [e_s(x)]^{l_s},$$

The irreducible divisors $e_1(x), e_2(x), \ldots, e_s(x)$ are distinct (with highest coefficient 1) and occur in $i_1(x), i_2(x), \ldots, i_n(x)$. All powers $[e_1(x)]^{c_1}, \ldots, [e_s(x)]^{l_s}$, which are different from 1, are called the *elementary divisors* of A in $\mathbf{F}$

Note that the coefficients of the characteristic polynomial and the invariant factors of an integer matrix are all integers. Furthermore, the set of eigenvalues of A is the same as the set of all roots of $\chi_A(x)$ which, in turn, is the set of all roots of $\mu_A(x)$.

Next, we define some natural problems in linear algebra we are looking at. If nothing else is said, our domain for the algebraic problems are the integers.

1. POWERELEMENT
   Input: an $n \times n$-matrix A and i, j, and m, $(1 \le i, j, m \le n)$.
   Output: $(A^m)_{i,j}$, the (i, j)-th element of A^m.
2. DETERMINANT
   Input: an $n \times n$-matrix A.
   Output: $\det(A)$, the determinant of A.
3. CHARPOLYNOMIAL
   Input: an $n \times n$-matrix A.
   Output: $(c_0, c_1, \ldots, c_{n-1})$, the coefficients of the characteristic polynomial $\chi_A(x) = x^n + c_{n-1}x^{n-1} + \cdots + c_0$ of the matrix A.
4. MINPOLYNOMIAL
   Input: an $n \times n$-matrix A.
   Output: $(c_0, c_1, \ldots, c_{m-1})$, the coefficients of the minimal polynomial $\mu_A(x) = x^m + c_{m-1}x^{m-1} + \cdots + c_1 x + c_0$ of the matrix A.
5. INVSYSTEM
   Input: an $n \times n$-matrix A.
   Output: the system of invariant factors of the matrix A.

The first three problems are complete for **GapL** [ABO99, HT00, ST98]. MINPOLYNOMIAL and INVSYSTEM are in $\mathbf{RNC}^2$ [KS87], and in $\mathbf{NC}^2$ for integer matrices [Vil97].

For each of them, we define the corresponding *verification problem* as the graph of the corresponding function: for a fixed function $f(x)$, define V-f as the set all pairs (x, y) such that $f(x) = y$. This yields the verification problems V-POWERELEMENT, V-DETERMINANT, V-CHARPOLYNOMIAL, V-MINPOLYNOMIAL and V-INVSYSTEM. The first three problems are known to be complete for $\mathbf{C_=L}$ [HT00]. We note that a special case of V-DETERMINANT is SINGULARITY where one has to decide whether the determinant of a matrix is zero. SINGULARITY is complete for $\mathbf{C_=L}$ as well.

Related problems are computing the rank of a matrix, RANK, or deciding whether a system of linear equations is feasible, FSLE for short. FSLE is many-one complete for $\mathbf{AC^0(C_=L)}$ [ABO99].

SIMILARITY is another many-one complete problem for $\mathbf{AC^0(C_=L)}$ [HT00]. Two square matrices A and B are *similar*, if there exists a nonsingular matrix P such that $A = P^{-1}BP$. It is well known that A and B are similar iff they have the same invariant factors or, what is the same, the same elementary divisors (see for example [Gan77]). Another characterization of similarity is based on tensor products. This was used by Byrnes and Gauger [BG77] to get the $\mathbf{AC^0(C_=L)}$ upper bound on SIMILARITY.

3   The Minimal Polynomial

In this section we show that MINPOLYNOMIAL is in $\mathbf{AC^0(GapL)}$ and is hard for $\mathbf{GapL}$.

3.1   Upper Bound

We mentioned in the previous section that the minimal polynomial of an integer matrix can be computed in $\mathbf{NC^2}$ [Vil97]. We take a different approach and show that MINPOLYNOMIAL is in $\mathbf{AC^0(GapL)}$, a subclass of $\mathbf{NC^2}$.

Let $m(x) = x^m + c_{m-1}x^{m-1} + \cdots + c_0$ be a monic polynomial. Then $m(x)$ is the minimal polynomial of A iff 1) m is an annihilating polynomial of A, i.e., $m(A) = A^m + c_{m-1}A^{m-1} + \cdots + c_0 I = \mathbf{0}$, and 2) for every monic polynomial $p(x)$ of degree smaller than $m(x)$, we have $p(A) \neq \mathbf{0}$.

Define vectors $\boldsymbol{a}_i = vec(A^i)$ for $i = 0, 1, 2, \ldots, n$, where $vec(A^i)$ is the vector of length n^2 obtained by putting the columns of A^i below each other. The equation $m(A) = \mathbf{0}$ can be rewritten as

$$\boldsymbol{a}_m + c_{m-1}\boldsymbol{a}_{m-1} + \cdots + c_0\boldsymbol{a}_0 = \mathbf{0}. \tag{1}$$

In other words, the vectors $\boldsymbol{a}_m, \ldots, \boldsymbol{a}_0$ are linearly dependent. Consequently, for some polynomial p with degree $k < m$, the inequation $p(A) \neq \mathbf{0}$ means that the vectors $\boldsymbol{a}_k, \ldots, \boldsymbol{a}_0$ are linearly independent.

In summary, the coefficients of $\mu_A(x)$ are the solution $(c_{m-1}, \ldots, c_0)$ of the system (1), for the smallest m where this system has a solution. Hence we have the following algorithm to compute $\mu_A(x)$:

MINPOLYNOMIAL(A)
1   compute vectors $\boldsymbol{a}_i = vec(A^i)$ for $i = 0, \ldots, n$
2   determine m such that $\boldsymbol{a}_0, \boldsymbol{a}_1, \ldots, \boldsymbol{a}_{m-1}$ are linearly independent and $\boldsymbol{a}_0, \boldsymbol{a}_1, \ldots, \boldsymbol{a}_m$ are linearly dependent
3   solve the linear system $\boldsymbol{a}_m + c_{m-1}\boldsymbol{a}_{m-1} + \cdots + c_0\boldsymbol{a}_0 = \mathbf{0}$
4   **return** $(1, c_{m-1}, \ldots, c_0)$, the coefficients of $\mu_A(x)$.

Step 1 and 3 in the above algorithm can be computed in **GapL** (see [ABO99]). In Step 2, checking linear independence of given vectors is in **coC$_=$L** and linear dependence is in **C$_=$L** [ABO99]. Hence we end up in the **AC^0**-closure of **GapL**, namely **AC^0(GapL)**. Recall that **AC^0(GapL)** $\subseteq$ **NC^2**. We conclude:

Theorem 1. MINPOLYNOMIAL *is in* **AC^0(GapL)**.

3.2   Lower Bound

Our main result is to show the hardness of the computation of the minimal polynomial of a matrix. Namely, we show that it is hard for **GapL**.

A problem known to be complete for **GapL** is POWERELEMENT where one has to compute the entry (i, j) of A^m, for a $n \times n$ integer matrix A. W.l.o.g. we can focus on entry $(1, n)$ of A, i.e. $(A^m)_{1,n}$.

In order to reduce POWERELEMENT to MINPOLYNOMIAL, we construct a matrix C such that the value $(A^m)_{1,n}$ occurs as one of the coefficients of the minimal polynomial of C.

The reduction build on the techniques from Toda [Tod91], Valiant [Val92], and Hoang and Thierauf [HT00] to reduce matrix powering to the determinant, and the latter to the characteristic polynomial. We give the proof of this result here because we need the matrices constructed there. We follow the presentation from [ABO99] and [HT00].

Theorem 2. [HT00] POWERELEMENT $\leq_m^{AC^0}$ CHARPOLYNOMIAL.

Proof. Let A be an $n \times n$ matrix and $1 \leq m \leq n$. W.l.o.g. we fix $i = 1$ and $j = n$ in the definition of POWERELEMENT. In **AC^0** we construct a matrix C such that all the coefficients of its characteristic polynomial can be easily computed from the value $(A^m)_{1,n}$.

Interpret A as representing a directed bipartite graph on $2n$ nodes and e edges. That is, the nodes are arranged in two columns of n nodes each. In both columns, nodes are numbered from 1 to n. If entry $a_{k,l}$ of A is not zero, then there is an edge labeled $a_{k,l}$ from node k in the first column to node l in the second column. The number of non-zero entries in A is exactly e. Now, take m copies of this graph, put them in a sequence and identify each second column of nodes with the first column of the next graph in the sequence. Call the resulting graph G'.

Graph G' has $m + 1$ columns of nodes. The *weight* of a path in G' is the product of all labels on the edges of the path. The crucial observation now is that the entry at position $(1, n)$ in A^m is the sum of the weights of all paths in G' from node 1 in the first column to node n in the last column. Call these two nodes s and t, respectively.

Graph G' is further modified: for each edge (k, l) with label $a_{k,l}$, introduce a new node u and replace the edge by two edges, (k, u) with label 1 and (u, l) with label $a_{k,l}$. Now all paths from s to t have *even* length, but still the same weight. Add an edge labeled 1 from t to s. Call the resulting graph G. Let C be the adjacency matrix of G. Graph G has $N = m(n + e) + n$ nodes and therefore C is a $N \times N$ matrix.

From combinatorial matrix theory we know that the coefficient c_i in $\chi_C(x)$ equals the sum of the disjoint weighted cycles that cover $N - i$ nodes in G, with appropriate sign (see [BR91] or [CDS80] for more details). In the graph G, all edges go from a layer to the next layer. The only exception is the edge (t, s). So any cycle in G must use precisely this edge (t, s), and then trace out a path from s to t. Therefore each cycle in G have exactly the length $2m + 1$, and the weighted sum of all these cycles is precisely $(A^m)_{1,n}$ with the sign -1. Hence $c_{N-(2m+1)} = -(A^m)_{1,n}$ and all other coefficients must be zero. That is,

$$\chi_C(x) = x^N - ax^{N-(2m+1)},$$

is the characteristic polynomial of C, where $a = (A^m)_{1,n}$.   □

Theorem 3. POWERELEMENT $\leq_m^{AC^0}$ MINPOLYNOMIAL.

Proof. We consider the $N \times N$ matrix C from the previous proof in more detail.

Except for the edge from t to s, graph G is acyclic. Thus we can put the nodes of G in such an order, that adjacency matrix C is upper triangular for the first $N - 1$ rows with zeros along the main diagonal. The last row of C has a one in the first position (representing edge (t, s)), and the rest is zero.

We also consider the upper triangle in C. Each column of graph G' was split in our construction into two columns and we got a new node on every edge. The first part we describe by the $n \times e$ matrix F:

$$F = \begin{pmatrix} 1 \cdots 1\, 0 \cdots 0 \cdots 0 \cdots 0 \\ 0 \cdots 0\, 1 \cdots 1 \cdots 0 \cdots 0 \\ \vdots \quad \vdots \quad \ddots \quad \vdots \\ 0 \cdots 0\, 0 \cdots 0 \cdots 1 \cdots 1 \end{pmatrix}$$

The number of ones in the k-th row of F is the number of edges leaving node k in the first column of G'.

From each of the newly introduced nodes there is one edge going out. Hence this second part we can describe by the $e \times n$-matrix S, which has precisely one non-zero entry in each row. The value of the non-zero entry is the weight of the corresponding edge in G'. With the construction of graph G it is not hard to see that $FS = A$. Now we can write C as a block matrix as follows:

$$C = \begin{pmatrix} \begin{array}{c|c} \begin{matrix} F \\ & S \\ & & \ddots \\ & & & F \\ & & & & S \end{matrix} & \\ \hline L & \end{array} \end{pmatrix}$$

There is m-times matrix F, alternating with m-times matrix S. L is the $n \times n$ matrix with a one at position $(n,1)$ and zero elsewhere. Hence C is a $(2m+1) \times (2m+1)$ block matrix. The empty places in C are all zero matrix.

Let a denote the element $(A^m)_{1,n}$. We claim that the minimal polynomial of C is $\mu_C(x) = x^{4m+2} - ax^{2m+1}$.

First, we observe that $d_{N-1}(x) = x^l$ for some l, because the minor of order $N-1$ of the matrix $xI - C$ at the position $(1,1)$ is x^{N-1}. Therefore the minimal polynomial must have the form

$$\mu_C(x) = \chi_C(x)/d_{N-1}(x) = x^{N-l} - ax^{N-(2m+1)-l}.$$

Define polynomials $p_k(x) = x^{(2m+1)+k} - ax^k$ for $0 \le k \le N - (2m+1)$. To prove our claim, we have to show that $p_{2m+1}(C) = 0$ and $p_k(C) \ne 0$ for $k < 2m+1$. To do so, we explicitly construct all the powers of C. The general form of C^i for $i \le 2m$ is as follows:

The entry $(C^i)_{j,i+j}$ for $1 \le j \le 2m-i+1$ and $i \le 2m$ lies on the sub-diagonal $(* \cdots *)$ and has the following form:

$$(C^i)_{j,i+j} = \begin{cases} S^{(j-1) \bmod 2}(FS)^{\frac{i-1}{2}} F^{j \bmod 2}, & \text{for odd } i, \\ (FS)^{j \bmod 2} S^{(j-1) \bmod 2}(FS)^{\frac{i-2}{2}} F^{(j-1) \bmod 2}, & \text{otherwise.} \end{cases}$$

The entry $(C^i)_{2m+1-i+k,k}$ for $1 \le k \le i$ and $i \le 2m$ lies on the sub-diagonal $(+ \cdots +)$ and has the following form:

$$(C^i)_{2m+1-i+k,k} = S^{(i+k) \bmod 2}(FS)^{\lfloor \frac{i-k}{2} \rfloor} L(FS)^{\lfloor \frac{k-1}{2} \rfloor} F^{(j-1) \bmod 2}.$$

From this we get in particular

$$C^{2m+1} = diag\{A^m L, SA^{m-1}LF, A^{m-1}LA, \ldots, LA^m\},$$
$$C^{4m+2} = diag\{A^m LA^m L, SA^{m-1}LA^m LF, A^{m-1}LA^m LA, \ldots, LA^m LA^m\}$$

Since $LA^m L = aL$, we have $p_{2m+1}(C) = C^{4m+2} - aC^{2m+1} = 0$. It remains to prove that $p_k(C) = C^{2m+1+k} - aC^k \ne 0$ for all $k \le 2m$. Note that it suffices to prove this for $k = 2m$, because $p_k(C) = 0$ for some k implies $p_{k+1}(C) = 0$.

For technical reasons we assume that the input matrix A is a nonsingular upper triangular matrix. The following lemma says that we can w.l.o.g. make this assumption.

Lemma 4. *Suppose A is an $n \times n$ matrix. Then there is a nonsingular upper triangular $p \times p$ matrix B such that $(B^m)_{1,p} = (A^m)_{1,n}$.*

Proof. We define B as an $(m+1) \times (m+1)$ block matrix in which all the elements of the principal diagonal are $n \times n$ identity matrices, all the elements of the first super-diagonal are matrices A and all the the other elements are zero-matrices. For $p = (m+1)n$ we have $(B^m)_{1,p} = (A^m)_{1,n}$ as claimed.   □

We compute C^{4m+1} as the product $C^{2m+1}C^{2m}$. Now we have $p_{2m}(C) = 0$ iff $C^{4m+1} = aC^{2m}$ iff $A^m L A^m = aA^m$. However, the latter equation cannot hold: by Lemma 4 we can assume that A is nonsingular. Therefore $rank(A^m L A^m) = rank(L) = 1$, whereas $rank(aA^m) \neq 1$. We conclude that $p_{2m}(C) \neq 0$.

   In summary, we have $\mu_C(x) = x^{4m+2} - ax^{2m+1}$, where $a = (A^m)_{1,n}$. Since the construction of graph G can be done in $\mathbf{AC}^0$, we have POWERELEMENT $\leq_m^{AC^0}$ MINPOLYNOMIAL as claimed.   □

3.3   The Invariant Factors

The system of all invariant factors of a matrix can be computed in $\mathbf{NC}^2$ [Vil97]. Since the minimal polynomial is one of the invariant factors, it follows from Theorem 3 that these are hard for **GapL** as well.

   In the *verification versions* of the above problems we have given A and coefficients of one, respectively several polynomials and have to decide whether these coefficients represent in fact the minimal polynomial, respectively the invariant factors of A.

   Note that in the case of the invariant factors we get potentially more information with the input than in the case of the minimal polynomial. Therefore, it could be that the invariant factors are easier to verify than the minimal polynomial. Interestingly we locate in fact the verification of the invariant factors in a seemingly smaller complexity class.

   To verify the minimal polynomial we can simplify the above algorithm for MINPOLYNOMIAL as follows:

V-MINPOLYNOMIAL$(A, c_{m-1}, \ldots, c_0)$
1   compute vectors $a_i = vec(A^i)$ for $i = 0, \ldots, m$
2   **if** $a_m + c_{m-1}a_{m-1} + \cdots + c_0 a_0 = 0$ **and**
        $a_0, a_1, \ldots, a_{m-1}$ are linearly independent
3   **then** accept **else** reject.

   Hence we get the same upper bound as for MINPOLYNOMIAL, namely $\mathbf{AC}^0(\mathbf{GapL})$. Since MINPOLYNOMIAL is hard for **GapL**, V-MINPOLYNOMIAL must be hard for $\mathbf{C}_=\mathbf{L}$. We summarize:

Corollary 5. V-MINPOLYNOMIAL *is in* $\mathbf{AC}^0(\mathbf{GapL})$ *and hard for* $\mathbf{C}_=\mathbf{L}$.

Next we show that the verification of the invariant factors is hard for $\mathbf{C}_=\mathbf{L}$ as well. However, as an upper bound we get the seemingly smaller class $\mathbf{AC}^0(\mathbf{C}_=\mathbf{L})$.

Theorem 6. V-INVSYSTEM *is in* $\mathbf{AC^0(C_=L)}$ *and hard for* $\mathbf{C_=L}$.

Proof. **Inclusion.** Let $\mathcal{S} = \{i_1(x), \ldots, i_n(x)\}$ be the system of n given monic polynomials and let A be an $n \times n$ matrix. We construct the companion matrices that correspond to the non-constant polynomials in $\mathcal{S}$. Let B denote the diagonal block matrix of all these companion matrices. Recall that $\mathcal{S}$ is the system of invariant factors of A iff A is similar to B. Testing similarity can be done in $\mathbf{AC^0(C_=L)}$ [HT00], therefore V-INVSYSTEM is in $\mathbf{AC^0(C_=L)}$ too.

Hardness. We continue with the setting from the proof of Theorem 3, in particular with matrix C. Our goal is to determine the system of all invariant factors of C. We have already shown that $i_1(x) = \mu_C(x) = x^{4m+2} - ax^{2m+1}$, where $(A^m)_{1,n} = a$. Next, we compute the invariant factors $i_2(x), \ldots, i_N(x)$.

It follows from the proof of Theorem 3 that $d_{N-1}(x) = x^{N-(4m+2)}$. Since $d_{N-1}(x) = i_2(x) \cdots i_N(x)$, each of the invariant factors must have the form x^l for some number l. Note that all non-constant invariant factors of the form x^l are already elementary divisors.

Define g_l to be the *number of occurrences of the elementary divisor* x^l. Clearly, if we have all numbers g_l, we can deduce the invariant factors. Numbers g_l can be determined from the ranks of matrices C^j (see [Gan77]). More precisely, let r_j denote *the rank of* C^j. The following formula relates the ranks to numbers g_j:

$$g_j = r_{j-1} + r_{j+1} - 2r_j, \tag{2}$$

for $j = 1, \ldots, t$, where $r_0 = N$ and t is the smallest index such that $r_{t-1} > r_t = r_{t+1}$. We can actually compute all the ranks r_j from the expressions we already have for matrices C^j.

Let us consider the blocks of C^j. By Lemma 4 we may assume that A is nonsingular, that is $\text{rank}(F) = \text{rank}(S) = \text{rank}(A) = n$. Therefore $\text{rank}(A^k) = \text{rank}(A^k F) = \text{rank}(A^k S) = n$ for any k. Hence blocks in C^j of the form $(FS)^k$, $(FS)^k F$, $(SF)^k$, or $(SF)^k S$ all have rank n (recall that $FS = A$). In all other blocks occurs matrix L. Recall that matrix L is all-zero except for the entry at the lower left corner, which is 1. Therefore, for any matrix M, we have $\text{rank}(ML) = 1$ iff the n-th column of M is a non-zero column. Analogously, $\text{rank}(LM) = 1$ iff the first row of M is a non-zero row. We conclude that all blocks that contain matrix L have rank 1.

Since the non-zero blocks of C^j are in pairwise different lines and columns, we can simply add up their ranks to obtain the rank of C^j. That way we get

$$r_j = \begin{cases} (2m + 1 - j)n + j, & \text{for } j = 1, \ldots, 2m, \\ 2m + 1, & \text{for } 2m + 1 \leq j. \end{cases}$$

The ranks don't change any more from $j = 2m+1$ on. Hence $t = 2m+1$. Plugged into the formula (2) we get

$$g_j = \begin{cases} N - n(2m + 1), & \text{for } j = 1, \\ 0, & \text{for } j = 2, \ldots, 2m, \\ n - 1, & \text{for } j = 2m - 1. \end{cases} \tag{3}$$

From equations (3) we can deduce the invariant factors:

$$i_k(x) = \begin{cases} x^{2m+1}, & \text{for } k = 2, \ldots, n, \\ x, & \text{for } k = n+1, \ldots, N - 2nm, \\ 1, & \text{for } k = N - 2nm + 1, \ldots, N. \end{cases} \qquad (4)$$

In summary, $(A^m)_{1,n} = a$ iff $i_1(x) = x^{4m+2} + ax^{2m+1}$, and $i_2(x), \ldots, i_N(x)$ are as in (4). This completes the proof of Theorem 6. $\qquad\square$

With the proof for the hardness result of v-INVSYSTEM we remark that computing the system of invariant factors is hard for **GapL**.

4 Diagonalization

If a matrix A is similar to a diagonal matrix then we say for short that A is *diagonalizable*. That is, the Jordan normal form of A is a diagonal matrix, called J, where all the entries on the diagonal of J are the eigenvalues of A. We ask for the complexity to check whether a given matrix is diagonalizable.

An obvious way is to compute the Jordan normal form of A and then decide whether it is in diagonal form. However, in general, the eigenvalues of an integer matrix are in the complex field. That is, we run into the problem of dealing with real-arithmetic.

We use another characterization: matrix A is diagonalizable iff the minimal polynomial of A can be factored into pairwise different linear factors.

Theorem 7. DIAGONALIZABLE *is in* $\mathbf{AC}^0(\mathbf{GapL})$ *and hard for* $\mathbf{AC}^0(\mathbf{C_=L})$.

Proof. To decide whether a matrix A is diagonalizable we use the following algorithm:

DIAGONALIZABLE(A)
1 compute the minimal polynomial $m(x)$ of A
2 construct from $m(x)$ the companion matrix B
3 **if** B has pairwise different eigenvalues
4 **then** accept **else** reject.

We have already seen that step 1 is in $\mathbf{AC}^0(\mathbf{GapL})$. We argue below (see Corollary 9) that the condition in Step 3 can be decided in $\mathbf{AC}^0(\mathbf{C_=L})$. Therefore DIAGONALIZABLE $\in \mathbf{AC}^0(\mathbf{GapL})$.

For the hardness result provide a reduction from FSLE, the set of *feasible linear equations*. That is FSLE is the set of pairs (A, b) such that the linear system $Ax = b$ has a solution $x \in \mathbf{Q}^n$, where A is $m \times n$ integer matrix and b a integer vector of length m. FSLE is complete for $\mathbf{AC}^0(\mathbf{C_=L})$ [ABO99].

Define the symmetric matrix $B = \begin{pmatrix} \mathbf{0} & A \\ A^T & \mathbf{0} \end{pmatrix}$ and vector $c = (b^T, \mathbf{0})^T$ of length $m + n$. The reduction goes as follows:

$$(A, b) \in \text{FSLE} \iff (B, c) \in \text{FSLE} \qquad (5)$$

$$\Longleftrightarrow \quad C = \begin{pmatrix} B & \mathbf{0} \\ 0 \cdots 0 \end{pmatrix} \text{ is similar to } D = \begin{pmatrix} B & c \\ 0 \cdots 0 \end{pmatrix} \tag{6}$$

$$\Longleftrightarrow \quad D \in \text{Diagonalizable.} \tag{7}$$

Equivalence (5) holds, since the system $A^T y = \mathbf{0}$ is always feasible.

To show equivalence (6), let x_0 be a solution of the system $Bx = c$. Define the nonsingular matrix $T = \begin{pmatrix} I & x_0 \\ 0 & -1 \end{pmatrix}$. It is easy to check that $CT = TD$, therefore C is similar to D. Conversely, if the above system is not feasible, then C and D have different ranks and can therefore not be similar.

To show equivalence (7), observe that matrix C is symmetric. Therefore C is always diagonalizable, i.e., similar to a diagonal matrix, say C'. Now, if C is similar to D, then D is similar to C' as well, because the similarity relation is transitive. Hence D is diagonalizable as well.

Conversely, if D is diagonalizable then all of its elementary divisors are linear of the form $(\lambda - \lambda_i)$ where λ_i is any of its eigenvalues. Since C is diagonalizable, its elementary divisors are linear too. Note furthermore that C and D have the same characteristic polynomial. Therefore they must have the same system of elementary divisors. This implies that they are similar.   □

To complete the proof of Theorem 7, we show how to test whether all eigenvalues of a given matrix are pairwise different.

Lemma 8. *All eigenvalues of the matrix A are pairwise different iff the matrix $B = A \otimes I - I \otimes A$ has 0 as an eigenvalue of multiplicity n (here, $\otimes$ denotes the tensor product (see [Gra81]))*.

Proof. Just note that if $\lambda_1, \ldots, \lambda_n$ are the eigenvalues of the $n \times n$ matrix A, then $(\lambda_i - \lambda_j)$, for all $1 \leq i, j \leq n$, are the eigenvalues of matrix B.   □

Corollary 9. *Whether all eigenvalues of a matrix A are pairwise different can be decided in $\mathbf{AC}^0(\mathbf{C_=L})$.*

Proof. Let $B = A \otimes I - I \otimes A$. The matrix B has 0 as an eigenvalue of multiplicity n iff $\chi_B(x) = x^{n^2} + c_{n^2-1}x^{n^2-1} + \cdots + c_n x^n$ such that $c_n \neq 0$. Recall that the coefficients of the characteristic polynomial can be computed in **GapL**. Therefore the test whether $c_0 = c_1 = \cdots = c_{n-1} = 0$ and $c_n \neq 0$ is in $\mathbf{AC}^0(\mathbf{C_=L})$.   □

Open Problems

The coefficients of the characteristic polynomial of a matrix can be computed in **GapL**. We do not know whether the minimal polynomial of a matrix can be computed in **GapL** as well. In other words, we want to close the gap between the upper bound (Theorem 1) and the lower bound (Theorem 3) we have for the minimal polynomial.

Analogously, we ask to close the gaps for the verification of the minimal polynomial (Corollary 5), the invariant factors (Theorem 6), and the diagonalization problem (Theorem 7). Note that *if one could show* that the minimal polynomial can be computed in **GapL** (or just in $\mathbf{AC}^0(\mathbf{C_=L})$) then the minimal polynomial can be verified in $\mathbf{AC}^0(\mathbf{C_=L})$ and it follows that DIAGONALIZABLE is complete for $\mathbf{AC}^0(\mathbf{C_=L})$.

An important question not directly addressed here is whether $\mathbf{C_=L}$ is closed under complement. An affirmative answer would solve many open problems in this area.

References

[ABO99] E. Allender, R. Beals, and M. Ogihara. The complexity of matrix rank and feasible systems of linear equations. *Computational Complexity*, 8:99 –126, 1999. 410, 411, 412, 417

[BG77] C. Byrnes and M. Gauger. Characteristic free, improved decidability criteria for the similarity problem. *Linear and Multilinear Algebra*, 5:153–158, 1977. 411

[BR91] R. Brualdi and H. Ryser. *Combinatorial Matrix Theory*, volume 39 of *Encyclopedia of Mathematics and its Applications*. Cambridge University Press, 1991. 413

[CDS80] D. Cvetković, M. Doob, and H. Sachs. *Spectra of Graphs, Theory and Application*. Academic Press, 1980. 413

[Gan77] F. Gantmacher. *The Theory of Matrices*, volume 1 and 2. AMS Chelsea Publishing, 1977. 411, 416

[Gra81] A. Graham. *Kronnecker Products and Matrix Calculus With Applications*. Ellis Horwood Ltd., 1981. 418

[HT00] T. M. Hoang and T. Thierauf. The complexity of verifying the characteristic polynomial and testing similarity. In *15th IEEE Conference on Computational Complexity (CCC)*, pages 87–95. IEEE Computer Society Press, 2000. 409, 410, 411, 412, 416

[KS87] E. Kaltofen and B. Saunders. Fast parallel computation of hermite and smith forms of polynomial matrices. *SIAM Algebraic and Discrete Methods*, 8:683–690, 1987. 408, 410

[ST98] M. Santha and S. Tan. Verifying the determinant in parallel. *Computational Complexity*, 7:128–151, 1998. 410

[Sto98] A. Storjohann. An $O(n^3)$ algorithm for frobenius normal form. In *International Symposium on Symbolic and Algebraic Computation (ISSAC)*, 1998. 408

[Tod91] S. Toda. Counting problems computationally equivalent to the determinant. Technical Report CSIM 91-07, Dept. of Computer Science and Information Mathematics, University of Electro-Communications, Chofu-shi, Tokyo 182, Japan, 1991. 412

[Val92] L. Valiant. Why is boolean complexity theory difficult. In M. S. Paterson, editor, *Boolean Function Complexity*, London Mathematical Society Lecture Notes Series 169. Cambridge University Press, 1992. 412

[Vil97] G. Villard. Fast parallel algorithms for matrix reduction to normal forms. *Applicable Algebra in Engineering Communication and Computing (AAECC)*, 8:511–537, 1997. 408, 409, 410, 411, 415

[vzGG99]   J. von zur Gathen and J. Gerhard. *Modern Computer Algebra*. Cambridge University Press, 1999.   408

Note on Minimal Finite Automata

Galina Jiráskova*

Mathematical Institute, Slovak Academy of Sciences
Grešákova 6, 040 01 Košice, Slovakia
jiraskova@duro.upjs.sk

Abstract. We show that for all n and α such that $1 \leq n \leq \alpha \leq 2^n$ there is a minimal n-state nondeterministic finite automaton whose equivalent minimal deterministic automaton has exactly α states.

1 Introduction

Finite automata belong among the few fundamental computing models that were intensively investigated for more then four decades. Despite on this fact there are still several fundamental open questions about them. These open questions [3] are mainly related to the estimation of the power of nondeterminism in finite automata. One of these questions is considered in this paper.

The subset construction [8] shows that any nondeterministic finite automaton (NFA) can be simulated by a deterministic finite automaton (DFA). This theorem is often stated as "NFA's are not stronger than DFA's". But we have to be careful. The general simulation is only possible by increasing the number of states (the size of the automata). It is known [6] that there is an NFA of n states which needs 2^n states to be simulated by a DFA. Thus, in this case NFA's are exponentially smaller than DFA's. But on the other hand the DFA which counts the number of 1's modulo k needs k states and all equivalent NFA's need the same number of states. So, nondeterminism does not help in this case.

The question that arises is how many states may the minimal DFA equivalent to a given minimal n-state NFA have? This question was first considered by Iwama, Kambayashi and Takaki [4]. They showed that for $\alpha = 2^n - 2^k$ or $\alpha = 2^n - 2^k - 1$, $k \leq n/2 - 2$, there is an n-state NFA which needs α deterministic states. The question has been studied also by Iwama, Matsuura and Paterson [5]. In their paper they called an integer Z, $n < Z < 2^n$, a "magic number" if no DFA of Z states can be simulated by any NFA of n states. They proved that for all integers $n \geq 7$ and α such that $5 \leq \alpha \leq 2n - 2$, with some coprimality condition, $2^n - \alpha$ cannot be a magic number. It seemed more likely for them that there is no such magic number.

In this paper we prove their conjecture and we show that there are no such magic numbers at all. For each integers n, α such that $n \leq \alpha \leq 2^n$ we give an example of minimal n-state NFA whose equivalent minimal DFA has exactly α states. However, the size of the input alphabet of these NFA's is very big,

* This research was supported by VEGA grant No. 2/7007/2.

J. Sgall, A. Pultr, and P. Kolman (Eds.): MFCS 2001, LNCS 2136, pp. 421–431, 2001.

namely $2^{n-1}+1$. In the second part of this paper we show that it can be reduced to $2n$. In the third part of the paper we discuss the situation for one and two-letters alphabet. For $O(n^2)$ values of α we construct a minimal n-state NFA over the alphabet $\{0,1\}$ which needs α deterministic states. The situation for unary automata is sligtly different because of known result of Chrobak [1] who proved that any n-state unary NFA can be simulated by a DFA having $O(e^{\sqrt{n \ln n}})$ states.

The paper is organized as follows. In Section 2 the basic definitions and notations are given. The main results are presented in Section 3. Section 4 is devoted to the automata over one and two-letters input alphabet.

2   Definitions and Notations

A *finite automaton* M is determined by giving the following five items: (i) a finite set Q of states, (ii) a finite set Σ of input symbols, (iii) an initial state $(\in Q)$, (iv) a set $F(\subseteq Q)$ of accepting states, (v) a state transition function δ.

If δ is a mapping from $Q \times \Sigma$ into Q then M is said to be *deterministic*. If δ is a mapping from $Q \times \Sigma$ into 2^Q then M is said to be *nondeterministic*. The domain of δ can be naturally extended from $Q \times \Sigma$ to $Q \times \Sigma^*$. The definition of the language accepted by M is as usual and we omit it. Two finite automata are said to be *equivalent* if they accept the same language. A DFA (NFA) M is said to be *minimal* if there is no DFA (NFA) M' that is equivalent to M and has fewer states than M. It is well known [8] that a DFA M is minimal if (i) all its states are reachable from the initial state and (ii) there are no two equivalent states (two states q_1 and q_2 are said to be equivalent if for all $x \in \Sigma^*, \delta(q_1, x) \in F$ iff $\delta(q_2, x) \in F$).

For an NFA M of n states $\Delta(M, n)$ denotes the number of states of the minimal DFA that is equivalent to M. NFA's should also be minimal. Note that in [4,5] the values of $\Delta(M, n)$ were large enough to guarantee the minimality of M. This is not the case in this paper and the minimality of NFA's has to be proved. For this purpose let us define the notion of a generalized 1-fooling set in the same way as in [2].

Definition 1. *A set $\mathcal{A}$ of pairs of strings is called a generalized 1-fooling set for a language L if for each (x, y) from $\mathcal{A}$ the string xy is in L and for every different pairs (x_1, x_2) and (y_1, y_2) from $\mathcal{A}$ at least one of the strings $x_1 y_2, y_1 x_2$ is not in L.*

Example 2. Let $L = \{w \mid$ the number of 1's in w is $k\}$. Then the set $\mathcal{A} = \{(\epsilon, 1^k), (1, 1^{k-1}), (1^2, 1^{k-2}), \ldots, (1^k, \epsilon)\}$ is a generalized 1-fooling set for this language, since for all i the string $1^i 1^{k-i}$ is in L and for all $i \neq j$ the string $1^i 1^{k-j}$ is not in L.

Now, we formulate a lemma and since it will be used throughout the rest of the paper we give the proof of it although it follows from the considerations in the 5th chapter of [2].

Lemma 3. *Let $\mathcal{A}$ be a generalized 1-fooling set for a regular language L. Then any NFA for L has at least $|\mathcal{A}|$ states (here, $|\mathcal{A}|$ denotes the cardinality of $\mathcal{A}$).*

Proof. Let $\mathcal{A} = \{(x_1^1, x_2^1), (x_1^2, x_2^2), \ldots, (x_1^n, x_2^n)\}$ be a generalized 1-fooling set for L. That means that for all $i = 1, 2, \ldots, n$ the string $x_1^i x_2^i$ is in L. Let M be an NFA for L and let us consider an accepting computation of M on $x_1^i x_2^i$. Denote q^i the state which M enters during this accepting computation after reading x_1^i. For $i \neq j$ it has to be $q^i \neq q^j$, otherwise there would be an accepting computation of M on $x_1^i x_2^j$ and also on $x_1^j x_2^i$ which contradicts to the assumption that $\mathcal{A}$ is a generalized 1-fooling set for L. So, we have obtained at least n different states of M which proves the lemma.   □

According to Lemma 3 to prove that an NFA M of n states is minimal it is sufficient to find a generalized 1-fooling set of the cardinality n for the language accepted by M.

Let M be an NFA of n states $Q = \{q_1, q_2, \ldots, q_n\}$. Then an equivalent DFA M_{det} can be constructed as follows. First, all 2^n subsets of Q are introduced; each of them can be a state of M_{det}. The initial state of M_{det} is $\{q_1\}$ if q_1 is the initial state of M. A state $X \subseteq Q$ is an accepting state if it includes at least one accepting state of M. The transition function δ_{det} is defined using the transition function δ of M as follows: $\delta_{det}(X, a) = \cup_{q \in X} \delta(q, a)$. After determining this δ_{det}, we remove all sets which cannot be reached from the initial state $\{q_1\}$ of M_{det}. This procedure is usually called the "subset construction" [8]. Note that this DFA M_{det} may still not be minimal since some two states might be equivalent.

In the following lemma we give a sufficient condition for an NFA which guarantees that no two states obtained by the subset construction are equivalent.

Lemma 4. *Let M be an n-state NFA over an alphabet $\Sigma, 1 \in \Sigma$, such that*

(1) q_1 is the initial state and q_n is the only accepting state,
(2) $\delta(q_i, 1) = \{q_{i+1}\}$ for $i = 1, 2, \ldots, n-1$ and $\delta(q_n, 1) = \emptyset$ (i.e., transitions on reading 1 look like in Fig. 1, the other transitions may be arbitrary).

Then (i) M is a minimal NFA for the language it accepts, (ii) no two reachable states in the subset construction of M_{det} are equivalent.

Fig. 1. Transitions on reading 1 of the NFA M

Proof. (i) It is not difficult to see that the set $\mathcal{A} = \{(\epsilon, 1^{n-1}), (1, 1^{n-2}), \ldots, (1^{n-1}, \epsilon)\}$ is a generalized 1-fooling set of the cardinality n for the language $L(M)$. So, according to Lemma 3 M is a minimal NFA for the language it accepts.

(ii) Note that for all $i = 1, 2, \ldots, n$ the string 1^{n-i} is accepted starting at the state q_i but it is not accepted starting at any other state. This immediately implies that two different reachable sets in the subset construction are unequivalent (since there is an i such that q_i is in one of them, say X, but not in the other, say Y, and so $\delta(X, 1^{n-i})$ involves q_n, the only accepting state, and $\delta(Y, 1^{n-i})$ does not involve q_n).   □

Thus, by the lemma above if a NFA M fulfils the conditions (1) and (2), then M is a minimal NFA for its language and to obtain $\Delta(M, n)$ i.e., the size of the minimal DFA equivalent to M, it is sufficient to find the number of reachable sets in the subset construction for M.

3   Main Results

In this section we show that for any integers n, α such that $n \le \alpha \le 2^n$ there is a minimal n-state nondeterministic automaton whose equivalent minimal deterministic automaton has exactly α states.

First, we give a minimal k-state NFA A_k over the alphabet $\{0, 1\}$ such that $\Delta(A_k, k) = 2^k$ and A_k fulfils the conditions (1) and (2) in Lemma 4 (note that the automaton in [6] does not fulfil these conditions, see Fig. 2, "missing" arcs on reading 0 from the accepting state correspond to transitions to the state "empty set").

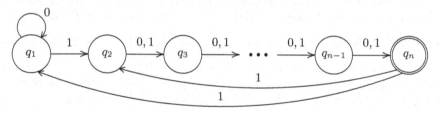

Fig. 2. Moore's n-state NFA whose equivalent minimal DFA has 2^n states

Using k-state automata A_k, $1 \le k \le n$, we construct n-state automata B_k over the alphabet $\{0, 1\}$ with $\Delta(B_k, n) = 2^k + n - k$. Finally, the n-state NFA M_α with $\Delta(M_\alpha, n) = \alpha$ will be constructed from an automaton B_k using further letters of $2^{n-1} + 1$ letters input alphabet.

Lemma 5. *For each integer k there is a minimal k-state NFA A_k over the alphabet $\{0, 1\}$ such that A_k fulfils the conditions (1) and (2) in Lemma 4 and $\Delta(A_k, k) = 2^k$ i.e., its equivalent minimal DFA has 2^k states.*

Proof. Let the set of states of A_k be $\{q_1, \ldots, q_k\}$ with the initial state q_1 and the only accepting state q_k. Define transition function δ as follows (see Fig.3)

$$\delta(q_i, 1) = \{q_{i+1}\} \qquad i = 1, 2, \ldots, k - 1$$

$$\delta(q_i, 0) = \{q_1, q_{i+1}\} \qquad i = 1, 2, \ldots, k-1$$
$$\delta(q_k, 1) = \delta(q_k, 0) = \emptyset$$

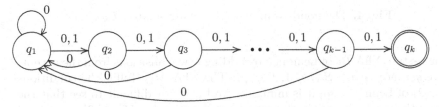

Fig. 3. The nondeterministic finite automaton A_k

Since q_k is the only accepting state of A_k and transitions on reading 1 are as in Lemma 4, A_k is minimal and to prove that $\Delta(A_k, k) = 2^k$ it is sufficient to show that each subset of $\{q_1, q_2, \ldots, q_k\}$ is reachable in the subset construction for A_k. It can be easily seen that $\emptyset, \{q_1\}, \{q_2\}, \ldots, \{q_k\}$ are all reachable. Suppose $X = \{q_{i_1}, q_{i_2}, \ldots, q_{i_l}\}$, where $l \geq 2$ and $1 \leq i_1 < i_2 < \ldots < i_l \leq k$. We show that X is reachable from a set Y with the smaller cardinality than X. Let $Y = \{q_{i_2-i_1}, q_{i_3-i_1}, \ldots, q_{i_l-i_1}\}$ i.e., $|Y| = |X| - 1$. Since $1 \leq i_j - i_1 \leq k-1$, $j = 2, 3, \ldots, l$, $\delta(Y, 01^{i_1-1})$ is exactly X. Using induction on cardinality we have proved that all subsets of $\{q_1, q_2, \ldots, q_k\}$ are reachable in the subset construction for A_k and so, $\Delta(A_k, k) = 2^k$. □

Lemma 6. *For all n, k such that $1 \leq k \leq n$ there is a minimal n-state NFA B_k over the alphabet $\{0, 1\}$ such that $\Delta(B_k, n) = 2^k + n - k$.*

Proof. We construct the n-state NFA B_k from the k-state NFA A_k described in Lemma 5 by adding new states $q_{k+1}, q_{k+2}, \ldots, q_n$ and new transitions on reading 1: $\delta(q_{k+1}, 1) = q_1$ and $\delta(q_i, 1) = q_{i-1}$ for $i = k+2, \ldots, n$ (see Fig. 4). The initial state of B_k is q_n, the only accepting state is q_k. Since the conditions of Lemma 4 are fulfilled, the NFA B_k is minimal and to find $\Delta(B_k, n)$ it is sufficient to count the number of reachable sets in the subset construction for B_k. Note that $\{q_n\}$, $\{q_{n-1}\}, \ldots, \{q_{k+1}\}$ and $\{q_1\}$ are reachable. Further, by Lemma 5 all subsets of $\{q_1, q_2, \ldots, q_k\}$ are reachable and it is not difficult to see that no other set is reachable. Thus, there are $2^k + n - k$ reachable sets in the subset construction for B_k which implies that $\Delta(B_k, n) = 2^k + n - k$. □

Lemma 7. *For all n, k and j such that $1 \leq k \leq n-1$, $1 \leq j \leq 2^k - 1$, there is a minimal n-state NFA $M_{k,j}$ such that $\Delta(M_{k,j}, n) = 2^k + n - k + j$.*

Proof. Let a_i, $i = 1, 2, \ldots, 2^{n-1} - 1$ be pairwise different symbols of the input alphabet, $a_i \neq 0, a_i \neq 1$. Let $S_1, S_2, \ldots, S_{2^k-1}$ be all subsets of $\{q_1, q_2, \ldots, q_{k+1}\}$ involving the state q_{k+1}, except for $\{q_{k+1}\}$. We construct the n-state NFA $M_{k,j}$

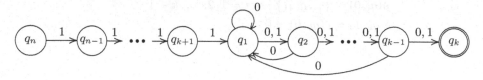

Fig. 4. The nondeterministic finite automaton B_k

from the n-state NFA B_k of Lemma 6 by adding new transitions from the accepting state q_k: $\delta(q_k, a_i) = S_i, i = 1, 2, \ldots, j$. The NFA $M_{k,j}$ fulfils the conditions (1) and (2) of Lemma 4 so, it is minimal. And it is not difficult to see that the number of the reachable sets in the subset construction for $M_{k,j}$ is $2^k + n - k + j$ (all subsets of $\{q_1, q_2, \ldots, q_k\}$, $\{q_{k+1}\}$, $\{q_{k+2}\}$, $\ldots, \{q_n\}$, S_1, S_2, $\ldots$, S_j) which implies (again according to Lemma 4) that $\Delta(M_{k,j}, n) = 2^k + n - k + j$. $\qquad\square$

Now we are ready to prove the main result.

Theorem 8. *For all n and α such that $n \leq \alpha \leq 2^n$ there is a minimal n-state NFA whose equivalent minimal DFA has exactly α states.*

Proof. The case $\alpha = n$ is trivial (the automaton accepting all strings of length at least $n-1$ can be taken). For $\alpha \geq n+1$ let k be such integer that $2^k + n - k \leq \alpha < 2^{k+1} + n - (k+1)$. If $\alpha = 2^k + n - k$ then the automaton B_k from Lemma 6 can be taken. In the other case $\alpha = 2^k + n - k + j$, where $k \leq n-1$, $1 \leq j < 2^k - 1$, and the automaton $M_{k,j}$ from Lemma 7 can be taken. $\qquad\square$

Thus, we have proved that nondeterminism can be helpful "step by step" for finite automata. Howewer, the automata which we used to prove it were over many-letters alphabet (namely $2^{n-1} + 1$). The following theorem shows that the size of input alphabet can be decreased to $2n$. Note that we only prove the existence of such automata but we do not give the construction of them. Before proving this we formulate a lemma which is similar to Lemma 4.

Lemma 9. *Let an NFA M have the states $q_1, q_2, \ldots, q_n$, where q_1 is the initial and q_n is the only accepting state of M. Let $a_1, a_2, \ldots a_{n-1}$ be pairwise different letters of the input alphabet. Let (see Fig. 5)*

$$\delta(q_i, a_i) = \{q_{i+1}\} \qquad i = 1, 2, \ldots, n-1,$$
$$\delta(q_i, a_j) = \emptyset \qquad for \qquad i \neq j,$$
$$\delta(q_n, a_i) = \emptyset \qquad for \qquad i = 1, 2, \ldots, n-1.$$

Then M is minimal and $\Delta(M, n)$ is equal to the number of reachable sets in the subset construction for M.

Proof. The proof of this lemma is similar as the proof of Lemma 4. The set $\mathcal{A} = \{(\epsilon, a_1 a_2 \ldots a_{n-1}), (a_1, a_2 a_3 \ldots a_{n-1}), \ldots, (a_1 a_2 \ldots a_{n-1}, \epsilon)\}$ is a generalized

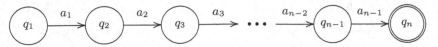

Fig. 5. Transitions on reading a_i of the NFA M

1-fooling set for $L(M)$ of the cardinality n which implies that M is minimal. For all $i = 1, 2, \ldots, n - 1$ the string $a_i a_{i+1} \ldots a_{n-1}$ is accepted starting at the state q_i but it is not accepted starting at any other state. This implies that no two reachable sets in the subset construction for M are equivalent. □

Theorem 10. *For all n and α, $n \le \alpha \le 2^n$, there is a minimal n-state NFA over $2n$ letters input alphabet whose equivalent minimal DFA has exactly α states.*

Proof. The case $\alpha = n$ is trivial. For $n + 1 \le \alpha \le 2^n$ we prove this theorem by induction on n. The automata for $n = 2$ are depicted in Fig. 6.

Fig. 6. The two-state NFA's whose equivalent minimal DFA's have 3 resp. 4 states

Suppose that for all α, $n + 1 \le \alpha \le 2^n$, there is a minimal n-state NFA over the input alphabet $\{a_1, b_1, a_2, b_2, \ldots, a_n, b_n\}$, with the initial state q_n, the only accepting state q_1 and the transition function δ such that

$$\delta(q_i, a_i) = \{q_{i-1}\} \qquad i = n, n - 1, \ldots, 2$$
$$\delta(q_j, a_i) = \emptyset \qquad \text{for} \qquad i \ne j$$
$$\delta(q_1, a_i) = \emptyset \qquad \text{for} \qquad i = n, n - 1, \ldots, 2$$

We show that than for all α between $n + 2$ and 2^{n+1} there is a minimal $(n + 1)$-state NFA over $2(n + 1)$ letters alphabet with a_i-transitions as above which needs α deterministic states.

Let D be the n-state NFA from the induction assumption with $\Delta(D, n) = \alpha$. We construct the $(n + 1)$-state NFA D' (see Fig. 7) from the n-state NFA D by adding a new initial state q_{n+1} connected with q_n through a_{n+1} and new transitions on reading b_{n+1}: $\delta(q_i, b_{n+1}) = \{q_i, q_{n+1}\}$ for $i = 1, 2, \ldots, n$. Note that if X is reachable in the subset construction for D, then $X \cup \{q_{n+1}\}$ is reachable in the subset construction for D' and moreover, all reachable sets for D' are either equal to a reachable set for D or can be written as $X \cup \{q_{n+1}\}$, where X is reachable for D. This implies that $\Delta(D', n + 1) = 2 \cdot \Delta(D, n)$.

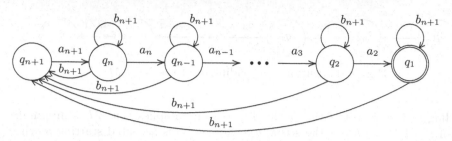

Fig. 7. Transitions on reading a_i and b_{n+1} of the NFA D'

Further, if we remove the transition on b_{n+1} from q_n to q_{n+1}, the set $\{q_n, q_{n+1}\}$ becomes unreachable and for the automaton D'' obtained in this way it holds $\Delta(D'', n+1) = 2 \cdot \Delta(D, n) - 1$.

If $\Delta(D, n) = \alpha$, $n + 1 \leq \alpha \leq 2^n$, we have constructed the automata D' and D'' whose equivalent minimal deterministic automata have 2α and $2\alpha - 1$ states. Note that these automata fulfil the conditions of Lemma 9 and also the induction assumption concerning transitions on a_i. It remains to prove that there is a minimal $(n + 1)$-state NFA (fulfilling the induction assumption on a_i-transitions) which needs α deterministic states for $n + 2 \leq \alpha < 2n + 1$. But this is not difficult. For $\alpha = n + 2$ it is the automaton in Fig. 8 with transitions $\delta(q_i, a_i) = \{q_{i-1}\}$ for $i = 2, 3, \ldots, n+1$. For $n + 2 < \alpha < 2n + 1$ denote $\beta = \alpha - (n + 2)$ (i.e., $1 \leq \beta < n - 1$) and consider the automaton obtained from the previous one by adding the transitions $\delta(q_1, b_i) = \{q_{i+1}, q_1\}$, $i = 1, 2, \ldots, \beta$. For this automaton the number of reachable sets is exactly $n + 2 + \beta = \alpha$.

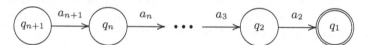

Fig. 8. The $(n + 1)$-state NFA which needs $n + 2$ deterministic states

□

4   Automata over the Alphabet $\{0, 1\}$ and $\{1\}$

In the first part of this section we will consider automata over the input alphabet $\{0, 1\}$. Note that only such automata were considered in [4] and [5]. In [4] it is shown that if $\alpha = 2^n - 2^k$ or $\alpha = 2^n - 2^k - 1, 0 \leq k \leq n/2 - 2$, then there is a minimal n-state NFA which needs α deterministic states. In [5] the same is shown for $\alpha = 2^n - k, n \geq 7$, $5 \leq k \leq 2n - 2$ (with some coprimality condition). In this section we prove this property for all α such that $n \leq \alpha \leq 1 + n(n+1)/2$ (the case $\alpha = n$ is again trivial). While in [4,5] the $\Delta(M, n)$ value was large enough to guarantee the minimality of the NFA M in this case $\Delta(M,n)$ value is small and so, the minimality of the NFA M should be proved. But we will construct nondeterministic automata in such a way that the conditions (1) and (2) in Lemma 4 will be fulfilled and so, the automata will be minimal. Moreover, by Lemma 4 $\Delta(M,n)$ will be equal to the number of reachable sets in the subset construction for M.

The following lemma shows that for any k, $0 \le k \le n-1$, there is a minimal n-state NFA over the alphabet $\{0,1\}$ whose equivalent minimal DFA has exactly $1+n+(n-1)+(n-2)+\ldots+(n-k)$ states. In the second lemma it is shown that for any k and j, $0 \le k \le n-1$, $1 \le j < n-(k+1)$ there is a minimal n-state NFA over the alphabet $\{0,1\}$ which needs $1+n+(n-1)+\ldots+(n-k)+j$ deterministic states.

Lemma 11. *For all n and k, $0 \le k \le n-1$, there is a minimal n-state NFA M_k over the aplhabet $\{0,1\}$ such that $\Delta(M_k,n) = 1+n+(n-1)+\ldots+(n-k)$.*

Proof. Let $q_1, q_2, \ldots, q_n$ be the states of M_k, let q_1 be the initial state and q_n be the only accepting state. Define transitions on reading 1 by $\delta_k(q_i,1) = \{q_{i+1}\}$, $i = 1,2,\ldots,n-1$ and $\delta_k(q_n,1) = \emptyset$. The conditions (1) and (2) of Lemma 4 are fulfilled and therefore M_k is minimal and $\Delta(M_k,n)$ is equal to the number of reachable sets in the subset construction. Further, define transitions on reading 0 by $\delta_k(q_i,0) = \{q_1, q_2, \ldots, q_i, q_{i+1}\}$ for $i = 1,2,\ldots,k$ and $\delta_k(q_j,0) = \emptyset$ for $j = k+1, k+2, \ldots, n$. So, if $k=0$ there are no 0-transitions and for $k \ge 1$ the NFA M_k looks like in Fig. 9

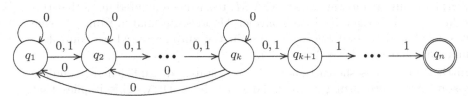

Fig. 9. The nondeterministic finite automaton M_k

It is easy to see that the following sets are reachable in the subset construction for M_k:
$$\emptyset, \{q_1\}, \{q_2\}, \ldots, \{q_n\},$$
$$\{q_1, q_2\}, \{q_2, q_3\}, \ldots, \{q_{n-1}, q_n\},$$
$$\{q_1, q_2, q_3\}, \{q_2, q_3, q_4\}, \ldots, \{q_{n-2}, q_{n-1}, q_n\},$$
$$\vdots$$
$$\{q_1, q_2, \ldots, q_{k+1}\}, \{q_2, q_3, \ldots, q_{k+2}\}, \ldots, \{q_{n-k}, q_{n-k+1}, \ldots, q_n\}.$$
It is not difficult to prove that no other set is reachable. Thus, $\Delta(M_k,n) = 1+n+(n-1)+(n-2)+\ldots+(n-k)$. $\square$

Lemma 12. *For all k and j, $0 \leq k \leq n-1$, $1 \leq j < n-(k+1)$ there is a minimal n-state NFA $M_{k,j}$ over the alphabet $\{0,1\}$ such that $\Delta(M_{k,j},n) = 1+n+(n-1)+(n-2)+\ldots+(n-k)+j$.*

Proof. We construct the automaton $M_{k,j}$ from the automaton M_k in Lemma 11 by adding the transition on reading 0 from the state q_{k+1} to the set $\{q_1,q_2,\ldots,q_k,q_{k+1},q_{n-j+1}\}$ (note that $k+2 < n-j+1 \leq n$). All the sets which were reachable in the subset construction for M_k are also reachable in the subset construction for $M_{k,j}$. Moreover, the following sets (and no other) are reachable: $\{q_1,q_2,\ldots,q_k,q_{k+1},q_{n-j+1}\}$, $\{q_2,q_3,\ldots,q_{k+1},q_{k+2},q_{n-j+2}\}$, $\ldots$, $\{q_j,q_{j+1},\ldots,q_{j+k},q_n\}$. Since $M_{k,j}$ fulfils the conditions (1) and (2) of Lemma 4 it is minimal and $\Delta(M_{k,j},n) = 1+n+(n-1)+(n-2)+\ldots+(n-k)+j$. $\square$

Corollary 13. *For all n and α such that $n \leq \alpha \leq 1+n(n+1)/2$ there is a minimal n-state NFA over the input alphabet $\{0,1\}$ whose equivalent minimal DFA has exactly α states.*

Proof. The case $\alpha = n$ is again trivial. Since $1+n+(n-1)+(n-2)+\ldots+2+1 = 1+n(n+1)/2$, the proof for $\alpha \geq n+1$ follows from the two lemmata above. $\square$

We have $O(n^2)$ values of α for which there is a minimal n-state NFA over the alphabet $\{0,1\}$ which needs α deterministic states. We are not able to prove it for all α, $n \leq \alpha \leq 2^n$, although it seems to be true for us. We have verified it for all $n \leq 8$, but for larger n the question remains open.

The last part of the paper we devote to a few notes on the automata over one-letter alphabet (unary automata). Again, we are interested in the value of $\Delta(M,n)$ for a minimal n-state unary NFA M. It has been pointed in [6] that this value has to be less than 2^n. If we consider Moore's automaton from [6] using input symbol 1 alone (see Fig. 2) we obtain the automaton whose equivalent minimal DFA has $(n-1)^2+2$ states [6].

Further, in [1] it is shown that $O(e^{\sqrt{n\ln n}})$ states are sufficient to simulate an n-state NFA recognizing a unary language by a DFA. This implies that $\Delta(M,n)$ cannot be between (approximately) $e^{\sqrt{n\ln n}}$ and 2^n for any n-state unary NFA M. On the other hand in [1] it is shown that there is a unary n-state NFA which needs $\Omega(e^{\sqrt{n\ln n}})$ deterministic states. For $n \leq 10$ we have verified that for all α such that $n \leq \alpha \leq (n-1)^2+2$, there is a minimal n-state unary NFA whose equivalent minimal DFA has exactly α states. Whether it is true for larger n and all α between n and $e^{\sqrt{n\ln n}}$ remains open.

Acknowledgement

I would like to thank Juraj Hromkovič for his comments concerning this work. I am also grateful to Jozef Jirásek for his helpful computer programs which enable us to verify some of our conjectures.

References

1. M. Chrobak: Finite automata and unary languages. Theoretical Computer Science 47(1986), 149-158 422, 430
2. J. Hromkovič: Communication Complexity and Parallel Computing. Springer 1997 422
3. J. Hromkovič, J. Karhumäki, H. Klauck, S. Seibert, G. Schnitger: Measures on nondeterminism in finite automata. In: ICALP'00, Lecture Notes in Computer Science 1853, Springer-Verlag 2000, pp.199-210 421
4. K. Iwama, Y. Kambayashi and K. Takaki: Tight bounds on the number of states of DFA's that are equivalent to n-state NFA's. Theoretical Computer Science 237(2000) 485-494 421, 422, 428
5. K. Iwama. A. Matsuura and M. Paterson: A family of NFA's which need $2^n - \alpha$ deterministic states. Proc. MFCS'00, Lecture Notes in Computer Science 1893, Springer-Verlag 2000, pp.436-445 421, 422, 428
6. F. Moore: On the bounds for state-set size in proofs of equivalence between deterministic, nondeterministic and two-way finite automata. IEEE Trans. Comput. C-20, pp.1211-1214, 1971 421, 424, 430
7. O. B. Lupanov: Uber der Vergleich zweier Typen endlicher Quellem. Probleme der Kybernetik, Vol.6, pp. 329-335, Akademie-Verlag, Berlin, 1966
8. M. Rabin and D. Scott: Finite automata and their decision problems. IBM J. Res. Develop, 3, pp. 114-129, 1959 421, 422, 423

Synchronizing Finite Automata on Eulerian Digraphs

Jarkko Kari*

Department of Computer Science, 15 MLH, University of Iowa
Iowa City, IA, 52242 USA
jjkari@cs.uiowa.edu

Abstract. Černý's conjecture and the road coloring problem are two open problems concerning synchronizing finite automata. We prove these conjectures in the special case that the underlying digraph is Eulerian, that is, if the in- and out-degrees of all vertices are equal.

1 Introduction

Let us call a directed graph $G = (V, E)$ admissible (k-admissible, to be precise) if all vertices have the same out-degree k. A deterministic finite automaton (without initial and final states) is obtained if we color the edges of a k-admissible digraph with k colors in such a way that all k edges leaving any node have distinct colors. Let $\Sigma = \{1, 2, \ldots, k\}$ be the coloring alphabet. We use the standard notation Σ^* for the set of words over Σ. Every word $w \in \Sigma^*$ defines a transformation $f_w : V \longrightarrow V$ on the vertex set V: the vertex $f_w(v)$ is the endpoint of the unique path starting at v whose labels read w. For a set $S \subseteq V$ we define $f_w(S) = \{f_w(v) \mid v \in S\}$. Word w is called synchronizing if $f_w(V)$ is a singleton set, and the automaton is called synchronizing if a synchronizing word exists. A coloring of an admissible graph is synchronizing if the corresponding automaton is synchronizing.

We investigate the following two natural questions:

- Which admissible digraphs have synchronizing colorings ?
- What is the length of a shortest synchronizing word on a given synchronizing automaton ?

The road coloring problem and the Černý conjecture are two open problems related to these questions. The road-coloring problem [1,2] asks whether a synchronizing coloring exists for every admissible digraph that is strongly connected and aperiodic. A digraph is called aperiodic if the gcd of the lengths of its cycles is one. This is clearly a necessary condition for the existence of a synchronizing coloring. The road-coloring problem has been solved in some special cases but the general case remains open. In particular, it is known to be true if the digraph has a simple cycle of prime length and there are no multiple edges in the digraph [6].

---

* Research supported by NSF Grant CCR 97-33101

J. Sgall, A. Pultr, and P. Kolman (Eds.): MFCS 2001, LNCS 2136, pp. 432–438, 2001.

In this work we prove the road coloring conjecture in the case the digraph is Eulerian, that is, also the in-degrees of all nodes are equal to k. (We assume k-admissibility throughout this paper.) This partial solution is interesting because such digraphs can also be colored in a completely non-synchronizing way: there exist labelings where every color specifies a permutation of the vertex set so that no input word can synchronize any vertices.

Proposition 1. *Let $G = (V, E)$ be a digraph with all in- and out-degrees equal to some fixed k. Then the edges can be colored with k colors in such a way that at every vertex all entering edges have distinct colors and all leaving edges have distinct colors.*

Proof. Let us start by splitting each vertex into two parts. One part gets all entering edges and the other part all leaving edges. As a result we get a digraph with a vertex set $V_1 \cup V_2$ and all edges are from V_1 into V_2. Moreover, all vertices in V_1 have in-degree zero and out-degree k, and all vertices in V_2 have out-degree zero and in-degree k. We need to color the edges with k colors in such a way in every node all k incident edges have different colors.

Let us show how any incorrect coloring can be improved in the sense that the total number of missing colors over all vertices is reduced. Without loss of generality assume that there are at least two edges with color a and no edges with color b leaving node $x \in V_1$. Construct a new digraph with the same vertex set $V_1 \cup V_2$ and all original edges of color a, as well as all edges of color b inverted. In the new digraph node x has positive out-degree and in-degree equal to zero. Consequently there exists a path P starting at x that ends in a node whose in-degree is greater than its out-degree. (To find such a path just follow edges, without repeating the same edge twice, until the path can no longer be extended.) If in the original graph we swap the a vs. b coloring of the edges along the selected path we get an improved coloring: Node x has now leaving edges with both colors a and b, and no new missing colors were introduced at any node. The improvement process can be repeated until a balanced coloring is reached.

We also prove the Černý conjecture if the underlying digraph is Eulerian. Černý's conjecture [3] states that if an n-state automaton is synchronizing then there always exists a synchronizing word of length at most $(n-1)^2$. Currently the best known bounds for the shortest synchronizing word are cubic in n.

2   Maximum Size Synchronizable Subsets

Our proofs rely on a result by J.Friedman [5]. We only need this result in the Eulerian case, which we include here for the sake of completeness. All results in this section were originally presented in the Friedman paper [5].

Given an automaton A, let us say that set $S \subseteq V$ is synchronizable if there exists a word w such that $f_w(S)$ is a singleton set. Clearly, for every word u, if S is synchronizable then also $f_u^{-1}(S)$ is synchronizable, synchronized by word uw.

Let us only consider finite automata that are based on Eulerian digraphs, and let k be the in- and out-degree of every node. We are interested in synchronizable sets of maximal cardinality, called maximum size synchronizable sets. Let m be the largest cardinality of any synchronizable set. If n-state automaton A is synchronizing then $m = n$, but in general $1 \leq m \leq n$. In a moment we'll see that m has to divide n, and that the vertex set V can be partitioned into non-overlapping maximum size synchronizable subsets.

Because in the Eulerian case every vertex has k incoming edges we have

$$\sum_{a \in \Sigma} |f_a^{-1}(S)| = k \cdot |S| \tag{1}$$

for every $S \subseteq V$. If set S is a maximum size synchronizable set then $|f_a^{-1}(S)| \leq |S|$ for every color $a \in \Sigma$, which combined with (1) gives $|f_a^{-1}(S)| = |S| = m$. More generally we have $|f_u^{-1}(S)| = m$ for every word u. In other words, predecessors of maximum size synchronizable sets have also maximum size.

Consider a collection of i maximum size synchronizable sets $S_1, S_2, \ldots, S_i$ that are synchronized by the same word w, and that are hence disjoint. Let y_j denote the unique element of $f_w(S_j)$ for every $j = 1, 2, \ldots, i$. If $im < n$ then the collection is not yet a partitioning of V and there exists a vertex x that does not belong to any S_j. Because the graph is strongly connected there exists a word u such that $f_u(y_1) = x$. Consider subsets that are synchronized by word wuw. These include S_1 (which is already synchronized by the first w) as well as $f_{wu}^{-1}(S_j)$ for all $j = 1, 2, \ldots, i$. All these $i + 1$ sets are maximum size synchronizable sets, and they are different from each other because they are synchronized by wuw into distinct vertices $f_w(x)$ and $u_j = f_w(S_j)$, for $j = 1, 2, \ldots, i$.

The reasoning above can be repeated until we reach a partitioning of V into maximum size synchronizable sets, all synchronized by the same word. We have proved

Proposition 2. *In any automaton based on an Eulerian digraph $G = (V, E)$ there exists a word w such that subsets of vertices synchronized by w form a partitioning of V into maximum size synchronizable sets.*

As a corollary we see that the maximum size m of synchronizable sets must divide n, the number of vertices. It is also clear that if w satisfies the property of the proposition then so does uw for every $u \in \Sigma^*$.

In [5] a more general approach is taken, applicable also in non-Eulerian cases. In this approach the notion of the cardinality of a subset is replaced by its weight (defined below) and instead of maximum size sets we consider maximum weight synchronizable sets. To define weights, notice that the adjacency matrix of the digraph has a positive left eigenvector e with eigenvalue k. The eigenvector is chosen with relatively prime integer components. Components of e are the weights of the corresponding vertices, and the weight of a set of vertices is the sum of the weights of its elements. In Eulerian cases the weight of a set is simply its cardinality because $e = (1, 1, \ldots, 1)$.

3   Synchronizing Colorings of Eulerian Digraphs

To prove the road coloring property of Eulerian digraphs we use the notion of stability from [4]. Nodes x and y of an automaton are called stable if

$$(\forall u \in \Sigma^*)(\exists w \in \Sigma^*) \qquad f_{uw}(x) = f_{uw}(y),$$

that is, nodes x and y are synchronizable, and all pairs of states reachable from x and y are synchronizable as well. It is easy to see that stability is an equivalence relation, and that all forward transformations f_a respect stability, i.e., they map equivalence classes into equivalence classes. Let us use the term stability class for these equivalence classes.

It was pointed out in [4] that the road coloring problem is equivalent to the conjecture that there always exists a coloring with at least one stable pair of vertices. The idea is that if a coloring with non-trivial stability classes exists then a factor automaton can be considered whose nodes are the stability classes. The factor automaton is strongly connected and aperiodic if the original automaton has these properties. The factor automaton has fewer nodes, so we use mathematical induction and assume that the factor automaton can be relabeled into a synchronizing automaton. The relabeling can be lifted to the original automaton, providing it with a synchronizing coloring because a relabeling does not break the stability of any nodes that were in the same stability class.

In the following we adapt this reasoning to the Eulerian case. Given an admissible Eulerian digraph $G = (V, E)$ with $|V| > 1$ we start with a fully non-synchronizing coloring, where all vertices have different colors on all entering and leaving edges (Proposition 1). Let x be a vertex and a and b two colors such that $f_a(x) \neq f_b(x)$. These must exist because otherwise the digraph would be periodic. Let $y = f_a(x)$ and $z = f_b(x)$. Let us swap the colors of edges $x \longrightarrow y$ and $x \longrightarrow z$. Then two edges labeled a enter z, and two edges labeled b enter node y. Because all other nodes have one edge of each color entering them, it is clear that for any set S of vertices

$$z \in S, y \notin S \Longrightarrow |f_a^{-1}(S)| > |S|, \text{ and}$$
$$y \in S, z \notin S \Longrightarrow |f_b^{-1}(S)| > |S|.$$

Therefore any maximum size synchronizable set of vertices must either contain both y and z or neither of them.

Let us prove that y and z are stable: Let $u \in \Sigma^*$ be arbitrary, and let w be a word satisfying Proposition 2. Then also uw satisfies Proposition 2, which means that $f_{uw}(y) = f_{uw}(z)$.

Because there are non-trivial stability classes the factor automaton has fewer states than the original automaton. It is trivial that (the underlying digraph of) the factor automaton is admissible, strongly connected and aperiodic. To prove that it is also Eulerian it is enough to show that there are edges of all colors entering all stability classes. This is trivially true for any class that contains any node different from y and z as all these nodes have incoming edges of all colors.

And since y and z are in the same stability class, also that class has entering edges of all colors.

The factor automaton has fewer states, so we may assume that there exists a synchronizing re-coloring of the factor automaton. This induces a synchronizing coloring of the original automaton. We have proved:

Proposition 3. *The road coloring conjecture is true for Eulerian digraphs.*

The reasoning fails on non-Eulerian digraphs. It is however easy to see that even in the non-Eulerian case nodes x and y are stable if and only if every maximum weight synchronizable set either contains both x and y or neither of them. In other words, stability classes are intersections of maximum weight synchronizable sets.

4   The Černý Conjecture

We have proved that Eulerian digraphs have synchronizing colorings. In this section we show that any such coloring admits a synchronizing word of length at most $(n-1)^2$, where $n = |V|$.

Let us view the vertices as basis vectors of an n-dimensional vector space, and subsets of V as elements of this vector space obtained as linear combinations with coefficients 0 and 1: Set $S \subseteq V$ is the vector

$$\sum_{x \in S} x.$$

It is useful to view functions f_a^{-1} as linear transformations of this vector space, for all $a \in \Sigma$. To make an explicit distinction to the function f_a^{-1} that acts on sets of vertices, we introduce a new notation φ_a for the linear transformation. For each color $a \in \Sigma$ we define

$$(\forall x \in V) \qquad \varphi_a(x) = \sum_{f_a(y)=x} y.$$

In other words, the transformation matrix of φ_a is the transpose of the adjacency matrix for edges colored with a. Compositions of functions φ_a give linear functions φ_w for all $w \in \Sigma^*$:

$$\varphi_\varepsilon = id,$$
$$\varphi_{aw} = \varphi_w \circ \varphi_a, \text{ for every } a \in \Sigma \text{ and } w \in \Sigma^*.$$

This means that φ_w is the linear extension of the set function f_w^{-1}:

$$(\forall w \in \Sigma^*)(\forall x \in V) \qquad \varphi_w(x) = \sum_{f_w(y)=x} y.$$

It follows from (1) that if the automaton is based on an Eulerian digraph then for every $S \subseteq V$, either

$$(\forall a \in \Sigma) \qquad |f_a^{-1}(S)| = |S|$$

or

$$(\exists a \in \Sigma) \qquad |f_a^{-1}(S)| > |S|.$$

Let us construct an $k \times n$ matrix M whose element (a, x) is

$$M_{ax} = |f_a^{-1}(x)| - 1,$$

that is, the change in the cardinality if set $\{x\}$ is replaced by its predecessor $f_a^{-1}(x)$ under color a. A set $S \subseteq V$ (viewed as a vector) is in $\ker M$, the kernel of M, if and only if for every color $|f_a^{-1}(S)| = |S|$ (where S is viewed as a set). So if S is not in the kernel then $|f_a^{-1}(S)| > |S|$ for some color a.

For any given proper subset S of V we would like to find a word u such that $|f_u^{-1}(S)| > |S|$, and we would like to find a shortest possible such u. Equivalently, $u = aw$ where w is a shortest word such that $\varphi_w(S)$ is not in $\ker M$. Because $\ker M$ is a linear subspace, we can use the following simple lemma.

Lemma 4. *Let U be a d-dimensional subspace, and let $x \in U$. If there exists a word w such that $\varphi_w(x) \notin U$ then there exists such word w of length at most d.*

Proof. Consider vector spaces $U_0 \subseteq U_1 \subseteq \ldots$ where U_i is generated by

$$\{\varphi_w(x) \mid |w| \leq i\}.$$

Clearly, if $U_{i+1} = U_i$ for some i then $U_j = U_i$ for every $j \geq i$. Namely, $U_{i+1} = U_i$ means that $\varphi_a(U_i) \subseteq U_i$ for all $a \in \Sigma$.

Let i be the smallest number such that $\varphi_w(x) \notin U$ for some w of length i, that is, the smallest i such that $U_i \not\subseteq U$. This means that in $U_0 \subset U_1 \subset \ldots \subset U_i$ all inclusions are proper. In terms of dimensions of the vector spaces:

$$1 = \dim U_0 < \dim U_1 < \ldots < \dim U_{i-1} < \dim U_i,$$

which means that $\dim U_{i-1} \geq i$. But $U_{i-1} \subseteq U$ so that we also have $\dim U_{i-1} \leq \dim U = d$, which means that $i \leq d$.

If the automaton is synchronizing then $\ker M$ is at most $n - 1$ dimensional subspace. Consequently, for any proper subset S of V there exists a word u of length at most $1 + (n - 1) = n$ such that $|f_u^{-1}(S)| > |S|$. After $n - 2$ repeated applications of this observation we conclude that for any subset S of cardinality 2 or more there exists a word u of length at most $n(n - 2)$ such that $f_u^{-1}(S) = V$. Because there must exist a vertex x and a color a such that the cardinality of the set $f_a^{-1}(x)$ is greater than one, we conclude that word ua synchronizes the automaton, and the length of the synchronizing word ua is at most $n(n-2)+1 = (n - 1)^2$. We have proved

Proposition 5. *If the underlying digraph of a synchronizing automaton is Eulerian then there exists a synchronizing word of length at most $(n - 1)^2$.*

Also this reasoning fails for non-Eulerian digraphs. While it is always true that for every proper subset S of V there is a word u of length at most n such that $|f_u^{-1}(S)| \neq |S|$, in the non-Eulerian case it is possible that $|f_u^{-1}(S)| < |S|$.

On the other hand, replacing cardinalities with weights is possible, and the same reasoning as above shows that for every proper subset S there is a word u of length at most n such that the weight of $f_u^{-1}(S)$ is greater than the weight of S. Unfortunately this result does not give the desired bound for the length of the shortest synchronizing word because the total weight of V can be larger than n.

References

1. R. L. Adler, L. W. Goodwyn, B. Weiss. Equivalence of topological Markov shifts. Israel Journal of Mathematics 27 (1977), 49–63. 432
2. R. L. Adler, B. Weiss. Similarity of automorphisms of the torus. Memoires of the American Mathematical Society, n.98. 1970. 432
3. J. Černý. Poznámka k. homogénnym experimenton s konečnými automatmi. Mat. fyz. čas SAV 14 (1964), 208–215. 433
4. K. Culik, J. Karhumäki, J. Kari. Synchronized automata and the road coloring problem. TUCS technical report 323. Turku Centere for Computer Science, University of Turku, 1999. 435
5. J. Friedman. On the road coloring problem. Proceedings of the American Mathematical Society 110 (1990), 1133–1135. 433, 434
6. G. L. O'Brien. The road coloring problem. Israel Journal of Mathematics 39 (1981), 145–154. 432

A Time Hierarchy for Bounded One-Way Cellular Automata

Andreas Klein and Martin Kutrib

Institute of Informatics, University of Giessen
Arndtstr. 2, D-35392 Giessen, Germany

Abstract. Space-bounded one-way cellular language acceptors (OCA) are investigated. The only inclusion known to be strict in their time hierarchy from real-time to exponential-time is between real-time and linear-time! We show the surprising result that there exists an infinite hierarchy of properly included OCA-language families in that range. A generalization of a method in [10] is shown that provides a tool for proving that languages are not acceptable by OCAs with small time bounds. By such a language and a translation result the hierarchies are established.

Keywords Models of Computation, computational complexity, cellular automata, time hierarchies.

1 Introduction

Linear arrays of interacting finite automata are models for massively parallel language acceptors. Their advantages are simplicity and uniformity. It has turned out that a large array of not very powerful processing elements operating in parallel can be programmed to be very powerful.

One type of system is of particular interest: the cellular automata whose homogeneously interconnected deterministic finite automata (the cells) work synchronously at discrete time steps obeying one common transition function. Here we are interested in a very simple type of cellular automata. The arrays are real-space bounded, i.e., the number of cells is bounded by the number of input symbols, and each cell is connected to its immediate neighbor to the right only. Due to the resulting information flow from right to left such devices are called one-way cellular automata (OCA). If the cells are connected to their both immediate neighbors the information flow becomes two-way and the device is a (two-way) cellular automaton (CA).

Although parallel language recognition by (O)CAs has been studied for more than a quarter of a century some important questions are still open. In particular, only little is known of proper inclusions in the time hierarchy. Most of the early languages known not to be real-time but linear-time OCA-languages are due to the fact that every unary real-time OCA-language is regular [4]. In [9] and [10] a method has been shown that allows proofs of non-acceptance for non-unary

---

J. Sgall, A. Pultr, and P. Kolman (Eds.): MFCS 2001, LNCS 2136, pp. 439–450, 2001.

languages in real-time OCAs. Utilizing these ideas the non-closure of real-time OCA-languages under concatenation could be shown.

Since for separating the complexity classes in question there are no other general algebraic methods available, specific languages as potential candidates are of particular interest. In [5] several positive results have been presented. Surprisingly, so far there was only one inclusion known to be strict in the time hierarchy from real-time to exponential-time. It is the inclusion between real-time and linear-time languages. In [2] the existence of a non-real-time OCA-language that is acceptable in $n + \log(n)$-time has been proved yielding a lower upper bound for the strict inclusion. Another valuable tool for exploring the OCA time hierarchy is the possible linear speed-up [7] from $n + r(n)$ to $n + \epsilon \cdot r(n)$ for $\epsilon > 0$.

The main contribution of the present paper is to show that there exists an infinite time hierarchy of properly included language families. These families are located in the range between real-time and linear-time. The surprising result covers the lower part of the time hierarchy in detail.

The paper is organized as follows: In Section 2 we define the basic notions and the model in question. Since for almost all infinite hierarchies in complexity theory the constructibility of the bounding functions is indispensable, in Section 3 we present our notion of constructibility in OCAs and prove that it covers a wider range of functions than the usual approach. Section 4 is devoted to a generalization of the method in [10] to time complexities beyond real-time. This key tool is utilized to obtain a certain language not acceptable with a given time bound. Finally, in Section 5 the corresponding proper inclusion is extended to an infinite time hierarchy by translation arguments.

2   Basic Notions

We denote the positive integers $\{1, 2, ...\}$ by $\mathbb{N}$ and the set $\mathbb{N} \cup \{0\}$ by $\mathbb{N}_0$. The empty word is denoted by λ and the reversal of a word w by w^R. For the length of w we write $|w|$. We use $\subseteq$ for inclusions and $\subset$ if the inclusion is strict. For a function $f : \mathbb{N}_0 \to \mathbb{N}$ we denote its i-fold composition by $f^{[i]}$, $i \in \mathbb{N}$. If f is increasing then its inverse is defined according to $f^{-1}(n) = \min\{m \in \mathbb{N} \mid f(m) \geq n\}$. As usual we define the set of functions that grow strictly less than f by $o(f) = \{g : \mathbb{N}_0 \to \mathbb{N} \mid \lim_{n \to \infty} \frac{g(n)}{f(n)} = 0\}$. In terms of orders of magnitude f is an upper bound of the set $O(f) = \{g : \mathbb{N}_0 \to \mathbb{N} \mid \exists n_0, c \in \mathbb{N} : \forall n \geq n_0 : g(n) \leq c \cdot f(n)\}$. Conversely, f is a lower bound of the set $\Omega(f) = \{g : \mathbb{N}_0 \to \mathbb{N} \mid f \in O(g)\}$.

A one-way resp. two-way cellular array is a linear array of identical deterministic finite state machines, sometimes called cells, which are connected to their nearest neighbor to the right resp. to their both nearest neighbors. The array is bounded by cells in a distinguished so-called boundary state. For convenience we identify the cells by positive integers. The state transition depends on the current state of each cell and the current state(s) of its neighbor(s). The transition function is applied to all cells synchronously at discrete time steps. Formally:

Definition 1. *A one-way cellular automaton (OCA) is a system* $\langle S, \delta, \#, A, F \rangle$*,
where*

1. S *is the finite, nonempty set of* cell states,
2. $\# \notin S$ *is the* boundary state,
3. $A \subseteq S$ *is the nonempty set of* input symbols,
4. $F \subseteq S$ *is the set of* accepting (or final) states, *and*
5. $\delta : (S \cup \{\#\})^2 \to S$ *is the* local transition function.

If the flow of information is extended to two-way the resulting device is
a *(two-way) cellular array* (CA) and the local transition function maps from
$(S \cup \{\#\})^3$ to S.

A *configuration* of a cellular automaton at some time $t \geq 0$ is a description
of its global state, which is actually a mapping $c_t : [1, \ldots, n] \to S$ for $n \in \mathbb{N}$.

The configuration at time 0 is defined by the initial sequence of states. For
a given input $w = a_1 \cdots a_n \in A^+$ we set $c_{0,w}(i) = a_i$ for $1 \leq i \leq n$. During
a computation the (O)CA steps through a sequence of configurations whereby
successor configurations are computed according to the global transition function Δ:

Let c_t for $t \geq 0$ be a configuration, then its successor configuration is as
follows:

$$c_{t+1} = \Delta(c_t) \iff$$
$$c_{t+1}(1) = \delta\big(\#, c_t(1), c_t(2)\big)$$
$$c_{t+1}(i) = \delta\big(c_t(i-1), c_t(i), c_t(i+1)\big), i \in \{2, \ldots, n-1\}$$
$$c_{t+1}(n) = \delta\big(c_t(n-1), c_t(n), \#\big)$$

for CAs and correspondingly for OCAs. Thus, Δ is induced by δ.

If the state set is a Cartesian product of some smaller sets $S = S_0 \times S_1 \times
\cdots \times S_r$, we will use the notion *register* for the single parts of a state.

Fig. 1. A one-way cellular automaton

An input w is accepted by an (O)CA if at some time i during its course of
computation the leftmost cell enters an accepting state.

Definition 2. *Let* $\mathcal{M} = \langle S, \delta, \#, A, F \rangle$ *be an (O)CA.*

1. *An input* $w \in A^+$ *is accepted by* $\mathcal{M}$ *if there exists a time step* $i \in \mathbb{N}$ *such
that* $c_i(1) \in F$ *holds for the configuration* $c_i = \Delta^{[i]}(c_{0,w})$.
2. $L(\mathcal{M}) = \{w \in A^+ \mid w$ *is accepted by* $\mathcal{M}\}$ *is the language accepted by* $\mathcal{M}$.
3. *Let* $t : \mathbb{N} \to \mathbb{N}$, $t(n) \geq n$, *be a mapping. If all* $w \in L(\mathcal{M})$ *can be accepted
with at most* $t(|w|)$ *time steps, then* L *is said to be of time complexity* t.

The family of all languages that are acceptable by some OCA (CA) with time complexity t is denoted by $L_t(\text{OCA})$ ($L_t(\text{CA})$). If t equals the identity function $id(n) = n$, acceptance is said to be in *real-time*, and if t is equal to $k \cdot id$ for an arbitrary rational number $k > 1$, then acceptance is carried out in *linear-time*. Correspondingly, we write $L_{rt}((\text{O})\text{CA})$ and $L_{lt}((\text{O})\text{CA})$.

In this article we prove:

Theorem 3. *Let $r_1, r_2 : \mathbb{N} \to \mathbb{N}$ be two functions such that $r_2 \cdot \log^2(r_2) \in o(r_1)$ and r_1^{-1} is constructible, then*

$$L_{n+r_2(n)}(\text{OCA}) \subset L_{n+r_1(n)}(\text{OCA})$$

Example 4. Let $0 \leq p < q \leq 1$ be two rational numbers. Clearly, $n^p \cdot \log^2(n^p)$ is of order $o(n^q)$. In the next section the constructibility of the inverse of n^q will be established. Thus, an application of Theorem 3 yields the strict inclusion

$$L_{n+n^p}(\text{OCA}) \subset L_{n+n^q}(\text{OCA})$$

3   Constructible Functions

For the proof of Theorem 3 it will be necessary to control the lengths of words with respect to some internal substructures. The following notion of constructibility expresses the idea that the length of a word relative to the length of a subword should be computable.

Definition 5. *A function $f : \mathbb{N} \to \mathbb{N}$ is* constructible *if there exists an λ-free homomorphism h and a language $L \in L_{rt}(\text{OCA})$ such that*

$$h(L) = \{a^{f(n)-n}b^n \mid n \in \mathbb{N}\}$$

Since constructible functions describe the length of the whole word dependent on the length of a subword it is obvious that each constructible function must be greater than or equal to the identity. At a first glance this notion of constructibility might look somehow unusual or restrictive. But λ-free homomorphisms are very powerful so the family of (in this sense) constructible functions is very rich, and is, in fact, a generalization of the usual notion. The remainder of this section is devoted to clarify the presented notion and its power.

The next lemma states that we can restrict our considerations to length preserving homomorphisms. The advantage is that for length preserving homomorphisms each word in L is known to be of length $f(m)$ for some $m \in \mathbb{N}$.

Lemma 6. *Let $f : \mathbb{N} \to \mathbb{N}$ be a constructible function. Then there exists a length preserving λ-free homomorphism h and a language $L \in L_{rt}(\text{OCA})$ such that*

$$h(L) = \{a^{f(n)-n}b^n \mid n \in \mathbb{N}\}$$

The proof follows immediately from a proof in [1] where the closure of $L_{rt}(OCA)$ under λ-free homomorphisms is characterized by OCAs with limited nondeterminism.

Given an increasing constructible function $f : \mathbb{N} \to \mathbb{N}$ and a language $L_a \subseteq A^+$ acceptable by some OCA with time complexity $n + r(n)$, where $r : \mathbb{N} \to \mathbb{N}$, we now define a language that plays an important role in the sequel. Let the language $L_f \subseteq B^+$ be a witness for the constructibility of f, i.e., $L_f \in L_{rt}(OCA)$ and $h(L_f) = \{a^{f(n)-n}b^n \mid n \in \mathbb{N}\}$ for a length preserving λ-free homomorphism h. The language $L_1(L_a, L_f) \subseteq (A \cup \{\sqcup\}) \times B)^+$ is constructed as follows

1. The second component of each word w in $L_1(L_a, L_f)$ is a word of L_f that implies that w is of length $f(m)$ for some $m \in \mathbb{N}$.
2. The first component of w contains exactly $f(m) - m$ blank symbols and m non-blank symbols.
3. The non-blank symbols in the first component of w form a word in L_a.

The following proposition is used in later sections. Besides, it is an example that demonstrates how to use constructible functions. In Lemma 13 we will prove that the shown bound for the time complexity of L_1 is minimal.

Proposition 7. *The language $L_1(L_a, L_f)$ is acceptable by some OCA with time complexity $n + r(f^{-1}(n))$.*

Proof. We construct an OCA $\mathcal{A}$ with three registers that accepts L_1 obeying the time complexity $n + r(f^{-1}(n))$.

In its first register $\mathcal{A}$ verifies that the second component of each word in L_1 is a word of L_f. By definition of L_f this can be done in real-time.

In its second register $\mathcal{A}$ checks that the first component of L_1 contains exactly $f(m) - m$ blank symbols. Because it can be verified that the second component of L_1 belongs to L_f, we know that the first $f(m) - m$ symbols of the second component are mapped to a's and the last m symbols of the second component are mapped to b's. The task is to check that the number of a's in the second component is equal to the number of blank symbols in the first component. Therefore, $\mathcal{A}$ shifts the blank symbols from right to left. Each symbol a in the second component consumes one blank symbol. A signal that goes from the right to the left with full speed can check that no blank symbol has reached the leftmost cell and that each letter a has consumed one blank symbol, i.e., that the number of a's is equal to the number of blank symbols. The test can be done in real-time.

In order to verify that the non-blank symbols in the first component form a word of L_a the automaton $\mathcal{A}$ simulates the OCA that accepts L_a. But for every blank-symbol $\mathcal{A}$ has to be delayed for one time step, until it receives the necessary information for the next simulation step. Therefore, $\mathcal{A}$ needs $m + r(m) + (f(m) - m)$ steps for the simulation ($m + r(m)$ time steps for the simulation itself and $f(m) - m$ time steps delaying time). Substituting $m = f^{-1}(n)$ completes the proof. $\qquad\square$

Now we prove that the family of constructible functions is very rich. In particular, all Fischer-constructible functions are constructible in the sense of Definition 5. A function f is said to be *Fischer-constructible* if there exists an unbounded two-way CA such that the initially leftmost cell enters a final state at time $i \in \mathbb{N}$ if and only if $i = f(m)$ for some $m \in \mathbb{N}$. Thus, the Fischer-constructibility is an important notion that meets the intuition of constructible functions. For a detailed study of these functions see [8] where also the name has been introduced according to the author of [6].

For example, n^k for $k \in \mathbb{N}$, 2^n, $n!$, and p_n, where p_n is the nth prime number, are Fischer-constructible. Moreover, the class is closed under several operations.

Lemma 8. *If a function $f : \mathbb{N} \to \mathbb{N}$ is Fischer-constructible, then it is constructible in the sense of Definition 5.*

Proof. In [3] it is shown that for every language $L' \in \mathrm{L}_{lt}(\mathrm{CA})$ there exists a language $L \in \mathrm{L}_{rt}(\mathrm{OCA})$ and an λ-free homomorphism h such that $h(L) = L'$. Therefore, it is sufficient to prove that for Fischer-constructible functions f the languages $\{a^{f(n)-n}b^n \mid n \in \mathbb{N}\}$ are linear-time CA-languages:

The initially leftmost cell of an appropriate CA starts the construction of f, i.e., it enters a final state exactly at the time steps $f(1), f(2), \ldots$ In addition, the rightmost cell sends initially a signal to the left that runs with full speed. The CA accepts a word of the form a^+b^+ if and only if this signal arrives at the leftmost cell at a time step $f(m)$ for some $m \in \mathbb{N}$ and the number of b's is equal to m. The details of the easy CA construction are straightforward.   □

Without proof we mention that the class of constructible functions is closed under several operations such as addition, multiplication or composition.

4   Equivalence Classes

In order to prove lower bounds for the time complexity we generalize a lemma that gives a necessary condition for a language to be real-time acceptable by an OCA. At first we need the following definition:

Definition 9. *Let L be a language and X and Y be two sets of words. Two words w and w' are equivalent with respect to L, X and Y (in short (L, X, Y)-equivalent) if and only if $xwy \in L \iff xw'y \in L$ for all $x \in X$ and $y \in Y$.*

Let $L_d \subset \{0, 1, (,), |\}^+$ be a language whose words are of the form

$$x\,(\,x_1 \mid y_1\,) \cdots (\,x_n \mid y_n\,)\,y$$

where $x, x_i, y, y_i \in \{0, 1\}^*$ for $1 \le i \le n$, and $(\,x \mid y\,) = (\,x_i \mid y_i\,)$ for at least one $i \in \{1, \ldots, n\}$.

The language L_d can be thought as a dictionary. The task for the OCA is to check whether the pair $(\,x \mid y\,)$ appears in the dictionary or not.

Proposition 10. *Let $X = Y = \{0,1\}^*$. Two words $w = (x_1 \mid y_1) \cdots (x_n \mid y_n)$ and $w' = (x_1' \mid y_1') \cdots (x_m' \mid y_m')$ are equivalent with respect to L_d, X and Y if and only if $\{(x_1 \mid y_1), \ldots, (x_n \mid y_n)\} = \{(x_1' \mid y_1'), \ldots, (x_m' \mid y_m')\}$.*

Proof. First assume that the two sets are equal. Let $x \in X$ and $y \in Y$, then $xwy \in L_d$ implies $(x \mid y) = (x_i \mid y_i)$ for some i. Since the two sets are equal we have $(x \mid y) = (x_j' \mid y_j')$ for some j. Therefore, $xwy \in L_d$ implies $xw'y \in L_d$ and vice versa, i.e., w and w' are (L_d, X, Y)-equivalent.

Now assume the two sets are not equal. Without loss of generality we can assume that there exist $x \in X$ and $y \in Y$ with $(x \mid y) = (x_i \mid y_i)$ for some i, but $(x \mid y) \neq (x_j' \mid y_j')$ for all $j = 1, \ldots, m$. Then $xwy \in L_d$ but $xw'y \notin L_d$ and, thus, w and w' are not (L_d, X, Y)-equivalent. $\quad\square$

Now we are prepared to formulate the lemma we are going to use in order to prove lower bounds for the time complexities. For the special case $L \in \mathrm{L}_{rt}(\mathrm{OCA})$ the lemma has been shown in [10].

Lemma 11. *Let $r : \mathbb{N} \to \mathbb{N}$ be an increasing function, $L \in \mathrm{L}_{n+r(n)}(\mathrm{OCA})$ and X and Y be two sets of words. Let s be the minimal number of states needed by an OCA to accept L in $n + r(n)$ time steps.*

If all words in X are of length m_1 and all words in Y are of length m_2, then the number N of (L, X, Y)-equivalence classes of the words at most of length $n - m_1 - m_2$ is bounded by

$$N \leq s^{m_1 |X|} s^{(m_2 + r(n))|Y|}$$

Proof. Let $\mathcal{A}$ be an OCA with s states that accepts L in $n + r(n)$ time steps. Let S be the state set of $\mathcal{A}$.

We consider the computation of $\mathcal{A}$ on the word xwy for some $x \in X$ and $y \in Y$. After $|w|$ time steps the interesting part of the configuration of $\mathcal{A}$ can be described by $f_w(x) f_w'(y)$ where (cf. Figure 2)

1. $f_w(x) \in S^*$ and $f_w'(y) \in S^*$.
2. $|f_w(x)| = |x|$ and $|f_w'(y)| = y + r(n)$. During the remaining $|xwy| + r(|xwy|) - |w| \leq |x| + |y| + r(n)$ time steps the result of the computation of $\mathcal{A}$ depends only on the states of the $|x| + |y| + r(n)$ leftmost cells.
3. $f_w'(y)$ depends only on w and y since no information can move from left to right.
4. $f_w(x)$ depends only on w and x since during $|w|$ time steps only the leftmost $|x| + |w|$ cells can influence the states of the leftmost $|x|$ cells.

If $f_w(x) = f_{w'}(x)$ and $f_w'(y) = f_{w'}'(y)$ for all $x \in X$ and $y \in Y$, then w and w' are equivalent with respect to L, X and Y. Thus, if w and w' are not equivalent, then $f_w \neq f_{w'}$ or $f_w' \neq f_{w'}'$.

Now we count the number of functions f_w and f_w'. Since f_w maps X into the set S^{m_1} which contains s^{m_1} elements, the number of different functions f_w is bounded by $(s^{m_1})^{|X|}$. Analogously, it follows that the number of different functions f_w' is bounded by $(s^{m_2 + r(n)})^{|Y|}$.

Since each upper bound on the number of pairs (f_w, f_w') is also an upper bound on the number of (L, X, Y)-equivalence classes the lemma follows. $\quad\square$

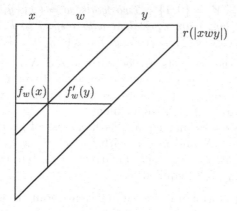

Fig. 2. OCA computation in the proof of Lemma 11

Now we apply the lemma to the language L_d.

Proposition 12. *Let* $r : \mathbb{N} \to \mathbb{N}$ *be a function. If* $r(n) \log^2(r(n)) \in o(n)$, *then* L_d *is not acceptable by an OCA with time complexity* $n + r(n)$ *but* L_d *belongs to* $L_{lt}(\text{OCA})$.

Proof. For fixed $m_1 \in \mathbb{N}$ and $m_2 \in \mathbb{N}$ we investigate all words of the form $(x_1 \mid y_1) \cdots (x_k \mid y_k)$ with $x_i \in \{0,1\}^{m_1}$ and $y_i \in \{0,1\}^{m_2}$ for all $i \in \{1,\ldots,k\}$ and $(x_i \mid y_i) \neq (x_j \mid y_j)$ for $i \neq j$. We call this words of type (m_1, m_2).

Since there are at most $2^{m_1+m_2}$ different pairs (x_i, x_j) the length of words of type (m_1, m_2) is at most $2^{m_1+m_2} \cdot (m_1 + m_2 + 3)$.

As has been shown in Proposition 10 two words are equivalent iff the sets of subwords are equal. Thus, there are $2^{2^{m_1+m_2}}$ words of type (m_1, m_2) which belong to different equivalence classes with respect to L_d, $X = \{0,1\}^{m_1}$ and $Y = \{0,1\}^{m_2}$. (For each subset of $X \times Y$ there exists one equivalence class.)

Assume L_d belongs to $L_{n+r(n)}(\text{OCA})$, then an accepting OCA must be able to distinguish all these equivalence classes. By Lemma 11 there must exist a number of states s such that

$$2^{2^{m_1+m_2}} \leq s^{m_1 2^{m_1}} s^{(m_2+r(n))2^{m_2}}$$

for $n = 2^{m_1+m_2} \cdot (m_1 + m_2 + 3) + m_1 + m_2$.

In order to obtain a contradiction let $m_1 = c \cdot 2^{m_2}$ for some arbitrary rational number c. Approximating the order of n we obtain

$$n \in O(2^{m_1+m_2} \cdot (m_1 + m_2)) = O(2^{m_1+m_2} \cdot m_1) = O(2^{m_1} 2^{m_2} \cdot m_1) = O(2^{m_1} m_1^2)$$

Since $r(n) \log^2(r(n)) \in o(n)$ it holds $r(2^{m_1} m_1^2) \log^2(r(2^{m_1} m_1^2)) \in o(2^{m_1} m_1^2)$. Observe $2^{m_1} \log^2(2^{m_1}) = 2^{m_1} m_1^2$ and, therefore, $r(2^{m_1} m_1^2)$ and, hence, $r(n)$ must be of order $o(2^{m_1})$. It follows

$$s^{m_1 2^{m_1}} s^{(m_2+r(n))2^{m_2}} = s^{m_1 2^{m_1}} s^{(m_2+o(2^{m_1}))2^{m_2}} = s^{m_1 2^{m_1}} s^{o(2^{m_1})2^{m_2}}$$

$$= s^{m_1 2^{m_1}+o(2^{m_1} 2^{m_2})} = s^{c 2^{m_2} 2^{m_1}+o(2^{m_1} 2^{m_2})} = s^{c 2^{m_2+m_1}+o(2^{m_1+m_2})}$$

Since c has been arbitrarily chosen we can let it go to 0 and obtain

$$= s^{o(2^{m_1+m_2})} = o(2^{2^{m_1+m_2}})$$

This is a contradiction, thus, L is not acceptable by an OCA in $n + r(n)$ time.

To see that L is acceptable in linear-time, we construct an appropriate OCA. Starting with an input word of the form $x(x_1|y_1)\cdots(x_m|y_m)y$ the OCA shifts the subword y with full speed to the left. During the first n time steps the OCA marks all pairs $(x_i|y_i)$ with $y_i = y$. Each marked pair starts moving to the left with half speed. Each time a pair $(x_i|y)$ reaches the left hand side the OCA checks whether $x_i = x$. The pairs of the form $(x_i|y)$ reach the leftmost cell sequentially because y moves with full speed but the pairs of form $(x_i|y)$ with half speed only. This guarantees that the OCA has sufficient time to check whether $x = x_i$. Figure 3 illustrates the computation. The basic task for the OCA is to check whether $y = y_i$. This is equivalent to the acceptance of the real-time OCA-language $\{w \bullet w \mid w \in \{0,1\}^+\}$.   □

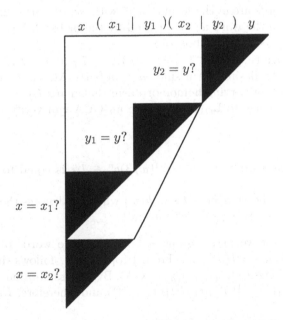

Fig. 3. Linear-time acceptance of L_d. The black triangles mark the areas where a check of the form $y_i = y$ takes place. It is easy to see that the black triangels are disjoint, i.e., the checks can be done one after the other with a finite number of states

5   Time Hierarchies

The last section is devoted to the proof of the main result which has been stated in Theorem 3. The next step towards the proof is a translation lemma that allows to extend a single proper inclusion to a time hierarchy.

Lemma 13. *Let $t_1, t_2 : \mathbb{N} \to \mathbb{N}$ be two functions and let L_a be a language that belongs to $L_{n+t_1(n)}(\text{OCA})$ but is not acceptable by any OCA within $n + o(t_2(n))$ time as can be shown by applying Lemma 11. Further let $f : \mathbb{N} \to \mathbb{N}$ be a constructible function and $r_1, r_2 : \mathbb{N} \to \mathbb{N}$ be two functions such that $r_1(f(n)) \in \Omega(t_1(n))$ and $r_2(f(n)) \in o(t_2(n))$. Then*

$$L_{n+r_2(n)}(\text{OCA}) \subset L_{n+r_1(n)}(\text{OCA}).$$

Proof. For $f(n) \in O(n)$ we have $r_2(O(n)) \in o(t_2(n))$ what implies $r_2(n) \in o(t_2(n))$ and, thus, $L_a \notin L_{n+r_2(n)}(\text{OCA})$. Conversely, $r_1(O(n)) \in \Omega(t_1(n))$ and, therefore, $r_1(n) \in \Omega(t_1(n))$. It follows $L_a \in L_{n+r_1(n)}(\text{OCA})$ and, hence, the assertion.

In order to prove the lemma for $n \in o(f(n))$ let L_f be a language that proves the constructibility of f in Lemma 6. At first we show that we can always find such an L_f whose words are of the form $a^n w b^n$ with $|w| = f(n) - 2n$.

By $w/3$ we denote the word w compressed by the factor 3, i.e., one symbol of $w/3$ is interpreted as three symbols of w.

Now define $\bar{L}_f$ such that $a^{f(n)-n-\frac{1}{3}f(n)}(w/3)b^n \in \bar{L}_f$ iff $w \in L_f$. Clearly, the words of $\bar{L}_f$ are of the desired form since $n \in o(f(n))$. Moreover, there exists a trivial λ-free, length preserving homomorphism that maps $\bar{L}_f$ to $\{a^{f(n)-n}b^n \mid n \in \mathbb{N}\}$. Also, $\bar{L}_f$ belongs to $L_{rt}(\text{OCA})$ since an OCA can verify in real-time that an input w

- belongs to L_f,
- the length of the word $a^{f(n)-n-\frac{1}{3}f(n)}(w/3)b^n \in \bar{L}_f$ is equal to the length of w, and
- that n is equal to the number of b's in $h(w)$ where h denotes the homomorphism that maps L_f to $\{a^{f(n)-n}b^n \mid n \in \mathbb{N}\}$.

Thus, from now on we may assume w.l.o.g. that the words of L_f are of the form $a^n w b^n$ with $|w| = f(n) - 2n$. From Proposition 7 follows that the language $L_1(L_a, L_f)$ belongs to $L_{n+t_1(f^{-1}(n))}(\text{OCA})$. By the assumption on $r_1(f(n))$ we obtain $r_1(n) = r_1(f(f^{-1}(n))) \in \Omega(t_1(f^{-1}(n)))$ and, therefore, $L_1(L_a, L_f) \in L_{n+r_1(n)}(\text{OCA})$.

It remains to show that $L_1(L_a, L_f) \notin L_{n+r_2(n)}(\text{OCA})$.

Since $r_2(f(n)) \in o(t_2(n))$ and L_a is not acceptable within $n + o(t_2(n))$ time by any OCA, the language L_a is not acceptable within $n + r_2(f(n))$ time by any OCA, either. Due to the assumption, by Lemma 11 for every $s \in \mathbb{N}$ there must exist sets X and Y and an $n \in \mathbb{N}$ such that all words in X are of length m_1, all words in Y are of length m_2, and the number of (L_a, X, Y)-equivalence classes of the words at most of length $n - m_1 - m_2$ is not bounded by $s^{m_1|X|}s^{(m_2+r_2(f(n)))|Y|}$.

Define
$$X' = \{(x_1, a) \cdots (x_{m_1}, a) \mid x = x_1 \ldots x_{m_1} \in X\}$$

and
$$Y' = \{(y_1, b) \cdots (y_{m_2}, b) \mid y = y_1 \ldots y_{m_2} \in Y\}$$

and for every word $v = v_1 \cdots v_{n-m_1-m_2}$ a word v' by

$$v' = (v_1, w_1) \cdots (v_{n-m_1-m_2}, w_{n-m_1-m_2})(\sqcup, w_{n-m_1-m_2+1}) \cdots (\sqcup, w_{f(n)-m_1-m_2})$$

where $a^{m_1} w_1 \cdots w_{f(n)-m_1-m_2} b^{m_2}$ is a word of L_f. (Remember that each word in L_f starts with n symbols a and ends with n symbols b and $m_1 + m_2 \leq n$.)

For $x \in X$ and $y \in Y$ let x' and y' denote the corresponding words in X' and Y'.

By construction $xvy \in L_a$ iff $x'v'y' \in L_1$. (The word $x'v'y'$ belongs to L_1 if the second component of $x'v'y'$ is a word in L_f, which is always true, and the first component of $x'v'y'$ is a word in L_a concatenated with some blank symbols, i.e., $xvy \in L_a$.) Thus, the (L_a, X, Y)-equivalence classes have corresponding (L_1, X', Y')-equivalence classes and the number of (L_1, X', Y')-equivalence classes under the words whose length is at most $f(n) - m_1 - m_2$ is not bounded by $s^{m_1|X|} s^{(m_2 + r_2(f(n)))|Y|}$.

Applying Lemma 11 with L_1, X', Y' and $f(n)$ in place of L_a, X, Y and n yields that $L_1 \notin L_{n+r_2(n)}(\text{OCA})$. This completes the proof. □

Finally the main Theorem 3 is just a combination of the preceding lemmas:

Theorem 3. Let $r_1, r_2 : \mathbb{N} \to \mathbb{N}$ be two functions such that $r_2 \cdot \log^2(r_2) \in o(r_1)$ and r_1^{-1} is constructible, then

$$L_{n+r_2(n)}(\text{OCA}) \subset L_{n+r_1(n)}(\text{OCA})$$

Proof. Proposition 12 shows that the previously defined language L_d is acceptable in linear-time but is not acceptable in $n + r(n)$ time if $r(n) \log^2(r(n)) \in o(n)$. Now set $t_1(n) = n$ and t_2 such that $t_2(n) \log^2(t_2(n)) = n$.

Inserting yields $r(n) \log^2(r(n)) \in o(t_2(n) \log^2(t_2(n)))$. We conclude $r(n) \in o(t_2(n))$ and, thus, L_d is not acceptable in $n + o(t_2(n))$ time. In order to apply Lemma 13 we consider the constructible function $f = r_1^{-1}$.

Clearly, $r_1(f(n)) = n \in \Omega(n) = \Omega(t_1(n))$.

Since $r_2(n) \log^2(r_2(n)) \in o(r_1(n))$ we have

$$r_2(f(n)) \log^2(r_2(f(n))) \in o(r_1(f(n))) = o(n) = o(t_2(n) \log^2(t_2(n)))$$

We conclude $r_2(f(n)) \in o(t_2(n))$.

Now all conditions of Lemma 13 are satisfied and an application proves the assertion $L_{n+r_2(n)}(\text{OCA}) \subset L_{n+r_1(n)}(\text{OCA})$. □

References

1. Buchholz, Th., Klein, A., and Kutrib, M. *One guess one-way cellular arrays.* Mathematical Foundations of Computer Science 1998, LNCS 1450, 1998, pp. 807–815. 443
2. Buchholz, Th., Klein, A., and Kutrib, M. *On tally languages and generalized interacting automata.* Developments in Language Theory IV. Foundations, Applications, and Perspectives, 2000, pp. 316–325. 440
3. Buchholz, Th., Klein, A., and Kutrib, M. *On interacting automata with limited nondeterminism.* Fund. Inform. (2001), to appear. 444
4. Choffrut, C. and Čulik II, K. *On real-time cellular automata and trellis automata.* Acta Inf. 21 (1984), 393–407. 439
5. Dyer, C. R. *One-way bounded cellular automata.* Inform. Control 44 (1980), 261–281. 440
6. Fischer, P. C. *Generation of primes by a one-dimensional real-time iterative array.* J. Assoc. Comput. Mach. 12 (1965), 388–394. 444
7. Ibarra, O. H. and Palis, M. A. *Some results concerning linear iterative (systolic) arrays.* J. Parallel and Distributed Comput. 2 (1985), 182–218. 440
8. Mazoyer, J. and Terrier, V. *Signals in one dimensional cellular automata.* Theoret. Comput. Sci. 217 (1999), 53–80. 444
9. Terrier, V. *On real time one-way cellular array.* Theoret. Comput. Sci. 141 (1995), 331–335. 439
10. Terrier, V. *Language not recognizable in real time by one-way cellular automata.* Theoret. Comput. Sci. 156 (1996), 281–287. 439, 440, 445

Checking Amalgamability Conditions for CASL Architectural Specifications

Bartek Klin[1], Piotr Hoffman[2], Andrzej Tarlecki[2,3],
Lutz Schröder[4], and Till Mossakowski[4]

[1] BRICS, Århus University
[2] Institute of Informatics, Warsaw University
[3] Institute of Computer Science, Polish Academy of Sciences
[4] BISS, Department of Computer Science, University of Bremen

Abstract. CASL, a specification formalism developed recently by the CoFI group, offers architectural specifications as a way to describe how simpler modules can be used to construct more complex ones. The semantics for CASL architectural specifications formulates static *amalgamation conditions* as a prerequisite for such constructions to be well-formed. These are non-trivial in the presence of subsorts due to the failure of the amalgamation property for the CASL institution. We show that indeed the static amalgamation conditions for CASL are undecidable in general. However, we identify a number of practically relevant special cases where the problem becomes decidable and analyze its complexity there. In cases where the result turns out to be **PSPACE**-hard, we discuss further restrictions under which polynomial algorithms become available. All this underlies the static analysis as implemented in the CASL tool set.

Keywords: formal specification and program development; CASL; architectural specifications; amalgamation; algorithms; decidability.

Introduction

Architectural specifications are a mechanism to support formal program development [2,11]. They have been introduced in the CASL specification language, which provides them in addition to structured specifications [4,3]. While the latter facilitate building global requirement specifications in a modularized fashion, the former aim at formalizing the development process: an architectural specification describes how an implementation of the global specification may be built from implementations of other (simpler) specifications. One of the operations which may be used in this process is *amalgamation*, i.e., one may build a bigger unit simply by combining (amalgamating) smaller ones. (Modules in CASL architectural specifications are called *units*; parametricity aside, they correspond semantically to models in the underlying CASL institution.)

Whenever such an amalgamation occurs in an architectural specification, the semantics requires the amalgamated unit to exist and be unique. Whether or not this holds in concrete cases is undecidable, since the CASL underlying logic is undecidable. However, in [11] we developed a somewhat stricter but

J. Sgall, A. Pultr, and P. Kolman (Eds.): MFCS 2001, LNCS 2136, pp. 451–463, 2001.

entirely static *amalgamation condition* that ensures amalgamability of units. This condition is certainly decidable for CASL without subsorts. The problem of its (un)decidability for arbitrary CASL specifications with subsorts is investigated here, and the tractability of important special cases is studied.

After introducing a simplified version of the CASL institution in Section 1, we recall the institution independent amalgamation conditions as defined in [11] in Section 2 and, in Section 3, reformulate them in the simplified CASL institution. This is followed by a sound and complete *cell calculus* for discharging them. In Section 4, we show that the cell calculus and, hence, the amalgamation conditions in CASL are undecidable. This motivates the search for practically relevant special cases for which an algorithm to check amalgamation conditions can be given. First, we indicate that when the number of cells to consider is finite, the problem becomes decidable but **PSPACE**-hard. Moreover, we discuss a simple situation where the amalgamation conditions can be checked in polynomial time. In Section 5, we investigate another special case which arises naturally in the categorical framework we use, but can also be interpreted as a methodological directive on the use of subsorting in architectural specifications. Essentially, it corresponds to the requirement that we can amalgamate all units in the architectural specification, even with hidden components. No surprise here: the problem remains undecidable. However, if we once again assume that the number of cells is finite, then this problem admits a polynomial time algorithm. This algorithm is currently used by the CASL toolset, CATS[1]. In Section 6 we discuss the possibility of restricting the cell calculus to a finite domain and observe that checking the resulting amalgamation condition is decidable but **PSPACE**-hard.

1   The CASL Institution

The following considerations rely on the notion of institution [6]. An *institution I* consists of a category **Sign** of *signatures*, a *model functor* **Mod** : **Sign**^{op} → **CAT**, and further components which formalize sentences and satisfaction. In this context, we use only the category of signatures and the model functor. The functor **Mod** maps a signature to a category of *models* over this signature, and (contravariantly) a signature morphism to a *reduct functor* between model categories.

We say that an institution I has the *amalgamation property* if **Mod** preserves limits, i.e., maps colimits of signature diagrams to limits in **CAT**.

The underlying logic of CASL is formalized by the institution *SubPCFOL*$^=$ ('subsorted partial first order logic with equality and sort generation constraints', see [3]). Signatures in this institution are the usual many- and subsorted algebraic signatures with partial and total operations and predicates. In the following, for clarity of the presentation, we ignore operations and predicates and view CASL signatures simply as preorders of sorts. Although the presence of operations and predicates adds some technical complication to the issues considered, it does not change the presented results in an essential way.

---

[1] Accessible by WWW, at http://www.tzi.de/cofi/CASL/CATS

Formally, our simplified CASL institution consists of the category **PreOrd** of all finite preorders as the signature category. A model over a signature $\Sigma = (S, \leq)$ is an S-sorted set $\{X_s\}_{s \in S}$ together with an injective function $e_{(s,t)} : X_s \to X_t$ for all $s \leq t$ in Σ, satisfying the commutativity condition $e_{(t,u)}e_{(s,t)} = e_{(s,u)}$ for all $s \leq t \leq u$ in Σ, and such that $e_{(s,s)}$ is the identity for $s \in S$. With the usual homomorphisms, models over a signature Σ form a category **Mod**(Σ). Together with a mapping of signature morphisms (i.e., order-preserving functions) to the obvious reduct functors, this defines the model functor **Mod** : **PreOrd**$^{op} \to$ **CAT**. Unfortunately, **Mod** does not preserve limits, i.e., the CASL institution does not have the amalgamation property (see the example given below).

We will often speak of CASL signatures as (preorder, or thin) categories. Their objects, in turn, will be called *sorts*, and their morphisms *sort embeddings*. *Signature morphisms*, or shortly *morphisms* (in **PreOrd**) are functors between signatures as thin categories.

2   Amalgamation Conditions

In [11], a semantics for architectural specifications as defined in CASL [4] has been proposed. The semantics is institution independent, i.e., it is parametrized by the institution of the underlying specification formalism. We now briefly describe a problem that occurs in checking the correctness of an architectural specification with respect to the proposed semantics, and then go on to analyse this problem in the case of our simplified CASL institution.

For an architectural specification to be correct, it has to satisfy certain *amalgamation conditions*. Informally, these conditions ensure that whenever two units are combined (amalgamated) into one unit, they must 'share' components from the intersection of their signatures, to allow the amalgamation to be defined unambiguously. Moreover, in the presence of subsorts, natural compatibility requirements on subsort embeddings must be ensured. One would expect the amalgamation conditions to be checked during static analysis of a specification, as in the analogous case of 'sharing conditions' in module systems of programming languages like Standard ML [8]. However, the most natural formulation of the amalgamation conditions for architectural specifications does not suggest any method of checking them statically, as it involves classes of models, possibly defined by sets of axioms of an undecidable underlying logic.

Because of this, an *extended static semantics* of architectural specifications, with the amalgamation conditions reformulated statically, has been proposed in [11] . In this approach, each part of an architectural specification is evaluated to a diagram $D : \mathbf{I} \to \mathbf{Sign}$, where $\mathbf{I}$ is a finite category, and **Sign** is the signature category of the underlying institution. A family of models $\langle M_i \rangle_{i \in \mathbf{Ob(I)}}$ is *compatible* with D if for all $i \in \mathbf{Ob(I)}$, $M_i \in \mathbf{Ob(Mod}(D(i)))$, and for all $m : i \to j$ in $\mathbf{I}$, $M_i = \mathbf{Mod}(D(m))(M_j)$. For two diagrams $D : \mathbf{I} \to \mathbf{Sign}$, $D' : \mathbf{I}' \to \mathbf{Sign}$, where D' extends D, we say that D *ensures amalgamability for* D', if any model family compatible with D can be uniquely extended to a family compatible with D'.

spec NUM1 = **sorts** *Number, List[Digit], List[Number]*
                    ops ... **axioms** ...
spec NUM2 = **sorts** *Number <List[Digit], List[Number]*
spec NUM3 = **sorts** *Number <List[Number], List[Digit]<List[Number]*

arch spec *A* =
units   *U* : NUM1;
             F : NUM1 → NUM2;
             G : NUM1 → NUM3;
result *F[U]* **and** *G[U]*
end

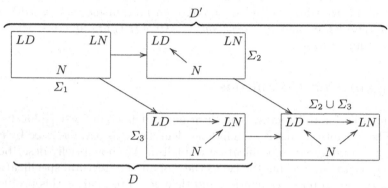

Fig. 1. An architectural specification in CASL and its corresponding signature diagram

Example. Consider the architectural specification *A* presented in Fig. 1, with the simplified CASL institution as the underlying institution. To implement this specification, one has to implement a unit *U* — a model consisting of three sorts (we omit the expected operations and axioms here). Moreover, one has to provide two parametrized units *F* and *G*, extending any such unit with some injective functions between sorts, according to specifications NUM2 and NUM3. The resulting unit of the specification *A* is the amalgamation of both extensions applied to *U*.

The extended static semantics maps the specification *A* to the signature diagram *D'* shown in Fig. 1, while the diagram *D* corresponds to the specification *A* 'before' the amalgamation *F[U]* **and** *G[U]* ($\Sigma_1, \Sigma_2, \Sigma_3$ are the signatures of the CASL basic specifications NUM1, NUM2, NUM3, respectively).

It is intuitively clear that amalgamation in this example is not always possible. If the units *F* and *G* of Fig. 1 are implemented independently, the injective functions they introduce may not commute (in fact, they do not commute under the informally expected realization of lists in *U* and of the embeddings in *F* and *G*). This is formalized in the extended static semantics by the fact that *D* does not ensure amalgamability for *D'*. This is connected to the lack of amalgamation property in the CASL institution: indeed, the model functor **Mod** does not map the pushout of D (i.e., the signature $\Sigma_2 \cup \Sigma_3$) to a pullback in **CAT**.

Now, the amalgamation conditions can be reformulated as conditions requiring that some diagrams $D : \mathbf{I} \to \mathbf{Sign}$ (corresponding to certain parts of the evaluated specification) ensure amalgamability for diagrams $D' : \mathbf{I}' \to \mathbf{Sign}$ such that $\mathbf{I}'$ results from $\mathbf{I}$ by adding a new object l and a morphism $m : k \to l$ for some $k \in \mathbf{Ob}(\mathbf{I})$, and D' extends D and maps m to an extremal epimorphism in **Sign** (the general situation can be easily reduced to this case, see [11] and the example below). In the following, the image of $m : k \to l$ under D will be denoted by $\tau : D(k) \to \Delta$.

Example (ctd.). In our example, the diagram D is extended by $\Delta = \Sigma_2 \cup \Sigma_3$ and a morphism τ from the disjoint union of Σ_2 and Σ_3 (silently added to D) to Δ. The morphism τ is an extremal epi in **PreOrd**, i.e., it is surjective on sorts and generates the preorder on its codomain.

The reformulated amalgamation conditions are 'static', because the model classes involved are not restricted by any axioms of the underlying logic. Unfortunately, in the CASL institution, even such static conditions cannot be checked directly due to the failure of the amalgamation property. As suggested in [11], this problem can be circumvented by a translation of the CASL institution to an institution with the amalgamation property:

Consider two institutions $I = (\mathbf{Sign}, \mathbf{Mod})$, $I' = (\mathbf{EnrSign}, \mathbf{Mod}_e)$ (the sentence and satisfaction components are omitted), and a functor $\Phi : \mathbf{Sign} \to \mathbf{EnrSign}$ such that $\mathbf{Mod}_e \circ \Phi$ is naturally isomorphic to **Mod**. Then one can show that, given two diagrams $D : \mathbf{I} \to \mathbf{Sign}$ and $D' : \mathbf{I}' \to \mathbf{Sign}$, D ensures amalgamability for D' if and only if ΦD ensures amalgamability for $\Phi D'$. Moreover, if I' satisfies the amalgamation property and certain additional conditions (see [11] for details), then ΦD ensures amalgamability for $\Phi D'$ if and only if a colimit cocone for ΦD extends to a cocone for $\Phi D'$. The latter condition is entirely 'static', in the sense that it does not directly involve models at all.

The above construction is institution independent. To reformulate the amalgamation conditions in the specific case of CASL, we need an appropriate institution $(\mathbf{EnrSign}, \mathbf{Mod}_e)$ together with a representation functor Φ. In [11], we proposed the category of left cancellable small categories **lcCat** (i.e., small categories with all morphisms being monic) as the category **EnrSign**, and the usual interpretation of preorders as thin categories as the functor Φ (when operations and predicates are present, the details are a bit more complicated, but the main idea is the same, cf. [12]). With the natural model functor $\mathbf{Mod}_e$, this construction ensures the amalgamation property and all additional requirements. Thus the amalgamation conditions for CASL architectural specifications have been reformulated as conditions involving extendability of colimit cocones for 'enriched' signature diagrams that correspond, via the extended static semantics, to architectural specifications.

In the following we present a detailed analysis of the problem of checking the latter conditions.

3   Cell Calculus

As announced in the previous section, we work with the enriched signature category **EnrSign** being the category **lcCat** of left cancellable small categories, the representation functor $\Phi : \mathbf{PreOrd} \to \mathbf{lcCat}$ interpreting preorders as thin categories, and the natural model functor $\mathbf{Mod}_e : \mathbf{lcCat}^{op} \to \mathbf{CAT}$. Note that Φ preserves extremal epimorphisms. A morphism in **lcCat** is an extremal epimorphism iff it is surjective on sorts and every embedding in its codomain is a composite of images of embeddings in its domain. To simplify the notation, we will in the following regard **PreOrd** as an actual subcategory of **lcCat**, i.e. identify preorders and thin categories (which are left cancellable), so that Φ becomes an inclusion and we can omit its application whenever convenient.

In this setting, the amalgamation conditions have the following form: given a diagram $D : \mathbf{I} \to \mathbf{lcCat}$ of thin categories, a thin category Δ, and an extremal epimorphism $\tau : D(k) \to \Delta$ for some $k \in \mathbf{Ob(I)}$, let $D' : \mathbf{I}' \to \mathbf{lcCat}$ be the result of adding l as a new object and $m : k \to l$ to $\mathbf{I}$, and extending D by mapping k to Δ and m to τ. The amalgamation condition now is that the colimit cocone for D extends to a cocone for D'. We can, of course, assume that $\mathbf{I}$ is finite (as it represents the structure of an architectural specification), as well as Δ and the $D(i)$ (as they represent CASL signatures).

Example (ctd.). The static incorrectness of specification A in Fig. 1 corresponds to the fact that the colimit cocone in **lcCat** for D does not extend to a cocone for D'. Indeed, fewer equations between paths of sort embeddings hold in the colimit object presented below, than in the object $\Delta = \Sigma_2 \cup \Sigma_3$ shown in Fig. 1.

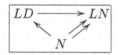

Informally, such an amalgamation condition states that the signature morphism τ does not 'glue' more sorts and embeddings than the colimit cocone for D. Verifying this condition requires some way of constructing colimits in **lcCat**: we introduce a *cell calculus* for this purpose. Our approach is related to the use of generalized congruences as worked out in [1] (see also [9,10]). To construct a colimit of a signature diagram D, we collect all sorts and embedding paths occuring in D, and divide the two resulting sets by two suitable equivalence relations $\simeq$ and $\cong$, ensuring that the resulting object forms a left cancellable category and a colimit for D.

Let $Sorts(D) = \biguplus_{i \in \mathbf{Ob(I)}} \mathbf{Ob}(D(i))$ and $Embs(D) = \biguplus_{i \in \mathbf{Ob(I)}} \mathbf{Mor}(D(i))$. Thus, elements of $Sorts(D)$ are of the form (i, s), where $i \in \mathbf{Ob(I)}$ and s is a sort in $D(i)$, and elements of $Embs(D)$ are of the form (i, e), where $i \in \mathbf{Ob(I)}$ and e is a sort embedding in $D(i)$.

Now let $\simeq$ be the least equivalence on $Sorts(D)$ such that for all $m : i \to j$ in $\mathbf{I}$ and $s \in \mathbf{Ob}(D(i))$, $(i, s) \simeq (j, D(m)(s))$.

Consider the set of non-empty finite words over the alphabet $Embs(D)$. For a word $\omega = \langle (i_n, e_n), \ldots, (i_1, e_1) \rangle$, we define $\mathrm{dom}(\omega) = (i_1, \mathrm{dom}(e_1))$ and $\mathrm{cod}(\omega) = (i_n, \mathrm{cod}(e_n))$. A word $\langle (i_n, e_n), \ldots, (i_1, e_1) \rangle$ is $\simeq$-*admissible* if, for $1 \leq k < n$, $(i_k, \mathrm{cod}(e_k)) \simeq (i_{k+1}, \mathrm{dom}(e_{k+1}))$. Let $Adm_\simeq$ be the set of all $\simeq$-admissible words over $Embs(D)$. A pair (ω, v) of $\simeq$-admissible words such that $\mathrm{dom}(\omega) \simeq \mathrm{dom}(v)$ and $\mathrm{cod}(\omega) \simeq \mathrm{cod}(v)$ is called a *cell over D*. For a cell $c = (\omega, v)$, we define $\mathrm{dom}(c) = \mathrm{dom}(\omega)$ and $\mathrm{cod}(c) = \mathrm{cod}(\omega)$.

Let $\cong$ be the least (equivalence) relation between $\simeq$-admissible words that satisfies the following rules:

$$\frac{m : i \to j \text{ is a morphism in } \mathbf{I}}{\langle (i, e) \rangle \cong \langle (j, D(m)(e)) \rangle} \textbf{(Diag)} \qquad \frac{ed \text{ is defined in } D(i)}{\langle (i, e), (i, d) \rangle \cong \langle (i, ed) \rangle} \textbf{(Comp)}$$

$$\frac{\omega \text{ is } \simeq\text{-admissible}}{\omega \cong \omega} \textbf{(Refl)} \qquad \frac{\omega \cong v}{v \cong \omega} \textbf{(Symm)} \qquad \frac{\omega \cong v \quad v \cong \psi}{\omega \cong \psi} \textbf{(Trans)}$$

$$\frac{\omega \cong v \quad \psi \cong \phi \quad \mathrm{cod}(\omega) \simeq \mathrm{dom}(\psi)}{\psi\omega \cong \phi v} \textbf{(Cong)} \qquad \frac{\psi\omega \cong \psi v}{\omega \cong v} \textbf{(Lc)}$$

Given these rules, *derivations* are defined in the usual way as minimal sequences of pairs of words such that every pair in the sequence follows from the previous ones by one of the rules.

It follows by induction on the length of the derivation that if $\omega \cong v$, then (ω, v) forms a cell over D. In particular, if the conditions for the rule **(Cong)** hold, then both $\psi\omega$ and ϕv are $\simeq$-admissible. Thus, we will think of $\cong$ as a set of cells over D defined inductively by the above rules, which we therefore call the *cell calculus*; a derivation is essentially a proof that a certain cell is in $\cong$.

Now, the category $\mathbf{C}$ with $Sorts(D)/_\simeq$ and $Adm_\simeq/_\cong$ as the sets of objects and morphisms respectively, and with the domain, codomain, identity and composition functions defined in the obvious way, is a colimit for D in $\mathbf{lcCat}$.

Equipped with this construction, we come back to the amalgamation condition formulated as above. Consider a relation $\simeq_\tau$ on $\mathbf{Ob}(D(k))$, defined as the kernel of τ on sorts: $s \simeq_\tau t$ iff $\tau(s) = \tau(t)$. Then consider the set of all non-empty finite words over the alphabet $\mathbf{Mor}(D(k))$ and define the set of $\simeq_\tau$-admissible words in the same manner as the set of $\simeq$-admissible words. Next, define the relation $\cong_\tau$ on the set of $\simeq_\tau$-admissible words: $\omega \cong_\tau v$ if $\mathrm{dom}(\omega) \simeq_\tau \mathrm{dom}(v)$ and $\mathrm{cod}(\omega) \simeq_\tau \mathrm{cod}(v)$. Finally, consider the obvious injections $\iota_o : \mathbf{Ob}(D(k)) \to Sorts(D)$ and $\iota_e : \mathbf{Mor}(D(k)) \to Embs(D)$. We will abuse the names ι_o, ι_e to denote also the obvious extensions to words over $\mathbf{Mor}(D(k))$ and to relations on these words.

Recall that, informally, amalgamation condition requires that the morphism τ does not 'glue' more objects and embedding paths than 'glued' in the colimit of D. This is captured by the following theorem.

Theorem 1. *In the above notation, a colimit cocone for D extends to a cocone for D' iff*

(*) $\iota_o(\simeq_\tau) \subseteq \simeq$ and
(**) $\iota_e(\cong_\tau) \subseteq \cong$.

The condition (**) is called the *cell condition*. If the inclusion (*) holds, then ι_e maps $\simeq_\tau$-admissible words to $\simeq$-admissible words, so (**) concerns an inclusion between two sets of cells.

Thinness of the colimit of D is a stronger property than the cell condition:

Lemma 2. *In the above notation, if (*) holds and the colimit C of D is a thin category, then (**) holds as well.*

Conversely, if, for all $i \in \mathbf{Ob(I)}$, there is a unique morphism $D(i) \to \Delta$ in the image of D', and if both (*) and (**) hold, then C is isomorphic to Δ (hence thin).

Theorem 1 reformulates the amalgamation condition using four relations: $\simeq$, $\simeq_\tau$, $\cong$, and $\cong_\tau$. The former two relations always have finite domains and hence can be represented explicitly, which makes (*) decidable.

However, the relations $\cong$ and $\cong_\tau$ have possibly infinite domains. One can always choose a finite subrelation $\cong'_\tau$ of $\cong_\tau$ such that if $\iota_m(\cong'_\tau) \subseteq \cong$ then $\iota_m(\cong_\tau) \subseteq \cong$ as well (roughly, $\cong_\tau$ can be restricted to 'loopless' words, see Sect. 6 below for a formal definition). This does not solve all problems though, since there is no obvious finite representation of $\cong$, more explicit than the cell calculus.

4   Checking the Cell Condition

We now discuss the issue of checking the cell condition, relying on the notation introduced in its formulation above and assuming that (*) holds.

In the following we will sometimes treat $\cong_\tau$ and $\cong$ as sets of cells over D (this is legal for $\cong_\tau$ because (*) holds). Moreover, we assume that $\cong_\tau$ can be represented explicitly, so the lack of a more explicit representation for $\cong$ (which is only defined inductively by the cell calculus) is the main difficulty.

4.1   Simple Case

First we discuss a simple condition that implies the cell condition and holds in many practical examples of architectural specifications.

Consider D, Δ and τ given as above, with (*) satisfied. Let $\cong_0$ be the least equivalence relation on the set of $\simeq$-admissible words that satisfies the rules **(Diag)** and **(Comp)** of the cell calculus. Obviously $\cong_0 \subseteq \cong$, so if $\iota_e(\cong_\tau) \subseteq \cong_0$, then the cell condition (**) holds.

Even though $\cong_0$ has an infinite domain, it has only finitely many non-reflexive elements, since $\mathbf{I}$ and the $D(i)$ are finite and the $D(i)$ are thin. Hence it can be explicitly represented during static analysis. Together with a method of explicit representation of $\cong_\tau$, this gives an algorithm (in fact, a polynomial time algorithm) for checking the condition $\iota_e(\cong_\tau) \subseteq \cong_0$.

Intuitively speaking, this captures the situation where each embedding cell present in the signature Δ can be traced to a cell in one of the signatures in D. This condition holds in many practical cases, including all cases where the architectural specification does not contain any subsort embeddings (then the problem is even simpler: if all signatures in D are discrete categories, then the cell condition (**) follows from (*)).

4.2   General Case

Unfortunately, the cell condition turns out to be undecidable in general:

Theorem 3. *In the above notation, and under the assumption that (*) holds, (**) is undecidable.*

The proof is by reduction from the problem of deciding whether the monoid with left cancellation presented by a given finite set of equations is trivial. For any finite set R of equations over a finite alphabet, we construct a diagram D, a category Δ and a morphism τ such that (*) holds, and such that (**) holds if and only if the monoid with left cancellation presented by R is trivial. There is a direct correspondence between the rules of the cell calculus and the axioms of monoids with left cancellation.

4.3   Special Case

Since the cell conditions are undecidable in general, we need to detect special cases of practical relevance in which the conditions are decidable. In fact, one such case is presented in Subsection 4.1.

Consider a situation where $Adm_\simeq$, the domain of $\cong$, is finite (leaving identity embeddings aside). In other words, there is no $\simeq$-admissible word ω such that $dom(\omega) \simeq cod(\omega)$, except when all letters of ω are identities. It follows that in the colimit $\mathbf{C}$ of D identities are the only embeddings with domain equal to codomain. This condition can be effectively checked, since the relation $\simeq$ can be represented explicitly, and it is enough to check the condition $dom(\omega) \not\simeq cod(\omega)$ for words of length smaller than the number of equivalence classes of $\simeq$.

Requiring finiteness of $Adm_\simeq$ rules out architectural specifications declaring mutual subsorting relations, for example, when there is an embedding $Num < List[Digit]$ (encoding a number as a list of digits) in one unit, and an embedding $List[Digit] < Num$ (computing a number from a list of digits) in another unit. It also disallows declaring two sorts to be isomorphic.

It is easy to see that if the set of $\simeq$-admissible words is finite, then the cell condition is decidable, because then the relation $\cong$ has a finite domain and hence can be computed. However, it is very unlikely that one can design a polynomial time algorithm for checking the cell condition in this case:

Theorem 4. *In the above notation, it is **PSPACE**-hard to check, knowing that (*) holds and that the set $Adm_\simeq$ is finite, whether (**) holds.*

The proof idea is to encode any linear bounded deterministic Turing machine $\mathcal{M}$ [7] in a signature diagram D together with its extension D' so that the cell condition holds for if and only if there is a two-way computation between two given configurations of $\mathcal{M}$.

5   Checking Colimit Thinness

In this section we discuss the following problem: given a finite diagram $D : \mathbf{I} \to$ lcCat of finite thin categories, is the colimit $\mathbf{C}$ of the diagram thin? Lemma 1 shows that this condition (together with the easily decidable (*)) implies the cell condition (and is equivalent to it under reasonable further conditions).

Unfortunately, colimit thinness is again undecidable in general. Indeed, one can construct the diagram in the proof of Theorem 2 in such a way that the assumptions of both parts of Lemma 1 are satisfied, so that the cell condition is equivalent to colimit thinness there.

However, unlike the cell condition, colimit thinness is decidable by a polynomial time algorithm if the set of $\simeq$-admissible words is finite. This algorithm works roughly as follows:

We think of $\cong$ as a set of cells over D and ask whether it contains all cells, i.e., whether every cell over D is derivable in the cell calculus. Since $Adm_\sim$ is finite, one can define a partial order $\leq$ on $\mathbf{Ob}(\mathbf{C})$ as follows: $\mathbf{s} \leq \mathbf{t}$ if there exists a $\simeq$-admissible word ω with $\mathrm{dom}(\omega) = (i, s)$ and $\mathrm{cod}(\omega) = (j, t)$ such that $\mu_i(s) = \mathbf{s}$ and $\mu_j(t) = \mathbf{t}$, where μ_i and μ_j are the respective colimit injections. We say that a cell (ω, v) is an $\mathbf{s}$-cell if $\mathrm{dom}(\omega)$ is mapped to $\mathbf{s}$ by the respective colimit injection. The algorithm relies on the observation that for any $\mathbf{s}$-cell c, all derivations of c involve only $\mathbf{s}'$-cells for which $\mathbf{s} \leq \mathbf{s}'$. This implies that one can sort the set $\mathbf{Ob}(\mathbf{C})$ topologically and, for each $\mathbf{s} \in \mathbf{Ob}(\mathbf{C})$, try to prove that all $\mathbf{s}$-cells are in $\cong$ provided that for any $\mathbf{s}' > \mathbf{s}$, all $\mathbf{s}'$-cells are in $\cong$. This can be checked in polynomial time, exploiting the fact that the set of all derivable $\mathbf{s}$-cells is closed with respect to a suitable connectedness relation.

Implemented in a slightly different setting, this algorithm is used in the static analysis of CASL architectural specifications performed by the system CATS. It has polynomial time complexity, but its practical efficiency turns out to be satisfactory only for specifications of moderate size.

6   Imposing Restrictions on the Cell Calculus

There is yet another way of trying to check the cell condition, namely by simply limiting the length of words in the cells considered.

Consider D, Δ and τ as usual. We call a word $\langle (i_n, e_n), \dots, (i_1, e_1) \rangle \in Adm_\sim$ loopless if there are no j, l such that $1 \leq j < l \leq n$ and $(i_j, e_j) = (i_l, e_l)$. We denote the (finite) set of all loopless $\simeq$-admissible words over $Embs(D)$ by $Loopless_\sim$.

Let $\cong^R$ be the least relation on $Loopless_\sim$ that satisfies all the rules of the cell calculus. Obviously $\cong^R \subseteq \cong$, so the *restricted cell condition*

$$(**R) \quad \iota_e(\cong_\tau) \subseteq \cong^R$$

implies the original cell condition (**). Since $\cong^R$ has a finite domain, it can be effectively computed. Hence the restricted cell condition is decidable, even if the set $Adm_\sim$ is infinite. However, one can see that it is **PSPACE**-hard in the general case. This follows from Theorem 3, since if $Adm_\sim$ is finite, then the conditions (**R) and (**) are equivalent.

The two properties, colimit thinness and the restricted cell condition, are mutually independent: for either of them one can give a CASL architectural specification for which it fails while the other property holds. However, the restricted cell condition seems to cover all practically relevant architectural specifications. All known examples of architectural specifications that satisfy (**) but not (**R) are complicated and unlikely to occur in practice.

7   Conclusions and Future Work

In [11], a semantics of CASL architectural specifications has been proposed, with static amalgamation conditions formulated in an institution independent way. However, no method of checking the conditions in CASL is given there.

In the present paper, we discuss the problem of checking amalgamation conditions for a simplified version of the CASL institution. We show that the conditions are undecidable in general, and we discuss some important variations of the problem. The first special situation, covering many practical examples of architectural specifications, occurs when the amalgamation condition is ensured by tracing each cell to be considered to a particular signature of a given unit. This implies the cell condition and is easily decidable. Another special case we present is that the set of all cells considered is finite: here the problem is again decidable, although **PSPACE**-hard. Two further sufficient conditions for the amalgamation condition are considered. One of them is colimit thinness, which is undecidable in general, but checkable in polynomial time in an important special case. The other one is given by the restricted cell calculus, which is always decidable, but **PSPACE**-hard.

All results are presented in a somewhat simplified setting, dealing only with the subsort relation in CASL signatures and leaving aside all operations and predicates. When operations are present, the details become somewhat more involved. One extends the definition of an admissible word by adding single operation names as letters, and presents a modified version of the cell calculus dealing with extended words. This adds, for instance, some more complication to the algorithm for checking colimit thinness, but does not change its polynomial complexity. Obviously, this does not affect the negative results presented here (Theorems 2 and 3).

As a practical result of the paper, we propose a method of checking the amalgamation conditions for CASL architectural specifications. Given such a specification, we propose the following steps to establish its (static) correctness, for each of the amalgamation conditions emerging in the static analysis:

1. Compute the relations $\simeq, \simeq_\tau$ and $\cong_\tau$, as described in Section 3.
2. Check the condition (*) as described in Section 3. If this fails, the specification is incorrect.

3. Check the condition described in Section 4.1. If this succeeds, the specification is correct.
4. Check if the set $Adm_{\approx}$, as described is Section 3, is finite. If not, go directly to step 7.
5. Check the colimit thinness condition, using the algorithm sketched in Section 5. If it succeeds, the specification is correct.
6. Check the cell condition in its full generality (it is decidable, as shown in Section 4.3). The specification is correct if and only if the check succeeds.
7. Check the restricted cell condition (it is decidable, as argued in Section 6). If the check succeeds, the specification is correct.
8. Invoke interaction with the user, generating proof obligations possibly supported by some heuristics, to prove the cell condition as described in Section 3. The specification is correct whenever a proof is found.

In practice, as it is unlikely that polynomial time algorithms for Steps 6 and 7 exist, one should go to Step 8 immediately after unsuccesful completion of Step 5. Steps 1–5 can be performed in polynomial time, keeping in mind that Step 5 is practically efficient only for specifications of moderate size. Step 8 above indicates perhaps the most important direction for our future work on this topic: developing methods to generate proof obligations for checking the amalgamation conditions, and to use theorem provers to assist the user in proving them.

References

1. M. Bednarczyk, M. Borzyszkowski, and W. Pawłowski. Generalized congruences – epimorphisms in Cat. *Theory and Applications of Categories* **5** (1999), 266–280. 456
2. M. Bidoit, D. Sannella and A. Tarlecki. Architectural specifications in CASL. *Algebraic Methodology and Software Technology, LNCS* **1548**, 341–357. Springer, 1999. 451
3. CoFI Semantics Task Group. CASL – The CoFI Algebraic Specification Language – Semantics. Note S-9 (v. 0.96), July 1999. Accessible by WWW[2] and FTP[3]. 451, 452
4. CoFI Language Design Task Group. CASL – The CoFI Algebraic Specification Language – Summary, version 1.0.1, June 2000. Accessible by WWW[2] and FTP[3]. 451, 453
5. M. Davis. Unsolvable Problems. in: J. Barwise (ed.), *Handbook of Mathematical Logic*, North Holland, Amsterdam, 1977.
6. J.Goguen and R.Burstall. Institutions: Abstract model theory for specifications and programming. *J. ACM* **39** (1992), 95–146. 452
7. J. E. Hopcroft and J. D. Ullman. *Introduction to Automata Theory, Languages, and Computation.* Addison-Wesley, Reading, 1979. 459
8. L. C. Paulson. *ML for the Working Programmer* (2nd ed.). Cambridge, 1996. 453

---

[2] http://www.brics.dk/Projects/CoFI
[3] ftp://ftp.brics.dk/Projects/CoFI

9. L. Schröder. *Composition Graphs and Free Extensions of Categories* (in German). Ph.D. thesis, University of Bremen, 1999; also: Logos, Berlin, 1999. 456
10. L. Schröder and H. Herrlich. Free adjunction of morphisms. *Applied Categorical Structures* **8** (2000), 595–606. 456
11. L. Schröder, T. Mossakowski, A. Tarlecki, B. Klin, and P. Hoffman. Semantics of Architectural Specification in CASL. *Fundamental Approaches to Software Engineering, LNCS* **2029**, 253-268. Springer, 2001. 451, 452, 453, 455, 461
12. L. Schröder, T. Mossakowski, A. Tarlecki. Amalgamation via enriched CASL signatures. ICALP 2001, to appear. 455

On-Line Scheduling with Tight Deadlines

Chiu-Yuen Koo, Tak-Wah Lam, Tsuen-Wan Ngan, and Kar-Keung To

Department of Computer Science, The University of Hong Kong
Hong Kong
{cykoo,twlam,twngan,kkto}@cs.hku.hk

Abstract. This paper is concerned with the on-line problem of scheduling jobs with tight deadlines in a single-processor system. It has been known for long that in such a setting, no on-line algorithm is optimal (or 1-competitive) in the sense of matching the optimal off-line algorithm on the total value of jobs that meet the deadlines; indeed, no algorithm can be $\Omega(k)$-competitive, where k is the importance ratio of the jobs. Recent work, however, reveals that the competitive ratio can be improved to $O(1)$ if the on-line scheduler is equipped with a processor $O(1)$ times faster [8]; furthermore, optimality can be achieved when using a processor $O(\log k)$ times faster [12]. This paper presents a new on-line algorithm for scheduling jobs with tight deadlines, which can achieve optimality when using a processor that is only $O(1)$ times faster.

1 Introduction

This paper is concerned with on-line firm deadline scheduling in a single-processor system. Jobs are released in an unpredictable fashion, each requests a certain amount of processing on the processor. The processing time, deadline, and value of a job are known only when the job is released. Deadlines are firm in the sense that completing a job after its deadline gives zero value. Notice that a system may be overloaded and there is no way to schedule every job released to meet the deadline. The aim of a scheduler is to maximize the total value of jobs meeting the deadlines. Preemption is allowed at no cost (i.e., a preempted job can be restarted from the point of preemption at any time).

The design of a good scheduler is further complicated by the fact that jobs may have different value densities, i.e., different ratios of value to processing time. The importance ratio k of a system is defined as the ratio of the largest possible value density to the smallest possible value density. When $k = 1$, maximizing the total value is equivalent to maximizing the processor utilization on jobs that meet the deadlines.

Consider any number $c \geq 1$. An on-line algorithm for firm deadline scheduling is said to be c-competitive if it guarantees to achieve at least a fraction $1/c$ of the total value obtained by the optimal off-line algorithm. In this case, the number c is called the competitive ratio of the on-line algorithm. Furthermore, the on-line algorithm is said to be *optimal* or *1-competitive* if it can always match the optimal off-line algorithm on the total value obtained. It has been known for

J. Sgall, A. Pultr, and P. Kolman (Eds.): MFCS 2001, LNCS 2136, pp. 464–473, 2001.
© Springer-Verlag Berlin Heidelberg 2001

long that no on-line algorithm for firm deadline scheduling [5] can be optimal, and the best known algorithm achieves a competitive ratio of $(\sqrt{k}+1)^2$, where k is the importance ratio [9].

In recent years, there are a number of exciting results on improving the performance guarantee without making assumption on future inputs; the basic idea is to allow the on-line scheduler to have more resources than the adversary (e.g., [3,6,8,13,7,11,12]). For the single-processor firm deadline scheduling problem, Kalyanasundaram and Pruhs [8] showed that the competitive ratio can be reduced significantly if the on-line scheduler is given a faster processor. For instance, with a processor that is 32 times faster, the competitive ratio can be improved from $(\sqrt{k}+1)^2$ [9] to roughly 2. Recently, Lam and To [12] proved that even optimality can be achieved. Precisely, they showed that the earliest deadline first (EDF) algorithm is optimal when given a processor that is $4\lceil \log k \rceil$ times faster; for the special case where $k = 1$, a processor that is two times faster suffices to guarantee optimality. Note that when k is big, demanding a processor that is $4\lceil \log k \rceil$ times faster may not be practical. A natural open question is whether the speed requirement for optimality in the case of general k can be improved to $o(\log k)$ or even $O(1)$.

In this paper, we address the above open problem with a focus on jobs with tight deadlines, i.e., the deadline of each job is equal to its release time plus processing time. Roughly speaking, scheduling jobs with tight deadlines is no easier than the general problem. In fact, the best lower bound results on firm deadline scheduling are based on the tight deadline setting. Even with the tight deadline assumption, no on-line algorithm (using a processor of ordinary speed) can be optimal and Baruah et al. [2,1] actually showed that no algorithm can achieve a competitive ratio less than $(\sqrt{k}+1)^2$. Note that when $k = 1$, $(\sqrt{k}+1)^2 = 4$. If deadlines are known to be not tight, only a weaker lower bound has been known – DasGupta and Palis [4] showed that if all jobs have a stretch factor $\alpha \geq 1$[1], then no algorithm is $4\lceil \alpha \rceil / (4\lceil \alpha \rceil - 3)$-competitive even when $k = 1$. For example, when $\alpha = 2$, this lower bound is 1.6, which is smaller than the lower bound of 4 in the tight deadline setting. Regarding optimality via a faster processor, Lam et al. [10] also gave a lower bound result that even with the tight deadline assumption, any optimal on-line algorithm for the case of $k = 1$ must use a processor at least ϕ times faster, where $\phi = (1 + \sqrt{5})/2 \approx 1.618$.

The contribution of this paper is a new on-line algorithm for scheduling jobs with tight deadlines. With this algorithm, we show that a processor that is $O(1)$ times faster is sufficient to guarantee optimality even if jobs have general value densities. Note that this algorithm, unlike the algorithms in [8,12], cannot handle jobs whose deadlines are not tight. Nevertheless, we believe that jobs with tight deadlines are the most difficult to handle, and an $O(1)$-times-faster processor should also be sufficient to guarantee optimality for the general case.

Organization: The remainder of this paper is divided into three sections. Section 2 describes the new on-line algorithm for firm deadline scheduling. Sec-

---

[1] The stretch factor of a job is defined to be the ratio of its span (i.e., deadline minus release time) to its processing time.

tion 3.1 gives some basic properties related to this algorithm, and Section 3.2 proves that this algorithm, when given a processor 14 times faster, is optimal for scheduling jobs with tight deadlines.

Notations: For any job J, let $r(J)$, $p(J)$, $d(J)$, and $v(J)$ denote the release time, processing time, deadline and value of J, respectively. Define the value density $\rho(J)$ to be $v(J)/p(J)$, and the *span* of J to be the period $[r(J), d(J)]$.

When we say that an on-line algorithm is using a speed-s processor, where $s \geq 1$, it is meant that the algorithm can schedule one unit of work to complete in $1/s$ unit of time. That is, the processor is s times as fast as the processor used by the off-line algorithm.

2   Algorithm

In this section, we present a firm deadline scheduling algorithm that uses two speed-s processors, where s is an integer greater than 1. Since two speed-s processors can be simulated by one speed-$2s$ processor, this algorithm can be considered as an algorithm using a single processor.

Definition 1. *Let* $w = \frac{1}{2} + \frac{1}{2s} \leq 1$. *A job* J, *once released, is said to be* fresh *up to the time* $d(J) - w \cdot p(J)$. *(Roughly speaking, a job is no longer fresh when the time is close to its deadline.)*

The following lemma states an important property of a fresh job.

Lemma 2. *While a job* J *is fresh, it is feasible to complete* J *on or before its deadline using a speed-s processor.*

Proof. Consider any time t when a job J is fresh. By definition, $d(J) - t \geq wp(J) = (\frac{1}{2} + \frac{1}{2s})p(J)$. Note that a speed-$s$ processor needs $p(J)/s$ time to complete J. Since $s \geq 1$, we have $\frac{1}{2} \geq \frac{1}{2s}$, or equivalently, $\frac{1}{2} + \frac{1}{2s} \geq \frac{1}{s}$. Thus, starting from time t, a speed-s processor must be able to complete J on or before its deadline. □

Figure 1 gives the details of the new on-line scheduling algorithm. The algorithm maintains a pool P of jobs that will be given priority for scheduling. Intuitively, if a job is still fresh and has sufficiently large value density, the algorithm will put it into the pool for possible scheduling. Among the jobs in the pool, the algorithm always schedules the two most dense jobs available. Once a job is scheduled, it can be preempted by a newly released job with sufficiently high density. Note that there is no guarantee that a job, once put into pool or scheduled, will complete.

In the following section, we will show that if all jobs have tight deadlines, this algorithm, when given a speed-14 processor (or two speed-7 processors), is optimal, i.e., it can match the optimal off-line algorithm which uses a speed-1 processor regarding the total value of jobs that meet their deadlines.

Consider a job J that the on-line algorithm fails to meet its deadline, but the optimal off-line algorithm can meet its deadline. During the span of J, the

(1) Initialization:
(2) $P \leftarrow \emptyset$
(3)
(4) When Job J is released:
(5)   **if** $|P| < 2$ **or** $\rho(J) \geq 2\max_{T \in P} \rho(T)$
(6) $P \leftarrow P \cup \{J\}$
(7)   Schedule the two most dense jobs, if available, in P
(8)
(9) When Job J completes:
(10) $P \leftarrow P - \{J\}$
(11)   **if** there exists a fresh job not in P
(12)   Denote J_0 as the most dense fresh job not in P
(13)   **if** $|P| < 2$ **or** $\rho(J_0) \geq 2\max_{T \in P} \rho(T)$
(14) $P \leftarrow P \cup \{J_0\}$
(15)   Schedule the two most dense jobs, if available, in P
(16)
(17) When Job $J \in P$ can't be completed by its deadline:
(18) $P \leftarrow P - \{J\}$

Fig. 1. A new scheduling algorithm

on-line algorithm must process other jobs for a considerably long period. These jobs should have reasonably high value density, yet the on-line algorithm may not complete them and generate any value. The optimality of the on-line algorithm is proven by a non-trivial amortization scheme, showing that the on-line algorithm will complete some extra jobs (compared to the off-line algorithm) within or beyond the span of J, which can be used to pay off the value of J.

3   Analysis

3.1   Basic Properties

Let $\mathcal{A}$ and $\mathcal{O}$ denote the schedules produced by our algorithm (using two speed-s processors) and the optimal off-line algorithm (using one speed-1 processor) for any particular job set I. Denote $\mathcal{J}(\mathcal{A})$ as the set of jobs that has ever scheduled by $\mathcal{A}$, and similarly for $\mathcal{J}(\mathcal{O})$. Without loss of generality, we assume that every job that has ever scheduled by $\mathcal{O}$ can be completed by its deadline. Note that such an assumption is not valid for $\mathcal{A}$.

We construct schedules $\mathcal{A}'$ and $\mathcal{O}'$ from $\mathcal{A}$ and $\mathcal{O}$ respectively by removing all jobs that are completed by both $\mathcal{A}$ and $\mathcal{O}$. An example is shown in Fig. 2.

By definition, for every job $J \in \mathcal{J}(\mathcal{O}')$, both $\mathcal{O}$ and $\mathcal{O}'$ schedule J to completion by its deadline, but neither $\mathcal{A}$ nor $\mathcal{A}'$ can complete J by its deadline.

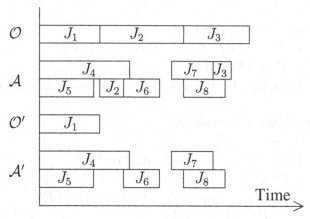

Fig. 2. A sample schedule for $\mathcal{O}$ and $\mathcal{A}$ and the respective $\mathcal{O}'$ and $\mathcal{A}'$. Note that J_2 and J_3 are removed in schedule $\mathcal{O}'$ and $\mathcal{A}'$

Lemma 3. *For any two jobs scheduled by $\mathcal{O}$, their span must be disjoint.*

Proof. Recall that any job scheduled by $\mathcal{O}$ can be completed by its deadlines. Since all jobs have tight deadlines, $\mathcal{O}$ or any off-line algorithm can complete a job only by scheduling it throughout its entire span. Thus, any two jobs that can be completed by $\mathcal{O}$ must have non-overlapping span. □

Lemma 4. *If $\mathcal{A}$ misses the deadline of a job J, then at any time while J is fresh, $\mathcal{A}$ schedules at least one job with value density greater than $\rho(J)/2$.*

Proof. Let J be a job for which $\mathcal{A}$ misses the deadline. Consider any time t when J is fresh. If J is in P, $\mathcal{A}$ is executing some job(s) of value density at least $\rho(J)$. It remains to consider the case where J is not in P at time t. In this case, J is not in P during the entire period $[r(J), t]$ (otherwise, by Lemma 2, even at time t, it is still feasible for $\mathcal{A}$ to complete J by its deadline and J, once put into P, will not been removed up to time t). Thus, $\mathcal{A}$ must be executing some job(s) of value density greater than $\rho(J)/2$ during the entire period $[r(J), t]$. In either case, $\mathcal{A}$ is executing a job with value density greater than $\rho(J)/2$. □

Lemma 5. *At any time, let J_i and J_{i+1} be the i-th and $(i+1)$-st most dense jobs in P. If J_{i+1} is not the least dense job in P, then $\rho(J_i) \geq 2\rho(J_{i+1})$.*

Proof. When a job J is added to P, either there is at most one job in P, or the density of J is at least double of the most dense job in P. Removing a job from P does not invalidate this property. □

3.2   Optimality

Denote $\|\mathcal{A}\|$ as the value obtained by schedule $\mathcal{A}$, and similarly for $\|\mathcal{A}'\|$, $\|\mathcal{O}\|$, and $\|\mathcal{O}'\|$. This section shows that if $s \geq 7$, then $\|\mathcal{A}'\| \geq \|\mathcal{O}'\|$, or equivalently, $\|\mathcal{A}\| \geq \|\mathcal{O}\|$. It follows that the algorithm shown in Fig. 1, when given two speed-7 processors or a speed-14 processor, can always match the off-line algorithm on the total value obtained.

Consider the schedule $\mathcal{A}'$. For any job $J \in \mathcal{J}(\mathcal{A}')$, let $T(J)$ be the total time $\mathcal{A}'$ schedules J. At any particular time, J is said to be the primary job scheduled by $\mathcal{A}'$ if $\mathcal{A}'$ schedule J at that time and either there is no other job scheduled at the same time, or J has a higher job density than the other job scheduled (we break the tie by requiring J to be the job with smaller job identity). For a job J_0 scheduled by $\mathcal{O}'$, define $T_{J_0}(J)$ to be the total time $\mathcal{A}'$ schedules J as the primary job during the period of time when J_0 is fresh.

To prove that $\|\mathcal{A}'\| \geq \|\mathcal{O}'\|$, we imagine that for each job J that has scheduled by $\mathcal{A}'$, we can withdraw some credits at a certain rate whenever $\mathcal{A}'$ schedules J. Precisely, we define the rate $\sigma(J)$ to be $\frac{4s}{s-1}\rho(J)$. I.e., we can withdraw a total of $\sigma(J)T(J)$ credits due to J.

The rest of this section is divided into two parts. The first part, comprising Lemmas 6 and 7 and Corollary 8, shows that $\|\mathcal{O}'\|$ is no more than the total amount of credits withdrawn due to the jobs in $\mathcal{J}(\mathcal{A}')$, which is exactly $\sum_{J \in \mathcal{J}(\mathcal{A}')} \sigma(J)T(J)$. In the second part, we show that $\|\mathcal{A}'\| \geq \sum_{J \in \mathcal{J}(\mathcal{A}')} \sigma(J)T(J)$. Combining both parts, we can conclude that $\|\mathcal{A}'\| \geq \|\mathcal{O}'\|$.

Lemma 6. *Consider any job J_0 scheduled by $\mathcal{O}'$. For any job $J \in \mathcal{J}(\mathcal{A}')$ such that $T_{J_0}(J) \geq 0$, $\rho(J) \geq \rho(J_0)/2$.*

Proof. Let J_0 be any job scheduled by $\mathcal{O}'$. By definition, J_0 cannot be completed by $\mathcal{A}$. Consider any job $J \in \mathcal{J}(\mathcal{A}')$ such that $T_{J_0}(J) \geq 0$. Let t be any time when J_0 is fresh and $\mathcal{A}'$ schedules J as the primary job. By Lemma 4, at time t, $\mathcal{A}$ is scheduling a job J' with density at least $\rho(J_0)/2$. Note that J may not be equal to J'. Nevertheless, we can argue that J' must be scheduled by $\mathcal{A}'$ at time t.

- If $J' = J_0$, then J' is not completed by $\mathcal{A}$. By the definition of $\mathcal{A}'$, J' is left in $\mathcal{A}'$.
- Suppose $J' \neq J_0$. By Lemma 3, J' is also not equal to any other job scheduled by $\mathcal{O}$. Again, by the definition of $\mathcal{A}'$, J' is left in $\mathcal{A}'$.

Since J is the primary job scheduled by $\mathcal{A}'$ at time t, we have $\rho(J) \geq \rho(J') \geq \rho(J_0)/2$. $\qquad\square$

Lemma 7. *For any job J_0 scheduled by $\mathcal{O}'$,*

$$\sum_{J \in \mathcal{J}(\mathcal{A}')} \sigma(J)T_{J_0}(J) \geq v(J_0) \ .$$

Proof. Note that J_0, being a job scheduled by $\mathcal{O}'$, cannot be completed by $\mathcal{A}$. By Lemma 4, $\mathcal{A}$ must schedule at least one job at any time while J_0 is fresh. By Lemma 3, every such job is not scheduled by $\mathcal{O}$ and must be found in $\mathcal{A}'$. Therefore, $\sum_{J \in \mathcal{J}(\mathcal{A}')} T_{J_0}(J)$ is equal to the length of the fresh period of J_0. By definition, J_0 is fresh for a period of length $(1-w)p(J_0) = \frac{1}{2}(1 - \frac{1}{s})p(J_0)$. Thus, $\sum_{J \in \mathcal{J}(\mathcal{A}')} T_{J_0}(J) = \frac{1}{2}(1 - \frac{1}{s})p(J_0)$. Furthermore,

$$
\sum_{J \in \mathcal{J}(\mathcal{A}')} \sigma(J) T_{J_0}(J) = \sum_{J \in \mathcal{J}(\mathcal{A}')} \frac{4s}{s-1} \rho(J) T_{J_0}(J)
$$

$$
\geq \frac{4s}{s-1} \sum_{J \in \mathcal{J}(\mathcal{A}')} \frac{\rho(J_0)}{2} T_{J_0}(J) \qquad \text{(by Lemma 6)}
$$

$$
= \frac{2s}{s-1} \rho(J_0) \sum_{J \in \mathcal{J}(\mathcal{A}')} T_{J_0}(J)
$$

$$
= \frac{2s}{s-1} \rho(J_0) \cdot \frac{1}{2}\left(1 - \frac{1}{s}\right) p(J_0)
$$

$$
= \rho(J_0) p(J_0)
$$

$$
= v(J_0) .
$$

$\square$

Corollary 8. $\displaystyle \sum_{J \in \mathcal{J}(\mathcal{A}')} \sigma(J) T(J) \geq \|\mathcal{O}'\|.$

Proof. At any time t, Lemma 3 implies that among all jobs scheduled by $\mathcal{O}$, there is at most one which is fresh at t. Thus, for each job $J \in \mathcal{J}(\mathcal{A}')$, $T(J) \geq \sum_{J_0 \in \mathcal{O}'} T_{J_0}(J)$. We have

$$
\sum_{J \in \mathcal{J}(\mathcal{A}')} \sigma(J) T(J) \geq \sum_{J \in \mathcal{J}(\mathcal{A}')} \sigma(J) \sum_{J_0 \in \mathcal{O}'} T_{J_0}(J) \geq \sum_{J_0 \in \mathcal{O}'} \sum_{J \in \mathcal{J}(\mathcal{A}')} \sigma(J) T_{J_0}(J)
$$

By Lemma 7, we conclude that

$$
\sum_{J_0 \in \mathcal{O}'} \sum_{J \in \mathcal{J}(\mathcal{A}')} \sigma(J) T_{J_0}(J) \geq \sum_{J_0 \in \mathcal{O}'} v(J_0) = \|\mathcal{O}'\| .
$$

$\square$

To prove $\|\mathcal{A}'\| \geq \sum_{J \in \mathcal{J}(\mathcal{A}')} \sigma(J) T(J)$, we consider the following amortization scheme on $\mathcal{A}'$. We associate an account with each job in $\mathcal{J}(\mathcal{A}')$, all having zero balance initially. Credits are put into or removed from these accounts in accordance to the way $\mathcal{A}'$ schedules the jobs.

- When a job J is completed, $v(J)$ is deposited into the account of J.
- Whenever $\mathcal{A}'$ schedules J, we withdraw credits from the account of J at rate $\sigma(J)$.

- Whenever $\mathcal{A}'$ schedules J, credits are transferred from the account of J to the account of each idling job J' in the pool P at the rate $I(J')$, where $I(J) = \frac{2s}{s-1}\rho(J)$.

Denote $\Psi(J)$ as the final balance of the account of each job J in $\mathcal{J}(\mathcal{A}')$. Note that the sum of all deposits is exactly $\|\mathcal{A}'\|$, and the sum of all withdrawals is $\sum_{J \in \mathcal{J}(\mathcal{A}')} T(J)\sigma(J)$. That is,

$$\|\mathcal{A}'\| = \sum_{J \in \mathcal{J}(\mathcal{A}')} \Psi(J) + \sum_{J \in \mathcal{J}(\mathcal{A}')} \sigma(J)T(J).$$

Corollary 8 states that $\sum_{J \in \mathcal{J}(\mathcal{A}')} \sigma(J)T(J) \geq \|\mathcal{O}'\|$. In the rest of this section, we will show that if $s \geq 7$, every account has a non-negative final balance and $\sum_{J \in \mathcal{J}(\mathcal{A}')} \Psi(J) \geq 0$. Then we can conclude that $\|\mathcal{A}'\| \geq \|\mathcal{O}'\|$ (see Theorem 11).

The following observation is crucial to the proof of the fact that all accounts have non-negative final balance. By definition, each job completed by $\mathcal{A}'$ will receive credits equal to the value of the job, which are enough to pay off for all withdrawals and transfers (see Lemma 10). The nontrivial part is concerned with those jobs $J \in \mathcal{J}(\mathcal{A}')$ that miss deadline. Note that J must idle for a long time while it is in the pool P. Note that whenever J is idle after being added into P, $\mathcal{A}$ schedules two other jobs. By Lemma 3, $\mathcal{O}$ cannot schedule both of these two jobs. In other words, at least one of these two jobs, say, J_a, is not completed by $\mathcal{O}$. By the definition of $\mathcal{A}'$, J_a is left in $\mathcal{A}'$. When $\mathcal{A}'$ schedules J_a and J is idle, credits are transferred from the account of J_a to the account of J. J thus receives credits transferred from other jobs that are scheduled while J is idle in P. More precisely, the following lemma shows that J receives at least $v(J)$ credits.

Lemma 9. *For each job J that is scheduled by $\mathcal{A}'$, at least $v(J)$ credits have been put into the account of J.*

Proof. If J is completed by $\mathcal{A}'$, the lemma holds by the definition. Now suppose J entered P but is eventually discarded. Then J has been idle in P for a period of at least $\frac{1}{2}(1-\frac{1}{s})p(J)$. At these times it receives credits from at least one running job at the rate $I(J)$, so it eventually receives at least $\frac{1}{2}(1-\frac{1}{s})p(J)I(J) = v(J)$ credits. $\square$

We then bound the amount of credits that is removed, (i.e., withdrawn or transferred) from the account of each job J. When $s \geq 7$, we show that the remaining balance is non-negative.

Lemma 10. *Assume that $s \geq 7$. Then for any job J, $\Psi(J) \geq 0$.*

Proof. We only consider jobs that have been scheduled since credits are only removed from the account of a job when the job is being scheduled. We first give an upper bound to the rate of transfer from J to idling jobs. The algorithm ensures that whenever J is scheduled, it is denser than all idling jobs in P. By Lemma 5, each job in P is at least twice as dense as the next dense job in P, except that the second least dense job needs only be at least as dense as the least dense job. Therefore, if there are i idling jobs, the rate of transfer is at most

$$\frac{2s}{s-1}\frac{\rho(J)}{2} + \frac{2s}{s-1}\frac{\rho(J)}{2^2} + \cdots + \frac{2s}{s-1}\frac{\rho(J)}{2^{i-1}} + \frac{2s}{s-1}\frac{\rho(J)}{2^{i-1}} = \frac{2s}{s-1}\rho(J) .$$

The total rate of transfer and withdrawal is thus at most $\sigma(J) + \frac{2s}{s-1}\rho(J)$. Now consider the final balance $\Psi(J)$.

$$\Psi(J) \geq v(J) - T(J)\left[\sigma(J) + \frac{2s}{s-1}\rho(J)\right]$$

$$\geq p(J)\rho(J) - \frac{p(J)}{s}\left[\frac{4s}{s-1}\rho(J) + \frac{2s}{s-1}\rho(J)\right]$$

$$= p(J)\rho(J) \cdot \frac{s-7}{s-1}$$

$$\geq 0 \quad \text{(assume } s \geq 7\text{)}$$

$\square$

We thus have the following main theorem.

Theorem 11. *Assume that the on-line algorithm is using two speed-s processors, where $s \geq 7$. Then $\|\mathcal{A}'\| \geq \|\mathcal{O}'\|$.*

Proof. $\|\mathcal{A}'\| = \sum_{J\in\mathcal{J}(\mathcal{A}')} \Psi(J) + \sum_{J\in\mathcal{J}(\mathcal{A}')} \sigma(J)T(J) \geq \sum_{J\in\mathcal{J}(\mathcal{A}')} \Psi(J) + \|\mathcal{O}'\| \geq \|\mathcal{O}'\|.$

$\square$

References

1. S. Baruah, G. Koren, D. Mao, B. Mishra, A. Raghunathan, L. Rosier, D. Shasha, and F. Wang. On the competitiveness of on-line task real-time task scheduling. *Journal of Real-Time Systems*, 4(2):124–144, 1992. 465
2. S. Baruah, G. Koren, B. Mishra, A. Raghunathan, L. Rosier, and D. Shasha. On-line scheduling in the presence of overload. In *Proceedings of the IEEE Thirty-second Annual Symposium on the Foundations of Computer Science*, pages 101–110, San Juan, Porto Rico, October 1991. 465
3. P. Berman and C. Coulston. Speed is more powerful than clairvoyance. In *Proceedings of the Sixth Scandinavian Workshop on Algorithm Theory*, pages 255–263, July 1998. 465
4. Bhaskar DasGupta and Michael A. Palis. Online real-time preemptive scheduling of jobs with deadlines. In *Proceedings of the Third International Workshop on Approximation Algorithms for Combinatorial Optimization*, volume 1913 of *Lecture Notes in Computer Science*, pages 96–107. Springer, September 2000. 465
5. Michael L. Dertouzos and Aloysius Ka-Lau Mok. Multiprocessor on-line scheduling of hard-real-time tasks. *IEEE Transactions on Software Engineering*, 15(12):1497–1506, December 1989. 465

6. Jeff Edmonds. Scheduling in the dark. In *Proceedings of the Thirty-First Annual ACM Symposium on Theory of Computing*, pages 179–188, 1999. 465
7. Bala Kalyanasundaram and Kirk Pruhs. Maximizing job completions online. In *Proceedings of the Sixth European Symposium on Algorithms*, pages 235–246, 1998. 465
8. Bala Kalyanasundaram and Kirk Pruhs. Speed is as powerful as clairvoyance. *Journal of the ACM*, 47(4):617–643, July 2000. 464, 465
9. Gilad Koren and Dennis Shasha. D^{over}: An optimal on-line scheduling algorithm for overloaded real-time systems. *SIAM Journal of Computing*, 24(2):318–339, April 1995. 465
10. Tak-Wah Lam, Tsuen-Wan Ngan, and Kar-Keung To. On the speed requirement for optimal deadline scheduling in overloaded systems. In *Proceedings of the Fifteenth International Parallel and Distributed Processing Symposium*, page 202, 2001. 465
11. Tak-Wah Lam and Kar-Keung To. Trade-offs between speed and processor in hard-deadline scheduling. In *Proceedings of the Tenth Annual ACM-SIAM Symposium on Discrete Algorithms*, pages 623–632, 1999. 465
12. Tak-Wah Lam and Kar-Keung To. Performance guarantee for online deadline scheduling in the presence of overload. In *Proceedings of the Twelfth Annual ACM-SIAM Symposium on Discrete Algorithms*, pages 755–764, 2001. 464, 465
13. Cynthia A. Phillips, Cliff Stein, Eric Torng, and Joel Wein. Optimal time-critical scheduling via resource augmentation. In *Proceedings of the Twenty-Ninth Annual ACM Symposium on Theory of Computing*, pages 140–149, 1997. 465

Complexity Note on Mixed Hypergraphs

Daniel Král'[1*], Jan Kratochvíl[1**], and Heinz-Jürgen Voss[2]

[1] Department of Applied Mathematics and Institute
for Theoretical Computer Science[***] Charles University
Malostranské nám. 25, 118 00 Prague, Czech Republic
{kral,honza}@kam.mff.cuni.cz
[2] Institute of Algebra, Technische Universität Dresden
Germany
voss@math.tu-dresden.de

Abstract. A mixed hypergraph H is a triple $(V, \mathcal{C}, \mathcal{D})$ where V is its vertex set and $\mathcal{C}$ and $\mathcal{D}$ are families of subsets of V, $\mathcal{C}$–edges and $\mathcal{D}$–edges. The degree of a vertex is the number of edges in which it is contained. A vertex coloring of H is proper if each $\mathcal{C}$–edge contains two vertices with the same color and each $\mathcal{D}$–edge contains two vertices with different colors. The feasible set of H is the set of all k's such that there exists a proper coloring using exactly k colors. The lower (upper) chromatic number of H is the minimum (maximum) number in the feasible set. We prove that it is NP–complete to decide whether the upper chromatic number of mixed hypergraphs with maximum degree two is at least a given k. We present polynomial time algorithms for mixed hypergraphs with maximum degree two to decide their colorability, to find a coloring using the number of colors equal to the lower chromatic number and we present a 5/3–aproximation algorithm for the upper chromatic number. We further prove that it is coNP-hard to decide whether the feasible set of a given general mixed hypergraph is an interval of integers.

1 Introduction

Hypergraphs are well established combinatorial objects, see [1]. A hypergraph is a pair $(V, \mathcal{E})$ where $\mathcal{E}$ is a family of subsets of V of size at least 2; the members of V are called vertices and the members of $\mathcal{E}$ are called edges. A mixed hypergraph G is a triple $(V, \mathcal{C}, \mathcal{D})$ where $\mathcal{C}$ and $\mathcal{D}$ are families of subsets of V; the members of $\mathcal{C}$ are called $\mathcal{C}$–edges and the members of $\mathcal{D}$ are called $\mathcal{D}$–edges. A mixed hypergraph is *reduced* if it does not contain a $\mathcal{C}$–edge of size two. A *proper k–coloring* c of G is a mapping $c : V \rightarrow \{1, \dots, k\}$ such that there are two vertices with $\mathcal{D}$ifferent colors in each $\mathcal{D}$–edge and there are two vertices with a $\mathcal{C}$ommon color in each $\mathcal{C}$–edge. The coloring c is a *strict k–coloring* if it uses all k colors. The number of edges which contain a vertex v is called the *degree*

* The author acknowledges partial support by GAČR 201/99/0242.
** The author acknowledges support by GAUK 158/99 and KONTAKT 338/99.
*** supported by Ministry of Education of Czech Republic as project LN00A056

J. Sgall, A. Pultr, and P. Kolman (Eds.): MFCS 2001, LNCS 2136, pp. 474–485, 2001.

of v; each edge which is both a $\mathcal{C}$–edge and a $\mathcal{D}$–edge is counted twice. The mixed hypergraph is colorable iff G has a proper coloring. Mixed hypergraphs were introduced in [13]; they find their application in different areas: coloring block designs (see [2,3,4,10,11,12]), list–coloring of graphs and others ([9]).

The *feasible set* $F(G)$ of a mixed hypergraph G is the set of all k's such that there exists a strict k–coloring of G. The *(lower) chromatic number* $\chi(G)$ of G is the minimum number in $F(G)$ and the *upper chromatic number* $\bar{\chi}(G)$ of G is the maximum number in $F(G)$. The feasible set of G is *gap–free* (unbroken) iff $F(G) = [\chi(G), \bar{\chi}(G)]$; we write $[a, b]$ for all the integers between a and b (inclusively). If the feasible set of G contains a gap, we say it is *broken*. An example of a mixed hypergraph with a broken feasible set was first given in [5]. On the other hand, it was proved in [8] that feasible sets of mixed hypertrees, special class of mixed hypergraphs, are gap–free. Other definitions and basic properties of mixed hypergraphs, previously used also in [6,8,13], are in Section 2.

We focus our attention on complexity of coloring problems for mixed hypergraphs and in particular of those with maximum vertex degree two; they are in nice correspondence with mixed multigraphs (Observation 3). Mixed hypergraphs with vertex degree bounded by three (or more) are not interesting from the complexity point of view: The coloring problems for them are equally difficult as the coloring problems for all mixed hypergrahs (Observation 2).

Properties of mixed hypergraphs with maximum degree two are studied and mixed multigraphs, natural representation of such mixed hypergraphs, are introduced in Section 3. We can often restrict our attention to reduced mixed hypergraphs (cf. Lemma 1). We prove that the feasible sets of mixed hypergraphs with maximum degree two are gap–free and their lower chromatic numbers are at most three (Theorem 5). We state several properties of colorings using the maximum possible number of colors in Lemma 7. We conclude Section 3 with the proof that it is NP–complete to decide whether the upper chromatic number of a given mixed hypergraph with maximum degree two is at least a given k (Theorem 8).

In Section 4, we present some polynomial algorithms for mixed hypergraphs with maximum degree two. We find a polynomial algorithm to decide their colorability; this problem is NP–complete for general mixed hypergraphs. We find a polynomial algorithm to determine the lower chromatic number of a mixed hypergraph with maximum degree two which even finds a coloring using that number of colors. We find a 5/3–aproximation algorithm for the upper chromatic number; the NP–completeness of this problem is established in Theorem 8 in Section 3 and there is no aproximation algorithm (unless $P = NP$) with a constant aproximation ratio for general mixed hypergraphs (thus also for those with maximum degree three) due to the results of [7]. The aproximation ratio can be improved to 3/2 by taking more care of triangular color subgraphs (see Section 4 for definitions) of mixed multigraphs in Lemma 9 and in Lemma 10, but we omit this due to space limitations.

We show in Section 5 that it is NP–hard to decide whether the feasible set of a given mixed hypergraph contains a gap (Theorem 14). (We also know that this problem is coNP–hard but we do not provide the proof of this fact due to space limitations and we refer the reader to [7].)

2   Basic Properties of Mixed Hypergraphs

We state a lemma about reducing mixed hypergraphs with maximum degree 2:

Lemma 1. *There is a polynomial time algorithm which for a given mixed hypergraph with maximum degree two either outputs that it is uncolorable or finds a reduced one with maximum degree two such that its proper colorings one–to–one correspond to proper colorings of the given mixed hypergraph.*

Proof. Let H be the given mixed hypergraph. If H does not contain a C–edge of size two, then we are done. Otherwise let $\{u, v\}$ be such a C–edge. If H contains a D–edge $\{u, v\}$, then H is uncolorable and we stop. If $\{u, v\}$ is a subset of another C–edge of H, we remove the super–C–edge from H (this clearly does not affect proper colorings of H). Next, we proceed as follows: Let w be a new vertex; we remove the C–edge $\{u, v\}$ from H, replace (at most one other) occurrence of u by w and another (at most one) occurrence of v by w. The vertex w is contained in at most two edges of the new mixed hypergraph and thus the new mixed hypergraph has the maximum degree at most two. Moreover, its proper colorings one–to–one correspond to proper colorings of the original one (the colors of u and v have to be the same, and thus we can color w with their common color and vice versa). We stop when either we find that the mixed hypergraph is uncolorable or we obtain a reduced one.

We made use of the fact that the maximum degree of the original mixed hypergraph was at most two. If we slighty weaken this condition, we can perform the following construction: Let H be any mixed hypergraph with n vertices $v_1, \ldots, v_n$ and m edges $E_1, \ldots, E_m$. We create the following mixed hypergraph H' with nm vertices v_{ij} where $1 \le i \le n$ and $1 \le j \le m$. We create C–edges $\{v_{ij}, v_{ij+1}\}$ for $1 \le i \le n$ and $1 \le j \le m - 1$; these C–edges assure that the vertices v_{ij} and $v_{ij'}$ are colored with the same color. We create new edges E'_k such that $E'_k = \{v_{ik} | v_i \in E_k\}$ and we assign the same type to E'_k as E_k has. It is clear that proper colorings of the constructed mixed hypergraph one–to–one correspond to the proper colorings of the original one and moreover, its maximum degree is at most three; thus we can conclude:

Observation 2. For each mixed hypergraph H, one can construct in polynomial time a mixed hypergraph with maximum degree three whose proper colorings one–to–one correspond to the proper colorings of H.

Let H be a mixed hypergraph with maximum degree two. We create a multigraph G as follows: Each edge of H corresponds a vertex of G; two vertices of G are joined by as many edges as they share vertices in H and these edges correspond to the shared vertices. A vertex of H contained only in one (no) edge of H is represented by an edge of G whose one end–vertex (both of them) do not correspond to an edge of H. G contains vertices of three types: C–vertices, D–vertices and other vertices. We color edges of G and we demand that there are two edges with the same color incident to each C–vertex and two edges with different colors incident to each D–vertex. This allows us to introduce the notion of *mixed multigraphs* which contain C and D–vertices (all these special vertices must have degree at least two) and their edges are to be colored as described above. All the definitions for mixed hypergraphs translate to mixed multigraphs, in particular, a mixed multigraph is reduced iff it does not contain a C–vertex of degree two. We conclude this paragraph with the following observation:

Observation 3. For each mixed hypergraph H with maximum degree two, there is the mixed multigraph G whose proper edge–colorings one–to–one correspond to the proper vertex–colorings of H. On the other hand, for each mixed multigraph G, there is a mixed hypergraph H with maximum degree two, such that their proper colorings are also in one–to–one correspondence.

3   Colouring Bounded Degree Mixed Hypergraphs

We use the correspondence between mixed hypergraphs of maximum degree two and mixed multigraphs established in Observation 3; any statement for mixed hypergraphs holds also for mixed multigraphs and vice versa.

Lemma 4. *Let H be a mixed hypergraph with maximum degree two. Let c be a proper coloring of H using $k \geq 4$ colors. Then there exists a proper coloring c' of H using exactly $k - 1$ colors.*

Proof. Assume that c uses colors $1, \ldots, k$. We replace each C–edge of H containing two or more vertices colored with the color k with a C–edge containing just two of its vertices colored by the color k and containing no other vertices. Let H' be this new mixed hypergraph. Clearly c is a proper coloring of H' and the maximum degree of H' is two. We contract the C–edges of H' of size two as in Lemma 1. Let $\tilde{H}$ be the mixed hypergraph after performing the contractions and let $\tilde{c}$ be its coloring using k colors corresponding to the coloring c of H. It is clear that $\mathrm{F}(H') = \mathrm{F}(\tilde{H}) \subseteq \mathrm{F}(H)$ and thus it is enough to construct a strict $(k-1)$–coloring of $\tilde{H}$.

Let $v_1, \ldots, v_l$ be all the vertices of $\tilde{H}$ colored by $\tilde{c}$ with the color k; no two vertices of $v_1, \ldots, v_l$ are contained together in a C–edge of $\tilde{H}$. We recolor these vertices one by one by colors different from the color k. We distinguish four cases:

- If v_i is contained in no D–edges of $\tilde{H}$, then we can assign to v_i any color, e.g. the color 1.
- If v_i is contained in exactly one D–edge D, we color v_i as follows: Let c_0 be a color of another vertex of D and we color v_i with a color different from c_0.
- If v_i is contained in two D–edges D_1 and D_2 of $\tilde{H}$, let c_1 be a color used on some other vertex of D_1 and let c_2 be a color used on some other vertex of D_2. We color v_i with a color different from both c_1 and c_2 (note that $k - 1 \geq 3$).

This construction gives a proper coloring of $\tilde{H}$ using exactly $k-1$ colors and this finishes the proof due to the above mentioned inclusions of the feasible sets.

The immediate corollary of this lemma is the following theorem:

Theorem 5. *The feasible set of any mixed hypergraph H with maximum degree two is gap–free. If H is colorable, then $[3, \overline{\chi}(H)] \subseteq F(H)$. Moreover, the chromatic number of H is at most 3.*

Proof. If H is not colorable or $\chi(H) = 1$ (and thus H contains no $\mathcal{D}$–edges), then its feasible set is gap–free. In other cases, it holds that $[3, \overline{\chi}(H)] \subseteq F(H)$ due to Lemma 4. This assures that $F(H)$ is gap–free and $\chi(H) \le 3$.

Let G be a mixed multigraph and let c be a proper coloring of G. The edges colored with the same color form a *color–subgraph* of G. If it is a path/cycle, we call it a *color–path/color–cycle*. The length of a color–path/cycle is the number of edges it contains. The following two lemmas give some information about the structure of colorings using the maximal possible number of colors:

Lemma 6. *If c is a proper coloring of a mixed multigraph G using k colors and if there exists a vertex adjacent to three or more edges colored by c with the same color, then there exists a proper coloring c' using $k+1$ colors.*

Proof. Let C_0 be a color–subgraph containing a vertex of degree at least three and let c_0 be the color to which C_0 corresponds. We distinguish two cases:

– C_0 contains a cycle with a vertex of degree at least three.
  Let C' be a cycle which C_0 contains; let c' be the coloring obtained from c by recoloring the edges contained in C' with a new color. We claim that c' is a proper coloring. It is clear that all the $\mathcal{D}$–vertices are colored properly; let C be a $\mathcal{C}$–vertex. If C is not colored properly, it has to be adjacent to exactly two edges colored by c with the color c_0 (otherwise it would be still properly colored) — but either both these edges are recolored (if they are in C_0) or both of them keep the color c_0 and thus C is colored properly.
– C_0 does not contain a cycle with a vertex of degree at least three.
  Let v be a vertex of degree at least three in C_0. Since v is not contained in any cycle of C_0, it is an articulation with respect to C_0. Let C' be one of the components of $C_0 \setminus v$. Let c' be the coloring obtained from c by recoloring the edges of C' and the edge joining C' to v with a new color. We claim that c' is a proper coloring. The only affected vertex through this procedure is v (since the edges colored with c_0 incident to other vertices were either all recolored or none of them was recolored). But v still contains at least two vertices colored with c_0 and thus v is colored properly.

Lemma 7. *If c is a coloring of a mixed multigraph G using $\overline{\chi}(G)$ colors, then no vertex of G is incident to three or more edges colored with the same color. Moreover, the color–subgraphs are only color–paths and color–cycles.*

Proof. All the vertices of the color–subgraphs have degree at most two due to Lemma 6. If there is a disconnected color subgraph, recoloring its components with mutually different colors gives a proper coloring using more colors. Thus any color subgraph is connected and its vertices have degree at most two.

We are ready to prove our NP–completeness result on the upper chromatic number of mixed hypergraphs with maximum degree two.

Theorem 8. *It is NP–complete to decide whether $\overline{\chi}(H) \geq k$ for a given mixed hypergraph H with maximum degree two and a given k.*

Proof. We present a reduction from a well–known NP–complete graph problem of triangle–covering. This problem is to decide whether a given graph G with $3n$ vertices can be covered by n triangles in such way that each vertex of G is contained in precisely one triangle.

Let G be an instance of the problem of triangle–covering; let $3n$ be the number of vertices of G and let m be the number of its edges. We construct a mixed multigraph from G by setting all its vertices to be C–vertices. We claim that $\overline{\chi}(G) = m - 2n$ iff it admits a triangle–covering.

We first prove that if G admits a triangle–covering, then it can be colored with $m - 2n$ colors: Let $T_1, \ldots, T_n$ be the triangles of the triangle–covering. We color the edges of T_i with the color i and the remaining $m - 3n$ edges with colors $n+1, \ldots, m - 2n$. This gives a proper coloring using exactly $m - 2n$ colors.

Let c be a coloring using $\overline{\chi}(H)$ colors. The color subgraphs of G are only color–paths and color–cycles due to Lemma 7. Let P be the number of the color–paths and Q be the number of the color–cycles; let p_i be the lengths of the color–paths and q_j be the lengths of the color–cycles. The coloring of H uses exactly the following number of colors:

$$P + Q = m - \sum_i p_i - \sum_j q_j + P + Q =$$

$$m - \left(\sum_i p_i + \sum_j q_j - P\right)\left(1 - \frac{Q}{\sum_i p_i + \sum_j q_j - P}\right)$$

Since each C–vertex has to be either an inner vertex of a color–path or a vertex of a color–cycle, it has to hold that $\sum_i p_i + \sum_j q_j - P \geq 3n$. If the coloring c uses at least $m - 2n$ colors, than it has to hold that $\frac{Q}{\sum_i p_i + \sum_j q_j - P} \geq \frac{1}{3}$. On the other hand, it holds that $P \leq \sum_i p_i$ and $Q \leq \frac{1}{3}\sum_j q_j$, since each color–path has length at least one and each color–cycle has length at least three; the equality is only in the case that all p_i's are ones and all q_j's are threes. Thus the inequality $\frac{Q}{\sum_i p_i + \sum_j q_j - P} \geq \frac{1}{3}$ implies that $p_i = 1$ for all i's (thus $\sum_i p_i = P$) and $q_j = 3$ for all j's. Since $\frac{Q}{\sum_i p_i + \sum_j q_j - P} = \frac{1}{3}$, it has to hold that $\sum_i p_i + \sum_j q_j - P = \sum_j q_j = 3n$. Thus each vertex is contained in exactly one color–cycle and all the color–cycles have length three; the color–cycles are triangles of a covering of G. This finishes the reduction.

4   Algorithms

We focus our attention to algorithms for mixed hypergraphs with maximum degree two in this section. We say that a vertex u of a mixed hypergraph of maximum degree two is *doubled* iff there is another vertex v such that u and v are in the same two C–edges of H; the *doubling number* of H is the maximum number of its doubled vertices such that no two of them are in the same C–edge. The doubling number of a mixed hypergraph can be easily obtained by a matching algorithm; this can be easily seen if we translate the doubling number to mixed multigraphs: The doubling number of the original mixed hypergraph is equal to the maximum number of disjoint pairs of C–vertices of the mixed multigraph such that the vertices of each pair are joined by at least two edges.

We first prove an upper bound on the upper chromatic number:

Lemma 9. *Let H be a reduced mixed hypergraph with maximum degree at most two. Then $\overline{\chi}(H) \leq n - 2m/3 + d/3$ where n is the number of the vertices of H, m is the number of the C–edges of H and d is the doubling number of H.*

Proof. Let G be the mixed multigraph from Observation 3; G contains exactly m C–vertices and n edges. Let c be its coloring using $\overline{\chi}(G) = \overline{\chi}(H)$ colors. The color subgraphs are only color–paths or color–cycles due to Lemma 7. Let P be the number of the color–paths and Q the number of the color–cycles; let p_i be the lengths of the color–paths and q_j the lengths of the color–cycles.

We can assume w.l.o.g. that the color–cycles of length two are vertex–disjoint: Otherwise, let uv and vw be two (non–disjoint) color-cycles of length two; we create a new vertex v' which is neither a C–vertex nor a D–vertex and we replace one of the edges between v and w by an edge between v' and w. The same coloring (which assigns the color of the replaced edge to the edge between v' and w) is a proper coloring of the new mixed multigraph and thus its upper chromatic number is at least the upper chromatic number of the original one. Neither n nor m is affected and d can only be decreased. Repeating this procedure yields a mixed multigraph such that one of its coloring using the upper chromatic number of colors does not contain two non–disjoint color–cycles of length two.

The coloring c of G uses exactly the following number of colors:

$$P + Q = n - \sum_{i} p_i - \sum_{j,q_j=2} q_j - \sum_{j,q_j>2} q_j + P + Q =$$

$$n - \left(\sum_{i} p_i + \sum_{j,q_j>2} q_j - P\right)\left(1 - \frac{Q - \sum_{j,q_j=2} 1}{\sum_{i} p_i + \sum_{j,q_j>2} q_j - P}\right) - \sum_{j,q_j=2}(q_j - 1) \leq$$

$$n - \frac{2}{3}\left(\sum_{i} p_i + \sum_{j,q_j>2} q_j - P\right) - \sum_{j,q_j=2}(q_j - 1) =$$

$$n - \frac{2}{3}\left(\sum_{i} p_i + \sum_{j} q_j - P\right) - \sum_{j,q_j=2}(q_j/3 - 1) =$$

$$n - \frac{2}{3}\left(\sum_i p_i + \sum_j q_j - P\right) + \frac{1}{6}\sum_{j,q_j=2} q_j \le n - \frac{2}{3}m + \frac{1}{3}d$$

The first inequality is due to the trivial facts that $P \le \sum_i p_i$ and $Q - \sum_{j,q_j=2} 1 \le 1/3 \sum_{j,q_j>2} q_j$ (the length of each summed cycle on each side is at least three). The last inequality holds, since each C–vertex is either an inner vertex of a color–path or a vertex of a color–cycle and thus $\sum_i p_i + \sum_j q_j - P \ge m$, and since the number of the disjoint color cycles of length two in the coloring using the maximum possible number of colors is at most the doubling number.

We start with the promised algorithmic issues in the following lemmas:

Lemma 10. *Let H be a reduced mixed hypergraph with maximum degree two. There is a polynomial algorithm which outputs a coloring of H using at least $n - m + d$ colors where n is the number of the vertices of H, m is the number of the C–edges of H and d is the doubling number of H.*

Proof. Let us deal with the case that $d = 0$, first. Let M be the number of edges of H. We first find a matching of size M between the vertices and the edges of H such that each edge is matched to one of its vertices and no two edges are matched to the same vertex. This matching exists, since each vertex of H is contained in at most two edges and each edge of H contains at least two vertices. Let V_C be the set of the vertices of H matched to the C–edges of H. We color the remaining $n - m$ vertices of H with mutually distinct colors. We extend this coloring to the remaining m vertices of H. If there exists an edge such that the vertex matched to it is its only yet uncolored vertex, we color it as follows: Let C be such an edge and let v be its uncolored vertex. We distinguish two cases:

- The vertex v is contained only in C–edges.
  We color v with any color used on some other vertex of C.
- The vertex v is contained in the C–edge C and in a D–edge D.
  Let c_0 be a color used on some vertex of D (D contains a colored vertex, since at least the vertex matched to D is colored). If the vertices of C different from v are all colored by c_0, we color v with any color different from c_0. If there are more colors used to color other vertices of C, we color v with any color used to color a vertex of C different from the color c_0.

We repeat this procedure as long as there exists an edge whose only uncolored vertex is the vertex matched to it. Note that when we finish coloring the vertices of the C–edge C, the edge C is colored properly (recall that H is reduced and thus each C–edge of H contains at least three vertices). We don't violate coloring of D–edges of H during the extension of the coloring.

We have extended the coloring to some vertices of H, but possibly not to all of them yet. Each C–edge contains either no or at least two uncolored vertices. There may also be some D–edges with uncolored vertices, but each D–edge contains at least one colored vertex (that one which was matched to it). We color all the uncolored vertices with a completely new color — this finishes properly coloring of all the yet not fully colored C–edges and D–edges. Thus we have just obtained a proper coloring of H using at least $n - m$ colors.

We deal with the case that $d > 0$. Let $u_1, \ldots, u_d$ be a set of its doubled vertices such that no two of them are contained together in a C–edge and let $v_1, \ldots, v_d$ be the copies of these vertices (i.e. u_i and v_i are contained in the same two C–edges), obtained by a matching algorithm as described earlier. We color each pair of vertices u_i and v_i with the same color and we remove these vertices together with the C–edges, which they are contained in, from the mixed hypegraph. We obtain a reduced mixed hypergraph with $n - 2d$ vertices, $m - 2d$ C–edges and with no doubled vertices. We proceed as described above and we obtain its coloring using at least $(n - 2d) - (m - 2d) = n - m$ colors. We extend this coloring to the $2d$ removed vertices (using d new colors) and we get a proper coloring of the original mixed hypergraph using at least $n - m + d$ colors.

Lemma 11. *Let H be a reduced mixed hypergraph with maximum degree at most two. H is two–colorable iff it does not contain an odd cycle consisting of vertices and D–edges of size two. Moreover there exists a polynomial algorithm which finds the coloring using exactly $\chi(H)$ colors.*

Proof. If H contains an odd cycle consisting of vertices and D–edges of size two, then it is clearly not two–colorable. Let us prove the opposite implication. Since H is reduced, all the C–edges of H have size at least three and they are colored properly in any (not necessarily proper) two–coloring of H. Thus we can assume w.l.o.g. that H contains only D–edges. We further replace any D–edge D of H with four or more vertices by any three–vertex subset D' of D. This new mixed hypergraph H' has still maximum degree at most two, it does not contain an odd cycle consisting of D–edges of size two and the sizes of its D–edges are only two or three. We replace further its three–vertex edges with their two–vertex subsets, but this step needs some care. Let $D = \{u, v, w\}$ be a D–edge of H' consisting of three vertices. If replacing D with $\{u, v\}$ introduces an odd cycle consisting of D–edges of size two, then there is a path between u and v consisting of an even number D–edges of size two; each vertex of this path is contained in exactly two D–edges, those edges used by this path; it cannot be contained in more, since the maximum degree of H_0 is two. Thus there is not such a path between u and w and replacing D with $\{u, w\}$ will not introduce an odd cycle consisting of D–edges of size two. This assures that we can replace D with $\{u, v\}$ or $\{u, w\}$ without introducing an odd cycle consisting of D–edges of size two. We proceed in this manner as long as there are some edges of size three. Let H'' be the obtained mixed hypergraph. This is actually a simple bipartite graph which is clearly two–colorable. Its coloring is also a proper coloring of H' and thus a proper coloring of the original mixed hypergraph H.

The algorithm for finding the coloring with $\chi(H)$ colors proceeds as follows: If H contains no D–edges, we color all the vertices of H with the same color. If H contains D–edges, the algorithm checks whether it contains an odd cycle consisting of D–edges of size two. If not, we use the procedure described in the previous paragraph to get a two–coloring. If it contains an odd cycle consisting of D–edges of size two, we find a proper coloring of H as described in Lemma 10 and then we find a three–coloring as described in Lemma 4.

We are now ready to state and prove the main theorem of this section:

Theorem 12. *There exist the following polynomial algorithms for mixed hypergraphs with maximum degree two:*

- *to decide colorability;*
- *to determine the lower chromatic number and to find a coloring using $\chi(H)$ colors;*
- *to find a coloring using at least $3\overline{\chi}(H)/5$ colors. This algorithm is a 5/3–aproximation algorithm for the upper chromatic number of H.*

Proof. The first statement follows from Lemma 1. The second statement follows from Lemma 11. As to the last statement: We first reduce the given mixed hypergraph; this does not affect its feasible set, in particular it does not change its upper chromatic number, and the proper colorings of the reduced mixed hypergraph one–to–one correspond to the proper colorings of the original one (Lemma 1). Let H be the obtained mixed hypergraph. We can find a proper coloring using at least $n - m + d$ colors due to Lemma 10 where n is the number of the vertices of H, m is the number of its $\mathcal{C}$–edges and d is its doubling number; let k be the actual number of colors this coloring uses. Since each $\mathcal{C}$–edge contains at least three vertices, it holds that $m \leq 2n/3$. On the other hand, we know that $\overline{\chi}(H) \leq n - 2m/3 + d/3$ (cf. Lemma 9). Easy calculation gives the desired estimate:

$$\frac{k}{\overline{\chi}(H)} \geq \frac{n - m + d}{n - 2m/3 + d/3} \geq 1 - \frac{m/3 - 2d/3}{n - 2m/3 + d/3} \geq$$

$$1 - \frac{m/3}{n - 2m/3} \geq 1 - \frac{2n/9}{n - 4n/9} = \frac{3}{5}$$

Thus the algorithm really finds a coloring using at least $3\overline{\chi}(H)/5$ colors.

5   NP-Hardness of Gaps in Feasible Sets

We focus our attention on the following algorithmic problem: "Is the feasible set of an input mixed hypergraph gap–free?" We first deal with NAE–satisfiability to prepare for the proof of the main theorem in this section.

A formula is called *necessarily unbalanced* if in every NAE–satisfying assignment at least 80% of variables have the same value.

Lemma 13. *Not-all-equal satisfiability of necessarily unbalanced formulae without negations is NP-complete.*

Proof. Not-all-equal satisfiability is known to be NP-complete for formulae without negations. Given such a formula Φ with n variables, construct Φ' by introducing $4n + 5$ new variables $y_0, y_1, \ldots, y_{4n+4}$ and adding to Φ clauses (y_0, y_i), $i = 1, 2, \ldots, 4n + 4$. Obviously, Φ' is NAE-satisfiable iff Φ is, and in any NAE-satisfying assignment, all variables y_i, $i = 1, 2, \ldots, 4n + 4$ have the same value; i.e., at least $4n + 4$ variables out of $5n + 5$ variables of Φ' have the same value.

Theorem 14. *It is NP-hard to decide whether the feasible set of a given mixed hypergraph contains a gap.*

Proof. For a necessarily unbalanced formula Φ on n variables without negations, we create a mixed hypergraph $H = (V, \mathcal{C}, \mathcal{D})$ with $v = 4n$ vertices such that:

1. H has nonempty feasible set iff Φ is NAE-satisfiable
2. $F(H) \subset [1..\frac{11v}{20}] \cup [\frac{14v}{20}..v]$
3. $F(H) \cap [1..\frac{11v}{20}] \neq \emptyset$ and $F(H) \cap [\frac{14v}{20}..v] \neq \emptyset$ if Φ is NAE-satisfiable (and hence $F(H)$ contains a gap in this case)

Let X be the set of variables of Φ and let C be the set of its clauses. For every variable $x \in X$, we introduce four vertices x_1, x_2, x_3, x_4, and we set $V = \{x_1, x_2, x_3, x_4 | x \in X\}$. On the four vertices arising from a variable x we create the following edges:

$$\mathcal{C}_x = \{(x_1, x_3, x_4), (x_2, x_3, x_4)\} \text{ and } \mathcal{D}_x = \{(x_1, x_2), (x_1, x_3, x_4), (x_2, x_3, x_4)\}$$

and we call $H_x = (\{x_1, x_2, x_3, x_4\}, \mathcal{C}_x, \mathcal{D}_x)$ the *variable gadget*.

For any two distinct variables $x, y \in X$, we introduce the following $\mathcal{D}$-edges:

$$\mathcal{D}_{xy} = \{(x_i, y_j) | 1 \leq i, j \leq 4\}$$

Thus the vertices of different variable gadgets must be colored by different colors. Finally, for every clause $\phi \in C$, we introduce the following $\mathcal{C}$-edges:

$$\mathcal{C}_\phi = \{(x_3, x_4 | x \in \phi), (x_1, x_3 | x \in \phi)\}$$

Our hypergraph H is then constructed as follows:

$$H = (V, \mathcal{C} = \bigcup_{x \in X} \mathcal{C}_x \cup \bigcup_{\phi \in C} \mathcal{C}_\phi, \mathcal{D} = \bigcup_{x \in X} \mathcal{D}_x \cup \bigcup_{x \neq y \in X} \mathcal{D}_{xy})$$

Suppose H admits a proper coloring c. Since colors used on different variable gadgets are distinct, we may assume that colors used on vertices of H_x are x_1^*, x_2^*, x_3^* (since each H_x has at least one $\mathcal{C}$-edge, the four vertices x_1, x_2, x_3, x_4 of H_x cannot be all colored by distinct colors). We may further assume without loss of generality that $c(x_1) = x_1^*, c(x_2) = x_2^*$. If $c(x_3) \notin \{x_1^*, x_2^*\}$ then $c(x_3) = c(x_4) = x_3^*$ (since both (x_1, x_3, x_4) and (x_2, x_3, x_4) must contain two vertices colored by the same color). We call this coloring of Type 1.

On the other hand, if $c(x_3) \in \{x_1^*, x_2^*\}$ (and thus $c(x_4) \in \{x_1^*, x_2^*\}$ due to symmetry of H_x), the $\mathcal{D}$-edges $(x_1, x_3, x_4), (x_2, x_3, x_4)$ imply that either $c(x_3) = x_1^*, c(x_4) = x_2^*$ or $c(x_3) = x_2^*, c(x_4) = x_1^*$. We call such a coloring of Type 2.

We define a truth assignment $\varphi : X \longrightarrow \{T, F\}$ so that $\varphi(x) = T$ iff $c[H_x]$ is of Type 1. We claim that φ NAE-satisfies Φ. Consider a clause ϕ. Since $(x_3, x_4 | x \in \phi) \in \mathcal{C}$, for some variable occuring in ϕ, say y, it must be $c(y_3) = c(y_4)$, i.e. $\varphi(y) = T$. Further, since $(x_1, x_3 | x \in \phi) \in \mathcal{C}$, for some variable occuring in ϕ, say z, it must be $c(z_1) = c(z_3)$, i.e. $\varphi(z) = F$. Hence each clause has at least one True and at least one False variable.

Suppose on the other hand that $\varphi : X \longrightarrow \{T, F\}$ NAE-satisfies Φ. Define a coloring c of H as follows:

$$c(x_1) = x_1^*, c(x_2) = x_2^*$$

$$c(x_3) = c(x_4) = x_3^* \text{ if } \varphi(x) = T$$

$$c(x_3) = x_1^*, c(x_4) = x_2^* \text{ if } \varphi(x) = F$$

for every variable $x \in X$. This coloring is proper on every variable gadget, it uses different colors for vertices of different gadgets and so it is proper in the hyperedges $\mathcal{D}_{xy}, x, y \in X$. Also the $\mathcal{C}$-edges arising from the clause gadgets are colored properly, because all clauses are NAE-satisfied.

Let c be a proper coloring and φ the corresponding NAE-satisfying truth assignment. The actual number of colors used by c is $k = 2n + |\{x : \varphi(x) = T\}|$. Since Φ was necessarily unbalanced, either $k \geq 2n + \frac{4}{5}n$ or $k \leq 2n + \frac{n}{5}$, and substituting $n = \frac{v}{4}$ gets the desired bounds. The claim that $F(H)$ has nonempty intersection with both $[1..\frac{11}{20}v]$ and $[\frac{14}{20}v..v]$ follows from the fact that the negation of a NAE-satisfying truth assignment is a NAE-satisfying assignment as well.

References

1. C. Berge: Graphs and Hypergraphs, North Holland, 1973. 474
2. C. Colbourn, J. Dinitz, A. Rosa: Bicoloring Triple Systems, Electronic J. Combin. 6# 1, paper 25, 16 pages. 475
3. Ch. J. Colbourn, A. Rosa: Triple Systems, Clarendon Press, Oxford, 1999, sect. 18.6. Strict colouring and the upper chromatic number, 340–341. 475
4. T. Etzion and A. Hartman: Towards a large set of Steiner quadruple systems, SIAM J. Discrete Math. 4 (1991), 182–195. 475
5. T. Jiang, D. Mubayi, Zs. Tuza, V. Voloshin and D. B. West: Chromatic spectrum is broken, 6th Twente Workshop on Graphs and Combinatorial Optimization, 26–28, May, 1999, H. J. Broersma, U. Faigle and J. L. Hurink (eds.), University of Twente, May, 1999, 231–234. 475
6. D. Král': On Complexity of Colouring Mixed Hypertrees, to appear in Procced-ings 13th International Symposium on Fundamentals of Computing Theory, 1st International Workshop on Efficient Algorithms, LNCS, 2001. 475
7. D. Král': On Feasible Sets of Mixed Hypergraphs, in preparation. 475, 476
8. D. Král', J. Kratochvíl, A. Proskurowski, H.-J. Voss: Coloring mixed hypertrees, Proceedings 26th Workshop on Graph-Theoretic Concepts in Computer Science, LNCS vol. 1928, 2000, p. 279–289. 475
9. H. Lefmann, V. Rödl, and R. Thomas: Monochromatic vs. multicolored paths, Graphs Combin. 8 (1992), 323–332. 475
10. L. Milazzo: On upper chromatic number for SQS(10) and SQS(16), Le Matematiche L (Catania, 1995), 179–193. 475
11. L. Milazzo and Zs. Tuza: Upper chromatic number of Steiner triple and quadruple systems, Discrete Math. 174 (1997), 247–259. 475
12. L. Milazzo and Zs. Tuza: Strict colorings for classes of Steiner triple systems, Discrete Math. 182 (1998), 233–243. 475
13. V. Voloshin: The mixed hypergraphs, Computer Science Journal of Moldova 1, 1993, 45–52. 475

News from the Online Traveling Repairman

Sven O. Krumke[1]*, Willem E. de Paepe[2], Diana Poensgen[1], and Leen Stougie[3]**

[1] Konrad-Zuse-Zentrum für Informationstechnik Berlin
Department Optimization, Takustr. 7, 14195 Berlin-Dahlem, Germany.
{krumke,poensgen}@zib.de
[2] Department of Technology Management, Technical University of Eindhoven
P. O. Box 513, 5600MB Eindhoven, The Netherlands
w.e.d.paepe@tm.tue.nl
[3] Department of Mathematics, Technical University of Eindhoven
P. O. Box 513, 5600MB Eindhoven, The Netherlands
and Centre for Mathematics and Computer Science (CWI)
P. O. Box 94079, NL-1090 GB Amsterdam, The Netherlands
leen@win.tue.nl

Abstract. In the traveling repairman problem (TRP), a tour must be found through every one of a set of points (cities) in some metric space such that the weighted sum of completion times of the cities is minimized. Given a tour, the completion time of a city is the time traveled on the tour before the city is reached.

In the online traveling repairman problem OLTRP requests for visits to cities arrive online while the repairman is traveling. We analyze the performance of algorithms for the online problem using competitive analysis, where the cost of an online algorithm is compared to that of an optimal offline algorithm.

Feuerstein and Stougie [8] present a 9-competitive algorithm for the OLTRP on the real line. In this paper we show how to use techniques from online-scheduling to obtain a 6-competitive deterministic algorithm for the OLTRP on any metric space. We also present a randomized algorithm with competitive ratio of $\frac{3}{\ln 2} < 4.3281$ against an oblivious adversary. Our results extend to the "dial-a-ride" generalization L-OLDARP of the OLTRP, where objects have to be picked up and delivered by a server. We supplement the deterministic lower bounds presented in [8] with lower bounds on the competitive ratio of any randomized algorithm against an oblivious adversary. Our lower bounds are $\frac{\ln 16 + 1}{\ln 16 - 1} > 2.1282$ for the L-OLDARP on the line, $\frac{4e-5}{2e-3} > 2.41041$ for the L-OLDARP on general metric spaces, 2 for the OLTRP on the line, and 7/3 for the OLTRP on general metric spaces.

* Research supported by the German Science Foundation (DFG, grant Gr 883/5-3)
** Supported by the TMR Network DONET of the European Community ERB TMRX-CT98-0202

J. Sgall, A. Pultr, and P. Kolman (Eds.): MFCS 2001, LNCS 2136, pp. 487–499, 2001.
© Springer-Verlag Berlin Heidelberg 2001

1   Introduction

In the *traveling repairman problem* (TRP) [1] a server must visit a set of m points $p_1, \ldots, p_m$ in a metric space. The server starts in a designated point 0 of the metric space, called the *origin*, and travels at most at unit speed. Given a tour through the m points, the completion time C_j of point p_j is defined as the time traveled by the server on the tour until it reaches p_j $(j = 1, \ldots, m)$. Each point p_j has a weight w_j, and the objective of the TRP is to find the tour that minimizes the total weighted completion time $\sum_{j=1}^{m} w_j C_j$. This objective is also referred to as the *latency*.

In this paper we consider an online version of the TRP called the *online traveling repairman problem* (OLTRP). Requests for visits to points are released over time while the repairman (the server) is traveling. In the online setting the completion time of a request r_j at point p_j with release time t_j is the first time at which the repairman visits p_j after the release time t_j. The online model allows the server to wait. However, waiting yields an increase in the completion times of the points still to be served. Decisions are revocable as long as they have not been executed.

In the dial-a-ride generalization L-OLDARP (for "latency online dial-a-ride problem") each request specifies a ride from one point in the metric space, its *source*, to another point, its *destination*. The server can serve only one ride at a time, and preemption of rides is not allowed: once a ride is started it has to be finished without interruption.

An online algorithm does not know about the existence of a request before its release time. It must base its decisions solely on past information. A common way to evaluate the quality of online algorithms is *competitive analysis* [6]: An algorithm is called *c-competitive* if its cost on any input sequence is at most c times the cost of an optimal offline algorithm.

In [8] a 9-competitive algorithm for the OLTRP on the real line is presented and a 15-competitive algorithm is given for L-OLDARP on the real line in the case that the server is allowed to serve any number of rides simultaneously. In the same paper lower bounds of $1 + \sqrt{2}$ and 3 on the competitive ratio of any deterministic algorithm for OLTRP and L-OLDARP, respectively, are proven.

The online traveling salesman problem where the objective is to minimize the makespan (the time at which the tour is completed) is studied in [4]. Online dial-a-ride problems with this objective have been studied in [3,8].

Dial-a-ride problems can be viewed as generalized scheduling problem with setup costs and order dependent execution times (see [3]). The main difference to scheduling problems is that the "execution time" of jobs depends on their execution order and also on the position of the server.

The offline problem TRP is known to be NP-hard even on trees [1]. Approximation algorithms for TRP have been studied in [9,5,2].

Our main results are competitive algorithms and randomized lower bounds for OLTRP and L-OLDARP. Our algorithms improve the competitive ratios given in [8], and, moreover, the results are valid for any metric space and not just the real line. For the case of the L-OLDARP our algorithms are the first competitive

algorithms. The randomized lower bounds are the first ones for OLTRP and
L-OLDARP.

Our algorithms are adaptions of the GREEDY-INTERVAL algorithm for online-
scheduling presented in [10,11] and of the randomized version given in [7]. Our
lower bound results are obtained by applying Yao's principle in conjunction with
a technique of Seiden [13]. An overview of the results is given in Tables 1 and 2.

Table 1. Deterministic upper and lower bounds

	Deterministic UB	Previous best UB	Deterministic LB
OLTRP	6 (Corollary 8)	9 [8] (real line)	$1 + \sqrt{2}$ [8]
L-OLDARP	6 (Theorem 7)	15 [8] (real line, server capacity ∞)	3 [8]

In Section 2 we define the problems formally. Section 3 contains the deterministic algorithm INTERVAL and the proof of its competitive ratio. In Section 4
we present the randomized algorithm RANDINTERVAL which achieves a better
competitive ratio. In Section 5 we prove lower bounds on the competitive ratio
of randomized algorithms against an oblivious adversary.

Table 2. Randomized upper and lower bounds

	Randomized UB	Randomized LB
OLTRP	$3/\ln 2 < 4.3281$ (Corollary 12)	general: $\frac{7}{3}$ (Theorem 16)
		real line: 2 (Theorem 17)
L-OLDARP	$3/\ln 2 < 4.3281$ (Theorem 11)	general: $\frac{4e-5}{2e-3} > 2.4104$ (Theorem 14)
		real line: $\frac{\ln 16 + 1}{\ln 16 - 1} > 2.1282$ (Theorem 15)

2 Preliminaries

An instance of the online dial-a-ride problem OLDARP consists of a metric space
$M = (X, d)$ with a distinguished origin $0 \in X$ and a sequence $\sigma = r_1, \ldots, r_m$ of
requests. We assume that M has the property that for all pairs of points $(x, y) \in$
M there is a continuous path $p \colon [0, 1] \to X$ in X with $p(0) = x$ and $p(1) = y$
of length $d(x, y)$ (see [4] for a thorough discussion of this model). Examples of
metric spaces that satisfy the above condition are the Euclidean space $\mathbb{R}^p$ and a
metric space induced by a connected undirected edge-weighted graph.

Each request $r_j = (t_j; a_j, b_j, w_j)$ specifies a ride, defined by two points: $a_j \in$
X, the ride's *source*, and $b_j \in X$, its *destination*, and a weight $w_j \geq 0$. Each
request r_j is released at a nonnegative time $t_j \geq 0$, its *release time*. For $t \geq 0$
we denote by $\sigma_{\leq t}$ ($\sigma_{=t}$) the set of requests in σ released no later than time t

(exactly at time t). A server is located at the origin $0 \in X$ at time 0 and can move at most at unit speed. We consider the case in which the server can serve only one ride at a time, and in which preemption of rides is not allowed.

An online algorithm learns from the existence of request r_j only at its release time t_j. In particular, it has neither information about the release time of the last request nor about the total number of requests. Hence, at any moment in time t, an online algorithm must make its decisions only knowing the requests in $\sigma_{\leq t}$. An offline algorithm has complete knowledge about the sequence σ already at time 0.

Given a sequence σ of requests, a *feasible route* for σ is a sequence of moves of the server such that the following conditions are satisfied: (a) The server starts in the origin 0, (b) each ride requested in σ is served, and (c) the repairman does not start to serve any ride (in its source) before its release time. Let C_j^S denote the completion time of request r_j on a feasible route S. The *length* of a route S, denoted by $l(S)$, is defined as the time difference between the time when S is completed and its start time.

Definition 1 (L-OlDarp, OlTrp). *Given a request sequence* $\sigma = r_1, \ldots, r_m$ *the problem* L-OLDARP *is to find a feasible route* S *minimizing* $\sum_{j=1}^{m} w_j C_j^S$. *The online traveling repairman problem* (OLTRP) *is the special case of* L-OLDARP *in which for each request* r_j *source and destination coincide, i.e.,* $a_j = b_j$.

We denote by $\text{ALG}(\sigma)$ the objective function value of the solution produced by an algorithm ALG on input σ. We use OPT to denote an optimal offline algorithm.

Definition 2 (Competitive Algorithm). *A deterministic online algorithm* ALG *for* L-OLDARP *is* c-competitive, *if there exists a constant* c *such that for any request sequence* σ: $\text{ALG}(\sigma) \leq c \cdot \text{OPT}(\sigma)$.

A randomized online algorithm is a probability distribution over a set of deterministic online algorithms. The objective value produced by a randomized algorithm is therefore a random variable. In this paper we analyze the performance of randomized online algorithms only against an *oblivious adversary*. An oblivious adversary does not see the realizations of the random choices made by the online algorithm and therefore has to generate a request sequence in advance. We refer to [6] for details on the various adversary models.

Definition 3 (Competitive Randomized Algorithm). *A randomized online algorithm* RALG *is* c-competitive *against an oblivious adversary if for any request sequence* σ: $\mathbb{E}[\text{RALG}(\sigma)] \leq c \cdot \text{OPT}(\sigma)$.

3   A Deterministic Algorithm for L-OlDarp

Our deterministic strategy is an adaption of the algorithm GREEDY-INTERVAL presented in [10,11] for online scheduling. The proof of performance borrows concepts of the proofs in [10,11].

Strategy interval *Phase 0*: In this phase the the algorithm is initialized.
Set L to be the earliest time when a request could be completed by OPT (we can assume that $L > 0$ since $L = 0$ means that there are requests released at time 0 with source and destination 0. These requests are served at no cost). Until time L remain in 0. For $i = 0, 1, 2, \ldots$, set $B_i := 2^{i-1}L$.

Phase i, $i = 1, 2, \ldots$: At time B_i compute a route S_i for the set of yet unserved requests released up to time B_i with the following properties:

(i) S_i starts at the endpoint x_{i-1} of route S_{i-1} (we set $x_0 := 0$).

(ii) S_i ends at a point x_i such that $d(0, x_i) \leq B_i$.

(iii) The length of route S_i denoted by $l(S_i)$ satisfies

$$l(S_i) \leq \begin{cases} B_1 & \text{if } i = 1 \\ \frac{3}{2}B_i & \text{if } i \geq 2 \end{cases}$$

(iv) S_i maximizes the sum of the weights of requests served among all routes satisfying (i)–(iii).

If $i = 1$, then follow S_1 starting at time L until time $\frac{3}{2}B_2$. If $i \geq 2$, follow S_i starting at time $\frac{3}{2}B_i$ until $\frac{3}{2}B_{i+1}$.

To justify the correctness of the algorithm notice that by definition route S_i computed at time B_i can actually be finished before time $\frac{3}{2}B_{i+1}$, the time when route S_{i+1}, computed at time B_{i+1} needs to be started.

Lemma 4. *Let R_i be the set of requests served on route S_i computed at time B_i, $i = 1, 2, \ldots$ and let R_i^* be the set of requests in the optimal offline solution which are completed in the time interval $(B_{i-1}, B_i]$. Then*

$$\sum_{i=0}^{k} w(R_i) \geq \sum_{i=0}^{k} w(R_i^*) \quad \text{for } k = 1, 2, \ldots \, .$$

Proof. We first argue that for any $k \geq 1$ we can obtain from the optimal offline solution S^* a route S which starts in the origin, has length at most B_k, ends at a point with distance at most B_k from the origin, and which serves all requests in $\bigcup_{i=1}^{k} R_i^*$.

Consider the optimal offline route S^*. Start at the origin and follow S^* for the first B_k time units with the modification that, if a ride is started in S^* before time B_k but not completed before time B_k, omit this ride. Observe that there is at most one such ride. We thereby obtain a route S of length at most B_k which serves all requests in $\bigcup_{i=1}^{k} R_i^*$. Since the server moves at unit speed, it follows that S ends at a point with distance at most B_k from the origin.

We now consider phase k and show that by the end of phase k, at least requests of weight $\sum_{i=0}^{k} w(R_i^*)$ have been scheduled by INTERVAL. If $k = 1$, the route S obtained as outlined above satisfies already all conditions (i)–(iii) required by INTERVAL. If $k \geq 2$, then condition (i) might be violated, since S starts in the origin. However, we can obtain a new route S' from S starting at the endpoint x_{k-1} of the route from the previous phase by starting in x_{k-1}, moving the empty server from x_{k-1} to the origin and then following S. Since $d(x_{k-1}, 0) \leq B_{k-1} = B_k/2$, the new route S' has length at most $B_k/2 + l(S) \leq B_k/2 + B_k = 3/2 \cdot B_k$ which means that it satisfies all the properties (i)–(iii) required by INTERVAL.

Recall that route S and thus also S' serves all requests in $\bigcup_{i=1}^{k} R_i^*$. Possibly, some of the requests from $\bigcup_{i=1}^{k} R_i^*$ have already been served by INTERVAL in previous phases. As omitting requests can never increase the length of a route, in phase k, INTERVAL can schedule at least all requests from $\bigcup_{i=1}^{k} R_i^* \setminus \bigcup_{i=1}^{k-1} R_i$. Consequently, the weight of all requests served in routes $S_1, \ldots, S_k$ of INTERVAL is at least $\bigcup_{i=1}^{k} R_k^*$ as claimed.   □

The previous lemma gives us the following bound on the number of phases that INTERVAL uses to process a given input sequence σ.

Corollary 5. *Suppose that the optimum offline route is completed in the interval $(B_{p-1}, B_p]$ for some $p \geq 1$. Then, the number of phases of Algorithm INTERVAL is at most p. The route S_p computed at time B_p by INTERVAL is completed no later than time $\frac{3}{2}B_{p+1}$.*   □

To prove competitiveness of INTERVAL we need an elementary lemma which can be proven by induction.

Lemma 6. *Let $a_i, b_i \in \mathbb{R} \geq 0$ for $i = 1, \ldots, p$, for which (i) $\sum_{i=1}^{p} a_i = \sum_{i=1}^{p} b_i$; and (ii) $\sum_{i=1}^{p'} a_i \geq \sum_{i=1}^{p'} b_i$ for all $1 \leq p' \leq p$. Then $\sum_{i=1}^{p} \tau_i a_i \leq \sum_{i=1}^{p} \tau_i b_i$ for any nondecreasing sequence $0 \leq \tau_1 \leq \cdots \leq \tau_p$.*   □

Theorem 7. *Algorithm INTERVAL is 6-competitive for L-OLDARP.*

Proof. Let $\sigma = r_1, \ldots, r_m$ be any sequence of requests. By definition of INTERVAL, each request on route S_i completes no later than time $\frac{3}{2}B_{i+1}$. Summing over all phases $1, \ldots, p$ yields

$$\text{INTERVAL}(\sigma) \leq \frac{3}{2} \sum_{i=1}^{p} B_{i+1} w(R_i) = 6 \sum_{i=1}^{p} B_{i-1} w(R_i). \tag{1}$$

From Lemma 4 we know that $\sum_{i=0}^{k} w(R_i) \geq \sum_{i=0}^{k} w(R_i^*)$ for $k = 1, 2, \ldots$, and from Corollary 5 we know that $\sum_{i=0}^{p} w(R_i) = \sum_{i=0}^{p} w(R_i^*)$. Therefore, application of Lemma 6 to the sequences $w(R_i)$ and $w(R_i^*)$ with the weighing sequence $\tau_i := B_{i-1}$, $i = 1, \ldots, p$, gives

$$6 \sum_{i=1}^{p} B_{i-1} w(R_i) \leq 6 \sum_{i=1}^{p} B_{i-1} w(R_i^*) \tag{2}$$

Denote by C_j^* the completion time of request r_j in the optimal solution $\text{OPT}(\sigma)$. For each request r_j denote by $(B_{\phi_j}, B_{\phi_j+1}]$ the interval that contains C_j^*. Then

$$6 \sum_{i=1}^{p} B_{i-1} w(R_i^*) = 6 \sum_{j=1}^{m} B_{\phi_j} w_j \leq 6 \sum_{j=1}^{m} w_j C_j^*. \tag{3}$$

(1), (2), and (3) together complete the proof.   □

Corollary 8. *Algorithm* INTERVAL *is 6-competitive for* OLTRP.   □

4   An Improved Randomized Algorithm

In this section we use techniques from [7] to devise a randomized algorithm RANDINTERVAL. At the beginning, RANDINTERVAL chooses a random number $\delta \in (0,1]$ according to the uniform distribution. From this moment on, the algorithm is completely deterministic, working in the same way as the deterministic algorithm INTERVAL presented in the previous section. For $i \geq 0$ define $B_i' := 2^{i-1-\delta}L$, where again L is the earliest time that a request could be completed by OPT. As stated before in the case of INTERVAL we can assume that $L > 0$. The difference to INTERVAL is that all phases are defined using $B_i' := 2^{i-1-\delta}L$ instead of B_i, $i \geq 1$. Phase 1 is still started at time L.

Lemma 9. *Let R_i be the set of requests scheduled in phase $i \geq 1$ of Algorithm* RANDINTERVAL *and denote by R_i^* the set of requests that are completed by* OPT *in the time interval $(B_{i-1}', B_i']$. Then*

$$\sum_{i=1}^{k} w(R_i) \geq \sum_{i=0}^{k} w(R_i^*) \quad \text{for } k = 1, 2, \dots .$$

Proof. We only have to make sure that the route S_1 is finished before time $\frac{3}{2}B_2'$. The rest of the proof is the same as that for Lemma 4. The proof for the first phase follows from the fact that $\frac{3}{2}B_2' - L = (3 \cdot 2^{-\delta} - 1)L > 2^{-\delta}L = B_1'$.   □

Using the proof of Theorem 7 with Lemma 4 replaced by Lemma 9, we can conclude that for a sequence $\sigma = r_1, \dots, r_m$ of requests the expected objective function value of RANDINTERVAL satisfies

$$\mathbb{E}\left[\text{RANDINTERVAL}(\sigma)\right] \leq \mathbb{E}\left[6\sum_{i=1}^{m} B_{\phi_j}' w_j\right] = 6\sum_{i=1}^{m} w_j \mathbb{E}\left[B_{\phi_j}'\right], \qquad (4)$$

where $(B_{\phi_j}', B_{\phi_{j+1}}']$ is the interval containing the completion time C_j^* of request r_j in the optimal solution OPT(σ). To prove a bound on the performance of RANDINTERVAL we compute $\mathbb{E}\left[B_{\phi_j}'\right]$. Notice that B_{ϕ_j}' is the largest value $2^{k-\delta}L$, $k \in \mathbb{N}$, which is strictly smaller than C_j^*.

Lemma 10. *Let $z \geq L$ and $\delta \in (0,1]$ be a random variable uniformly distributed on $(0,1]$. Define B by $B := \max\{ 2^{k-\delta}L : 2^{k-\delta}L < z \text{ and } k \in \mathbb{N} \}$. Then, $\mathbb{E}[B] = \frac{z}{2\ln 2}$.*

Proof. Suppose that $2^k L \leq z < 2^{k+1}L$ for some $k \geq 0$. Observe that

$$B = \begin{cases} 2^{k-\delta}L & \text{if } \delta \leq \log_2 \frac{2^{k+1}L}{z} \\ 2^{k+1-\delta}L & \text{otherwise} \end{cases}$$

Hence

$$E\left[B\right] = \int_0^{\log_2 \frac{2^{k+1}L}{z}} 2^{k-\delta} L \, d\delta + \int_{\log_2 \frac{2^{k+1}L}{z}}^1 2^{k+1-\delta} L \, d\delta = \frac{z}{2 \ln 2}.$$

This completes the proof.   □

From Lemma 10 we can conclude that $E\left[B'_{\phi_j}\right] = \frac{1}{2 \ln 2} C_j^*$. Using this result in inequality (4) yields the following theorem:

Theorem 11. *Algorithm* RANDINTERVAL *is c-competitive with $c = \frac{3}{\ln 2} < 4.3281$ for* L-OLDARP *against an oblivious adversary.*   □

Corollary 12. *Algorithm* RANDINTERVAL *is c-competitive with $c = \frac{3}{\ln 2} < 4.3281$ for* OLTRP *against an oblivious adversary.*   □

5   Lower Bounds

In this section we show lower bounds for the competitive ratio of any randomized algorithm against an oblivious adversary for the problem L-OLDARP. The basic method for deriving such a lower bound is Yao's principle (see also [6,12,13]). Let X be a probability distribution over input sequences $\Sigma = \{ \sigma_x : x \in \mathcal{X} \}$. We denote the expected cost of the deterministic algorithm ALG according to the distribution X on Σ by $\mathbb{E}_X\left[\text{ALG}(\sigma_x)\right]$. Yao's principle can now be stated as follows.

Theorem 13 (Yao's principle). *Let $\{ \text{ALG}_y : y \in \mathcal{Y} \}$ denote the set of deterministic online algorithms for an online minimization problem. If X is a distribution over input sequences $\{ \sigma_x : x \in \mathcal{X} \}$ such that*

$$\inf_{y \in \mathcal{Y}} \mathbb{E}_X\left[\text{ALG}_y(\sigma_x)\right] \geq \bar{c}\,\mathbb{E}_X\left[\text{OPT}(\sigma_x)\right]$$

for some real number $\bar{c} \geq 1$, then $\bar{c}$ is a lower bound on the competitive ratio of any randomized algorithm against an oblivious adversary.   □

We use a method explained in [13] to compute a suitable distribution once our ground set of request sequences has been fixed.

5.1   A General Lower Bound for L-OlDarp

We provide a general lower bound where the metric space is a star, which consists of three rays of length 2 each. The center of the star is the origin, denoted by 0.

Let the rays of the star be named A, B and C. Let x_A be the point on A with distance x to 0, $0 < x \leq 2$. Points on B and C will be denoted in the same manner. Let $k \in \mathbb{N}$, $k \geq 2$ be arbitrary but fixed. At time 0 there is one request from 0 to 1_A with weight 1, denoted by $r_1 = (0; 0, 1_A, 1)$. With probability

α there are no further requests. With probability $1 - \alpha$ there are k requests at time $2x$: with probability $\frac{1}{2}(1-\alpha)$ from $2x_B$ to $2x_B$ and with probability $\frac{1}{2}(1-\alpha)$ from $2x_C$ to $2x_C$, where $x \in (0,1]$ is chosen according to the density function p, where $\alpha + \int_0^1 p(x)\,dx = 1$ (each of these requests has also weight 1). This yields a probability distribution X over the set $\Sigma = \{\, \sigma_{x,R} : x \in [0,1], R \in \{B,C\}\,\}$ of request sequences where $\sigma_{0,R} = r_1$ for $R \in \{B,C\}$, and

$$\sigma_{x,R} = r_1, \underbrace{(2x; 2x_R, 2x_R, 1), \dots, (2x; 2x_R, 2x_R, 1)}_{k \text{ times}} \text{ for } 0 < x \le 1 \text{ and } R \in \{B,C\}.$$

We first calculate the expected cost $\mathbb{E}_X\left[\text{OPT}(\sigma_{x,R})\right]$ of an optimal offline algorithm with respect to the distribution X on Σ. With probability α there is only request r_1 to be served and in this case the offline cost is 1. Now consider the situation where there are k additional requests at position $2x_B$ or $2x_C$. Clearly the optimal offline cost is independent of the ray on which the requests arrive. First serving request r_1 and then the k requests yields the objective function value $1 + k(2 + 2x)$, whereas first serving the set of k requests and then r_1 results in a total cost of $2kx + 4x + 1$. Hence, for $k \ge 2$, we have

$$\mathbb{E}_X\left[\text{OPT}(\sigma_{x,R})\right] = \alpha + \int_0^1 (2kx + 4x + 1)p(x)\,dx. \tag{5}$$

The strategy of a generic deterministic online algorithm ALG_y can be cast into the following framework: ALG_y starts serving request r_1 at time $2y$ where $y \ge 0$, unless further requests are released before time $2y$. If the sequence ends after r_1, the online costs are $2y + 1$. Otherwise, two cases have to be distinguished.

If $2x > 2y$, that is, the set of k requests is released after the time at which ALG_y starts the ride requested in r_1, the algorithm must first finish r_1 before it can serve the k requests. In this case, the cost of ALG_y is at least $2y + 1 + k(2y + 2 + 2x)$.

If $2x \le 2y$, the server of ALG_y has not yet started r_1 and can serve the k requests before r_1. To calculate the cost it incurs in this case, let l denote the distance of ALG_y's server to the origin at time $2x$. Then $0 \le l \le y$ since otherwise, the server cannot start r_1 at time $2y$. We may assume that ALG_y is either on ray B or C, since moving onto ray A without serving r_1 is clearly not advantageous.

Now, with probability $\frac{1}{2}$, ALG_y's server is on the "wrong" ray. This yields cost of at least $(2x + (l + 2x))k + 6x + l + 1$ for ALG_y. Being on the "right" ray will cost $(2x + (2x - l))k + 6x - l + 1$. Putting this together, we get that for $y \le 1$

$$\mathbb{E}_X\left[\text{ALG}_y(\sigma_{x,R})\right] \ge \alpha(2y + 1) + \int_y^1 (2y + 1 + k(2y + 2 + 2x))\,p(x)\,dx$$

$$+ \frac{1}{2}\int_0^y (4kx + 6x + kl + l + 1)p(x)\,dx$$

$$+ \frac{1}{2}\int_0^y (4kx + 6x - kl - l + 1)p(x)\,dx.$$

This results in

$$\mathbb{E}_X\left[\mathrm{ALG}_y(\sigma_{x,R})\right] \geq F(y) := \alpha(2y+1) + \int_y^1 (2y+1+k(2y+2+2x))\,p(x)\,dx$$

$$+ \int_0^y (4kx+6x+1)p(x)\,dx.$$

(6)

Observe that for $y \geq 1$ we have that

$$\mathbb{E}_X\left[\mathrm{ALG}_y(\sigma_x)\right] \geq \alpha(2y+1) + \int_0^1 (2y+1+k(2y+2+2x))\,p(x)\,dx \geq F(1).$$

Hence in what follows it suffices to consider the case $y \leq 1$. To maximize the expected cost of any deterministic online algorithm on our randomized input sequence, we wish to choose α and a density p such that $\min_{y\in[0,1]} F(y)$ is maximized. We use the following heuristic approach (cf.[13]): Assume that α and the density p maximizing the minimum has the property that $F(y)$ is constant on $[0,1]$. Hence $F'(y) = 0$ and $F''(y) = 0$ for all $y \in (0,1)$. Differentiating we find that

$$F'(y) = 2\left(\alpha + (1+k)\int_y^1 p(x)\,dx - (k-2y)p(y)\right)$$

and $\quad F''(y) = -2(k-1)p(y) - 2(k-2y)p'(y).$

From the condition $F''(y) = 0$ for all $y \in (0,1)$ we obtain the differential equation

$$-2(k-1)p(x) - 2(k-2x)p'(x) = 0,$$

which has the general solution

$$p(x) = \beta(k-2x)^{\frac{1}{2}(k-1)}.$$

(7)

The value of $\beta > 0$ is obtained from the initial condition $\alpha + \int_0^1 p(x)\,dx = 1$ as

$$\beta = \frac{1-\alpha}{\int_0^1 (k-2x)^{\frac{1}{2}(k-1)}\,dx} = \frac{(1+k)\,(\alpha-1)}{(k-2)^{\frac{1+k}{2}} - k^{\frac{1+k}{2}}}.$$

(8)

It remains to determine α. Recall that we attempted to choose α and p in such a way that F is constant over the interval $[0,1]$. Hence in particular we must have $F(0) = F(1)$. Using

$$F(0) = \alpha + \int_0^1 (1+k(2+2x))p(x)\,dx \quad \text{and} \quad F(1) = 3\alpha + \int_0^1 (4kx+6x+1)p(x)\,dx$$

and substituting p and β from (7) and (8), respectively, we obtain

$$\alpha = \frac{(k-2)^{\frac{1+k}{2}}(1+k)}{(k-2)^{\frac{1+k}{2}}k + k^{\frac{1+k}{2}}}.$$

We now use the distribution obtained this way in (6) and (5). This results in

$$
\mathbb{E}_X\left[\mathrm{ALG}_y(\sigma_{x,R})\right] \geq \frac{(1+k)\left(-5(k-2)^{\frac{2+k}{2}}\sqrt{k}+\sqrt{k-2}\,k^{\frac{k}{2}}\,(3+4k)\right)}{\sqrt{k-2}\,(3+k)\left((k-2)^{\frac{1+k}{2}}\sqrt{k}+k^{\frac{k}{2}}\right)}
$$

and

$$
\mathbb{E}_X\left[\mathrm{OPT}(\sigma_{x,R})\right] = \frac{\sqrt{k-2}\,k^{\frac{1+k}{2}}(1+k)(3+2k)-(k-2)^{\frac{2+k}{2}}(1+k)(4+3k)}{\sqrt{k-2}\left((k-2)^{\frac{1+k}{2}}k(3+k)+k^{\frac{1+k}{2}}(3+k)\right)}.
$$

Hence we conclude that

$$
\frac{\mathbb{E}_X\left[\mathrm{ALG}_y(\sigma_{x,R})\right]}{\mathbb{E}_X\left[\mathrm{OPT}(\sigma_{x,R})\right]} \geq \frac{-5(k-2)^{\frac{2+k}{2}}k+\sqrt{k-2}\,k^{\frac{1+k}{2}}(3+4k)}{\sqrt{k-2}\,k^{\frac{1+k}{2}}(3+2k)-(k-2)^{\frac{2+k}{2}}(4+3k)}. \tag{9}
$$

For $k \to \infty$, the right hand side of (9) converges to $\frac{4e-5}{2e-3} > 2.4104$. Hence by Yao's principle we obtain the following result:

Theorem 14. *Any randomized algorithm for* L-OLDARP *has competitive ratio greater or equal to* $\frac{5-4e}{3-2e}$ *against an oblivious adversary.*   □

5.2   A Lower Bound on the Real Line

The lower bound construction on the star uses the fact that the online server does not know on which ray to move if he wants to anticipate on the arrival of the k requests at time $2x$. If the metric space is the real line, there are only two rays, and this argument is no longer valid. The server can move towards the point $2x$ (of course still at the risk that there will be no requests at all on this ray) in anticipation of the set of k requests. The same construction therefore leads to a slightly worse lower bound. Using exactly the same techniques, we obtain the following theorem:

Theorem 15. *Any randomized algorithm for* L-OLDARP *on the real line has competitive ratio greater or equal to* $\frac{\ln 16+1}{\ln 16-1} > 2.1282$ *against an oblivious adversary.*   □

5.3   Lower Bounds for the OlTrp

For the OLTRP we provide a general lower bound, again using a star with three rays of length 2 as a metric space. With probability $\alpha = \frac{k+1}{k+2}$, the input sequence consists of only one request r_1 at distance 1 from the origin, released at time 1, and with weight 1. The ray on which this request occurs is chosen uniformly at random among the three rays. With probability $1 - \alpha$ there will be an additional series of k requests at distance 2 from the origin, each with weight equal to 1.

These requests are released at time 2, and the ray on which they occur is chosen uniformly among the two rays that do not contain r_1.

The cost for the adversary is given by

$$\mathbb{E}_X \left[\text{OPT}(\sigma_x) \right] = \alpha + (1 - \alpha)(2k + 5) = \frac{3k + 6}{k + 2}.$$

It is easy to show that no online algorithm can do better than one whose server is in the origin at time 1 and, at time 2, at distance $0 \leq y \leq 1$ from the origin on the ray where r_1 is located. Using this fact, we get

$$\mathbb{E}_X \left[\text{ALG}_y(\sigma_x) \right] \geq \alpha(3 - y) + (1 - \alpha)((4 + y)k + 7 + y) = \frac{7k + 10}{k + 2}.$$

This leads to

$$\frac{\mathbb{E}_X \left[\text{ALG}_y(\sigma_x) \right]}{\mathbb{E}_X \left[\text{OPT}(\sigma_x) \right]} \geq \frac{7k + 10}{3k + 6}. \tag{10}$$

By letting $k \to \infty$ and applying Yao's principle once more, we obtain the following result.

Theorem 16. *Any randomized algorithm for* OLTRP *has competitive ratio greater or equal to 7/3 against an oblivious adversary.*   □

On the real line, a lower bound is obtained by the following very simple randomized request sequence. With probability $1/2$ we give a request at time 1 in -1, and with probability $1/2$ we give a request at time 1 in $+1$. This leads to the following theorem.

Theorem 17. *Any randomized algorithm for* OLTRP *on the real line has competitive ratio greater or equal to 2 against an oblivious adversary.*   □

References

1. F. Afrati, C. Cosmadakis, C. Papadimitriou, G. Papageorgiou, and N. Papakostantinou, *The complexity of the traveling repairman problem*, Informatique Theorique et Applications **20** (1986), no. 1, 79–87. 488
2. S. Arora and G. Karakostas, *Approximation schemes for minimum latency problems*, Proceedings of the 31st Annual ACM Symposium on the Theory of Computing, 1999, pp. 688–693. 488
3. N. Ascheuer, S. O. Krumke, and J. Rambau, *Online dial-a-ride problems: Minimizing the completion time*, Proceedings of the 17th International Symposium on Theoretical Aspects of Computer Science, Lecture Notes in Computer Science, vol. 1770, Springer, 2000, pp. 639–650. 488
4. G. Ausiello, E. Feuerstein, S. Leonardi, L. Stougie, and M. Talamo, *Algorithms for the on-line traveling salesman*, Algorithmica (2001), To appear. 488, 489
5. A. Blum, P. Chalasani, D. Coppersmith, B. Pulleyblank, P. Raghavan, and M. Sudan, *The minimum latency problem*, Proceedings of the 26th Annual ACM Symposium on the Theory of Computing, 1994, pp. 163–171. 488
6. A. Borodin and R. El-Yaniv, *Online computation and competitive analysis*, Cambridge University Press, 1998. 488, 490, 494

7. S. Chakrabarti, C. A. Phillips, A. S. Schulz, D. B. Shmoys, C. Stein, and J Wein, *Improved scheduling algorithms for minsum criteria*, Proceedings of the 23rd International Colloquium on Automata, Languages and Programming, Lecture Notes in Computer Science, vol. 1099, Springer, 1996, pp. 646–657. 489, 493

8. E. Feuerstein and L. Stougie, *On-line single server dial-a-ride problems*, Theoretical Computer Science (2001), To appear. 487, 488, 489

9. M. Goemans and J. Kleinberg, *An improved approximation ratio for the minimum latency problem*, Proceedings of the 7th Annual ACM-SIAM Symposium on Discrete Algorithms, 1996, pp. 152–158. 488

10. L. Hall, D. B. Shmoys, and J. Wein, *Scheduling to minimize average completion time: Off-line and on-line algorithms*, Proceedings of the 7thAnnual ACM-SIAM Symposium on Discrete Algorithms, 1996, pp. 142–151. 489, 490

11. L. A. Hall, A. S. Schulz, D. B. Shmoys, and J. Wein, *Scheduling to minimize average completion time: Off-line and on-line approximation algorithms*, Mathematics of Operations Research **22** (1997), 513–544. 489, 490

12. R. Motwani and P. Raghavan, *Randomized algorithms*, Cambridge University Press, 1995. 494

13. S. Seiden, *A guessing game and randomized online algorithms*, Proceedings of the 32nd Annual ACM Symposium on the Theory of Computing, 2000, pp. 592–601. 489, 494, 496

Word Problems for 2-Homogeneous Monoids and Symmetric Logspace

Markus Lohrey

Universität Stuttgart, Institut für Informatik
Breitwiesenstr. 20–22, 70565 Stuttgart, Germany
lohreyms@informatik.uni-stuttgart.de

Abstract. We prove that the word problem for every monoid presented by a fixed 2-homogeneous semi-Thue system can be solved in log-space, which generalizes a result of Lipton and Zalcstein for free groups. The uniform word problem for the class of all 2-homogeneous semi-Thue systems is shown to be complete for symmetric log-space.

1 Introduction

Word problems for finite semi-Thue systems, or more precisely word problems for monoids presented by finite semi-Thue systems, received a lot of attention in mathematics and theoretical computer science and are still an active field of research. Since the work of Markov [18] and Post [22] it is known that there exists a fixed semi-Thue system with an undecidable word problem. This has motivated the search for classes of semi-Thue systems with decidable word problems and the investigation of the computational complexity of these word problems, see e.g. [9,7,16]. In [1] Adjan has investigated a particular class of semi-Thue systems, namely n-homogeneous systems, where a semi-Thue system is called n-homogeneous if all rules are of the form $s \to \epsilon$, where s is a word of length n and ϵ is the empty word. Adjan has shown that there exists a fixed 3-homogeneous semi-Thue system with an undecidable word problem and furthermore that every 2-homogeneous semi-Thue system has a decidable word problem. Book [8] has sharpened Adjan's decidability result by proving that the word problem for every 2-homogeneous semi-Thue system can be solved in linear time.

In this paper we will continue the investigation of 2-homogeneous semi-Thue systems. In the first part of the paper we will prove that the word problem for every 2-homogeneous semi-Thue system can be solved in logarithmic space. This result improves Adjan's decidability result in another direction and also generalizes a result of Lipton and Zalcstein [15], namely that the word problem for a finitely generated free group can be solved in logarithmic space. Furthermore our log-space algorithm immediately shows that the word problem for an arbitrary 2-homogeneous semi-Thue system can be solved in DLOGTIME-uniform NC^1 if the word problem for the free group of rank 2 is solvable in DLOGTIME-uniform NC^1. Whether the later holds is one of the major open questions concerning the class DLOGTIME-uniform NC^1. In the second part of this paper we will consider the uniform word problem for 2-homogeneous semi-Thue systems. In this

J. Sgall, A. Pultr, and P. Kolman (Eds.): MFCS 2001, LNCS 2136, pp. 500–512, 2001.

decision problem the 2-homogeneous semi-Thue system is also part of the input. Building on the results from the first part, we will show that the uniform word problem for the class of all 2-homogeneous semi-Thue systems is complete for symmetric log-space. This result is in particular interesting from the viewpoint of computational complexity, since there are quite few natural and nonobvious SL-complete problems in formal language theory, see [2].

2   Preliminaries

We assume some familiarity with computational complexity, see e.g. [21], in particular with circuit complexity, see e.g. [27]. L denotes deterministic logarithmic space. SL (symmetric log-space) is the class of all problems that can be solved in log-space on a symmetric (nondeterministic) Turing machine, see [14] for more details. Important results for SL are the closure of SL under log-space bounded Turing reductions, i.e., $SL = L^{SL}$ [19], and the fact that problems in SL can be solved in deterministic space $O(\log(n)^{\frac{4}{3}})$ [3]. A collection of SL-complete problems can be found in [2]. For the definition of DLOGTIME-uniformity and DLOGTIME-reductions see e.g. [10,5]. DLOGTIME-uniform NC^1, briefly uNC^1, is the class of all languages that can be recognized by a DLOGTIME-uniform family of polynomial-size, logarithmic-depth, fan-in two Boolean circuits. It is well known that uNC^1 corresponds to the class ALOGTIME [24]. An important subclass of uNC^1 is DLOGTIME-uniform TC^0, briefly uTC^0. It is characterized by DLOGTIME-uniform families of constant depth, polynomial-size, unbounded fan-in Boolean circuits with majority-gates. Using the fact that the number of 1s in a word over $\{0,1\}$ can be calculated in uTC^0 [5], the following result was shown in [4].

Theorem 1. *The Dyck-language over 2 bracket pairs is in uTC^0.*

By allowing more than one output gate in circuits we can speak of functions that can be calculated in uTC^0. But with this definition only functions $f : \{0,1\}^* \to \{0,1\}^*$ that satisfy the requirement that $|f(x)| = |f(y)|$ if $|x| = |y|$ could be computed. In order to overcome this restriction we define for a function $f : \{0,1\}^* \to \{0,1\}^*$ the function $\mathrm{pad}(f) : \{0,1\}^* \to \{0,1\}^*\{\#\}^*$ by $\mathrm{pad}(f)(x) = y\#^n$, where $f(x) = y$ and $n = \max\{|f(z)| \mid z \in \{0,1\}^{|x|}\} - |y|$. Then we say that a function f can be calculated in uTC^0 if the function $\mathrm{pad}(f)$ can be calculated by a family of circuits that satisfy the restrictions for uTC^0, where the alphabet $\{0,1,\#\}$ has to be encoded into the binary alphabet $\{0,1\}$. Hence we also have a notion of uTC^0 many-one reducibility. More generally we say that a language A is uTC^0-*reducible* to a language B if A can be recognized by a DLOGTIME-uniform family of polynomial-size, constant-depth, unbounded fan-in Boolean circuits containing also majority-gates and oracle-gates for the language B. This notion of reducibility is a special case of the NC^1-reducibility of [11]. In particular [11, Proposition 4.1] immediately implies that L is closed under uTC^0-reductions. Moreover also uNC^1 and uTC^0 are closed under uTC^0-reducibility and uTC^0-reducibility is transitive. The following inclusions are known between the classes introduced above: $uTC^0 \subseteq uNC^1 \subseteq L \subseteq SL$.

For a binary relation $\rightarrow$ on some set we denote by $\xrightarrow{*}$ the reflexive and transitive closure of $\rightarrow$. In the following let Σ be a finite alphabet. An *involution* $\bar{\ }$ on Σ is a function $\bar{\ } : \Sigma \rightarrow \Sigma$ such that $\bar{\bar{a}} = a$ for all $a \in \Sigma$. The empty word over Σ is denoted by ϵ. Let $s = a_1 a_2 \cdots a_n \in \Sigma^*$ be a word over Σ, where $a_i \in \Sigma$ for $1 \leq i \leq n$. The *length* of s is $|s| = n$. For $1 \leq i \leq n$ we define $s[i] = a_i$ and for $1 \leq i \leq j \leq n$ we define $s[i,j] = a_i a_{i+1} \cdots a_j$. If $i > j$ we set $s[i,j] = \epsilon$. Every word $s[1,i]$ with $i \geq 1$ is called a *non-empty prefix* of s. A *semi-Thue system* $\mathcal{R}$ over Σ, briefly STS, is a finite set $\mathcal{R} \subseteq \Sigma^* \times \Sigma^*$. Its elements are called *rules*. See [13,6] for a good introduction into the theory of semi-Thue systems. The length $\|\mathcal{R}\|$ of $\mathcal{R}$ is defined by $\|\mathcal{R}\| = \sum_{(s,t) \in \mathcal{R}} |st|$. As usual we write $x \rightarrow_{\mathcal{R}} y$ if there exist $u, v \in \Sigma^*$ and $(s,t) \in \mathcal{R}$ with $x = usv$ and $y = utv$. We write $x \leftrightarrow_{\mathcal{R}} y$ if ($x \rightarrow_{\mathcal{R}} y$ or $y \rightarrow_{\mathcal{R}} x$). The relation $\leftrightarrow_{\mathcal{R}}$ is a congruence with respect to the concatenation of words, it is called the *Thue-congruence* generated by $\mathcal{R}$. Hence we can define the quotient monoid $\Sigma^* / \leftrightarrow_{\mathcal{R}}$, which is briefly denoted by $\Sigma^*/\mathcal{R}$. A word t is a $\mathcal{R}$-*normalform* of s if $s \xrightarrow{*}_{\mathcal{R}} t$ and t is $\mathcal{R}$-*irreducible*, i.e., there does not exist a u with $t \rightarrow_{\mathcal{R}} u$. The STS $\mathcal{R}$ is *confluent* if for all $s,t,u \in \Sigma^*$ with ($s \xrightarrow{*}_{\mathcal{R}} t$ and $s \xrightarrow{*}_{\mathcal{R}} u$) there exists a v with ($t \xrightarrow{*}_{\mathcal{R}} v$ and $u \xrightarrow{*}_{\mathcal{R}} v$). It is well-known that $\mathcal{R}$ is confluent if and only if $\mathcal{R}$ is *Church-Rosser*, i.e., for all $s,t \in \Sigma^*$ if $s \leftrightarrow_{\mathcal{R}} t$ then ($s \xrightarrow{*}_{\mathcal{R}} u$ and $t \xrightarrow{*}_{\mathcal{R}} u$) for some $u \in \Sigma^*$, see [6, p 12]. For a morphism $\phi : \Sigma^* \rightarrow \Gamma^*$ we define the STS $\phi(\mathcal{R}) = \{(\phi(\ell), \phi(r)) \mid (\ell,r) \in \mathcal{R}\}$. Let $n \geq 1$. A STS $\mathcal{R}$ is n-*homogeneous* if all rules of $\mathcal{R}$ have the form (ℓ, ϵ) with $|\ell| = n$. An important case of a confluent and 2-homogeneous STS is the STS $\mathcal{S}_n = \{c_i \bar{c}_i \rightarrow \epsilon, \bar{c}_i c_i \rightarrow \epsilon \mid 1 \leq i \leq n\}$ over $\Gamma_n = \{c_1, \ldots, c_n, \bar{c}_1, \ldots, \bar{c}_n\}$. The monoid $\Gamma_n^*/\mathcal{S}_n$ is the *free group* F_n of rank n.

A decision problem that is of fundamental importance in the theory of semi-Thue systems is the uniform word problem. Let $\mathcal{C}$ be a class of STSs. The *uniform word problem*, briefly UWP, for the class $\mathcal{C}$ is the following decision problem:

INPUT: An $\mathcal{R} \in \mathcal{C}$ (over some alphabet Σ) and two words $s,t \in \Sigma^*$.

QUESTION: Does $s \leftrightarrow_{\mathcal{R}} t$ hold?

Here the length of the input is $\|\mathcal{R}\| + |st|$. The UWP for a singleton class $\{\mathcal{R}\}$ is called the *word problem*, briefly WP, for $\mathcal{R}$. In this case we also speak of the word problem for the monoid $\Sigma^*/\mathcal{R}$ and the input size is just $|st|$. In [15] the word problem for a fixed free group was investigated and the following theorem was proven as a special case of a more general result on linear groups.

Theorem 2. *The WP for the free group F_2 of rank 2 is in L.*

This result immediately implies the following corollary.

Corollary 3. *The UWP for the class $\{\mathcal{S}_n \mid n \geq 1\}$ is uTC^0-reducible to the WP for F_2, and therefore is also in L.*

Proof. The group morphism $\varphi_n : F_n \rightarrow F_2$ defined by $c_i \mapsto \bar{c}_1^i c_2 c_1^i$ is injective, see e.g. [17, Proposition 3.1]. Furthermore $\varphi_n(w)$ can be calculated from w and $\mathcal{S}_n$ in uTC^0. The second statement of the theorem follows with Theorem 2. □

Finally let us mention the following result, which was shown in [23].

Theorem 4. *The WP for the free group F_2 of rank 2 is uNC^1-hard under DLOGTIME-reductions.*

3   The Confluent Case

In this section we will investigate the UWP for the class of all confluent and 2-homogeneous STSs. For the rest of this section let $\mathcal{R}$ be a confluent and 2-homogeneous semi-Thue system over an alphabet Σ. It is easy to see that w.l.o.g. we may assume that every symbol in Σ appears in some rule of $\mathcal{R}$.

Lemma 5. *There exist pairwise disjoint sets $\Sigma_\ell, \Sigma_r, \Gamma \subseteq \Sigma$, an involution $^{-}$: $\Gamma \to \Gamma$, and a STS $\mathcal{D} \subseteq \{(ab, \epsilon) \mid a \in \Sigma_\ell,\ b \in \Sigma_r\}$ such that $\Sigma = \Sigma_\ell \cup \Sigma_r \cup \Gamma$ and $\mathcal{R} = \mathcal{D} \cup \{(a\bar{a}, \epsilon) \mid a \in \Gamma\}$. Furthermore given $\mathcal{R}$ and $a \in \Sigma$ we can decide in uTC^0 whether a belongs to Σ_ℓ, Σ_r, or Γ.*

Proof. Define subsets $\Sigma_1, \Sigma_2 \subseteq \Sigma$ by $\Sigma_1 = \{a \in \Sigma \mid \exists b \in \Sigma : (ab, \epsilon) \in \mathcal{R}\}$, $\Sigma_2 = \{a \in \Sigma \mid \exists b \in \Sigma : (ba, \epsilon) \in \mathcal{R}\}$, and let $\Sigma_\ell = \Sigma_1 \backslash \Sigma_2$, $\Sigma_r = \Sigma_2 \backslash \Sigma_1$, and $\Gamma = \Sigma_1 \cap \Sigma_2$. Obviously Σ_ℓ, Σ_r, and Γ are pairwise disjoint and $\Sigma = \Sigma_\ell \cup \Sigma_r \cup \Gamma$. Now let $a \in \Gamma$. Then there exist $b, c \in \Sigma$ with $(ab, \epsilon), (ca, \epsilon) \in \mathcal{R}$. It follows $cab \to_{\mathcal{R}} b$ and $cab \to_{\mathcal{R}} c$. Since $\mathcal{R}$ is confluent we get $b = c$, i.e, $(ab, \epsilon), (ba, \epsilon) \in \mathcal{R}$ and thus $b \in \Gamma$. Now assume that also $(ab', \epsilon) \in \mathcal{R}$ for some $b' \neq b$. Then $bab' \to_{\mathcal{R}} b$ and $bab' \to_{\mathcal{R}} b'$ which contradicts the confluence of $\mathcal{R}$. Similarly there cannot exist a $b' \neq b$ with $(b'a, \epsilon) \in \mathcal{R}$. Thus we can define an involution $^{-} : \Gamma \to \Gamma$ by $\bar{a} = b$ if $(ab, \epsilon), (ba, \epsilon) \in \mathcal{R}$. The lemma follows easily.   □

Note that the involution $^{-} : \Gamma \to \Gamma$ may have fixed points. For the rest of this section it is helpful to eliminate these fixed points. Let $a \in \Gamma$ such that $\bar{a} = a$. Take a new symbol a' and redefine the involution $^{-}$ on the alphabet $\Gamma \cup \{a'\}$ by $\bar{a} = a'$ and $\bar{a'} = a$. Let $\mathcal{R}' = (\mathcal{R} \cup \{(aa', \epsilon), (a'a, \epsilon)\}) \backslash \{(aa, \epsilon)\}$. Furthermore for $w \in \Sigma^*$ let $w' \in (\Sigma \cup \{a'\})^*$ be the word that results from s by replacing the ith occurrence of a in w by a' if i is odd and leaving all other occurrences of symbols unchanged. Then it follows that for $s, t \in \Sigma^*$ it holds $s \overset{*}{\leftrightarrow}_{\mathcal{R}} t$ if and only if $s' \overset{*}{\leftrightarrow}_{\mathcal{R}'} t'$. Note that s', t', and $\mathcal{R}'$ can be calculated from s, t, and $\mathcal{R}$ in uTC^0. In this way we can eliminate all fixed points of the involution $^{-}$. Thus for the rest of the section we may assume that $a \neq \bar{a}$ for all $a \in \Gamma$. Let $\mathcal{S} = \{(a\bar{a}, \epsilon) \mid a \in \Gamma\} \subseteq \mathcal{R}$. Then $\Gamma^*/\mathcal{S}$ is the free group of rank $|\Gamma|/2$.

Define the morphism $\pi : \Sigma^* \to \{(,)\}^*$ by $\pi(a) = ($ for $a \in \Sigma_\ell$, $\pi(b) =)$ for $b \in \Sigma_r$, and $\pi(c) = \epsilon$ for $c \in \Gamma$. We say that a word $w \in \Sigma^*$ is *well-bracketed* if the word $\pi(w)$ is well-bracketed. It is easy to see that if $w \overset{*}{\to}_{\mathcal{R}} \epsilon$ then w is well-bracketed. Furthermore Theorem 1 implies that for a word w and two positions $i, j \in \{1, \ldots, |w|\}$ we can check in uTC^0 whether $w[i, j]$ is well-bracketed. We say that two positions $i, j \in \{1, \ldots, |w|\}$ are *corresponding brackets*, briefly $co_w(i, j)$, if $i < j$, $w[i] \in \Sigma_\ell$, $w[j] \in \Sigma_r$, $w[i, j]$ is well-bracketed, and $w[i, k]$ is not well-bracketed for all k with $i < k < j$. Again it can be checked in uTC^0, whether two positions are corresponding brackets. If w is well-bracketed then we can factorize w uniquely as $w = s_0 w[i_1, j_1] s_1 \cdots w[i_n, j_n] s_n$, where $n \geq 0$, $co_w(i_k, j_k)$ for all $k \in \{1, \ldots, n\}$ and $s_k \in \Gamma^*$ for all $k \in \{0, \ldots, n\}$. We define $\mathcal{F}(w) = s_0 \cdots s_n \in \Gamma^*$.

Lemma 6. *The partial function $\mathcal{F} : \Sigma^* \to \Gamma^*$ (which is only defined on well-bracketed words) can be calculated in uTC^0.*

Proof. First in parallel for every $m \in \{1, \ldots, |w|\}$ we calculate in uTC^0 the value $f_m \in \{0, 1\}$, where $f_m = 0$ if and only if there exist positions $i \leq m \leq j$ such that $co_w(i, j)$. Next we calculate in parallel for every $m \in \{1, \ldots, |w|\}$ the sum $F_m = \sum_{i=1}^{m} f_i$, which is possible in uTC^0 by [5]. If $F_{|w|} < m \leq |w|$ then the m-th output is set to the binary coding of $\#$. If $m \leq F_{|w|}$ and $i \in \{1, \ldots, |w|\}$ is such that $f_i = 1$ and $F_i = m$ then the m-th output is set to the binary coding of $w[i]$. □

Lemma 7. *Let $w = s_0 w[i_1, j_1] s_1 \cdots w[i_n, j_n] s_n$ be well-bracketed, where $n \geq 0$, $co_w(i_k, j_k)$ for all $1 \leq k \leq n$, and $s_k \in \Gamma^*$ for all $0 \leq k \leq n$. Then $w \xrightarrow{*}_{\mathcal{R}} \epsilon$ if and only if $\mathcal{F}(w) = s_0 \cdots s_n \xrightarrow{*}_{\mathcal{S}} \epsilon$, $(w[i_k]w[j_k], \epsilon) \in \mathcal{R}$, and $w[i_k + 1, j_k - 1] \xrightarrow{*}_{\mathcal{R}} \epsilon$ for all $1 \leq k \leq n$.*

Proof. The if-direction of the lemma is trivial. We prove the other direction by an induction on the length of the derivation $w \xrightarrow{*}_{\mathcal{R}} \epsilon$. The case that this derivation has length 0 is trivial. Thus assume that $w = w_1 \ell w_2 \to_{\mathcal{R}} w_1 w_2 \xrightarrow{*}_{\mathcal{R}} \epsilon$. In case that the removed occurrence of ℓ in w lies completely within one of the factors s_k ($0 \leq k \leq n$) or $w[i_k + 1, j_k - 1]$ ($1 \leq k \leq n$) of w we can directly apply the induction hypothesis to $w_1 w_2$. On the other hand if the removed occurrence of ℓ contains one of the positions i_k or j_k ($1 \leq k \leq n$) then, since $co_w(i_k, j_k)$, we must have $\ell = w[i_k]w[j_k]$, $w[i_k + 1, j_k - 1] = \epsilon$, and $w_1 w_2 = s_0 w[i_1, j_1] s_1 \cdots w[i_{k-1}, j_{k-1}](s_{k-1} s_k) w[i_{k+1}, j_{k+1}] s_{k+1} \cdots w[i_n, j_n] s_n \xrightarrow{*}_{\mathcal{R}} \epsilon$. We can conclude by using the induction hypothesis. □

Lemma 8. *For $w \in \Sigma^*$ it holds $w \xrightarrow{*}_{\mathcal{R}} \epsilon$ if and only if w is well-bracketed, $\mathcal{F}(w) \xrightarrow{*}_{\mathcal{S}} \epsilon$, and for all $i, j \in \{1, \ldots, |w|\}$ with $co_w(i, j)$ it holds $((w[i]w[j], \epsilon) \in \mathcal{R}$ and $\mathcal{F}(w[i + 1, j - 1]) \xrightarrow{*}_{\mathcal{S}} \epsilon)$.*

Proof. The only if-direction can be shown by an induction on $|w|$ as follows. Let $w \xrightarrow{*}_{\mathcal{R}} \epsilon$. Then w must be well-bracketed, thus we can factorize w as $w = s_0 w[i_1, j_1] s_1 \cdots w[i_n, j_n] s_n$, where $n \geq 0$, $co_w(i_k, j_k)$ for all $k \in \{1, \ldots, n\}$, and $s_k \in \Gamma^*$ for all $k \in \{0, \ldots, n\}$. By Lemma 7 above we obtain $\mathcal{F}(w) \xrightarrow{*}_{\mathcal{S}} \epsilon$, $(w[i_k]w[j_k], \epsilon) \in \mathcal{R}$, and $w[i_k + 1, j_k - 1] \xrightarrow{*}_{\mathcal{R}} \epsilon$ for all $k \in \{1, \ldots, n\}$. Since $|w[i_k + 1, j_k - 1]| < |w|$ we can apply the induction hypothesis to each of the words $w[i_k + 1, j_k - 1]$ which proves the only if-direction. For the other direction assume that w is well-bracketed, $\mathcal{F}(w) \xrightarrow{*}_{\mathcal{S}} \epsilon$, and for all $i, j \in \{1, \ldots, |w|\}$ with $co_w(i, j)$ it holds $((w[i]w[j], \epsilon) \in \mathcal{R}$ and $\mathcal{F}(w[i + 1, j - 1]) \xrightarrow{*}_{\mathcal{S}} \epsilon)$. We claim that for all $i, j \in \{1, \ldots, |w|\}$ with $co_w(i, j)$ it holds $w[i, j] \xrightarrow{*}_{\mathcal{R}} \epsilon$. This can be easily shown by an induction on $j - i$. Together with $\mathcal{F}(w) \xrightarrow{*}_{\mathcal{S}} \epsilon$ we get $w \xrightarrow{*}_{\mathcal{R}} \epsilon$. □

The previous lemma implies easily the following partial result.

Lemma 9. *The following problem is uTC^0-reducible to the WP for F_2.*
INPUT: A confluent and 2-homogeneous STS $\mathcal{R}$ and a word $w \in \Sigma^*$.
QUESTION: Does $w \xrightarrow{*}_{\mathcal{R}} \epsilon$ (or equivalently $w \xleftrightarrow{*}_{\mathcal{R}} \epsilon$) hold?

Proof. A circuit with oracle gates for the WP for F_2 that on input $w, \mathcal{R}$ determines whether $w \xrightarrow{*}_{\mathcal{R}} \epsilon$ can be easily built using Lemma 8. The quantification over all pairs $i, j \in \{1, \ldots, |w|\}$ in Lemma 8 corresponds to an and-gate of unbounded fan-in. In order to check whether $\mathcal{F}(w[i, j]) \xrightarrow{*}_{\mathcal{S}} \epsilon$ for two positions i and j, we first calculate in uTC^0 the word $\mathcal{F}(w[i, j])$ using Lemma 6. Next we apply Corollary 3, and finally we use an oracle gate for the WP for F_2. $\square$

For $w \in \Sigma^*$ we define the set $\Pi(w)$ as the set of all positions $i \in \{1, \ldots, |w|\}$ such that $w[i] \in \Sigma_\ell \cup \Sigma_r$ and furthermore there does not exist a position $k > i$ with $w[i, k] \xrightarrow{*}_{\mathcal{R}} \epsilon$ and there does not exist a position $k < i$ with $w[k, i] \xrightarrow{*}_{\mathcal{R}} \epsilon$. Thus $\Pi(w)$ is the set of all positions in w whose corresponding symbols are from $\Sigma_\ell \cup \Sigma_r$ but which cannot be deleted in any derivation starting from w. The following lemma should be compared with [20, Lemma 5.4] which makes a similar statement for arbitrary special STSs, i.e., STSs for which it is only required that each rule has the form (s, ϵ) with s arbitrary.

Lemma 10. *For $u, v \in \Sigma^*$ let $\Pi(u) = \{i_1, \ldots, i_m\}$ and $\Pi(v) = \{j_1, \ldots, j_n\}$, where $i_1 < i_2 < \cdots < i_m$ and $j_1 < j_2 < \cdots < j_n$. Define $i_0 = j_0 = 0$, $i_{m+1} = |u| + 1$, and $j_{n+1} = |v| + 1$. Then $u \xleftrightarrow{*}_{\mathcal{R}} v$ if and only if $m = n$, $u[i_k] = v[j_k]$ for $1 \le k \le n$ and $\mathcal{F}(u[i_k + 1, i_{k+1} - 1]) \xleftrightarrow{*}_{\mathcal{S}} \mathcal{F}(v[j_k + 1, j_{k+1} - 1])$ for $0 \le k \le n$.*

Proof. First we show the following statement:

$$\text{Let } w \in \Sigma^*. \text{ If } \Pi(w) = \emptyset \text{ then } w \text{ is well-bracketed and } w \xrightarrow{*}_{\mathcal{R}} \mathcal{F}(w). \tag{1}$$

The case that there does not exist an $i \in \{1, \ldots, |w|\}$ with $w[i] \in \Sigma_\ell \cup \Sigma_r$ is clear. Otherwise there exists a smallest $i \in \{1, \ldots, |w|\}$ with $w[i] \in \Sigma_\ell \cup \Sigma_r$. Thus $w = s\,w[i]\,t$ for some $s \in \Gamma^*, t \in \Sigma^*$. Since $\Pi(w) = \emptyset$ we must have $w[i] \in \Sigma_\ell$ and there exists a minimal $j > i$ with $w[i, j] \xrightarrow{*}_{\mathcal{R}} \epsilon$. Lemma 7 implies $\text{co}_w(i, j)$. Let u be such that $w = s\,w[i, j]\,u$. Since $\Pi(w) = \emptyset$ we must have $\Pi(u) = \emptyset$. Inductively it follows that u is well-bracketed and $u \xrightarrow{*}_{\mathcal{R}} \mathcal{F}(u)$. Thus w is well-bracketed and $w \xrightarrow{*}_{\mathcal{R}} s\mathcal{F}(u) = \mathcal{F}(w)$, which proves (1).

Now we prove the lemma. Consider a factor $u_k := u[i_{k-1} + 1, i_k - 1]$ of u. Let $i_{k-1} < i < i_k$ such that $u[i] \in \Sigma_\ell$. Then $i \notin \Pi(u)$, hence there exists a $j > i$ such that $u[i, j] \xrightarrow{*}_{\mathcal{R}} \epsilon$. But since $i_k \in \Pi(u)$ we must have $j < i_k$. A similar argument holds if $u[i] \in \Sigma_r$, hence $\Pi(u_k) = \emptyset$ and thus $u_k \xrightarrow{*}_{\mathcal{R}} \mathcal{F}(u_k)$ by (1). We obtain $u \xrightarrow{*}_{\mathcal{R}} \mathcal{F}(u_1)u[i_1]\mathcal{F}(u_2)u[i_2] \cdots \mathcal{F}(u_m)u[i_m]\mathcal{F}(u_{m+1}) =: u'$ and similarly $v \xrightarrow{*}_{\mathcal{R}} \mathcal{F}(v_1)v[j_1]\mathcal{F}(v_2)v[j_2] \cdots \mathcal{F}(v_n)v[j_n]\mathcal{F}(v_{n+1}) =: v'$. Thus $u \xleftrightarrow{*}_{\mathcal{R}} v$ if and only if $u' \xleftrightarrow{*}_{\mathcal{R}} v'$ if and only if u' and v' can be reduced to a common word. But only the factors $\mathcal{F}(u_k)$ and $\mathcal{F}(v_k)$ of u' and v', respectively, are reducible. The lemma follows easily. $\square$

With the previous lemma the following theorem follows easily.

Theorem 11. *The UWP for the class of all confluent and 2-homogeneous STSs is uTC^0-reducible to the WP for F_2.*

Proof. Let two words $u, v \in \Sigma^*$ and a confluent and 2-homogeneous STS $\mathcal{R}$ be given. First we calculate in parallel for all $i, j \in \{1, \ldots, |u|\}$ with $i < j$ the Boolean value $e_{i,j}$, which is false if and only if $u[i, j] \xrightarrow{*}_{\mathcal{R}} \epsilon$. Next we calculate in parallel for all $i \in \{1, \ldots, |u|\}$ the number $g_i \in \{0, 1\}$ by

$$
g_i = \begin{cases} 1 & \text{if } u[i] \in \Sigma_\ell \cup \Sigma_r \wedge \bigwedge_{k=1}^{i-1} e_{k,i} \wedge \bigwedge_{k=i+1}^{|u|} e_{i,k} \\ 0 & \text{else} \end{cases}
$$

Thus $g_i = 1$ if and only if $i \in \Pi(u)$. Similarly we calculate for all $j \in \{1, \ldots, |v|\}$ a number $h_j \in \{0, 1\}$, which is 1 if and only if $j \in \Pi(v)$. W.l.o.g. we assume that $g_1 = g_{|u|} = h_1 = h_{|v|} = 1$, this can be enforced by appending symbols to the left and right end of u and v. Now we calculate in parallel for all $i \in \{1, \ldots, |u|\}$ and all $j \in \{1, \ldots, |v|\}$ the sums $G_i = \sum_{k=1}^{i} g_k$ and $H_j = \sum_{k=1}^{j} h_k$, which can be done in uTC^0 by [5]. Finally by Lemma 10, $u \xleftrightarrow{*}_{\mathcal{R}} v$ holds if and only if $G_{|u|} = H_{|v|}$ and furthermore for all $i_1, i_2 \in \{1, \ldots, |u|\}$ and all $j_1, j_2 \in \{1, \ldots, |v|\}$ such that $(g_{i_1} = g_{i_2} = h_{j_1} = h_{j_2} = 1, G_{i_1} = H_{j_1}, \text{ and } G_{i_2} = H_{j_2} = G_{i_1} + 1)$ it holds $(u[i_1] = v[j_1], u[i_2] = v[j_2], \text{ and } \mathcal{F}(u[i_1 + 1, i_2 - 1]) \xleftrightarrow{*}_S \mathcal{F}(v[j_1 + 1, j_2 - 1]))$. Using Corollary 3, Lemma 6, and Lemma 9 the above description can be easily converted into a uTC^0-reduction to the WP for F_2. □

Corollary 12. *The UWP for the class of all 2-homogeneous and confluent STSs is in L. Furthermore if the WP for F_2 is in uNC^1 then the UWP for the class of all 2-homogeneous and confluent STSs is in uNC^1.*

4   The Nonuniform Case

In this section let $\mathcal{R}$ be a fixed 2-homogeneous STS over an alphabet Σ which is not necessarily confluent. W.l.o.g. we may assume that $\Sigma = \{0, \ldots, n - 1\}$. The following two lemmas are easy to prove.

Lemma 13. *Let $a, b \in \Sigma$ such that $a \xleftrightarrow{*}_{\mathcal{R}} b$ and define a morphism $\phi : \Sigma^* \to \Sigma^*$ by $\phi(a) = b$ and $\phi(c) = c$ for all $c \in \Sigma \backslash \{a\}$. Then for all $s, t \in \Sigma^*$ we have $s \xleftrightarrow{*}_{\mathcal{R}} t$ if and only if $\phi(s) \xleftrightarrow{*}_{\phi(\mathcal{R})} \phi(t)$.*

Lemma 14. *Let $\phi : \Sigma^* \to \Sigma^*$ be the morphism defined by $\phi(a) = \min\{b \in \Sigma \mid a \xleftrightarrow{*}_{\mathcal{R}} b\}$. Then for all $u, v \in \Sigma^*$ it holds $u \xleftrightarrow{*}_{\mathcal{R}} v$ if and only if $\phi(u) \xleftrightarrow{*}_{\phi(\mathcal{R})} \phi(v)$. Furthermore the STS $\phi(\mathcal{R})$ is confluent.*

Proof. All critical pairs of $\phi(\mathcal{R})$ can be resolved: If $\phi(a) \leftarrow \phi(a)\phi(b)\phi(c) \to \phi(c)$ then $a \xleftrightarrow{*}_{\mathcal{R}} b$ and thus $\phi(a) = \phi(b)$. The second statement of the lemma follows immediately from Lemma 13.

Theorem 15. *Let $\mathcal{R}$ be a fixed 2-homogeneous STS over an alphabet Σ. Then the WP for $\Sigma^*/\mathcal{R}$ is in L. Furthermore if the WP for F_2 is in uNC^1 then also the WP for $\Sigma^*/\mathcal{R}$ is in uNC^1.*

Proof. Let $\mathcal{R}$ be a fixed 2-homogeneous STS over an alphabet Σ and let ϕ be the fixed morphism from Lemma 14. Since the morphism ϕ can be calculated in uTC^0, the result follows from Corollary 12. □

The next theorem gives some lower bounds for word problems for 2-homogeneous STSs. It deals w.l.o.g. only with confluent and 2-homogeneous STSs. We use the notations from Lemma 5.

Theorem 16. *Let $\mathcal{R}$ be a confluent and 2-homogeneous STS over the alphabet $\Sigma = \Sigma_\ell \cup \Sigma_r \cup \Gamma$. Let $|\Gamma| = 2 \cdot n + f$, where f is the number of fixed points of the involution $^- : \Gamma \to \Gamma$. If $n + f \geq 2$ but not $(n = 0$ and $f = 2)$ then the WP for $\mathcal{R}$ is uNC^1-hard under DLOGTIME-reductions. If $n + f < 2$ or $(n = 0$ and $f = 2)$ then the WP for $\mathcal{R}$ is in uTC^0.*

Proof. If we do not remove the fixed points of the involution $^- : \Gamma \to \Gamma$ then the considerations from Section 3 imply that the WP for $\mathcal{R}$ is uTC^0-reducible to the WP for $G = F_n * \mathbb{Z}_2 * \cdots * \mathbb{Z}_2$, where $*$ constructs the free product and we take f copies of $\mathbb{Z}_2$ (each fixed point of $^-$ generates a copy of $\mathbb{Z}_2$, and the remaining $2n$ many elements in Γ generate F_n). The case $n + f = 0$ is clear. If $n + f = 1$, then either $G = \mathbb{Z}$ or $G = \mathbb{Z}_2$. Both groups have a word problem in uTC^0. If $n = 0$ and $f = 2$ then $G = \mathbb{Z}_2 * \mathbb{Z}_2$. Now $\mathbb{Z}_2 * \mathbb{Z}_2$ is a solvable group, see [23, Lemma 6.9]. Furthermore if we choose two generators a and b of G, where $a^2 = b^2 = 1$ in G, then the number of elements of G definable by words over $\{a, b\}$ of length at most n grows only polynomially in n, i.e, G has a polynomial growth function. Now [23, Theorem 7.6] implies that the WP for G is in uTC^0. Finally let $n + f \geq 2$ but not $(n = 0$ and $f = 2)$. Then $G = G_1 * G_2$, where either $G_1 \neq \mathbb{Z}_2$ or $G_2 \neq \mathbb{Z}_2$, hence G has F_2 as a subgroup, see e.g. the remark in [17, p 177]. Theorem 4 implies that the WP for G and thus also the WP for $\mathcal{R}$ are uNC^1-hard under DLOGTIME-reductions. □

5   The General Uniform Case

In this section let $\mathcal{R}$ be an arbitrary 2-homogeneous STS over an alphabet Σ which is not necessarily confluent. We start with some definitions. A word $w = a_1 a_2 \cdots a_n \in \Sigma^*$, where $n \geq 1$ and $a_i \in \Sigma$ for $i \in \{1, \ldots, n\}$, is an $\mathcal{R}$-*path from* a_1 *to* a_n if for all $i \in \{1, \ldots, n-1\}$ we have $(a_i a_{i+1}, \epsilon) \in \mathcal{R}$ or $(a_{i+1} a_i, \epsilon) \in \mathcal{R}$. Let $\triangleright$ and $\triangleleft$ be two symbols. For an $\mathcal{R}$-path $w = a_1 \cdots a_n$ the set $D_{\mathcal{R}}(w) \subseteq \{\triangleright, \triangleleft\}^*$ contains all words of the form $d_1 \cdots d_{n-1}$ such that for all $i \in \{1, \ldots, n-1\}$ if $d_i = \triangleright$ (respectively $d_i = \triangleleft$) then $(a_i a_{i+1}, \epsilon) \in \mathcal{R}$ (respectively $(a_{i+1} a_i, \epsilon) \in \mathcal{R}$). Since $\mathcal{R}$ may contain two rules of the form (ab, ϵ) and (ba, ϵ), the set $D_{\mathcal{R}}(u)$ may contain more than one word. We define a confluent and 2-homogeneous STS over $\{\triangleright, \triangleleft\}$ by $\mathcal{Z} = \{(\triangleright\triangleright, \epsilon), (\triangleleft\triangleleft, \epsilon)\}$. Finally let $[\epsilon]_{\mathcal{Z}} = \{s \in \{\triangleright, \triangleleft\}^* \mid s \xrightarrow{*}_{\mathcal{Z}} \epsilon\}$. Note that every word in $[\epsilon]_{\mathcal{Z}}$ has an even length.

Lemma 17. *Let $a, b \in \Sigma$. Then $a \overset{*}{\leftrightarrow}_{\mathcal{R}} b$ if and only if there exists an $\mathcal{R}$-path w from a to b with $D_{\mathcal{R}}(w) \cap [\epsilon]_{\mathcal{Z}} \neq \emptyset$.*

Proof. First assume that $w = a_1 \cdots a_n$ is an $\mathcal{R}$-path such that $a_1 = a$, $a_n = b$, and $d_1 \cdots d_{n-1} \in D_{\mathcal{R}}(w) \cap [\epsilon]_{\mathcal{Z}}$. The case $n = 1$ is clear, thus assume that $n \geq 3$, $s = d_1 \cdots d_{i-1} d_{i+2} \cdots d_{n-1} \in [\epsilon]_{\mathcal{Z}}$, and $d_i = \triangleright = d_{i+1}$ (the case $d_i = \triangleleft = d_{i+1}$ is analogous). Thus $(a_i a_{i+1}, \epsilon), (a_{i+1} a_{i+2}, \epsilon) \in \mathcal{R}$ and $a_i \leftarrow a_i a_{i+1} a_{i+2} \rightarrow a_{i+2}$. Define a morphism φ by $\varphi(a_{i+2}) = a_i$ and $\varphi(c) = c$ for all $c \in \Sigma \backslash \{a_{i+2}\}$. Then $w' = \varphi(a_1) \cdots \varphi(a_i) \varphi(a_{i+3}) \cdots \varphi(a_n)$ is a $\varphi(\mathcal{R})$-path such that $s \in D_{\varphi(\mathcal{R})}(w')$. Inductively we obtain $\varphi(a) \overset{*}{\leftrightarrow}_{\varphi(\mathcal{R})} \varphi(b)$. Finally Lemma 13 implies $a \overset{*}{\leftrightarrow}_{\mathcal{R}} b$.

Now assume that $a \overset{*}{\leftrightarrow}_{\mathcal{R}} b$ and choose a derivation $a = u_1 \leftrightarrow_{\mathcal{R}} u_2 \leftrightarrow_{\mathcal{R}} \cdots u_{n-1} \leftrightarrow_{\mathcal{R}} u_n = b$, where n is minimal. The case $a = b$ is clear, thus assume that $a \neq b$ and hence $n \geq 3$. First we will apply the following transformation step to our chosen derivation: If the derivation contains a subderivation of the form $u v \ell_2 w \leftarrow u \ell_1 v \ell_2 w \rightarrow u \ell_1 v w$, where $(\ell_1, \epsilon), (\ell_2, \epsilon) \in \mathcal{R}$ then we replace this subderivation by $u v \ell_2 w \rightarrow u v w \leftarrow u \ell_1 v w$. Similarly we proceed with subderivations of the from $u \ell_2 v w \leftarrow u \ell_2 v \ell_1 w \rightarrow u v \ell_1 w$. Since the iterated application of this transformation step is a terminating process, we finally obtain a derivation $\mathcal{D}$ from a to b which does not allow further applications of the transformation described above. We proceed with the derivation $\mathcal{D}$. Note that $\mathcal{D}$ is also a derivation of minimal length from a to b. Since a and b are both $\mathcal{R}$-irreducible, $\mathcal{D}$ must be of the form $a \overset{*}{\leftrightarrow}_{\mathcal{R}} u \leftarrow v \rightarrow w \overset{*}{\leftrightarrow}_{\mathcal{R}} b$ for some u, v, w. The assumptions on $\mathcal{D}$ imply that there exist $s, t \in \Sigma^*$ and $(a_1 a_2, \epsilon), (a_2 a_3, \epsilon) \in \mathcal{R}$ such that $u = s a_1 t$, $v = s a_1 a_2 a_3 t$, and $w = s a_3 t$ (or $u = s a_3 t$, $v = s a_1 a_2 a_3 t$, and $w = s a_1 t$, this case is analogous). Thus $a_1 \overset{*}{\leftrightarrow}_{\mathcal{R}} a_3$. Define the morphism φ by $\varphi(a_3) = a_1$ and $\varphi(c) = c$ for all $c \in \Sigma \backslash \{a_3\}$. Lemma 13 implies $\varphi(a) \overset{*}{\leftrightarrow}_{\varphi(\mathcal{R})} \varphi(b)$ by a derivation which is shorter then $\mathcal{D}$. Inductively we can conclude that there exists a $\varphi(\mathcal{R})$-path w' from $\varphi(a)$ to $\varphi(b)$ with $D_{\varphi(\mathcal{R})}(w') \cap [\epsilon]_{\mathcal{Z}} \neq \emptyset$. By replacing in the path w' some occurrences of a_1 by one of the $\mathcal{R}$-paths $a_1 a_2 a_3$, $a_3 a_2 a_1$, or $a_3 a_2 a_1 a_2 a_3$, we obtain an $\mathcal{R}$-path w from a to b. For instance if w' contains a subpath of the form $c a_1 d$, where $(c a_1, \epsilon), (a_1 d, \epsilon) \notin \mathcal{R}$ but $(c a_3, \epsilon), (a_3 d, \epsilon) \in \mathcal{R}$, then we replace $c a_1 d$ by $c a_3 a_2 a_1 a_2 a_3 d$. Since for all $v \in \{a_1 a_2 a_3, a_3 a_2 a_1, a_3 a_2 a_1 a_2 a_3\}$ we have $D_{\mathcal{R}}(v) \cap [\epsilon]_{\mathcal{Z}} \neq \emptyset$ it follows $D_{\mathcal{R}}(w) \cap [\epsilon]_{\mathcal{Z}} \neq \emptyset$. □

Define the set $\mathcal{I}$ by $\mathcal{I} = \{s \in [\epsilon]_{\mathcal{Z}} \backslash \{\epsilon\} \mid \forall p, q \in \Sigma^* \backslash \{\epsilon\} : s = pq \Rightarrow p \notin [\epsilon]_{\mathcal{Z}}\}$. The following lemma follows immediately from the definition of $\mathcal{I}$.

Lemma 18. *It holds $\mathcal{I} \subseteq \triangleright \{\triangleright, \triangleleft\}^* \triangleright \cup \triangleleft \{\triangleright, \triangleleft\}^* \triangleleft$ and $[\epsilon]_{\mathcal{Z}} = \mathcal{I}^*$.*

Define a binary relation $T \subseteq \Sigma \times \Sigma$ by $(a, b) \in T$ if and only if there exists an $\mathcal{R}$-path w from a to b with $|w|$ odd and furthermore there exist $c, d \in \Sigma$ such that either $(ac, \epsilon), (db, \epsilon) \in \mathcal{R}$ or $(ca, \epsilon), (bd, \epsilon) \in \mathcal{R}$. Note that T is symmetric.

Lemma 19. *Let $a, b \in \Sigma$. Then $a \overset{*}{\leftrightarrow}_{\mathcal{R}} b$ if and only if $(a, b) \in T$.*

Proof. For the if-direction it suffices to show that $a \overset{*}{\leftrightarrow}_{\mathcal{R}} b$ if $(a, b) \in T$. Thus assume that there exists an $\mathcal{R}$-path w from a to b with $|w|$ odd and furthermore there exist $c, d \in \Sigma$ such that $(ac, \epsilon), (db, \epsilon) \in \mathcal{R}$ (the case that $(ca, \epsilon), (bd, \epsilon) \in \mathcal{R}$ is analogous). Let $s \in D_{\mathcal{R}}(w)$. Since $(ac, \epsilon), (db, \epsilon) \in \mathcal{R}$, also the word $w_i = (ac)^{|s|} w (db)^i$ is an $\mathcal{R}$-path for every $i \geq 0$. It holds $s_i = (\triangleright \triangleleft)^{|s|} s (\triangleleft \triangleright)^i \in D_{\mathcal{R}}(w_i)$. Since $|s|$ is even and $|s| < |(\triangleright \triangleleft)^{|s|}|$, the (unique) $\mathcal{Z}$-normalform of the prefix $(\triangleright \triangleleft)^{|s|} s$ of s_i has the form $(\triangleright \triangleleft)^j$ for some $j \geq 0$. Thus $s_j \in [\epsilon]_{\mathcal{Z}}$ and $D_{\mathcal{R}}(w_j) \cap [\epsilon]_{\mathcal{Z}} \neq \emptyset$. By Lemma 17 we have $a \stackrel{*}{\leftrightarrow}_{\mathcal{R}} b$.

Now let $a \stackrel{*}{\leftrightarrow}_{\mathcal{R}} b$. By Lemma 17 there exists an $\mathcal{R}$-path w from a to b and a word $s \in D_{\mathcal{R}}(w) \cap [\epsilon]_{\mathcal{Z}}$. Let $s = s_1 \cdots s_m$ where $m \geq 0$ and $s_i \in \mathcal{I}$. Let w_i be a subpath of w which goes from a_i to a_{i+1} such that $s_i \in D_{\mathcal{R}}(w_i)$ and $a_1 = a$, $a_{m+1} = b$. It suffices to show that $(a_i, a_{i+1}) \in T$. Since $s_i \in \mathcal{I} \subseteq [\epsilon]_{\mathcal{Z}}$, the length of s_i is even. Thus $|w_i|$ is odd. Next $s_i \in \triangleright\{\triangleright, \triangleleft\}^* \triangleright \cup \triangleleft \{\triangleright, \triangleleft\}^* \triangleleft$ by Lemma 18. Let $s_i \in \triangleright\{\triangleright, \triangleleft\}^* \triangleright$, the other case is symmetric. Hence there exist rules $(a_i c, \epsilon), (d a_{i+1}, \epsilon) \in \mathcal{R}$. Thus $(a_i, a_{i+1}) \in T$. $\qquad\square$

The preceding lemma is the key for proving that the UWP for the class of 2-homogeneous STSs is in SL. In general it is quite difficult to prove that a problem is contained in SL. A useful strategy developed in [12] and applied in [25,26] is based on a logical characterization of SL. In the following we consider finite structures of the form $\mathcal{A} = (\{0, \ldots, n-1\}, 0, \max, s, R)$. Here $\max = n - 1$, and R and s are binary relations on $\{0, \ldots, n-1\}$, where $s(x, y)$ holds if and only if $y = x + 1$. The logic FO+posSTS is the set of all formulas build up from the constant 0 and $\max$, first-order variables $x_1, x_2, \ldots$, the binary relations s and R, the equality $=$, the Boolean connectives $\neg, \wedge$, and $\vee$, the quantifiers $\forall$ and $\exists$, and the symmetric transitive closure operator STC, where STC is not allowed to occur within a negation $\neg$. The semantic of STC is the following. Let $\varphi(x, y)$ be a formula of FO+posSTS with two free variables x and y, and let $\mathcal{A} = (\{0, \ldots, n-1\}, 0, \max, s, R)$ be a structure. Assume that $\varphi(x, y)$ describes the binary relation S over $\{0, \ldots, n-1\}$, i.e, $\mathcal{A} \models \varphi(i, j)$ if and only if $(i, j) \in S$ for all $i, j \in \{0, \ldots, n-1\}$. Then $[\mathrm{STC}x, y \, \varphi(x, y)]$ is a formula of FO+posSTS with two free variables, and for all $i, j \in \{0, \ldots, n-1\}$ it holds $\mathcal{A} \models [\mathrm{STC}x, y \, \varphi(x, y)](i, j)$ if and only if (i, j) belongs to the symmetric, transitive, and reflexive closure of S, i.e, $(i, j) \in (S \cup S^{-1})^*$. In [12] it was shown that for every fixed variable-free formula φ of FO+posSTC the following problem belongs to SL:

INPUT: A binary coded structure $\mathcal{A} = (\{0, \ldots, n-1\}, 0, \max, s, R)$
QUESTION: Does $\mathcal{A} \models \varphi$ hold?

Theorem 20. *The following problem is SL-complete:*
INPUT: A 2-homogeneous STS $\mathcal{R}$ over an alphabet Σ and $a, b \in \Sigma$.
QUESTION: Does $a \stackrel{*}{\leftrightarrow}_{\mathcal{R}} b$ hold?

Proof. First we show containment in SL. Let $\mathcal{R}$ be a 2-homogeneous STS over an alphabet Σ and let $a, b \in \Sigma$. W.l.o.g. we may assume that $\Sigma = \{0, \ldots, n-1\}$ and $a = 0, b = n-1$. If $a = 0$ and $b = n-1$ does not hold then it can be enforced by relabeling the alphabet symbols. This relabeling can be done in deterministic log-space and we can use the fact that $\mathrm{L}^{\mathrm{SL}} = \mathrm{SL}$. We identify the input $\mathcal{R}$, a, b with the structure $\mathcal{A} = (\Sigma, 0, \max, s, R)$, where $R = \{(i, j) \mid (ij, \epsilon) \in \mathcal{R}\}$. Now define formulas $S(x, y)$ and $T(x, y)$ as follows:

$$S(x, y) \quad :\Leftrightarrow \quad \exists z \{(R(x, z) \vee R(z, x)) \wedge (R(y, z) \vee R(z, y))\}$$

$$T(x, y) \quad :\Leftrightarrow \quad [STCu, v\, S(u, v)](x, y) \wedge \exists x', y' \begin{Bmatrix} (R(x, x') \wedge R(y', y)) \vee \\ (R(x', x) \wedge R(y, y')) \end{Bmatrix}$$

By Lemma 19, $a \overset{*}{\leftrightarrow}_{\mathcal{R}} b$ if and only if $\mathcal{A} \models [STCu, v\, T(u, v)](0, \max)$. Thus containment in SL follows from [12]. In order to show SL-hardness we use the SL-complete undirected graph accessibility problem (UGAP), see also [14]:

INPUT: An undirected graph $G = (V, E)$ and two nodes $a, b \in V$.

QUESTION: Does there exist a path in G from a to b?

Let $G = (V, E), a, b$ be an instance of UGAP, where $E \subseteq \{\{v, w\} \mid v, w \in V\}$ and of course $V \cap E = \emptyset$. We define a 2-homogeneous STS $\mathcal{R}$ over $V \cup E$ by $\mathcal{R} = \{(ce, \epsilon), (ec, \epsilon) \mid c \in V, e \in E, c \in e\}$. We claim that there exists a path in G from a to b if and only if $a \overset{*}{\leftrightarrow}_{\mathcal{R}} b$. First assume that there exists a path $a = a_1, a_2, \cdots, a_n = b$ with $\{a_i, a_{i+1}\} = e_i \in E$. The case $n = 1$ is clear. If $n > 1$ then by induction $a \overset{*}{\leftrightarrow}_{\mathcal{R}} a_{n-1}$. Thus $a \overset{*}{\leftrightarrow}_{\mathcal{R}} a_{n-1} \leftarrow a_{n-1}e_{n-1}a_n \rightarrow a_n = b$. Conversely assume that a and b belong to different connected components of G. Let V_a and E_a be the set of all nodes and edges, respectively, that belong to the connected component of a. Define a projection $\pi : V \cup E \rightarrow V_a \cup E_a$ by $\pi(x) = \epsilon$ if $x \notin V_a \cup E_a$ and $\pi(x) = x$ if $x \in V_a \cup E_a$. If $a \overset{*}{\leftrightarrow}_{\mathcal{R}} b$ then $a = \pi(a) \overset{*}{\leftrightarrow}_{\pi(\mathcal{R})} \pi(b) = \epsilon$, which is impossible since $u \overset{*}{\leftrightarrow}_{\pi(\mathcal{R})} \epsilon$ implies $|u|_V = |u|_E$, where $|u|_X$ is the number of occurrences of symbols from X in u. $\quad\square$

Theorem 21. *The UWP for the class of all 2-homogeneous STSs is SL-complete.*

Proof. By Theorem 20 it remains to show containment in SL. W.l.o.g. let $\Sigma = \{0, \ldots, n-1\}$. Let ϕ be the morphism from Lemma 14. We check whether $u \overset{*}{\leftrightarrow}_{\mathcal{R}} v$ by essentially running the log-space algorithm for the UWP for confluent and 2-homogeneous STSs from Section 3, but each time we read from the input-tape (the binary coding of) a symbol $a \in \Sigma$, we replace a by $\phi(a)$. Since $\phi(a) = \min\{b \in \Sigma \mid a \overset{*}{\leftrightarrow}_{\mathcal{R}} b\}$, Theorem 20 implies that we can find $\phi(a)$ by at most n queries to an SL-oracle. Since $L^{SL} = SL$, the theorem follows. $\quad\square$

Acknowledgments I would like to thank Klaus Wich for valuable comments.

References

1. S. I. Adjan. *Defining relations and algorithmic problems for groups and semigroups*, volume 85 of *Proceedings of the Steklov Institute of Mathematics*. American Mathematical Society, 1967. 500
2. C. Alvarez and R. Greenlaw. A compendium of problems complete for symmetric logarithmic space. *Electronic Colloquium on Computational Complexity*, Report No. TR96-039, 1996. 501

3. R. Armoni, A. Ta-Shma, A. Widgerson, and S. Zhou. An $O(\log(n)^{4/3})$ space algorithm for (s,t) connectivity in undirected graphs. *Journal of the Association for Computing Machinery*, 47(2):294–311, 2000. 501

4. D. A. M. Barrington and J. Corbet. On the relative complexity of some languages in NC^1. *Information Processing Letters*, 32:251–256, 1989. 501

5. D. A. M. Barrington, N. Immerman, and H. Straubing. On uniformity within NC^1. *Journal of Computer and System Sciences*, 41:274–306, 1990. 501, 504, 506

6. R. Book and F. Otto. *String–Rewriting Systems*. Springer, 1993. 502

7. R. V. Book. Confluent and other types of Thue systems. *Journal of the Association for Computing Machinery*, 29(1):171–182, 1982. 500

8. R. V. Book. Homogeneous Thue systems and the Church–Rosser property. *Discrete Mathematics*, 48:137–145, 1984. 500

9. R. V. Book, M. Jantzen, B. Monien, C. P. O'Dunlaing, and C. Wrathall. On the complexity of word problems in certain Thue systems. In J. Gruska and M. Chytil, editors, *Proceedings of the 10rd Mathematical Foundations of Computer Science (MFCS'81), Štrbské Pleso (Czechoslovakia)*, number 118 in Lecture Notes in Computer Science, pages 216–223. Springer, 1981. 500

10. S. R. Buss. The Boolean formula value problem is in ALOGTIME. In *Proceedings of the 19th Annual Symposium on Theory of Computing (STOC 87)*, pages 123–131. ACM Press, 1987. 501

11. S. A. Cook. A taxonomy of problems with fast parallel algorithms. *Information and Control*, 64:2–22, 1985. 501

12. N. Immerman. Languages that capture complexity classes. *SIAM Journal on Computing*, 16(4):760–778, 1987. 509, 510

13. M. Jantzen. Confluent string rewriting. In *EATCS Monographs on theoretical computer science*, volume 14. Springer, 1988. 502

14. H. R. Lewis and C. H. Papadimitriou. Symmetric space-bounded computation. *Theoretical Computer Science*, 19(2):161–187, 1982. 501, 510

15. R. J. Lipton and Y. Zalcstein. Word problems solvable in logspace. *Journal of the Association for Computing Machinery*, 24(3):522–526, 1977. 500, 502

16. M. Lohrey. Word problems and confluence problems for restricted semi-Thue systems. In L. Bachmair, editor, *Proceedings of the 11th International Conference on Rewrite Techniques and Applications (RTA 2000), Norwich (UK)*, number 1833 in Lecture Notes in Computer Science, pages 172–186. Springer, 2000. 500

17. R. C. Lyndon and P. E. Schupp. *Combinatorial Group Theory*. Springer, 1977. 502, 507

18. A. Markov. On the impossibility of certain algorithms in the theory of associative systems. *Doklady Akademii Nauk SSSR*, 55, 58:587–590, 353–356, 1947. 500

19. N. Nisan and A. Ta-Shma. Symmetric logspace is closed under complement. *Chicago Journal of Theoretical Computer Science*, 1995. 501

20. F. Otto and L. Zhang. Decision problems for finite special string-rewriting systems that are confluent on some congruence class. *Acta Informatica*, 28:477–510, 1991. 505

21. C. H. Papadimitriou. *Computational Complexity*. Addison Wesley, 1994. 501

22. E. Post. Recursive unsolvability of a problem of Thue. *Journal of Symbolic Logic*, 12(1):1–11, 1947. 500

23. D. Robinson. *Parallel Algorithms for Group Word Problems*. PhD thesis, University of California, San Diego, 1993. 502, 507

24. W. L. Ruzzo. On uniform circuit complexity. *Journal of Computer and System Sciences*, 22:365–383, 1981. 501

25. I. Stewart. Complete problems for symmetric logspace involving free groups. *Information Processing Letters*, 40:263–267, 1991. 509

26. I. Stewart. Refining known results on the generalized word problem for free groups. *International Journal of Algebra and Computation*, 2:221–236, 1992. 509

27. H. Vollmer. *Introduction to Circuit Complexity*. Springer, 1999. 501

Variations on a Theorem of Fine & Wilf

Filippo Mignosi[1]*, Jeffrey Shallit[2]**, and Ming-wei Wang[2]

[1] Dipartimento di Matematica ed Applicazioni, Universitá di Palermo
Via Archirafi 34, 90123 Palermo, Italy
mignosi@math.unipa.it
http://dipinfo.math.unipa.it/~mignosi/
[2] Department of Computer Science, University of Waterloo
Waterloo, Ontario, Canada N2L 3G1
shallit@uwaterloo.ca, m2wang@math.uwaterloo.ca
http://www.math.uwaterloo.ca/~shallit

Abstract. In 1965, Fine & Wilf proved the following theorem: if $(f_n)_{n \geq 0}$ and $(g_n)_{n \geq 0}$ are periodic sequences of real numbers, of periods h and k respectively, and $f_n = g_n$ for $0 \leq n < h + k - \gcd(h, k)$, then $f_n = g_n$ for all $n \geq 0$. Furthermore, the constant $h + k - \gcd(h, k)$ is best possible. In this paper we consider some variations on this theorem. In particular, we study the case where $f_n \leq g_n$ instead of $f_n = g_n$. We also obtain a generalization to more than two periods.

1 Introduction

Periodicity is an important property of words that has applications in various domains. For instance, it has applications in string searching algorithms (cf. [4]), in formal languages (cf. for instance the pumping lemmas in Salomaa [11]), and it is an important part of combinatorics on words (cf. [3,1].

The first significants results on periodicity are the theorem of Fine & Wilf and the critical factorization theorem. The main result of this paper is a variation on the theorem of Fine & Wilf in which equality is replaced by inequality.

Besides its intrinsic interest, we started our researches on this subject by trying to solve the still-open "decreasing length conjecture" on $D0L$-sequences. This conjecture states that if $\varphi : \Sigma^* \to \Sigma^*$ is a morphism on the free monoid, $|\Sigma| = n$, and $w \in \Sigma^*$, then

$$|w| > |\varphi(w)| > \cdots > |\varphi^k(w)|$$

implies that $k \leq n$.

We soon realized that the Theorem of Fine & Wilf was not an appropriate tool for approaching this conjecture. But, by using Theorem 11, we were able to prove this conjecture when w consists only of non-growing letters, a result that seems impossible with Fine & Wilf. Details will appear in the full paper.

We say a sequence $(f_n)_{n \geq 0}$ is periodic with period $h \geq 1$ if $f_n = f_{n+h}$ for all $n \geq 0$. The following is a classical "folk theorem":

* Research supported in part by CNR-NATO fellowship no. 215.31
** Research supported in part by a grant from NSERC

J. Sgall, A. Pultr, and P. Kolman (Eds.): MFCS 2001, LNCS 2136, pp. 512–523, 2001.
© Springer-Verlag Berlin Heidelberg 2001

Theorem 1. *If* $(f_n)_{n\geq 0}$ *is a sequence of real numbers which is periodic with periods* h *and* k*, then it is periodic with period* $\gcd(h,k)$*.*

Proof. By the extended Euclidean algorithm, there exist integers $r, s \geq 0$ such that $rh - sk = \gcd(h,k)$. Then we have

$$f_n = f_{n+rh} = f_{n+rh-sk} = f_{n+\gcd(h,k)}$$

for all $n \geq 0$. □

The 1965 theorem of Fine & Wilf [6] is the following:

Theorem 2. *Let* $(f_n)_{n\geq 0}$*,* $(g_n)_{n\geq 0}$ *be two periodic sequences of real numbers, of periods* h *and* k *respectively.*

(a) If $f_n = g_n$ *for* $0 \leq n < h + k - \gcd(h,k)$*, then* $f_n = g_n$ *for all* $n \geq 0$*.*
(b) The conclusion in (a) would be false if $h+k-\gcd(h,k)$ *were replaced by any smaller number.*

As mentioned previously, in this paper we consider a variation on the theorem of Fine & Wilf in which equality is replaced by inequality. It is easy to see that we must necessarily impose some additional condition in order for the result to hold, and so we add a "normality" condition that the periods sum to 0.

Our main theorem follows:

Theorem 3. *Let* $(f_n)_{n\geq 0}$*,* $(g_n)_{n\geq 0}$ *be two periodic sequences of real numbers, of periods* h *and* k*, respectively, such that*

$$\sum_{0 \leq i < h} f_i = \sum_{0 \leq j < k} g_j = 0.$$

Let $d = \gcd(h,k)$*.*

(a) If

$$f_n \leq g_n \quad \text{for } 0 \leq n < h + k - d \tag{1}$$

then
(i) $f_n = g_n$ *for all* $n \geq 0$*; and*
(ii) $\sum_{j \leq i < j+d} f_i = \sum_{j \leq i < j+d} g_i = 0$ *for all integers* $j \geq 0$*.*

(b) The conclusion (a)(i) would be false if in the hypothesis $h + k - d$ *were replaced by any smaller integer.*

Remarks. Although Theorem 3 is superficially similar to that of Fine & Wilf (Theorem 2), it is not trivially implied by Theorem 2. This can be seen most easily by comparing the generalization to three or more periods found in Section 3 to the work of Castelli, Mignosi, and Restivo [2] on Fine & Wilf for three periods: for three periods h_1, h_2, h_3 their periodicity bound is approximately $\frac{1}{2}(h_1 + h_2 + h_3)$ while ours is approximately $h_1 + h_2 + h_3$. (Also see Justin [8].)

Theorem 3 is reminiscent of some classical theorems on trigonometric polyno-mials. For example, Fejér [5] proved that a real trigonometric polynomial with 0 constant term

$$\lambda_1 \cos\theta + \mu_1 \sin\theta + \lambda_2 \cos(2\theta) + \mu_2 \sin(2\theta) + \cdots + \lambda_r \cos(r\theta) + \mu_r \sin(r\theta)$$

cannot have the same sign for all real θ unless it is identically zero. Also see Pólya and Szegö [10, pp. 80, 263] and Gilbert and Smyth [7].

2   Proof of Theorem 3

Proof. Let $d = \gcd(h,k)$. First, observe that any periodic sequence $(b_n)_{n\geq0}$ of period p can be written in the form

$$b_n = \sum_{\omega^p=1} c_\omega\,\omega^n$$

where the sum is over all the complex p'th roots of unity.

Thus there are constants c_i and d_j such that

$$f_n = \sum_{0\leq i<h} c_i\omega_h^{in};$$

$$g_n = \sum_{0\leq j<k} d_j\omega_k^{jn};$$

for all $n \geq 0$, where ω_p is a primitive p'th root of unity.

Let

$$r = [r_1, r_2, \ldots, r_m]$$
$$= [1, \omega_h, \omega_h^2, \ldots, \omega_h^{h-1}, 1, \omega_k, \omega_k^2, \ldots, \omega_k^{k-1}],$$

where $m = h + k$.

Let $B = h + k - d$, and define a $B \times m$ Vandermonde-type matrix $T = (t_{i,j})$ by

$$t_{i,j} = r_j^i$$

for $0 \leq i < B$, $1 \leq j \leq m$. Define a column vector

$$v = [c_0, c_1, \ldots, c_{h-1}, -d_0, -d_1, \ldots, -d_{k-1}]^T.$$

Then the hypothesis (1) of the theorem states that $Tv \leq 0$, and the desired conclusion in part (a)(i) is equivalent to $Tv = 0$.

Now, since the period of f sums to 0, we have

$$0 = \sum_{0\leq n<h} f_n = \sum_{0\leq n<h}\sum_{0\leq i<h} c_i\omega_h^{in} = \sum_{0\leq i<h} c_i \sum_{0\leq n<h} \omega_h^{in} = hc_0,$$

and so it follows that
$$c_0 = 0. \tag{2}$$

Similarly, we have $d_0 = 0$.

Now some of the elements of r will appear more than once in r, and hence some of the columns of T are identical. Let T' be the matrix obtained from T by deleting the extra identical columns. Then T' is a $B \times B$ matrix. Let v' be the vector obtained from v by summing the entries corresponding to the identical columns of T. Then $Tv = T'v'$.

Let T'' be the $B \times (B-1)$ matrix obtained from T' by deleting the single column of 1's. Let v'' be the vector obtained from v' by deleting the entry corresponding to the column of 1's in T', which by (2) must be 0. Then we have $Tv = T''v''$.

Define
$$y(X) = \frac{(X^h - 1)(X^k - 1)}{(X^d - 1)(X - 1)} = \sum_{0 \le i < B} a_i X^i.$$

We now claim that all the coefficients a_i of y are all strictly positive. To see this, note that

$$y(X) = \frac{X^h - 1}{X^d - 1} \cdot \frac{X^k - 1}{X - 1}$$
$$= (1 + X^d + X^{2d} + \cdots + X^{h-d})(1 + X + X^2 + \cdots + X^{k-1}).$$

Now y is a polynomial of degree $h + k - d - 1$. If $i < h$, write $i = qd + r$ where $0 \le r < d$, and choose the term X^{qd} from the left factor and X^r from the right factor to show $a_i > 0$. If $h \le i < h + k - d$, choose X^{h-d} from the left factor and X^{i-h+d} from the right factor to show $a_i > 0$.

Now let
$$z = [a_0 \; a_1 \; a_2 \; \cdots \; a_{B-1}].$$

Then $zT'' = 0$, since multiplying z by column i of T'' is just evaluating the polynomial $y(X)$ at its root r_i.

Suppose there exists a column vector v such that $Tv \le 0$ but $Tv \ne 0$. Then from above there exists v'' such that $T''v'' \le 0$ but $T''v'' \ne 0$. But since the entries of z are all strictly positive we would have $zT''v'' \le 0$ and $zT''v'' \ne 0$. But $zT'' = 0$, a contradiction. This proves (a)(i).

Now we prove (a)(ii). Since $f_n = g_n$ for all $n \ge 0$, it follows that f is periodic of period h and k, and hence by Theorem 1, of period d.

Then $f_j + f_{j+1} + \cdots + f_{j+d-1}$ is just a cyclic permutation of $f_0 + f_1 + \cdots + f_{d-1}$, which equals 0. A similar argument applies to g.

We now turn to the proof of part (b). To prove the claim we will construct, for each pair of integers (h, k) with $0 < h \le k$ and $(h, k) \ne (1, 1)$, strings $v_{h,k}$ and $w_{h,k}$ of length h and k respectively, over the alphabet $\{-1, 0, +1\}$. The intent is that the terms of $v_{h,k}$ represent the values of $f_0, f_1, \ldots, f_{h-1}$ and the terms of $w_{h,k}$ represent the values of $g_0, g_1, \ldots, g_{k-1}$.

We introduce some notation. First, we let an overbar denote negation, so that $\overline{1} = -1$ (and $\overline{0} = 0$). Second, we define a function $A(x)$ as follows: if $x = a_1 a_2 \cdots a_i$ for some $i \geq 1$, then $A(x) = (a_1 + 1)a_2 \cdots a_i$. Thus A has the effect of adding one to the first term of its argument. Similarly we have $A^{-1}(x) = (a_1 - 1)a_2 \cdots a_i$.

The definition of $v_{h,k}$ and $w_{h,k}$ is as follows: write $k = qh+r$, with $0 \leq r < k$. Then we have

$$v_{h,k} = \begin{cases} 0^h, & \text{if } r = 0; \\ -w_{r,h}, & \text{otherwise.} \end{cases}$$

$$w_{h,k} = \begin{cases} 1\,0^{k-2}\,\overline{1}, & \text{if } r = 0; \\ A(-w_{r,h})\,(-w_{r,h})^{q-1}\,A^{-1}(-v_{r,h}), & \text{otherwise.} \end{cases}$$
(3)

Figure 1 gives some examples of $v_{h,k}$ and $w_{h,k}$:

h	k	$v_{h,k}$	$w_{h,k}$
1	2	0	$1\overline{1}$
2	2	00	$1\overline{1}$
2	3	$\overline{1}1$	$01\overline{1}$
2	4	00	$100\overline{1}$
3	3	000	$10\overline{1}$
3	5	$0\overline{1}1$	$1\overline{1}10\overline{1}$
4	6	$\overline{1}001$	$0001\overline{1}0$
5	8	$\overline{1}1\overline{1}01$	$01\overline{1}01\overline{1}1\overline{1}$
5	12	$0\overline{1}1\overline{1}1$	$1\overline{1}1\overline{1}10\overline{1}1\overline{1}10\overline{1}$
8	13	$0\overline{1}10\overline{1}1\overline{1}1$	$1\overline{1}10\overline{1}1\overline{1}10\overline{1}10\overline{1}$

Fig. 1. Some sample values of $v_{h,k}$ and $w_{h,k}$

As it turns out, $v_{h,k}$ and $w_{h,k}$ are closely related to the so-called standard Sturmian words [9], which can be defined as follows:

$$\sigma(h,k) = \begin{cases} 0, & \text{if } (h,k) = (0,1); \\ 0^{k-1}1, & \text{if } h = 1; \\ \sigma(r,h)^q\,\sigma(r',r), & \text{if } h > 1 \text{ and } k = qh + r,\ h = q'r + r'. \end{cases}$$
(4)

We introduce some notation dealing with continued fractions. For any rational number u, express u as a finite simple continued fraction $u = [a_0, a_1, a_2, \ldots, a_n]$, with the additional restriction that $a_n \neq 1$ if $u \in \mathbb{Q} - \mathbb{Z}$. Then we define the length of u, $\ell(u)$, to be n. Note that $\ell(u) = 0$ if $u \in \mathbb{Z}$, and further that $\ell(u/v) = \ell((v \bmod u)/u)$ for integers $0 < u < v$.

The standard Sturmian words possess a number of beautiful properties, including the following [9]:

Theorem 4. *Suppose* $0 < h \leq k$ *and* $\gcd(h, k) = 1$. *Write* $k = qh + r$. *Then* $\sigma(h, k)\sigma(r, h)$ *coincides with* $\sigma(r, h)\sigma(h, k)$ *except for the last two symbols. If* $\ell(h, k)$ *is even, then* $\sigma(h, k)\sigma(r, h)$ *ends in* 01, *while* $\sigma(r, h)\sigma(h, k)$ *ends in* 10, *while if* $\ell(h, k)$ *is odd, the situation is reversed.*

Lemma 5. *Let* h, k *be integers with* $0 < h \leq k$. *Then*

$$v_{h,k} \in (0^* \; \bar{1} \; 0^* \; 1)^* \; 0^*;$$
$$w_{h,k} \in (0^* \; 1 \; 0^* \; \bar{1})^* \; 0^*.$$

Proof. By induction on $\ell(h/k)$. If $\ell(h/k) = 0$, then we have $v_{h,k} = 0^h$ and $w_{h,k} = 1\;0^{k-2}\;\bar{1}$.

Otherwise, assume the result is true for $\ell(h/k) = n$. We prove it for $\ell(h/k) = n + 1$. Write $k = qh + r$. Then $\ell(r/h) = n$, and $w_{r,h} \in (0^* \; 1 \; 0^* \; \bar{1})^* \; 0^*$ by induction. Hence $v_{h,k} = -w_{r,h} \in (0^* \; \bar{1} \; 0^* \; 1)^* \; 0^*$, as desired.

Also $v_{r,h} \in (0^* \; \bar{1} \; 0^* \; 1)^* \; 0^*$ by induction, so

$$
\begin{aligned}
w_{h,k} &= A(-w_{r,h}) \; (-w_{r,h})^{q-1} \; A^{-1}(-v_{r,h}) \\
&\in (1 \; 0^* \; \bar{1} \; 0^* \; 1 + 0^* \; 1)(0^* \; \bar{1} \; 0^* \; 1)^* \; 0^*((0^* \; \bar{1} \; 0^* \; 1)^* 0^*)^{q-1} \\
&\quad (\bar{1} \; 0^* \; 1 \; 0^* \; \bar{1} + 0^* \; \bar{1})(0^* \; 1 \; 0^* \; \bar{1})^* 0^* \\
&\subseteq (0^* \; 1 \; 0^* \; \bar{1})^* \; 0^*,
\end{aligned}
$$

as desired. $\qquad\square$

We also define the sum map s and the running sum map S as follows: if $x = b_1 b_2 \cdots b_i$, then $s(x) = b_1 + b_2 + \cdots + b_i$ and $S(x) = b_1(b_1 + b_2) \cdots (b_1 + b_2 + \cdots + b_i)$.

Lemma 6. *Let* $0 \leq h \leq k$ *with* $(h, k) \neq (1, 1)$. *We have*

(a) $s(v_{h,k}) = s(w_{h,k}) = 0$;
(b) $|v_{h,k}| = h$;
(c) $|w_{h,k}| = k$.

Proof. By induction on $\ell(h/k)$. $\qquad\square$

We introduce one more piece of notation: τ is a coding which maps $0 \to 1$ and $1 \to 0$.

Lemma 7. *Let* $x \in (0^* \; 1 \; 0^* \; \bar{1})^* 0^*$. *Then* $\tau(S(x)) = S(A(-x))$.

Proof. If $x \in (0^* \ 1 \ 0^* \ \bar{1})^* 0^*$, then we can write

$$x = 0^{a_1} \ 1 \ 0^{b_1} \ \bar{1} \ 0^{a_2} \ 1 \ 0^{b_2} \ \bar{1} \cdots 0^{a_n} \ 1 \ 0^{b_n} \ \bar{1} \ 0^{a_{n+1}}$$

for some non-negative integers $n, a_1, a_2, \ldots, a_n, a_{n+1}, b_1, b_2, \ldots, b_n$. Hence

$$S(x) = 0^{a_1} \ 1^{b_1+1} \ 0^{a_2+1} \ 1^{b_2+1} \cdots 0^{a_n+1} \ 1^{b_n+1} \ 0^{a_{n+1}+1}$$

and

$$\tau(S(x)) = 1^{a_1} \ 0^{b_1+1} \ 1^{a_2+1} \ 0^{b_2+1} \cdots 1^{a_n+1} \ 0^{b_n+1} \ 1^{a_{n+1}+1}.$$

On the other hand, we have

$$A(-x) = \begin{cases} 1 \ 0^{a_1-1} \ \bar{1} \ 0^{b_1} \ 1 \ 0^{a_2} \ \bar{1} \ 0^{b_2} \ 1 \cdots 0^{a_n} \ 1 \ 0^{b_n} \ \bar{1} \ 0^{a_{n+1}}, & \text{if } a_1 > 0; \\ 0^{b_1+1} \ 1 \ 0^{a_2} \ \bar{1} \ 0^{b_2} \ 1 \cdots 0^{a_n} \ 1 \ 0^{b_n} \ \bar{1} \ 0^{a_{n+1}}, & \text{if } a_1 = 0; \end{cases}$$

and so

$$S(A(-x)) = \begin{cases} 1^{a_1} \ 0^{b_1+1} \ 1^{a_2+1} \ 0^{b_2+1} \cdots 0^{a_n+1} \ 1^{b_n+1} \ 0^{a_{n+1}}, & \text{if } a_1 > 0; \\ 0^{b_1+1} \ 1^{a_2+1} \ 0^{b_2+1} \cdots 0^{a_n+1} \ 1^{b_n+1} \ 0^{a_{n+1}}, & \text{if } a_1 = 0. \end{cases}$$

$\square$

We introduce a family of morphisms, φ_d, for integers $d \geq 1$. We define $\varphi_d(0) = 0^d$ and $\varphi_d(1) = 0^{d-1}1$.

Theorem 8. *Suppose* $0 < h \leq k$ *with* $\gcd(h, k) = d$. *Then*

$$S(w_{h,k}) = \tau^{\ell(h/k)}(\varphi_d(\sigma(h/d, k/d))).$$

Proof. By induction on $\ell(h/k)$.

If $\ell(h/k) = 1$ then, using Eq. (4), we find

$$\begin{aligned} \tau(\varphi_d(\sigma(h/d, k/d))) &= \tau(\varphi_d(0^{k/d} - 1)) \\ &= \tau(0^{k-d} \ 0^{d-1} \ 1) \\ &= 1^{k-1} \ 0. \end{aligned}$$

On the other hand, $w_{h,k} = 1 \ 0^{k-2} \ \bar{1}$. Hence $S(w_{h,k}) = 1^{k-1}0$.
If $\ell(h/k) = 2$ then write $k = qh + r$ with $r = d$. Then

$$\begin{aligned} \tau^2(\varphi_d(\sigma(h/d, k/d))) &= \varphi_d(\sigma(h/d, k/d)) \\ &= \varphi_d(\sigma(r/d, h/d)^q \ \sigma(0, r/d)) \\ &= \varphi_d((0^{h/d-1} \ 1)^q \ 0) \\ &= (0^{h-d} \ 0^{d-1} \ 1)^q \ 0^d \\ &= (0^{h-1} \ 1)^q \ 0^d. \end{aligned}$$

On the other hand,

$$w_{h,k} = A(-w_{d,h}) \, (-w_{d,h})^{q-1} \, A^{-1}(-v_{d,h})$$
$$= A(\bar{1} \, 0^{h-2} \, 1)(\bar{1} \, 0^{h-2} \, 1)^{q-1} A^{-1}(0^d)$$
$$= 0^{h-1} \, 1(\bar{1} \, 0^{h-2}1)^{q-1} \, \bar{1} \, 0^{d-1}$$

so that $S(w_{h,k}) = (0^{h-1}1)^q \, 0^d$.

Now assume $n > 2$ and the result is true for $\ell(h/k) < n$; we prove it for $\ell(h/k) = n$. Write $k = qh + r$, $h = q'r + r'$. Then

$$S(w_{h,k}) = S(A(-w_{r,h})(-w_{r,h})^{q-1}A^{-1}(w_{r',r})) \quad \text{(by Eq. (3))}$$
$$= S(A(-w_{r,h}))^q S(w_{r',r})$$
$$= \tau(S(w_{r,h}))^q \, S(w_{r',r}) \quad \text{(by Lemma 7)}$$
$$= \tau(\tau^{\ell(r/h)}(\varphi_d(\sigma(r/d, k/d))))^q \tau^{\ell(r'/r)}(\varphi_d(\sigma(r'/d, r/d))) \quad \text{(by induction)}$$
$$= \tau^{\ell(h/k)}(\varphi_d(\sigma(r/d, k/d)^q \sigma(r'/d, r/d)))$$
$$= \tau^{\ell(h/k)}(\varphi_d(\sigma(h/d, k/d))).$$

This completes the proof. □

Corollary 9. *Let* $0 < h \le k$, *and* $d = \gcd(h, k)$. *Then* $D_1 := w_{h,k}A(v_{h,k})$
coincides with $D_2 := A(v_{h,k})A^{-1}(w_{h,k})$ *except for the last* $d + 1$ *symbols.*

(a) If $\ell(h, k)$ *is even, then* D_1 *ends in* $0^d \, 1$ *while* D_2 *ends in* $1 \, \bar{1} \, 0^{d-1}$.
(b) If $\ell(h, k)$ *is odd, then* D_1 *ends in* $\bar{1} \, 1 \, 0^{d-1}$ *while* D_2 *ends in* $0^d \, \bar{1}$.

Proof. We prove only the case where $\ell(h, k)$ is even. The case where $\ell(h, k)$ is odd is similar and is left to the reader.

By Theorem 4 we know that

$$E_1 := \sigma(h/d, k/d)\sigma(r/d, k/d)$$

coincides with

$$E_2 := \sigma(r/d, h/d)\sigma(h/d, k/d)$$

except that E_1 ends in 01 while E_2 ends in 10. It follows that $\varphi_d(E_1)$ coincides with $\varphi_d(E_2)$ except that $\varphi_d(E_1)$ ends in $0^{2d-1} \, 1$ while $\varphi_d(E_2)$ ends in $0^{d-1} \, 1 \, 0^d$.

By Theorem 8 we know

$$\varphi_d(E_1) = S(w_{h,k})\tau(S(w_{r,h}));$$
$$\varphi_d(E_2) = \tau(S(w_{r,h}))S(w_{h,k}).$$

By Lemma 7 we have $\tau(S(w_{r,h})) = S(A(-w_{r,h})) = S(A(v_{h,k}))$. Hence

$$S(w_{h,k})\tau(S(w_{r,h})) = S(w_{h,k})S(A(v_{h,k})) = S(w_{h,k}A(v_{h,k})),$$

where we have used Lemma 6 (a). Similarly

$$\tau(S(w_{r,h}))S(w_{h,k}) = S(A(v_{h,k}))S(w_{h,k}) = S(A(v_{h,k})A^{-1}(w_{h,k})).$$

It now follows that $S(w_{h,k}A(v_{h,k}))$ coincides with $S(A(v_{h,k})A^{-1}(w_{h,k}))$ except that the first ends in $0^{2d-1}\,1$ while the second ends in $0^{d-1}\,1\,0^d$. Hence $w_{h,k}A(v_{h,k})$ coincides with $A(v_{h,k})A^{-1}(w_{h,k})$ except that the first ends in $0^d\,1$ while the second ends in $1\,\bar{1}\,0^{d-1}$, as desired.   □

We can now complete the proof of Theorem 3 (b). If x is a finite sequence, then by x^ω we mean the infinite sequence $xxx\cdots$. We will show that if $(f_n)_{n\geq 0} = v_{h,k}^\omega$ and $(g_n)_{n\geq 0} = w_{h,k}^\omega$, then $f_i \leq g_i$ for $0 \leq i < h+k-d-1$, while $f_{h+k-d-1} > g_{h+k-d-1}$.

For $\ell(h/k) = 1$ we have $k = qh$, so $d = h$. Then $v_{h,k} = 0^h$ and $w_{h,k} = 1\,0^{k-2}\,\bar{1}$, and so $f_i \leq g_i$ for $0 \leq i < k-1 = h+k-d-1$, while $f_{k-1} = 0 > g_{k-1} = -1$.

Now assume $\ell(h/k) = 2$. Write $k = qh+r$. The following diagram illustrates a prefix of $v_{h,k}^\omega$ compared to $w_{h,k}^\omega$:

$$
\begin{aligned}
v_{h,k}^\omega &= \quad v_{h,k} \quad \overbrace{v_{h,k}\ v_{h,k}\ \cdots\ v_{h,k}}^{q-1} \quad v_{h,k} \qquad v_{h,k} \quad \cdots \\
w_{h,k}^\omega &= A(v_{h,k})\ \overbrace{v_{h,k}\ v_{h,k}\ \cdots\ v_{h,k}}^{q-1}\ A^{-1}(-v_{r,h})\ A(v_{h,k})\ \cdots
\end{aligned}
\tag{5}
$$

From this picture we see that $f_0 < g_0$, and $f_i = g_i$ for $1 \leq i < qh$. For the range $qh \leq i \leq h+k+d-1$, we must compare $v_{h,k}v_{h,k}$ to $A^{-1}(-v_{r,h})A(v_{h,k})$.

However, since $v_{h,k} = -w_{r,h}$, and $w_{r,h}$ begins with $A(v_{r,h})$, it suffices to compare $F_1 := v_{h,k}A^{-1}(-v_{r,h})$ to $F_2 := A^{-1}(-v_{r,h})A(v_{h,k})$.

From Corollary 9, it follows that

(a) if $\ell(h,k)$ is even, then F_1 and F_2 coincide, except that F_1 ends in $1\,\bar{1}\,0^{d-1}$ while F_2 ends in $0^d\,1$;
(b) if $\ell(h,k)$ is odd, then F_1 and F_2 coincide, except that F_1 ends in $0^d\,\bar{1}$ while F_2 ends in $\bar{1}\,1\,0^{d-1}$.

Thus $f_i \leq g_i$ for $0 \leq i < h+k-d-1$, but $f_{h+k-d-1} > g_{h+k-d-1}$. This completes the proof of Theorem 3.

Example. Suppose $h = 8$ and $k = 13$. Then from Figure 1 we have

n	0	1	2	3		4	5		6	7	8		9	10	11		12	13		14	15	16		17	18		19
f_n	1	-1	1	0		-1	1		-1	1	0		-1	1	0		-1	1		-1	1	0		-1	1		0
g_n	1	-1	1	0		-1	1		-1	1	0		-1	2	0		-1	1		-1	1	0		-1	1		-1

so we find $f_n \leq g_n$ for $0 \leq n < 19$.

There is another construction for proving Theorem 3 (b), under the additional restriction that $h \neq k$. This construction gives strict inequality for $h + k - \gcd(h, k) - 1$ consecutive terms. We describe it next.

First, we need a slight generalization of the standard Sturmian words to the case where $\gcd(h, k) > 1$. To do so, we modify the construction given in Eq. (4) as follows:

$$
\sigma(h, k) = \begin{cases} 0, & \text{if } h = 0; \\ 0^{k-1}1, & \text{if } h \mid k; \\ \sigma(r, h)^q \, \sigma(r', r), & \text{if } h > 1 \text{ and } k = qh + r, \, h = q'r + r'. \end{cases}
$$

It is now not difficult to see that a periodic function with period $\sigma(h, k)$ and a periodic function of period $\sigma(r, h)$ (with $k = qh + r$) coincide for the first $h + k - \gcd(h, k) - 1$ terms, but disagree at term $h + k - \gcd(h, k)$.

Now let A and B be two integers to be specified later, and define two codings τ_0 and τ_1 as follows: $\tau_0(0) = A$, $\tau_0(1) = B$, $\tau_1(0) = A + 1$, $\tau_1(1) = B + 1$.

Theorem 10. *There exist choices for A and B such that if $k = qh + r$, and $0 < h < k$, then $f := \tau_0(\sigma(r, h))$ and $g := \tau_1(\sigma(h, k))$ both have periods that sum to 0 and satisfy $f_i < g_i$ for $0 \le i < h + k - \gcd(h, k) - 1$, but not at $i = h + k - \gcd(h, k) - 1$.*

Proof. First, we observe that the number of 1's in $\sigma(h, k)$ is is equal to u, where $hu/d \equiv (-1)^{\ell(h,k)+1} \pmod{k/d}$, $0 \le u < k/d$, and $d = \gcd(h, k)$. See, for example, [1]. (Strictly speaking this was proved only for the case $d = 1$ but the proof in the more general case is not difficult.) Similarly, the number of 1's in $\sigma(r, h)$ is equal to t, where $th/d \equiv (-1)^{\ell(r,h)+1} \pmod{h/d}$ and $0 \le t < h/d$. It follows that the sum of the period for $\tau_0(\sigma(r, h))$ is $A(h - t) + Bt$, and the sum of the period for $\tau_1(\sigma(h, k))$ is $(A + 1)(k - u) + (B + 1)u$. We want both these sums to be 0. Such A and B exist provided

$$
\det \begin{bmatrix} h - t & t \\ k - u & u \end{bmatrix} \ne 0.
$$

In other words, we want $(h - t)u - (k - u)t = hu - kt \ne 0$. It suffices to show that $(h/d)u - (k/d)t$ is nonzero. Consider this expression modulo k/d; we get $(h/d)u \pmod{k/d}$. But from above this is ± 1, and hence $\ne 0$ provided $k/d > 1$. But since $0 < h < k$, we have $k = d$ if and only if $h = k$. Since $h < k$, the required A and B exist.

Now, by the construction, f and g have periods that sum to 0, and $f_i < g_i$ for $0 \le i < h + k - \gcd(h, k) - 1$, but $f_i > g_i$ for $i = h + k - \gcd(h, k) - 1$. $\square$

Example.

Take $h = 5$, $k = 8$. Then we find $\sigma(r, h) = 01001$ and $\sigma(h, k) = 01001010$. Solving the linear system $3A + 2B = 0$, $5(A+1) + 3(B+1) = 0$, we find $A = -15$, $B = 24$. Thus if f is the periodic sequence $(-16, 24, -16, -16, 24)^\omega$ and g is the periodic sequence $(15, 25, -15, -15, 25, 15, -25, 15)^\omega$, we see from the following table

n	0	1	2	3	4	5	6	7	8	9	10	11
f_n	−16	24	−16	−16	24	−16	24	−16	−16	24	−16	24
g_n	−15	25	−15	−15	25	−15	25	−15	−15	25	−15	−15

that $f_n < g_n$ for $0 \le n \le 10$, but not at $n = 11$.

3   More than Two Periods

In this section we consider results analogous to Theorem 3 for more than two periods.

Theorem 11. *Let $(f_1(n))_{n\geq 0}$, $(f_2(n))_{n\geq 0}$, ..., $(f_r(n))_{n\geq 0}$ be r periodic real-valued sequences of periods $h_1, h_2, \ldots, h_r$, respectively. Suppose that the periods of the f_j sum to 0, that is, for all j, $1 \leq j \leq r$, we have*

$$\sum_{0\leq n<h_j} f_j(n) = 0.$$

If

$$\sum_{1\leq j\leq r} f_j(n) \leq 0$$

for $0 \leq n < h_1 + h_2 + \cdots + h_r - r + 1$, then

$$\sum_{1\leq j\leq r} f_j(n) = 0$$

for all $n \geq 0$.

Proof. The proof is very similar to the proof of Theorem 3, and we indicate only what needs to be changed. The argument goes through as stated, except that we define the polynomial y as follows:

$$y(X) = \prod_{1\leq i\leq r} \frac{X^{h_i} - 1}{X - 1}.$$

Then $\deg y = h_1 + h_2 + \cdots + h_r - r$, and the coefficients of y are strictly positive. □

We note that the bound $h_1 + h_2 + \cdots + h_r - r + 1$ is not, in general, optimal, although the bound is optimal if the period lengths $h_1, h_2, \ldots, h_r$ are relatively prime. The optimal bound can be computed using the following algorithm.

Define $s(X) = \prod_{1\leq i\leq r}(X^{h_i} - 1)$ and

$$t(X) = \frac{s(X)}{(X - 1)\gcd(s(X), s'(X))}.$$

Let $a_0, a_1, \ldots, a_m$ be defined such that $t(X) = \sum_{0\leq i\leq m} a_i X^i$, with $m = \deg t$. Let $v_0, v_1, \ldots$ be defined as follows:

$$v_0 = [a_0, ..., a_m, 0, ...]$$
$$v_1 = [0, a_0, ..., a_m, 0, ...]$$
$$v_2 = [0, 0, a_0, ..., a_m, 0, ...].$$

$$\vdots$$

Let k be the largest integer such that the orthogonal complement of $V_k :=$ $\{v_0, \ldots, v_k\}$ intersects the non-negative quadrant nontrivially in the first $m+k+1$ coordinates. Then the bound is $m + k + 2 = (\deg t) + k + 2$.

Example. Consider the case where $h_1 = 6$, $h_2 = 10$, and $h_3 = 15$. Then

$$t(X) = X^{21} + X^{19} + X^{18} + X^{17} + 2X^{16} + X^{15} + 2X^{14} + 2X^{13} + 2X^{12} +$$
$$2X^{11} + 2X^{10} + 2X^9 + 2X^8 + 2X^7 + X^6 + 2X^5 + X^4 + X^3 + X^2 + 1.$$

Then we have $m = 21$ and $k = 0$. Hence if $\sum_{1 \le j \le r} f_j(n) \le 0$ for $0 \le n < 23$, then $\sum_{1 \le j \le r} f_j(n) = 0$ for all $n \ge 0$. However, the following is an example of three periodic sequences of period lengths 6, 10, and 15, respectively, whose periods sum to 0 and such that the sum is ≤ 0 for the first 22 terms:

$$f_1 = (0,0,0,-1,1,0)^\omega;$$
$$f_2 = (0,0,0,0,0,1,-1,0,-1,1)^\omega;$$
$$f_3 = (0,0,0,1,-1,-1,1,0,1,0,-1,0,0,0,0)^\omega.$$

References

1. J. Berstel. Sturmian words. In M. Lothaire, editor, *Algebraic Combinatorics on Words*. Cambridge University Press, to appear, 2001. Preliminary version available at http://www-igm.univ-mlv.fr/~berstel/Lothaire/ 512, 521
2. M. G. Castelli, F. Mignosi, and A. Restivo. Fine and Wilf's theorem for three periods and a generalization of Sturmian words. *Theoret. Comput. Sci.* **218** (1999), 83–94. 513
3. C. Choffrut and J. Karhumäki. Combinatorics of Words. In G. Rozenberg and A. Salomaa, eds., *Handbook of Formal Languages*, 1997, pp. 329–438. 512
4. M. Crochemore and W. Rytter. *Text Algorithms*. Oxford University Press, 1994. 512
5. L. Fejér. Sur les polynomes harmoniques quelconques. *C. R. Acad. Sci. Paris* **157** (1913), 506–509. 514
6. N. J. Fine and H. S. Wilf. Uniqueness theorems for periodic functions. *Proc. Amer. Math. Soc.* **16** (1965), 109–114. 513
7. A. D. Gilbert and C. J. Smyth. Zero-mean cosine polynomials which are non-negative for as long as possible. Preprint, 2000. 514
8. J. Justin. On a paper by Castelli, Mignosi, Restivo. *Theoret. Inform. Appl.* **34** (2000), 373–377. 513
9. D. E. Knuth, J. Morris, and V. Pratt. Fast pattern matching in strings. *SIAM J. Comput.* **6** (1977), 323–350. 516, 517
10. G. Pólya and G. Szegö. *Problems and Theorems in Analysis II*. Springer-Verlag, 1976. 514
11. A. Salomaa. *Formal Languages*. Academic Press, 1973. 512

Upper Bounds on the Bisection Width of 3- and 4-Regular Graphs

Burkhard Monien and Robert Preis

Department of Mathematics and Computer Science, University of Paderborn
D-33098 Paderborn, Germany
{bm,robsy}@uni-paderborn.de

Abstract. We derive new upper bounds on the bisection width of graphs which have a regular vertex degree. We show that the bisection width of large 3-regular graphs with $|V|$ vertices is at most $\frac{1}{6}|V|$. For the bisection width of large 4-regular graphs we show an upper bound of $\frac{2}{5}|V|$.

Keywords: graph partitioning; bisection width; regular graphs; local improvement;

1 Introduction

There are graph-partitioning problems in a wide range of applications. The task is to divide the set of vertices of a graph equally into a given number of parts while keeping the number of crossing edges between vertices belonging to different parts, called the *cut size* of the partition, as small as possible. The special case of a partition of the graph into 2 parts is called a *bisection*, and the minimal cut size of all balanced bisections of a graph is called its *bisection width*. Its calculation is **NP**-complete for arbitrary graphs [11] and remains **NP**-complete for regular graphs [5].

There are several results on bounds on the bisection width of regular graphs (discussed below). Results for 3- and 4-regular graphs are of special interest because these are the lowest non-trivial degrees. Some previous results for small degrees have been generalized to results for larger degrees.

It is a general theoretical interest to improve previous upper bounds on 3- and 4-regular graphs. Moreover, there are some direct applications of these results. As a motivating example, upper bounds on the bisection width of 4-regular graphs have successfully been applied to the configuration of transputer systems [13].

Definitions and Previous Results Let $G = (V, E)$ be a simple undirected graph with vertex set V of cardinality $n := |V|$ and edge set E. A graph is d-regular if for all $v \in V$ it is $|\{w \in V; \{v, w\} \in E\}| = d$. Let $\pi : V \rightarrow \{0, 1\}$ be a **bisection** of G. It distributes the vertices among parts V_0 and V_1. We focus on **balanced** bisections, i. e. the number of vertices in the parts differ by at most 1. Let $cut(\pi) := |\{\{v, w\} \in E; \pi(v) \neq \pi(w)\}|$ be the **cut size** of π. The **bisection width** of a graph G is $bw(G) := \min\{cut(\pi); \pi \text{ is a balanced bisection of G}\}$.

J. Sgall, A. Pultr, and P. Kolman (Eds.): MFCS 2001, LNCS 2136, pp. 524–536, 2001.
© Springer-Verlag Berlin Heidelberg 2001

The bisection width is known for some graph classes with regular degree such as tori, cube-connected-cycles [18] or butterflies [4].

There are several results on bounds on the bisection width of arbitrary regular graphs. Clark and Entringer [7] present an upper bound of $\frac{n+138}{3}$ for the bisection width of 3-regular graphs. Kostochka and Melnikov improve this asymptotically and show an upper bound of $\frac{n}{4} + O(\sqrt{n}\log n)$ [15]. Recently, an upper bound of $0.198n + O(\log(n))$ has been proved in [24]. Hromkovic and Monien [13] proved an upper bound of $\frac{n}{2}+1$ for the bisection width of 4-regular graphs with $n \geq 350$. A general upper bound of $\frac{n}{2}+5$ for 4-regular graphs with any number of vertices is proven in [24]. The result of [15] for 3-regular graphs above is a corollary of an upper bound of $\frac{d-2}{4}n + O(d\sqrt{n}\log n)$ for the bisection width of d-regular graphs in the same paper. An upper bound of $\frac{d-2}{4}n + 1$ for $n \geq n_0(d)$ with some function $n_0(d)$ is shown in [20,21] by generalizing the techniques of [13]. Alon [1] uses probabilistic arguments to show that the bisection width is at most $(\frac{d}{2} - \frac{3\sqrt{d}}{16\sqrt{2}})\frac{n}{2}$ for d-regular graphs with $n > 40d^9$.

Bollobas [3] shows that for $d \to \infty$ the bisection width of almost every d-regular graph is at least $(\frac{d}{2} - \sqrt{\ln(2) \cdot d})\frac{n}{2}$. For $d = 4$ he shows that almost all 4-regular graphs have a bisection width of at least $\frac{11}{50}n = 0.22n$. Furthermore, Kostochka and Melnikov show that almost every 3-regular graph has a bisection width of at least $\frac{1}{9.9}n \approx 0.101n$ [16]. There are some (slightly weaker) results for explicitly constructible infinite graph classes with high bisection width. The *Ramanujan Graphs* (see e. g. [6,17,19,22]) have a regular degree d and a bisection width of at least $(\frac{d}{2} - \sqrt{d-1})\frac{n}{2}$. This value is derived by the use of the well-known spectral lower bound $\frac{\lambda_2 \cdot n}{4}$ with λ_2 being the second smallest eigenvalue of the Laplacian of the graph (cf. [10]). This implies lower bounds of $0.042n$ and $0.133n$ for the bisection widths of 3-regular and 4-regular Ramanujan graphs. The spectral lower bound has been improved in [2] to a lower bound of $0.082|V|$ for the bisection width of large 3-regular Ramanujan graphs and a lower bound of $0.176|V|$ for the bisection width of large 4-regular Ramanujan graphs.

There are many heuristics for graph partitioning which are successfully being used in applications. Furthermore, efficient software implementations of the most relevant methods are available by using software tools like e. g. CHACO [12], JOSTLE [25], METIS [14], SCOTCH [23] or PARTY [24]. These heuristics try to calculate a bisection with a small cut size. However, they do not guarantee an approximation of the bisection width. Recently, it has been shown that the bisection width can be approximated by a polynomial time algorithm within a factor of $O(\log^2(|V|))$ [9].

New Results and Outline of the Paper In this paper we improve previous upper bounds on the bisection width of large 3- and 4-regular graphs. In Section 2 we prove for any $\epsilon > 0$ an upper bound of $(\frac{1}{6} + \epsilon)n$ on the bisection width of large 3-regular graphs. We are able to prove an upper bound of $(\frac{2}{5} + \epsilon)n$ on the bisection width of large 4-regular graphs. This proof is omitted in this version of the paper due to space limitations and in favour of a detailed description of the 3-regular case. The proof will be published in the full version. As discussed above, there are large 3-regular graphs with a bisection width of at least $0.101n$

and large 4-regular graphs with a bisection width of at least $0.22n$. Thus, the results are optimal up to constant factors and our results improve these factors.

Iterative Local Improvement with Helpful Sets The proofs in this paper are constructive and follow an iterative local improvement scheme. It starts with an arbitrary balanced bisection. If the cut size of it does not fulfill the stated upper bound, it performs two steps to improve the bisection as illustrated in Fig. 1. In the first step, a small set $S_0 \subset V_0$ is moved to V_1. S_0 is chosen such that this move decreases the cut size. In the second step, a set $S_1 \subset V_1 \cup S_0$ with $|S_1| = |S_0|$ is moved to V_0. S_1 is chosen such that the cut size does not increase too much, i. e. such that the increase is less than the decrease in the first step. Thus, the resulting bisection is balanced and has a smaller cut size. These steps are repeated until the cut size drops below the upper bound. The proofs in this paper ensure that there are sets S_0 and S_1 with the desired property as long as the cut size is higher than the stated upper bound.

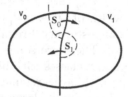

Fig. 1. One iteration of a local improvement

This local improvement scheme has successfully been used to derive upper bounds on the bisection width of 4-regular graphs [13,20]. Furthermore, it is the basis for the *Helpful-Set* heuristic which is able to calculate bisections with low cut sizes for very large graphs in a short time [8,21]. An implementation of the Helpful-Set heuristic can be found in the software tool PARTY [24].

A move of a set of vertices from one part to the other changes the cut size of the bisection. The helpfulness of the set is the amount of this change.

Definition 1. *Let* π *be a bisection of a graph* $G = (V, E)$. *For* $S \subset V_p(\pi)$, $p \in \{0, 1\}$, *let*

$$H(S) = |\{\{v, w\} \in E; v \in S, w \in V\backslash V_p(\pi)\}| - |\{\{v, w\} \in E; v \in S, w \in V_p(\pi)\backslash S\}|$$

be the **helpfulness** *of* S. S *is called* $H(S)$-**helpful.**

2   Upper Bound on the Bisection width of 3-Regular Graphs

In this section we derive a new upper bound on the bisection width of 3-regular graphs. The proof is based on the iterative local improvement scheme described in the previous section. We will use Lemma 5 for the first step of the improvement scheme and Lemma 6 for the second step. These lemmas will be used to prove the theorem at the end of this section. Lemma 3 is the main lemma of this section and may be of interest on its own. It will be used to prove Lemma 5.

Lemma 2. *Let $T = (V, E)$ be a tree with weights $w : V \rightarrow \mathbf{N}$. Let $A = (\sum_{v \in V} w(v))/(|V| - 1)$. Let $w(l) \geq A$ for all leaves l. Then there are adjacent vertices u and v with $w(u) + w(v) \leq 2 \cdot A$.*

Proof. By induction on $|V|$. The case if T is a path is trivial.

Otherwise consider an arbitrary leaf v_0. Let $v_1, \ldots, v_{k+1}$ be the path connecting v_0 to a vertex v_{k+1} of degree at least 3, i. e. the degree of v_i, $1 \leq i \leq k$, is 2. Let T' be the subtree (path) $v_0, \ldots, v_k$.

By considering the vertex pairs $\{v_1, v_2\}, \{v_3, v_4\}, \ldots, \{v_{k-1}, v_k\}$ if k is even or $\{v_0, v_1\}, \{v_2, v_3\}, \ldots, \{v_{k-1}, v_k\}$ if k is odd we deduce that we either found the desired pair or it is $w(T') \geq A \cdot |V(T')|$. But in this case the tree $T \backslash T'$ satisfies $w(T \backslash T') \leq A \cdot |V(T \backslash T')|$ and we can apply induction on $T \backslash T'$.   □

Lemma 3. *Let $G = (V, E)$, $E = B \uplus R$, be a 3-regular graph with black edges B and red edges R. Let each vertex be adjacent to at least one black edge. Let $|R| > (\frac{1}{2} + \epsilon)|V|$ for an $\epsilon > 0$. Then there is a set $S \subset V$ of size $O(\frac{1}{\epsilon^2})$ such that the number of red edges between vertices of S is larger than the number of black edges between S and $V \backslash S$.*

Proof. Let e be the number of vertices which are adjacent to 3 black edges, d_1 be the number of vertices which are adjacent to 2 black edges and d_2 be the number of vertices which are adjacent to 1 black edge, i. e. $|V| = e + d_1 + d_2$ and $2|R| = 2d_2 + d_1$. It is

$$\frac{|R|}{|V|} = \frac{d_1 + 2d_2}{2(d_1 + d_2 + e)} \ . \tag{1}$$

The fact $|V| < 2|R|$ leads us to

$$e = |V| - d_2 - d_1 < 2|R| - d_2 - d_1 = d_2 \ . \tag{2}$$

Call a set $S \subset V$ **positive** if it has more internal red edges than external black edges, **negative** if it has more external black edges than internal red edges and **neutral** if the numbers are equal.

Consider the graph consisting of black edges only and let F be the family of its connected components. Clearly, the elements of F are neutral or positive. As a simple example, a positive set $I \in F$ of size $O(\frac{1}{\epsilon})$ fulfills the lemma. In the following we manipulate the sets in F. However, the number of sets in F remains constant and the sets in F remain neutral or positive.

The number of vertices of degree 1 in a connected component exceeds the vertices of degree 3 by at most 2. Thus, for each $I \in F$ it is $e(I) \geq d_2(I) - 2$. More precisely, it is $e(I) = d_2(I) + 2r(I) - 2$ with $r(I)$ being the number of edges which can be removed from I without splitting it into disconnected components. Let $r = \sum_{I \in F} r(I)$. It is

$$e = d_2 + 2r - 2|F| \ . \tag{3}$$

Let $\delta > 0$ be a constant. The value of δ will be assigned below. A set $I \in F$ is called **small** if $|I| \leq \frac{1}{\delta}$ and **large** otherwise. Denote with $\alpha(S)$, $S \subset V$, the number of red edges between vertices of S and vertices of small sets. Call a black path P a **thin path** if it has the following property. P contains only vertices with black degree of 2, P has a maximal path length, i. e. both end vertices are adjacent via a black edge to a vertex of black degree 1 or 3 (or, as described below, to a marked vertex of degree 2) and it is $|P| \leq \frac{\alpha(P)}{\delta}$. Call a black path P a **thick path** if it has all these properties except the last one.

Let $B(I)$, $I \in F$, be the family of disconnected components of I if all red edges and thin paths of I are removed. Let $\beta(B) = |B(I)|$. Let $s(I)$ be the size of the union of I and all small sets which are connected to I via a red edge. A set $I \in F$ is called **thin** if $s(I) < \frac{\beta(I)-1}{\delta}$ and **thick** otherwise.

The outline of the proof is the following. We state an algorithm below which manipulates the elements of F. Especially, we will disconnect certain thin paths from the rest of their set. The disconnection of a thin path P will shade it and mark both vertices previously connected to the ends of P as well as mark the small sets connected to P via a red edge. We disconnect thin paths in each set $I \in F$ until there is no thin path in any cycle of I. For such a set $I \in F$ we construct the tree $T(I) = (X, Y)$ with the vertex set X being the disconnected components from $B(I)$ and with the edges connecting the components if there is an unshaded thin path between them. It is $\beta(I) = |X| = |Y| + 1$. Let the weight $w(v)$, $v \in X$, be the size of the subset represented by v.

The algorithm may terminate with a positive set of size $O(\frac{1}{\delta^2})$. Since we will set later $\delta = \frac{\epsilon}{6(1+2\epsilon)}$, we can say that any positive set of size at most $O(\frac{1}{\delta^2})$ fulfills the lemma. Otherwise, the algorithm terminates with a family F consisting of sets of the following types only.

(i) $I \in F$ is large.

(ii) $I \in F$ is small and has at least one red edge to a thick set or to a thick path.

(iii) $I \in F$ is small and has at least 2 vertex marks or there is a cycle of black edges in I.

We will finish the proof by showing that if there are only sets of these types, we get a contradiction. Thus, the algorithm was successful in finding a set fulfilling the lemma. The following algorithm manipulates the elements of F.

Step 1: If there is a positive small set $I \in F$, then I fulfills the lemma. If there is a red edge between two different small sets $I_a, I_b \in F$, then $I_a \cup I_b$ fulfills the lemma because I_a and I_b are neutral themselves and the union has an additional internal red edge. Both kind of fulfilling sets have a size of $O(\frac{1}{\delta})$.

Step 2: If there is a thin path P with $\alpha(P) \geq 3$, we show that there is a set fulfilling the lemma. We will take 3 small sets which are connected to P via a red edge and unify them with a subpath of P such that the subpath connects the 3 small sets. This union is a positive set. We will show that there are 3 small sets such that the connecting subpath is not too long, i. e. such that the size of the union is at most $O(\frac{1}{\delta})$.

P can be divided into subpaths by deleting the vertices in P which are connected to a small set via a red edge. There are $\alpha(P)$ such vertices, i. e. $\alpha(P)+1$ such (possibly empty) subpaths P_i, $1 \leq i \leq \alpha(P)+1$. Let $x_i = |P_i|$.

It is $|P| = \sum_{i=1}^{\alpha(P)+1} x_i + \alpha(P)$. In the following we do not consider the subpaths P_1 and $P_{\alpha(P)+1}$ on both ends of the path.

If $\alpha(P) = 3$, it is $x_2 + x_3 = 2\frac{1}{\alpha(P)-1}\sum_{j=2}^{\alpha(P)} x_j$. If $\alpha(P) > 3$, there is an i, $2 \leq i < \alpha(P)$ such that $x_i + x_{i+1} \leq 3\frac{1}{\alpha(P)-1}\sum_{j=2}^{\alpha(P)} x_j$. Otherwise we get $2\sum_{j=2}^{\alpha(P)} x_j = \sum_{j=2}^{\alpha(P)-1}(x_j + x_{j+1}) + x_2 + x_{\alpha(P)} > 3\frac{\alpha(P)-2}{\alpha(P)-1}\sum_{j=2}^{\alpha(P)} x_j + x_2 + x_{\alpha(P)} \geq 2\sum_{j=2}^{\alpha(P)} x_j + x_2 + x_{\alpha(P)}$, which is a contradiction.

Thus, there is an i, $2 \leq i < \alpha(P)$, with $x_i + x_{i+1} \leq 3\frac{|P|}{\alpha(P)-1} \leq 3\frac{\alpha(P)}{(\alpha(P)-1)\delta} \leq 3\frac{3}{2\delta} = O(\frac{1}{\delta})$. The union of P_i, P_{i+1}, the three vertices connecting the paths P_{i-1} with P_i, P_i with P_{i+1} and P_{i+1} with P_{i+2} and the three small sets connected to these vertices is a positive set with a size of $O(\frac{1}{\delta})$.

From now on it is $\alpha(P) \leq 2$ for each thin path P.

Step 3: If there is a thin path P in a large set $I \in F$ and there is a cycle in I which includes P, remove the edges connecting P with $I\backslash P$ and shade P. The cycle in I ensures that both vertices which are connected to the ends of P had a black degree of 3 or (as we will see below) they had a degree of 2 and were marked once.

It is $\alpha(P) \leq 2$ due to step 2. If $\alpha(P) = 1$, let $S_1 \in F$ be the small set connected to P via a red edge and assign 2 marks to S_1. If $\alpha(P) = 2$, let $S_1, S_2 \in F$ be the small sets connected to P via a red edge and assign 1 mark to S_1 and 1 mark to S_2. The vertices which were connected to both ends of P get a mark, too. Their black degree is reduced by one due to the removal of the connecting edges. Although a vertex with black degree of 2 may be generated, it is not to be taken as part of a thin or thick path.

The removal of the connecting edges changes the graph. However, only two black edges are removed which are internal to I, i. e. I remains neutral or positive. Although P is disconnected from the rest of I and shaded, we still consider it to be a part of I. Therefore, neither the value $s(I)$ nor the number of elements in F does change.

The changes of the graph in this step come into account again when a positive set S of size $O(\frac{1}{\delta})$ is found in a future part of the algorithm. S is only positive with respect to the graph with removed edges. As we will see in step 4, S may include a subset of $I\backslash P$. Especially, S may include a marked vertex v which is connected to the end of P. However, this external edge can be compensated by enlarging S with P, S_1 and, if existing, S_2. Thus, the black edge between v and P is now internal and the possible external black edge between P and the vertex connected to the other end of P is compensated by the internal edge between P and S_1. Thus, the enlarged set S is positive after adding the formerly removed edges. S is getting enlarged for each vertex mark of a vertex in S. Each enlargement adds at most $|P| + |S_1| + |S_2| \leq \frac{4}{\delta}$ vertices. All enlargements lead to a positive set S of size $O(\frac{1}{\delta^2})$ fulfilling the lemma.

Repeat step 3 until there are no such thin paths.

Step 4: Let $I \in F$ be a thin set. Let $T(I) = (X, Y)$ be the graph of I with edges Y representing the unshaded thin paths in I as described above. $T(I)$

is a tree due to step 3. For each leaf l in $T(I)$ let $L(l)$ be the union of the vertices represented by l, all adjacent (possibly shaded) thin paths and all small sets which are adjacent to these vertices via a red edge. Clearly, such a set $L(l)$ is neutral or positive. We defined the weight $w(v)$ of a vertex $v \in T$ above to be the number of vertices represented by v. We redefine the weight $w(l)$ for each leaf l in $T(I)$ to be $|L(l)|$. Call a leaf l *small* if $w(l) < \frac{1}{\delta}$. If there is a shaded thin path between two small leaves l_1 and l_2, then $L(l_1) \cup L(l_2)$ is positive of size $O(\frac{1}{\delta})$ and the lemma is fulfilled. In the remainder we can assume that there is no such path and, therefore, for the new definition of the weights it holds $\sum_{v \in X} w(v) \le s(I)$.

If there is a thin set $I \in F$ with $w(l) \ge \frac{1}{\delta}$ for each leaf l in $T(I)$, we can fulfill the lemma. It is $w(l) \ge \frac{1}{\delta} \ge \frac{s(I)}{\beta(I)-1} \ge \frac{\sum_{v \in X} w(v)}{|X|-1}$. $T(I)$ fulfills the requirements of Lemma 2. Thus, there are adjacent vertices $u, v \in T$ with

$$w(u) + w(v) \le 2 \cdot \left(\sum_{v \in X} w(v) \right) / (|X| - 1) \le 2 \cdot \frac{s(I)}{\beta(I) - 1} \le \frac{2}{\delta} \, .$$

The union S of the vertices represented by u, v and the edge between u and v has a size of $O(\frac{1}{\delta})$. However, we have to insert the edges which were removed in step 3 and 4. For each removed edge of step 4 (see below) and for each vertex mark of step 3 we enlarge S with the adjacent thin path and a small set connected to this path via a red edge. Each enlargement increases S by at most $O(\frac{1}{\delta})$ and compensates an external black edge as discussed in step 3. All enlargements together result in a positive set S of size $O(\frac{1}{\delta^2})$ and the lemma is fulfilled.

If we could not fulfill the lemma, consider all thin sets $I \in F$ which have a leaf l in $T(I)$ with $w(l) < \frac{1}{\delta}$. We will manipulate two elements of F. Let P be the thin path which connects l with its neighbor v in the tree and let S be the small set connected to P. Remove I and S from F and construct some new sets I_1 and I_2. Let I_1 be the union of the vertices represented by l, the path P and the small set S. Let I_2 be the set I reduced by P and the vertices represented by l. If $|I_2| \le \frac{1}{\delta}$, then the union $L(l) \cup I$ fulfills the lemma. Otherwise, remove the edge which connects P with v. Now I_1 and I_2 are disconnected and both are neutral or positive with I_1 being a small set and I_2 being a large set. Unlike before, let the value $|L(l)|$ be the size of I_1. To keep our notations correctly we exchange the red color of the edge between P and S and the black color of the edge between P and v. This does not change the neutral status of I_1.

We now add I_1 and I_2 to F. Thus, this step does not change the number of elements in F.

Note that, like in step 3, we will find a positive set of size $O(\frac{1}{\delta})$ in a future part of the algorithm. And, again, we have to insert the removed edges and we compensate the external edge with enlarging the set by the path P and the small set S. Thus, each single enlargement will increase the size of the set by at most $O(\frac{1}{\delta})$ leading to a positive set of size $O(\frac{1}{\delta^2})$ fulfilling the lemma. Go back to step 1 if any new small set was generated in step 4.

In the remaining we show the contradiction if only sets of types (i)-(iii) remain. Let z be the number of sets of type (i) or (ii). Let z_2 be the number of sets of type (iii) with at least two vertex marks and let z_1 be the number of sets of type (iii) with an internal black cycle.

It is $e = d_2 + 2r - 2|F|$ due to equation (3). At most r thin paths were shaded, leading to at most $2(r - z_2)$ vertex marks. Notice that only thin paths in large sets were shaded. It is $2(z_1 + z_2) \leq 2r$. It follows $e = d_2 + 2r - 2z - 2(z_1 + z_2) \geq d_2 - 2z$.

Each set of type (ii) is adjacent to a set of type (i) via a red edge. Let $I \in F$ be a set of type (i). We will show that $s(I)$ is large enough to reserve an average size of $\frac{1}{6\delta}$ for I itself and for all small sets of type (ii) connected to thick paths and unshaded thin paths of I. Each thick path P is connected to $\alpha(P)$ sets of type (ii) and has a size of at least $\frac{\alpha(P)}{\delta}$. We reserve a value of $\frac{1}{2\delta}$ for each set of type (ii) which is connected to a thick path. It remains a value of at least $\frac{s(I)}{2}$ for I and all small sets connected to unshaded thin paths of I. There are at most $\beta(I) - 1$ unshaded thin paths in I, the others were shaded in step 3. Thus, there are at most $2\beta(I) - 2$ small sets of type (ii) which are adjacent to unshaded thin paths of I via a red edge. Together with I these are at most $2\beta(I) - 1$ sets of type (i) or (ii). If $\beta(I) = 1$ it is $\frac{s(I)}{2} \geq \frac{1}{2\delta}$. If $\beta(I) \geq 2$ it is $\frac{s(I)}{2} \geq \frac{\beta(I)-1}{2\delta} \geq \frac{2\beta(I)-1}{6\delta}$. Thus, we can reserve at least $\frac{1}{6\delta}$ for each set of type (i) and (ii) and it is $e + d_1 + d_2 \geq \frac{z}{6\delta}$. With $e < d_2$ (equation (2)) we get $z \leq 6\delta(d_1 + 2d_2)$.

With $e \geq d_2 - 2z$ and $z \leq 6\delta(d_1 + 2d_2)$ the equation (1) leads us to

$$\frac{|R|}{|V|} = \frac{d_1 + 2d_2}{2(d_1 + d_2 + e)} \leq \frac{d_1 + 2d_2}{2(d_1 + 2d_2 - 2z)} \leq \frac{d_1 + 2d_2}{2(d_1 + 2d_2 - 12\delta(d_1 + 2d_2))} = \frac{1}{2(1 - 12\delta)}.$$

However, it is $\frac{|R|}{|V|} > \frac{1}{2} + \epsilon$. This leads to contradiction for $\delta \leq \frac{\epsilon}{6(1+2\epsilon)}$. This shows that during the algorithm we have found a set fulfilling the lemma. $\square$

In the following we state the lemmas 5 and 6 which are used for the two steps of the local improvement scheme. Before we do so, we classify the vertices.

Definition 4. *The vertices of V_0 (or V_1) are classified according to their distance to the cut. It is $V_0 = C \uplus D \uplus E$ with C vertices being at a distance of 1 to the cut, i. e. they are incident to a cut edge. D vertices are at a distance of 2 and E vertices at a distance of at least 3. D vertices are further classified with respect to the number of adjacent C vertices. I. e. $D = D_3 \uplus D_2 \uplus D_1$ and each D_x vertex is adjacent to x vertices in C. Overall, it is $V_0 = C \uplus D_3 \uplus D_2 \uplus D_1 \uplus E$.*

The values c, d_3, d_2, d_1 and e specify the number of vertices of each type and the values $c(X)$, $d_3(X)$, $d_2(X)$, $d_1(X)$ and $e(X)$ denote the number of the according vertices in a set $X \subset V$.

Lemma 5. *Let π be a bisection of a 3-regular graph $G = (V, E)$ with $V = V_0 \uplus V_1$. If $cut(\pi) > (\frac{1}{3} + 2\epsilon)|V_0|$, $\epsilon > 0$, then there is an at least 1-helpful set of size $O(\frac{1}{\epsilon^2})$ in V_0.*

Proof. We focus on part V_0 of the bisection only. Let $m := |V_0|$.

There are some structures which would directly lead to small 1-helpful sets.

(i) If a C vertex is incident to two or three cut edges, it is an at least 1-helpful set by itself.

(ii) A set of three connected C vertices is 1-helpful.

(iii) If there are two adjacent C vertices and one of them is adjacent to a D_3 vertex, then the union of the adjacent C vertices, the D_3 vertex and its two other adjacent C vertices form a 1-helpful set of size 5.

(iv) Another 1-helpful set can be formed if two D_3 vertices are adjacent to a common C vertices. Then, the union of both D_3 vertices and their adjacent C vertices is a 1-helpful set of size 7.

(v) Let v be a vertex which is adjacent to a C vertex which itself is adjacent to another C vertex or a D_3 vertex. If both other neighbors of v are from $C \cup D_2 \cup D_3$, the union of the mentioned vertices and their adjacent C vertices forms an at least 1-helpful set of size at most 11.

In the remainder we can assume that these types of structures do not exist. Especially, since (i) is excluded, it is $c = cut(\pi)$.

We manipulate the edges between vertices of V_0 such that G is transformed into a 3-regular graph $\bar{G} = (V, \bar{E})$ with the following properties.

- V_0 of $\bar{G}$ does not contain D_3 vertices or adjacent C vertices.
- If there is a 1-helpful set $\bar{H} \subset V_0$ of $\bar{G}$, then there is an at least 1-helpful set H of G, $\bar{H} \subset H \subset V_0$, with $\frac{|H|}{|\bar{H}|}$ bounded from above by a constant value.

We omit the proof of the transformation of G into $\bar{G}$ and the reverse transformation of $\bar{H}$ to H due to space limitations.

$\bar{G} = (V, \bar{E})$ has no D_3 vertices and no adjacent C vertices, i. e. it is now $d_3 = 0$ and $2c = 2d_2 + d_1$. Furthermore, because of $m < 3c$ it is (similar to equation (2))

$$e = m - c - d_2 - d_1 < 2c - d_2 - d_1 = 2d_2 + d_1 - d_2 - d_1 = d_2 . \qquad (4)$$

Construct a new graph K consisting of the D and E vertices of $\bar{G}$, i. e. $K = (U, F)$ with $U = D \uplus E$. Let $F = B \uplus R$ with black edges B and red edges R. The black edges are the edges between the D and E vertices as in $\bar{G}$. Furthermore, there is a red edge between two vertices if they are adjacent to a common C vertex in $\bar{G}$, i. e. $|R| = c$. Thus, K is 3-regular with a maximum red degree of 2, due to the fact that there are no D_3 vertices. It is

$$|R| = c > (\frac{1}{3} + 2\epsilon)(c + d_2 + d_1 + e) > (\frac{1}{3} + 2\epsilon)\frac{3}{2}(d_2 + d_1 + e) = (\frac{1}{2} + 3\epsilon)|U| .$$

Thus, K fulfills the requirements of Lemma 3 for $\bar{\epsilon} = 3\epsilon$. We use Lemma 3 to derive a set S of D and E vertices with size $O(\frac{1}{\bar{\epsilon}^2}) = O(\frac{1}{\epsilon^2})$.

The number b_{ext} of external black edges of S with respect to K is equal to the number of edges between S and other D and E vertices in V_0. The number r_{int} of internal red edges of S with respect to K is equal to the number of C vertices in V_0 which are connected to two vertices in S. Lemma 3 ensures $r_{int} > b_{ext}$.

Let $\bar{S}$ be the union of S with all adjacent C-vertices. It is $|\bar{S}| = O(\frac{1}{\epsilon^2})$. Each external black edge connects $\bar{S}$ with $V_0 \backslash \bar{S}$. The external red edges are neutral, because they connect S via a C vertex to $V_0 \backslash \bar{S}$. Thus, such a C vertex has one edge to V_1 and one edge to $V_0 \backslash \bar{S}$. Each internal red edge is a C vertex which is connected to two vertices in S. Thus, such a vertex has one edge to V_1 and no edges to $V_0 \backslash \bar{S}$. Overall, there are r_{int} edges between $\bar{S}$ and V_1 and b_{ext} edges between $\bar{S}$ and $V_0 \backslash \bar{S}$. This leads to $H(\bar{S}) = r_{int} - b_{ext} > 0$ and $\bar{S}$ fulfills the lemma. $\qquad \square$

In the following, 'log(x)' denotes the logarithm of x to the basis 2.

Lemma 6. *Let $G = (V, E)$ be a connected 3-regular graph and let π be a bisection of G. If $|V_1(\pi)| < 3 \cdot cut(\pi)$ and $0 < x < |V_1(\pi)|$, then there is a set $S \subset V_1(\pi)$ with $|S| = x$ and $H(S) \geq -1 - \log(|S|)$.*

Proof. We first discuss the following cases.

(i) If $x \leq 2$, any set S of x vertices which are incident to a cut edge has the desired property $H(S) \geq -1 - \log(|S|)$.
(ii) If we find a set $Z \subset V_1(\pi)$ with $|Z| \leq x$ and $H(Z) \geq 0$, we can move Z from V_1 to V_0 without increasing the cut size. It remains to apply the lemma again with $\bar{x} = x - |Z|$. Notice that in the case $H(Z) > 0$ the move may result in $|V_1| \geq 3 \cdot cut$. In this case vertices which are incident to a cut edge can be moved from V_1 to V_0 until we either moved a total of x vertices or until it holds $|V_1| < 3 \cdot cut$. In the latter case we apply the lemma again.
(iii) If we find a set $Z \subset V_1(\pi)$ with $\frac{x}{2} \leq |Z| \leq x$ and $H(Z) \geq -1$, we can move Z from V_1 to V_0 with increasing the cut size by at most 1. It remains to apply the lemma again with $\bar{x} = x - |Z| < \frac{x}{2}$. This will construct a set $\bar{S}$ with $|\bar{S}| = x - |Z|$ and $H(\bar{S}) \geq -1 - \log(|\bar{S}|)$, and a unified set $S = Z \cup \bar{S}$ with $|S| = x$ and $H(S) \geq -1 - 1 - \log(|\bar{S}|) \geq -1 - \log(|S|)$.

In the following we can exclude the existence of certain small 0-helpful sets. One example are C vertices incident to two or three cut edges and any set of two adjacent C vertices. A D_3 vertex, together with its adjacent C vertices, also forms a 0-helpful set. In the remainder there are no such sets, i. e. it is $cut = c$, $d_3 = 0$ and $2c = 2d_2 + d_1$. Because of $|V_1| < 3c$ it holds equation (4).

Consider the graph induced by the vertex set $D \cup E$ and its connected components. Let F be the family of these components. For a set $I \subset V_1(\pi)$ define the enlarged set $Z(I) = I \cup \{v \in C; \exists w \in I \text{ with } \{v, w\} \in E\}$ which includes the adjacent C-vertices. Clearly, each set $Z(I)$ for an $I \in F$ is at least 0-helpful. If there is a set $Z(I)$, $I \in F$, with $|Z(I)| \leq x$, we proceed as discussed in case (ii).

Consider a connected component $I \in F$ and let $K = (I, J)$ be the subgraph of G induced by I. The E vertices in K have degree 3, D_1 vertices have degree 2 and D_2 vertices have degree 1. It is easy to see that $e(I) \geq d_2(I)$ iff K contains a cycle and $e(I) = d_2(I) - 2$ otherwise. Because of equation (4) there is an $I \in F$ for which the induced subgraph is a tree.

Let $I \in F$ be a connected component with the induced subgraph $T = (I, J)$ being a tree. Assign a weight $w(v)$ to each vertex v in the tree with $w(v) = |Z(\{v\})|$. For each vertex v this is one higher than the number of C vertices adjacent to v. Thus, each leaf has a weight of 3, each vertex of degree 2 has a weight of 2 and each vertex with a degree of 3 has a weight of 1. It is $\sum_{v \in L} w(v) = |Z(L)|$ for an $L \subset I$ if there are no C vertices which are connected to two vertices of L. It is $\sum_{v \in L} w(v) > |Z(L)|$ for an $L \subset I$ if there is at least one such vertex.

With $|Z(I)| > x$ it is $\sum_{v \in T} w(v) > x$. Clearly, for this type of weight distribution there is an edge in T which separates T into T_1 and T_2 with $\frac{x}{2} \le \sum_{v \in T_1} w(v) \le x$. If $|Z(T_1)| < \sum_{v \in T_1} w(v)$, it is $|Z(T_1)| < x$ and $H(Z(T_1)) \ge 0$ and we proceed with case (ii) above. If $|Z(T_1)| = \sum_{v \in T_1} w(v)$, it is $\frac{x}{2} \le |Z(T_1)| \le x$ and $H(Z(T_1)) \ge -1$. We proceed with case (iii) above. □

Theorem 7. *For any $\epsilon > 0$ there is a value $n(\epsilon)$ such that the bisection width of any 3-regular graph $G = (V, E)$ with $|V| > n(\epsilon)$ is at most $(\frac{1}{6} + \epsilon)|V|$.*

Proof. We start with an arbitrary bisection and follow the iterative local improvement scheme described in Section 1. As long as the cut is above the bound, we repeatedly use Lemma 5 and 6 to calculate a new bisection with a lower cut. Thus, we can limit our focus on one iteration of the two lemmas. Let π_0 be a balanced bisection at the start of the iteration with $cut(\pi_0) > (\frac{1}{6} + \epsilon)|V|$.

Step 1: We construct a small helpful set $S \subset V_0$. Set $k = 4 \cdot \log(\frac{1}{\epsilon})$. The value of k is discussed below. We apply Lemma 5 several times. Each time we find an at least 1-helpful set. We proceed until we reach a total helpfulness of at least k, i. e. we apply the lemma k' times with $k' \le k$. Let $S_i \subset V_0$, $1 \le i \le k'$, with $|S_i| = O(\frac{1}{\epsilon^2})$ be the sets constructed with Lemma 5. After a 1-helpful set S_i is constructed, it is moved from V_0 to V_1 and the next set S_{i+1} is constructed. Let $S = \biguplus_{1 \le i \le k'} S_i$. It is $|S| = k' \cdot O(\frac{1}{\epsilon^2}) = k \cdot O(\frac{1}{\epsilon^2})$ and $H(S) \ge k$.

It remains to show that the requirement of Lemma 5 is fulfilled before each construction of a helpful set. Let $\bar{\epsilon} = \frac{\epsilon}{2}$. It is $|V| \ge 2|V_0| - 1$ and $cut(\pi_0) > (\frac{1}{3} + 2\bar{\epsilon})|V_0(\pi_0)| - (\frac{1}{6} + \bar{\epsilon}) + \bar{\epsilon}|V|$ at the beginning. Let $n(\epsilon)$ be large enough such that $\bar{\epsilon}|V| \ge k + (\frac{1}{6} + \bar{\epsilon})$ for all $|V| > n(\epsilon)$. Thus, it is $cut(\pi_0) > (\frac{1}{3} + 2\bar{\epsilon})|V_0(\pi_0)| + k$. Each application of Lemma 5 decreases the size of the cut. We perform the lemma as long as $cut(\pi) > cut(\pi_0) - k > (\frac{1}{3} + 2\bar{\epsilon})|V_0(\pi_0)| \ge (\frac{1}{3} + 2\bar{\epsilon})|V_0(\pi)|$ with π being the current bisection. Thus, the condition $cut(\pi) > (\frac{1}{3} + 2\bar{\epsilon})|V_0(\pi)|$ is true before each application.

Let π_1 be the new bisection with $cut(\pi_1) = cut(\pi_0) - H(S)$.

Step 2: If $H(S) = k$, it is $cut(\pi_1) = cut(\pi_0) - k$. If $H(S) > k$, it is $cut(\pi_1) < cut(\pi_0) - k$ and we change π_1 by iteratively moving border vertices from V_1 to V_0 until we either get to $cut(\pi_1) = cut(\pi_0) - k$ or to a balanced bisection (in this case we are already finished). Each move of a border vertex decreases the imbalance of the bisection and increases the cut by at most one.

Let $i := |V_1(\pi_1)| - \frac{n}{2}$ be the imbalance of π_1. It is $i \le k \cdot O(\frac{1}{\epsilon^2})$. We use Lemma 6 to find a balancing set $\bar{S} \subset V_1(\pi_1)$ with $|\bar{S}| = i$.

Lemma 6 can only be applied if $|V_1(\pi_1)| < 3 \cdot cut(\pi_1)$. The fact $cut(\pi_0) > (\frac{1}{6} + \epsilon)n$ implies $|V_1(\pi_1)| = \frac{n}{2} + i < 3cut(\pi_0) - 3\epsilon \cdot n + i = 3cut(\pi_1) + 3k - 3\epsilon \cdot n + i \le 3cut(\pi_1)$ if $3k + i \le 3\epsilon \cdot n$. Clearly, there is a value $n(\epsilon)$ such that this equation holds for all graphs with $n > n(\epsilon)$.

We use Lemma 6 to get a set $\bar{S} \subset V_1(\pi_1)$ with $|\bar{S}| = i$ and $H(\bar{S}) \ge -1 - \log(i)$. The move of $\bar{S}$ from V_1 to V_0 results in a balanced bisection π_2 with $cut(\pi_2) \le cut(\pi_1) + 1 + \log(i)$.

We need to ensure $cut(\pi_2) < cut(\pi_0)$ in order to show a decrease of the cut size. It is $cut(\pi_2) \le cut(\pi_0) - k + 1 + \log(i)$ and $i \le k \cdot x \frac{1}{\epsilon^2}$. for some constant x. Choosing $k = 4 \cdot \log(\frac{1}{\epsilon})$ fulfills $k > 1 + \log(k \cdot x \frac{1}{\epsilon^2})$ for $\frac{1}{\epsilon} \ge 2^8$ and $\frac{1}{\epsilon} \ge x$. □

Acknowledgement

The authors would like to thank Thomas Lücking and the anonymous reviewers for their helpful comments.

References

1. N. Alon. On the edge-expansion of graphs. *Combinatorics, Probability and Computing*, 6:145–152, 1997. 525
2. S. L. Bezroukov, R. Elsässer, B. Monien, R. Preis, and J.-P. Tillich. New spectral lower bounds on the bisection width of graphs. In *Workshop on Graph-Theoretic Concepts in Computer Science (WG)*, LNCS 1928, pages 23–34, 2000. 525
3. B. Bollobas. The isoperimetric number of random regular graphs. *Europ. J. Combinatorics*, 9:241–244, 1988. 525
4. C. F. Bornstein, A. Litman, B. M. Maggs, R. K. Sitaraman, and T. Yatzkar. On the bisection width and expansion of butterfly networks. In *Proc. Int. Parallel Processing Symp. (IPPS)*, pages 144–150, 1998. 525
5. T. N. Bui, S. Chaudhuri, F. T. Leighton, and M. Sisper. Graph bisection algorithms with good average case behaviour. *Combinatorica*, 7(2):171–191, 1987. 524
6. P. Chiu. Cubic ramanujan graphs. *Combinatorica*, 12(3):275–285, 1992. 525
7. L. H. Clark and R. C. Entringer. The bisection width of cubic graphs. *Bulletin of th Australian Mathematical Society*, 39:389–396, 1988. 525
8. R. Diekmann, B. Monien, and R. Preis. Using helpful sets to improve graph bisections. In *Interconnection Networks and Mapping and Scheduling Parallel Computations*, pages 57–73. AMS, 1995. 526
9. U. Feige and R. Krauthgamer. A polylogarithmic approximation of the minimum bisection. In *Symp. on Found. of Computer Science (FOCS)*, pages 105–115, 2000. 525
10. M. Fiedler. A property of eigenvectors of nonnegative symmetric matrices and its application to graph theory. *Czechoslovak Math. J., Praha*, 25(100):619–633, 1975. 525
11. M. R. Garey, D. S. Johnson, and L. Stockmeyer. Some simplified NP-complete graph problems. *Theoretical Computer Science*, 1:237–267, 1976. 524
12. B. Hendrickson and R. Leland. Chaco 2.0. Technical Report 94-2692, Sandia, 1994. 525
13. J. Hromkovič and B. Monien. The bisection problem for graphs of degree 4. In *Festschrift zum 60. Geburtstag von Günter Hotz*, pages 215–234. Teubner, 1992. 524, 525, 526
14. G. Karypis and V. Kumar. *METIS Manual 4.0*. University of Minnesota, 1998. 525
15. A. V. Kostochka and L. S. Melnikov. On bounds of the bisection width of cubic graphs. In *Czech. Symp. on Comb., Graphs and Complexity*, pages 151–154, 1992. 525
16. A. V. Kostochka and L. S. Melnikov. On a lower bound for the isoperimetric number of cubic graphs. In *Probabilistic Methods in Discrete Math.*, pages 251–265, 1993. 525

17. A. Lubotzky, R. Phillips, and P. Sarnak. Ramanujan graphs. *Combinatorica*, 8(3):261–277, 1988. 525
18. Y. Manabe, K. Hagihara, and N. Tokura. The minimum bisection widths of the cube-connected cycles graph and cube graph. *Transactions of the IEICE*, J67-D(6):647–654, 1994. (in Japanese). 525
19. G. A. Margulis. Explicit group-theoretical constructions of combinatorial schemes. *Probl. Inf. Transm.*, 24(1):39–46, 1988. 525
20. B. Monien and R. Diekmann. A local graph partitioning heuristic meeting bisection bounds. In *8th SIAM Conf. on Parallel Processing for Scientific Computing*, 1997. 525, 526
21. B. Monien, R. Preis, and R. Diekmann. Quality matching and local improvement for multilevel graph-partitioning. *Parallel Computing*, 26(12):1609–1634, 2000. 525, 526
22. M. Morgenstern. Existence and explicit constructions of $q + 1$ regular ramanujan graphs for every prime power q. *J. Comb. Theory, Ser. B*, 62(1):44–62, 1994. 525
23. F. Pellegrini. SCOTCH 3.1. Technical Report 1137-96, University of Bordeaux, 1996. 525
24. R. Preis. *Analyses and Design of Efficient Graph Partitioning Methods*. Heinz Nixdorf Institut Verlagsschriftenreihe, 2000. Dissertation, Universität Paderborn. 525, 526
25. C. Walshaw. *The Jostle user manual: Version 2.2.* University of Greenwich, 2000. 525

Satisfiability of Systems of Equations over Finite Monoids

Cristopher Moore[1], Pascal Tesson[2], and Denis Thérien[2]*

[1] University of New Mexico, Albuquerque and The Santa Fe Institute
moore@santafe.edu
[2] School of Computer Science, McGill University
{ptesso,denis}@cs.mcgill.ca

Abstract. We study the computational complexity of determining whether a systems of equations over a fixed finite monoid has a solution. In [6], it was shown that in the restricted case of groups the problem is tractable if the group is Abelian and NP-complete otherwise. We prove that in the case of an arbitrary finite monoid, the problem is in P if the monoid divides the direct product of an Abelian group and a commutative idempotent monoid, and is NP-complete otherwise. In the restricted case where only constants appear on the right-hand side, we show that the problem is in P if the monoid is in the class $\mathbf{R}_1 \vee \mathbf{L}_1$, and is NP-complete otherwise.

Furthermore interesting connections to the well known CONSTRAINT SATISFIABILITY PROBLEM are uncovered and exploited.

1 Introduction

Using ideas and tools from algebraic automata theory, a number of algebraic characterizations of complexity classes have been uncovered ([1,7] among many others). This has increased the importance of the study of problems whose computational complexity is parametrized by the properties of an underlying algebraic structure [2,6,12].

In [6], Goldmann and Russell studied the computational complexity of solving equations and systems of equations over a fixed finite group. These investigations were partly completed in [2] where we considered the more general problem of checking satisfiability of a single equation over a fixed finite monoid. Formally, an equation over the monoid M is given as: $c_0 X_1 c_1 \ldots X_k c_k = d_0 Y_1 d_1 \ldots Y_l d_l$ where the $c_i, d_i \in M$ are constants and the X_i's and Y_i's are variables.

The SYSTEM OF EQUATIONS SATISFIABILITY problem for the finite monoid M (denoted EQN_M^*) is to determine whether a given system of equations over M has a solution. We also consider the restriction of EQN_M^* where the right-hand side of each equation is a constant, i.e. variables occur only on the left, and denote it $\mathrm{T}-\mathrm{EQN}_M^*$ (where "T" stands for "target"). For groups these problems are equivalent, and it is shown in [6] that they are in P for Abelian groups but NP-complete otherwise.

* Research supported by NSERC, FCAR and the Von Humboldt Foundation.

J. Sgall, A. Pultr, and P. Kolman (Eds.): MFCS 2001, LNCS 2136, pp. 537–547, 2001.
© Springer-Verlag Berlin Heidelberg 2001

Since this problem is understood for groups, we examine the complementary case where M contains no non-trivial groups, i.e. when it is aperiodic. We prove that dichotomies exist for aperiodic M similar to that for groups, namely that EQN_M^* is in P if M is idempotent and commutative, and is NP-complete otherwise. For the case with only constants on the right, we show that $\mathrm{T-EQN}_M^*$ is tractable when M belongs to the class $\mathbf{R}_1 \vee \mathbf{L}_1$ (defined below), and is NP-complete when M is aperiodic but outside $\mathbf{R}_1 \vee \mathbf{L}_1$. For the case of arbitrary finite monoids, we show that EQN_M^* is tractable if M belongs to the class $\mathbf{J}_1 \vee \mathbf{G}_{\mathrm{com}}$ (where $\mathbf{G}_{\mathrm{com}}$ is the variety of Abelian groups) and is NP-complete otherwise. Furthermore, we explore (and exploit) the relationship between this work and the extensively studied CONSTRAINT SATISFACTION PROBLEM [5,10,3].

2   Background and Notation

2.1   Finite Monoids

We review here some basic notions of finite semigroup theory. A detailed introduction on the material sketched here can be found in [4,11].

A monoid M is a set with a binary associative operation and an identity. This operation defines a canonical surjective morphism $eval_M : M^* \to M$ by

$$eval_M(m_1 m_2 \ldots m_k) = m_1 \cdot m_2 \cdots m_k$$

which just maps a sequence of monoid elements to their product in M. Throughout this paper, M denotes a finite monoid.

We say that a monoid N divides M and write $N \prec M$ if N is the homomorphic image of a submonoid of M. A class of finite monoids $\mathcal{M}$ is said to be a (pseudo)-variety if it is closed under direct product and division. Varieties are the natural unit of classification of finite monoids.

A monoid M is *aperiodic* or group-free if no subset of it forms a non-trivial group or, equivalently, if it satisfies $m^{t+1} = m^t$ for some integer t and all $m \in M$. Any finite monoid divides a wreath product[1] of aperiodics and simple groups. Thus aperiodics and groups act as "building blocks" for all monoids, so they are the two special cases to consider first.

An element $m \in M$ is said to be *idempotent* if $m^2 = m$ and if this holds for all $m \in M$, we say that M is idempotent.

The *left (right) ideal* generated by an element x is the set Mx (resp. xM), and its *two-sided ideal* is the set MxM. The following equivalence relations on M, called Green's relations, ask whether two elements generate the same ideals.
- $x\mathcal{J}y$ if $MxM = MyM$
- $x\mathcal{L}y$ if $Mx = My$
- $x\mathcal{R}y$ if $xM = yM$
- $x\mathcal{H}y$ if both $x\mathcal{R}y$ and $x\mathcal{L}y$
It is known that $\mathcal{L}$ and $\mathcal{R}$ commute and that for finite monoids $\mathcal{L} \circ \mathcal{R} = \mathcal{J}$.

---

[1] See e.g. [11] for a formal definition

If we instead ask whether element's ideals are contained in one another, we can define natural pre-orders $\leq_{\mathcal{J}}, \leq_{\mathcal{R}}, \leq_{\mathcal{L}}$ on M with e.g. $x \leq_{\mathcal{J}} y$ iff $MxM \subseteq MyM$. We will say that "x is (strictly) $\mathcal{J}$-above y" if $x \geq_{\mathcal{J}} y$ (resp. $x >_{\mathcal{J}} y$), and so on. Note that $x \leq_{\mathcal{J}} y$ iff there exists u, v such that $x = uyv$ and similarly $x \leq_{\mathcal{R}} y$ iff there is u with $x = yu$. We will make extensive use of the following fact:

Lemma 1. *If $x\mathcal{J}y$ and x, y are $\mathcal{R}$-comparable (resp. $\mathcal{L}$-comparable), then $x\mathcal{R}y$ (resp. $x\mathcal{L}y$). Thus if $x\mathcal{J}y$ and x, y are both $\mathcal{R}$- and $\mathcal{L}$-comparable, then $x\mathcal{H}y$.*

Monoid M is said to be $\mathcal{J}$-trivial (resp. $\mathcal{R}$-, $\mathcal{L}$-, or $\mathcal{H}$-trivial) if $x\mathcal{J}y$ (resp. $x\mathcal{R}y$, $x\mathcal{L}y$ or $x\mathcal{H}y$) implies $x = y$; that is, if each of the relevant equivalence classes contain exactly one element. A $\mathcal{J}$-class (resp. $\mathcal{R}, \mathcal{L}, \mathcal{H}$-class) is *regular* if it contains an idempotent.

An $\mathcal{H}$ class contains at most one idempotent and is regular iff it forms a group whose identity element is the idempotent of the class. Conversely, any maximal subgroup of M is precisely a regular $\mathcal{H}$ class. Therefore, M is aperiodic iff it is $\mathcal{H}$-trivial. More generally, if all its $\mathcal{H}$-classes are regular then M is the union of its maximal subgroups, in which case it is called a *union of groups*. We denote by $\mathcal{R}_s$ (and similarly for $\mathcal{J}, \mathcal{H}, \mathcal{L}$) the $\mathcal{R}$-class of s. One can prove the following:

Lemma 2. *Let $s\mathcal{J}t$, then $st \in \mathcal{R}_s \cap \mathcal{L}_t$ if and only if $\mathcal{L}_s \cap \mathcal{R}_t$ contains an idempotent. Otherwise, we have $st <_{\mathcal{J}} s$.*

2.2   CSP

Let D be a finite domain and Γ be a finite set of relations on D. To each pair D, Γ corresponds a CONSTRAINT SATISFACTION PROBLEM (CSP). An instance of CSP is a list of constraints, i.e. of pairs $R_i(S_i)$ where $R_i \in \Gamma$ is a k-ary relation and S_i, the scope of R_i, is an ordered list of of k-variables (with possible repetitions) and we want to determine whether the variables can be assigned values in D such that each constraint is satisfied.

This class of combinatorial decision problems has received a lot of attention because of the wide variety of problems which it encompasses. CSP is NP-complete in general, and one important question concerning CSP is to identify restrictions on Γ which make the problem tractable for a given domain D. Recently, tools from universal algebra [3], relational database theory [5], and group theory [5] have been used to identify tractable classes of CSP when the relations in Γ are closed under certain families of operations. In our context, the domain is the monoid and Γ is the set of constraints definable as equations over M.

The CSP problem on boolean domains is also known as GENERALIZED SATISFIABILITY and was studied by Schaefer in [13]. He proved a complete dichotomy and showed the problem to be NP-complete unless it was one of six tractable special cases: 2-SAT, 0-valid SAT, 1-valid SAT, affine-SAT, Horn-SAT and anti-Horn SAT. Affine-SAT is the case where each relation is the solution set of a system of equations over the cyclic group $\mathbb{Z}_2$. There exists a second 2-element monoid, namely U_1, the two element aperiodic having elements 1 and 0 and multiplication given by $0 \cdot 1 = 1 \cdot 0 = 0 \cdot 0 = 0$ and $1 \cdot 1 = 1$. The following provides a connection to our concern in this paper.

Lemma 3. *A boolean relation is Horn or anti-Horn, i.e. expressible as tuples satisfying a conjunction of disjuncts containing each at most one un-negated (resp. negated) variable, iff it is the set of solutions of a system of equations over U_1.*

Proof. Identify the element 1 of U_1 with TRUE and 0 with FALSE. Then the Horn clause $X_1 \wedge X_2 \wedge \ldots X_n \to Y$ is satisfied when one of the X_i's is FALSE or when all X_i's and Y are TRUE. These are exactly the tuples which satisfy the equation

$$X_1 X_2 \ldots X_n = X_1 \ldots X_n Y$$

over U_1.

Conversely, the equation $X_1 \ldots X_n = Y_1 \ldots Y_m$ corresponds to the Horn formula:

$$\bigwedge_{1 \le i \le m} (X_1 \wedge \ldots \wedge X_n \to Y_i) \wedge \bigwedge_{1 \le i \le n} (Y_1 \wedge \ldots \wedge Y_n \to X_i)$$

If on the other hand we choose to identify 1 with FALSE and 0 with TRUE, a similar argument shows the relationship of U_1 systems to anti-Horn formulas.

3   Hardness Results

In this section we show several hardness results for EQN_M^*. Note that the problem trivially lies in NP.

Theorem 4. *If M is aperiodic but not idempotent, then EQN_M^* is NP-complete.*

Proof. Let $m \ne m^2$ be a $\mathcal{J}$-maximal non-idempotent element of M.

We use the following reduction from 1-3SAT, where each clause consists of three literals, exactly one of which must be true. For each Boolean variable X_i, we use two variables x_i, $\overline{x}_i$ in M. For each i we have equations

$$(1) \ x_i \overline{x}_i = m \text{ and} \qquad (2) \ \overline{x}_i x_i = m$$

and for each clause of the formula, e.g. $(X_1 \vee \overline{X_2} \vee X_3)$ we add the equation

$$(3) \ x_1 \overline{x}_2 x_3 = m$$

Suppose that the original 1-in-3 SAT formula is satisfiable. Then for each i we set $x_i = m$ and $\overline{x}_i = 1$ whenever X_i is TRUE, and $x_i = 1$ and $\overline{x}_i = m$ whenever X_i is FALSE. It is easy to see that this satisfies the sets of Equations (1), (2) and (3), so these equations are satisfiable if the 1-in-3 SAT formula is.

Conversely, suppose that this system of equations is satisfiable. Let us first look at the Equations (1) and (2). Note that x_i (resp. $\overline{x}_i$) is both $\mathcal{R}$-above and $\mathcal{L}$-above m. So if x_i (say) lies in m's $\mathcal{J}$-class, it must be $\mathcal{R}$ and $\mathcal{L}$ equivalent to m, hence, by Lemma 1, $\mathcal{H}$-equivalent and thus equal to m by aperiodicity. It follows that at least one of x_i or $\overline{x}_i$ lies strictly $\mathcal{J}$-above m; otherwise we would have $x_i = \overline{x}_i = m$, and since m is not idempotent this would violate Equation (1). Moreover, since m is $\mathcal{J}$-maximal among the non-idempotent elements of M, whichever one of x_i, $\overline{x}_i$ is strictly $\mathcal{J}$-above m must be some idempotent e.

Therefore, suppose $x_i = e$ where $e >_{\mathcal{J}} m$ is idempotent. Then Equation (1) gives us $m = e\overline{x}_i = ee\overline{x}_i = em$, and similarly (2) gives $m = me$. We cannot also have $\overline{x}_i >_{\mathcal{J}} m$, since then $\overline{x}_i$ would also be idempotent, and this leads to the contradiction $m^2 = e\overline{x}_i\overline{x}_i e = e\overline{x}_i e = em = m$. Thus $\overline{x}_i \mathcal{J} m$; by Lemma 1 this gives $\overline{x}_i \mathcal{H} m$, and so $\overline{x}_i = m$ from aperiodicity.

Similarly, if $\overline{x}_i >_{\mathcal{J}} m$ then $x_i = m$. So if we set X_i to TRUE when $x_i = m$ and FALSE when $\overline{x}_i = m$, Eqs (1) and (2) insure that our mapping between Boolean variables and variables in M is consistent, in the sense that for all i, exactly one of x_i, $\overline{x}_i$ is m and the other is an idempotent in a higher $\mathcal{J}$-class.

Finally, suppose that all 3 variables in Equation (3) have idempotent values. By a previous argument, these values fix m and so $m^2 = x_1\overline{x}_2 x_3 m = x_1\overline{x}_2 m = x_1 m = m$, a contradiction. We get a similar contradiction if two or more of the variables are set to m. Therefore (3) insures that exactly one literal in each clause is true, and the 1-in-3 SAT formula is satisfiable.

Theorem 5. *If M contains some irregular $\mathcal{H}$-class, i.e. if M is not a union of groups, then* EQN^*_M *is NP-complete.*

Proof. Let $m \neq m^2$ be an element of a $\mathcal{J}$-maximal irregular $\mathcal{H}$-class, i.e. such that any monoid element u with $u >_{\mathcal{J}} m$ lies in a regular $\mathcal{H}$-class. In particular, since regular $\mathcal{H}$-classes form groups, there must exist $q \geq 1$ such that for all such u we have u^q is idempotent and $u^{q+1} = u$.

We use the same reduction from 1-3SAT, namely: for each literal x_i and its complement $\overline{x}_i$, we add equations $x_i\overline{x}_i = m$ and $\overline{x}_i x_i = m$ and for each clause we add the equation e.g.

$$x_1\overline{x}_2 x_3 = m$$

Once again, a satisfiable instance of 1-3SAT trivially gives rise to a satisfiable system of equations.

For the converse, Eqs. (1) and (2) show that if $x_i \mathcal{J} m$ then in fact $x_i \mathcal{H} m$, just as in Theorem 4. Since the $\mathcal{H}$-class of m contains no idempotent, Lemma 2 shows that the product of any two elements of $\mathcal{H}_m$ lies strictly $\mathcal{J}$-below m. Eqs. (1) and (2) thus force at least one of x_i, $\overline{x}_i$ to be strictly $\mathcal{J}$-above m.

Suppose both x_i and $\overline{x}_i$ are strictly $\mathcal{J}$-above m. Then x_i^q and $\overline{x}_i^q$ are idempotents, and $m = x_i\overline{x}_i = x_i\overline{x}_i^{q+1} = m\overline{x}_i^q$. Similarly $m = mx_i^q$. Moreover, x_i and $\overline{x}_i$ commute by Eqs. (1) and (2), so we get $m = mx_i^q\overline{x}_i^q = m(x_i\overline{x}_i)^q = m^{q+1} <_{\mathcal{J}} m$, a contradiction. Therefore, at least one of $x_i, \overline{x}_i$ must be $\mathcal{H}$-equivalent to m and the other fixes the $\mathcal{H}$-class of m, so if we identify true literals with variables taking a value $\mathcal{H}$-equivalent to m, we obtain a consistent truth assignment, and the same argument as before shows that exactly one literal in each clause corresponding to Equation (3) must be true.

Note that the reductions used to prove Theorems 4 and 5 actually show that $T-EQN^*$ is NP-complete in both these cases.

Theorem 6. *If M is aperiodic and idempotent but is not commutative, then EQN^*_M is NP-complete.*

Proof. Let a and b be two non-commuting elements, $ab \neq ba$, such that a is a $\mathcal{J}$-maximal element which is not *central* in M (i.e. which does not commute with every element) and b is a $\mathcal{J}$-maximal element which does not commute with a. We reduce from 3-SAT. For each Boolean variable X_i, we create variables $x_i, \bar{x}_i, y_i, \bar{y}_i$ and equations

$$
\begin{array}{ll}
(1) \quad x_i \bar{x}_i = a & (2) \quad \bar{x}_i x_i = a \\
(3) \quad y_i \bar{y}_i = b & (4) \quad \bar{y}_i y_i = b \\
(5) \quad x_i \bar{y}_i = \bar{y}_i x_i & (6) \quad \bar{x}_i y_i = y_i \bar{x}_i
\end{array}
$$

Also, for each 3-SAT clause, e.g. $X_1 \vee \overline{X_2} \vee X_3$, we add an equation

$$x_1 \bar{x}_2 x_3 = a$$

It is easy to verify that if the 3-SAT instance is satisfiable, the above system is satisfiable by setting $x = a$, $\bar{x} = 1$, $y = b$, and $\bar{y} = 1$ whenever X is TRUE, and $x = 1$, $\bar{x} = a$, $y = 1$, $\bar{y} = b$ whenever X is FALSE.

Conversely, suppose the system of equations is satisfiable. Eq. (1) shows that $x_i, \bar{x}_i \geq_{\mathcal{J}} a$. Since a and b don't commute, a cannot be the product of two elements commuting with b. However, any element strictly $\mathcal{J}$-above a is central and commutes with everything, so at least one of $x_i, \bar{x}_i$ must be $\mathcal{J}$-equivalent to a. Moreover, Eqs. (1) and (2) insure that $x_i, \bar{x}_i$ are both $\mathcal{L}$-above and $\mathcal{R}$-above a, so if $x_i \mathcal{J} a$ (say) we must also have $x \mathcal{H} a$ by Lemma 1 and so $x = a$ by aperiodicity. Thus at least one of $x_i, \bar{x}_i$ must be a, Similarly at least one of $y_i, \bar{y}_i$ must be b, since any elements strictly $\mathcal{J}$-above b commute with a.

If $x_i = a$, then $\bar{y}_i$ commutes with a by Eq. (5). Thus $\bar{y}_i \neq b$ must be strictly $\mathcal{J}$-above b. If $y_i = b$, then $\bar{x}_i$ commutes with b by Eq. (6), so $\bar{x}_i \neq a$ is strictly $\mathcal{J}$-above a. If we identify true literals with a and b and false ones with elements strictly $\mathcal{J}$-above a and b, this shows that our truth assignment is consistent.

Since every element strictly $\mathcal{J}$-above a is central but a is not, a cannot be a product of such elements. Therefore, at least one of the variables in $x_1 \bar{x}_2 x_3 = a$ must be a, so the 3-SAT clause is satisfiable.

Goldmann and Russell already showed that $T-EQN^*$ is NP-complete for a non-Abelian group. We can extend this result to any M having an $\mathcal{H}$-class that forms a non-Abelian group G. Indeed, let e be the idempotent of this $\mathcal{H}$-class and consider a system S of equations over G with no variables on the right-hand side. By replacing each occurrence of a variable x in S by exe, we obtain a system S' which is satisfiable over M if and only if S is satisfiable over G since for any $x \in M$, either exe is in G or exe is strictly $\mathcal{J}$-below the equation's target.

We can combine the results of this section into one algebraic condition on M insuring NP-completeness of EQN^*_M. Let $\mathbf{J}_1 \vee \mathbf{G}_{\mathrm{com}}$ be the variety generated by idempotent commutative monoids and Abelian groups. It can be shown that M belongs to $\mathbf{J}_1 \vee \mathbf{G}_{\mathrm{com}}$ if and only if it is a union of groups and is commutative.

Theorem 7. *If M is not in $\mathbf{J}_1 \vee \mathbf{G}_{\mathrm{com}}$, then EQN^*_M is NP-complete.*

Proof. By Theorem 5 and remarks following its proof, this is certainly true unless M is a union of maximal Abelian subgroups. Such a monoid can easily be shown to be commutative iff all four of Green's relations coincide in M.

If this is not the case, however, there must exist a $\mathcal{J}$-maximal $\mathcal{J}$-class containing two idempotents a, b which are not $\mathcal{H}$-equivalent and thus, by Lemma 2, do not commute. Moreover, because each $\mathcal{H}$-class is a subgroup, there exists some q such that for all $m \in M$, m^q is idempotent. We can therefore show NP-completeness using the reduction in the proof of Theorem 6 where we replace all occurrences of a variable x by x^q.

4   Tractable Cases

It is shown in [6] that EQN^* is in P for Abelian groups. This could also be proved using Theorem 33 of [5] which insures tractability of CSP over a finite group G when each relation of Γ is a coset[2] of G^k. For aperiodics, we have:

Theorem 8. *If M is idempotent and commutative, then EQN^*_M is in P.*

Proof. It is known that idempotent and commutative finite monoids are $\mathcal{J}$-trivial [4]. In particular, they always have a unique $\mathcal{J}$-minimal element 0 such that $0 \cdot x = x \cdot 0 = 0$ for all $x \in M$. Our algorithm looks for a "$\mathcal{J}$-minimal" solution to the system.

For each equation of our system in which the variable x_i appears only on one side, $u x_i v = E$ where x_i does not appear in E, we write an inequality $x_i \geq_{\mathcal{J}} E$. Note that any satisfying assignment must satisfy all these $\mathcal{J}$-inequalities.

Initially, we set all our variables to 0. We update these values in the following way: for each variable x_i, we set its new value to the least value satisfying all inequalities $x_i \geq_{\mathcal{J}} E$ where E is evaluated using the current values for $x_1, \ldots, x_n$. We iterate this until we reach a fixed point. It should be clear that no satisfying assignment of the system assigns to x_i a value which is smaller than this one in the $\mathcal{J}$-ordering. Moreover, this process terminates in at most $n \cdot |M|$ steps where n is the number of variables, since the new values are increasing in the $\mathcal{J}$-order.

If the assignment resulting from this process does not satisfy the system, then some equation is violated. For this to happen, there must be a constant appearing on only one side of the equation such that all the variables appearing on the same side are $\mathcal{J}$-above it, and all variables on the other side are strictly $\mathcal{J}$-above it. Thus, if this assignment does not satisfy our system, the system is unsatisfiable because any assignment $\mathcal{J}$-above it will also violate this inequality.

---

[2] Note that, in general, the solution set of an equation over G can be assumed to have this property only when G is Abelian.

Note once more that when the arity of the equations is bounded, this can be obtained by appealing to [8] which shows that CSP is tractable when the relations are preserved by an associative, commutative and idempotent operation. In our case, this operation is simply multiplication in M.

As a corollary, EQN^*_M is tractable if M is a direct product of an Abelian group and a commutative idempotent. On the other hand, when M is not a divisor of such a direct product, i.e. when M is not in $\mathbf{J}_1 \vee \mathbf{G}_{\text{com}}$, then by Theorem 5, EQN^*_M is NP-complete. Very recently, these authors and, independently, Klima [9] proved that EQN^*_M is in fact tractable for any $M \in \mathbf{J}_1 \vee \mathbf{G}_{\text{com}}$, thus completing the dichotomy for EQN^*_M.

Unfortunately, time and space constraints prevented these results (and a proof of Lemma 12) from being included in our final submission to MFCS. They should however be available from one of the authors' web pages by the time these proceedings are published.

5   T−EQN*: Equations with only Constants on the Right-Hand Side

If we restrict our equations to have only constants on the right-hand side, we can do better. Note that we can assume that the left-hand side contains no constants by adding to our system equations $X_m = m$ for all $m \in M$ and replacing constants on the left side by the appropriate variable.

We denote as $\mathbf{R}_1 \vee \mathbf{L}_1$ the variety generated by idempotent $\mathcal{R}$-trivial and idempotent $\mathcal{L}$-trivial monoids. A monoid M is in $\mathbf{R}_1 \vee \mathbf{L}_1$ iff it satisfies $m^2 = m$ for all $m \in M$ and $abaca = abca$ for all $a, b, c \in M$ (See e.g. [4]).

Theorem 9. *If M belongs to $\mathbf{R}_1 \vee \mathbf{L}_1$, then T−EQN^*_M can be solved in polynomial time.*

Proof. Our algorithm works by shrinking a list of possible values for each variable and implicitly uses the fact that the relations defined by equations over M are closed under a set function [5]. We use the following lemma:

Lemma 10. *Let $M \in \mathbf{R}_1 \vee \mathbf{L}_1$ and suppose $x_1 \ldots x_k = m$ and $y_1 \ldots y_l = m$. For all shuffles K of $x_1 \ldots x_k$ with $y_1 \ldots y_l$, we have $K = m$.*

To see this, recall [4] that in idempotent monoids, the product of two elements $\mathcal{J}$-above some $u \in M$ is also $\mathcal{J}$-above u. Hence we have $K \geq_{\mathcal{J}} m$ since K is a product of elements $\mathcal{J}$-above m. On the other hand, since all x_i's appear in K, we can use the relation $abaca = abca$ to get $KmK = Kx_1 \cdots x_k K = K^2 = K$. Thus, $m \geq_{\mathcal{J}} K$ and so $m \mathcal{J} K$. Note also that every prefix of $x_1 \ldots x_k$ is $\mathcal{R}$-above m and so $x_1 \ldots x_i m = m$ for all $i \leq k$.

We claim that $K \geq_{\mathcal{R}} m$. Indeed, we have $Km = Kmx_1 \ldots x_k y_1 \ldots y_l$. Using again the relation $abaca = abca$, we can replace the occurrence of x_i in K on the right-hand side of this equation with the prefix $x_1 \ldots x_i$ since all the x_j with $j \leq i$ appear both before and after x_i. Hence Km can be written as a product of prefixes of $x_1 \cdots x_k$ or $y_1 \ldots y_l$ times m. Thus $Km = m$ and $K \geq_{\mathcal{R}} m$.

By a symmetric argument, $K \geq_{\mathcal{L}} m$. Since $m \mathcal{J} K$, Lemma 1 implies that $m \mathcal{H} K$ and $m = K$ by aperiodicity and we obtain the lemma.

Now consider the following algorithm. For each variable X_i, $1 \leq i \leq n$, initialize a set $A_i = M$ of "possible values." Then repeat the following until either the A_i are fixed or some $A_i = \emptyset$: for all i from 1 to n, for each equation E involving X_i, and each $a_i \in A_i$, if there exists no n-tuple $(a_1, \ldots, a_i, \ldots a_n)$ with $a_j \in A_j$ that satisfies E, then set $A_i \Rightarrow A_i - \{a_i\}$.

If some A_i is empty, the system clearly has no solution. Conversely, we are left with sets A_i such that for all $a_i \in A_i$ and all equations E in the system, there are $a_j \in A_j$ for all $i \neq j$ such that the n-tuple $(a_1, \ldots, a_n)$ satisfies E. We claim that this guarantees the existence of a solution to the system.

Indeed, let t_i be the product in M of all elements of $A_i = \{a_i^{(1)}, \ldots, a_i^{(s_i)}\}$ in some arbitrary order. Then $(t_1, \ldots, t_n)$ satisfies all equations in the system. To see this, consider some equation $E = X_1 X_2 \ldots X_k = m$. The product $t_1 t_2 \ldots t_k$ is a shuffle of solutions to this equation by definition of the A_i's, so by the above lemma $(t_1, \ldots, t_n)$ is also a solution.

It remains to show that the algorithm has a polynomial running time. It is sufficient to show that we can test in polynomial time for the existence of a tuple $\bar{a} = (a_1, \ldots, a_i, \ldots, a_n)$, $a_j \in A_j$ satisfying a given equation, say $X_1 \ldots X_k = m$.

For each $m \in M$, we will denote M_m the subset of M^* given by $M_m = \{m_1 m_2 \ldots m_k : eval_M(m_1, m_2 \ldots m_k) = m\}$. Because $\mathbf{R}_1 \vee \mathbf{L}_1$ is contained in the aperiodic variety $\mathbf{DA}$ (see [14]), each M_m can be expressed as the disjoint union of languages of the form[3] $B_0^* b_1 B_1^* \ldots b_r B_r^*$ with $b_j \in M$ and $B_j \subseteq M$. So to test for the existence of an $\bar{a}$ with the right properties, it is sufficient to consider, for each of the concatenations $B_0^* b_1 B_1^* \ldots b_r B_r^*$ that compose M_m, the $\binom{k}{r}$ possibilities of placing the b_i's among the k variables. To validate this choice, it is enough to check that $b_i \in A_j$ if we set $X_j = b_i$, and that for all other variables the corresponding A_i contains at least one element which belongs to the right B_j's.

Here again, we seek a dichotomy result for aperiodic monoids. By our proof of Theorem 4, $\mathrm{T-EQN}^*$ is NP-complete for non-idempotent aperiodics. So we only need to consider idempotent aperiodics outside $\mathbf{R}_1 \vee \mathbf{L}_1$ that is idempotent monoids that fail to satisfy the relation $abaca = abca$.

Theorem 11. *If M is idempotent aperiodic outside $\mathbf{R}_1 \vee \mathbf{L}_1$ then $\mathrm{T-EQN}^*$ is NP-complete.*

Proof. We start by stating a technical lemma about idempotent monoids outside $\mathbf{R}_1 \vee \mathbf{L}_1$ whose proof is due to Klima [9].

Lemma 12. *Suppose M is an idempotent aperiodic monoid outside of $\mathbf{R}_1 \vee \mathbf{L}_1$. There exists $a, b, c \in M$ such that $abaca \neq abca$ and moreover for all e, f lying strictly $\mathcal{J}$-above b (resp. lying strictly $\mathcal{J}$-above c) the product ef is also strictly $\mathcal{J}$-above b (resp. strictly $\mathcal{J}$-above c).*

---

[3] We do not need this in our proof, but it can also be assumed that these concatenations are unambiguous, i.e. that if $w \in B_0^* b_1 B_1^* \ldots b_r B_r^*$, its factorization is unique.

Choose $a, b, c \in M$ as in the preceding lemma. We now reduce from Monotone NAE-3SAT, whose NP-hardness is insured by Schaefer's theorem.

For each variable X_i, create variables s_i, t_i and equations:

$$(1) \quad bs_i = b \qquad\qquad (2) \; s_ib = b$$
$$(3) \quad t_ic = c \qquad\qquad (4) \; ct_i = c$$
$$(5) \; abas_it_iaca = abaca$$

For each clause e.g. $X_1 \lor X_2 \lor X_3$ we further add two equations:

$$(6) \; s_1s_2s_3 = b \qquad\qquad (7) \; t_1t_2t_3 = c$$

If we have a satisfying NAE-3SAT assignment, we can satisfy this system by setting $s_i = b$ and $t_i = 1$ if X_i is TRUE and $s_i = 1$ and $t_i = c$ if X_i is FALSE.

Conversely, Equations (1-4) show that s_i and t_i are $\mathcal{R}$- and $\mathcal{L}$-above b and c respectively. By Lemma 1 and aperiodicity, $s_i\mathcal{J}b$ forces $s_i = b$ and $t_i\mathcal{J}c$ forces $t_i = c$. Thus either $s_i = b$ or $s_i >_\mathcal{J} b$, and similarly either $t_i = c$ or $t_i >_\mathcal{J} c$.

Moreover, if we have both $s_i = b$ and $t_i = c$ then Eq. (5) is violated since $(ab)(ab)(ca)(ca) = abca \neq abaca$. Thus if we interpret $s_i = b$ as X_i is TRUE and $t_i = c$ as X_i is FALSE, we are guaranteed that no variable has both truth values.

If neither $s_i = b$ nor $t_i = c$, then we view X_i as indeterminate, and unable to contribute to the satisfaction of a NAE-3SAT clause: that is, among the remaining two variables there must be one true and one false. This corresponds to the fact that if (say) $s_i >_\mathcal{J} b$, then it cannot help satisfy Eq. (6), and one of the remaining two variables must be b.

Indeed, by Lemma 12 neither b nor c is the product of elements $\mathcal{J}$-above them, so if Eqs. (6) and (7) are satisfied, at least one of the s's must be set to b and one of the t's to c. Since we are considering *monotone* formulas, if a satisfying assignment exists with some variables indeterminate, then one exists with these variables set to true. So the Monotone NAE-3SAT formula is also satisfiable.

6   Conclusion

We have proved that EQN^*_M is tractable for M in $\mathbf{J}_1 \lor \mathbf{G}_{\mathrm{com}}$ and is NP-complete otherwise. One might expect that a similar dichotomy also exists for $\mathrm{T-EQN}^*_M$. Arguments analogous to the ones mentioned at the end of Section 4 can be used to show that $\mathrm{T-EQN}^*_M$ is tractable whenever M belongs to the variety $\mathbf{R}_1 \lor \mathbf{L}_1 \lor \mathbf{G}_{\mathrm{com}}$ and we conjecture that outside this class $\mathrm{T-EQN}^*_M$ is in fact NP-complete.

The connection to CSP can also be further explored. For instance, it would be interesting to relate, as in Lemma 3, tractable sets of constraints with solution sets to equations over the domain viewed as a monoid.

We wish to thank Ondrej Klima for sending us a proof of Lemma 12 and many helpful comments and suggestions.

References

1. D. A. M. Barrington and D. Thérien. Finite monoids and the fine structure of NC^1. *Journal of the ACM*, 35(4):941–952, Oct. 1988. 537
2. D. M. Barrington, P. McKenzie, C. Moore, P. Tesson, and D. Thérien. Equation satisfiability and program satisfiability for finite monoids. In *MFCS'00*, pages 172–181, 2000. 537
3. V. Dalmau. *Computational Complexity of Problems over Generalized Formula*. PhD thesis, Universita Politécnica de Catalunya, 2000. 538, 539
4. S. Eilenberg. *Automata, Languages and Machines*, volume B. Academic Press, 1976. 538, 543, 544
5. T. Feder and M. Y. Vardi. The computational structure of monotone monadic SNP and constraint satisfaction: A study through datalog and group theory. *SIAM Journal on Computing*, 28(1):57–104, 1999. 538, 539, 543, 544
6. M. Goldmann and A. Russell. The complexity of solving equations over finite groups. In *Proc. 14th Conf. on Computational Complexity*, pages 80–86, 1999. 537, 543
7. Hertrampf, Lautemann, Schwentick, Vollmer, and Wagner. On the power of polynomial time bit-reductions. In *Conf. on Structure in Complexity Theory*, 1993. 537
8. P. Jeavons, D. Cohen, and M. Gyssens. Closure properties of constraints. *J. ACM*, 44(4):527–548, 1997. 544
9. O. Klima. Private communication, 2001. 544, 545
10. J. Pearson and P. Jeavons. A survey of tractable constraint satisfaction problems. Technical Report CSD-TR-97-15, University of London, 1997. Available at http://www.dcs.rhbnc.ac.uk/research/. 538
11. J.-E. Pin. *Varieties of formal languages*. North Oxford Academic Publishers Ltd, London, 1986. 538
12. J.-F. Raymond, P. Tesson, and D. Thérien. An algebraic approach to communication complexity. *Lecture Notes in Computer Science*, 1443:29–40, 1998. 537
13. T. J. Schaefer. The complexity of satisfiability problems. In *Proc. 10th ACM STOC*, pages 216–226, 1978. 539
14. M. P. Schützenberger. Sur le produit de concaténation non ambigu. *Semigroup Forum*, 13:47–75, 1976. 545

Rational Graphs Trace Context-Sensitive Languages

Christophe Morvan[1] and Colin Stirling[2]

[1] IRISA, Campus de Beaulieu
35042 Rennes, France
christophe.morvan@irisa.fr
[2] Division of Informatics, University of Edinburgh
cps@dcs.ed.ac.uk

Abstract. This paper shows that the traces of rational graphs coincide with the context-sensitive languages.

1 Introduction

Infinite transition graphs have become a focus of attention recently. For example, the behaviour of an infinite state system is an infinite transition graph, and researchers have examined when property checking and equivalence checking such graphs is decidable [3]. Muller and Schupp pioneered the study of infinite transition graphs by examining the graphs of pushdown automata [10]. They were interested in extending Cayley graphs to structures that are more general than groups. Courcelle defined a more extensive class, the equational graphs, using deterministic graph grammars [5]. More recently, Caucal constructed a richer class, the prefix-recognisable graphs [4], using transformations of the complete binary tree [4]: such a graph is characterised by an inverse rational substitution followed by a rational restriction of the complete binary tree. Caucal also provided a mechanism for generating the prefix-recognisable graphs, using finite sets of rewrite rules. The first author produced an even richer family, the rational graphs, again using transformations of the complete binary tree [9]: the characterisation involves an inverse linear substitution followed by a rational restriction of the complete binary tree. He also showed that rational graphs are generated by finite-state transducers whose words are the vertices of a graph.

There is a natural relation between transition graphs and languages, namely *the trace*. A trace of a path in a transition graph is its sequence of labels. Relative to designated initial and final sets of vertices, the trace of a transition graph is the set of all path traces which start at an initial vertex and end at a final vertex. For example, the regular languages are the traces of finite transition graphs. For richer families of languages one needs to consider infinite transition graphs. The context-free languages are the traces of the transition graphs of pushdown automata. This remains true for both equational and prefix-recognisable graphs. The next family of languages in the Chomsky hierarchy is the context-sensitive languages. Although they have been studied from a language theoretic point

J. Sgall, A. Pultr, and P. Kolman (Eds.): MFCS 2001, LNCS 2136, pp. 548–559, 2001.
© Springer-Verlag Berlin Heidelberg 2001

of view, (see [8], for example), only recently have their canonical graphs been examined [7].

In this paper we prove that the traces of rational graphs (relative to regular initial and final vertex sets) coincide exactly with the context-sensitive languages. In Section 2 we describe the rational graphs. In Section 3 we prove that the traces of rational graphs are context-sensitive, using linear bounded Turing machines. Finally, in Section 4 we prove the converse inclusion, using a normal form due to Penttonen for context-sensitive languages [11].

2 Rational Graphs

In this section we concentrate on a presentation of rational graphs using finite state transducers that generate them. For a more detailed introduction to rational graphs, their basis using partial semigroups and rational relations, and their characterization in terms of transformations of the complete binary tree, see [9].

Assume a finite alphabet $\mathcal{A}$ and a finite set of symbols X. A vertex of a transition graph is a word $u \in X^*$, and a transition has the form (u, a, v), which we write as $u \xrightarrow{a} v$, where $a \in \mathcal{A}$ and u, v are vertices. A transition graph $G \subseteq X^* \times \mathcal{A} \times X^*$ is a set of transitions.

A transducer is a finite state device that transforms an input word into an output word in stages, see [1,2]. At each stage, it reads a subword, transforms it, and changes state. A transition of a transducer has the form $p \xrightarrow{u/v} q$ where p and q are states and u is the input subword and v is its transformation. For our purposes, both u and v are elements of X^*, and final states of the transducer are labelled by subsets of $\mathcal{A}$.

Definition 2.1 *A labelled transducer $T = (Q, I, F, E, L)$ over X and $\mathcal{A}$, is a finite set of states Q, a set of initial states $I \subseteq Q$, a set of final states $F \subseteq Q$, a finite set of transitions $E \subseteq Q \times X^* \times X^* \times Q$ and a labelling $L : F \to 2^{\mathcal{A}}$.*

A transition $u \xrightarrow{a} v$ is *recognized* by a labelled transducer T if there is a path in T, $p_0 \xrightarrow{u_1/v_1} p_1 \xrightarrow{u_2/v_2} \cdots \xrightarrow{u_{n-1}/v_{n-1}} p_{n-1} \xrightarrow{u_n/v_n} p_n$, with $p_0 \in I$ and $p_n \in F$ and $u = u_1 \ldots u_n$ and $v = v_1 \ldots v_n$ and $a \in L(p_n)$. Labelled transducers provide a simple characterization of rational graphs, see [9]. Here we treat the characterization as a definition.

Definition 2.2 *A graph $G \subseteq X^* \times \mathcal{A} \times X^*$ is rational if, and only if, G is recognized by labelled transducer.*

A transducer is *normalised* if all its transitions have the form, $p \xrightarrow{u/v} q$, where $|u| + |v| = 1$. It is straightforward to show that any rational graph is generable from a normalised transducer.

For each a in $\mathcal{A}$, the subgraph G_a is the restriction of G to transitions labelled a. If G is rational then so is G_a, and we let T_a be the transducer that recognises G_a. If u is a vertex of G then $G_a(u)$ is the set of vertices $\{v \mid u \xrightarrow{a} v\}$.

Example 2.3 *The transition graph, below on the right, is generated by the transducer, below on the left.*

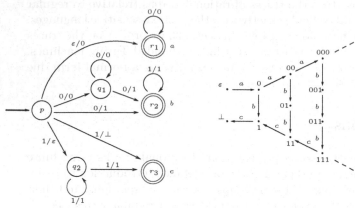

For example, $001 \xrightarrow{b} 011$ is recognized because of the following path, $p \xrightarrow{0/0} q_1 \xrightarrow{0/1} r_2 \xrightarrow{1/1} r_2$ and b is associated with the final state r_2.

A path in a graph G is a sequence of transitions $u_0 \xrightarrow{a_0} u_1 \xrightarrow{a_1} \dots \xrightarrow{a_n} u_n$. If $x = a_0 a_1 \dots a_n$ we write $u_0 \xrightarrow{x} u_n$ and we say that x is the trace of this path. The trace of a graph G, relative to a set I of initial vertices and a set F of final vertices, is the set of traces of all paths between vertices in I and vertices in F.

$$L(G, I, F) := \{x \mid \exists s \in I \; \exists t \in F, s \xrightarrow{x} t\}$$

In other words, the trace of a graph is "the language of its labels". In this paper, we are interested in the trace of a rational graph relative to regular vertex sets I and F.

In Section 4 we shall appeal to rational relations $R \subseteq X^* \times X^*$. A relation R is rational if it is induced by an unlabelled transducer over X: that is, a transducer that does not have a labelling function L, see [1,2].

3 Traces of Rational Graphs Are Context-Sensitive

In this section, we prove that the trace of a rational graph G, relative to regular vertex sets I and F, is a context-sensitive language. In fact, we only need to consider the trace of G relative to $I = \{\$\}$ and $F = \{\#\}$, where $\$$ and $\#$ are two new symbols. If G is rational then the following graph G',

$$G \cup (\bigcup_{a \in \mathcal{A}} \{\$\} \times \{a\} \times G_a(I)) \cup (\bigcup_{a \in \mathcal{A}} G_a^{-1}(F) \times \{a\} \times \{\#\}) \cup (\underbrace{\bigcup \{\$\} \times \{a\} \times \{\#\}}_{C})$$

where C is the constraint $a \in \mathcal{A}$, $G_a(I) \cap F \neq \emptyset$, is also rational and has the property, $L(G', \{\$\}, \{\#\}) = L(G, I, F)$.

The two most common characterizations of a context-sensitive (cs) language are that it is generated by a cs grammar and that it is recognized by a linear bounded machine, LBM, see for instance [8].

Definition 3.1 *A linear bounded machine, LBM, is a Turing machine such that the size of its tape is bounded, linearly, by the length of the input.*

At a first glance, one might expect that the trace of a rational graph is recognisable by a LBM because it can store information on its tape (namely the current vertex of the graph), and use it to compute new information (the next vertex). However, the linear bound is a problem as the next example illustrates.

Example 3.2 *The very simple transducer,* $p \xrightarrow{A/AA} p$ *(where p is both an initial and a final state labelled with a), produces the following graph G.*

$$A \xrightarrow{a} AA \xrightarrow{a} A^4 \xrightarrow{a} \cdots \xrightarrow{a} A^{2^n} \cdots$$
$$A^3 \xrightarrow{a} A^6 \xrightarrow{a} A^{12} \xrightarrow{a} \cdots \xrightarrow{a} A^{3.2^n} \cdots$$
$$A^5 \xrightarrow{a} A^{10} \xrightarrow{a} A^{20} \xrightarrow{a} \cdots \xrightarrow{a} A^{5.2^n} \cdots$$

$$\vdots \qquad \vdots \qquad \vdots$$

The trace of G relative to $I = \{A\}$ and $F = A^*$ is the language a^*. But, the length of vertices is exponential in the length of the word that is recognized. For example, the path recognizing a^3 is $A \xrightarrow{a} AA \xrightarrow{a} A^4 \xrightarrow{a} A^8$.

Our solution[1] is to work on transitions in parallel. If $u \xrightarrow{a} v \xrightarrow{b} w$ and the first "output" in the transducer T_a involves a transition $q \xrightarrow{\epsilon/X} q'$ then X will be the first element of v and therefore we can also activate T_b (and remember that q' is then the current state of T_a). In this way, as we shall see, we only ever need to be working with current head elements which may activate subsequent transitions.

Proposition 3.3 *Traces of rational graphs are cs-languages.*

Proof (Sketch). Assume a rational graph G that is generated by a normalised transducer T. We now show that $L(G, \{\$\}, \{\#\})$ is recognised by a LBM, M. Let $w \in \mathcal{A}^*$, and assume that w_k is the kth letter of w. The tape of M has length $2|w| + 3$, and it has a left-end marker X. The initial configuration of M is,

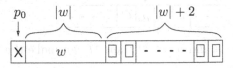

where the tape symbols are the left-end marker, the letters of $\mathcal{A}$, and then blank symbols. M then transforms this configuration into

---

[1] One might believe that it is possible to encode vertices in such a way that their lengths are linear in the length of the recognized word. But, the problem is then how to deduce the "next vertex function". In particular, if some branches of the transducer have a linear growth and others have an exponential growth, we are unable to construct the machine.

$$p_1$$
$$\downarrow$$

| X | i_{w_1} | \$ | i_{w_2} | ε | – | – | – | i_{w_k} | ε | – | – | – | i_{w_n} | ε | % | ε |

where ε and % are new tape-symbols, and i_{w_k} are symbols corresponding to an initial state of T_{w_k}. We assume that there are tape symbols for each state of the transducer T.

Now, we consider transitions where the head moves to the right. Assume that the current configuration is,

$$p_1$$
$$\downarrow$$

| – | – | – | p_a | A | p_b | C | – | – | – | – |

where A and C both belong to $\mathcal{A} \cup \{\$, \#, \varepsilon\}$, p_a is any state of T_a and p_b is either a state of T_b or %. There are two possible moves. First, if there is a transition $p_a \xrightarrow{A/\varepsilon} q_a$ in T_a then M can make the following step.

$$p_1$$
$$\downarrow$$

| – | – | – | q_a | ε | p_b | C | – | – | – | – | Move (a)

A is removed because the transition $p_a \xrightarrow{A/\varepsilon} q_a$ consumes this input. C is unchanged because ε is the output. We also assume that M remains in state p_1. Second, if there is a transition $p_a \xrightarrow{\varepsilon/B} q'_a$ in T_a and C is ε then M can make the following step.

$$p_1$$
$$\downarrow$$

| – | – | – | q'_a | A | p_b | B | – | – | – | – | Move (b)

A is not removed because the transition $p_a \xrightarrow{\varepsilon/B} q'_a$ consumes nothing. $C = \varepsilon$ is changed to B. Again, we assume M remains in state p_1.

Next, we consider transitions where the head moves to the left. If the current configuration is

$$p_1$$
$$\downarrow$$

| – | – | – | p_b | B | p_a | C | – | – | – | – |

and either $C = \varepsilon$ or ($C = \#$ and $p_a = \%$) then M can make the following step.

$$p_1$$
$$\downarrow$$

| – | – | – | p_b | B | p_a | C | – | – | – | – | Move (c)

Finally, there is the case when M checks for success. This happens either when the head is scanning % and the next symbol is # or the head is scanning f_{w_i} and the next symbol is ε.

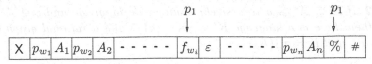

M now checks that each p_{a_i} is a final state f_{a_i} of T_{a_i} and that A_i is ε, and if this is true, then M enters a final accepting state. It is now not difficult to show that M recognizes the language $L(G, \{\$\}, \{\#\})$. $\qquad\qquad\qquad\square$

4 Cs-Languages Are Traces of Rational Graphs

In this section we prove the converse inclusion that every cs-language is the trace of a rational graph relative to regular initial and final sets of vertices. The proof is subtle, for reasons that we shall explain, and uses Penttonen's characterization of a cs-language as the image of a linear language under a length preserving left cs transformation [11].

A cs-language is generated by a cs grammar. A cs grammar Γ consists of a finite set of nonterminals N, a finite set of terminals T and a finite set of productions P, each of which has the form $UAW \to UVW$, where U and W are in $(N \cup T)^*$, A in N and V in $(N \cup T)^+$. An application of the production $UAW \to UVW$ in Γ to U_1UAWW_1 produces U_1UVWW_1, which we write as $U_1UAWW_1 \underset{\Gamma}{\to} U_1UVWW_1$. Moreover, we assume that $\underset{\Gamma}{\overset{*}{\to}}$ is the reflexive and transitive closure of $\underset{\Gamma}{\to}$. Sometimes, we shall use the notation $\underset{\Gamma}{\overset{*}{\to}}(U)$ to be the set $\{V \mid U \underset{\Gamma}{\overset{*}{\to}} V\}$. The language of $U \in (N \cup T)^+$, denoted by $\mathsf{L}(U)$, is the set of words $\left\{u \in T^* \mid U \underset{\Gamma}{\to}{}^* u\right\}$. Usually, there is a start symbol $S \in N$ of Γ, in which case the language generated by the grammar, $\mathsf{L}(\Gamma)$, is $\mathsf{L}(S)$. There are various normal forms for cs grammars, including the "left form" due to Penttonen [11].

Theorem 4.1 (Penttonen 74) *Every cs-language can be generated by a grammar, whose productions have the form*

$$A \to BC, AB \to AC, A \to a$$

where A, B, C are nonterminals and a is a terminal.

If ε-transitions are allowed in rational graphs and Γ is a cs grammar, then it is easy to construct a rational graph whose trace is $\mathsf{L}(\Gamma)$. One merely simulates the derivation $S \underset{\Gamma}{\to}{}^* v$ in the graph. Each production can be simulated by a transducer whose final state is labelled ϵ (for instance, $AB \to AC$ is captured by the transducer $p \xrightarrow{D/D} p$, $p \xrightarrow{AB/AC} q$ and $q \xrightarrow{D/D} q$ for each $D \in N$) and the graph outputs v at the end, $v \xrightarrow{v} \varepsilon$. However, the traces of rational graphs with ε-transitions coincide with the recursively enumerable languages. Indeed, following Knapik and Payet's exposition in [7], it is straightforward to prove the next result. It appeals to the standard notion of projection π with respect to a subalphabet: if $J \subseteq \mathcal{B}^*$ and $\mathcal{C} \subseteq \mathcal{B}$ then $\pi_\mathcal{C}(J)$ is the language that is the result of erasing all occurrences of letters in $\mathcal{B} - \mathcal{C}$ from words in J.

Corollary 4.2 *If $K \subseteq \mathcal{A}^*$ is a recursively enumerable language and $c \notin \mathcal{A}$ is a new letter, then there is a language $K' \subseteq (\mathcal{A} \cup \{c\})^*$ and a rational graph G whose trace is K' and $K = \pi_{\mathcal{A}}(K')$.*

This result is also a well known property of cs-languages (see, for example, [6]). However, the trace of a rational graph without ε-transitions is a recursive language, [9], and therefore the rational graphs with ε-transitions are more expressive than rational graphs.

What is the real problem of being able to generate a cs-language from a rational graph? As we can see simulating an application of a single production does not create a problem. And therefore simulating sequences of productions is also not problematic. The main issue is generating each word of length n in "real time", that is, in precisely n steps. This is in stark contrast with derivations in cs grammars. The number of applications of productions in a derivation $S \underset{\Gamma}{\to}^* u$ can be exponential in the length of the word u, as the next example illustrates.

Example 4.3

$$\Gamma \begin{cases} S \to AT & AB \to AC & AC \to AB \\ T \to RT \mid BD & BE \to BF & BG \to BD \\ R \to BD & CD \to CE & CF \to CG \\ DC \to DB & EB \to EC & GC \to GB \end{cases}$$

There is the derivation $S \underset{\Gamma}{\to}^* A(BD)^n$ just by applying the S, T and R rules. However, consider the number of steps it takes to replace an occurrence of D with G in $A(BD)^n$. For instance, when $n = 2$, $\boxed{AB}DBD \to A\boxed{CD}BD \to AC\boxed{EB}D \to ACE\boxed{CD} \to \boxed{AC}ECE \to A\boxed{BE}CE \to \boxed{AB}FCE \to A\boxed{CF}CE \to ACGCE$. To replace D at position $2n + 1$ with G requires 2^n applications of productions involving the initial letter A in the first position, as it requires 2^{n-1} applications of both $AB \to AC$ and $AC \to AB$. This also means that the initial A cannot be removed before the end of the computation.

Consequently, we use a more refined characterization of the cs-languages than that they are generable by cs grammars, that is due to Penttonen (Theorem 3 in [11]).

Theorem 4.4 (Penttonen 74) *There is a linear language L_{lin} such that every cs-language K can be represented in the form,*
$K = \left\{ u \in \mathcal{A}^* \mid \exists v \in L_{lin} \land \exists n \in \mathbb{N} \land v \underset{\tau}{\to}^n u \right\}$, *where τ is a length preserving left cs transformation.*

A *linear language* is generated from a context-free grammar where each production contains at most one nonterminal on its right hand side: for example, if $S \to aSb$, $S \to ab$ then $\mathsf{L}(S)$ is linear. In Theorem 4.4, a length preserving left cs transformation is a set of rewrite rules of the form $ua \to ub$ where $u \in \mathcal{A}^*$ and $a, b \in \mathcal{A}$. However, we shall work with linear languages whose alphabet is a set of nonterminals that, for us, will be vertex symbols. Clearly, Theorem 4.4 can be easily recast with respect to such an alphabet just by assuming each element of $\mathcal{A}$ is coded as a nonterminal A. However, we also need rules to transform nonterminals into terminals, which will just be of the form $A \to a$. A length preserving left cs transformation τ therefore is a set of productions of the form $UA \to UB$. The relation $\overset{*}{\underset{\tau}{\to}}$, therefore, has the important property that if $U \overset{*}{\underset{\tau}{\to}} V$ then the length of U is the same as the length of V. In fact, the proof of Theorem 4.4 in [11] shows that it is only necessary to consider productions of length at most 2.

Definition 4.5 *A 2-left-length-preserving transformation is a set of productions of the form, $AB \to AC$ and $A \to a$, where $A, B, C \in N$ and $a \in \mathcal{A}$.*

It is clear how to define a rational graph whose trace is a linear language, L_{lin}. However, we then need to "filter" it through a 2-left-length-preserving transformation τ. But, there is no obvious transducer associated with τ. In particular, the relation $\overset{*}{\underset{\tau}{\to}}$ may not be rational (that is, definable by a finite transducer), as the following example illustrates.

Example 4.6 *If τ is $\{AB \to AC, AC \to AA, CA \to CB\}$, then its rewriting relation is not rational.* $\overset{*}{\underset{\tau}{\to}}((AB)^*) \cap A^*B^* = \{A^n B^m \mid n \geqslant m\}$. $(AB)^*$ *is a regular language and, therefore, if $\overset{*}{\underset{\tau}{\to}}$ were rational then $\overset{*}{\underset{\tau}{\to}}((AB)^*)$ and $\{A^n B^m \mid n \geqslant m\}$ would also be regular sets.*

A more subtle construction is needed to rationally capture τ. Consider a derivation $U(1) \ldots U(n) \overset{*}{\underset{\tau}{\to}} V(1) \ldots V(n)$ where $n > 1$ and each $U(i)$ and $V(i)$ is a nonterminal. We can represent such a derivation in the following 2-dimensional form

$$
\begin{array}{ccc}
U_0(1) & \cdots & U_0(n) \\
U_1(1) & \cdots & U_1(n) \\
\vdots & & \vdots \\
U_m(1) & \cdots & U_m(n)
\end{array}
$$

where $U_0(i) = U(i)$ and $U_m(i) = V(i)$ and $U_i(1) \ldots U_i(n) \underset{\tau}{\to} U_{i+1}(1) \ldots U_{i+1}(n)$. As noted previously, m can be exponential in n. We wish to capture this derivation in $(n-1)$ steps with a rational relation R_τ in one left to right swoop. First, notice that in the first column there is just one element because $U(1) = V(1)$. In the second column there may be a number of different, and repeating, elements. For instance, if $U_i(2) \neq U_{i+1}(2)$ then $U(1)U_i(2) \underset{\tau}{\to} U(1)U_{i+1}(2)$. Consider the subsequence of this column with initial element $U_0(2)$ and subsequent elements when it changes, $[U_0(2), U_{i_1}(2), \ldots, U_{i_{k_2}}(2)]$, so the final element $U_{i_{k_2}}(2)$ is $V(2)$. Next, consider the third column starting with $U_0(3)$. Changes to this element may depend on elements in the second column and not just on $U(2)$. For example $U_{i_j}(2)U_\ell(3) \underset{\tau}{\to} U_{i_j}(2)U_{\ell+1}(3)$. And so on. What we now define is a rational relation R_τ which will include

$$[U(1)]U(2) \ldots U(n) \; R_\tau \; [U_1(2)U_{i_1}(2) \ldots U_{i_{k_2}}(2)]U(3) \ldots U(n)$$

and by composition, it will have the property

$$[U(1)]U(2)\ldots U(n) \; R_\tau^{j-1} \; [U_1(j)U_{i_1}(j)\ldots U_{i_{k_j}(j)}]U(j+1)\ldots U(n)$$

where R_τ^k is the $k$fold composition of R_τ. Consequently, this relation transforms a word of the form $[U]AV$ into a word $[AU']V$ where $U \in N^+$ and $V, U' \in N^*$ and $A \in N$ which says that A may be rewritten to elements of U' in turn, depending on the changing context represented by U. We now define R_τ.

Definition 4.7 *If τ is a 2-left-length-preserving transformation and $X = N \cup \{[,]\}$, then $R_\tau \subseteq X^* \times X^*$ is the rational relation recognized by the following transducer T_τ.*

$$T_\tau \begin{cases} I \xrightarrow{[X/[A} (A, X, A) & \forall A, X \in N & \textit{Type 1} \\ (A, X, Y) \xrightarrow{\varepsilon/Z} (A, X, Z) & \forall A, X, Y, Z \in N \text{ and } XY \xrightarrow{\tau} XZ & \textit{Type 2} \\ (A, X, Y) \xrightarrow{Z/\varepsilon} (A, Z, Y) & \forall A, X, Y, Z \in N & \textit{Type 3} \\ (A, X, Y) \xrightarrow{]A/]} F & \forall A, X, Y \in N & \textit{Type 4} \\ F \xrightarrow{X/X} F & \forall X \in N & \textit{Type 5} \end{cases}$$

States of this transducer are: (X, Y, Z) for all $X, Y, Z \in N$, the initial state I and the final state F.

Example 4.8 *Assume $\tau = \{AB \to AC, AC \to AB, CB \to CE, BE \to BF\}$.*

$$[A]B \; R_\tau \; [BCB]: \begin{array}{c|c} A & B \\ A & C \\ A & B \end{array}$$

$$[BCB]BAB \; R_\tau \; [BEF]AB: \begin{array}{c|c|c} AB & B & AB \\ AC & B & AB \\ AC & E & AB \\ AB & E & AB \\ AB & F & AB \end{array}$$

The relation R_τ underlies a rational graph whose trace is the corresponding language. Before proving this, we need a technical lemma.

Lemma 4.9 *If $U, V \in N^*$ and $|U| = |V| = n$, then the following statements are equivalent.*

(i) $U \xrightarrow{\tau}^* V$, using only productions of the form $AB \to AC \in \tau$

(ii) $[U(1)]U(2) \cdots U(n) \; R_\tau^{n-1} \; [U(n)WV(n)]$, for some W in N^*

Proof. (Sketch) Let U and V belong to N^*, and assume that they have equal length n. Moreover, assume that $\{r_1, r_2, \cdots, r_k\}$ are all the productions in τ of the form $AB \to AC$. First, we show $(i) \Rightarrow (ii)$. If $U \xrightarrow{\tau}^* V$, then there exists $m \geqslant 1$ such that $U = U_1 \xrightarrow{r_{\ell_1}} U_2 \cdots \to \xrightarrow{r_{\ell_{m-1}}} U_m = V$. For each j between 1 and $m-1$ assume that I_j is the set $\{i \mid U_i(j) \neq U_{i+1}(j)\}$: that is, it is the set of indices i such that the rule r_{ℓ_i} of τ changes the jth letter of U_i. Notice that $I_1 = \emptyset$, and if $i \neq j$ then $I_i \cap I_j = \emptyset$ and, moreover, $\bigcup_{j=1}^{n} I_j = [m]$. The following holds, for all $j \geq 2$.

$$[U_1(j-1) \prod_{i \in I_{j-1}} U_{i+1}(j-1)]U_1(j) \; R_\tau \; [U_1(j) \prod_{i \in I_j} U_{i+1}(j)]$$

Therefore, there is a path in T_τ which starts in I and ends in F that is labelled $[U_1(j-1) \prod_{i \in I_{j-1}} U_{i+1}(j-1)]U_1(j)$ on the left and $[U_1(j) \prod_{i \in I_j} U_{i+1}(j)]$ on the right. By Definition 4.7, therefore, there is the following transition in T_τ.

$$I \xrightarrow{[U_1(j-1)/[U_1(j)} (U_1(j), U_1(j-1), U_1(j))$$

Using a type 4 transition, there is a transition to the final state.

$$(U_m(j), U_m(j-1), U_m(j)) \xrightarrow{]U_m(j)/]} F$$

Thus, for all $j \geqslant 2$

$$[U_1(j-1) \prod_{i \in I_{j-1}} U_{i+1}(j-1)]U_1(j) \; R_\tau \; [U_1(j) \prod_{i \in I_j} U_{i+1}(j)]$$

By induction on j it follows that

$$[U(1)]U(2) \cdots U(n) \; R_\tau{}^{j-1} \; [U_1(j) \prod_{i \in I_j} U_{i+1}(j)]U_1(j+1) \cdots U_1(n)$$

The proof $(ii) \Rightarrow (i)$ is easier. Associated with any $k$fold composition of R_τ is a sequence of applications of productions of τ.   □

This lemma is used in the proof of the next result.

Proposition 4.10 *Cs-languages are traces of rational graphs.*

Proof. Assume K is a cs-language, and let τ be a 2-left-length-preserving transformation, by Theorem 4.4, such that $\rightarrow_\tau^*(L_{lin}) = K$. We use the relation R_τ of Definition 4.7 to construct a rational graph. For each letter a in $\mathcal{A}$, N_a is the set of nonterminals A such that $A \rightarrow_\tau a$. Let R_a be the relation $R_\tau \cap [N^* N_a]N^+ \times [N^*]N^*$ which is rational, see [2]. Therefore, the following graph G_0 is rational, $G_0 = \bigcup_{a \in \mathcal{A}} \{(x, a, y) \mid (x, y) \in R_a\}$. Therefore, the following graph G_1 is also rational.

$$G_1 = G_0 \cup \bigcup_{a \in \mathcal{A}} [N^* N_a] \times \{a\} \times \{\varepsilon\}$$

Let $[L_{lin}]$ be the language $\{[A]U \mid A \in N \wedge U \in N^* \wedge AU \in L_{lin}\}$. Therefore, $L(G_1, [L_{lin}], \{\varepsilon\}) = K$.

$$u \in L(G_1, [L_{lin}], \{\varepsilon\}), \text{and the length of } u \text{ is } n$$

$$\Leftrightarrow \quad [V(1)]V(2) \cdots V(n) \xrightarrow[G_1]{u} \varepsilon$$

$$\Leftrightarrow \quad [V(1)]V(2) \cdots V(n) \; R_{u(1)} \; [V(2) \cdots V'(2)]V(3) \cdots V(n) \; R_{u(2)} \cdots$$
$$\cdots R_{u(n-1)}[V(n) \cdots V'(n)] \; R_{u(n)} \; \varepsilon$$

$$\overset{\text{Lemma } 4.9}{\Longleftrightarrow} \quad V(1)V(2) \cdots V(n) \underset{\tau}{\to}^* V(1)V'(2) \cdots V'(n)$$

$$\text{and } V(1) \underset{\tau}{\to} u(1) \wedge \forall i \in [2 \ldots n], \; V'(i) \underset{\tau}{\to} u(i)$$

$$\Leftrightarrow \quad u \in K$$

The problem is now that $[L_{lin}]$ is not a regular language. To get a proper converse of Proposition 3.3 we need to find a rational graph whose trace between two regular sets is K. However, $[L_{lin}]$ is generable by a context-free grammar Γ in Greibach normal form, where every production has a single nonterminal (belonging to N) on its right hand side. The relation R_τ transforms a word in L_{lin} from left to right, and therefore G_1 can be transformed into G in such a way that it starts with extended words in Γ that begin with precisely two terminal letters. Let I be this finite set of words. The transducer generating G will also apply productions of Γ to obtain more letters of N so that the computation continues. Consequently, it now follows that $K = L(G, I, \{\varepsilon\})$. $\qquad\square$

The proof of Proposition 4.10 constructs a graph that may have infinite degree. However, there can only be infinite degree if there are cycles in the transformation τ. It is possible to remove such cycles: for any 2-left-length-preserving transformation τ, there is an equivalent 2-left-length-preserving transformation τ' which is acyclic. Therefore, it follows that for any cs-language there is a rational graph of finite, but not necessarily bounded, degree whose trace coincides with the language.

Proposition 4.11 *If $K \subseteq \mathcal{A}^*$ is a cs-language, then there is a rational graph G of finite degree such that $K = L(G, I, F)$ where I and F are finite sets of vertices.*

Combining Propositions 3.3 and 4.10, we thereby obtain the main result of the paper.

Theorem 4.12 *Traces of rational graphs are all the cs-languages.*

5   Conclusion

We have shown that the traces of rational graphs relative to regular initial and final sets of vertices coincide with the cs-languages. This new characterization of the cs-languages may offer new insights into these languages. For instance, an interesting question is: is it possible to define a subfamily of rational graphs whose traces are the deterministic cs-languages?

Acknowledgements

The authors are grateful to Thierry Cachat, Didier Caucal and Jean-Claude Raoult for their helpful comments on this paper.

References

1. AUTEBERT, J.-M. and BOASSON, L. *Transductions rationelles.* MASSON, 1988. 549, 550
2. BERSTEL, J. *Transductions and context-free languages.* Teubner, 1979. 549, 550, 557
3. MOLLER, F. BURKART, O., CAUCAL, D. and STEFFEN B. *Handbook of Process Algebra*, chapter Verification on infinite structures, pages 545–623. Elsevier, 2001. 548
4. CAUCAL, D. On transition graphs having a decidable monadic theory. In Icalp 96, volume 1099 of *LNCS*, pages 194–205, 1996. 548
5. COURCELLE, B. *Handbook of Theoretical Computer Science*, chapter Graph rewriting: an algebraic and logic approach. Elsevier, 1990. 548
6. GINSBURG, S. and ROSE, G. F. Preservation of languages by transducers. *Information and control*, 9:153–176, 1966. 554
7. KNAPIK, T. and PAYET, E. Synchronization product of linear bounded machines. In *FCT*, volume 1684 of *LNCS*, pages 362–373, 1999. 549, 553
8. MATEESCU, A., SALOMAA, A. *Handbook of Formal Languages*, volume 1, chapter Aspects of classical language theory, pages 175–252. Springer-Verlag, 1997. 549, 550
9. MORVAN, C. On rational graphs. In Fossacs 00, volume 1784 of *LNCS*, pages 252–266, 2000. 548, 549, 554
10. MULLER, D. and SCHUPP, P. The theory of ends, pushdown automata, and second-order logic. *Theoretical Computer Science*, 37:51–75, 1985. 548
11. PENTTONEN, M. One-sided and two-sided context in formal grammars. *Information and Control*, 25:371–392, 1974. 549, 553, 554, 555

Towards Regular Languages over Infinite Alphabets

Frank Neven[1]*, Thomas Schwentick[2], and Victor Vianu[3]**

[1] Limburgs Universitair Centrum
[2] Friedrich-Schiller-Universität Jena
[3] U.C. San Diego

Abstract. Motivated by formal models recently proposed in the context of XML, we study automata and logics on strings over infinite alphabets. These are conservative extensions of classical automata and logics defining the regular languages on finite alphabets. Specifically, we consider register and pebble automata, and extensions of first-order logic and monadic second-order logic. For each type of automaton we consider one-way and two-way variants, as well as deterministic, non-deterministic, and alternating control. We investigate the expressiveness and complexity of the automata, their connection to the logics, as well as standard decision problems.

1 Introduction

One of the significant recent developments related to the World Wide Web (WWW) is the emergence of the Extensible Markup Language (XML) as the standard for data exchange on the Web [1]. Since XML documents have a tree structure (usually defined by DTDs), XML queries can be modeled as mappings from trees to trees (tree transductions), and schema languages are closely related to tree automata, automata theory has naturally emerged as a central tool in formal work on XML [13,14,15,16,17,18,19]. The connection to logic and automata proved very fruitful in understanding such languages and in the development of optimization algorithms and static analysis techniques. However, these abstractions ignore an important aspect of XML, namely the presence of *data values* attached to leaves of trees, and comparison tests performed on them by XML queries. These data values make a big difference – indeed, in some cases the difference between decidability and undecidability (e.g., see [3]). It is therefore important to extend the automata and logic formalisms to trees with data values. In this initial investigation we model data values by infinite alphabets, and consider the simpler case of strings rather than trees. Understanding such automata on strings is also interesting for the case of trees, as most formalisms allow reasoning along paths in the tree. In the case of XML, it would be more

---

* Post-doctoral researcher of the Fund for Scientific Research, Flanders.
** This author supported in part by the National Science Foundation under grant number IIS-9802288.

J. Sgall, A. Pultr, and P. Kolman (Eds.): MFCS 2001, LNCS 2136, pp. 560–572, 2001.

accurate to consider strings labeled by a finite alphabet and attach data values to positions in the string. However, this would render the formalism more complicated and has no bearing on the results. Although limited to strings, we believe that our results provide a useful starting point in investigating the more general problem. In particular, our lower-bound results will easily be extended to trees.

We only consider models which accept precisely the regular languages when restricted to finite alphabets. It is useful to observe that for infinite alphabets it is no longer sufficient to equip automata with states alone. Indeed, automata should at least be able to check equality of symbols. There are two main ways to do this: (1) store a finite set of positions and allow equality tests between the symbols on these positions; (2) store a finite set of symbols only and allow equality tests with these symbols. The first approach, however, leads to multi-head automata, immediately going beyond regular languages. Therefore, we instead equip automata with a finite set of *pebbles* whose use is restricted by a stack discipline. The automaton can test equality by comparing the pebbled symbols. In the second approach, we follow Kaminski and Francez [11,10] and extend finite automata with a finite number of *registers* that can store alphabet symbols. When processing a string, an automaton compares the symbol on the current position with values in the registers; based on this comparison it can decide to store the current symbol in some register. In addition to automata, we consider another well-known formalism: monadic second-order logic (MSO). To be precise, we associate to strings first-order structures in the standard way, and consider the extensions of MSO and FO denoted by MSO* and FO*, as done by Grädel and Gurevich in the context of meta-finite models [6].

Our results concern the expressive power of the various models, provide lower and upper complexity bounds, and consider standard decision problems. For the above mentioned automata models we consider deterministic (D), non-deterministic (N), and alternating (A) control, as well as one-way and two-way variants. We denote these automata models by dC-X where $d \in \{1, 2\}$, $C = \{D, N, A\}$, and $X \in \{RA, PA\}$. Here, 1 and 2 stand for one- and two-way, respectively, D, N, and A stand for deterministic, non-deterministic, and alternating, and PA and RA for pebble and register automata. Our main results are the following (the expressiveness results are also represented in Figure 1).

Registers. We pursue the investigation of register automata initiated by Kaminski and Francez [11]. In particular, we investigate the connection between RAs and logic and show that they are essentially incomparable. Indeed, we show that MSO* cannot define 2D-RA. Furthermore, there are even properties in FO* that cannot be expressed by 2A-RAs. Next, we consider the relationship between the various RA models. We separate 1N-RAs, 2D-RAs, 2N-RAs, and 2A-RAs, subject to standard complexity-theoretic assumptions.

Pebbles. We consider two kinds of PAs: one where every new pebble is placed on the first position of the string and one where every new pebble is placed on the position of the current pebble. We refer to them as strong and weak PAs, respectively. Clearly, this pebble placement only makes a difference in the case of one-way PAs. In the one-way case, strong 1D-PA can simulate FO* while weak 1N-PA cannot (whence the names). Furthermore, we show that all pebble automata variants can be defined in MSO*. Finally, we provide more evidence that strong PAs are a robust notion by showing that the power of strong 1D-PA, strong 1N-PA, 2D-PA, and 2N-PA coincide.

Decision Problems. Finally, we consider decision problems for RAs and PAs, and answer some open questions from Kaminsky and Francez We show that universality and containment of 1N-RAs and non-emptiness of 2D-RA are undecidable and that non-emptiness is undecidable even for weak 1D-PAs.

As RAs are orthogonal to logically defined classes one might argue that PAs are better suited to define the notion of regular languages over infinite alphabets. Indeed, they are reasonably expressive as they lie between FO* and MSO*. Furthermore, strong PAs form a robust notion. Adding two-wayness and nondeterminism does not increase expressiveness and the class of languages is closed under Boolean operations, concatenation and Kleene star. Capturing exactly MSO* most likely requires significant extensions of PAs, as in MSO* one can express complete problems for every level of the polynomial hierarchy, while computations of 2A-PAs are in P.

Related work. Kaminski and Francez were the first to consider RAs (which they called *finite-memory automata*) to handle strings over infinite alphabets. They showed that 1N-RAs are closed under union, intersection, concatenation, and Kleene star. They further showed that non-emptiness is decidable for 1N-RAs and that containment is decidable for 1N-RAs when the automaton in which containment has to be tested has only two registers.

When the input is restricted to a finite alphabet, PAs recognize the regular languages, even in the presence of alternation [14]. We point out that the pebbling mechanism we employ is based on the one of Milo, Suciu, and Vianu [14] and is more liberal than the one used by Globerman and Harel [7]: indeed, in our case, after a pebble is placed the automaton can still walk over the whole string and sense the presence of the other pebbles. Globerman and Harel prove certain lower bounds in the gap of succinctness of the expressibility of their automata.

Overview. This paper is organized as follows. In Section 2, we provide the formal framework. In Section 3, we study register automata. In Section 4, we examine pebble automata. In Section 5, we compare the register and pebble models. In Section 6, we discuss decision problems. We conclude with a discussion in Section 7. Due to space limitations proofs are only sketched.

2   Definitions

We consider strings over an infinite alphabet $\mathbf{D}$. Formally, a $\mathbf{D}$-*string* w is a finite sequence $d_1 \cdots d_n \in \mathbf{D}^*$. As we are often dealing with 2-way automata we delimit input strings by two special symbols, $\triangleright, \triangleleft$ for the left and the right end of the string, neither of which is in $\mathbf{D}$. I.e., automata always work on strings of the form $w = \triangleright v \triangleleft$, where $v \in \mathbf{D}^*$. By $\mathrm{dom}(w)$ we denote the set $\{1, \ldots, |w|\}$ with $|w|$ the length of w. For $i \in \mathrm{dom}(w)$, we also write $\mathrm{val}_w(i)$ for d_i.

2.1 Register Automata

Definition 1 ([11,10]). A *k-register automaton* $\mathcal{B}$ is a tuple (Q, q_0, F, τ_0, P) where Q is a finite set of states; $q_0 \in Q$ is the initial state; $F \subseteq Q$ is the set of final states; $\tau_0 : \{1, \ldots, k\} \rightarrow \mathbf{D} \cup \{\triangleright, \triangleleft\}$ is the initial register assignment; and, P is a finite set of transitions of the forms $(i, q) \rightarrow (q', d)$ or $q \rightarrow (q', i, d)$. Here, $i \in \{1, \ldots, k\}$, $q, q' \in Q$ and $d \in \{\text{stay}, \text{left}, \text{right}\}$.

Given a string w, a *configuration of* $\mathcal{B}$ *on* w is a tuple $[j, q, \tau]$ where $j \in \mathrm{dom}(w)$, $q \in Q$, and $\tau : \{1, \ldots, k\} \rightarrow \mathbf{D} \cup \{\triangleright, \triangleleft\}$. The *initial* configuration is $\gamma_0 := [1, q_0, \tau_0]$. A configuration $[j, q, \tau]$ with $q \in F$ is *accepting*. Given $\gamma = [j, q, \tau]$, the transition $(i, p) \rightarrow \beta$ (respectively, $p \rightarrow \beta$) *applies* to γ iff $p = q$ and $\mathrm{val}_w(j) = \tau(i)$ (respectively, $\mathrm{val}_w(j) \neq \tau(i)$ for all $i \in \{1, \ldots, k\}$).

Given $\gamma = [j, q, \tau]$ and $\gamma' = [j', q', \tau']$, we define the one step transition relation $\vdash$ on configurations as follows: $\gamma \vdash \gamma'$ iff there is a transition $(i, q) \rightarrow (q', d)$ that applies to γ, $\tau' = \tau$, and $j' = j$, $j' = j - 1$, $j' = j + 1$ whenever $d = \text{stay}$, $d = \text{left}$, or $d = \text{right}$, respectively; or there is a transition $q \rightarrow (q', i, d)$ that applies to γ, j' is as defined in the previous case, and τ' is obtained from τ by setting $\tau'(i)$ to $\mathrm{val}_w(j)$. We denote the transitive closure of $\vdash$ by $\vdash^*$. Intuitively, transitions $(i, q) \rightarrow (q', d)$ can only be applied when the value of the current position is in register i. Transitions $q \rightarrow (q', i, d)$ can only be applied when the value of the current position differs from all the values in the registers. In this case, the current value is copied into register i.

We require that the initial register assignment contains the symbols $\triangleright$ and $\triangleleft$, so automata can recognize the boundaries of the input. Furthermore, from a $\triangleright$ only right-transitions and from a $\triangleleft$ only left-transitions are allowed.

As usual, a string w is accepted by $\mathcal{B}$, if $\gamma_0 \vdash^* \gamma$, for some accepting configuration γ. The language $L(\mathcal{B})$ accepted by $\mathcal{B}$, is defined as $\{v \mid \triangleright v \triangleleft$ is accepted by $\mathcal{B}\}$.

The automata we defined so far are in general *non-deterministic*. An automaton is *deterministic*, if in each configuration at most one transition applies. If there are no left-transitions, then the automaton is *one-way*. Alternating automata are defined in the usual way. As explained in the introduction, we refer to these automata as dC-RA where $d \in \{1, 2\}$ and $C = \{\mathrm{D}, \mathrm{N}, \mathrm{A}\}$. Clearly, when the input is restricted to a finite alphabet, RAs accept only regular languages.

2.2 Pebble Automata

We borrow some notation from Milo, Suciu, and Vianu [14].

Definition 2. A *k-pebble automaton* $\mathcal{A}$ *over* $\mathbf{D}$ is a tuple (Q, q_0, F, T) where

- Q is a finite set of *states*; $q_0 \in Q$ is the *initial state*; $F \subseteq Q$ is the set of *final states*; and,
- T is a finite set of *transitions* of the form $\alpha \to \beta$, where α is of the form (i, s, P, V, q) or (i, P, V, q), where $i \in \{1, \ldots, k\}$, $s \in \mathbf{D} \cup \{\triangleright, \triangleleft\}$, $P, V \subseteq \{1, \ldots, i-1\}$, and β is of the form (q, d) with $q \in Q$ and $d \in \{$stay, left, right, place-new-pebble, lift-current-pebble$\}$.

Given a string w, a *configuration of $\mathcal{A}$ on w* is of the form $\gamma = [i, q, \theta]$ where $i \in \{1, \ldots, k\}$, $q \in Q$ and $\theta : \{1, \ldots, i\} \to \mathrm{dom}(w)$. We call θ a *pebble assignment* and i the *depth* of the configuration (and of the pebble assignment). Sometimes we denote the depth of a configuration γ (pebble assignment θ) by depth(γ) (depth(θ)). The *initial* configuration is $\gamma_0 := [1, q_0, \theta_0]$ where $\theta_0(1) = 1$. A configuration $[i, q, \theta]$ with $q \in F$ is *accepting*.

A transition $(i, s, P, V, p) \to \beta$ *applies to* a configuration $\gamma = [j, q, \theta]$, if

1. $i = j$, $p = q$,
2. $P = \{l < i \mid \mathrm{val}_w(\theta(l)) = \mathrm{val}_w(\theta(i))\}$,
3. $V = \{l < i \mid \theta(l) = \theta(i)\}$, and
4. $\mathrm{val}_w(\theta(i)) = s$.

A transition $(i, P, V, q) \to \beta$ *applies* to γ if (1)-(3) hold and no transition $(i', s', P', V', q') \to \beta$ applies to γ. Intuitively, $(i, s, P, V, p) \to \beta$ applies to a configuration, if i is the current number of placed pebbles, p is the current state, P is the set of pebbles that see the same symbol as the top pebble, V is the set of pebbles that are at the same position as the top pebble, and the current symbol seen by the top pebble is s. We define the transition relation $\vdash$ as follows: $[i, q, \theta] \vdash [i', q', \theta']$ iff there is a transition $\alpha \to (p, d)$ that applies to γ such that $q' = p$ and $\theta'(j) = \theta(j)$, for all $j < i$, and

- if $d = $ stay, then $i' = i$ and $\theta'(i) = \theta(i)$,
- if $d = $ left, then $i' = i$ and $\theta'(i) = \theta(i) - 1$,
- if $d = $ right, then $i' = i$ and $\theta'(i) = \theta(i) + 1$,
- if $d = $ place-new-pebble, then $i' = i + 1$, $\theta'(i+1) = \theta'(i) = \theta(i)$,
- if $d = $ lift-current-pebble then $i' = i - 1$.

The definitions of the *accepted language, deterministic, alternating* and *one-way* are analogous to the case of register automata. We refer to these automata as dC-RA where d and C are as before.

In the above definition, new pebbles are placed at the position of the most recent pebble. An alternative would be to place new pebbles at the beginning of the string. While the choice makes no difference in the two-way case, it is significant in the one-way case. We refer to the model as defined above as *weak* pebble automata and to the latter as *strong* pebble automata. Strong pebble automata are formally defined by setting $\theta'(i+1) = 1$ (and keeping $\theta'(i) = \theta(i)$) in the *place-new-pebble* case of the definition of the transition relation.

2.3   Logic

We consider first-order and monadic second-order logic over **D**-strings. The representation as well as the logics are special instances of the meta-finite structures and their logics as defined by Grädel and Gurevich [6]. A string w is represented by the logical structure with domain $\text{dom}(w)$, the natural ordering $<$ on the domain, and a function $\text{val} : \text{dom}(w) \to \mathbf{D}$ instantiated by val_w. An atomic formula is of the form $x < y$, $\text{val}(x) = \text{val}(y)$, or $\text{val}(x) = d$ for $d \in \mathbf{D}$, and has the obvious semantics. The logic FO* is obtained by closing the atomic formulas under the boolean connectives and first-order quantification over $\text{dom}(w)$. Hence, no quantification over **D** is allowed. The logic MSO* is obtained by adding quantification over sets over $\text{dom}(w)$; again, no quantification over **D** is allowed.

2.4   Complexity Classes over Infinite Alphabets

Some of our separating results are relative to complexity-theoretic assumptions. To this end, we assume the straightforward generalization of standard complexity classes (like LOGSPACE, NLOGSPACE, and PTIME) to the case of infinite alphabets that can be defined, e.g., by using multi-tape Turing machines which are able to compare and move symbols between the current head positions. It should be clear that the collapse of two of these generalized classes immediately implies the collapse of the respective finite alphabet classes.

3   Register Automata

We start by investigating RAs. In particular, we compare them with FO* and MSO*. Our main conclusion is that RAs are orthogonal to these logics as they cannot even express all FO* properties but can express properties not definable in MSO*. Further, we separate the variants of RAs subject to standard complexity-theoretic assumptions.

3.1   Expressiveness

Theorem 3. *MSO* cannot define all 2D-RA.*

Proof. (sketch) Consider strings of the form $u\#v$ where $u, v \in (\mathbf{D} - \{\#\})^*$. Define N_u and N_v as the set of symbols occurring in u and v, respectively. Denote by n_u and n_v their cardinalities. We show that there is a 2D-RA $\mathcal{A}$ that accepts $u\#v$ iff $n_u = n_v$ while there is no such MSO* sentence.

A 2D-RA can recognize this property by checking that the number of leftmost occurences of symbols is the same in u and v. To this end it moves back and forth between leftmost occurences of symbols in u and v.   □

We next show that RAs cannot capture FO*, even with alternation. The proof is based on communication complexity[9].

Theorem 4. *2A-RA cannot express FO*.*

Proof. We start with some terminology. Let D be a finite or infinite set. A *1-hyperset over D* is a finite subset of D. For $i > 1$, an *i-hyperset over D* is a finite set of $(i-1)$-hypersets over D. For clarity, we will often denote i-hypersets with a superscript i, as in $S^{(i)}$. Let us assume that $\mathbf{D}$ contains all natural numbers and let, for $j > 0$, $\mathbf{D}_j$ be $\mathbf{D} - \{1, \ldots, j\}$. Let $j > 0$ be fixed. We inductively define *encodings* of i-hypersets over $\mathbf{D}_j$. A string $w = 1d_1d_2 \cdots d_n 1$ over $\mathbf{D}_j$ is an encoding of the 1-hyperset $H(w) = \{d_1, \ldots, d_n\}$ over $\mathbf{D}_j$. For each $i \le j$, and encodings $w_1, \ldots, w_n$ of $(i-1)$-hypersets, $iw_1iw_2 \cdots iw_ni$ is an encoding of the i-hyperset $\{H(w_i) \mid i \le n\}$. Define $L_=^m$ as the language $\{u\#v \mid u \text{ and } v \text{ are encodings of } m\text{-hypersets over } \mathbf{D}_m - \{\#\} \text{ and } H(u) = H(v)\}$.

Lemma 5. *For each m, $L_=^m$ is definable in FO^*.*

Next, we show that no 2A-RA can recognize $L_=^m$ for $m > 4$. The idea is that for large enough m, a 2A-RA simply cannot communicate enough information between the two sides of the input string to check whether $H(u)$ equals $H(v)$. Our proof is inspired by a proof of Abiteboul, Herr, and Van den Bussche [2]. To simulate 2A-RA, however, we need a more powerful protocol where the number of messages depends on the number of different data values in u and v. This protocol is defined next. Let $\exp_0(n) := n$ and $\exp_i(n) := 2^{\exp_{i-1}(n)}$, for $i > 0$.

Definition 6. Let P be a binary predicate on i-hypersets over $\mathbf{D}$ and let $k, l \ge 0$. We say that P can be *computed by a (k,l)-communication protocol* between two parties (denoted by I and II) if there is a polynomial p such that for all i-hypersets $X^{(i)}$ and $Y^{(i)}$ over a finite set D there is a finite alphabet Δ of size at most $p(|D|)$ such that $P(X^{(i)}, Y^{(i)})$ can be computed as follows: I gets $X^{(i)}$ and II gets $Y^{(i)}$; both know D and Δ; they send k-hypersets over Δ back and forth; and, after $\exp_l(p(|D|))$ rounds of message exchanges, both I and II have enough information to decide whether $P(X^{(i)}, Y^{(i)})$ holds. We refer to strings of the form $u\#v$, where u and v do not contain $\#$, as *split strings*. A communication protocol computes on such strings by giving u to I and v to II.

Lemma 7. *For $m > 4$, $L_=^m$ cannot be computed by a $(2,2)$-communication protocol.*

Proof. Suppose there is a protocol computing $L_=^m$. For every finite set D with d elements, the number of different possible messages is the number of 2-hypersets which is at most $\exp_2(p(d))$. Call a complete sequence of exchanged messages $a_1b_1a_2b_2\ldots$ a *dialogue*. Every dialogue has at most $\exp_2(p(d))$ rounds. Hence, there are at most $\exp_2(2d \cdot \exp_2(p(d)))$ different dialogues. However, the number of different m-hypersets over D is $\exp_m(d)$. Hence, for $m > 4$ and D large enough there are m-hypersets $X^{(m)} \ne Y^{(m)}$ such that the protocol gives the same dialogue for $P(X^{(m)}, X^{(m)})$ and $P(Y^{(m)}, Y^{(m)})$, and therefore also on $P(X^{(m)}, Y^{(m)})$ and $P(Y^{(m)}, X^{(m)})$. This leads to the desired contradiction. □

Lemma 8. *On split strings, the language defined by a 2A-RA can be recognized by a $(2,2)$-communication protocol.*

Proof. (sketch) Let $\mathcal{B}$ be a 2A-RA working on split strings over $\mathbf{D}$. On an input string $w := u\#v$, $[e, q, \tau]$ is a #-configuration when e is the position of # in w. Define $p(n) := |Q|n^k$, where Q and k are the set of states and number of registers of $\mathcal{B}$, respectively. Then the number of #-configurations is $|Q|m^k$ where m is the number of different symbols in w. We assume w.l.o.g. that there are no transitions possible from final configurations. Further, we assume that $\mathcal{B}$ never changes direction or accepts at the symbol #. Hence, on w, when $\mathcal{B}$ leaves u to the right it enters v and vice versa. In essence, both parties compute partial runs where they send the #-configurations in which $\mathcal{B}$ walks off their part of the string to the other party. We omit further details.   □

Theorem 4 now follows from Lemmas 5, 7, and 8.   □

When restricted to one-way computations, RAs can only express "regular" properties. The next proposition is easily shown using standard techniques.

Proposition 9. *MSO* can simulate every 1N-RA.*

3.2   Control

On strings of a special shape, RAs can simulate multi-head automata. These strings are of even length where the odd positions contain pairwise distinct elements and the even positions carry an a or a b. By storing the unique ids preceding the a's and b's, the RA can remember the positions of the heads of a multi-head automaton. Note that 2D-RAs can check whether the input string is of the desired form. As deterministic, nondeterministic, and alternating multi-head automata recognize precisely LOGSPACE, NLOGSPACE, and PTIME languages, respectively, membership for 2D-RA, 2N-RA, and 2A-RA is hard for these classes, respectively [20,12]. Furthermore, it is easy to see that the respective membership problems also belong to the infinite alphabet variants of these classes. Thus, we can show the following proposition in which all complexity classes are over infinite alphabets.

Proposition 10.   *1. Membership of 2D-RA is complete for LOGSPACE;*
2. Membership of 2N-RA is complete for NLOGSPACE; and
3. Membership of 2A-RA is complete for PTIME.

The indexing technique above cannot be used for 1N-RAs, because they cannot check whether the odd positions form a unique index. However, we can extend (2) to 1N-RAs using a direct reduction from an NLOGSPACE-complete problem: ordered reachability.

Proposition 11. *Membership of 1N-RA is complete for NLOGSPACE.*

From Theorem 3 and Proposition 9 we get immediately that the class accepted by 1N-RA is different from those accepted by 2D-RA and 2N-RA. It follows from Proposition 11 that all four classes defined by the mentioned automata models are different unless the corresponding complexity classes collapse.

4   Pebble Automata

In this section we show that PAs are better behaved than RAs with regard to the connection to logic. In a sense, PAs are more "regular" than RAs. Indeed, we show that strong 1D-PAs can simlate FO* and that even the most liberal pebble model, 2A-PA, can be defined in MSO*. 1D-PAs are clearly more expressive than FO*; furthermore, we can separate 2A-RAs from MSO* under usual complexity-theoretic assumptions. Next, we show that weak one-way PAs do not suffice to capture FO*. Again, the proof is based on communication complexity. Finally, we prove that for strong PAs, the one-way, two-way, deterministic and nonde-terministic variants collapse. Together with the straightforward closure under Boolean operations, concatenation and Kleene star, these results suggest that strong PAs define a robust class of languages.

4.1   Expressiveness

Proposition 12. *FO* is strictly included in strong 1D-PA.*

PAs are subsumed by MSO*. Thus, they behave in a "regular" manner.

Theorem 13. *MSO* can simulate 2A-PA.*

It is open whether the above inclusion is strict. However, we can show the following.

Proposition 14. *For every $i \in \mathbb{N}$, there are MSO* formulas φ_i and ψ_i such that the model checking problem for φ_i and ψ_i is hard for Σ_i^P and Π_i^P, respectively. In contrast, membership for 2A-PAs is PTIME-complete.*

Here, the model checking problem for a logical formula consists in determining, given a string w, whether $w \models \varphi$. Since the first part of the proposition is already known for graphs, it suffices to observe that graphs can readily be encoded as strings. We end this section by considering weak PAs. Recall that this notion only makes a difference for one-way PAs. Unlike their strong counterparts, we show that they cannot simulate FO*, which justifies their name.

Theorem 15. *Weak 1N-PA cannot simulate FO*.*

Proof.(sketch) The proof is similar to the proof of Theorem 4. We show by a communication complexity argument that the FO*-expressible language $L_=^2$ defined in that proof cannot be recognized by a weak 1N-PA. In the current proof, however, we use a different kind of communication protocol which better reflects the behaviour of a weak 1N-PA. We define this protocol next. Again we use strings of the form $u\#v$ where u and v encode 2-hypersets. Let k be fixed and let S_1, S_2 be finite sets. The protocol has only one agent which has arbitrary access to the string u but only limited access to the string v. On u, its computational power is unlimited. The access to v is restricted as follows. There is a fixed function $f : \mathbf{D}^* \times \mathbf{D}^k \times S_1 \to S_2$ and the agent can evaluate f on all arguments $(v, \mathbf{d}, s)$, where $\mathbf{d}$ is a tuple of length k of symbols from u and $s \in S_1$. Based on this information and on u the agent decides whether $u\#v$ is accepted.   □

4.2   Control

Our next result shows that all variants of strong pebble automata without alternation collapse. Hence, strong PAs provide a robust model of automata.

Theorem 16. *The following have the same expressive power: 2N-PA, 2D-PA, strong 1N-PA and strong 1D-PA.*

Proof. We show that, for each 2N-PA $\mathcal{A}$, there is a strong 1D-PA $\mathcal{B}$ which accepts the same language. Actually, $\mathcal{B}$ will use the same number k of pebbles as $\mathcal{A}$. Very briefly, we show by an induction on i that, for all states q and pebble placements θ, $\mathcal{B}$ can compute all states q' such that $[i, q, \theta] \vdash^*_{>i} [i, q', \theta]$, where the subscript $> i$ indicates that only subcomputations are considered in which, at every step, more than i pebbles are present. $\qquad\square$

5   Registers versus Pebbles

The known inclusions between the considered classes are depicted in Figure 1. The pebble and register models are rather incomparable. From the connection with logic we can deduce the following. As 2D-RA can already express non-MSO* definable properties, no two-way register model is subsumed by a pebble model. Conversely, as strong 1D-PAs can already express FO*, no strong pebble model is subsumed by any two-way register model. Some open problems about the relationships between register and pebble automata are given in Section 7.

6   Decision Problems

We briefly discuss the standard decision problems for RAs and PAs. Kaminski and Francez already showed that non-emptiness of 1N-RAs is decidable and that it is decidable whether for a 1N-RA $\mathcal{A}$ and a 1N-RA $\mathcal{B}$ with 2-registers $L(\mathcal{A}) \subseteq L(\mathcal{B})$. We next show that *universality* (does an automaton accept every string) of 1N-RAs is undecidable, which implies that containment of *arbitry* 1N-RAs is undecidable. Kaminski and Francez further asked whether the decidability of non-emptiness can be extended to 2D-RAs: we show it cannot. Regarding PAs, we show that non-emptiness is already undecidable for weak 1D-PAs. This is due to the fact that, when PAs lift pebble i, the control is transferred to pebble $i - 1$. Therefore, even weak 1D-PAs can make several left-to-right sweeps of the input string.

Theorem 17. *1. It is undecidable whether a 1N-RA is universal.*
  2. Containment of 1N-RAs is undecidable.
  3. It is undecidable whether a 2D-RA is non-empty.
  4. It is undecidable whether a weak 1D-PA is non-empty.

7   Discussion

We investigated several models of computations for strings over an infinite alphabet. One main goal was to identify a natural notion of regular language and corresponding automata models. In particular, the extended notion should agree in the finite alphabet case with the classical notion of regular language. We considered two plausible automata models: RAs and PAs. Our results tend to favor PAs as the more natural of the two. Indeed, the expressiveness of PAs lies between FO* and MSO*. The inclusion of FO* provides a reasonable expressiveness lower bound, while the MSO* upper bound indicates that the languages defined by PAs remain regular in a natural sense. Moreover, strong PAs are quite robust: all variants without alternation (one or two-way, deterministic or non-deterministic) have the same expressive power.

Some of the results in the paper are quite intricate. The proofs bring into play a variety of techniques at the confluence of communication complexity, language theory, and logic. Along the way, we answer several questions on RAs left open by Kaminski and Francez.

Several problems remain open: (*i*) can weak 1D-PA or weak 1N-PA be simulated by 2D-RAs? (*ii*) are 1D-RA or 1N-RA subsumed by any pebble model? (We know that they can be defined in MSO*. As 1N-RAs are hard for NLOGSPACE they likely cannot be simulated by 2A-PAs.) (*iii*) are weak 1N-PAs strictly more powerful than weak 1D-PAs? (*iv*) are 2A-PAs strictly more powerful than 2N-PAs?

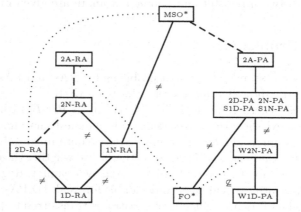

Fig. 1. Inclusions between the classes under consideration. Solid lines indicate inclusion (strictness shown as $\neq$), dotted lines indicate that the classes are incomparable. Dashed lines indicate strict inclusion subject to complexity-theoretic assumptions.

References

1. S. Abiteboul, P. Buneman, and D. Suciu. *Data on the Web : From Relations to Semistructured Data and XML.* Morgan Kaufmann, 1999. 560

2. S. Abiteboul, L. Herr, and J. Van den Bussche. Temporal connectives versus explicit timestamps to query temporal databases. *Journal of Computer and System Sciences*, 58(1):54–68, 1999. 566

3. N. Alon, T. Milo, F. Neven, D. Suciu, and V. Vianu. XML with Data Values: Typechecking Revisited. To be presented at LICS 2001. 560

4. J. Clark XSL Transformations (XSLT) specification, 1999. http://www.w3.org/TR/WD-xslt.

5. G. Rozenberg and A. Salomaa, editors. *Handbook of Formal Languages*, volume 3. Springer, 1997. 572

6. E. Grädel and Y. Gurevich. Metafinite model theory. *Information and Computation*, 140(1):26–81, 1998. 561, 565

7. N. Globerman and D. Harel. Complexity results for two-way and multi-pebble automata and their logics. *Theoretical Computer Science*, 169(2):161–184, 1996. 562

8. J. E. Hopcroft and J. D. Ullman. *Introduction to Automata Theory, Languages, and Computation.* Addison-Wesley, 1979.

9. J. Hromkovič. *Communication Complexity and Parallel Computing.* Springer-Verlag, 2000. 565

10. M. Kaminski and N. Francez. Finite-memory automata. In *Proceedings of 31th IEEE Symposium on Foundations of Computer Science (FOCS)*, pages 683–688, 1990. 561, 563

11. M. Kaminski and N. Francez. Finite-memory automata. *Theoretical Computer Science*, 134(2):329–363, 1994. 561, 563

12. A. K. Chandra, L. J. Stockmeyer. Alternation. In *Proceedings of 17th IEEE Symposium on Foundations of Computer Science (FOCS 1976)*, 98–108, 1976. 567

13. S. Maneth and F. Neven. Structured document transformations based on XSL. In R. Connor and A. Mendelzon, editors, *Research Issues in Structured and Semistructured Database Programming (DBPL'99)*, volume 1949 of *Lecture Notes in Computer Science*, pages 79–96. Springer, 2000. 560

14. T. Milo, D. Suciu, and V. Vianu. Type checking for XML transformers. In *Proceedings of the Nineteenth ACM Symposium on Principles of Database Systems*, pages 11–22. ACM Press, 2000. 560, 562, 563

15. F. Neven. Extensions of attribute grammars for structured document queries. In R. Connor and A. Mendelzon, editors, *Research Issues in Structured and Semistructured Database Programming (DBPL'99)*, volume 1949 of *Lecture Notes in Computer Science*, pages 97–114. Springer, 2000. 560

16. F. Neven and T. Schwentick. Query automata. In *Proceedings of the 18th ACM Symposium on Principles of Database Systems (PODS 1999)*, pages 205–214. ACM Press, 1999. 560

17. F. Neven and T. Schwentick. Expressive and efficient pattern languages for tree-structured data. In *Proceedings of 19th ACM Symposium on Principles of Database Systems (PODS 2000), Dallas*, pages 145–156, 2000. 560

18. F. Neven and J. Van den Bussche. Expressiveness of structured document query languages based on attribute grammars. In *Proceedings of 17th ACM Symposium on Principles of Database Systems (PODS 1998)*, pages 11–17. ACM Press, 1998. 560

19. Y. Papakonstantinou and V. Vianu. DTD inference for views of XML data. In *Proceedings of 19th ACM Symposium on Principles of Database Systems (PODS 2000)*, pages 35–46. ACM, Press 2000. 560

20. I. H. Sudborough. On tape-bounded complexity classes and multihead finite automata. *Journal of Computer and System Sciences*, 10(1):62–76, 1975. 567

21. W. Thomas. Languages, automata, and logic. In Rozenberg and Salomaa [5], chapter 7.

Partial Information and Special Case Algorithms

Arfst Nickelsen

Technische Universität Berlin, Fakultät für Informatik und Elektrotechnik
10623 Berlin, Germany
nicke@cs.tu-berlin.de

Abstract. The connection is investigated between two well known notions which deal with languages that show polynomial time behaviour weaker than membership decidability. One notion is polynomial time bi-immunity (p-bi-immunity). The other one is polynomial time $\mathcal{D}$-verboseness which captures p-selectivity, p-cheatability, p-verboseness and similar notions, where partial information about the characteristic function is computed. The type of partial information is determined by a family of sets of bitstrings $\mathcal{D}$.
A full characterization of those $\mathcal{D}$ for which there are p-bi-immune polynomially $\mathcal{D}$-verbose languages is given. Results of the same type for special cases of polynomial $\mathcal{D}$-verboseness were already given by Goldsmith, Joseph, Young [GJY93], Beigel [Bei90], and Amir, Gasarch [AG88].

1 Introduction

If a language A is not decidable in polynomial time ($A \notin$ P), one may ask whether it nevertheless exhibits some polynomial time behaviour. One then does not expect the polynomial time algorithm for the language A to actually answer the question "$x \in A$?" for all inputs x. Instead one weakens this demand in different ways. One way of weakening is to look for a *special case algorithm*. Such an algorithm decides membership in A not for all input words, but only for an infinite set of special cases. The existence of such a special case algorithm implies that A or $\overline{A}$, the complement of A, has an infinite polynomial time decidable subset. Languages for which such polynomial time algorithms do *not* exist are called polynomially bi-immune or p-bi-immune [BS85].

Alternatively, one may look for an algorithm which works for all instances instead of a small (though infinite) part of the instances, but does not completely solve the membership question. Instead it gives only some *partial information* on membership. To allow intermediate levels of information on membership in A, the algorithm is run on tuples of input words $(x_1, \ldots, x_n)$. To yield partial information then means to narrow the range of possibilities for values of $\chi_A(x_1, \ldots, x_n)$ (where χ_A is the characteristic function for A). The algorithm outputs a set $D \subseteq \{0,1\}^n$ such that $\chi_A(x_1, \ldots, x_n) \in D$. The sets D which are allowed as possible outputs of the algorithm specify the type of information that the algorithm computes.

This approach is not new. Many types of partial information have been studied, e.g., verboseness (or approximability), cheatability, frequency computations,

J. Sgall, A. Pultr, and P. Kolman (Eds.): MFCS 2001, LNCS 2136, pp. 573–584, 2001.

easily countable languages, multiselectivity, sortability etc. (for detailed information see [AG88], [ABG90], [BKS95b], [HN93], [BKS95a], [HW97], [HJRW97]). To get a more unified picture Beigel, Gasarch and Kinber introduced $\mathcal{D}$-verboseness and strong $\mathcal{D}$-verboseness [BGK95], where the type of partial information is specified by a family $\mathcal{D}$ of sets of bitstrings. The definitions for the time bounded version, namely polynomially $\mathcal{D}$-verbose languages, are given below. Basic properties of these polynomial $\mathcal{D}$-verboseness classes are presented in [Nic97]. The facts needed for this paper are given in Section 2.

How do the two approaches of defining "some polynomial time behaviour" relate? The existence of a special case algorithm does not imply the existence of a partial information algorithm. To see this, just consider a language A that is the join of a hard language (one that has no partial information algorithm – such languages exist) and an infinite easy language (a language in P). Then A has a special case algorithm, but no partial information algorithm.

For the other direction, the situation is much more intricate. If the partial information algorithm narrows the range of possible values for $\chi_A(x_1, \ldots, x_n)$ sufficiently, then there is a special case algorithm, i.e. A is not p-bi-immune. On the other hand, for weaker types of partial information there are languages which allow a partial information algorithm and nevertheless are *hard almost everywhere*, i.e. any polynomial time decision algorithm decides "$x \in A$?" only on a finite set of instances. The achievement of this paper is to fully characterize those $\mathcal{D}$ for which languages can be at the same time polynomially $\mathcal{D}$-verbose and p-bi-immune. This characterization is given in Section 4, Theorem 19. Three important results (presented in Section 3) had already been obtained for special types of polynomial $\mathcal{D}$-verboseness on their relation to p-bi-immunity. Our theorem fills the gap left open by these results and gives a uniform proof for them.

The relationship between partial information and special case solutions can of course also be studied for algorithms without resource bounds. When Jockusch [Joc66] introduced semirecursive sets, the purpose of this new definition was to separate many-one reductions from Turing reductions in the recursion theoretic setting. One of the main steps to reach this goal was to show that there are semirecursive languages that are bi-immune. On the other hand, if for a given $\mathcal{D}$ the class of strongly $\mathcal{D}$-verbose languages does not contain all semirecursive languages, then this class only contains recursive languages [Nic97]. Therefore the relation between strong $\mathcal{D}$-verboseness and bi-immunity is completely solved (and the solution is easier to state than in the polynomial time case): For a family $\mathcal{D}$ there are bi-immune strongly $\mathcal{D}$-verbose languages iff there are non-recursive $\mathcal{D}$-verbose languages iff all semirecursive languages are strongly $\mathcal{D}$-verbose.

2   Definitions

Languages are subsets of $\Sigma^* = \{0,1\}^*$. The natural numbers (without 0) are denoted by $\mathbb{N}$. Let $\mathbb{B} := \{0,1\}$. For a language A the characteristic function

$\chi_A : \Sigma^* \to \mathbb{B}$ is defined by $\chi_A(x) = 1 \Leftrightarrow x \in A$. We extend χ_A to tuples of words by $\chi_A(x_1, \ldots, x_n) := \chi_A(x_1) \cdots \chi_A(x_n)$.

In the following, elements of $\{0,1\}^*$ for which membership in languages is of interest are called words, elements of $\mathbb{B}^*$ which are considered as possible values of characteristic functions are called bitstrings. For a bitstring b the number of 1's in b is denoted $\#_1(b)$, $b[i]$ is the i-th bit of b, and $b[i_1, \ldots, i_j]$ is the string formed by the i_1-th to the i_j-th bit of b; i.e. $b[i_1, \ldots, i_j] = b[i_1] \cdots b[i_j]$. We extend this to sets of bitstrings by $P[i_1, \ldots, i_j] := \{b[i_1, \ldots, i_j] \mid b \in P\}$.

We will first define p-bi-immunity and polynomial $\mathcal{D}$-verboseness.

Definition 1 (p-Bi-immunity). *A language A is p-bi-immune if neither A nor $\overline{A}$ has an infinite subset $B \in P$.*

Definition 2 (n-Families). *Let $n \in \mathbb{N}$ and let $\mathcal{D} = \{D_1, \ldots, D_r\}$ be a set of sets of bitstrings of length n, i.e. $D_i \subseteq \mathbb{B}^n$ for each i. We call $\mathcal{D}$ an n-family if*

1. *$\mathcal{D}$ covers $\mathbb{B}^n$, i.e. $\bigcup_{i=1}^r D_i = \mathbb{B}^n$, and*
2. *$\mathcal{D}$ is closed under subsets, i.e. $D \in \mathcal{D}$ and $D' \subseteq D$ implies $D' \in \mathcal{D}$.*

Definition 3 (Polynomially $\mathcal{D}$-Verbose). *For a given n-family $\mathcal{D}$ a language A is in $\mathrm{P}\,[\mathcal{D}]$ iff there is a polynomially time-bounded deterministic Turing machine M that on input $(x_1, \ldots, x_n)$ outputs a $D \in \mathcal{D}$ such that $\chi_A(x_1, \ldots, x_n) \in D$. Such languages are called polynomially $\mathcal{D}$-verbose.*

We now present additional definitions that make it easier to deal with polynomial $\mathcal{D}$-verboseness and then state some known facts which will be applied in the following. For more details on polynomial $\mathcal{D}$-verboseness see [Nic97], [Nic99].

Definition 4 (Operations on Bitstrings).

1. *Let S_n be the group of permutations of $\{1, \ldots, n\}$. For $\sigma \in S_n$ and $b_1 \cdots b_n \in \mathbb{B}^n$ we define $\sigma(b_1 \cdots b_n) := b_{\sigma(1)} \cdots b_{\sigma(n)}$.*
2. *Let $i \in \{1, \ldots, n\}$ and $c \in \mathbb{B}$. Define projections $\pi_i^c \colon \mathbb{B}^n \to \mathbb{B}^n$ by $\pi_i^c(b_1 \cdots b_n) := b_1 \cdots b_{i-1} c\, b_{i+1} \cdots b_n$.*
3. *Let $i, j \in \{1, \ldots, n\}$. Define a replacement operation $\rho_{i,j} \colon \mathbb{B}^n \to \mathbb{B}^n$ by $\rho_{i,j}(b_1 \cdots b_n) := b'_1 \cdots b'_n$ where $b'_k = b_k$ for $k \neq j$ and $b'_j = b_i$.*

We extend such operations ω from bitstrings to sets of bitstrings by $\omega(D) = \{\omega(b) \mid b \in D\}$. An n-family $\mathcal{D}$ is closed under permutations, projections and replacements iff $\omega(D) \in \mathcal{D}$ for all $D \in \mathcal{D}$ and for all such operations ω.

Definition 5 (Normal Form). *An n-family $\mathcal{D}$ is in normal form if it is closed under permutations, projections and replacements.*

Fact 6 (Normal Form). *For every n-family $\mathcal{D}$ there is a unique n-family $\mathcal{D}'$ in normal form with $\mathrm{P}\,[\mathcal{D}] = \mathrm{P}\,[\mathcal{D}']$.*

Fact 7 (Class Inclusion Reduces to Family Inclusion). *Let $\mathcal{D}_1, \mathcal{D}_2$ be n-families in normal form. Then $\mathrm{P}\,[\mathcal{D}_1] \subseteq \mathrm{P}\,[\mathcal{D}_2]$ iff $\mathcal{D}_1 \subseteq \mathcal{D}_2$.*

Definition 8 (Change of Tuple-Length). *For an n-family $\mathcal{D}$ define $\lceil \mathcal{D} \rceil_m$, the translation of $\mathcal{D}$ to length m, by*

$$\lceil \mathcal{D} \rceil_m = \{D \subseteq \mathbb{B}^m \mid \text{for all } i_1, \ldots, i_n \in \{1, \ldots, m\} \colon D[i_1, \ldots, i_n] \in \mathcal{D}\}.$$

Fact 9 (Change of Tuple-Length).   *For $m, n \in \mathbb{N}$ and an n-family $\mathcal{D}$ in normal form, $\lceil \mathcal{D} \rceil_m$ is in normal form, and $\mathrm{P}[\mathcal{D}] \subseteq \mathrm{P}[\lceil \mathcal{D} \rceil_m]$. Moreover, if $m \geq n$, then $\mathrm{P}[\mathcal{D}] = \mathrm{P}[\lceil \mathcal{D} \rceil_m]$.*

Fact 10 (Intersection).   *For n-families $\mathcal{D}_1$, $\mathcal{D}_2$ in normal form $\mathrm{P}[\mathcal{D}_1] \cap \mathrm{P}[\mathcal{D}_2] = \mathrm{P}[\mathcal{D}_1 \cap \mathcal{D}_2]$.*

Definition 11 (Generated Family).   *Let $D_1, \ldots, D_r \subseteq \mathbb{B}^n$ be nonempty sets of bitstrings. Then $\langle D_1, \ldots, D_r \rangle$ denotes the minimal n-family $\mathcal{D}$ in normal form with $\{D_1, \ldots, D_r\} \subseteq \mathcal{D}$. This means that $\langle D_1, \ldots, D_r \rangle$ is the closure of $\{D_1, \ldots, D_r\}$ under permutations, projections and replacements. We say that $\langle D_1, \ldots, D_r \rangle$ is generated by $D_1, \ldots, D_r$.*

Some *n*-families are of special interest. We define SEL_n, $k\text{-}\mathrm{SIZE}_n$, and $k\text{-}\mathrm{CARD}_n$.

Definition 12 (SEL, SIZE, CARD).

1. $\mathrm{SEL}_n := \langle \{0^i 1^{n-i} \mid 0 \leq i \leq n\} \rangle$.
2. *For $1 \leq k \leq 2^n$*: $k\text{-}\mathrm{SIZE}_n := \{D \subseteq \mathbb{B}^n \mid |D| \leq k\}$.
3. *For $1 \leq k \leq (n+1)$*: $k\text{-}\mathrm{CARD}_n := \{D \subseteq \mathbb{B}^n \mid |\{\#_1(b) \mid b \in D\}| \leq k\}$.

Languages in $\mathrm{P}[n\text{-}\mathrm{CARD}_n]$ are called easily countable. The class $\mathrm{P}[\mathrm{SEL}_2]$ equals the class P-sel of p-selective languages. Languages in $\mathrm{P}[(2^n - 1)\text{-}\mathrm{SIZE}_n]$ are called *n*-approximable or non-*n*-p-superterse or *n*-p-verbose. Languages in $\mathrm{P}[n\text{-}\mathrm{SIZE}_n]$ may be called *n*-cheatable but the term cheatability is not used in exactly this sense everywhere in the literature. We state two important facts; the one on SEL_n was implicitly shown in [Sel79], the one on $k\text{-}\mathrm{SIZE}_n$ in [Bei91]:

Fact 13 (SEL, SIZE).

1. P-sel $= \mathrm{P}[\mathrm{SEL}_2] = \mathrm{P}[\mathrm{SEL}_n]$ *for $n \geq 2$.*
2. $\mathrm{P}[k\text{-}\mathrm{SIZE}_k] = \mathrm{P}[k\text{-}\mathrm{SIZE}_n]$ *for $n \geq k$.*

3   Previous Results

For polynomial $\mathcal{D}$-verboseness three results connecting it to p-bi-immunity are known. Goldsmith, Joseph, and Young [GJY87][GJY93] showed that – in their nomenclature – there are 4-for-3-cheatable sets that are bi-immune (and hence also p-bi-immune). Inspection of their proof shows that in fact they construct a p-selective set that is bi-immune. If we rephrase their result we get:

Fact 14 (Goldsmith, Joseph, Young).
$\mathrm{P}[\mathrm{SEL}_2]$ *contains bi-immune languages.*

In the same paper, Goldsmith, Joseph, and Young [GJY93] also construct a language that is 4-for-4-cheatable and p-bi-immune. This result was independently obtained by Beigel [Bei90], where the 4-for-4-cheatable languages are called 2-cheatable. In fact, the constructions of [GJY93] and [Bei90] even yield a better result. This result, implicit in their proofs, can be stated as follows:

Fact 15 (Beigel; Goldsmith, Joseph, Young).
$P\left[\langle\{000, 001, 010, 011^n\}\rangle\right]$ *contains a p-bi-immune language.*

On the other hand Amir and Gasarch [AG88] showed for a polynomial verboseness class that it does not contain p-bi-immune languages. Stated in our nomenclature they proved:

Fact 16 (Amir, Gasarch).
$P\left[2\text{-}\mathrm{SIZE}_2\right]$ *does not contain p-bi-immune languages.*

By Fact 15 and 16 the question of p-bi-immunity was settled for $P\left[4\text{-}\mathrm{SIZE}_4\right]$ and for $P\left[2\text{-}\mathrm{SIZE}_2\right]$, but it was left open for $P\left[3\text{-}\mathrm{SIZE}_3\right]$ (which turns out to contain p-bi-immune languages, see Corollary 20). For tuple-length 2, Fact 14 and Fact 16 settle the question for seven out of ten non-trivial 2-families. See [Nic97] for an overview of these families. One of the three families for which the question was left open was $2\text{-}\mathrm{CARD}_2$, for which the corresponding class turns out to contain p-bi-immune languages, see Corollary 20. The other two classes $P\left[\mathrm{TOP}_2\right]$ and $P\left[\mathrm{BOTTOM}_2\right]$, see Definition 17, do not contain p-bi-immune languages, see Theorem 25.

4   Stating the Main Result

To characterize the classes $P\left[\mathcal{D}\right]$ that contain p-bi-immune languages, we partition the set of n-families in normal form into an upper and a lower part. Families in the upper part, which consists of families above two specific families, yield classes which contain p-bi-immune languages. The lower part consists of families which are below (i.e. are subfamilies of) two other specific families. Languages in $P\left[\mathcal{D}\right]$ with $\mathcal{D}$ from this lower part allow special case solutions (i.e. are not p-bi-immune).

We first define the two families which will turn out to be maximal for the lower part. The generating sets of bitstrings for these families express the information "at most one input word is outside the language" (respectively "... in the language").

Definition 17 (TOP_n, BOTTOM_n). *For $n \geq 2$, define sets of bitstrings Top_n and $Bottom_n$. Then use these sets of bitstrings to define for every n two n-families:*

1. *$Top_n := \{b \in \mathbb{B}^n \mid \#_1(b) \geq n - 1\}$ and $\mathrm{TOP}_n := \langle Top_n \rangle$.*
2. *$Bottom_n := \{b \in \mathbb{B}^n \mid \#_1(b) \leq 1\}$ and $\mathrm{BOTTOM}_n := \langle Bottom_n \rangle$.*

Now we define sets of bitstrings that generate the minimal families in the upper part:

Definition 18. *For $n \geq 2$ define sets of bitstrings S_n, T_n, and B_n as follows:*

1. $S_n = \{000^{n-2}, 010^{n-2}, 110^{n-2}\}$
2. $T_n = \{011^{n-2}, 101^{n-2}, 111^{n-2}\}$
3. $B_n = \{000^{n-2}, 010^{n-2}, 100^{n-2}\}$

The notation is meant to remind of selectivity, top and bottom. For $n = 2$, the set S_2 is a generating set for SEL_2, hence $\langle S_2 \rangle = \text{SEL}_2$. For the other two sets we have $T_2 = \text{Top}_2$ and $B_2 = \text{Bottom}_2$ and thus $\langle T_2 \rangle = \text{TOP}_2$, $\langle B_2 \rangle = \text{BOTTOM}_2$, and $\langle T_2, B_2 \rangle = 2\text{-CARD}_2$. But for $n \geq 3$ such equalities do not hold anymore.

Now we are ready to state our main result. It gives a complete characterization of the polynomial verboseness classes that contain p-bi-immune languages.

Theorem 19 (Main Theorem). *Let $\mathcal{D}$ be an n-family in normal form, $n \geq 2$. Then the following five statements are equivalent:*

1. $\text{P}[\mathcal{D}]$ *contains p-bi-immune languages.*
2. $\mathcal{D} \not\subseteq \text{BOTTOM}_n$ *and* $\mathcal{D} \not\subseteq \text{TOP}_n$.
3. $\text{P}[\mathcal{D}] \not\subseteq \text{P}[\text{BOTTOM}_2]$ *and* $\text{P}[\mathcal{D}] \not\subseteq \text{P}[\text{TOP}_2]$.
4. $\langle S_n \rangle \subseteq \mathcal{D}$ *or* $\langle T_n, B_n \rangle \subseteq \mathcal{D}$.
5. $\text{P}[\langle S_n \rangle] \subseteq \text{P}[\mathcal{D}]$ *or* $\text{P}[\langle T_n, B_n \rangle] \subseteq \text{P}[\mathcal{D}]$.

Proof. We put together several results that are proved subsequently.

The equivalence of the last four statements is treated in Section 5. Statement 2 and 3 are equivalent by Lemma 21. Statement 4 and 5 are equivalent by Lemma 22 and by Lemma 23. Statement 3 and 5 are equivalent by Theorem 24. To show that 1 implies 3, we have to prove that $\text{P}[\text{TOP}_2]$ and $\text{P}[\text{BOTTOM}_2]$ do not contain p-bi-immune languages. This is done in Section 6, see Theorem 25. To show that 5 implies 1, in Section 6 we prove that $\text{P}[\langle T_n, B_n \rangle]$ and $\text{P}[\langle S_3 \rangle]$ contain p-bi-immune languages, see Theorem 26 and Theorem 27.   □

Specializing the Main Theorem to tuple-length 2 this solves the three open cases for this length, specializing it to length 3 solves the case 3-SIZE_3, as $\langle T_3, B_3 \rangle \subseteq 3\text{-SIZE}_3$.

Corollary 20 (Tuple-Length 2 and 3).

1. $\text{P}[2\text{-CARD}_2]$ *contains p-bi-immune languages.*
2. $\text{P}[\text{TOP}_2]$ *and* $\text{P}[\text{BOTTOM}_2]$ *do not contain p-bi-immune languages.*
3. $\text{P}[3\text{-SIZE}_3]$ *contains p-bi-immune languages.*

5   Splitting the Set of Families into Two Parts

Lemma 21. *For $n \geq 2$:* $\text{P}[\text{BOTTOM}_n] = \text{P}[\text{BOTTOM}_2]$ *and* $\text{P}[\text{TOP}_n] = \text{P}[\text{TOP}_2]$.

Proof. The class P [TOP$_n$] is the complement class of P [BOTTOM$_n$] and P [TOP$_2$] is the complement class of P [BOTTOM$_2$]. Therefore it suffices to show that P [BOTTOM$_n$] $\subseteq$ P [BOTTOM$_2$] and P [BOTTOM$_2$] $\subseteq$ P [BOTTOM$_n$]. But this can be seen by applying Fact 9 in both directions. □

Lemma 22. *For $n \geq 3$: P $[\langle S_n \rangle] = $ P $[\langle S_3 \rangle]$.*

Proof. For $n \geq 3$, we have $\langle S_n \rangle = $ SEL$_n \cap 3$-SIZE$_n$. By Fact 13, for $n \geq 3$ we have P [SEL$_3$] $=$ P [SEL$_n$] and P [3-SIZE$_3$] $=$ P [3-SIZE$_n$]. By applying Fact 10 we get P $[\langle S_n \rangle] = $ P [SEL$_n \cap 3$-SIZE$_n$] $=$ P [SEL$_n$]$\cap$P [3-SIZE$_n$] $=$ P [SEL$_3$]$\cap$P [3-SIZE$_n$] $=$ P [SEL$_3 \cap 3$-SIZE$_n$] $=$ P $[\langle S_3 \rangle]$. □

The proofs of Lemma 23 and Theorem 24 can be found in [Nic99] and are omitted due to lack of space. Theorem 24 states that with the families TOP$_n$, BOTTOM$_n$, $\langle S_n \rangle$, and $\langle T_n, B_n \rangle$ we really partition the set of all n-families in normal form.

Lemma 23. *For $n \geq 3$: P $[\langle T_n, B_n \rangle] = $ P $[\langle T_3, B_3 \rangle]$.*

Theorem 24. *For n-families $\mathcal{D}$ in normal form exactly one of these cases holds:*

1. $\mathcal{D} \subseteq$ TOP$_n$ *or* $\mathcal{D} \subseteq$ BOTTOM$_n$
2. $\langle S_n \rangle \subseteq \mathcal{D}$, *or* $\langle T_n, B_n \rangle \subseteq \mathcal{D}$.

6   No p-Bi-immune Languages in the Lower Part

The next theorem also yields that P [TOP$_2$] does not contain p-bi-immune languages because P [TOP$_2$] contains the complements of languages in P [BOTTOM$_2$].

Theorem 25. *If $A \in$ P [BOTTOM$_2$], then A is not p-bi-immune.*

Proof. We can restrict ourselves to tally languages A. Let $A \subseteq \{1\}^*$ be in P [BOTTOM$_2$] via a machine M_A. Consider the following procedure that stepwise tries to compute as much information about χ_A as possible. After step n the procedure has constructed disjoint sets $A_n, \overline{A}_n, B_1^n, \ldots, B_{r_n}^n$ such that

- $A_n \cup \overline{A}_n \cup B_1^n \cup \cdots \cup B_{r_n}^n = \{1^i \mid i \leq n\}$,
- $A_n \subseteq A$ and $\overline{A}_n \subseteq \overline{A}$,
- $B_j^n \subseteq A$ or $B_j^n \subseteq \overline{A}$ for all $j \leq r_n$,
- $B_j^n \subseteq A$ for at most one $j \leq r_n$.

Let this construction be already completed up to step $n-1$. Now we have to add the word 1^n to this structure. Compute $M_A(x, 1^n)$ for all $x \in B_1^n \cup \cdots \cup B_{r_n}^n$. Four cases can occur:

1. $M_A(x, 1^n) \subseteq \{01, 11\}$ for some x. This means $1^n \in A$. Let $A_n = A_{n-1}\cup\{1^n\}$. Leave the other sets unchanged.
2. Not case 1 and $M_A(x, 1^n) \subseteq \{10, 11\}$ for some x, say $x \in B_j^{n-1}$. Then $B_j^{n-1} \subseteq A$. Let $A_n = A_{n-1}\cup B_j^{n-1}$, $\overline{A}_n = \overline{A}_{n-1}\cup\bigcup_{i \neq j} B_i^{n-1}$ and $B_1^n = \{1^n\}$. This means that for all words up to 1^{n-1} membership in A is decided.

3. Neither case 1 nor case 2 and $M_A(x, 1^n) \subseteq \{00, 11\}$ for some x, say $x \in B_j^{n-1}$. Let $B_j^n = B_j^{n-1} \cup \{1^n\}$ and leave the other sets unchanged.

4. Neither case 1, 2, nor 3. Then $M_A(x, 1^n) = \{00, 01, 10\}$ for all $x \in B_j^{n-1}$ for all j. Add a new set B_i, i.e. let $r_n = r_{n-1} + 1$ and $B_{r_n}^n = \{1^n\}$ and leave the other sets unchanged.

It is clear, that this construction can be done in time bounded by a polynomial in n. – How can we use this construction to exhibit an infinite polynomial time decidable subset of A or $\overline{A}$? Four different cases can occur:

a. In the procedure case 1 occurs infinitely often. Then infinitely often words 1^n are put into A. We get an infinite polynomial time subset of A.

b. In the procedure case 2 occurs infinitely often. Then infinitely often words 1^n are put into $\overline{A}$. We get an infinite polynomial time subset of $\overline{A}$.

c. There is an n_0 such that after step n_0 cases 1 and 2 do not occur anymore. Suppose the number of B_i produced is finite. Then there is at least one index i_0 such that infinitely often a word is put into $B_{i_0}^n$. The set $B_{i_0} = \{1^n \mid 1^n$ is put into $B_{i_0}^n, n \geq n_0\}$ then is in P and $B_{i_0} \subseteq A$ or $B_{i_0} \subseteq \overline{A}$.

d. Neither case a, b, nor c. Then after n_0 infinitely many B_i are introduced during the procedure. Only one of them, say B_{i_0}, is a subset of A. Therefore $\bigcup_{i \neq i_0, n \geq n_0} B_i^n$ is infinite, is in P and a subset of $\overline{A}$.   □

7   Constructing p-Bi-immune Languages

Theorem 26. *The class* $\mathrm{P}\left[\langle T_3, B_3\rangle\right]$ *contains a p-bi-immune language.*

Proof. We construct a set $A \subseteq \Sigma^*$ that is in $\mathrm{P}\left[\langle T_3, B_3\rangle\right]$ and p-bi-immune. We will diagonalize against every Turing machine that could possibly decide an infinite subset of A or $\overline{A}$. Let $\langle M_k\rangle_{k \in \mathbb{N}}$ be a standard enumeration of polynomial time Turing machines such that the running time of M_k is bounded by a polynomial p_k, say $n^{\log k} + \log k$. We will ensure that for every M_k that accepts infinitely many words, there are words $w_1, w_2 \in L(M_k)$ with $w_1 \in A$ and $w_2 \in \overline{A}$. For every $n \in \mathbb{N}$ we define stepwise approximations to A and to $\overline{A}$. For this purpose we use a sequence of natural numbers $\langle n_k\rangle_{k \in \mathbb{N}}$ that grows fast enough such that on inputs of length $\geq n_i$ we can simulate all M_j with $j \leq i - 1$ on all words of length $\leq n_{i-1}$. It is sufficient to define $n_0 := 1$, $n_{i+1} := 2^{n_i}$. During the construction of A and $\overline{A}$ we keep track of a bunch of parameters. After step n we will have constructed a partition of $\Sigma^{\leq n}$ into four disjoint sets IN_n, OUT_n, OLD_n and NEW_n.

For all n it will hold that $\mathrm{IN}_n \subseteq \mathrm{IN}_{n+1}$ and $\mathrm{OUT}_n \subseteq \mathrm{OUT}_{n+1}$. In the end we define $A = \bigcup_n \mathrm{IN}_n$, the construction then yields $\overline{A} = \bigcup_n \mathrm{OUT}_n$.

The sets OLD_n and NEW_n contain those words up to length n for which membership in A or $\overline{A}$ is not yet decided after step n. We also maintain a finite list $L_n \subseteq \mathbb{N}$ of indices of Turing machines. An index k enters the list at construction step n_k. It is removed from the list at a later step n only if both requirements for M_k are fulfilled, i.e., there are $w_1 \in \text{IN}_n$ and $w_2 \in \text{OUT}_n$ with $M_k(w_1) = \text{accept}$ and $M_k(w_2) = \text{accept}$.

Which requirements are still unfulfilled after step n is expressed by a function req_n that maps every k to a subset of $\{in, out\}$. For example, $req_n(k) = \{in\}$ means that it is still required to put a w_1 with $M_k(w_1) = \text{accept}$ into A at some later stage, but there already is a $w_2 \in \text{OUT}_n$ with $M_k(w_2) = \text{accept}$. We also need a parameter $status_n$ with possible values t and b. If $status_n = b$ then at most one of the sets OLD_n and NEW_n will become part of A in a later step, if $status_n = t$ then at least one of the sets OLD_n and NEW_n will enter A later on. As a last parameter we need $urgent_n \in L_n \cup \{\infty\}$. If after a construction step $urgent_n = k \in L_n$, then there is a requirement for M_k that we would like to fulfill by putting a $w \in \text{OLD}_{n-1}$ into A or $\overline{A}$, but at the moment we are hindered to do this by the current value of $status_n$. Therefore we have to wait to do so until the status has changed or until the requirement is overruled by a requirement with higher priority, i.e. by a requirement for a machine $M_{k'}$ with $k' < k$. Now let IN_{n-1}, OUT_{n-1}, OLD_{n-1}, NEW_{n-1}, L_{n-1}, req_{n-1}, $status_{n-1}$ and $urgent_{n-1}$ be already constructed. At step n consider the words of length n. Check whether $n = n_i$ for some i.

Case 1: $n \neq n_i$ for all i. Add Σ^n to the set NEW, leave everything else unchanged.

$$\text{IN}_n = \text{IN}_{n-1} \qquad\qquad \text{OUT}_n = \text{OUT}_{n-1}$$
$$\text{OLD}_n = \text{OLD}_{n-1} \qquad \text{NEW}_n = \text{NEW}_{n-1} \cup \Sigma^n$$
$$L_n = L_{n-1} \qquad\qquad req_n = req_{n-1}$$
$$status_n = status_{n-1} \qquad urgent_n = urgent_{n-1}$$

Case 2: $n = n_i$ for some i. We check wether a requirement can be fulfilled. If the current status is b, we only fulfill out-requirements, if the status is t we only fulfill in-requirements. Suppose that $status_{n-1} = b$. (In case $status_{n-1} = t$ we have analogous subcases. See explanation below.) For every $k \in L_{n-1}$ with $k < urgent_{n-1}$ and for every $x \in \text{OLD}_{n-1}$ compute $M_k(x)$. We check only requirements with higher priority than k, i.e. $k < urgent_{n-1}$, because if $urgent_{n-1} \in L_{n-1}$ then there is a word in OLD_{n-1} that can serve to fulfill an in-requirement for $M_{urgent_{n-1}}$. Three different cases can occur:

Case 2.1: $M_k(x) = \text{reject}$ for all k and x. No requirement for $k < urgent_{n-1}$ can be fulfilled at this stage. On the other hand, there is nothing wrong with putting OLD_{n-1} into $\overline{A}$. Therefore we add OLD_{n-1} to OUT_{n-1}, change the status from b to t and give up the restriction on indices possibly imposed by $urgent_{n-1}$.

$$\text{IN}_n = \text{IN}_{n-1} \qquad\qquad \text{OUT}_n = \text{OUT}_{n-1} \cup \text{OLD}_{n-1}$$
$$\text{OLD}_n = \text{NEW}_{n-1} \qquad\quad \text{NEW}_n = \Sigma^n$$
$$L_n = L_{n-1} \cup \{i\} \qquad\quad req_n = req_{n-1}$$
$$status_n = t \qquad\qquad\quad urgent_n = \infty$$

Case 2.2: $M_k(x) = \text{accept}$ for some k and x. Choose k_0 as the minimal k for which this happens. Let x_0 be the smallest word in OLD_{n-1} with $M_{k_0}(x_0) = \text{accept}$. We distinguish two subcases depending on the value of $req_{n-1}(k_0)$.

Case 2.2.1: $out \in req_{n-1}(k_0)$. We can directly fulfill the requirement for k_0 by putting x_0 (and the whole set OLD_{n-1}) into $\overline{A}$. If both requirements for k_0 are then fulfilled we remove k_0 from the list of machine indices; else we remove the requirement *out* from $req_{n-1}(k_0)$. We also change the status to t.

$$\text{IN}_n = \text{IN}_{n-1} \qquad\qquad \text{OUT}_n = \text{OUT}_{n-1} \cup \text{OLD}_{n-1}$$
$$\text{OLD}_n = \text{NEW}_{n-1} \qquad\qquad \text{NEW}_n = \Sigma^n$$
$$req_n(k_0) = req_{n-1}(k_0) \setminus \{out\} \qquad req_n(k) = req_{n-1}(k) \text{ for } k \neq k_0$$
$$status_n = t \qquad\qquad urgent_n = \infty$$
$$\text{if } req_{n-1}(k_0) = \{in, out\} \text{ then } L_n = L_{n-1} \cup \{i\}$$
$$\text{if } req_{n-1}(k_0) = \{out\} \text{ then } L_n = (L_{n-1} \setminus \{k_0\}) \cup \{i\}$$

Case 2.2.2: $req_{n-1}(k_0) = \{in\}$. We would like to put OLD_{n-1} including x_0 into A, but we cannot do this at this stage because then we would have to put NEW_{n-1} into $\overline{A}$ (remember that in status b at most one of the sets OLD_{n-1} and NEW_{n-1} may enter A). Maybe we need words from NEW_{n-1} to fulfill *in*-requirements with higher priority. Therefore we only change *urgent* to k_0 without fulfilling any requirement. We let the sets OLD and NEW change their roles.

$$\text{IN}_n = \text{IN}_{n-1} \qquad\qquad \text{OUT}_n = \text{OUT}_{n-1}$$
$$\text{OLD}_n = \text{NEW}_{n-1} \qquad\qquad \text{NEW}_n = \text{OLD}_{n-1} \cup \Sigma^n$$
$$L_n = L_{n-1} \cup \{i\} \qquad\qquad req_n = req_{n-1}$$
$$status_n = status_{n-1} \qquad\qquad urgent_n = k_0$$

How do we deal with case 2 if $status_{n-1} = t$? Essentially the roles of IN and OUT are exchanged.

Case 2.1: $M_k(x) = $ reject for all k and x. We add OLD_{n-1} to IN_{n-1}, change *status* from t to b and put $urgent_n = \infty$.

Case 2.2.1: $M_{k_0}(x_0) = $ accept and $in \in req_{n-1}(k_0)$. Put OLD_{n-1} into A, set $req_n(k_0) = req_{n-1}(k_0) \setminus \{in\}$ and change the status. **Case 2.2.2:** $M_{k_0}(x_0) = $ accept and $req_{n-1}(k_0) = \{out\}$. As in the case $status_{n-1} = b$ change the role of OLD and NEW and set $urgent_n = k_0$.

This ends the description of the procedure. We now state some properties of the construction.

Claim 1 Suppose for some n for words x, y we have $x \in \text{OLD}_n$ and $y \in \text{NEW}_n$. If $status_n = b$, then x and y are not both put into A. If $status_n = t$, then x and y are not both put into $\overline{A}$ (i.e. into OUT_m for some m).

This claim is easily verified by considering the different cases that can happen during the procedure after step n.

Claim 2 implies that $\overline{A} = \bigcup_n \text{OUT}_n$.

Claim 2 For every n there is a step m in the construction where Σ^n enters IN_m or OUT_m.

The set Σ^n enters NEW at step n. Suppose $n_{i-1} \leq n < n_i$. Then at step n_i, Σ^n enters OLD. At step n_{i+1} one of the cases 2.1, 2.2.1 or 2.2.2 occurs. In case 2.1 and case 2.2.1, Σ^n is moved to IN or OUT. Suppose case 2.2.2 occurs and *urgent* is set to k_0. Then together with Σ^n there is a word x_0 in OLD with $M_{k_0}(x_0) = $ accept. Now Σ^n and x_0 can possibly oscillate from OLD to NEW and back again. But each time Σ^n and x_0 are in NEW it holds that *urgent* $< k_0$. Therefore the open requirement for k_0 cannot be fulfilled in a situation where $\Sigma^n \subseteq$ NEW.

How many oscillations can happen without putting Σ^n into IN or OUT? If at a step n_j the set Σ^n is in OLD_{n_j-1}, then either Σ^n is put into IN or OUT, or *urgent* is changed from ∞ to a $k \leq k_0$, or *urgent* is lowered by at least 1, or a requirement is fulfilled for a $k < k_0$. Therefore after at most $k_0 \cdot 2(k_0 - 1)$ oscillations, Σ^n finally moves to IN or OUT.

Claim 3 implies that A indeed is p-bi-immune:

Claim 3 For each k where $L(M_k)$ is infinite, there is some step of the procedure where $req(k)$ is empty.

Assume that k_0 is the smallest k where $L(M_k)$ is infinite and $req_n(k) \neq \emptyset$ for all n. Let m_0 be such that for all $k < k_0$ with $L(M_k)$ infinite it holds that $req_{m_0}(k) = \emptyset$, and for all $k < k_0$ with $L(M_k)$ finite it holds that $L(M_k) \subseteq \Sigma^{\leq m_0-1}$, and $m_0 > n_{k_0}$.

Consider $m_1 = \min\{m \mid 1^m \in L(M_k), m > m_0\}$. Assume w.l.o.g. that $req_{m_1} = \{in\}$. If $n_{i-1} \leq m_1 < n_i$, then in step n_i the word 1^{m_1} enters OLD, k_0 is in the requirement list and there is no requirement left with higher priority. This means that in step n_{i+1} either case 2.2.1 or 2.2.2 occurs. In case 2.2.1 where *status* $= t$ the *in*-requirement is fulfilled which contradicts the assumption. In case 2.2.2 where *status* $= b$ *urgent* is set to k_0. Then at step n_{i+2} case 2.1 occurs (because there are no k in list $L_{n_{i+2}}$ with $k <$ *urgent*). The *status* is changed to t, the word 1^{m_1} is in OLD again and at step n_{i+3} it will be moved to IN because case 2.2.1 occurs; so again the *in*-requirement is fulfilled, contradicting the assumption.

Claim 4 The construction of IN_n, OUT_n, OLD_n, NEW_n, and *status*$_n$ can be done in time polynomial in n.

Claim 5 A is in $\mathrm{P}\,[\langle T_3, B_3 \rangle]$.

Consider an input (x_1, x_2, x_3). If $n = \max_i |x_i|$, compute IN_n, OUT_n, OLD_n, NEW_n, and *status*$_n$. Suppose *status*$_n = b$ (the case *status*$_n = t$ is treated analogously). Suppose that there are i and j, $i < j$ with $x_i \in \mathrm{OLD}_n$ and $x_j \in \mathrm{NEW}_n$ (or $x_i \in \mathrm{NEW}_n$ and $x_j \in \mathrm{OLD}_n$). Because of Claim 1 we know that $\chi_A(x_1, x_2, x_3)[i,j] \in \{00, 01, 10\}$. For x_l with $l \neq i, j$ different cases can occur.

- If $x_l \in \mathrm{IN}_n$ then $\chi_A(x_l) = 1$.
- If $x_l \in \mathrm{OUT}_n$ then $\chi_A(x_l) = 0$.
- If x_l and x_i are both in OLD_n (or both in NEW_n) then $\chi_A(x_l) = \chi_A(x_i)$.
- If x_l and x_j are both in NEW_n (or both in OLD_n) then $\chi_A(x_l) = \chi_A(x_j)$.

Hence we can apply projections π_l^1 or π_l^0 or replacements $\rho_{i,l}$ or $\rho_{j,l}$ to determine the bits of $\chi_A(x_1, x_2, x_3)[l]$ depending on $\chi_A(x_1, x_2, x_3)[i,j]$. Thus we get a set D of three bitstrings containing $\chi_A(x_1, x_2, x_3)$ and $D \in \langle B_n \rangle$. □

Theorem 27. *The class* $\mathrm{P}\,[\langle S_3 \rangle]$ *contains a p-bi-immune language.*

Proof. The proof is by a priority technique similar to that in the proof of Theorem 26, and is omitted due to lack of space. □

References

[ABG90]   A. Amir, R. Beigel, and W. Gasarch. Some connections between bounded query classes and non-uniform complexity. In *Proc. 5th Structure in Complexity Theory*, pages 232–242, 1990. 574

[AG88]   A. Amir and W. Gasarch. Polynomial terse sets. *Information and Computation*, 77:27–56, 1988. 573, 574, 577

[Bei90]   R. Beigel. Bi-immunity results for cheatable sets. *TCS*, 73(3):249–263, 1990. 573, 577

[Bei91]   R. Beigel. Bounded queries to SAT and the boolean hierarchy. *Theoretical Computer Science*, 84(2):199–223, 1991. 576

[BGK95]   R. Beigel, W. Gasarch, and E. Kinber. Frequency computation and bounded queries. In *Proc. 10th Structure in Complexity Theory*, pages 125–132, 1995. 574

[BKS95a] R. Beigel, M. Kummer, and F. Stephan. Approximable sets. *Information and Computation*, 120(2):304–314, 1995. 574

[BKS95b] R. Beigel, M. Kummer, and F. Stephan. Quantifying the amount of verboseness. *Information and Computation*, 118(1):73–90, 1995. 574

[BS85]   J. Balcázar and U. Schöning. Bi-immune sets for complexity classes. *Mathematical Systems Theory*, 18:1–10, 1985. 573

[GJY87]   J. Goldsmith, D. Joseph, and P. Young. Self-reducible, P-selective, near-testable, and P-cheatable sets: The effect of internal structure on the complexity of a set. In *Proceedings 2nd Structure in Complexity Theory Conference*, pages 50–59. IEEE Computer Society Press, 1987. 576

[GJY93]   J. Goldsmith, D. Joseph, and P. Young. A note on bi-immunity and p-closeness of p-cheatable sets in *P/Poly*. *JCSS*, 46(3):349–362, 1993. 573, 576, 577

[HJRW97] L. A. Hemaspaandra, Z. Jiang, J. Rothe, and O. Watanabe. Polynomial-time multi-selectivity. *J.UCS: Journal of Universal Computer Science*, 3(3):197–229, 1997. 574

[HN93]   A. Hoene and A. Nickelsen. Counting, selecting, and sorting by query-bounded machines. In *Proc. STACS 93*, pages 196–205. LNCS 665, 1993. 574

[HW97]   M. Hinrichs and G. Wechsung. Time bounded frequency computations. *Information and Computation*, 139(2):234–257, 1997. 574

[Joc66]   C. G. Jockusch, Jr. *Reducibilities in recursive function theory*. PhD thesis, Massachusetts Institute of Technology, Cambridge, Mass., 1966. 574

[Nic97]   A. Nickelsen. On polynomially $\mathcal{D}$-verbose sets. In *Proc. STACS 97*, pages 307–318, 1997. 574, 575, 577

[Nic99]   A. Nickelsen. *Polynomial Time Partial Information Classes*. PhD thesis, Technical University of Berlin, 1999. 575, 579

[Sel79]   A. Selman. P-selective sets, tally languages and the behaviour of polynomial time reducibilities on NP. *Mathematical Systems Theory*, 13:55–65, 1979. 576

The Complexity of Computing the Number of Self-Avoiding Walks in Two-Dimensional Grid Graphs and in Hypercube Graphs

Mitsunori Ogihara[1*] and Seinosuke Toda[2]

[1] Department of Computer Science, University of Rochester
Box 270226, Rochester, NY 14627-0226, USA
ogihara@cs.rochester.edu
[2] Department of Applied Mathematics, Nihon University
3-25-40 Sakurajyou-shi, Setagaya-ku, Tokyo 156, Japan
toda@am.chs.nihon-u.ac.jp

Abstract. Valiant (*SIAM Journal on Computing* 8, pages 410–421) showed that the problem of counting the number of s-t paths in graphs (both in the case of directed graphs and in the case of undirected graphs) is complete for #P under polynomial-time one-Turing reductions (namely, some post-computation is needed to recover the value of a #P-function). Valiant then asked whether the problem of counting the number of self-avoiding walks of length n in the two-dimensional grid is complete for $\#P_1$, i.e., the tally-version of #P. This paper offers a partial answer to the question. It is shown that a number of versions of the problem of computing the number of self-avoiding walks in two-dimensional grid graphs (graphs embedded in the two-dimensional grid) is polynomial-time one-Turing complete for #P.
This paper also studies the problem of counting the number of self-avoiding walks in graphs embedded in a hypercube. It is shown that a number of versions of the problem is polynomial-time one-Turing complete for #P, where a hypercube graph is specified by its dimension, a list of its nodes, and a list of its edges. By scaling up the completeness result for #P, it is shown that the same variety of problems is polynomial-time one-Turing complete for #EXP, where the post-computation required is right bit-shift by exponentially many bits and a hypercube graph is specified by: its dimension, a boolean circuit that accept its nodes, and one that accepts its edges.

1 Introduction

Self-avoiding walks are random walks that do not intersect themselves. Computing the exact number of self-avoiding walks of a given length n on the two-dimensional lattice is well-known for its difficulty and has been studied extensively (see Madras and Slade [MS93] for the details of the problem, and also

---

* Supported in part by NSF grants CCR-9701911, CCR-9725021, DUE-9980943, INT-9726724, an NIH/NIA grant RO1-AG18231 and a DARPA grant F30602-98-2-013.

J. Sgall, A. Pultr, and P. Kolman (Eds.): MFCS 2001, LNCS 2136, pp. 585–597, 2001.
© Springer-Verlag Berlin Heidelberg 2001

Welch [Wel93] for its connection to other related problems). No simple recurrences are known for the counting problem. The exact number has been calculated only for small lengths n. The largest n for which the exact number is computed is 51, by Conway and Guttman [CG96].

Valiant [Val79b] is the first to find connections between self-avoiding walks and computational complexity theory. He showed that the problem of counting the number of all simple s-t paths in graphs is #P-complete both for the directed graphs and for the undirected graphs. He then asked whether the exact counting problem for the two-dimensional grid is complete for #P_1, the "tally" version of #P, provided that the length of walks is specified by a tally string.

While it is very easy to prove the membership of the problem in #P_1, settling on the question of its hardness is difficult. The difficulty comes from the fact that the two-dimensional grid has a very rigid structure. A straightforward approach for proving the hardness would be to embed the computation of a nondeterministic Turing machine in the grid, by defining a correspondence between its configurations and the grid points. However, since every line segment of the grid can be traversed in either direction and the configuration graph of a nondeterministic Turing machine is of high dimension, such an embedding seems impossible.

This observation leads to the question of whether this counting problem is complete if one is allowed to eliminate some nodes and edges, namely counting the number of self-avoiding walks in two-dimensional grids, those composed of the nodes and the edges of the two-dimensional grid. Since elimination of edges and nodes is dependent on the input, here one should perhaps think of #P-completeness rather than #P_1-completeness of the problem. So, the question asked is:

Is the problem of computing the number of self-avoiding walks in two-dimensional grid graphs #P-complete?

This paper provides a positive answer to the question. The problem is #P-complete under polynomial-time one-Turing reductions if the kind of self-avoiding walks to be counted is: (1) restricted to those between two specific nodes, (2) restricted to those starting from a specific node, or (3) totally unrestricted. However, it is unknown whether the problem is #P-complete under any of these restrictions if the length of the walks to be counted is specified. The kind of one-Turing reduction imposed here is quite simple; the only post-computation performed is to divide the number by a power of two; i.e., to shift the bits of the binary representation of the number.

Another question asked in this paper is whether the counting problem is complete if the dimension of the grid is dependent on the input; i.e., whether it is #P-complete to count the number of self-avoiding walks in hypercube graphs, the graphs consisting of nodes and edges of a hypercube. This paper gives a positive answer to the question, too. The problem is #P-complete under polynomial-time one-Turing reductions if the kind of self-avoiding walks to be counted is: (1) restricted to those between two specific nodes, (2) restricted to those starting from a specific node, and (3) totally unrestricted. Furthermore, the problem is complete even when the length of the walks to be counted is specified. When we reduce a #P-function to the counting problem of self-avoiding walks in a hypercube, the dimension of the hypercube in which the graph is embedded is $\mathcal{O}(\log n)$. It is natural to ask about the complexity of the counting problem when the dimension of the hypercube graphs increases as a polynomial in the input size. Since there are 2^n nodes in the n-dimensional hypercube, it is stipulated here that, instead of an enumeration of the nodes and edges, boolean circuits are used to describe the structure of a hypercube graph, one circuit for the nodes and the other for the edges. It is shown here that under that stipulation the counting problem is #EXP-complete, the exponential-time version of #P.

2   Preliminaries

This section sets down some concepts and notation. The reader's familiarity with the basic notions in computational complexity theory is assumed; a reader unfamiliar with the subject may consult [Pap94].

Let M be a nondeterministic Turing. $\#acc_M$ denotes the function that maps each string x to the number of accepting computation paths of M on input x. Then $\#P = \{\#acc_M \mid M$ is a polynomial-time nondeterministic Turing machine $\}$ and $\#EXP = \{\#acc_M \mid M$ is an exponential-time nondeterministic Turing machine $\}$.[1]

Let f and g be two functions from Σ^* to $\mathbf{N}$. The function f is *polynomial-time one-Turing reducible* to g, denoted by $f \leq^p_{1\text{-}T} g$, if there is a pair of polynomial-time computable functions, $R_1 : \Sigma^* \to \Sigma^*$ and $R_2 : \Sigma^* \times \mathbf{N} \to \mathbf{N}$, such that for all x, $f(x) = R_2(x, g(R_1(x)))$. The function f is *polynomial-time right-shift reducible* to g, denoted by $f \leq^p_{r\text{-}shift} g$, if there is a pair of polynomial-time computable functions, $R_1 : \Sigma^* \to \Sigma^*$ and $R_2 : \Sigma^* \to \mathbf{N}$, such that for all x, $f(x) = g(R_1(x)) \operatorname{div} 2^{R_2(x)}$, where div is integer division.

3SAT is the problem of testing satisfiability of CNF formulas with exactly three literals per clause. #SAT is the function that counts the number of satisfying assignments of a 3CNF formula. By the standard reduction from nondeterministic Turing machine computations to 3CNF formulas, #SAT is complete for #P [Val79b].

Proposition 1. *For every #P function f, there is a polynomial-time computable function R such that for all x, $R(x)$ is a 3CNF formula and satisfies $\#SAT(R(x)) = f(x)$.*

Hamilton Path (HP, for short) is the problem of deciding, given a graph G with two specified nodes s and t, whether there is a Hamilton path from s to t in G. #HP is the problem of counting, given a graph G and a node pair (s, t), the number of Hamilton paths from s to t in G.

---

[1] The reader should be cautioned that Valiant defined #EXP as $\#P^{EXP}$ in [Val79a].

3   Reducing #SAT to #HP

#P-completeness of the self-avoiding walk problems is proven based upon a reduction from #SAT to #HP, where the instances in #HP generated by the reduction have a low degree. More precisely, for the completeness in grid graphs, the graphs have to be planar and the maximum degree is ≤ 4, and for the hypercube graph results, the maximum degree is $\mathcal{O}(\log n)$.

To meet the requirement, modifications are given to the reduction by Garey, Johnson, and Tarjan [GJT76], which reduces SAT to HP of planar, 3-regular graphs (thereby showing that this special case of HP is NP-complete).[2] This reduction is referred to by the term GJT-reduction.

One cannot directly use their reduction to prove #P-completeness of #HP, because the number of Hamilton paths representing a satisfying assignment depends on how the assignment satisfies each clause. More precisely, let φ be a satisfiable 3CNF formula and (G, s, t) be the instance of HP generated by the GJT-reduction. Let A be the set of all satisfying assignments of φ and P be the set of all Hamilton paths from s to t in G. The GJT-reduction defines an onto mapping from P to A so that from each element path in P an element in A can be recovered in polynomial time. For each $a \in A$, let P_a denote the set of all Hamilton paths in P that map to a by this onto mapping. Clearly, $\|P\| = \sum_{a \in A} \|P_a\|$. It holds that $\|P_a\| = 2^p 3^q$, where p as well as q is a linear combination of the following three quantities: the number of clauses such that a satisfies all the three literals, the number of clauses such that a satisfies exactly two literals, and the number of clauses such that a satisfies exactly one literal.

Not-All-Equal-3-SAT [Sch78] (NAE3SAT, for short) is a special case of 3SAT in which it is asked to test whether a given CNF formula can be satisfied by an assignment that dissatisfies at least one literal per clause. Schaefer [Sch78] shows that NAE3SAT is NP-complete. In this paper the formulas are prohibited to have clauses with only two literals but permitted to have clauses with only one literal. The following lemma states that there is a polynomial-time many-one reduction from 3SAT to NAE3SAT that preserves the number of satisfying assignments such that for all formulas ϕ produced by the reduction, for all satisfying assignments a of ϕ, and for all three-literal clauses C of ϕ, a dissatisfies at least one literal of C.

Lemma 2. *There is a polynomial-time computable mapping f from the set of all 3CNF formulas to the set of all CNF formulas consisting of clauses with either one or three literals such that for every 3CNF formula φ, the following conditions hold for $\psi = f(\varphi)$:*

1. *The number of variables of ψ is $n + 2m$ and the number of clauses of ψ is $9m$, out of which $8m$ are three-literal clauses and m are single-literal clauses, where n and m are respectively the number of variables and the number of clauses of φ.*

---

[2] Garey, Johnson, and Tarjan reduce 3SAT to the Hamilton Cycle problem. The same reduction can be used to show the completeness of HP, since for every graph generated by their reduction, there is an edge e that each Hamilton cycle has to traverse.

2. $\#\text{SAT}(\varphi) = \#\text{SAT}(f(\varphi))$.
3. *Every satisfying assignment a of ψ satisfies all the m single-literal clause of ψ, exactly two literals for exactly $4m$ three-literal clauses, and exactly one literal for exactly $4m$ three-literal clauses.*

Proof.   Let a 3CNF formula $\varphi = C_1 \wedge \cdots \wedge C_m$ be given. Let $1 \leq i \leq m$ and let x, y, z be the literals of C_i. Introduce a new variable u_i that is equivalent to $x \vee y$. The equivalence can be written as a CNF formula: $(\overline{u_i} \vee x \vee y) \wedge (u_i \vee \overline{x}) \wedge (u_i \vee \overline{y})$. Furthermore, introduce another new variable w_i that needs to be false by the clause $(\overline{w_i})$. Let

$$C_i' = (\overline{u_i} \vee x \vee y) \wedge (u_i \vee \overline{x} \vee w_i) \wedge (u_i \vee \overline{y} \vee w_i) \wedge (u_i \vee z \vee w_i)$$
$$\wedge (u_i \vee \overline{x} \vee \overline{y}) \wedge (\overline{u_i} \vee x \vee \overline{w_i}) \wedge (\overline{u_i} \vee y \vee \overline{w_i}) \wedge (\overline{u_i} \vee \overline{z} \vee \overline{w_i}) \wedge (\overline{w_i}).$$

Then, every satisfying assignment a of C_i' satisfies exactly two literals for four three-literal clauses and exactly one literal for four three-literal clauses. Let $\psi = C_1' \wedge \ldots \wedge C_m'$. Then ψ has $n + 2m$ variables, consists of $8m$ three-literal clauses, m single-literal clauses, and has as many satisfying assignment as φ. Furthermore, for every satisfying assignment a of ψ, there are precisely $4m$ three-literal clauses such that a dissatisfies exactly one literal and precisely $4m$ three-literal clauses such that a dissatisfies exactly two literals. This proves the lemma.
□

Let φ be a 3CNF formula for which the number of satisfying assignments needs to be evaluated. Suppose φ is defined over n variables and has m clauses. Let ψ be the formula generated by applying the transformation in Lemma 2 to φ. The GJT-reduction with slight modifications is applied to ψ.

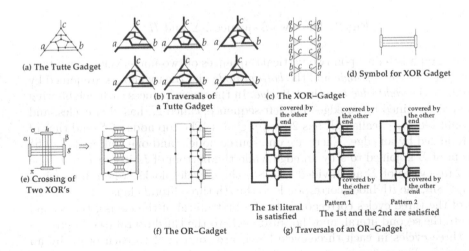

(a) The Tutte Gadget

(b) Traversals of a Tutte Gadget

(c) The XOR–Gadget

(d) Symbol for XOR Gadget

(e) Crossing of Two XOR's

(f) The OR–Gadget

covered by the other end   covered by the other end   covered by the other end

covered by the other end   covered by the other end

The 1st literal is satisfied

Pattern 1   Pattern 2
The 1st and the 2nd are satisfied

(g) Traversals of an OR–Gadget

Fig. 1. The Gadgets

Basic components of the construction are the Tutte-gadget (Figure 1 (a)), the XOR-gadget (Figure 1 (c)), and the OR-gadget (Figure 1 (f)). Here the first two gadgets are inherited from the GJT-reduction while the OR-gadget is new. The Tutte-gadget forces branching; to visit c without missing a node, one has to either enter from a and visit b on its way or enter from b and visit a on its way. There are four ways to do the former and two ways to do the latter (see Figure 1 (b)). The XOR-gadget is a ladder built using eight copies of the Tutte-gadget. In order to go through all the nodes in an XOR-gadget one has to enter and exit on the same vertical axis. For each of the two vertical axes there are $(4 \cdot 2)^4 = 2^{12}$ Hamilton paths. XOR-gadgets can be crossed without losing planarity by inflating the number of Hamilton paths (see Figure 1 (b) and (c)). Let g be an XOR-gadget connecting two vertical lines, α and β, and h be an XOR-gadget connecting two horizontal lines, σ and π. In order to cross these two XOR-gadgets, a two-node cycle is inserted in each of the four horizontal lines of g, the edges of the inserted cycles together with σ and π, and then each pair is connected by an XOR-gadget. Since four XOR-gadgets are added, the number of Hamilton paths is increased by a multiplicative factor of 2^{48}.

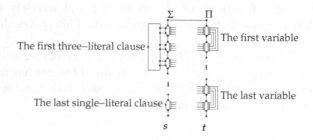

Fig. 2. The Two Sequences, Σ and Π

The graph is built upon two vertical sequences of two-node cycles. Each two-node cycle has the *top node* and the *bottom node* and the two nodes are joined by two edges, the *right edge* and the *left edge*. In the two sequences each neighboring cycle-pair is joined by an edge. The left sequence, called Σ, has $25m$ cycles, and the right sequence, called Π, has $2n + 4m$ cycles. The top node of Σ and that of Π are joined by an edge. There are one source node s and one sink node t. The bottom of Σ is joined to s by an edge while the bottom of Π is joined to t. The first $24m$ cycles of Σ are divided into $8m$ three-cycle blocks, where for each i, $1 \le i \le 8m$, the ith block corresponds to the ith three-literal clause of ψ, i.e., the first of the three cycles corresponds to the first literal of the clause, the second cycle to the second literal, and the third cycle to the third literal (see Figure 2). The three cycles in each three-cycle block are connected to each other by an OR-gadget. The remaining m cycles correspond to the m single-literal clauses of ψ. Each of these m cycles has a node in the middle of the left edge of the cycle. The sequence Π is divided into $n + 2m$ blocks of cycle-pairs, where for each i,

$1 \leq i \leq n + 2m$, the ith pair corresponds to x_i, the ith variable. For each pair, the top cycle corresponds to $\overline{x_i}$ and the bottom to x_i. To these sequences add a number of XOR-gadgets:

- For each i, $1 \leq i \leq n + 2m$, add an XOR-gadget to connect the right edge of the top cycle and that of the bottom cycle in the ith cycle-pair in Π.
- For each i, $1 \leq i \leq 2(n + 2m)$, and each position j, $1 \leq j \leq 25j$, if the ith cycle of Π and the jth cycle correspond to the same literal, then join them by an XOR-gadget.
- If two XOR-gadgets connecting Σ and Π need to be crossed, it is done so by addition of an XOR-gadget as described earlier.

For every i, $1 \leq i \leq n + 2m$, traversing the XOR-gadget within the ith cycle-pair in Π from the top cycle corresponds to assigning 1 to x_i, since it frees up the left edge of the bottom cycle; traversing it from the bottom corresponds to assigning 0 to x_i. Then for every i, $1 \leq i \leq n + 2m$, from the "free" left edge of the ith cycle-pair, all the XOR-gadgets satisfied by the assignment correspond to the "free" left edge must be traversed. So, for every assignment a to ψ, and for every j, $1 \leq j \leq 8m$, the number of "free" left edges in the jth cycle-triple in Σ is the number of literals that a satisfies in the jth three-literal clause of ψ. In each OR-gadget, the number of ways to traverse all of its nodes is the number of literals that are satisfied (see Figure 1 (g)). Let Ξ denote the number of crossings of XOR-gadgets. Then, for every satisfying assignment a of ψ, the number of Hamilton paths (between s and t) corresponding to a is: $2^{4m} \cdot 2^{12(n+29m+4\Xi)} = 2^{12n+352m+48\Xi}$. The maximum degree of the resulting graph is 3, and the graph is planar. Combining these observations and Proposition 1 and Lemma 2 yields the following lemma.

Lemma 3. *Every #P function is $\leq^p_{r\text{-}shift}$-reducible to the problem of computing the number of Hamilton paths in planar graphs of maximum degree 3.*

4 Self-Avoiding Walks in Two-Dimensional Grid Graphs

Let f be an arbitrary #P function. Let x be a string for which $f(x)$ needs to be evaluated. Let g be the reduction from f to #HP stated in Lemma 3. Let $(G, s, t) = g(x)$. Let N be the number of nodes in G.

It is known that planar graphs can be embedded in the two-dimensional grid in polynomial time (for example, see [CP95]). In the case when the maximum degree is 3, the embedding can be made so that there is no edge contention. Pick one such method and apply it to G so that s is the origin. Then the resulting embedding, call it $\mathcal{E}$, is a grid graph. Let $\alpha = 8N^2$ and $\beta = N^2$. Based on $\mathcal{E}$ construct a new grid graph $\mathcal{F}$ as follows:

First, enlarge the embedding by 5α. This is done by moving each corner (a, b) of the embedding to $(5\alpha \cdot a, 5\alpha \cdot b)$. Then it is possible to allocate to each edge e of G an area of size $\alpha \times \alpha$ that is adjacent to a straight-line segment realizing e in the embedding so that no two edges share points in the allocated area.

Second, for each edge of G, attach to the path realizing it a block of β unit-size rectangles so that the rectangles do not meet with those for another edge. Such a block of rectangles is considered by $S(p, q, r)$ in Figure 3, where $p, q, r \geq 0$. In short, $S(p, q, r)$ is a path of length $p + r + 2q - 1$ with q rectangles attached. This second step of the conversion replaces each path π realizing an edge of G

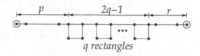

Fig. 3. $S(p, q, r)$

by $S(p, \beta, r)$ for some $p, r \geq 1$, such that $p + r + 2\beta - 1 = |\pi|$. Let L be an integer such that every path in $\mathcal{F}$ realizing an edge of G has length at most L. Then $p + r + 2\beta - 1 \leq L \leq N^c$ for some fixed constant $c > 0$. Call the left and the right ends of the structure u and v, respectively. For all $p, q, r \geq 1$, such that $p + 2q - 1 + r \leq L$, the simple paths within $S(p, q, r)$ have the following properties:

1. The number of those that connect u and v is 2^q.
2. The number of those that touch u but not v is $p+1+4(2^q+\cdots+2)+2^q(r-1) = 2^q(r+7)+p-7 < L \cdot 2^q$. Similarly, the number of those that touch v but not u is less than $L \cdot 2^q$.
3. The number of those that touch neither u nor v is less than $L^2 \cdot 2^{q+1}$. This can be seen as follows:
   - The number of such paths of length zero is at most $L + 2q$.
   - The number of such paths of length one is at most $L + 3q$.
   - The number of paths within the rectangles of length two is $4q$.
   - The number of paths within the rectangles of length three is $4q$.
   - The number of paths π such that either the left end of π is a point between u and the leftmost rectangle or the right end of π is a point between v and the rightmost rectangle is less than $2(p \cdot r \cdot 2^q) < L^2 2^q$.
   - All the paths that are yet to be counted sit between the leftmost rectangle and the rightmost rectangle. For every i, $2 \leq i \leq q$, the number of such paths that touch precisely i rectangles is less than $7^2 \cdot 2^{i-2}$. The sum of the quantity for all i, $2 \leq i \leq q$, is less than $49 \cdot 2^{q-1}$.

   The grand total is less than $2L+13q+L^2 2^q+49 \cdot 2^{q-1}$. This is less than $L^2 \cdot 2^{q+1}$ for all but finitely many L.

In the final step of the conversion, add two new nodes s' and t' and join s and s' by $S(7, 2\beta, 1)$ and join t and t' by $S(7, 2\beta, 1)$. For each simple path π in G, let $M(\pi)$ denote the set of all simple paths $\sigma = [v_1, \ldots, v_k]$ such that eliminating from σ all the nodes not corresponding to the nodes of G produces π. Also, for each simple path π in G, let $\mu(\pi)$ denote the size of $M(\pi)$. Let $B = 2^{(N+3)\beta+6}$. The value of $\mu(\pi)$ is evaluated as follows:

(Case 1) π is a Hamilton path: $\mu(\pi) = 2^{\beta(N-1)}(8 \cdot 2^{2\beta})^2 = 2^{(N+3)\beta+6}$.

(Case 2) π is nonempty, non-Hamiltonian, and connects s and t: Here $1 \le |\pi| \le N - 2$. So, $\mu(\pi) = 2^{\beta|\pi|}(8 \cdot 2^{2\beta})^2 \le 2^{(N+2)\beta+6}$.

(Case 3) π touches exactly one of s and t: Here $1 \le |\pi| \le N - 2$. There are at most two edges attached to the end node of π that is neither s nor t. By (2) in the above, there are $2L \cdot 2^\beta$ paths that can grow out of the end points of π along each of the two edge, so $\mu(\pi) \le 2^{|\pi|\beta} \cdot (8 \cdot 2^{2\beta}) \cdot (2L \cdot 2^\beta) < 16L \cdot 2^{\beta(N+1)}$.

(Case 4) π touches neither s nor t: Here $1 \le |\pi| \le N - 3$. There are at most two edges attached to each end node of π. So, $\mu(\pi) \le 2^{|\pi|\beta} \cdot (2L \cdot 2^\beta)^2 < 4L^2 \cdot 2^{\beta(N-1)}$.

(Case 5) π is empty: Since G has maximum degree 3. There are at most $3N/2$ edges in N. By (3) in the above, $\mu(\pi) < L^2 2^{\beta+1}$.

Note that the largest of the upper bounds in Cases 2 through 5 is $2^{(N+2)\beta+6}$. The number of non-Hamiltonian paths in G is bounded by $3N \cdot (2^{N-2} + 2^{N-3} + \cdots + 1) < 3N \cdot 2^{N-1}$. This is because there are N choices for one end point, there are at most three possible directions for the first edge, there are at most two possibilities there after, and the length may vary from 1 to $N - 2$. Thus, the sum of $\mu(\pi)$ for all non-Hamiltonian paths is less than $3N \cdot 2^{N-1} \cdot 2^{(N+2)\beta+6} < B$. So:

Claim. The number of simple paths in $\mathcal{F}$ is $\#HP(G, s, t) \cdot B + R$, where $0 \le R \le B - 1$.

Let $B' = 2^{(N+1)\beta+3}$. By an analysis similar to the above the following fact can be proven.

Claim. The number of simple paths in $\mathcal{F}$ starting from the origin (the image of s) is $\#HP(G, s, t) \cdot B' + R$, where $0 \le R \le B' - 1$.

Also, let $B'' = 2^{(N-1)\beta}$. By an analysis similar to the above, the following fact holds.

Claim. The number of simple paths in $\mathcal{F}$ between s' and t' is $\#HP(G, s, t) \cdot B'' + R$, where $0 \le R \le B'' - 1$.

Hence, the following theorem has been proven.

Theorem 4. *The problem of counting the number of self-avoiding walks in two-dimensional grid graphs is $\#P$-complete under $\le^p_{r\text{-}shift}$-reductions in each of the three possible definitions of the counting problem:*

- *the number of self-avoiding walks between two specified nodes needs to be counted;*
- *the number of self-avoiding walks from a specified node needs to be counted;*
- *the number of all self-avoiding walks needs to be counted.*

5   Self-Avoiding Walks in Hypercube Graphs

5.1   #P-Completeness

Let f be an arbitrary #P function. Let x be a string for which $f(x)$ needs to be evaluated. Let g be the reduction from f to the #HP as stated in Lemma 3. Let $(G, s, t) = g(x)$. Let $\alpha = 8N^2$ and $\beta = N^2$ as in the previous proof. Construct from G a graph H by replacing each edge by $S(1, \beta, \alpha-2\beta)$ and attach $S(1, 2\beta, 7)$ to s and t. Call the end of the two copies of $S(1, 2\beta, 7)$ by s' and t'. Let $L = \alpha$. This graph H is still planar and has maximum degree 3. Exactly the same analysis that holds for the two-dimensional grid graph can be applied to H. In particualr, the three claims for $\mathcal{F}$ hold with H in place of $\mathcal{F}$. Furthermore, let $\Lambda = 2\alpha(N - 1) + 2(8 + 2\beta)$. Then the number of paths in H having length Λ is equal to the number of Hamilton paths in G. Since the maximum degree of the graph is 3, H can be embedded as a hypercube graph $\mathcal{E}$ of dimension $\mathcal{O}(\log N)$ so that all paths realizing an edge of G have length d, where $d = 2^e$ for some integer e. Let $\mathcal{H}$ be the resulting embedding. Since $d = \mathcal{O}(\log N)$, the analysis is not affected much. The three facts hold with $\mathcal{H}$ in place of $\mathcal{F}$. Also, the following fact holds.

Claim. The total number of simple paths in $\mathcal{H}$ having length Λ is #HP(G, s, t).

Thus, the following theorem has been proven.

Theorem 5. *The problem of counting the number of self-avoiding walks in hypercube graphs is #P-complete under $\leq^p_{r\text{-}shift}$-reductions in each of the following four cases:*

- *the number of self-avoiding walks of a specified length needs to counted;*
- *the number of self-avoiding walks between two specified nodes needs to be counted;*
- *the number of self-avoiding walks from a specified node needs to be counted;*
- *the number of all self-avoiding walks needs to be counted.*

Now it remains to show that the embedding is indeed possible.

Lemma 6. *Let $d \geq 1$ be an integer. There exists a polynomial-time computable embedding $\mathcal{E}$ and a pair of polynomial-time computable functions $s, \ell : \mathbf{N} \to \mathbf{N}$ such that the following properties hold:*

1. *$s(n) = \mathcal{O}(\log n)$ and $\ell(n) = \mathcal{O}(\log \log n)$.*
2. *For every $n \geq 1$, and every undirected graph $G = (V, E)$ of n nodes having maximum degree d, $\mathcal{E}(G)$ is an embedding of G in the $s(n)$-dimensional hypercube $HC^{(s(n))}$ such that*
   - *for all $u \neq v$ in G, $\mu(u) \neq \mu(v)$, and*
   - *for every edge $e = (u, v)$ in G, $\nu(e)$ has dilation $2^{\ell(n)}$ and visits no $\mu(w)$ between $\mu(u)$ and $\mu(v)$.*

   Here μ and ν are respectively the embedding of nodes and the embedding of edges specified by $\mathcal{E}(G)$.

Proof. Let $d, n, m \geq 1$ be integers. Let G be a degree d graph with n nodes and m edges. Note that $m < n^2/2$. Identify the nodes in G with the set of integers from 0 to $n - 1$ and identify the edges in G with the set of integers from 0 to $m - 1$. For each node u of G, fix an enumeration of its neighbors. For every edge (u, v) of G, let $I_u(v)$ denote the order of v in the enumeration of the neighbors of u. Let q be the smallest integer such that $3q + 4$ is a power of 2 and such that $q \geq \lceil \log n \rceil$. Let $s = 6q + d + 2$. Let H denote the s-dimensional hypercube. Each node of G will be viewed as a q-bit binary number and each edge of G as a $2q$-bit binary number. For a binary string u, let $\bar{u}$ denote the string constructed from u by flipping all of its bits.

The s-bit representation of a node in H is divided into the following five components.

- The Node Part: This part has length $2q$. Here for each node $u = u_1 \cdots u_q$ of G, $\bar{u}u$ encodes u. Note that this encoding has exactly q many 1s in it.
- The Edge Part: This part has length $4q$ and is used to encode the edge to be traversed. Here each edge $e = e_1 \cdots e_{2q}$ is encoded as $\bar{e}e$.
- The Neighbor Part: This part has length d and is used to encode the value of $I_u(v)$ or the value of $I_v(u)$ when the edge (u, v) is traversed.
- The Switch Part: This part has length 2.

Let u, v, w, y be binary strings of length $2q$, $4q$, d, and 2, respectively. Then, $\langle u, v, w, y \rangle$ denotes the node $uvwy$ in $HC^{(s)}$.

The embedding $\mathcal{E}(G) = (\mu, \nu)$ is defined as follows: For each node $u = u_1 \cdots u_q$,

$$\mu(u) = \langle \bar{u}u, 0^{2q}, 0^d, 000 \rangle.$$

As for ν, let $e = (u, v)$ be an edge in G such that $u < v$. Let $A = a_1 \cdots a_{2q} = \bar{u}u$, $B = b_1 \cdots b_{2q} = \bar{v}v$, $C = c_1 \cdots c_{4q} = \bar{e}e$, $W_1 = 0^{I_u(v)-1}10^{d-I_u(v)}$, and $W_2 = 0^{I_v(u)-1}10^{d-I_v(u)}$. Let $i_1, \ldots, i_q$ be the enumeration of all the positions at which A has a bit 1 in increasing order. Let $j_1, \ldots, j_q$ be the enumeration of all the positions at which B has a bit 1 in increasing order. Let $k_1, \ldots, k_{2q}$ be the enumeration of all the positions at which C has a bit 1 in increasing order. For each t, $1 \leq t \leq q$, let $A_t = 0^{i_t}a_{i_t+1} \cdots a_{2q}$ and $B_t = 0^{j_t}b_{j_t+1} \cdots b_{2q}$. Also, for each t, $1 \leq t \leq 2q$, let $C_t = c_1 \cdots c_{k_t}0^{4q-k_t}$. Note that $A_q = B_q = 0^q$ and $C_{2q} = C$. The edge e is represented by a path from $\langle A, 0^{4q}, 0^d, 00 \rangle$ to $\langle 0^q, C, 0^d, 11 \rangle$ and a path $\langle B, 0^{4q}, 0^d, 00 \rangle$ to $\langle 0^q, C, 0^d, 11 \rangle$, each of length $3q+4$. The first path is defined by:

$$\langle A, 0^{4q}, 0^d, 00 \rangle, \langle A, 0^{4q}, W_1, 00 \rangle, \langle A, 0^{4q}, W_1, 10 \rangle, \langle A, C_1, W_1, 10 \rangle, \cdots,$$
$$\langle A, C_{2q}, W_1, 10 \rangle, \langle A_1, C, W_1, 10 \rangle, \cdots, \langle A_q, C, W_1, 10 \rangle, \langle 0^q, C, 0^d, 10 \rangle,$$
$$\langle 0^q, C, 0^d, 11 \rangle.$$

The second path is defined with B in place of A. The total length of the join of the two paths is $6q + 8$ and this is a power of 2 by assumption. Note that every node in the entire path contains one of the following: (i) C in the edge part, (ii) A in the node part and W_1 in the neighbor part, or (iii) B in the node part and W_2 in the neighbor part. So, no two edges share internal nodes in their path representation. It is not hard to see that the embedding can be computed in logarithmic space. This proves the lemma.   □

5.2   Scaling Up to Exponential Time

Let f be a function belonging to #EXP. Then there is a one-tape nondeterministic exponential-time machine M such that $f = \#acc_M$. It can be assumed that there is a polynomial p such that for all strings x, M on input x halts at step $2^{p(|x|)}$. By applying the tableau method to the computation of M and then by reducing the number of occurrences of each variable by introducing its copies, it can be shown that the computation of f can be transformed to #SAT by a polynomial-time local computation.

Lemma 7. *There is a reduction R from the evaluation problem of f to #SAT that satisfies the following conditions.*

1. *For every string x, $R(x)$ is a 3CNF formula.*
2. *For every string x, $f(x) = \#SAT(R(x))$.*
3. *For every string x, $R(x)$ can be locally computed in polynomial time in the following sense:*
   (a) For each variable y of $R(x)$, y or $\bar{y}$ appears in exactly three clauses.
   (b) For each string x, let $\nu(x)$ denote the number of variables in $R(x)$. Then ν is polynomial-time computable.
   (c) For each string x, let $\mu(x)$ denote the number of variables in $R(x)$. Then ν is polynomial-time computable.
   (d) For each string x and each i, $1 \le i \le \mu(x)$, let $C(x,i)$ be the ith clause in $R(x)$. Then C is polynomial-time computable.
   (e) For each string x and each i, $1 \le i \le \nu(x)$, let $S(x,i)$ be the set of all indices j such that the ith variable appears in $C(x,j)$. Then S is polynomial-time computable.

Proof.   Due to the space limitation the proof of the lemma is omitted.   □
    Now apply conversion of Lemma 2 to each formula generated by R. Denote the resulting transformation from the set of strings to NAE3SAT by R'. Since R is locally polynomial-time computable, so is R'. Modify the construction for Lemma 3 so that crossing of XOR-gadgets is allowed. Then the maximum degree remains three and the scaling factor becomes $2^{12n+352m}$. The connectivity of the gadgets in the graph essentially depends only on the occurrences of corresponding variables, so the mapping can be computed locally in polynomial time. Now apply the embedding described above. Denote the resulting transformation from the set of strings to the set of hypercube graphs by H. The length of the path is proportional to the logarithm of the number of nodes, so they are polynomially bounded. For each string x, let $d(x)$ be the dimension of $H(x)$, $N(x)$ denote the set of all strings of length $d(x)$ that encode a node in $H(x)$, and $E(x)$ denote the set of all strings ww' having length $2d(x)$ such that (w, w') is an edge of $H(x)$.

Lemma 8. *d is polynomial-time computable. There are polynomial-time computable functions C_N and C_E such that $C_N(x)$ and $C_E(x)$ are both polynomial-size boolean circuits and accept $N(x)$ and $E(x)$, respectively.*

This lemma gives the following theorem.

Theorem 9. *The problem of counting the number of simple paths in hypercube graphs is #EXP-complete under $\leq^p_{r\text{-}shift}$-reductions if the graphs are specified by circuits.*

Acknowledgement

The authors would like to thank Joerg Rothe for careful reading of the manuscript.

References

[CG96]   A. R. Conway and A. J. Guttmann. Square lattice self-avoiding walks and corrections-to-scaling. *Physical Review Letters*, 77:5284–5287, 1996. 586

[CP95]   M. Chrobak and T. H. Payne. A linear time algorithm for drawing a planar graph on a grid. *Information Processing Letters*, pages 241–246, 1995. 591

[GJT76] M. Garey, D. Johnson, and E. Tarjan. The planar Hamiltonian circuit problem is NP-complete. *SIAM Journal on Computing*, 5(4):704–714, 1976. 588

[MS93]   N. Madras and G. Slade. *The Self-Avoiding Walk*. Birkhäuser, Boston, MA, 1993. 585

[Pap94] C. Papadimitriou. *Computational Complexity*. Addison-Wesley, 1994. 587

[Sch78] T. Schaefer. The complexity of satisfiability problem. In *Proceedings of 10th Symposium on Theory of Computing*, pages 216–226. ACM Press, 1978. 588

[Val79a] L. Valiant. The complexity of computing the permanent. *Theoretical Computer Science*, 8:189–201, 1979. 587

[Val79b] L. Valiant. The complexity of enumeration and reliability problems. *SIAM Journal on Computing*, 8(3):410–421, 1979. 586, 587

[Wel93] D. Welsh. *Complexity: Knots, Colourings and Counting*. Cambridge University Press, 1993. 586

From Bidirectionality to Alternation

Nir Piterman[1] and Moshe Y. Vardi[2*]

[1] Weizmann Institute of Science, Department of Computer Science
Rehovot 76100, Israel
nirp@wisdom.weizmann.ac.il
http://www.wisdom.weizmann.ac.il/~nirp
[2] Rice University, Department of Computer Science
Houston, TX 77251-1892, U.S.A.
vardi@cs.rice.edu
http://www.cs.rice.edu/~vardi

Abstract. We describe an explicit simulation of 2-way nondeterministic automata by 1-way alternating automata with quadratic blow-up. We first describe the construction for automata on finite words, and extend it to automata on infinite words.

1 Introduction

The theory of finite automata is one of the fundamental building blocks of theoretical computer science. As the basic theory of finite-state systems, this theory is covered in numerous textbooks and in any basic undergraduate curriculum in computer science. Since its introduction in the 1950's, the theory had numerous applications in practically all branches of computer science, from the construction of electrical circuits [Koh70], to the design of lexical analyzers [JPAR68], and to the automated verification of hardware and software designs [VW86].

From its very inception, one fundamental theme in automata theory is the quest for understanding the relative power of the various constructs of the theory. Perhaps the most fundamental result of automata theory is the robustness of the class of regular languages, the class of languages definable by means of finite automata. Rabin and Scott showed in their classical paper that neither nondeterminism nor bidirectionality changes the expressive power of finite automata; that is, nondeterministic 2-way automata and deterministic 1-way automata have the same expressive power [RS59]. This robustness was later extended to alternating automata, which can switch back and forth between existential and universal modes (nondeterminism is an existential mode) [BL80, CKS81, LLS84].

In view of this robustness, the concept of relative expressive power was extended to cover also succinctness of description. For example, it is known that nondeterministic automata and two-way automata are exponentially more succinct than deterministic automata. The language $L_n = \{uv : u, v \in \{0,1\}^n$ and

* Supported in part by NSF grants CCR-9700061 and CCR-9988322, by BSF grant 9800096, and by a grant from the Intel Corporation.

J. Sgall, A. Pultr, and P. Kolman (Eds.): MFCS 2001, LNCS 2136, pp. 598–610, 2001.
© Springer-Verlag Berlin Heidelberg 2001

$u \neq v\}$ can be expressed using a 1-way nondeterministic automaton or a 2-way deterministic automaton of size polynomial in n, but a 1-way deterministic automaton accepting L_n must be of exponential size (cf. [SS78]). Alternating automata, in turn, are doubly exponentially more succinct than deterministic automata [BL80, CKS81].

Consequently, a major line of research in automata theory is establishing tight simulation results between different types of automata. For example, given a 2-way automaton with n states, Shepherdson showed how to construct an equivalent 1-way automaton with $2^{O(n \; log(n))}$ states [She59]. Birget showed how to construct an equivalent 1-way automaton with 2^{3n} states [Bir93] (see also [GH96]). Vardi constructed the *complementary* automaton, an automaton accepting the words rejected by the 2-way automaton, with 2^{2n} states [Var89]. Birget also showed, via a chain of reductions, that a 2-way nondeterministic automaton can be converted to a 1-way alternating automaton with quadratic blow-up [Bir93]. As the converse efficient simulation is impossible [LLS84], alternation is more powerfull than bidirectionality.

Our focus in this paper is on simulation of bidirectionality by alternation. The interest in bidirectionality and alternation in not merely theoretical. Both constructs have been shown to be useful in automated reasoning. For example, reasoning about modal μ-calculus with past temporal connectives requires alternation and bidirectionality [Str82, Var88, Var98]. Recently, model checking of specifications in μ-calculus on context-free and prefix-recognizable systems has been reduced to questions about 2-way automata [KV00]. In a different field of research, 2-way automata were used in query processing over semistructured data [CdGLV00].

We found Birget's construction, simulating bidirectionality by alternation with quadratic blow-up, unsatisfactory. As noted, his construction is indirect, using a chain of reductions. In particular, it uses the reverse language and, consequently, can not be extended to automata on infinite words. The theory of finite automata on infinite objects was established in the 1960s by Büchi, Mc-Naughton and Rabin [Büc62, McN66, Rab69]. They were motivated by decision problems in mathematical logic. More recently, automata on infinite words have shown to be useful in computer-aided verification [Kur94, VW86]. We note that bidirectionality does not add expressive power also in the context of automata on infinite words. Vardi has already shown that given a 2-way nondeterministic Büchi automaton with n states one can construct an equivalent 1-way nondeterministic Büchi with $2^{O(n^2)}$ states [Var88].

Our main result in this paper is a direct quadratic simulation of bidirectionality by alternation. Given a 2-way nondeterministic automaton with n states, we construct an equivalent 1-way alternating automaton with $O(n^2)$ states. Unlike Birget's construction, our construction is explicit. This has two advantages. First, one can see exactly how alternation can efficiently simulate bidirectionality. (In order to convert the nondeterministic automaton into an alternating automaton we use the fact that the run of the 2-way nondeterministic automaton looks like a tree of "zigzags" [1]. We analyze the form such a tree can take and recognize, using an alternating automaton, when such a tree exists.) Second, the explicitness of the construction enables us to extend it to Büchi automata. (In the full version we also give a construction for 2-way nondeterministic Rabin and parity automata.) Since it is known how to simulate alternating Büchi automata by nondeterministic Büchi automata with exponential blow-up [MH84], our construction provides another proof of the result that a 2-way nondeterministic Büchi automaton with n states can be simulated by a 1-way nondeterministic Büchi with $2^{O(n^2)}$ states [Var88].

2   Preliminaries

We consider finite or infinite sequences of symbols from some finite alphabet Σ. Given a *word* w, an element in $\Sigma^* \cup \Sigma^\omega$, we denote by w_i the ith letter of the word w. The *length* of w is denoted by $|w|$ and is defined to be ω for infinite words.

A *2-way nondeterministic automaton* is $A = \langle \Sigma, S, S_0, \rho, F \rangle$, where Σ is the finite alphabet, S is the finite set of states, $S_0 \subseteq S$ is the set of initial states, $\rho : S \times \Sigma \to 2^{S \times \{-1,0,1\}}$ is the transition function, and F is the acceptance set. We can run A either on finite words (*2-way nondeterministic finite automaton* or *2NFA* for short) or on infinite words (*2-way nondeterministic Büchi automaton* or *2NBW* for short). In the full version we show that we can restrict our attention to automata whose transition function is of the form $\rho : S \times \Sigma \to 2^{S \times \{-1,1\}}$.

A *run* on a finite word $w = w_0, ..., w_l$ is a finite sequence of states and locations $(q_0, i_0), (q_1, i_1), ..., (q_m, i_m) \in (S \times \{0, ..., l+1\})^*$. The pair (q_j, i_j) represents the automaton is in state q_j reading letter i_j. Formally, $q_0 = s_0$ and $i_0 = 0$, and for all $0 \leq j < m$, we have $i_j \in \{0, ..., l\}$ and $i_m \in \{0, ..., l+1\}$. Finally, for all $0 \leq j < m$, we have $(q_{j+1}, i_{j+1} - i_j) \in \delta(q_j, w_{i_j})$. A run is *accepting* if $i_m = l+1$ and $q_m \in F$.

A *run* on an infinite word $w = w_0, w_1, ...$ is defined similarly as an infinite sequence. The restriction on the locations is removed (for all j, the location i_j can be every number in $\mathbb{N}$). In 2NBW, a run is *accepting* if it visits $F \times \mathbb{N}$ infinitely often. A word w is *accepted* by A if it has an accepting run over w. The *language* of A is the set of words accepted by A, denoted by $L(A)$.

In the finite case we are only interested in runs in which the same state in the same position do no repeat twice during the run. In the infinite case we minimize the amount of repetition to the unavoidable minimum. A run $r = (s_0, 0), (s_1, i_1), (s_2, i_2), ..., (s_m, i_m)$ on a finite word is *simple* if for all j and k such that $j < k$, either $s_j \neq s_k$ or $i_j \neq i_k$. A run $r = (s_0, 0), (s_1, i_1), (s_2, i_2), ...$ on an infinite word is *simple* if one of the following holds (1) For all $j < k$, either $s_j \neq s_k$ or $i_j \neq i_k$. (2) There exists $l, m \in \mathbb{N}$ such that for all $j < k < l + m$, either $s_j \neq s_k$ or $i_j \neq i_k$, and for all $j \geq l$, $s_j = s_{j+m}$ and $i_j = i_{j+m}$.

---

[1] The analysis of the form of the "zigzags" is similar to the analysis of runs of pushdown-automata done in [Ruz80, Ven91].

In the full version we show that there exists an accepting run iff there exists a simple accepting run. Hence, it is enough to consider simple accepting runs.

Given a set S we first define the set $B^+(S)$ as the set of all positive formulas over the set S with 'true' and 'false' (i.e., for all $s \in S$, s is a formula and if f_1 and f_2 are formulas, so are $f_1 \wedge f_2$ and $f_1 \vee f_2$). We say that a subset $S' \subseteq S$ satisfies a formula $\varphi \in B^+(S)$ (denoted $S' \models \varphi$) if by assigning 'true' to all members of S' and 'false' to all members of $S \setminus S'$ the formula φ evaluates to 'true'. Clearly 'true' is satisfied by the empty set and 'false' cannot be satisfied.

A *tree* is a set $T \subseteq \mathbb{N}^*$ such that if $x \cdot c \in T$ where $x \in \mathbb{N}^*$ and $c \in \mathbb{N}$, then also $x \in T$. The elements of T are called *nodes*, and the empty word ϵ is the *root* of T. For every $x \in T$, the nodes $x \cdot c$ where $c \in \mathbb{N}$ are the *successors* of x. A node is a *leaf* if it has no successors. A *path* π of a tree T is a set $\pi \subseteq T$ such that $\epsilon \in \pi$ and for every $x \in \pi$, either x is a leaf or there exists a unique $c \in \mathbb{N}$ such that $x \cdot c \in \pi$. Given an alphabet Σ, a Σ-*labeled tree* is a pair (T, V) where T is a tree and $V : T \to \Sigma$ maps each node of T to a letter in Σ.

A *1-way alternating automaton* is $B = \langle \Sigma, Q, s_0, \Delta, F \rangle$ where Σ, Q and F are like in nondeterministic automata. s_0 is a unique initial state and $\Delta : S \times \Sigma \to B^+(Q)$ is the transition function. Again we may run A on finite words (*1-way alternating automata on finite words* or *1AFA* for short) or on infinite words (*1-way alternating Büchi automata* or *1ABW* for short).

A *run* of A on a finite word $w = w_0...w_l$ is a labeled tree (T, r) where $r : T \to Q$. The maximal depth in the tree is $l + 1$. A node x labeled by s describes a copy of the automaton in state s reading letter $w_{|x|}$. The labels of a node and its successors have to satisfy the transition function Δ. Formally, $r(\epsilon) = s_0$ and for all nodes x with $r(x) = s$ and $\Delta(s, w_{|x|}) = \varphi$ there is a (possibly empty) set $\{s_1, ..., s_n\} \models \varphi$ such that the successors of x, $\{x \cdot 0, ..., x \cdot (n-1)\}$ are labeled by $\{s_1, ..., s_n\}$. The run is *accepting* if all the leaves in depth $l + 1$ are labeled by states from F.

A run of A on an infinite word $w = w_0 w_1...$ is defined similarly as a (possibly) infinite labeled tree. A run of a 1ABW is *accepting* if every infinite path visits the accepting set infinitely often. As before, a word w is *accepted* by A if it has an accepting run over the word. We similarly define the language of A, $L(A)$.

3 Automata on Finite Words

We start by transforming 2NFA to 1AFA. We analyze the possible form of an accepting run of a 2NFA and using a 1AFA check when such a run exists over a word.

Theorem 1. *For every 2NFA $A = \langle \Sigma, S, s_0, \rho, F \rangle$ with n states, there exists an 1AFA $B = \langle \Sigma, Q, s_0, \Delta, F \rangle$ with $O(n^2)$ states such that $L(B) = L(A)$.*

Given a 2NFA $A = \langle \Sigma, S, s_0, \delta, F \rangle$, let $B = \langle \Sigma, Q, s_0, \Delta, F \rangle$ denote its equivalent 1AFA. Note that B uses the acceptance set and the initial state of A.

Recall that a run of A is a sequence $r = (s_0, 0), (s_1, i_1), (s_2, i_2), ..., (s_m, i_m)$ of pairs of states and locations, where s_j is the state and i_j is the location of

the automaton in the word w. We refer to each state as a *forward* or *backward*
state according to its predecessor in the run. If it resulted from a backward
movement it is a *backward* state and if from a forward movement it is a *forward*
state. Formally, (s_j, i_j) is a forward state if $i_j = i_{j-1} + 1$ and backward state
if $i_j = i_{j-1} - 1$. The first state $(s_0, 0)$ is defined to be a forward state.

Given the 2NFA A our goal is to construct the 1AFA B recognizing the
same language. In Figure 1a we see that a run of A takes the form of a tree of
'zigzags'. Our one-way automaton reads words moving forward and accepts if
such a tree exists. In Figure 1a we see that there are two transitions using a_1.
The first $(s_2, 1) \in \delta(s_1, a_1)$ and the second $(s_4, 1) \in \delta(s_3, a_1)$. In the one-way
sweep we would like to make sure that s_3 indeed resulted from s_2 and that the
run continuing from s_3 to s_4 and further is accepting. Hence when in state s_1
reading letter a_1 we guess that there is a part of the run coming from the future
and spawn two processes. The first checks that s_1 indeed results in s_3 and the
second ensures that the part $s_3, s_4, \dots$ of the run is accepting.

Hence the state set of the alternating automaton is $Q = S \cup (S \times S)$. A
singleton state $s \in Q$ represents a part of the run that is only looking forward (s_4
in Figure 1a). In fact, we use singleton states to represent only the last forward
state in the run of A that visits a letter. A *pair state* $(s_1, s_3) \in Q$ represents a
part of the run that consists of a forward moving state and a backward moving
state (s_1 and s_3 in Figure 1a). Such a pair ensures that there is a run segment
linking the forward state to the backward state. We introduce one modification,
since s_3 is a backward state (i.e. $(s_3, -1) \in \delta(s_2, a_2)$) it makes sense to associate
it with a_2 and not with a_1. As the alternating automaton reads a_1 (when in
state s_1), it guesses that s_3 comes from the future and changes direction. The
alternating automaton then spawns two processes: the first, s_4 and the second,
(s_2, s_3), and both read a_2 as their next letter. Then it is easier to check that
$(s_3, -1) \in \delta(s_2, a_2)$.

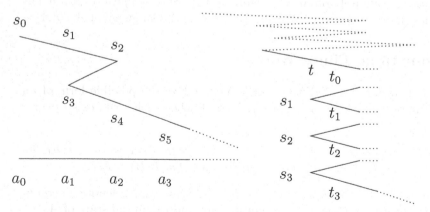

Fig. 1. (a) A zigzag run (b) The transition at the singleton state t

3.1 The Construction

The Transition at a Singleton State We define the transitions of B in two stages. First we define transitions from a singleton state. When in a singleton state $t \in Q$ reading letter a_j (See Figure 1b) the alternating automaton guesses that there are going to be k more visits to letter a_j in the rest of the run (as the run is simple k is bounded by the number of states of the 2NFA A). We refer to the states reading letter a_j according to the order they appear in the run as $s_1, ..., s_k$. We assume that all states that read letters prior to a_j have already been taken care of, hence $s_1, ..., s_k$ themselves are backward states (i.e. $(s_i, -1) \in \delta(p_i, a_{j+1})$ for some p_i). They read the letter a_j and move forward (there exists some t_i such that $(t_i, 1) \in \delta(s_i, a_j)$). Denote the successors of $s_1, ..., s_k$ by $t_1, ..., t_k$. The alternating automaton verifies that there is a run segment connecting the successor of t (denoted before a_j have been taken care of, this run segment should not go back to letters before a_j). Similarly the alternating automaton verifies that a run segment connects t_1 to s_2, etc. In general the alternating automaton checks that there is a part of the run connecting t_i to s_{i+1}. Finally, from t_k the run has to read the rest of the word and reach location $|w|$ in an accepting state.

Given a state t and an alphabet letter a, consider the set R_a^t of all possible sequences of states of length at most $2n - 1$ where no two states in an even place (forward states) are equal and no two states in an odd place (backward states) are equal. We further demand that the first state in the sequence be a successor of t ($(t_0, 1) \in \delta(t, a)$) and similarly that t_i be a successor of s_i ($(t_i, 1) \in \delta(s_i, a)$). Formally

$$R_a^t = \left\{ \langle t_0, s_1, t_1, ..., s_k, t_k \rangle \;\middle|\; \begin{array}{l} 0 \leq k < n \\ (t_0, 1) \in \delta(t, a) \\ \forall i < j, \; s_i \neq s_j \text{ and } t_i \neq t_j \\ \forall i, \; (t_i, 1) \in \delta(s_i, a) \end{array} \right\}$$

The transition of B chooses one of these sequences and ensures that all promises are kept, i.e. there exists a run segment connecting t_{i-1} to s_i.

$$\Delta(t, a) = \bigvee_{\langle t_0, ..., t_k \rangle \in R_a^t} (t_0, s_1) \wedge (t_1, s_2) \wedge ... \wedge (t_{k-1}, s_k) \wedge t_k$$

The Transition at a Pair State When the alternating automaton is in a pair state (t, s) reading letter a_j it tries to find a run segment connecting t to s using only the suffix $a_j...a_{|w|-1}$. We view t as a forward state reading a_j and s as a backward state reading a_{j-1} (Again $(s, -1) \in \delta(p, a_j)$). As shown in Figure 2a, the run segment connecting t to s might visit letter a_j but should not visit a_{j-1}.

Figure 2b provides a detailed example. The automaton in state (t, s) guesses that the run segment linking t to s visits a_2 twice and that the states reading letter a_2 are s_1 and s_2. The automaton further guesses that the predecessor of s is s_3 ($(s, -1) \in \delta(s_3, a_2)$) and that the successors of t, s_1 and s_2 are t_0, t_1 and t_2 respectively. The alternating automaton spawns three processes: $(t_0, s_1), (t_1, s_2)$

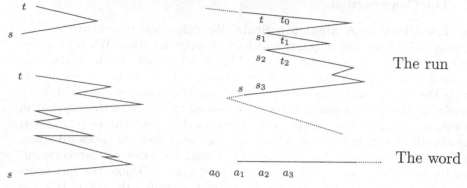

Fig. 2. (a) Different connecting segments (b) The transition at the pair state (t, s)

and (t_2, s_3) all reading letter a_{j+1}. Each of these pair states has to find a run segment connecting the two states.

We now define the transition from a state in $S \times S$. Given a state (t, s) and an alphabet letter a, we define the set $R_a^{(t,s)}$ of all possible sequences of states of length at most $2n$ where no two states in an even position (forward states) are equal and no two states in an odd position (backward states) are equal. We further demand that the first state in the sequence be a successor of t $((t_0, 1) \in \delta(t, a))$, that the last state in the sequence be a predecessor of s $((s, -1) \in \delta(s_{k+1}, a))$ and similarly that t_i be a successor of s_i $((t_i, 1) \in \delta(s_i, a))$.

$$
R_a^{(t,s)} = \left\{ \langle t_0, s_1, t_1, ..., s_k, t_k, s_{k+1} \rangle \; \middle| \; \begin{array}{l} 0 \le k < n \\ (t_0, 1) \in \delta(t, a) \\ (s, -1) \in \delta(s_{k+1}, a) \\ \forall i < j, \; s_i \ne s_j \text{ and } t_i \ne t_j \\ \forall i, \; (t_i, 1) \in \delta(s_i, a) \end{array} \right\}
$$

The transition of B chooses one sequence and ensures that all pairs meet:

$$
\Delta((t, s), a) = \begin{cases} true & \text{If } (s, -1) \in \\ & \delta(t, a) \\[2ex] \bigvee\limits_{\langle t_0,...,s_{k+1}\rangle \in R_a^{(t,s)}} (t_0, s_1) \wedge (t_1, s_2) \wedge ... \wedge (t_k, s_{k+1}) & \text{Otherwise} \end{cases}
$$

Claim. $L(A) = L(B)$

Proof. Given an accepting simple run of A on a word w of the form $(s_0, 0)$, $(s_1, i_1), \ldots, (s_m, i_m)$, we annotate each pair by the place it took in the run of A.

Thus the run takes the form $(s_0, 0, 0), (s_1, i_1, 1), \ldots, (s_m, i_m, m)$. We build a run tree (T, V) of B by induction. In addition to the labeling $V : T \to S \cup S \times S$, we attach a single tag to a singleton state and a pair of tags to a pair state. The tags are triplets from the annotated run of A. For example the root of the run tree of B is labeled by s_0 and tagged by $(s_0, 0, 0)$. The labeling and the tagging conforms to the following:

- Given a node x labeled by state s tagged by (s', i, j) from the run of A we build the tree so that $s = s'$, $i = |x|$ and furthermore all triplets in the run of A whose third element is larger than j have their second element at least i.
- Given a node x labeled by state (t, s) tagged by (t', i_1, j_1) and (s', i_2, j_2) in the run of A we build the tree so that $t = t'$, $s = s'$, $i_1 = i_2 + 1 = |x|$, $j_1 < j_2$ and that all triplets in the run of A whose third element is between j_1 and j_2 have their second element be at least i_1.

We start with the root labeling it by s_0 and tagging it by $(s_0, 0, 0)$. Obviously this conforms to our demands.

Given a node x labeled by t, tagged by (t, i, j), and adhering to our demands (see state t in Figure 1b). If (t, i, j) has no successor in the run of A, it must be the case that $i = |w|$ and that $t \in F$. Otherwise we denote the triplets in the run of A whose third element is larger than j and whose second element is i by $(s_1, i, j_1), \ldots, (s_k, i, j_k)$. By assumption there is no point in the run of A beyond j visiting a letter before i. Since the run is simple, $k < n$. Denote by $(t_0, i + 1, j + 1)$ the successor of (t, i, j) and by $(t_1, i + 1, j_1 + 1), \ldots, (t_k, i + 1, j_k + 1)$ the successors of $s_1, \ldots, s_k$. We add $k + 1$ successors to x, label them $(t_0, s_1), (t_1, s_2), \ldots, (t_{k-1}, s_k), t_k$, to a successor of x labeled by (t_{l-1}, s_l) we add the tags $(t_{l-1}, i + 1, j_l + 1)$ and (s_l, i, j_l), and to the successor of x labeled by t_k we add the tag $(t_k, i + 1, j_k + 1)$. We now show that the new nodes added to the tree conform to our demands. By assumption there are no visits beyond the jth step in the run of A to letters before a_i and $s_1, \ldots, s_k$ are all the visits to a_i after the jth step of A.

Let y be the successor of x labeled t_k (tagged $(t_k, i + 1, j_k + 1)$). Since $|x| = i$, we conclude $|y| = i + 1$. All the triplets in the run of A appearing after $(t_k, i + 1, j_k + 1)$ do not visit letters before a_{i+1} (We collected all visits to a_i).

Let y be a successor of x labeled by (t_l, s_{l+1}) (tagged $(t_l, i + 1, j_l + 1)$ and (s_{l+1}, i, j_{l+1})). We know that $i = |x|$ hence $i + 1 = |y|$, $j_l + 1 < j_{l+1}$ and between the $j_l + 1$ element in the run of A and the j_{l+1} element letters before a_{i+1} are not visited.

We turn to continuing the tree below a node labeled by a pair state. Given a node x labeled by (t, s) tagged (t, i, j) and and $(s, i - 1, k)$. By assumption there are no visits to a_{i-1} in the run of A between the jth triplet and k^{th} triplet. If $k = j + 1$ then we are done and we leave this node as a leaf. Otherwise we denote the triplets in the run of A whose third element is between j and k and whose second element is i by $(s_1, i, j_1), \ldots, (s_m, i, j_m)$ (see Figure 2b). Denote by $(t_1, i + 1, j_1 + 1), \ldots, (t_m, i + 1, j_m + 1)$ their successors, by $(t_0, i + 1, j + 1)$ the successor of t and by $(s_{k+1}, i, k - 1)$ the predecessor of s. We add $k + 1$

successors to x and label them $(t_0, s_1), (t_1, s_2), ..., (t_k, s_{k+1})$. To a successor of x labeled by (t_{l-1}, s_l) we add the tags $(t_{l-1}, i+1, j_{l-1}+1)$ and (s_l, i, j_l). As in the previous case when we combine the assumption with the way we chose $t_0, ...t_k$ and $s_1, ..., s_{k+1}$, we conclude that the new nodes conform to the demands.

Clearly, all pair-labeled paths terminate with 'true' before reading the whole word w and the path labeled by singleton states reaches the end of w with an accepting state.

In the other direction we stretch the tree run of B into a linear run of A. In the full version, we give a recursive algorithm that starts from the root of the run tree and constructs a run of A. When first reaching a node x labeled by pair-state (s, t), we add s to the run of A. Then we handle recursively the sons of x. When we return to x we add t to the run of A. When reaching a node x labeled by a singleton state s we simply add s to the run of A and handle the sons of x recursively.

$\square$

4   Automata on Infinite Words

We may try to run the 1AFA from Section 3 on infinite words. We demand that pair-labeled paths be finite and that the infinite singleton-labeled path visit F infinitely often. Although an accepting run of A visited F infinitely often we cannot ensure infinitely many visits to F on the infinite path. The visits may be reflected in the run of B in the pair-labeled paths. Another problem is when the run ends in a loop.

Theorem 2. *For every 2NBW $A = \langle \Sigma, S, s_0, \rho, F \rangle$ with n states, there exists an 1ABWs $B = \langle \Sigma, Q, s_0', \Delta, F' \rangle$ with $O(n^2)$ states such that $L(B) = L(A)$.*

We have to record hidden visits to F. This is done by doubling the set of states. While in the finite case the state set is $S \cup S \times S$, this time we also annotate the states by $\perp$ and $\top$. Hence $Q = (S \cup S \times S) \times \{\perp, \top\}$. A pair state labeled by $\top$ is a promise to visit the acceptance set. The state $(s, t, \top)$ means that in the run segment linking s to $(s, \top)$ is displaying a visit to F in the zigzags connecting s to the previous singleton state. The initial state is $s_0' = (s_0, \perp)$.

With the same notation we solve the problem of a loop. We allow a transition from a singleton state to a sequence of pair states. One of the pairs promises a visit to F. The acceptance set is $F' = (S \times \{\top\})$ and the transition function Δ is defined as follows.

The Transition at a Singleton State Just like in the finite case we consider all possible sequences of states of length at most $2n - 1$ with same demands.

$$
R_a^t = \left\{ \langle t_0, s_1, t_1, ..., s_k, t_k \rangle \; \middle| \; \begin{array}{l} 0 \le k < n \\ (t_0, 1) \in \delta(t, a) \\ \forall i < j, \; s_i \ne s_j \text{ and } t_i \ne t_j \\ \forall i, \; (t_i, 1) \in \delta(s_i, a) \end{array} \right\}
$$

Recall that a sequence $(t_0, s_1), (t_1, s_2), ..., (t_{k-1}, s_k), t_k$ checks that there is a zigzag run segment linking t_0 to t_k. We mentioned that t_k is annotated with $\top$ in case this run segment has a visit to F. If t_k is annotated with $\top$, at least one of the pairs has to be annotated with $\top$. Although more than one pair might visit F we annotate all other pairs by $\bot$. Hence for a sequence $\langle t_0, s_1, t_1, ..., s_k, t_k \rangle$ we consider the sequences of $\bot$ and $\top$ of length $k+1$ in which if the last is $\top$ so is another one. Otherwise all are $\bot$.

$$\alpha_k^R = \left\{ \langle \alpha_0, ..., \alpha_k \rangle \in \{\bot, \top\}^{k+1} \, \middle| \, \begin{array}{l} \text{If } \alpha_k = \top \text{ then } \exists! i \text{ s.t. } 0 \leq i < k \text{ and } \alpha_i = \top \\ \text{If } \alpha_k = \bot \text{ then } \forall\, 0 \leq i < k, \; \alpha_i = \bot \end{array} \right\}$$

This is, however, not enough. We have to consider also the case of a loop. The automaton has to guess that the run terminates with a loop when it reads the first letter of w that is read inside the loop. The only states reading this letter inside the loop are backward states. We consider pairs of sequences of at most $2n$ states, where the last state in the two sequences is equal. This repetition closes the loop. In both sequences no two states in an even/odd position are equal. For example, in Figure 3, we see that in state t reading letter a_1, the alternating automaton guesses the sequence $(t_0, s_1), (t_1, s_2)$ and the sequence $(t_2, s_3), (t_3, s_2)$. The last state in both sequences is s_2.

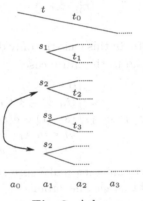

Fig. 3. A loop

More formally, we demand that the first state in the first sequence be a successor of t $((t_0^1, 1) \in \delta(t, a))$, that the first state in the second sequence be a successor of the last state in the first sequence $((t_0^2, 1) \in \delta(s_{k+1}^1, a))$, that t_i^p be a successor of s_i^p for $p \in \{1, 2\}$ $((t_i^p, 1) \in \delta(s_i^p, a))$ and that the last state in the first sequence be equal to the last state in the second sequence $(s_{k+1}^1 = s_{l+1}^2)$.

$$L_a^t = \left\{ \left\langle \begin{array}{c} \langle t_0^1, s_1^1, t_1^1, ..., s_k^1, t_k^1, s_{k+1}^1 \rangle, \\ \langle t_0^2, s_1^2, t_1^2, ..., s_l^2, t_l^2, s_{l+1}^2 \rangle \end{array} \right\rangle \,\middle|\, \begin{array}{l} 0 \le k < n, \ 0 \le l < n \\ (t_0^1, 1) \in \delta(t, a), \ (t_0^2, 1) \in \delta(s_{k+1}^1, a) \\ \forall i < j, \ s_i^1 \ne s_j^1 \text{ and } t_i^1 \ne t_j^1 \\ \forall i < j, \ s_i^2 \ne s_j^2 \text{ and } t_i^2 \ne t_j^2 \\ \forall i, \ \forall p, \ (t_i^p, 1) \in \delta(s_i^p, a) \\ s_{k+1}^1 = s_{l+1}^2 \end{array} \right\}$$

It is obvious that a visit to F has to occur within the loop. Hence we have to make sure that the run segment connecting one of the pairs in the second sequence visits F. Hence we annotate one of the pairs $(t_0^2, s_1^2), ..., (t_l^2, s_{l+1}^2)$ with $\top$. One visit to F is enough hence all other pairs are annotated by $\bot$.

$$\alpha_l^L = \{ \langle \alpha_0, ..., \alpha_l \rangle \in \{\bot, \top\}^{l+1} \mid \exists ! i \text{ s.t. } \alpha_i = \top \}$$

The transition of B chooses a sequence in $R_a^t \cup L_a^t$ and a sequence of $\bot$ and $\top$.

$$\Delta((t, \bot), a) = \Delta((t, \top), a) = \bigvee \begin{cases} \displaystyle\bigvee_{R_a^t, \alpha_k^R} (t_0, s_1, \alpha_0) \wedge ... \wedge (t_{k-1}, s_k, \alpha_{k-1}) \wedge (t_k, \alpha_k) \\[2ex] \displaystyle\bigvee_{L_a^t, \alpha_l^L} \left(\begin{array}{l} (t_0^1, s_1^1, \bot) \wedge ... \wedge (t_k^1, s_{k+1}^1, \bot) \wedge \\ (t_0^2, s_1^2, \alpha_0) \wedge ... \wedge (t_l^2, s_{l+1}^2, \alpha_l) \end{array} \right) \end{cases}$$

The Transition at a Pair State In this case the only difference is the addition of $\bot$ and $\top$. The set $R_a^{(t,s)}$ is equal to the finite case.

$$R_a^{(t,s)} = \left\{ \langle t_0, s_1, t_1, ..., s_k, t_k, s_{k+1} \rangle \,\middle|\, \begin{array}{l} 0 \le k < n \\ (t_0, 1) \in \delta(t, a) \\ (s, -1) \in \delta(s_{k+1}, a) \\ \forall i < j, \ s_i \ne s_j \text{ and } t_i \ne t_j \\ \forall i, \ (t_i, 1) \in \delta(s_i, a) \end{array} \right\}$$

In the transition of 'top' states we have to make sure that a visit to F indeed occurs. If the visit occured in this stage the promise ($\top$) can be removed ($\bot$). Otherwise the promise must be passed to one of the successors.

$$\alpha_{s,t,k}^R = \left\{ \langle \alpha_0, ..., \alpha_k \rangle \in \{\bot, \top\}^{k+1} \,\middle|\, \begin{array}{l} \text{If } s \notin F \text{ and } t \notin F \text{ then } \exists ! i \text{ s.t. } \alpha_i = \top \\ \text{Otherwise } \forall\, 0 \le i \le k, \ \alpha_i = \bot \end{array} \right\}$$

The transition of B chooses a sequence of states and a sequence of $\bot$ and $\top$.

$$\Delta((t, s, \bot), a) = \begin{cases} true & \text{If } (s, -1) \in \delta(t, a) \\ \displaystyle\bigvee_{R_a^{(t,s)}} (t_0, s_1, \bot) \wedge ... \wedge (t_k, s_{k+1}, \bot) & \text{Otherwise} \end{cases}$$

$$\Delta((t, s, \top), a) = \begin{cases} true & \begin{array}{l} \text{If } (s, -1) \in \delta(t, a) \\ \text{and } (s \in \text{Fort} \in F) \end{array} \\ \displaystyle\bigvee_{R_a^{(t,s)}, \alpha_{s,t,k}^R} (t_0, s_1, \alpha_0) \wedge ... \wedge (t_k, s_{k+1}, \alpha_k) & \text{Otherwise} \end{cases}$$

Claim. L(A)=L(B)

The proof is just an elaboration on the proof of the finite case.

Remark: In both the finite and the infinite cases, we get a 1-way alternating automaton with $O(n^2)$ states and transitions of exponential size. Birget's construction also results in exponential-sized transitions [Bir93]. Globerman and Harel use 0-steps in order to reduce the transition to polynomial size [GH96]. Their construction uses the reverse language and can not be applied to infinite words. If we use 0-steps, it is quite simple to change our construction so that it uses only polynomial-sized transitions. We note that the transition size does not effect the conversion from 1ABW to 1NBW.

5   Acknowledgments

We would like to thank Orna Kupferman for her remarks on the manuscript and an anonymous referee for pointing out the works of Ruzzo and Venkateswaran.

References

[Bir93]   J. C. Birget. State-complexity of finite-state devices, state compressibility and incompressibility. *Mathematical Systems Theory*, 26(3):237–269, 1993. 599, 609

[BL80]   J. A. Brzozowski and E. Leiss. Finite automata and sequential networks. *Theoretical Computer Science*, 10:19–35, 1980. 598, 599

[Büc62]   J. R. Büchi. On a decision method in restricted second order arithmetic. In *Proc. Internat. Congr. Logic, Method. and Philos. Sci. 1960*, pages 1–12, Stanford, 1962. Stanford University Press. 599

[CdGLV00]   D. Calvanese, G. de Giacomo, M. Lenzerini, and M. Y. Vardi. View-based query processing for regular path queries with inverse. In *19th PODS*, 58–66, ACM, 2000. 599

[CKS81]   A. K. Chandra, D. C. Kozen, and L. J. Stockmeyer. Alternation. *Journal of the Association for Computing Machinery*, 28(1):114–133, January 1981. 598, 599

[GH96]   N. Globerman and D. Harel. Complexity results for two-way and multi-pebble automata and their logics. *TCS*, 143:161–184, 1996. 599, 609

[JPAR68]   W. L. Johnson, J. H. Porter, S. I. Ackley, and D. T. Ross. Automatic generation of efficient lexical processors using finite state techniques. *Communications of the ACM*, 11(12):805–813, 1968. 598

[Koh70]   Z. Kohavi. *Switching and Finite Automata Theory*. McGraw-Hill, New York, 1970. 598

[Kur94]   R. P. Kurshan. *Computer Aided Verification of Coordinating Processes*. Princeton Univ. Press, 1994. 599

[KV00]   O. Kupferman and M. Y. Vardi. Synthesis with incomplete informatio. In *Advances in Temporal Logic*, pages 109–127. Kluwer Academic Publishers, January 2000. 599

[LLS84]   Richard E. Ladner, Richard J. Lipton, and Larry J. Stockmeyer. Alternating pushdown and stack automata. *SIAM Journal on Computing*, 13(1):135–155, 1984. 598, 599

[McN66]   R. McNaughton. Testing and generating infinite sequences by a finite automaton. *Information and Control*, 9:521–530, 1966. 599

[MH84]   S. Miyano and T. Hayashi. Alternating finite automata on ω-words. *Theoretical Computer Science*, 32:321–330, 1984. 600

[Rab69]   M. O. Rabin. Decidability of second order theories and automata on infinite trees. *Transaction of the AMS*, 141:1–35, 1969. 599

[RS59]   M. O. Rabin and D. Scott. Finite automata and their decision problems. *IBM Journal of Research and Development*, 3:115–125, 1959. 598

[Ruz80]   W. Ruzzo. Tree-size bounded alternation. *JCSS*, 21:218–235, 1980. 600

[She59]   J. C. Shepherdson. The reduction of two-way automata to one-way automata. *IBM Journal of Research and Development*, 3:198–200, 1959. 599

[SS78]   W. J. Sakoda and M. Sipser. Nondeterminism and the size of two way finite automata. In *10th STOC*, 275–286, 1978. ACM. 599

[Str82]   R. S. Streett. Propositional dynamic logic of looping and converse. *Information and Control*, 54:121–141, 1982. 599

[Var88]   M. Y. Vardi. A temporal fixpoint calculus. In *Proc. 15th ACM Symp. on Principles of Programming Languages*, 250–259, 1988. 599, 600

[Var89]   M. Y. Vardi. A note on the reduction of two-way automata to one-way automata. *Information Processing Letters*, 30(5):261–264, March 1989. 599

[Var90]   M. Y. Vardi. Endmarkers can make a difference. *Information Processing Letters*, 35(3):145–148, July 1990.

[Var98]   M. Y. Vardi. Reasoning about the past with two-way automata. In *25th ICALP*, LNCS 1443, 628–641. Springer-Verlag, July 1998. 599

[Ven91]   H. Venkateswaran. Properties that characterize logCFL. *JCSS*, 43(2):380–404, 1991. 600

[VW86]   M. Y. Vardi and P. Wolper. An automata-theoretic approach to automatic program verification. In *1st LICS*, 332–344, 1986. 598, 599

Syntactic Semiring of a Language
(Extended Abstract)

Libor Polák*

Department of Mathematics, Masaryk University
Janáčkovo nám 2a, 662 95 Brno, Czech Republic
polak@math.muni.cz
http://www.math.muni.cz/~polak

Abstract. A classical construction assigns to any language its (ordered) syntactic monoid. Recently the author defined the so-called syntactic semiring of a language. We discuss here the relationships between those two structures. Pin's refinement of Eilenberg theorem gives a one-to-one correspondence between positive varieties of rational languages and pseudovarieties of ordered monoids. The author's modification uses so-called conjunctive varieties of rational languages and pseudovarieties of idempotent semirings. We present here also several examples of our varieties of languages.

Keywords: syntactic semiring, rational languages

1 Introduction

The syntactic monoid is a monoid canonically attached to each language. Certain classes of rational languages can be characterized by the syntactic monoids of their members. The book [3] and the survey [5] by Pin are devoted to a systematic study of this correspondence. One also speaks about the algebraic theory of finite automata.

Recently the author defined in [6] the so-called syntactic semiring of a language. Basic definitions and properties are stated in Sect. 2. In Sect. 3 we recall the main result of [6]. It is an Eilenberg-type theorem relating the so-called conjunctive varieties of rational languages and pseudovarieties of idempotent semirings.

Section 4 is devoted to examples. The first two varieties of languages arise when considering identities satisfied by syntactic semirings. Certain operators on classes of languages yield further examples.

The relationships between the (ordered) syntactic monoid and the syntactic semiring of a given language are studied in Sect. 5 and the last section presents a method for computing the syntactic semiring of a given language.

* The author acknowledges the support of the Grant no. 201/01/0323 of the Grant Agency of the Czech Republic.

J. Sgall, A. Pultr, and P. Kolman (Eds.): MFCS 2001, LNCS 2136, pp. 611–620, 2001.

All the background needed is thoroughly explored in above quoted Pin's sources or in Almeida's book [1]. Our contribution is meant as an extended abstract, we present full proofs only in Sect. 5.

2 Ordered Syntactic Monoid and Syntactic Semiring

A structure $(O, \cdot, \leq)$ is called an *ordered monoid* if

(i) $(O, \cdot)$ is a monoid with the neutral element 1,
(ii) $(O, \leq)$ is an ordered set,
(iii) $a, b, c \in O$, $a \leq b$ implies $ac \leq bc$ and $ca \leq cb$.

A structure $(S, \cdot, \vee)$ is called a *semilattice-ordered monoid* if

(i) $(S, \cdot)$ is a monoid,
(ii) $(S, \vee)$ is a semilattice,
(iii) $a, b, c \in S$ implies $a(b \vee c) = ab \vee ac$ and $(a \vee b)c = ac \vee bc$.

An alternative name is an *idempotent semiring* which we will prefer here. A semilattice $(S, \vee)$ becomes an ordered set with respect to the relation $\leq$ defined by $a \leq b \Leftrightarrow a \vee b = b$, $a, b \in S$. Moreover, in this view, any idempotent semiring is an ordered monoid. For an ordered set $(O, \leq)$, a subset H of O is *hereditary* if $a \in H$, $b \in O$, $b \leq a$ implies $b \in H$.

We also denote

2^B the set of all subsets of a set B,
$\mathsf{F}(B)$ the set of all non-empty finite subsets of a set B,
$\Delta_B = \{(b, b) \mid b \in B\}$ the diagonal relation on a set B,
$(B] = \{a \in O \mid a \leq b \text{ for some } b \in B\}$ the hereditary subset of $(O, \leq)$ generated by a given $B \subseteq O$,
$\mathsf{H}(O, \leq)$ the set of all non-empty finitely generated hereditary subsets of an ordered set $(O, \leq)$,
$\mathsf{c}(u)$ the set of all letters of A in a word $u \in A^*$.

Recall that an *ideal* of a semilattice $(S, \vee)$ is a hereditary subset closed with respect to the operation $\vee$. *Morphisms* of monoids are semigroup homomorphisms with $1 \mapsto 1$ and morphisms of ordered monoids (idempotent semirings) are isotone monoid morphisms (monoid morphisms which respect also the operation $\vee$).

We say that a monoid $(M, \cdot)$ *divides* a monoid $(N, \cdot)$ if $(M, \cdot)$ is a morphic image of a submonoid of $(N, \cdot)$; similarly for ordered monoids and semirings.

A hereditary subset H of an ordered monoid $(O, \cdot, \leq)$ defines a relation $\approx_H$ on O by

$$a \approx_H b \text{ if and only if } (\forall p, q \in O)(paq \in H \Leftrightarrow pbq \in H).$$

This relation is a congruence of $(O, \cdot)$ and the corresponding factor-structure is called the *syntactic monoid* of H in $(O, \cdot, \leq)$. It is ordered by

$$a\approx_H \ \leq \ b\approx_H \text{ if and only if } (\forall p, q \in O)(pbq \in H \Rightarrow paq \in H)$$

and we speak about the *ordered syntactic monoid*. We also write $a \preceq_H b$ instead of $a \approx_H\ \leq\ b \approx_H$. Let σ_H denote the mapping $a \mapsto a \approx_H$, $a \in O$. It is a surjective ordered monoid morphism.

Similarly, an ideal I of an idempotent semiring $(S, \cdot, \vee)$ defines a relation $\sim_I$ on the set S by

$$a \sim_I b \text{ if and only if } (\forall p, q \in S)(paq \in I \Leftrightarrow pbq \in I).$$

This relation is a congruence of $(S, \cdot, \vee)$ and the corresponding factor-structure is called the *syntactic semiring* of I in $(S, \cdot, \vee)$. Let ρ_I denote the mapping $a \mapsto a \sim_I$, $a \in S$. It is a surjective semiring morphism.

Any language L over A is a hereditary subset of the trivially ordered monoid $(A^*, \cdot, =)$ and the set $\mathsf{F}(L)$ is an ideal of $(\mathsf{F}(A^*), \cdot, \cup)$ where

$$U \cdot V = \{u \cdot v \mid u \in U,\ v \in V\}.$$

The above constructions gives the *ordered syntactic monoid* and *syntactic semiring of the language* L; we denote them by $\mathsf{O}(L)$ and $\mathsf{S}(L)$. We also put $\mathsf{o}(L) = \sigma_L(L)$ and $\mathsf{s}(L) = \rho_{\mathsf{F}(L)}(\mathsf{F}(L))$.

For $L \subseteq A^*$, the congruence $\sim_{\mathsf{F}(L)}$ can be also expressed as

$$\{u_1, \ldots, u_k\} \sim_{\mathsf{F}(L)} \{v_1, \ldots, v_l\} \text{ if and only if}$$

$$(\forall x, y \in A^*)(xu_1y, \ldots, xu_ky \in L \Leftrightarrow xv_1y, \ldots, xv_ly \in L).$$

Notice that $(A^*, \cdot)$ is a free monoid over the set A and that $(\mathsf{F}(A^*), \cdot, \cup)$ is a free idempotent semiring over A. Further, any ideal of $(\mathsf{F}(A^*), \cdot, \cup)$ is of the form $\mathsf{F}(L)$ for $L \subseteq A^*$.

A language $L \subseteq A^*$ is *recognizable* by a monoid $(M, \cdot)$ with respect to its subset B (by an ordered monoid $(O, \cdot, \leq)$ with respect to its hereditary subset H, by an idempotent semiring $(S, \cdot, \vee)$ with respect to its ideal I) if there exists a monoid morphism $\alpha : A^* \to M$ such that $L = \alpha^{-1}(B)$ ($\alpha : A^* \to O$ such that $L = \alpha^{-1}(H)$, $\alpha : A^* \to S$ such that $L = \alpha^{-1}(I)$).

Clearly, the last notion can be rephrased as follows: A language $L \subseteq A^*$ is recognizable by an idempotent semiring $(S, \cdot, \vee)$ with respect to its ideal I if there exists a semiring morphism $\beta : (\mathsf{F}(A^*), \cdot, \cup) \to (S, \cdot, \vee)$ such that $\mathsf{F}(L) = \beta^{-1}(I)$.

We say that a language L is *recognizable* by a monoid $(M, \cdot)$ (by an ordered monoid $(O, \cdot, \leq)$, by an idempotent semiring $(S, \cdot, \vee)$) if it is recognizable with respect to some subset (hereditary subset, ideal). Clearly, the recognizability by $(M, \cdot)$ gives the recognizability by $(M, \cdot, =)$ and conversely the recognizability by $(M, \cdot, \leq)$ yields the recognizability by $(M, \cdot)$. Furthermore, the *recognizability* means the recognizability by a finite monoid.

Proposition 1. *Let $(O, \cdot, \leq)$ be an ordered monoid. Then $(\mathsf{H}(O, \leq), \circ, \cup)$, where*

$$H \circ K = \{c \in O \mid \text{ there exist } a \in H,\ b \in K \text{ such that } c \leq a \cdot b\},$$

is an idempotent semiring and the mapping $\iota : a \mapsto (a]$, $a \in O$, is an injective monoid morphism of $(O, \cdot, \leq)$ to $(\mathsf{H}(O, \leq), \circ, \cup)$ satisfying $a \leq b$ if and only if

$\iota(a) \subseteq \iota(b)$, $a, b \in O$. *Moreover, the latter structure is a free idempotent semiring over* $(O, \cdot, \leq)$ *with respect to* ι, *that is, for any idempotent semiring* $(S, \cdot, \vee)$ *and an ordered monoid morphism*

$$\alpha : (O, \cdot, \leq) \rightarrow (S, \cdot, \vee)$$

there exists exactly one semiring morphism

$$\beta : (\mathsf{H}\,(O, \leq), \circ, \cup) \rightarrow (S, \cdot, \vee)$$

such that $\beta \circ \iota = \alpha$.

Proof. Obviously $d \in (H \circ K) \circ L$ iff there exist $a \in H$, $b \in K$, $c \in L$ such that $d \leq abc$ and the same holds for $H \circ (K \circ L)$. Also the validity of the distributive laws is immediate. Further, $a \leq b$ iff $(a] \subseteq (b]$ (in particular, ι is injective) and $(ab] = (a] \cdot (b]$ for any $a, b \in O$.

Given $(S, \cdot, \vee)$ and α, define $\beta((a_1, \ldots, a_k]) = \alpha(a_1) \vee \ldots \vee \alpha(a_k)$ which yields the rest of the proposition. □

Proposition 2. *A language* $L \subseteq A^*$ *is recognizable by a finite idempotent semiring if and only if it is recognizable.*

Proof. If L is recognizable by a finite ordered monoid $(O, \cdot, \leq)$ with respect to its hereditary subset H then L is also recognizable by the idempotent semiring $(\mathsf{H}\,(O, \leq), \circ, \cup)$ with respect to its ideal $(H]$ of all hereditary subsets of H.

The opposite implication is trivial. □

The following result is classical (see [5]).

Proposition 3. *(i) Let A be a finite set, let $(O, \cdot, \leq)$ be an ordered monoid, H its hereditary subset, $\alpha : (A^*, \cdot) \rightarrow (O, \cdot)$ a surjective monoid morphism, and let $L = \alpha^{-1}(H)$. Then*

$$((O, \cdot)/ \approx_H, \leq) \text{ is isomorphic to } (\mathsf{O}\,(L), \cdot, \leq), \; \mathsf{o}\,(L) \text{ being the image of } \sigma_H(H)$$

$$\text{and } \preceq_{\mathsf{o}\,(L)} \; = \; \leq \text{ on } \mathsf{O}\,(L) \; .$$

(ii) A language L over a finite set is recognizable by a finite ordered monoid $(O, \cdot, \leq)$ if and only if $(\mathsf{O}\,(L), \cdot, \leq)$ divides $(O, \cdot, \leq)$. □

An analogy of the last result follows.

Proposition 4. *(i) Let A be a finite set, let $(S, \cdot, \vee)$ be an idempotent semiring, I its ideal, $\beta : (\mathsf{F}\,(A^*), \cdot, \cup) \rightarrow (S, \cdot, \vee)$ a surjective semiring morphism, and let*

$$L = \{\, u \in A^* \mid \{u\} \in \beta^{-1}(I) \,\} \; .$$

Then

$$(S, \cdot, \vee)/ \sim_I \text{ is isomorphic to } (\mathsf{S}\,(L), \cdot, \vee), \; \mathsf{s}\,(L) \text{ being the image of } \rho_I(I)$$

$$\text{and } \sim_{\mathsf{s}\,(L)} \; = \; \Delta_{\mathsf{S}\,(L)} \text{ on } \mathsf{S}\,(L) \; .$$

(ii) A language L over a finite set is recognizable by a finite idempotent semiring $(S, \cdot, \vee)$ if and only if $(\mathsf{S}\,(L), \cdot, \vee)$ divides $(S, \cdot, \vee)$.

Proof. (i) Here $\{u_1, \ldots, u_k\} \sim_{F(L)} \; \mapsto \; \beta(\{u_1, \ldots, u_k\}) \sim_I$ is the desired isomorphism and the second part is in Lemma 2 of [6]. The item (ii) is Lemma 10 of [6]. $\qquad\square$

3   Eilenberg-Type Theorems

For languages $K, L \subseteq A^*$ we define

$$K \cdot L = \{ \; uv \mid u \in K, \; v \in L \; \}, \quad K^* = \{ \; u_1 \cdot \ldots \cdot u_k \mid k \geq 0, \; u_1, \ldots, u_k \in K \; \} \; .$$

Recall that the set of all *rational* languages over a finite alphabet A is the smallest family of subsets of A^* containing the empty set, all singletons $\{u\}$, $u \in A^*$, closed with respect to binary unions and the operations $\cdot$ and $*$.

As well-known, the rational languages are exactly the recognizable ones.

A class of finite monoids (ordered monoids, idempotent semirings) is called a *pseudovariety of monoids (ordered monoids, idempotent semirings)* if it is closed under forming of products of finite families, substructures and morphic images.

For sets A and B, a semiring morphism

$$\psi : (F(A^*), \cdot, \cup) \to (F(B^*), \cdot, \cup)$$

and $L \subseteq B^*$ we define

$$\psi^{[-1]}(L) = \{ \; u \in A^* \mid \psi(\{u\}) \subseteq L \; \} \text{ and}$$

$$\psi^{(-1)}(L) = \{ \; u \in A^* \mid \psi(\{u\}) \cap L \neq \emptyset \; \} \; .$$

A *class* of rational languages is an operator $\mathcal{L}$ assigning to every finite set A a set $\mathcal{L}(A)$ of rational languages over the alphabet A containing both $\emptyset$ and A^*. Conditions which such a class of languages can satisfy follow.

($\cap$) : for every A, the set $\mathcal{L}(A)$ is closed with respect to finite intersections,

($\cup$) : for every A, the set $\mathcal{L}(A)$ is closed with respect to finite unions,

(Q) : for every A, $a \in A$ and $L \in \mathcal{L}(A)$ we have $a^{-1}L$, $La^{-1} \in \mathcal{L}(A)$,

($^{-1}$) : for every sets A and B, a monoid morphism $\phi : A^* \to B^*$ and $L \in \mathcal{L}(B)$ we have $\phi^{-1}(L) \in \mathcal{L}(A)$,

($^{[-1]}$) : for every sets A and B, a semiring morphism $\psi : (F(A^*), \cdot, \cup) \to (F(B^*), \cdot, \cup)$ and $L \in \mathcal{L}(B)$ we have $\psi^{[-1]}(L) \in \mathcal{L}(A)$,

($^{(-1)}$) : for every sets A and B, a semiring morphism $\psi : (F(A^*), \cdot, \cup) \to (F(B^*), \cdot, \cup)$ and $L \in \mathcal{L}(B)$ we have $\psi^{(-1)}(L) \in \mathcal{L}(A)$,

(C) : for every A, the set $\mathcal{L}(A)$ is closed with respect to complements.

For a class $\mathcal{L}$ of languages we define its *complement* $\mathcal{L}^C$ by

$$\mathcal{L}^C(A) = \{ \; A^* \setminus L \mid L \in \mathcal{L}(A) \; \} \text{ for every finite set } A \; .$$

Clearly, $\mathcal{L}$ satisfies the condition ($\cap$) if and only if $\mathcal{L}^c$ satisfies ($\cup$). Similarly, $\mathcal{L}$ satisfies the condition ($^{[-1]}$) if and only if $\mathcal{L}^c$ satisfies ($^{(-1)}$). Further, either of the conditions ($^{[-1]}$) and ($^{(-1)}$) implies ($^{-1}$).

A class $\mathcal{L}$ is called a *conjunctive variety of languages* if it satisfies the conditions ($\cap$), ($^{[-1]}$) and (Q). Similarly, it is called a *disjunctive variety of languages* if it satisfies ($\cup$), ($^{(-1)}$) and (Q).

Further, $\mathcal{L}$ is called *a positive variety of languages* if it satisfies the conditions ($\cap$), ($\cup$), ($^{-1}$), (Q) and such a variety is called *a boolean variety of languages* if it satisfies in addition the condition (C).

We can assign to any class of languages $\mathcal{L}$ the pseudovarieties

$$\mathsf{M}(\mathcal{L}) = \langle \{ (\mathsf{O}(L), \cdot) \mid A \text{ a finite set}, L \in \mathcal{L}(A) \} \rangle_\mathsf{M} ,$$
$$\mathsf{O}(\mathcal{L}) = \langle \{ (\mathsf{O}(L), \cdot, \le) \mid A \text{ a finite set}, L \in \mathcal{L}(A) \} \rangle_\mathsf{O} ,$$
$$\mathsf{S}(\mathcal{L}) = \langle \{ (\mathsf{S}(L), \cdot, \vee) \mid A \text{ a finite set}, L \in \mathcal{L}(A) \} \rangle_\mathsf{S}$$

of monoids (ordered monoids, idempotent semirings) generated by all syntactic monoids (ordered syntactic monoids, syntactic semirings) of members of $\mathcal{L}$.

Conversely, for pseudovarieties $\mathcal{M}$ of monoids, $\mathcal{P}$ of ordered monoids and $\mathcal{V}$ of idempotent semirings and a finite set A, we put

$$(\mathsf{L}(\mathcal{M}))(A) = \{ L \subseteq A^* \mid (\mathsf{O}(L), \cdot) \in \mathcal{M} \} ,$$
$$(\mathsf{L}(\mathcal{P}))(A) = \{ L \subseteq A^* \mid (\mathsf{O}(L), \cdot, \le) \in \mathcal{P} \} ,$$
$$(\mathsf{L}(\mathcal{V}))(A) = \{ L \subseteq A^* \mid (\mathsf{S}(L), \cdot, \vee) \in \mathcal{V} \} .$$

Theorem 5 (Eilenberg [2], Pin [4], Polák [6]). *(i) The operators* M *and* L *are mutually inverse bijections between boolean varieties of languages and pseudovarieties of monoids.*
(ii) The operators O *and* L *are mutually inverse bijections between positive varieties of languages and pseudovarieties of ordered monoids.*
(iii) The operators S *and* L *are mutually inverse bijections between conjunctive varieties of languages and pseudovarieties of idempotent semirings.*

4   Examples

Let $X = \{x_1, x_2, \ldots\}$ be the set of variables. Any identity of idempotent semirings is of the form

$$f_1 \vee \ldots \vee f_k = g_1 \vee \ldots \vee g_l \text{ where } f_1, \ldots, f_k, g_1, \ldots, g_l \in X^* .$$

It is equivalent to the inequalities

$$f_1 \vee \ldots \vee f_k \ge g_1 \vee \ldots \vee g_l, \quad f_1 \vee \ldots \vee f_k \le g_1 \vee \ldots \vee g_l .$$

The last inequality is equivalent to k inequalities

$$f_i \le g_1 \vee \ldots \vee g_l, \; i = 1, \ldots, k .$$

Thus when dealing with varieties of idempotent semirings, it is enough to consider identities of the form

$$f \leq g_1 \vee \ldots \vee g_l .$$

Notice that identities of the form $f \leq g$ are not sufficient since, for instance, the identity $xy \leq x \vee yz$ is not equivalent to any set of identities of the previous form. In fact, the variety (pseudovariety) of all (finite) idempotent semirings $(S, \cdot, \vee)$ such that also $(S, \cdot)$ is a semilattice has precisely three proper non-trivial subvarieties (subpseudovarieties). These are given by additional identities $x \leq 1$, $1 \leq x$ and $xy \leq x \vee yz$, respectively.

For idempotent semirings the identities $x^2 \leq x$ and $xy \leq x \vee y$ are equivalent. The syntactic semiring of a language L over an alphabet A satisfies the identity $xy \leq x \vee y$ if and only if all its quotients $p^{-1}Lq^{-1}$ $(p, q \in A^*)$ satisfy

$$u, v \in p^{-1}Lq^{-1} \text{ implies } u \cdot v \in p^{-1}Lq^{-1} .$$

This can be generalized to the identities of the form

$$(I_k) \quad x_1 y_1 \ldots x_k y_k \leq x_1 \ldots x_k \vee y_1 \ldots y_k .$$

Now the syntactic semiring of a language L satisfies (I_k) if and only if every quotient K of L satisfies

$$u_1 \ldots u_k, \ v_1 \ldots v_k \in K \text{ implies } u_1 v_1 \ldots u_k v_k \in K .$$

Recall that the *shuffle* of words $u, v \in A^*$ is the set $u \sqcup v =$

$$\{ u_1 v_1 \ldots u_k v_k \mid k \in \mathbb{N}, \ u = u_1 \ldots u_k, \ v = v_1 \ldots v_k, \ u_1, \ldots, u_k, v_1, \ldots, v_k \in A^* \} .$$

Thus the system of all (I_k), $k \in \mathbb{N}$ characterizes languages all quotients of which are shuffle-closed.

Consider now the identity $x \leq 1 \vee xy$. The syntactic semiring of a language L satisfies this identity if and only if every quotient K of L satisfies

$$1, \ uv \in K \text{ implies } u \in K .$$

It is equivalent to the following condition on right quotients Lq^{-1} of L: if a deterministic automaton for Lq^{-1} accepts a word p and does not accept pa for a letter a, it does not need to continue reading of the input since any par $(r \in A^*)$ is not accepted.

In certain aspects the disjunctive varieties of languages are more natural than the conjunctive ones. For languages K and L over the alphabet A we define their *shuffle product* by

$$K \sqcup L = \bigcup_{u \in K, v \in L} u \sqcup v .$$

Let $\mathcal{L}(A)$ consist of finite unions of

$$\{v\} \sqcup C^*, \ vA^* \sqcup C^*, \ A^* vA^* \sqcup C^*, \ A^* v \sqcup C^*, \text{ where } v \in A^*, \ C \subseteq A .$$

One can show that $\mathcal{L}$ is a disjunctive variety of languages. $\mathcal{L}$ is not a positive variety since, for instance, the language $aA^* \cap A^*b$ over $A = \{a, b\}$ is not from $\mathcal{L}(A)$.

Another method for obtaining a disjunctive variety of languages is to close a positive variety of languages first with respect to $((^{-1}))$ and then to finite unions. On the other hand, starting from a disjunctive variety $\mathcal{L}$ of languages one gets a positive one closing every $\mathcal{L}(A)$ with respect to finite intersections.

5   Relationships between Syntactic Ordered Monoid and Syntactic Semiring of a Language

Proposition 6 ([6], Lemma 7). *Let L be a language over an alphabet A. The mapping*

$$\iota : u \approx_L \; \mapsto \; \{u\} \sim_{F(L)}, \; u \in A^*$$

is an injective monoid morphism of $(O(L), \cdot, \leq)$ into $(S(L), \cdot, \vee)$ satisfying $a \leq b$ if and only if $\iota(a) \leq \iota(b)$, $a, b \in O(L)$. Moreover, $\iota(o(L)) = s(L) \cap \iota(O(L))$ and $\iota(O(L))$ contains all join-irreducible elements of $(S(L), \vee)$.

Proposition 7. *Let L be a recognizable language over a finite alphabet A. Then*

$$(S(L), \cdot, \vee) \; \text{is isomorphic to} \; (H(O(L), \leq), \circ, \cup)/ \sim_{(o(L)]} \; .$$

Proof. This is a consequence of Prop. 4 (i) since, as mentioned in the proof of Prop. 2, the idempotent semiring $(H(O(L), \leq), \circ, \cup)$ recognizes L with respect to $(o(L)]$. □

Proposition 8. *Let $(S, \cdot, \vee)$ be a finite idempotent semiring and let I be its ideal such that $\sim_I \; = \Delta_S$. Let O be a submonoid of $(S, \cdot)$ containing all join-irreducible elements of $(S, \vee)$. Then there exists a recognizable language L over a finite set A such that*

$$(O(L), \cdot, \leq) \; \text{is isomorphic to} \; (O, \cdot, \leq) \; \text{and} \; (S(L), \cdot, \vee) \; \text{is isomorphic to} \; (S, \cdot, \vee) .$$

In particular, it is the case for $O = S$.

Proof. Let $a, b \in O$. We show that

$$a \preceq_{O \cap I} b \; \text{if and only if} \; a \vee b = b \; \text{in} \; (S, \vee) .$$

Really, $a \preceq_{O \cap I} b$ if and only if

$$(\forall p, q \in O)(\, pbq \in O \cap I \; \Rightarrow \; paq \in O \cap I \,) ,$$

that is, iff $(\forall p, q \in O)(\, pbq \in I \; \Rightarrow \; paq \in I \,) ,$

that is, iff $(\forall p, q \in S)(\, pbq \in I \; \Rightarrow \; paq \in I \,) ,$

(since O generates $(S, \vee)$) that is, iff

$$(\forall\, p, q \in S\,)(\, p(a \vee b)q \in I \quad \Leftrightarrow \quad pbq \in I\,)\ ,$$

that is, iff $a \vee b \approx_I b$.

Let $\alpha : (A^*, \cdot) \to (O, \cdot)$ be a surjective monoid homomorphism (for an appropriate finite set A). By Prop. 3 (i), $(O, \cdot, \leq)$ is isomorphic to the ordered syntactic monoid of $L = \alpha^{-1}(O \cap I)$.

By Prop. 1, the inclusion $O \subseteq S$ extends to a semiring morphism

$$\beta : (\mathsf{H}\,(O, \leq), \circ, \cup) \to (S, \cdot, \vee),\ (a_1, \ldots, a_k] \mapsto a_1 \vee \ldots \vee a_k\ .$$

We show that the kernel of β is $\sim_{(O \cap I]}$.

Really,

$$(a_1, \ldots, a_k]\ \sim_{(O \cap I]}\ (b_1, \ldots, b_l]$$

if and only if

$$(\forall\, p, q \in O\,)(\, pa_1q, \ldots, pa_kq \in O \cap I \quad \Leftrightarrow \quad pb_1q, \ldots, pb_lq \in O \cap I\,)\ ,$$

that is, iff $(\forall\, p, q \in O\,)(\, pa_1q, \ldots, pa_kq \in I \quad \Leftrightarrow \quad pb_1q, \ldots, pb_lq \in I\,)\ ,$

that is, iff $(\forall\, p, q \in O\,)(\, p(a_1 \vee \ldots \vee a_k)q \in I \quad \Leftrightarrow \quad p(b_1 \vee \ldots \vee b_l)q \in I\,)\ ,$

which is equivalent to $(\forall\, p, q \in S\,)(\, p(a_1 \vee \ldots \vee a_k)q \in I \quad \Leftrightarrow \quad p(b_1 \vee \ldots \vee b_l)q \in I\,)$

since every element of S is a join of elements of O.

Thus, by Prop. 6, $(S, \cdot, \vee)$ is isomorphic to the syntactic semiring of L.   □

Consider the language $L = \{u \in A^* \mid \mathsf{c}\,(u) \neq A\}$ over a finite set A.

The ordered syntactic monoid of L is isomorphic to $(2^A, \cup, \subseteq)$. Direct calculations yield that the syntactic semiring of L is isomorphic to $(\mathsf{H}\,(2^A, \subseteq), \circ, \cup)$. Denote these structures by $(O_n, \cdot, \leq)$ and by $(S_n, \cdot, \vee)$ in the case of an n-element alphabet A $(n \in \mathbb{N})$. Clearly, any $(O_n, \cdot, \leq)$ generates the pseudovariety J_1^-. Note that J_1^- is given by the identities $x^2 = x$, $xy = yx$, $1 \leq x$ and that it is an atom in the lattice of pseudovarieties of finite ordered monoids. One can calculate that $(S_n, \cdot, \vee)$ generates the variety of idempotent semirings given by $x^{n+1} = x^n$, $xy = yx$, $1 \leq x$.

Using Prop. 7, we can get languages with their ordered syntactic monoids outside of J_1^- having some of $(S_n, \cdot, \vee)$ as syntactic semiring. For instance, take K consisting of all words over $A = \{a, b, c\}$ having none of $ab, ac, ba, bc, ca, cb, cc$ as a segment has the syntactic semiring isomorphic to $(S_2, \cdot, \vee)$.

We can comment the situation as follows: the ordered syntactic monoid and the syntactic semiring of a language are equationally independent.

6   A Construction of the Syntactic Semiring

We extend here a well-known construction of the syntactic monoid of a rational language to the case of the syntactic semiring.

Let a rational language L over a finite alphabet A be given. Classically one assigns to L its minimal automaton $\mathcal{A}$ using left quotients; namely

$Q = \{\, u^{-1} \cdot L \mid u \in A^* \,\}$ is the (finite) set of states,

$a \in A$ acts on $u^{-1}L$ by $(u^{-1}L) \cdot a = a^{-1} \cdot (u^{-1}L)$,

$q_0 = L$ is the initial state and $u^{-1}L$ is a final state if and only if $u \in L$.

Now $(\mathrm{O}\,(L), \cdot)$ is the transition monoid of $\mathcal{A}$.

The order on $\mathrm{O}\,(L)$ is given by

$$f \leq g \text{ if and only if for every } q \in Q \text{ we have } q \cdot f \supseteq q \cdot g \ .$$

We extend the set of states to

$$\overline{Q} = \{\, q_1 \cap \ldots \cap q_m \mid m \in \mathbb{N}, q_1, \ldots, q_m \in Q \,\} \ .$$

The action of a letter $a \in A$ is now given by

$$(q_1 \cap \ldots \cap q_m) \cdot a = q_1 \cdot a \, \cap \ldots \cap \, q_m \cdot a \ .$$

It can be extended to transitions induced by non-empty finite sets of words by

$$q \cdot \{u_1, ..., u_k\} = q \cdot u_1 \cap \ldots \cap q \cdot u_k \text{ for } q \in \overline{Q}, \ u_1, ..., u_k \in A^* \ .$$

It can be shown that this transition monoid with $\vee$ being the union is the syntactic semiring of the language L.

To make the computation finite, one considers instead of words from A^* their representatives in $\mathrm{O}\,(L)$. Moreover, it suffices to take only the actions of hereditary subsets of $(\mathrm{O}\,(L), \leq)$. At present stage we are far from a concrete implementation.

References

1. Almeida, J.; Finite Semigroups and Universal Algebra, World Scientific, 1994   612
2. Eilenberg, S.; Automata, Languages and Machines, Vol. B, Academic Press, 1976   616
3. Pin, J.-E.; Varieties of Formal Languages, Plenum, 1986   611
4. Pin, J.-E.; A variety theorem without complementation, Izvestiya VUZ Matematika **39** (1995) 80–90. English version: Russian Mathem. (Iz. VUZ) **39** (1995) 74–83   616
5. Pin, J.-E.; Syntactic semigroups, Chapter 10 in Handbook of Formal Languages, G. Rozenberg and A. Salomaa eds, Springer, 1997   611, 614
6. Polák, L.; A classification of rational languages by semilattice-ordered monoids, http://www.math.muni.cz/~polak   611, 615, 616, 618

On Reducibility and Symmetry of Disjoint NP-Pairs

Pavel Pudlák*

Mathematical Institute, AV ČR and ITI
Prague, Czech Republic

Abstract. We consider some problems about pairs of disjoint NP sets. The theory of these sets with a natural concept of reducibility is, on the one hand, closely related to the theory of proof systems for propositional calculus, and, on the other, it resembles the theory of NP completeness. Furthermore, such pairs are important in cryptography. Among others, we prove that the Broken Mosquito Screen pair of disjoint NP-sets can be polynomially reduced to Clique-Coloring pair and thus is polynomially separable and we show that the pair of disjoint NP-sets canonically associated with the Resolution proof system is symmetric.

1 Introduction

The subject of study of this paper is the concept of pairs of disjoint NP-sets. Thus instead of studying sets (or in other words, languages), the most common object in complexity theory, we study pairs of sets and we require, moreover, that they are disjoint and that they belong to NP. The research of such pairs was initiated by Razborov in [13]. He studied them in connection with some formal systems, in particular, proof systems for propositional calculus and systems of bounded arithmetic.

There is a natural concept of polynomial reducibility between pairs of disjoint sets. We say that a pair (A, B) is *polynomially reducible* to (C, D) if there is a polynomial time computable function f defined on all strings such that f maps A into C and B into D. (Note that polynomial reducibility does not imply that the corresponding sets are polynomially (Karp) reducible.) We say that a pair (A, B) is *polynomially separable*, if there exists a function f computable in polynomial time time such that f is 0 on A and it is 1 on B.

A related concept is the concept of a *propositional proof system*. A general propositional proof system, as defined by Cook and Reckhow [5], is simply a nondeterministic algorithm for the set of propositional tautologies. There are several well-studied concrete systems, coming from logic, automated reasoning and others. Proof systems can be compared using the relation of polynomial simulation (see Section 3 for definitions). It has been conjectured that there is no strongest propositional proof system.

---

* Partially supported by grant A1019901 of the AV ČR.

J. Sgall, A. Pultr, and P. Kolman (Eds.): MFCS 2001, LNCS 2136, pp. 621–632, 2001.

Razborov [13] associated a pair of disjoint NP sets in a natural way to each proof system: roughly speaking, one set is the set of tautologies that have short proofs in the given system, the other is the set of non-tautologies. This relation gives a reason to believe that in the lattice of the degrees of pairs there is no biggest element. It seems that the lattice of degrees of pairs reflects the strength of the systems, hence there should not be the biggest degree of a pair (unless we define it as the degree of pairs that are *not* disjoint), but we are not able to derive this statement from the standard complexity theoretical conjectures such as $P \neq NP$. Most people believe that $P \neq NP \cap coNP$, which implies that that there are pairs of disjoint NP sets that are not polynomially separable. The only concrete sets in $NP \cap coNP$ that are conjectured not to belong to P come from cryptography. (In cryptography one assumes even more, namely, that there exists a set $A \in NP \cap coNP$ such that a random element of A cannot be distinguished from a random non-element of A using a probabilistic algorithm with probability significantly larger than $1/2$.)

In this paper we show that a pair called Broken Mosquito Screen, introduced by A. Haken [4] is polynomially separable. Pairs similar to BMS have been proposed for bit commitment schemas in cryptography. The polynomial separability implies that such schemas are not secure. Furthermore, we show simple monotone reductions between BMS and the Clique-Coloring pair. Hence one can deduce exponential lower bounds on monotone boolean circuits for BMS [4] and Clique-Coloring [12] one from the other. Note that all lower bounds on monotone computation models, with the exception of Andrejev's, are in fact lower bounds on devices separating two NP sets.

In section 3 we consider some basic relations between proof systems and disjoint NP-pairs. This section contains some new observations, but mostly it is a survey of simple basic facts. It is mainly intended as a brief introduction into the subject for those who are not experts in it.

In section 4 we shall show a symmetry property of the pair associated to the Resolution proof system. This is not a surprising result, as such properties have been already established in Razborov's original paper for stronger systems. The reason for presenting the reduction explicitly is that Resolution is relatively weak, so it does not share all good properties of strong systems. Furthermore, we would like to understand this pair and, possibly, to find a simpler combinatorial characterization of its degree.

2   The Broken Mosquito Screen Pair

Definition 1. *The BMS pair is a pair of sets of graphs (BMS_0, BMS_1) such that*

- *BMS_0 is the set of graphs such that for some $k > 2$ the graph has $k^2 - 2$ vertices and contains k disjoint cliques with $k - 1$ cliques of size k and one of size $k - 2$,*

- and BMS_1 is the set of graphs such that for some $k > 2$ the graph has $k^2 - 2$ vertices and contains k disjoint independent sets with $k - 1$ independent sets of size k and one of size $k - 2$.

Clearly $BMS_0, BMS_1 \in NP$. To prove that the two sets are disjoint, suppose that a graph G satisfies both conditions at the same time. Each independent set of size k must contain a vertex that is not contained in any of the cliques of size k, since there are only $k - 1$ such cliques and an independent set can have at most one vertex in common with a clique. But then we get $k - 1$ vertices outside of the $k - 1$ cliques of size k, so the graph has at least $(k-1)k + k - 1 > k^2 - 1$ vertices, which is a contradiction. Thus $BMS_0 \cap BMS_1 = \emptyset$. This pair was introduced by A. Haken along with his new method for proving exponential lower bounds on the size of monotone boolean circuits. Then, in a joint paper with Cook [4], it was used to prove an exponential lower bound on the size of cutting planes proofs. We define a modification of the pair, denoted by BMS', by relaxing the conditions a little. In the BMS' pair (BMS_0', BMS_1') we ask for only $k - 1$ cliques of size k, respectively, $k - 1$ independent sets of size k. A very important pair is the following Clique-Coloring pair.

Definition 2. *The CC pair is a pair of sets (CC_0, CC_1) such that CC_0 and CC_1 are sets of pairs (G, k) with G a graph and $k \geq 2$ an integer such that*

- *CC_0 is the set of pairs (G, k) such that G contains a clique of size k*
- *and CC_1 is the set of pairs (G, k) such that G can be colored by $k - 1$ colors.*

It is well-known that the CC pair is polynomially separable; the function that separates CC is the famous θ function of Lovász [10]. We will show a reduction of BMS' to CC, hence BMS' and BMS are also polynomially separable.

Proposition 3. *BMS' is polynomially reducible to CC.*

Proof. Let $G = (V, E)$ be a graph on $k^2 - 2$ vertices. We assign a graph H to G as follows. The vertices of H are (i, v), $1 \leq i \leq k - 1$, $v \in V$; $((i, v), (j, u))$ is an edge in H, if $i = j$ and $(v, u) \in E$, or $i \neq j$ and $v \neq u$. If G contains $k - 1$ disjoint cliques of size k, we can take one such clique in each copy, different cliques in different copies, and thus get a clique of size $k^2 - k$ in H. Now suppose G contains $k - 1$ disjoint independent sets of size k. Let X be the union of these sets. Thus the graph induced on X by G can be colored by $k - 1$ colors and the size of X is $k(k - 1)$. Hence we can color the vertices $[1, k - 1] \times X$ of H by $(k - 1)^2$ colors. The remaining vertices can be colored by $|V \setminus X| = k - 2$ colors (by coloring (i, v) by v). Thus we need only $(k - 1)^2 + k - 2 = k^2 - k - 1$ colors. Hence $G \mapsto (H, k^2 - k)$ is a reduction of BMS' to CC.

Corollary 4. *The BMS pair is polynomially separable.*

If a pair is polynomially separable, then, trivially, it can be polynomially reduced to any other pair. The algorithm for separation of the CC pair is, however, highly nontrivial, therefore the next proposition gives us additional information. Recall that a function f is a *projection*, if for every fixed input size n, the output size is a fixed number m and each bit of $f(x)$ is either constant or depends on only one bit of x. In other words, $f(x)$ is computed by depth 0 circuit. So far it was irrelevant in what form we represent the integers in the pairs. In the following we shall need that they are represented in unary.

Proposition 5. *CC is reducible to BMS' using a polynomial time computable projection.*

Proof. Let (G, k) be given, let $G = (V, E)$, $n = |V|$. We can assume w.l.o.g. that n is even $n \geq 4$ and $k = n/2$. We construct a graph H from $2k - 2$ copies of G and some additional vertices. The edges connecting the copies and the edges connecting the additional vertices do not depend on G. Thus H is defined as a projection of G. The set of vertices of H is $\{0, 1\} \times [1, k - 1] \times V$ plus a set U of n elements and a set W of $n - 2$ elements. A pair $((i, r, v), (j, s, u))$ is an edge in H, if either $i = j$ and $r = s$ and $(v, u) \in E$, or $i = j$ and $r \neq s$ and $v = u$, or $i \neq j$. On U we put a matching and W will be an independent set. Every vertex of $\{0, 1\} \times [1, k - 1] \times V$ will be connected with every vertex of U and W, and there will be no edges between U and W. The number of vertices of H is $2(k - 1)n + n + n - 2 = 2(n/2 - 1)n + n + n - 2 = n^2 - 2$.

Assume that G has a clique K of size k. Then H has $k - 1 = n/2 - 1$ disjoint cliques of size $2k = n$ of the form $\{0, 1\} \times \{r\} \times K$. Furthermore we get $n - k = n/2$ disjoint cliques of size n by taking $\{0, 1\} \times [1, k - 1] \times \{v\}$, $v \in V \setminus K$ together with a pair from U. Thus H contains $n - 1$ disjoint cliques of size n.

Assume that $\chi(G) \leq k - 1$. Then we can cover each of the two sets $\{i\} \times [1, k - 1] \times V$ by $k - 1$ independent sets of size n by uniting the independent sets diagonally. Thus we get $2(k - 1) = n - 2$ disjoint independent sets of size n. On $U \cup W$ we have another independent set of size $n/2 + n - 2 \geq n$.

Note that the reduction of BMS' to CC presented above is also projection, thus the two pairs are very close to each other. We believe, though we do not have a proof yet, that a refinement of the proof will give the same for the original BMS. Furthermore, these projections are monotone (hence computable by linear size monotone circuits), thus one can get exponential lower bounds on the size monotone boolean circuits for one pair from the other.

What are the pairs that we still believe that they are not polynomially separable? As noted above, the most likely inseparable pairs are from cryptography. Any bit commitment schema that we believe is secure gives such a pair. For instance, the encryption schema RSA can be used to encode a single bit by using the parity of the encoded number. Thus one set is the set of codes of odd numbers and the other consists of the codes of even numbers. Every one-way permutation can be used to define a inseparable pair. All these pairs are based on number theory. Pairs based on pure combinatorics are rather scarce. A somewhat combinatorial pair of disjoint NP sets is implicit in the lower bound on monotone span programs of [2]. This pair is based on bipartite graphs with special properties. There are two known constructions of such graphs. The first construction uses deep results from commutative algebra, the second uses deep results from number theory. We do not know a polynomial time separation algorithm for these pairs, but also we do not have any particular reasons to believe that they are not separable. Here is another pair that we do not know how to separate.

Definition 6. *The MMMT (Monotone-Min-Max-Term) pair is the pair of sets* $(MMMT_0, MMMT_1)$ *in which both sets are sets of some pairs* (C, k), C *a monotone circuit and* k *a number and*

- $MMMT_0$ *is the set of pairs such that* C *has* $k + 1$ *disjoint minterms,*
- $MMMT_1$ *is the set of pairs such that* C *has a maxterm of size* k.

We suspect, however, that $MMMT$ can be reduced to CC, since to prove the disjointness of the sets in the pair, essentially, only the pigeon hole principle is needed.

3   Propositional Proof Systems

In 1970's Cook initiated systematic study of the complexity of propositional proofs. In a joint paper with Reckhow [5] they defined a general concept of a propositional proof system: a *propositional proof system* is a polynomial time computable function S mapping all strings in a finite alphabet *onto* the set of all tautologies $TAUT$. To be precise one has to specify in what language the tautologies are. In this paper we will need only tautologies in DNF. The meaning of the definition is: x *is a proof of* $S(x)$. The fact that every string is a proof seems strange at first glance, but, clearly, it is only a technicality. The crucial property is that one can test in polynomial time whether a given string is a proof of a given formula.

Propositional proof systems are quasi-ordered by the relation of polynomial simulation. We say that P *polynomially simulates* S, if there exists a polynomial time computable function f such that $P(f(x)) = S(x)$ for all x. Thus given an S proof x of a formula ϕ (i.e. $\phi = S(x)$), f finds a P proof $f(x)$ of this formula (i.e. $\phi = P(f(x))$).

As in the next section we will consider the resolution proof system, which is a refutation system, we shall often talk about refutations, i.e., proofs of contradiction from a given formula, rather than direct proofs. Again, this is only *façon de parler*.

Disjoint NP pairs are closely related to propositional proof systems. Following [13] we define, for a proof system S, $REF(S)$ to be the set of pairs $(\phi, 1^m)$, where ϕ is a CNF formula that has a refutation of length $\leq m$ in S and 1^m is a string of 1's of length m. Furthermore, SAT^* is the set of pairs $(\phi, 1^m)$ where ϕ is a satisfiable CNF. We say that $(REF(S), SAT^*)$ is *the canonical NP-pair* for the proof system S.

The polynomial reducibility quasi-ordering of canonical pairs reflects the polynomial simulation quasi-ordering of proof systems.

Proposition 7. *If P polynomially simulates S, then the canonical pair of S is polynomially reducible to the canonical pair of P.*

Proof. The reduction is given by $(\phi, 1^m) \mapsto (\phi, 1^{p(m)})$, where p is a polynomial bound such that $|f(x)| \leq p(|x|)$ for all x.

It is possible, however, to give an example of two systems that are not equivalent with respect to polynomial simulation, but still have canonical pairs mutually polynomially reducible. We will give the example a few lines below.

The main problem about canonical pairs is, how hard it is to distinguish elements of one of the sets from the elements of the other set, in particular, is the pair polynomially separable? This question is related to the so called automatizability of a proof system. A proof system S is *automatizable*, if there exists a polynomial time algorithm that for a given formula ϕ and a number m finds a refutation of ϕ in time polynomial in m, provided a refutation of length at most m exists. The following is trivial:

Lemma 8. *If S is automatizable, then the canonical pair of S polynomially separable.*

The converse may be not true, but a the following weaker statement is true.

Lemma 9. *If the canonical pair of S polynomially separable, then there exists a proof system S' which polynomially simulates S and which is automatizable.*

Proof. Let f be a polynomial time computable function that is 0 on $REF(S)$ and 1 on SAT^*. In the proof system S' a refutation of ϕ is a sequence 1^m such that $f(\phi, 1^m) = 0$. Formally, we define S' by $S'(w) = \phi$, if $w = (\phi, 1^m)$ and $f(\phi, 1^m) = 0$; $S'(w) = x_1 \vee \neg x_1$ otherwise. A polynomial simulation of S by S' is the function $w \mapsto (S(w), 1^{|w|})$.

Corollary 10. *The canonical pair of a proof system S is polynomially separable iff there exists an automatizable proof system S' that polynomially simulates S.*

The last corollary shows that from the point of view of proof search the problem of the polynomial separation of the canonical pair is more important than automatizability. For example, assuming a reasonable complexity theoretical conjecture, it has been established that Resolution is not automatizable [1]. But this does not exclude the possibility that an extension of Resolution is automatizable. To show that the latter possibility is excluded means to prove that the canonical pair of Resolution is not polynomially separable. (Thus the relation of these two concepts is similar to *undecidability* and *essential undecidability* of first order theories in logic.)

We shall mention two more concepts that are connected with disjoint NP-pairs. The first is the feasible interpolation property. We say that a system S has *the feasible interpolation property* if, given a proof of a formula

$$\phi(\bar{x}, \bar{y}) \vee \psi(\bar{x}, \bar{z}), \tag{1}$$

in which $\bar{x}, \bar{y}, \bar{z}$ are strings of distinct propositional variables, one can construct in polynomial time a boolean circuit C with the property

$$C(\bar{x}) = 0 \Rightarrow \phi(\bar{x}, \bar{y}) \quad \text{and} \quad C(\bar{x}) = 1 \Rightarrow \phi(\bar{x}, \bar{z}).$$

The meaning of this is that the sets $\{\bar{x} \; ; \; \exists \bar{y} \neg \phi(\bar{x}, \bar{y})\}$ and $\{\bar{x} \; ; \; \exists \bar{z} \neg \psi(\bar{x}, \bar{z})\}$, which have polynomial size nondeterministic boolean circuits, can be separated by a polynomial size (deterministic) circuit. If we had a sequence of formulas of the form above given uniformly in polynomial time and also their proofs given in this way, we would get, from the feasible interpolation property, a pair of disjoint NP-sets and a polynomial time separation algorithm for them. On the other hand, given an NP set A, we can construct (in fact, generate uniformly in polynomial time) a sequence of formulas α_n such that for $|\bar{x}| = n$, $\bar{x} \in A$ iff $\exists \bar{y} \, \alpha_n(\bar{x}, \bar{y})$. So the statement that two NP sets are disjoint can be expresses as a sequence of formulas of the form 1.

Consequently: *feasible interpolation means that whenever we have short proofs that two NP sets are disjoint, then they can be polynomially separated.*

Now we sketch the promised example of the two nonequivalent proof systems with essentially the same canonical pair. In [11] we have shown (using the feasible interpolation property) that in the cutting planes proof system CP the tautology expressing the disjointness of sets of the pair CC has only exponentially long proofs. Note that the disjointness of the CC pair is based on the pigeon hole principle: it is not possible to color a k-clique by $k - 1$-colors. This may seem paradoxical, as the pigeon hole principle has polynomial size proofs in CP. The explanation is that in order to use the pigeon hole principle we need to define a mapping and the mapping from the clique to the colors cannot be defined using the restricted means of CP. In CP one can use only *linear* inequalities with propositional variables. To define the mapping we need quadratic terms, namely, terms of the form $x_i y_j$ for x_i coding a vertex of the clique and y_j coding a color. So let us define an extension of CP, denoted by CP^2 that allows quadratic terms. What it means precisely is the following. Given a formula with variables $x_1, \ldots, x_n$ we allow in its proofs inequalities with terms of the form x_i and $x_i x_j$ for $i < j$ (and, of course, constants). On top of the axioms and rules of CP the proofs of CP^2 may use the following axioms about the quadratic terms:

$$0 \leq x_i x_j \leq 1, \quad x_i x_j \leq x_i, \quad x_i x_j \leq x_j, \quad x_i + x_j \leq x_i x_j + 1.$$

One can show that in this system the CC tautology has polynomial size proofs. To prove that the canonical pair of CP^2 is polynomially reducible to the one of CP, use the following mapping: $(\phi, 1^m) \mapsto (\phi', 1^{p(m)})$ with ϕ' expressing that the above axioms for quadratic terms imply ϕ. Since CP is a refutation system, we can think of ϕ as a set of inequalities from which we want to derive a contradictory inequality and then ϕ' is the union of this set with the inequalities for the quadratic terms. $p(m)$ is a suitable polynomial overhead.[1]

---

[1] Note for Experts. The Lovász-Schrijver system combined with CP that we considered in [11] seems not to be strong enough to polynomially simulate CP^2, as it does not allow to apply the rounding up rule to quadratic inequalities.

The last property of proof systems that we mention in this paper is the feasible reflection. We say that a system S has *the feasible reflection property* if the formulas

$$\neg \pi_{S,n,m}(\bar{x}, \bar{y}) \vee \neg \sigma_n(\bar{x}, \bar{z})$$

have polynomial size proofs, where $\pi_{S,n,m}(\bar{x}, \bar{y})$ is a propositional encoding of '*y is an S refutation of length m of formula x of length n*' and $\sigma_n(\bar{x}, \bar{z})$ is an encoding of '*z is a satisfying assignment of formula x of length n*'. Furthermore, we will assume that the proofs of these formulas are given uniformly by a polynomial time algorithm. The meaning of the formulas, actually tautologies, is that either the formula x has no refutation of length m or it is not satisfiable. Thus feasible reflection of S means that we can generate in polynomial time proofs of propositional instances of the statement $REF(S) \cap SAT^* = \emptyset$.

Proposition 11. *If a proof system has both feasible interpolation and feasible reflection properties, then its canonical pair is polynomially separable.*

Proof. Feasible reflection means that one can efficiently generate proofs of the tautologies expressing the disjointness of the canonical pair. Feasible interpolation property means that any NP-pair that has such proofs is polynomially separable. Hence the canonical pair is polynomially separable.

We know of strong systems that have feasible reflection property (see [6] Thms 9.1.5 and 9.3.4),[2] we also know of weak systems that have feasible interpolation property, but we have no example of a proof system that has both properties. In fact we do not know of any natural proof system the canonical pair of which is polynomially separable. Let us conclude by noting that the last proposition can be refined. Thus to prove polynomial separation of the canonical pair of a system S we only need to have short P proofs of the reflection principle for S in some, possibly stronger, system P that has the feasible interpolation property.

4   The NP-Pair of Resolution

We shall consider the canonical pair of the Resolution proof system. Resolution uses only formulas that are disjunctions of variables and negated variables (called *literals*); these formulas are called *clauses*. The only rule of Resolution is the cut in which we combine two clauses with a complementary literal into one, omitting the complementary literal. A proof is a sequence of clauses such that at the beginning we have the clauses that we want to refute and then a sequence of clauses follows such that each of these clauses follows by an application of the resolution rule from two clauses before it. In general, the length of a proof is the

---

[2] For a logician this may look surprising, since reflection principles for first order theories are stronger than consistency and even the latter is unprovable by Gödel theorem. Furthermore, reflection for strong enough propositional proof systems is equivalent to their consistency [6].

length of a binary sequence that encodes the proof. In Resolution the size of each step of the proof, which is a clause, is bounded by the number of propositional variables that appear in the clauses to be refuted. Hence we can assume w.l.o.g. that the length is simply the number of clauses in the proof.

To get more information on the pair $(REF(R), SAT^*)$, where R stands for the resolution proof system, we prove the following symmetry property of it.

Definition 12. *A pair (A, B) is symmetric, if (A, B) is polynomially reducible to (B, A).*

This property has been shown for some stronger systems using first order theories associated to the proof systems [13]. The symmetry of the canonical pairs of such systems can also be derived from the feasible reflection property. Resolution is weaker than such systems, in particular it is unlikely that it possesses the feasible reflection property. Therefore we give a direct proof of the symmetry of the canonical pair of Resolution. The idea of the proof is to show a property that is a little weaker than feasible reflection.

Theorem 13. *The canonical pair of Resolution is symmetric.*

Proof. We need, for a given CNF ϕ and a number m, to construct in polynomial time a CNF ψ such that if ϕ is refutable by a resolution refutation of length m then ψ is satisfiable and if ϕ is satisfiable, then ψ is refutable by a refutation with length polynomial in m. Let ϕ be the conjunction of clauses $C_1, \ldots, C_r$, let the variables in the clauses be $x_1, \ldots, x_n$. We shall represent a refutation of ϕ of length m by a $2n \times m$ matrix, plus some additional information. The columns of the matrix will encode the clauses of the refutation. The additional information will specify for each clause that is not an assumption, from which two clauses it has been derived. Furthermore we shall specify the variable that was resolved in this step of the refutation.

It will be clear from the construction of ψ that the formula is a correct description of the refutation, ie., if a refutation exists, then ψ is satisfiable. Thus the assignment $(\phi, 1^m) \mapsto (\psi, 1^{m'})$ maps $REF(R)$ into SAT^*, (whatever m' we choose).

The nontrivial part is to show that if ϕ is satisfiable, then there is a resolution refutation of ψ that is polynomial in the size of ϕ and m (the size of this proof will determine the m' and then we get that SAT^* is mapped to $REF(R)$). This will be proved as follows. We take a satisfying assignment and derive gradually, for each $j = r, r+1, \ldots, m$, the clause that says that the j-th clause of the proof agrees with the satisfying assignment at least in one literal. The contradiction is obtained by using the clauses of ψ that express that the last clause C_m should be empty (clauses (1) below). Here is a detailed proof.

Variables $y_{e,i,j}$, $e = 0, 1$, $i = 1, \ldots, n$, $j = 1, \ldots, m$ encode clauses. Namely, $y_{0,i,j}$ (resp. $y_{1,i,j}$) means that $\neg x_i$ (resp. x_i) is present in the clause C_j. Variables $p_{j,k}$ (resp. $q_{j,k}$) $1 \leq j < k$, $r < k \leq m$ say that C_k was obtained from C_j and C_j contains negated (resp. positively) the resolved variable. Finally, variables $v_{i,j}$ determine that C_j was obtained by resolving variable x_i. The following are the clauses of ψ.

(0) $y_{0,i,j}$ or $y_{1,i,j}$ for all i and all $j \leq r$, according to which literal occurs in C_j (recall that for $j \leq r$ the clauses are given by ϕ);

(1) $\neg y_{e,i,m}$, for all e, i (the last clause is empty);

(2) $\neg y_{0,i,j} \vee \neg y_{1,i,j}$, for all i, j (C_j does not contain x_i and $\neg x_i$ at the same time);

(3a) $\bigvee_{j<k} p_{j,k}$, (3b) $\bigvee_{j<k} q_{j,k}$, for $k > r$;

(4) $\neg p_{j,k} \vee \neg q_{j,k}$, for $j < k$, $r < k$;

(5) $\neg p_{j,k} \vee \neg p_{j',k}$, $\neg q_{j,k} \vee \neg q_{j',k}$ for $j, j' < k$, $j \neq j'$, $r < k$ ((3-5) say that there are exactly two clauses that are assigned to C_k);

(6a) $\neg p_{j,k} \vee \neg v_{i,k} \vee y_{0,i,j}$ (the C_j assigned to C_k contains literal $\neg x_i$);

(6b) $\neg q_{j,k} \vee \neg v_{i,k} \vee y_{1,i,j}$ (the C_j assigned to C_k contains literal x_i);

(7a) $\neg p_{j,k} \vee v_{i,k} \vee \neg y_{e,i,j} \vee y_{e,i,k}$ (C_k contains C_j except for $\neg x_i$);

(7b) $\neg q_{j,k} \vee v_{i,k} \vee \neg y_{e,i,j} \vee y_{e,i,k}$ (C_k contains C_j except for $\neg x_i$);

(8a) $\bigvee_i v_{i,k}$ for $r < k$;

(8b) $\neg v_{i,k} \vee \neg v_{i',k}$ for $i \neq i'$, $r < k$ (the resolution variable x_i is uniquely assigned to C_k).

This finishes the description of the ψ that is assigned to $(\phi, 1^m)$. Now, given a satisfying assignment $(e_1, \ldots, e_n)$ for $C_1, \ldots, C_r$ we construct a polynomial size refutation from (0)-(8). We shall use the weakening rule, which is superfluous, but it simplifies notation. Put

$$D_k := y_{e_1,1,k} \vee \ldots \vee y_{e_n,n,k}.$$

We shall gradually derive clauses $D_1, \ldots, D_m$. Once we have D_m, a contradiction follows immediately using clauses (1).

Clauses $D_1, \ldots, D_r$ follow immediately from (0) using weakening. To derive D_k, assuming D_j for $j < k$, we first derive clauses

(9) $\neg p_{j,k} \vee \neg q_{l,k} \vee D_k$

for $j \neq l$, $j, l < k$. Fix i and l and assume w.l.o.g. $e_i = 0$. From (6b) and (2) we get $\neg q_{l,k} \vee \neg v_{i,k} \vee \neg y_{0,i,l}$. Resolving with D_l we get

(10) $\neg q_{l,k} \vee \neg v_{i,k} \vee (D_l \setminus \{y_{0,i,l}\})$.

From (7b) and (8b) we get

(11) $\neg q_{l,k} \vee \neg v_{i,k} \vee \neg y_{e_{i'},i',l} \vee \neg y_{e_{i'},i',k}$,

for all $i' \neq i$. Resolving (10) with clauses in (11) we get $\neg q_{l,k} \vee \neg v_{i,k} \vee (D_k \setminus \{y_{0,i,k}\})$. Using weakening we get

(12) $\neg p_{j,k} \vee \neg q_{l,k} \vee \neg v_{i,k} \vee D_k$.

Having these for all i, we can resolve with (8a) and get (9). To get D_k from (9), first resolve with (3a) to get $\neg p_{l,k} \vee \neg q_{l,k} \vee D_k$. Then resolve with (4) to get $\neg q_{l,k} \vee D_k$. Finally resolve with (3b) and get D_k.

We have shown that if ϕ is satisfiable, then there exists a proof of ψ the size of which is polynomial in the size of ϕ and m. Let m' be the polynomial bound on this proof; we can compute this bound *without* having the proof of ψ. Thus, if we define the reduction by $(\phi, 1^m) \mapsto (\psi, 1^{m'})$, the set SAT^* will be mapped into $REF(R)$.

The following operation on pairs, clearly, defines the the meet in the lattice of degrees of pairs,

$$(A, B) \wedge (C, D) = (A \times C, B \times D).$$

Given a pair (A, B) we can thus form a symmetric pair by taking $(A, B) \wedge (B, A)$. This symmetrization satisfies the following stronger property: there exists a polynomial time computable isomorphism that transposes the sets in the pair. An example of a concrete pair that has this property is BMS. We observe that the symmetry of a pair implies that there is an equivalent pair, namely $(A, B) \wedge (B, A)$, with the stronger property.

Proposition 14. *If (A, B) is symmetric, then (A, B) and $(A, B) \wedge (B, A)$ are polynomially equivalent.*

Proof. $(A, B) \wedge (B, A)$ is always reducible to (A, B) by the projection on the first coordinate. Let f be a polynomial reduction of (A, B) to (B, A). Then $x \mapsto (x, f(x))$ is a polynomial reduction of (A, B) to $(A, B) \wedge (B, A)$.

Consequently, there is a pair of disjoint NP sets that has this stronger symmetry property and that is polynomially equivalent to the canonical pair of Resolution.

5   Open Problems and Further Research Topics

Our first result shows that seemingly different pairs may be in fact equivalent. Our second result shows that the canonical pair of Resolution is equivalent to a very symmetric pair. This gives some hope that a nice combinatorial characterization of the degree of the canonical pair of Resolution and other systems may be found. If the systems are natural and robust, there should be simple combinatorial principles on which they are based. Ideally, we would like to prove that some canonical pair is polynomially equivalent to some combinatorially defined pair. At present we only have reductions of cryptographic pairs to canonical pairs of proof systems [8,3], but we do not have converse reductions. We do not have any reductions from canonical pairs to pairs defined in another way.

An important problem is to decide if the canonical pairs of weak systems are polynomially separable. In particular, prove or disprove (using plausible complexity theoretical assumptions) that the canonical pair of Resolution is polynomially separable. If it were polynomially separable, it might have practical consequences for automated theorem proving (see Lemma 9).

In this paper we have considered a concept of reduction between pairs that corresponds to many one reductions between sets. One can define also the concept corresponding to Turing reductions:

Definition 15. *(A, B) is polynomially Turing reducible to (C, D), if there exists a polynomial time oracle Turing machine M such that M^A separates (A, B) for every oracle A that separates (C, D).*

With this definition of reduction, it should be possible to show the equivalence of more pairs. Eg., every (A, B) is polynomially Turing equivalent to (B, A). Furthermore, given a pair (A, B), define $P_0^{A,B}, P_1^{A,B}$ by $(x_1, \ldots, x_n) \in P_i^{A,B}$ iff $x_1, \ldots, x_n \in A \cup B$ and the parity of the number of x_j's such that $x_j \in A_j$ is i. It is an easy exercise to show that this pair is polynomially Turing equivalent to (A, B).

It would be interesting to learn more about the lattice of degrees of disjoint NP pairs. We know about this structure even less than we know about the degrees of proof systems. Does there exist the biggest element in it? How is this question related to the same question about the degrees of proof systems? Etc.

References

1. M. Alekhnovich and A. A. Razborov, Resolution is Not Automatizable Unless W[P] is Tractable, http://genesis.mi.ras.ru/~razborov/. 626
2. L. Babai, A. Gál and A. Wigderson, Superpolynomial lower bounds for monotone span programs, Combinatorica 19 (3), 1999, 301-319. 624
3. M. Bonet, T. Pitassi and R. Raz, On interpolation and automatization for Frege systems. SIAM J. on Computing 29(6), (2000), 1939-1967. 631
4. S. A. Cook and A. Haken, An exponential lower bound for the size of monotone real circuits, JCSS 58, (1999), 326-335. 622, 623
5. S. A. Cook and Reckhow, The relative efficiency of propositional proof systems, J. of Symbolic Logic, 44(1), 36-50. 621, 625
6. J. Krajíček, Bounded arithmetic, propositional logic, and complexity theory, Cambridge Univ. Press, 1995. 628
7. J. Krajíček and P. Pudlák, Propositional proof systems, the consistency of first order theories and the complexity of computations, J. of Symbolic Logic, 54(3), 1063-1079.
8. J. Krajíček and P. Pudlák, Some consequences of cryptographical conjectures for S_2^1 and EF. Information and Computation 142, (1998), 82-94 631
9. L. Kučera, Cryptography and random graphs, preprint.
10. L. Lovász, On the Shannon capacity of graphs, *IEEE Trans. Inform. Theory* 25, 1979, 1-7. 623
11. P. Pudlák, On the complexity of propositional calculus, Sets and Proofs, Invited papers from Logic Colloquium'97, Cambridge Univ. Press, 1999, 197-218. 627
12. A. A. Razborov, Lower bounds on the monotone complexity of some Boolean functions, *Soviet Mathem. Doklady* 31, 354-357. 622
13. A. A. Razborov, On provably disjoint NP-pairs, BRICS Report Series RS-94-36, 1994, http://www.brics.dk/RS/94/36/index.html. 621, 622, 625, 629

Hierarchy of Monotonically Computable Real Numbers

Extended Abstract

Robert Rettinger[1] and Xizhong Zheng[2*]

[1] FernUniversität Hagen
58084-Hagen, Germany
[2] BTU Cottbus
03044 Cottbus, Germany
zheng@informatik.tu-cottbus.de

Abstract. A real number x is called *h-monotonically computable* (*h*-mc), for some function h, if there is a computable sequence $(x_s)_{s \in \mathbb{N}}$ of rational numbers such that $h(n)|x - x_n| \geq |x - x_m|$ for any $m \geq n$. x is called *ω-monotonically computable* (*ω*-mc) if it is h-mc for some recursive function h and, for any $c \in \mathbb{R}$, x is c-mc if it is h-mc for the constant function $h \equiv c$. In this paper we discuss the properties of c-mc and ω-mc real numbers. Among others we will show a hierarchy theorem of c-mc real numbers that, for any constants $c_2 > c_1 \geq 1$, there is a c_2-mc real number which is not c_1-mc and that there is an ω-mc real number which is not c-mc for any $c \in \mathbb{R}$. Furthermore, the class of all ω-mc real numbers is incomparable with the class of weakly computable real numbers which is the arithmetical closure of semi-computable real numbers.

1 Introduction

The effectiveness of a real number x are usually described by a computable sequence $(x_s)_{s \in \mathbb{N}}$ of rational numbers which converges to x with some further restrictions on the error-estimation of the approximation. The optimal situation is that we have full control over errors of the approximation. Namely, there is an unbounded recursive function $e : \mathbb{N} \to \mathbb{N}$ such that $|x - x_s| \leq 2^{-e(s)}$ for any $s \in \mathbb{N}$, or equivalently, there is a computable sequence $(x_s)_{s \in \mathbb{N}}$ of rational numbers which converges to x effectively in the sense that $|x_n - x_m| \leq 2^{-n}$ if $n \leq m$. Such real numbers are called by A. Turing [16] *computable*. The class of all computable real numbers is denoted by **E**. There are a lot of equivalent ways to characterize this class **E**. In fact, corresponding to any classical definition of real numbers in mathematics there is an effectivization which induces an equivalent definition of the computable real numbers (see [9,8,17]). Typically, besides the fast converging Cauchy sequence definition mentioned above, computable real numbers can also be described by effective version of Dedekind cuts, binary or decimal expansions, and so on.

---

* Corresponding author

J. Sgall, A. Pultr, and P. Kolman (Eds.): MFCS 2001, LNCS 2136, pp. 633–644, 2001.

The optimal error-estimation of an approximation to some real number is not always available. E. Specker [15] has found an easy example of such real numbers. Let $(A_s)_{s \in \mathbb{N}}$ be an effective enumeration of a non-recursive r.e. set $A \subseteq \mathbb{N}$. That is, $(A_s)_{s \in \mathbb{N}}$ is a computable sequence of finite sets of natural numbers such that $A_s \subset A_{s+1}$ and $A = \bigcup_{s \in \mathbb{N}} A_s$. Define $x_A := \sum_{i \in A} 2^{-(i+1)}$ and $x_s := \sum_{i \in A_s} 2^{-(i+1)}$. Then $(x_s)_{s \in \mathbb{N}}$ is a computable sequence of rational numbers which converges to x_A. But an effective error-estimation for this approximation is impossible, because x_A is not a computable real number (it has a non-computable binary expansion). Namely, the sequence $(x_s)_{s \in \mathbb{N}}$ converges to x non-effectively. On the other hand, since $(x_s)_{s \in \mathbb{N}}$ is an increasing sequence, we get always better and better approximations x_s to x with increasing index s. Such kind of real numbers have the effectiveness only weaker than that of the computable real numbers and thay are called *left computable*[1]. The class of all left computable real numbers is denoted by **LC**. The class **LC** has been widely discussed in literature (see e.g., [1,2,3,12,13,18]). Unlike the case of computable real numbers, effectivizations of classical definitions of real numbers (to the level of recursive enumerabiliy) do not induce always the same notions of left computability. Let's say that $x \in \mathbb{N}$ has a r.e. binary expansion if there is a r.e. set $A \subseteq \mathbb{N}$ such that $x = x_A$. Then, C.G. Jockush (cf. [12]) pointed out that the real number[2] $x_{A \oplus \bar{A}}$ has a r.e. Dedekind cut but it has no r.e. binary expansion, if $A \subseteq \mathbb{N}$ is a non-recursive r.e. set, since $A \oplus \bar{A}$ is not r.e. In fact, as shown by Specker [15], the effectivizations of classical definitions of real numbers to the level of primitive recursiveness are also not equivalent.

Left computable real numbers are not equally difficult (or easy) to be computed. Solovay [14] introduced a "domination relation" on the class **LC** to compare them. For any $x, y \in$ **LC**, x *dominates* y if for any computable increasing sequence $(y_s)_{s \in \mathbb{N}}$ of rational numbers which converges to y, there is a computable increasing sequence $(x_s)_{s \in \mathbb{N}}$ of rational numbers which converges to x such that the sequence $(x_s)_{s \in \mathbb{N}}$ dominates $(y_s)_{s \in \mathbb{N}}$ in the sense that $c(x - x_s) \geq y - y_s$ for some constant c and all $s \in \mathbb{N}$. Intuitively, if x dominates y, then it is not more difficult to approximate y than x, thus x contains at least so much information as y. If a left computable x dominates all $y \in$ **LC**, then x is called Ω-like by Solovay. Solovay shows that the Chaitin Ω numbers (i.e., the halting probability of a universal self-delimiting Turing machine, cf [5]) are in fact Ω-like. That is, the Chaitin Ω numbers are, in some sense, the most complicated left computable real numbers. This is also confirmed by a nice result, which is proved in stages by Chaitin [5], Solovay [14], Calude, Hertling, Khoussainov and Wang [3] and Slaman [10] (see also [2] for a good survey), that a real number x is Ω-like iff it is a Chaitin Ω number and iff it is a left computable random real number, where x is random means that its binary expansion is a random sequence of $\{0,1\}^\omega$ in the sense of Martin-Löf [6].

---

[1] Some authors call these real numbers effectively enumerable or computably enumerable. See e.g. [2,3,12,13]

[2] $A \oplus B := \{2n : n \in A\} \cup \{2n+1 : n \in B\}$ is the join of the sets A and B and $\bar{A}$ is the complement of set A.

The arithmetical closure of the left computable real number class **LC** (under the operations $+, -, \times, \div$) is denoted by **WC**, the class of so-called *weakly computable* real numbers. Interestingly, Weihrauch and Zheng [18] shown that x is weakly computable iff there is a computable sequence $(x_s)_{s \in \mathbb{N}}$ of rational numbers which converges to x weakly effectively in the sense that the sum of the jumps $\sum_{s \in \mathbb{N}} |x_s - x_{s+1}|$ is finite. The class **WC** is strictly between **LC** and the class **RA** of recursively approximable real numbers which are simply the limits of some computable sequences of rational numbers.

Symmetrically, we define also the class **RC** of right computable real numbers by the limits of computable decreasing sequences of rational numbers. Left and right computable real numbers are all called semi-computable whose class is denoted by $\mathbf{SC} := \mathbf{LC} \cup \mathbf{RC}$. The semi-computability can also be characterized by the monotone convergence of the computable sequences (see [7]). A sequence $(x_s)_{s \in \mathbb{N}}$ converges to x monotonically means that $|x - x_n| \geq |x - x_m|$ holds for any $m \geq n$. Then, x is semi-computable iff there is a computable sequence $(x_s)_{s \in \mathbb{N}}$ of rational numbers which converges to x *monotonically*. Notice that, if a sequence converges monotonically, then a later element of the sequence is always a better approximation to the limit. However, since the improvement of the approximation changes during the procedure, it is still impossible to give an exact error estimation.

More generally, Calude and Hertling [4] discussed c-monotone convergence for any positive constant c in the sense of $c|x - x_n| \geq |x - x_m|$ for any $m \geq n$. They have shown that, although there are computable sequences of rational numbers which converge to computable real numbers very slowly, c-convergent computable sequences converge always fast, if their limits are computable. It is easy to see that, for different constants c, c-monotonically convergent sequences converge in different speeds. More precisely, if $(x_s)_{s \in \mathbb{N}}$ and $(y_s)_{s \in \mathbb{N}}$ converge c_1- and c_2-monotonically, for $c_1 < c_2$, to x and y, respectively, then, $(x_s)_{s \in \mathbb{N}}$ seems to be a better approximation to x than $(y_s)_{s \in \mathbb{N}}$ to y. Let's call a real number x *c-monotonically computable* (c-mc for short) if there is a computable sequence of rational numbers which converges to x c-monotonically and denote by c-**MC** the class of all c-mc real numbers. It is not difficult to see that 1-**MC** = **SC**, c-**MC** = **E** if $c < 1$ and c_1-**MC** $\subseteq c_2$-**MC** if $c_1 \leq c_2$. Furthermore, it is also shown in [7] that c-mc real numbers are weakly computable for any constant c, but there is a weakly computable real number which is not c-mc for any constant c, i.e., $\bigcup_{c \in \mathbb{R}^+} c$-**MC** $\subsetneq$ **WC**. To separate the different c-mc classes, only a very rough hierarchy theorem was shown in [7] that, for any constant c_1, there is a $c_2 > c_1$ such that c_1-**MC** $\neq c_2$-**MC**. However, it remains open in [7] whether c_1-**MC** $\neq c_2$-**MC** holds for any different c_1 and c_2. A positive answer will be given in this paper (Theorem 7).

For any $n \in \mathbb{N}$ and any sequence $(x_s)_{s \in \mathbb{N}}$ which converges c-monotonically to x, we know that the error of the approximation x_m, for any $m \geq n$, is always bounded by $c|x - x_n|$. Although we do not know exactly the current error $|x - x_n|$ itself, we do know that later errors are bounded by the current error some how. This requirement can be further weakened in the way that the constant c is replaced by a function h whose value depends on the index n. Namely we can define the h-monotone convergence of the sequence $(x_s)_{s\in\mathbb{N}}$ to x by the condition that $h(n)|x - x_n| \geq |x - x_m|$ for any $m \geq n$ and define the h-mc real numbers accordingly. Especially we call a real number x ω-mc if it is h-mc for some recursive function h. The class of ω-mc real numbers is denoted by ω-**MC**. Of course, all c-mc real numbers are ω-mc. But we will show that there is an ω-mc real number which is not c-mc for any constant c. In fact we can prove that the classes ω-**MC** and **WC** are incomparable. Hence ω-**MC** is not covered completely by **WC** while all c-**MC** are contained in **WC**.

Notice that, the classes $\mathbf{E}, \mathbf{LC}, \mathbf{RC}, \mathbf{WC}$ and $\mathbf{RA}$ can be defined by the existence of some computable sequences of rational numbers with some special properties which do not relate directly to their limits. These definitions are purely *procedural*. The classes c-**MC** and ω-**MC**, on the other hand, are defined in a different way. For example, the property that a sequence $(x_s)_{s\in\mathbb{N}}$ converges c-monoconically to x involves inevitably the limit x itself. These definitions are called *abstract*. Some classes, e.g. $\mathbf{E}$ and $\mathbf{SC}$, can be defined either by procedural or abstract definitions. Classes ω-**MC** and **WC** supply a first example of incomparable classes which are defined by procedural and abstract definitions respectively.

The outline of this paper is as follows. In the next section we will explain notion and notation we need and recall some related known results. The hierarchy theorem about c-mc classes is proved in Section 3. The last section discusses the relationship between the classes ω-**MC** and **WC**.

2   Preliminaries

In this section we explain some notions and notations which are needed for later sections and recall some related results. We suppose that the reader know the very basic notions and results of classic computability theory. But no knowledge from computable analysis is assumed.

We denote by $\mathbb{N}, \mathbb{Q}$ and $\mathbb{R}$ the sets of all natural, rational and real numbers, respectively. For any sets A and B, $f :\subseteq A \to B$ is a partial function with $\text{dom}(f) \subseteq A$ and $\text{range}(f) \subseteq B$. If f is a total function, i.e., $\text{dom}(f) = A$, then it is denoted by $f : A \to B$. The computability notions like computable (or recursive) function, recursive and r.e. (recursively enumerable) set, etc., on $\mathbb{N}$ are well defined and developed in classic computability theory. For example, the pairing function $\langle \cdot, \cdot \rangle : \mathbb{N}^2 \to \mathbb{N}$ defined by $\langle m, n \rangle := (n + m)(n + m + 1)/2 + m$ is a computable function. Let $\pi_1, \pi_2 : \mathbb{N} \to \mathbb{N}$ be its inverse functions, i.e., $\pi_1 \langle n, m \rangle = n$ and $\pi_2 \langle n, m \rangle = m$ for any $n, m \in \mathbb{N}$. Then π_1 and π_2 are obviously computable too. Let $\sigma : \mathbb{N} \to \mathbb{Q}$ be a coding function of $\mathbb{Q}$ using $\mathbb{N}$ defined by $\sigma(\langle n, m \rangle) := n/(m + 1)$. By this coding, the computability notions on $\mathbb{N}$ can be easily transferred to that of $\mathbb{Q}$. For example, a function $f :\subseteq \mathbb{N} \to \mathbb{Q}$ is computable if there is a computable function $g :\subseteq \mathbb{N} \to \mathbb{N}$ such that $f(n) = \sigma(g(n))$ for any $n \in \text{dom}(f)$, $A \subseteq \mathbb{Q}$ is recursive if $\{n \in \mathbb{N} : \sigma(n) \in A\}$ is recursive, a sequence $(x_n)_{n \in \mathbb{N}}$ of rational numbers is computable if there is a computable total function $f : \mathbb{N} \to \mathbb{Q}$ such that $x_n = f(n)$ for all n, and so on.

Frequently, we would like to diagonalize against all computable sequences of rational numbers or some subset of such sequences. In this case, an effective enumeration of all computable sequences of rational numbers would be useful. Unfortunately, the set of computable total functions are not effectively enumerable and hence there is no effective enumeration of computable sequences of rational numbers. Instead we consider simply the effective enumeration $(\varphi_e)_{e \in \mathbb{N}}$ of all computable partial functions $\varphi_e :\subseteq \mathbb{N} \to \mathbb{Q}$. Of course, all computable sequences of rational numbers (i.e., total computable functions $f : \mathbb{N} \to \mathbb{Q}$, more precisely) appear in this enumeration. Thus it suffices to carry out our diagonalization against this enumeration. Concretely, the enumeration $(\varphi_e)_{e \in \mathbb{N}}$ can be defined from some effective enumeration $(M_e)_{e \in \mathbb{N}}$ of Turing machines. Namely, $\varphi_e :\subseteq \mathbb{N} \to \mathbb{Q}$ is the function computed by the Turing machine M_e. Furthermore, let $\varphi_{e,s}$ be the approximation of φ_e computed by M_e until the stage s. Then $(\varphi_{e,s})_{e,s \in \mathbb{N}}$ is an uniformly effective approximation of $(\varphi_e)_{e \in \mathbb{N}}$ such that $\{(e, s, n, r) : \varphi_{e,s}(n) \downarrow = r\}$ is a recursive set and $\varphi_{e,t}(n) \downarrow = \varphi_{e,s}(n) = \varphi_e(n)$ for any $t \geq s$, if $\varphi_{e,s}(n) \downarrow = r$, where $\varphi_{e,s}(n) \downarrow = r$ means that $\varphi_{e,s}(n)$ is defined and it has the value r.

In the last section we have introduced the classes $\mathbf{E}, \mathbf{LC}, \mathbf{RC}, \mathbf{SC}, \mathbf{WC}$ and $\mathbf{RA}$ of real numbers. Some important properties about these classes are summarized again in the following theorem.

Theorem 1 (Weihrauch and Zheng [18]).

1. The classes $\mathbf{E}, \mathbf{LC}, \mathbf{RC}, \mathbf{SC}, \mathbf{WC}$ and $\mathbf{RA}$ are all different and have the following relationships $\mathbf{E} = \mathbf{LC} \cap \mathbf{RC} \subsetneq \mathbf{SC} = \mathbf{LC} \cup \mathbf{RC} \subsetneq \mathbf{WC} \subsetneq \mathbf{RA}$;

2. $x \in \mathbf{WC}$ iff there is a computable sequence $(x_s)_{s \in \mathbb{N}}$ of rational numbers such that the sum $\sum_{s \in \mathbb{N}} |x_s - x_{s+1}|$ is finite; and

3. The classes $\mathbf{E}, \mathbf{WC}$ and $\mathbf{RA}$ are algebraic fields. That is, they are closed under the arithmetic operations $+, -, \times$ and $\div$.

Now let's give a precise definition of c-monotonically computable real numbers.

Definition 2 (Rettinger, Zheng, Gengler and von Braunmühl [7]). Let x be any real number and $c \in \mathbb{R}$ a positive constant.

1. A sequence $(x_n)_{n \in \mathbb{N}}$ of real numbers converges to x c-monotonically if the sequence converges to x and satisfies the following condition

$$\forall n, m \in \mathbb{N} \, (m \geq n \implies c \cdot |x_n - x| \geq |x_m - x|). \tag{1}$$

2. The real number x is called c-monotonically computable (c-mc, for short) if there is a computable sequence $(x_n)_{n \in \mathbb{N}}$ of rational numbers which converges to x c-monotonically. The classes of all c-mc real numbers is denoted by $c\text{-}\mathbf{MC}$. Furthermore, we denote also the union $\bigcup_{c \in \mathbb{R}^+} c\text{-}\mathbf{MC}$ by $*\text{-}\mathbf{MC}$.

The next proposition follows easily from the definition.

Proposition 3. *Let x be any real number and c_1, c_2, c positive constants. Then*
1. *If $c_1 \leq c_2$ and x is c_1-mc, then it is c_2-mc too, i.e., c_1-**MC** $\subseteq c_2$-**MC**;*
2. *For any $c < 1$, x is c-mc iff it is computable, i.e, c-**MC** $= \mathbf{E}$; and*
3. *x is 1-mc iff it is semi-computable. Thus, $\mathbf{SC} = 1$-**MC**.*

Some further results about c-mc real numbers are summarized in the next theorem.

Theorem 4 (Rettinger, Zheng, Gengler and von Braunmühl [7]).
   1. *A c-mc real number is also weakly computable for any constant c. But there is a weakly computable real number which is not c-mc for any constant c. That is, $*$-**MC** $\subsetneq$ **WC**.*
   2. *For any constant c_1, there is a constant $c_2 > c_1$ such that c_1-**MC** $\subsetneq c_2$-**MC**. Therefore there is an infinite hierarchy of the class $*$-**MC**.*

Notice that, for any class $\mathbf{C} \subseteq \mathbb{R}$ discussed in this paper, $x \in \mathbf{C}$ iff $x \pm n \in \mathbf{C}$ for any $x \in \mathbb{R}$ and $n \in \mathbb{N}$. Therefore we can assume, without loss of generality, that any real number and corresponding sequence of rational numbers discussed in this paper is usually in the interval $[0; 1]$ except for being pointed out explicitly otherwise.

Here are some further notations: For any alphabet Σ, let Σ^* and Σ^ω be the sets of all finite strings and infinite sequences of Σ, respectively. For $u, v \in \Sigma^*$, denote by uv the concatenation of v after u. If $w \in \Sigma^* \cup \Sigma^\omega$, then $w[n]$ denotes its n-th element. Thus, $w = w[0]w[1]\cdots w[n-1]$, if $|w|$, the length of w, is n, and $w = w[0]w[1]w[2]\cdots$, if $|w| = \infty$. The unique string of length 0 is always denoted by λ (so-called empty string). For any finite string $w \in \{0; 1\}^*$, and number $n < |w|$, the restriction $w \restriction n$ is defined by $(w \restriction n)[i] := w[i]$ if $i < n$ and $(w \restriction n)[i] := \uparrow$, otherwise.

3   A Hierarchy Theorem of c-MC Real Numbers

From Theorem 4 of last section we know that there is an infinite hierarchy on the c-mc real numbers. Unfortunately, this hierarchy shown in [7] is very rough. In fact, it is only shown that, for any $c_1 \geq 1$, there is a, say, $c_2 := (c_1 + 8)^2$ such that c_1-**MC** $\neq c_2$-**MC**. In this section we will prove a dense hierarchy theorem on c-mc real numbers that c_1-**MC** $\neq c_2$-**MC** for any real numbers $c_2 > c_1 \geq 1$. To this end we need the following technical lemmas.

Lemma 5. *For any rational numbers c_1, c_2, a and b with $1 < c_1 < c_2$ and $a < b$, there are positive rational numbers ϵ, δ_1 and δ_2 with $\epsilon \leq \min\{\delta_1, \delta_2\}$ such that*

1. *$\delta_1 + \delta_2 + 5\epsilon \leq b - a$;*
2. *$c_1(\delta_1 + 2\epsilon) < \delta_1 + \delta_2 + \epsilon$;*
3. *$\delta_1 + \delta_2 + 3\epsilon \leq c_2\delta_1$; and*
4. *$\delta_2 + 2\epsilon \leq c_2\delta_2$.*

Lemma 6. *Let $c_1, c_2, a, b \in \mathbb{Q}$ be any rational numbers with $1 < c_1 < c_2$ & $a < b$ and $\epsilon, \delta_1, \delta_2 \in \mathbb{Q}$ satisfy the conditions 1. – 4. of Lemma 5. Then the interval $[a; b]$ can be divided into seven subintervals $I_i := [a_i; a_{i+1}]$, for $i < 7$, by*

$$
\begin{aligned}
a_0 &:= a & a_2 &:= a_1 + \epsilon & a_4 &:= a_3 + \epsilon & a_6 &:= a_5 + \epsilon \\
a_1 &:= a + \epsilon & a_3 &:= a_2 + \delta_1 & a_5 &:= a_4 + \delta_2 & a_7 &:= b
\end{aligned}
\tag{2}
$$

such that the following hold

$$
x \in I_1 \;\&\; x_1 \in I_3 \;\&\; x_2 \in I_5 \implies c_1 \cdot |x - x_1| < |x - x_2|, \; and \tag{3}
$$

$$
\left.
\begin{aligned}
&(x \in I_1 \;\&\; x_1 \in I_3 \;\&\; x_2 \in I_5) \\
&\text{or } (x \in I_1 \;\&\; x_1, x_2 \in I_3) \\
&\text{or } (x \in I_1 \;\&\; x_1, x_2 \in I_5) \\
&\text{or } (x \in I_5 \;\&\; x_1, x_2 \in I_3)
\end{aligned}
\right\}
\implies c_2 \cdot |x - x_1| \geq |x - x_2|. \tag{4}
$$

Now we prove our hierarchy theorem at first for the case of rational constants c_1 and c_2. The reason is, that we have to divide some interval I into subintervals according to Lemma 6. To make sure this procedure effective, the interval I and numbers a_i, hence also the numbers c_1, c_2, a, b, should be rational.

Theorem 7. *For any rational numbers c_1, c_2 with $1 < c_1 < c_2$, there is a c_2-mc real number x which is not c_1-mc. Thus, c_1-**MC** $\subsetneq c_2$-**MC**.*

Proof. (sketch) Let c_1, c_2 be any rational numbers such that $1 \leq c_1 < c_2$, $(\varphi_i)_{i \in \mathbb{N}}$ an effective enumeration of all computable functions $\varphi_i :\subseteq \mathbb{N} \to \mathbb{Q}$ and $(\varphi_{i,s})_{s \in \mathbb{N}}$ the uniformly effective approximation of φ_i. We will construct a computable sequence $(x_s)_{s \in \mathbb{N}}$ of rational numbers which satisfies, for any $e \in \mathbb{N}$, the following requirements:

$$
N : (x_s)_{s \in \mathbb{N}} \text{ converges } c_2\text{-monotonically to some } x, \text{ and}
$$
$$
R_e : (\varphi_e(n))_{n \in \mathbb{N}} \text{ converges } c_1\text{-monotonically to } y_e \implies x \neq y_e.
$$

The strategy to satisfy a single requirement R_e is an application of Lemma 6. Let $a := 0, b := 1$. We fix at first three rational numbers $\epsilon, \delta_1, \delta_2$ which satisfy the conditions 1. – 4. of Lemma 5. Furthermore, for $i < 7$, let a_i be defined by (2) and $I_i := [a_i; a_{i+1}]$ whose length is denoted by $l_i := a_{i+1} - a_i$. We denote by I^o the open interval which consists of all inner points of I for any interval $I \subseteq \mathbb{R}$.

We take the interval $(0; 1)$ as our base interval and will try to find out a so-called witness interval $I_w \subseteq (0; 1)$ such that any $x \in I_w$ satisfies the requirement R_e. Let I_3^o be our first (default) candidate of the witness interval. If no element of the sequence $(\varphi_e(n))_{n \in \mathbb{N}}$ appears in this interval, then it is automatically a correct witness interval, since the limit $\lim_{n \to \infty} \varphi_e(n)$, if exists, will not be in this interval. Otherwise, if there are some $s_1, n_1 \in \mathbb{N}$ such that $\varphi_{e,s_1}(n_1) \in I_3^o$, then we choose I_5^o as our new candidate of witness interval. In this case we denote n_1 by $c_e(s_1)$. Again, if there is no $n_2 > n_1$ such that $\varphi_e(n_2)$ comes into this interval, then any element from this interval witnesses the requirement R_e. Otherwise, if $\varphi_{e,s_2}(n_2) \in I_5^o$ for some $s_2 > s_1$ and some $c_e(s_2) := n_2 > n_1 = c_e(s_1)$, then we choose the interval I_1^o as the new candidate of the witness interval. By the implication (3) of Lemma 6, this interval turns out to be really a correct witness interval, since $c_1|x - \varphi_e(n_1)| < |x - \varphi_e(n_2)|$, for any $x \in I_1^o$, and hence $(\varphi_e(n))_{n \in \mathbb{N}}$ does not converge to x c_1-monotonically.

To satisfy all requirements R_e simultaneously, we implement the above strategies for different requirements R_e on different base intervals. More precisely, we need a tree of intervals on which the above strategy can be implemented. Let $\mathbb{I}$ be the set of all rational subintervals of $[0;1]$ and $\Sigma_7 := \{0, 1, \cdots, 6\}$. We define a tree I on $\mathbb{I}$ as a function $I : \Sigma_7^* \to \mathbb{I}$ such that, for all $w \in \Sigma_7^*$, $I(w) := [a_w; b_w]$ for $a_w, b_w \in [0;1]$ defined inductively by $(a_\lambda := 0, a_{wi} := a_w + a_i \cdot l_w)$ and $(b_\lambda := 1, b_{wi} := a_w + a_{i+1} \cdot l_w)$, where a_i is defined by (2) and l_w is the length of interval $I(w)$ and can be defined inductively by $l_\lambda := 1$ and $l_{wi} := l_w \cdot l_i$. Notice that we do not distinguish a_i for $i \in \Sigma_7$ from a_i for $i \in \Sigma_7^*$ with $|i| = 1$, because their value are simply the same.

Now for any $w \in \Sigma_7^*$ of length e, the intervals $I^o(w)$ are reserved exclusively for the base intervals of the requirement R_e. Suppose that $I^o(w)$ is a current base interval for R_e. We will try to find some subinterval $I^o(wi) \subset I^o(w)$, for some $i \in \{1, 3, 5\}$, as a witness interval for R_e such that any real number of this interval witnesses the requirement R_e. More precisely, the limit $\lim_{n \to \infty} \varphi_e(n)$, if exists, will not be in the interval $I(wi)$, if the sequence $(\varphi_e(n))_{n \in \mathbb{N}}$ converges c_1-monotonically. At the same time we take this witness interval $I^o(wi)$ of R_e as our current base interval for the requirement R_{e+1} and will try to find out a subinterval $I^o(wij) \subset I^o(wi)$ for $j \in \{1, 3, 5\}$ as a witness interval of R_{e+1}, and so on. This means that, if $I^o(w_1)$ and $I^o(w_2)$ are witness intervals for R_{i_1} and R_{i_2}, respectively, and $i_1 > i_2$, then $I^o(w_{i_2}) \subset I^o(w_{i_1})$. Thus, the sequence of all witness intervals for all requirements form a nested interval sequence whose common point x satisfies all requirements R_e for $e \in \mathbb{N}$.

Of course, the choices of base and witness intervals have to be corrected continuously according to the behaviors of the sequences $(\varphi_{e,s}(n))_{n \in \mathbb{N}}$ for different $s \in \mathbb{N}$. The choice of the witness intervals for the requirements $R_0, R_1, \cdots, R_{e-1}$ corresponds to a string $w \in \{1, 3, 5\}^*$ of length e. Namely, for any $e < |w|$, the interval $I^o(w \upharpoonright (e+1))$ is the base interval for R_{e+1} and at the same time the witness interval of requirement R_e. We denote by w_s our choice of such string at stage s, which seems correct at least for the s-th approximation sequences $(\varphi_{e,s}(n))_{n \in \mathbb{N}}$ instead of $(\varphi_e(n))_{n \in \mathbb{N}}$ for all $e < |w_s|$. As the limit, $w := \lim_{s \to \infty} w_s \in \Sigma^\omega$ describes a correct sequence of witness intervals $(I(w \upharpoonright e))_{e \in \mathbb{N}}$ for all requirements $(R_e)_{e \in \mathbb{N}}$, i.e., x_w is not c_1-mc. Let $x_s := a_{w_s 3}$ for any $s \in \mathbb{N}$. Then $\lim_{s \to \infty} x_s = x_w$. By 2.–3. of the Lemma 5 and the condition (4), the sequence $(x_s)_{s \in \mathbb{N}}$ converges in fact c_2-monotonically to x_w.

By the density of $\mathbb{Q}$ in $\mathbb{R}$, our hierarchy theorem for the real constants c_1 and c_2 follows immediately from Theorem 7

Corollary 8 (Hierarchy Theorem). *For any real numbers c_1, c_2 with $1 \leq c_1 < c_2$, there is a c_2-mc real number which is not c_1-mc. Therefore c_1-**MC** $\subsetneq c_2$-**MC**.*

4 ω-Monotone Computability and Weak Computability

In this section we will extend the c-monotone computability of real numbers to a more general one, namely, ω-monotone computability. We discuss the relationships between this computability and the weak computability discussed in [1] and show that they are in fact incomparable in the sense that ω-**MC** $\not\subseteq$ **WC** and **WC** $\not\subseteq \omega$-**MC**.

Definition 9. Let $h : \mathbb{N} \to \mathbb{Q}$ be any total function.

1. A sequence $(x_s)_{s \in \mathbb{N}}$ of real numbers *converges to x h-monotonically* if the sequence converges to x and the following condition holds

$$\forall n \, \forall m \geq n \, (h(n)|x - x_n| \geq |x - x_m|). \tag{5}$$

2. A real number x is called *h-monotonically computable* (h-mc, for short) if there is a computable sequence of rational numbers which converges to x h-monotonically. The class of all h-mc real numbers is denoted by h-**MC**.

3. A real number x is called ω-monotonically computable (ω-mc, for short) if it is h-monotonically computable for some recursive function h. The class of all ω-mc real numbers is denoted by ω-**MC**.

By definition, a c-mc real number is h-mc for the constant function $h \equiv c$. Therefore c-mc real numbers are also ω-mc. But we will see that not every ω-mc real number is a c-mc for some constants c.

From Theorem 4, we know that not every weakly computable real number is a c-mc real number for some constant c. Next theorem shows that even the class of ω-mc real numbers does not cover the class of weakly computable real numbers.

Theorem 10. *There is a weakly computable real number x which is not ω-mc.*

Proof. (sketch) Let $(\varphi_i)_{i \in \mathbb{N}}$ be an effective enumeration of computable functions $\varphi_i :\subseteq \mathbb{N} \to \mathbb{Q}$. $(\varphi_{i,s})_{s \in \mathbb{N}}$ is the uniformly effective approximation of φ_i. We will construct effectively a computable sequence $(x_s)_{s \in \mathbb{N}}$ of rational numbers which converges to x such that $\sum_{n \in \mathbb{N}} |x_n - x_{n+1}| \leq c$, for some $c \in \mathbb{N}$, and x satisfies, for all $e := \langle i, j \rangle \in \mathbb{N}$, the following requirements

$$R_e : \left. \begin{array}{l} \varphi_i, \varphi_j \text{ are total, } \lim_{n \to \infty} \varphi_i(n) = y_i \text{ exists and} \\ \forall n \forall m \geq n \, (\varphi_j(n) \cdot |y_i - \varphi_i(n)| \geq |y_i - \varphi_i(m)|) \end{array} \right\} \implies x \neq y_i \tag{6}$$

The strategy for satisfying a single requirement R_e is as follows. Fix a nonempty interval (a, b) as default "witness" interval and wait for $t_1, t_2 \in \mathbb{N}$ with $t_1 < t_2$ and $s \in \mathbb{N}$ such that $\varphi_{i,s}(t_1) \downarrow := \varphi_i(t_1)$ and $\varphi_{i,s}(t_2) \downarrow := \varphi_i(t_2)$

are both defined and different such that $\varphi_i(t_1), \varphi_i(t_2) \in (a, b)$ (R_e requires attention). Then define $(a'; b') := (\varphi_i(t_1) - \delta; \varphi_i(t_1) + \delta) \subset (a; b)$ for $\delta := \min\{|\varphi_i(t_1) - \varphi_i(t_2)|/(\varphi_j(t_1)+1), |a - \varphi_i(t_1)|/2, |b - \varphi_i(t_1)|/2\}$ (R_e receives attention). In this case, for any $x \in (a'; b')$, the sequence $(\varphi_i(t))_{t\in\mathbb{N}}$ does not converge φ_j-monotonically to x, since $\varphi_j(t_1) \cdot |x - \varphi_i(t_1)| \geq |x - \varphi_i(t_2)|$. Namely, any $x \in (a'; b')$ witnesses the requirement $R_{\langle i,j\rangle}$ and the interval $(a'; b')$ will be called a (correct) witness interval of $R_{\langle i,j\rangle}$. Otherwise, if there are no such s, t_1 and t_2, then the limit $y_i := \lim_{n\to\infty} \varphi_i(n)$, if it exists, cannot be in the interval (a, b). Thus the interval $(a; b)$ is itself a witness interval of the requirement $R_{\langle i,j\rangle}$. Note that, although this strategy always succeeds, we have no effective way to decide which of the above two possible approaches will be eventually applied.

To satisfy all requirements simultaneously, we will try to construct a nested rational interval sequences $((a_e; b_e))_{e\in\mathbb{N}}$ such that $(a_{e+1}; b_{e+1}) \subset (a_e; b_e)$ and the interval $(a_e; b_e)$ is a witness interval of R_e for any $e \in \mathbb{N}$. In this case, any real number $x \in \bigcap_{e\in\mathbb{N}}(a_e; b_e)$ satisfies all requirement R_e. Unfortunately, as pointed above, it is not uniformly effectively to define the witness intervals for all requirements. Therefore we construct its approximation $((a_{e,s}; b_{e,s}))_{s\in\mathbb{N}}$ instead. Namely, at any stage s, we define finite many intervals $(a_{e,s}; b_{e,s})$ for $e \leq d_s$, where d_s is some natural number such that $\lim_{s\to\infty} d_s = \infty$. These intervals should be nested in the sense that $(a_{e+1,s}; b_{e+1,s}) \subset (a_{e,s}; b_{e,s})$. As witness interval for R_e, $(a_{e,s}; b_{e,s})$ seems "correct" at least for the approximation sequences $(\varphi_{i,s}(n))_{n\in\mathbb{N}}$ ($e = \langle i,j\rangle$) instead of the sequence $(\varphi_i(n))_{n\in\mathbb{N}}$ itself. Of course we have to correct continuously our approximations according to the behaviors of $(\varphi_{i,s}(n))_{n\in\mathbb{N}}$. To this end, a priority argument is applied.

We say that R_e has a higher priority than R_{e_1} if $e < e_1$. If several requirements require attention at the same stage, then only the requirement of highest priority receives attention. If R_e receives attention at some stage, then a new witness interval for R_e will be defined according to above strategy. In this case, all (possible) old witness intervals for $R_{e'}$ may be destroyed (set to be undefined) for $e' > e$, i.e., the requirement $R_{e'}$ is injured at this stage. The witness intervals for $R_{e'}$ has to be redefined at a later stage again. On the other hand, if some correct witness interval I for R_e is defined by the above strategy, then I witnesses the requirement R_e and we don't need to do anything more for R_e unless it is destroyed again. To avoid any unnecessary multiple action, we will set the requirement R_e into the state of "satisfied" if it receives attention. If it is injuried, then set it back to the state "unsatisfied". Only the requirement of state "unsatisfied" can receive attention. In this way we guarantee that any requirement R_e can be injured at most $2^e - 1$ times and can receive attention at most 2^e times. This means also that the limit $(a_e; b_e) := \lim_{s\to\infty}(a_{e,s}; b_{e,s})$ exists and it is a correct witness interval for R_e.

At any stage s, we define x_s to be the middle point of the smallest witness interval defined at stage s. This guarantees that x_s locates in all current defined witness intervals. On the other hand, we choose the interval $(a_{e,s}; b_{e,s})$ small enough that its length not longer as, say, 2^{-2e}. Then we have also that $b_e - a_e \leq 2^{-2e}$, hence the nested interval sequence $((a_e; b_e))_{e\in\mathbb{N}}$ has a unique common point x which is in fact the limit $\lim_{s\to\infty} x_s$. The limit x satisfies all requirement R_e, hence it is not ω-mc. Furthermore, by the construction, we have $|x_s - x_{s+1}| \leq 2^{-2e}$ if R_e receives attention at stage $s + 1$, because $x_s, x_{s+1} \in (a_{e,s}; b_{e,s})$. This implies that $\sum_{s\in\mathbb{N}} |x_s - x_{s+1}| \leq \sum_{e\in\mathbb{N}} 2^{-2e} \cdot 2^e \leq 2$ since R_e receives attention at most 2^e times. This means that x is weakly computable.

By Theorem 4, any c-monotonically computable real number is also weakly computable. Next theorem shows that this is not the case any more for ω-mc real number.

Theorem 11. *There is an ω-mc real number which is not weakly computable.*

Proof. (sketch) Let $(\varphi_i)_{i\in\mathbb{N}}$ be an effective enumeration of computable partial functions $\varphi_i :\subseteq \mathbb{N} \to \mathbb{Q}$ and $(\varphi_{i,s})_{s\in\mathbb{N}}$ its uniform approximation. We will define a computable sequence $(x_s)_{s\in\mathbb{N}}$ of rational numbers and a recursive function $h : \mathbb{N} \to \mathbb{N}$ such that the limit $x := \lim_{s\to\infty} x_s$ exists and satisfies, for all $i \in \mathbb{N}$, the following requirements

$$N: \quad (x_s)_{s\in\mathbb{N}} \text{ converges to } x \text{ } h\text{-monotonuically, and}$$

$$R_i: \quad \sum_{s\in\mathbb{N}} |\varphi_i(s) - \varphi_i(s+1)| \leq i \text{ \& } y_i := \lim_{s\in\mathbb{N}} \varphi_i(s) \Longrightarrow y_i \neq x.$$

Since every sequence $(\varphi_i(s))_{s\in\mathbb{N}}$ have infinitely many different indices i by the Padding Lemma, the real number x can't be weakly computable, if all requirements R_i are satisfied. The strategy to satisfy a single requirement R_i is quite straightforward. We take the interval $[0; 1]$ as the base interval and define at first $x_0 := 1/7$. If there is some m_1 and s_1 such that $\varphi_{i,s_1}(m_1) \leq 3/7$, then define $x_{s_1} := 5/7$. Again, at stage $s_2 > s_1$, if there is an $m_2 > m_1$ such that $\varphi_{i,s_2}(m_2) \geq 4/7$, then define $x_{s_2} := 1/7$, and so on. Otherwise, x_{s+1} takes simply the same value as x_s. Now, if $\sum_{s\in\mathbb{N}} |\varphi_i(s) - \varphi_i(s+1)| \leq i$, then x_s can be redefined at most $7 \cdot i$ times (suppose that $\varphi_i(s) \in [0; 1]$) and the last value $x := \lim_{s\to\infty} x_s$ has at least the distant $1/7$ from $y_i := \lim_{s\to\infty} \varphi_i(s)$, if this limit exists.

Of coures the sequence $(x_s)_{s\in\mathbb{N}}$ defined in this way does not satisfy the requirement N, because some elements x_s of the sequence are simply equal to the limit x which destroies the monotone convergence. Fourtunately, this problem disappears if we implement a stragegy to satisfy all requirement R_i simultanuously. In this case, we need an interval tree $I : \Sigma_7^* \to \mathbb{I}$ which is defined by $I(w) := [a_w; b_w]$, where $a_w := \sum_{i<|w|} w[i] \cdot 7^{-(i+1)}$ and $b_w := a_w + 7^{-|w|}$, for all $w \in \Sigma_7^*$. The interval $I(w)$ for $w \in \Sigma_7^*$ of length i is reserved for the requirement R_i. We will define $x_s := a_{w1}$ or $x_s := a_{w5}$ in order to guarantee that the limit $y_i := \lim_{s\to\infty} \varphi_i(s)$, if exists, has at least the distant $7^{-(i+1)}$ from x_s, hence the limit $\lim_{s\to\infty} x_s$ is also different from y_i. On the other hand, this construction guarantees also that the element x_s defined at stage s for satisfying R_i has at least a distant $7^{-(i+2)}$ from the limit $\lim_{s\to\infty} x_s$. Therefore the sequence $(x_s)_{s\in\mathbb{N}}$ converges h-monotonically for $h(s) := 7^{(i+2)}$.

Since every c-mc real number is weakly computable, the Theorem 11 implies that there is an ω-mc real number which is not c-mc for any constant c, i.e., $*$-MC $\subsetneq$ ω-MC. Besides, Theorem 11 and Theorem 10 implies directly the following corollary.

Corollary 12. *The classes of weakly computable and ω-monotonically computable real numbers are incomparable, i.e.,* WC $\not\subseteq$ ω-MC & ω-MC $\not\subseteq$ WC.

References

1. K. Ambos-Spies, K. Weihrauch and X. Zheng Weakly computable real numbers. *J. of Complexity.* 16(2000), 676–690. 634, 641
2. C. Calude. A characterization of c.e. random reals. CDMTCS Research Report Series 095, March 1999. 634
3. C. Calude, P. Hertling, B. Khoussainov, and Y. Wang, Recursive enumerable reals and Chaitin's Ω-number, in *STACS'98*, pp596–606. 634
4. C. Calude and P. Hertling Computable approximations of reals: An information-theoretic analysis. *Fundamenta Informaticae* 33(1998), 105–120. 635
5. G. J. Chaitin A theory of program size formally identical to information theory, *J. of ACM.*, 22(1975), 329–340. 634
6. P. Martin-Löf The definition of random sequences, *Information and Control*, 9(1966), 602–619. 634
7. R. Rettinger, X. Zheng, R. Gengler and B. von Braunmühl Monotonically computable real numbers. *DMTCS'01*, July 2-6, 2001, Constanţa, Romania. 635, 637, 638
8. H. G. Rice Recursive real numbers, *Proc. Amer. Math. Soc.* 5(1954), 784–791. 633
9. R. M. Robinson Review of "R. Peter: 'Rekursive Funktionen', Akad. Kiado. Budapest, 1951", *J. Symb. Logic* 16(1951), 280. 633
10. T. A. Slaman Randomness and recursive enumerability, preprint, 1999. 634
11. R. Soare *Recursively Enumerable Sets and Degrees*, Springer-Verlag, Berlin, Heidelberg, 1987.
12. R. Soare Recursion theory and Dedekind cuts, *Trans, Amer. Math. Soc.* 140(1969), 271–294. 634
13. R. Soare Cohesive sets and recursively enumerable Dedekind cuts, *Pacific J. of Math.* 31(1969), no.1, 215–231. 634
14. R. Solovay. Draft of a paper (or series of papers) on Chaitin's work ... done for the most part during the period of Sept. –Dec. 1975, unpublished manuscript, IBM Thomas J. Watson Research Center, Yorktoen Heights, New York, May 1975, 215pp. 634
15. E. Specker Nicht konstruktive beweisbare Sätze der Analysis, *J. Symbolic Logic* 14(1949), 145–158 634
16. A. M. Turing. On computable number, with an application to the "Entscheidungsproblem". *Proceeding of the London Mathematical Society*, 43(1936), no.2, 230–265. 633
17. K. Weihrauch. *An Introduction to Computable Analysis*. Texts in Theoretical Computer Science, Springer-Verlag, Heidelberg 2000. 633
18. K. Weihrauch & X. Zheng A finite hierarchy of the recursively enumerable real numbers, *MFCS'98* Brno, Czech Republic, August 1998, pp798–806. 634, 635, 637

On the Equational Definition of the Least Prefixed Point

Luigi Santocanale*

BRICS**, University of Aarhus
luigis@brics.dk

Abstract. We show how to axiomatize by equations the least prefixed point of an order preserving function and discuss the domain of application of the proposed method. Thus, we generalize the well known equational axiomatization of Propositional Dynamic Logic to a complete equational axiomatization of the Boolean Modal μ-Calculus. We show on the other hand that the existence of a term which does not preserve the order is an essential condition for the least prefixed point to be definable by equations.

Introduction

The least and the greatest fixed points of an order preserving function have shown to be basic ingredients of logics of programs. In several cases adding a fixed point constructor to a logic increases its expressive power without turning its model checking into an intractable problem. This observation is among the motivations behind the intensive study of equational properties of fixed points, leading to the framework of iteration theories and their models [6,7]. However, if we focus on canonical fixed points, we observe that a natural way to axiomatize them is by means of equational implications. For example the least fixed point $g(y)$ of a function $f(x, y)$, order preserving in the variable x, has the natural axiomatization

$$f(g(y), y) \leq g(y) \tag{1}$$

$$f(z, y) \leq z \quad \Rightarrow \quad g(y) \leq z \tag{2}$$

stating that $g(y)$ is the least *prefixed* point of $f(x, y)$. The second clause, usually called Park induction rule [9,17], is an equational implication if the order relation is expressible by equations between terms. The main aim of this paper is to understand when this equational implication can be eliminated in favor of an axiomatization that uses only equations. Equivalently we want to understand when equational logic is a complete tool to reason about programs.

---

* Current affiliation: Department of Computer Science, University of Calgary.
** Basic Research in Computer Science,
Centre of the Danish National Research Foundation.

J. Sgall, A. Pultr, and P. Kolman (Eds.): MFCS 2001, LNCS 2136, pp. 645–656, 2001.

We present here a method to substitute Park induction rule by an equation. This method applies as soon as a host theory has an order preserving binary term $x \otimes y$ with a unit 1 and a right adjoint $x \multimap z$. In order to apply the method we should be interested in axiomatizing the least prefixed point of $f(x, \boldsymbol{y}) \otimes z$ as well, and not just that of $f(x, \boldsymbol{y})$. Then, for every theory which extends one among: the theory of groups, the theory of Boolean algebras, the theory of Heyting algebras, the theory of Girard quantales or of implicative quantales (that is, Classical or Intuitionistic Linear Logic), the least fixed point of an order preserving function turns out to be definable by equations.

As a corollary of our considerations we show that the well known equational axiomatization of Propositional Dynamic Logic [13,19,23] can be generalized to a complete axiomatization of the Boolean Modal μ-Calculus. We provide a simple list of equations forming an equational base for the theory and develop some considerations from the point of view of universal algebra: we argue that it is not possible to find finite bases (i.e. both a finite signature and a finite equational base) for the μ-Calculus. The fact that the Boolean Modal μ-Calculus has an equational axiomatization has the consequence that its algebraic models are very well behaved: they form a variety of algebras, in the usual sense of universal algebra [10], or an exact category [3]. The main characteristic of these models is that quotients are in a bijective correspondence with congruences. From a logical point of view, what arises as a surprising fact from our considerations is that the μ-Calculus, a powerful logic with quantifier-like operators μ, ν, can be reduced to the realm of equational logic. This implies that Kozen's axiomatization of the μ-Calculus [12] can be given purely in term of axiom schemes and does not require additional inference rules for those two operators.

In the last section of this paper we give a partial converse to our definability by equations theorem. The main feature of a right adjoint to a covariant binary operator is its contravariancy in one of its two variables, that is, it reverses the order. We consider a theory extending the theory of bounded lattices and show that if the theory contains only order preserving operators, then the least prefixed point of an order preserving function cannot be axiomatized by equations, unless an equation stating that the least prefixed point is at a finite distance from the bottom of the lattice holds in the theory.

1   Equationally Definable Least Prefixed Points

A signature Ω is a collection of pairwise disjoint sets $\{\Omega_n\}_{n \geq 0}$. By writing $f \in \Omega_n$ we mean that f is a function symbol of arity n from the signature. Terms over Ω are generated from a countable set of variables $\{x_1, \ldots, x_n, \ldots\}$ by substitution of previously defined terms as argument of the function symbols. By $\mathcal{T}(\Omega, X)$ we shall denote the set of terms over Ω whose free variables are contained in the finite subset of variables X.

We shall restrict our attention to Horn theories $\mathbb{T}$ of the form $\langle \Omega, I \rangle$ where Ω is a signature and I is a set of equational implications of the form

$$\bigwedge_{i=1,\ldots,k} s_i = t_i \;\Rightarrow\; s_0 = t_0 \,,$$

with $s_i, t_i \in T(\Omega, X)$, X being some finite subset of variables. We allow k to be equal to 0 in which case such an implication is simply an equation. The notions of a model of the theory and of a homomorphism of models are standard from model theory and we let $\mathcal{M}(\mathbb{T})$ denote the category of models of $\mathbb{T}$. By $\mathcal{I}(\mathbb{T})$ we shall denote the set of equational implications holding in every model of $\mathbb{T}$ and by $\mathcal{E}(\mathbb{T})$ the set of equations holding in every model of $\mathbb{T}$. We say that an equational implication (resp. an equation) holds if it is in $\mathcal{I}(\mathbb{T})$ (resp. in $\mathcal{E}(\mathbb{T})$). We shall use the notations $x, y, \ldots l, m, \ldots$ to range over vectors of variables and of elements respectively.

Definition 1. *A theory* $\mathbb{T} = \langle \Omega, I \rangle$ *is* equational *if every implication in I is an equation and it is* algebraic *if $\mathcal{M}(\mathbb{T})$ is equivalent to a category $\mathcal{M}(\mathbb{T}')$ for an equational theory $\mathbb{T}'$.*

Definition 2. *A theory* $\mathbb{T}$ *is* ordered *if it comes with two terms $f(x,y)$ and $g(x,y)$ such that the relation $\leq$, defined as $l \leq m$ if and only if $f(l,m) = g(l,m)$, is a partial order in every model of the theory.*

The theory of lattices is the main example of an ordered theory. If the theory $\mathbb{T}$ is ordered, the equations of the theory are determined by relations of the form $s \leq t$ and vice-versa so that we shall refer to those relations simply as equations. We shall say that a term $f(x, \boldsymbol{z})$ is covariant in x – or that x occurs positive in it – if the implication

$$x \leq y \Rightarrow f(x, \boldsymbol{z}) \leq f(y, \boldsymbol{z})$$

holds. Similarly, we shall say that a term $f(x, \boldsymbol{z})$ is contravariant in x – or that x occurs negative in it – if the implication obtained from the previous one by exchanging x and y on the left holds. We shall say that a term $g(\boldsymbol{y})$ is the least prefixed point of a term $f(x, \boldsymbol{y})$, covariant in x, if an instance of the fixed point equation (1) at page 645 belongs to $\mathcal{E}(\mathbb{T})$ and an instance of the Park induction rule (2) belongs to $\mathcal{I}(\mathbb{T})$.

Theorem 3. *Suppose that in the ordered theory* $\mathbb{T}$ *we can find a binary term $x \otimes y$, covariant in x and y, with a right adjoint $x \multimap z$, that is, such that the relation*

$$x \otimes y \leq z \; \text{iff} \; y \leq x \multimap z$$

holds. Let $f(x, \boldsymbol{y})$ be a term covariant in x. A term $g(\boldsymbol{y}, z)$, covariant in z, is the least prefixed point of the term $f(x, \boldsymbol{y}) \otimes z$ - parameterized in $\boldsymbol{y}$ and z- if and only if the equations

$$f(g(\boldsymbol{y}, z), \boldsymbol{y}) \otimes z \leq g(\boldsymbol{y}, z) \tag{3}$$
$$g(\boldsymbol{y}, f(x, \boldsymbol{y}) \multimap x) \leq x \tag{4}$$

hold. *In this case, if* $x \otimes 1 = x$ *holds for some term* 1, *then* $g(\boldsymbol{y}, 1)$ *is the least prefixed point of* $f(x, \boldsymbol{y})$.

Proof. Suppose that for each $\boldsymbol{y}$ and z $g(\boldsymbol{y}, z)$ is the desired least prefixed point. Then equation (3) is simply stating that $g(\boldsymbol{y}, z)$ is a prefixed point and equation (4) follows from the relation $f(x, \boldsymbol{y}) \otimes (f(x, \boldsymbol{y}) \multimap x) \leq x$ and the Park induction rule. On the other hand, if the above relations hold, then $g(\boldsymbol{y}, z)$ is a prefixed point and if $f(x, \boldsymbol{y}) \otimes z \leq x$, then $z \leq f(x, \boldsymbol{y}) \multimap x$ and

$$g(\boldsymbol{y}, z) \leq g(\boldsymbol{y}, f(x, \boldsymbol{y}) \multimap x) \leq x.$$

The last claim follows since $f(x, \boldsymbol{y}) \leq x$ if and only if $f(x, \boldsymbol{y}) \otimes 1 \leq x$.   □

Remark 4. Observe that the adjointness condition is equational since it is equivalent to the two equations $x \otimes (x \multimap y) \leq y$ and $y \leq x \multimap (x \otimes y)$. The equational characterization of the transitive closure in Action Logic [18] has suggested to use adjoints (or residuation) to axiomatize the least prefixed point. We can use theorem 3 to provide an alternative axiomatization of the transitive closure as well as a general method to axiomatize the least prefixed point. This method has an "inconvenience" which we exemplify now. Suppose that we want to axiomatize the operator $p \operatorname{Until} q$ of temporal logic, which is the least fixed point of $h(x, q, p) = q \vee (p \wedge \Diamond x)$. It might be asked whether it is possible to do this without introducing an additional operator $p \operatorname{Until}_z q$, axiomatized as the least prefixed point of $h(x, q, p) \wedge z$, and by imposing the equation $p \operatorname{Until} q = p \operatorname{Until}_\top q$. The answer is positive since in this case it is possible to axiomatize the least prefixed point of $p \wedge \Diamond(q \vee x)$ and deduce the existence of a least prefixed point of $q \vee (p \wedge \Diamond x)$. However this idea is not generally available and similar questions remain open.

2 μ-Theories, μ-Algebras and the Modal μ-Calculus

The goal of this section is to present a complete equational axiomatization of the Boolean Modal μ-Calculus, illustrating in this way the domain of application of theorem 3. We do this by introducing μ-theories of algebraic theories, a syntactic counterpart of the μ-algebras defined in [15]. Even if the role of the syntax is explicit also there, we are interested here in discussing the possibility that some operator of theory is not order preserving; moreover, we need to represent μ-terms as terms constructible from a signature by substitution.

Henceforth, we shall fix an algebraic theory $\mathbb{T} = \langle \Omega, E \rangle$ and make the following assumptions on it. We shall assume that $\mathbb{T}$ extends the theory of bounded lattices, which means that it comes with given terms $\bot, x \vee y, \top, x \wedge y$ so that the group of equations (1) in figure 1 belong to $\mathcal{E}(\mathbb{T})$. Such a theory is then ordered by the lattice theoretic order. We shall also assume that each function symbol $f \in \Omega_n$ induces an order preserving function

$$f : L^l \times (L^{op})^k \longrightarrow L,$$

for every model L of $\mathbb{T}$, where $l + k = n$: f is order preserving in its first l variables and order reversing in its second set of variables. We shall write in this case $f \in \Omega_{l,k}$.

As if we were working with a two sorted theory, we construct now a polarized signature $\{\mu\Omega_{l,k}\}_{l \geq 0, k \geq 0}$ and sets of polarized terms $\mu\mathcal{T}(X, Y)$ representing μ-terms. A polarized term comes with two disjoint sets of free variables $X, Y \subseteq \{x_1, \ldots, x_n, \ldots\}$ reminding us which variable occurs positive and which variable occurs negative. In forming terms, we always suppose that X, Y are disjoint and apply rules whenever this property is preserved.

Definition 5. *The polarized signature $\mu\Omega$ and the sets of polarized terms $\mu\mathcal{T}(X, Y)$ are defined by induction as follows:*

1. *If $f \in \Omega_{l,k}$, then $f \in \mu\Omega_{l,k}$.*
2. *For each $x \in X$, $x \in \mu\mathcal{T}(X, Y)$.*
3. *If $f \in \mu\Omega_{l,k}$, $t_i \in \mu\mathcal{T}(X, Y)$ for $i = 1, \ldots, l$ and $t_i \in \mu\mathcal{T}(Y, X)$ for $i = l+1, \ldots, n$, then $f(t_1, \ldots, t_n) \in \mu\mathcal{T}(X, Y)$.*
4. *If $t \in \mu\mathcal{T}(X, Y)$ and $x \in X$, then $\mu_x.t, \nu_x.t \in \mu\Omega_{l,k}$, where l is the cardinality of $X \setminus \{x\}$ and k is the cardinality of Y.*

The single sorted signature $\mu\Omega$ is defined by letting $\mu\Omega_n$ be the disjoint union of the $\mu\Omega_{l,k}$ with $l + k = n$.

Definition 6. *The μ-theory of $\mathbb{T}$ is the Horn theory $\langle \mu\Omega, I \rangle$ where I contains the equations of $\mathbb{T}$ and for each $t \in \mu\mathcal{T}(X, Y)$ and $x \in X$ the equations*

$$t[\mu_x.t(\boldsymbol{z}, \boldsymbol{y})/x, \boldsymbol{z}, \boldsymbol{y}] \leq \mu_x.t(\boldsymbol{z}, \boldsymbol{y})$$
$$t[\nu_x.t(\boldsymbol{z}, \boldsymbol{y})/x, \boldsymbol{z}, \boldsymbol{y}] \leq \mu_x.t(\boldsymbol{z}, \boldsymbol{y}),$$

as well as the Park induction rules

$$t[x, \boldsymbol{z}, \boldsymbol{y}] \leq x \Rightarrow \mu_x.t(\boldsymbol{z}, \boldsymbol{y}) \leq x$$
$$x \leq t[x, \boldsymbol{z}, \boldsymbol{y}] \Rightarrow x \leq \nu_x.t(\boldsymbol{z}, \boldsymbol{y}),$$

where $\boldsymbol{z}$ is the sorted list of variables in $X \setminus \{x\}$ and $\boldsymbol{y}$ is the sorted list of variables in Y.

The reader should be aware that the one given above is just a representation of μ-terms. For example, given usual μ-terms $\mu_x.t_1$ and t_2 with the only free variable $y \neq x$, the equation $(\mu_x.t_1)[t_2/y] = \mu_x.(t_1[t_2/y])$ is an actual equality, where if we are working with their representation this is only a derivable equality of the theory. The above discussion has shown that ideas related to μ-calculi can be casted in the framework of Horn theories. In the following we shall make informal use of μ-terms, for example by writing $\mu_x.t(x, \boldsymbol{y})$ even if, in our representation of μ-terms, the variable x is not strictly speaking a subterm of this term.

Example 7. If $\mathbb{T}$ is the theory of lattices, then the models of the μ-theory of $\mathbb{T}$ are the μ-lattices studied in [21,22].

Example 8. A theory $\mathbb{T}$ is said to be distributive if $\mathcal{E}(\mathbb{T})$ contains equation (2) of figure 1. If $\mathbb{T}$ is the theory of distributive lattices, then a model of the μ-theory is simply a distributive lattice. This follows from the fact that the equations

$$\mu_x.(a \vee (b \wedge x)) = a\,,$$
$$\mu_x.(a \wedge (b \vee x)) = a \wedge b\,,$$

and their dual hold in any μ-lattice and that any term $f(x)$ of the theory of distributive lattices is equivalent to one of the form $a \vee (b \wedge x)$. This is essentially Kozen's observation [12] that every formula of the μ-Calculus is equivalent to a guarded one and tells that in order to obtain an interesting μ-theory a distributive theory $\mathbb{T}$ should properly extend the theory of distributive lattices.

Example 9. Let $\mathbb{T}$ be the distributive theory $\langle \Omega, E \rangle$ where Ω contains the lattice operators $\top, \wedge, \bot, \vee$ as well as the modal operators $\{\langle a \rangle, [a]\}_{a \in A}$ and E contains also the axiom schemes (3) of figure 1. Following [1,11], we call a model of the μ-theory of $\mathbb{T}$ a modal μ-algebra. We say that a modal μ-algebra is a Boolean modal μ-algebra if its underlying distributive lattice is a boolean algebra. Boolean modal μ-algebras are the models of the μ-theory of $\mathbb{T}'$, where $\mathbb{T}'$ is obtained from the previous $\mathbb{T}$ by adding to Ω a complement operator $\neg$ and the axioms (4) of figure 1.

Boolean modal μ-algebras are the algebraic models of the Boolean Modal μ-Calculus. According to theorem 3 the μ-Calculus has the equational axiomatization of figure 1: we let $x \otimes y$ be $x \wedge y$, 1 be $\top$, $x \multimap y$ be $\neg x \vee y$, and use dual operators to axiomatize the greatest postfixed point. This axiomatization is complete with respect to Kripke structures because it is equivalent to Kozen's axiomatization [12] and because of Walukiewicz theorem [24]. We add some comments on the group of axioms (5). The last two axioms are the equations (3) and (4) of theorem 3. We need the second axiom in order to turn the least prefixed point of $f(x, y) \wedge z$ into a term covariant in z: if we let $g(y, z)$ be this least prefixed point, then this axiom has the form $g(y, z_1) \leq g(y, z_1 \vee z_2)$. Observe that we cannot argue that $\mu_x.t_1 \leq \mu_x.t_2$ from $t_1 \leq t_2$ since this is not a law of equational logic. This inference will be deducible as soon as we deduce the Park induction rule, along the lines of 3. For a similar reason we cannot infer $\mu_x.f(x, y) = \mu_x.(f(x, y) \wedge \top)$ from $f(x, y) = f(x, y) \wedge \top$ and this explains why we need the first axiom.

Definition 10. *A theory $\mathbb{T}$ is* finitely based *if there exists a theory $\mathbb{T}' = \langle \Omega, I \rangle$ such that $\mathcal{M}(\mathbb{T})$ and $\mathcal{M}(\mathbb{T}')$ are equivalent concrete categories and moreover the disjoint sum of the Ω_n is a finite set.*

As a consequence of the strictness of the alternation hierarchy of the μ-Calculus, the theory of modal μ-algebras is not finitely based. According to [16], we let $\Sigma_0 = \Pi_0$ be the closure under substitution of modal and boolean operators; we let Σ_{n+1} be the closure of $\Sigma_n \cup \Pi_n$ under substitution and the operation of taking the least prefixed point, and let Π_{n+1} be the closure of the same set

(1) Lattice axioms:

$$x \vee \bot = x \qquad\qquad x \wedge \top = x$$
$$x \vee y = y \vee x \qquad\qquad x \wedge y = y \wedge x$$
$$(x \vee y) \vee z = x \vee (y \vee z) \qquad\qquad (x \wedge y) \wedge z = x \wedge (y \wedge z)$$
$$x \vee x = x \qquad\qquad x \wedge x = x$$
$$x \vee (x \wedge y) = x \qquad\qquad x \wedge (x \vee y) = x$$

(2) Distributive lattice axiom:

$$x \wedge (y \vee z) = (x \wedge y) \vee (x \wedge z)$$

(3) Modal axiom schemes:

$$\langle a \rangle \bot = \bot \qquad\qquad [a]\top = \top$$
$$\langle a \rangle x \vee \langle a \rangle y = \langle a \rangle (x \vee y) \qquad\qquad [a]a \wedge [a]y = [a](x \wedge y)$$
$$\langle a \rangle x \wedge [a]y \le \langle a \rangle (x \wedge y) \qquad\qquad [a](x \vee y) \le [a]x \vee \langle a \rangle y$$

(4) Boolean algebra axioms:

$$\neg x \vee x = \top \qquad\qquad \neg x \wedge x = \bot$$

(5) Fixed point axiom schemes:

$$\mu_x . f(x, \boldsymbol{y}) = \mu_x . (f(x, \boldsymbol{y}) \wedge \top)$$
$$\mu_x . (f(x, \boldsymbol{y}) \wedge z_1) \le \mu_x . (f(x, \boldsymbol{y}) \wedge (z_1 \vee z_2))$$
$$f(\mu_x . (f(x, \boldsymbol{y}) \wedge z), \boldsymbol{y}) \wedge z \le \mu_x . (f(x, \boldsymbol{y}) \wedge z)$$
$$\mu_x . (f(x, \boldsymbol{y}) \wedge (\neg f(u, \boldsymbol{y}) \vee u)) \le u$$

$$\nu_x . f(x, \boldsymbol{y}) = \nu_x . (f(x, \boldsymbol{y}) \vee \bot)$$
$$\nu_x . (f(x, \boldsymbol{y}) \vee z_1) \le \nu_x . (f(x, \boldsymbol{y}) \vee (z_1 \vee z_2))$$
$$\nu_x . (f(x, \boldsymbol{y}) \vee z) \le f(\nu_x . (f(x, \boldsymbol{y}) \vee z), \boldsymbol{y}) \vee z$$
$$u \le \nu_x . (f(x, \boldsymbol{y}) \vee (\neg f(u, \boldsymbol{y}) \wedge u))$$

Fig. 1. (1)–(5) Boolean modal μ-algebra axioms

under substitution and the greatest postfixed point operation. The inclusions $\Sigma_n \cup \Pi_n \subseteq \Sigma_{n+1} \cap \Pi_{n+1}$ are shown to be strict in the semantics: for each $n \ge 0$, we can find a boolean modal μ-algebra L and a term $t \in \Sigma_{n+1} \cap \Pi_{n+1}$ the interpretation of which, as a function of its free variables, is different from the interpretation of every term $t' \in \Sigma_n \cup \Pi_n$ [2,8,14]. Suppose that we can find a finite signature Ω that is equivalent to the infinite signature $\mu\Omega$. Henceforth, we can find an integer $k \ge 0$ such that every function symbol of Ω is expressible as a term from $\Sigma_k \cap \Pi_k$. Similarly, every μ-term is equivalent to a term built up from the signature Ω. It follows that every term is equivalent to a term from $\Sigma_k \cap \Pi_k$ since this class is closed under substitution. For related considerations see also [4].

Definition 11. *An algebraic theory* $\mathbf{T}$ *is equationally finitely based if there exists a theory* $\mathbf{T}' = \langle \Omega, E \rangle$ *such that* $\mathcal{M}(\mathbf{T})$ *and* $\mathcal{M}(\mathbf{T}')$ *are equivalent concrete categories and* E *is a finite set of equations.*

If a theory is not finitely based, then it might be expected that it is not equationally finitely based. We can argue that the theory of Boolean modal μ-

algebras is not equationally finitely based as follows. Suppose that a presentation of the theory, say $\langle \mu\Omega, E \rangle$ where E is a finite set of equations, is available. Let $f_1, \ldots, f_n, \ldots$ be an enumeration of the symbols of $\mu\Omega$ and find an integer $k \geq 0$ such that all the equations in E involve terms constructed from the function symbols $f_1, \ldots, f_k$. Choose a term t in the class Π_m such that $\mu_x.t$, as a function symbol of the signature $\mu\Omega$, has the form f_l with $l > k$ and moreover $\mu_x.t$, as a term of the class Σ_{m+1}, is hard: this means that we can find a model L such that the interpretation of $\mu_x.t$ is different from the interpretation of any term in $\Sigma_m \cup \Pi_m$. We transform L into a model L' of $\langle \mu\Omega, E \rangle$ such that at least one of the relations axiomatizing f_l as a canonical fixed point of t – the relations of group (5) in figure 1 – does not hold. The underlying set of L' is the same as the one of L. The interpretation of the function symbols is the same as in L apart from the fact that we interpret $\mu_x.t$ as the greatest postfixed point of t. Since the interpretation of $f_1, \ldots, f_k$ is as in L, all the equations from E hold in L'. If all the four axioms (5) in figure 1 for $\mu_x.t$ hold, then $\mu_x.t$ is determined as the least prefixed point of t and the new model L' coincides the old model L. But this means that the interpretation of $\mu_x.t$ in L is equal to the interpretation of the term $\nu_x.t$ which belongs to the class Π_m, since Π_m is closed under the greatest postfixed point. This contradicts the choice of the term t.

The argument can be generalized so that it does not depend on the choice of the signature $\mu\Omega$. Observe that we have done essential use of the strictness of the alternation hierarchy of the μ-Calculus while it can be conjectured that analogous results of non existence of finite bases hold for the theory determined by the μ-terms in the class Σ_1.

3   Equationally Undefinable Least Prefixed Points

In this section we prove a partial converse to theorem 3. There we have implicitly used the fact that the term $x \multimap y$ is not order preserving in x.

Definition 12. *An ordered theory* $\mathbb{T}$ *is* positive *if for each term* $f \in T(\Omega, x)$ *the implication*

$$\bigwedge_{i=1}^{n} x_i \leq y_i \Rightarrow f(x) \leq f(y)$$

belongs to $\mathcal{I}(\mathbb{T})$.

In the following we let $f^{(0)}(\bot, y) = \bot$ and $f^{(n+1)}(\bot, y) = f(f^{(n)}(\bot, y), y)$.

Theorem 13. *Let* $\mathbb{T}$ *be a positive algebraic theory which extends the theory of bounded lattices. Suppose that we can find a pair of terms* $f(x, y)$ *and* $g(y)$ *such that, for each* $\mathbb{T}$-*model* L *and* $m \in L^y$, $g(m)$ *is the least prefixed point of the order preserving function* $x \longmapsto f(x, m)$. *Then an equation of the form*

$$g(y) = f^{(n)}(\bot, y)$$

holds.

Example 14. Let $\mathbb{T}$ be the theory of distributive lattices and consider its μ-theory, as discussed in example 8. This theory is positive and moreover it is algebraic since we have seen that its category of models is exactly the category of distributive lattices. It is easily seen that the equation

$$\mu_x . f(x, \boldsymbol{y}) = f(\bot, \boldsymbol{y})$$

holds in 2, the distributive lattice with two elements. We can argue that this equation holds in every distributive lattice since such a lattice is a sublattice of a power of 2.

From now on our goal is to prove theorem 13. We shall consider a positive Horn theory $\mathbb{T} = \langle \Omega, I \rangle$ which extends the theory of lattices and which comes with terms $f(x, \boldsymbol{y})$ and $g(\boldsymbol{y})$, so that $g(\boldsymbol{y})$ is the least prefixed point of $f(x, \boldsymbol{y})$.

Definition 15. *Let L be a $\mathbb{T}$-model, a congruence on L is an equivalence relation $\sim \subseteq L \times L$ such that, for all $f \in T(\Omega, X)$ and $\boldsymbol{l}, \boldsymbol{m} \in L^X$, if $l(x) \sim m(x)$ for all $x \in X$, then also $f(\boldsymbol{l}) \sim f(\boldsymbol{m})$.*

Given a $\mathbb{T}$-model L and a congruence $\sim$ on L, we can construct the quotient $L/\sim$. Its elements are equivalence classes under $\sim$; if $\boldsymbol{l} \in L^X$ we shall denote by $[\boldsymbol{l}]$ the vector of classes of $\boldsymbol{l}$ under the equivalence relation $\sim$. The quotient $L/\sim$ is a $\mathbb{T}$-algebra, meaning that we can define $f([\boldsymbol{l}]) = [f(\boldsymbol{l})]$ so that all the equations of $\mathcal{E}(\mathbb{T})$ holds. Observe that $L/\sim$ need not to be a $\mathbb{T}$-model, since some equational implication in $\mathcal{I}(\mathbb{T})$ could be false in $L/\sim$. However, it is a poset where $[l] \leq [m]$ if and only if $l \vee m \sim m$.

Definition 16. *An order congruence on L is a preorder $\preceq \subseteq L \times L$ which extends the partial order of L and such that for all $f \in T(\Omega, X)$, $\boldsymbol{l}, \boldsymbol{m} \in L^X$, if $\boldsymbol{l} \preceq \boldsymbol{m}$, then $f(\boldsymbol{l}) \preceq f(\boldsymbol{m})$.*

The above concept, which we will need for calculations, was introduced in [5] for varieties of ordered algebras. The following lemma shows that if the theory $\mathbb{T}$ extends the theory of lattices, then the point of view of varieties of ordered algebras coincides with the usual one of universal algebra.

Lemma 17. *There is a bijection between congruences and order congruences on a $\mathbb{T}$-model L.*

Proof. The bijection maps an order congruence to its antisymmetric closure. On the other hand, if $\sim$ is a congruence on L, then we obtain a preorder $\preceq$ by saying that $l \preceq m$ if and only if $l \vee m \sim m$. This is an order congruence: let $f \in T(\Omega, X)$ and $\boldsymbol{l}, \boldsymbol{m} \in L^X$, and suppose that $\boldsymbol{l} \vee \boldsymbol{m} \sim \boldsymbol{m}$; by usual lattice theoretic reasoning it follows that $\boldsymbol{l} \wedge \boldsymbol{m} \sim \boldsymbol{l}$, so that

$$f(\boldsymbol{l}) \vee f(\boldsymbol{m}) \quad \sim \quad f(\boldsymbol{l} \wedge \boldsymbol{m}) \vee f(\boldsymbol{m}) \quad = \quad f(\boldsymbol{m}),$$

using the fact that f is order preserving. □

We shall freely use both representations. Even if $g(l)$ is the least prefixed point of the correspondence $x \longmapsto f(x, l)$, $g([l])$ need not to be the least prefixed point of the order preserving function $[x] \longmapsto f([x, l])$; this would certainly be true if $\mathbb{T}$ were algebraic, by Birkhoff's theorem on varieties of algebras. We are motivated to give the following definition:

Definition 18. *Let L be a model of $\mathbb{T}$ and let $\sim$ be a congruence on L. We say that $\sim$ is (f, g)-effective if for every $l \in L^{y}$, $[g(l)]$ is the least prefixed point of the order preserving function*

$$[x] \longmapsto [f(x, l)].$$

The following is an equivalent formulation of theorem 13.

Theorem 19. *Suppose we can find a model L of $\mathbb{T}$, $l \in L^{y}$ such that the chain*

$$\bot \le f^{(1)}(\bot, l) \le \ldots \le f^{(n)}(\bot, l) \le \ldots$$

is strict. Then there exists a model L^{ω} of $\mathbb{T}$ and a congruence on L^{ω} which is not (f, g)-effective. Hence the theory $\mathbb{T}$ is not algebraic.

Proof. Consider such a $\mathbb{T}$-model L and form its power $L^{\mathbb{N}}$. Being the product $\prod_{n \ge 0} L$, $L^{\mathbb{N}}$ is a $\mathbb{T}$-model. Its elements are sequences of the form $\{l_n\}_{n \ge 0}$, where $l_n \in L$. If $l \in L^{\mathbb{N}}$, then l_n shall denote the n-th element of the sequence. Hence, for a term $f \in T(\Omega, X)$, $f(l)$ is calculated by the formula $f(l)_n = f(l_n)$. Proof of the following two lemmas can be found in [5].

Lemma 20. *Let $L^{\omega} \subseteq L^{\mathbb{N}}$ be the subset of weakly increasing sequences $\{l_n\}$, i.e. such that if $n \le m$ then $l_n \le l_m$: this set is a sub-$\mathbb{T}$-model of $L^{\mathbb{N}}$.*

Lemma 21. *Define the relation $\preceq \subseteq L^{\omega} \times L^{\omega}$ by saying that $l \preceq m$ if and only if for all $n \ge 0$ there exists $k(n) \ge 0$ such that $l_n \le m_{k(n)}$: the relation $\preceq$ is an order congruence on L^{ω}.*

We claim that the congruence $\sim$ on L^{ω} arising from the order congruence $\preceq$ defined above is not (f, g)-effective. For each $y \in y$, let $\Delta l(y) \in L^{\omega}$ be the constant sequence such that $\Delta l(y)_n = l(y)$. Let ψ be the sequence such that $\psi_n = f^{(n)}(\bot, l)$ and $L(\psi)$ be the sequence $L(\psi)_n = \psi_{n+1}$. Observe that $L(\psi) \preceq \psi$ and therefore $[\psi] = [L(\psi)]$. Moreover

$$f(\psi, \Delta l)_n \;=\; f(\psi_n, \Delta l_n) \;=\; f(f^n(\bot, l), l) \;=\; \psi_{n+1} \;=\; L(\psi)_n.$$

Therefore $[\psi]$ is a fixed point of $f(x, [\Delta l])$:

$$f([\psi], [\Delta l]) \;=\; [f(\psi, \Delta l)] \;=\; [L(\psi)] \;=\; [\psi].$$

However $g([\Delta l]) \not\preceq [\psi]$: suppose on the contrary that for all $n \ge 0$ there exists $k(n) \ge 0$ such that $g(\Delta l)_n \le \psi_{k(n)}$. Let $n = 0$, so that $g(l) \le \psi_{k(0)}$, we obtain

$$\psi_{k(0)+1} \;\le\; g(l) \;\le\; \psi_{k(0)}$$

which contradicts the assumption that the ascending chain is strict. This completes the proof of theorem 19.   □

Corollary 22. *The theory of μ-lattices is not algebraic. The theory of modal μ-algebras is not algebraic.*

Proof. The polynomial

$$\psi(x) = a \wedge (b \vee (c \wedge (a \vee (b \wedge (c \vee x)))))$$

has no fixed point in the free lattice on three generators $\{a, b, c\}$, cf. [25], and this lattice embeds in the free μ-lattice over three generators, cf. [21]. Also, a Kripke frame such that the polynomial $[a]x$ does not converge in a finite number of steps is easily constructed. Other polynomials with similar properties are constructed using results on the alternation hierarchy [2,8,14,20]. □

Proposition 23. *Let L be a countable $\mathbb{T}$-model, then the $\mathbb{T}$-algebra $L^\omega / \sim$ defined above is order isomorphic to the set of all directed ideals of L.*

Proof. Let $l \in L^\omega$, we associate to l the ideal $\bigcup_{n \geq 0} \downarrow l_n$, where $\downarrow l_n$ is the principal ideal generated by l_n. The correspondence is well defined from $L^\omega / \sim$, since if $l \preceq m$, then $\bigcup_{n \geq 0} \downarrow l_n \subseteq \bigcup_{n \geq 0} \downarrow m_n$. The correspondence is easily seen to be an embedding, since the relation $\bigcup_{n \geq 0} \downarrow l_n \subseteq \bigcup_{n \geq 0} \downarrow m_n$ states exactly that $l \preceq m$. Eventually the correspondence is surjective: let I be a directed ideal of L – which we can suppose not to be principal, otherwise the result is obvious – and let $i_0, i_1, \ldots, i_n, \ldots$ be an enumeration of its elements. We define an ascending chain of elements of I in this way: let $k_0 = 0$ and define k_{n+1} to be the least index for which $i_j < i_{k_{n+1}}$ for all $j = 0, \ldots, k_n$. Such an index exists since I is a non principal filtered ideal. By construction we have $k_n < k_{n+1}$, $i_{k_n} < i_{k_{n+1}}$, and eventually

$$I \subseteq \bigcup_{n \geq 0} \downarrow i_{k_n}$$

since if $i \in I$, then $i = i_h$ for some h, so that $h < k_n$ for some n, and therefore $i_h \in \downarrow i_{k_{n+1}}$. The reverse inclusion is obvious. □

The above proposition can be used to show that, whenever L is countable, the set of ideals of L carries a canonical structure of a $\mathbb{T}$-algebra. For our purposes, we observe that if L is a countable μ-lattice, then the set of its ideals belongs to the equational hull of the category of μ-lattices in two different ways. Indeed, it carries this structure as a quotient of a μ-lattice, as shown in lemmas 20 and 21, and, on the other hand, it is a μ-lattice since it is a complete lattice. We conclude with the following interesting fact:

Corollary 24. *The underlying order of an object in the equational hull of the category of μ-lattices does not determine the algebraic structure of the object.*

References

1. S. Ambler, M. Kwiatkowska, and N. Measor. Duality and the completeness of the modal μ-calculus. *Theoret. Comput. Sci.*, 151(1):3–27, 1995. 650
2. A. Arnold. The μ-calculus alternation-depth hierarchy is strict on binary trees. *Theor. Inform. Appl.*, 33(4-5):329–339, 1999. 651, 655
3. M. Barr. Catégories exactes. *C. R. Acad. Sci. Paris Sér. A-B*, 272:A1501–A1503, 1971. 646
4. D. Beauquier and A. Rabinovich. Monadic logic of order over naturals has no finite base. To appear in *Journal of Logic and Computation*, 2001. 651
5. S. L. Bloom. Varieties of ordered algebras. *J. Comput. System Sci.*, 13(2):200–212, 1976. 653, 654
6. S. L. Bloom and Z. Ésik. Iteration algebras. *Internat. J. Found. Comput. Sci.*, 3(3):245–302, 1992. 645
7. S. L. Bloom and Z. Ésik. *Iteration theories*. Springer-Verlag, Berlin, 1993. 645
8. J. C. Bradfield. The modal mu-calculus alternation hierarchy is strict. *Theoret. Comput. Sci.*, 195(2):133–153, 1998. 651, 655
9. Z. Ésik. Completeness of Park induction. *Theoret. Comput. Sci.*, 177(1):217–283, 1997. 645
10. G. Grätzer. *Universal algebra*. Springer-Verlag, New York, second edition, 1979. 646
11. C. Hartonas. Duality for modal μ-logics. *Theoret. Comput. Sci.*, 202(1-2):193–222, 1998. 650
12. D. Kozen. Results on the propositional μ-calculus. *Theoret. Comput. Sci.*, 27(3):333–354, 1983. 646, 650
13. D. Kozen and R. Parikh. An elementary proof of the completeness of PDL. *Theoret. Comput. Sci.*, 14(1):113–118, 1981. 646
14. G. Lenzi. A hierarchy theorem for the μ-calculus. In *Automata, languages and programming*, volume 1099, pages 87–97, 1996. 651, 655
15. D. Niwiński. Equational μ-calculus. In *Computation theory (Zaborów, 1984)*, pages 169–176. Springer, Berlin, 1985. 648
16. D. Niwiński. On fixed-point clones (extended abstract). In *Automata, languages and programming (Rennes, 1986)*, pages 464–473. Springer, Berlin, 1986. 650
17. D. Park. Fixpoint induction and proofs of program properties. In *Machine intelligence, 5*, pages 59–78. American Elsevier, New York, 1970. 645
18. V. Pratt. Action logic and pure induction. In *Logics in AI (Amsterdam, 1990)*, pages 97–120. Springer, Berlin, 1991. 648
19. V. Pratt. Dynamic algebras: examples, constructions, applications. *Studia Logica*, 50(3-4):571–605, 1991. 646
20. L. Santocanale. The alternation hierarchy for the theory of μ-lattices. Research Series RS-00-29, BRICS, Department of Computer Science, University of Aarhus, November 2000. 655
21. L. Santocanale. Free μ-lattices. Research Series RS-00-28, BRICS, Department of Computer Science, University of Aarhus, November 2000. 649, 655
22. L. Santocanale. *Sur les μ-treillis libres*. PhD thesis, Université du Québec à Montréal, July 2000. 649
23. K. Segerberg. A completeness theorem in the modal logic of programs. In *Universal algebra and applications (Warsaw, 1978)*, pages 31–46. PWN, Warsaw, 1982. 646
24. I. Walukiewicz. Completeness of Kozen's axiomatisation of the propositional μ-calculus. *Inform. and Comput.*, 157(1-2):142–182, 2000. 650
25. P. M. Whitman. Free lattices. II. *Ann. of Math. (2)*, 43:104–115, 1942. 655

On the Periods of Partial Words

Arseny M. Shur* and Yulia V. Konovalova

Dept. of Algebra and Discrete Math., Ural State University
Ekaterinburg, Russia
Arseny.Shur@usu.ru

Abstract. In [1], partial words were defined to be partial mappings of a set $\{1, ..., n\}$ to a finite alphabet. We continue the research of periodic partial words started in that paper. The main goal is to clarify the interaction of different periods of a word. In particular, we exhibit some propeties of periodic partial words, which are quite similar to classical Theorem of Fine and Wilf.

Keywords: formal languages, combinatorics of words.

1 Introduction

Partial words appear in a very natural way in the fields of both pure mathematics and computer science. In mathematics they are the result of generalizing of usual, "total", words to partial ones, while for computer science they allow to simulate data whose valuable part is lost or unknown; such situations are common in studying genes in biology, in data communication, and so on.

In the paper [1], a *partial word* is defined as a partial mapping W from the set $\{1, \dots, n\}$ for some n to a finite alphabet A. If the mapping is defined everywhere, then it corresponds to a usual word. Instead of the partial word W over the alphabet A one can consider its *companion*, that is a usual word U over an alphabet $A \cup \{\diamond\}$, defined as follows:

$$U(i) = \begin{cases} W(i), \text{ if exists,} \\ \diamond \text{ otherwise.} \end{cases}$$

The expression $U(i)$ always stands for the i-th symbol of a word U (total or partial). Throughout the paper partial words *are always represented by their companions*, and so "partial word" means a finite string of symbols from the set $A \cup \{\diamond\}$, which are placed in positions numbered starting with 1. In [1], the symbol $\diamond$ was called a *hole*. We find that its role is better expressed by the term *joker*, used here. We also use the term "letter" for an alphabetical symbol, different from the joker.

We recall that a positive integer p is called a period of a (usual) word W, if $W(i) = W(i+p)$ for $i = 1, \dots, |W| - p$, where $|W|$ stands for the length of W. Any long enough word with two given periods p and q must also have the period $\gcd(p, q)$ according to the classical Theorem of Fine and Wilf.

* supported by the INTAS grant no.99-1224

J. Sgall, A. Pultr, and P. Kolman (Eds.): MFCS 2001, LNCS 2136, pp. 657–665, 2001.

Theorem I ([2]). *Let p and q be positive integers. Then every word of length at least $p + q - \gcd(p, q)$ with periods p and q has the period $\gcd(p, q)$.*

In particular, the theorem claims that a word with two coprime periods p and q having the length at least $p + q - 1$ must be unary, and thus, is completely determined by its periods. The following generalisation of this fact was derived in [3] as a corollary from the proof of the above theorem.

Corollary I ([3]). *Let p, q, r be positive integers, $\gcd(p, q) = 1$, $1 \leq r \leq q < p$. Then every word of length at least $p + q - r$ with periods p and q has at most r different letters.*

Now it is natural to ask whether partial words behave similarly to usual ones from the periodicity standpoint. Namely, given a generalisation of the notion of a period to partial words, some conditions similar to that shown above have to be found. There are two reasonable generalisations of this notion to partial words (see [1]). A positive integer p is called
– a *period* of a partial word U, if $U(i)=U(j)$ for all i, j such that $1 \leq i, j \leq |U|$, $U(i) \neq \diamond$, $U(j) \neq \diamond$, and $i \equiv j \,(\mathrm{mod}\,p)$;
– a *local period* of a partial word U, if $U(i) = U(j)$ for all i, j such that $1 \leq i, j \leq |U|$, $U(i) \neq \diamond$, $U(j) \neq \diamond$, and $|i - j| = p$.
So, if p is a period of a partial word U, then jokers in U can be replaced with letters such that p is a period of the resulted (usual) word. The notion of a local period is weaker; for example, the partial word $a\diamond bc$ has the local period 2, but has no period 2 provided $a \neq c$.
In [1], a variant of Fine-Wilf's theorem for partial words with one joker was proved.

Theorem II ([1]). *Let p and q be positive integers. Then every partial word of length at least $p + q$ with one joker and local periods p and q has the period $\gcd(p, q)$.*

It is quite easy to see that there exist partial words of any length, with given local periods p and q, but without the (local) period $\gcd(p, q)$ (where $\gcd(p, q) < \min(p, q)$). Moreover, such an example can be constructed with only 2 jokers. Suppose U has local periods p, q and $U(q + 1) = U(p + 1) = \diamond$. Now we can change the letter $U(1)$ to obtain $U(1) \neq U(\gcd(p, q) + 1)$. Since the local periods p and q do not depend on the value of $U(1)$, any such word gives the desired example.

So, to obtain results analogous to Theorem of Fine and Wilf for a wide class of partial words, one should impose restrictions on periods of partial words, while conditions on local periods seem to be useless. The general problem we are interested in can be formulated as follows. Given positive integers p, q, k, and L, where L is strictly greater then the others, find a maximal possible number of different letters in a partial word of length L with k jokers and periods p and q. The aim of this paper is to find some conditions under which this number is equal to $\gcd(p, q)$. The following observation shows why we may restrict ourselves to the case of coprime p and q.

Observation ([1]). If $\gcd(p, q) = d > 1$, replace U with a set of d partial words $U_1, \ldots, U_d$, where $U_i = U(i)U(d+i)U(2d+i)\ldots$. Each of these words have periods $p/d, q/d$, which are coprime. Thus, the word U will have the period d provided each word U_i has the period 1.

2   Relations and Graphs

In our proofs we use a technique based on graphs of binary relations. Let a partial word U of length L have k jokers in positions $i_1, \ldots, i_k$ and periods p_1, p_2. We associate with this word a binary relation R on the set $D = \{1, \ldots, L\}\backslash\{i_1, \ldots, i_k\}$. The relation consists of all pairs (s, t), where $s \equiv t \pmod{p_j}$ for some $j = 1, 2$ and there is no number n between s and t such that $n \in D$ and $s \equiv n \pmod{p_j}$. As is easily seen, R is symmetric and so the pair (D, R) can be considered as an undirected graph. Let us consider the equivalence relation R^*, which is the reflexive and transitive closure of R. Then, positions with equivalent numbers are occupied with equal letters in U. Thus, the number of equivalence classes (or the number of connected components in corresponding graph) is just the maximal possible number of different letters in U.

Example. Let $n = 9, k = 3, i_1 = 2, i_2 = 4, i_3 = 6, p_1 = 3, p_2 = 5$. Then $U = a_1 \diamond a_3 \diamond a_5 \diamond a_7 a_8 a_9$. The graph of a usual word with such periods and the graph of U look as follows:

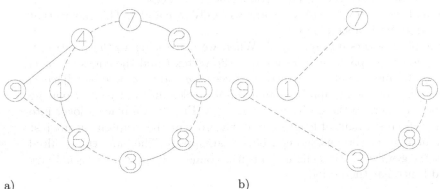

a)   b)
Fig. 1 Examples of graphs of usual (a) and partial (b) words

So, U can have at most 2 different letters, which correspond to connected components $\{1, 7\}$ and $\{3, 5, 8, 9\}$.

3   Fine-Wilf's Type Theorem for Partial Words

Throughout this section we suppose that $2 \leq q < p$ and $\gcd(p, q) = 1$.

Theorem 3. *Let the partial word U with periods p and q have k jokers. Then U has the period 1 whenever $|U| \geq p + (k+1)q - 1$. This bound is sharp provided $q = 2$ and p divides k; otherwise it can be improved.*

Remark 1. It follows from Theorem 1 that a partial word of length $p + (k+1)q - 2$ with periods p and q, but without the period 1, exists for any $k \neq 2^t, t \geq 0$, and suitable p and q. Namely, k has an odd divisor in this case, which can be taken as p together with $q = 2$.

Remark 2. For usual words ($k = 0$) Theorem 1 gives the same bound that Theorem of Fine and Wilf for coprime periods.

To prove the first statement of the theorem we need two lemmas.

Lemma 1. *Let a partial word U of length $p + q$ have periods p and q, a joker in exactly one of the positions $1, \ldots, q, p+1, \ldots, p+q$, and any number of jokers in positions $q+1, \ldots, p$. Then U has the period 1.*

Proof. Let us construct the graph of the word U. First, consider a word of length $p + q - 1$ with periods p, q and without jokers. It has the period 1 by Fine-Wilf's Theorem; thus, its graph is connected. What vertices are adjacent in this graph to the vertex t? Since the length of the word is $p + q - 1$, we find the following possibilities: 1)$t + q, t + p$; 2)$t + q$; 3)$t + q, t - q$; 4)$t - q$; 5)$t - q, t - p$. The cases 2) and 4) correspond to the vertices q and p respectively; any other vertex has two adjacent ones. Since the graph is connected, it is a chain.

Let us add the vertex $(p + q)$. It is adjacent only to p and q. This means that the obtained graph is a simple cycle.

Now we place jokers into our word. When we place joker to the position t, $q + 1 \leq t \leq p$, the graph changes as follows: the vertex t and the edges $(t - q, t)$ and $(t, t+q)$ disappear, but instead, the vertices $t - q$ and $t + q$ become adjacent. As a result, the obtained graph is again a simple cycle, and so it is after placing any number of jokers to the positions $q + 1, \ldots, p$. Finally, we place a joker into the initial or final segment of length q of our word. For the graph it means just a deletion of a vertex, because no new edges can appear. Therefore the resulted graph, i.e. the graph of U, is a chain. Thus it is connected; therefore U is a unary word, and hence has the period 1.

Corollary 2. *A partial word U with one joker and periods p, q has the period 1 whenever $|U| \geq p + q$. This bound is sharp.*

Proof. The given bound immediately follows from Lemma 1. To prove that it is sharp, consider in a similar way the procedure of placing jokers into the word of length $p + q - 1$ (whose graph is a chain, as we know). A joker in any position of $1, \ldots, q - 1, p + 1, \ldots, p + q - 1$ will break this chain into two pieces.

From now on, we call a joker *essential* to a word U of length $p + q$ with periods p, q, if it is situated in the initial or final segment of length q of U.

Lemma 2. *If a partial word U with periods p and q has a segment of length $p + q$ with at most one essential joker, then U has the period 1.*

Proof. Let such a segment be situated in U in positions $t + 1, \ldots, t + p + q$. By Lemma 1, all letters among $U(t + 1), \ldots, U(t + p + q)$ are equal; denote this letter by a. It suffices to show that any other letter in U is equal to a. Take a position s which is not in the chosen segment and mark all the positions equal to s modulo q. Then one of the positions $t + 1, \ldots, t + q$ and one of the positions $t + p + 1, \ldots, t + p + q$ get marked. At most one of them is occupied with a joker by the condition of the lemma; hence one of them is surely occupied with the letter a. However, all letters in the marked positions are equal. Hence $U(s) = a$.

Proof (of Theorem 1). Let us prove that a partial word, satisfying the conditions of the theorem but without the period 1, is shorter that the given bound. By Lemma 2, it has to contain at least two essential jokers in every segment of length $p + q$. Since there are $2q$ positions for essential jokers in each such segment, any joker can be essential for at most $2q$ segments. Further, the first joker in the word cannot occupy the last position in the segment with two essential jokers; hence, it can be essential only for $2q - 1$ segments. Symmetrically, the same is true for the last joker in the word.

Since two essential jokers are needed for any segment of length $p + q$, the number of such segments in our word does not exceed $(2qk - 2)/2$, i.e. $qk - 1$. On the other hand, the word of length L contains $L - p - q + 1$ such segments. So we have

$$L - p - q + 1 \le qk - 1,$$

and hence $L \le p + (k + 1)q - 2$. Therefore, a partial word of length at least $p + (k + 1)q - 1$ has the period 1. The first statement of the theorem is proved.

Now verify the second statement. If $k = 1$ then the bound can be decreased to $p + q$ by Corollary 1. So, let $k \ge 2$. First, we prove that the bound can be sharp only if $q = 2$. Suppose the contrary: $q \ge 3$ and there exists a partial word W with k jokers, periods p and q, length $p + (k + 1)q - 2$, but without the period 1.

Let us examine the segments of length $p + q$ in W. Note that each segment should contain exactly 2 essential jokers. In addition, any joker in W should be essential for maximal possible number of such segments. In particular, the first joker should be essential for $2q - 1$ segments. Hence, this joker occupies second from the right position in the initial segment of length $p + q$ of W; and the last position of this segment is also occupied by a joker. So, the first two jokers stay in the positions $p + q - 1$ and $p + q$ in W.

One can imagine that we look at the word W through a window which is $p + q$ symbols wide and is divided into three parts, which are q, $p - q$ and q symbols wide respectively (see Figure 2a,b). At the beginning the window shows the initial segment of W. Further, the window shifts to the right step by step, one position each time. On every step there are exactly two jokers in the lateral parts of the window.

The first two jokers in the word are initially located in two right positions of the window and move to the left as we move the window to the right. The third joker will appear in the window at the moment when the first one moves to the central part of the window (Figure 2a). Therefore, the 3rd and 4th jokers occupy positions $p+2q-1, p+2q$ in W. Note that these positions exist, because $k \geq 2$, and hence $|W| \geq p+3q-2$. Furthermore, we see now that $k \geq 4$, and hence $|W| \geq p+5q-2$, and so on. This way we can show that all positions in the consideration below are in W. Note also, that there is at least one letter between the 2nd and 3rd jokers, because $q \geq 3$.

While the first joker moves through the central part of the window, the situation above can repeat several times: when the current pair of jokers moves to the central part, another pair appears in the right with the positions greater by q than the ones of the current pair.

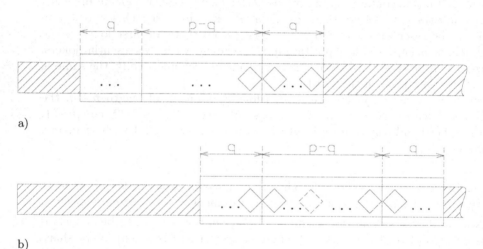

a)

b)

Fig. 2 Considering a partial word through a "window"

At the moment when the first pair of jokers enter the left part of the window, another pair should move to the central part (Figure 2b). Denote the word the window shows at that moment by V. On the one hand, V has jokers in positions $q, q+1$ and $p, p+1$. On the other hand, the distance between the first jokers of these pairs is a multiple of q. Hence q divides $p - q$, which contradicts the assumption $\gcd(p, q) = 1$.

Now let $q = 2$ and use our window again. The first two jokers occupy the positions $p+1$ and $p+2$. In the previous case pairs of jokers in W were separated by one or more letters. But now the right part of the window holds only two symbols; thus, the 3rd joker appears immediately after the 2nd one. Moreover,

it is not hard to see that W contains p consequent jokers in the positions $p +$
$1, \ldots, 2p$. They are followed by p letters, another p jokers, and so on, until the
end of W. Thus, if $k = sp$ for some s, then the length of W can be equal to
$(2s+1)p$, which coincides with the desired bound $p + (k+1)q - 2$. On the other
hand, if $k \neq sp$, then the symbol, following the kth joker in W, should also be a
joker. So, this symbol does not exist, and hence, the last symbol in W is a joker.
This joker is essential for only one segment of length $p + q$, instead of $2q - 1$, as
is required. Therefore, in this case $|W| < p + (k+1)q - 2$.

Up to this moment we verified only necessary conditions and find that these
conditions are satisfied only if $q = 2$ and $k = sp$ for some s. Now we have to show
that the partial word W of length $(2s+1)p$ with sp jokers placed as above and
periods 2 and p actually can have at least two different letters. As we proved,
the jokers in W should be placed as follows:

$$W = a_1 a_2 \ldots a_p \underbrace{\diamond \diamond \ldots \diamond}_{p} a_{2p+1} \ldots a_{(2s-1)p} \underbrace{\diamond \diamond \ldots \diamond}_{p} a_{2sp+1} a_{2sp+2} \ldots a_{(2s+1)p}.$$

Recall that an edge in its graph can connect only vertices equal modulo 2 or
modulo p. Any two positions in W, which are different by p, contain one letter
and one joker. Thus, the vertices in its graph, equal modulo p, are also equal
modulo $2p$. It means that the graph has two connected components, consisting
from odd and even numbers respectively. Hence, W can have two different letters.
The theorem is proved.

As it follows from Remark 1, for 2^t jokers the maximal length of a nonunary
word with two given coprime periods is less than the theorem shows. Of course,
the case of two jokers is most interesting. In the following statement we find the
sharp bound for the length of nonunary words which have coprime periods and
contain two jokers. Remind that if we consider local periods instead of periods,
then there is no finite bound in this case at all.

Proposition 1. *Let a partial word U with periods p and q contain 2 jokers.
Then U has the period 1 whenever $|U| \geq p + 2q - 1$. This bound is sharp for
any p and q.*

Proof. Prove the first statement. Suppose first, that U has a joker in one of the
positions $q+1, \ldots, p, p+q+1, \ldots, p+2q-1$. Then the initial segment of length
$p + q$ in U contains at most one essential joker. Symmetrically, there is at most
one essential joker in the final segment of U, if a joker is located in one of the
positions $1, \ldots, q-1, 2q, \ldots, p+q-1$. Applying Lemma 2 to these cases we
obtain that U has the period 1.

Further, suppose that U has a joker in a position t with $p+1 \leq t < 2q$ (this
case is impossible if $p+1 \geq 2q$). Then for the segment of length $p+q$ in U,
starting in the position $t - p + 1$, this joker is not essential, as is located in pth
position. Again, by Lemma 2, we obtain the period 1 for U.

Only one possibility remains: when the jokers occupy the qth and $(p+q)$th
positions. Prove that in this case the graph of U is connected. Recall that the

graph of a usual word of length $p+q$ with periods p and q is a simple cycle (see the proof of Lemma 1). The vertices q and $p+q$ are adjacent in this graph. So, if we remove these vertices from the graph (placing jokers in the positions q and $p+q$), the graph will remain connected. Thus, the graph of the initial segment from U of length $p+q$ is connected. By symmetric argument, so is the graph of the final segment. These segments have $p+1$ common symbols and hence have at least one common letter. This implies that the graph of U is connected.

Now prove the second statement. Take a word W of length $p+2q-2$ with periods p and q and two jokers in positions $q-1,q$. The graph of such a word for $q=5$ and $p=7$ is presented on Figure 3b. The length $p+2q-1$ is equal to 16 here; removing of the vertex 16 splits the graph into two components.

On the Figure 3a one may see how the graph of a usual word with the same periods looks like. A graph with 24 vertices is presented in order to show the cyclic structure of such graphs.

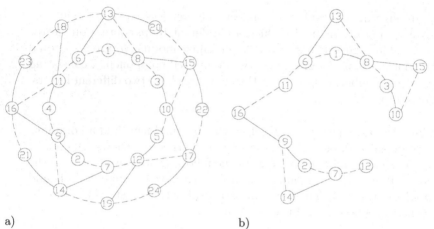

a)   b)

Fig. 3 Graphs of words with periods 5 and 7: usual word (a), unary partial word (b)

In the general case, the graph of the initial segment of length $p+q$ from W is obtained from a simple cycle by removing two nonadjacent vertices. So, it consists of two disconnected chains. Now begin to add next vertices. The vertex $p+q+1$ is adjacent to vertices $p+1$ and $q+1$. These vertices are adjacent to the vertex 1; therefore they are already in the same component, and the obtained graph still have two components. By similar argument, adding the vertices $p+q+2,\ldots,p+2q-2$ does not make the graph connected. Thus, the graph of W is disconnected, and W may be nonunary (namely, binary). Note then the vertex $p+2q-1$ will connect the vertices $p+q-1$ and $2q-1$, which were not connected since $W(q-1)=\diamond$.

We conclude the paper with a few words about the computational aspect of our results. First, by Theorem 1, there are only finitely many nonunary partial words with k jokers and given periods q and p over a fixed finite alphabet. Moreover, the number of such words is not too big, since the bound in Theorem 1 is linearly dependent on each of parameters.

Second, the maximal possible number of different letters in a partial word of length L with given periods and positions of jokers can be found effectively using graphs (namely, in the $O(L)$ time). Let us prove this. Constructing the graph of a usual word with the same length and periods requires $O(L)$ operations: we have to verify for each vertex t whether the vertices $t-p, t-q, t+q, t+p$ belong to the graph (see Figure 3a for example of such graph). After this we remove vertices corresponding to the positions of jokers. Removing one such vertex takes a constant time. Hence, the graph of a partial word with given parameters can be constructed in a linear time. The connected components in a graph with m edges can be determined in $O(m)$ operations. In our case $m < 2L$, so, we again obtain the bound $O(L)$.

References

1. J. Berstel, L. Boasson, *Partial words and a theorem of Fine and Wilf*, Theor. Comp. Sci., **218** (1999), 135-141. 657, 658, 659
2. N. J. Fine, H. S. Wilf, *Uniqueness theorem for periodic functions*, Proc. Amer. Math. Soc., **16** (1965), 109-114. 658
3. C. Choffrut, J. Karhumäki, *Combinatorics on words*, in: G.Rosenberg, A.Salomaa (Eds.), Handbook of formal languages, v.1, Ch.6, Springer, Berlin, 1997. 658

The Size of Power Automata

Klaus Sutner

Computer Science Department, Carnegie Mellon University
Pittsburgh, PA 15213
sutner@cs.cmu.edu

Abstract. We describe a class of simple transitive semiautomata that exhibit full exponential blow-up during deterministic simulation. For arbitrary semiautomata we show that it is PSPACE-complete to decide whether the size of the accessible part of their power automata exceeds a given bound. We comment on the application of these results to the study of cellular automata.

1 Motivation

Consider the following semiautomaton $\mathcal{A} = \langle [n], \Sigma, \delta \rangle$ where $[n] = \{1, \ldots, n\}$, $\Sigma = \{a, b, c\}$ and the transition function is given by

δ_a a cyclic shift on $[n]$,

δ_b the transposition that interchanges 1 and 2,

δ_c sends 1 and 2 to 2, identity elsewhere.

It is well-known that $\mathcal{A}$ has a transition semigroup of maximal size n^n, see [13]. In other words, every function $f : [n] \to [n]$ is already of the form δ_w for some word w. Note that δ_a, δ_b can be replaced by any other pair of generators for the symmetric group on n points, and δ_c can be replaced by any function whose range has cardinality $n - 1$. It was shown by Vorobev and Salomaa that, for a three-letter alphabet Σ, those are the only choices that produce a maximal transition semigroup, see [15,16,20]. If we think of the transition function as operating on sets of states, it follows that for all $A, B \subseteq [n]$ such that $|A| \geq |B| \geq 1$, there is a word w such that $\delta_w(A) = B$.

Now suppose we reverse all transitions in $\mathcal{A}$. In rev($\mathcal{A}$), for any $B \subseteq [n]$, there is a word w such that $\delta_w(\{1\}) = B$. Indeed, if we augment the semiautomaton rev($\mathcal{A}$) by selecting 1 as initial and final state, the corresponding power automaton has size 2^n, see [19,3,24], and this power automaton turns out to be reduced. The same is true if we select all states to be initial and final. However, the semiautomaton rev($\mathcal{A}$) is transitive (i.e., the underlying digraph is strongly connected), whereas its power automaton has 3 strongly connected components: $[n]$, $\emptyset$ and the remaining subsets of $[n]$. The question arises whether there are transitive nondeterministic semiautomata on n states whose corresponding power

J. Sgall, A. Pultr, and P. Kolman (Eds.): MFCS 2001, LNCS 2136, pp. 666–677, 2001.
© Springer-Verlag Berlin Heidelberg 2001

automata have size 2^n, and where deletion of the sink $\emptyset$ produces a transitive machine. Equivalently, can we find a semiautomaton $\langle Q, \Sigma, \delta \rangle$ such that for all $\emptyset \neq A, B \subseteq Q$ there is a word w such that $\delta_w(A) = B$? As is customary, we assume that the set of initial states as well as final states of a semiautomaton is the whole state set.

Let us fix some notation. For a nondeterministic automaton $\mathcal{A}$ (with or without initial and final states), we write $\text{pow}(\mathcal{A})$ for the accessible part of the full power automaton of $\mathcal{A}$, $\pi(\mathcal{A})$ for the size of $\text{pow}(\mathcal{A})$, and $\mu(\mathcal{A})$ for the size of the minimal automaton of $\mathcal{A}$. Hence, we have the obvious bounds

$$1 \leq \mu(\mathcal{A}) \leq \pi(\mathcal{A}) \leq 2^n$$

Also, $\mu(\mathcal{A})$ can be computed in polynomial time from $\text{pow}(\mathcal{A})$. But can $\pi(\mathcal{A})$ be computed efficiently, without having to construct $\text{pow}(\mathcal{A})$ first? When are $\mu(\mathcal{A})$ and $\pi(\mathcal{A})$ equal to or close to the upper bound? In particular, what happens for semiautomata, and for transitive semiautomata? When is the sink-free version of $\text{pow}(\mathcal{A})$ again transitive?

These questions are originally motivated by the study of discrete dynamical systems, see [12,1] for details and references on the topic. Briefly, let Σ be an alphabet, and denote by Σ^∞ the collection of all biinfinite words over Σ, usually referred to as *configurations* in this context. We can associate every configuration X with its *cover* $\text{cov}(X) \subseteq \Sigma^*$, the set of all finite factors of X. For a set $\mathcal{X}$ of configurations define its cover $\text{cov}(\mathcal{X})$ to be the union of all the covers $\text{cov}(X)$ where $X \in \mathcal{X}$. A *shift space* is a subset $\mathcal{X}$ of Σ^∞ that is topologically closed and invariant under the shift map σ: $\sigma(X)_i = X_{i+1}$. By compactness, a shift space can be reconstructed from its cover: a configuration X is in $\mathcal{X}$ iff it is the limit of a sequence of words in the cover. Of particular interest are *sofic systems*, subshifts $\mathcal{X}$ where $\text{cov}(\mathcal{X})$ is a regular language. A notable subclass of all sofic systems are the *subshifts of finite type*: shifts whose cover is a finite complement language: there is a finite set F of words over Σ^* such that $\text{cov}(\mathcal{X}) = \Sigma^* - \Sigma^* F \Sigma^*$, see Weiss [21].

The proper morphisms for shift spaces are given by continuous maps that commute with the shift. By the Curtis-Lyndon-Hedlund theorem [9], these maps are precisely the global maps of one-dimensional *cellular automata*. For our purposes, a (one-dimensional) cellular automaton is simply a local map $\rho : \Sigma^w \to \Sigma$. The local map naturally extends to a global map $\Sigma^\infty \to \Sigma^\infty$ that we also denote by ρ: $\rho(X)_i = \rho(X_{i-w+1} \ldots X_i)$. Weiss [21] showed that every sofic system is the homomorphic image of a subshift of finite type.

By repeatedly applying the global map ρ to the full shift Σ^∞, we obtain a descending sequence of sofic shifts. Let $L_t = \text{cov}(\rho^t(\Sigma^\infty))$ denote the corresponding regular cover languages. Clearly, all these languages are factorial, extensible and transitive (i.e., $uv \in L \implies u, v \in L$, $u \in L \implies \exists a, b \in \Sigma(aub \in L)$, and $u, v \in L \implies \exists x(uxv \in L)$). For languages of this type there is an alternative notion of minimal automaton, first introduced by Fischer [5] and discovered independently by Beauquier [2] in the form of the 0-minimal ideal in the syntactic semigroup of L. A *Fischer* automaton is a deterministic transitive semiautomaton. For each factorial, extensible and transitive language there is a unique Fischer automaton with the minimal number of states. Actually, the minimal Fischer automaton naturally embeds into the ordinary minimal DFA, see [17].

Each cellular automaton $\rho : \Sigma^{w+1} \to \Sigma$ is associated with a natural semiautomaton $B(\rho)$ whose underlying digraph is a de Bruijn graph $B(\Sigma, w) = \langle \Sigma^w, E \rangle$ where $E = \{ (ax, xb) \mid a, b \in \Sigma, x \in \Sigma^{w-1} \}$ If one labels edge (ax, xb) by $\rho(axb)$ one obtains a semiautomaton that accepts $\text{cov}(\rho(\Sigma^\infty))$. We will write $\mu(\rho)$ [$\mu_F(\rho)$, $\pi(\rho)$] for the size of the minimal DFA [minimal Fischer automaton, power automaton of $B(\rho)$, respectively] for $\text{cov}(\rho(\Sigma^\infty))$. In the mid eighties, Wolfram performed extensive calculations in a effort to understand the behavior of the sequence $(\mu(\rho^t))_t$, see [23]. In [22] he posed the question whether these sequences are generally increasing, and what can be said about the complexity of the limit set $L_\infty = \bigcap L_t$. It appears that for automata in Wolfram's classes *III* and *IV*, the sequence grows exponentially. For the lower classes the μ sequence appears to be bounded (and therefore eventually constant), or grow polynomially. The limit languages may be undecidable, though they are trivially co-recursively enumerable [10]. Surprisingly, there are examples of computationally universal cellular automata whose limit languages are regular [7], so one should not expect a simple answer to Wolfram's questions.

Recent work in computational mechanics also focuses on finite state machines as a tool to describe interesting behavior of cellular automata, see [8] for further references. One of the main objectives here is to find so-called *domains*, extensible, factorial, transitive, regular languages $L \subseteq \Sigma^*$ that are invariant or periodic under ρ (with cyclic boundary conditions). There are several examples where the typical dynamics of a cellular automaton can be described concisely on these domains, but not on the space of all configurations. In the reference, the authors use heuristic methods to construct candidate machines for domains. In general, one would expect a careful analysis of the Fischer automaton of the cellular automaton to provide a more systematic approach to the construction of domains.

Unfortunately, gathering computational data for the parameters $\mu(\rho)$ and $\mu_F(\rho)$ is rather difficult. For the sake of simplicity, let us only consider binary cellular automata, i.e., cellular automata over a two-symbol alphabet. The only obvious bounds are

$$\mu_F(\rho) \le \mu(\rho) \le \pi(\rho) \le 2^{2^{w-1}}$$

where w is the width of the local map of ρ. Hence, even for elementary cellular automata ($w = 3$), the computations for the iterates easily get out of hand since ρ^t has width $t(w - 1) + 1$.

We will first show some examples that demonstrate that full or nearly full exponential blow-up occurs quite often. In section 3, we prove that it is in general PSPACE-hard to determine even the size of the power automaton of a given semiautomaton. Our argument does not directly apply to the automata one encounters in the context of cellular automata, but it is not clear how the hardness obstruction could be avoided. Lastly, in section 4 we comment on some more related results in the study of cellular automata.

2 1-Permutation Automata and Blow-Up

2.1 Near Permutation Automata

The example given in the introduction shows that a transitive semiautomaton on a three-letter alphabet and n states may have a minimal automaton of size 2^n. In the following, we will limit our discussion to automata on two-letter alphabets. A *permutation automaton* is an automaton where each symbol $s = a, b$ induces a permutation δ_s of the state set. In other words, both the automaton and its reverse are deterministic. It is clear that the underlying graph of a permutation automaton has to be $(2, 2)$-regular: every node has indegree 2 as well as outdegree 2. On the other hand, any $(2, 2)$-regular graph admits labelings that produce a permutation automaton.

The following simple construction produces semiautomata with full blow-up. Start with a permutation automaton, and switch the label of one transition. These machines will be called 1-*permutation automata* to indicate that the labeling has Hamming distance 1 to a permutation labeling. If the switched label is in particular a self-loop, we refer to the semiautomaton as a *loop-1-permutation automaton*. We have the following result, see [17] for a proof.

Theorem 1. *Let A be a loop-1-permutation automaton of size n. Then the corresponding power automaton has 2^n states. Moreover, the power automaton is already reduced.*

The result applies in particular to cellular automata, since $B(\rho)$ is based on a de Bruijn graph. In fact, permutation automata are a standard way to construct cellular automata that have global maps that are open (in the sense of the usual product topology) and therefore surjective, but fail to be injective, see [9,18]. The one-bit change moves the cover from being trivial, to having maximum possible complexity as a regular language.

It is shown in [17] that the minimal Fischer automaton of any factorial, extensible and transitive language L can be described as the uniquely determined strongly connected component of the minimal automaton of L that has transitions only to the sink. Minimal Fischer automata are *synchronizing*: there is a word w such that $\delta_w(Q) = \{p\}$ for any state p. Hence, in the case where L is given by a 1-permutation automaton A for which pow(A) is reduced, the construction of the minimal Fischer automaton can be construed as yet another power automaton problem: the minimal Fischer automaton is (isomorphic to) the kernel automaton pow($S, \{p\}$) (the power automaton obtained by selecting $\{p\}$ as set of initial states), except for the sink. As we will see, this observation often provides a better way to compute the minimal Fischer automaton.

2.2   The Zig-zag Decomposition

To construct a permutation automaton start with an arbitrary transitive $(2,2)$-regular graph G. A *zig-zag* in G is an alternating cycle of the form

$$x_1 \to x_2 \leftarrow x_3 \to \ldots \leftarrow x_{s-1} \to x_s \leftarrow x_1$$

where all edges are distinct, but the vertices $x_1, \ldots, x_n$ need not be distinct. A *k-zig-zag* is a zig-zag containing $2k$ edges, and therefore between k and $2k$ vertices, depending on the number of self-loops in the zig-zag. Since every edge belongs to exactly one zig-zag we can form a zig-zag decomposition of the graph. We denote by $\zeta(G)$ the number of zig-zags in G. For example, for binary de Bruijn graphs, all the zig-zags have size 4. Hence, $\zeta(B(\mathbf{2}, w)) = 2^{w-1}$.

Our interest in zig-zags comes from the fact that in a permutation automaton, the label of a single edge on a zig-zag determines the labels of all the other edges. Hence, for any $(2,2)$-regular graph G there are $2^{\zeta(G)}$ permutation automata based on G. Note, though, that many of these automata may be isomorphic.

To produce a 1-permutation automaton $\mathcal{A}$ we select one of the $2^{\zeta(G)}$ permutation automata on G, choose an edge in one zig-zag, and flip the label of that edge. This will usually produce a nondeterministic machine, but there is one exception: when the selected edge belongs to a 1-zig-zag $x \to y \leftarrow x$. In this case, we effectively remove the edge whose label was flipped, so that, say, δ_a is still a permutation, but δ_b is now a partial function. It follows that $|\delta_w(A)| \le |A|$ for any word w and $A \subseteq Q$. The power automaton of $\mathcal{A}$ can not be transitive in this case, even after removal of the sink.

Before we establish full blow-up for a number of 1-permutation automata in the next section, we briefly comment on 2-permutation automata, automata whose labeling has Hamming distance 2 from the nearest permutation automaton.

Lemma 2. *Let $\mathcal{A}$ be a 2-permutation automaton, based on some $(2,2)$-regular graph of size n. Then the power automaton of $\mathcal{A}$ has size strictly less than 2^n.*

The proof of the lemma shows that indeed $\pi(\mathcal{A}) < c \cdot 2^n$ for some constant $c < 1$ in all cases except the one commented on earlier: the switched labels belong to 1-zig-zags.

2.3   Circulants

In contrast to the last lemma, we will now exhibit a class of 1-permutation automata that do exhibit full blow-up. Obvious candidates are machines based on *circulant graphs* since the latter are vertex-transitive. Recall that the circulant graph $C(n; d_1, d_2)$ has vertex set $\{0, 1, \ldots, n-1\}$ and edges $(i, i + d_s \bmod n)$, $s = 1, 2$. We will insist that $\gcd(n, d_1, d_2) = 1$ so that the graphs are always strongly connected and we can obtain a transitive semiautomaton, say, over the alphabet $\{a, b\}$, by attaching a label to each edge. It is easy to see that the number of zig-zags is

$$\zeta(C(n; d_1, d_2)) = \gcd(n, d_1 - d_2)$$

and they are all isomorphic as subgraphs of the circulant. Thus, there are $2^{\gcd(n, d_1 - d_2)}$ permutation automata based on a circulant graph, though many of them will be isomorphic.

In the following we will study a number of 1-permutation automata based on circulants $C(n; d_1, d_2)$ and compute their π, μ and μ_F values. We may safely assume that $0 \le d_1 \le d_2 < n$. Suppose e and n are coprime. Then $C(n; d_1, d_2)$ is isomorphic to $C(n; ed_1 \bmod n, ed_2 \bmod n)$, so we can cover a great many cases focusing on $d_1 = 0$ and $d_1 = 1$. E.g., for n prime all circulants are isomorphic to one of those two types.

One can establish the following results.

Lemma 3. *Let S be a 1-permutation automaton based on the circulant graph $C(n; 0, 1)$, $n \ge 2$, where the switched edge is a self-loop. Then $\pi(S) = \mu(S) = 2^n$ and $\mu_F(S) = 2^n - 1$.*

Lemma 4. *Let S be a 1-permutation automaton based on the circulant graph $C(n; 0, 1)$, $n \ge 2$, where the switched edge lies on the main cycle. Then $\pi(S) = \mu(S) = 3 \cdot 2^{n-2}$ and $\mu_F(S) = n$.*

Lemma 5. *Let S be a 1-permutation automaton based on the circulant graph $C(n; 1, 1)$, $n \ge 2$. Then $\pi(S) = \mu(S) = 2^n$ and $\mu_F(S) = n$.*

Lemma 6. *Let S be a 1-permutation automaton based on a bi-cycle where one of the secondary edges has been switched to label a. Then $\pi(S) = \mu(S) = 2^n$, and $\mu_F(S) = m(2^{n/m} - 1)$ where $m = \gcd(n, d - 1)$.*

We briefly indicate how to prove these lemmata. To avoid confusion, we refer to the states of the semiautomaton as *points*, and denote the set of all points by Q. In the following, we write the transition function as a right action $P' = P \cdot x$ of Σ^* on the semimodule $\mathrm{pow}(Q)$. *State* then always refers to a state in the power automaton, i.e., a set $P \subseteq Q$ of points. P' is *reachable* from $P \subseteq Q$ if there is some word $x \in \Sigma^*$ such that $P' = P \cdot x$. State P is reachable if P is reachable from Q, in which case a word x such that $Q \cdot x = P$ is a *witness* for P. See [16] for on overview of results concerning the lengths of such witnesses in the special case where P has cardinality 1, and the automaton is deterministic.

Thus $\mathrm{pow}(\mathcal{A})$ consists of all reachable states and is natural to consider *representable operators*, maps $f : \mathcal{P}(Q) \rightarrow \mathcal{P}(Q)$, such that for each subset P of Q, there is a word x such that $Pf = P \cdot x$. Every word w gives rise to a representable operator $[w]$ where $P[w] = P \cdot w$. Representable operators are obviously closed under composition and thus form a monoid that acts naturally on $\mathcal{P}(Q)$; the state set of $\mathrm{pow}(\mathcal{A})$ is the orbit of Q under this monoid. In some cases one can show that the monoid of representable operators contains certain shift and delete operators that are sufficient to generate the whole power set of Q. If only a part of the full power set can be generated, one may succeed in giving simple membership conditions, which provide a count of the reachable states.

As an example, suppose that S is obtained from the permutation automaton on $C(n; 0, 1)$ by switching the label at a loop, say, the loop at node 0. We adopt the convention that the nodes are numbered $0, 1, \ldots, n - 1$, and will assume that mods are taken whenever necessary. Thus, -1 is an alternative notation for node $n - 1$, and so forth. The following operators σ and κ_0 are representable:

$$P\sigma = \{ i + 1 \bmod n \mid i \in P \},$$
$$P\kappa_0 = P - \{0\}.$$

Here, σ is a cyclic shift and κ_0 corresponds to deletion of point 0. To see that both are representable, note that $P\sigma = P \cdot a$ if $0 \notin P$ or $-1 \in P$. Otherwise, $P\sigma = P \cdot ab$. And $P\kappa_0 = P \cdot b$ for all P. By a sequence of shift and delete operations we can generate an arbitrary set $P \subseteq Q$ from Q. More precisely, consider an arbitrary state $P \subseteq Q$ and define the representable operator

$$f = \sigma \circ \tau_{-1} \circ \sigma \circ \tau_{-2} \circ \ldots \circ \sigma \circ \tau_1 \circ \sigma \circ \tau_0.$$

Here τ_i is κ_0 if $i \notin P$, and the identity operator otherwise. It is easy to check that $Qf = P$.

Thus, $\pi(S) = 2^n$. Moreover, $\mu(S)$ must also be equal to 2^n. For suppose $p \notin P \subseteq Q$. Then there is a representable operator f similar to the one just used such that $pf \neq \emptyset$ but $Pf = \emptyset$: rotate and successively delete all the states in P. But then any two different states $P, P' \subseteq Q$ in $\mathrm{pow}(S)$ must have distinct behavior and it follows that $\mathrm{pow}(S)$ is the minimal automaton.

It remains to determine the minimal Fischer automaton. We claim that $\mathrm{pow}(S)$ has exactly two strongly connected components: the sink $\emptyset$ and the Fischer automaton. To see that other than $\emptyset$ there is only one component, note that the "sticky shift" operator σ_0 is representable:

$$P\sigma_0 = \begin{cases} P\sigma & \text{if } 0 \notin P, \\ P\sigma \cup \{0\} & \text{otherwise.} \end{cases}$$

It follows immediately that $P\sigma_0^{2n} = Q$ for any $P \neq \emptyset$.

Summarizing, we have established lemma 3 above. What happens if we switch the label of an edge belonging to the big cycle of length n in $C(n; 0, 1)$? Without loss of generality, we may assume that the label of edge $(-1, 0)$ is switched to b. Then symbol a induces the operator $\sigma\kappa_0$ but symbol b induces a new operator

$$P_{-1}\alpha_0 = \begin{cases} P & \text{if } -1 \notin P, \\ P \cup \{0\} & \text{otherwise.} \end{cases}$$

Thus, $_{-1}\alpha_0$ adds point 0 to P provided that P contains point -1. As a consequence, the modified shift operator

$$P\sigma' = \begin{cases} P\sigma & \text{if } -2 \in P, \\ P\sigma\kappa_0 & \text{otherwise,} \end{cases}$$

is representable. This suffices to apply the argument from the previous case and show that all states P can be reached from Q as long as $-1 \in P$ or $0 \notin P$. On the other hand, it is easy to see that the property "$-1 \notin P$ & $0 \in P$" is preserved under all representable operators. Hence $\pi(S) = 2^n - 2^{n-2} = 3 \cdot 2^{n-2}$. As in the previous case, we can show that all states in the power automaton have distinct behavior, so that $\mu(S) = 3 \cdot 2^{n-2}$. However, the minimal Fischer automaton is now smaller: its state set has cardinality n and has the form $\{\{-1\}, \{-1, 0\}, \{1\}, \{2\}, \ldots\}$. Thus, lemma 4 describes one of the cases where the minimal Fischer automaton can be generated cheaply.

3   The Hardness Argument

We now turn to the *Power Automaton Size problem*, or PAS for short. More precisely, given a nondeterministic automaton $\mathcal{A}$, we wish to determine the size $\pi(\mathcal{A})$ of the accessible part of the power automaton associated with $\mathcal{A}$. To see that counting the number of states in $\mathrm{pow}(\mathcal{A})$ can be done in nondeterministic linear space, it is best to consider a slightly more general problem: the succinct version of the Weak Connected Component Size problem. In the ordinary version of the problem, we are given a digraph G and a source node s, and we have to determine the size of the weakly connected component of s. In other words, we have to count the number of nodes of G reachable from s by some path. By the celebrated Immerman-Szelepsényi theorem, the Weak Connected Component Size problem is in NLOG. The corresponding reachability problem, where we are given an additional target node t and we have to decide if t is reachable from s, is NLOG-complete, see [6].

In the succinct version of the problem, the graph in question has $N = 2^n$ nodes and the adjacencies between the nodes are not given by a standard data structure, such as adjacency lists, but by a boolean circuit A. The circuit has $2n$ inputs and $A(x_1, \ldots, x_n, y_1, \ldots, y_n) = 1$ iff there is an edge from node $x_1 \ldots x_n$ to node $y_1 \ldots y_n$. The counting problem is now solvable in $\mathrm{NSPACE}(n)$, and the corresponding reachability problem is PSPACE-complete.

But PAS is clearly a special case of Weak Connected Component Size in its succinct form: the graph has as vertex set the full power set of Q, the source vertex is $I \subseteq Q$, the set of initial states of the nondeterministic machine, and the adjacencies are given by the condition

$$(P, P') \in E \iff \exists a \in \Sigma (P \cdot a = P').$$

The latter condition is easily expressed as a boolean circuit of size $O(kn)$ which codes the transition relation of the nondeterministic automaton.

To establish hardness, it is convenient to consider a decision version of PAS. For the next theorem we have chosen $\mathrm{PAS}_{\geq}$ where the input is a nondeterministic semiautomaton $\mathcal{A}$ together with a positive bound B, and one has to determine if $\pi(\mathcal{A}) \geq B$. Note that since $\mathrm{NSPACE}(n)$ is closed under complementation by Immerman-Szelepsényi, it does not matter whether the query is phrased as

"$\pi(\mathcal{A}) \geq m$", "$\pi(\mathcal{A}) < m$", or "$\pi(\mathcal{A}) = m$"; the problem always remains in NSPACE(n).

Theorem 7. *PAS$_\geq$, the problem of determining whether the size of the power automaton of a given nondeterministic semiautomaton exceeds a given bound, is* Γ*SPACE-complete.*

Proof. To establish PSPACE-hardness of the problem, we use Kozen's result that it is PSPACE-hard to determine whether a given collection of DFAs determines an empty intersection language, see [11] or [6]. The construction in the first reference produces machines with unique final states.

So suppose we have a list $\mathcal{A}_1, \ldots, \mathcal{A}_r$ of DFAs over some alphabet Σ. We assume that a, b, c are three new symbols not occurring in Σ, and set $\Gamma = \Sigma \cup \{a, b, c\}$. Let n_i be the size of machine $\mathcal{A}_i$. We may safely assume that $r \geq 4$ and that $n_i \leq n_{i+1}$. Let p_i be the least prime larger than 7, r, n_i and p_{i-1}, for $i = 1, \ldots, r$.

Construct new machines $\mathcal{A}'_i$ by attaching a cycle of length p_i to $\mathcal{A}_i$. The transitions on the cycle are all labeled by b. Furthermore, at each node of the cycle except for one, there is a self-loop labeled by the new symbol c. The one node without a self-loop will be called the base-point of the cycle. There are transitions labeled b from the unique final state of $\mathcal{A}_i$ to all the points on the cycle C_i. Lastly, we attach a self-loop labeled a to the initial state of $\mathcal{A}$.

The semiautomaton $\mathcal{A}$ also contains an auxiliary cycle C of length $m = 7$. Again, the cycle edges are labeled b. There are self-loops at all nodes of C labeled by all symbols in Σ as well as c. Furthermore, the loop at a selected base-point q_0 is in addition labeled a. There are no transitions from or to C from any other part of $\mathcal{A}$.

Hence, $\mathcal{A} = \bigoplus \mathcal{A}'_i \oplus C$ is the disjoint union of all the DFAs with their attached cycles plus the auxiliary cycle C. We have to count the number of states in pow($\mathcal{A}$) that are reachable from $Q = \bigcup Q_i \cup \bigcup C_i \cup C$, the state set of $\mathcal{A}$. Again, we will refer to the states of $\mathcal{A}$ as points, so that state from now on always refers to pow($\mathcal{A}$), i.e., a set of points. We will write Q' for $\bigcup Q_i$ and C' for $\bigcup C_i$.

Let us say that an input string of the form $w = uaxbv$ is proper, where $u, x \in \Sigma^*$ and $v \in \Gamma^*$. Correspondingly, all states obtained from proper inputs are proper. First consider the case $u = v = \varepsilon$. After the first symbol a, all the Kozen automata are in their respective initial states and the base-point on C is also active. Let

$$A(x) = \{\, i \in [r] \mid \mathcal{A}_i \text{ accepts } x \,\}.$$

The next input symbol b will then generate the state $\bigcup_{i \in A(x)} C_i$ plus one point on the auxiliary cycle C. As there is no self-loop at the base-points of the cycles C_i, the symbol c induces a delete operation on these cycles (delete all base-points). Symbol b, on the other hand, induces a cyclic shift. Since the lengths of these cycles as well as the auxiliary cycle C are relatively prime, it follows from the Chinese remainder theorem that we can generate $m \, \alpha(A(x))$ states from $Q \cdot axb$. Here for any subset A of $[r]$: $\alpha(A) = \prod_{i \in A} 2^{p_i}$. Hence, from any proper input $uaxbv$ we can generate at least $m \, \alpha(A(x))$ states.

Improper inputs are somewhat more tedious to deal with. First, we have the reachable states Q and $\emptyset$. Then, there are at most $\prod_{i \in [r]} 2^{n_i}$ states due to inputs of the form Σ^*, and at most $\alpha([r])$ states due to inputs of the form $\Sigma^*(b+c)(\Gamma - a)^*$. Either of these states contain C as a subset. Then there are m states due to inputs of the form $\Sigma^*(b+c)(\Gamma - a)^* a \Gamma^*$, namely all the single points on C. Lastly, inputs of the form $\Sigma^* a \Gamma^*$ which fail to be proper can produce at most an additional $\prod_{i \in [r]} (p_i + 1)$ states, consisting of the base-point in C and at most one point in Q_i.

Since $n_i \leq p_i$ and $p_1 + 1 \leq 2^{p_i}$, the number of states reachable from Q is bounded from above by

$$K = 3\alpha([r]) + m\left(1 + \sum_x \alpha(A(x))\right),$$

where the summation is over suitably chosen factors $x \in \Sigma^*$ in proper inputs. On the other hand, the number of reachable states is bounded from below by

$$m \sum_x \alpha(A(x)).$$

Now consider the bound $B = m \cdot \alpha([r])$.

First suppose that the acceptance languages of the Kozen automata have non-empty intersection. Then there is some input $x \in \Sigma^*$ such that all DFAs are in their accepting state after scanning x. Hence, at least B states are reachable, as required.

On the other hand, suppose that the intersection language of the Kozen automata is empty. Then $A(x)$ has cardinality at most $r - 1$ and we have

$$
\begin{aligned}
K/B &\leq \frac{3\alpha([r]) + m(1 + \sum_x \alpha(A(x)))}{m \cdot \alpha([r])} \\
&= 3/m + \sum_x \alpha(A(x))/\alpha([r]) + 1/\alpha([r]) \\
&< 1/2 + \sum_i 2^{-p_i} + 2^{-\sum p_i} \quad < \quad 1.
\end{aligned}
$$

Thus, $K < B$, and we are done.   □

A minor modification of the last argument also shows that it is hard to determine the size of the *kernel automaton*. Recall that the (full) kernel automaton of a semiautomaton $\mathcal{A} = \langle Q, \Sigma, \tau \rangle$ is the subautomaton of the full power automaton of $\mathcal{A}$ that is induced by the singleton states $\{p\}$ for $p \in Q$.

Corollary 8. *The problem of determining whether the size of the kernel automaton of a given nondeterministic semiautomaton exceeds a given bound, is PSPACE-complete.*

Proof. Modify the machine $\mathcal{A}$ from the last theorem by attaching a new state q_0 and transitions labeled a from q_0 to the initial states of the Kozen automata as well as to the base point of the auxiliary cycle. The old self-loops labeled a are removed. The argument then proceeds almost verbatim as in the last proof. □

4 Conclusion

It was pointed out by Ravikumar [14] that the uniqueness of the final states in the Kozen automata implies that the related State Reachability Problem for the power automaton of a nondeterministic machine is also PSPACE-complete. The input for State Reachability is a nondeterministic machine with initial states, say, $I \subseteq Q$, a target set $P \subseteq Q$, and one has to determine whether $P = I \cdot x$ for some input x. It is not hard to modify the last construction in order to show that State Reachability is hard even for semiautomata.

The semiautomaton $\mathcal{A}$ in the proof of the the theorem is of course nowhere near transitive. We do not know if PAS remains hard for transitive semiautomata, but it appears likely that even for this restricted class of machines there is essentially no other way to determine $\pi(\mathcal{A})$ than to construct the accessible part of the power automaton. In fact, we do not know how to compute $\pi(\mathcal{A})$ efficiently even if $\mathcal{A}$ is a 1-permutation automaton based on a circulant, or on a de Bruijn graph.

With respect to cellular automata, theorem 1 can be strengthened to show that any loop-1-permutation automaton of size n whose underlying graph is a de Bruijn graph, has a minimal Fischer automaton of size $2^n - 1$. However, this result is only the tip of an iceberg. For binary cellular automata of width $w \geq 2$, there are 2^{w-2} zig-zags and $2^w \cdot 2^{2^{w-2}}$ 1-permutation automata. The following table shows the frequencies of the differences $2^{16} - \pi(\mathcal{A})$ for all binary 1-permutation automata of width 5.

Δ	0	1	2	4	8	16	32	64
freq.	4096	512	896	480	240	208	328	352
Δ	124	128	170	256	512	1024	1052	2048
freq.	8	296	8	224	160	120	8	256

Another automata construct that is of importance in symbolic dynamics is the kernel automaton. Surjectivity of a cellular automaton ρ is equivalent to $B(\rho)$ being unambiguous. As a consequence, for a surjective cellular automaton, the states of maximal size in the kernel automaton of $B(\rho)$ form a subautomaton, the so-called Welch automaton. The size of the states in this subautomaton is an important parameter, e.g., one can show that this so-called Welch index is a homomorphism from the monoid of all epimorphisms of Σ^∞ to the multiplicative monoid of the positive natural numbers, see [9]. As we have seen, determining the size of the kernel automaton of a semiautomaton is also PSPACE-hard.

References

1. M.-P. Beal and D. Perrin. Symbolic dynamics and finite automata. In G. Rozenberg and A. Salomaa, editors, *Handbook of Formal Languages*, volume 2, chapter 10. Springer Verlag, 1997. 667
2. D. Beauquier. Minimal automaton for a factorial, transitive, rational language. *Theoretical Computer Science*, 67:65–73, 1989. 667
3. W. Brauer. On minimizing finite automata. *EATCS Bulletin*, 39:113–116, 1988. 666
4. M. Delorme and J. Mazoyer. *Cellular Automata: A Parallel Model*, volume 460 of *Mathematics and Its Applications*. Kluwer Academic Publishers, 1999. 677
5. R. Fischer. Sofic systems and graphs. *Monatshefte für Mathematik*, 80:179–186, 1975. 667
6. M. R. Garey and D. S. Johnson. *Computers and Intractability*. Freeman, 1979. 673, 674
7. E. Goles, A. Maass, and S. Martinez. On the limit set of some universal cellular automata. *Theoretical Computer Science*, 110:53–78, 1993. 668
8. J. E. Hanson and J. P. Crutchfield. Computational mechanics of cellular automata. Technical Report 95-10-095, Santa Fe Institute, 1995. 668
9. G. A. Hedlund. Endomorphisms and automorphisms of the shift dynamical system. *Math. Systems Theory*, 3:320–375, 1969. 667, 669, 676
10. L. Hurd. Formal language characterizations of cellular automata limit sets. *Complex Systems*, 1(1):69–80, 1987. 668
11. D. Kozen. Lower bounds for natural proof systems. In *Proc. 18-th Ann. Symp. on Foundations of Computer Science*, pages 254–266. IEEE Computer Society, 1977. 674
12. D. Lind and B. Marcus. *Introduction to Symbolic Dynamics and Coding*. Cambridge University Press, 1995. 667
13. J. E. Pin. *Varieties of Formal Languages*. Foundations of Computer Science. Plenum Publishing Corporation, 1986. 666
14. B. Ravikumar. Private communication, 1994. 676
15. A. Salomaa. On the composition of functions of several variables ranging over a finite set. *Ann. Univ. Turkuensis*, 41, 1960. 666
16. A. Salomaa. Many-valued truth functions, černý's conjecture and road coloring. *Bulletin EATCS*, (68):134–150, June 1999. 666, 671
17. K. Sutner. Linear cellular automata and Fischer automata. *Parallel Computing*, 23(11):1613–1634, 1997. 668, 669
18. K. Sutner. *Linear Cellular Automata and De Bruijn Automata*, pages 303–320. Volume 460 of *Mathematics and Its Applications* [4], 1999. 669
19. R. A. Trakhtenbrot and Y. M. Barzdin. *Finite Automata: Behavior and Sythesis*. North-Holland, 1973. 666
20. N. Vorobev. On symmetric associative systems. *Leningrad Gos. Ped. Inst., Uch. Zap.*, 89:161–166, 1953. 666
21. B. Weiss. Subshifts of finite type and sofic systems. *Monatshefte für Mathematik*, 77:462–474, 1973. 667
22. S. Wolfram. Twenty problems in the theory of cellular automata. *Physica Scripta*, T9:170–183, 1985. 668
23. S. Wolfram. *Theory and Applications of Cellular Automata*. World Scientific, 1986. 668
24. Sheng Yu. Regular languages. In G. Rozenberg and A. Salomaa, editors, *Handbook of Formal Languages*, volume 1, chapter 2. Springer Verlag, 1997. 666

On the Approximability of the Steiner Tree Problem

Martin Thimm *

Humboldt Universität zu Berlin, Institut für Informatik
Lehrstuhl für Algorithmen und Komplexität
D-10099 Berlin
thimm@informatik.hu-berlin.de

Abstract. We show that it is not possible to approximate the minimum Steiner tree problem within $\frac{136}{135}$ unless $co-RP = NP$. This improves the currently best known lower bound by about a factor of 3. The reduction is from Håstad's nonapproximability result for maximum satisfiability of linear equation modulo 2. The improvement on the nonapproximability ratio is mainly based on the fact that our reduction does not use variable gadgets. This idea was introduced by Papadimitriou and Vempala.

Keywords: Minimum Steiner tree, Approximability, Gadget reduction, Lower bounds.

1 Introduction

Suppose that we are given a graph $G = (V, E)$, a metric given by edge weights $c : E \to \mathbb{R}$, and a set of required vertices $T \subset V$, the *terminals*. The minimum Steiner tree problem consists of finding a subtree of G of minimum weight that spans all vertices in T.

The Steiner tree problem appears in many different kinds of applications. Examples are the computation of phylogenetic trees in biology or the routing phase in VLSI-design.

The Steiner tree problem is well known to be NP-complete even in the very special cases of Euclidian or rectilinear metric. Arora [1] has shown that Euclidian and rectilinear Steiner tree problems admit a polynomial time approximation scheme, i.e. they can be approximated in polynomial time up to a factor of $1 + \epsilon$ for any constant $\epsilon > 0$. In contrast to these two special cases the Steiner tree problems is known to be APX-complete [2,5] which means that unless $P = NP$ there does not exist a polynomial time approximation scheme for this problem.

During the last ten years a lot of work has been done on designing approximation algorithms for the Steiner tree problem [16,4,14,10,9]. The currently best approximation ratio is 1.550 and is due to Robins and Zelikovsky [15]. For more details on approximation algorithms for the Steiner tree problem see [7].

---

* Graduate School "Algorithmische Diskrete Mathematik", supported by Deutsche Forschungsgemeinschaft, grant 219/3.

J. Sgall, A. Pultr, and P. Kolman (Eds.): MFCS 2001, LNCS 2136, pp. 678–689, 2001.

But very little is known about lower bounds. The presently best known lower bound is 1.0025 and follows from a nonapproximability result for $VERTEX$-$COVER$ in graphs of bounded degree [3]. We improve this bound by about a factor of 3.

The improvement on the nonapproximability ratio is mainly based on the fact that our reduction does not use variable gadgets. This idea was introduced by Papadimitriou and Vempala [13]. They prove that the (symmetric) traveling salesman problem cannot be approximated within $\frac{129}{128}$, unless $P = NP$.

We reduce from Håstad's nonapproximability result for maximum satisfiability of linear equations modulo 2 with three variables per equation, MAX-$E3$-LIN-2, [8]. Our construction uses two types of gadgets. There is one "equation gadgets" for each equation in the MAX-$E3$-LIN-2 instance. We also have "edge gadgets" which connect the nodes of the equation gadgets corresponding to literals x and $\bar{x}$ in a special way that is induced by some d-regular bipartite graph. We prove that if this graph is an expander then every optimal Steiner tree for this instance has a special structure which allows us to derive a legal truth assignment for the MAX-$E3$-LIN-2 instance (without relying on variable gadgets).

Our nonapproximability ratio heavily depends on the parameters of the above mentioned expanders, namely the degree d and the expansion coefficient c. In [17] the existence of such graphs is proved by counting arguments. We adapt the proof of Sarnak's result [17] for our purpose.

The rest of the paper is organized as follows: In Section 2 we describe the reduction in detail and in Section 3 we prove our main theorem. The existence of the expander graphs needed in the reduction is proved in Section 4. In the last Section we shortly discuss some ideas how to possibly improve our main result.

2   The Reduction

2.1   The Graph

We reduce from Håstad's nonapproximability result for maximum satisfiability of linear equations modulo 2 with three variables per equation, MAX-$E3$-LIN-2, [8].

As already mentioned in [13] we can state Håstad's result [8] as follows:

Theorem 1. *[8] For every $\epsilon > 0$ there is an integer k such that it is NP-hard to tell whether a set of n linear equations modulo 2 with three variables per equation and with $2k$ occurences of each variable has an assignment that satisfies $n(1 - \epsilon)$ equations, or has no assignment that satisfies more than $n(\frac{1}{2} + \epsilon)$ equations.*

We start with such an instance of MAX-$E3$-LIN-2, namely a set of n linear equations modulo 2, where each equation has exactly three literals and with exactly $2k$ occurences of each variable. We may also assume that each variable appears exactly k times negated and k times unnegated and also that all equations are of the form $x + y + z = 1$. The latter condition can be enforced by flipping some literals, the former by adding three copies of each equation with all possible pairs of literals negated. (e.g. $x + y + z = 0$ may be transformed to $x + y + \bar{z} = 1$ and we add $\bar{x} + \bar{y} + \bar{z} = 1$, $\bar{x} + y + z = 1$ and $x + \bar{y} + z = 1$.)

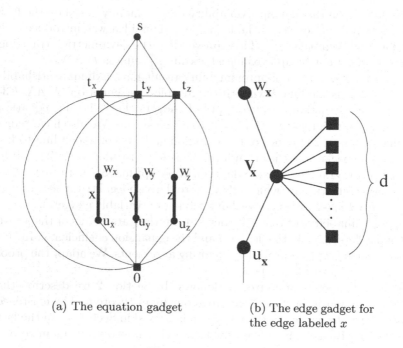

(a) The equation gadget

(b) The edge gadget for the edge labeled x

Fig. 1. The gadgets

We construct an instance for the Steiner tree problem as follows:

The graph consists of several gadgets. For each equation there is an *equation gadget* (see Figure 1(a)). The nodes drawn as boxes are terminals. All equation gadgets share the node 0. The bold edges labeled with x, y and z correspond to the variables in an equation of the form $x + y + z = 1$ and are not really edges of the graph but represent whole substructures as shown in Figure 1(b). (The edge weights and the value of d in the edge gadget will be specified later.)

As in [13] we do not use "variable gadgets" to assure a correct assignment of the literals and its opposite. The construction will enforce this implicitly. By assumption each variable appears exactly k times negated and k times unnegated. Consider one variable, say x. In our graph we then have k edge gadgets that correspond to the occurences of x and k edge gadgets for the occurences of $\bar{x}$. We now connect these edge gadgets in the following way: Suppose we are given a bipartite d-regular graph $H = (A \cup B, E)$, $|A| = |B| = k$. Now identify a terminal in the i-th edge gadget corresponding to an occurence of x with a terminal in the j-th edge gadget corresponding to a occurence of $\bar{x}$ iff the i-th vertex in A is connected to the j-th vertex in B ($1 \le i, j \le k$) (see Figure 2).

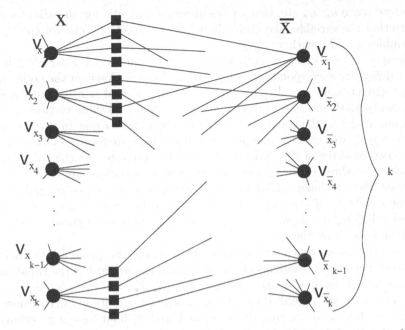

Fig. 2. Edge gadgets of variable x after identification of the terminals

We will see in what follows that the only thing we need to know about H is that it is an expander, i.e. with $V(H) = A \cup B$, $|A| = |B| = k$ we have that for all $S \subset A$ with $|S| \leq \frac{k}{2}$: $|\Gamma(S)| \geq c|S|$ for some $c > 1$ (where $\Gamma(s)$ denotes the set of neighbours of S).

If we now define the edge weights appropriately, our graph will have some useful properties. First, a truth assignment for our $MAX\text{-}E3\text{-}LIN\text{-}2$ instance directly yields a solution for the Steiner tree instance which has a "nice" structure: All subtrees of the resulting Steiner tree which correspond to satisfied equations have the same length l_s and a special structure. Also the subtrees which correspond to the equations that are not satisfied have some special structure and the same length l_{ns}. Additionally we have that $l_{ns} = l_s + const$. By this the length of the Steiner tree reflects the number of equations that are not satisfied.

Second, we will see that we always may assume that an optimal Steiner tree has a simple special structure which makes it easy to define the corresponding truth assignment.

2.2   The Edge Weights

We have already defined the graph for the Steiner tree instance. Let us now define the edge weights. First, we want to guarantee that in any optimal Steiner tree all the terminals in the edge gadgets have degree 1. We define the weight of any edge in any edge gadget which is incident to a terminal to be b. Then it suffices to have that the weight of a path $(0, u_\bullet, v_\bullet)$ and $(v_\bullet, w_\bullet, t_\bullet)$ is at most b.

(Whenever we write $u_\bullet$, $w_\bullet$ etc. this always means u_x, w_x etc. for all variables x. To indicate that the variables should be distict we write $c(t_\bullet, t_\star)$ etc. for $c(t_x, t_y)$ for all varaibles x and y with $x \neq y$.)

Suppose now we are given a truth assignment to our $MAX\text{-}E3\text{-}LIN\text{-}2$ instance. We define the corresponding Steiner tree as follows: Consider the nodes $v_\bullet$ in the edge gadgets corresponding to variables with assigned truth value 1 and connect them to all their adjacent terminals. Now connect all these vertices $v_\bullet$ to the remaining nodes in the equation gadgets to get a Steiner tree in the cheapest possible way. If we consider an equation gadget there are four possible cases: None, one, two or three of the nodes $v_\bullet$ have to be connected to the tree. The cases with one or three of these nodes correspond to equations that are satisfied. We want them to have partial solutions (subtrees of the equation gadgets) of the same weight l_s. If none or two nodes have to be connected to the tree, the corresponding equations were not satisfied. The subtrees in these equation gadgets should also have the same weight, l_{ns}.

For technical reasons which will become clear during the proof of the main theorem we add up the edge weights of the edge gadgets and those of the equation gadgets separatly. If we now define $c(0, u_\bullet) = a$, $c(u_\bullet, v_\bullet) = b - a$, $c(v_\bullet, w_\bullet) = 0$ and $c(w_\bullet, t_\bullet) = b$ we have fullfilled the above mentioned condition. Furthermore, connecting a vertex $v_\bullet$ to the tree always costs b and without loss of generality we may assume that connecting $v_\bullet$ to the tree is the same as connecting $u_\bullet$ to the tree (that is to pay a in the equation gadget and $b - a$ in the edge gadget). Let us now define the remaining edge weights: $c(0, t_\bullet) = f$, where $f = 2a + b$, $c(s, t_\bullet) = e$ where $e = \frac{2}{3}(a + b)$ and $c(t_\bullet, t_\star) = f$ (see Figure 3).

It is easy to find the optimal partial solution in all the four cases. They are displayed in Figure 4. The subtrees corresponding to satisfied equations (one or three variables have truth value 1) have weight $3a + 3b$ (Figure 4, right hand side), those corresponding to equations that are not satisfied have weight $4a + 3b$ (Figure 4, left hand side). Remember that we count the weights in the edge gadgets separately.

The way these numbers a and b are related will become clear at the end of the proof of the main theorem. We will also see how d, the number of terminals in the edge gadgets, has to be chosen.

We have now fully described our Steiner tree instance and prove our main theorem.

3   Proof of the Main Theorem

Theorem 2. *No polynomial time approximation algorithm for the minimum Steiner tree problem can have a peformance ratio below* 1.0074, *unless* $co - RP = NP$.

Proof. Given a truth assignment to our $MAX\text{-}E3\text{-}LIN\text{-}2$ instance. We have already seen in the last Section how to define the corresponding Steiner tree. The weight of this Steiner tree consists of the weight of the edge gadgets and the weight of the equation gadgets.

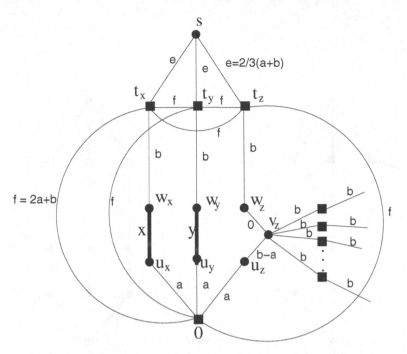

Fig. 3. Edge weights (edge gadget for z inserted)

There are $3n$ edge gadgets, half of them correspond to literals with truth value 1 (since each variable appears the same number of times and also the same number of times negated and unnegated). The weight of an edge gadget is $b - a + db$ so that the total weight of all edge gadgets sums up to $\frac{3}{2}n(b - a + db)$.

The subtrees in the equation gadgets corresponding to satisfied equations have weight $3a + 3b$, those corresponding to unsatisfied equations have weight $4a + 3b$.

Suppose the truth assignment satisfies all but M equations. Then the Steiner tree constructed above has weight $\frac{3}{2}n(b - a + db) + n(3a + 3b) + Ma$.

We call such a tree *standard* (for this assignment), i.e. for the collection of edge gadget of any variable, say x, it is true that either all the nodes v_x corresponing to the occurences of the literal x are Steiner nodes or all the nodes $v_{\bar{x}}$ corresponing to the occurences of the literal $\bar{x}$ are of that kind.

To prove the theorem we have to show that an optimal Steiner tree is standard (for some assignment). This is done in Lemma 3.

Now we use Håstad's result: for every $\epsilon > 0$ it is NP-hard to decide whether a set of n linear equations modulo 2 with three variables per equation has an assignment that satisfies $n(1 - \epsilon)$ equations, or has no assignment that satisfies more than $n(\frac{1}{2} + \epsilon)$ equations. By our reduction the same is true for standard trees of length $n(\frac{3}{2}(db + (b-a)) + (3a + 3b) - \epsilon)$ and $n(\frac{3}{2}(db + (b-a)) + (3a + 3b) + \frac{1}{2}a + \epsilon)$.

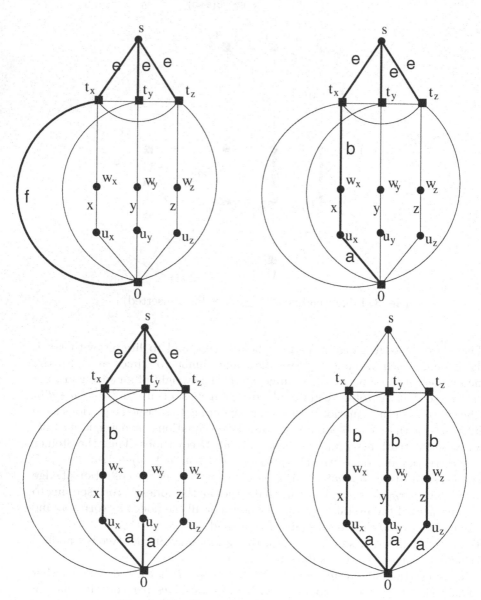

Fig. 4. The four cases of optimal subtrees

We get an nonapproximability ratio of

$$r = \frac{\frac{3}{2}n(db + (b - a)) + n(3a + 3b) + \frac{1}{2}na}{\frac{3}{2}n(db + (b - a)) + n(3a + 3b)}.$$

If we set $a = tb$ and cancel n and b, this reads as

$$r = \frac{\frac{3}{2}(d + 1 - t) + 3(1 + t) + \frac{1}{2}t}{\frac{3}{2}(d + 1 - t) + 3(1 + t)}.$$

With $a = tb$ and $c = \frac{b+a}{b-a}$ (see Lemma 3 below), we also get $c = \frac{1+t}{1-t}$ ($t \leq \frac{1}{3}$, since $t = \frac{c-1}{c+1}$ and $c \leq 2$). Plugging this in, we finally get

$$r = \frac{\frac{3}{2}(d + 1 - \frac{c-1}{c+1}) + 3(1 + \frac{c-1}{c+1}) + \frac{1}{2}\frac{c-1}{c+1}}{\frac{3}{2}(d + 1 - \frac{c-1}{c+1}) + 3(1 + \frac{c-1}{c+1})}$$

$$= 1 + \frac{\frac{c-1}{c+1}}{3(d + 3 + \frac{c-1}{c+1})}. \tag{1}$$

With $d = 6$ and $c = 1.5144$ the theorem follows (see Lemma 4). $\qquad\square$

Lemma 3. *Every optimal Steiner tree can be transformed into a standard tree without increasing the weight.*

Proof. Given an optimal Steiner tree. Remember that we may assume that all terminals in the edge gadgets have degree 1.

Consider all edge gadgets corresponding to one variable, say x and $\bar{x}$ and the graph H_x which shows the identification of the terminals of the edge gadgets. We partition the node set $V(H_x)$ into three classes A_x, T_x and $B_{\bar{x}}$ (see Figure 5). We then partition A_x into $C1$, which are the nodes in A_x that are Steiner nodes of the tree, and $U1$, all other nodes in A_x. In the same way we partition $B_{\bar{x}}$ into $C2$ and $U2$ (see Figure 5).

Without loss of generality let $|U1| \leq |U2|$. Consider the following modification of the Steiner tree: (see Figure 5) All nodes in A_x and none in $B_{\bar{x}}$ are Steiner nodes after this step, all terminals in the edge gadgets are linked to these Steiner nodes. The subtrees of those equation gadgets which contain the nodes lying in $U1$ and $C2$ are changed. We construct them according to the new conditions (number of vertices $v_\bullet$ to be connected to the tree) completely new (in the cheapest possible way).

We claim that if we modify the Steiner tree in this way , the weight of the resulting new tree does not increase. It is easy to see that if we have done this for all variables, one after another, the result is a tree in standard form.

It remains to show that this modification does not increase the weight of the tree. Consider H_x. In order to connect the Steiner nodes $v_\bullet$ in H_x to the tree we have to pay a weight of b for each of them. So the cost to do this in the old tree is $b(|C1| + |C2|)$. In the new tree we only need kb with $k = |U1| + |C1|$, hence we gain $b(|C2| - |U1|)$.

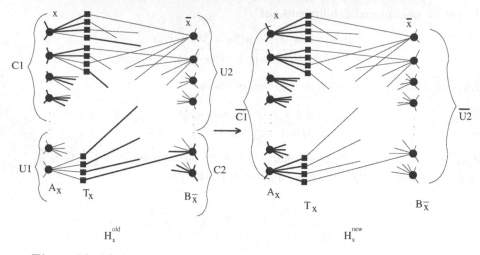

Fig. 5. Modification of the edges of the Steiner tree in the edge gadgets

On the other hand we have to look at the newly constructed subtrees of the equation gadgets, which contain nodes out of $U1$ and $C2$. Since the number of nodes $v_\bullet$ which have to be connected to the subtrees has changed, in each such case we may have to pay an extra cost of a (see Figure 4). So the total extra cost can be bounded by $a(|U1| + |C2|)$.

We are done if we can show that

$$(|C2| - |U1|)b \geq (|U1| + |C2|)a .$$

This inequality can be rearranged to stand as

$$|C2| \geq \frac{b+a}{b-a}|U1| . \tag{2}$$

Let $\Gamma(U1)$ be the set of nodes in $B_{\bar{x}}$ which are reachable from $U1$ by paths of length 2. We get that $\Gamma(U1) \subset C2$. If we can show that

$$\Gamma(U1) \geq \frac{b+a}{b-a}|U1|, \tag{3}$$

we have proven (2).

But if we consider H_x as a bipartite graph with node sets A_x and $B_{\bar{x}}$ and recall that $|U1| \leq \frac{k}{2}$, (3) is just a typical expander condition with expansion constant $c = \frac{b+a}{b-a}$.

It is known that such bipartite regular expanders exist, as well as how to construct them probabilistically. The following Section will deal with this question in detail. To get our nonapproximability result we use 6-regular expanders with $c = 1.5144$ (see Lemma 4). $\qquad\square$

4 Expanders

Let $G = (I \cup O, E)$ a bipartite d-regular graph, $d \geq 3$, $|I| = |O| = n$. We call G a (n, d, c)-*expander*, if for all $A \subset I$, $|A| \leq \frac{n}{2}$ and some $2 \geq c > 1$

$$\Gamma(A) \geq c|A|. \tag{4}$$

Explicit constructions of such expanders seem to be very hard. This was first done by Margulis in [12]. He could prove that the constant c arising in his construction is bounded away from 1. In [6] such a constant was explicitly calculated by Gabber and Galil, before Lubotzky, Philips and Sarnak [11] came up with a different construction and a better constant c.

Compared to this it is quite easy to prove the existence of linear expanders by counting arguments.

Following the ideas of [17] we will do this in greater detail. The main result of this Section is

Lemma 4. *For sufficiently large n there exist (n, d, c)-expanders for*

$$d > max\left\{c + \frac{3}{2}, \frac{2}{2-c}, \frac{\frac{c}{2}\ln(\frac{c}{2}) + (1 - \frac{c}{2})\ln(1 - \frac{c}{2}) - \ln(2)}{\frac{c}{2}\ln(c) - \frac{c-1}{2}\ln(c-1) - \ln(2)}\right\}. \tag{5}$$

Sketch of the proof. Let $I = O = \{1, 2, \ldots, n\}$ and X be a bipartite graph with $V(X) = I \cup O$. Consider d permutations $\pi_1, \pi_2, \ldots, \pi_d$ of I and connect each $j \in I$ with all $\pi_r(j) \in O$, $1 \leq r \leq d$. We get a bipartite d-regular (multi)graph. Let us call $\pi = (\pi_1, \pi_2, \ldots, \pi_d)$ *bad*, if there is a set $A \subset I$, $|A| \leq \frac{n}{2}$ and a set $B \subset O$, $|B| = c|A|$ such that $\pi_j(A) \subset B$ for all $j = 1, 2, \ldots, d$.

We will bound the number of bad π's from above. For some given A and B, $|A| = t \leq \frac{n}{2}$, $|B| = ct$, the number of bad π's is

$$(ct(ct-1)(ct-2)\ldots(ct-t+1)(n-t)!)^d = \left(\frac{(ct)!(n-t)!}{(ct-t)!}\right)^d.$$

So the total number of bad π's, BAD, over all A and B is given by

$$BAD = \sum_{t \leq \frac{n}{2}} \binom{n}{t}\binom{n}{ct}\left(\frac{(ct)!(n-t)!}{(ct-t)!}\right)^d. \tag{6}$$

Since there are $(n!)^d$ π's in total, we are interested in the function

$$R(t) = \binom{n}{t}\binom{n}{ct}\left(\frac{(ct)!(n-t)!}{(ct-t)!n!}\right)^d, 1 \leq t \leq \frac{n}{2}, \tag{7}$$

especially for $n \to \infty$.

$R(t) \to 0$ for $n \to \infty$ means that if we pick a π randomly to construct our graph X, we will get a (n, d, c)-expander with very high probability.

We proceed as follows: To analyse the function $R(t)$ we consider its continuous version and its first derivative. We show that $R'(t) \approx R(t)\ln(f(t))$ for some function f. By looking at f we can prove that $R(t)$ decreases in the interval $1 \leq t \leq \alpha$ for some α that only depends on c and d and inreases for all other values of t up to $\frac{n}{2}$. We bound the values of $R(1)$, $R(\sqrt{n})$ and $R(\frac{n}{2})$ to give upper bounds on (6) by $\sqrt{n}R(1) + \alpha n R(\sqrt{n}) + \frac{n}{2}R(\frac{n}{2})$.

$\square$

5   Discussion

Table 1 may help to see the numbers hidden in (5). The general form (1) of our main result – the nonapproximability ratio is just a function of c and d – immediately gives better results if better expanders are found. For example, an $(n, 5, \frac{7}{4})$-expander would give a ratio of 1.011.

We believe that using this method a nonapproximability ratio of about 1.01 is within reach.

Table 1. Some values for c and d in (5)

d	3	4	5	6	7	8	10	15	50	100
c	1.162	1.310	1.425	1.514	1.583	1.637	1.716	1.821	1.954	1.978

References

1. S. Arora, Polynomial time approximation schemes for the Euclidian TSP and other geometric problems, Proceedings of the 37th Annual Symposium on Foundations of Computer Science (1996), pp.2–11. 678
2. S. Arora, C. Lund, R. Motwani, M. Sudan, M. Szegedy, Proof verification an hardness of approximation problems, Proceedings of the 33rd Annual Symposium on Foundations of Computer Science (1992), pp. 14–23. 678
3. P. Berman, M. Karpinski , On Some Tighter Inapproximability Results, Further Improvements , Electronic Colloquium on Computational Complexity, Report No.65 (1998) 679
4. P. Berman, V. Ramaiyer, Improved approximations for the Steiner tree problem, Journal of Algorithms 17 (1994), pp.381–408. 678
5. M. Bern, P. Plassmann, The Steiner Problem with edge lengths 1 an 2, Information Processing Letters 32 (1989), pp. 171-176. 678
6. O. Gabber, Z. Galil, Explicit construction of linear-sized superconcentrators, J. Comput. System Sci. 22 (1981) pp.407-420 687
7. C. Gröpl, S. Hougardy, T. Nierhoff, H. J. Prömel, Approximation algorithms for the Steiner tree problem in graphs, technical report, Humboldt-Universität zu Berlin, 2000. 678
8. J. Håstad, Some optimal inapproximability results, Proceedings of the 28rd Annual Symposium on Theory of Computing, ACM, 1997. 679

9. S. Hougardy, H. J. Prömel, A 1.598 approximation algorithm for the Steiner Problem in graphs, In: Proceeding of the Tenth Annual ACM-SIAM Symposium on Discrete Algorithms 1999, pp. 448–453. 678

10. M. Karpinski, A. Zelikovsky, New approximation algorithms for the Steiner tree problems, Journal of Combinatorial Optimization 1 (1997), pp.47–65. 678

11. A. Lubotzy, P. Philips, P. Sarnak, Ramanujan graphs, Combinatorica, 8, (1988) pp.261–277. 687

12. G. A. Margulis, Explicit construction of concentrators, Problemy Inf. Trans. 9 (1973), pp.325-332 687

13. C. H. Papadimitriou, S. Vempala, On the Approximability of the Traveling Salesman Problem, Proceedings of the 32nd ACM Symposium on the theory of computing, Portland, 2000. 679, 680

14. H. J. Prömel, A. Steger, A new approximation algorithm for the Steiner tree problem with perfomance ratio 5/3, Journal of Algorithms 36 (2000), pp.89-101. 678

15. G. Robins, A. Zelikovsky, Improved Steiner tree approximation in graphs, In: Proceedings of the Eleventh Annual ACM-SIAM Symposium on Discrete Algorithms 2000, pp.770-779. 678

16. A. Zelikovsky, An 11/6-approximation algorithm for the network Steiner problem, Algorithmica 9 (1993), pp. 463–470. 678

17. P. Sarnak, Some Applications of Modular Forms, Cambridge Tracts in Mathematics 99, Cambridge University Press, 1990 679, 687

Alignment between Two RNA Structures[*]

Zhuozhi Wang and Kaizhong Zhang

Dept. of Computer Science, University of Western Ontario
London, Ont. N6A 5B7, Canada
zzwang@csd.uwo.ca,kzhang@csd.uwo.ca

Abstract. The primary structure of a ribonucleic acid (RNA) molecule can be represented as a sequence of nucleotides (bases) over the four-letter alphabet $\{A, C, G, U\}$. The RNA secondary and tertiary structures can be represented as a set of nested base pairs and a set of crossing base pairs, respectively. These base pairs form bonds between $A - U$, $C - G$, and $G - U$.

This paper considers alignment with affine gap penalty between two RNA molecule structures. In general this problem is Max SNP-hard for tertiary structures. We present an algorithm for the case where aligned base pairs are non-crossing. Experimental results show that this algorithm can be used for practical application of RNA structure alignment.

1 Introduction

Ribonucleic Acid (RNA) is an important molecule which performs a wide range of functions in biological systems. In particular it is RNA (not DNA) that contains genetic information of viruses such as HIV and therefore regulates the functions of such viruses. RNA has recently become the center of much attention because of its catalytic properties, leading to an increased interest in obtaining structural information.

It is well known that secondary and tertiary structural features of RNAs are important in the molecular mechanism involving their functions. The presumption, of course, is that to a preserved function there corresponds a preserved molecular conformation and, therefore, a preserved secondary and tertiary structure. Hence the ability to compare RNA structures is useful [8,16,6,9,1,5]. In many problems involving RNAs [3,13], it is actually required to have an alignment between RNA structures in addition to a similarity measure [14].

RNA secondary and tertiary structures are represented as a set of bonded pairs of bases. A bonded pair of bases (base pair) is usually represented as an edge between the two complementary bases involved in the bond. It is assumed that any base participates in at most one such pair. If there is no crossing in the set of edges representing a structure, the structure is considered as secondary structure. Otherwise, the structure is considered as tertiary structure.

---

[*] Research supported partially by the Natural Sciences and Engineering Research Council of Canada under Grant No. OGP0046373.

J. Sgall, A. Pultr, and P. Kolman (Eds.): MFCS 2001, LNCS 2136, pp. 690–703, 2001.
© Springer-Verlag Berlin Heidelberg 2001

In [16], edit distance, a similarity measure, between RNA structures is proposed. This measure takes into account the primary, the secondary and the tertiary information of RNA structures. This measure treats a base pair as a unit and does not allow it to match to two unpaired bases. In general this is a reasonable model since in RNA structures when one base of a pair changes, we usually find that its partner also changes so as to conserve that base pair. Computing edit distance for tertiary structures is proved to be NP-hard [16].

An efficient algorithm has also been developed to compute edit distance when at least one of the structures involved is a secondary structure [16]. In [4], a method was proposed to deal with the practical problem of comparing RNA tertiary structures. The edit distance between two RNA structures corresponds to an optimal mapping between them: in fact, this mapping induces an alignment between RNA structures. The problem with this induced alignment is that it may contain some small gaps. In general, if possible, longer gaps are preferred. The reason is that it is hard to delete the first element, but after that to continue deleting is much easier. A better alignment can be achieved by introducing a gap opening penalty. This kind of affine gap penalty has long been used in sequence alignment [7]. In this paper we consider RNA structural alignment with affine gap penalty.

Another line of works are primary structure based where the comparison is basically done on the primary structure while trying to incorporate secondary structure data [1,5]. The weakness of this approach is that it does not treat a base-pair as a whole entity. For example, in the comparison of two RNAs, a base pair from one RNA can have one nucleotide deleted while the other nucleotide matched to a nucleotide (unpaired or even paired) in the other RNA. Our method treats base-pairs as units, and is closer to the spirit of the comparative analysis method currently being used in the analysis of RNA secondary structures either manually or automatically.

Recently an improved edit distance model was proposed [10] to measure the similarity between RNA structures. Unfortunately, even for secondary structures, optimal solution exists only for special scoring schemes. Therefore this paper is based on the model defined in [16].

Results

We show that computing alignment between RNA tertiary structures is Max SNP-hard. This means that there is no polynomial time approximation scheme (PTAS) for this problem unless P=NP. We present an algorithm for the case where aligned base pairs are non-crossing. With experiments on real RNA data, we then show that this algorithm can be used to compute alignment between RNA tertiary structures in practical applications.

2   RNA Structural Alignment

An RNA structure is represented by $R(P)$, where R is a sequence of nucleotides with $r[i]$ representing the ith nucleotide, and $P \subset \{1, 2, \cdots, |R|\}^2$ is a set of pairs of which each element (i, j), $i < j$, represents a base pair $(r[i], r[j])$ in R.

We use $R[i,j]$ to represent the subsequence of nucleotides from $r[i]$ to $r[j]$. We assume that base pairs in $R(P)$ do not share participating bases. Formally for any (i_1,j_1) and (i_2,j_2) in P, $j_1 \neq i_2$, $i_1 \neq j_2$, and $i_1 = i_2$ if and only if $j_1 = j_2$.

Let $s = r[k]$ be an unpaired base and $p = (r[i],r[j])$ be a base pair in $R(P)$. We define the relation between s and p as follows. We say s is *before* p if $k < i$. We say s is *inside* p if $i < k < j$. We say s is *after* p if $j < k$.

Let $s = (r[i],r[j])$ and $t = (r[k],r[l])$ be two base pairs in $R(P)$, we define the relation between s and t as follows. We say s is *before* t if $j < k$. We say s is *inside* t if $k < i$ and $j < l$. We say s and t are *crossing* if $i < k < j < l$ or $k < i < l < j$.

For an RNA $R(P)$, we define $p_r(\ )$ as follows.

$$p_r(i) = \begin{cases} i & \text{if } r[i] \text{ is an unpaired base} \\ j & \text{if } (r[i],r[j]) \text{ or } (r[j],r[i]) \text{ is a base pair in } P \end{cases}$$

By this definition $p_r(i) \neq i$ if and only if $r[i]$ is a base in a base pair of $R(P)$ and $p_r(i) = i$ if and only if $r[i]$ is an unpaired base of $R(P)$. If $p_r(i) \neq i$, then $p_r(i)$ is the base paired with base i. When there is no confusion, we use R, instead of $R(P)$, to represent an RNA structure assuming that there is an associated function $p_r(\ )$.

Following the tradition in sequence comparison [11,15], we define three edit operations, substitute, delete, and insert, on RNA structures. For a given RNA structure R, each operation can be applied to either a base pair or an unpaired base. To substitute a base pair is to replace one base pair with another. This means that at the sequence level, two bases may be changed at the same time. To delete a base pair is to remove the base pair. At the sequence level, this means to delete two bases at the same time. To insert a base pair is to insert a new base pair. At the sequence level, this means to insert two bases at the same time. Note that there is no relabel operation that can change a base pair to an unpaired base or vice versa.

We represent an edit operation as $a \rightarrow b$, where a and b are either λ, the null label, or labels of base pairs from $\{A,C,G,U\} \times \{A,C,G,U\}$, or unpaired bases from $\{A,C,G,U\}$.

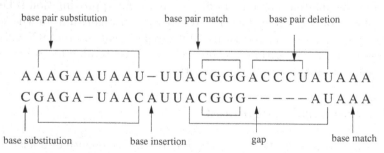

Fig. 1. An RNA structural alignment and the edit operations

We call $a \rightarrow b$ a substitute operation if $a \neq \lambda$ and $b \neq \lambda$; a delete operation if $b = \lambda$; and an insert operation if $a = \lambda$. (see Figure 1 as a simple illustration). Let Γ be a cost function which assigns to each edit operation $a \rightarrow b$ a nonnegative real number $\Gamma(a \rightarrow b)$. We constrain Γ to be a distance metric.

Given two RNA structures R_1 and R_2, a structural alignment of R_1 and R_2 is represented by (R_1', R_2') satisfying the following conditions.

1) R_1' is R_1 with some new symbols $'-'$ inserted and R_2' is R_2 with some new symbols $'-'$ inserted such that $|R_1'| = |R_2'|$.
2) If $r_1'[i]$ is an unpaired base in R_1', then either $r_2'[i]$ is an unpaired base in R_2' or $r_2'[i] = '-'$. If $r_2'[i]$ is an unpaired base in R_2', then either $r_1'[i]$ is an unpaired base in R_1' or $r_1'[i] = '-'$.
3) If $(r_1'[i], r_1'[j])$ is a base pair in R_1', then either $(r_2'[i], r_2'[j])$ is a base pair in R_2' or $r_2'[i] = r_2'[j] = '-'$. If $(r_2'[i], r_2'[j])$ is a base pair in R_2', then either $(r_1'[i], r_1'[j])$ is a base pair in R_1' or $r_1'[i] = r_1'[j] = '-'$.

From this definition, it is clear that alignments preserve the order of unpaired bases and the topological relationship between base pairs.

A gap in an alignment (R_1', R_2') is a consecutive subsequence of $'-'$ in either R_1' or R_2' with maximal length. More formally $[i \cdots j]$ is a gap in (R_1', R_2') if either $r_1'[k] = '-'$ for $i \leq k \leq j$, $r_1'[i-1] \neq '-'$, and $r_1'[j+1] \neq '-'$, or $r_2'[k] = '-'$ for $i \leq k \leq j$, $r_2'[i-1] \neq '-'$, and $r_2'[j+1] \neq '-'$. For each gap in an alignment, in addition to the insertion/deletion costs, we will assign a constant, *gap_cost*, as the gap cost. This means that longer gaps are preferred since for a longer gap the additional cost distributed to each base is relatively small.

Given an alignment (R_1', R_2'), we define single base match SM, single base deletion SD, single base insertion SI, base pair match PM, base pair deletion PD, and base pair deletion PI, as follows.

$$SM = \{\, i \mid r_1'[i] \text{ and } r_2'[i] \text{ are unpaired bases in } R_1 \text{ and } R_2 \}.$$
$$SD = \{\, i \mid r_1'[i] \text{ is an unpaired base in } R_1 \text{ and } r_2'[i] = \; '-'\}.$$
$$SI = \{\, i \mid r_2'[i] \text{ is an unpaired base in } R_2 \text{ and } r_1'[i] = \; '-'\}.$$
$$PM = \{\, (i,j) \mid (r_1'[i], r_1'[j]) \text{ and } (r_2'[i], r_2'[j]) \text{ are base pairs in } R_1 \text{ and } R_2 \}.$$
$$PD = \{\, (i,j) \mid (r_1'[i], r_1'[j]) \text{ is a base pair in } R_1 \text{ and } r_2'[i] = r_2'[j] = \; '-'\}.$$
$$PI = \{\, (i,j) \mid (r_2'[i], r_2'[j]) \text{ is a base pair in } R_2 \text{ and } r_1'[i] = r_1'[j] = \; '-'\}.$$

The cost of an alignment (R_1', R_2') is defined as follows, where $\#gap$ is the number of gaps in (R_1', R_2').

$$cost((R_1', R_2')) = gap\_cost \times \#gap$$
$$+ \sum_{i \in SM} \Gamma(r_1'[i] \rightarrow r_2'[i]) + \sum_{i \in SD} \Gamma(r_1'[i] \rightarrow \lambda) + \sum_{i \in SI} \Gamma(\lambda \rightarrow r_2'[i])$$
$$+ \sum_{(i,j) \in PM} \Gamma((r_1'[i], r_1'[j]) \rightarrow (r_2'[i], r_2'[j])) + \sum_{(i,j) \in PD} \Gamma((r_1'[i], r_1'[j]) \rightarrow \lambda)$$
$$+ \sum_{(i,j) \in PI} \Gamma(\lambda \rightarrow (r_2'[i], r_2'[j]))$$

Given two RNA structures R_1 and R_2, our goal is to find the alignment with minimum cost:

$$A(R_1, R_2) = \min_{(R_1', R_2')} \{cost((R_1', R_2'))\}.$$

When $gap\_cost = 0$, this measure is the same as the edit distance between RNA structures proposed in [16].

3   Max-SNP Hard Result

We now consider the problem of alignment between RNA structures with affine gap penalty where both structures are tertiary structures. We show that this is in general Max-SNP hard.

We L-reduce the problem Max-Cut to this problem. Max-Cut is Max-SNP hard even when the degrees of the vertices in the graph are bounded by 3 [12]. Since by attaching a small constant size graph to the nodes with degree 1 or 2, we can make them to be degree 3, it is easy to see that Max-Cut problem is still Max-SNP hard when the degrees of the vertices are exactly 3. Suppose that we are given such a graph $G = \langle V, E \rangle$, where $|V| = \{v_1, v_2, \ldots, v_n\}$, $|E| = m$, $deg(v) = 3$ for any $v \in V$, and K is the value of maximum cut of G.

For any $v_i \in V$, construct two structures p_i and q_i as in Figure 2. p_i^{in} consists of the three inner most base pairs and the five bases inside these base pairs. p_i^{out} consists of these base pairs enclosing p_i^{in} and the number of base pairs in p_i^{out} is a fixed constant. q_i^{out} is identical to p_i^{out} and q_i^{in1} and q_i^{in2} are identical to p_i^{in}. q_i^{in1} is the left piece and q_i^{in2} is the right piece.

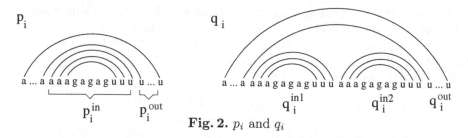

Fig. 2. p_i and q_i

Let S_1 and S_2 be the following structures.

$$S_1 = p_1\, p_2 \; \cdots \; p_n \qquad S_2 = q_1\, q_2 \; \cdots \; q_n.$$

For any $e = \langle v_i, v_j \rangle \in E$, we arbitrarily choose an unpaired g from each of p_i^{in} and p_j^{in} and pair them up. At the same positions where we choose g from p_i and p_j, we pair the two g of q_i^{in1} with q_j^{in2}, and pair the two g of q_i^{in2} with q_j^{in1}. Figure 3 illustrates the case. Note that we use (g, g) for convenience, for real RNAs we can use (c, g).

We use the following cost function: $gap\_cost = 1$, $\Gamma((a, u) \to \lambda) = 1$, $\Gamma((g, g) \to \lambda) = 1$, $\Gamma((a, u) \to (g, g)) = 2$, and $\Gamma(a \to \lambda) = 2$.

Lemma 1. *Given an alignment between S_1 and S_2, we can construct in polynomial time another alignment having equal or smaller cost with the following properties:*

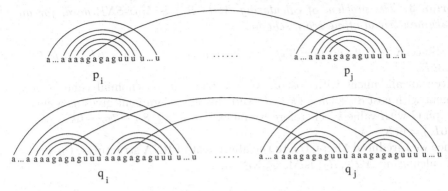

Fig. 3. The construction corresponding the edge $\langle v_i, v_j \rangle$

1. All (a, u) pairs in S_1 are aligned with (a, u) pairs in S_2,
2. p_i^{in} is aligned to either $q_i^{in_1}$ or $q_i^{in_2}$,
3. If p_i^{in} and p_j^{in} are aligned to $q_i^{in_1}$ and $q_j^{in_2}$ or $q_i^{in_2}$ and $q_j^{in_1}$, then any (g, g) pair between p_i and p_j is aligned with a (g, g) pair.

Proof (sketch). When the number of base pairs in p_i^{out} is sufficiently large (≥ 15), one can prove that it is better to align p_i^{out} to q_i^{out}. Then it is easy to show that it is better to align p_i^{in} to either $q_i^{in_1}$ or $q_i^{in_2}$. By the construction of S_2, see Figure 3, there is no edge between $q_i^{in_1}$ and $q_j^{in_1}$ or $q_i^{in_2}$ and $q_j^{in_2}$. Therefore only when p_i^{in} and p_j^{in} are aligned to $q_i^{in_1}$ and $q_j^{in_2}$ or $q_i^{in_2}$ and $q_j^{in_1}$, any (g, g) pair between p_i and p_j can be aligned with (g, g) pair between $q_i^{in_1}$ and $q_j^{in_2}$ or $q_i^{in_2}$ and $q_j^{in_1}$. ☐

Lemma 2. *Let A be an alignment satisfying properties in Lemma 1 and let V' contain the v_i's such that p_i^{in} is aligned to $q_i^{in_1}$ and V'' contain the v_i's such that p_i^{in} is aligned to $q_i^{in_2}$. Then $cost(A) = 8n + 7m - 6k$, where k is the number of edges between V' and V''.*

Proof (sketch). Since only one of $q_i^{in_1}$ and $q_i^{in_2}$ is aligned, we need to delete $3n$ (a, u) pairs and $2n$ a bases from S_2. There are also n gap costs. The total cost is $3n + 4n + n = 8n$. There are m (g, g) pairs in S_1 and there are $2m$ (g, g) pairs in S_2. All (g, g) pairs corresponding to edges between V' and V'' are aligned. And only these (g, g) pairs can be aligned. Therefore we need to delete $m - k$ (g, g) pairs from S_1 with a cost of $3(m - k)$, $m - k$ for deleting these pairs and $2(m - k)$ for gap cost. We also need to delete $2m - k$ (g, g) pairs from S_2 with a cost of $2m - k + 2(m - k)$, $2m - k$ for deleting these pairs and $2(m - k)$ for gap cost. Hence $cost(A) = 8n + 7m - 6k$. ☐

From Lemma 1 and 2, it is clear that $A(S_1, S_2) = 8n + 7m - 6K$.

Theorem 3. *The problem of calculating $A(R_1, R_2)$ is Max-SNP hard, for an arbitrary non-trivial affine score scheme.*

Proof. Since $K \geq n/2$ and $m = 3n/2$, we have $d_{opt} = 8n + 7m - 6K = 37n/2 - 6K \leq 31K$.

Given an alignment with cost d, we construct, in polynomial time, a new alignment with cost $8n + 7m - 6k \leq d$. And this new alignment gives us a cut of the graph G with value k. Therefore we have, $d - d_{opt} \geq 8n + 7m - 6k - (8n + 7m - 6K) = 6(K - k)$.

Since we L-reduced Max-CUT to the alignment of RNA structures, the problem of calculating $A(R_1, R_2)$ is Max-SNP hard.   □

4   Algorithms

Since aligning crossing base pairs is difficult, we add one more condition in defining the structural alignment.

4) If $(r_1'[i], r_1'[j])$ and $(r_1'[k], r_1'[l])$ are base pairs in R_1' and $(r_2'[i], r_2'[j])$ and $(r_2'[k], r_2'[l])$ are base pairs in R_2', then $(r_1'[i], r_1'[j])$ and $(r_1'[k], r_1'[l])$ are non-crossing in R_1' and $(r_2'[i], r_2'[j])$ and $(r_2'[k], r_2'[l])$ are non-crossing in R_2'.

Therefore even though the input RNA structures may have crossing base pairs, the aligned base pairs are non-crossing. We present an algorithm which computes the optimal alignment based on this new alignment definition. We show that our algorithm can be used for aligning tertiary structures in practical application, though the alignment may not be the optimal one according to the original definition.

In extending techniques of Gotoh [7] from sequence alignment to structural alignment, the main difficulty is that with the deletion of a base pair two gaps might be created simultaneously.

We use a bottom up dynamic programming algorithm to find the optimal alignment between R_1 and R_2. We consider the smaller substructures first and eventually consider the whole structures R_1 and R_2.

4.1   Property of Optimal Alignments

Consider two RNA structures R_1 and R_2, we use $\Gamma(\ )$ to define $\gamma(i, j)$ for $0 \leq i \leq |R_1|$ and $0 \leq j \leq |R_2|$.

$$\gamma(i, 0) = \Gamma(r_1[i] \to \lambda) \qquad\qquad \text{if } i = p_{r_1}(i),$$
$$\gamma(0, i) = \Gamma(\lambda \to r_2[i]) \qquad\qquad \text{if } i = p_{r_2}(i),$$
$$\gamma(i, j) = \Gamma(r_1[i] \to r_2[j]) \qquad\qquad \text{if } i = p_{r_1}(i) \text{ and } j = p_{r_2}(j),$$
$$\gamma(i, 0) = \gamma(j, 0) = \Gamma((r_1[i], r_1[j]) \to \lambda)/2 \quad \text{if } i = p_{r_1}(j) < j,$$
$$\gamma(0, i) = \gamma(0, j) = \Gamma(\lambda \to (r_2[i], r_2[j]))/2 \quad \text{if } i = p_{r_2}(j) < j,$$
$$\gamma(i, j) = \Gamma((r_1[i_1], r_1[i]) \to (r_2[j_1], r_2[j])) \quad \text{if } i_1 = p_{r_1}(i) < i \text{ and } j_1 = p_{r_2}(j) < j.$$

From this definition, if $r_1[i]$ is a single base, then $\gamma(i,0)$ is the cost of deleting this base and if $r_1[i]$ is a base of a base pair, then $\gamma(i,0)$ is half of the cost of deleting this base pair. Therefore we distribute evenly the deletion cost of a base pair to its two bases. The meaning of $\gamma(0,i)$ is similar. When $i > 0$ and $j > 0$, $\gamma(i,j)$ is the cost of aligning base pairs $(r_1[i_1], r_1[i])$ and $(r_2[j_1], r_2[j])$.

We now consider the optimal alignment between $R_1[i_1, i_2]$ and $R_2[j_1, j_2]$. We use $A(i_1, i_2 \; ; \; j_1, j_2)$ to represent the optimal alignment cost between $R_1[i_1, i_2]$ and $R_2[j_1, j_2]$. We use $D(i_1, i_2 \; ; \; j_1, j_2)$ to represent the optimal alignment cost such that $r_1[i_2]$ is aligned to $'-'$. If $i_1 \leq p_{r_1}(i_2) < i_2$, then by the definition of alignment, in the optimal alignment of $D(i_1, i_2 \; ; \; j_1, j_2)$, $r_1[p_{r_1}(i_2)]$ has to be aligned to $'-'$. We use $I(i_1, i_2 \; ; \; j_1, j_2)$ to represent the optimal alignment cost such that $r_2[j_2]$ is aligned to $'-'$. If $j_1 \leq p_{r_2}(j_2) < j_2$, then in the optimal alignment of $I(i_1, i_2 \; ; \; j_1, j_2)$, $r_2[p_{r_2}(j_2)]$ has to be aligned to $'-'$.

In computing $A(i_1, i_2 \; ; \; j_1, j_2)$, $D(i_1, i_2 \; ; \; j_1, j_2)$ and $I(i_1, i_2 \; ; \; j_1, j_2)$, for any $i_1 \leq i \leq i_2$, if $p_{r_1}(i) < i_1$ or $i_2 < p_{r_1}(i)$, then $r_1[i]$ will be forced to be aligned to $'-'$; for any $j_1 \leq j \leq j_2$, if $p_{r_2}(j) < j_1$ or $j_2 < p_{r_2}(j)$, then $r_2[j]$ will be forced to be aligned to $'-'$. It will be clear (see Lemma 6, 7 and 8) that this is used to deal with two situations: aligning one base pair among crossing base pairs and deleting a base pair.

We can now consider how to compute the optimal alignment between $R_1[i_1, i_2]$ and $R_2[j_1, j_2]$. The first two lemmas are trivial, so we omit their proofs.

Lemma 4.
$$A(\emptyset \; ; \; \emptyset) = 0$$
$$D(\emptyset \; ; \; \emptyset) = gap\_cost$$
$$I(\emptyset \; ; \; \emptyset) = gap\_cost$$

Lemma 5. For $i_1 \leq i \leq i_2$ and $j_1 \leq j \leq j_2$,

$$D(i_1, i \; ; \; \emptyset) = D(i_1, i-1 \; ; \; \emptyset) + \gamma(i,0) \qquad I(\emptyset \; ; \; j_1, j) = I(\emptyset \; ; \; j_1, j-1) + \gamma(0,j)$$
$$A(i_1, i \; ; \; \emptyset) = D(i_1, i \; ; \; \emptyset) \qquad A(\emptyset \; ; \; j_1, j) = I(\emptyset \; ; \; j_1, j)$$
$$I(i_1, i \; ; \; \emptyset) = D(i_1, i \; ; \; \emptyset) + gap\_cost \qquad D(\emptyset \; ; \; j_1, j) = I(\emptyset \; ; \; j_1, j) + gap\_cost$$

Lemma 6. For $i_1 \leq i \leq i_2$ and $j_1 \leq j \leq j_2$,

$$D(i_1, i \; ; \; j_1, j) = \min \begin{cases} D(i_1, i-1 \; ; \; j_1, j) + \gamma(i,0) \\ A(i_1, i-1 \; ; \; j_1, j) + \gamma(i,0) + gap\_cost \end{cases}$$

Proof. If $D(i_1, i \; ; \; j_1, j)$ is from $D(i_1, i-1 \; ; \; j_1, j)$, then aligning $r_1[i]$ to $'-'$ does not open a gap. Therefore there is no gap penalty. If $D(i_1, i \; ; \; j_1, j)$ is from either $I(i_1, i-1 \; ; \; j_1, j)$ or an alignment such that $r_1[i]$ aligned to $r_2[j]$, then aligning $r_1[i]$ to $'-'$ opens a gap. Therefore there is a gap penalty.

Notice that if $i_1 \leq p_{r_1}(i) < i$, then aligning $r_1[i]$ to $'-'$ means aligning $r_1[p_{r_1}(i)]$ to $'-'$. Therefore with the deletion of a base pair, two gaps may be opened. However aligning $r_1[p_{r_1}(i)]$ to $'-'$ is indeed true in both

$D(i_1, i-1 \; ; \; j_1, j)$ and $A(i_1, i-1 \; ; \; j_1, j)$. The reason for this is that for the base pair $(r_1[p_{r_1}(i)], r_1[i])$, one base, $r_1[p_{r_1}(i)]$, is inside the interval $[i_1, i-1]$ and one base, $r_1[i]$, is outside the interval $[i_1, i-1]$. This means that $r_1[p_{r_1}(i)]$ is forced to be aligned to $'-'$. $\qquad\square$

Lemma 7. *For $i_1 \le i \le i_2$ and $j_1 \le j \le j_2$,*

$$I(i_1, i \; ; \; j_1, j) = \min \begin{cases} I(i_1, i \; ; \; j_1, j-1) + \gamma(0, j) \\ A(i_1, i \; ; \; j_1, j-1) + \gamma(0, j) + gap\_cost \end{cases}$$

Proof. Similar to Lemma 6. $\qquad\square$

Lemma 8. *For $i_1 \le i \le i_2$ and $j_1 \le j \le j_2$,*

if $i = p_{r_1}(i)$ and $j = p_{r_2}(j)$, then

$$A(i_1, i \; ; \; j_1, j) = \min \begin{cases} D(i_1, i \; ; \; j_1, j) \\ I(i_1, i \; ; \; j_1, j) \\ A(i_1, i-1 \; ; \; j_1, j-1) + \gamma(i, j) \end{cases}$$

if $i_1 \le p_{r_1}(i) < i$ and $j_1 \le p_{r_2}(j) < j$, then

$$A(i_1, i \; ; \; j_1, j) = \min \begin{cases} D(i_1, i \; ; \; j_1, j) \\ I(i_1, i \; ; \; j_1, j) \\ A(i_1, p_{r_1}(i) - 1 \; ; \; j_1, p_{r_2}(j) - 1) + \\ + A(p_{r_1}(i) + 1, i-1 \; ; \; p_{r_2}(j) + 1, j-1) + \gamma(i, j) \end{cases}$$

otherwise,

$$A(i_1, i \; ; \; j_1, j) = \min \begin{cases} D(i_1, i \; ; \; j_1, j) \\ I(i_1, i \; ; \; j_1, j) \end{cases}$$

Proof. Consider the optimal alignment between $R_1[i_1, i]$ and $R_2[j_1, j]$. There are three cases: 1. $i = p_{r_1}(i)$ and $j = p_{r_2}(j)$, 2. $i_1 \le p_{r_1}(i) < i$ and $j_1 \le p_{r_2}(j) < j$, and 3. all the other cases.

For case 1, since $i = p_{r_1}(i)$ and $j = p_{r_2}(j)$, both $r_1[i]$ and $r_2[j]$ are unpaired bases. In the optimal alignment, $r_1[i]$ may be aligned to $'-'$, $r_2[j]$ may be aligned to $'-'$, or $r_1[i]$ may be aligned to $r_2[j]$. Therefore we take the minimum of the three cases.

For case 2, since $i_1 \le p_{r_1}(i) < i$ and $j_1 \le p_{r_2}(j) < j$, both $(r_1[p_{r_1}(i)], r_1[i])$ and $(r_2[p_{r_2}(j)], r_2[j])$ are base pairs. In the optimal alignment, $(r_1[p_{r_1}(i)], r_1[i])$ may be aligned to $('-', '-')$, $(r_2[p_{r_2}(j)], r_2[j])$ may be aligned to $('-', '-')$, or $(r_1[p_{r_1}(i)], r_1[i])$ may be aligned to $(r_2[p_{r_2}(j)], r_2[j])$.

If $(r_1[p_{r_1}(i)], r_1[i])$ is aligned to $('-', '-')$, then $A(i_1, i \; ; \; j_1, j) = D(i_1, i \; ; \; j_1, j)$. If $(r_2[p_{r_2}(j)], r_2[j])$ is aligned to $('-', '-')$ then $A(i_1, i \; ; \; j_1, j) = I(i_1, i \; ; \; j_1, j)$.

If $(r_1[p_{r_1}(i)], r_1[i])$ is aligned to $(r_2[p_{r_2}(j)], r_2[j])$, then the optimal alignment between $R_1[i_1, i]$ and $R_2[j_1, j]$ is divided into three parts: 1. the optimal alignment between $R_1[i_1, p_{r_1}(i) - 1]$ and $R_2[j_1, p_{r_2}(j)] - 1]$, 2. the optimal alignment between $R_1[p_{r_1}(i) + 1, i-1]$ and $R_2[p_{r_2}(j)] + 1, j-1]$, and 3. the alignment of $(r_1[p_{r_1}(i)], r_1[i])$ to $(r_2[p_{r_2}(j)], r_2[j])$. This is true since any base pair across $(r_1[p_{r_1}(i)], r_1[i])$ or $(r_2[p_{r_2}(j)], r_2[j])$ should be aligned to $'-'$ and the cost of such an alignment has already been included in part 1 and part 2. Hence we have $A(i_1, i \; ; \; j_1, j) = A(i_1, p_{r_1}(i) - 1 \; ; \; j_1, p_{r_2}(j) - 1) + A(p_{r_1}(i) + 1, i - 1 \; ; \; p_{r_2}(j) + 1, j - 1) + \gamma(i, j)$.

In case 3, we consider all the other possibilities in which we cannot align $r_1[i]$ to $r_2[j]$. We examine several sub-cases involving base pairs.

- sub-case 1: $p_{r_1}(i) > i$. This means that $r_1[p_{r_1}(i)]$ is outside the interval $[i_1, i]$ and we have to align $r_1[i]$ to $'-'$.
- sub-case 2: $p_{r_2}(j) > j$. This is similar to sub-case 1. Together with sub-case 1, this implies that when $p_{r_1}(i) > i$ and $p_{r_2}(j) > j$, even if $r_1[i]=r_2[j]$, we cannot align them to each other.
- sub-case 3: $p_{r_1}(i) < i_1$. This is similar to sub-case 1. Together with sub-case 1, we know that if a base pair is across an aligned base pair, then it has to be aligned to $'-'$.
- sub-case 4: $p_{r_2}(j) < j_1$. This is similar to sub-case 3.   □

4.2   Basic Algorithm

From the above lemmas, we can compute $A(R_1, R_2)=A(1, |R_1| \; ; \; 1, |R_2|)$ using a bottom up approach. Moreover, it is clear that we do not need to compute all $A(i_1, i_2 \; ; \; j_1, j_2)$. From Lemma 8, we only need to compute those $A(i_1, i_2 \; ; \; j_1, j_2)$ such that $(r_1[i_1 - 1], r_1[i_2 + 1])$ is a base pair in R_1 and $(r_2[j_1 - 1], r_2[j_2 + 1])$ is a base pair in R_2.

Given R_1 and R_2, we can first compute sorted base pair lists L_1 for R_1 and L_2 for R_2. This sorted order is in fact a bottom up order since, for two base pairs s and t, if s is before or inside t, then s is before t in the sorted list. For each pair of base pairs $L_1[i] = (i_1, i_2)$ and $L_2[j] = (j_1, j_2)$, we use Lemma 4 through Lemma 8 to compute $A(i_1 + 1, i_2 - 1 \; ; \; j_1 + 1, j_2 - 1)$.

Let R_1 and R_2 be the two given RNA structures and P_1 and P_2 be the number of base pairs in R_1 and R_2 respectively. The time to compute $A(i_1, i_2 \; ; \; j_1, j_2)$ is $O((i_2 - i_1)(j_2 - j_1))$ which is bounded by $O(|R_1| \times |R_2|)$. The time complexity of the algorithm in worst case is $O(P_1 \times P_2 \times |R_1| \times |R_2|)$. We can improve our algorithm so that the worst case running time is $O(S_1 S_2 |R_1||R_2|)$ where S_1 and S_2 are the number of stems, i.e. stacked pairs, in R_1 and R_2 respectively. The space complexity of the algorithm is $O(|R_1| \times |R_2|)$.

Notice that when one of the RNAs is a secondary structure, this algorithm computes the optimal solution of the problem. Also, since the number of tertiary interactions is relatively small compared with the number of secondary interactions, we can use this algorithm to compute the alignment between RNA tertiary structures. Essentially the algorithm tries to find the best sets of non-crossing base pairs to align and delete tertiary interactions. Although this is not an optimal solution, in practice it would produce a reasonable result by aligning most of the base pairs [4].

We can add another step to align tertiary base pairs if they are not in conflict with the base pairs already aligned by our algorithm. This can be considered as a constrained alignment problem where our goal is to find the optimal alignment using these aligned base pairs as the constraints. Due to the page limitation, we omit the details of the constrained alignment.

5   Experimental Results

We performed extensive experiments of our alignment algorithm on real RNA structures. The results show that our algorithm can be used to produce high quality alignments between RNA tertiary structures.

In our experiments, we compute alignments between RNA tertiary structures. Figure 4 shows a 2-D drawing of two RNA structures where secondary bondings are represented by a dash or a dot between two bases and tertiary bondings are represented by a solid line between distant bases.

These RNA structures are taken from the RNase P Database [2]. Ribonuclease P is the ribonucleoprotein endonuclease that cleaves transfer RNA precursors, removing 5' precursor sequences and generating the mature 5' terminus of the tRNA. Alcaligenes eutrophus is from the beta purple bacteria group and Streptomyces bikiniensis is from the high G+C gram positive group. Notice that both RNAs are tertiary structures.

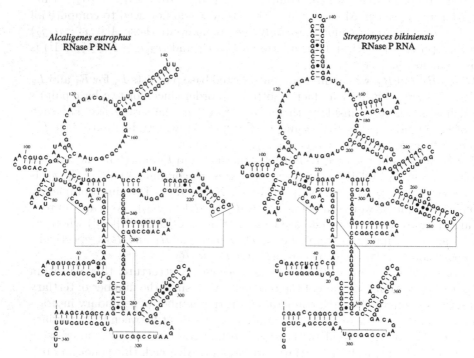

Fig. 4. Alcaligenes eutrophus and Streptomyces bikiniensis from the RNase P database. These images are taken from http://www.mbio.ncsu.edu/RNaseP/

We deal with tertiary structures in the following way. Given two RNA tertiary structures, we first apply our alignment algorithm to produce an alignment where aligned base pairs are non-crossing and then, using constrained alignment, we align tertiary base pairs if they are not in conflict with the base pairs already aligned. The reason for this two step procedure is that in real RNA data, the number of tertiary base pairs is relatively small compared with the number of secondary base pairs. Therefore the first step will handle the majority secondary base pairs and the second step will handle the minority tertiary base pairs.

In our experiment we compare the alignment produced by our procedure with the alignment induced by edit distance mapping defined in [16] using Alcaligenes eutrophus and Streptomyces bikiniensis. Figure 5 shows the two alignments. The top one is produced by using the optimal mapping from the edit distance between two RNAs [16]. The bottom one is produced by our alignment algorithm. It is clear that our alignment is better. Our alignment reduces the number of small gaps and can align tertiary base pairs. In addition, it also corrected some wrong alignments of unpaired bases, i.e. these aligned unpaired bases scattered in the second and third rows of the first alignment (marked by * in Figure 5).

Acknowledgements

We thank Greg Collins, Shuyun Le, Bin Ma and Lusheng Wang for helpful discussions concerning this work. We also thank the anonymous referees for their informative comments.

References

1. V. Bafna, S. Muthukrishnan, and R. Ravi, 'Comparing similarity between RNA strings', Proc. Combinatorial Pattern Matching Conf. 95, LNCS 937, pp.1-14, 1995 690, 691
2. J. W. Brown, 'The Ribonuclease P Database', Nucleic Acids Research, 27:314, 1999. 700
3. J. H. Chen, S. Y. Le, and J. V. Maizel, 'Prediction of common secondary structures of RNAs: a genetic algorithm approach', Nucleic Acids Research, 28:4, pp.991-999, 2000. 690
4. G. Collins, S. Y. Le, and K. Zhang, 'A new method for computing similarity between RNA structures', Proceedings of the Second International workshop on Biomolecular Informatics, pp.761-765, Atlantic City, March 2000. 691, 699
5. F. Corpet and B. Michot, 'RNAlign program: alignment of RNA sequences using both primary and secondary structures', Comput. Appl. Biosci vol. 10, no. 4, pp.389-399, 1995. 690, 691
6. P. A. Evans, Algorithms and Complexity for Annotated Sequence Analysis, PhD thesis, University of Victoria, 1999. 690
7. O. Gotoh, 'An improved algorithm for matching biological sequences', J. Mol. Biol., 162, pp.705-708, 1982. 691, 696
8. T. Jiang, G.-H. Lin, B. Ma, and K. Zhang, 'The longest common subsequence problem for arc-annotated sequences', Proceedings of the 11th Annual Symposium on Combinatorial Pattern Matching (CPM 2000), LNCS 1848, pages 154–165, 2000. 690
9. H. Lenhof, K. Reinert, and M. Vingron. 'A polyhedral approach to RNA sequence structure alignment'. Proceedings of the Second Annual International Conference on Computational Molecular Biology (RECOMB'98), pages 153–159, 1998. 690

```
Alcaligenes eutrophus, Streptomyces bikiniensis: edit distance, gap_cost=0, score= 203.0

((((((((((-------((((((( (( ((((((( - )))))))) ))   [[[ [[[[[[((( ----[[[[ -(((((- ((((    ))))- (((((    )))) )
((((((((((((((((((------- (( ((((((- --)))))))) ))   [[[ [[[[[[((( [[[[---- (((-(((-((((    )))))-((((( ))))) )
AAAGCAGGCCA-------GGCAACCGCUGCCUGCACCG-CAAGGUGCAGGGGGAGGAAAGUCCGGACUCCACA----GGGCA-GGGUG-UUGGCUAACAGCCA-UCCACGGCAACGUGCG
CGAGCCGGGCGGGCGGCC-------GCGUGGGGGUC-UUCG--GACCUCCCCGAGGAACGUCCGGGCUCCACAGAGC----AGGG-UGG-UGGCUAACGGCCAC-CCGGGGUGACCCGCG

(( ((         - -   (((( ((((((( ))))))) )--)))          ------ ---------------------- --------- - --- -))))))))
(( ((         --(((-(((((((( )))))))-) ))-  (((((     ))))) (((((((( )))))))     ))))))-
GAAUAGGGCCACAGAGA-C-GAGUCUUGCCGCCGGGUUCGCCCGGCGGGA--AGGGUGAAAC------G----------------------CG---------G-UA---A-CCUCCACC
GGACAGUGCCACAGAAAACAGA--CCG-CCGGGGACCUCGGUCCUCGG-UAAGG-GUGAAACGGUGGUGUAAGAGACCACCAGCGCCUGAGGCGACUCAGGCGGCUAGGUAAACCCCA-C
                                                      *          **     * **  *

)-))))  ((((  - ------- - --------(((((( -((((-( ----]]]])-)))))))))    ))))    -(((((((( ))))))))-  -
))))))  ((((  ((((((( )))))))) ((((((  ((((( ]]]----)))))))))))    ))))      (((((((-( )-)))))))
U-GGAGCAAUCCCAA-A--------UA-G----------GCAGGCGAU-GAAG-CG----GCCCG-CUGAGUCUGCGGGUAGGGAGCUGGA-GCCGGCUGGUAACAGCCGGC-CUAGA-G
UCGGAGCAAGGUCAAGAGGGGACACCCGGUGUCCCUGCGCGGAUGUUCGAGGGCUGCUC----GCCCGAGUCCGCGGGUAGACCGCACGAGGCCGGC-GGCAAC-GCCGGCCCUAGAUG
       *   ** *

-------))))))))  (( (((--((((((   ))))-))))) ))   ]]]]]]]] ) ))))))))))
- ))))))))------- --((-((( --(((( )))) --)))-))--   ]]]]]]] ) ))))))))))
GAAU-------GGUUGUCACGCACCG--UUUGCCGCAAGGCG-GGCGGGGCGCACAGAAUCCGGCUUUAUCGGCCUGCUUUGCUU
GA-UGGCCGUC-------G-CC-CCGAC--GACCGCGAGGUCC--CGG-GG--ACAGAACCCGGCGUACAGCCCGACUCGUCUG


Alcaligenes eutrophus, Streptomyces bikiniensis: gap_cost=4, score= 242.0

((((((((((((((((((( (( ((((((( ))))))) ))   [[[ [[[[[[((( [[[[ (((((- ((((    ))))- (((( ))))  )(( ((
((((((((((((((((((( (( ((((((- -))))))) ))   [[[ [[[[[[((( [[[[ (((((-(((( )))))-(((( )))) )(( ((
AAAGCAGGCCAGGCAACCGCUGCCUGCACCGCAAGGUGCAGGGGGAGGAAAGUCCGGACUCCACAGAGGGCAGGGUG-UUGGCUAACAGCCA-UCCACGGCAACGUGCGGAAUAGGGCCACA
CGAGCCGGGCGGGCGGCCGCGUGGGGGUC-UUCG-GACCUCCCCGAGGAACGUCCGGGCUCCACAGAGCAGGGUGG-UGGCUAACGGCCAC-CCGGGGUGACCCGCGGGGACAGUGCCACA

-(((( ((((( ))))))) ))-))          ------------------------------------------   ))))))))))))) ((((
(((--((((((( )))))))--) ))   ((((((     ))))) (((((((( ))))))))                  ))))))))))))) ((((
GAGACGAG-UCUUGCCGCCGGGUUCGCCCGGCGGGAA--GGGUGAAAC----------------------------------------------GCGGUAACCUCCACCUGGAGCAAUCCCAA
GAAAACAGACCG--CCGGGGACCUCGGUCCUCGG--UAAGGGUGAAACGGUGGUGUAAGAGACCACCAGCGCCUGAGGCGACUCAGGCGGCUAGGUAAACCCCACUCGGAGCAAGGUCAA

-------------------- ((((((  -(((((- ]]]]-)))))))))))    ))))    (((((((( ))))))))   )))))))  (( (((((-
((((((((  )))))))) (((((    ((((( ]]])))))))))))    ))))      (((((((( ))))))))   )))))))  ----(((((
AU--------------------AGGCAGGCGAU-GAAGC-GGCCC-GCUGAGUCUGCGGGUAGGGAGCUGGAGCCGGCUGGUAACAGCCGGCCUAGAGGAAUGGUUUGUCACGCACCGUU-
GAGGGGACACCCGGUGUCCCUGCGCGGAUGUUCGAGGGCUGCUCCGCCCGAGUCCGCGGGUAGACCGCACGAGGCCGGCGGCAACGCCGGCCCUAGAUGGAUGGCCGUCG----CCCCGA

-(((( ))))-)))) ))   ]]]]]]]] ) ))))))))))
(((( )))) )))))----- ]]]]]]] ) ))))))))))
-UGCCGCAAGGCG-GGCGGGGCGCACAGAAUCCGGCUUUAUCGGCCUGCUUUGCUU
CGACCGCGAGGUCCCGGGG-----ACAGAACCCGGCGUACAGCCCGACUCGUCUG
```

Fig. 5. Comparison of edit distance mapping and alignment

10. G.-H. Lin, B. Ma, and K. Zhang, 'Edit distance between two RNA structures', to appear *Proceedings of the Fifth Annual International Conference on Computational Molecular Biology (RECOMB'2001)*, pp. 200-209, 2001. 691

11. S. E. Needleman and C. D. Wunsch, 'A general method applicable to the search for similarities in the amino-acid sequences of two proteins', *J. Mol. Bio.*, 48, pp.443-453, 1970 692

12. C. H. Papadimitriou and M. Yannakakis, 'Optimization, approximation, and complexity classes', *J. Comput. System Sciences*, vol. 43, pp.425-440, 1991. 694

13. Y. Sakakibara, M. Brown, R. Hughey, I. S. Mian, K. Sjölander, R. Underwood, and D. Haussler, 'Stochastic context-free grammar for tRNA modeling', *Nucleic aids research*, vol. 22, no. 23, pp. 5112-5120, 1994. 690

14. D. Sankoff, 'Simultaneous solution of the RNA folding, alignment and protosequence problems', *SIAM J. Appl. Math* vol. 45, no. 5, pp.810-824, 1985. 690

15. T. F. Smith and M. S. Waterman, 'Comparison of biosequences', *Adv. in Appl. Math.* 2, pp.482-489, 1981 692

16. K. Zhang, L. Wang, and B. Ma, 'Computing similarity between RNA structures', *Proceedings of the Tenth Symposium on Combinatorial Pattern Matching. LNCS 1645*, pp. 281-293, 1999. 690, 691, 694, 701

Characterization of Context-Free Languages with Polynomially Bounded Ambiguity

Klaus Wich

Institut für Informatik, Universität Stuttgart,
Breitwiesenstr. 20-22, 70565 Stuttgart
wich@informatik.uni-stuttgart.de

Abstract. We prove that the class of context-free languages with polynomially bounded ambiguity (*PCFL*) is the closure of the class of unambiguous languages (*UCFL*) under projections which deletes Parikh bounded symbols only. A symbol a is Parikh bounded in a language L if there is a constant c such that no word of L contains more than c occurrences of a. Furthermore *PCFL* is closed under the formally stronger operation of Parikh bounded substitution, i.e., a substitution which is the identity for non Parikh bounded symbols. Finally we prove that the closure of *UCFL* under union and concatenation is a proper subset of *PCFL*.

1 Introduction

A context-free (for short cf) grammar is unambiguous if each word has at most one derivation tree. Otherwise it is ambiguous. There are cf languages which cannot be generated by unambiguous cf grammars [8]. These languages are called (inherently) ambiguous. A cf grammar has a k-bounded ambiguity if no word has more than k derivation trees. It is k-ambiguous if it is k-bounded but not $(k-1)$-bounded. A language L is ambiguous of degree k if it is generated by a k-ambiguous grammar, but it cannot be generated by a $(k-1)$-bounded grammar. There are examples for k-ambiguous languages for each $k \in \mathbb{N}$ [6]. But even languages with infinite degree of ambiguity exist [3]. We can distinguish these languages by the asymptotic behavior of their ambiguity with respect to the length of the words. The ambiguity function of a cf grammar is a monotonous function which yields the maximal number of derivation trees for words up to a given length. By an undecidable criterion which is related to the Pumping Lemma it can be shown that each cycle-free cf grammar is either exponentially ambiguous or its ambiguity is bounded by a polynomial [10,11]. Consequently there is a gap between polynomial and exponential ambiguity. We introduce (inherent) asymptotic ambiguity for languages similarly as above. There are cf languages with exponential ambiguity and with $\Theta(n^k)$ ambiguity for each $k \in \mathbb{N}$ [7]. Even a linear cf language with logarithmic ambiguity is known [12].

A symbol a is Parikh bounded in a language L if there is a constant c such that no word of L contains more than c occurrences of a. Thus c is an upper

J. Sgall, A. Pultr, and P. Kolman (Eds.): MFCS 2001, LNCS 2136, pp. 703–714, 2001.
© Springer-Verlag Berlin Heidelberg 2001

bound for the corresponding component in the Parikh mapping. A substitution is called Parikh bounded if it is the identity for non Parikh bounded symbols. It is easily seen that $PCFL$ is closed under Parikh bounded substitution. We prove that $PCFL$ is the closure of unambiguous context-free languages ($UCFL$) under Parikh bounded projection, i.e., a substitution which deletes Parikh bounded symbols only. The construction is effective and can be computed in polynomial time. Note that Parikh bounded projection is a special case of Parikh bounded substitution for $UCFL$ languages since singletons are unambiguous. Finally we prove that the closure of $UCFL$ under union and concatenation ($UCFL[\cdot, \cup]$) is a proper subset of $PCFL$. This class is interesting since each language in $UCFL[\cdot, \cup]$ is generated by a cf grammar with quadratic Earley parsing time.

2   Preliminaries

Let Σ be a finite alphabet. Let $u = a_1 \cdots a_n \in \Sigma^*$ be a word over Σ, where $a_i \in \Sigma$ for $1 \leq i \leq n$. The *length* of u is $|u| := n$. For $1 \leq i \leq n$ we define $u[i] = a_i$. The *empty word* is denoted by ε. Formal languages and their concatenation are defined as usual [1]. Let Σ and Γ be two alphabets. A *substitution* σ is the homomorphic extension of a mapping $\sigma : \Sigma \to 2^{\Gamma^*}$. For $a \in \Sigma$, $L \subseteq \Sigma^*$, and $\tilde{L} \subseteq \Gamma^*$, the *single symbol substitution* defined by $\sigma(a) = \tilde{L}$ and $\sigma(b) = \{b\}$ for each $b \in \Sigma \setminus \{a\}$ is denoted $[a/\tilde{L}]$. We write $L[a/\tilde{L}]$ for $[a/\tilde{L}](L)$. For $L \subseteq \Sigma^*$ and a substitution σ we define $\sigma(L) := \{u \mid u \in \sigma(w) \text{ for some } w \in L\}$. For $\Gamma \subseteq \Sigma$ the *projection* π_Γ is the homomorphism given by $\pi_\Gamma(b) = b$ for all $b \in \Gamma$ and $\pi_\Gamma(b) = \varepsilon$ for $b \in \Sigma \setminus \Gamma$. For each $w \in \Sigma^*, a \in \Sigma$, and $\Gamma \subseteq \Sigma$ we define $|w|_\Gamma := |\pi_\Gamma(w)|$ and $|w|_a := |w|_{\{a\}}$.

A *context-free grammar* is a triple $G = (V, P, S)$, where V is a finite alphabet, $P \subseteq V \times V^*$ is a finite set of *productions*, and $S \in V$ is the *start symbol*. $N_G := \{A \in V \mid A \times V^* \cap P \neq \emptyset\}$ is called the set of *nonterminals*. $\Sigma_G := V \setminus N_G$ is the set of *terminals*. A production $p = (A, \alpha)$ is denoted by $A \to \alpha$ or $(A \to \alpha)$. The *left-*, and *right-hand sides* of p are $\ell(p) := A$, and $r(p) := \alpha$, respectively. We call p an *ε-production* if $r(p) = \varepsilon$.

The usual way to introduce the ambiguity of a word w over Σ_G is by the number of its leftmost derivations, which are in one to one correspondence with the derivation trees for w. This correspondence does not hold for trees which contain nonterminals in their frontier. But in our consideration trees with this property, especially so called pumping trees, play a crucial rule. On the other hand we do not need the full term rewriting formalism to handle derivation trees. Therefore we use an intermediate formalism.

Let $X \in V \cup P$ and $\tau \in (V \cup P)^*$. The *root* of $X\tau$ is X if $X \in V$ and $\ell(X)$ if $X \in P$. The root of $X\tau$ is denoted by $\uparrow X\tau$. The projection $\pi_V(\tau)$ is the *yield* of τ denoted by $\downarrow\tau$. A *derivation tree* ρ is either an element of $V \cup \{p\, r(p) \mid p \in P\}$ or it can be decomposed as $\rho = \tau_1 \rho' \tau_2$ such that ρ' and $\tau_1 (\uparrow\rho') \tau_2$ are derivation trees. The set of derivation trees of G is denoted by $\triangle(G)$. If ρ is a derivation tree then the *interface* of ρ is the pair $\updownarrow\rho := (\uparrow\rho, \downarrow\rho)$. An element from $\{1, \ldots, |\rho|\}$ is said to be a *node* of ρ. A node ν is an *internal node* if $\rho[\nu] \in P$ and we say that the *P-label* of ν is $\rho[\nu]$. The node ν is a *leaf* if $\rho[\nu] \in V$. The *label of* ν is $\rho[\nu]$ if ν is a leaf and it is $\ell(\rho[\nu])$ if ν is an internal node. Note that $|\downarrow\rho|_\Gamma = |\rho|_\Gamma$ holds for each $\Gamma \subseteq V$. Thus $|\rho|_\Gamma$ is the number of leaves in ρ labeled with an element of Γ. A node μ is an *ancestor* of ν if $\rho = \tau_1 \cdots \tau_5$, $\mu = |\tau_1| + 1$, $\nu = |\tau_1\tau_2| + 1$, and $\tau_2\tau_3\tau_4$, $\tau_3 \in \triangle(G)$ for some $\tau_1, \ldots, \tau_5 \in (V \cup P)^*$. The node ν is a *descendant* of μ if μ is an ancestor of ν .The node μ is a *proper ancestor (descendant)* of ν if μ is an ancestor (a descendant) such that $\mu \neq \nu$. Let $\mathcal{P}$ be a property of nodes. The node μ is the *first ancestor* of ν with property $\mathcal{P}$ if no proper descendant of μ which satisfies $\mathcal{P}$ is an ancestor of ν. The *first common ancestor* of two nodes ν_1 and ν_2 is the first ancestor of ν_1 which is an ancestor of ν_2. A node ν is a *son* of a node μ if μ is the first proper ancestor of ν. A *tree format* $\triangle$ is a mapping that assigns to each cf grammar G a subset of $\triangle(G)$. The tree format defined by $\triangle(G)$ for each cf grammar G is $\triangle$. The set of *compressed derivation trees* is defined by $comp(\triangle(G)) := \{\pi_{V \cup P}(\rho) \mid \rho \in \triangle(G)\}$. Note that $\pi_{V \cup P}$ restricted to $\triangle(G)$ is a bijection from $\triangle(G)$ to $comp(\triangle(G))$. If $\downarrow\rho \in \Sigma^*$ then $\pi_{V \cup P}(\rho)$ coincides with the well known left-parse. The *standard derivation tree format* $\triangle_\Sigma^S$ is defined by $\triangle_\Sigma^S(G) := \{\rho \in \triangle(G) \mid \updownarrow\rho \in \{S\} \times \Sigma_G^*\}$. The *language generated by* G is $L(G) := \{\downarrow\rho \mid \rho \in \triangle_\Sigma^S(G)\}$. The set of *sentential forms* is $\zeta(G) := \{\downarrow\rho \mid \rho \in \triangle(G)\}$. *In the sequel the notions language and grammar represent context-free languages and context-free grammars, respectively.* A grammar $G = (V, P, S)$ is *cycle-free* if $\uparrow\rho = \downarrow\rho$ implies $\rho \in V$. It is *reduced* if either for each $p \in P$ there is a $\rho \in \triangle_\Sigma^S(G)$, such that p appears in ρ, or $P = \{S \to S\}$ in case $L(G) = \emptyset$. A reduced grammar $G = (V, P, S)$ is *strongly reduced* if either $L((V, P, A)) \neq \{\varepsilon\}$ for each $A \in N_G$ or $P = \{S \to \varepsilon\}$ in case $L(G) = \{\varepsilon\}$. For each grammar G there is an equivalent strongly reduced grammar which is not larger than G. If G is a grammar, then its canonical strongly reduced equivalent grammar is denoted by $red(G)$. *If not stated otherwise* $G = (V, P, S)$ *is an implicitly "for all" quantified cycle-free strongly reduced grammar and* $N = N_G$, $\Sigma = \Sigma_G$, $A, B, C \in N$; $a, b, c \in \Sigma$; $X, Y \in V$. The *ambiguity function* $am_G : \mathbb{N}_0 \to \mathbb{N}_0$ is defined by $am_G(n) := \max\{|\{\rho \in \triangle_\Sigma^S(G) \mid \downarrow\rho = w\}| \mid w \in \Sigma^* \wedge |w| \leq n\}$. The grammar G is *unambiguous* if $am_G(n) \leq 1$ for each $n \in \mathbb{N}$, it is of *polynomially bounded ambiguity* if $am_G(n) = \mathcal{O}(n^k)$ for some $k \in \mathbb{N}$, and it is *exponentially ambiguous* if $am_G(n) = 2^{\Omega(n)}$. The set of grammars with the corresponding asymptotic ambiguities are denoted by *UCFG*, *PCFG*, and *ECFG*, respectively. The class of languages generated by unambiguous grammars, and grammars with polynomially bounded ambiguity are called *UCFL*, and *PCFL*, respectively. The class of languages which require an exponentially ambiguous grammar to be generated is called *ECFL*. If X is a set of languages or grammars then $\overline{X}$ denotes the set of cycle-free strongly reduced cf grammars or cf languages which do not belong to X. Let $\triangle$ be a tree format. The set $\triangle(G)$ is unambiguous if for all $\rho_1, \rho_2 \in \triangle(G)$ we have $\updownarrow\rho_1 = \updownarrow\rho_2 \Rightarrow \rho_1 = \rho_2$. The set of grammars which are unambiguous w.r.t. $\triangle$ is defined by $U(\triangle) = \{G$ cycle-free and strongly reduced $\mid \triangle(G)$ *is unambiguous*$\}$. Note that $U(\triangle) = UCFG = U(\triangle_\Sigma^S)$.

3   Closure Properties of *PCFL*

Lemma 1. *PCFL is not closed under length preserving homomorphisms.*

Proof. Let $L := (\{a^i \$ b^{i-1} c^j \mid i,j \geq 1\} \cup \{a^i b^{j-1} \$ c^j \mid i,j \geq 1\})^*$ and $h : \{a,b,c,\$\}^* \to \{a,b,c\}^*$ be defined by $h(x) = x$ for $x \neq \$$ and $h(\$) = b$. It is easily seen that $L \in UCFL$. But $h(L) = \{a^i b^j c^k \mid i = j \vee j = k\}^*$ is an inherently exponentially ambiguous language [7].

Definition 2. *We extend $\mathbb{N}$ to a complete lattice by adding a maximal element ω. Thus each subset of $\mathbb{N} \cup \{\omega\}$ has a supremum. Let L be a language over Σ. Let $\Gamma \subseteq \Sigma$. The Parikh supremum is defined by $\Psi_{\sup} : 2^{\Sigma^*} \to (2^\Sigma \to \mathbb{N} \cup \{\omega\})$ where $\Psi_{\sup}(L)(\Gamma) := \sup\{|w|_\Gamma \mid w \in L\}$. For each grammar $G = (V,P,S)$ and each $\Gamma \subseteq V$, the set Γ is called Parikh bounded (Pb) if $\Psi_{\sup}(G)(\Gamma) := \Psi_{\sup}(\zeta(G))(\Gamma) < \omega$. A symbol $X \in V$ is Pb if $\{X\}$ is Pb.*

The Parikh supremum is the maximum number of Γ symbols which can occur in a word of L. We write $\Psi_{\sup}(L)(a)$ for $\Psi_{\sup}(L)(\{a\})$. $\Psi_{\sup}(L)(a)$ can be considered as the supremum of the corresponding component over all Parikh vectors for L. See [9] for the definition of Parikh vectors.

Definition 3. *Let Σ and Γ be finite alphabets. A substitution $\sigma : \Sigma^* \to 2^{\Gamma^*}$ is Parikh bounded for a language $L \subseteq \Sigma^*$ if for each $a \in \Sigma$ which is not Parikh bounded $\sigma(a) = \{a\}$ holds. The projection $\pi_\Gamma(L)$ is a Parikh bounded projection if $\Sigma \setminus \Gamma$ contains Parikh bounded symbols only.*

Lemma 4. *The language class PCFL is closed under Parikh bounded substitution, Parikh bounded projection, concatenation and union.*

Proof. Let $L_1, L_2 \in PCFL$ generated by $G_1, G_2 \in PCFG$ with disjoint sets of nonterminals. Let a be a Parikh bounded symbol in L_1 with the Parikh supremum k. We can construct a grammar for $L_1[a/L_2]$ by replacing each occurrence of an a by the start symbol of G_2. Then each word of $L_1[a/L_2]$ consists of a word from L_1, where all occurrences of a in w (at most k many) have been replaced by words of L_2. Each inserted word is specified by the positions where it starts and ends. Therefore the number of possible factorizations is bounded by $\mathcal{O}(n^{2k})$. Since ambiguity functions are monotonous the ambiguity of a word of length n in $L_1[a/L_2]$ is bounded by $\mathcal{O}(am_{G_1}(n) \cdot (am_{G_2}(n))^k \cdot n^{2k})$, which is polynomially bounded. Each Parikh bounded substitution can be written as a sequence of single symbol substitutions. For the Parikh bounded projection $\pi_{\Sigma_{G_1} \setminus \{a\}}(L_1)$, concatenation, and union we obtain the better upper bounds $\mathcal{O}(am_{G_1}(n) \cdot n^k)$, $\mathcal{O}(am_{G_1}(n) \cdot am_{G_2}(n) \cdot n)$, and $\mathcal{O}(am_{G_1}(n) + am_{G_2}(n))$, respectively. □

4   Types of Symbols, Productions, Trees and Tree Formats

Definition 5. *We say A derives B, denoted by $A \vdash B$, if there exists $\rho \in \triangle(G)$ such that $\uparrow \rho = A$ and $|\rho|_B > 0$. We define an equivalence relation $\equiv$ by: $A \equiv B$ if $A \vdash B \wedge B \vdash A$. The equivalence class of A is denoted by $[A]$.*

Lemma 6. *Let* $X \in V$, $\Gamma_X = \{A \in N \mid A \vdash X\}$ *and* $\Psi := \Psi_{\sup}((V, P, S))$. *Then* $\Psi(\Gamma_X) = \Psi(X)$.

Proof. Since each Element of Γ_X can generate at least one occurrence of X, we obtain $\Psi(\Gamma_X) \leq \Psi(X)$. On the other hand $X \in \Gamma_X$. Therefore $\Psi(X) \in \Psi(\Gamma_X)$. Thus the claim follows. □

Definition 7. *A production* $(A \to \alpha) \in P$ *is called*

- pumping production *if* $|\alpha|_{[A]} > 0$.
- descending production *if it is not a pumping production.*
- bounded production *if it is descending and* A *is Parikh bounded.*
- unbounded production *if it is not bounded.*

The sets of pumping, descending, bounded, and unbounded productions are denoted by, $P_=$, $P_<$, $P_{<\omega}$, and P_ω, respectively.

On a path from a leaf to the root a pumping production can occur arbitrarily often, while a descending production can occur only once. Hence the maximal number of occurrences of a descending production in an arbitrary derivation tree is bounded by the Parikh supremum of its left-hand side. Bounded productions will play a particularly crucial role in the sequel.

Definition 8. *A derivation tree* $\rho \in \Delta(G)$ *is a pumping tree if* $|\rho|_{\uparrow\rho} > 0$, *it is a partial pumping tree if* $|\rho|_{[\uparrow\rho]} > 0$, *and it is an unbounded production tree if it doesn't contain a bounded production, i.e., $|\rho|_{P_{<\omega}} = 0$. The set of pumping, partial pumping, and unbounded production trees, are denoted by $\Delta(G)$, $\text{ǁ}(G)$, and $\Delta(G)$, respectively. The corresponding tree formats are Δ, ǁ, and Δ. For an arbitrary tree format Δ the tree formats Δ^ω and $\Delta^{<\omega}$ are defined by the restriction of Δ to trees with non Parikh bounded and Parikh bounded roots, respectively.*

Note that each tree obtained from a pumping tree by cutting off some subtrees is a partial pumping tree.

Definition 9. *A symbol X is pumpable, if there is a $\rho \in \Delta(G)$ such that* $|\rho|_X > |{\uparrow}\rho|_X$.

Lemma 10. *A symbol X is pumpable if and only if it is not Parikh bounded.*

Proof. If X is pumpable then by definition we have a pumping tree, which pumps occurrences of X. Hence X is not Parikh bounded. If X is not Parikh bounded, we choose a word with sufficiently many occurrences of X. We mark them according to Ogden's iteration Lemma [1] for context-free grammars, and obtain a pumping tree which is appropriate to show the pumpability of X. □

In the sequel we will use the notion pumpable symbol synonymous for non Parikh bounded symbols, without explicitly referencing the previous lemma.

5   Computation of Parikh Suprema

There is a polynomial time algorithm which takes a pair, consisting of a reduced grammar $G = (V, P, S)$ and an alphabet $\Gamma \subseteq V$ (in a binary encoding), as the input and computes $\Psi_{\text{sup}}(G)(\Gamma)$. The algorithm works as follows: First we compute $directlyPumpable := \{X \mid \exists p \in P_= : |r(p)|_X > |\ell(p)|_X\}$. Then we compute $Pumpable := \{Y \mid X \in directlyPumpable \land X \vdash Y\}$. It is easily seen that $Pumpable$ is the set of pumpable symbols. The set $\Gamma \subseteq V$ is not Parikh bounded if and only if at least one $X \in \Gamma$ is not Parikh bounded. Hence we can determine by the algorithm above whether Γ is Parikh bounded or not. If Γ is not Parikh bounded then $\Psi_{\text{sup}}(G)(\Gamma) = \infty$. Otherwise we proceed as follows: First we construct G' by erasing all productions p where $\ell(p)$ is a pumpable symbol and by erasing all occurrences of pumpable symbols in right-hand sides of productions. By Lemma 6 no Parikh bounded symbol can ever be generated by a pumpable symbol. Hence G' is reduced. Again by Lemma 6 the Parikh bound is an invariant w.r.t. the equivalence $\equiv$. Hence we can replace each occurrence of a symbol by its equivalence class and obtain grammar G''. The remaining pumping productions all have the form $[X] \to [X]$ and can be eliminated to obtain G'''. It is easily seen that $\Psi_{\text{sup}}(G)(\Gamma) = \Psi_{\text{sup}}(G''')(\Gamma')$ for $\Gamma' := \{[X] \mid X \in \Gamma\}$. Since the grammar G''' has descending productions only $\triangle(G''')$ is finite. Obviously we can explore all the trees in $\triangle(G''')$ to compute the Parikh supremum in exponential time. But we can use dynamic programming methods to compute the Parikh bound efficiently. Let $G''' = (V, P, S)$. We can allow productions as start symbols with the semantic that this production has to be the first one in each derivation tree. Obviously this definition does not increase the generative power of cf grammars. We compute for each $\mu \in P \cup V$ the Parikh supremum of Γ' in the grammar (V, P, μ). For $\mu \in \Sigma_{G'''}$ the task is trivial. Starting with the terminals we compute $\Psi_{\text{sup}}(G''')(\Gamma')$ bottom up by the use of the equation:

$$\Psi_{\text{sup}}((V, P, X))(\Gamma') = \begin{cases} \max\{\Psi_{\text{sup}}((V, P, \nu))(\Gamma') \mid \nu \in P \cap X \times V^*\} \text{ if } X \in V \\ \sum_{i=1}^{k} \Psi_{\text{sup}}((V, P, X_k))(\Gamma') \text{ if } X = A \to X_1 \cdots X_k \in P \end{cases}$$

6   Characterization of $PCFL$

Theorem 11 ([11]). $\overline{U(\triangle)} \subseteq ECFG$.

The idea of the proof is as follows. If $G \notin U(\triangle)$ then there exist two different pumping trees ρ_1, ρ_2 with a common interface. Thus we can construct trees which contain chains of n such pumping trees. Their interfaces do not depend on the chosen pumping trees at an arbitrary position in the chain. Hence we have at least 2^n ways to construct derivation trees of that kind, all having the same interface. To be sure that this idea yields exponential ambiguity we need the fact that the grammar is *cycle-free* and *strongly reduced*. Even $U(\triangle) \subseteq PCFG$ and the undecidability of $U(\triangle)$ have been proved in [11]. But we will use Theorem 11 only. Included in our final result we will get an alternative proof for $U(\triangle) \subseteq PCFG$. As an immediate consequence of the definitions we obtain the following Lemma:

Lemma 12. *Let Δ_1 and Δ_2 be tree formats. Then $(\forall G : \Delta_1(G) \subseteq \Delta_2(G)) \Rightarrow U(\Delta_2) \subseteq U(\Delta_1)$.*

Lemma 13. *Let Δ, Δ_1, and Δ_2 be tree formats such that: $\forall G : \Delta(G) \subseteq \Delta_1(G) \cup \Delta_2(G)$ and $\{\updownarrow \rho \mid \rho \in \Delta_1(G)\} \cap \{\updownarrow \rho \mid \rho \in \Delta_2(G)\} = \emptyset$. Then $U(\Delta_1) \cap U(\Delta_2) \subseteq U(\Delta)$.*

Proof. Let $G \in U(\Delta_1) \cap U(\Delta_2)$ and $\rho_1, \rho_2 \in \Delta(G)$ such that $\updownarrow \rho_1 = \updownarrow \rho_2$. In case $\rho_1 \in \Delta_1(G)$, the assumption $\rho_2 \in \Delta_2(G)$ would lead to a nonempty intersection of the interfaces of $\Delta_1(G)$ and $\Delta_2(G)$, contradicting our requirements. Hence $\rho_2 \in \Delta_1(G)$ too. Thus $\rho_1 = \rho_2$ follows from $G \in U(\Delta_1)$. The assumption $\rho_1 \in \Delta_2(G)$ yields in an analogous manner $\rho_1 = \rho_2$. Hence $G \in U(\Delta)$. □

Lemma 14. $U(\Delta) = U(\underset{\Delta}{\mp})$.

Proof. Since $\Delta(G) \subseteq \underset{\Delta}{\mp}(G)$ for each grammar G the inclusion $U(\underset{\Delta}{\mp}) \subseteq U(\Delta)$ is a consequence of Lemma 12. Let $G \in U(\Delta)$. Choose arbitrary $\rho_1, \rho_2 \in \underset{\Delta}{\mp}(G)$ such that $\updownarrow \rho_1 = \updownarrow \rho_2$. We have to show that this implies $\rho_1 = \rho_2$. By definition of $\underset{\Delta}{\mp}$ there is a $B \in N$ such that B is contained in the yield and $\uparrow \rho_1 \equiv B$. By definition of $\equiv$ this implies that there is a derivation tree $\rho = \tau_1(\uparrow \rho_1)\tau_2 \in \Delta(G)$ such that $\uparrow \rho = B$. Then we obtain that $\tau_1 \rho_1 \tau_2$ and $\tau_1 \rho_2 \tau_2$ are pumping trees with a common interface. Since $G \in U(\Delta)$ this implies $\tau_1 \rho_1 \tau_2 = \tau_1 \rho_2 \tau_2$. By left- and right cancellation we obtain $\rho_1 = \rho_2$. Thus $G \in U(\underset{\Delta}{\mp})$. □

Lemma 15. $U(\Delta) \subseteq U(\Delta^\omega)$.

Proof. Let $G \in U(\Delta)$. Choose arbitrary $\rho_1, \rho_2 \in \Delta^\omega(G)$ such that $\updownarrow \rho_1 = \updownarrow \rho_2$. Then $X := \uparrow \rho_1$ is pumpable. Hence by definition there is a derivation tree $\rho = \tau_1 X \tau_2 \in \Delta(G)$, such that $|\rho|_X - |\uparrow \rho|_X > 0$. Now either $X = \uparrow \rho$ then $|\tau_1 \tau_2|_{\uparrow \rho} = |\tau_1(\uparrow \rho)\tau_2|_{\uparrow \rho} - |\uparrow \rho|_{\uparrow \rho} = |\rho|_X - |\uparrow \rho|_X > 0$ or $\uparrow \rho \neq X$ then $|\tau_1 \tau_2|_{\uparrow \rho} = |\tau_1 X \tau_2|_{\uparrow \rho} = |\rho|_{\uparrow \rho} > 0$ since $\rho \in \Delta(G)$. Hence in both cases $|\tau_1 \tau_2|_{\uparrow \rho} > 0$. This implies $\tau_1 \rho_1 \tau_2, \tau_1 \rho_2 \tau_2 \in \Delta(G)$. Now $G \in U(\Delta)$ implies $\tau_1 \rho_1 \tau_2 = \tau_1 \rho_2 \tau_2$. By left- and right cancellation we obtain $\rho_1 = \rho_2$. Hence $G \in U(\Delta^\omega)$. □

Lemma 16. *Let $\rho \in \Delta(G)$. If ρ contains a node ν labeled with a Pb symbol A then each ancestor of ν is labeled with a Pb symbol.*

Proof. If ν' is an ancestor of ν labeled with B then B generates A (i.e. $B \vdash A$). Thus by Lemma 6 we have $\Psi_{\sup}(B) \leq \Psi_{\sup}(A)$. Hence B is a Pb symbol. □

Lemma 17. *Let $\rho \in \Delta(G)$. Let ν_1 and ν_2 be two nodes labeled with a Pb symbol where neither one is an ancestor of the other. Then there is a node $\nu < \min\{\nu_1, \nu_2\}$ such that the P-label of ν is a bounded production.*

Proof. Let ν be the first common ancestor of ν_1 and ν_2. Then $\nu \leq \min\{\nu_1, \nu_2\}$. Since neither one is an ancestor of the other $\nu < \min\{\nu_1, \nu_2\}$. By Lemma 16 the node ν and two distinct sons of ν are labeled with Pb symbols. But among the descendants of a pumping production there is at most one Pb symbol. Therefore the P-label of ν must be a descending production. Thus we obtain that the P-label of ν is a bounded production. □

Corollary 18. *Let $\rho \in \triangle(G)$. If ρ contains more than one leaf labeled with a Pb symbol then $\rho \notin \triangleq(G)$.*

Proof. Let ν_1, ν_2 be two distinct leaves of ρ labeled with a Pb symbol. Obviously neither is an ancestor of the other. Hence by Lemma 17 the tree ρ contains a bounded production. □

Lemma 19. *Let $\rho \in \triangle^{<\omega}(G)$. If $|\rho|_{[\uparrow\rho]} = 0$ then $\rho \notin \triangleq(G)$.*

Proof. By definition $\uparrow\rho$ is a Pb symbol. Hence ρ contains a node ν, labeled with a symbol in $[\uparrow\rho]$, which has no descendants in $[\uparrow\rho]$. If $|\rho|_{[\uparrow\rho]} = 0$ then this node cannot be a leaf, i.e., there is a production p which is the P-label of ν. This production is descending by the choice of ν. Moreover $\ell(p) \in [\uparrow\rho]$ is Parikh bounded. Thus p is a bounded production, which implies $\rho \notin \triangleq(G)$. □

Immediately by Corollary 18 and Lemma 19 we obtain:

Lemma 20. *If $\rho \in \triangleq^{<\omega}(G)$ then ρ contains exactly one leaf labeled with a Pb symbol and this symbol is in $[\uparrow\rho]$.*

Lemma 21. $\triangleq^{<\omega}(G) = ⩘^{<\omega}(G)$.

Proof. The inclusion $\triangleq^{<\omega}(G) \subseteq ⩘^{<\omega}(G)$ is an immediate consequence of Lemma 20. Let ρ be an arbitrary element of $⩘^{<\omega}(G)$. Then ρ contains a leaf ν labeled with a symbol $A \in [\uparrow\rho]$. Since $\uparrow\rho$ is Pb it suffices to show that an arbitrary internal node μ of ρ must not have a bounded production as its P-label. First assume μ is an ancestor of ν labeled B. Then $\uparrow\rho \vdash B \vdash A \vdash \uparrow\rho$ since $A \in [\uparrow\rho]$. Hence $B \in [\uparrow\rho]$. But then the first ancestor of ν which is a son of μ is in $[\uparrow\rho]$ as well. This implies that the P-label of μ is a pumping production p. Thus p is not bounded. Now assume μ is no ancestor of ν, then the first common ancestor of μ and ν is a pumping production capable to pump the label of μ and again the P-label of μ must not be a bounded production. □

Theorem 22. $PCFG \subseteq U(\triangle) \subseteq U(\triangleq)$.

Proof. $PCFG \subseteq_{T11} U(\triangle) \subseteq_{L14} U(⩘) \subseteq_{L12} U(⩘^{<\omega})$. Moreover $U(\triangle) \subseteq_{L15} U(\triangle^{\omega})$. Hence $U(\triangle) \subseteq U(\triangle^{\omega}) \cap U(⩘^{<\omega})$. Obviously $\{\updownarrow\rho \mid \rho \in \triangle^{\omega}\} \cap \{\updownarrow\rho \mid \rho \in ⩘^{<\omega}\} = \emptyset$ and $\triangleq(G) \subseteq \triangle^{\omega}(G) \cup \triangleq^{<\omega}(G) \subseteq_{L21} \triangle^{\omega}(G) \cup ⩘^{<\omega}(G)$. Thus by Lemma 13 we obtain $U(\triangle^{\omega}) \cap U(⩘^{<\omega}) \subseteq U(\triangleq)$. □

Obviously $U(\triangleq)$ and $\overline{U(\triangleq)}$ are closed under deletion and insertion of terminals in bounded productions. Now Theorem 22 states that for the generation of polynomially bounded ambiguity, bounded productions must be essential. If we insert sufficiently many markers in bounded productions to destroy their capacity to cause ambiguity then the resulting grammar should be unambiguous. This is exactly what we are about to do. Note that we will use productions of the original grammar as marker symbols in the constructed grammar.

Lemma 23. *Each $\rho \in (\triangle^{<\omega}(G) \cap (\triangle(G) \setminus \mathbb{\triangle}(G)))$ has a unique decomposition $\rho = \xi p \tau \chi$ where $p \in P_{<\omega}$ is a bounded production, such that $p\tau \in \triangle(G)$, and $\xi \ell(p) \chi \in \triangle^{<\omega}$.*

Proof. Let ρ be an element of $(\triangle^{<\omega}(G) \cap (\triangle(G) \setminus \mathbb{\triangle}(G)))$. Then ρ contains at least on bounded production. For each internal node μ in ρ there is a uniquely defined decomposition $\rho = \xi p \tau \chi$ such that $|\xi| = \mu - 1$ and $p\tau$, $\xi \ell(p) \chi \in \triangle(G)$. Our task is to find the appropriate μ. Since ξ must not contain a bounded production but p is a bounded production, the only possible candidate for μ is the smallest integer i such that $\rho[i]$ is a bounded production. Now we choose the uniquely defined $\xi \in (V \cup P_\omega)^*$, $\tau, \chi \in (V \cup P)^*$, and $p \in P_{<\omega}$ with the property $\rho = \xi p \tau \chi$ and $\rho' := \xi \ell(p) \chi$, $p\tau \in \triangle(G)$. Now assume that there is a node μ' in the χ portion of ρ' which is P-labeled by a bounded production. Since $\mu < \mu'$ the node μ' cannot be an ancestor of μ. On the other hand μ is a leaf and is therefore no ancestor of μ'. Thus by Lemma 17 there is a node $\nu < \mu$ which is P-labeled by a bounded production, which is a contradiction to our choice of μ. That implies χ does not contain bounded productions. Therefore $\xi \ell(p) \chi \in \mathbb{\triangle}(G)$. Finally, since $\ell(p)$ is Pb, Lemma 16 implies that $\xi \ell(p) \chi \in \triangle^{<\omega}(G)$. □

Definition 24. *Let $h_G : (N \cup P)^* \to (N \cup (N \times (V \cup P_{<\omega})^*))^*$ be the homomorphism defined by $h_G(X) := X$ for $X \in N \cup P_\omega$ and $h_G(p) := (A \to p u_0 A_1 \cdots p u_{k-1} A_k p u_k)$ for $p = (A \to u_0 A_1 \cdots u_{k-1} A_k u_k) \in P_{<\omega}$. The skeleton grammar for $G = (V, P, S)$ is defined by $s(G) := (V \cup P_{<\omega}, P', S)$ where $P' := \{h_G(p) \mid p \in P\}$.*

Note that restricted to compressed derivation trees the mapping h_G is a bijection from $comp(\triangle(G))$ to $comp(\triangle(s(G)))$.

Lemma 25. $G \in U(\mathbb{\triangle}) \Rightarrow s(G) \in UCFG.$

Proof. Observe that $p \in P$ is an unbounded production of G if and only if $h_G(p)$ is an unbounded production of $s(G)$. Moreover $P_\omega = P'_\omega$ which implies $\mathbb{\triangle}(G) = \mathbb{\triangle}(s(G))$. For $G \in U(\mathbb{\triangle})$ we must show that arbitrary $\rho_1, \rho_2 \in \triangle(s(G))$ have common interfaces only if $\rho_1 = \rho_2$. We prove this by induction on $|\rho_1|_{P_{<\omega}}$. The basis is that $\rho_1 \in \mathbb{\triangle}(s(G))$. Now $\downarrow \rho_1$ does not contain any symbols in $P_{<\omega}$. Since each production in $h(P_{<\omega})$ generates symbols in $P_{<\omega}$ and $\updownarrow \rho_1 = \updownarrow \rho_2$ we obtain that ρ_2 is in $\mathbb{\triangle}(s(G))$ too. By the observation above $\rho_1, \rho_2 \in \mathbb{\triangle}(G)$. Hence $G \in U(\mathbb{\triangle})$ implies $\rho_1 = \rho_2$. Assume the claim has been proved for all $\rho \in \triangle(s(G))$ with at most n bounded productions. Let ρ_1 contain $n+1$ bounded productions. By Lemma 23 for $i \in \{1, 2\}$ we can uniquely decompose $\rho_i = \xi_i \, h(p_i) \, \tau_i \chi_i$ such that $\rho'_i := \xi_i l(h(p_i)) \chi_i \in \mathbb{\triangle}(s(G))$, $h(p_i) \in P'_{<\omega}$, and $h(p_i)\tau_i \in \triangle(s(G))$. Since all bounded productions happen to occur in $h(p_i)\tau_i$ it follows that p_1 and p_2 generate the leftmost and rightmost occurrences of symbols from $P_{<\omega}$ in $\downarrow \rho_1$ and $\downarrow \rho_2$, respectively. By $\updownarrow \rho_1 = \updownarrow \rho_2$ this implies $p := p_1 = p_2$ and $\updownarrow \rho'_1 = \updownarrow \rho'_2$. Now $\rho'_1, \rho'_2 \in \triangle(s(G))$ implies $\rho'_1, \rho'_2 \in \mathbb{\triangle}(G)$. Since $G \in U(\mathbb{\triangle})$ this implies $\rho'_1 = \rho'_2$. Now p is both a terminal of $s(G)$ and a bounded production of G. Thus $h(p) = (A \to p u_0 A_1 p u_1 \cdots A_k p u_k)$ for some $k \in \mathbb{N}$, and for each $j \in \{1, \ldots, k\}$ we have

$A, A_j \in N$ and $u_j \in \Sigma^*$. Then $\tau_i = \tau_{i,1} \cdots \tau_{i,k}$ has, for each $i \in \{1,2\}$, a unique decomposition in k derivation trees $\tau_{i,1}, \ldots, \tau_{i,k} \in \triangle(s(G))$ such that $A_j = \uparrow\tau_{i,j}$. Since $h(p)$ is a descending production it cannot occur in any $\tau_{i,j}$. Hence their yields cannot contain a p. Therefore we can uniquely retrieve the yield of each $\tau_{i,j}$ from $\downarrow\rho_i$, i.e., for each $j \in \{1, \ldots, k\}$ we have $\downarrow\tau_{1,j} = \downarrow\tau_{2,j}$. Hence $\updownarrow\tau_{1,j} = \updownarrow\tau_{2,j}$ for each $j \in \{1, \ldots, k\}$. But since they must not contain $h(p)$ they contain at most n bounded productions. Therefore by the inductive hypothesis $\tau_{1,j} = \tau_{2,j}$ for each $j \in \{1, \ldots, k\}$. This finally implies $\rho_1 = \rho_2$.   □

By elementary combinatorial considerations we obtain:

Lemma 26. $s(G) \in UCFG \Rightarrow am_G(n) = \mathcal{O}(n^k)$, where $k = \psi(s(G))(P_{<\omega})$.

Lemma 27. $PCFG = U(\triangle) = U(\text{⚔}) = U(\text{⚔})$.

Proof. $PCFG \subseteq_{T11} U(\triangle) \subseteq_{L22} U(\text{⚔})$. By Lemma 14 we have $U(\triangle) = U(\text{⚔})$. Finally let $G \in U(\text{⚔})$ then $s(G) \in UCFG$ by Lemma 25. Thus $G \in PCFG$ by Lemma 26.   □

Theorem 28. $am_G = 2^{\otimes(n)}$ or $am_G = \mathcal{O}(n^k)$, where $k = \psi(s(G))(P_{<\omega})$.

Proof. If $G \in \overline{U(\triangle)}$ then $G \in ECFG$ by Theorem 11, i.e., $am_G = 2^{\otimes(n)}$. If $G \in U(\triangle)$ then $G \in U(\text{⚔})$ follows by Lemma 27. Thus $s(G) \in UCFG$ by Lemma 25 and by Lemma 26 we obtain $am_G(n) = \mathcal{O}(n^k)$.   □

Note that the value of k in the theorem above can be computed in polynomial time w.r.t. the size of the grammar.

Theorem 29. *The closure of UCFL under Parikh bounded projection coincides with PCFL*

Proof. By Lemma 4 the closure of $UCFL$ under Parikh bounded projection is a subset of $PCFL$. Let $L \in PCFL$ and let $G = (V, P, S) \in PCFG$ such that $L = L(G)$. Then $G \in U(\text{⚔})$ by Lemma 27. By Lemma 25 this implies $L(s(G)) \in UCFL$. Obviously $L = \pi_V(L(s(G)))$. Finally we observe that π_V is Parikh bounded for $L(s(G))$.   □

Corollary 30. *A grammar* $G = (V, P, S)$ *is in PCFG if and only if* (V, P_ω, S) *is in* $U(\triangle)$.

Proof. Obviously $\triangle((V, P_\omega, S)) = \triangle(G)$. Hence by Theorem 27 the claim follows.   □

In general $PCFG$ is undecidable, which has been shown in [11]. But in special cases Corollary 30 can help us to decide whether G belongs to $PCFG$ since there are fewer derivation trees to consider. Furthermore the constructed grammar is not necessarily reduced. This can be used to break it into a bunch of grammars which are likely to be much smaller and easier to handle than the original one:

Corollary 31. *Let* $G = (V, P, S)$ *be a grammar and* $N_r := \{B \in N \mid \exists p \in P_{<\omega} : |r(p)|_B > 0\} \cup \{S\}$. *Then* $G \in PCFG$ *if and only if for all* $A \in N_r$ *we have* $red((V \cup \tilde{N}_G, P_\omega \cup P', A)) \in UCFG$, *where* $\tilde{N}_G := \{\tilde{X} \mid X \in N\}$ *is a copy of the nonterminals disjoint from* V, *and* $P' := \{A \to \tilde{A} \mid A \in \tilde{N}_G\}$.

7 The Semiring Closure of *UCFL*

Definition 32. *If X is a language class then $X[\cup]$ and $X[\cdot]$ denote the closure of X under union and concatenation, respectively. $X[\cdot, \cup]$ denotes the closure under union and concatenation. $X[\cdot, \cup]$ is called the semiring closure of X.*

Lemma 33. *Each language in $UCFL[\cdot, \cup]$ can be parsed by the Earley algorithm in $\mathcal{O}(n^2)$.*

The previous lemma can be proved analogously to the proof for the quadratic parsing time of metalinear languages [4].

Definition 34. *We define $L_p := \{u \in \{b, c\}^* \mid u = u^R\}$, where u^R is the reversal of u. Thus L_p is the set of palindromes. Now we define $L_\diamond := \{w \in \{a, b, c, \#\}^* \mid \exists i \in \mathbb{N} : \exists u, v \in L_p : w = a^i \# uv \# a^i\}$.*

Lemma 35. $L_\diamond \in PCFL$.

Proof. Let $L_1 := \{a^i \# \$\$ \# a^i \mid i \in \mathbb{N}\}$. Obviously $L_1, L_p \in UCFL$ and the substitution $[\$/L_p]$ is Pb. By Lemma 4 this implies $L_\diamond = L_1[\$/L_p] \in PCFL$. □

Lemma 36. $L_\diamond \notin UCFL[\cup, \cdot]$.

Proof. Assume $L_\diamond \in UCFL[\cup, \cdot]$. By the distributive laws this is equivalent to $L_\diamond \in UCFL[\cdot][\cup]$. Thus for some $k, \ell \in \mathbb{N}$, $U_1, \dots, U_\ell \in UCFL$, and $L_1, \dots, L_k \in UCFL[\cdot] \setminus UCFL$ we have $L_\diamond = (\cup_{i=1}^{\ell} U_i) \cup (\cup_{i=1}^{k} L_i)$. Let us consider L_i for an arbitrary $i \in \{1, \dots, k\}$. Now for some minimal $m \in \mathbb{N}$ we can write $L_i = \tilde{U}_1 \cdots \tilde{U}_m$ where $\tilde{U}_1, \dots, \tilde{U}_m \in UCFL$. Since L_i is ambiguous we have $m > 1$. Each word in $L_\diamond$ contains exactly two $\#$'s. Therefore $\forall j \in \{1, \dots, m\} : \forall u, v \in \tilde{U}_j : |u|_\# = |v|_\#$. Assume the words in $\tilde{U}_1$ do not contain a $\#$ then $\tilde{U}_1$ only contains words of the form a^*. Recall that for each $w \in L_i = \tilde{U}_1 \cdots \tilde{U}_m$ the number of a's to the left of the first $\#$ must match the number of a's to the right of the second $\#$. Therefore $\tilde{U}_1$ must be a singleton. But then $\tilde{U}_1 \cdot \tilde{U}_2$ is unambiguous contradicting the minimal choice of m. Thus each word in $\tilde{U}_1$ must contain the first $\#$. Similarly we obtain that each word in $\tilde{U}_m$ must contain the second $\#$. This implies that the words in $\tilde{U}_1$ and $\tilde{U}_m$ consist of words of the forms $a^* \# \{b, c\}^*$ and $\{b, c\}^* \# a^*$, respectively. Again, if the number of a's would not be fixed we could compose words with non matching "a" blocks. Hence $L_i \subseteq a^{n_i} \# \{b, c\}^* \# a^{n_i}$ for some $n_i \in \mathbb{N}$. We define $n = \max\{n_i \mid i \in \{1, \dots, k\}\} + 1$. Let $R := a^n \# \{b, c\}^* \# a^n$. Then $L_\diamond \cap R = ((\cup_{i=1}^{\ell} U_i) \cup (\cup_{i=1}^{k} L_i)) \cap R = ((\cup_{i=1}^{\ell} U_i) \cap R) \cup ((\cup_{i=1}^{k} L_i) \cap R) = (\cup_{i=1}^{\ell} U_i) \cap R = \cup_{i=1}^{\ell} (U_i \cap R)$. Since unambiguous languages are closed under intersection with regular sets [5], this implies $L_\diamond \cap R \in UCFL[\cup]$. Moreover, unambiguous languages are closed under cancellation of singletons [5]. By cancellation of $a^n \#$ from the left-hand side and $\# a^n$ from the right-hand side, we obtain $L_p L_p \in UCFL[\cup]$. But this is false since in [3] it is proved that $L_p L_p$ has infinite ambiguity. Therefore $L_\diamond \notin UCFL[\cup, \cdot]$. □

As an immediate consequence of Lemmas 4, 35 and 36 we obtain:

Theorem 37. $UCFL[\cup, \cdot] \subsetneq PCFL$.

8   Conclusion

We have shown that *PCFL* is the closure of unambiguous languages under Parikh bounded projection. Even if we use the formally stronger operation of Parikh bounded substitution we cannot leave *PCFL*. There is another nontrivial characterization of *PCFL* which proves a gap between polynomial and exponential ambiguity [11]. By Corollary 30 we know that exponential ambiguity is independent from bounded productions. On the other hand ambiguity which is polynomially bounded crucially depends on bounded productions. Recall that in the construction of the skeleton grammar we have inserted terminals in bounded productions only. By Lemma 27 and Lemma 25 this is sufficient to destroy subexponential ambiguity. The class *PCFL* is not only interesting for structural research, but also for applications. For example, the concept of polynomially bounded ambiguity has been recently applied in [2].

Acknowledgments: Thanks to Markus Lohrey and Gundula Niemann for proofreading and valueable discussions, and to Horst Prote, for some LaTeX tips.

References

1. J. Berstel. *Transductions and context-free languages*. Teubner, Stuttgart, 1979. 704, 707
2. A. Bertoni, M. Goldwurm, and M. Santini. Random generation and approximate counting of ambiguously described combinatorical structures. In H. Reichel and S. Tison, editors, *Proc. STACS 2000*, LNCS 1770, pp. 567–580, Berlin-Heidelberg-New York, 2000. Springer. 714
3. J. Crestin. Un langage non ambigu dont le carré est d'ambiguité non bornée. In M. Nivat, editor, *Automata, Languages and Programming*, pp. 377–390. Amsterdam, North-Holland, 1973. 703, 713
4. J. C. Earley. *An efficient context-free parsing algorithm*. PhD thesis, Carnegie-Mellon Uni., 1968. 713
5. M. A. Harrison. *Introduction to Formal Language Theory*. Addison-Wesley, Reading, 1978. 713
6. H. Maurer. The existence of context-free languages which are inherently ambiguous of any degree. Research series, Dept. of Mathematics, Uni. of Calgary, 1968. 703
7. M. Naji. Grad der Mehrdeutigkeit kontextfreier Grammatiken und Sprachen, 1998. Diplomarbeit, FB Informatik, JWG-Universität Frankfurt/M. 703, 706
8. R. J. Parikh. Language–generating devices. In *Quarterly Progress Report*, volume 60, pp. 199–212. Research Laboratory of Electronics, M. I.T, 1961. 703
9. A. Salomaa and M. Soittola. *Automata theoretic aspects of formal power series*. Springer, 1978. 706
10. K. Wich. Kriterien für die Mehrdeutigkeit kontextfreier Grammatiken, 1997. Diplomarbeit, FB Informatik, JWG-Universität Frankfurt/M. 703
11. K. Wich. Exponential ambiguity of context-free grammars. In G. Rozenberg and W. Thomas, editors, *Proc. DLT, 1999*, pp. 125–138. World Scientific, Singapore, 2000. 703, 708, 712, 714
12. K. Wich. Sublinear ambiguity. In M. Nielsen and B. Rovan, editors, *Proc. MFCS 2000*, LNCS 1893, pp. 690–698, Berlin-Heidelberg-New York, 2000. Springer. 703

Author Index

Lecture Notes in Computer Science

For information about Vols. 1–2048
please contact your bookseller or Springer-Verlag

Vol. 2096: J. Kittler, F. Roli (Eds.), Multiple Classifier Systems. Proceedings, 2001. XII, 456 pages. 2001.

Vol. 2097: B. Read (Ed.), Advances in Databases. Proceedings, 2001. X, 219 pages. 2001.

Vol. 2098: J. Akiyama, M. Kano, M. Urabe (Eds.), Discrete and Computational Geometry. Proceedings, 2000. XI, 381 pages. 2001.

Vol. 2099: P. de Groote, G. Morrill, C. Retoré (Eds.), Logical Aspects of Computational Linguistics. Proceedings, 2001. VIII, 311 pages. 2001. (Subseries LNAI).

Vol. 2100: R. Küsters, Non-Standard Inferences in Description Logocs. X, 250 pages. 2001. (Subseries LNAI).

Vol. 2101: S. Quaglini, P. Barahona, S. Andreassen (Eds.), Artificial Intelligence in Medicine. Proceedings, 2001. XIV, 469 pages. 2001. (Subseries LNAI).

Vol. 2102: G. Berry, H. Comon, A. Finkel (Eds.), Computer-Aided Verification. Proceedings, 2001. XIII, 520 pages. 2001.

Vol. 2103: M. Hannebauer, J. Wendler, E. Pagello (Eds.), Balancing Reactivity and Social Deliberation in Multi-Agent Systems. VIII, 237 pages. 2001. (Subseries LNAI).

Vol. 2104: R. Eigenmann, M.J. Voss (Eds.), OpenMP Shared Memory Parallel Programming. Proceedings, 2001. X, 185 pages. 2001.

Vol. 2105: W. Kim, T.-W. Ling, Y-J. Lee, S.-S. Park (Eds.), The Human Society and the Internet. Proceedings, 2001. XVI, 470 pages. 2001.

Vol. 2106: M. Kerckhove (Ed.), Scale-Space and Morphology in Computer Vision. Proceedings, 2001. XI, 435 pages. 2001.

Vol. 2107: F.T. Chong, C. Kozyrakis, M. Oskin (Eds.), Intelligent Memory Systems. Proceedings, 2000. VIII, 193 pages. 2001.

Vol. 2108: J. Wang (Ed.), Computing and Combinatorics. Proceedings, 2001. XIII, 602 pages. 2001.

Vol. 2109: M. Bauer, P.J. Gymtrasiewicz, J. Vassileva (Eds.), User Modeling 2001. Proceedings, 2001. XIII, 318 pages. 2001. (Subseries LNAI).

Vol. 2110: B. Hertzberger, A. Hoekstra, R. Williams (Eds.), High-Performance Computing and Networking. Proceedings, 2001. XVII, 733 pages. 2001.

Vol. 2111: D. Helmbold, B. Williamson (Eds.), Computational Learning Theory. Proceedings, 2001. IX, 631 pages. 2001. (Subseries LNAI).

Vol. 2116: V. Akman, P. Bouquet, R. Thomason, R.A. Young (Eds.), Modeling and Using Context. Proceedings, 2001. XII, 472 pages. 2001. (Subseries LNAI).

Vol. 2117: M. Beynon, C.L. Nehaniv, K. Dautenhahn (Eds.), Cognitive Technology: Instruments of Mind. Proceedings, 2001. XV, 522 pages. 2001. (Subseries LNAI).

Vol. 2118: X.S. Wang, G. Yu, H. Lu (Eds.), Advances in Web-Age Information Management. Proceedings, 2001. XV, 418 pages. 2001.

Vol. 2119: V. Varadharajan, Y. Mu (Eds.), Information Security and Privacy. Proceedings, 2001. XI, 522 pages. 2001.

Vol. 2120: H.S. Delugach, G. Stumme (Eds.), Conceptual Structures: Broadening the Base. Proceedings, 2001. X, 377 pages. 2001. (Subseries LNAI).

Vol. 2121: C.S. Jensen, M. Schneider, B. Seeger, V.J. Tsotras (Eds.), Advances in Spatial and Temporal Databases. Proceedings, 2001. XI, 543 pages. 2001.

Vol. 2123: P. Perner (Ed.), Machine Learning and Data Mining in Pattern Recognition. Proceedings, 2001. XI, 363 pages. 2001. (Subseries LNAI).

Vol. 2124: W. Skarbek (Ed.), Computer Analysis of Images and Patterns. Proceedings, 2001. XV, 743 pages. 2001.

Vol. 2125: F. Dehne, J.-R. Sack, R. Tamassia (Eds.), Algorithms and Data Structures. Proceedings, 2001. XII, 484 pages. 2001.

Vol. 2126: P. Cousot (Ed.), Static Analysis. Proceedings, 2001. XI, 439 pages. 2001.

Vol. 2129: M. Goemans, K. Jansen, J.D.P. Rolim, L. Trevisan (Eds.), Approximation, Randomization, and Combinatorial Optimization. Proceedings, 2001. IX, 297 pages. 2001.

Vol. 2130: G. Dorffner, H. Bischof, K. Hornik (Eds.), Artificial Neural Networks – ICANN 2001. Proceedings, 2001. XXII, 1259 pages. 2001.

Vol. 2132: S.-T. Yuan, M. Yokoo (Eds.), Intelligent Agents. Specification. Modeling, and Application. Proceedings, 2001. X, 237 pages. 2001. (Subseries LNAI).

Vol. 2136: J. Sgall, A. Pultr, P. Kolman (Eds.), Mathematical Foundations of Computer Science 2001. Proceedings, 2001. XII, 716 pages. 2001.

Vol. 2138: R. Freivalds (Ed.), Fundamentals of Computation Theory. Proceedings, 2001. XIII, 542 pages. 2001.

Vol. 2139: J. Kilian (Ed.), Advances in Cryptology – CRYPTO 2001. Proceedings, 2001. XI, 599 pages. 2001.

Vol. 2141: G.S. Brodal, D. Frigioni, A. Marchetti-Spaccamela (Eds.), Algorithm Engineering. Proceedings, 2001. X, 199 pages. 2001.

Vol. 2143: S. Benferhat, P. Besnard (Eds.), Symbolic and Quantitative Approaches to Reasoning with Uncertainty. Proceedings, 2001. XIV, 818 pages. 2001. (Subseries LNAI).

Vol. 2146: J.H. Silverman (Eds.), Cryptography and Lattices. Proceedings, 2001. VII, 219 pages. 2001.

Vol. 2147: G. Brebner, R. Woods (Eds.), Field-Programmable Logic and Applications. Proceedings, 2001. XV, 665 pages. 2001.

Vol. 2149: O. Gascuel, B.M.E. Moret (Eds.), Algorithms in Bioinformatics. Proceedings, 2001. X, 307 pages. 2001.

Vol. 2150: R. Sakellariou, J. Keane, J. Gurd, L. Freeman (Eds.), Euro-Par 2001 Parallel Processing. Proceedings, 2001. XXX, 943 pages. 2001.

Vol. 2154: K.G. Larsen, M. Nielsen (Eds.), CONCUR 2001 – Concurrency Theory. Proceedings, 2001. XI, 583 pages. 2001.

Vol. 2161: F. Meyer auf der Heide (Ed.), Algorithms – ESA 2001. Proceedings, 2001. XII, 538 pages. 2001.

Vol. 2164: S. Pierre, R. Glitho (Eds.), Mobile Agents for Telecommunication Applications. Proceedings, 2001. XI, 292 pages. 2001.